Sensory Mechanisms of the Spinal Cord

Third Edition

Volume 1

Primary Afferent Neurons and the Spinal Dorsal Horn

Sensory Mechanisms of the Spinal Cord
Third Edition

Volume 1
Primary Afferent Neurons and the Spinal Dorsal Horn

William D. Willis, Jr.

and

Richard E. Coggeshall

The University of Texas Medical Branch
Galveston, Texas

Kluwer Academic / Plenum Publishers
New York, Boston, Dordrecht, London, Moscow

ISBN 0-306-47729-7 (set)
ISBN 0-306-48033-6 (vol. 1)

233 Spring Street, New York, New York 10013

http://www.wkap.nl/

10 9 8 7 6 5 4 3 2 1

A C.I.P. record for this book is available from the Library of Congress

PREFACE

The third edition of this monograph continues to have the goal of providing an overview of current thought about the spinal cord mechanisms that are responsible for sensory processing. We hope that the book is of value to both basic and clinical neuroscientists.

Several changes have been made in the presentation, as well as additions because of the research advances that have been made during the past decade. Chapters 3 and 4 in the previous edition have been subdivided, and now the morphology of primary afferent neurons of the dorsal root ganglia is described in Chapter 3 and the chemical neuroanatomy of these neurons in Chapter 4. The description of the dorsal horn in the previous Chapter 4 is now included in Chapter 5, and the chemical neuroanatomy of the dorsal horn in Chapter 6. Furthermore, discussions of the descending control systems have now been consolidated at the end of Chapter 12.

The authors would like to express their appreciation for the help provided by several individuals. R.E.C. wishes to acknowledge the many things he learned about primary afferent neurons from conversations with Dr S. N. Lawson. He also thanks Lyn Shilling for her assistance with the typing. WDW thanks Dr Nada Lawand for her critical reading of parts of the manuscript, Rosaline Leigh for help with the manuscript, and Griselda Gonzales for preparing the illustrations.

William D. Willis and Richard E. Coggeshall

CONTENTS

Volume 1

Chapter 1

Introduction

SENSORY MODALITIES AND CHANNELS

The term "sensory modality" was introduced by Helmholtz (Boring, 1942). A sensory modality is a class of sensations. That is, sensations of two different modalities are qualitatively different (e.g., hearing versus vision), whereas sensations within a single modality are only quantitatively different (e.g., different frequencies of two sounds). Müller (1840–2) regarded the sensation derived from skin stimulation as a single sensory modality ("touch"), following Aristotle's classification of the five senses (touch, vision, hearing, smell, and taste). However, contemporary investigators consider such sensations as touch, pain, cold, and warm to be discrete modalities. A given sensory modality would have a number of characteristics or attributes, such as quality, intensity, duration, and extension.

On the basis of subjective awareness and also of the results of clinical and psychophysical testing, a substantial number of sensory modalities pertaining to the sensory experiences that occur following stimulation of the skin or of subcutaneous tissue may be recognized. These include touch-pressure, flutter-vibration, tickle, warmth, cold, pain, itch, position sense, and kinesthesia. Some of these may be further subdivided. For example, cutaneous pain may be considered to have several submodalities, including pricking pain and burning pain (Lewis, 1942), whereas deep somatic structures, such as muscle, give rise to aching (Lewis, 1942) or cramping pain (cf. Marchettini et al., 1996). More complicated somatic sensations, perhaps involving combinations of modalities, are also of interest, especially in clinical testing. These include two-point discrimination, stereognosis, graphesthesia, and the abnormal sensations called paresthesias. Visceral sensations include the awareness of distention, hunger, nausea, and visceral pain. Geldard (1953) provides an even more extensive list of somatovisceral sensory modalities.

Each modality of sensation depends upon information that is transmitted along one or more sensory pathways. Transection of the spinal cord completely abolishes any awareness of sensation from regions of the body below the transection. Interruption of one or more of the sensory pathways may partially or totally eliminate a particular kind of sensory experience. The neural circuitry for transmission of information concerning a sensory modality can be defined as a "sensory channel." For sensations originating from the body, a sensory channel would include one or more sets of sensory receptors, one or more of the sensory pathways that ascend in the spinal cord, and the particular regions of the brain that are involved in receiving and processing the information. Activity in a sensory channel would typically be under centrifugal control by way of pathways that descend from the brain to the spinal cord.

This monograph is concerned with the portions of the somatosensory channels that involve the primary afferent neurons of the peripheral nervous system that supply the body and the processing circuits and ascending pathways of the spinal cord. A discussion of somatosensory processing at thalamic and cerebral cortical levels can be found in works by Mountcastle (1984; 1998), Creutzfelt (1995), Steriade et al. (1997), Price (1999), and Casey and Bushnell (2000).

SPECIFICITY VERSUS PATTERN THEORIES OF SENSATION

The concept of a sensory channel implies that the neural circuits concerned with somatosensory processing can distinguish between different sensory modalities. A number of theories about how this might be done have been proposed and two major theories have emerged to account for the processing and differentiation of various forms of sensory experience: the specificity theory and the pattern theory (Sinclair, 1981). A mixed theory involving specificity of primary afferent fibers but patterned responses of central neurons was later developed to account for several aspects of pain sensation (Melzack and Wall, 1965). This has been called the Gate Theory of Pain.

Historical Perspective

Specificity Theory

The manner in which the nervous system differentiates between the various forms of sensory experience has been a central issue since the beginning of sensory physiology. A fundamental discovery was the separation of sensory and motor functions in the spinal cord dorsal and ventral roots, later termed the Law of Bell and Magendie (see Cranefield, 1974). The notion of specificity of cutaneous sensation is attributed by Sinclair to Bell (1811), although it was actually Magendie (1822) who demonstrated the sensory function of the spinal cord dorsal roots. The idea of specificity of cutaneous sensation was forwarded by Müller's doctrine of specific nerve energies (Müller, 1840–2). Müller had in mind the Aristotelean five senses and, as already mentioned, lumped together the sensations derived from the body surface under the category of "touch." According to Sinclair (1981), Volkmann, Natanson, and others in the 1840s extended the specificity concept to include the postulate of separate nerve endings for each variety of sensation arising from cutaneous stimulation. However, Sinclair emphasizes the distinction between "specific nerve energies" (i.e., particular nerves evoke particular sensations) and "specific irritability" (i.e., particular stimuli activate particular sense organs). It is one thing to show that specific stimuli preferentially activate certain sensory receptors; it is another to show that the same sensory receptors are responsible for a particular quality of sensory experience.

Evidence supporting the notion that the doctrine of specific nerve energies applies to the different cutaneous senses came from the observations of Blix (1884), who discovered that stimulation of separate localized points on the skin gave rise to distinct sensations of pressure, warmth, cold, or pain. Numerous other investigators, including Goldscheider (1884) and Donaldson (1885), confirmed this observation. Specificity was shown by the observation that a cold spot, for example, can be stimulated by cold, but not by heat, whereas a warm spot responds to heat but not cold (Blix, 1884). Even if a cold (or warm) spot is stimulated by an electric current, the sensation that results is still cold (or warm). Mapping of marked areas of skin proved that the cold and warm spots are in fixed positions that can be identified on subsequent days with a high degree of accuracy (Fig. 1.1A), provided that care is taken to avoid a number of sources of technical error (Dallenbach, 1927).

Careful mapping showed that pressure and pain spots (Fig. 1.1B) are distinct from each other and from cold and warm spots (Blix, 1884; Goldscheider, 1884; von Frey, 1896). The density of the different kinds of sensory spots was found to vary from one part of the body surface to another. For example, von Frey (1896) estimated (using von Frey hairs as stimulating devices, Fig. 1.1C) that there are about 100–200 pain spots/cm^2 on the back of the hand; this is about eight times the density of pressure points in the same region. By contrast, he found 100–200 pressure spots/cm^2 on the palm. In an area of 29 mm^2 on the conjunctiva, von Frey (1896) could locate 35 pain spots and 120 cold spots. However, in general, pain spots are the most common. Cold spots are somewhat more frequent than warm spots (Donaldson, 1885).

The spot-like distribution of mechanically and thermally sensitive areas suggested that a specific type of sensory receptor organ would be found beneath each sensory spot. If so, there would be a causal linkage between individual types of sensory receptors and specific sensations. Histologists had begun to describe a variety of types of cutaneous sense organs (Krause, 1859; Meissner, 1859; Ruffini, 1894), and so it was reasonable for

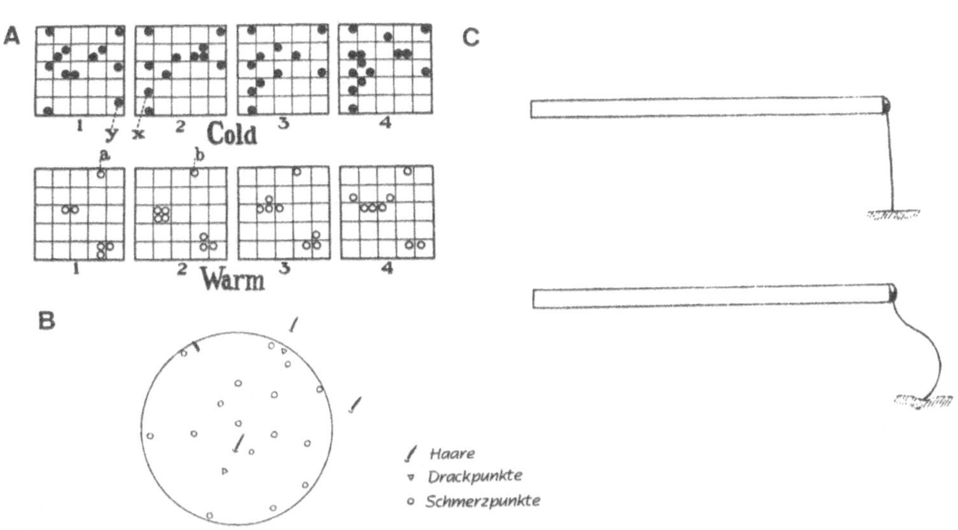

FIGURE 1.1. A, maps of the distribution of cold and warm spots within an area of 1 cm^2 in a single subject. The maps were made on four different days (1–4). The cold spots were mapped before the warm spots in a given session. (From Dallenbach, 1927.) B, map of pressure (Druckpunkte) and pain (Schmerzpunkte) spots in an area of 12.5 mm^2 on the back of the hand. The locations of hairs (Haare) are also indicated. (From von Frey, 1896.) C, use of a von Frey hair to stimulate a pressure point. The hair is calibrated in terms of its diameter and bending force. Once the hair bends, the pressure applied is constant. (From von Frey, 1896.)

von Frey (1906; 1910) to suggest that each kind of sensory spot is associated with a particular type of sensory organ. On the basis of his understanding of the distribution of sense organs, von Frey proposed the following relationships between sensation and activation of particular sense organs: he suggested that touch is associated with hair follicle endings in hairy skin and Meissner's corpuscles in glabrous skin; cold with Krause's end bulbs; warmth with Ruffini endings; and pain with free endings.

The match suggested by von Frey between particular sensory receptors and sensory spots was correct for touch and pain, but was incorrect with respect to the proposed association between Ruffini endings and Krause's end bulbs and the warm and cold spots (Donaldson, 1885; Dallenbach, 1927; Weddell and Sinclair, 1953; Sinclair, 1955; Weddell, 1955; Lele and Weddell, 1956; 1959; Weddell and Miller, 1962). Ruffini endings have now been identified with SA II slowly adapting mechanoreceptors (Chambers et al., 1972), and Krause's end bulbs are a type of rapidly adapting mechanoreceptor (Iggo and Ogawa, 1977). Thermal sensibility, like pain, depends upon the activation of "free" (i.e., unencapsulated) endings, although the nature of the receptor membrane is different for different types of sensory endings (Andres and von Düring, 1973; Hensel et al., 1974; Kruger et al., 1981; 1985; Heppelmann et al., 1995).

Pattern Theory

The criticisms by Weddell and colleagues of von Frey's theory were directed not so much at the notion of sensory spots as at the histological correlates of the spots. However, even the existence of sensory spots was called into question by some investigators (Guilford and Lovewell, 1936; Jenkins, 1941a, b; see Sinclair, 1955). Furthermore, Weddell and Miller (1962) point out that it is unlikely that a naturally occurring stimulus applied to a sensory spot would activate only a single afferent fiber, since a stimulus applied to just 1 mm^2 of skin could activate more than 100 underlying endings. This criticism loses some force if the differential sensitivity of the endings is taken into account. Nevertheless, Burgess et al. (1974) estimate that the application of a weak stimulus (less than 100 mg) by a probe with a diameter of 0.5–1 mm will excite 10–15 rapidly adapting mechanoreceptive afferent fibers that supply several of the types of receptors that are found in cat hairy skin. Stronger stimuli, if favorably placed, would also activate slowly adapting mechanoreceptors. Similarly, Johansson, R. S. and Vallbo (1976) provide evidence "that practically all naturally occurring tactile stimuli to the human hand excite a large number of sensory units" and that "even the smallest variation of a suprathreshold stimulus... will appreciably change this pattern." Thus, pressure spots are likely to be complex entities and to involve several afferent fibers that innervate more than one receptor type.

Because of these difficulties with the specificity theory, another theory of cutaneous sensation seemed to be required. An alternative was the pattern theory. This theory was based on the observation that sensory stimuli are encoded by trains of nerve impulses in the nerve fibers that supply sensory receptor organs (Adrian, 1946). Nafe (1927; 1929) suggested that a sensation results from a patterned input from sense organs of the skin that is usually, but not necessarily always, associated with a particular kind of stimulus. Learning provides the name for the sensation associated with the stimulus and the resultant patterned responses in sensory afferent fibers. Specific sensory channels are not needed, just particular spatial and temporal patterns of nerve impulses that reach and are recognized by the central nervous system. Sinclair (1955) and Weddell (1955) supported this theory.

However, a difficulty for the pattern theory was the finding that large myelinated afferent fibers play a special role in touch and small myelinated and unmyelinated fibers in pain and temperature sensations (Heinbecker et al., 1933; 1934; Lewis and Pochin, 1938a, b; Torebjörk and Hallin, 1973; Hallin and Torebjörk, 1976). For instance, pressure applied to a mixed peripheral nerve in a human subject results in a progressive conduction block of large myelinated afferent fibers and then of small myelinated fibers before unmyelinated fibers cease conducting (Fig. 1.2, Panel I). Touch and other mechanically evoked sensations are lost first and then cold and pricking pain, leaving warm and burning pain intact so long as C fibers continue to conduct (MacKenzie et al., 1975; Campbell et al., 1988; Torebjörk et al., 1992). Conversely, small fibers are blocked first when a local anesthetic is injected near a human peripheral nerve (Fig. 1.2, Panel II; Hallin and Torebjörk, 1976; see Campbell et al., 1988). The result is a loss of pain and temperature sensations, with retention of touch. Furthermore, single-unit recording studies demonstrated that mechanoreceptors are often supplied by large myelinated fibers and nociceptors and thermoreceptors by small myelinated and unmyelinated fibers (Hensel and Zotterman, 1951a; Dodt, 1952; Dodt and Zotterman, 1952a; Iggo, 1959; 1969; Hensel et al., 1960; Hunt, C. C. and McIntyre, 1960c; Iriuchijima and Zotterman, 1960), although some mechanoreceptors are innervated by small myelinated (Burgess et al., 1968; Perl, 1968) or unmyelinated (Douglas and Ritchie, 1957; Iggo, 1960; Vallbo et al., 1999) fibers and some nociceptors by large myelinated fibers (Burgess and Perl, 1967; Georgopoulos, 1976). See section below on "Large- versus small-fiber systems."

Gate Theory of Pain

With the strengthening of the evidence that sensory receptors play specific sensory roles, a compromise theory was proposed in which specific information from the sense organs generates patterned activity centrally (Melzack and Wall, 1965). The model of Melzack and Wall is known as the Gate Theory of Pain. However, this theory does not attempt to account for modalities of cutaneous sensation other than pain. The details of the original neural circuit suggested by Melzack and Wall have been challenged, as discussed in Chapter 7.

Clinical evidence suggests that sensory pathways in the central nervous system are specific, since dissociated sensory loss can be produced in such disorders as the Wallenberg syndrome and after therapeutic interventions such as anterolateral cordotomy for the relief of pain (Willis, 1985; Gybels and Sweet, 1989). Melzack and Wall (1962) caution that a lesion placed in the spinal cord white matter can affect sensory transmission in a number of ways in addition to the possible interruption of a specific sensory pathway. For example, a lesion will reduce the total number of axons of ascending tract neurons that convey sensory input to the brain; furthermore, the pattern of activity in the ascending pathways will be altered. Another important consideration is that pathways descending from the brain that control the activity in sensory pathways may be interrupted, with a consequent change in the operation of the sensory pathways. However, a case report by Noordenbos and Wall (1976) provides additional evidence that sensory information concerning pain, temperature, and certain aspects of tactile sense is conveyed to the brain by pathways that ascend to the brain in the anterolateral quadrant of the spinal cord, whereas other mechanical senses depend upon pathways that ascend in other parts of the spinal cord (see discussion in Chapter 8 and Fig. 8.5). These observations are consistent with the notion that there are specific somatosensory pathways in the central nervous system.

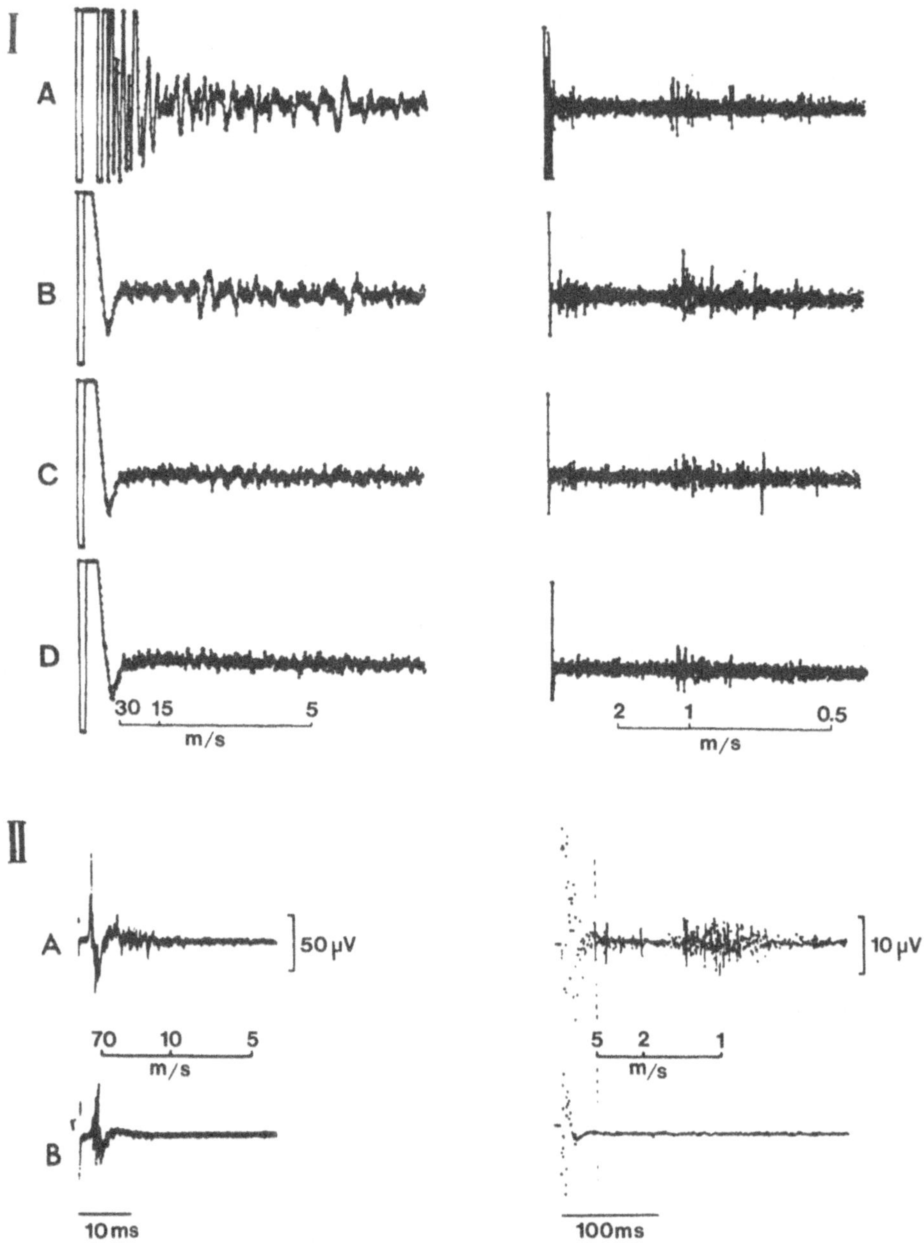

FIGURE 1.2. I, changes in the afferent volley and in evoked sensation during a progressive nerve block produced by pressure. The records in the left column show the compound action potential recorded from myelinated axons of the superficial radial nerve in a human subject using a microneurography electrode. The column at the right shows the activity in unmyelinated axons (near the middle of the traces), as well as in myelinated fibers (at the far left of the traces). A, activity was not yet blocked and perception was normal except for some dysesthesia. B, the Aαβ volley was blocked, and the Aδ volley was reduced. Light touch was absent, and pinprick and cord were impaired. C, most or all of the myelinated fibers were blocked, and pinprick and cold were absent. However, warm and pain sensations were still felt. D, the unmyelinated fiber volley was reduced, and warm and pain sensations were impaired. (From Mackenzie et al., 1975.) II, activity recorded from the human radial nerve in response to electrical stimulation of the dorsum of the hand. A, the volleys are seen to include Aδ and C fibers (note the two different sweeps and the bars indicating conduction velocity). The stimulus was painful. B, the Aδ and C-fiber activity was abolished by local anesthetic, but some Aαβ activity remained. The subject no longer felt pain, but could still feel touch. (From Hallin and Torebjörk, 1976; reproduced with permission.)

Supporting Evidence

Molecular Basis of Specificity of Primary Afferent Neurons

Ultimately, it should be possible to relate functional differences in the various kinds of sensory receptors and their primary afferent fibers to the phenotypes of these neurons. Specificity will then be defined at the molecular level. For example, it should be possible to correlate sensory function and such markers of genetic expression as particular membrane receptors and intracellular protein or peptide content to provide a means to distinguish between primary afferent neurons of different functional types (Nagy and Hunt, 1982; Lawson et al., 1984; 1985; Alvarez et al., 1988; Kruger et al., 1989; Caterina et al., 1997; Tominaga et al., 1998; see Chapter 2). Other markers include cell surface carbohydrates (Chou et al., 1988), which can be recognized by monoclonal antibodies (Dodd and Jessell, 1985; Jessell and Dodd, 1985; Lawson et al., 1985). The antibodies label not only the cell bodies of dorsal root ganglion neurons but also the axons and terminals of these sensory neurons (Alvarez et al., 1989a, b). Globoseries carbohydrates identify dorsal root ganglion cells that project to laminae III and IV (Dodd et al., 1984; Jessell and Dodd, 1985), whereas lactoseries oligosaccharides recognize dorsal root ganglion cells that have terminations in laminae I and II (Dodd and Jessell, 1985; Alvarez et al., 1989a). These surface carbohydrates are likely to play a crucial role in the cell–cell interactions required for the development of primary afferent connections to the spinal cord dorsal horn (Dodd and Jessell, 1985).

Descriptions of several primary afferent phenotypes are currently being developed. Based on immunocytochemical evidence, Averill et al. (1995) suggest that there are three major groups of dorsal root ganglion cells (with some overlap).

1. About 40% of lumbar DRG cells are immunoreactive to an antibody, RT97, to the phosphorylated form of the neurofilament protein NF200 (cf. Lawson et al., 1984; Lawson and Waddell, 1991; Perry et al., 1991). These DRG cells belong to the "light" cell class of DRG cells. These neurons were formerly called "large light cells," but some are in fact medium or small in size (Lawson and Waddell, 1991; see Chapter 3 for a discussion of the classification of DRG cells as "large light" or "small dark" neurons). They are nonpeptidergic and do not bind the lectin IB4. However, they do bind antibodies to globoseries carbohydrates (Dodd et al., 1984). Many contain tyrosine kinase B (trkB) and/or trkC receptors (McMahon et al., 1994; Wright and Snider, 1995). Those with trkB receptors are presumably regulated by brain-derived neurotrophic factor (BDNF) and/or neurotrophic factor-4/5 (NT-4/5), whereas those with trkC receptors are probably regulated by neurotrophic factor-3 (NT-3). The "light" DRG cells have myelinated axons, many of which are Aβ fibers that innervate mechanoreceptors. However, some RT97-positive DRG cells appear to have Aδ-sized axons, some of which could be nociceptors (Lawson and Waddell, 1991). Many of the axons of "light" DRG cells belong to cutaneous receptors and project to laminae III and IV of the dorsal horn, but others are large muscle afferents that project to the intermediate nucleus and ventral horn. The latter have trkC receptors (Wright and Snider, 1995).

2. About 30% of DRG cells are "small dark" neurons that are immunoreactive to an antibody known as LA4 that reacts with an oligosaccharide located on the surface of the neurons (Dodd and Jessell, 1985), but not to the antibody to RT97. These DRG cells have unmyelinated axons (Alvarez et al., 1989b). These cells also contain binding sites for the lectin IB4 and express fluoride-resistant acid phosphatase (FRAP) (Alvarez et al., 1991),

and they do not contain neuropeptides (Guo, A. et al., 1999). Dorsal root ganglion cells that bind IB4 may express P2X3 purinoreceptors, and these receptors are regulated by glial-derived neurotrophic factor (GDNF) (Bradbury et al., 1998). The synaptic endings of this class of DRG cells form a narrow band in the inner part of the substantia gelatinosa, lamina II_{inner} (Bradbury et al., 1998; Guo, A. et al., 1999).

3. The remaining group of "small dark" DRG cells contains CGRP and often other peptides, such as substance P, somatostatin, or galanin (see Chapter 3). Many of these can be labeled with the LD2 antibody, which recognizes α-galactose extended gangliosides (Chou et al., 1988). Most substance P-containing dorsal root ganglion cells have neurokinin 1 receptors (von Banchet and Schaible, 1999). This third population of DRG cells also contains tyrosine kinase A (trkA) receptors and is regulated by nerve growth factor (NGF; Averill et al., 1995; Bennett et al., 1996b). These neurons lack IB4 binding sites, $P2X_3$ receptors and FRAP expression (Bradbury et al., 1998), and they do not label with RT97. This group of generally small DRG cells have unmyelinated axons, although some are medium-sized and presumably have small, myelinated axons. The synaptic terminals of these neurons are concentrated in laminae I and II_{outer} (Bradbury et al., 1998), although some also project to deeper layers of the dorsal horn and to lamina X (Alvarez et al., 1989a).

Many of the small DRG cells that belong to the second two phenotypes are nociceptors (Hunt and Rossi, 1985). Cutaneous Aδ and C mechano-heat nociceptors and groups III and IV muscle nociceptors respond to capsaicin (Szolcsanyi, 1987; Szolcsanyi et al., 1988; Baumann et al., 1991; LaMotte et al., 1992; Marchettini et al., 1996). This responsiveness to capsaicin can be attributed to the presence of VR-1 vanilloid receptors in their surface membranes. VR-1 receptors make these neurons responsive to the application of capsaicin and also to heat and to lowered pH (Caterina et al., 1997; Helliwell et al., 1998; Tominaga et al., 1998; Nagy and Rang, 1999a, b). However, some heat-sensitive afferents have a high heat threshold and are insensitive to capsaicin (Nagy and Rang, 1999a). These neurons may express VRL-1 receptors (Caterina et al., 1999). Curiously, experiments on single channels in isolated membrane patches from DRG cells show that a given channel may respond to either capsaicin or to heat but rarely to both (Nagy and Rang, 1999b). As mentioned, many of the capsaicin-sensitive afferents also contain neuropeptides, such as calcitonin gene-related peptide and/or substance P (Alvarez et al., 1988; Hanesch et al., 1991; 1997), whereas others have binding sites for IB4 (Guo et al., 1999). Thus, there are at least two phenotypes for heat-responsive nociceptors.

Molecular markers may eventually also help distinguish between dorsal root ganglion cells that belong to cutaneous, muscle, and visceral nerves (Perry and Lawson, 1998). For example, an antibody to soybean agglutinin labels a similar proportion of cutaneous, muscle, and visceral afferents, whereas one to peanut agglutinin labels more cutaneous and visceral than muscle afferents (Perry and Lawson, 1998). A high proportion of cutaneous (37%) and of visceral afferents (31%) belong to the phenotype that expresses $P2X_3$ receptors, whereas only a small fraction (2%) of muscle afferents are in this category (Bradbury et al., 1998). Furthermore, a much higher proportion of visceral than of cutaneous DRG cells contains CGRP and trkA receptors and a much lower proportion has IB4 binding sites (Bennett et al., 1996b). Most muscle afferents (70%) have trkC receptors, but only 10% of cutaneous afferents and no visceral afferents express these receptors (McMahon et al., 1994).

Sinclair's Challenge and Evidence for Specificity from Microneurography

As mentioned above, specificity of responsiveness of sense organs must be distinguished from specificity of the sensations that their activity elicits. Response specificity does not necessarily lead to sensory specificity. For example, baroreceptor afferents respond in a rather specific way to changes in blood pressure, but there is no corresponding sensory experience. A similar question is whether activation of a cold receptor would necessarily lead to a sensation of cold or perhaps just contribute to thermoregulation. Is the receptor both a cold detector and a provoker of cold sensation? In 1981, Sinclair stated that "There is as yet no unassailable evidence that stimulation of a single fibre in isolation can give rise to a sensation of any kind. To stimulate a single fibre in an intact human subject, to prove satisfactorily that only that fibre and no other has been stimulated, and to record a simultaneous meaningful sensory judgement is an almost incredibly difficult technical and psychophysiological feat, and it may be a long time before unequivocal evidence can be obtained."

Sensory Role of Single Tactile Afferents in Human Subjects

Experiments that appear to meet the stringent criteria set by Sinclair have now been done using the intraneural microstimulation technique in human subjects (Hagbarth and Vallbo, 1967; Torebjörk and Ochoa, 1980; 1990; Konietzny et al., 1981; Vallbo, 1981; Ochoa and Torebjörk, 1983; 1989; Schady and Torebjörk, 1983; Schady et al., 1983; Torebjörk et al., 1984a, b; 1987; Macefield et al., 1990; Marchettini et al., 1996). For microneurography, a metal microelectrode is introduced into a peripheral nerve. The usual search procedure for intraneural microstimulation is to stimulate through the electrode while finding a site within the nerve that evokes a unitary sensation. The sensation is investigated psychophysically, and then the electrode is switched to a recording mode. There is consistently a sensory unit within recording range whose receptive field corresponds to the area of sensory referral (Fig. 1.3A; Torebjörk and Ochoa, 1980; Schady and Torebjörk, 1983). The response properties of the unit are then examined (Fig. 1.3B), and the conduction velocity is determined (Fig. 1.3C). Verification that the same nerve fiber is responsible for both the unitary sensation and the recorded response is obtained by application of a high-frequency train of stimuli. For a large myelinated fiber, this causes a state of hyperexcitability, as shown by increases in its responses to stimulation of the receptive field, either mechanically or electrically (Fig. 1.3D and E). However, fine myelinated and unmyelinated fibers do not become hyperexcitable, but instead their conduction velocity slows when they are "marked" by repetitive stimulation (Nordin, 1990; Torebjörk and Ochoa, 1990; Schmelz et al., 1995). The reason for this is that the relatively refractory period of fine fibers, especially C fibers, is long-lasting. This property is very useful for distinguishing between a particular unmyelinated fiber and others in the same recording, since stimulation of the receptive field to activate a given fine afferent fiber will cause just that fiber to show a slowed conduction velocity (see Fig. 2.3).

In some experiments, a second electrode has been inserted into the nerve at a more proximal location to show that the unitary sensation corresponded in an all-or-nothing fashion to the activation of a single large myelinated sensory unit (Ochoa and Torebjörk, 1983). That is, a recording is made from the same axon that is stimulated! These observations are good evidence that only a single afferent fiber is being stimulated when a unitary

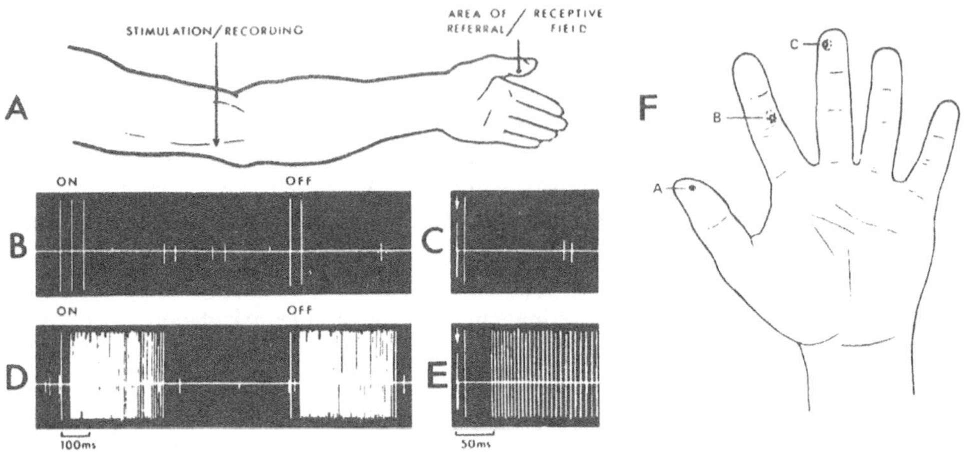

FIGURE 1.3. A, method used for stimulating single afferent nerve fibers during microneurography. Recordings are made from a nerve fiber by using a metal microelectrode inserted percutaneously into a peripheral nerve. The receptive field is mapped, and then the fiber is stimulated through the recording electrode. The area to which sensation is referred ("projected receptive field") is compared with the mapped receptive field. The recordings in panels B–E were made from the axon of a rapidly adapting receptor, whose stimulation evoked a sensation of flutter-vibration. B, brief indentation of the skin produced on- and off-responses of the unit. C, conduction delay following electrical stimulation of the skin; the conduction velocity was 45 m/s. D and E, responses to mechanical and electrical stimulation were increased after the receptor was sensitized by high-frequency stimulation. (From Torebjörk and Ochoa, 1980.) F, correspondence between the mapped and projected receptive fields for three units in the median nerve. One unit was a rapidly adapting receptor (A), and the other two were Merkel cell (SAI) receptors (B and C). (From Ochoa and Torebjörk, 1983.)

sensation is experienced. Unfortunately, this approach is impractical in the case of small myelinated and unmyelinated afferent fibers.

The correspondence between the projected and receptive fields is shown for three different units in Fig. 1.3F (Ochoa and Torebjörk, 1983). If the stimulus intensity is raised, activation of an adjacent sensory unit is signaled by the development of an additional area of sensory referral. For example, in a recording from the median nerve, the receptive field of an afferent fiber might be on a digit, and stimulation of the axon might result in a projected receptive field in the same location. Raising the stimulus intensity slightly might cause no change in the projected receptive field, but another sensation may be detected with higher stimulus intensities. This sensation is typically referred to a different receptive field, for example on another digit (Torebjörk and Ochoa, 1980; Ochoa and Torebjörk, 1983), although sometimes the original area of referral enlarges (Vallbo, 1981; Schady et al., 1983).

A single electrical stimulus applied to a single cutaneous afferent fiber usually does not evoke any sensation, although for Meissner's corpuscles it may. Generally, repetitive stimulation is required. However, when a sensation is produced (by using a suprathreshold stimulus intensity lower than the intensity required to cause recognition of a second sensory experience), it is always of the same modality, whatever the frequency of stimulation. The modality appears to be more "pure" than that experienced using a natural stimulus. As mentioned above, a natural stimulus would activate several kinds of sensory receptors, and the mixed input would produce mixed sensations. The particular modality of sensation elicited by stimulation of a single sensory axon depends on the receptor type, and the intensity of the sensation depends upon the frequency of stimulation. For rapidly adapting

TABLE 1.1. Sensations Evoked by Stimulation of Single Identified Cutaneous Afferent Fibers during Intraneural Microneurography in Human Subjects[a]

Receptor	Sensation
FA I (RA)	Tapping at 1 Hz; flutter at 10 Hz; vibration at 50 Hz
FA II (PC)	Tickling or vibration over 20–50 Hz
SA I	Sustained pressure over 5–10 Hz
SA II	Usually no sensation (occasionally position)
Aδ mechanical nociceptor	Sharp pain
C polymodal nociceptor	Dull pain, burning pain, or itch

[a]From Torebjörk and Ochoa (1980); Ochoa and Torebjörk (1983); Schady et al. (1983); Torebjörk et al. (1984a, b); Ochoa and Torebjörk (1989); Torebjörk and Ochoa (1990); and Macefield et al. (1990). For C fibers, small groups of axons rather than individual fibers may have been stimulated.

receptors, increases in frequency at first affect just the quality of the sensation, but above some frequency the perceived intensity increases (Ochoa and Torebjörk, 1983). Stimulation of afferents from most of the commonly observed receptor types produces the sensations listed in Table 1.1. Of particular note is that pain is produced only when nociceptors are stimulated; no pain results from high-frequency stimulation of mechanoreceptors. It is possible that the negative results from stimulation of SA II afferents from Ruffini endings reflect the need for activation of more than one fiber to meet the requirements for spatial summation in the central pathways.

The general conclusion that can be reached from these experiments is that there is specificity of primary afferent fibers supplying particular cutaneous sense organs with respect both to the adequate stimulus and to the sensation evoked. Although it can be questioned whether only single afferent fibers were excited in all of these experiments (Calancie and Stein, 1988), particularly those involving small fibers, the conclusion about specificity appears to hold even if a small number of afferent fibers were sometimes being stimulated. As already mentioned, in some cases recordings more proximally in a nerve confirmed the correspondence between activity in a single large afferent fiber and a unitary sensation.

Criticism of Microneurography Results

Wall and McMahon (1985) have contested the interpretation of these experiments using microneurography. They suggest that the large microelectrodes used in such studies (Fig. 1.4) are unlikely to be able to record and stimulate single axons selectively unless adjacent fibers are blocked by the mechanical trauma produced by the search procedure. If a large number of axons are blocked, the sensory experience reported might be abnormal. Furthermore, a number of observations from the microneurography experiments suggest that the sensations reported might depend upon central processing of input over a number of afferent fibers rather than a single fiber, and so the identification of sensory modalities may depend upon a patterned input to the central nervous system in these experiments. Experimental results in both rats and humans were consistent with at least a temporary block of many axons following introduction of a microneurography electrode into a peripheral nerve (Rice et al., 1993; 1994). However, recordings with concentric electrodes indicate that conduction is maintained in many axons despite the presence of a microneurography electrode (Hallin et al., 1986).

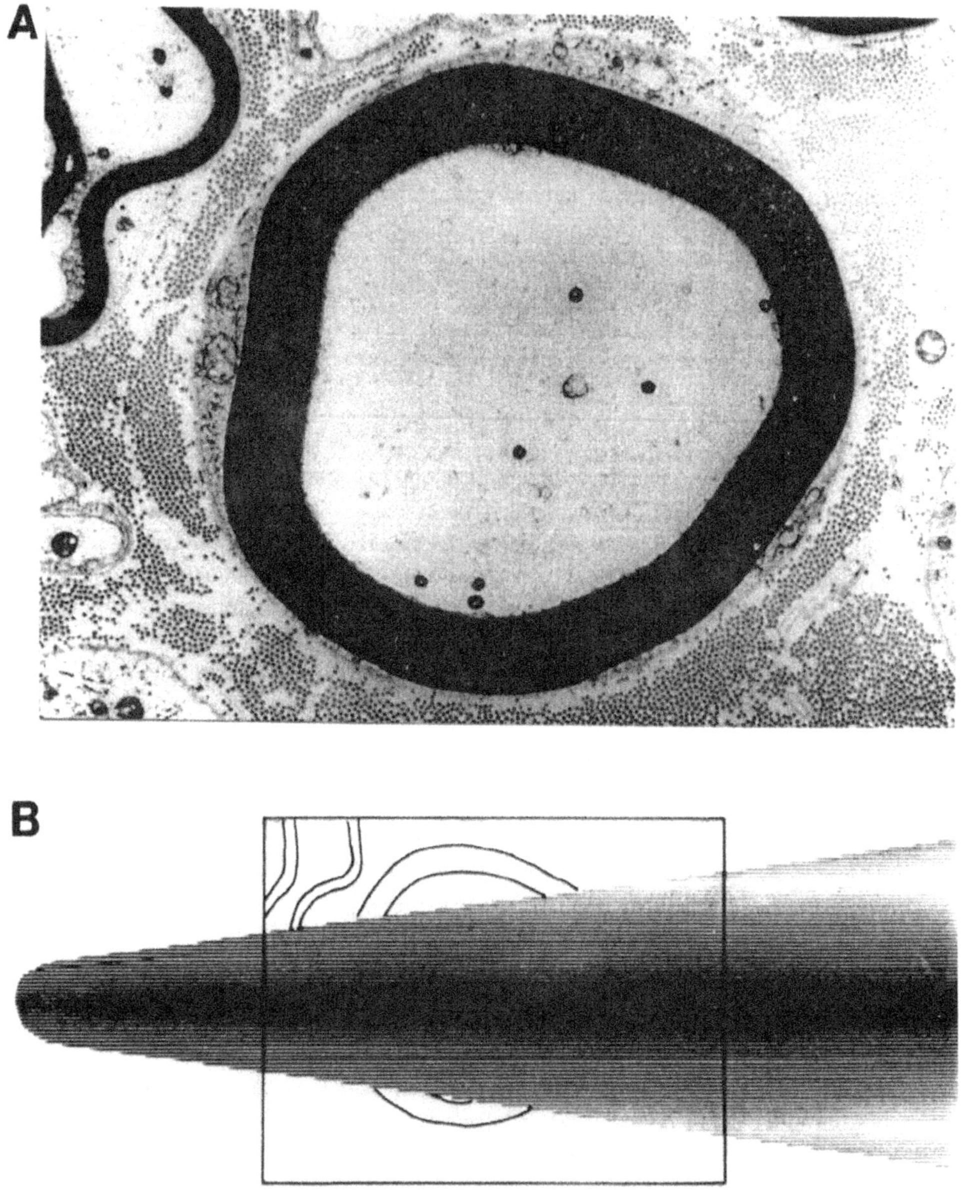

FIGURE 1.4. A, electron micrograph of a cross-section of a myelinated axon in the human sural nerve. The axon had a diameter of 9.5 μm. B, drawing of the same axon, with a representation of the tip of a microneurography recording electrode superimposed. (From Wall and McMahon, 1985.)

Rebuttal of this Criticism

Although Wall and McMahon (1985) present a reasonable case for caution in interpreting the results of microneurography experiments, it is important to recognize that if a single afferent fiber of a particular type can be activated and can be shown to evoke a specific sensation, the evidence in favor of the specificity theory is strengthened. Although one can never be absolutely certain that only one axon is excited in experiments on humans, it is still hard to dismiss many of the observations obtained in the microneurography experiments. Furthermore, in animal experiments, in which it is possible to obtain better stimulus control, stimulation of an individual sensory receptor can result in the activation of neurons in the spinal cord dorsal horn and the dorsal column nuclei, cortical evoked potentials, and behavioral responses (McIntyre et al., 1967; Tapper and Mann, 1968; Tapper, 1970; Mann et al., 1972; P. B. Brown et al., 1973; Ferrington et al., 1987a; Vickery et al., 1994; Gynther et al., 1995). For example, a cat can be trained to recognize whether or not there is activity in an afferent that supplies a single Merkel cell ending (Tapper, 1970). Therefore, it certainly seems plausible that stimulation through a microneurography electrode can activate a single afferent fiber, which in turn evokes a unitary sensation.

Torebjörk et al. (1987) deal carefully with the issues raised by Wall and McMahon (1985). The microneurography technique is reviewed, and evidence is presented that there is no severe damage to the nerve with consequent block of conduction in a large number of nerve fibers. Torebjörk et al. (1987) state that there is no reason to believe that conduction is blocked in most of the fibers being recorded from, although there is a change in the action potential configuration as a result of the presence of the recording electrode. The sensory experience is argued not to depend upon changes due to a loss of afferent barrage because of the damage, since the damage was limited and since the only cutaneous mechanoreceptive afferent units that are normally active are SA II receptors, which make up only 19% of the mechanoreceptor population. The intensities of current that were required to produce threshold sensations averaged only 0.81 μA, which agrees with the current needed to activate single axons in animal experiments. The currents used were anodal. Since the microelectrode tip was thought to be within the myelin sheath or partially intracellular in the axon, as indicated by the configuration of the action potential, fibers adjacent to the one stimulated should not be excited. Torebjörk et al. (1987) conclude that "specificity is a key principle for tactile systems."

Torebjörk et al. (1987) list a series of proposals that were made by Melzack and Wall (1962) and by Wall and McMahon (1985) and for which there is contrary evidence from microneurography. These proposals and the contrary findings from microneurogaphy include the following. (1) "More than one nerve impulse from a single afferent fibre, or more than one fibre carrying single nerve impulses, is essential for central cells to detect the characteristics of a sensory stimulus" (Melzack and Wall, 1962). This does not hold for FAI receptors in the finger, since a single impulse in the axon of at least some of these receptors can produce a specific sensation. (2) "Every discriminably different somaesthetic perception is produced by a unique pattern of nerve impulses" (Melzack and Wall, 1962). However, changing the temporal pattern of impulses in an SAI unit does not change the quality of pressure sensation that is experienced, and stimulating an FAI unit or an SAI unit with the same pattern evokes quite different sensations. (3) "Perceived sensation is not determined in quality and location and time course by the presence of action potentials in uniquely specialized fibres" (Wall and McMahon, 1985). This does not hold for FAI and SAI units. (4) "Threshold of perceived sensation depends not on the presence or absence of impulses in a particular type of fibre but on their achieving certain levels of firing" (Wall and McMahon, 1985). This does not hold for FAI units in the fingers, but does apply to SAI units. (5) "The time course of the afferent barrage is poorly related to the time course of perception" (Wall and McMahon, 1985). This does not apply to FAI units, whose activity reproduces the time course of the sensation. (6) "Wide spatial gradients of stimulus and afferent activity are abstracted by the brain to achieve a perception of apparent elementary sensations with a spot location" (Wall and McMahon, 1985). This holds in part, since surround inhibition may sharpen receptive fields, but spatial summation is not needed for accurate localization of the fields of FAI and SAI units. However, spatial summation may be required for SA II receptors to evoke sensation.

Two "essential conclusions" reached by Torebjörk et al. (1987) are that "physiological specificity of low-threshold mechanoreceptors in the hand is linked to distinct attributes of very simple tactile sensations" and that "afferent inputs are processed differently in tactile subsystems."

Microneurography and Pain

Ochoa and Torebjörk (1989) performed an extensive investigation of the sensations associated with the activation of C polymodal nociceptors during microneurography. They used microstimulation in a peripheral nerve as a search stimulus. The sensory quality and localization of the subjective responses to the stimuli were then correlated with the ability to record from one or more C units at the same electrode position. They found that stimulation of C fibers supplying the glabrous skin of the hand more often provoked a sensation of dull pain than of burning pain, but that stimulation of C fibers from hairy skin always caused a burning pain sensation. By contrast, stimulation of group III and IV muscle nociceptors elicits cramping pain (Marchettini et al., 1996).

Commonly, the activity of more than one C fiber could be recorded from a stimulation site that evoked pain. Furthermore, the projected receptive fields for pain were often larger than the receptive fields of single C fibers, and the areas of the receptive fields grew larger as the stimulus intensity was increased. These observations indicate that several C fibers were probably activated simultaneously by the stimuli. Nevertheless, a pure sensation of dull or burning pain resulted (although, in some instances, stimulation of C fibers produced itch). It was suggested that the difference between the glabrous and hairy skin with respect to the sensations of dull and burning pain might relate to the number of C fibers that were simultaneously activated, since it appeared that fewer glabrous- than hairy-skin C fibers were stimulated at a given site. This would be consistent with the idea that burning pain might require spatial facilitation.

An important observation was that the quality of the pain was unchanged when the myelinated fibers in the nerve were blocked by pressure, indicating that the sensory quality depended strictly on the input rather than on modulation by activity in more rapidly conducting afferent fibers.

Signals from the C-fiber volleys were relatively well localized if the projected receptive fields were compared with those mapped. The correspondence was generally within 10 mm. This can be compared with the psychophysical finding that a hot stimulus applied to glabrous skin on the hand is localized within 7–10 mm (tactile stimuli are localized within 4–7 mm).

SOMATOSENSORY SYSTEM DUALITIES

Epicritic versus Protopathic Sensations

Brief consideration should be given to the proposal of Head (1920) that the cutaneous senses can be divided into two broad categories—protopathic and epicritic. The protopathic system, according to Head's view, mediates pain and the extremes of temperature sensation, whereas the epicritic system is responsible for touch, size, and two-point discrimination, and the detection of small thermal gradients. The experimental basis for the proposal was the introspective analysis of sensory changes in an area of skin after denervation and during regeneration of the cut nerve. Head himself was the experimental subject. Many of the observations and most of the interpretations of this experiment have been contested (Trotter and Davies, 1909; Boring, 1916; Walshe, 1942; Weddell et al., 1948), and the hypothesis should be laid to rest. However, perhaps because of the appeal of the terminology (cf. Walshe, 1942), the protopathic–epicritic dichotomy lingers. In fact, this terminology is sometimes applied to central nervous system pathways, such as the spinothalamic tract and the dorsal column–medial lemniscus pathway, even though Head clearly stated that the information conveyed by the two proposed systems in the periphery became intermingled centrally.

The notion of Head (1920) that somatic sensory functions can be described in terms of a dual system of epicritic and protopathic sensory mechanisms has been superceded by two other dualities: lemniscal versus non-lemniscal systems and large- versus small-fiber systems.

Lemniscal versus Non-lemniscal Systems

The lemniscal versus non-lemniscal nomenclature was developed by Poggio and Mountcastle (1960; 1963) when describing the responses of neurons in the thalamus to cutaneous stimulation. Lemniscal responses were defined as those with the following attributes: the receptive fields of the thalamic neurons were small and contralateral; the kinds of effective stimuli were restricted, indicating that only one or a few types of sensory receptors were involved in the activation of the thalamic neuron; and synaptic transmission was secure, so the neurons could follow relatively high rates of stimulation. Non-lemniscal neurons were defined by the following traits: the receptive fields of the thalamic neurons were large and often bilateral; the neurons received convergent input from different types of sensory receptors; and the cells were unable to follow repetitive stimulation accurately. The assumption was made that the lemniscal neurons of the ventrobasal complex of the thalamus were activated by the dorsal column–medial lemniscus pathway (or its trigeminal equivalent), whereas the non-lemniscal neurons, whose activity was recorded from other parts of the thalamus, such as the posterior nuclear complex, were excited by the spinothalamic tract (Poggio and Mountcastle, 1960). This nomenclature has been criticized (Boivie and Perl, 1975). Both the medial lemniscus and the spinothalamic tract project to both the ventrobasal complex and to the posterior complex. Furthermore, neurons of both the dorsal column–medial lemniscus system and the spinothalamic tract can be found that have properties that are consistent with each category of thalamic neuron (as do neurons in other somatosensory pathways that access the thalamus). Boivie and Perl (1975) suggest that the terms "lemniscal" and "non-lemniscal" be replaced by "specified" and "unspecified," although they warn that "unspecified" may simply reflect ignorance of the functional role of the neuron. Neither of these latter terms has gained general acceptance.

Large- versus Small-Fiber Systems

The other duality that is often cited in the literature on pain is the dichotomy between large-fiber versus small-fiber systems (Noordenbos, 1959; Melzack and Wall, 1965).

The implication of this terminology is that the large-fiber system has to do with innocuous forms of mechanoreception, whereas the small-fiber system is associated with pain. Interactions occur between inputs conveyed by the large- and small-fiber systems to central neurons, and the central nervous system determines from the outcome of these interactions whether or not a stimulus is painful. The difficulty with the nomenclature is that it tends to obscure the fact that some large (Aβ) afferent fibers supply nociceptors and many small (Aδ and C) afferent fibers in cutaneous nerves supply mechanoreceptors. Most nociceptors have small axons, but there are some that have axons that conduct at Aβ velocities (see Chapter 2, Table 2.1). In addition, many cutaneous mechanoreceptors have small axons. These include down hair afferents (supplied by Aδ fibers) and C mechanoreceptors (see Chapters 2 and 10; it should be noted that C mechanoreceptors have now been described in the skin of the human forearm, as well as in the face; Nordin, 1990; Vallbo et al., 1999). Thermoreceptors and itch receptors are left out of consideration by this terminology, but these are supplied by small afferents. Consequently, there is a danger when employing this terminology that the results of electrical stimulation of large fibers in cutaneous nerves can be falsely attributed to the activation of just mechanoreceptors and that the results of electrical stimulation of fine afferents can be falsely attributed to activation of just nociceptors. Furthermore, mixed somatic nerves contain muscle and joint, as well as cutaneous, afferents, and so more receptor types are represented in mixed nerves than in purely cutaneous nerves. Therefore, the interpretation of the results of stimulation of mixed nerves is even more problematic.

***

The first part of this monograph will be concerned with the organization of the peripheral nervous system and spinal cord. Then the sensory pathways in the spinal cord will be discussed, along with the kinds of information they carry. Portions of several of the sensory channels and their descending controls will be discussed in the last chapter.

CONCLUSIONS

1. Sensory modalities are classes of sensation. A number of cutaneous, deep tissue, and visceral sensory modalities can be recognized.
2. A sensory channel is the sensory mechanism responsible for conveying the information needed for recognition of a sensory modality. A sensory channel would include one or more sets of sensory receptors, one or more ascending pathways, certain regions of the thalamus and cerebral cortex, and also the descending pathways that can modify the ascending activity.
3. The specificity theory of cutaneous sensation was originally based on the discovery of localized sensory "spots" that respond specifically to tactile, cold, warm, or painful stimuli.
4. The proposed correlations between cold spots and Krause's end bulbs and between warm spots and Ruffini endings proved to be incorrect. Krause's end bulbs and Ruffini endings have been shown to be mechanoreceptors. Specific thermoreceptors appear to have free endings with specialized membranes, as do nociceptors.

5. Stimulation of a sensory spot is likely to activate a number of sensory receptors, although activation of a single afferent fiber can evoke behavioral or sensory events (see below).
6. A pattern theory of cutaneous sensation does not require a high degree of specificity of sensory receptors or of central nervous system pathways to elicit a particular sensation. It is also possible to develop theories of sensation that accept an admixture of specific and nonspecific elements. However, the weight of the evidence favors the specificity theory. This theory presumes that specific sensory receptors are responsible for activating sensory channels that are responsible for specific sensory modalities.
7. Lesions that affect sensation can do so not only by interruption of specific sensory pathways but also by reducing the total number of neurons conveying information to the brain, by altering the pattern of activity in ascending pathways, and by changing the operation of control systems that originate in the brain and that regulate activity in sensory pathways.
8. Evidence is being developed that should allow a description of the phenotypes of different specific primary afferent somatosensory neurons. There are at least three partially overlapping phenotypes, one for large and two for medium-sized and small dorsal root ganglion cells. These are based on such markers as antigens recognized by antibodies to neurofilament protein or surface carbohydrates, peptide content, expression of fluoride-resistant acid phosphatase, the presence of binding sites for the lectin, IB4, and of particular membrane receptors (vanilloid receptors, purinergic receptors, trk receptors).
9. Evidence from microneurography experiments in humans indicates that stimulation of the individual axons of particular types of sensory fibers can lead to specific sensory experiences.
10. Head's theory of the epicritic and protopathic divisions of the peripheral nervous system should be discarded.
11. It is preferable not to use the terms "lemniscal" and "non-lemniscal" to distinguish between activity evoked via the dorsal column–medial lemniscus pathway and that evoked via the spinothalamic tract. Better terms are needed to classify somatosensory neurons in the brain that have such contrasting response properties as small, contralateral receptive fields, restricted convergence of inputs, and secure synaptic coupling on the one hand and large, bilateral receptive fields, wide convergence, and weak synaptic coupling on the other hand.
12. The subdivision of afferent fibers into large- and small-fiber systems is misleading, since sensory receptors of a variety of types fall into each of these subdivisions. Small fibers cannot be equated with nociceptive afferents.

Chapter 2

Sensory Receptors and Peripheral Nerves

COMPOSITION OF PERIPHERAL NERVES

Peripheral nerves are composed of the axons of somatic and visceral sensory neurons and of somatic and autonomic motor neurons, along with the investing connective-tissue sheaths (epineurium, perineurium, and endoneurium; see Landon, 1976; Ross et al., 1995). The axons may be myelinated or unmyelinated (Fig. 2.1; Peters et al., 1976; Willis and Grossman, 1981). Since the emphasis will be on sensory mechanisms, the characteristics of motor axons will not be considered.

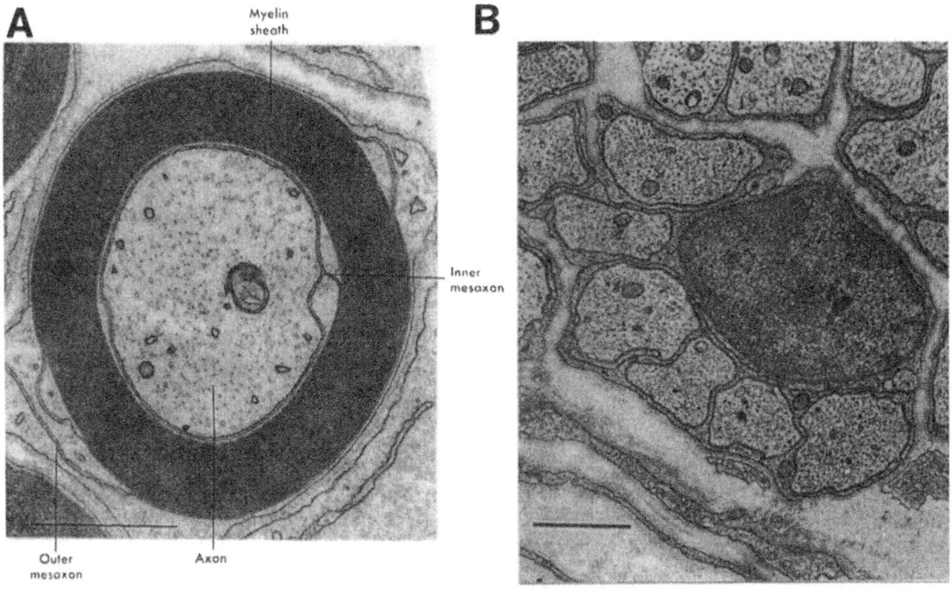

FIGURE 2.1. Myelinated and unmyelinated axons in peripheral nerve. A, cross-section of a large myelinated axon examined in an electron microscope. Cytoplasm of the Schwann cell that forms the myelin sheath is seen outside and inside the myelin. Outer and inner mesaxons are indicated. The axon proper is in the center. B, cross-section of a peripheral nerve that contains a large number of unmyelinated axons. In the center, a group of unmyelinated axons (bundle of Remak) is seen imbedded in the cytoplasm of a Schwann cell whose nucleus is the prominent dark structure. Parts of other bundles of Remak are nearby. Calibration bars = 1 μm. (Electron micrographs by Coggeshall, R. E.; from Willis and Grossman, 1981.)

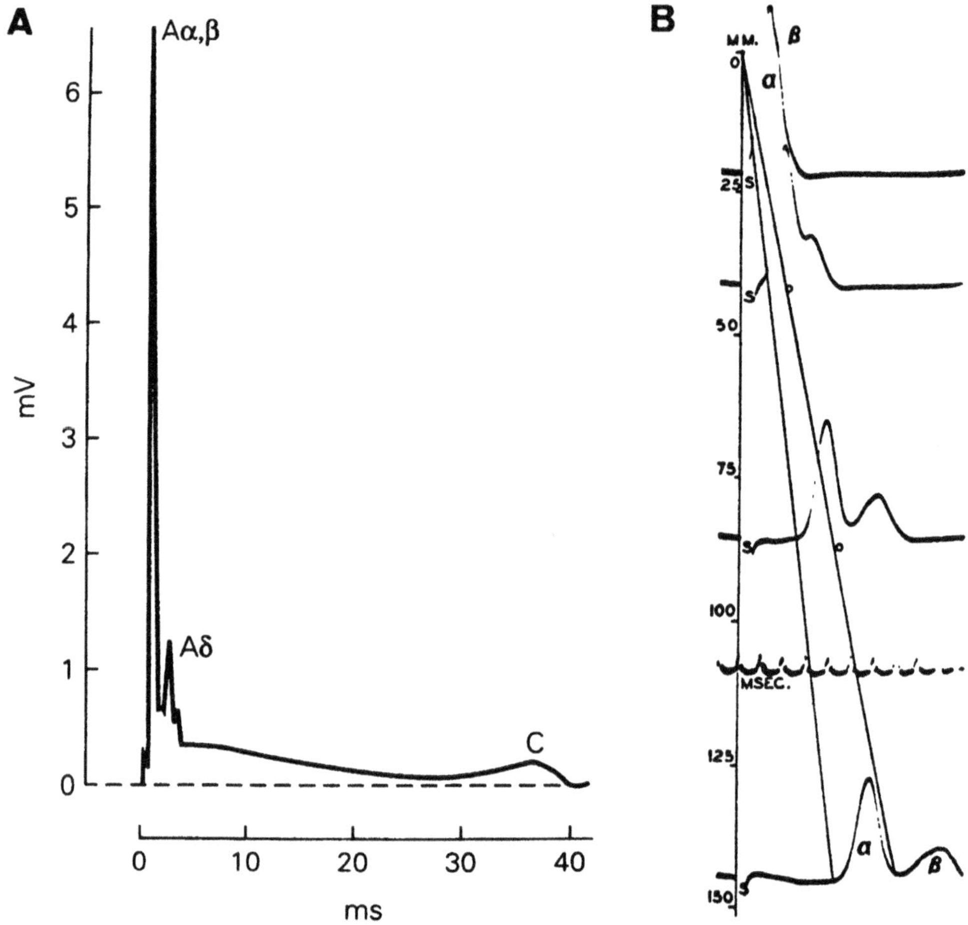

FIGURE 2.2. Conduction velocities of peripheral nerve fibers. A, compound action potential recorded from a peripheral nerve following an electrical stimulus that activated axons of all sizes. The abscissa gives the latency of the arrival of different components of the compound action potential at the recording electrode following the stimulus. Conduction velocity is calculated from the ratio of distance between the stimulating and recording electrodes and the latency (mm/ms = m/s). The earliest deflection is the summed action potential in the largest nerve fibers, the Aαβ group of axons. After a slight delay, the compound action potential in Aδ fibers appears. The compound action potential in C fibers is seen after a much greater delay. The C fiber potential is smaller than the A fiber potential, even though there are more C fibers than A fibers. The reasons for this are that the extracellular action potentials of individual unmyelinated axons are much smaller than those of myelinated axons and because the C-fiber action potentials are spread out more in time. (From Gasser, 1941.) B, Aαβ component of the compound action potential as it is conducted progressively farther along the nerve from the stimulating electrode. There is a progressive shift in the relative positions of the Aα and Aβ potentials because of the difference in the conduction velocities in these two populations of axons. Whereas these potentials completely overlap near the stimulating electrode (0 mm), by the time they conduct 150 mm, the potentials are almost completely separate. (From Erlanger and Gasser, 1937.) The same reasoning applies to the separation of the various components of the compound action potential seen in A, except that the separation of the slowest potentials can be detected nearer the stimulating electrode.

The sensory axons in cutaneous and visceral nerves are subdivided into three classes. The largest myelinated sensory axons belong to the Aαβ class, whereas the small myelinated fibers are in the Aδ group of axons (Fig. 2.2; Erlanger and Gasser, 1937; Gasser, 1941). In cats, Aαβ fibers conduct at 30–100 m/s and Aδ fibers at 4–30 m/s (Boivie and Perl, 1975).

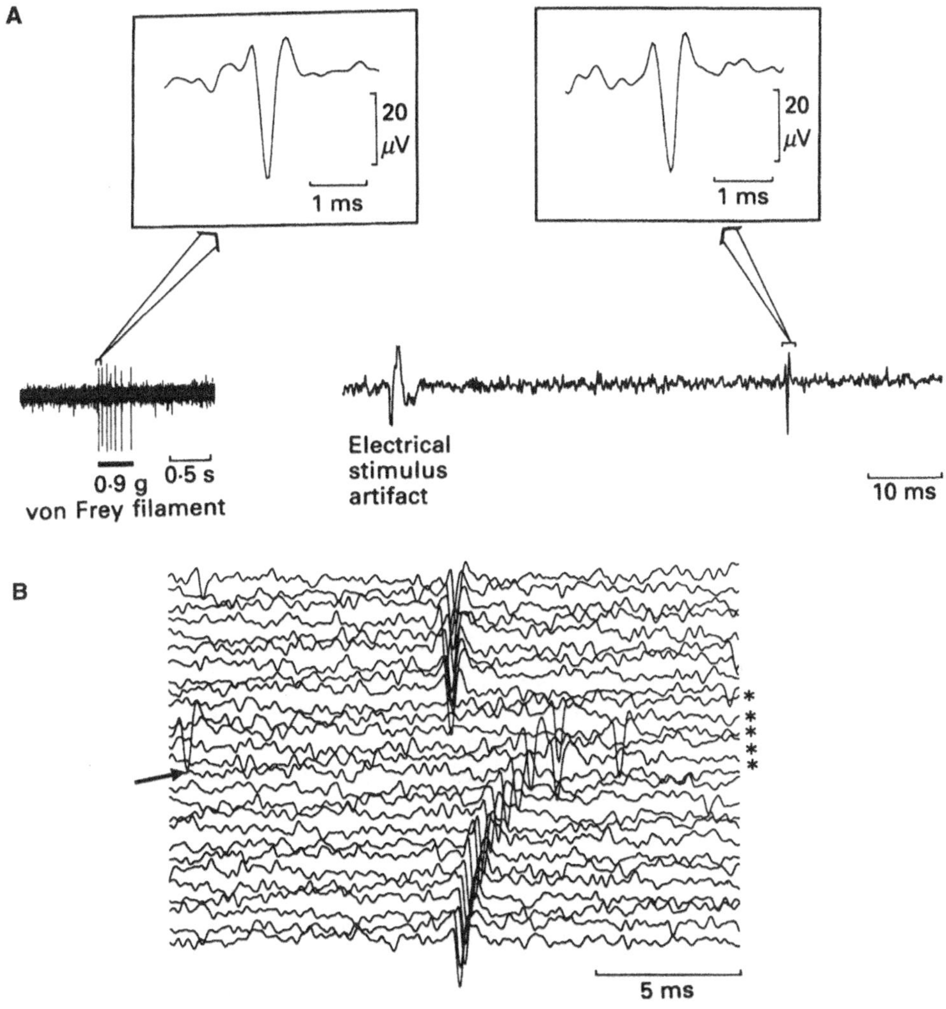

FIGURE 2.3. Slowed conduction velocity of a C fiber belonging to a C mechanoreceptor recorded from the human supraorbital nerve following evoked activity. A, action potentials of the unit after mechanical (left) or electrical stimulation of the receptive field. In the latter case, the latency of the spike was 53 ms, corresponding to a conduction velocity of 1.3 m/s. B, the unit was activated by an electrical stimulus repeated at a rate of 0.9/s. A separate oscilloscope sweep is shown for each response; successive sweeps are shifted progressively downward. During the sweeps indicated by the asterisks, a mechanical stimulus was applied to the receptive field of the unit. The arrow shows one of the action potentials that were evoked by the mechanical stimulus. Note the increase in latency of the action potential evoked by the electrical stimulus and its gradual recovery. (From Nordin, 1990.)

Unmyelinated fibers are often designated as C fibers; they conduct at less than 2.5 m/s (Gasser, 1950). Of the total number of sensory axons in the cat sural nerve, a cutaneous nerve, 61% are unmyelinated (Langford, 1983).

A different terminology is used for muscle and joint nerves. The myelinated afferent fibers of muscle nerves are subdivided into groups I, II, and III (Lloyd and Chang, 1948; Rexed and Therman, 1948). In cats, these groups of axons conduct at 72–120, 24–71, and 6–23 m/s, respectively (Hunt, C. C., 1954). Muscle nerves also contain numerous unmyelinated, or group IV, afferent fibers (Stacey, 1969) that conduct at less than 2.5 m/s. Unmyelinated sensory axons account for 58% of the total axons in the medial gastrocnemius nerve, a muscle nerve (Langford, 1983).

Joint nerves resemble muscle nerves, except that they have only a few group I fibers, which appear to arise from Golgi tendon organs in tendons around the joint or from muscle spindles of nearby muscles (Gardner, 1944; 1948; Skoglund, 1956; Andersen, H. T. et al., 1967; Burgess and Clark, 1969; Gregory et al., 1989; Hines et al., 1996). The sensory axons in joint nerves belong mainly to groups II, III, and IV. Nearly half (45–49%) of cat knee joint afferents are group IV (unmyelinated) axons (Langford, 1983; Langford and Schmidt, 1983). Of the nerve fibers in branches of the nerve that innervates the posterior part of the knee joint capsule in humans, 70–78% are unmyelinated (Hines et al., 1996). However, this figure would include sympathetic postganglionic axons.

For at least the large myelinated fibers in cats, it is possible to predict the conduction velocity in meters per second by multiplying the axon diameter (in micrometers), including the myelin sheath, by a factor of 6 (Hursh, 1939) or 5.7 (Arbuthnott et al., 1975). A factor of 4.5 is more accurate for group II and III myelinated fibers (Arbuthnott et al., 1975). Conduction velocities vary with species. For example, the conduction velocities of axons of a particular functional category in the peripheral nerves of mice are considerably slower than are those of comparable fibers in cats (Cain et al., 2001).

The conduction velocity of a C fiber decreases markedly during repetitive firing, because nerve impulse transmission is considerably slowed during the prolonged relative refractory periods of these small axons (Nordin, 1990; Raymond et al., 1990; Torebjörk and Ochoa, 1990; Schmelz et al., 1995). In microneurography studies (see Chapters 1 and 12), this slowing of conduction velocity can be used to "mark" the action potentials of a particular axon following activation of the axon by stimulation of the receptive field to test the conduction delay (see Fig. 2.3).

SENSORY RECEPTORS

The structural and functional characteristics of somatic and visceral sensory receptors have been reviewed elsewhere (Matthews, 1964; 1972; 1981; Neil, 1972; Burgess and Perl, 1973; Hensel, 1973; 1974; Boivie and Perl, 1975; Price and Dubner, 1977; Burgess et al., 1982; Darian-Smith, 1984; Willis, 1985; Halata, 1993; Cervero, 1994). However, it is necessary to provide a brief description of them here as background for later discussion of the effects evoked by the different receptors in central neurons and proposals about their contributions to sensation.

The somatovisceral sensory receptors can be grouped by their location in the skin, skeletal muscles, joints, or viscera. The receptor types will be discussed in this same order. A summary of the main types of somatic and visceral sensory receptors and their response properties in a variety of mammals is given in Tables 2.1, 2.2, and 2.3. In addition, several recent reports have provided information about cutaneous receptors in mice that will be useful in studies employing transgenic animals (Koltzenburg et al., 1997; Cain et al., 2001). In general, the receptor types that were found in mouse skin are quite similar to those in other mammals. However, as mentioned, the conduction velocities of the myelinated afferent fiber groups in mice are slower than in larger animals, such as the cat. Some receptor types, such as Pacinian corpuscles and field receptors were not found in the

TABLE 2.1. Characteristics of Cutaneous Mechanoreceptors.

Receptor type		
Cutaneous displacement and velocity detectors		
SAI (Merkel cell)		
Adequate stimulus	Indentation of tactile dome	
Signals	Displacement of skin, velocity of displacement	
Background activity	None	
Conduction velocity (m/s)		
Rat	38.9 ± 5.6	Leem et al. (1993)
Cat	57.2 ± 0.99 (33–95)	Brown, A. G. and Iggo (1967)
	65 (47–84)	Burgess et al. (1968)
	63.6 ± 2.3	Tapper et al. (1973)
Rabbit	47.3 ± 1.77 (16–96)	Brown, A. G. and Iggo (1967)
Monkey	46 ± 4; 51 ± 2	Perl (1968)
Human	58.7 ± 2.3	Knibestöl (1975)
SAII (Ruffini)		
Adequate stimulus	Skin indentation, skin stretch	
Signals	Displacement of skin, velocity of displacement	
Background activity	Sometimes	
Conduction velocity (m/s)		
Rat	42.78 ± 4.85	Leem, et al. (1993)
Cat	53.6 ± 2.21 (20–100)	Brown, A. G. and Iggo (1967)
	54 (39–68)	Burgess et al. (1968)
Rabbit	31 ± 2.4 (24–45)	Brown, A. G. and Iggo (1967)
Monkey	34 ± 12; 40 ± 10	Perl (1968)
Human	45.3 ± 3.6	Knibestöl (1975)
Cutaneous velocity detectors		
FAI (RA, Krause's end bulb, Meissner corpuscle)		
Adequate stimulus	Skin movement	
Signals	Velocity	
Background activity	None	
Conduction velocity (m/s)		
Rat	45.37 ± 5.58	Leem et al. (1993)
Cat	>60 (50%)	Jänig et al. (1968a)
	36.1 ± 6.2	Iggo and Ogawa (1977)
Monkey	(40–80)	Talbot, W. H. et al. (1968)
Human	55.3 ± 3.4 (26–91)	Knibestöl (1973)
G_2 (and T) hair follicle		
Adequate stimulus	G2 hair or skin movement	
Signals	Velocity	
Background activity	None	
Conduction velocity (m/s)		
Cat	68 ± 2.72 (44–72)	Brown, A. G. and Iggo (1967)
	53 (39–73)	Burgess et al. (1968)
Rabbit	35.6 ± 1.7 (8–53)	Brown, A. G. and Iggo (1967)
Monkey	36 ± 14; 42 ± 16	Perl (1968)

(*Continued*)

TABLE 2.1. Continued

Receptor type		
D hair follicle		
Adequate stimulus	D hair or skin movement	
Signals	Velocity (very low threshold)	
Background activity	None (unless hair is moving)	
Conduction velocity (m/s)		
Cat	17.9 ± 0.23 (15–24)	Brown, A. G. and Iggo (1967)
	18.3 (11–30)	Burgess and Perl (1967)
	21 (15–32)	Burgess et al. (1968)
Rabbit	9 ± 0.2 (5–16)	Brown, A. G. and Iggo (1967)
Monkey	16 ± 6; 19 ± 6	Perl (1968)
Field		
Adequate stimulus	Skin movement	
Signals	Velocity; also, to some extent, displacement	
Background activity	None	
Conduction velocity (m/s)		
Cat	55 (36–72)	Burgess et al. (1968)
Monkey	39 ± 13; 39 ± 8	Perl (1968)
C mechanoreceptor		
Adequate stimulus	Skin movement	
Signals	Velocity (slow); also, to some extent, displacement	
Background activity	None	
Conduction velocity (m/s)		
Rat	0.74 ± 0.12	Leem et al. (1993)
Cat	(0.55–1.25)	Iggo (1960)
	(0.5–1.0)	Bessou et al. (1971)
	(0.5–2.0)	
Human	(0.6–1.3)	Nordin (1990)
Cutaneous transient detectors		
FAII (Pacinian corpuscle)		
Adequate stimulus	Vibration	
Signals	Transients	
Background activity	None	
Conduction velocity (m/s)		
Rat	46.64 ± 8.03	Leem et al. (1993)
Cat	(54–84)	Hunt, C. C. and McIntyre (1960b)
	65 (54–82)	Burgess et al. (1968)
	>57 (50%)	Jänig et al. (1968a)
Monkey	41 ± 8; 56 ± 11	Perl (1968)
Human	46.9 ± 3.6 (34–61)	Knibestöl (1973)
G_1 hair follicle		
Adequate stimulus	Rapid movement of G_1 hair or of skin	
Signals	Transients	
Background activity	None	
Conduction velocity (m/s)		
Cat	75 (56–85)	Burgess et al. (1968)
Monkey	47 ± 12; 49 ± 13	Perl (1968)

TABLE 2.2. Characteristics of Cutaneous Nociceptors and Thermoreceptors.

Receptor type		
Nociceptors		
Aδ mechanical (A–M) nociceptor		
Adequate stimulus	Mechanical damage to skin	
Signals	Skin damage or threat of skin damage	
Background activity	None	
Conduction velocity (m/s)		
Rat	14.17 ± 5.86	Leem et al. (1993)
Cat	27 (5.5–49)	Fitzgerald and Lynn (1977)
	(6–51)	Burgess and Perl (1967)
	(6–65)	Burgess et al. (1968)
Rabbit	15 (5–32.5)	Fitzgerald and Lynn (1977)
Monkey	25 ± 11; 21 ± 13	Perl (1968)
	(4–44)	Georgopoulos (1976)
Aδ mechanoheat (A–MH; A polymodal) nociceptor		
Adequate stimulus	Noxious mechanical, heat and often chemical stimuli	
Signals	Skin damage or threat of skin damage; Type II: first pain	
Background activity	None	
Conduction velocity (m/s)		
Rat	(6.2–12.9)	Szolcsanyi et al. (1988)
	(1.7–14.9)	Kirchhoff et al. (1990)
Monkey	(3.9–6.8)	Iggo and Ogawa (1971)
	(4–40)	Georgopoulos (1976)
Type I	25.4 ± 15.7	Treede et al. (1998)
Type II	14.2 ± 5.2	Treede et al. (1998)
Aδ mechanically insensitive (A–MIA) nociceptor		
Adequate stimulus	Heat, chemical stimuli or none of those tested	
Signals	Skin damage or threat of skin damage; first pain	
Background activity		
Conduction velocity (m/s)		
Monkey	13.3 ± 0.8	Meyer et al. (1991)
Aδ cold or mechano-heat-cold and mechano-cold nociceptor		
Adequate stimulus	Severe cold	
Signals	Skin damage to threat of skin damage	
Background activity	None	
Conduction velocity (m/s)		
Rat AMC	14.5 ± 1.25	Simone and Kajander (1997)
AMHC	11.3 ± 2.56 and 11.1 ± 4.8	Simone and Kajander (1997)
Monkey	(4–40)	Georgopoulos (1976)

(*Continued*)

TABLE 2.2. Continued

Receptor type		
C mechanoheat (C-MH; polymodal) nociceptor		
Adequate stimulus	Noxious mechanical, heat, chemical, and sometimes cold stimuli	
Signals	Skin damage or threat of skin damage	
Background activity	None	
Conduction velocity (m/s)		
Rat	0.77 ± 0.14	Leem et al. (1993)
	0.82 ± 0.03	White et al. (1990)
	(0.22–0.7)	Steen et al. (1995)
Cat	(0.4–1.1)	Bessou and Perl (1969)
	(0.4–1.8)	Beck et al. (1974)
Monkey	(0.6–1.1)	Croze et al. (1976)
Human	0.89 (0.66–1.1)	Van Hees and Gybels (1972)
	0.97	Schmidt et al. (1995)
	(0.79–1.27)	Weidner et al. (1999)
C mechanical (C–M) nociceptor		
Adequate stimulus	Mechanical damage to skin	
Signals	Skin damage or threat of skin damage	
Background activity	None	
Conduction velocity (m/s)		
Rat	(0.47–0.66)	Steen et al. (1995)
Cat	(0.6–1.4)	Bessou and Perl (1969)
Monkey	(0.8–1 mostly)	Georgopoulos (1976)
Human	0.84	Schmidt et al. (1995)
C mechanically insensitive (C–MIA; C–H and CM_iH_i) nociceptor		
Adequate stimulus	None known	
Signals	Inflammation?	
Background activity		
Conduction velocity (m/s)		
Monkey	0.7 ± 0.06	Davis et al. (1993)
Human	0.8	Schmidt et al. (1995)
C–H	(0.59–1.19)	
CM_iH_i	(0.49–1.24)	Weidner et al. (1999)
C cold or mechano-heat-cold (C–MHC) nociceptor		
Adequate stimulus	Severe cold	
Signals	Skin damage or threat of skin damage	
Background activity	None	
Conduction velocity (m/s)		
Rat	<2	Simone and Kajander (1996)
Monkey	(0.8–1)	Georgopoulos (1976)

(*Continued*)

TABLE 2.2. Continued

Receptor type		
Itch receptors		
C chemoreceptors		
Adequate stimulus	Itch-provoking chemicals, such as histamine; C–H or CM_i H_i units	
Signals	Itch	
Background activity	None	
Conduction velocity (m/s)		
Human	0.52 ± 0.05	Schmelz et al. (1997)
Thermoreceptors		
Cold receptor		
Adequate stimulus	Reduced skin temperature	
Signals	Skin cooling	
Background activity	At ambient temperatures	
Conduction velocity (m/s)		
Rat	(0.5–1.0)	Iriuchijima and Zotterman (1960)
	(<1.5)	Iggo (1969)
	1.74 ± 3.59	Leem et al. (1993)
Cat	(0.6–1.5)	Hensel et al. (1960)
Dog	(9–18)	Iggo (1969)
Monkey	8 (4–14)	Perl (1968)
	(<1.5–15.3)	Iggo (1969)
	0.7 ± 0.3 (0.3–1.3) and 5.8 ± 2.5 (2.2–9.5)	Hensel and Iggo (1971)
	14.5 (5–31)	Darian-Smith et al. (1973)
	13.3 ± 2.8 (glabrous)	
	7.0 ± 2.8 (hairy)	
	10.8 ± 3.1 (junction)	Long, R. R. (1977)
Warm receptor		
Adequate stimulus	Increased temperature	
Signals	Skin warming	
Background activity	At ambient temperatures	
Conduction velocity (m/s)		
Rat	(0.7–1.5)	Iriuchijima and Zotterman (1960)
Cat	(0.84 for 1 fiber)	Hensel et al. (1960)
Monkey	0.7 ± 0.2 (0.4–0.9)	Hensel and Iggo (1971)
	(0.8 ± 0.09)	Duclaux and Kenshalo (1980)
Human	(0.5–0.75)	Konietzny and Hensel (1975)

TABLE 2.3. Characteristics of Muscle, Joint, and Visceral Receptors.

Receptor type		
Muscle mechanoreceptors		
Muscle spindle primary ending		
Adequate stimulus	Stretch of muscle	
Signals	Length and rate of change of length of muscle	
Background activity	Depends on limb position	
Conduction velocity (m/s)		
Cat	(72–120)	Hunt, C. C. (1954)
Muscle spindle secondary ending		
Adequate stimulus	Stretch of muscle	
Signals	Length of muscle	
Background activity	Depends on limb position	
Conduction velocity (m/s)		
Cat	(24–72)	Hunt, C. C. (1954)
Golgi tendon organ		
Adequate stimulus	Contraction or stretch of muscle	
Signals	Tension	
Background activity	Often, depending on limb position	
Conduction velocity (m/s)		
Cat	(72–120)	Hunt, C. C. (1954)
Pacinian corpuscle (see Table 2.1 on cutaneous receptors)		
Muscle nociceptors		
Group III muscle afferents		
Adequate stimulus	Pressure; damage to muscle (some may be ergoreceptors)	
Signals	Pressure; damage or threat of damage to muscle	
Background activity	None or slight	
Conduction velocity (m/s)		
Cat	(6–91), mostly (10–15)	Paintal (1960)
Human	6.7 ± 1.38	Simone et al. (1994)
Group IV muscle afferents		
Adequate stimulus	Pressure; damage to muscle	
Signals	Damage or threat of damage to muscle	
Background activity	Little	
Conduction velocity (m/s)		
Rat	(0.4–2.2)	Diehl et al. (1993)
Cat	<2.5	Mense and Schmidt (1974)
	<2.5	Franz and Mense (1975)
Human	0.85 ± 0.19	Simone et al. (1994)
Joint mechanoreceptors		
Ruffini ending		
Adequate stimulus	Flexion or extension to extremes	
Signals	Joint pressure	
Background activity	Only at extreme joint positions	
Conduction velocity (m/s)		
Cat	(20–70)	Burgess and Clark (1969)
	(9–77)	Clark, F. J. (1975)
Golgi tendon organ (see above under muscle mechanoreceptors)		

(*Continued*)

TABLE 2.3. Continued

Receptor type		
Paciniform endings (see above under cutaneous mechanoreceptors)		
Joint nociceptors		
Group III		
Adequate stimulus	Extreme bending of joint; probing joint capsule	
Signals	Damage or threat of damage to joint	
Background activity	Little or none	
Conduction velocity (m/s)		
Cat	<30	Clark, F. J. and Burgess (1975)
	(2.5–20)	Schaible and Schmidt (1983a, b)
Group IV		
Adequate stimulus	Extreme bending of joint; probing joint capsule	
Signals	Damage or threat of damage to joint	
Background activity	Little or none	
Conduction velocity (m/s)		
Cat	(0.3–1.3)	Schaible and Schmidt (1983a, b)
Visceral mechanoreceptors		
Esophageal mechanoreceptors		
Adequate stimulus	Distension, peristaltic contractions, longitudinal stretch	
Signals	Damage	
Background activity	Little or none	
Conduction velocity (m/s)		
Opposum	(2.7–7.8)	Sengupta et al. (1990)
	(0.32–2.5)	Sengupta et al. (1990)
Intestinal mechanoreceptors		
Adequate stimulus	Distension; tension on mesentery or blood vessels	
Signals	Movements	
Background activity	May have respiratory, cardio-vascular, or gastrointestinal rhythm	
Conduction velocity (m/s)		
Cat	(2–21)	Bessou and Perl (1966)
	(0.6–5)	Ranieri et al. (1973)
	Aδ or <2.5	Clifton et al. (1976)
	(0.6–30)	Morrison (1977)
Dog	(0.5–36)	Morrison (1977)
Mesenteric pacinian corpuscles (see above under cutaneous mechanoreceptors)		
Adequate stimulus	Movements of intestine	
Signals	Vibration; may have cardiac rhythm	
Background activity	Can have some background activity	

(*Continued*)

TABLE 2.3. Continued

Receptor type		
Conduction velocity (m/s)		
Cat	(61–70)	Ranieri et al. (1973)
Anal mucosa mechanoreceptors		
Adequate stimulus	Movement across mucosa	
Signals	Possibly distinguish between feces and flatus	
Background activity	None	
Conduction velocity (m/s)		
Rat	Myelinated axons	Sengupta and Gebhart (1994)
Reproductive tract mechanoreceptors		
Adequate stimulus	Mechanical stimulation, distension, chemical stimulation	
Signals	Reproductive activity	
Background activity	Little or none	
Conduction velocity (m/s)		
Rat vagina, cervix, uterus	(0.36–1.7)	Berkeley et al. (1990)
Urinary bladder mechanoreceptors		
Adequate stimulus	Bladder distension or contraction	
Signals	Tension	
Background activity	Can have some background activity	
Conduction velocity (m/s)		
Cat	(<2–22)	Winter (1971)
	Aδ	Clifton et al. (1976)
Ureter mechanoreceptors		
Adequate stimulus	Ureter distension or contraction	
Signals	Peristalsis	
Background activity	None	
Conduction velocity (m/s)		
Guinea-pig	0.6 ± 0.02	Cervero and Sann (1989)
Visceral nociceptors		
Cardiac nociceptors		
Adequate stimulus	Myocardial ischemia, chemical stimuli (bradykinin, prostaglandins)	
Signals	Angina	
Background activity	Can have some background activity	
Conduction velocity (m/s)		
Cat	<2.5	Tjen-A-Looi et al. (1998)
	(2.5–30)	Tjen-A-Looi et al. (1998)
Gastrointestinal nociceptors		
Adequate stimulus	Intense mechanical, thermal or chemical stimuli	

(*Continued*)

TABLE 2.3. Continued

Receptor type		
Signals	Damage or threat of damage to viscera; some are high threshold, others respond to innocuous and noxious intensities of stimulation; many respond to or are sensitized by chemical stimuli (bradykinin, prostaglandins, histamine, adenosine, serotonin, acetylcholine)	
Background activity	Little or none	
Conduction velocity (m/s)		
Rat jejunum	Multiunit recordings	Brunsden and Grundy (1999)
Rat ileum	0.7 ± 0.03	Cervero and Sharkey (1988)
Rat colon	1.52 ± 0.11	Sengupta and Gebhart (1994)
	3.56 ± 1.58	Sengupta and Gebhart (1994)
Opossum esophagus	<2.5	Sengupta et al. (1990)
	(2.5–7.8)	Sengupta et al. (1990)
Cat anal canal	<2.5	Clifton et al. (1976)
Reproductive tract nociceptors		
Adequate stimulus	Intense mechanical or chemical stimuli	
Signals	Reproductive function, labor pain	
Background activity	Little or none	
Conduction velocity (m/s)		
Rat vagina, cervix, uterus	(0.36–1.7)	Berkley et al. (1990)
Ureter nociceptors		
Adequate stimulus	Intense compression or distension of ureter	
Signals	Ureteric pressure	
Background activity	Some afferents discharge spontaneously, but these may be sensitized	
Conduction velocity (m/s)		
Guinea-pig	0.37 ± 0.21	Cervero and Sann (1989)

mouse skin-nerve preparation because of the way the preparation is made (Koltzenburg et al., 1997).

CUTANEOUS RECEPTORS

Receptors in the skin (and the adjacent subcutaneous connective tissue) include mechanoreceptors, nociceptors, thermoreceptors, and chemoreceptors. Some of the cutaneous sense organs are encapsulated, whereas others are unencapsulated (Chouchkov, 1978; cf. Halata, 1975). Encapsulated receptors in the glabrous skin include Meissner corpuscles, Ruffini endings, and Pacinian corpuscles. These are located in the dermis and, in the case of Pacinian corpuscles, also in subcutaneous tissue. In hairy skin, the hair follicle receptors replace the Meissner corpuscles. Unencapsulated receptors in glabrous and hairy skin include Merkel cell endings and free nerve endings that terminate either in the epidermis (epidermal nerve endings) or in the dermis (penicillate and papillary dermal nerve endings). Penicillate endings are widely distributed in the dermis, and the papillary endings are found around the openings of hair follicles in the dermis.

Cutaneous Mechanoreceptors

Weak mechanical forces applied to the skin provide the adequate stimuli for cutaneous mechanoreceptors (Table 2.1). Some are also excited by cold stimuli (Hunt, C. C. and McIntyre, 1960b; Iggo, 1963; Simone and Kajander, 1997), but the thermal responses are considered too weak to be significant in thermoreception (see Chapter 12; Werner and Mountcastle, 1965; Duclaux and Kenshalo, 1972; Johnson, K. O. et al., 1973; however, cf. Burton et al., 1972), and they do not encode cold intensity (Simone and Kajander, 1997). Noxious mechanical stimuli produce no greater excitation of mechanoreceptors than do innocuous ones (Burgess and Perl, 1973).

Cutaneous mechanoreceptors can be classified in several ways. An important parameter is the adaptation rate to a maintained stimulus (Adrian and Zotterman, 1926a, b; Adrian, 1928). Slowly adapting receptors tend to continue to discharge repetitively so long as a stimulus is maintained (Fig. 2.4A and B, top pairs of traces). On the other hand, rapidly adapting receptors discharge only when a stimulus is applied (and sometimes again when the stimulus is removed; Fig. 2.4A and B, second pairs of traces). Nerve fibers generally discharge only once when stimulated mechanically (Fig. 2.4A, fifth trace), unless they are damaged.

Another approach to the classification of mechanoreceptors is based on the most likely property of the stimulus for which they encode information (Burgess and Perl, 1973; see also Looft, 1996a, b). For slowly adapting cutaneous mechanoreceptors, this may be the amount of indentation of the skin. The discharge rate of such receptors would then be a function of skin displacement (position; Fig. 2.5A). Rapidly adapting receptors signal the rate at which the skin is displaced (stimulus velocity; Fig. 2.5B, top response) or some higher derivative of skin position (acceleration, jerk; Fig. 2.5B, bottom response). Thus, the number of discharges that occur when such receptors are activated is a function of stimulus velocity, acceleration, or jerk. Receptors that signal velocity can be termed velocity detectors, and those that signal acceleration or jerk can be grouped as transient detectors. Often a given receptor will behave as if it signaled two properties of the stimulus. For instance, the slowly adapting (or static) response of a displacement detector might be preceded by a velocity-sensitive (dynamic) response. Such receptors would signal both displacement and the velocity of displacement.

A key that is useful for the classification of cutaneous mechanoreceptors in mammals has been published (Horch et al., 1977; reprinted in Willis and Coggeshall, 1991). Whitehorn et al. (1974) have estimated the proportions of different classes of myelinated sensory fibers that supply the part of the hairy skin that is innervated by a branch of the sural nerve in cats.

Mechanoreceptors in the Glabrous Skin of the Primate Hand

The hand of primates serves as a sense organ that rivals the eye in importance (Darian-Smith, 1984). The glabrous surface of the hand in humans, monkeys, and apes is well

A

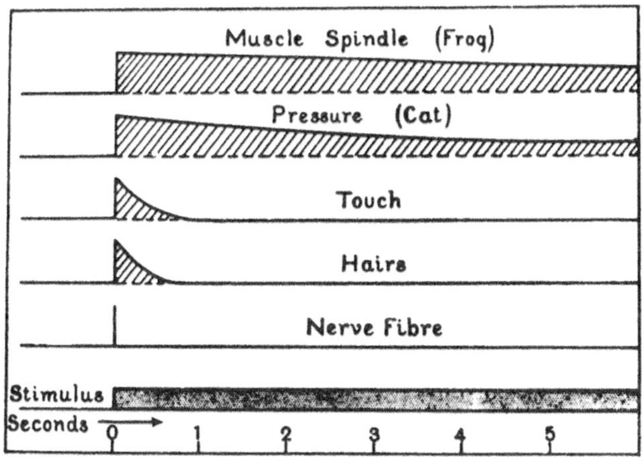

B

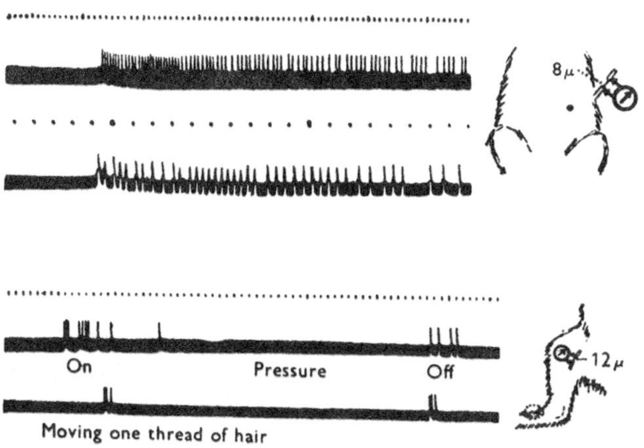

FIGURE 2.4. A, diagram of the durations of the responses of slowly and rapidly adapting receptors in relation to a long-lasting stimulus. The responses from slowly adapting receptors include that recorded from a muscle spindle afferent in a frog and that from a pressure afferent in a cat. Rapidly adapting responses are from touch and hair follicle afferents. The nerve fiber discharges just once at the start of a stimulus applied directly to the fiber. (From Adrian, 1928, with permission.) B, discharges of an afferent fiber from a slowly adapting tactile receptor (top traces, which were recorded with different time bases) and of an afferent fiber from a rapidly adapting hair follicle receptor (bottom traces). Time marks indicate 10 ms. The stimulus was maintained for the duration of the discharges. (From Maruhashi et al., 1952.)

adapted for the analysis of the size, shape, weight, and texture of objects. Sensory discrimination is especially good when there is movement of the hand in relation to an object with which it is in contact. "Active" touch is performed when the hand moves over an object, and "passive" touch occurs when an object is moved over the surface of the hand (Gordon, 1978). Discrimination is equally good for active and passive touch. In contrast to primates, many non-primate mammals utilize the mystacial vibrissae, snout region, and mouth for tactile exploration, rather than the forepaws.

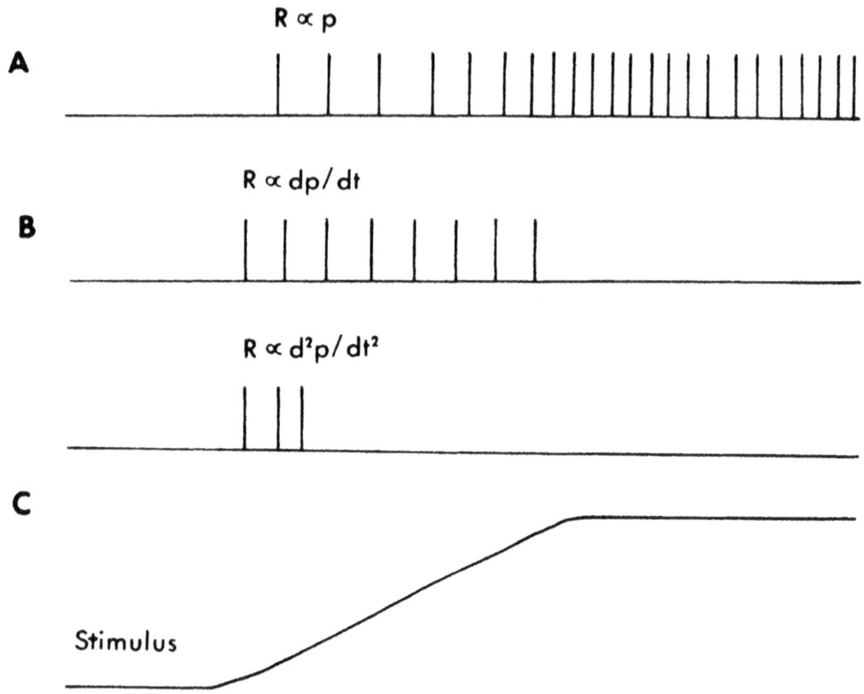

FIGURE 2.5. Responses (R) of different types of cutaneous mechanoreceptors to a ramp displacement of the skin. A, receptor whose discharge signals skin displacement or position (p). B, response in the upper trace is in proportion to the first derivative of position (velocity). That in the lower trace is proportional to the second derivative of position (acceleration). C, ramp and hold stimulus. (From Willis and Grossman, 1981.)

The tactile discriminative capability of the primate hand depends in large part on the presence of four types of mechanoreceptors: these are the Merkel disks (SA I receptors), Ruffini endings (SA II receptors), Meissner corpuscles (FA I receptors) and Pacinian corpuscles (FA II receptors). These receptor types and their location in primate glabrous skin are illustrated in Fig. 2.6. During active touch, there is also an important sensory contribution from receptors located in the joints and muscles of the hand (Darian-Smith, 1984).

Freeman, A. W. and Johnson (1982a, b) have developed a mathematical model that accounts for the responses to vibratory stimuli of slowly adapting, rapidly adapting, and Pacinian corpuscle receptors in monkeys.

Cutaneous Displacement and Velocity Detectors

Two kinds of slowly adapting cutaneous mechanoreceptors have been described, the Merkel (SA I) and Ruffini (SA II) endings (Table 2.1). Each signals both displacement and velocity.

SA I Receptors

SA I mechanoreceptors are associated with Merkel cell complexes (Fig. 2.7A). Merkel cells are specialized cells in the basal epidermis adjacent to the basement membrane. About 5–10 Merkel cells and the region of epidermis that contains them form a "tactile dome." Tactile domes are innervated by large myelinated fibers that can branch to supply as many as seven

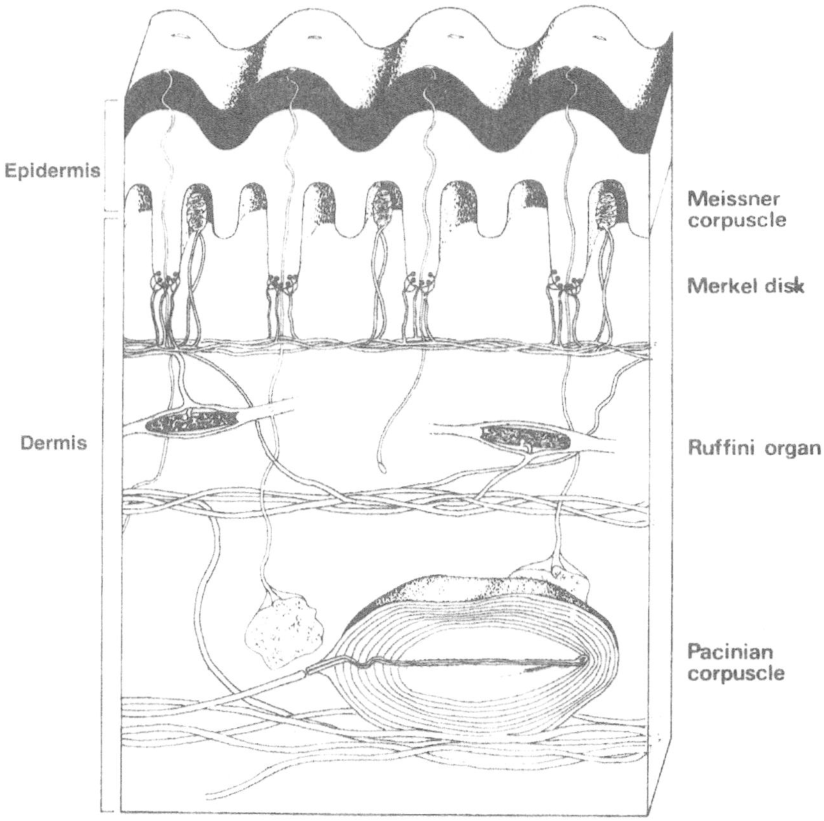

FIGURE 2.6. Drawing showing the arrangement of tactile sensory receptors in the glabrous skin of the primate hand. The epidermis of the finger pad has a series of surface ridges along which sweat glands open. Beneath the epidermis are a series of dermal papillae that protrude toward the surface of the skin between different types of epidermal ridges. Meissner corpuscles are found in many of the dermal papillae. Merkel disks are in the basal epidermis of the ridges through which the ducts of sweat glands pass. Ruffini endings are found in the dermis, and cutaneous Pacinian corpuscles are mainly in the deep dermis and subcutaneous tissue. (From Darian-Smith, 1984.)

separate domes, although more commonly there are only two or three branches and domes per fiber. Two or more axons supply a given Merkel cell complex (Paré et al., 2002).

The tactile domes are easily recognized in the skin of some animals, including cats, rabbits, mice, rats, and guinea-pigs. Demarcation of tactile domes in the glabrous skin of primates (including humans) is difficult. However, according to Darian-Smith (1984), tactile domes can be recognized in the hairy skin of humans if the skin is viewed with a dissecting microscope. The domes are slightly elevated, they have some pigmentation, and there are sometimes blood vessels in the central region of the dome. In non-primate animals, tactile domes are often associated with a tylotrich hair follicle (see below; Smith, K. R., 1968). These morphological features give rise to alternative names for this receptor type: Merkel cell ending, tactile dome, touch corpuscle, and Haarscheibe. (For details about the structural features of SA I receptors, see Merkel, 1875; Iggo, 1963; Pinkus, 1964; Munger, 1965; Smith, K. R., 1968; 1970; Iggo and Muir, 1969; Straile, 1969; Jänig, 1971a, b; Halata, 1975; English, 1977; Rice and Rasmusson, 2000; Paré et al., 2002.)

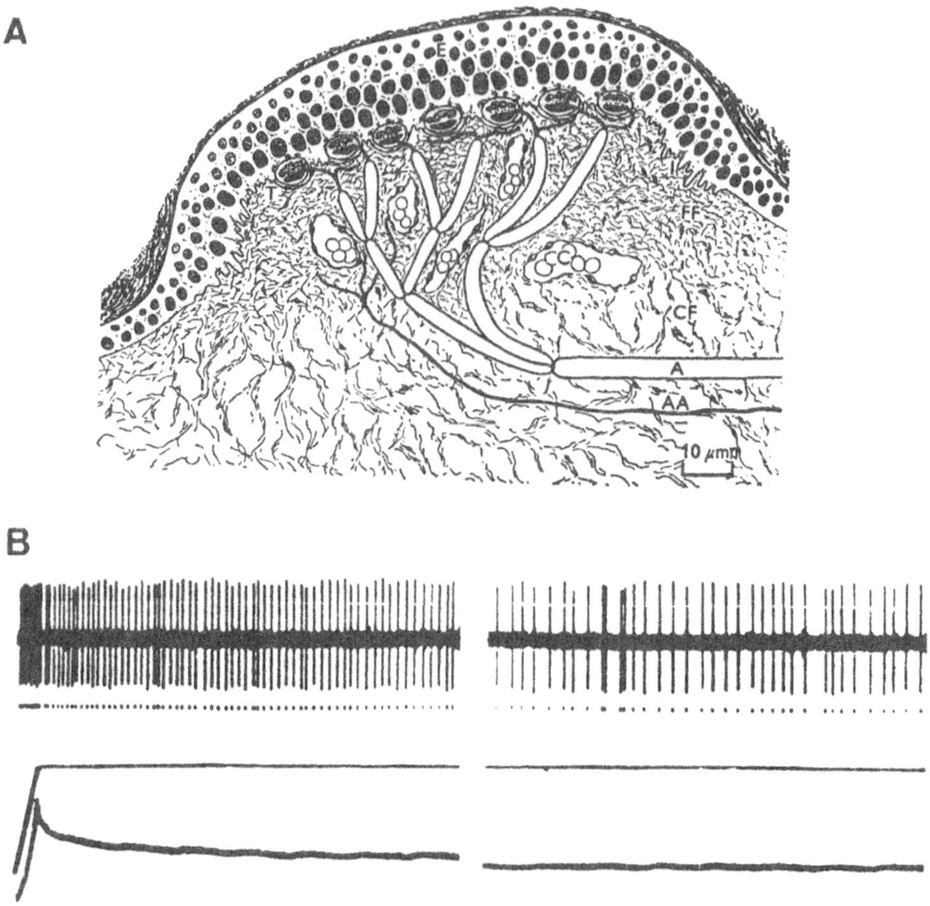

FIGURE 2.7. Structure and function of an SA I receptor (Merkel cell ending). A, tactile dome supplied by an afferent fiber that divides repeatedly. Each final branch ends in relation to a Merkel cell in the basal layer of the epidermis. (From Iggo and Muir, 1969.) B, discharge of an SA I receptor located in the glabrous skin of the hand of a squirrel monkey in response to a constant displacement stimulus. Position and pressure monitors are below. The left trace was recorded during the first 2 s of stimulation. Note the dynamic and static responses. The trace on the right shows the response during the last 2 s of the 10-s stimulus. Note the irregularity of the discharges (From Pubols, B. H. and Benkich, 1986.)

Merkel cells contain numerous electron-dense, membrane bound secretory granules, and antibodies to neuron-specific enolase can be used to label Merkel cells (Gu et al., 1981; López et al., 1998). Merkel cells contact the expanded afferent terminals of the axon by junctions that have many of the ultrastructural characteristics of synapses, including presynaptic densities and postsynaptic membrane thickenings; however, they lack a concentration of small, clear vesicles near the presynaptic densities (Hartschuh and Weihe, 1980). The Merkel cells of several species contain vasoactive intestinal polypeptide (VIP), which may be released to modulate the activity of the afferent endings (Hartschuh et al., 1983). At least in some species, Merkel cells also contain CGRP (Fundin et al., 1997; Rice and Rasmusson, 2000). However, it should be emphasized that the afferent nerve endings on Merkel cells contain neither VIP nor CGRP. Recordings from isolated Merkel cells using the patch-clamp technique have allowed the demonstration of an influx of Ca^{2+} through L-type high threshold voltage-gated Ca^{2+} channels (Yamashita et al., 1992). It was suggested that the Ca^{2+} may play a role in transmitter release from Merkel cells onto the terminals of SA I afferent fibers. A role has also been proposed for intracellular Ca^{2+} release in the operation of Merkel cell complexes (Senok and Baumann, 1997).

The response properties of SA I receptors have been well characterized (Table 2.1; Fig. 2.7B, 2.9A; Werner & Mountcastle, 1965; Iggo and Muir, 1969; Harrington and Merzenich, 1970; Mountcastle et al., 1972; Knibestöl, 1975; Knibestöl and Vallbo, 1980; Coleman et al., 2001). Without a stimulus, SA I endings are normally silent. However, they discharge following indentation of the skin directly over the ending, and they have a low threshold (displacement of less than 15 μm). They are relatively insensitive to displacement of the skin immediately adjacent to the dome and to skin stretch. A step-displacement applied to the dome evokes a dynamic response followed by a static response; the discharges during the static response tend to occur at irregular intervals (Figs. 2.7B and 2.9A). Sensitivity is chiefly to stimulus position, although there is also a response to stimulus velocity (Looft, 1996b).

For more information about the functional properties of SA I endings, see Frankenhaeuser (1949); Maruhashi et al. (1952); Witt and Hensel (1959); Hunt, C. C. and McIntyre (1960b); Tapper (1965); Burgess et al. (1968); Perl (1968); Iggo and Muir (1969); Hagbarth et al. (1970); Kenton and Kruger (1971); Pubols et al. (1971); Horch et al. (1974); Whitehorn et al. (1974); Horch and Burgess (1975; 1976); Pubols, B. H. and Pubols, L. M. (1976; 1983); Konietzny and Hensel (1977); Aitken and Lal (1982a); Lynn and Carpenter (1982); Pubols, B. H. and Benkich (1986); Turnbull and Rasmusson (1986); Goodwin and Morley (1987a, b); Handwerker et al. (1987); Morley and Goodwin (1987); Pubols, B. H. (1990); Edin and Abbs (1991); Cohen and Vierck (1993); Koltzenburg et al. (1997); Senok and Baumann (1997); Guinard et al. (1998); Rice and Rasmussen (2000). See also Figs. 12.1–12.6, 12.8.

SA II Receptors

SA II slowly adapting mechanoreceptors are identified with Ruffini endings (Fig. 2.8A; Ruffini, 1894; Chambers, M. R. et al., 1972; Halata, 1977). These large endings (dimensions of up to 200 by 1,000 μm) are located in the dermis (Fig. 2.6). Beneath the capsule, there is a fluid-filled subcapsular space that surrounds a core region that contains the nerve endings, as well as Schwann cells and collagen. As in the Golgi tendon organ, which Ruffini endings closely resemble (Darian-Smith, 1984), the collagen fibers apply mechanical force to the nerve endings when the Ruffini corpuscle is stretched. Each SA II receptor is supplied by a large myelinated axon. One axon may innervate several Ruffini endings, but there is generally only one low-threshold spot on the skin for each SA II fiber. Paré et al. (2002) were unable to find Ruffini endings in the fingerpads of monkeys, although they did find such endings under the base of fingernails. The absence of Ruffini endings in this study is consistent with reports that the glabrous skin in monkeys lacks innervation by SA II endings (Lindblom, 1965; Perl, 1968; Darian-Smith and Kenins, 1980; Darian-Smith and Oke, 1980; Lamb, 1983a, b).

The primary afferents that supply SA II endings may show a background discharge in the absence of an overt stimulus, and they respond to small displacements of the skin either directly over the receptor (threshold of about 15 μm) or as a result of stretch of the skin from a distance (Table 2.1; Figs. 2.8B, 2.9B, C, E; Knibestöl and Vallbo, 1970;

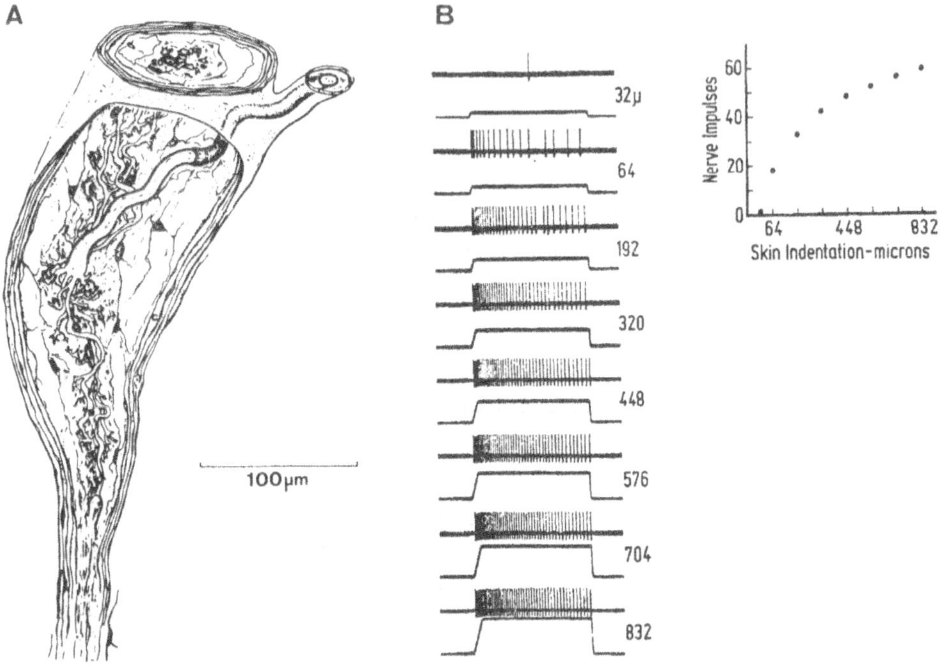

FIGURE 2.8. Structure and function of an SA II mechanoreceptor. A, Ruffini or SA II ending. (From Chambers, M. R. et al., 1972.) B, series of responses of an SA II ending to different amounts of displacement of the skin and a graph of the stimulus-response function. (From Harrington and Merzenich, 1970.)

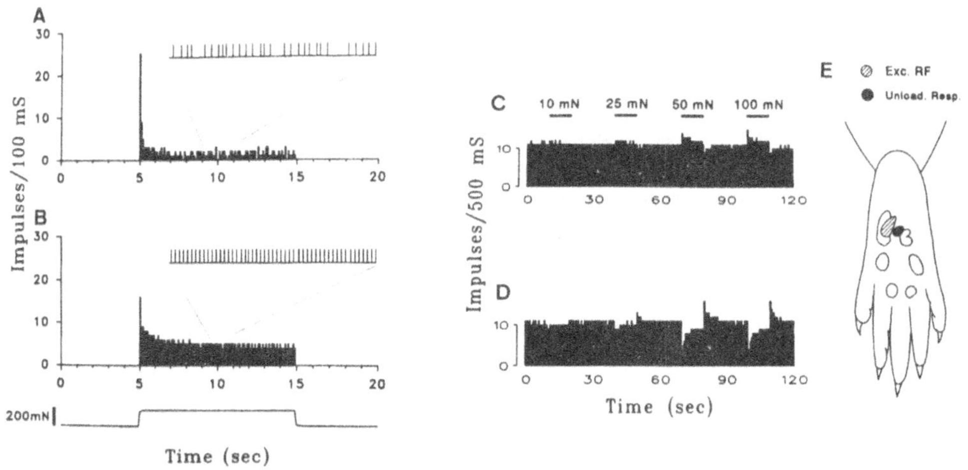

FIGURE 2.9. Responses of SA units in the rat foot. A, response of an SA I unit, and B, responses of an SA II unit to application of a force of 200 mN to the receptive field. The histograms plot the activity during a 10 s stimulus. The insets show the irregularity of the tonic discharges of the SA I unit and the regularity of the tonic discharges of the SA II unit. The lower trace in B is the stimulus monitor. C, excitation of another SA II unit to graded applications of force to the receptive field. Note the background activity of this unit. D, reduction in activity (unloading effect) when stimuli were applied in an area adjacent to the receptive field. E, hatched region on the drawing of the foot was the receptive field; the unloading effect was produced when the black area was stimulated. (From Leem et al., 1993.)

Chambers et al., 1972; Johannson, R. S., 1978; Johansson and Vallbo, 1979a, b; Leem et al., 1993). Stretching the skin at right angles to the orientation of the lines of force that activate SA II endings can lead to a reduction in afferent activity, presumably by unloading the tension on the ending (Fig. 2.9D, E). SA II endings are also located near nails (Paré et al., 2002) and can respond to pressure on the nail. Following a step-displacement of the skin, the discharge has both a dynamic and a static component (Figs. 2.8B, 2.9B), but the dynamic response is smaller than is typical of SA I receptors (compare Fig. 2.9B with Fig. 2.9A). The discharges of SA II endings are quite regular during the static response, in contrast to those of SA I receptors (Figs. 2.8B, 2.9B; cf. Fig. 2.9A). The dynamic responses of SA II receptors may provide a velocity signal that is useful for the control of movements (Grill and Hallett, 1995).

For additional details about SA II receptors, see Burgess et al. (1968); Perl (1968); Harrington and Merzenich (1970); Hagbarth et al. (1970); Kenton and Kruger (1971); Horch and Burgess (1975; 1976); Halata and Munger (1981); Aitken and Lal (1982b); Lynn and Carpenter (1982); Rasmusson and Turnbull (1986); Turnbull and Rasmusson (1986); Handwerker et al. (1987); Edin and Abbs (1991); Grill and Hallett (1995); Wu, G. et al. (1998); Paré et al. (2002). See also Figs. 12.4, 12.6–12.8.

Cutaneous Velocity Detectors

There are at least five kinds of cutaneous mechanoreceptors that detect stimulus velocity, but not position (Table 2.1). One of these is found only in the glabrous skin, and four are found only in hairy skin. The velocity detector in the glabrous skin of primates, including humans, is the FA I (for fast adapting, type I) receptor. The velocity detectors in hairy skin include G_2 hair follicle receptors (and a subpopulation of T hair follicle receptors), D hair follicle receptors, field receptors, and C mechanoreceptors (Table 2.1).

FA I receptors

The morphology of the FA I receptor varies, even within a single species (Malinovský, 1966). In primate (including human) glabrous skin, this receptor can be identified with Meissner's corpuscle (Fig. 2.10; Wagner and Meissner, 1852; Meissner, 1859; Cauna, 1956; Cauna and Ross, 1960; Munger, 1971; Quillium, 1975; Paré et al., 2002). The comparable receptor in the foot and toe pads of cats is Krause's end bulb (Krause, 1859; Lynn, 1969; Jänig, 1971a, b; Iggo and Ogawa, 1977). Meissner's corpuscles and Krause's end bulbs may be evolutionarily derived from a common type of ending, and both may be morphological equivalents in glabrous skin to hair follicle endings in hairy skin (Malinovský, 1966; Munger, 1971).

Meissner's corpuscles in the fingerpads of the monkey or human hand are supplied by up to six large myelinated axons, and a given myelinated axon can supply more than one Meissner's corpuscle (Fig. 2.10A; Cauna, 1956; Halata, 1975; Johansson, O. et al., 1999; Paré et al., 2001; 2002). Meissner's corpuscles in the human hand are most common in the fingerprint skin and less common in the skin of the palm and plantar foot (Bolton et al., 1966). Their size is about 100 by 50 μm. Collagen fibers connect the corpuscle to tonofibrils of the basal epidermis, providing a mechanical linkage for delivery of mechanical forces to the nerve endings (Fig. 2.10B; Andres and von Düring, 1973). The nerve terminals form disk-like expansions at right angles to the long axis of the Meissner corpuscles (Fig. 2.10C; Castano and Ventura, 1978). These are interleaved with capsular cells.

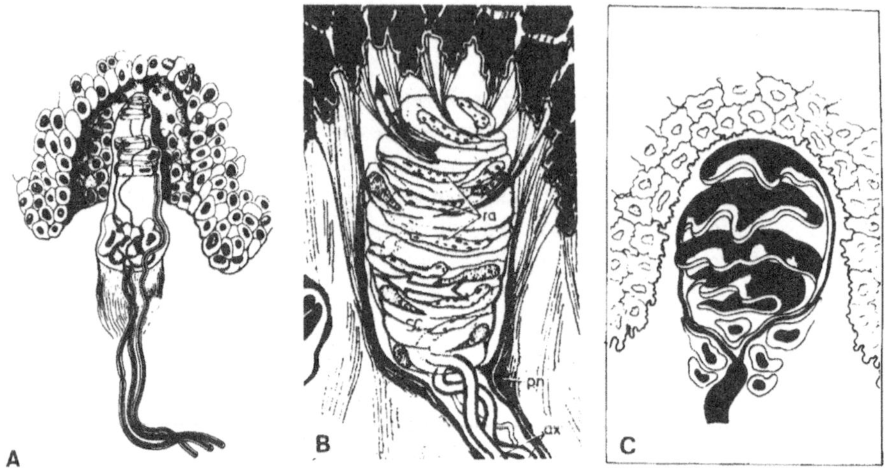

FIGURE 2.10. Structure of the Meissner corpuscle. A, Meissner corpuscle in a dermal papilla. The corpuscle is supplied by three large myelinated nerve fibers. (From Quilliam, 1975.) B, collagen fibers connecting the basal epidermis to a Meissner corpuscle. (From Andres and von Düring, 1973.) C, flattened terminals of the nerve fibers that supply a Meissner's corpuscle. (From Castano and Ventura, 1978.)

The endings of the Aβ fibers that innervate Meissner corpuscles may contain NK-1 receptors, as well as adrenoreceptors, δ-opiate receptors, P2X receptors, and VRL-1 (Paré et al., 2001).

Besides their innervation by large myelinated afferents, Meissner's corpuscles also receive accessory innervation from two different kinds of unmyelinated afferent axons (Dogiel, 1892; Cauna, 1956; Johansson, O. et al., 1999; Paré et al. 2001). These C fibers can be immunoreactive for calcitonin gene-related peptide or for substance P (Ishida-Yamamoto et al., 1988; 1989; Johansson, O. et al., 1999; Paré et al. 2001). The peptidergic C fibers may also contain μ- and δ-opiate receptors, as well as P2X receptors. Other, non-peptidergic C fibers are immunoreactive for VR-1 receptors (Paré et al., 2001). The function of these accessory C fibers in human, monkey, and rat Meissner's corpuscles is unclear, but a role in nociception has been proposed (Paré et al., 2001). It is unclear what this role might be.

FA I receptors in primates have localized receptive fields, which are smallest on the fingerpads (Talbot et al., 1968; Knibestöl, 1973; Johansson, R. S., 1978). The thresholds of the receptors in the skin of the cat are low (often less than 10 μm of indentation; ramp thresholds average 2 μm/ms), and there is no discharge during a maintained stimulus (Jänig et al., 1968a; Jänig, 1971b; Iggo and Ogawa, 1977). The FA I receptors in primate glabrous skin have similar properties, although their thresholds seem to be higher than in the cat. They are activated best by repetitive stimuli at rates of 30–40 Hz (Fig. 2.11A, C; Talbot et al., 1968). When FA I receptors are activated by low-intensity stimuli, the receptors are sensitive to stimulus position; however, with higher-intensity stimuli, they encode velocity (Looft, 1996b).

For further details about FA I receptors, see Lindblom, 1965; Pubols, L. M. et al., 1971; Johnson, K. O., 1974; Pubols, B. H. and Pubols, L. M. 1976; Iggo and Ogawa, 1977; Vallbo and Johansson, 1984; Goodwin and Morley, 1987a, b; Morley and Goodwin, 1987; Rumio et al., 1995; Johansson, O. et al., 1999; Coleman et al., 2001; Paré et al., 2002. See also Figs. 12.4, 12.6, 12.10–12.13.

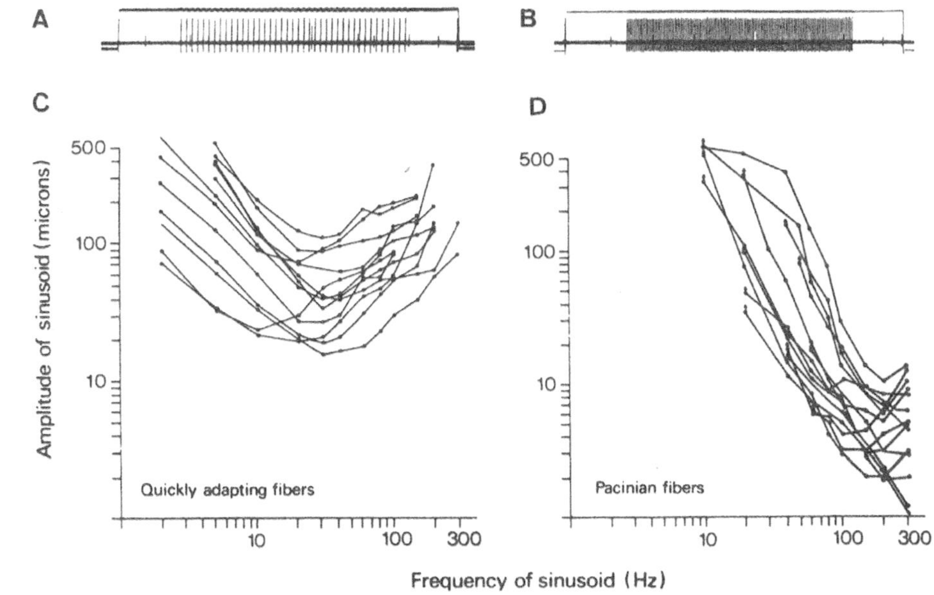

FIGURE 2.11. A, recording from the axon of an FA I receptor (presumably a Meissner's corpuscle) in the finger of a monkey. The stimulus was an 18-μm sinusoidal indentation of the skin at 40 Hz superimposed on a maintained step indentation. B, recording from an axon of an FA II receptor (presumably a Pacinian corpuscle) in the palm of a monkey's hand. The stimulus in this case was a 19-μm sinusoidal indentation at 150 Hz superimposed on a step indentation. C–D, tuning curves for FA I and FA II fibers to sinusoidal stimuli of different frequencies. The tuning curves for FA I units are U-shaped with a best frequency of about 30–40 Hz. Those for FA II units are similar but with best frequencies of 250–350 Hz. (From Talbot, W. H. et al., 1968.)

Hair Follicle Receptors

Hair follicles in mammals can be subdivided into three main types: tylotrichs, guard hairs, and down hairs (Table 2.1; Noback, 1951; Straile, 1960; 1961; Brown, A. G. and Iggo, 1967; Brown, A. G. et al., 1967; Millard and Woolf, 1988). A fourth type, the sinus hair, is found only in certain locations and is not represented in humans. Sinus hairs will be discussed separately below.

Tylotrichs are the largest and least numerous of the three main types of hairs. Both tylotrichs and guard hairs arise individually from hair follicles. Down hairs are the smallest hairs; they are wavy; and they arise in groups from single follicles. Each hair is innervated by several myelinated nerve fibers that join the follicle just below the sebaceous glands, lose their myelin, and terminate in a circumferential array around the follicle (pilo-Ruffini endings; Fig. 2.12A). Other afferents end as lanceolate endings within the dermal sheath (Fig. 2.12B; Halata, 1975; Carpenter and Sutin, 1983). Guard hairs receive a denser innervation than do down hairs. A given hair follicle afferent supplies about ten guard hair follicles, and a given hair follicle may receive the terminations of several afferents (Burgess et al., 1968; Halata, 1993). In addition, Aδ and C fibers may end in the connective tissue between the hair follicle and the sebaceous gland; these are likely to be nociceptors (Halata, 1993).

Hair movement activates many of the myelinated fibers that supply the skin (Hunt and McIntyre, 1960c; Ray et al., 1985). The afferent fibers that innervate guard hair follicles can be subdivided into two groups based on their responses to mechanical stimulation.

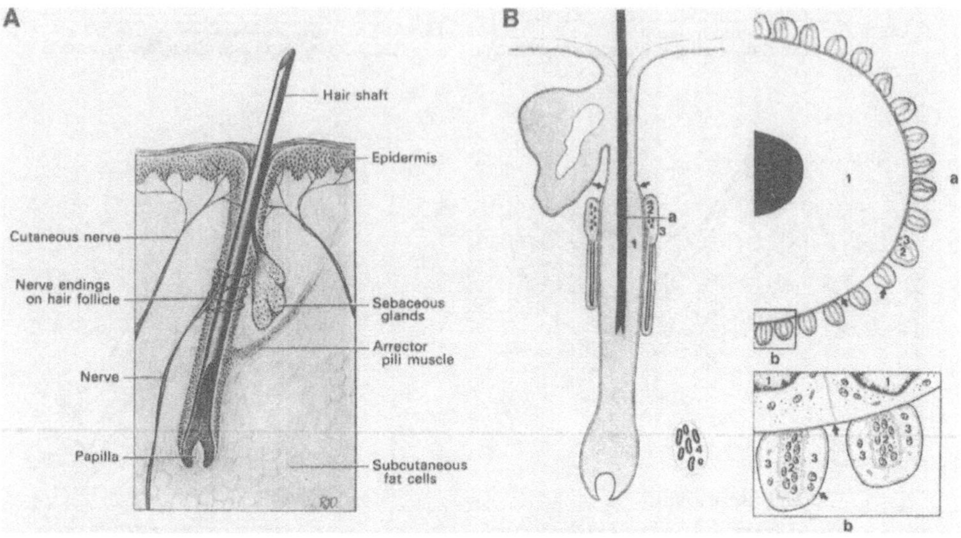

FIGURE 2.12. Innervation of hair follicles. A, section through the hairy skin. The shaft of the hair follicle is supplied by several myelinated nerve fibers that terminate just below the connection of sebaceous glands with the follicle. (From Carpenter and Sutin, 1983.) B, left, diagram of a longitudinal section through a hair follicle; right, a cross-section taken at the level indicated by the lead-line labeled "a." Detailed drawing below (b). The arrows on the left indicate the juncture between the epidermal (1) and dermal components of the follicle. Lanciform nerve endings (2) are covered by Schwann cells (3). (From Halata, 1975.)

Most G_1 hair follicle receptors are transient detectors, as will be discussed below. G_2 hair follicle receptors are generally velocity detectors. However, Hahn and Wall (1975) failed to find a clear separation between G_1 and G_2 receptors. Some T (for tylotrich) hair follicle receptors appear to behave like G_1 receptors and others like G_2 receptors. Therefore, different T hair follicle receptors are grouped with the appropriate type of G hair follicle receptor (cf. Brown, A. G. and Iggo, 1967; Brown, A. G. et al., 1967; Burgess et al., 1968; Perl, 1968; Tuckett et al., 1978; Tuckett, 1982).

G_2 hair follicle receptors and some T hair follicle receptors respond to the velocity of hair movement (Fig. 2.13). That is, higher rates of discharge are produced when the velocity at which the stimulus occurs is increased. Ramp and hold stimuli are useful for determining responses to stimuli that have onsets with different velocities. The skin or a hair is displaced at a constant velocity until a particular displacement is achieved. The rate at which the displacement occurs can be varied systematically for each of a series of displacements of a constant amount. Thresholds to ramp movements of individual hairs in cats are in the range of 0.5–1.5 μm/ms. The threshold for a stimulus applied to the skin is less than 0.05 μm/ms. If the hair is moved gently, there is a steady discharge. However, no discharge occurs during a maintained stimulus.

In the case of the responses of the tylotrich unit illustrated in Fig. 2.13, the threshold for discharge was a displacement of the hair at a velocity of less than 0.26 μm/ms, and the responses increased as the velocity of displacement of the hair was increased to 6.24 μm/ms (Brown, A. G. and Iggo, 1967).

D hair follicle receptors also signal stimulus velocity, but their thresholds are lower than those of G_2 receptors (that is, they are more sensitive). For hair movements, threshold responses to ramp stimuli are 0.1–1 μm/ms. D hair follicle receptors are so sensitive that

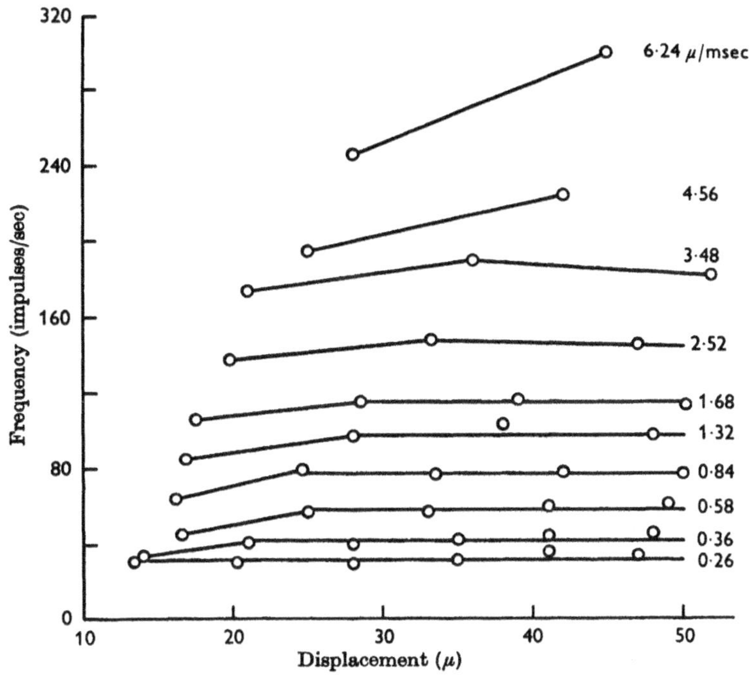

FIGURE 2.13. Input–output curves for a rapidly adapting receptor (T hair follicle unit) that signaled stimulus velocity. The stimuli were constant-velocity displacements of the hair. The ordinate shows the instantaneous frequency of the discharge, and the abscissa shows the amount of displacement. The stimuli were applied over a range of velocities. The family of curves indicates that the discharge rate was constant for a given velocity and was a function of stimulus velocity. (From Brown, A. G. and Iggo, 1967.)

they can be activated indirectly by movement of nearby down or guard hairs by mechanical spread of the stimulus. There is no discharge in the absence of an overt stimulus, but very small movements of the skin will activate these receptors (Maruhashi et al., 1952; Brown, A. G. and Iggo, 1967; Burgess et al., 1968; Perl, 1968; Merzenich and Harrington, 1969). They may appear to be slowly adapting when activated by a hand-held stimulator, perhaps because of repeated stimulation by physiological tremor. D hair mechanoreceptors respond vigorously to air jet stimuli (Ray et al., 1985).

Field Receptors

Field receptors are distinguished from hair follicle receptors by the fact that field receptors cannot be activated by the movement of single hairs, whereas hair follicle receptors can (Table 2.1). Field receptors are excited when large numbers of hairs are brushed, probably by spread of the mechanical stimulus to the underlying skin. Direct stimulation of the skin, along with hair movement, is the best stimulus. The morphology of field receptors is unknown. Field receptors signal stimulus velocity and have low thresholds to ramp stimuli (0.1–1 μm/ms). Some field receptors have a tendency to continue to discharge during a maintained stimulus. Several categories of field receptors (F1, intermediate, F2) can be distinguished on the basis of gradations between rapidly and slowly adapting responses (Burgess et al., 1968; Perl, 1968; Burgess and Perl, 1973; Tuckett et al., 1978).

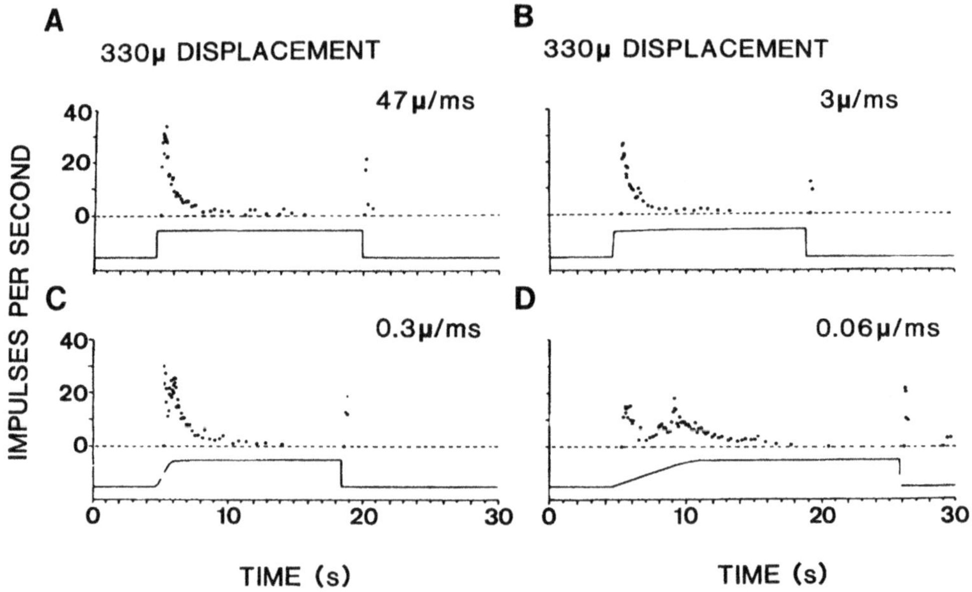

FIGURE 2.14. A–D, responses of a C-mechanoreceptor to indentation of the skin at different velocities. (From Bessou et al., 1971.)

C Mechanoreceptors

C mechanoreceptors are sensitive mechanoreceptors that have unmyelinated afferent fibers (Table 2.1; Fig. 2.14). They respond to skin indentation, but they are especially sensitive to stimuli that move slowly across the receptive field ("caress detectors"). They signal chiefly velocity, although there is a tendency for their discharge to continue during a maintained stimulus. Ramp thresholds can be less than 0.1 μm/ms (Fig. 2.14; Iggo, 1960; Bessou et al., 1971). Following cessation of stimulation, there may be an after-discharge. C mechanoreceptors are not distributed universally. For example, they are found in the proximal hairy skin in cats, but not in the pad of the cat's foot (Bessou et al., 1971; Beck et al., 1974), nor in the glabrous skin of the monkey hand (Georgopoulos, 1976). They occur, but are uncommon, in the hairy skin of monkeys (Kumazawa and Perl, 1977). C mechanoreceptors have recently been identified in recordings from forelimb nerves in human subjects, using the microneurography technique (Vallbo et al., 1999), as well as in nerves to the face (Nordin, 1990).

For additional details about hair follicle receptors, field receptors, and C mechanoreceptors, see Zotterman (1939); Douglas, W. W. and Ritchie (1957); Iggo (1960); Iriuchijima and Zotterman (1960); Bessou and Perl (1969); Bessou et al. (1971); Iggo and Kornhuber (1977); Lynn and Carpenter (1982); Ray et al. (1985); Millard and Woolf (1988); Halata (1993). See also Figs. 12.14 and 12.25.

Cutaneous Transient Detectors

Two types of cutaneous receptors are designed to detect mechanical transients: Pacinian corpuscles and G_1 hair follicle receptors (Table 2.1).

FA II Receptors

FA II receptors or Pacinian corpuscles (Fig. 2.15A) are large receptors (0.3 to 1.5 mm long, 0.2 to 0.7 mm wide) located in the dermis, subcutaneous tissue, and deep tissue (Fig. 2.15B). They are very sensitive to mechanical transients that result from cutaneous stimulation. However, they do not respond to maintained pressure because the capsule of the Pacinian corpuscle (Fig. 2.15A) serves as a mechanical filter that allows mechanical transients to affect the terminal but prevents lower-frequency mechanical events from doing so (Hubbard, 1958). The structure of the Pacinian corpuscle that produces the mechanical filtering action is an outer, fluid-filled capsule, within which are a subcapsular space and an inner core. The core is composed of a bilayered structure formed by laminar cells separated by a cleft, in the center of which is the nerve fiber (Halata, 1975; 1977). The axon is oval in cross-section, and extensions of the axon called axonal spines or filopodia protrude into the cleft (Malinovský et al., 1982; Bolanowski et al., 1994). It has been speculated that the filopodia are sites for mechanical transduction, where the receptor potential is produced (Ide et al., 1988; Bolanowski et al., 1994). Freeze-fracture experiments demonstrate a large number of intramembranous particles in the membrane of the axon terminal (Ide and Hayashi, 1987). The function of these intramembranous particles is unclear. One speculation is that they might represent mechanically sensitive channels. An accessory innervation of Pacinian corpuscles by sympathetic axons has been described (Santini et al., 1971). The function of this accessory innervation is unknown.

The distribution of Pacinian corpuscles has been described in the forefoot of the cat and the hand of the monkey (Kumamoto et al., 1993a, b), as well as in the human hand (Fig. 2.15B; Cauna and Mannan, 1959; Stark et al., 1998). They are less common than other cutaneous receptors. In the hand, none are found dorsally (Cauna and Mannan, 1958), and many are in the deep dermis and subcutaneous fat of the thenar and hypothenar eminences (Bushong, 1963, quoted in Darian-Smith, 1984). However, the densest concentration is in the fingers, especially near the metacarpophalageal joints (Stark et al., 1998).

A B

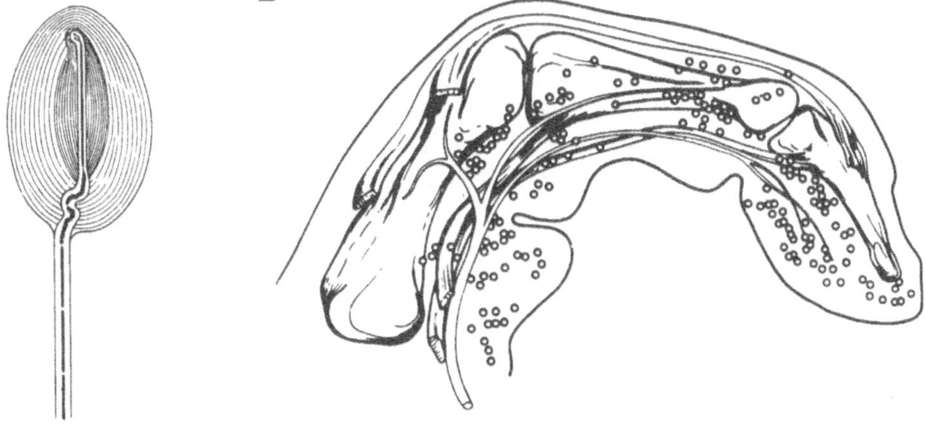

FIGURE 2.15. A, Pacinian corpuscle cut longitudinally. The afferent nerve fiber is seen entering the corpuscle, losing its myelin sheath and then extending along the inner core before ending. The capsule consists of numerous fluid-filled lamellae. (From Quillium and Sato, 1955.) B, distribution of Pacinian corpuscles (open circles) in the index finger of a human fetus. (From Cauna and Mannan, 1959.)

Afferent axons from Pacinian corpuscles tend to cluster within human peripheral nerves (Wu, G. et al., 1999). Similarly, the axons of SA II afferents in human peripheral nerves are clustered (Wu, G. et al., 1998), and Pacinian corpuscle afferents are clustered with SA II afferents (Wu, G. et al., 1999; cf. Roberts, W. J. and Elardo, 1986, who find that several classes of cutanous afferents in cats are also not dispersed randomly in the saphenous nerve, but rather are clustered). Pacinian corpuscles do not appear to project by way of cutaneous nerves in monkeys, although they do so in cats (Merzenich and Harrington, 1969).

Pacinian corpuscles adjacent to the interosseous membrane or beneath the pads of the cat's foot are extraordinarily sensitive. They can be activated by hair movement, as well as by stimuli applied to the skin or bony prominences. In cats they can be excited by movements of the floor at some distance from the experimental preparation, by dropping a paper match folder on the experimental table, or even by a loud sound (Hunt, C. C. and McIntyre, 1960a). The threshold to indentation of the skin is often less than 1 μm (Hunt, C. C. 1961; Jänig et al., 1968a; Lynn, 1969; 1971). In humans, the median threshold is 9.2 μm, slightly more than the diameter of a red blood cell (Johansson, R. S. and Vallbo, 1979b).

Pacinian corpuscles are activated best by frequencies of vibration in the range of 60–300 Hz (Fig. 2.11B, C; Talbot, W. H. et al., 1968; Iggo and Ogawa, 1977), but they can follow frequencies of 500 Hz or even higher (Hunt, C. C., 1961; Burgess et al., 1968). They are sensitive to velocity when low-intensity stimuli are used and to acceleration with high-intensity stimuli (Looft, 1996b). The excitability of these sensory organs depends on the concentrations of sodium and potassium within the fluid content of the corpuscle (Ilyinsky et al., 1976). However, divalent cations do not seem to contribute to the transduction mechanism (Akoev, 1982).

Mathematical models of the Pacinian corpuscle (Freeman, A. W. and Johnson, 1982a, b; Grandori and Pedotti, 1982; Bell and Holmes, 1992) and of the mechanical properties of the tissue in which Pacinian corpuscles are imbedded (Van Doren, 1989) have been proposed. Such models can predict a number of the functional characteristics of Pacinian corpuscles, such as the off-response and the frequency characteristics of the action potentials.

For details about the structure, distribution, and mechanical properties of Pacinian corpuscles, see Vater (1741); Pacini (1840); Henle and Kolliker (1845); Quillium and Sato (1955); Pease and Quillium (1957); Cauna and Mannan (1958; 1959); Hubbard (1958); Loewenstein and Skalak (1966); Malinovský (1966); Malinovský et al. (1982); Lynn (1969); Nishi et al. (1969); Chouchkov, H. N. (1971); Halata (1975; 1977); Chouchkov, C. (1978); Zelena (1978); Ide and Hayashi (1987); Munger et al. (1988); Ide et al. (1988); Bell et al. (1994); Vega et al. (1994a; 1996); Dubovy and Bednarova (1999). For more information about the functional properties of Pacinian corpuscles, see Hulliger et al. (1979); Hämäläinen and Pertovaara (1984); Bolanowski (1984); Bolanowski and Zwislocki (1984a, b); Greenstein et al. (1987); Goodwin and Morley (1987a, b); Morley and Goodwin (1987); Palmer and Gardner (1990); Checkosky, C. M. and Bolanowski, S. J. (1992; 1994); Looft (1996a, b); Coleman et al. (2001). See also Figs. 12.4, 12.6, 12.10–13, and 12.15.

G_1 Hair Follicle Receptors

Rapid movements of guard hairs activate G_1 hair follicle receptors. The thresholds to ramp stimuli exceed 80 μm/ms when single hairs are moved and 5–20 μm/ms when the skin is stimulated (Burgess et al., 1968; cf. Tuckett et al., 1978). As mentioned, some T hair follicle endings behave in a manner similar to G_1 hair follicle receptors.

Sinus Hairs

A special sensory apparatus is associated with the sinus hairs of animals. However, sinus hairs do not occur in humans (Halata, 1993). These hairs form the mystacial vibrissae on the snout, and they are also present in other locations, such as the carpal hairs in the wrist region in cats. The innervation of sinus hairs is complex and includes several types of sensory endings (Vincent, 1913; Nilsson and Skoglund, 1965; Andres, 1966; Nilsson, 1969a, b; Halata, 1975; Halata and Munger, 1980; Halata, 1993; Fundin et al., 1997). Recordings from the afferent fibers that innervate the vibrissae show that these sinus hairs are supplied by two kinds of slowly adapting receptors that behave like SA I and SA II endings and also by two types of rapidly adapting receptors, one with a high threshold and one with a low threshold (Gottschaldt et al., 1973; cf. Nilsson and Skoglund, 1965). Carpal hairs are also supplied by two kinds of slowly adapting receptors, but the rapidly adapting receptors are missing. Instead, movements of carpal hairs excite adjacent Pacinian corpuscles (Gottschaldt et al., 1973).

The Role of Growth Factors in the Regulation of Mechanoreceptors

SA I (Merkel Cell) Endings

Observations of the innervation of the mystacial vibrissae in mice provide an opportunity to examine the role of neurotrophins on different types of mechanoreceptor endings, using transgenic animals (Fundin et al., 1997; Cronk et al., 2002). For example, Merkel cell endings fail to develop in knock-out mice that lack either NT-3 or trkC receptors. Curiously, there is an increase in Merkel cell complexes in NT-4, BDNF, and trkB-null animals. Fundin et al. (1997) conclude that each of the trk receptors has a major role in the formation of at least one type of mechanoreceptor, and the absence of the related neurotrophin has an effect similar to that of the lack of the receptor(s) it acts on.

Observations have also been made on SA I endings in other parts of the skin than in sinus hairs. After denervation, tactile domes lose many of their Merkel cells, apparently through phagocytosis by Schwann and Langerhans cells in the dome (English, 1977). Studies have been done to determine the possible role of neurotrophins in the maintenance of Merkel cells. Nerve growth factor (NGF) is found in keratinocytes of tactile domes, but not in the surrounding epidermis, and the low-affinity NGF receptor (p75) is found in Merkel cells and in the nerve endings on these cells (English et al., 1994). Kinkelin et al. (1999) observed that Merkel cell numbers are reduced in mice that lack p75. The presence of NGF in tactile domes could suggest that NGF regulates Merkel cells. However, several neurotrophins other than NGF also bind to p75, and NT-3 seems more likely to be required for the regulation of Merkel cells than is NGF. For example, transgenic mice that overexpress NT-3 have enlarged tactile domes with increased numbers of Merkel cells (Albers et al., 1996). Heterozygous and null mutants for NT-3 (Airaksinen et al., 1996) and rats treated with antibody to NT-3 have a reduced number of Merkel cells (Zhou, X. F. et al., 1998). Experiments on mice with null mutations indicate that NGF and NT-3, as well as the receptors for trk A, trkC, and p75, all promote the survival of Merkel cell complexes during development. However, NT-3 and trkC are crucial determinants for their survival in the adult (Fundin et al., 1997; Cronk et al., 2002; see also López et al., 1998), either because of actions on the afferent innervation or because of direct effects on Merkel cells. Interestingly, BDNF and NT-4 appear to suppress Merkel cell complexes through an action on trkB receptors. On the other hand, BDNF-null mice show a reduction in the mechanical sensivity of slowly adapting mechanoreceptors (Carroll et al., 1998). Heterozygous mice also have a reduced mechanical sensitivity, and this reduction can be reversed by treatment with BDNF.

An important observation derived from studies of neurotrophic actions on Merkel cells is that despite a substantial loss of Merkel cells in p75 knock-out mice, the apparent numbers and the response properties of SA I receptors in these animals are unchanged (Kinkelin et al., 1999). This suggests that Merkel cells have some function other than transduction of mechanical stimuli, since it appears that the terminals of the nerve fibers that normally end on Merkel cells can manage this role in the absence of Merkel cells.

FA I (Meissner Corpuscle) Receptors

The axon and lamellar cells of Meissner corpuscles have been shown to demonstrate immunoreactivity for epidermal growth factor receptors and the lamellar cells for trkA and p75 receptors (Vega et al., 1993; 1994b; López et al., 1998). Thus, components of Meissner corpuscles appear to be regulated by several growth factors. Many of the lamellar cells of Meissner's corpuscles have been found to contain Bcl-2, but not Bax protein (Ortiz-Rey et al., 2002). It was proposed that this favorable ratio of Bcl-2 to Bax protects Meissner's corpuscles from apoptosis.

FA II (Pacinian Corpuscle) Receptors

The axons of Pacinian corpuscles, the inner core, and the capsule have been reported to label immunohistochemically for receptors for nerve growth factor (trkA and p75 receptors) and epidermal growth factor, as well as for neurofilament protein and S-100 protein (Vega et al., 1993; 1994b; 1996; López et al., 1998).

Hair Follicle Receptors

The periodic growth and regression of hair follicles appear to depend on interactions with a number of growth factors. These include epidermal growth factor (Hansen et al., 1997), NGF, NT-3, NT-4, BDNF (Botchkarev et al., 1998; 1999; 2000), and insulin-like growth factor (Su et al., 1999). There is an increase in the density of the circumferential nerve endings on hair follicles in transgenic mice that overexpress NT-3 (Albers et al., 1996), suggesting that this neurotrophic factor influences the innervation of hair follicles, as well as the hair follicles themselves. In mice, NT-4 is required for the survival of D-hair receptors (Stucky et al., 1998). The absence of BDNF or trkB affects the innervation of hair follicles by lanceolate endings and also prevents the development of Ruffini endings.

Transduction in Mechanoreceptors

The mechanism by which cutaneous mechanoreceptors are activated is controversial. One hypothesis is that stretch-evoked tension within the membrane of a mechanoreceptor might directly activate mechanically sensitive membrane channels (Hamill and Martinac, 2001; Welsh et al., 2002). A variant of this concept is that the channels may be tethered to the extracellular matrix and the cytoskeleton and that this allows mechanical forces to open them (Garcia-Añoveros and Corey, 1997; Garcia-Añoveros et al., 2001; Welsh et al., 2002).

Another proposal is that mechanical forces might cause the release of a ligand, such as ATP, into the extracellular space, and that the ligand would then produce a receptor potential by an action on purinergic receptors (Hamill and Martinac, 2001).

McCarter et al. (1999) showed that dorsal root ganglion neurons, but not sympathetic ganglion cells, were depolarized by activation of non-selective cation channels in response to mechanical stimuli. The current evoked by mechanical stimuli could be partially blocked by gadolinium or by an amiloride analog. These observations are consistent with the suggestion that the mechanoreceptor channel protein belongs to the degenerin/epithelial sodium channel (DEG/ENaC) family of proteins, which are typically blocked by gadolinium or amiloride (Corey and García-Añoveras, 1996; McCarter et al., 1999; see reviews by Morris, 1990; García-Añoveros and Corey, 1997; Welsh et al., 2002). These proteins are known to form voltage-insensitive, but mechanically sensitive, cation channels in the nematode, *Caenorhabditis elegans* (García-Añoveros et al., 1997; McCarter et al., 1999; Drummond et al., 2000; Price, M. P. et al., 2000). Branches of this family of proteins are also found in insects and mammals and include the acid-sensitive ion channels (ASICs; Waldmann and Lazdunski, 1998). ENaC protein is also found in the terminals of baroreceptors, where it is thought to have a mechanoreceptive function (Drummond et al., 1998).

The mRNA for the β- and γ- subunits, but not the α-subunits, of BNaC1 protein, were initially reported in rat dorsal root ganglion cells and in the neural components of several types of mechanoreceptors, including Merkel cell complexes, and Meissner and Paciniform corpuscles (Drummond et al., 2000). However, the mRNA for α-subunits has since been observed in most large DRG cells, as well as in a few smaller, presumably nociceptive DRG cells (Price et al., 2000; García-Añoveraos et al., 2001). This isoform is transported distally in the peripheral process of the DRG cells, but apparently not centrally. It was identified immunohistochemically in Meissner's corpuscles, Merkel cell endings, and hair follicle afferents, as well as in some free nerve endings (Price, M. P. et al., 2000; García-Añoveraos et al., 2001; Welsh et al., 2002). García-Añoveraos et al. (2001) proposed that the BNaC1α molecule might serve as a mechanical transducer in mechanoreceptors and also in Aδ mechanical nociceptors.

The functional role of BNaC1 (also called BNC1, MDEG, or ASIC2) has been compared in wild-type and knock-out mice (Price, M. P. et al., 2000). The absence of the gene for BNaC1 results in a reduced sensitivity of both rapidly adapting (FA I) and slowly adapting (S I) cutaneous mechanoreceptors. However, there was no apparent change in the sensitivity of D hair endings or of Aδ or C mechanical nociceptors. Furthermore, responses to lowered pH were unaffected. The threshold for activating cultured dorsal root ganglion cells by current injection was unchanged in the knock-out animals, indicating that the reduced sensitivity was more likely to be related to a change in the receptor potential than a change in threshold for action potential generation. Immunostaining showed that BNaC1 was present in the lanceolate nerve endings of hair follicles in wild-type mice but not in the knock-out mice. The fact that knocking out BNaC1 did not abolish mechanical sensitivity led to the suggestion that other proteins are also involved in the formation or tethering of mechanically sensitive channels (Price, M. P. et al., 2000).

The excitability of mechanoreceptors is also determined in part by their potassium ion channel composition. Large diameter dorsal root ganglion cells and their large myelinated axons, many of which belong to cutaneous mechanoreceptors, contain delayed rectifier potassium channels. The subunit composition of these slowly inactivating channels includes dendrotoxin-sensitive Kv1.1, Kv1.2, and/or Kvβ2.1 molecules (Rasband et al., 2001). Another slowly inactivating potassium channel in large sensory neurons includes Kv1.1,

Kv1.2, Kv1.4, and Kvβ2.1 subunits. Large DRG cells may have a rapidly inactivating A-type current, for which the channel includes Kv1.1, Kv1.4, and Kvβ2.1 subunits (Rasband et al., 2001).

Cutaneous Nociceptors

Nociceptors can be defined as sensory receptors that respond to stimuli that threaten or actually damage tissue (Sherrington, 1906). Besides receptors that respond only to damaging stimuli, receptors that respond at threshold to moderately intense stimuli are included in the nociceptor category if they discharge progressively more as the stimulus intensity is increased, and the greatest response occurs when the stimulus becomes damaging.

Some nociceptors in the skin are unresponsive to mechanical stimuli unless they are first sensitized ("silent nociceptors"; Lynn and Carpenter, 1982; Meyer et al., 1991; Kress et al., 1992; Davis, K. D. et al., 1993; Treede et al., 1998; Schmidt, R. et al., 1995; 2000; cf. Coggeshall et al., 1983; Schaible and Schmidt, 1983a, b). These receptors may become sensitized following the application of long-lasting noxious mechanical stimuli (Schmidt, R. et al., 2000) or during inflammation (see section on **Sensitization of Nociceptors**). Silent nociceptors can be regarded as chemoreceptors, since their responses change after exposure to chemical mediators of inflammation (Baumann et al., 1991; Handwerker et al., 1991; Meyer et al., 1991; Davis, K. D. et al., 1993; cf. Schaible and Schmidt, 1985; Lang et al., 1990). Since mechanical search stimuli cannot be used to locate and identify such receptors and a chemical search stimulus is not practical, Meyer et al. (1991) introduced the use of an electrical search stimulus. Strong electrical stimuli activate the axons of all types of receptors, including those that do not respond to mechanical stimuli. Once a sensory receptor is identified by its response to an electrical stimulus, other stimuli, such as application of thermal stimuli or chemical agents to the skin, can then be employed to characterize the receptive field properties of the receptor.

A convenient in vitro skin-nerve preparation for studying mammalian cutaneous nociceptors has been described by Reeh (1986). This has now been adapted for the study of cutaneous receptors in mice to take advantage of the availability of transgenic animals (Koltzenburg et al., 1997).

There are several types of cutaneous nociceptors (Table 2.2). Most of these are supplied by Aδ or C fibers. However, some so-called Aδ nociceptors have axons that actually conduct at Aβ velocity (Burgess and Perl, 1967; Georgopoulos, 1976; Treede et al., 1998; Djouhri and Lawson, 1999). Nociceptors are often given subtype names that reflect the adequate stimuli that activate them, as well as the conduction velocity of the afferent fibers that innervate them. For example, Aδ mechanical nociceptors are nociceptors that have finely myelinated afferent axons and that respond to noxious mechanical stimuli. On the other hand, C polymodal nociceptors have unmyelinated afferents and respond to noxious mechanical, thermal, and chemical stimuli. Cutaneous nociceptors may respond well to just one type of intense stimulus (mechanical, thermal, or chemical), to any combination of two of these stimuli, or to all three forms of stimulus energy.

Aδ Nociceptors

The morphology of Aδ mechanical nociceptors has been described (Fig. 2.16A; Kruger et al., 1981; Perl, 1984). The terminals extend into the epidermis and are largely ensheathed

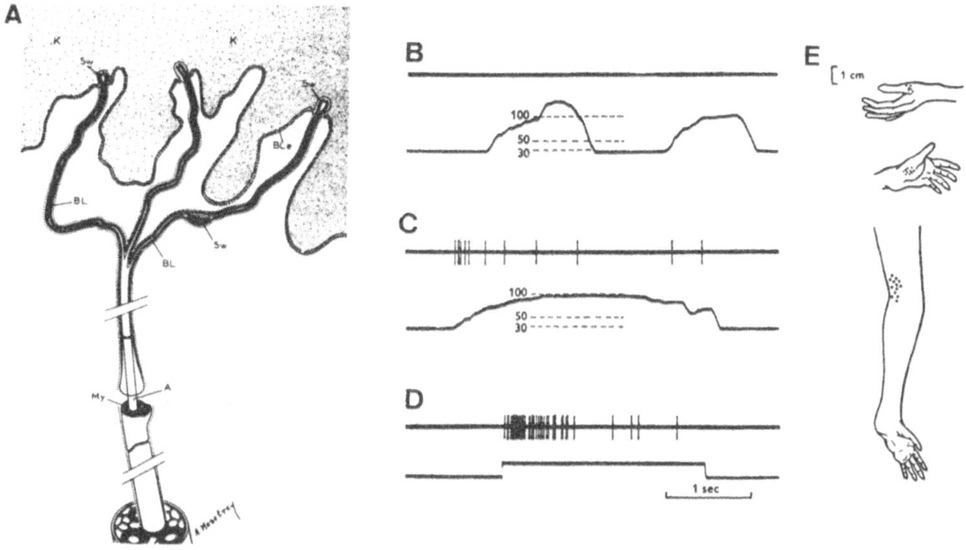

FIGURE 2.16. Aδ mechanical nociceptors. A, structure of a cutaneous Aδ mechanical nociceptor. After losing their myelin sheaths, several branches penetrate into the basal epidermis. Most of the axonal tips, however, remain enveloped by Schwann cells (Sw). BL = basal lamina; My = myelin; A = axon. (From Perl, 1984.) B–D, responses of an Aδ mechanical nociceptor to mechanical stimulation of the glabrous skin of a monkey. In B there was no response to pressure by a probe with a 2.2-mm tip diameter (numbers indicate force applied), whereas in C there was a response when the skin was indented with the tip of a needle. In D the skin was pinched with a serrated forceps. E shows the receptive fields of three different Aδ mechanical nociceptors in monkey skin. The receptive fields consisted of sets of punctate spots separated by insensitive zones. (From Perl, 1968.)

by Schwann cells. This feature suggests that the term "free nerve ending" is inaccurate (Kruger et al., 1981), although there are bare areas of the surface membrane. The receptive field of a single afferent fiber in monkey skin consists of a set of small spots (Fig. 2.16E) distributed over an area of about 2 cm^2.

Aδ mechanical nociceptors are excited best by mechanical stimuli that damage the skin (Fig. 2.16B–D; Hunt and McInytre, 1960c; Burgess and Perl, 1967; Perl, 1968; Handwerker et al., 1987). Thresholds for these receptors vary. The lowest thresholds are in the innocuous range, although the receptors discharge progressively more as the stimulus intensity extends into the noxious range. Repeated strong mechanical stimulation of these receptors can sensitize them to later mechanical stimuli; when sensitized, the receptors develop spontaneous activity and expanded receptive fields (Reeh et al., 1987). It has been thought that these receptors do not respond to noxious heat or to algesic chemicals under normal conditions (Burgess and Perl, 1967; Burgess et al., 1968; Perl, 1968; Beck et al., 1974; Georgopoulos, 1976; Lynn and Carpenter, 1982). However, Aδ mechano-heat nociceptors have been described (see Iggo and Ogawa 1971; Beck et al., 1974; Georgopoulos, 1976; 1977; Kumazawa and Perl, 1977; LaMotte, R. H. et al., 1982; Treede et al., 1998). These respond well to both noxious mechanical and heat stimuli, although the heat threshold of these is higher than that of C polymodal nociceptors (LaMotte, R. H. et al., 1983).

Aδ mechano-heat nociceptors respond best to high rates of heat stimulation, in contrast to C mechano-heat nociceptors, which preferentially respond to low rates of heating (Yeomans and Proudfit, 1996; Yeomans et al., 1996). Aδ mechano-heat nociceptors also respond to intense cold stimuli (thresholds in the range of 14 to −18 °C; Simone and Kajander, 1996; 1997). Previous reports had indicated that less intense cold stimuli were ineffective, but in the experiments of Simone and Kajander (1996; 1997), the cold stimuli ranged down to well below 0 °C, and in these experiments all Aδ nociceptors that were responsive to mechanical stimulation were also responsive to cold. Therefore, a better name for these receptors would presumably be mechano-thermal, mechano-cold, or mechano-heat-cold nociceptors. Aδ and C mechano-cold nociceptors respond both to intense mechanical stimuli and to extreme cold (Iggo, 1959; Bessou and Perl, 1969; Georgopoulos, 1976; 1977; Simone and Kajander, 1996; 1997).

Recent studies using an electrical search stimulus to identify Aδ nociceptors in monkey skin (see above) have demonstrated that there are many more of these receptors than had previously been recognized. This is because a large fraction of these receptors are insensitive to mechanical stimuli and so are not observed when mechanical search stimuli are used. Such receptors have been termed "mechanically insensitive afferents" (Meyer et al., 1991). Most of the mechanically sensitive and insensitive Aδ nociceptors in monkey skin respond to noxious heat. However, such units fall into two categories. Prolonged heat pulses activate Type I Aδ mechano-heat nociceptors, but the peak discharge is delayed until well after the stimulus onset (Fig. 2.17A; Treede et al., 1998). These heat responses were undoubtedly missed in earlier work on these afferents (Perl, 1968), leading investigators to call the receptors Aδ mechanical nociceptors. The thresholds of these nociceptors

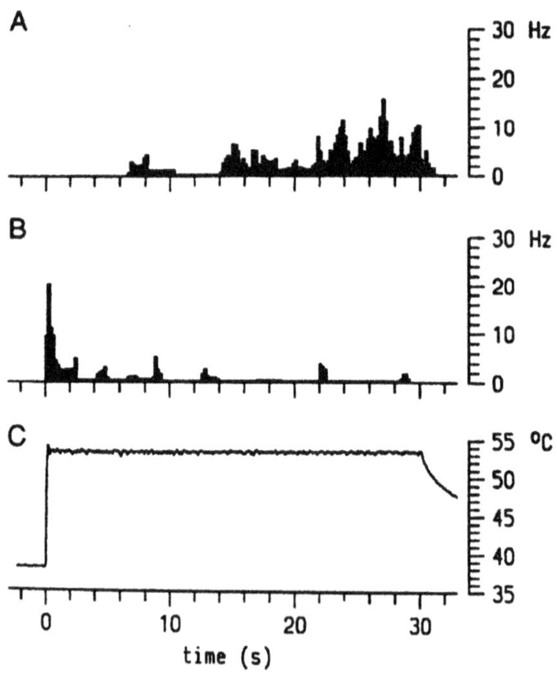

FIGURE 2.17. Responses typical of Type I and Type II Aδ mechano-heat nociceptors found in monkey hairy skin. A, response of a Type I Aδ nociceptor (mechanical threshold 2.8 bar; heat threshold >53 °C). The latency of the heat response was 6.8 s, and the latency of the peak response was 27.1 s. B, response of a Type II Aδ nociceptor (mechanical threshold 7.3 bar, heat threshold 47 °C). The latency of the response was 136 ms, and the peak latency was 0.3 s. C is the temperature monitor. (From Treede et al., 1998.)

to brief heat pulses are unchanged, and so the late response to heat does not reflect sensitization (cf. Fitzgerald and Lynn, 1977). Type II Aδ mechano-heat nociceptors respond promptly to heat stimuli, but the response adapts quickly (Fig. 2.17B; Treede et al., 1998). Because of the early and vigorous responses of Type II receptors to heat, it was suggested that these nociceptors account for the first pain sensation that is produced by heat stimuli that have a rapid onset (Treede et al., 1998).

Administration of a mixture of inflammatory mediators ("inflammatory soup" containing bradykinin, serotonin, prostaglandin, and histamine) activates a large fraction of the Aδ nociceptors in monkey skin (Fig. 2.18A, C; Davis, K. D. et al., 1993). This often results in sensitization of the afferents to mechanical stimuli, as shown by a reduced threshold for mechanical stimuli and enlarged receptive fields (Fig. 2.19; see section below on **Sensitization of Nociceptors**). Another substance that can activate and sensitize Aδ nociceptors is glutamate (Kinkelin et al., 2000; Du et al., 2001). Some Aδ nociceptors have VR1 receptors and so can respond to capsaicin (Nagy, J. I. et al., 1983; Michael and Priestley, 1999). Thus, many cutaneous Aδ nociceptors can be characterized as polymodal nociceptors, since they respond to mechanical, thermal, and chemical stimuli (Treede et al., 1998). Only a few are actually high-threshold mechanical nociceptors of the type originally proposed (Perl, 1968).

C Nociceptors

C nociceptors are an abundant receptor type, especially in primates (including humans). These receptors are formed by free nerve endings, many of which extend into the epidermis. In humans, the maximum receptive field diameters for mechanical stimulation of human C mechanoheat receptors average 22.3 mm on the lower leg, 15.3 mm on the foot, and 7.7 mm on the toes (Schmidt, R. et al., 1997). The sizes of the cutaneous receptive fields of different nociceptors accounts for the accuracy of the localization that is observed, and the reduced sizes of the receptive fields of nociceptors in the distal limb corresponds to the better discrimination that is possible at more distal sites (Schmidt, R. et al., 1997).

Jørum et al. (1989) observed that the axons of C nociceptors of humans that supply a given area of skin tended to cluster in peripheral nerves. In the same study, the ability of the subjects to estimate the location of a heat stimulus was tested after conduction in large afferents was blocked by pressure. The average distance between the estimated location of the painful stimulus and its actual location was 17.1 mm. This is only slightly less accurate than the localization of tactile stimuli, which was in error by an average of 11.8 mm. This shows that the localization of painful stimuli does not depend on simultaneous tactile input. In other experiments, painful stimuli applied to skin could be localized to within about 9.5 mm under normal conditions and 8.9 mm when conduction in A fibers was blocked (Koltzenburg et al., 1993). In the same study, tactile stimuli applied with a von Frey filament were localized on average within 5.5 mm. The fact that blockade of large afferents did not alter the ability of the subjects to localize painful stimuli indicates not only that the localization of painful stimuli does not depend on tactile cues but also that noxious inputs reduce the capability of the subjects to localize tactile stimuli (Koltzenburg et al., 1993).

C polymodal nociceptors respond well to noxious mechanical, thermal, and chemical stimuli (Fig. 2.20A–E; Bessou and Perl, 1969; Beck et al., 1974; see also Handwerker and Neher, 1976; Fleischer et al., 1983; Martin and Murphy, 1995; Sorkin et al., 1997; and historical review by Handwerker, 1996). However, their chemical sensitivity is often not

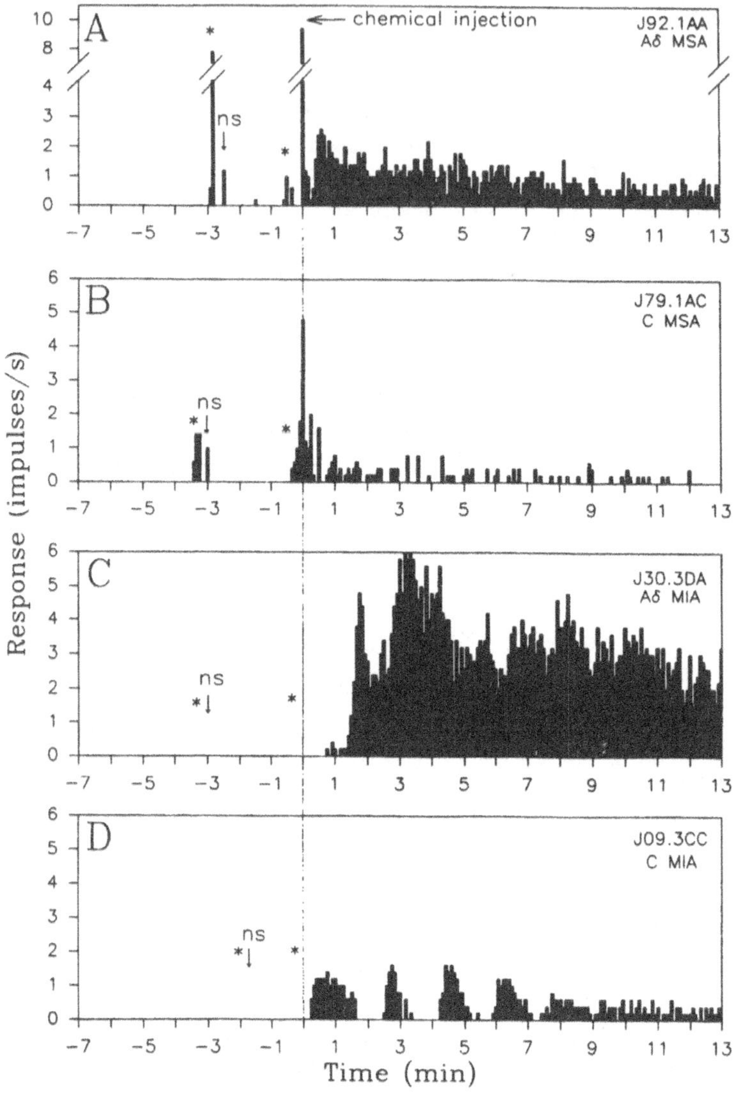

FIGURE 2.18. Effects of chemical stimulation, using "inflammatory soup," on the responses of Aδ nociceptors in monkey hairy skin. A–D, responses of four different Aδ nociceptors to injection of a mixture of bradykinin, histamine, serotonin, and prostaglandin E_1 into the skin.* Shows time of needle insertions; ns, normal saline. The receptors in A and B were mechanically sensitive (thresholds of 4.8 and 4.2 bar), and the one in B was also responsive to heat. The receptors in C and D were mechanically insensitive (thresholds of 17.5 and 10.1 bar), and the one in C did not respond to heat. Note the oscillatory response in D. (From Davis et al., 1993.)

tested; in such cases these receptors are generally called C mechano-thermal nociceptors. As for many other nociceptors, threshold mechanical stimuli can be in the innocuous range. The effective thermal stimuli are noxious heat (greater than 45 °C; Bessou and Perl, 1969; Handwerker and Neher, 1976) or intense cold (Simone and Kajander, 1996). As already mentioned, C nociceptors are preferentially activated by low rates of skin heating, whereas Aδ nociceptors are preferentially activated by high rates of skin heating

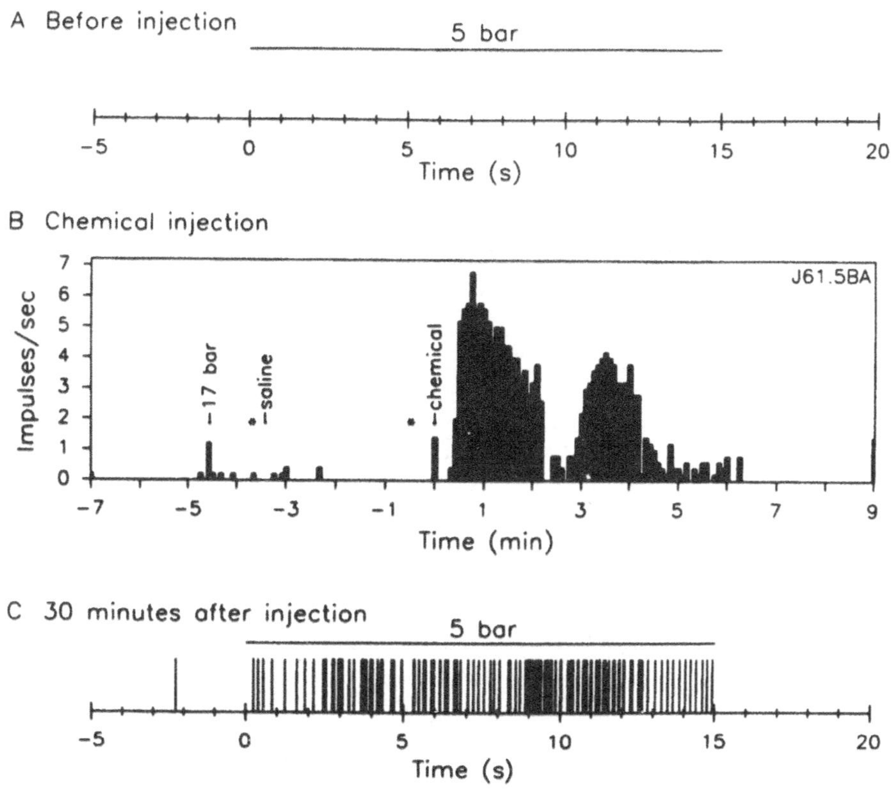

FIGURE 2.19. Sensitization of an Aδ mechanically insensitive nociceptor to mechanical stimulation following injection of inflammatory soup into the receptive field. A, the receptor did not respond to a 5.1-bar mechanical stimulus applied to the receptive field; it did respond to a 10.1-bar stimulus. B, responses to a 17-bar stimulus, injection of saline, and injection of inflammatory soup. C, after the injection, the unit responded to the 5.1-bar stimulus. The threshold to the von Frey filament dropped to 4.2 bar, and the receptive field size increased. There was no response to heat before or after the injection. (From Davis et al., 1993.)

(Yeomans and Proudfit, 1996; Yeomans et al., 1996). Yeomans and Proudfit (1996) suggest that C mechano-heat nociceptors evoke burning pain, whereas Aδ mechano-heat nociceptors produce pricking pain. Effective chemical stimuli include topical application of acid solutions or injections of algesic chemicals. These nociceptors can be sensitized by inflammatory mediators (Kessler et al., 1992; Kress et al., 1992), including low pH (Steen et al., 1992), and by glutamate (Du et al., 2001).

C mechanical nociceptors do not respond to heat (Iggo, 1960; Bessou and Perl, 1969; Beck et al., 1974; Georgopoulos, 1976; Kumazawa and Perl, 1977; LaMotte, R. H. and Campbell, 1978; Fleischer et al., 1983; Schmidt, R. et al., 1997). It is possible that some of these receptors are also C cold nociceptors, since sufficiently cold stimuli may not always have been used in characterizing them (Bessou and Perl, 1969). Some could thus be C mechano-cold nociceptors (Simone and Kajander, 1996).

Some C nociceptors in humans respond to heat but not to mechanical stimuli; these have been termed C-heat units (Weidner et al., 1999). These mechanically insensitive C nociceptors can be sensitized by prolonged intense mechanical stimuli, which may account for the incrementing nature of pain induced in humans by long-lasting noxious mechanical

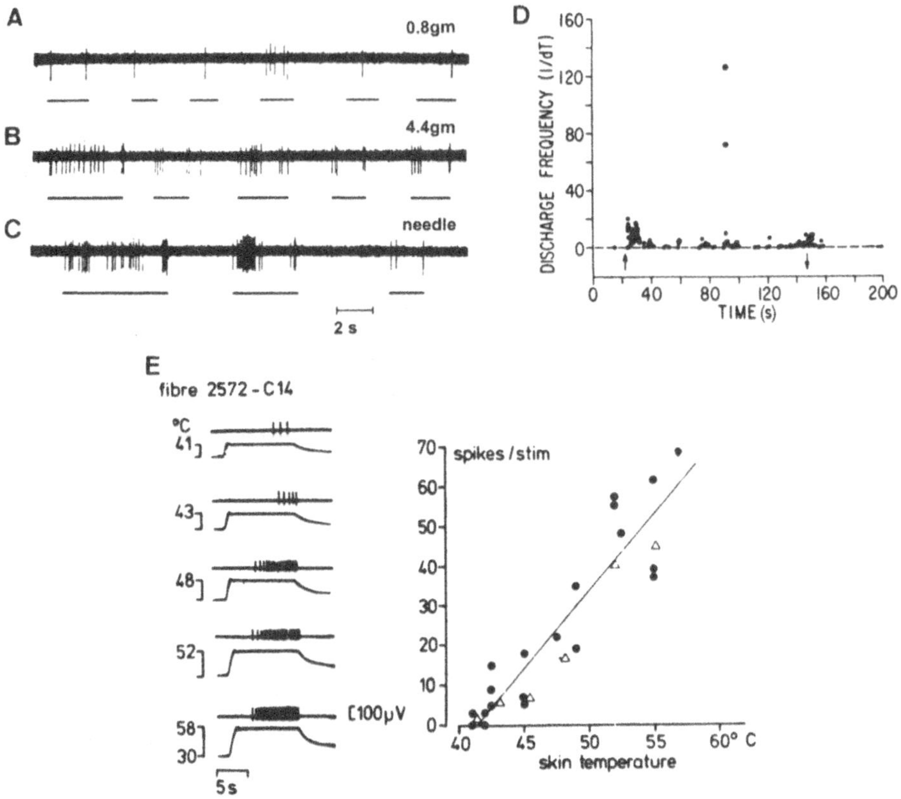

FIGURE 2.20. A, responses of a C-polymodal nociceptor to stimulation with a von Frey filament that bent with a force of 0.8 g. The horizontal bars under the recordings show the times when the stimulus was applied. B, greater response to a 4.4-g von Frey filament. C, response to stimulation with a needle that penetrated the skin. D, discharge produced in a C-polymodal nociceptor by the application of dilute acid to the skin. (From Bessou and Perl, 1969.) E, responses of a C-polymodal nociceptor to graded heat stimuli. The relationship between the number of discharges evoked by each stimulus and skin temperature is shown by the graph at the right. (From Beck et al., 1974.)

stimuli (Schmidt, R. et al., 2000). Some C nociceptors in humans are initially insensitive to both mechanical and heat stimuli. These do respond to injection of capsaicin into the skin and so are chemosensitive (Weidner et al., 1999). Activity-dependent slowing of the action potentials of mechanically insensitive C fibers is much greater than that in mechanically sensitive C fibers (Weidner et al., 1999; 2000).

For more detail about cutaneous nociceptors, see Iggo (1959); Iriuchijima and Zotterman (1960); Fjällbrant and Iggo (1961); Bessou and Perl (1969); Iggo and Ogawa (1977); Van Hees and Gybels (1972); Beck and Handwerker (1974); Beck et al. (1974); Torebjörk (1974); Georgopoulos (1976; 1977); Handwerker and Neher (1976); Croze et al. (1976); Kumazawa and Perl (1977); LaMotte, R. H. and Campbell (1978); Lynn and Carpenter (1982); Adriaensen et al. (1983; 1984); Fleischer et al. (1983); LaMotte, R. H. et al. (1983); Handwerker et al. (1987); Kenins (1988); Lang et al. (1990); Kress et al. (1992); Koltzenburg and Handwerker (1994); Schmidt, R. et al. (1995; 1997; 2000); Khalsa et al. (1997); Weidner et al. (1999; 2000; 2002); Schmelz et al. (2000). See also Figs. 12.20–5.

Transduction in Nociceptors

Since nociceptors can respond to various combinations of mechanical, thermal, and chemical stimuli, it should not be surprising to find that nociceptors utilize a variety of transduction mechanisms (Fig. 2.21; see review by Woolf and Salter, 2000). The transduction mechanisms that are operative at a given time depend on the type of nociceptor and on whether or not it has been sensitized.

The activation of nociceptors in rat skin has been found to signal tensile and compressive strains in the local environment of the nerve endings (Khalsa et al., 1997). It is unclear what transducer protein is responsible for sensitivity to noxious mechanical stimuli. Woolf and Salter (2000) and García-Añoveros et al. (2001) suggest that it is possible that this protein may belong to the DEG/ENaC ion channel family described in *Caenorhabditis elegans* (Fig. 2.21; Waldmann and Lazdunski, 1998; see section on **Transduction in mechanoreceptors**). As already mentioned, the mRNA for BNaC1α subunits has been observed in a few small, presumably nociceptive DRG cells (Price, M. P. et al., 2000; García-Añoveraos et al., 2001). This isoform was identified immunohistochemically in some free nerve endings. It was proposed that this molecule might serve as a mechanical transducer in Aδ mechanical nociceptors, as well as in mechanoreceptors (García-Añoveraos et al., 2001).

Responses to noxious heat depend in part on the presence of VR1 or VRL1 vanilloid receptors or other heat-sensitive molecules (Tominaga et al., 1998; Caterina and Julius, 2001;

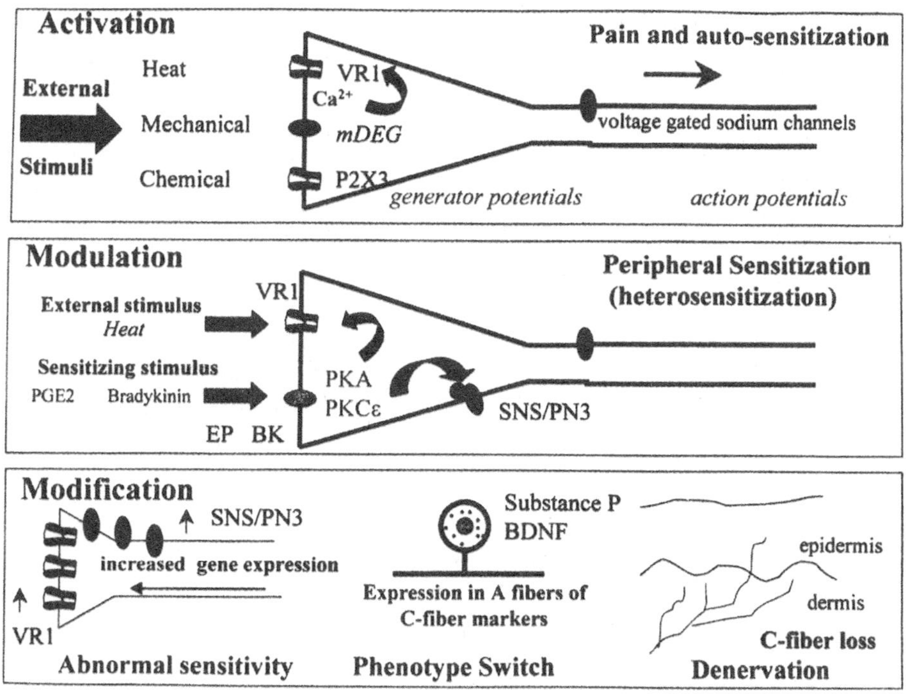

FIGURE 2.21. Sensitization and other plastic changes in primary afferent neurons in response to noxious stimuli. Some of the molecules are shown that are involved in the activation of nociceptors, their modulation during sensitization, and their long-term modification in response to changes in gene expression. (From Woolf and Salter, 2000.)

Caterina et al., 1999; 2000; Davis, J. B. et al., 2000). Detection of changes in pH depends both on VR1 receptors (Tominaga et al., 1998; Caterina et al., 1999; cf. Kress et al., 1996a) and on other acid-sensing (ASIC) channels (Habelt et al., 2000; Immke and McCleskey, 2001; Sutherland et al., 2001). Noxious chemical stimuli are detected by the activation of any of a large number of receptors: VR1, purinergic (especially $P2X_3$ or heteromeric $P2X_{2/3}$ receptors), bradykinin, glutamate, and many other receptor types (Szolcsanyi et al., 1988; Wood et al., 1988; Dray et al., 1990; Lang et al., 1990; Cook, S. P. and McCleskey, 2002; Cook, S. P. et al., 1997; Burgard et al., 1999; see reviews by Coggeshall and Carlton, 1997; Bradbury et al., 1998; McCleskey and Gold, 1999; Burnstock, 2000; Kinkelin et al., 2000).

VR1 receptors are found in some Aδ nociceptors (Nagy, J. I. et al., 1983; Michael and Priestley, 1999), but mRNA for VR1 is mostly expressed in C nociceptors (Michael and Priestley, 1999). Of DRG cells that fail to stain for neurofilament protein and are therefore likely to have unmyelinated axons, 83% express VR1. These VR1-containing neurons include 65% of DRG cells that also label for trkA and 75% of DRG cells that bind IB4. A small population of DRG cells that contain immunoreactivity for RET have a particularly high concentration of mRNA for VR1 (Michael and Priestley, 1999). These neurons do not stain for trkA or IB4 and are likely to be GDNF-sensitive (cf. Bennett et al., 1998a). Most DRG cells that contain substance P or CGRP also express VR1, except for large CGRP-immunoreactive cells, which do not (Michael and Priestley, 1999).

Nociceptors that are activated by capsaicin at low doses can be desensitized or the terminals killed by high doses of capsaicin or by administration of resiniferotoxin, and the entire primary afferent nociceptive neuron can be destroyed by neonatal capsaicin treatment (see reviews by Lynn et al., 1984; Buck and Burks, 1986; Holzer, 1991).

A variant of the vanilloid receptor called the stretch-inactivated channel (SIC) has been found on dorsal root ganglion cells. This channel is activated by cell shrinkage, and so it has been suggested that it mediates the pain that is caused by exposure of tissue to hypertonic solutions (Schumacher et al., 2000).

The responsiveness of nociceptors to adequate stimuli depends in part on the voltage-gated ion channels that are involved in action potential generation. Important ion channels in nociceptors include voltage-gated sodium channels, including tetrodotoxin-resistant sodium channels, and potassium channels (Gold et al., 1996a; 1998; Khasar et al., 1998). Nociceptors express voltage-gated potassium channel Kvl.4 subunits in transient (A-type) potassium channels, in contrast to mechanoreceptors, which express Kv1.1, Kv1.2, and Kvβ2.1 subunits (Rasband et al., 2001). Activation of these channels would interfere with the excitability of nociceptive axons. Conversely, a reduction in their activity should lead to hyperexcitability. This may account in part for the increased spontaneous activity of nociceptors in neuropathic pain, since Kv1.4 subunits are greatly reduced in ligated spinal nerves in a model of neuropathic pain (Rasband et al., 2001).

Sensitization of Nociceptors

Nociceptors can be sensitized by exposure to irritant chemicals, inflammatory mediators, and/or low pH (Figs. 2.18 and 2.19), or following noxious heat stimulation (Fig. 2.22; Bessou and Perl, 1969; Fitzgerald and Lynn, 1977; Meyer and Campbell, 1981; LaMotte, R. H. et al., 1982; 1983; Kocher et al., 1987; Reeh et al., 1987; Kirchhoff et al., 1990; Handwerker et al., 1991; Kessler et al., 1992; Kress et al., 1992; Steen et al., 1992; 1995; 1996; Davis, K. D. et al., 1993; Schmelz et al., 1994; 1996; 2000; Schuligoi et al., 1994; Schmidt, R. et al.,

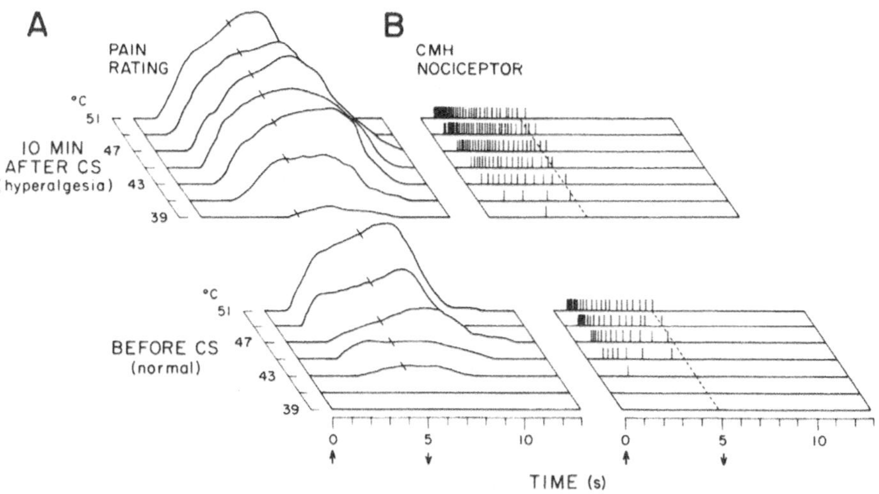

FIGURE 2.22. Thermal hyperalgesia and sensitization of the responses of a cutaneous C nociceptor. A, human magnitude estimates of pain in response to graded intensities of heat stimuli before and after a conditioning stimulus (CS) that produced a mild burn (50 °C for 100 s). The pain responses for any given thermal stimulus increased following the burn. B, responses of a C polymodal nociceptor recorded from a monkey following the same heat stimuli before and after a comparable burn. The responses of the C nociceptor to any given stimulus increased following the burn. Presumably, the increased responses of such nociceptors to heat stimuli led to the primary heat hyperalgesia reported by the human subject. (From LaMotte et al., 1983.)

1995; Kress et al., 1997; Sorkin et al., 1997; Chen, X. et al., 1999; Du et al., 2001; Liang et al., 2001; see review by Woolf and Salter, 2000). Different mechanisms may lead to selective sensitization of the responses to mechanical, thermal, or chemical stimuli.

Sensitization of a C mechano-heat nociceptor in a monkey and the development of heat hyperalgesia in a human subject after a mild burn are illustrated in Fig. 2.22 (LaMotte, R. H. et al., 1983). Evidence of this kind has led to the concept that sensitized primary afferent nociceptors can help account for primary heat hyperalgesia (Meyer and Campbell, 1981; LaMotte, R. H. et al., 1983; Carlton et al., 1998a). Primary mechanical hyperalgesia after injury seems more likely to be due to sensitization of Aδ mechanical nociceptors (Fig. 2.19) than of C mechano-heat nociceptors, since the latter show fewer signs of sensitization to mechanical stimuli (Reeh et al., 1987).

The interactions between inflammatory mediators and low pH are not reciprocal. Inflammatory mediators sensitize the responses of nociceptors to low pH (Steen et al., 1995; 1996), but low pH does not sensitize the responses to inflammatory agents (Steen et al., 1995). However, low pH does sensitize responses to mechanical stimuli (Steen et al., 1992).

If a portion of the axon of a C mechano-heat nociceptor is sensitized to heat (but not to mechanical stimuli) by application of an irritant, such as mustard oil, the remainder of the terminal arborization does not necessarily become sensitized (Schmelz et al., 1994; 1996; see also Martin, H. A. et al., 1987). However, the entire terminal arborization can be sensitized, for example, by widespread application of mustard oil or capsaicin (Schmelz et al., 1994). This indicates that sensitization can be very localized within a single axon if the sensitizing stimulus is confined to just a part of the axon, whereas if the sensitizing stimulus is applied simultaneously to a number of branches of the axon, each can become sensitized. These observations suggest that the signal transduction mechanisms underlying peripheral sensitization are confined within a restricted length of an axon. One implication is that secondary hyperalgesia is unlikely to depend on a peripheral mechanism that involves the spread of peripheral sensitization to branches of axons that were unaffected by the irritant but rather is due to central sensitization (Schmelz et al., 1996; see Chapter 7).

Intradermal capsaicin injections in humans have quite different effects on mechanically sensitive versus mechanically insensitive C nociceptors (Schmelz et al., 2000). Most C nociceptors are activated by capsaicin. However, mechanically sensitive nociceptors produce only a brief discharge, and then they become desensitized to mechanical stimuli. By contrast, mechanically insensitive C nociceptors are activated for as long as 15–20 minutes, and then they develop mechanical and heat responsiveness. The heat threshold can fall to as low a temperature as 35 °C. The latter units seem likely to account for the primary heat and mechanical hyperalgesia that develops following intradermal injection of capsaicin. Mechanically insensitive C nociceptors in human subjects become active during a long-lasting noxious mechanical stimulus, in contrast to mechanically sensitive ones, which adapt to a similar prolonged mechanical stimulus (Schmidt et al., 2000). It was proposed that sensitized C nociceptors that had originally been classified as mechanically insensitive help account for the primary mechanical hyperalgesia that develops after such stimuli. Neither type of unit becomes responsive to stimuli applied in the area of secondary hyperalgesia (Schmelz et al., 2000).

Sensitization of nociceptors is triggered by the action of bradykinin, prostaglandins and other arachidonic acid products, serotonin, catecholamines, ATP, adenosine, histamine, excitatory amino acids, and/or neurokinins on surface membrane receptors. For more details, see Martin, H. A. et al. (1987); Cohen and Perl (1990); Martin, H. A., (1990); Taiwo and Levine (1990); White, D. M. et al. (1990); Nakamura-Craig and Gill (1991); Kessler et al. (1992); Davis, K. D. et al. (1993); Rueff and Dray (1993a, b); Ferreira and Lorenzetti (1994); Gold et al. (1994); Khasar et al. (1994); Schuligoi et al. (1994); Kress et al. (1997); Chen, X. et al., (1999); Khasar et al. (1999a); Kress and Guenther (1999); Du et al. (2001); Liang et al. (2001); Shin, J. et al. (2002); Zeitz et al. (2002); and reviews by Walker et al. (1995); Coggeshall and Carlton (1997); and Burnstock (2000). Nerve growth factor plays an important role in the sensitization of nociceptors that occurs in inflammation (Koltzenburg et al., 1999).

A key step in the initiation of sensitization is an increase in intracellular Ca^{++} concentration (Guenther et al., 1999; Kress and Guenther, 1999). This leads to the activation of second messenger cascades, which can then alter the responsiveness of the afferents through actions on ion channels (Fitzgerald, E. M. et al., 1999; England, S. et al., 1996; Gold et al., 1996b; 1998; Nicol et al., 1997; Khasar et al., 1998; Djouhri and Lawson, 1999; cf. Akopian et al., 1999) or on membrane receptors, such as VR1 (Lopshire and Nicol, 1998; see Guenther et al., 1999).

The second messenger systems that are involved in sensitization of primary afferent nociceptors include the protein kinase C, protein kinase A, and protein kinase G pathways (Dray et al., 1988; Taiwo and Levine, 1991; Rueff and Dray, 1993a; Cui and Nicol, 1995; Ouseph et al., 1995; England, S. et al., 1996; Kress et al., 1996b; Wang, J. F. et al., 1996; Gold et al., 1998; Lopshire and Nicol, 1998; Aley and Levine, 1999; Cesare et al., 1999; Fitzgerald, E. M. et al., 1999; Khasar et al., 1999a, b; Aley et al., 1998; 2000).

An example of the role of signal transduction pathways in the sensitization of nociceptors is provided by studies of the sensitization of nociceptors by PGE_2, 5-HT, or adenosine. Activation of the appropriate membrane receptors results in an increase in voltage-gated currents through a tetrodotoxin-resistant sodium channel (Gold et al., 1996b;

Khasar et al., 1998). Intracellular release of a diffusible second messenger is responsible for this action. Both PKA and PKC are involved in the action of PGE_2 in increasing the TTX-resistant sodium currents (Gold et al., 1998).

In a more detailed study, the sensitization of nociceptors that is produced in rats by injection of PGE_2, an A_2 adenosine receptor agonist or a $5HT_{1A}$ agonist into the paw was found to require the sequential activation of adenylyl cyclase and then PKA. The initiating event involving adenylyl cyclase is transient, but activation of PKA leads to the maintenance of hyperalgesia for several hours (Aley and Levine, 1999). A longer lasting hyperalgesia can be produced if carrageenan is injected first and then later PGE_2. The prostaglandin injection that follows the carrageenan injection results in hyperalgesia that lasts for more than a day. Whereas the early phase of the hyperalgesia depends on PKA, the later, more prolonged component depends on PKCϵ (Aley et al., 2000). As mentioned, a key factor in the sensitization of primary afferent neurons by prostaglandins and some other agents is phosphorylation of tetrodotoxin-resistant sodium channels (Aley et al., 1998).

Nitric oxide release may also contribute to the sensitization process, although there is controversy about this and about the role of the cGMP/protein kinase G system (Duarte et al., 1992; Ferreira and Lorenzetti, 1994; Ouseph et al., 1995; Kress et al., 1996b; Aley et al., 1998; Lopshire and Nicol, 1998; Chen and Levine, 1999; Nakamura and Shiomi, 1999).

Epinephrine has been found to produce sensitization of the responses of nociceptors to mechanical stimuli in rats, both by the production of mechanical hyperalgesia and by enhanced excitability of small DRG cells (Khasar et al., 1999a). Signal transduction cascades that are involved include the PKA and PKC pathways. This effect is mediated by β-adrenoreceptors. Since elevated circulating epinephrine levels occur during stress, a role for epinephrine in hyperalgesia is reasonable. Another signal transduction pathway that contributes to the sensitization of nociceptors by epinephrine acting on β_2 adrenoreceptors involves extracellular signal-regulated kinases (ERKs; Aley et al., 2001).

Cesare et al. (1999) provide evidence that bradykinin causes heat sensitization of dorsal root ganglion cells by causing a translocation of PKCϵ to the cell membrane. A similar translocation of this isoform was produced by phorbol ester. Intracellular release of a PKCϵ analog from a patch-clamp pipette increased the current produced in the DRG neurons by heat, and an inhibitor of PKCϵ translocation prevented the effect of bradykinin on the heat-activated current (Cesare et al., 1999).

Bradykinin also sensitizes nociceptors during inflammation in part by activation of the 12-lipogenase pathway (Shin et al., 2002). This pathway produces an eicosanoid, 12-HPETE, which is an agonist of VR1.

Liang et al. (2001) have proposed a theory concerning the mechanism of sensitization of nociceptors by bradykinin that would involve heat-sensitive membrane receptors, such as VR1. They find an apparent dissociation between the excitation of nociceptors by bradykinin and sensitization of these neurons to heat stimuli. Repeated responses of afferents that are excited by bradykinin show tachyphylaxis, and yet sensitization is maintained (Fig. 2.23). Their hypothesis is that each bradykinin administration lowers the heat threshold to a level below the bath temperature. Although there is desensitization of this effect with repeated doses, the heat threshold still dips below the bath temperature, although to a lesser extent, leading to less excitation (tachyphylaxis). The same process would cause sensitization of the responses of the unit to heat. This idea is consistent with the known interactions between temperature and VR1 (Tominaga et al., 1998; Vyklicky et al., 1999).

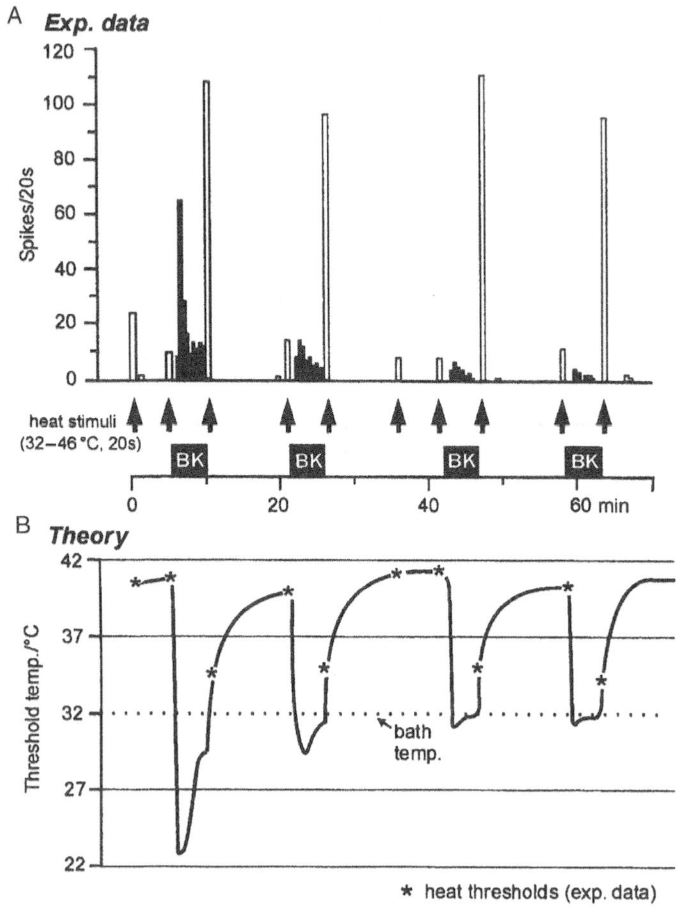

FIGURE 2.23. Theory of bradykinin action. A, responses of a C polymodal nociceptor to repeated applications of bradykinin (BK) and heat stimuli (arrows). Although the responses to BK showed tachyphylaxis, the sensitization of the responses to heat remained stable. B, theoretical model suggesting that bradykinin lowers the heat threshold of the unit to a level below ambient temperature (32 °C), and so the tissue temperature determines the reponsiveness to bradykinin. Desensitization reduces the excitatory effect of bradykinin. (From Liang et al., 2001.)

Strong or persistent noxious stimuli can trigger changes in gene expression in primary afferent fibers through the activation of transcription factors (see review by Woolf and Salter, 2000). The result can be an enduring change in the expression of ion channels, receptors, neuropeptides (such as substance P), and/or growth factors (such as nerve growth factor and brain-derived neurotrophic factor) (Djouhri and Lawson, 1999; Kerr, B. J. et al., 1999; Mannion et al., 1999).

The plastic changes that occur in primary afferent fibers after nerve injury can be quite different from those that occur as a result of inflammation and that lead to sensitization of nociceptive afferents. The changes produced by nerve injury can occur not only in nociceptors but also in mechanoreceptors. Axotomy may result in sprouting of the axons of primary afferent fibers and rewiring of dorsal horn circuits (Woolf et al., 1992; Koerber et al., 1994; 1999; Kohama et al., 2000 however see Tong et al., 1999; Bao et al., 2002) and in down-regulation of peptides and receptors, perhaps secondarily to a reduction in the availability of neurotrophic factors, such as NGF or GDNF (Michael and Priestley, 1999). Many dorsal root ganglion cells that label with RT97 or with IB4 express *raf* serine/threonine protein kinases; the *raf* protein kinase is translocated to the surface membrane following transection of the sciatic nerve (Mihaly et al., 1996). This suggests a role for this protein kinase in the changes observed following nerve injury.

Neurogenic Inflammation

Peptidergic and Non-Peptidergic Nociceptors

Some primary afferent neurons that have terminals in the dermis and epidermis and that may be nociceptors are immunoreactive to antibodies against calcitonin gene-related peptide, substance P, and/or other peptides (O'Shaughnessy et al., 1983; Dalsgaard et al., 1984; Gibbins et al., 1985; 1987a; Alvarez et al., 1988; Ishida-Yamamoto et al., 1989; Kruger et al., 1989; O'Brien et al., 1989). A few of these peptidergic neurons contain co-localized fluoride resistant acid phosphatase (FRAP), although most small DRG cells contain either a peptide or FRAP but not both (Dalsgaard et al., 1984). In a study in which the receptive field properties of individual dorsal root ganglion cells labeled for substance P in the guinea-pig were examined, all of the substance P-positive neurons were nociceptors (Lawson et al., 1997).

The presence of peptides in these nociceptive neurons may reflect their synthesis by the sensory neurons for release centrally as neurotransmitters or modulators. However, peptidergic primary afferent neurons also have an effector role in the periphery that involves the release of peptides. For instance, capsaicin and also noxious heat cause the release of CGRP in the skin (Kilo et al., 1997; Kessler et al., 1999). Much of the release is Ca^{2+}-dependent. Such a peripheral release of peptides is known to play an important role in neurogenic inflammation (see next section). Furthermore, substance P has been shown to potentiate the effect of inflammatory mediators in sensitizing nociceptor terminals (Kessler et al., 1992). NK1 receptors have been found on substance P-positive dorsal root ganglion cells but on very few IB4-containing dorsal root ganglion neurons, suggesting that NK1 receptors could serve as autoreceptors (von Banchet and Schaible, 1999).

Other, non-peptidergic nociceptors have binding sites for the lectin, IB4 (Guo, A. et al., 1999). These primary sensory neurons form a population of nociceptors that is distinctly different from the peptidergic population. For example, peptidergic nociceptors terminate in laminae I and IIo of the superficial dorsal horn (as well as in laminae V and X); by contrast, IB4-positive primary afferent neurons synapse in lamina II_i. NaN tetrodotoxin-resistant sodium channels are preferentially expressed in IB4-positive neurons (Fjell et al., 1999a; 2000). Many of the same neurons also contain the SNS sodium channel, another tetrodotoxin-resistant sodium channel that is expressed preferentially in small dorsal root ganglion neurons (Fjell et al., 1999b). About half of IB4+ and half of IB4− neurons express SNS channels (Fjell et al., 1999a; 2000).

Nociceptors and Neurogenic Inflammation

Neurogenic inflammation depends on intact innervation (Bruce, 1910; Jancsó, N. et al., 1967) and involves plasma extravasation from postcapillary venules and arteriolar vasodilation ("flare") that spreads well beyond the area of damage (reviewed in Lewis, 1927;

Holzer, 1988; Geppetti and Holzer, 1996; Holzer and Maggi, 1998). Neurogenic inflammation can be initiated by antidromic activation of the terminals of sensory axons (Bayliss, 1901; reviewed by Willis, 1999). The responsible afferents include C polymodal nociceptors but not C mechanoreceptors or warm receptors (Kenins, 1981). Although the effective terminals are mainly those of unmyelinated axons (Low and Westerman, 1989), some Aδ fibers also contribute (Jänig and Lisney, 1989). In humans, cutaneous stimuli of sufficient strength to evoke vasodilation often cause pain, although weaker, non-painful stimuli can also result in vasodilation (Magerl et al., 1987; Koltzenburg and Handwerker, 1994). The initial events in neurogenic inflammation depend on the release of peptides (substance P, calcitonin gene-related peptide) and possibly other substances (such as excitatory amino acids) from sensory nerve terminals (Lembeck and Holzer, 1979; Kenins, 1981; Brain et al., 1985; Gamse and Saria, 1987; Magerl et al., 1987; Szolcsanyi, 1988; 1996; Delay-Goyet et al., 1992; Kessler et al., 1999). Michaelis et al. (1996b) speculate on the possibility that "silent" nociceptors may be especially involved in an efferent, instead of or in addition to an afferent function, by participating in peripheral release of inflammatory substances in response to tissue damage (see also Holzer and Maggi, 1998). Another candidate set of dorsal root ganglion neurons that were considered as possibly having just an efferent function are those that project through the ventral roots (Michaelis et al., 1996b; Holzer and Maggi, 1998; see Coggeshall et al., 1973; 1974; Clifton et al., 1976; Häbler et al., 1990a). However, stimulation of ventral root axons does not cause antidromic vasodilation, and so this possibility was discounted.

Although low pH contributes to inflammation, in part by increasing the release of CGRP from nerve terminals in the skin, low pH by itself does not enhance SP release (Averbeck and Reeh, 2001). Consequently, low pH causes vasodilation but not plasma extravasation (Steen et al., 1995).

Interestingly, peptides released from antidromically activated nerve terminals do not appear to sensitize C nociceptors to thermal or mechanical stimuli, at least in rats and monkeys (Reeh et al., 1986; Meyer et al., 1988; however, cf. Lynn, 1988). Inflammation progresses as substances enter the tissue from the plasma or are released by immune or other cells. Bradykinin is synthesized from plasma proteins, and other inflammatory agents, such as prostaglandins, histamine, serotonin, and cytokines, are released from damaged cells, immune cells, platelets, and other sources. Interactions between some of these agents have been described (Kessler et al., 1999; Averbeck and Reeh, 2001).

The spread of flare and neurogenic edema in the skin is generally attributed to axon reflexes in branches of sensory neurons (Lewis, 1927; cf. Lynn, 1988; Forster et al., 1995; Schmelz et al., 1997; Schmidt, R. et al., 1997). The control of vasodilation seems to differ for superficial versus deeper blood vessels in the skin (Forster et al., 1995). An alternative explanation for the spread of flare and edema, at least in rats, cats, and monkeys, is the generation of dorsal root reflexes in fine afferent fibers under conditions of inflammation (Rees et al., 1994; 1995; 1996; Lin et al., 1999d; 2000; see review by Willis, 1999, and section on Joint Nociceptors; see also Chapter 7).

Interaction of Growth Factors with Nociceptors

The peptidergic, IB4-negative primary sensory neurons have TrkA receptors and are regulated by nerve growth factor (McMahon et al., 1994; Averill et al., 1995). Most of these neurons contain CGRP (Averill et al., 1995). The maintenance of tetrodotoxin-resistant

SNS channels in small primary afferent neurons depends on nerve growth factor, but maintenance of NaN channels does not (Fjell et al., 1999a, b). Glial cell line-derived neurotrophic factor (GDNF) is involved in the maintenance of both SNS and NaN channels in IB4-positive neurons (Fjell et al., 1999a).

Although IB4-containing neurons also have trkA receptors and are regulated by NGF during development, these neurons down-regulate trkA and become dependent on GDNF in adulthood (Bennett, D. L. H. et al., 1996a; Molliver et al., 1997). IB4-positive sensory neurons have the receptors necessary for GDNF signaling (RET, GFRα-1 and GFRα-2), and most lack trk receptors (Bennett, D. L. H. et al., 1998a). Of dorsal root ganglion cells that contain $P2X_3$ purinoceptors, over 95% contain IB4 binding sites, and these receptors are regulated by GDNF (Bradbury et al., 1998).

The p75 receptor is needed for normal responses of Aδ nociceptors to heat stimuli and of C nociceptors to mechanical stimuli (Stucky and Koltzenburg, 1997). Administration of recombinant human NGF to transgenic mice that lack the low affinity p75 receptor results in the development of heat and mechanical hyperalgesia to the same degree as in wild-type animals, suggesting that trkA receptors are sufficient to mediate the pronociceptive action of NGF (Bergmann et al., 1998). However, the p75 receptor is required for the upregulation of bradykinin binding sites on DRG cells by NGF (Petersen et al., 1998b).

Transgenic mice that lack NGF are analgesic in the tail-pinch and tail-flick tests, and they undergo a loss of most (70%) dorsal root ganglion cells at L4 and L5, as well as of a substantial number of sympathetic neurons (Crowley et al., 1994). Administration of NGF in neonatal rats reduces the thresholds of Aδ nociceptors. In adults, NGF administration does not have this effect, but it does result in mechanical and heat hyperalgesia, probably by different mechanisms (Lewin et al., 1993; 1994). Mast cells have to be present for this change to be observed (Rueff and Mendell, 1996). This suggests that hyperalgesia is secondary to activation of mast cells by the release of a neuropeptide, such as substance P. The mast cells produce the hyperalgesia by releasing 5-HT, since the hyperalgesia can be converted to hypalgesia when antagonists of 5-HT receptors are administered (Lewin et al., 1994). NGF administration in neonatal rats increases the proportion of C mechano-heat nociceptors but does not cause their sensitization (Lewin and Mendell, 1994). Neurotrophin-5 has a similar effect; it is unclear if this is mediated through trkA or trk B receptors (Rueff and Mendell, 1996).

Stucky et al. (1999) have shown that over-expression of NGF in the skin of transgenic mice results in dramatic changes in the cutaneous innervation by Aδ and C nociceptors, but not by Aβ fibers. The mice become hyperalgesic. Among the population of fine afferents, there is an increase in both the total numbers and the percentage of nociceptors. The mechanical sensitivity, but not the heat sensitivity, of Aδ nociceptors increases. By contrast, the heat responsiveness of C-fiber nociceptors increases, whereas the mechanical sensitivity decreases. It was proposed that NGF affects the regulation of transduction rather than general excitability.

Injection of NGF into the glabrous skin of rat hindpaws results in thermal hyperalgesia that can be reduced markedly by surgical or chemical sympathectomy (Andreev et al., 1995). This suggests that the action of NGF may be at least partially indirect.

Administration of antibody to nerve growth factor (NGF) to rats in the neonatal period reduces the number of Aδ nociceptors that can be found in the adults and increases the proportion of D-hair follicle afferents (Ritter et al., 1991; Lewin et al., 1992). However, there is no change in the proportion of Aδ fibers that supply structures deep to the skin, This argues against a simple loss of Aδ nociceptors and supports the idea of a shift in the phenotype of primary afferent neurons. Similarly, treatment of young rats with NGF antibodies reduces the proportion of C-mechano-heat nociceptors found in a cutaneous nerve in adults and results in an increase in the proportion of C-mechanical nociceptors (Lewin and Mendell, 1994). Administration of antibodies to NGF in neonatal rats also produces a reversible reduction in the proportion of adult dorsal root ganglion cells that label for calcitonin gene-related peptide (Tonra and Mendell, 1998).

Blocking the action of NGF in rats by injection of a fusion molecule into the skin results in a reduction in the sensitivity of the animals to stimulation with heat and or bradykinin administration (Bennett, D. L. H. et al., 1998b). There is also a reduction in the innervation of the epidermis by axon terminals that presumably belong to C-nociceptors (Bennett, D. L. H. et al., 1998b; cf. Hsieh et al., 1996). Responsiveness to mechanical and noxious cold stimuli is unchanged. This study indicates that endogenous NGF regulates the sensitivity of nociceptors to thermal and chemical stimuli in adult animals.

Inhibitory Interactions with Peripheral Cutaneous Nociceptor Endings

Activation of some membrane receptors on nociceptive terminals in the periphery reduces rather than increases nociception (Stein, 1995; Nakamura and Shiomi, 1999).

Opiate receptors have been found on peripheral nerve terminals in skin. These include μ-, δ-, and κ-opiate receptors (Stein et al., 1989; Stein, 1995; Coggeshall et al., 1997). During inflammation, but not in control animals, activation of these different classes of opiate receptors results in an antinociceptive action (Stein et al., 1989). Injection of the μ-opiate receptor agonist, DAMGO, into the skin reduces the nociceptive behaviors that result from the injection of glutamate (Coggeshall et al., 1997). Interleukin 1β and corticotropin-releasing factor relieve pain by causing the release of opioids from immune cells in inflamed tissue (Schäfer et al., 1994). During inflammation, μ-opiate receptors in primary afferent neurons, including their peripheral terminals, become upregulated (Mousa et al., 2002), and opioids that could activate these receptors are upregulated in immune cells (Cabot et al., 1997; 2001; Mousa et al., 2002). Endogenous ligands for opiate receptors include β-endorphin, endomorphin 1 and 2, enkephalin, and dynorphin. Interestingly, endomorphin 1 and 2 are not upregulated in primary afferents during inflammation, suggesting that the peripheral control of nociception depends on interactions between opiate receptors on neurons and opioids released by immune cells (Mousa et al., 2002).

Cannabinoid receptors are G-protein activated and include the CB1 and CB2 receptors (Matsuda et al., 1990; Munro, S. et al., 1993; see Di Marzo, 1998). CB1 receptors are found in a number of regions of the central nervous system, including the spinal cord (Howlett, 1995; Abood and Martin, 1996), whereas CB2 receptors are concentrated peripherally in the immune system (reviewed in Di Marzo, 1998). However, CB1 receptors have also been found in sensory neurons (Hohmann and Herkenham, 1999a, b). CB1 receptors mediate analgesic as well as anti-inflammatory actions (Richardson et al., 1998). Possible endogenous ligands for the cannabinoid receptors are anandamide, which acts on CB1 receptors, and palmitoyl-ethanolamide, which acts on CB2 receptors (Calignano et al., 1998; Di Marzo, 1998: Hirst et al., 1998; Jaggar et al., 1998; Richardson et al., 1998). One action of CB1 receptors is the inhibition of forskolin-induced cAMP formation; other actions include inhibition of Ca^{2+} channels and activation of a K^+ current (see Di Marzo, 1998). It has been suggested that the endocannabinoids cause presynaptic inhibition by blocking Ca^{2+} channels, as well as suppression of action potentials through activation of K^+ channels. Actions on adenylate cyclase would reduce neurotransmitter synthesis and neuronal plasticity. Such actions could help account for changes in cognition. With respect to nociception, CB1 receptor activation can inhibit the thermal hyperalgesia that is produced by carrageenan injection, as well as the release of CGRP by capsaicin (Richardson et al., 1998). The enzyme "fatty acid amide hydrolase" (FAAH) breaks down anandamide (Di Marzo, 1998).

Peripheral antinociceptive actions have also been described following the activation of somatostatin receptors (Carlton et al., 2001a, b; see also Heppelmann and Pawlak, 1997b; 1999) and $GABA_A$ receptors (Carlton et al., 1999). The mRNA for three different somatostatin receptors ($STTR_{2a}$, $STTR_{2b}$ and $STTR_3$) has been localized to dorsal root ganglia (Senaris et al., 1995; Schulz et al., 1998a, b; Schindler et al., 1999a, b). $SSTR_{2a}$ receptors have been demonstrated immunocytochemically on the terminals of C fibers in the dermal-epidermal junction (Carlton et al., 2001a). An agonist of SSTRs, octreotide, reduces the behavioral responses to formalin injection. The STTR antagonist, c-SOM, reverses this action. Similarly, activation of SSTR results in an inhibition of the responses of nociceptors in an in vitro skin-nerve preparation to noxious heat both in the non-sensitized state and after sensitization by bradykinin. Unlike opioid receptors, somatostatin receptors are tonically active, and so blocking them in normal animals results in nociceptive behaviors (Carlton et al., 2001b).

Following peripheral administration of a low dose of muscimol, a $GABA_A$ receptor agonist, into the skin of the paw, the behavioral responses to formalin injection are reduced, possibly because muscimol interferes with the release of inflammatory agents from nerve terminals by a process that resembles presynaptic inhibition centrally (Carlton et al., 1999). A high dose of muscimol, on the other hand, increases nociceptive behaviors, perhaps by triggering action potentials in a fashion similar to the production of dorsal root reflexes centrally (see Chapter 7).

Cutaneous Chemoreceptors Possibly Responsible for Itch

In animal experiments, some C fibers that are classified initially as polymodal nociceptors respond to cowhage, an itch-provoking substance (Tuckett and Wei, 1987b). By contrast, myelinated afferent fibers do not respond selectively to cowhage (Tuckett and Wei, 1987a). Microstimulation of some C afferents classified as C polymodal nociceptors in human subjects give rise to a sensation of itch, rather than of pain (Ochoa and Torebjörk, 1989).

Recently, a category of C fibers has been described in humans that responds to histamine with long-lasting discharges that parallel the time course of the itch sensation triggered by locally applied histamine (Schmelz et al., 1997). These C fibers are believed to serve as itch receptors (Table 2.2). The itch afferents are mechanically insensitive, although some do respond to heat. Others are insensitive to both mechanical and heat stimuli (CMiHi units). Their terminal branches extend as far as 85 mm on the lower extremity, and so if they release neuropeptides, they could account for the distribution of flare in humans by an axon reflex mechanism.

Cutaneous Thermoreceptors

Two kinds of receptors in the skin signal innocuous changes in temperature (Table 2.2). These specific thermoreceptors include cold receptors and warm receptors. Thermoreceptors respond very poorly or not at all to mechanical stimuli. They are involved not only in thermal sensation (Konietzny, 1984; Spray, 1986) but also in thermoregulation (Gordon and Heath, 1986). Simon (2000) has reviewed the concept of deep-body thermoreception.

Cold Receptors

The morphology of cold receptors has been described (Hensel, 1973; Hensel et al., 1974). The afferent fibers from cold receptors may be myelinated or unmyelinated; they are chiefly myelinated in humans (Fruhstorfer et al., 1974). In rats, C cold receptor afferents are especially prevalent in the coccygeal nerve (Fleischer et al., 1983), suggesting that they may play an important role in the thermosensitivity of the tail. Cold receptors are activated by changes in skin temperature as small as 0.1 °C. They show a background discharge at normal skin temperatures. With cooling, there is an increase in the discharge rate, with both a dynamic and a static component (Fig. 2.24; Darian-Smith et al., 1973). The dynamic response signals the rate at which the temperature is changed, whereas the static response signals the temperature level. Static discharges are seen at temperatures from 5 to 43 °C (Fig. 2.25; Hensel, 1973). The maximum static discharge occurs between about 18 and 34 °C. Cold receptors can also be activated chemically by exposing them to menthol (Hensel and Zotterman, 1951).

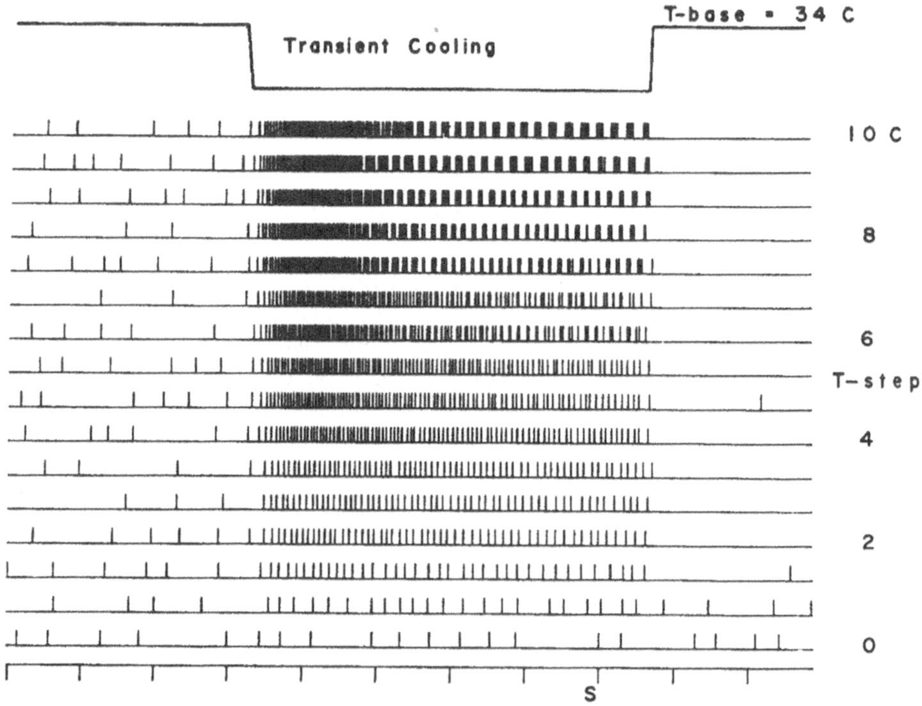

FIGURE 2.24. Responses of an afferent fiber that supplied a cold receptor to graded cooling pulses. The monitor trace at the top shows the timing of step-changes in skin temperature from a baseline temperature of 34 °C (T-base). The series of recordings of nerve impulses show the background activity and the responses of the receptor to cold pulses that were varied in intensity over a range of 0 to 10 °C below the T-base. Note that there was a dynamic and then a static response to large temperature steps. Furthermore, rhythmic bursting occurred following the strongest stimuli, 8–10 °C cold pulses. (From Darian-Smith et al., 1973.)

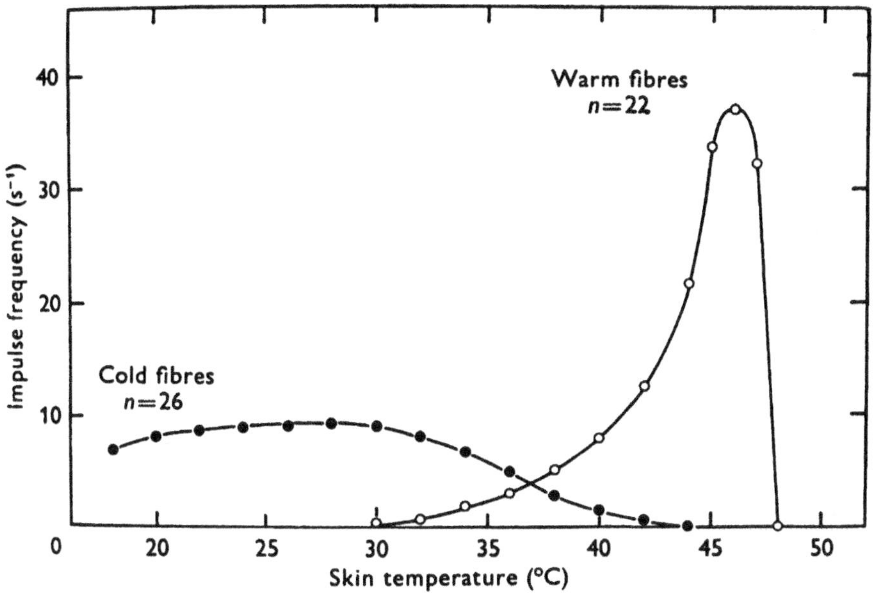

FIGURE 2.25. Average static discharge rates for populations of cold and warm receptors. Recordings were made from afferent fibers in the infra-orbital nerve in cats that supply warm and cold receptors, and the temperature was varied over a wide range. The maximum static firing rate of warm fibers occurred when the temperature was 46 °C and that of cold fibers was seen when the temperature was 27 °C. (From Hensel, 1973.)

It is of interest that the static discharge rate is the same for pairs of temperature readings above and below the optimum temperature (Fig. 2.25). Cold receptors tend to have a bursting pattern of discharge in the lower part of their temperature range (Fig. 2.24; Dodt, 1952; Darian-Smith et al., 1973), and it has been suggested that this bursting activity may permit the central nervous system to distinguish between temperatures above and below the optimum (Iggo, 1969; Poulos, 1971; Dykes, 1975). However, the coding for static temperature by bursts of discharges in cold fibers may be more important for thermoregulation than for thermal perception (Dykes, 1975). Noxious heat may also evoke a discharge in cold receptors, the "paradoxical cold response" (Dodt and Zotterman, 1952b); the presence of this response depends in part upon the core body temperature (Long, R. R., 1977).

The activity of cold receptors is sensitive to the extracellular calcium concentration. An increased concentration of calcium reduces the discharges of cold receptors at a given temperature, whereas a decrease causes enhanced activity. The effect of changes in the extracellular calcium concentration may depend on changes in a calcium-activated outward current in cold receptors (Schäfer, 1987).

LaMotte, R. H. and Thalhammer (1982) have described a class of high-threshold cold receptors with Aδ- and C-sized axons in the skin of monkeys. The receptors responded to colder, but innocuous, stimuli, below the range for maximal responses of low threshold cold receptors.

Warm Receptors

The morphology of warm receptors is unknown, but they are presumed to be free endings, and they have unmyelinated axons (Hallin et al., 1981). Warm receptors are distinctly separate from cold receptors. They are active at normal body temperature. Their discharge slows with cooling of the skin, and the activity accelerates with warming, reaching a maximum at temperatures of about 45 °C (Fig. 2.25; Handwerker and Neher, 1976). However, noxious levels of heat (above 48 °C) silence warm receptors. Warm receptors do not have a paradoxical response to cooling (Hensel, 1973).

For additional details about thermoreceptors, see Dodt and Zotterman, 1952a; Hensel and Boman, 1960; Hensel et al., 1960; Iriuchijima and Zotterman, 1960; Iggo, 1969; Perl, 1968; Martin, H. F. and Manning, 1969; 1972; Poulos and Lende, 1970a, b; Hensel and Iggo, 1971; Stolwijk and Wexler, 1971; Darian-Smith et al., 1973; Dykes, 1975; Hellon et al., 1975; Konietzny and Hensel, 1975; Pierau et al., 1975; Handwerker and Neher, 1976; Kenshalo and Duclaux, 1977; LaMotte and Campbell, 1978; Duclaux and Kenshalo, 1980; Sumino and Dubner, 1981; Lynn and Carpenter, 1982; Fleischer et al., 1983.

Transduction in Thermoreceptors

The transduction mechanism in cold receptors has been suggested to involve the TRPM8 (or CMR1) ion channel, which is one of the TRP family of excitatory ion channels (cf. McKemy et al., 2002; Peier et al., 2002a, b). These channels have been found in some dorsal root ganglion cells (Reid and Flonta, 2001). They respond to cooling and to menthol.

MUSCLE RECEPTORS

Receptors in and around skeletal muscles include the stretch receptors (muscle spindles, Golgi tendon organs), pressure-pain endings with either myelinated or unmyelinated afferent fibers, and paciniform corpuscles (which are located in fascial planes and near interosseus ligaments) (Fig. 2.26; Table 2.3; Barker, 1962). The paciniform corpuscles respond like the Pacinian corpuscles found in subcutaneous tissue and so will not be discussed further (see above). Many of the pressure-pain endings are considered to be nociceptors (Mense, 1977a; 1996; Kniffki, 1986).

Stretch Receptors

The structure and function of the stretch receptors have been reviewed thoroughly (Barker, 1962; Matthews, 1964; 1972; 1981; Barker et al., 1974). No attempt will be made here to provide an inclusive reference list for these important receptors (for a bibliography of the literature up to the mid-1970s, see Eldred et al., 1977).

Muscle stretch receptors are thought to contribute not only to reflex activity but also to proprioception (position sense and kinesthesia; see Chapter 12; Goodwin et al., 1972a, b; Simon et al., 1984; Wei et al., 1986). Although stimulation of the axons of individual stretch receptors by intraneural microstimulation fails to elicit any sensation (Gandevia et al., 1992), sensation is likely to result from the activation of a population of stretch receptors. This is shown by experiments in which illusions of movement were produced when muscle was vibrated or large muscle afferents were stimulated electrically in human subjects (Goodwin et al., 1972b; Feldman and Latash, 1982; Gandevia, 1985; Inglis and Frank, 1990; Inglis et al., 1991; see Chapter 12). Interestingly, vibration of the antagonist muscle during a movement is the more significant event with respect to signals arising from muscle spindles than vibration of the agonist muscle (Inglis and Frank, 1990). This

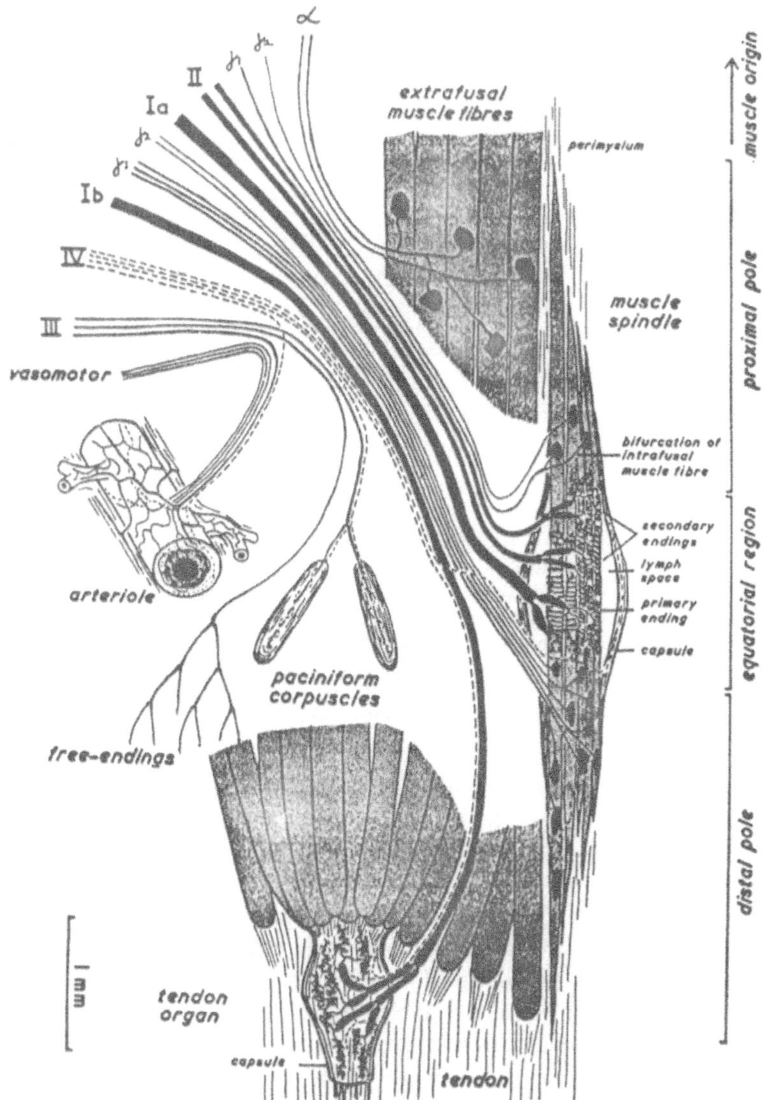

FIGURE 2.26. Innervation of mammalian skeletal muscle. (From Barker, 1962.)

is presumably because human muscle spindles in the agonist muscle are unloaded during contractions (Burke et al., 1978a; Prochazka et al., 1979).

Muscle Spindles

Muscle spindles have two types of sensory endings: a primary ending and one or more secondary endings (Fig. 2.27A and B; Boyd, 1962; Table 2.3). The primary ending of a muscle spindle is supplied by a large myelinated axon that is classified as a group Ia fiber (Fig. 2.27B). The primary ending is responsive both to the rate of stretch of the spindle (dynamic response) and to the new length (static response) (Fig. 2.27C; Matthews, 1972). The sensitivity of the dynamic response is set, in part, by the activity of dynamic

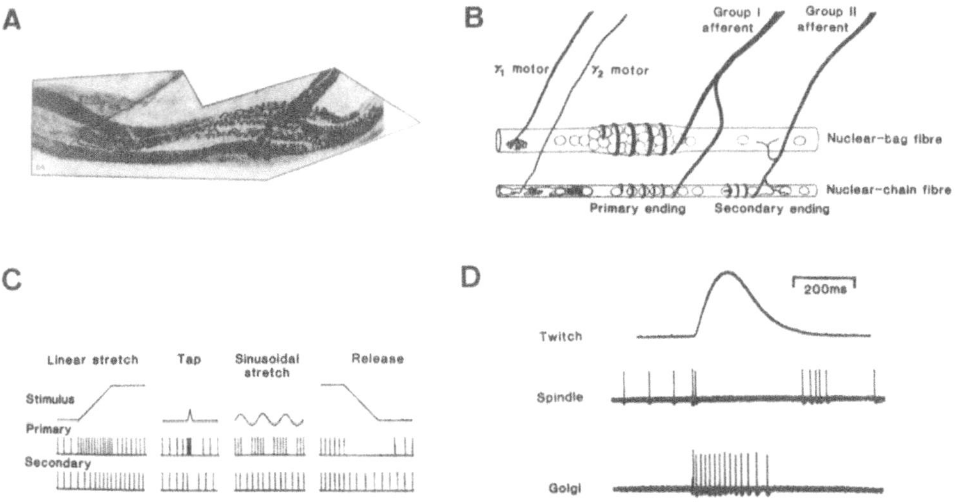

FIGURE 2.27. Structure and function of muscle spindle. A, innervation of an isolated muscle spindle. (From Boyd, 1962.) B, simplified diagram of the equatorial region of a muscle spindle. A nuclear bag fiber receives a primary ending from a group Ia afferent fiber and a plate ending from a dynamic fusimotor (γ) neuron. A nuclear chain fiber has both a primary ending from the same group Ia fiber and a secondary ending from a group II fiber. It also receives a trail ending from a static γ motor neuron. C, responses of typical primary and secondary endings to linear stretch, tap, sinusoidal stretch, and release. (B and C from Matthews, 1964.) D, unloading effect of a twitch on the discharges of a muscle spindle afferent and activation of a Golgi tendon organ afferent by the twitch. (D from Matthews, 1972, copyright Edward Arnold, Publishers.)

fusimotor fibers from a subpopulation of γ motoneurons (Bessou et al., 1968). The secondary endings are supplied by afferent axons of intermediate size, the group II fibers (Fig. 2.27B). Secondary endings show little dynamic responsiveness; instead, they signal chiefly muscle length (Fig. 2.27C). Static responsiveness is controlled in part by static fusimotor fibers from a second subpopulation of γ motoneurons (Appelberg et al., 1966). Some muscle spindles receive a motor supply from branches of α motor axons (termed β innervation by workers in the field). When a muscle nerve is stimulated at a strength that activates α motor axons, but not γ motor axons, muscle spindle afferent fibers stop discharging during the resultant twitch. By contrast, Golgi tendon organs discharge during the twitch (Fig. 2.27D).

Golgi Tendon Organs

Golgi tendon organs are located in the dense connective tissue of muscle tendons and aponeuroses, and their terminals are interleaved among bundles of collagen (Schoultz and Swett, 1972; 1974). The structure of the Golgi tendon organ is essentially identical to that of the Ruffini ending (Figs. 2.8A and 2.26). The main difference is that the location of Golgi tendon organs in the connective tissue of a muscle places them in series with skeletal muscle fibers and gives them a special function in relation to muscular contractions and muscle stretch (Table 2.3; Matthews, 1972; Darian-Smith, 1984). Their afferent fibers are large myelinated axons classified as group Ib. A Golgi tendon organ responds at a high threshold to passive muscle stretch, but at a low threshold to the contraction of muscle fibers ending on the tendon slip that contains the receptor (Fig. 2.26D; Houk and Henneman, 1967). Golgi tendon organs signal tension. When the muscle is stretched, there is only a small dynamic response, but a substantial static response.

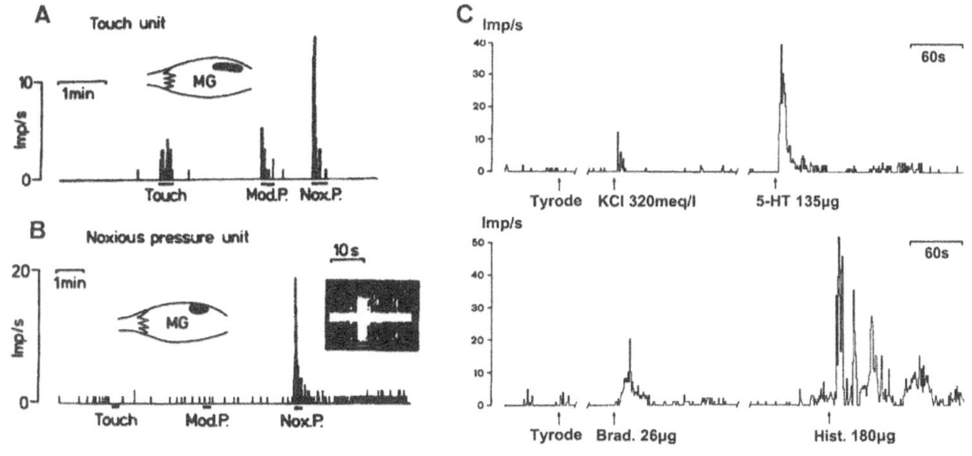

FIGURE 2.28. A and B, responses of group IV muscle afferent units to graded mechanical stimuli, ranging from tactile to moderate (Mod.P.) and noxious pressure (Nox.P.). (From Mense, 1996.) C, Responses of a group IV muscle afferent fiber to the intra-arterial injection of algesic chemicals (KCl, 5-HT, 5-hydroxytryptamine; Brad., bradykinin; Hist., histamine). (From Mense and Schmidt, 1974.)

Pressure-Pain Endings and Muscle Nociception

There are numerous unencapsulated endings in the fascia and in the adventitia of blood vessels of muscle. These form as much as 75% of the innervation of skeletal muscle (Stacey, 1969). Some endings are supplied by small myelinated group III afferent fibers, whereas others are supplied by group IV (unmyelinated) fibers. A few of the endings are connected with larger afferent fibers of group II size (Paintal, 1960; Stacey, 1969). Fine afferents that innervate muscle may contain neuropeptides, such as CGRP, SP, or somatostatin, and some contain fluoride-resistant acid phosphatase (Molander et al., 1987).

One of the functions of fine muscle afferents is to mediate muscle pain. However, some of these afferents may be ergoreceptors, whose input contributes to reflex adjustments of the circulatory and respiratory systems to exercise (see Mense, 1996). Intraneural microstimulation at sites in the human common peroneal nerve from which activity of group III or IV muscle afferents could be recorded elicits muscle pain that is described as "cramping" and that is well localized in the muscle (Simone et al., 1994). These muscle afferents can be activated by moderate or noxious pressure over the muscle, but not by muscle stretch. Examples of group III and IV muscle afferents that can be activated by intramuscular injection of capsaicin in human subjects have been reported (Marchettini et al., 1996; cf. Kaufman et al., 1982). Since intramuscular injections of capsaicin produce cramping pain, it was concluded that at least some group III and IV muscle afferents are nociceptors. However, as mentioned, it is likely that another population of group III and IV muscle receptors are ergoreceptors (exercise detectors; Mense, 1996). The amount of pain experienced increases when the stimulus rate is increased. The projected receptive field increases when the stimulation is prolonged (although the receptive field determined by natural stimulation in the muscle does not change).

Group III Muscle Afferents

Group III afferent fibers can be excited by mechanical stimulation of muscle. Several types of pressure receptors can be recognized: (1) some of these receptors respond with a slowly adapting discharge to pressure applied at the junction of muscle with tendon; (2) others with large receptive fields are excited by pressure on the belly of the muscle; (3) another group of rapidly adapting receptors have receptive fields in a small region of the muscle; (4) one group is activated by manipulations that affect the space between muscles or along the surface of a muscle; (5) the final group discharges in response to muscle stretch (Bessou and Laporte, 1961).

Mense and Meyer (1985) found that the most common type of group III muscle afferent fiber in their sample (44%) responded to innocuous mechanical stimuli. However, numerous group III fibers were classified as nociceptive (33%) or contraction-sensitive units (23%).

Group III afferent fibers can also be activated or sensitized by intra-arterial injections of algesic chemicals, such as bradykinin, serotonin, and histamine (Mense, 1977b; 1996; Mense and Schmidt, 1974; see also Hník et al., 1969) and by thermal stimuli applied to the muscle belly (Hertel et al., 1976; Kumazawa and Mizumura, 1976; 1977b; Mense and Meyer, 1988). Another effective stimulus is the injection of hypertonic sodium chloride solution into the muscle (Lewis, 1942; Paintal, 1960; Abrahams et al., 1984a).

Mense and Stahnke (1983) confirmed that many group III muscle afferent fibers are activated by muscle stretch. The same fibers often respond to muscle contractions as well. However, the responses to muscle contraction are reduced following a period of ischemia. The suggestion was made that these muscle afferent fibers would be activated during exercise and that they were more likely to function as ergoreceptors than as nociceptors.

Group IV Muscle Afferents

Group IV muscle afferent fibers are similar in some of their response properties to group III muscle afferent fibers, but their thresholds to mechanical stimuli tend to be high (Fig. 2.28A and B; Mense, 1996; see also Bessou and Laporte, 1961; Iggo, 1961). Many are readily activated or sensitized by algesic chemicals (Fig. 2.28C; Mense and Schmidt, 1974; Franz and Mense, 1975; Fock and Mense, 1976; Hiss and Mense, 1976; Mense, 1977b; 1981; Kumazawa and Mizumura, 1976; 1977b; Kniffki et al., 1978; Mense and Meyer, 1988), and they may also discharge in response to thermal stimuli (Hertel et al., 1976). Kniffki et al. (1978) proposed that group IV muscle afferent fibers that respond to algesic chemicals but not muscular activity are nociceptors, whereas those that respond to contractions but not to algesic chemicals are ergoreceptors. However, some group IV fibers have both response properties. Mense and Stahnke (1983) found that few group IV muscle afferent fibers respond to muscle stretch or contraction. However, some discharge vigorously when muscle contractions occur during ischemia. They suggest that such afferent fibers might be nociceptors that contribute to the pain of muscle ischemia.

Mense and Meyer (1985) observed that the most common type of group IV unit in their sample (43%) was nociceptive. However, there were also group IV units that responded to innocuous mechanical stimuli or to muscle contractions. A few group IV units behaved like specific thermoreceptors. It was suggested that these might participate in thermoregulation, rather than in thermal sensation.

Muscle afferent fibers have less effect than cutaneous ones in evoking plasma extravasation; this observation correlates with a smaller content of substance P in muscle nerves than in cutaneous nerves (McMahon et al., 1984).

Sensitization of High Threshold Muscle Afferents

Like cutaneous nociceptors, Group III and IV muscle receptors can by sensitized by exposure to inflammatory agents (Berberich et al., 1988; Diehl et al., 1988; 1993; Mense and Meyer, 1988). Since acetylsalicylic acid reduces the discharges of many sensitized fine muscle afferents, the release of prostaglandins is likely to contribute to the sensitization process (Diehl et al., 1993).

JOINT RECEPTORS

The innervation of joints has been well investigated. Encapsulated mechanoreceptors that are found in or around joints include Ruffini endings, Pacinian corpuscles, and Golgi tendon organs (Fig. 2.29; Table 2.3; Skoglund, 1956; see review by Darian-Smith, 1984). There are also reports of Golgi-Mazzoni endings in joint capsules (Polacek, 1961; Freeman, M. A. R. and Wyke, 1967; Grigg et al., 1982). The encapsulated joint receptors are mechanoreceptors. There are also many articular afferents with free nerve endings; these include nociceptors (see review by Schaible and Grubb, 1993).

For details about joint innervation, see Gardner, 1944; Samuel, 1952; Andrew, 1954; Boyd, 1954; Skoglund, 1956; Polacek, 1961; Freeman and Wyke, 1967; O'Conner and McConnaughey, 1978; Tracey, 1979; O'Conner and Woodbury, 1982; Langford and Schmidt, 1983; Langford, 1983; Schultz et al., 1984; Halata et al., 1985; Zimny et al., 1986; Heppelmann et al., 1988; 1990; Sjölander et al., 1989; Hildebrand et al., 1991; Marinozzi et al., 1991; Gomez-Barrena et al., 1996; Hines et al., 1996; Salo and Theriault, 1997.

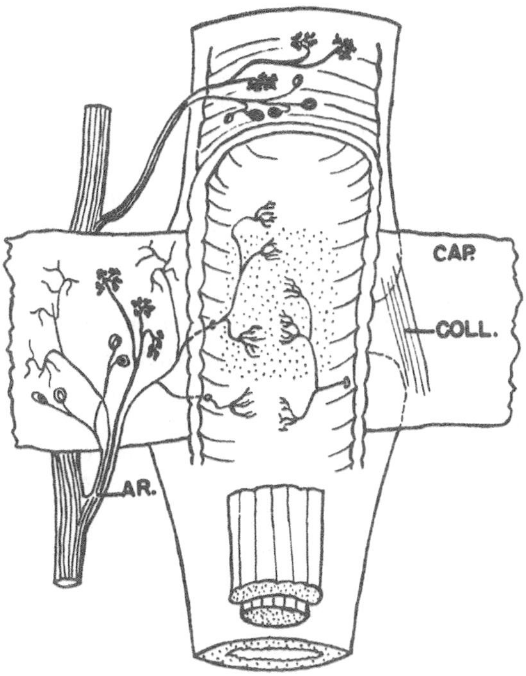

FIGURE 2.29. Innervation of the human metacarpophalageal joint. The joint is viewed from the palmar side. The flexor tendon sheath is opened and the tendon removed. The joint capsule has been incised dorsally and opened as flaps (CAP.). The collateral ligament is shown (COLL.). The articular nerve (AR.) supplies Ruffini, Pacinian and free nerve endings. (From Darian-Smith, 1984.)

Joint Mechanoreceptors

The types of mechanoreceptors that are associated with joints are classified based on the discharge properties of the afferent fibers. Joint receptors include both slowly adapting (Golgi tendon organs; Ruffini endings) and rapidly adapting (Paciniform) mechanoreceptors (Gardner, 1944; Boyd, I. A., 1954; Skoglund, 1956; Halata, 1977; Halata et al., 1984; 1985). The Golgi tendon organ and paciniform endings respond like comparable endings found elsewhere.

Slowly Adapting Joint Receptors

Slowly adapting discharges have been recorded from articular nerves in response to joint movements (Gardner, 1948; Andrew and Dodt, 1953; Andrew, 1954; Boyd and Roberts, 1953; Boyd, I. A., 1954; Cohen, 1955; Skoglund, 1956; Eklund and Skoglund, 1960). Several reports have analyzed the dynamic as well as the static responses of slowly adapting joint afferent fibers (McCall et al., 1974; Nade et al., 1987).

Golgi tendon organs seem to respond maximally when the joint is moved to one end of its range and are relatively insensitive; Ruffini endings may respond over a narrow range of joint angles, and they are very sensitive (Skoglund, 1973). For these reasons, it was initially proposed that Ruffini endings signal joint position (Boyd and Roberts, 1953; Boyd, 1954; Cohen, 1955; Skoglund, 1956; 1973; Carli et al., 1979; 1981). However, it has since been shown that Ruffini endings are seldom active at intermediate positions of joints (at least in proximal joints, such as the knee and hip joints of cats and dogs), and so it is unlikely that they signal joint position in proximal joints. Instead, they appear to signal the torque that develops as the joint is extended, flexed, or rotated to the extremes of its range (Burgess and Clark, 1969; Clark, 1975; Clark and Burgess, 1975; Grigg, 1975; 1976; Grigg and Greenspan, 1977; Grigg and Hoffman, 1982; 1984; Rossi and Grigg, 1982; Ferrell et al., 1986; Aloisi et al., 1988; see also the review by Burgess et al., 1982). The slowly adapting responses seen in early work when a joint was at midrange can be attributed to muscle afferent fibers that distribute through joint nerves (Burgess and Clark, 1969; McIntyre et al., 1978; Rossi and Grigg, 1982; Rossi and Rossi, 1985; Aloisi et al., 1988; Gregory et al., 1989). However, there are also reports of activity in slowly adapting joint afferents when the joint is held at mid-range (Ferrell, 1980; Ferrell et al., 1986). Of particular interest are recordings from the nerve supplying the elbow of the cat (Millar, 1973; 1975; Baxendale and Ferrell, 1983), since this nerve does not appear to be contaminated by muscle afferent fibers (Baxendale and Ferrell, 1983). In distal joints, proprioception seems to depend on a contribution from joint and cutaneous afferents, as well as from muscle stretch receptors (Gandevia et al., 1983; Ferrell and Craske, 1992; see Chapter 12).

Grigg et al. (1982) provide evidence for another type of slowly adapting joint receptor that responds to compression of the joint capsule but not to stretch. They argue that these are Golgi-Mazzoni corpuscles that might function to signal deep pressure. In vitro studies of the mechanical forces that activate afferents supplying the knee joint capsule of cats indicate that the responses are more likely to relate to local stress than to strain (Khalsa et al., 1996) and that they are not due to local compression of the afferent ending (Grigg and Hoffman, 1996). It was concluded that afferents supplying joint capsules are "stress detectors" (Khalsa et al., 1996).

Rapidly Adapting Joint Receptors

The Pacinian corpuscle afferents in joint nerves respond to vibration, like Pacinian corpuscles elsewhere (Aloisi et al., 1988). These receptors presumably signal mechanical transients in the joint (Skoglund, 1956; 1973).

Joint Nociceptors

Nociceptors in joints have been described in some detail (Table 2.3; Burgess and Clark, 1969; Clark, 1975; Clark and Burgess, 1975; Schaible and Schmidt, 1983a, b; see review by Schaible and Grubb, 1993).

Many group III and IV afferent fibers that supply the joint capsule respond to innocuous movements of the joint, although they may respond additionally to movements beyond the normal range of the joint, movements that are considered noxious. However, others respond best or only when the joint is moved beyond its normal range (Fig. 2.30). Some fine afferent fibers that supply a joint are not activated at all by joint movements unless the joint is inflamed (Coggeshall et al., 1983; Schaible and Schmidt, 1983a, b; 1985; Grigg et al., 1986). Some joint nociceptors are excited by capsaicin (He et al., 1990). These become at least temporarily unresponsive to passive movements of the joint and to bradykinin.

Sensitization of Joint Nociceptors

Initially silent articular nociceptors can often be activated even by weak mechanical stimuli after sensitization has developed during acute inflammation. Such silent nociceptors were initially observed in experiments on articular afferents (Coggeshall et al., 1983;

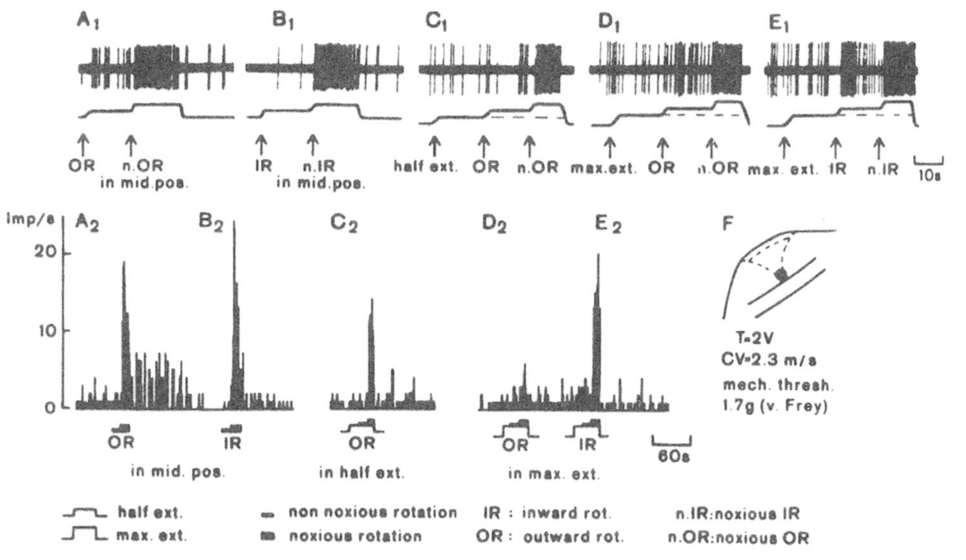

FIGURE 2.30. Responses of a group IV primary afferent fiber that innervated the knee joint of a cat. The fiber responded only weakly to innocuous movements of the joint, but strongly to noxious movements. The location of the receptive field is shown in panel F. (From Schaible and Schmidt, 1983b.)

Schaible and Schmidt, 1983a, b; 1985; 1988a), although similar silent nociceptors have now been observed to supply the skin and the viscera (see sections on **Cutaneous Nociceptors** and on **Visceral Nociceptors**). Administration of bradykinin, prostaglandins PGE_2 or PGI_2, or serotonin has been observed to excite and/or sensitize joint afferents (Schaible and Schmidt, 1988b; Neugebauer et al., 1989; Birrell et al., 1990; 1993; Herbert and Schmidt, 1992; Schepelmann et al., 1992), although bradykinin does not appear to be responsible for maintaining the sensitization seen after inflammation (Messlinger et al., 1993; 1994). Some joint afferents can be sensitized by substance P (Heppelmann and Pawlak, 1997a) or by excitatory amino acids (Lawand et al., 1997b). Inflammation of a joint sensitizes group III but not group II or IV joint afferents to serotonin (Herbert and Schmidt, 1992). Administration of phorbol ester results in sensitization of some joint afferents, especially many of those that were also responsive to bradykinin; this suggests a role for PKC in the sensitization process (Schepelmann et al., 1993). Nitric oxide release is also involved (Lawand et al., 1997a).

Peptide Content of Fine Joint Afferents

Some articular afferents in cats, rats, dogs, and humans have been found to contain neuropeptides, such as substance P, calcitonin gene-related peptide, neurokinin A, or somatostatin (O'Brien et al., 1989; Wojtys et al., 1990; Hanesch et al., 1991; 1992; 1995; Heppelmann et al., 1997; Salo and Theriault, 1997; Tamura et al., 1998). Quantification in immunohistochemical staining experiments has shown that about 16–17% of the articular afferents supplying the cat knee joint contain substance P and 32–35% contain calcitonin gene-related peptide (Hanesch et al., 1991). About half of the substance P-containing knee joint afferents in cats also contain NKA (Hanesch et al., 1992). Similar proportions of joint afferents in dogs were found to contain substance P (17%) and CGRP (29%) (Tamura et al., 1998). Higher proportions of these peptides were observed in joint afferents of rats by one group (29% contain substance P and 71% CGRP; O'Brien et al., 1989), but a similar proportion by another group (10% of joint afferents contained substance P and 33% CGRP; Salo and Theriault, 1997). Somatostatin and substance P are usually co-localized (Hanesch et al., 1995; however, cf. O'Brien et al., 1989, who found no somatostatin in rat joint afferents).

The functions of the peptides that are present in the terminals of knee joint afferents are unclear. It has been suggested that their release could help regulate blood vessels in the knee joint or provide a trophic action; another possibility is a contribution to neurogenic inflammation when this occurs (Heppelmann et al., 1997). Joint afferents only occasionally discharge after injection of substance P into the arterial circulation of the joint; however, some joint afferents become sensitized to joint movements (Heppelmann and Pawlak, 1997a). Somatostatin has an inhibitory action and so when it is released from the terminals of joint afferents it may have actions contrary to those of other peptides, such as substance P and CGRP (Heppelmann and Pawlak, 1997b; 1999).

Joint Inflammation

There have been numerous studies of the effects of inflammation of joints in experimental animals (see review by Schaible and Grubb, 1993). One model of acute arthritis involves the injection of kaolin and carrageenan into the knee joint in cats. As this type of inflammation develops, the joint swells because of plasma extravasation (neurogenic edema) and it becomes warm because of vasodilation (Schaible and Schmidt, 1985; Sluka and

Westlund, 1992). Many articular afferents develop spontaneous discharges, and the threshold for activation of fine articular afferents by joint movements is reduced dramatically in cats (Coggeshall et al., 1983; Schaible and Schmidt, 1983a, b; 1985; 1988a; Grigg et al., 1986) and rats (Sluka and Westlund, 1993b). However, the activity of large myelinated joint afferents is unchanged (Dorn et al., 1991).

Another model of arthritis is produced by injection of Freund's adjuvant into the foot or the base of the tail of rats (Colpaert, 1987). However, the inflammation that develops is systemic. An arthritis that is limited to a single joint can be produced by the injection of Freund's adjuvant directly into a joint (Butler et al., 1992). Freund's adjuvant monoarthritis develops over days and lasts for weeks. It is thus suitable for investigations of chronic arthritis and is often regarded as a model of rheumatoid arthritis. Injection of complete Freund's adjuvant near the ankle produces arthritis in that ankle and also in the contralateral ankle (Donaldson et al., 1995). Treatment of either sciatic nerve with capsaicin prevents the spread of the arthritis to the contralateral side. Recordings from afferent fibers in the dorsal roots show that dorsal root reflexes are produced bilaterally after injection of Freund's adjuvant into one knee joint (Rees et al., 1996). The dorsal root reflexes contralateral to the side of injection could be prevented by treatment of the sciatic nerve with capsaicin. It was proposed that the mirror-spread of the arthritis depend on the dorsal root reflexes, which presumably are conducted in part in peptidergic, capsaicin-sensitive C fibers, although this was not directly demonstrated (see the next section on the **Role of Dorsal Root Reflexes in the Neurogenic Inflammation of Arthritis**).

Inflammation results in the upregulation of NADPH-diaphorase activity in articular afferents (Bscheidl et al., 1994), as well as an increase in the content of substance P and calcitonin gene-related peptide (Sluka and Westlund, 1993b) and in mRNA for preprotachykinin and calcitonin gene-related peptide in dorsal root ganglion cells (Donaldson et al., 1995). Release of glutamate and/or of nitric oxide into a joint may be involved in joint inflammation (Lawand et al., 1997a, b).

The inflammation seen in experimental arthritis is associated with plasma extravasation into the knee joint. There is also a vasodilation, after an initial vasoconstriction due to activation of sympathetic postganglionic axons. These events can also be produced by antidromic stimulation of group IV fibers in articular nerves (Ferrell and Russell, 1985; 1986; Khoshbaten and Ferrell, 1990). Thus, the inflammation is at least initially neurogenic. The plasma extravasation and vasodilation depend on the release of substance P and calcitonin gene-related peptide from fine joint afferents (Brain and Williams, 1985; Brain et al., 1985; Ferrell and Russell, 1986; Khoshbaten and Ferrell, 1990). Ferrell and Russell found that sympathetic postganglionic efferents interfere with plasma extravasation (Ferrell and Russell, 1986), and Sluka et al. (1994) observed the same degree of joint inflammation in the presence of a sympathectomy and with an intact sympathetic nervous system. However, Levine et al. (1985) provide evidence that suggests a role for both substance P and sympathetic innervation in arthritis. Vasodilation also depends in part on release of substance P from articular afferents (Khoshbaten and Ferrell, 1990), although presumably calcitonin gene-related peptide would contribute more to the vasodilation.

Role of Dorsal Root Reflexes in the Neurogenic Inflammation of Arthritis

Sluka and Westlund (1993a) observed that administration of a non-NMDA glutamate receptor antagonist, CNQX, but not the NMDA receptor antagonist, AP7, into the lumbar dorsal horn by microdialysis resulted in a substantial reduction in the knee joint inflammation that would otherwise have been produced by intra-articular injection of kaolin and carrageenan. Microdialysis administration of bicuculline into the dorsal horn was then found to have the same effect (Sluka et al., 1993). These observations indicate that central processing is involved in the induction of arthritis and that this central processing depends on actions on non-NMDA glutamate and $GABA_A$ receptors, but not on NMDA receptors. Further investigation revealed that, during the development of arthritis, it becomes possible to elicit dorsal root reflexes in joint afferents, and it was shown that these dorsal root reflexes depend on dorsal horn circuits that utilize non-NMDA and $GABA_A$, but not NMDA, receptors (Rees et al., 1994; 1995). Dorsal root reflexes were conducted in fine articular afferents (Sluka et al., 1995). Dorsal root reflexes were also observed in the chronic monoarthritis produced by Freund's adjuvant, and these were shown to depend on conduction in capsaicin-sensitive afferent fibers (Rees et al., 1996).

Dorsal root reflexes are triggered by primary afferent depolarization (PAD), which normally results in presynaptic inhibition of the effects of volleys in primary afferent fibers (see Chapter 7; reviewed by Eccles, 1964; Rudomin and Schmidt, 1999). PAD is known to occur in joint afferents (Jankowska et al., 1993). The production of PAD depends on the release of glutamate at the first synapse of primary afferents in the dorsal horn onto a disynaptic pathway that ends on GABAergic inhibitory interneurons (Evans and Long, 1989). The administration of the non-NMDA glutamate receptor antagonist, CNQX, largely blocks primary afferent depolarization, whereas an NMDA antagonist is only partially effective. The inhibitory interneurons that cause PAD in turn release GABA, which activates $GABA_A$ receptors on the terminals of primary afferent fibers (Eccles et al., 1963b; see review by Willis, 1999). Bicuculline, which is a $GABA_A$ receptor antagonist, reduces primary afferent depolarization and blocks dorsal root reflex activity.

Inflammation causes an upregulation of GABAergic transmission in the dorsal horn (Castro-Lopes et al., 1992; 1994a). Presumably, this upregulation enables primary afferent depolarization to trigger dorsal root reflexes. Antidromic propagation of dorsal root reflexes in fine, peptidergic primary afferent fibers would cause release of peptides and glutamate in peripheral tissues, and this would have the same effect in inducing neurogenic edema and vasodilation as antidromic volleys evoked by stimulation of dorsal roots or peripheral nerves (Ferrell and Russell, 1985; 1986; Magerl et al., 1987; Low and Westerman, 1989; Jänig and Lisney, 1989). Similar events have now been shown to occur in the skin (see section on **Nociceptors and Neurogenic Inflammation** in the discussion of **Cutaneous Nociceptors**).

VISCERAL RECEPTORS

Although under ordinary circumstances some viscera do not contribute much to sensation, many visceral structures can produce sensations, including severe pain, given an adequate stimulus (Cervero, 1994; Hölzl et al., 1996; Jänig, 1996; Cervero and Laird, 1999). Even when afferent input from viscera to the central nervous system fails to reach consciousness, it may affect emotional state (Jänig, 1996; Mayer, E. A. et al., 2000). Some viscera signal non-painful sensations, such as distention or flow, under physiological conditions, as well as pain.

A variety of receptor types have been identified in viscera (see Table 2.3). Only those whose axons enter the spinal cord will be considered here. The afferent axons gain access to the spinal cord either by way of the splanchnic nerves and sympathetic chain or by way of the pelvic parasympathetic nerves (Cervero, 1994). Although the organization of the visceral afferents is complex, this organization can be analyzed with modern tracing techniques (Aldskogius et al., 1986; Morgan et al., 1986; Wang, H. F. et al., 1998a, b). The sensory receptor types include mechanoreceptors, nociceptors, and thermoreceptors. In addition, primary afferents that are involved in regulation, rather than in sensation, supply many viscera. These often access the central nervous system through the vagus nerve (see review by Cervero, 1994). Markers for different groups of visceral afferents include the antibody, RT97, which recognizes large, light DRG cells; antibodies for neuropeptides, such as CGRP and SP, which are contained in a group of small, dark DRG cells; and the lectin, IB4, which recognizes another group of small, dark DRG cells (Wang, S. et al., 1998). Interestingly, unmyelinated primary afferent fibers supplying the bladder were effectively labeled by cholera toxin B-conjugated HRP, in contrast to unmyelinated cutaneous afferents (Wang, H. F. et al., 1998b).

Visceral Mechanoreceptors

Visceral mechanoreceptors include Pacinian corpuscles located in the mesentery and in connective tissue associated with visceral organs (Sheehan, 1932). They behave like Pacinian corpuscles elsewhere, except that they sometimes discharge in phase with the heartbeat (Gammon and Bronk, 1935; Gernandt and Zotterman, 1946). It is not known whether signals from these receptors are of consequence in cardiovascular regulation.

Mechanoreceptors that have a low threshold to distention of the esophagus have been recorded from the vagus nerve of the opossum (Sengupta et al., 1989). Afferents from the esophagus that travel with the sympathetics are nociceptive (Sengupta et al., 1990), although some begin to discharge with low intensity stimuli (see below). Mechanoreceptors also supply the mucosa of the anal canal (Sengupta and Gebhart, 1994).

Other visceral mechanoreceptors have been found associated with the mesentery, with the serosal surface of the gut or other organs, smooth muscle, and along blood vessels (Paintal, 1957; Todd, J. K., 1964; Bessou and Perl, 1966; Morrison, 1973; 1977; Ranieri et al., 1973; Floyd and Morrison, 1974; Clifton et al., 1976; Floyd et al., 1976; 1977; Bahns et al., 1986; 1987; Cervero and Sharkey, 1988). They appear to respond to movements or to distention of hollow viscera. However, the discharges do not always signal intraluminal pressure or volume with precision, since some receptors are at a distance from the contractile region (Morrison, 1977). Another type of visceral mechanoreceptor that seems to be inserted in series with smooth muscle responds to distention or contraction of the muscle. Such receptors are typical of the bladder and ureter (Talaat, 1937; Iggo, 1955; Winter, 1971; Cervero and Sann, 1989; Häbler et al., 1993a; cf. Floyd et al., 1976).

Jänig and Koltzenburg (1991) have recorded from Aδ-mechanoreceptors of the colon that responded to distension of the colon but not to stimulation of the anal canal. Other Aδ mechanoreceptors had receptive fields in the anal canal and responded to shear forces and not to distension. Presumably these different populations of afferents subserve different aspects of colon function.

Berkley et al. (1990) have examined the responses of afferent fibers that supply the reproductive organs, as well as other pelvic organs, in female rats. Both mechanoreceptive and chemoreceptive responses were observed. The responses of some of the afferent fibers that supply reproductive organs vary during the estrous cycle.

Raybould et al. (1999) have demonstrated that dorsal root ganglion cells shown by retrograde labeling to innervate the stomach or colon respond to mechanical stimulation of the soma with an influx of calcium ions. This calcium influx could be blocked by gadolinium. They suggest that these stretch-sensitive ion channels might serve as a transducer for visceral mechanical stimuli.

Visceral Nociceptors

Many of the visceral afferent neurons that innervate the gastrointestinal tract and urinary bladder respond to both weak and strong mechanical stimuli (Fig. 2.31A, low threshold; Bahns et al., 1986; 1987; Jänig and Koltzenburg, 1990; 1991; Sengupta et al., 1990; Sengupta and Gebhart, 1994; see review by Jänig, 1996). The responses can be graded throughout the innocuous and noxious ranges. These afferents presumably contribute to visceral pain when the stimulus reaches sufficient intensity. This type of activity fits well with the "intensity" theory of visceral pain (see Cervero, 1994; Jänig, 1996). Other visceral afferents that have high thresholds are very likely to contribute to visceral pain (Fig. 2.31A, high threshold). Either type of visceral receptor can be supplied by a C fiber or by an Aδ fiber. In addition, there are other afferents that do not respond even to strong mechanical stimuli under normal conditions but that become quite responsive after visceral inflammation or ischemia (Haupt et al., 1983; Häbler et al., 1988b; 1990b; 1993b; Jänig and Koltzenburg, 1990; Sengupta and Gebhart, 1994).

Nociceptors have been found in visceral organs, including the heart and blood vessels, meninges, respiratory system, gastrointestinal tract, urinary system, and reproductive organs (Gernandt and Zotterman, 1946; Lim et al., 1962; Brown, A. M., 1967; Peterson and Brown, 1973; Uchida and Murao, 1974; Baker et al., 1980; Cervero, 1982b; Cervero and Sharkey, 1988; Cervero and Sann, 1989; Jänig and Koltzenburg, 1990; Sengupta et al., 1990; Tjen-A-Looi et al., 1998; Brunsden and Grundy, 1999). Sensations arising from some visceral organs appear to be restricted to pain, whereas other organs can give rise to non-painful as well as painful sensations (see review by Cervero, 1994). Examples of the former category include the heart and blood vessels; the respiratory tree; the biliary system; the pancreas; the kidney and ureter; the ovaries and uterus; as well as the epididymis, vas deferens, and testis. Examples of the latter include the esophagus, stomach, colon, and rectum; the urinary bladder and urethra.

One strategy for studying the mechanisms of visceral pain is to examine the responses of primary afferents supplying visceral organs that give rise only to pain sensation (although some of these organs may also produce afferent signals that are involved in regulation). For example, Cervero (1982b; 1983) has recorded the responses of afferents that supply the biliary system of the ferret to distention of the gallbladder and biliary ducts. Two types of response were observed. One population of afferents had low thresholds but responded throughout the physiological range of distention. These were proposed to be involved in reflex regulation. Because the other group of afferents responded only to noxious levels of distention, they were suggested to be nociceptors.

The renal pelvis and ureter are sources of severe pain when a renal stone obstructs the ureter. Cervero and Sann (1989) investigated the responses of afferents that innervate the ureter. Some of the afferents were sensitive to peristaltic contractions of the ureter (see section on **Visceral Mechanoreceptors**). However, most of the afferents examined responded

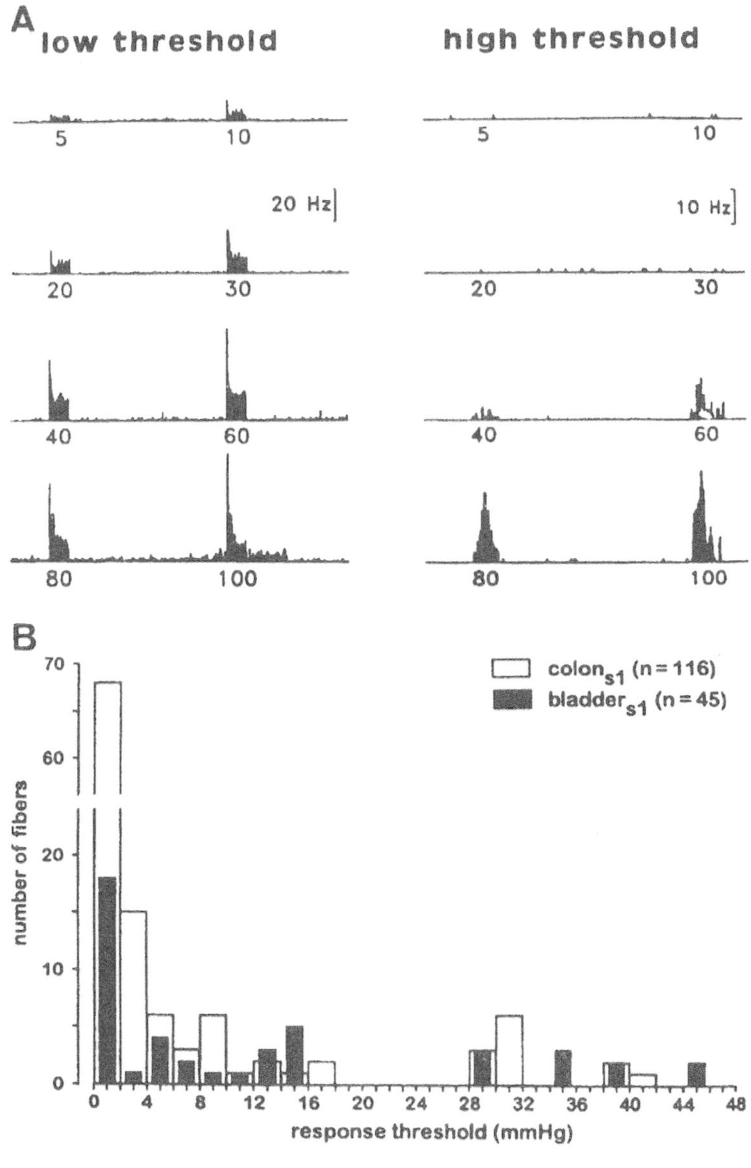

FIGURE 2.31. Nociceptive visceral afferents. A, responses of 2 visceral afferent fibers recorded from the pelvic nerve. The responses are to graded intensities of colorectal distention. The fiber whose responses are shown on the left had a threshold of 5 mmHg, but the responses increased as stimulus intensity increased into the noxious range. The fiber whose responses are shown on the right had a threshold of 40 mmHg. B, bar graph shows the distribution of thresholds of pelvic nerve afferents to distention of the colon (open bars) or urinary bladder (filled bars). Note the bimodal distribution of thresholds. (From Gebhart, 1996.)

only to strong stimuli, including distention, hypoxia, bradykinin, or capsaicin. It was concluded that these latter units were nociceptors. A rat model of the pain produced by a renal stone has been developed and used to study changes in behavioral responses to painful stimuli, such as the development of referred hyperalgesia in muscle (Giamberardino et al., 1995).

Afferent fibers innervating the reproductive organs of female rats respond to both mechanical and chemical stimuli (Berkley et al., 1988; 1990; 1993). Thresholds for activation of these are lowered during estrus (Berkley et al., 1988; 1990; Robbins et al., 1990; 1992). It was proposed that such afferents contribute to reproductive functions, including mating, conception, and parturition. However, it is likely that they also participate in labor pain in humans.

Kumazawa and Mizumura (1977a; 1979; 1980a, b; 1983), Kumazawa et al. (1987) and Sato et al. (1989) have described in some detail a class of polymodal nociceptors in the testis of the dog. Most have conduction velocities in the Aδ range, and they respond to mechanical, thermal, and chemical stimuli (Fig. 2.32; Gebhart, 1996). The ultrastructure of some of these endings has been described (Kruger et al., 1988a).

The lower anal canal is very sensitive to tactile and thermal sensations, resembling in this respect the perineal skin (Cervero, 1994). Some mechanoreceptors that innervate the anal canal mucosa respond with a rapidly adapting discharge following weak mechanical stimulation of the mucosa (Sengupta and Gebhart, 1994). A type of nociceptor has been described in the anal canal that has multiple spot-like receptive fields in the mucosal

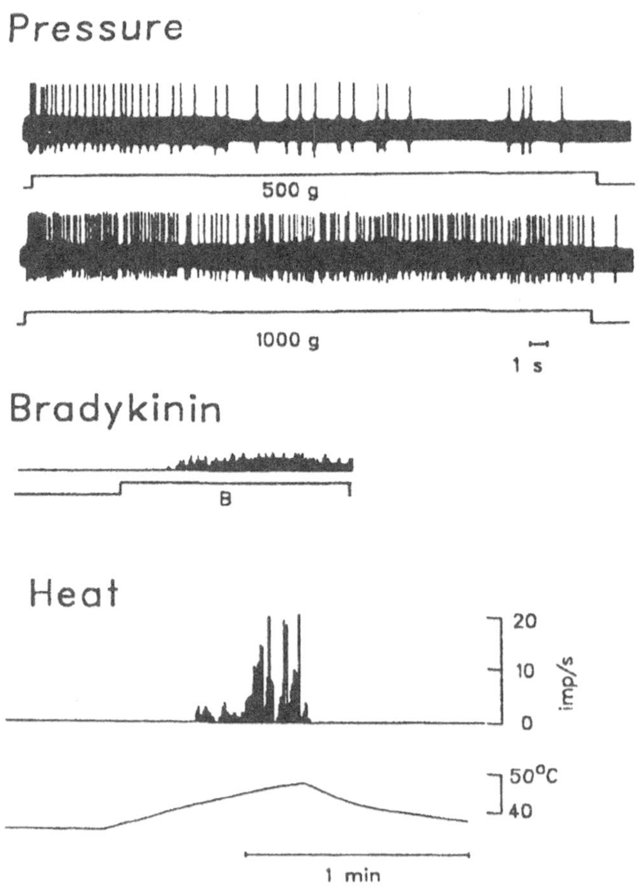

FIGURE 2.32. Polymodal visceral nociceptor. The receptor was located in the testicle. As shown, it responded to pressure, bradykinin, and heat. (From Kumazawa, 1986; modified by Gebhart, 1996.)

lining; this receptor has an unmyelinated axon, and it responds to noxious mechanical, thermal, and chemical stimuli, thus resembling C polymodal nociceptors of the skin (Clifton et al., 1976).

Peptides in Visceral Afferents

Substance P and calcitonin gene-related peptide are found in many visceral primary afferent neurons (Sharkey, 1987; Green and Dockray, 1988; Bennett, D. L. H. et al., 1996b; Goehler and Sternini, 1996; Traub et al., 1999). These peptides are often co-localized. Few pelvic afferents contain somatostatin and none either neuropeptide Y or vasoactive intestinal polypeptide (Wang, H. F. et al., 1998a; Traub et al., 1999). Other neurokinins than substance P may be involved in visceral signaling, since NK3 receptors are implicated (Julia et al., 1999). Peptides released from the terminals of primary afferent fibers in viscera could affect blood flow, secretion, and smooth muscle function (Green and Dockray, 1988). The peptide content of visceral primary afferents can be depleted during the early phase of inflammation (Traub et al., 1999) or following capsaicin administration (Sharkey, 1987). Some visceral afferents contain fluoride-resistant acid phosphatase (Molander et al., 1987), and some non-peptidergic visceral afferents express IB4 binding sites (Bennett, D. L. H. et al., 1996b; Wang, H. F. et al., 1998a, b). The proportion of afferent fibers that supply the urinary bladder and that express IB4 binding sites is smaller than the proportion of IB4-positive cutaneous afferents (Bennett et al., 1996b).

Electrical stimulation of afferent fibers supplying the urinary bladder in rats can elicit plasma extravasation, as can the introduction of mustard oil into the lumen of the bladder (Koltzenburg and McMahon, 1986). However, contractions of the bladder wall or distension of the bladder do not have this effect.

Growth Factors and Visceral Afferents

The effects of growth factors on visceral primary afferent neurons appear to be mediated by way of trkA and trkB receptors (McMahon et al., 1994; Bennett et al., 1996b).

Sensitization of Visceral Afferents

Like cutaneous, muscle, and joint afferents, mechanically insensitive visceral afferents can be sensitized under conditions of inflammation (Fig. 2.33; Häbler et al., 1988b; 1990b; 1993b; Jänig and Koltzenburg, 1990; Dmitrieva and McMahon, 1996; Gebhart, 1996; Michaelis et al., 1996b). Bradykinin activates visceral afferents by an effect on B2 receptors (Pan et al., 1994; Sengupta and Gebhart, 1994; Tjen-A-Looi et al., 1998). Although endogenous bradykinin release exerts an action on visceral C fibers that can be attenuated by aspirin (Pan et al., 1994), the action of bradykinin is at least partially independent of the activation of cyclo-oxygenase (Tjen-A-Looi et al., 1998). Visceral afferents are also activated by substance P and serotonin (Lew and Longhurst, 1986). Substances that cause sensitization of visceral afferents during ischemia include prostaglandins and other arachidonic acid products, histamine, adenosine, and serotonin (Longhurst and Dittman, 1987; Longhurst et al., 1991; Pan et al., 1994; 1995; Fu et al., 1997; Fu and Longhurst, 1998; Brunsden and Grundy, 1999). Sensitization can also be produced by cytokines, such as interleukin-1β (Fu and Longhurst, 1999) and by nerve growth factor (Dmitrieva and McMahon, 1996). Signal transduction pathways involved in sensitization of visceral

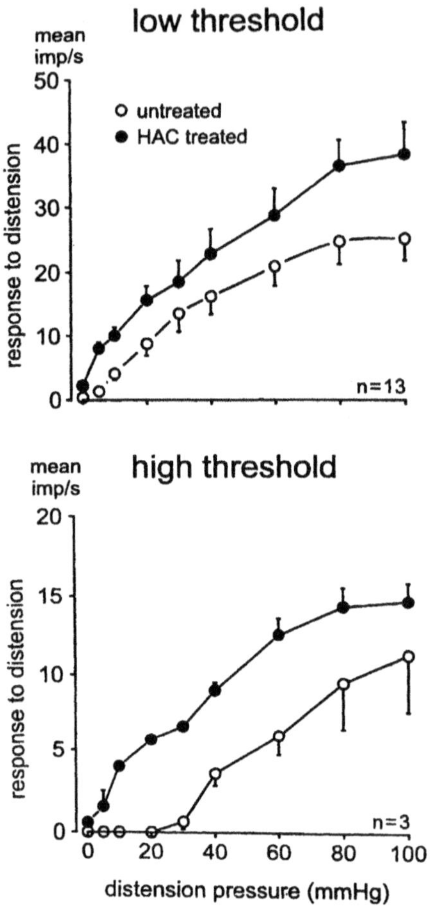

FIGURE 2.33. Sensitization of visceral afferents. The graphs show the responses of populations of pelvic nerve afferents that had low and high thresholds for colorectal distention. The responses before sensitization by intracolonic administration of 2.5% acetic acid are shown by the open circles and those after acetic acid administration are shown by the filled circles. (From Gebhart, 1996.)

afferents include the cAMP/protein kinase A (Brunsden and Grundy, 1999; Guo and Longhurst, 2000) and protein kinase C pathways (Mizumura et al., 1997; Guo, Z. L. et al., 1998; 1999).

Cervero (1994) argues that sensitization of previously silent afferents does not mean that such silent afferents are qualitatively different from other visceral afferents (or joint and cutaneous afferents). The afferents may initially have a high threshold, which is then lowered by the sensitization process. Other afferents with an initially lower threshold can also be sensitized.

Visceral Thermoreceptors

Recently, cold-sensitive C fibers that innervate the urinary bladder of cats have been described (Jiang et al., 2002). Activity in these afferent fibers was recorded from sacral dorsal roots. The afferents responded to cooling the bladder by filling it with cold water, and their activity decreased when the bladder was filled with warm water. The receptors were insensitive to mechanical stimuli, but they responded to menthol, as do cutaneous cold receptors. It was suggested that the transduction mechanism might involve the TRPM8 (or CMR1) ion channel, which is one of the TRP family of excitatory ion channels (cf. McKemy et al., 2002; Peier et al., 2002a, b) and is found in some dorsal root ganglion cells (Reid and Flonta, 2001; Peier et al., 2002a, b).

CONCLUSIONS

1. The sensory axons in peripheral nerves can be subdivided into myelinated and unmyelinated classes. There are two broad classes of myelinated cutaneous and visceral afferent fibers, the $A\alpha\beta$ fibers, which conduct at 30–100 m/s, and the $A\delta$ fibers, which conduct at 4–30 m/s. The unmyelinated or C fibers have conduction velocities of less than 2.5 m/s. Muscle and joint nerves have three classes of myelinated axons: group I, which conduct at 72–120 m/s; group II, which conduct at 24–71 m/s, and group III, which conduct at 6–23 m/s. There are also unmyelinated group IV axons that have conduction velocities of less than 2.5 m/s. These conduction velocities were measured in cats; the comparable fiber groups have slower conduction velocities in smaller animals, such as mice.
2. The conduction velocity of a myelinated axon in meters per second can be predicted by multiplying its total diameter in micrometers by 6 if it is a large fiber, or by 4.5 if it is medium-sized or small. The conduction velocities of fine afferent fibers slow during repetitive discharges.
3. Sensory receptors are found in skin, muscle, joints, and viscera.
4. Cutaneous sensory receptors in mammals can be classified as mechanoreceptors, nociceptors, thermoreceptors, and chemoreceptors. They may end in epidermis or dermis; some types are encapsulated and others unencapsulated; they can be subdivided into slowly adapting and rapidly adapting receptors.
5. Another way of classifying cutaneous receptors is by the most likely information that they encode. For example, cutaneous mechanoreceptors may encode skin indentation (position of the skin), velocity of skin indentation, or a higher derivative of indentation (acceleration, jerk).
6. Cutaneous slowly adapting mechanoreceptors (position detectors) include SA I and SA II receptors. The SA I receptors are associated with Merkel cells in tactile domes of the basal layer of the epidermis. They respond with a dynamic and then an irregular static discharge when the dome is indented, but they are insensitive to stimuli applied to adjacent skin or to skin stretch. They are regulated by the neurotrophin NT-3 acting on trkC receptors.
7. SA II receptors are associated with Ruffini endings. They may have a background discharge, and they respond with a dynamic and a regular static discharge to skin indentation or stretch over a broad area. Stretch at right angles to the orientation of the most effective line of stretch may unload these endings and decrease their discharge.
8. Cutaneous velocity detectors include FA I receptors (Meissner's corpuscles in primates and Krause's end bulbs in cats), G_2 hair follicle receptors, D hair follicle receptors, field receptors, and C mechanoreceptors.

9. FA I receptors are found in the glabrous (non-hairy) skin; they are very sensitive and respond best to repetitive stimulation at 5–40 Hz. G_2 hair follicle receptors supply guard hairs and are readily activated by movements of these hairs. D hair follicle receptors innervate down hairs; they are exquisitely sensitive to movements of down hairs by stimuli applied anywhere over a broad area. Field receptors are in the skin; they can be excited by movement of many hairs or by contact with the skin, but not by movement of single hairs. C mechanoreceptors are especially responsive to stimuli that move slowly across the hairy skin; they are much less common in primates than in cats.
10. The cycle of growth and regression of hair follicles depends on a number of growth factors. NT-3 regulates the nerve supply of hair follicles.
11. Cutaneous transient detectors include Pacinian corpuscles and G_1 hair follicle receptors. Pacinian corpuscles are found in subcutaneous tissue (as well as in fascial planes, along interosseous membranes, in tissue adjacent to joints, and in the mesentery). They are extraordinarily sensitive to mechanical transients, sometimes having a threshold of less than 1 μm. They respond best to vibratory stimuli at 60–300 Hz, but can often follow stimulation at 500 Hz or more. G_1 hair follicle receptors innervate guard hairs and respond only to rapid movements of these hairs.
12. Sinus hairs are not found in humans. Those found in animals include mystacial vibrissae and carpal hairs. These are richly innervated. Two types of slowly adapting and two types of rapidly adapting receptors supply vibrissae. Carpal hairs have two types of slowly adapting receptors, but no rapidly adapting receptors. Instead, movement of carpal hairs excites a cluster of nearby Pacinian corpuscles
13. Cutaneous nociceptors of several types have been described. Various combinations of noxious stimuli (mechanical, thermal, and/or chemical) can activate Aδ nociceptors. If sensitized, for example by "inflammatory soup," they become more responsive. Their terminals may invade the epidermis but remain ensheathed by Schwann cells; therefore, they should not be considered free nerve endings. Their activation causes first or "pricking" pain.
14. C nociceptors may also respond to noxious mechanical, thermal, and/or chemical stimuli. C-mechano-heat nociceptors are often called C-polymodal nociceptors. Their activation causes second or burning pain. They provide accurate information about the spatial location of a noxious stimulus.
15. Transduction by cutaneous and other nociceptors often involves special membrane molecules that make the nociceptive terminals sensitive to particular stimuli, such as heat or a reduction in pH. Examples include vanilloid (VR1) receptors, which respond to capsaicin, heat, and acidic conditions, and P2X3 receptors, which are activated by ATP.
16. Nociceptors can often be sensitized by exposure to strong noxious stimuli or inflammatory mediators. Sensitization involves an increase of intracellular Ca^{++} concentration and subsequent activation of signal transduction pathways. Phosphorylation of ion channels or of transduction molecules, such as VR1 receptors, presumably underlies the sensitization of primary afferent nociceptors.
17. Many of the fine afferent nociceptive fibers that innervate the skin contain peptides, such as calcitonin gene-related peptide, substance P, and others. Other cutaneous nociceptors have binding sites for the lectin, IB4. Peptidergic and IB4 nociceptors seem to form two major phenotypes of nociceptive primary afferents.

18. Peptidergic nociceptors release their peptides centrally, but in addition they may release them peripherally, triggering neurogenic inflammation. Substance P and calcitonin gene-related peptide result in neurogenic edema and vasodilation, respectively. In animals, the spread of neurogenic inflammation appears to depend chiefly on dorsal root reflexes in nociceptive afferents.
19. Nociceptive neurons are regulated by growth factors. Aδ nociceptors and peptidergic C fibers are dependent on nerve growth factor. In the adult, IB4-containing nociceptors depend on glial cell line-derived neurotrophic factor.
20. A class of C fibers has recently been described in human skin that is sensitive to histamine and can apparently account for the sensation of itch. The terminal fields of these afferents can be widespread and so they could account for the spread of flare in humans. However, the spread of flare in rats depends in part on dorsal root reflexes.
21. Cutaneous thermoreceptors include cold and warm receptors. Cold receptors may be innervated by Aδ or by C fibers (mostly Aδ fibers in humans). They respond with a dynamic and a static discharge to sudden cooling of the skin. The maximum static discharge occurs at 18–34 °C. At low temperatures, cold receptors show a bursting discharge, which may help distinguish between temperatures above and below the level of maximum responsiveness. Cold receptors may show a paradoxical cold response when exposed to noxious heat.
22. Warm receptors are supplied by unmyelinated fibers. They discharge when the skin is warmed, but they stop firing when noxious temperatures are reached.
23. Muscle contains stretch receptors and pressure-pain endings. The stretch receptors include muscle spindles and Golgi tendon organs.
24. Muscle spindles are complex sense organs with two types of sensory endings and at least two kinds of motor endings. The largest type of sensory axon, the group Ia afferent fiber, supplies the primary endings. They are sensitive to the rate of stretch of the muscle and to the length of the muscle. The sensitivity of the primary ending is set by activity in dynamic fusimotor fibers. Secondary endings are innervated by intermediate-sized group II axons and they signal muscle length. Their sensitivity is controlled by static fusimotor axons. β-innervation may also be provided by branches of α-motor axons.
25. Golgi tendon organs are innervated by group Ib axons and are located in tendons and aponeuroses. They respond to both muscle stretch and contractions, and they signal tension in the tendon or aponeurosis.
26. Group III muscle afferents have been called pressure-pain endings. Some group III afferents are activated by pressure on the junction between the muscle and its tendon or on the belly of the muscle; others are in fascial planes. Some can be excited by muscle stretch. Intra-arterial injection of algesic chemicals or intramuscular injection of hypertonic saline will activate many of these afferent fibers. Although many are excited by muscular contractions, their responses are decreased following ischemia. Therefore it has been proposed that these fibers are more likely to be ergoreceptors than nociceptors. Ergoreceptors provide information useful for cardiopulmonary adjustments to exercise. Activation of muscle nociceptors results in cramping pain.
27. Group IV muscle afferent fibers often have high thresholds for mechanical stimuli. They can be excited or sensitized by intra-arterial injection of algesic chemicals and often by thermal stimuli. It has been suggested that those that respond preferentially to chemical stimuli are nociceptors and those activated by muscle

contractions but not by algesic chemicals are ergoreceptors. Some group IV muscle afferents discharge vigorously during muscle ischemia and are suited for a role in ischemic muscle pain.

28. Muscle nociceptors can become sensitized, for example, by exposure to inflammatory mediators.
29. Joints are innervated by mechanoreceptors and nociceptors.
30. Joint mechanoreceptors include Ruffini endings, Golgi tendon organs, Pacinian corpuscles, and Golgi–Mazzoni endings. The Ruffini endings discharge best when the joint is brought to the limit of its range, and so they may signal joint pressure. Golgi tendon organs respond like those in muscle and Pacinian corpuscles like those in subcutaneous tissue. Golgi–Mazzoni endings respond with a slowly adapting discharge when they are compressed, and so they may provide deep pressure information.
31. There is controversy about the presence or absence of slowly adapting discharges when a joint is at mid-range. One opinion is that when such discharges are seen, they reflect the activity of muscle stretch receptors whose axons have entered the joint nerve. According to this view, joint position is signaled by muscle receptors rather than by joint receptors. The opposing view is that there are some slowly adapting joint receptors whose afferent fibers discharge best when the joint is at mid-range and that these can provide information to the central nervous system about joint position.
32. Joint nociceptors have been well characterized. They may have group III or IV afferent fibers. They respond best or only to noxious movements of the joint. Some do not respond normally to any joint movement, but these become very sensitive during inflammation.
33. Dorsal root reflexes were first implicated in neurogenic inflammation in experiments on experimental arthritis. Dorsal root reflexes can be triggered if the primary afferent depolarization that is normally responsible for presynaptic inhibition becomes large enough to exceed threshold. Impulses are then conduced antidromically from the spinal cord back out to the periphery. Sensory terminals may then release neuroactive substances, such as peptides. Substance P and calcitonin gene-related peptide are two of the peptides known to contribute to neurogenic inflammation. Substance P causes plasma extravasation and edema and CGRP produces vasodilation.
34. Viscera are supplied by mechanoreceptors and by nociceptors.
35. Mechanoreceptors associated with the gastrointestinal tract respond to movements of the bowel or of the mesentery. Some in the wall of the urinary bladder respond either to distention or to contraction of the bladder.
36. Visceral nociceptors have been described in the heart, esophagus, gallbladder, colon, ureter, testis, and female reproductive tract. Those in the testis have been particularly well characterized. Most are supplied by Aδ fibers, and they are responsive to noxious mechanical, thermal, and chemical stimuli. Many visceral nociceptors are supplied by C fibers. In some cases, the thresholds are in the innocuous range, but discharges of the afferents increase as stimulation is increased well into the noxious range. Other afferents have high thresholds. Nociceptive visceral afferents can be sensitized.

Chapter 3

Dorsal Root Ganglion Cells and their Processes

Somatic primary afferent neurons are important in spinal sensory mechanisms for they receive information from the periphery and transmit it to the cord. These neurons are housed in dorsal root ganglia (and cranial nerve equivalents), so they are called dorsal root ganglion (DRG) cells. These cells are unipolar, sometimes called pseudounipolar because they originate as bipolar cells (Dogiel, 1908; Ranson, 1912), with the main process splitting into peripheral and central processes (Fig. 3.1). The central processes are called dorsal root axons because they travel in the dorsal root on their way to the cord.

DORSAL ROOT GANGLION CELL BODIES

DRG cells have common features in that the cell bodies are in peripheral ganglia, their processes are organized in a similar way, and they have a common function, to transduce environmental stimuli and transmit the resulting information centrally (Lieberman, 1976; Scharf, 1958). Despite these similarities, DGR cells are diverse. This chapter will document these diversities.

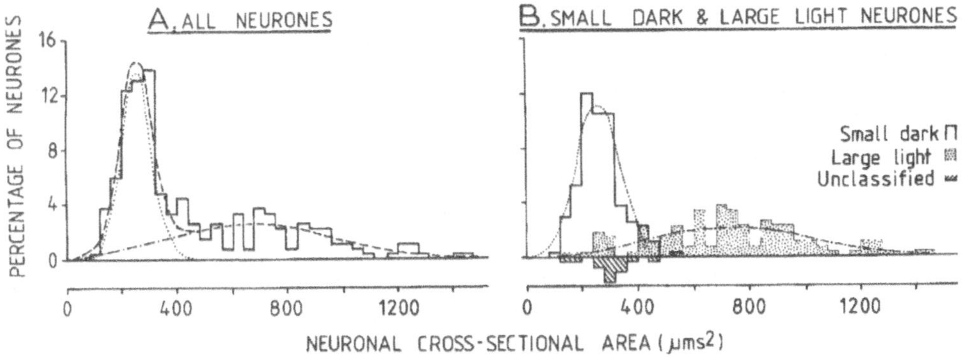

FIGURE 3.1. Histograms of dorsal root ganglion cell body sizes. A indicates all cells and B shows the cells split into large light and small dark categories. Note that there are some small large light cells. If the large light cells are labeled by the neurofilament antibody RT97, the histograms are essentially identical. (From Lawson et al., 1984; reprinted by permission of Wiley-Liss.)

DRG Cell Classifications

Cytologic and Size Classifications

The most widely used morphologic subdivision of DRG cells is by cell body size. Although occasional studies show unimodal or trimodal histograms (Warrington and Griffith, 1904; Dyck et al., 1978; 1979; Kawamura and Dyck, 1978; Sugimoto et al., 1988), bimodal histograms are almost always found (Fig. 3.1; Lawson, 1979; Lawson and Biscoe, 1979; Szarijanni and Rethelyi, 1979; Lawson et al., 1984; 1985; McDougal Jr. et al., 1985; Lee et al., 1986; Ohnishi and Ogawa, 1986; Schmalbruch, 1987) leading to the widely accepted classification of DRG cells into large and small categories (Lieberman, 1976; Lawson, 1979).

It was noted early that large DRG cells were clear or lightly stained and small cells were darkly stained with the basic analine dyes used so extensively in early neuron studies (Nissl, 1894; Dogiel, 1908; Ramón y Cajal, 1909; Scharf, 1958; Lieberman, 1976). Some did not accept this classification as representing the situation in the living, however, because distinctive cytologic differences were not seen or poor fixation produced small dark cells (summarized by Lieberman, 1976). However, these differences are seen in well-fixed material (Lawson et al., 1974; Lieberman, 1976), and the last birthdays of large cells are different from those of small cells (Lawson et al., 1974; Lawson and Biscoe, 1979; Sims and Vaughn, 1979; Altman and Bayer, 1984). In addition, large clear cells but not small dark cells are immunostained by RT97 antibodies, which recognize neurofilaments (Lawson et al., 1984). Thus the distinction between large clear cells and small dark cells is secure (Lieberman, 1976).

An important point is that not all clear cells are large. Thus cells stained by phosphorylated 200-kd neurofilament antibodies such as RT97, which have become standard for identifying large light cells (Lawson and Waddell, 1991), are of all sizes (Lawson et al., 1984). This contrasts with the small dark cells, which are always small (Fig. 3.2), and which in large part are stained for a 57-kd filament called peripherin (Ferri et al., 1990; Troy et al., 1990; Goldstein et al., 1991; 1996). Thus there are two populations of DRG cells, the first staining lightly with basic dyes but intensely for RT97 with sizes spanning the whole range and the second darkly staining and only small. This is a reason why large- and small-cell histograms are indistinctly separated in histograms.

DRG cells have been further subdivided on the basis of various cytologic criteria, for example, arrangements of granular endoplasmic reticulum (Nissl substance) (Clark, 1926; Orr and Rows, 1901), but they do not correlate with function or responses to various stimulus types (Clark, 1926; Duce and Keen, 1977). On the basis of numerous cytologoic criteria, Andres (1961a, b) divided the large light (A) cells into A1, A2, and A3 subgroups and small dark (B) cells into B1, B2, and B3 subgroups. Jacobs, J. et al. (1975) proposed two types of A and one type of B cells, Duce and Keen (1977) proposed three groups of A cells and four of B cells, Rambourg et al. (1983) divided DRG cells into A, B, and C cells, with the A type being subdivided into three and the B type into two subgroups, and Sommer et al. (1985) divided the cells into three A types, two B types, and a single C type. However, these cytologic subdivisions are not yet correlated with functional and biochemical variables. More recent classifications (below) use multiple criteria but within the context of large light and small dark categories.

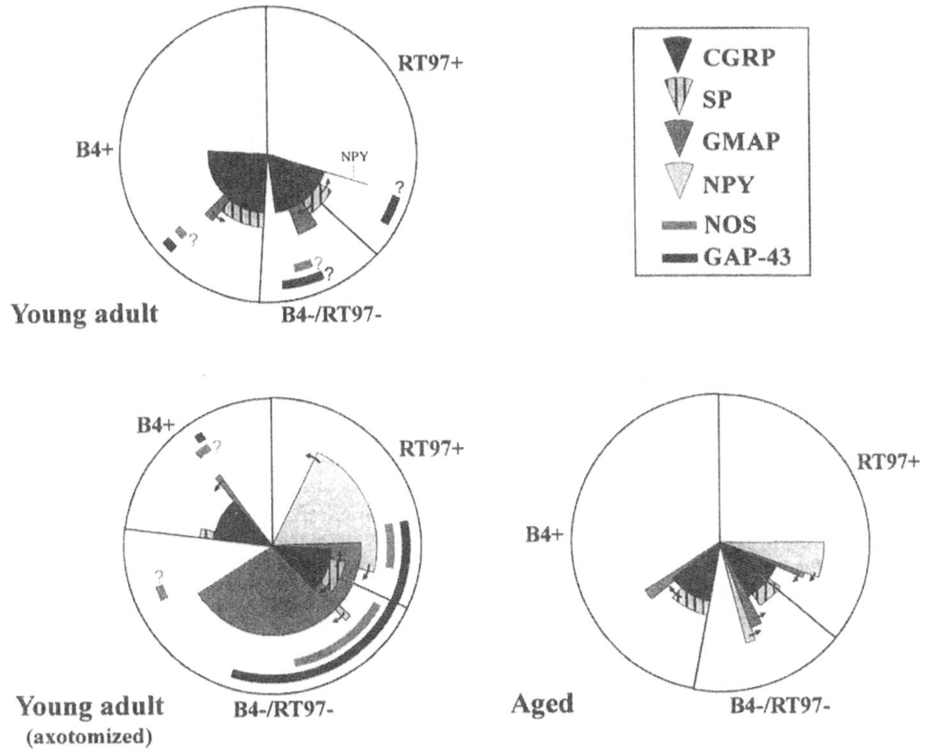

FIGURE 3.2. A pie chart summarizing expression of several peptides, NOS, GAP43, IB4 (+B4), and RT97 in normal and operated young adult animals and in aged animals. Note that there is little overlap of IB4 and RT97 populations but a considerable overlap of CGRP and IB4 populations. (From Bergman et al., 1999.)

Myelination

An obvious subdivision is between DRG cells with myelinated vs. unmyelinated axons, and for cells with myelinated axons, myelin thickness and length of internode are correlated with axon size (Thomas et al., 1993) and thus with conduction velocity and function (Chapter 2). Here we point out only that myelin sheaths are considerably thicker in peripheral nerves than they are in dorsal roots (Berthold and Carlstedt, 1977; Suh et al., 1984) and that peripheral myelinated axons can become unmyelinated as they cross the PNS–CNS interchange (Carlstedt, 1977; Thomas et al., 1993), so correlations depend on where you do your measurements.

Axon Conduction Velocity

Conduction velocities show three types of sensory axons, Aβ (sometimes split into Aα and Aβ), Aδ, and C fibers, and consequently one would expect three types of DRG cells. Large light cells are clearly associated with Aβ axons and small dark cells with C fibers (Lawson et al., 1984; Harper and Lawson, 1985a; Lawson and Waddell, 1991), but this leaves an intermediate group, the Aδ cells. Some studies describe intermediate-sized DRG cells (Matsuura, 1967; Matsuura et al., 1969; Honda et al., 1983; Lindh et al., 1983; Otten and Lorez, 1983; Kuo et al., 1985; Tuchscherer and Seybold, 1985; Hanko et al., 1986; Kuwayama et al., 1987), but only two peaks are seen in histograms (see above). Visualization of individually labeled Aδ cells shows that they distribute in the small or large groups and do not form a separate peak (Harper and Lawson, 1985a; Lawson and Waddell, 1991).

Using conduction velocity as an indicator of DRG cell subdivisions, a linear relationship between cell size and axon diameters is found for HRP-filled mouse DRG cells (Yoshida and Matsuda, 1979) and for HRP-filled cat L5-S3 dorsal root ganglion cells whose processes conducted impulses more rapidly than 2.5 m/s, but not for cells whose processes conducted at less than 2.5 m/s (Lee et al., 1986). Conduction velocity is an estimate of axon size, and conduction velocities almost always correlate with cell body size and function with the tightness of the correlations varying depending on the types of cells studied (Yoshida and Matsuda, 1979; Harper and Lawson, 1985a; Cameron et al., 1986; Koerber and Mendell, 1988; Lawson and Waddell, 1991; Lawson et al., 1996; McCarthy and Lawson, 1997). Exceptions are found, however, for similar sized sensory cells in the nodose ganglia conduct action potentials in A and C ranges (Gallego and Eyzaguirre, 1978), all sizes of dorsal root ganglion cells gave rise to group IV (unmyelinated) afferents from joints (Hoheisel and Mense, 1986), and there was no correlation between cell body size and conduction velocity of the axon in a study of unmyelinated fibers (Nagy et al., 1993a). In summary, with some exceptions, cell body size, axon conduction velocity, and type of information correlates lead to meaningful delineations of DRG cell types (Chapter 2).

Action Potentials and After Polarizations

Somatic potentials show remarkable differences between nociceptive and non-nociceptive neurons. Thus, DRG cells can be categorized as having broad or narrow action potentials (spikes), different-sized after hyperpolarizations and usually a hump on the falling phase of the action potential (Gallego and Eyzaguirre, 1978; Yoshida et al., 1978; Yoshida and Matsuda, 1979; Belmonte and Gallego, 1983; Harper and Lawson, 1985a, b; Cameron et al., 1986; Rose et al., 1986; Koerber and Mendell, 1988; Koerber et al., 1988; Waddell and Lawson, 1990; Mirnics and Koerber, 1997; Djouhri et al., 1998; Djouhri and Lawson, 2001; Lawson, 2002). The cells with wider action potentials and larger after hyperpolarizations are high-threshold responders or nociceptors whereas cells with the narrow action potentials and smaller hyperpolarizations are low-threshold responders or non-nociceptors (Maxwell and Jankowska, 1996). In correlative studies, there is an inverse correlation of axon conduction velocity and width of the somal spike (Mirnics and Koerber, 1997).

It is becoming apparent that neurons that respond to high-threshold inputs are organized differently from those that respond to low-threshold inputs. This is a truism in that many things (peptide and receptor expressions, etc.) differ in these populations, but it has not been fully appreciated that these differences may not be in terms of large light and small dark or Aβ, Aδ, and C DRG cells. Rather DRG cells are either nociceptors or low threshold responders and in each category are large light and small dark and Aβ, Aδ, and C DRG cells. A corollary to this is that there are large light Aβ nociceptors in considerable numbers (Koerber et al., 1988; Ritter and Mendell, 1992; Djouhri et al., 1998; Lawson, 2002). Thus, the dogma that large light cells respond to low-threshold inputs and small dark cells to high-threshold inputs is a considerable oversimplification.

Classifications by Chemical Neuroanatomical Criteria

Cytologic studies noted that large light cells were light because the somata contained many filaments that did not stain with basic dyes. Consequently the use of phosphorylated neurofilament markers such as RT97 gave an almost perfect map of the large light population (Lawson et al., 1984). A correlation of conduction velocities and RT97 staining showed that all A fiber DRG cells were RT97 positive and all C cells were RT97 negative (Lawson and Waddell, 1991). There is also almost no overlap of RT97 with IB4 staining, IB4 being a major marker for a large group of small dark cells (Wang et al., 1994). In addition, the RT97 population co-localized with the GM1 ganglioside receptor (for B-HRP), which is expressed by large DRG cells in normal animals (Robertson and Arvidsson, 1985) that transmit largely low-threshold mechanical information. Accordingly phosphorylated filament markers have become the standard for the large light population of DRG cells (Lawson and Waddell, 1991). Admittedly, the correlations are not absolute in that a few light or phosphorylated neurofilament stained cells are small (Lawson et al., 1984), there is a slight overlap of IB4 and GM1 with RT97 staining (Bergman et al., 1999; Wang et al., 1994), and some large DRG cells are nociceptors (Lawson et al., 1997). Thus the generalization is not absolute, but in the main the distinction is clear, and the phosphorylated neurofilament markers identify a population of large light DRG cells, comprising about one third of the total (Bergman and Ulfhake, 1998; Perry and Lawson, 1998; Bergman et al., 1999), that are distinctive from the various small cell populations (below).

Similar clarity has not been reached for the small dark population. It should be understood that this group of cells contains an enormous number of subpopulations in the sense that markers of many kinds (peptides, enzymes, receptors, lectins, trophic factors, etc.) are found in subgroups of these cells (Chapter 4). From the point of view of cell classification, attention has focused on fluoride-resistant acid phosphatase (FRAP), the peptide CGRP, the lectin IB4, and trk receptors. The first breakthrough came with the hypothesis that there are two populations of small dark neurons; the first that expresses acid phosphatase and whose axons project preferentially to lamina IIi, and the second that is peptide-containing and whose axons project preferentially to laminae I-IIo (Nagy and Hunt, 1982; Hunt and Rossi, 1985; Silverman and Kruger, 1988a, b; 1990). Lectin IB4 labeling, more adaptable than FRAP histochemistry, was then found to label the non-peptidergic population of DRG cells (Silverman and Kruger, 1990; Ambalavanar and Morris, 1992; Averill et al., 1995; Molliver et al., 1995). These findings were strongly supported by the observation that the NGF receptor trkA characterized the peptidergic (CGRP) population whereas GDNF receptors characterize the IB4 cells, which are also characterized by the tyrosine kinase Ret in adults (Molliver et al., 1995; 1997). The difficulty is that other studies report a considerable overlap of IB4 or FRAP with CGRP or trkA in DRG cells (Fig. 3.2; Carr et al., 1990; Wang et al., 1994; Bergman et al., 1999; Kashiba et al., 2001a), and there is a significant population of small DRG calls that do not express any trk receptors (Wright and Snider 1995). The distinction between trkA and trkC forms a meaningful subdivision; the cells with these receptors innervate different populations both by size and region, but the trkA cells contain both peptide and non-peptide small dark cells (McMahon et al., 1994). Accordingly completely satisfactory classifications are not yet available, but major progress is being made. A significant advance would be the demonstration of significant functional differences between populations of small dark cells, and in this regard IB4-positive cells respond differently than IB4-negative cells to sprouting after injury and thermal stimuli (Belyantseva and Lewin, 1999; Stucky and Lewin, 1999).

Topographic Organization

Another type of sensory cell organization would be functional categories of neurons in particular places in the ganglion. Usually, however, cells seem to be distributed randomly

(Norcio and DeSantis, 1976; McLachlan and Jänig, 1983; Peyronnard et al., 1986) particularly for visceral neurons (McLachlan and Jänig, 1983; Cervero et al., 1984a, b). Some organization is evident, however, in that the position of the perikaryon in cat DRGs predicts which dorsal root filament will carry the central axon to the spinal cord (Burton and McFarlane, 1973), a lamellar arrangement of cell bodies is seen following restricted injections of HRP into different parts of the periphery or spinal cord (Kausz and Rethelyi, 1985), and large neurons in cat sacrococcygeal ganglia cluster laterally (Szarijanni and Rethelyi, 1979). So there is some topographic organization within the ganglion, but the significance is not known.

Innervation of Ganglion Cell Bodies, Pericellular Baskets

There is a classic literature showing fibers entering normal dorsal root ganglia and forming pericellular baskets around DRG cell bodies (Ramón y Cajal, 1909; Scharf, 1958; Lieberman, 1976). Many of these fibers are adrenergic (Fig. 3.3; Owman and Santini, 1966; Lukas et al., 1970; Stevens et al., 1983), and some contain peptides such as TRH, SP, or CGRP (Kuwayama et al., 1987). In addition, synaptic terminals on cat cervical DRG cell bodies have been shown electron-microscopically (Kayahara et al., 1981; 1984; Kayahara 1986), but their function is not known and confirmation from other labs would be desirable. The above findings are a backdrop to the important demonstrations that peripheral nerve lesions cause a major increase in axon sprouting and pericellular baskets around DRG cell bodies (Chung et al., 1993a; McLachlan et al., 1993). Many of the sprouting fibers are adrenergic (Chung et al., 1993a; 1996; McLachlan et al., 1993), and stimulation of the sympathetic system leads to primary afferent activity (Jänig et al., 1996; Ramer et al., 1999), but the amount of sprouting does not correlate with sensory findings (Kim et al., 1996; 1998; Lee, B. H. et al., 1998). Many of the sprouting fibers are also GAP-43 positive

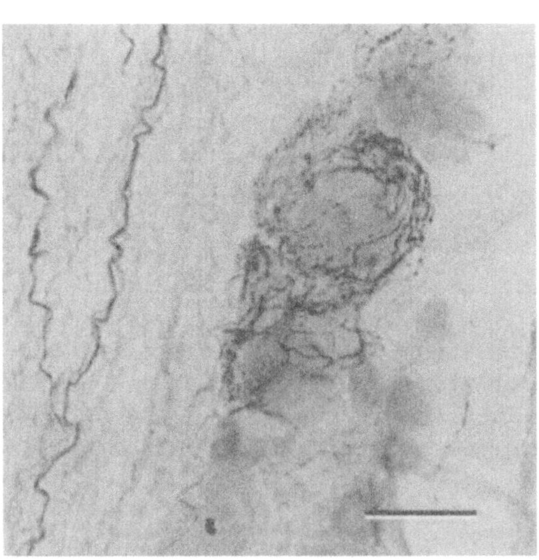

FIGURE 3.3. A picture of a sympathetic axon ramifying around a large DRG cell body to form a typical basket, bar = 20 microns. (Courtesy of K. Chung.)

(Chung and Chung, 2001), peptidergic (Michael et al., 1999; Hu and McLachlan, 2000; Holmberg et al., 2001; Lazarov, 2002), BDNF expressing (Michael et al., 1999), IB4 positive (Li, L. and Zhou, 2001), and NOS or 5-HT expressing (Lazarov, 2002). The baskets appear mainly around large cells (Chung et al., 1993a; McLachlan et al., 1993; Chung and Chung, 2001; Hu and McLachlan, 2001; Li, L. and Zhou, X. F. 2001), and in NGF over-expressing animals around trkA cells (Davis et al., 1998), but not around peptide containing DRG cells (Ramer and Bisby, 1998). Sprouting adrenergic fibers form synaptic varicosities without cytoplasmic thickenings on DRG somata (Chung et al., 1997), but in another study although varicosities were seen, none directly abutted DRG somata (Ramer et al., 1997). Adrenergic and some peptidergic fibers arise from sympathetic neurons (Davis et al., 1994; Chung et al., 1996; Chung and Chung, 2001; Holmberg et al., 2001), but some peptide and IB4 fibers arise from DRG cells (Hu and McLachlan, 2001; Li, L. and Zhou, X. F. 2001). The sprouting occurs no matter what type of lesion or species, but responses are different depending on such things as distance from the ganglion or type of lesion (Kim et al., 1996; Kim, H. J. et al., 2001; Ramer and Bisby, 1997a; Lee, B. H. et al., 1998; Ramer et al., 1999; Hu and McLachlan, 2000). Sprouting occurs principally around cells whose axons are transected by the lesion, but also around neighboring neurons with intact peripheral processes (Chung et al., 1996; Ma and Bisby, 1999a). Wallerian degeneration plays a role in the sympathetic sprouting (Ramer et al., 1997). The sprouting increases in the absence of C fibers, indicating that damage to the large axons is the primary stimulus (Abbadie and Basbaum, 1998).

To gain insight into mechanisms of basket formation, considerable attention has been paid to the role of trophic factors. One approach is to use transgenic mice that over-express NGF. They have considerably more adrenergic baskets than wild-type littermates (Davis et al., 1994; 1998; Ramer et al., 1998a) that largely co-localize with GAL (Holmberg et al., 2001), and these plexuses surround the large majority of trkA/CGRP-expressing DRG cells (Davis et al., 1998; Walsh et al., 1999), as well as cells that express NGF itself (Walsh and Kawaja, 1998). In NGF over-expressors that lack P75 receptors, the sprouting axons make few pericellular baskets (Ramer and Bisby, 1997b; Walsh et al., 1999). Unlike normal animals, there was no increase in baskets after nerve lesions, but GAL declined whereas TH was relatively unaffected (Holmberg et al., 2001). Antagonizing NGF retards basket development (Ramer and Bisby, 1999; Zhou et al., 1999; Deng et al., 2000), although in one study peripheral antiNGF had no effect (Jones et al., 1999), whereas intrathecal or intraventricular NGF increased the sprouting (Jones et al., 1999; Nauta et al., 1999). One difficulty with ascribing these changes to the influence of NGF is that transection of the nerve should cut off peripheral NGF (Ramer et al., 1999), but increased expression of NGF or P75 receptors by satellite cells may compensate (Zhou et al., 1996; Häbler et al., 2000).

Antagonizing NT3 and BDNF also reduces the sprouting (Zhou et al., 1999; Deng et al., 2000), and the expression of BDNF in both target DRG cells as well as sprouting baskets was greatly increased after nerve lesion (Michael et al., 1999). In addition, LIF increases adrenergic baskets in normals or after axotomy (Thompson and Majithia, 1998), and the sprouting is inhibited in interleukin-6 deficient mice (Ramer et al., 1998b). In addition, there may be changes in receptors, and there is no necessity for actual sympathetic synapses for DRG cell activation (Rubin et al., 1997; Shinder et al., 1999). It is difficult to summarize these findings concisely, but it does seem apparent that the sprouting fibers have a role in increasing primary afferent activity, but that it is not a simple role, and that the sprouting is controlled, at least in part, by multiple trophic factors, NGF being specially prominent.

DORSAL ROOT GANGLION CELL PROCESSES

Dermatomes

A dermatome is the area of skin innervated by the sensory axons from a single dorsal root. They have been delineated by: (1) surgical isolation of a root and mapping of the resulting area of projected sensation; (2) electrical stimulation of the root and determination of the area of sensation; (3) reflex vasodilation; (4) the distribution of blisters after herpes zoster infection; (5) dorsal root potentials; (6) sensory levels after spinal cord transection; and (7) sensory examination following disc impingement on the dorsal roots (Sherrington, 1893; 1898; Keegan and Garrett, 1948; Kirk and Denny-Brown, 1970). However, if a dermatome simply represents the distribution of the peripheral sensory axons from a root, it should be unchanging. But strychnine, cutting the lateral part of Lissauer's tract, or sectioning proximal peripheral nerves rather than dorsal roots to outline the dermatome, increased dermatome size, and cutting dorsal roots cranially and caudally decreased it (Denny-Brown et al., 1973). Thus, a dermatome, as measured clinically, is not static, and the above changes presumably result from interactions of primary afferents with intrinsic or other afferent axons. This cautions against taking a simplistic view of a dermatome and emphasizes that the way a dermatome is outlined will have a bearing on its size.

Branching

Branching of primary afferent axons is a suggested anatomical basis for referred pain (Sinclair et al., 1948). Some depictions show such axons (Dogiel, 1908; Hoheisel and Mense, 1985), but others do not (Ramón y Cajal, 1909). Also, some studies show considerably more dorsal root or peripheral sensory axons than DRG cells (Langford and Coggeshall, 1979; 1981a; Aldskogius and Risling, 1981; Chung and Coggeshall, 1984), but again others do not (Duncan and Keyser, 1936; Tandrup, 1995), the latter study being particularly noteworthy because of techniques for unbiased cell counts and visualization of unmyelinated axons with electron microscopy. However, double-labeled DRG cells can be found when markers are placed in different peripheral nerves or organs (Pierau et al., 1982; 1984; Taylor and Pierau, 1982; Borges and Moskowitz, 1983; Alles and Dom, 1985; Laurberg and Sorensen, 1985; McNeill and Burden, 1986; Atkinson and Kenyon, 1990; Wang, G. D. et al., 1996; Sameda et al., 2001), and action potentials can be recorded from DRG cell bodies when different nerves or organs are stimulated (Mense et al., 1980; Bahr et al., 1981; Pierau et al., 1982; Devor et al., 1984; McMahon and Wall, 1987; Häbler et al., 1988a; McCarthy et al., 1995; Nagy et al., 1995). The percentages of such cells vary widely but are usually quite small. Thus, it seems clear that branching of primary afferent axons occurs, but the percentages are usually low, so it is unlikely that this is a major mechanism for referred pain.

Ventral Root Afferent Fibers

The law of separation of function of the spinal roots postulates afferent fibers in the dorsal root and efferent fibers in the ventral root (Cranefield, 1974). However, stimulating ventral roots causes pain, myelinated and unmyelinated fibers survive distal to a ventral root

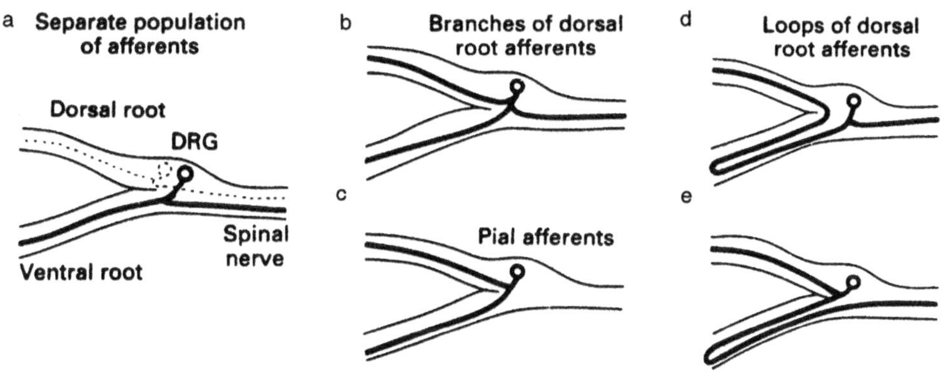

FIGURE 3.4. Diagrams of various arrangements of ventral root afferents. In (a) the ventral root afferent will either end blindly in the ventral root or enter the spinal cord directly; (b) is similar to (a) except that the DRG cell has a third branch in the dorsal root. The pial afferents are shown in (c). The two possibilities of looping afferents are shown in (d) and (e). (From Häbler et al., 1990.)

transection, these surviving fibers degenerate after DRG removal and they possess receptive fields (Magendie, 1822; Bernard, 1858; Sherrington, 1894; Windle, 1931; Frykholm, 1951; Frykholm et al., 1953; Dimsdale and Kemp, 1966; Kato and Hirata, 1968; Ryall and Piercey, 1970; Kato and Tanji, 1971; Coggeshall et al., 1974; Applebaum et al., 1976; Clifton et al., 1976; Coggeshall and Ito, 1977; Emery et al., 1977; Maynard et al., 1977; Yamamoto et al., 1977; Coggeshall, 1980; Hiura, 1982; Holmes, 1982; Wee et al., 1985; Fang, 1987; Phillips et al., 2000; Schenker and Birch, 2000). Thus, there are afferent fibers in the ventral root with possible arrangements diagramed (Fig. 3.4).

Spinal Cord

Some ventral root afferents enter the cord and go to the superficial dorsal horn or intermediolateral cell column (Light and Metz, 1978; Mawe et al., 1984; 1986; Beattie et al., 1987). They are few in number, however, and their function is not known, but they are an exception to the law of separation of function of the spinal roots.

Blind Fibers

Many primary afferent fibers end in the ventral root (Kim et al., 1987; Häbler et al., 1990). One group finds that most of these are exclusive projections (Häbler et al., 1990a) but others find that many have a third process in the dorsal root (Loeb, 1976; Kim et al., 1987).

Looping Fibers

Looping fibers have been postulated ever since Magendie (1822) discovered that stimulation of the ventral root caused pain relieved by section of the dorsal root (Bernard, 1858; Frykholm, 1951; Frykholm et al., 1953). Evidence for this is that numbers of unmyelinated axons decrease as one approaches the cord (Risling and Hildebrand, 1982; Risling et al., 1984a, b), and axonal U-turns can be demonstrated (Bostock, 1981; Risling et al., 1984b;

Azerad et al., 1986; Vergara et al., 1986; Baik-Han et al., 1989). The suggestion is that the central process loops (Vergara et al., 1986).

Pial Fibers

Numerous fibers enter the pia from the ventral root (Clark, 1931; Dalsgaard et al., 1982a; Risling et al., 1994). These fibers are characterized by peptides such as SP and CGRP, as well as GAP-43 and neurotrophin receptors (Fang, 1987; Risling et al., 1994), but in the root they are not sensitive to capsaicin (Karlsson and Hildebrand, 1994). They increase considerably after nerve lesions (Risling et al., 1984a).

Function

It has been known since the formulation of the law of separation of the spinal roots that stimulation of the ventral root causes pain (Magendie, 1822), and affects dorsal horn neurons as dorsal root stimulation does (Kim et al., 1988; Shin et al., 1985; 1986), although not when the stimulation is done proximal to a ventral root transection (Chung et al., 1983b; Chung et al., 1986; Endo et al., 1985). Also there are only minimal effects on blood pressure if the ventral root is stimulated when the dorsal root is cut (Longhurst et al., 1980; Chung et al., 1986b). Thus, as far as primary afferent input is concerned, most ventral root afferents seem to bring information to the cord through the dorsal root. Speculations are that because of the innervation of the meninges, they might have a role in migraine, or that they have trophic functions. A possible way to get insight into function would be to study animals with increased numbers of these axons, and numbers of these axons are markedly increased during development (Heath et al., 1986; Risling et al., 1981) or by neonatal sciatic nerve interruption or limb amputation (Risling et al., 1984a, c; Heath et al., 1986; Nam et al., 1989; Oh et al., 1989; Chung et al., 1993b; Park et al., 1994).

Segregation of Dorsal Root Fibers

Dorsal roots break into rootlets as they approach the cord, and large and small fibers in each rootlet segregate (Lissauer, 1886; Ingvar, 1927; O'Leary et al., 1932; Ranson, 1914; Szentagothai, 1964; Kerr, 1970a, b; Sindou et al., 1974), although the fine fibers may be in Lissauer's tract (Earle, 1952; Wall, 1962). To visualize unmyelinated fibers better, an electron-microscopic examination showed segregation of coarse and fine axons in monkey but not cat proximal rootlets (Snyder, 1977). In our experience, segregation is also not prominent in rat rootlets. Interest in this subject is because segregation would allow fine afferent fibers to be cut but large ones spared, thus relieving pain without making the segment anesthetic. This was done in the cat (Ranson and Billingsley, 1916), but the effects were later ascribed to interruption of the blood supply to the dorsal horn (Wall, 1962; Sindou et al., 1974). Nevertheless, such lesions have been used to interrupt fine sensory fibers selectively (LaMotte, 1977; Light and Perl, 1977), so the procedure is feasible. In the clinic, many patients get intractable pain with nerve and root avulsions, and sometimes the only effective method of treatment is a selective posterior rhizotomy or a dorsal root entry zone lesion (Nashold, 1987; Sindou et al., 1987). In addition, spasticity is a clinical problem sometimes amenable to selective primary afferent removal. The point here is that seemingly esoteric questions about the arrangements of the primary afferent fibers as they enter the cord has a bearing on relief of human suffering.

CONCLUSIONS

1. DRG cells are pseudounipolar neurons that convey information from the periphery to the spinal cord.
2. DRG cells are divided into two categories, small dark and large light. These cytologic characteristics are due to the fact that the large light cells contain many 200-kDa neurofilaments that stain poorly with most standard basic dyes so they appear pale as compared to the small dark cells that do not contain these filaments and thus stain much more intensely.
3. Axonal conduction velocities are subdivided into Aβ, Aδ, and C categories. The large light cells give rise to the Aβ fibers, the small dark cells give rise to the C fibers, and both cell types give rise to the Aδ fibers.
4. Characteristics such as the width and shape of the somatic action potential characterize different groups of DRG cells, and there is a general but not perfect correlation between axon conduction velocity, shape of the action potential, and cell size. A tighter correlation may be between nociceptors and non-nociceptors with the former having wider somatic action potentials and greater hyperpolarizations than the former. A corollary is that nociceptors can be of all sizes, so there are Aβ nociceptors.
5. There is controversy about subdivisions of the small dark category of cells. The present widely used construct is that small dark cells can be split into two groups, the first containing peptides such as CGRP and SP and the trophic factor trkA and projecting to laminae I-IIo, and the second being labeled by FRAP, IB4, retinin and responding to BDNF, and projecting to lamina IIi. The difficulty is that other studies find considerable overlap of, for example, FRAP and IB4 with various peptides. It is difficult to judge because findings in this area are often diametrically opposed. Thus, the construct will in our opinion have to stand with the caveats remembered until advancing knowledge provides clearer classifications that can be correlated with function.
6. Peripheral nerve lesions lead to the formation of pericellular baskets in DRG. These baskets arise from either sensory or sympathetic neurons. This implies a considerable control of primary afferent input at the level of the cell body after nerve injury.
7. Peripheral and central processes of DRG cells branch in the general vicinity of the ganglion. This branching is usually discussed as a possible mechanism for referred pain, but the number of branches is small so it is unlikely that this is the explanation of this important clinical phenomenon.
8. Ventral root afferent fibers, which were classically thought not to exist, are numerous and they are undoubtedly the explanation for the pain that accompanies ventral root manipulation. Few of these enter the cord through the ventral root, however, whereas most loop, end blindly in the root, or provide innervation from the nearby meninges.
9. The old question of the segregation of dorsal root fibers as they enter the cord has clinical relevance.

Chapter 4

Chemical Anatomy of Dorsal Root Ganglion Cells

The term "chemical anatomy" indicates cells that contain enough neuroactive material or express enough receptors to be identifiable by virtue of these markers. This field has exploded in the last few years.

AMINES

Catecholamines (CAs)

Catecholamine systems in the cord are regarded as arising almost exclusively from descending fibers. A few dorsal root ganglion (DRG) cells, however, contain high concentrations of the enzymes TH, DβH, or adenosine deaminase, or DA itself (Katz et al., 1983; Price and Mudge, 1983; Jonakait et al., 1984; Price, 1985; Katz and Erb, 1990; Vega et al., 1991; Weil-Fugazza et al., 1993); there are catecholamine fibers in the DRG (Owman and Santini, 1966; Stevens et al., 1983; Nagy et al., 1984; Nagy and Daddona, 1985; Weil-Fugazza et al., 1993), and sympathectomy does not decrease DOPA or dopamine in the DRG, indicating that DRG dopamine is not from sympathetic neurons (Bertrand and Weil-Fugazza, 1995). On the other hand, there is no evidence of DOPA decarboxylase in the DRG (Giacobini and Norre, 1971). So there is evidence that a few primary afferent neurons contain CAs, but their function is unknown.

Histamine

A few DRG cells are immunolabeled for histamine in adults and earlier in development (Nissinen and Panula, 1995).

Serotonin

DRG cells are not thought to be serotoninergic, as evidenced by lack of DRG immunostaining for 5-HT (Garry et al., 1989) and complete loss of spinal 5-HT after spinal transection (Arvidsson et al., 1990). On the other hand, autoradiography and immunochemistry (after tryptophan and an MAO inhibitor) show some 5-HT DRG cells (Calas et al., 1919; Kai-Kai and Keen, 1985) and some serotoninergic axons in the dorsal root (DiCarlo, 1983), and there is a decrease in spinal immunostaining after dorsal rhizotomy (Tuchscherer and Seybold, 1989). Accordingly the issue is not settled, but until further evidence for primary afferent 5-HT is available, we regard the dorsal horn 5-HT system as not having a significant primary afferent contribution.

AMINO ACIDS

Glutamate

High concentrations of glutamate, glutamine, glutamic oxaloacetic transaminase, glutaminase, and retrogradely transported aspartate are found in subpopulations of DRG cells (Barbaresi et al., 1985; Cangro et al., 1985; 1987; Jessell and Dodd, 1986; Battaglia et al., 1987; Inagaki et al., 1987; Wanaka et al., 1987; Battaglia and Rustioni, 1988; Kai-Kai and Howe, 1991; Miller et al., 1993; Keast and Stephensen, 2000). Early studies found large cells labeled (Wanaka et al., 1987) or relatively few cells labeled (Kai-Kai and Howe, 1991), but the other studies found numerous small cells labeled. There is considerable co-localization of glutamate or glutaminase with CGRP and SP (Battaglia et al., 1987; Battaglia and Rustioni, 1988; Miller et al., 1993; Keast and Stephensen, 2000), presumably implying dual release of amino acids and peptides during cell activity.

Approximately 8.5% of unmyelinated and 2% of myelinated axons in rat lumbosacral dorsal roots immunostain for glutamate (Westlund et al., 1989a), but considerably higher proportions of axons sequester glutamate in the medial articular nerve of the monkey with 28.6% myelinated and 12.1% unmyelinated axons labeled (Westlund et al., 1992). Peripheral inflammation approximately doubles proportions of labeled myelinated axons (Westlund et al., 1992).

Aspartate

Many neurons in the dorsal root ganglion, presynaptic terminals in laminae I–III and axons in the dorsal roots are labeled for aspartate (Westlund et al., 1989b; Merighi et al., 1991; Tracey et al., 1991; Valtschanoff et al., 1994; Keast and Stephensen, 2000; Larsson et al., 2001). Small cells and unmyelinated sensory fibers are preferentially stained (Westlund et al., 1989b; Tracey et al., 1991). Aspartate co-localizes with glutamate in many DRG cells (Tracey et al., 1991), but these amino acids are not in the same terminals in the dorsal horn (Merighi et al., 1991). There is some debate as to whether aspartate is a primary afferent neurotransmitter in the dorsal horn (Chapter 6).

GABA

Primary afferent neurons are not thought to use GABA as a transmitter. Many DRG cell profiles immunostain for GABA (Szabat et al., 1992; Stoyanova et al., 1998), however, and a small population of DRG cells retrogradely transports exogenous radioactive GABA, a technique used for locating GABAergic cells elsewhere (Barbaresi et al., 1985). There is no obvious flaw in these papers, but the absence of GABA labeling in central terminals of primary afferents and the lack of evidence for an inhibitory afferent input militate against the idea that there is GABAergic primary afferent input.

Glycine

There is no morphologic evidence for glycinergic primary afferent neurons or sensory terminals at the periphery.

CHANNELS

Calcium Channels

Calcium channel $\alpha 2\delta$-1 receptors are expressed by mostly small but all sizes of DRG neurons in naïve rats. After peripheral nerve injury the expression increases, and the increase is proportionately greatest in large DRG cells (Newton et al., 2001).

Sodium Channels

Early work considered a unitary voltage-dependent sodium channel that mediated Na+ shifts across membranes during action potentials. These channels consist of a pore forming α subunit, and three auxiliary β subunits. Finding multiple types of sodium currents indicated that a single channel concept is simplistic, however, and multiple isoforms are now known (Goldin et al., 2000). Voltage-gated channels are usually discussed in terms of the action potential and more recently as a mechanism for maintaining ectopic activity important in neuropathic pain. Non-voltage gated sodium channels have recently been discovered (see below).

Voltage-Gated Sodium Channels

The α isoforms are more than 50% identical in amino acid sequence, but the differences are functionally meaningful. The terminology was confusing, but recently a logical nomenclature has been proposed (Goldin et al., 2000). The terminology begins with the permeating ion (Na) with the principal physiological variable as a subscript (voltage, Na_V), then the gene subfamily (only one so far, Na_V1), followed by the isoform labeled numerically (e.g., $Na_V1.1$) followed by splice variants as lower case letters (e.g., $Na_V1.1a$). In addition, there are sodium channel-like proteins (Na_X). We will use this terminology but refer to the name used in the particular study where appropriate. Four isoforms are a family of tetrodotoxin (TTX) sensitive channels ($Na_V1.1$, $Na_V1.2$, $Na_V1.3$, and $Na_V1.7$), and three other closely related isoforms are a TTX resistant family ($Na_V1.5$, $Na_V1.8$, and $Na_V1.9$), whereas $Na_V1.4$ and $Na_V1.6$ are more distant genetically (Goldin et al., 2000). All these isoforms and Na_X are expressed in DRG cells to a greater or lesser extent except $Na_V1.4$ which so far is restricted to skeletal muscle (Fig. 4.1; Black et al., 1994; 1996; Felts et al., 1995; Akopian et al., 1996; Sangameswaran et al., 1996; Waxman and Black, 1996; Toledo-Aral et al., 1997; Dib-Hajj et al., 1998a; Novakovic et al., 1998a; Fjell et al.,

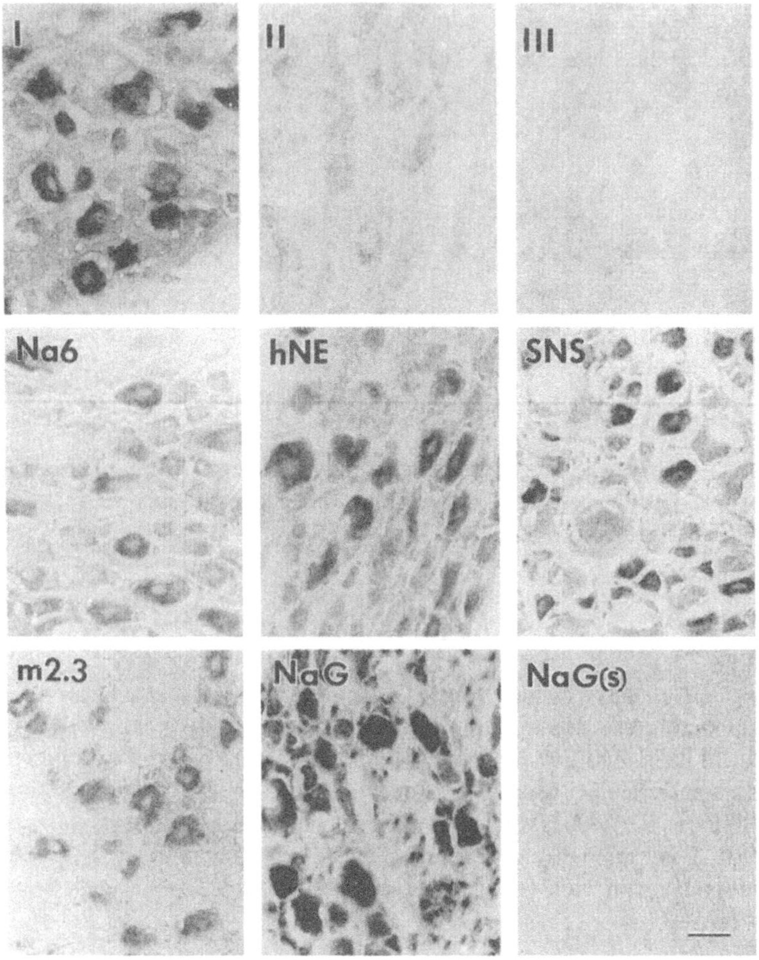

FIGURE 4.1. Sodium-channel α-subunit mRNAs in adult DRG cells. Note the numerous DRG cells expressing I ($Na_V1.1$), hNE ($Na_V1.7$), SNS ($Na_V1.7$), and NaG (Na_X). (From Black et al., 1996.)

1999a, b, c; 2000; Waxman et al., 1999; Amaya et al., 2000; Krzemien et al., 2000; Tzoumaka et al., 2000). $Na_V1.8$ (SNS) is essentially restricted to DRG cells, whereas $Na_V1.2$ (II, α2) is only seen at low levels in neurons and more strongly in glia (Black et al., 1996; Waxman and Black, 1996), and $Na_V1.3$ (III, α3) is expressed in embryo but not adult DRG neurons (Waxman et al., 1994). $Na_V1.8$ is expressed by a significant proportion but not all nociceptive DRG cells (Fang, X. et al., 2002)

Cell Size

$Na_V1.7$ (PN1/hNE) and its mRNA are expressed in virtually every DRG cell; $Na_V1.1$, $Na_V1.6$, and Na_X (α1, Na6, and NaG) are primarily in large neurons (Black et al., 1996; Fjell et al., 1999b; Tzoumaka et al., 2000; Waxman and Black, 1996), although Na_X is found in neurons of all sizes; and $Na_V1.8$ and $Na_V1.9$ (SNS and SNS2) are found primarily in small neurons (Dib-Hajj et al., 1998a, b; Novakovic et al., 1998a; Tanaka et al., 1998; Fjell et al., 1999b; 2000; Amaya et al., 2000; Bongenhielm et al., 2000), although there are medium-sized and an occasional large $Na_V1.8$ cells with a distinctive phenotype (Novakovic et al., 1998a; Amaya et al., 2000). $Na_V1.3$ is expressed in small cells after axotomy (Black et al., 1999). The profusion of different sodium channel isoforms with accompanying differences in sodium currents provide enormous possibilities for modulation.

Co-localizations

$Na_V1.8$ and $Na_V1.9$ (SNS and SNS2) are co-localized in many small DRG cells (Amaya et al., 2000). Many $Na_V1.8$ and $Na_V1.9$ (SNS and SNS2/NaN) neurons co-localize with IB4, VR1, and Trk-A (Amaya et al., 2000; Fjell et al., 2000). Trk-A and IB4 indicate responsiveness to NGF and GDNF respectively, so these factors presumably play roles in the sodium channel phenotypes of these cells.

β Subunits

The β subunits do not form sodium channel pores, but they influence pore function. All or almost all DRG neurons express β1 and β2 subunits with slightly greater expression in large cells, and all DRG neurons express a β1 variant called β1A (Black et al., 1996; Kazen-Gillespie et al., 2000). The more recently discovered β3 subunit is found almost exclusively in small DRG cells, and its mRNA increases in a chronic constriction neuropathic model and in medium diameter DRG cells following streptozocin diabetic neuropathy (Shah et al., 2000; 2001).

Development

$Na_V1.3$ (III) is expressed by E17 rat DRG cells but disappears by adulthood and is (re)expressed following nerve lesions in adult rats (Waxman et al., 1994). In the ganglion, $Na_V1.1$ is seen at P2 and increases slightly at P15 and P30, $Na_V1.2$ is seen in moderate numbers of cells at all ages, $Na_V1.3$ is high at E17 but steadily declines, NaG is in large neurons at all stages, $Na_V1.6$ starts at E17 and increases to high levels at P30 and $Na_V1.7$ starts at P2 and steadily increases in large cells (Felts et al., 1997). In NGF over-expressing rats, $Na_V1.2$ (α2) and β2 are strongly increased and $Na_V1.1$, $Na_V1.6$, $Na_V1.8$, and $Na_V1.9$ isoforms, Na_X sodium channel-like proteins (α1, NaG, Na6 SNS/PON3, NaN) and β1 subunits are moderately increased (Fjell et al., 1999c). Thus there is diversity of sodium channel expression during development.

Inflammation

Inflammation caused by intraplantar carrageenan or CFA administration causes a marked upregulation of the $Na_V1.8$ or generic α subtypes (Gould III et al., 1998; Tanaka et al., 1998) in both large and small DRG cells (Gould III et al., 1999). In contrast, expression of generic α subtypes decreased and then disappeared at nodes of Ranvier in an experimental neuritis, and this disappearance correlated with development of symptoms (Novakovic et al., 1998b). Bladder nociceptive responses are ameliorated after $Na_V1.8$ mRNA antisense treatment (Yoshimura et al., 2001).

Peripheral Axotomy

Peripheral axotomy leads to ectopic activity and thus central sensitization, a major factor in neuropathic pain. Voltage-gated sodium channels are candidates to underlie ectopic activity. Tetrodotoxin binding, which localizes TTX-sensitive sodium channels, accumulates on either side of a nerve ligation, indicating both retrograde and anterograde axonal transport of TTX-sensitive receptors (Lombet et al., 1985). After nerve transection, there is increased immunolabeling for pan-sodium channel receptors on axons in the neuroma and on small DRG cells (Devor et al., 1993; England et al., 1994; 1996a). TTX blocks ectopic activity at doses that usually have no effect on axon conduction (Matzner and Devor, 1994). Thus, peripheral axotomy increases sodium channels and so presumably plays an important role in the ectopic activity that follows nerve injury (Devor, 1993).

At present, the $Na_V1.3$ isoform is the major candidate underlying the above activity for it is expressed in small DRG neurons during development and (re)expressed after nerve lesions but not root lesions (central axotomy), and appears in nerve fibers in injured but not uninjured nerves (Waxman et al., 1994; 2000; Waxman, 1999; 2001; Coward et al., 2001a), the expression being predominantly in large cells in one study (Kim et al., 2001). By contrast, the TTX resistant subtypes $Na_V1.8$ and $Na_V1.9$, the TTX sensitive subtype $Na_V1.7$ and $Na_V1.6$ are attenuated or unchanged after peripheral or central axotomy (Dib-Hajj et al., 1996; 1998a, b; Novakovic et al., 1998a; Sleeper et al., 2000; Coward et al., 2001a; Decosterd et al., 2002; Kim et al., 2002). $Na_V1.8$ also has a role, however, in that antisense attenuation reduces pain behaviors without affecting non-noxious sensation after a neuropathic lesion (Lai et al., 2002), and the lesion-induced attenuation of $Na_V1.8$ and $Na_V1.9$ is only in damaged cells, neighboring DRG cells being unaffected (Decosterd et al., 2002). There is a rapid reduction in TTX-R and an increase in TTX-S currents after nerve lesion (Cummins and Waxman, 1997). $Na_V1.8$ mRNA expression is unchanged after axotomy but cell body mRNA declines whereas axonal protein increases implying peripheral translocation of the channel (Novakovic et al., 1998a). Na_X is unchanged after nerve lesions but prominent in supporting cells in one study (Coward et al., 2001b), and increased in another (Kim et al., 2002). In summary, sodium channels increase after nerve injury, and attention focuses on $Na_V1.3$, but $Na_V1.8$ and Na_X may also be involved.

Nerve Growth Factor (NGF)

The axotomy-induced attenuation of $Na_V1.8$ and $Na_V1.9$ (SNS and SNS2/NaN) can be prevented by NGF administration to the nerve (Dib-Hajj et al., 1998b). Systemic NGF upregulates $Na_V1.1$ (PN1) but not $Na_V1.3$ (PN3) in DRG cells, but the upregulation persists after the hyperalgesia recedes (Gould III et al., 2000).

Peripheral Sensory Axons

$Na_V1.6$ is found in the axon membrane at the nodes of Ranvier (Tzoumaka et al., 2000), and is distributed down the length of unmyelinated axons (Black et al., 2002). Thus this receptor presumably plays a key role in axonal conduction. $Na_V1.8$ and $Na_V1.9$ fibers are seen in various structures of the lumbar spine (Bucknill et al., 2002).

Non-voltage Gated Sodium Channels

The discovery of these channels has recently exploded in the field of somatic sensation. They belong to the degenerin epithelial sodium channel family (Waldmann and Lazdunski,

1998; Mano and Driscoll, 1999), and of particular interest are the epithelial and brain sodium channels, ENaCs and BNaCs (ASICS) respectively. The terminology is formidable and we will use the appellations in each study with some synonyms. Both ENaCs and BNaCs (ASICS) are found in DRG cells (Chen et al., 1998; Olson et al., 1998; Akopian et al., 2000; Drummond et al., 2000; Price et al., 2000; García-Añoveros et al., 2001; Alvarez et al., 2002). More particularly, the β and γ but not the α subunits of ENaC and a newly cloned BNaC1α are found in medium to large DRG cells and in peripheral sensory mechanoreceptors such as Meissner's corpuscles and Merkel receptor complexes, and disrupting the mouse BNC1 (BNaC1, ASIC2) gene markedly reduces responses of low-threshold mechanoreceptors (Drummond et al., 2000; Price et al., 2000; García-Añoveros et al., 2001). Thus these channels have an important role in mechanoreception. Other suggested functions are pH and cold sensing, and they are major factors in epithelial sodium transport (Waldmann and Lazdunski, 1998; Mano and Driscoll, 1999; Askwith et al., 2001; Alvarez et al., 2002). It is early times, however, and much more information will be available shortly.

ENZYMES

Alkaline Phosphatase

DRG cells are reported to contain alkaline phosphatase at the periphery of each cell body (Tewari and Bourne, 1962a, b; Shantha et al., 1967; Thakar and Tewari, 1967; Sethi et al., 1969; Sarrat, 1970; Tewari et al., 1970; Sethi and Tewari, 1971; Thomas, 1972; Manocha, 1973; Kopaczyk et al., 1974). However, electron microscopy revealed that this enzyme was found in the capsular or satellite cells (Matsuura et al., 1969), indicating that ganglion cell labeling is presumably artifactual. The function of satellite cell alkaline phosphatase is unknown but could play a role in surface activity such as ion transfer (Matsuura et al., 1970).

Carbonic Anhydrase (CA)

CA hydrates carbon dioxide and was found in DRG cells in an early histochemical study (Korhonen and Hyyppa, 1967). Modern emphasis began with the demonstration that a population of large and medium-sized DRG cells are intensely stained for CA (Wong et al., 1983; Riley et al., 1984; Sommer et al., 1985; Kazimierczak et al., 1986; Droz and Kazimierczak, 1987; Peyronnard et al., 1988a, b; Robertson and Grant, 1989). In accord with these observations, large myelinated CA fibers can be seen in peripheral nerves and centrally in dorsal columns (Riley et al., 1984; 1988). These axons appear predominantly in muscle nerves (Peyronnard et al., 1988b; Perry and Lawson, 1998), and large proprioceptive axons and sensory terminals in muscle spindles and Golgi tendon organs stain intensely (Szabolcs et al., 1989). Axon terminals in Pacinian and Meissner's corpuscles and terminals on Merkel cells also stain (Tohyama and Ide, 1987). Thus CA is thought to be involved in muscle sense and proprioception. However, large CA axons innervate tooth pulp where proprioception is minimal (Sugimoto et al., 1989a; Sugimoto and Takemura, 1993).

Almost all CA DRG cells are RT97 and GM1 ganglioside-positive (Robertson and Grant, 1989), so they belong to the large light population of DRG cells (Chapter 3), and many co-localize with calbindin and parvalbumin (Carr et al., 1989a). The CA cells are of all sizes, however, although large cells predominate, and they have distinctive action potentials (Prabhakar and Lawson, 1995). As previously mentioned, they predominate in muscle nerves, are sparse in cutaneous nerves and almost absent in visceral nerves (Perry and Lawson, 1998). CA is reduced by peripheral axotomy (Peyronnard et al., 1988a) and indeed by anything that interrupts contact with muscle or prevents muscle from contracting, including death of muscle itself, but CA is not affected by separation from the cord (Mayeaux and Valmier, 1995). In development, CA appears at the time that the muscle afferents attach to muscle (Mayeux et al., 1996). Thus, a peripheral trophic factor seems to be important in maintaining intense CA-expressing DRG cells, and CA expression is dependent on neuron–muscle interactions.

Cholinergic Enzymes

Cholinergic systems are outlined by histochemical procedures for two enzymes, choline acetyltransferase (ChAT) and acetylcholinesterase (ACHE).

ChAT

ChAT cells are usually not seen in the DRG (Barber et al., 1984; Borges and Iversen, 1986), and other work does not support the idea of cholinergic primary afferents, but ChAT DRG cells have been reported (Burt, 1971; Motavkin and Okhotin, 1980; Sann et al., 1995a). However, until other evidence for cholinergic primary afferent transmission is presented, we favor the idea that the cholinergic system in the dorsal horn is intrinsic.

ACHE

Unlike ChAT, there is a significant population of mostly small ACHE-expressing DRG cells (Coupland and Holmes, 1957; Gerebtzoff, 1959; Tewari and Bourne, 1962c; Cauna and Naik, 1963; Nandy and Bourne, 1964; Kokko, 1965; Novikoff et al., 1966; Matsuura, 1967; Shantha et al., 1967; Dixon, 1968; Schlaepfer, 1968; Kalina and Bubis, 1969; Matsuura et al., 1969; 1970; Kalina and Wolman, 1970; Lukas et al., 1970; Sarrat, 1970; Gwyn and Flumerfelt, 1971; Robain and Jardin, 1972; Manocha, 1973; Mazza et al., 1973; Hanker and Peach, 1976; Ambrose and McNeill, 1978; Homor and Kasa, 1978; McNeill and Norvell, 1978; Schoenen, 1978; Vega et al., 1989). The function of ACHE in these cell bodies is not known. Speculations are that the enzyme prevents "extraneous" activation by acetylcholine or may be necessary for a trophic or metabolic role unrelated to acetylcholine.

Cytochrome Oxidase

Cytochrome oxidase is a mitochondrial enzyme used to indicate oxidative activity (Wong-Riley, 1989). DRG cells of all sizes are stained at various intensities for this enzyme (Hyden et al., 1958; Burstone, 1961; Wong-Riley and Kageyama, 1986; Carr et al., 1989b; Karmy et al., 1991), presumably indicating the amount of oxidative activity in these cells. There is considerable co-localization with parvalbumin, calbindin, and carbonic anhydrase (Carr et al., 1989a, c; Karmy et al., 1991).

Fluoride-resistant Acid Phosphatase (FRAP)

Acid phosphatase (AP) is a lysosomal hydrolase important in general cell metabolism. Although all DRG cells have some label, many small DRG cells are intensely labeled (Fig 4.2; Colmant, 1959; Vitry et al., 1960; Kokko, 1965; Kalina and Bubis, 1968; Kalina and Wolman, 1970; Sarrat, 1970; Novikoff et al., 1971; Peach, 1972; Robain and Jardin, 1972; Kopaczyk et al., 1974). The intense labeling is due to an extralysosomal AP resistant to fluoride ions, fluoride-resistant acid phosphatase or FRAP (Gerebtzoff and Maeda, 1968; Schoenen et al., 1968; Knyihar, 1971; Knyihar and Gerebtzoff, 1973; Knyihar-Csillik and Csillik, 1981; Csillik and Knyihar-Csillik, 1986) that is used as a marker for a particular class of small dark nociceptive neurons (Silverman and Kruger, 1988a). FRAP localizations are particularly prominent in rodents (Sanyal and Rustioni, 1974), but they can be seen in DRG cells and cords of other mammals (Knyihar-Csillik and Csillik, 1981; Silverman and Kruger, 1988a). Some FRAP cells have cytoplasmic striations caused by heavy staining of cisternae of the granular endoplasmic reticulum (Knyihar-Csillik and Csillik, 1981; Csillik and Knyihar-Csillik, 1986). The function of FRAP is not known, and it is not a transmitter for sensory synaptic transmission is not obviously affected when FRAP is removed by nerve crush (Devor and Wall, 1981b). FRAP may be involved in transmitter metabolism or membrane turnover (Knyihar-Csillik and Csillik, 1981; Dodd et al., 1983; Csillik and Knyihar-Csillik, 1986; Silverman and Kruger, 1988a), but these remain only reasonable speculations. FRAP is stated to be involved in nociceptive transmission (Knyihar, 1971; Jancsó and Knyihar, 1975). Numerous AP axons are found in the cornea where nociceptive afferents predominate (Csillik and Knyihar-Csillik, 1986). Thiamine monophosphatase is thought to be a more specific marker than FRAP for extralysosomal acid phosphatase (Csillik et al., 1986; Knyihar-Csillik et al., 1986; Szabo et al., 1989). FRAP decreases and then increases in DRG cells after peripheral nerve section (Tenser, 1985).

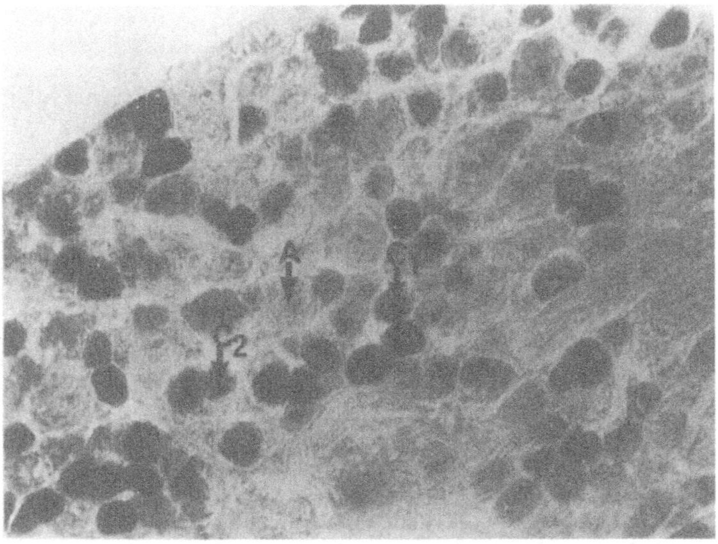

FIGURE 4.2. FRAP activity in dorsal root ganglion cells. The cell at A is a large essentially unreactive cell. Cell C1 shows an intensely labeled small cell with the enzyme reaction being dispersed homogeneously in the cytoplasm. Cell C2 is equally reactive but the reaction is more punctate. (From Knyihar-Csillik and Csillik, 1981.)

FRAP does not co-localize or only to a small extent with SP, somatostatin, CCK, adenosine deaminase, ACHE, or CGRP in most studies (Nagy and Hunt, 1982; Dalsgaard et al., 1984; Nagy and Daddona, 1985; Csillik et al., 1986; Vega et al., 1989), leading to a hypothesis that there are two fine primary afferent systems, the first peptide-containing (of which CGRP is the most prominent member) and the second, non-peptide containing, as exemplified by FRAP and IB4 populations (Hunt and Rossi, 1985). By contrast, however, FRAP was found to be extensively co-localized with the peptides oxytocin and vasopressin (Kai-Kai et al., 1986), although there is doubt about the presence of these peptides in DRG cells (see below), and with CGRP, and ACHE IB4, (Carr et al., 1990), and ChAT (Wang et al., 1993; 1994) in the rat. These findings raise questions about the peptide, non-peptide classification of DRG cells, but other investigators find a clear separation so this classification is still used (Chapter 3).

FRAP sensory axons are found in visceral and cutaneous but not in muscle nerves (McMahon et al., 1984; McMahon, 1986), although some FRAP cells project into the medial gastrocnemius nerve (Molander et al., 1987). Cross-anastomosing sural (cutaneous) and medial gastrocnemius (muscle) nerves reversed FRAP distribution patterns indicating that peripheral trophic factors control phenotypes of these cells (Molander et al., 1987).

Monoamine Oxidase

Monoamine oxidase breaks down monoamines, which are not supposed to be in DRG cells, but many DRG cells are faintly stained for this enzyme (Koelle and Valk, 1954; Nandy and Bourne, 1964; Matsuura, 1967; Shantha et al., 1967; Matsuura et al., 1969; Giacobini and Kerpel-Fronius, 1970; Kalina and Wolman, 1970; Hanker et al., 1973). There is no indication of strongly monoamine oxidase-positive DRG cells, however, but some primary afferent terminals in the superficial dorsal horn are immunostained for monoamine oxidase A (Westlund et al., 1993), and there are a few aminergic DRG cells (see amines), so the issue is unsettled.

Nitric Oxide (NO), Nitric Oxide Synthase (NOS) and NADPHd

The enzyme NADPHd is often used to indicate the presence of nitric acid synthase (NOS) and NOS or NADPHd indicate NO systems. Many NOS, NOS mRNA, or NADPHd cells are found in ganglia T5-L1 with fewer in the other ganglia in rats and guinea-pigs (Aimi et al., 1991; Verge et al., 1992; Aoki et al., 1993; Fiallos-Estrada et al., 1993; Terenghi et al., 1993; Zhang et al., 1993a; Ruda et al., 1994; Steel et al., 1994; Burnett et al., 1995; Farkas-Szallasi et al., 1995; Ren and Ruda, 1995; Vizzard et al., 1995; 1998; Rydh-Rinder et al., 1996; Doone et al., 1999) whereas cats, monkeys, and humans have relatively more ranging up to approximately 50% in human ganglia (Terenghi et al., 1993; Vizzard et al., 1993a; Zhang et al., 1993a). NOS cells are mostly small, but some are medium-sized (Terenghi et al., 1993; Vizzard et al., 1993a; Zhang et al., 1993a; Ruda et al., 1994; Steel et al., 1994; Blottner et al., 1995). Neonatal capsaicin removes NOS cells (Ruda et al., 1994; Ren and Ruda, 1995), but capsaicin in adults increases their numbers (Farkas-Szallasi et al., 1995; Vizzard et al., 1995). Most NOS cells co-localize CGRP, SP, and GAL after peripheral axotomy (Aimi et al., 1991; Zhang et al., 1993a; Hökfelt et al., 1994a). NOS axons can be traced to viscera and retrograde labeling shows NOS cells innervate pelvic viscera (McNeill et al., 1992a, b; Papka and McNeill, 1992; Vizzard et al., 1993b, c; 1994; 1996; Callsen-Cencic and Mense, 1997; Vizzard, 1997).

Numbers of rat NOS, NOS mRNA, or NADPHd neuron profiles are significantly increased by peripheral axotomy (Verge et al., 1992; Fiallos-Estrada et al., 1993; Zhang et al., 1993a; Steel et al., 1994; Zhang, X. Ji, R. R. et al., 1996; Vizzard et al., 1998), spinal cord injury (Vizzard, 1997), systemic capsaicin or resiniferatoxin in adults (Farkas-Szallasi et al., 1995; Vizzard et al., 1995), acute and chronic bladder inflammation (Vizzard et al., 1996), joint inflammation (Bscheidl et al., 1994), and AIDS (Nagano et al., 1996a; Callsen-Cencic and Mense, 1997). Thus this is a plastic population with considerable increases after neuropathic and inflammatory stimuli. This upregulation undoubtedly plays a role in the responses to the above perturbations, but this is not a universal response, in that peripheral axotomy does not significantly change numbers of NOS profiles in guinea-pig or monkey (Zhang et al., 1993a; Rydh-Rinder et al., 1996).

Ca^{++} Calmodulin Kinase II

Calmodulin kinase II is found in numerous DRG cells and co-localizes with MORs, IB4, CGRP, and VR1 (Bruggemann et al., 2000; Carlton and Hargett, 2002). It is transported into the central and peripheral processes of these cells and is increased in peripheral sensory axons after inflammation (Carlton, 2002).

Other Kinases

The serine/threonine kinase CDC2, the *c-kit* receptor tyrosine kinase, the cyclin-dependent kinase 5/p35, and the Src-suppressed kinase substrate are expressed in DRG cells (Ino et al., 1994; Hirata et al., 1995; Terada et al., 1998; Siegel et al., 2002).

HORMONES

Corticotrophin Releasing Factor (CRF)

CRF DRG cells are not seen (Puder and Papka, 2001a), but the loss of dorsal horn immunostaining after rhizotomy or capsaicin treatment indicates that some CRF primary afferent dorsal horn fibers exist (Skofitsch et al., 1984; Kawatani et al., 1996).

Insulin

Preproinsulin I and II and their mRNAs are present in DRG cells (Schechter et al., 1996).

PEPTIDES

Angiotensin

There are no morphologic demonstrations of angiotensin or its derivatives in DRG cells, but since angiotensin II concentrates in the superficial dorsal horn, it is a reasonable possibility that small DRG cells make this peptide.

Atrial and Brain Natriuretic Peptide

Atrial natriuretic peptide is found in a few DRG cells (Nohr et al., 1989). Large numbers of DRG cells contain brain natriuretic peptide (Saper et al., 1989).

Bombesin

Bombesin and related gastrin-releasing peptide and neuromedin B have GI and CNS effects (Ohki-Hamazaki, 2000), and are thought to have a role in nociceptive input. Although not always seen (Garry et al., 1989), a small population of DRG cells express bombesin-like activity (5% of DRG profiles) (Fuxe et al., 1983; Panula et al., 1983; Cameron et al., 1988). These cells are small (less than 40 microns in diameter) (Panula et al., 1983), which correlates with the approximately 3% bombesin labeled unmyelinated dorsal root axons (McNeill et al., 1989). Bombesin coexists to some extent with SP, CGRP, and SST (Fuxe et al., 1983; Cameron et al., 1988).

Cholycystokinin (CCK)

CCK is reported in small DRG cells in several species (Lundberg et al., 1978; Dalsgaard et al., 1982b; 1984; Gibson et al., 1982; Otten and Lorez, 1983; Leah et al., 1985a; Kuwayama and Stone, 1986; Clements and Beitz, 1987; Gibbins et al., 1987a; Kuwayama et al., 1987; McNeill and Burden, 1987). In the rat, however, the labeling seems to be due largely to cross-reactions with CGRP (Chapter 6). Thus in the rat, CCK in the cord would not appear to be presynaptic whereas in the cat, guinea-pig, and monkey, there should be a substantial amount of CCK on primary afferent terminals. It is not clear why the rat is specialized in this way.

CCK antagonizes morphine at the spinal level, and axotomy greatly increases numbers of CCK mRNA rat DRG cell profiles (Verge et al., 1993), but there is no increase in CCK itself in the dorsal horn (Bras et al., 1999). Nevertheless, the major increase in CCK mRNA in DRG cells after axotomy needs explanation (Hökfelt et al., 1994b, c). CCK fibers are found in various peripheral locations and CCK DRG cells project to both visceral and somatic regions (Rehfeld and Lundberg, 1983; Kawatani et al., 1986; Kuwayama and Stone, 1986; McNeill and Burden, 1987). It must be remembered, however, that the antibody may be recognizing a CGRP-like substance and there is considerable CCK in the peripheral autonomic system. Accordingly further work to sort out the organization of the peripheral sensory CCK system is warranted.

Calcitonin Gene-related Peptide (CGRP)

CGRP and its mRNA are expressed in a large proportion of DRG or trigeminal ganglion cells (Fig. 4.3). Estimates range from 20–80% with the figures clustering around 50% or 20% (Inagaki and Kito, 1986; Ju et al., 1987b; Cameron et al., 1988; Rethelyi et al., 1989a; Silverman and Kruger, 1989; Fabri and Conti, 1990; Ishida-Yamamoto and Senba, 1990; Noguchi et al., 1990a, b; Bonfanti et al., 1991; Hökfelt et al., 1992; Bergman et al., 1996;

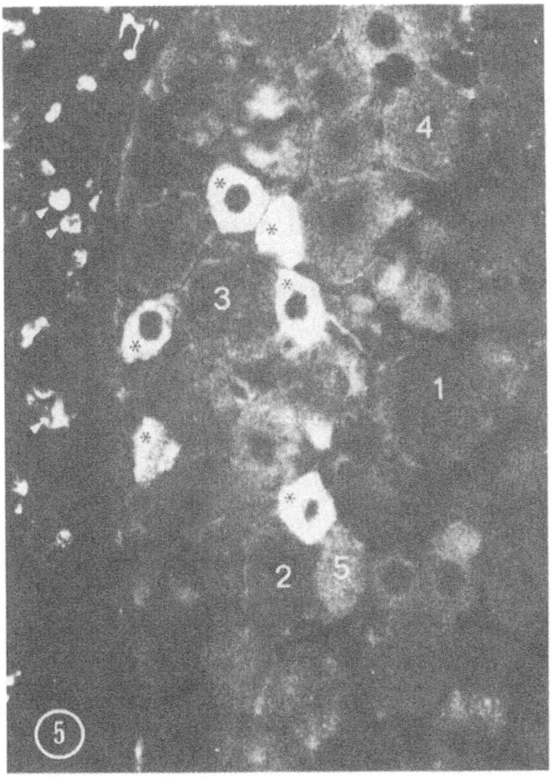

FIGURE 4.3. CGRP, an example of immunoreactive DRG cell bodies. In this case the reaction is for somatostatin. (From Hökfelt et al., 1976.)

Kashiba et al., 1996; Hall et al., 1997; Schicho and Donnerer, 1999; Tajti et al., 1999; Wiberg et al., 1999). It must be remembered, however, that most of these studies count profiles, so large cells tend to be over-counted (Lawson, 1992; Shi et al., 2001), that some studies use colchicine and some do not, and that there are species differences. Most CGRP DRG cells are small but some are medium-sized and a few are large. This is in concert with the observations that most CGRP afferent axons are unmyelinated but some are myelinated (Carlton et al., 1987; 1988; Harmann et al., 1988b; Ishida-Yamamoto et al., 1988; 1989; McNeill et al., 1988b) and that axon conduction velocities from individually identified CGRP DRG cells are in Aβ, Aδ, and C ranges (McCarthy and Lawson, 1990; Lawson et al., 1993; 1996). Individually identified A-fiber CGRP somata have longer action potentials, after-hyperpolarizations, and inflections in the falling phase than non-CGRP A fiber cells, but the action potentials on CGRP and non-CGRP C cells do not differ (Lawson et al., 1996), the inflections on the falling phase identifying nociceptors (Lawson, 1995; Gold et al., 1996c; Wang et al., 1998a), and action potential shapes differ between Aδ CGRP cells that do and do not co-localize with SP (Lawson, 1995; Lawson et al., 1996; McCarthy and Lawson, 1997). Finally, CGRP and its mRNA comes in α and β forms, and the α form is predominant in DRG cells (Gibson et al., 1988; Mulderry et al., 1988; Noguchi et al., 1990a, b; Sternini and Anderson, 1992).

Classification

DRG cells are divided into large light and small dark categories (Chapter 3). A proposal is that small dark cells be separated into peptidergic and non-peptidergic categories, with CGRP and SP being major peptide markers and IB-4 and acid phosphatases (FRAP and TMP) non-peptide markers (Hunt and Rossi, 1985; Silverman and Kruger, 1988a, b; 1990). A difficulty, however, is that there is co-localization of CGRP and non-peptide markers (Carr et al., 1990; McCarthy and Lawson, 1990; Ambalavanar and Morris, 1992; Wang et al., 1994; Bergman et al., 1999), and CGRP is found in DRG cells of all sizes (Ishida-Yamamoto and Senba, 1990; Bergman et al., 1999), although in some studies the distinction is clear (Molliver et al., 1997). It must also be remembered that phenotypes are plastic and change after peripheral axotomy or inflammation and during development. Thus, any categorization will be imperfect until we understand the rules governing phenotypes of primary afferent neurons.

Age

The CGRP content of DRG cells from normal and hypertensive rats decreases with age (Bergman et al., 1996; 1999; Yamaga et al., 2001).

Development

MRNA for β-CGRP appears at E14 and α-CGRP at E16 in rat DRG cells, but α-CGRP is more abundant in sensory systems (Terrado et al., 1999). In rats, CGRP itself appears early in development, E16 in one study and E19 in another, and then appears in the dorsal horn before it appears in the periphery (Marti et al., 1987; Groves et al., 1999). In dogs, the CGRP primary afferent system develops largely postnatally (Bonfanti et al., 1991). Mice that over-express NGF have approximately 100% more total neuron and CGRP neural profiles than normal littermates, interpreted as the result of prevention of normally occurring cell death (Mendelson et al., 1996).

Co-localizations

SP

SP is co-localized with CGRP in sensory neurons, in primary afferent terminals in the dorsal horn (often in the same large dense-core vesicles), and in peripheral cutaneous or visceral axons (Gibson et al., 1984b; Wiesenfeld-Hallin et al., 1984; Gibbins et al., 1985; 1987a, b; Lee et al., 1985a, b; Lundberg et al., 1985; Skofitsch and Jacobowitz, 1985a; Uddman et al., 1985a; Franco-Cereceda et al., 1986; Gulbenkian et al., 1986; Hanko et al., 1986; Matsuyama et al., 1986; Wanaka et al., 1986; Ju et al., 1987b; Cameron et al., 1988; Merighi et al., 1988; 1989; 1991; Garry et al., 1989; Tuchscherer and Seybold, 1989; Pereira da Silva and Carmo-Fonseca, 1990; Plenderleith et al., 1990; Bonfanti et al., 1991; Semenenko and Cervero, 1992; Ribeiro-da-Silva, 1995; Ribeiro-da-Silva and Cuello, 1995; Zhang, X. et al., 1995e; Tamura et al., 1998). Proportions of SP and CGRP and double-labeled cells differ from study to study, but the usual conclusion is that all SP-containing DRG cells express CGRP but only about half of CGRP cells express SP.

GAL

GAL is often co-localized with CGRP and SP in both primary afferent terminals and DRG cell bodies (Ju et al., 1987a, b; Villar et al., 1989; Zhang et al., 1993b).

SOM

SOM and CGRP coexist in DRG cells, central synapses, and peripheral axons (Gibbins et al., 1987b; Ju et al., 1987; Cameron et al., 1988; Merighi et al., 1988; Garry et al., 1989; Ribeiro-da-Silva, 1995; Kashiba et al., 1996).

CCK

CCK and CGRP are co-localized in DRG cells (Gibbins et al., 1987a), but there is evidence that CCK localizations may represent CGRP (Ju et al., 1986).

VIP

VIP is not strongly expressed in normal DRG cells but is expressed after peripheral axotomy, when it is often co-localized with CGRP (Ju et al., 1987a, b).

Bombesin, ENK, DYN, VIP, NT, and NPY

These peptides are occasionally co-expressed with CGRP, either in terminals or cell bodies (Gibbins et al., 1987a; Cameron et al., 1988; Zhang et al., 1993b).

FRAP

Although FRAP and CGRP are used as markers of non-peptide and peptide DRG cell populations respectively, and FRAP does not co-localize with CGRP (Doughty et al., 1991), some DRG cells are reported to express both (Carr et al., 1990).

Cysteine Proteases

CGRP and cysteine proteases are co-localized in lysosomes in DRG cells (Taniguchi et al., 1993). This presumably indicates a lysosomal degradation of CGRP.

Amino Acids

CGRP and glutamate coexist in primary afferent terminals with CGRP being localized in the large, dense-core vesicles and glutamate in the small, round, clear vesicles (Merighi et al., 1991).

IB4

Although IB4 and CGRP are often used to indicate non-peptidergic peptidergic and DRG populations respectively, they are sometimes co-localized (Ambalavanar and Morris, 1992; Averill et al., 1995).

Peripherin

More than 50% of peripherin profiles co-localize with CGRP mRNA (Goldstein et al., 1996).

BDNF

BDNF localizes with CGRP in DRG cells and synapses in the dorsal horn (Luo et al., 2001), and this proportion increases after NGF treatment (Michael et al., 1997a).

GDNF

There is considerable co-localization of CGRP and GDNF in human DRG cells (Quartu et al., 1999).

TrkA

Almost all CGRP DRG cells and many CGRP fibers in the dorsal horn coexpress trkA receptors (Verge et al., 1989; Averill et al., 1995; Michael et al., 1997b; Pezet et al., 2001). In addition, many trkA cells express α-CGRP mRNA (Kashiba et al., 1996).

P75

Of CGRP DRG cells, 46% coexpress the NP75 low affinity NGF receptor (Zhou et al., 1993; Pezet et al., 2001).

Experimental Manipulations or Disease

Dorsal Rhizotomy

CGRP in DRG cells increases after dorsal rhizotomy (Inaishi et al., 1992).

Peripheral Nerve Lesions

CGRP levels, the mRNA for CGRP, and numbers of cell profiles expressing CGRP decline in dorsal root ganglia after peripheral axotomy or axonal blockade (Dumoulin et al., 1990; Noguchi et al., 1990b; Doughty et al., 1991; Kashiba et al., 1992a; Hökfelt et al., 1994c; Ahmed et al., 1995a; Bergman et al., 1999; Elcock et al., 2001a; Hu and McLachlan, 2001). The decline in the mRNA is for both α and β forms (Noguchi et al., 1990) and precedes the decline in the peptide (Dumoulin et al., 1990), so presumably the primary cause of the decline is reduction in CGRP manufacture rather than increased release. On the other hand, transection or chronic constriction of a peripheral nerve did not lead to CGRP changes (Villar et al., 1989; Munglani et al., 1996), axotomy by corneal NaOH did not affect CGRP in corneal trigeminal ganglion cells (Felipe et al., 1999), and increases in CGRP or its mRNA were seen after blocking axon transport by cisplatin, axotomizing CGRP DRG cells that project in the dorsal column, or in a ganglion adjacent to a transected segmental nerve (Barajon et al., 1996; Fukuoka et al., 1998; Miki et al., 1998a). Thus, there may be a balance between reduction of mRNA followed by decreased production on one hand and "pile-up" because of blocked transport on the other, but peripheral axotomy leads to a decrease in CGRP mRNA and usually to a reduction in CGRP in DRG cells. The co-localization of

CGRP and VIP after a nerve lesion shows that cells that stop making CGRP begin making VIP (Villar et al., 1989; Doughty et al., 1991).

Inflammation

Inflammation increases the amount of CGRP mRNA, numbers of DRG cells expressing CGRP and the amount of CGRP in the ganglion with the mRNA increase usually preceding changes in the peptide (Kuraishi et al., 1989; Donaldson et al., 1992; 1994; Donnerer et al., 1992; 1993; Smith et al., 1992; Hanesch et al., 1993; Nahin and Byers, 1994; Ahmed et al., 1995a, b, c; Galeazza et al., 1995; Mulder et al., 1997; Calza et al., 1998; 2000; Keates et al., 1998; Nohr et al., 1999; Ohtori et al., 2001; Vizzard, 2001). Exceptions are that colonic and temporal artery inflammation cause a decrease (Saldanha et al., 1999; Traub et al., 1999), and polyarthritis results in no change (Weihe et al., 1988a) or an increase preceded by a decrease (Calza et al., 1998; 2000) in ganglionic CGRP. In addition, mRNA for CGRP is down at the height of ultraviolet irradiation inflammation (Gillardon et al., 1991). Thus, usually inflammation causes a rapid rise in CGRP mRNA in DRG cells followed by manufacture of the peptide and transport into DRG axons and terminals. Interestingly, the rise in CGRP and other peptides that follows a chronic experimental inflammation outlasts the inflammation suggesting permanent changes in peptide architecture that might be deleterious (Calza et al., 2000). In addition, there seems to be a phenotypic switch in that larger cells are labeled after inflammation than in the normal (Ohtori et al., 2001).

During chronic experimental or human rheumatoid arthritis or temporal arteritis, there is a considerable loss of CGRP axons, and the inflammation is often accompanied by tissue destruction (Konttinen et al., 1990; Mapp et al., 1990; Pereira da Silva and Carmo-Fonseca, 1990; Saldanha et al., 1999; Calza et al., 2000). In presumably earlier stages of polyarthritis, however, there is a proliferation of CGRP axons (Ahmed et al., 1995b) particularly around blood vessels (Weihe et al., 1988a). In other regions, inflammation leads to a major increase in CGRP in sciatic nerves, which seems to be related to the increase in NGF that accompanies inflammation, and administration of Clostridium toxin to isolated gut leads to a considerable increase in CGRP in gut and ganglion along with a reduction of the fluid loss that accompanies cholera (Keates et al., 1998). These data suggest acute CGRP axon sprouting followed by destruction, as the inflammation turns chronic. On the other hand, there was no increase in the CGRP content of bladder submucosa and detrusor muscle after interstitial cystitis (Hohenfellner et al., 1992).

The rise in CGRP or its mRNA after inflammation is prevented by nerve section, local anesthesia, or systemic capsaicin, indicating that neural activity, presumably in fine afferents, signals the CGRP increase (Donaldson et al., 1994; Ahmed et al., 1995a, c; Bulling et al., 2001), and by antibodies against NGF, indicating control in large part by NGF (Donnerer et al., 1992). Bowel inflammation after Clostridium toxin leads to a rise in CGRP, and pretreatment with a CGRP antagonist prevents this and much of the fluid loss, which might be therapeutic (Keates et al., 1998). Glucocorticoids prevent the increase in CGRP mRNA in polyarthritis (Nohr et al., 1999), and mineralocorticoids increase numbers of CGRP DRG cells (Covenas et al., 1994).

Diabetes

Experimental diabetes results in a loss of CGRP in sciatic nerves but no changes in mRNA in the DRG (Brewster et al., 1994) or CGRP labeling in the dorsal horn (Akkina et al., 2001).

In other studies, however, particularly longer term, there was downregulation of CGRP mRNA (Rittenhouse et al., 1996; Zochodne et al., 2001a). Loss of CGRP in the nerve may also relate to the preferential damage of distal sensory axons in diabetes.

Hypoxia

Hypoxia causes a reduction in CGRP in the sciatic nerve but no reduction in mRNA in the ganglion (Brewster et al., 1994).

Hereditary Sensory Neuropathy

There is a decrease in the excitatory peptides CGRP and SP and an increase in the opioid enkephalin in the dorsal horns of rats with a hereditary sensory neuropathy (Kar et al., 1989).

Capsaicin

When adult rats are treated with capsaicin, CGRP peptide and mRNA in DRG cells decrease approximately 50%. These decreases are averted by NGF (Schicho et al., 1999; Schicho and Donnerer, 1999).

Axonal Transport Blockers

Axonal transport blockers reduce numbers of CGRP DRG profiles, presumably because of lack of access of peripheral NGF to DRG cells (Kashiba et al., 1992a).

NGF Treatment

Prevention of NGF access to DRG cells, by systemic (Shadiack et al., 2001) or intrathecal (Christensen and Hulsebosch, 1997b) antiNGF, or nerve transection (see above), leads to reductions in CGRP, due presumably to a production decrease. In addition, lack of the NGF receptor P75 leads to sprouting of sympathetic axons around CGRP DRG cells (Walsh et al., 1999). On the other hand, increases in NGF, as in NGF over-expressing mice (Ma et al., 1995; Mendelson et al., 1996; Ritter et al., 2000), or by exogenous NGF, either in culture (Lindsay and Harmar, 1989) or in vivo (Donnerer et al., 1992; Amann et al., 1996; Supowit et al., 2001), the latter study showing a concomitant but transient fall in blood pressure, increase CGRP or its mRNA in DRG cells, peripheral nerves, and the cord. Also, local or systemic NGF reverses the CGRP lessening effects of axotomy (Verge et al., 1995; Wiberg et al., 1999) or capsaicin (Donnerer et al., 1996; Schicho et al., 1999; Schicho and Donnerer, 1999), and antiNGF significantly reduces the CGRP increase that accompanies inflammation (Donnerer et al., 1992; 1993). Thus, CGRP in DRG cells is largely under the control of NGF (Verge et al., 1995).

Female Steroid Hormones

An increase in female steroid hormones, either during pregnancy or by exogenous administration, increases CGRP in the DRG, and this increase accompanies an increase in NGF receptors on these cells (Gangula et al., 2000; Lanlua et al., 2001).

Sympathectomy

Surgical or chemical sympathectomy leads to a considerable increase in numbers and/or intensity of staining of CGRP axons in various viscera (Unger et al., 1985; 1988; Carvalho et al., 1986; Terenghi et al., 1986a; Aberdeen et al., 1990; 1992; Van Ranst and Lauweryns 1990; Bolden et al., 1997; but see Terenghi et al., 1985). This increase is thought to be the result of sprouting, and there is a concomitant increase in the growth associated protein, GAP-43 (Bolden et al., 1997), but there is still the possibility of phenotypic changes without actual axon growth or formation of new synapses. The functional significance of this presumed sprouting is not clear.

Dorsal Roots

There are numerous CGRP axons in dorsal roots, more numerous than SOM, BOMB, or SP, but fewer than GAL. The majority are unmyelinated but some are myelinated (McNeill et al., 1989; Klein et al., 1990a). CGRP primary afferent axons project surprisingly long distances (five segments) rostrally or caudally in the cord (Traub et al., 1989a, b)

Periphery

CGRP containing axons permeate every peripheral tissue that has been examined. These axons seem to have two major functions; to signal noxious events and to mediate neurogenic inflammation. Efferent functions may predominate in tissues with CGRP innervation that rarely signal pain (Kruger et al., 1989). Somatic CGRP innervation is extensive, but proportions of visceral CGRP axons are higher (Molander et al., 1987; Sternini and Anderson, 1992; Lawson, 1996; Perry and Lawson, 1998; Won et al., 1998). Listed below are peripheral regions that receive CGRP innervation as shown by direct demonstration of CGRP axons or retrograde labeling of CGRP DRG cells from a particular nerve or tissue. Evidence that all or the large majority of peripheral CGRP axons arise from sensory ganglion cells is that capsaicin, dorsal root, or Vth nerve ganglionectomy or nerve transection (the latter would not distinguish between sensory and sympathetic axons) removes these fibers (Terenghi et al., 1985; 1986a, b; Uddman et al., 1985a; Unger et al., 1985; 1988; Gibbins et al., 1987a; Green and Dockray, 1988; Ishida-Yamamoto et al., 1988; 1989; Van Ranst and Lauweryns, 1990; Ahmed et al., 1995a, c).

Somatic Innervation

1. Skin (Integument) or cutaneous nerves (Gibbins et al., 1985; 1987a, b; Molander et al., 1987; Weihe et al., 1988a; Ishida-Yamamoto et al., 1989; Kar et al., 1989; O'Brien et al., 1989; Kashiba et al., 1990a; Hosoi et al., 1993; Bennett et al., 1996b; Perry and Lawson, 1998).
2. Meissner's corpuscles (Ishida-Yamamoto et al., 1988).
3. General somatic nerves (Kar et al., 1990).
4. Joints or joint nerves (O'Brien et al., 1989; Konttinen et al., 1990; Mapp et al., 1990; 1993; Hanesch et al., 1991; Ahmed et al., 1995a, b, c; Salo and Theriault, 1997; Kallakuri et al., 1998; Tamura et al., 1998; Ohtori et al., 2001).
5. Muscle nerves (Molander et al., 1987; O'Brien et al., 1989).
6. Ventral roots (Risling et al., 1994).

7. Oral cavity, tongue, epiglottis, and pharynx (Terenghi et al., 1986a, b).
8. Dura and pia mater (Risling et al., 1994; Kallakuri et al., 1998; Yamada et al., 2001).
9. Eye (cornea, anterior chamber, iris) (Gibbins et al., 1985; Terenghi et al., 1985; 1986a, b; Unger et al., 1985; 1988; Matsuyama et al., 1986; Kuwayama et al., 1987; Silverman and Kruger, 1989; Bolden et al., 1997; Felipe et al., 1999).
10. Bone and teeth, especially dental pulp (Gazelius et al., 1987; Kruger et al., 1989).
11. Middle ear (Uddman et al., 1985b).

Visceral Innervation

1. Vascular tree including heart (Gibbins et al., 1985; Lundberg et al., 1985; Uddman et al., 1985a; Franco-Cereceda et al., 1986; Gulbenkian et al., 1986; Hanko et al., 1986; Inagaki and Kito, 1986; Matsuyama et al., 1986; Wanaka et al., 1986; Gibbins et al., 1987a; Tsai et al., 1988; Saldanha et al., 1999).
2. Lung and airway (Lundberg et al., 1985; Gibbins et al., 1987a; Van Ranst and Lauweryns, 1990).
3. Gastrointestinal (GI) tract and associated organs (Franco-Cereceda et al., 1986; Inagaki and Kito, 1986; Gibbins et al., 1987a; Green and Dockray, 1987; 1988; Su et al., 1987; Kashiba et al., 1990a; Sternini and Anderson, 1992; Won et al., 1998; Traub et al., 1999).
4. Visceral (splanchnic) nerves (Molander et al., 1987; Perry and Lawson, 1998).
5. Kidney and ureter (Semenenko and Cervero, 1992; Zheng and Lawson, 1997).
6. Pelvic viscera (bladder, uterus, vas deferens, etc.) (Gibbins et al., 1985; Franco-Cereceda et al., 1986; Inagaki and Kito, 1986; Inyama et al., 1986; Yokokawa et al., 1986; Silverman and Kruger, 1988b; Bennett et al., 1996b; Czaja, 2000; Vizzard, 2001).
7. Superior cervical and other autonomic ganglia (Inagaki and Kito, 1986; Yamamoto and Ohnishi, 1996).

Endothelin (ET)

Approximately 75% of DRG cell profiles, both large and small, express ET-1 mRNA (Giaid et al., 1989).

Galanin

Galanin was first isolated from the intestine and later found in DRG cells and superficial dorsal horn. It is markedly upregulated by peripheral axotomy (Bartfai et al., 1993), and levels correlate with behavioral hypersensitivity (Xu et al., 2000) although the effects of galanin are complicated (Kerr et al., 2000). Most studies report only a few small galanin cells in the normal (Chung et al., 1985; Skofitsch and Jacobowitz, 1985b, c; Hokfelt et al., 1987; Ju et al., 1987b; Arvidsson et al., 1991; Kashiba et al., 1992b; Corness et al., 1996; Ma and Bisby, 1997; Landry et al., 2000; Hu and McLachlan, 2001; Macdonald et al., 2001), except in one study 50% of small cell profiles labeled (DeLeon et al., 1994).

The low figures contrast with the relatively high percentages of dorsal root axons that express galanin (Klein et al., 1990a).

Nerve Lesions

Complete nerve transection, various types of partial transection, the latter being chronic pain models, nerve crush, or photochemical injury result in more galanin and its mRNA in DRGs and more small DRG cell profiles expressing these but also in some large and intermediate-size cells (Hökfelt et al., 1987; Arvidsson et al., 1991; Kashiba et al., 1992b; Nahin et al., 1994; Corness et al., 1996; Rydh-Rinder et al., 1996; Ma and Bisby, 1997; 1999b; Hao et al., 1999; Shi et al., 1999a; Landry et al., 2000; Holmberg et al., 2001; Hu and McLachlan, 2001; Macdonald et al., 2001), the only exception being a decline in the trigeminal ganglion of the ferret after infraorbital nerve lesions (Elcock et al., 2001a, b). Given the marked upregulation of galanin in the DRG, its increased release in the cord, and the general overall inhibitory action of this substance, it is tempting to speculate that this may have a role in damping neuropathic pain (Zhang, X. et al., 1998a; Wiesenfeld-Hallin and Xu, 2001). The action of galanin is complex, however, and in some instances galanin may have an algesic function (Ma and Bisby, 1997; 1999b; Ramer et al., 1998c; Kerr et al., 2000). On the other hand, there is relief of neuropathic pain behavior with galanin administration and inverse correlation of galanin with autotomy (Xu et al., 2000) so the jury is still out.

One question is whether galanin is upregulated in DRG cells whose peripheral axons have not been transected. No evidence was found for this in one study (Fukuoka et al., 1998) but a significant number of such cells were found after segmental nerve injury (Ma and Bisby, 1999b). The upregulation in undamaged sensory cells may have a role in secondary hyperalgesia.

Nerve injury results in the formation of axonal baskets around DRG cell bodies. These were first described for TH-expressing sympathetic axons, but galanin axonal baskets, some co-localized with CGRP and thus from DRG cells, and others with TH and thus from sympathetic neurons, form around DRG cells after peripheral axotomy (Hu and McLachlan, 2001). Thus, presumably both sensory and sympathetic activity could directly affect primary afferent somata after peripheral axotomy.

Since dorsal rhizotomy has little effect and peripheral axotomy great effect on galanin synthesis and expression, peripheral trophic factors presumably control the galanin increase after nerve lesions. One obvious candidate would be NGF, and antiNGF leads to a decrease in galanin in DRG cells (Shadiack et al., 2001), indicating NGF facilitation of galanin production. But NGF has no effect after nerve lesions (Verge et al., 1995), and there seems to be no fundamental difference in the DRG galanin response in NGF over-expressing compared to wild-type mice (Holmberg et al., 2001). In addition, withdrawal of NGF from cultured DRG cells led to an upregulation of galanin (Corness et al., 1998). Thus, after axotomy NGF has no or an inhibitory effect on galanin synthesis. By contrast, leukemia inhibitory factor (LIF) has a major influence on galanin expression (Corness et al., 1996; Sun and Zigmond, 1996), and indeed in culture a reduction in NGF and an increase in LIF produce maximal galanin increase in the DRG (Holmberg et al., 2001). IL-6 operates through the same receptor as LIF, is induced by CCI and exogenous application causes a rise in galanin in cord and ganglion (Murphy et al., 1999). The relation of IL-6 and LIF is not yet clear, but progress is being made on the mechanisms of galanin increase after nerve lesions.

Inflammation

After acute inflammation, one group finds more mRNA and immunostaining for galanin in the DRG (Ji et al., 1995a), and another less (Callsen-Cencic and Mense, 1997). The difference may reflect temporal factors in that in chronic inflammation, galanin and galanin mRNA DRG profile numbers decrease and then increase, particularly large profiles, and the increase persists after inflammation resolves (Munglani et al., 1996; Calza et al., 1998; 2000). The long-lasting upregulation may be neuropathic rather than inflammatory in that joint destruction occurs late in this condition.

Herpes simplex in footpad or cornea results in an increased number of galanin DRG cell profiles (Henken and Martin, 1992a, b). It would be of interest to see if the infected cells are the ones that begin to make galanin.

Resiniferatoxin

Systemic application of the potent capsaicin analog resiniferatoxin leads to a chronic increase of numbers of DRG profiles expressing galanin message associated peptide, which may underlie some of the chronic effects of this compound (Xu et al., 1997).

Axonal Transport Blockers

Local or systemic axonal transport blockade increases numbers of galanin profiles, many of which are large (Henken and Martin, 1992a, b; Kashiba et al., 1992a; Farkas-Szallasi, 1995; Barajon et al., 1996). Presumably the increase is due to loss of access to peripheral trophic factors.

Co-localizations

CGRP

In DRG cells and terminals in cord (Ju et al., 1987a, b; Arvidsson et al., 1991; Hu and McLachlan, 2001).

SP

In DRG cells and terminals in the cord, sometimes in the same dense-core vesicles (Ju et al., 1987; Tuchscherer and Seybold, 1989; Zhang, X. et al., 1995e).

VIP

(Kashiba et al., 1992b).

NPY

(Landry et al., 2000).

5-HT

In a few axons in the deep dorsal horn (Maxwell et al., 1996).

Glucocorticoid Receptors

In DRG and trigeminal ganglion cells (DeLeon et al., 1994).

Dorsal Roots

Between one quarter and one third of the myelinated and unmyelinated axons in normal rat lumbosacral dorsal roots immunostain for galanin (Klein et al., 1990a).

Neuropeptide FF (NPFF)

Although NPFF is not seen in DRG cells from normal animals, after colchicine a few small NPFF DRG cells are seen (Panula et al., 1987; Majane et al., 1989; Lee, J. H. et al., 1993). There is disagreement as to the effects of dorsal rhizotomy on dorsal horn immunostaining (Ferrarese et al., 1986; Majane et al., 1989).

Neuropeptide Y (NPY), Polypeptide Y (PPY), and Peptide YY

Normal adult DRG cells do not usually contain enough NPY to be immunostained (Allen et al., 1984; Hökfelt et al., 1998), although a few lightly stained cells are occasionally seen (Lindh et al., 1989). Numbers of NPY cells increase after axotomy (Zhang, X. and Ji, R. R. et al., 1996).

Peripheral Axotomy or Compression

After peripheral axotomy, many large and some small DRG cells express NPY and its mRNA (Wakisaka et al., 1991; 1992; 1993; Noguchi et al., 1993; Ma and Bisby, 1998; Marchand et al., 1999; Widerberg et al., 2001). Consonant with this, many NPY cells after axotomy are labeled with RT97, a marker for large (A fiber) primary afferent neurons (Frisen et al., 1992), in greater proportions than for galanin or VIP, which are other upregulated peptides (Noguchi et al., 1993; Hökfelt et al., 1998). NPY mRNA is not increased in undamaged DRG cells in ganglia adjacent to those cells whose peripheral processes were transected (Fukuoka et al., 1998), but some undamaged cells express NPY itself (Ma and Bisby, 1998). The lesion-induced increases are greater in middle-aged as opposed to young adult animals (Ma and Bisby, 2000). Similar changes in NPY and many other markers are seen if the nucleus gracilis is transected (Persson et al., 1995). Compression of a limb results in a transient increase in CPON, a marker for NPY, in DRG cells (Widerberg et al., 2001), indicating either that these changes do not depend on actual axon transection or that inflating a tourniquet (150 or 300 mmHg) damages peripheral nerves.

Trophic Factors

VIP in small DRG cells may be under the control of leukemia inhibitory factor (LIF) since the appearance of VIP after axotomy in small cells does not occur in LIF-deficient mice (Corness et al., 1996). Also chronic NT-3 administration attenuates the expression of NPY in large neurons, but not CGRP, SP, or VIP, which are predominately in small neurons (Sterne et al., 1998).

Development

In one study, no VIP cells are seen during development, although some PYY cells are seen in embryos (Jazin et al., 1993). Others find NPY DRG cells prenatally, some of which co-localize with SP, but these fade away after birth (Morris et al., 2001).

Co-localizations

Many post-axotomy NPY DRG cells express BDNF or BDNF mRNA (Kashiba and Senba, 1999; Michael et al., 1999), and many of the galanin cells that appear after axotomy also express NPY (Landry et al., 2000). Some small NPY DRG cells express histamine receptors (Kashiba et al., 2001a). In development, some DRG cells coexpress NPY and SP or NPY and CGRP (Morris et al., 2001).

Peripheral Innervation

Many VIP axons are found in viscera, but the distinction between sensory and sympathetic VIP fibers is not usually made.

Neurotensin (NT)

A very few large DRG cells express NT or its mRNA. Their numbers increase after peripheral axotomy and immunolabeled NT piles up 20 hours after the lesion on the ganglion side of a nerve crush indicating that DRG cells begin making the peptide in response to axotomy (Zhang et al., 1993c; 1996; Zhang, X. et al., 1995b). Since dorsal rhizotomies have little effect either in the normal or after nerve crush, it is unlikely that significant amounts of NT are present on primary afferent fibers in the cord, so NT in DRG cell bodies, being particularly prominent after axotomy, may activate DRG cell bodies that express these receptors (Zhang, X. et al., 1995b).

OPIOIDS

Dynorphin (DYN)

There is discrepancy in reports on DYN in the DRG. Thus, low levels of DYN are reported in DRG, there is no loss of DYN in cord after dorsal rhizotomy as assayed chemically (Botticelli et al., 1981; Przewlocki et al., 1983), and no DYN cells are seen in DRG, even with colchicine, in some studies (Basbaum et al., 1986a; Garry et al., 1989). On the other hand, DYN DRG cells are seen (Gibbins et al., 1987a) as are proDYN DRG cells (Weihe et al., 1985; 1988b), cells in DRG cultures express DYN (Sweetnam et al., 1986), ties on the dorsal roots, either in vivo (Basbaum et al., 1986a) or in vitro (Gibbins et al., 1987a) show DYN piling up on the ganglion side of the ligature and dorsal rhizotomy depletes dorsal horn DYN (Basbaum et al., 1986a; Tuchscherer and Seybold, 1989). Accordingly, the weight of evidence indicates DYN-containing primary afferent neurons, particularly at sacral levels where DYN immunostaining is most prominent. Some DYN cells contain SP (Gibbins et al., 1987a; Tuchscherer and Seybold, 1989), but none contain CGRP or CCK

(Gibbins et al., 1987a), the CGRP finding being puzzling since peripheral fibers co-localize DYN and CGRP (see below) and proDYN co-localizes with CGRP (Weihe et al., 1988b).

Many fine visceral and somatic regions contain fine sensory DYN and prodynorphin sensory fibers, as evidenced by co-labeling with CGRP or SP (Gibbins et al., 1987a; Weihe et al., 1988b; Steele et al., 1996), although these conclusions must be accepted with caution since the fibers in question are mostly unmyelinated and with LM resolution, it is difficult to see if the same or adjacent fibers are expressing the various labels.

Endomorphin

Endomorphin-2 fibers are seen in the dorsal root ganglion and skin and are numerous in the superficial dorsal horn (Martin-Schild et al., 1998; Pierce et al., 1998; Zhang, X., 1999).

Endorphin

To our knowledge, no POMC derivatives such as β-endorphin have been demonstrated morphologically in DRG cells, but numerous inflammatory cells containing β-endorphin migrate into inflamed regions where they can influence primary afferent input (Przewlocki et al., 1992), and chemical assays indicate that considerable β-endorphin exists caudal to a spinal transection, which must come from intrinsic or primary afferent neurons (Gutstein et al., 1992).

Enkephalin

Enkephalins are found in few or no adult DRG cells, although some are found in developing rats (Senba et al., 1982; Kummer and Heym, 1986; Marti et al., 1987; Weihe et al., 1988b; Garry et al., 1989; Pohl et al., 1994), and in support of this, dorsal rhizotomy or neonatal capsaicin has little effect on enkephalin immunostaining or content (Hökfelt et al., 1977a, b; Seybold and Elde, 1980; Priestley et al., 1982a; Micevych et al., 1986; Pohl et al., 1990), enkephalin does not pile up on the ganglion side of a dorsal root ligature (Bras et al., 2001) and enkephalin is not affected in long-term amputations in humans (Hunt et al., 1982). On the other hand, there is a significant amount of enkephalin and its mRNA in the DRG (Przewlocki et al., 1983; Pohl et al., 1997), 17% of unmyelinated cutaneous sensory axons are enkephalinergic (Carlton and Coggeshall, 1997), a few CGRP fibers co-localize with ENK (Zhang et al., 1993b), numerous enkephalinergic fibers can be seen in joints (Elhassan et al., 1998) or lungs (Shimosegawa et al., 1990), and enkephalin piles up on the proximal side of a ligature of the sciatic nerve (Bras et al., 2001). The different findings could be because a significant number of DRG cells manufacture enkephalin and send it rapidly only to the periphery, and if so, these cells might be examples of efferent sensory neurons (Holzer and Maggi, 1998) but releasing an inhibitory rather than an excitatory compound. In mice that over-express enkephalin, enkephalin neurons are obvious in the DRG, more enkephalin piles up against a sciatic nerve ligature, some enkephalin piles up against a dorsal root ligature, and the pain behavior and joint destruction that follows an experimental arthritis is greatly ameliorated (Bras et al., 1998; 2001; Braz et al., 2001).

Enkephalin has a considerable effect on spinal blood flow and enkephalinergic fibers are seen on spinal blood vessels, presumably from enkephalin interneurons but possibly from DRG or sympathetic cells (Zochodne et al., 2001b).

Nociceptin

No or few DRG cells express the mRNA for prepronociceptin, the precursor for nociceptin, in some studies (Andoh et al., 1997; Boom et al., 1999; Itoh et al., 2001), but numerous small DRG cells express this mRNA in another (Pettersson et al., 2002). After inflammation nociceptin mRNA appears in the ganglion (Andoh et al., 1997), and many small, some intermediate, and a few large cells that also express VR1 express prepronociceptin mRNA (Itoh et al., 2001).

Oxytocin and Vasopressin

Some investigators find no OXY, VP, or their mRNAs in DRG cells (Boehmer et al., 1989; Garry et al., 1989; Lawson, 1992; Lawson, et al., 1993), and preprovasopressin is not present in the DRG of the rat (Hallbeck et al., 1996). Others claim that approximately 50% of DRG cell profiles contain OXY and VP, that these compounds are removed from ganglion and cord by neonatal capsaicin, and that chromatographic and radioimmunoassays show OXY and VP in DRGs (Kai-Kai et al., 1985; 1986; Vecsernyes et al., 1987; Kai-Kai and Che, 1995). These findings are not reconciled, but the fact that OXY terminals do not form central glomerular endings (Chapter 6) speaks against them being in fine primary afferent fibers.

Secretoneurin and the Chromagranins

Chromagranins A and B, originally isolated from chromaffin tissue, and secretogranin II are contained in large dense core vesicles (chromaffin granules). Secretoneurin is the product of secretogranin II metabolism (Marksteiner et al., 1993). Up to 80% of DRG cell profiles express the RNA for secretogranin II (Kirchmair et al., 1993) and numerous cells express secretoneurin (Dun et al., 1997; Li, J. Y. et al., 1999). These are mostly small cells. Chromogranin A is found widely in all sizes of DRG cell and chromogranin B is found predominantly in large sensory neurons (Li, J. Y. et al., 1999).

Somatostatin

Approximately 10% of small DRG and trigeminal ganglion cell profiles contain somatostatin (Hökfelt et al., 1975a; 1976; Johansson, 1978; Lundberg et al., 1978; Dalsgaard et al., 1984; Leah et al., 1985a, b; Price, 1985; Skofitsch et al., 1985c; Tuchscherer and Seybold, 1985; Shehab and Atkinson, 1986a, b; Ju et al., 1987b; Marti et al., 1987; Molander et al., 1987; Cameron et al., 1988; Henken et al., 1988; Lazarov, 2002). Somatostatin is found in DRG cells that respond to either noxious or non-noxious stimuli (Leah et al., 1985a).

Co-localizations

Somatostatin coexists with SP, CGRP, leu-enkephalin, CCK, and bombesin in cat DRG cells (Leah et al., 1985a; Cameron et al., 1988; Garry et al., 1989), although there is no co-localization of somatostatin and SP in rat DRGs (Hökfelt et al., 1976; Tuchscherer and Seybold, 1985). Somatostatin either does (Dalsgaard et al., 1984) or does not co-localize with FRAP (Nagy and Hunt, 1982), the latter finding possibly related to the use of colchicine. Somatostatin does not co-localize with galanin in the normal (Ju et al., 1987b) or galanin and VIP after systemic capsaicin (Kashiba et al., 1997), although a few cells with both somatostatin and galanin have been seen (Garry et al., 1989). Most small DRG cells, including somatostatin cells, express latexin (Takiguchi-Hayashi et al., 1998). Cells expressing somatostatin mRNA are found exclusively in the IB4 population (Kashiba et al., 2001b), and somatostatin DRG cells express c-ret, related to the GDNF receptor (Kashiba and Senba, 2000), and no SST cells express trk receptors (Kashiba et al., 1996). Thus, unlike other peptides, DRG somatostatin seems to be under the control of GDNF (see below).

GDNF and Peripheral Axotomy

Total DRG somatostatin and numbers of somatostatin DRG profiles are reduced after peripheral axotomy (Noguchi et al., 1993; Bennett et al., 1998a), and NGF has no effect on this process or on somatostatin levels in general (Adler and Walker, 2000), whereas GDNF prevents the axotomy-induced loss (Bennett et al., 1998a). In addition, somatostatin is regulated by GDNF in DRG cells in culture (Adler, 1998), whereas NGF has no effect (Otten and Lorez, 1983; Mulderry, 1994). Accordingly somatostatin is not a typical primary afferent peptide in that, at least in adults, it is under the control not of NGF but rather of GDNF, and localizes in IB4 cells, IB4 being the marker often used for the "non-peptidergic" small DRG cell population (Chapter 3). Adrenalectomy increases numbers of somatostatin DRG cell profiles (Covenas et al., 1994).

Development

Somatostatin is expressed transiently in DRG cells before birth, and then a second expression occurs starting about one week after birth in rats (Katz et al., 1992; Maubert et al., 1992; Kashiba and Senba, 2000). It is the postnatal expression of somatostatin that is under the control of GDNF (Kashiba and Senba, 2000).

Periphery

Somatostatin fibers can be seen in various peripheral somatic and visceral targets, somatostatin containing DRG cells are retrogradely filled from somatic and visceral nerves and immunochemical assays show somatostatin passing from the DRG to the periphery (Rasool et al., 1981; Molander et al., 1987; Ositelu et al., 1987; Ackermann et al., 1999). Thus there are somatostatin-containing fine sensory fibers in the periphery, but somatostatin is also prominent in peripheral autonomics. There is a tonic inhibitory somatostatin influence on peripheral nociceptors (Carlton et al., 2001a).

Substance P (SP)

Approximately 20% of DRG cell profiles, mostly small cells, express SP or SP mRNA (Hökfelt et al., 1975b; Panula et al., 1983; Lee et al., 1985b; Henken et al., 1988; Tajti et al., 1999), although a few medium-sized cells also label (Chiba et al., 1986; Battaglia and Rustioni, 1988; Holtman, 1989; Hanesch et al., 1991; 1993; Holford et al., 1994; Airaksinen and Meyer, 1996; Ji et al., 1996), and in concert with these findings individually identified DRG SP cells have axons conducting in Aδ or C conduction velocity ranges (Lawson et al., 1997).

Potentials recorded from SP C cells have inflections on their falling phases and wider action potentials than A cells, and thus SP C cells are similar to C cells in general (Lawson, 1995; Gold et al., 1996b, c; McCarthy and Lawson, 1997). By contrast, SP Aδ cells have deeper and longer after-hyperpolarizations than non-SP Aδ cells (McCarthy and Lawson, 1997). In an impressive catalog of individually identified cells, all identified SP DRG cells were nociceptors (Lawson et al., 1997).

Segmental Levels

There are considerable differences in proportions of SP profiles in relation to segments with L2 having the highest proportion (Tuchscherer and Seybold, 1985).

Age and Development

SP starts being expressed at about the time that axons make their way to the periphery (Michael et al., 1997b), and preprotachykinin is expressed at adult levels at birth and remains constant from then on (Kashiba and Senba, 2000).

Co-localizations

Tachykinins

SP is co-localized with neurokinin A (Dalsgaard et al., 1985) or β-preprotachykinin (Henken et al., 1988).

CGRP

SP is co-localized with CGRP in sensory neurons, in primary afferent terminals in the dorsal horn, often in the same large dense-core vesicles, and in peripheral cutaneous or visceral axons (Gibson et al., 1984b; Wiesenfeld-Hallin et al., 1984; Gibbins et al., 1985; 1987a, b; Lee et al., 1985a, b; Lundberg et al., 1985; Skofitsch and Jacobowitz, 1985a; Uddman et al., 1985a; Franco-Cereceda et al., 1986; Gulbenkian et al., 1986; Hanko et al., 1986; Matsuyama et al., 1986; Wanaka et al., 1986; Ju et al., 1987a; Cameron et al., 1988; Merighi et al., 1988; 1989; 1991; Garry et al., 1989; Tuchscherer and Seybold, 1989; Pereira da Silva and Carmo-Fonseca, 1990; Plenderleith et al., 1990; Bonfanti et al., 1991; Semenenko and Cervero, 1992; Hökfelt et al., 1994c; Heym et al., 1995a, b; Ribeiro-da-Silva, 1995; Ribeiro-da-Silva and Cuello, 1995; Zhang, X. et al., 1995f). Proportions of SP and CGRP and double-labeled cells differ from study to study, but the usual conclusion is that all SP-containing DRG cells express CGRP but only about half of CGRP cells express SP.

GAL

SP and galanin are co-localized after injury (Ju et al., 1987a; Zhang et al., 1993b).

SOM

An early study showed that SP and SOM do not co-localize (Hökfelt et al., 1976), but later studies show considerable co-localization (Leah et al., 1985a; Ju et al., 1987a; Cameron et al., 1988; Garry et al., 1989; Hanesch et al., 1995; Heym et al., 1995b).

CCK

Many or all SP cells co-localize with CCK (Dalsgaard et al., 1982b; Leah et al., 1985a; Kuwayama and Stone, 1986), although it must be remembered that CCK and CGRP antibodies may recognize the same antigen (Ju et al., 1986). They are also co-localized in dorsal horn terminals (Tuchscherer and Seybold, 1985; Tuchscherer et al., 1987).

ENK and DYN

SP often co-localizes with these endogenous opioids (Weihe et al., 1986; Tashiro et al., 1987; Katoh et al., 1988a; Garry et al., 1989; Heym et al., 1995b).

Bombesin, VIP, NT, and NPY

These peptides are coexpressed with SP, either in terminals or DRG cell bodies (Fuxe et al., 1983; de Groat, 1986; Gibbins et al., 1987a; Cameron et al., 1988; Zhang et al., 1993b).

FRAP

There is no overlap of FRAP with SP (Nagy and Hunt, 1982), but it must be remembered that FRAP and CGRP are extensively co-localized in one study (Carr et al., 1990), and SP is completely co-localized with CGRP in DRG cells (see above), and SP and FRAP coexist in some DRG cells in another study (Dalsgaard et al., 1984).

NOS

(Heym et al., 1995b).

Glutamate

Glutamate and SP coexist in most small DRG cells (Battaglia and Rustioni, 1988).

IB-4

Some studies find that IB4 and SP essentially do not co-localize (Silverman and Kruger, 1988b; 1990), but others find a considerable co-localization (Ambalavanar and Morris, 1992; Wang et al., 1994; Kashiba et al., 2001b).

200 KD Neurofilament Protein

SP cells do not co-localize with 200 kD neurofilament protein in one study (Price, 1985), but some larger SP cells co-localize in another (Lawson et al., 1993). Thus, most SP cells fall in the small dark category but a few are large light cells.

Peripherin

There is more than 50% to almost complete co-localization of peripherin with SP mRNA in DRG cells (Holford et al., 1994; Goldstein et al., 1996).

GDNF

There is some coexistence of GDNF with SP in DRG cells (Quartu et al., 1999).

Opioid Receptors

SP and MORs co-localize in DRG cells and synaptic terminals in the dorsal horn (Ding et al., 1995a; Li, P. J. L. et al., 1998a).

Experimental Manipulations

Dorsal Rhizotomy

The effects of dorsal rhizotomy are described in Chapter 6.

Peripheral Nerve Lesions

Nerve crush or transection with or without regeneration depletes SP and SP mRNA cell profile numbers in DRG, dorsal roots, and peripheral nerves or trigeminal ganglion (Bisby and Keen, 1986; Nielsch and Keen, 1989; Villar et al., 1989; 1991; Kar et al., 1990; Verge et al., 1995; Zhang, Q. et al., 1995a; Zhang, X. et al., 1995a; Ji et al., 1996; Elcock et al., 2001a). There is return of SP in the DRG if regeneration is allowed, but a permanent depletion if regeneration is prevented (Bisby and Keen, 1986; Nielsch et al., 1987), suggesting that synthesis of SP is dependent on peripheral trophic factors. Increased preprotachynin mRNA in small DRG cells occurred early followed by marked decline, but synthesis then began in large cells, which may relate to the development of allodynia. By contrast, after infraorbital nerve section there was a relatively constant and long-lasting SP decline in the trigeminal ganglion (Elcock et al., 2001a), but there was no change if the infraorbital nerve was ligated (Elcock et al., 2001b). Thus the type of peripheral axotomy influences peptide expression. Neonatal crush of the infraorbital nerve causes a marked loss of SP DRG cells in the trigeminal ganglion, but there seems to be less loss than for other DRG cell types (Enfiejian et al., 1989).

Muscle nerves contain few SP afferent fibers and cutaneous nerves many, and when the nerves are transposed and regenerate, numerous SP axons are in the muscle nerve, that now innervates skin, and few are in the cutaneous nerve that innervates muscle, which may imply different peripheral trophic factors (McMahon and Gibson, 1987). Cisplatin applied to the nerve causes a slight increase in SP in ganglia and dorsal cord due presumably to interference with axonal transport (Barajon et al., 1996). Capsaicin administration prevents some of the SP increase in the DRG and periphery after hindpaw inflammation (Ahmed et al., 1995c) and also causes DR cell mitochondrial swelling (Chiba et al., 1986). Neonatal capsaicin essentially removes SP neurons in the adult (Jancso et al., 1981), and adult capsaicin reduces SP cell numbers (Kar et al., 1990). Diabetes and hypoxia cause nerve damage and hypoxia causes a decrease of SP in sciatic nerve with no change in the DRG (Brewster et al., 1994), and experimental diabetes causes a decrease in numbers of DRG cells expressing SP mRNA, especially long after the lesion (Zochodne et al., 2001a).

Inflammation

Preprotachykinin (the SP precursor) mRNA is increased after monoarthritis (Minami et al., 1989; Donaldson et al., 1992; 1994; Galeazza et al., 1995; Garrett et al., 1995) or intraplantar formalin (McCarson and Krause, 1995). For SP itself, there is a considerable increase in SP cell profiles or amount of SP in the DRG following various inflammations (Kar et al., 1991a; Donnerer et al., 1992; 1993; Smith et al., 1992; Hanesch et al., 1993; Ahmed et al., 1995a, b, c; Garrett et al., 1995; De Giorgio et al., 2001; Vizzard, 2001), although early in the course of inflammation, a reduction in the SP profiles is seen (Calza et al., 1998; Traub et al., 1999) or there is no change acutely and a slight reduction chronically (Mapp et al., 1993). On the assumption that a decline in SP precedes the increases, most findings are compatible with the idea of early release due to inflammation-induced activity followed by upregulation of preprotachynin mRNA and then increased SP in DRG SP cells. Supporting evidence is that nerve activity, presumably leading to depletion, is necessary because prevention by local anesthesia, capsaicin application to appropriate nerves, or nerve section stops the SP increases (Holland and Goldstein, 1990; Donaldson et al., 1994; Ahmed et al., 1995a, c).

The increase in SP cell profiles in most inflammatory situations presumably indicates that some DRG cells begin making SP. This phenotypic change in neurons giving rise to large myelinated axons could be important in allodynia (Neumann et al., 1996), but there is no increase in large SP DRG cell profiles after adjuvant inflammation of the foot (Hanesch et al., 1993). Inflammation upregulates NGF, and antiNGF prevents the SP changes in inflammation and administration of NGF in normals increases SP in appropriate nerves and ganglia (Donnerer et al., 1992; 1993). An opioid antagonist increases SP in the DRG after formalin-induced inflammation (McCarson and Krause, 1995). Thus opioids and NGF are factors in increased expression of SP. Some show SP changes correlated with the course of inflammation (Calza et al., 1998), but others do not (Hoheisel et al., 1998).

Release of SP from peripheral sensory axons is a major factor in neurogenic inflammation, which may play a role in adjuvant-induced or rheumatoid arthritis. Morphological findings are that SP and indeed all axons around joints are greatly reduced in chronic rheumatoid arthritis and adjuvant-induced arthritis (Konttinen et al., 1990; Mapp et al., 1990). On the other hand, others find an increase in SP axons around adjuvant inflamed joints (Weihe et al., 1988a; Ahmed et al., 1995a, b, c), or no change in SP levels (Garrett et al., 1995), and when hyperalgesia developed in experimental arthritis there was no change in peripheral SP fibers (Sluka and Westlund, 1993b). Thus, a coherent story has not yet emerged, but a possibility is that inflammation leads to release of SP peripherally, followed in turn by upregulation of SP mRNA in DRG cells, a peripheral increase in SP immunostaining and then as the arthritis turns chronic, a loss of SP and perhaps all sensory axons in and around the joint. There is also a report of increased numbers of SP fibers in human interstitial cystitis (Pang et al., 1995). An important point is that SP sensory systems change over time after inflammation, so findings at single time-points may be difficult to interpret (Calza et al., 1998). Systemic NGF prevents the capsaicin-induced cell loss of SP in sciatic nerve and skin and anti-NGF prevents the rise in SP after CFA inflammation (Donnerer et al., 1992; 1993; 1996).

Trophic Factors

NGF prevents the loss of SP or SP mRNA cells after nerve lesion (Wong and Oblinger, 1991; Verge et al., 1995; Zhang et al., 1995c; Ji et al., 1996), but in another study NGF has no effect on numbers of SP cell profiles (Wiberg et al., 1999). There is an increase in proportions of SP DRG profiles in transgenic mice that over-express NGF (Ma et al., 1995), and NGF given to 2-day rats increases SP in dorsal root ganglia and numbers of SP DRG cell profiles (Otten and Lorez, 1983). Neurotrophin-3 (NT-3) is necessary for survival of propriceptive afferents, and one report shows SP cells largely unaffected in NT-3 knockout mice, but another shows a major loss SP and CGRP DRG cell profiles in these knockouts (Airaksinen and Meyer, 1996). Neither BDNF nor NT-3 prevents the loss of DRG cells expressing SP mRNA after nerve section, but the combination does, as do high doses of LIF (Zhang, Q. et al., 1995a). Thus, the DGR cell populations supplied by different trophic factors are relatively but not absolutely distinct.

Adrenalectomy

Adrenalectomy increases the numbers of SP and SOM DRG cells, possibly indicating a tonic inhibition of SP synthesis by adrenal steroids (Covenas et al., 1994).

Resiniferatoxin

Resiniferatoxin does not change numbers of DRG profiles expressing the mRNA for SP but the signals are weaker, and the total mRNA in the ganglion is depleted. Thus presumably SP cells are not killed but SP production is down (Szallasi et al., 1999).

Periphery

Somatic Innervation

1. Skin or cutaneous nerves (Hökfelt et al., 1975b; Hunt and Rossi, 1985; Gibbins et al., 1987a; O'Brien et al., 1989; Perry and Lawson, 1998; Cuello et al., 1978).
2. General somatic nerves (Lundberg et al., 1978; Kawatani et al., 1986; Molander et al., 1987).
3. Muscle nerves (O'Brien et al., 1989; Perry and Lawson, 1998).
4. Meissner's corpuscles (Ishida-Yamamoto et al., 1988).
5. Nasal mucosa (Hökfelt et al., 1975b).
6. Dental pulp (Gazelius et al., 1987). Both CGRP and SP are found in dental pulp and they synergize when administered together.
7. Eye, iris, or cornea (Tervo et al., 1981; Lehtosalo 1984; Gibbins et al., 1985; Hill et al., 1996). Of SP cell profiles projecting to eye, 20% contain SP (Kuwayama and Stone, 1986; Matsuyama et al., 1986; Kuwayama et al., 1987; Felipe et al., 1999).

8. Joints or joint nerves (Weihe et al., 1988a; O'Brien et al., 1989; Konttinen et al., 1990; Mapp et al., 1990; Hanesch et al., 1991; Salo and Theriault, 1997; Tamura et al., 1998).
9. Ventral roots (Risling et al., 1994).
10. Diaphragm (Holtman, 1989).
11. Dura and pia maters (Risling et al., 1994; Kallakuri et al., 1998; Yamada et al., 2001).

Visceral Innervation

1. Vascular system including heart (Lundberg et al., 1985; Uddman et al., 1985a; Gulbenkian et al., 1986; Hanko et al., 1986; Matsuyama et al., 1986; Wanaka et al., 1986; Gibbins et al., 1987a; Vizzard, 2001).
2. GI tract and associated organs (Lindh et al., 1983; Green and Dockray, 1987; 1988; Won et al., 1998; Traub et al., 1999; De Giorgio et al., 2001).
3. Splanchnic nerves (Lundberg et al., 1978; Molander et al., 1987; Sharkey et al., 1987; Perry and Lawson, 1998).
4. Visceral afferent neurons, pelvic viscera or pelvic nerve (de Groat et al., 1983; Gibbins et al., 1985; 1987a; de Groat, 1986; Kawatani et al., 1986; McNeill and Burden, 1987).
5. Kidney, ureter, prostate, or bladder (Kuo et al., 1984; Semenenko and Cervero, 1992; Pang et al., 1995; Ishigooka et al., 2002).
6. Respiratory system (Lundberg et al., 1985; Gibbins et al., 1987a).
7. Adrenal gland (Heym et al., 1995a, b).

Some studies report partial or complete loss of SP staining after appropriate nerve lesions or capsaicin administration as evidence for the primary afferent nature of the peripheral SP axons, but the nerve lesions would also remove sympathetic axons containing SP (Barbut et al., 1981; McGregor et al., 1984; Green and Dockray, 1988; Ahmed et al., 1995a, c).

VIP

VIP and its mRNA are localized in a moderate number of small DRG cells, particularly in sacral segments and innervate pelvic viscera preferentially (Kawatani et al., 1983; 1985; Kuo et al., 1985; Kummer and Heym, 1986; Moller et al., 1993; Mulder et al., 1994; Dun, E. C. et al., 1996; Dun, N. J. et al., 1996; Zhang, Y. Z. et al., 1996; Zhang, X. and Ji, R.-R. et al., 1996). VIP cells appear early in development (Nielsen et al., 1998). VIP is co-localized with SP, CGRP, CCK, and α-neo-endorphin in DRG cells, but not with somatostatin (Leah et al., 1985a; Kummer and Heym, 1986; Moller et al., 1993). After peripheral axotomy, DRG cells with transected peripheral axons begin expressing VIP and appropriate mRNAs in much greater amounts than in the normal, and some of these cells are large (Shehab and Atkinson, 1986b; Nielsch and Keen, 1989; Noguchi et al., 1989b; Nahin et al., 1994; Zhang, Q. et al., 1995a; Zhang, Y. Z. et al., 1996; Zhang, X. and Bean, A. J. et al., 1995a, b). VIP cell numbers are also increased after systemic resiniferatoxin (Farkas-Szallasi et al., 1995). VIP axons are found in various viscera and some of these are sensory as they co-localize with CGRP (Moller et al., 1993). VIP is also found in autonomic neurons.

PROTEINS

Calcium Binding Proteins

The calcium binding proteins, calretinin, calbindin, and parvalbumin are localized in 4–33% of DRG cells. These cells are large but some reports find some small cells (Carr et al., 1989a; Antal et al., 1990; Celio, 1990; Kashiba et al., 1990b; Kuramoto et al., 1990; Zhang, J. H. et al., 1990; Duc et al., 1993; Heppelmann and Emson, 1993; Ichikawa et al., 1993; 1994; Ren et al., 1993a; Honda, 1995), and calbindin is found in essentially small cells in one study (Honda, 1995). These cells co-localize with cytochrome oxidase and carbonic anhydrase to a great extent so they are metabolically active and also proprioceptive in nature (Carr et al., 1989a, c; Duc et al., 1993). Large calcium protein binding DRG cells essentially do not co-localize with peptides such as CGRP, but many such trigeminal ganglion cells are small and these do co-localize with neurokinins, suggesting a dual population particularly with regard to innervation of the face and head (Ichikawa et al., 1992; 1993; 1994). Perhaps the small trigeminal ganglion cells are involved in migraine. Parvalbumin co-localizes with S100 in trigeminal ganglion cells (Ichikawa et al., 1998), and DRG cells transiently express calcium-binding protein S-100β following nerve lesions (Huang et al., 1998). Calbindin cells label almost exclusively viscera, whereas parvalbumin cells do not, but somatic tissues are innervated by both (Kashiba et al., 1990b; Honda, 1995). There is little co-localization of parvalbumin and calbindin in DRG cells (Zhang, J. H. et al., 1990; Honda, 1995), but very considerable co-localization of calretinin and parvalbumin (Ichikawa et al., 1994).

A calcium-binding peptide has been discovered that exists in approximately one third of DRG cells and immunolabels proprioceptive axons (Ichikawa et al., 1999a). The relation of this peptide to calcium-binding proteins is not known. Neurocalcin is a calcium-binding protein found in approximately 40% of DRG cells. Many of these cells also contain calbindin or parvalbumin (Okazaki et al., 1994; Junttila et al., 1995; Iino et al., 1998).

During development, parvalbumin and calbindin cells appear early and increase steadily until they reach the adult pattern (Zhang et al., 1990), whereas calretinin appears in almost 75% of early DRG cells, then disappears to reappear postnatally to reach adult levels (Ambrus et al., 1998). There is no discernible change in calbindin or parvalbumin cells after peripheral nerve section (Honda, 1995). Hypothyroidism, fetal trisomy 13, and acquired immunodeficiency reduce numbers of calcium-binding protein DRG cells, except that calbindin neurons are not affected by the immunodeficiency (Ninomiya et al., 1993; Nagano et al., 1996b; Barakat-Walter et al., 2000).

Calbindin and calretinin axons innervate muscle spindles, Pacinian corpuscles, subepidermal lamellar corpuscles, lanceolate endings around hairs and vibrassae, and intradermal endings in digital skin. This work, plus earlier work on chicks, shows that these calcium-binding proteins are concerned primarily with proprioceptive and exteroceptive afferent input (Duc et al., 1993).

Glutamate Transporters

The sodium-dependent high-affinity transporter EEAT1 but not EEAC2 is expressed by satellite cells in the dorsal root ganglion (Berger and Hediger, 2000), and EEAC1 is present early in development in DRG cells (Furuta et al., 1997). A vesicular glutamate transmitter is present in numerous DRG cells (Tong et al., 2001).

Reg-2

Reg-2, a pancreatitis-associated protein is not normally expressed in DRG cells, but it is rapidly upregulated in IB4 DRG cells after peripheral axotomy (Averill et al., 2002).

Bone Matrix Proteins

The bone matrix proteins osteopontin and osteocalcin are found in many, mostly large DRG cells (Ichikawa et al., 1999b; 2000).

β-arrestin

β-arrestins mediate desensitization of G-protein coupled receptors, and thus might mediate desensitization in DRG cells. β-arrestin is found in almost all small sensory neurons in rat lumbosacral DRGs (Komori et al., 1999).

RECEPTORS

Aminergic Receptors

Adrenergic Receptors

Adrenergic receptors expressed by DRG cells are of particular interest because changes in adrenergic sensitivity follow lesions that lead to neuropathic pain. Morphologic studies focus on the α_2 receptors. DRG α_{2A} mRNA was not seen in the DRG by polymerase chain technology (Gold et al., 1997; Ongjoco et al., 2000), but DRG cells express α_{2A} receptor mRNA (Nicholas et al., 1991; 1993; Cho et al., 1997; Shi et al., 1999b; 2000), and there is immunostaining of DRG cells for the α_{2A} receptor (Gold et al., 1997). DRG α_{2B} mRNA is abundant by polymerase chain technology (Gold et al., 1997; Ongjoco et al., 2000), but few DRG cells express these receptors (Cho et al., 1997; Shi et al., 2000). Two groups find the majority of DRG cells expressing α_{2C} receptor mRNA (Cho et al., 1997; Shi et al., 2000), but others find relatively few DRG cells expressing α_{2C} mRNA (Birder and Perl, 1999). We cannot explain these differences.

After peripheral axotomy, numbers of DRG cell profiles expressing α_{2A} receptor mRNA (Cho et al., 1997; Waldmann and Lazdunski, 1998) and immunostaining (Birder and Perl, 1999) increase. The increase is proportionately greatest in medium-large DRG cells and occurs in cells with either severed or intact peripheral processes (Birder and Perl, 1999). These findings, combined with the sprouting of adrenergic fibers into the ganglion after axotomy (Chapter 3), indicate a considerable increase in effectiveness of α_{2A} receptor ligands at the level of the DRG cell body, given that increases in receptor mRNA and immunostaining imply more active receptors. By contrast, α_{2C} receptor cells decrease slightly or do not change in numbers (Cho et al., 1997; Birder and Perl, 1999; Shi et al., 2000), so the assumption is that they do not take part in enhanced responses of DRG cells to adrenergic ligands after peripheral nerve injury. Also, however, $\alpha 1$ adrenergic ligands increase after axotomy so these may also play a role (Xie J. et al., 2001).

Dopamine Receptors

Polymerase chain reaction analyses show D_1, D_2, D_3, and D_4 mRNA in whole dorsal root ganglion (Xie et al., 1998), and DRG cells containing the mRNA for D_2 receptors have been shown (Peterfreund et al., 1995).

Histamine Receptors

Approximately 23% of monkey DRG cells, many of which contain SP, bind a histamine-1 ligand (Ninkovic and Hunt, 1985), and numerous small DRG cells in the rat and guinea-pig express the mRNA for this receptor and co-localize with IB-4 but not with CGRP or SP (Senba and Kashiba, 1996; Kashiba et al., 1999; Kashiba and Senba, 2001). After nerve crush, histamine-1 receptor-expressing DRG neuron profiles increase, and many of these express SP or CGRP (Kashiba et al., 1999). Thus, this receptor seems to be expressed by non-peptidergic DGR cells before injury and peptidergic after. These neurons are generally capsaicin-insensitive and co-localize with NPY (Kashiba et al., 2001a).

5-HT Receptors

There are seven 5-HT receptor classes (5-HT_1–7), some with many subtypes. The 5-HT_3 receptors are ligand gated ion channels whereas the others interact with G proteins and are linked to second messengers (Martin and Humphrey, 1994).

5-HT_{1A} Receptors

$5HT_{1A}$ receptors have not been described anatomically in DRG cells, but polymerase chain technology (Wu et al., 2001) and *in situ* hybridization (Morales and Wang, 2002) shows their mRNA in the DRG.

5-HT_{1B} Receptors

The mRNA for $5HT_{1B}$ receptors is found in a few (Bonaventure et al., 1998a) or many (Bruinvels et al., 1992; Longmore et al., 1997) DRG cells. There are conflicting reports on 5-HT_{1B} receptors in the cord (Chapter 6), but if none are there, this would suggest only peripheral transport of these receptors by sensory axons (Bruinvels et al., 1992). In quantitative work, 15% of trigeminal ganglion cell profiles express 5-HT_{1B} receptors, some of which co-localize with the 200-kd neurofilament protein (large light cells), some with CGRP and some with trkA, but none with SP or IB4 (Wotherspoon and Priestley, 2000), although other work shows co-localization with SP mRNA (Bonaventure et al., 1998a, b) or SP, CGRP, and NOS in medium-sized cells (Hou et al., 2001). Given the prominent innervation of meningeal vessels by $5HT_{1B}$ receptor-expressing axons, the suggestion has been made that activation of these receptors may lessen the vascular changes that play a role in migraine (Longmore et al., 1997; Wotherspoon and Priestley, 2000).

5-HT_{1C} Receptors

These are now known as $5HT_{2C}$ receptors (see below).

5-HT_{1D} Receptors

5-HT_{1D} receptors and 5-HT_{1D} receptor mRNA are found in trigeminal ganglion cells (Longmore et al., 1997; Bonaventure et al., 1998; Hou et al., 2001), with the peptide being predominantly the 5-$HT1_{D\alpha}$subtype in one study (Rebeck et al., 1994) and both subtypes in another (Bruinvels et al., 1992). 5-HT_{1D} receptor DRG cells also contain SP, CGRP, and NOS (Hou et al., 2001).

5-HT_{1E} Receptors

5HT_{1E} receptors have not been described anatomically in DRG cells or cord, but polymerase chain technology shows their mRNA in human DRG (Pierce et al., 1997).

5-HT_{1F} Receptors

5-HT_{1F} receptors or their mRNA are not found in the cord (Palacios et al., 1990; Adham et al., 1997), but the mRNA for these receptors is expressed in the majority of trigeminal ganglion cells of all sizes (Adham et al., 1997), again suggesting peripheral transport only.

5-HT_2 Receptors

5-HT_2 receptors are now split into three subtypes and the majority of large and small DRG cells express 5HT_{2A} receptors (Maeshima et al., 1998).

5-HT_3 Receptors

Dorsal rhizotomy or capsaicin administration removes 50–80% of 5-HT_3 ligand binding in the dorsal horn (Chapter 6), so these receptors are on primary afferent terminals. The mRNAs for 5-HT_{3A} and 5-HT_{3B} are expressed by many DRG cells, and DRG cells can be divided into those that express only the 5-HT_{3A} receptors and those that express 5-HT_{3A}/5-HT_{3B} heteromeric receptors (Morales et al., 2001; Morales and Wang, 2002). This contrasts with the CNS where only homomeric 5-HT_{3A} complexes are expressed (Morales and Wang, 2002). Activation of 5-HT_{3A} receptors is involved in acute but not persistent nociception, but these neurons do not take part in neurogenic inflammation as stimulating them does not lead to edema, and they do not contain SP (Zeitz et al., 2002).

5-HT_4, 5-HT_5, and 5-HT_7 Receptors

These have not been described anatomically on DRG cells, but polymerase chain technology shows their mRNA in the DRG (Pierce et al., 1996; 1997; Wu et al., 2001).

Amino Acid Receptors

Glutamate Receptors

Ionotropic Glutamate Receptors

Many, perhaps all dorsal root and trigeminal ganglion neurons express NR1 and/or its mRNA (Shigemoto et al., 1992a; Kiyama et al., 1993; Sato et al., 1993a; Petralia et al.,

1996a; Wang et al., 1999), including visceral afferents (McRoberts et al., 2001; Gougis et al., 2002). NR2A and NR2B receptors are expressed by many DRG cells of all sizes (Petralia et al., 1996a), but the NR2B receptors are found predominantly in the small cell population, both peptidergic and non-peptidergic (Ma and Hargreaves, 2000). For AMPA receptors, a few small cells express GluR1, and many large and small cells express GluR2/3, although the mRNA tends to be in small cells and none express GluR4 (Kiyama et al., 1993; Sato et al., 1993a, b; Tachibana et al., 1994). Kainate receptors are found mostly in small cells (Bettler et al., 1990; Petralia et al., 1994). NR1 receptors decline in appropriate DRG cells after localized inflammation (Wang et al., 1999). Thus, as would be expected, most or all DRG cells express ionotropic glutamate receptors with NR2B, GluR2/3 and kainate receptors predominantly in the small cell population.

Binding of radioactive glutamate and immunostaining for NR1 are seen on both sides of a ligated vagus nerve (Lewis et al., 1987; Cincotta et al., 1989) indicating transport of glutamate receptors bidirectionally. Immunologic demonstrations show that approximately 20% of unmyelinated cutaneous sensory axons in rats and humans express the three major types of ionotropic sensory receptors (Fig. 4.4; Carlton et al., 1995; 1998a; Coggeshall and Carlton, 1998; Kinkelin et al., 2000). The percentage of peripheral sensory axons is considerably less than the percentages of DRG cells expressing the Nr1 and GluR2/3 receptors, possibly indicating that many DRG cells do not send these receptors to their peripheral terminals.

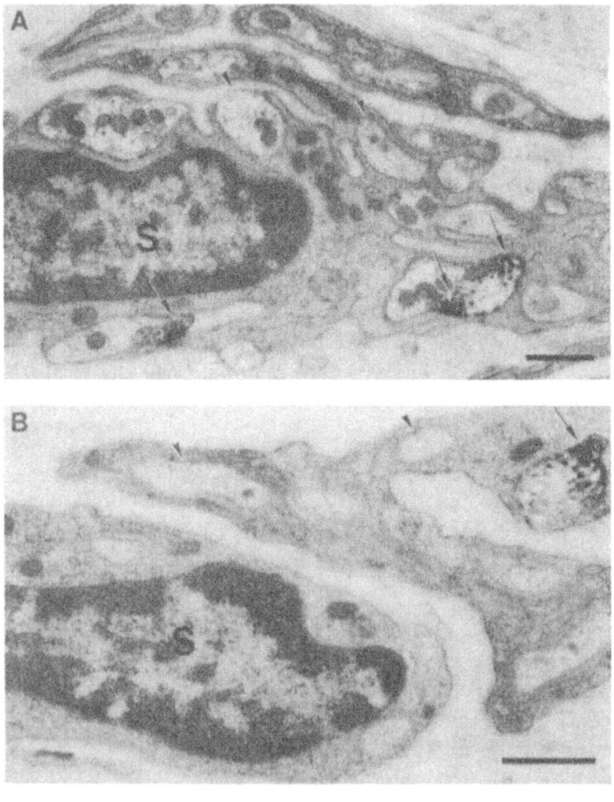

FIGURE 4.4. Peripheral unmyelinated sensory axons express NMDAR1 receptors. (From Carlton et al., 1995.)

Metabotropic Glutamate Receptors

Numerous DRG cells from rats and humans are immunolabeled for mGluR5 (Valerio et al., 1997a, b), and double labeling shows that many of these also express βIII tubulin and VR1 (Walker et al., 2001). These cells are important in inflammatory hyperalgesia (Walker et al., 2001). Group 2 mGluRs are found on 40% of predominantly small DRG cell profiles, and the majority of these are IB4 positive (Petralia et al., 1996b; Carlton et al., 2001b). The mGluR4 and mGluR7 mRNAs are intensely expressed in numerous, usually large, DRG cells (Belyantseva and Lewin, 1999), and many DRG cells are immunostained for mGluR4a or mGluR7 (Ohishi et al., 1995; Azkue et al., 2001), although one study could not find mGluR7 mRNA in DRG cells (Kinoshita et al., 1998). Dorsal rhizotomy significantly reduces spinal immunostaining (Ohishi et al., 1995a, c), and these receptors co-localize with primary afferent markers in spinal terminals (Li, H. et al., 1997), so they are located presynaptically.

GABA Receptors

More than a third of DRG cell profiles express the $GABA_{A\gamma}$ receptor subunit mRNA, and many of these co-localize with the glycine receptor β subunit (Furuyama et al., 1992). The $GABA_{A\gamma}2$ subunit is found in close to 50% of DRG cell profiles normally and is down-regulated after peripheral axotomy, whereas the $GABA_{A\alpha}$ subunit is not affected (Fukuoka et al., 1998). $GABA_A$ mRNA labels many DRG cells whereas $GABA_{A1b}$ labels hardly any (Towers et al., 2000). Immunostaining for the β2/β3 subunits of the $GABA_A$ receptor shows label in almost every DRG cell, but dorsal rhizotomy produces no change in central immunostaining (Alvarez et al., 1996), even though these receptors are transported peripherally to unmyelinated sensory axons (Carlton et al., 1999) possibly indicating that the receptor is only transported peripherally.

The mRNA for $GABA_{B2}$ receptors is expressed by numerous DRG cells of all sizes (Durkin et al., 1999). Immunostaining shows $GABA_{B1b}$ to be in large neurons, whereas $GABA_{B1a}$ and $GABA_{B2}$ neurons are of all sizes (Charles et al., 2001), but in another study $GABA_{B1a}$ staining is intense for essentially all DRG cells whereas $GABA_{B1a}$ is much lighter and evenly distributed (Poorkhalkali et al., 2000).

Glycine Receptors

Few studies report glycine receptors on primary afferent neurons, and there is no change in spinal glycine receptors after neonatal capsaicin, which removes fine primary afferents (Singer and Placheta, 1980). Also these receptors are not found on central terminals of glomeruli, which are fine primary afferent endings (Mitchell et al., 1993). However, almost half of rat DRG cells express the glycine β receptor subtype (Furuyama et al., 1993), which may imply an extensive primary afferent glycinergic input.

Cholinergic Receptors

Acetylcholine excites some primary afferent neurons, and cholinergic antagonists are antinociceptive. Cholinergic receptors are divided into nicotinic and muscarinic categories. Nicotinic receptors are ligand-gated ionic channels whose activation leads to rapid increases in permeability to sodium and potassium. Muscarinic receptors are G-protein coupled and activate second-messenger systems.

Nicotinic Receptors

Nicotinic receptors, as shown by α-bungarotoxin binding, are concentrated in 11–20% of mostly large DRG cells, and these receptors are transported bidirectionally in sensory axons (Polz-Tejera et al., 1980; Ninkovic and Hunt, 1983). More recently, nicotinic receptors, classically studied at the neuromuscular junction, have been subdivided into α, β, δ, γ, and ε subunits. So far, only α and β units are found in neurons, and there are nine α and four β subunit subtypes. The first immunolocalizations showed α and β subunits in DRG cells of all sizes (Swanson et al., 1987), and gel analyses or RT-PCR showed α1–7, α9, and β2–4 subunit transcripts in the trigeminal ganglion (Liu et al., 1998; Keiger and Walker, 2000). Almost all trigeminal primary sensory cells express α4 and α7 and 85% express β2 subunits (Liu et al., 1998). In another study, trigeminal ganglion cells express α3, α4, α5, β2, and β4 subunits, with α3β4 providing the most intense labeling, the two subunits being largely co-localized and found in 15% of large sensory cell profiles with α2β4 subunits also being found in considerable abundance (Flores et al., 1996). Developmentally, α3, α4, β2, and β4 receptors are found in considerable abundance in cells of the spinal cord and DRG early in development, but only α4 and β2 persist in the cord and α3 and β2 in the dorsal root ganglia after postnatal day 4 (Zoli et al., 1995). The subunits are thus more widespread in primary afferent neurons than previously thought, but ACh only excites a fraction of DRG cells in culture or an in vitro skin preparation (Steen and Reeh, 1993; Liu et al., 1998), although nicotinic ligands activate primary afferent fibers and cause pain, and nicotinic antagonists given intrathecally or locally relieve pain or pain-related behaviors (Steen and Reeh, 1993; Lawand et al., 1999). Thus, it remains to be determined if only ACh itself excites these cells or if other endogenous nicotinic ligands exist. Also, many DRG cells express acetylcholine esterase, perhaps to protect the DRG cell body from having nicotinic receptors stimulated.

Muscarinic Receptors

Autoradiographic localizations of muscarinic ligands show labeled dorsal root ganglion cells (Wamsley et al., 1981). Muscarinic receptors flow down the sciatic nerve, presumably in sensory fibers (Wamsley et al., 1981). Reductions in dorsal horn labeling are seen with dorsal rhizotomy, so these receptors are located on central primary afferent terminals (Gillberg and Wiksten, 1986; Gillberg and Askmark, 1991). There are five muscarinic receptor subtypes (M1–M5). Large numbers of small, IB4, M2 expressing cells are seen in rat thoracic dorsal root ganglia (Haberberger et al., 1999). Many medium-sized DRG cells in neonatal rats express the M2 subunit (Tata et al., 1999). The antinociceptive properties of intrathecal muscarinic ligands (Naguib and Yaksh, 1997) make this a likely area for further study.

Hormone Receptors

Androgen Receptors

No radioactive binding is seen in dorsal root ganglia (Sar and Stumpf, 1977), but approximately half of DRG cell nuclear profiles are immunolabeled for androgen receptors and approximately half of these express cytoplasmic CGRP (Keast and Gleeson, 1998). DRG cells express these receptors transiently during rat embryonic days 15 and 16 (Young and

Chang, 1998). Castration causes a marked diminution in nuclear androgen receptor immunostaining of DRG cells (Keast and Gleeson, 1998).

Estrogen Receptors

Estrogen is a gonadal steroid hormone with profound metabolic, developmental, and regenerative activities. Cloning has shown two forms of the estrogen receptor, α and β. A relatively surprising finding is that essentially every DRG cell from either normal or ovariectomized animals possessed estrogen receptor mRNA (Sohrabji et al., 1994). This finding was extended by the immunochemical observations that 87% of DRG cell profiles from normal animals and 96% from ovariectomized animals expressed the receptor (Yang, Y. et al., 1998). By contrast, however, another study found only moderate numbers of DRG cells labeled by in situ and immunologic localizations, and the *in situ* localizations were of large and medium-sized cells whereas the immunolocalizations were of small and medium-sized cells (Papka et al., 1997). In more recent work, immunolocalizations for both estrogen receptor-α and estrogen receptor-β showed predominantly small cells labeled, with some intermediate cells and almost no large cells. (Taleghany et al., 1999). In the same study, in situ localizations showed estrogen receptor-α localizations in predominantly small cells, although not all small cells, and in some supporting cells, and estrogen receptor-β localizations in all neurons (Taleghany et al., 1999). These investigators also showed that the mRNA expression of both estrogen receptor-α and estrogen receptor-β was more intense in L6–S1 than in L4–L5 levels (Taleghany et al., 1999). Thus, it is clear that large numbers of DRG cells express both the estrogen receptor-α and estrogen receptor-β and the mRNAs for making these proteins. It is not clear whether essentially all DRG neurons or only a subpopulation of mostly small neurons fall into these categories.

Gender Differences

Most of the above studies are on females, but one study did note that approximately half of dorsal root ganglion cells from males contained estrogen receptors, in contrast to essentially all neurons from females (Yang, Y. et al., 1998). By contrast, mRNAs for estrogen receptor expression are similar in male and female rats (Taleghany et al., 1999). Thus, estrogen receptors are found on significant numbers of rat DRG cells with a slight preponderance in females.

Development

During development, both estrogen receptor-α and estrogen receptor-β are found on dorsal root ganglion cells, and 17-B estradiol, which presumably acts through these receptors, ameliorates cell death in culture (Patrone et al., 1999).

Castration

Ovariectomy results in more cell profiles and/or more intense staining for these receptors or their mRNA (Sohrabji et al., 1994; Williams and Papka, 1996; Papka et al., 1997; Yang, Y. et al., 1998; Taleghany et al., 1999). Exogenous estrogen administration reverses the effects of ovariectomy (Yang, Y. et al., 1998) and upregulates estrogen receptor-β but downregulates estrogen receptor-α levels in rat DRG (Taleghany et al., 1999). Estrogen levels also correlate with changes in trophic factor expression. Thus, both p75 and trk A receptors, which are related to nerve growth factor, are upregulated during proestrus (high estrogen) and downregulated after ovariectomy (Sohrabji et al., 1994). The fact that changes in the normal estrous cycle correlate with changes in trophic factor and estrogen receptor expression in DRG cells suggests changes in nociceptive perception during the estrous cycle.

In summary, estrogen receptors and their mRNA are found in large numbers of neurons in both male and female dorsal root ganglia. There is some disagreement in that some report primarily small cells labeled and others essentially all cells labeled. Estrogen levels correlate inversely with the expression of the estrogen receptors or its mRNA and positively with expression of NGF receptors. Thus, although estrogen is not usually thought of in somatic sensation, these findings suggest that this hormone has important somatosensory functions and might play a role in pain perception during the estrous cycle.

Glucocorticoid Receptors

Glucocorticoid receptors are present in one third to one half of DRG cells expressing SP or CGRP (DeLeon et al., 1994).

Insulin Receptors

The IRR receptors coexist with TRK receptors and are intensely expressed by a subset of sensory neurons (Reinhardt et al., 1993; 1994; Sugimoto et al., 2002).

TRH Receptors

All DRG cells express thyroid NT_3Rs throughout life (Barakat-Walter et al., 1993; Walter-Barakat and Droz, 1995), particularly the TRα-common and TRβ-1 isoforms (Glauser and Walter, 1997), indicating that thyroid hormones could exert a powerful influence on primary afferent neurons.

Vitamin D Receptors

1,25-Dihydroxyvitamin D3 receptors are expressed by fetal DRG cells, and many coexpress calbindin (Johnson et al., 1996).

Peptide Receptors

Angiotensin Receptors

Binding of angiotensin II and angiotensin IV in the DRG shows these receptors presumably expressed by DRG cells (Oldfield et al., 1994; Moeller et al., 1995).

Bombesin Receptors

No bombesin receptor DRG cells are described.

Bradykinin Receptors

One way to show bradykinin receptors is to apply bradykinin in the presence of cobalt ions, which substitute for calcium, and then histochemically demonstrate the cobalt. When this was done, approximately 10% of small rat DRG neurons label (Nagy et al., 1993b). No B1 ligand binding was seen in cultured DRG cells even after the administration of interleukin-1β or prostaglandins, which are mediators of inflammation (Davis et al., 1996), but B1 and B2 receptors are expressed by DRG neurons in normal and B2 knock-out mice, although the administration of B1 agonists had no discernible effects (Seabrook et al., 1997; Ma et al., 2000a). B1 receptors are upregulated by DRG cells cultured at various times after unilateral sciatic nerve ligation, and the assumption is that this mirrors the *in vivo* condition (Petersen et al., 1998a). Small DRG and vagal nodose ganglion cells express the B2 receptor (Steranka et al., 1988; Lopes et al., 1993; Lopes and Couture, 1997; Krstew et al., 1998). Numerous B1-R cells co-localize with IB4 but not CGRP (Wotherspoon and Winter, 2000). B2 receptor ligands are strongly concentrated in the DRG (Murone et al., 1997). Since these receptors are found on DRG cells, they should have presynaptic influence in the cord.

Nerve growth factor but not BDNF or neurotrophin-3 upregulate bradykinin binding in cultured cells from normal animals but not in cells from animals lacking the p75 receptor (Petersen et al., 1998b). Peripheral axotomy prior to cell disassociation and culture upregulates bradykinin binding, which was ascribed to a *de novo* formation of B1 and an increase in B2 receptors (Petersen et al., 1998a), but cells whose axons had not been cut also had increased bradykinin binding, but only in neighboring ganglia for cells in more distant DRGs did not show an increase (Eckert et al., 1999).

Cannabinoid Receptors

Numerous DRG cells express CB1 receptors (Hohmann and Herkenham, 1999a; Salio et al., 2002). The cells are of all sizes, but most are medium-sized or large. CB1 receptors are transported distally in sensory fibers toward the periphery (Hohmann and Herkenham, 1999b).

CCK Receptors

Many DRG cells bind radioactive CCK, presumably representing predominantly the CCK_B subtype (Ghilardi et al., 1992), but only a few DRG cells express the mRNA for CCK_B receptors in the normal (Zhang, X. et al., 1994a). After peripheral axotomy or systemic resiniferatoxin there is a considerable increase in numbers of DRG cells that express CCK_BR mRNA (Zhang, X. et al., 1994a; Broberger et al., 2000; Hokfelt et al., 2001), but no increase in CCK_ARs after axotomy or inflammation (Broberger et al., 2001). The increase in CCK_BR mRNA cells, if it leads to an increase in spinal CCK, probably has a role in the ineffectiveness of opioids in neuropathic pain.

CGRP Receptors

CGRP receptors are not usually found on DRG cells, but a CGRP-receptor component protein is found on many DRG cells suggesting that this issue be revisited (Pokabla et al., 2002).

Endothelin (ET) Receptors

Endothelin-1 binding was reported principally over DRG satellite cells (Kar et al., 1991b), but further work shows ET_AR expressed by a subset of small peptidergic and non-peptidergic neurons whereas ET_BR was expressed by satellite cells and Schwann cells associated with unmyelinated axons (Pomonis et al., 2001). Thus endothelin could directly affect both neurons and supporting cells. Prenatal DRG cells express ET_BR suggesting a role for endothelin in development (Tsaur et al., 1997).

Galanin Receptors (GALR)

Monkey DRG cells bind radioactive galanin (Zhang et al., 1995c). GALR-1 and GALR-2 mRNA are expressed by approximately 25% of rat DRG cell profiles, with the GALR-1 population being small and medium-sized cells in one study (Xu et al., 1996a, b) and large cells in another (O'Donnell et al., 1999) whereas most GALR-2 cells are small (Shi et al., 1997; O'Donnell et al., 1999). The GAL-R1 neurons coexpress CGRP (Xu et al., 1996a, b). Rhizotomy and neonatal capsaicin increase central galanin binding so there is little evidence for galanin receptors on central primary afferent terminals (Kar and Quirion, 1995; Zhang et al., 1995c). Peripheral inflammation decreases GALR-1 and increases GALR-2 mRNAs in DRG and peripheral axotomy decreases both GALR-1 and GALR-2 mRNAs in the same ganglia (Xu et al., 1996a, b; Shi et al., 1997).

Neuropeptide Y Receptors (Y-Rs)

In rats, numerous, predominantly small, DRG cell profiles express or contain the mRNA for either Y1-Rs or Y2-Rs (Mantyh et al., 1994; Gustafson et al., 1997; Zhang, X. et al., 1997; Hökfelt et al., 1998; Migita et al., 2001), but monkeys have significantly fewer (Zhang, X. et al., 1995d). Y1-R receptors are not transmitted into the cord, as evidenced by lack of immunostaining in DRG cell processes and central elements of glomeruli and no change in spinal staining immunostaining (Zhang, X. et al., 1994b; Zhang et al., 1999a; Landry et al., 2000), although there are a few Y1-R afferent fibers, particularly large fibers, after peripheral axotomy, some of which contain CGRP (Landry et al., 2000).

Following peripheral axotomy, there is no change in binding of NPY ligands in the ganglion (Mantyh et al., 1994; Marchand et al., 1999), but Y1-R mRNA decreases in the ganglion after axotomy, even though many more large cells are labeled, indicating a shift in cells expressing this gene (Zhang et al., 1994c). Thus an increase in cells that project to laminae III–IV may balance a decrease in cells that project to laminae I–II (Zhang, X. et al., 1995c). In NPY-deficient mice, peripheral axotomy causes a decrease in both Y1-R and Y2-R mRNA in the DRG, but since the effect is the same in wild-type and knock-out mice, this implies that the expression and regulation of the receptor is not dependent on the presence of the ligand (Shi et al., 1998). One effect of peripheral axotomy is to induce expression of NPY and GAL in many DRG cells, and approximately half of the cells expressing these ligands also express Y1-R and Y2-Rs (Hökfelt et al., 1998). This suggests autoreception for many NPY DRG cells after axotomy (Mantyh et al., 1994; Hökfelt et al., 1998). Peripheral axotomy induces primarily distal transport of both Y1-Rs and Y2-Rs (Marchand et al., 1999; Brumovsky et al., 2002), with presumably less distribution to primary afferent terminals in the cord.

Peripheral inflammation leads to a considerable increase in the numbers of neuron profiles expressing Y1-R mRNA (Ji et al., 1994).

Both Y1-Rs and Y2-Rs are expressed early in embryonic life suggesting a role for these receptors in development of the DRG and other parts of the nervous system (Jazin et al., 1993; Naveilhan et al., 1998). Similarly there is a transient prenatal expression of Y1-R axons on vibrissal axons suggesting a developmental role for these receptors in this system (Ubink et al., 2001).

Neurotensin Receptors (NT-R)

Approximately 25% of DRG cell profiles express neurotensin mRNA, these cells being separate from other NPY-R, SP, and CGRP cells, and peripheral axotomy reduces the number of cells expressing this mRNA in DRG cells as well as similar cells and immunostaining in the dorsal horn (Zhang, X. et al., 1995b).

Neurokinin-1 Receptors (NKR1s)

DRG and trigeminal ganglion cells do not bind SP ligands (Mantyh et al., 1984a; Mantyh and Hunt, 1985b), transection of dorsal roots leads to an increase in SP binding in the dorsal horn (Massari et al., 1985; Brass et al., 1986; Helke et al., 1986; Yashpal et al., 1991a; Croul et al., 1995), NK-1Rs do not pile up on the proximal side of a nerve ligation (Brown et al., 1995), neonatal capsaicin has little effect on SP binding in the dorsal horn or the periphery (Mantyh and Hunt, 1985b; O'Flynn et al., 1989), and NK-1 mRNA cannot be found in dorsal root ganglia (McCarson, 1999). This speaks against NK-1Rs being expressed by primary afferent neurons. On the other hand, peptidergic DRG cells express the NK-I receptor (Li and Zhao, 1998), peripheral sensory axons express NK-1Rs (Carlton et al., 1996a; 1998a), mRNA for NK-1Rs is found in mouse dorsal root ganglia (Andoh et al., 1996), and NK1 receptor binding and immunocytochemical staining decrease after neonatal or perineural capsaicin (Rössler et al., 1993; Sann et al., 1995b; Ohtori et al., 2000). We cannot reconcile these differences.

NPFF Receptors

To our knowledge, no studies have shown NPFF receptors on DRG cells.

Opioid Receptors

MORs, DORs, and KORs

The main opioid receptors are MORs, DORs, and KORS. DRG cells expressing these receptors are small with occasional intermediate-sized and rare large DRG cells (Ninkovic et al., 1982; Dado et al., 1993; Schäfer, M. K. H. et al., 1994; Arvidsson et al., 1995a, b, c; Ji et al., 1995b; Li, J. L. et al., 1998a; Zhang et al., 1998c; Zhang, X. et al., 1998b), although mostly large cells reportedly label for opioid receptor mRNA (Mansour et al., 1994a, b; 1995a, b; Wang and Wessendorf, 2001). Difficulties with the latter conclusion are DRG cell co-localizations are consistent with small cell input, and MOR cells do not stain with RT97, an indicator of large light DRG cells (Arvidsson et al., 1995). Thus, until further evidence becomes available, we conclude that it is predominantly the small DRG cell population that expresses opioid receptors.

In an early binding study with a generic opioid ligand, only 8% of DRG cell profiles were labeled for opioid receptors (Ninkovic et al., 1982). This contrasts with the approximately 55% of DRG cells that express mRNA for MORs, 20% that express mRNA for DORs and 18% that express mRNA for KORs (Minami et al., 1995; Minami and Satoh, 1995). Later mRNA assays had a much lower proportion for MOR mRNAs, however (Hökfelt et al., 1998). In immunocytochemical studies, 20% of DRG cell profiles expressed MORs, 13.5% expressed DORs and 9% expressed KORs (Ji et al., 1995b). In later studies, however, the numbers were higher in that approximately 50% of DRG cell profile expressed MORs (Zhang et al., 1998c) and 40% expressed DORs (Zhang, X. et al., 1998b). These differences are ascribed to technical improvements and quality of different antibodies (Hökfelt et al., 1998), and it would seem that many small DRG cells express MORs and DORs.

Co-localizations

SP and CGRP are commonly found in cells expressing MORs and DORs (Ninkovic and Hunt, 1985; Dado et al., 1993; Elde et al., 1995; Lu et al., 1997; Li, J. L. et al., 1998a; Zhang, X. et al., 1998b; Zhang et al., 1998c). Estimates of the proportions of peptide-containing cells that also express MORs or DORs range from approximately 30% to almost all. These data indicate that opioid receptors are primarily associated with the small DRG cell population and also that activation of these receptors should influence the release of these important peptides. The opioid receptors do not co-localize with RT-97, a marker for large light cells, which strengthens the notion that opioid receptors are on small DRG cells (Arvidsson et al., 1995a, b, c). Many DRG cells coexpress two or all three opioid receptor subtypes (Ji et al., 1995b).

Peripheral Axotomy

MOR immunostaining and mRNA are considerably reduced in DRG cells ipsilateral to a sciatic nerve transection (Li et al., 1996a; Zhang et al., 1998c). For DOR staining, however, there is a reduction in stain intensity, but DOR cell profile numbers remain the same (Zhang, X. et al., 1998b). Accordingly, there does seem to be a reduction in both immunostaining and numbers of MOR cells, but for DORs the results are not as clear. The clear reduction for MORs may be one reason why morphine is relatively ineffective in neuropathic pain, and morphine has no effect on normal or axotomized DRG cell MORs (Li, J. L. et al., 1998a).

Inflammation

Numbers of MOR-expressing DRG cells are increased and numbers of DOR and KOR cells are decreased after unilateral Freund's adjuvant treatment (Ji et al., 1995b).

Development

More DRG cell profiles express MORs and DORs and the cells are proportionately larger in newborn than adolescent rats. This may play a role in opioid effects on innocuous reflexes in newborns as compared to the lack of effect in adults (Beland and Fitzgerald, 2001).

Periphery

Immunocytochemical localizations (Stein et al., 1990; Coggeshall et al., 1997; Wenk and Honda, 1999), ligand binding by dorsal roots (Fields et al., 1980), and receptor accumulation on either side of dorsal root (Zarbin et al., 1990) or peripheral nerve ligations (Young, W. S. et al., 1980; Laduron, 1984; Hassan et al., 1993; Ji et al., 1995b) show opioid receptors on peripheral sensory axons. Small DRG cells express these receptors (see above), capsaicin reduces the receptor binding (Laduron, 1984), unmyelinated axons are labeled in cutaneous sensory nerves (Coggeshall et al., 1997), and the opioid receptors co-localize with CGRP, a fine afferent marker and not with RT-97 and TH, myelinated and sympathetic markers respectively (Wenk and Honda, 1999), showing these receptors primarily on fine sensory axons. DOR sensory axons innervate cornea, hair follicles, glandular tissue, and blood vessels (Wenk and Honda, 1999). The demonstration of opioid receptors on peripheral sensory axons indicates that integration of afferent input could take place in the periphery (Coggeshall and Carlton, 1997). There are major changes in opioids and their receptors peripherally in non-neural tissues during inflammation (Stein, 1995).

Endomorphin Receptors

The endomorphin receptors are MORs.

Endorphin Receptors

β-endorphin binds preferentially to MORs and also to DORs, which are present on DRG cells and their peripheral processes and increase in inflammation. There is β-endorphin binding by sensory axons in peripheral nerves, with increased transport after inflammation (Hassan et al., 1993).

Opioid-like Receptor 1 (ORL1)

Homology cloning yielded a receptor with considerable sequence homology to known opioid receptors. It was referred to as an "orphan" receptor (ORL1) because classic opioid ligands do not bind with high affinity to this receptor and there was no known endogenous ligand. Recently, however, an endogenous compound with considerable sequence homology to prodynorphin has been found to bind to this receptor. The compound is referred to as orphaninFQ or nociceptin. Numerous DRG cells, particularly large ones, express ORL1 mRNA even though nociceptin binding could be demonstrated in one study but not in another (Neal et al., 1999; Pettersson et al., 2002), and only scattered DRG cells are weakly labeled immunochemically (Monteillet-Agius et al., 1998).

Sigma Receptors

Sigma receptor ligands bind strongly to DRG cells (Gundlach et al., 1986).

Oxytocin and Vasopressin Receptors

To our knowledge, DRG cells have not been examined for the presence of OXY and VP receptors.

Somatostatin Receptors

A significant number of medium-sized DRG cells and 11% of unmyelinated cutaneous sensory axons express the $sst2_a$ receptor (Schulz et al., 1998a; Carlton et al., 2001c). Interestingly, dorsal rhizotomy has no effect on spinal $sst2_a$ staining which may imply that DRG cells express the receptor but transport it only to the periphery. Every DRG cell is reported to express the sst_3 receptor (Señaris et al., 1995), which if they are active, implies that every DRG cell would be influenced by somatostatin.

Vasoactive Intestinal Polypeptide (VIP) Receptors

Although considerable evidence for VIP receptors on sensory cells exists (Chaudhary and Baumann, 2002), we are not aware of morphologic studies demonstrating the receptors on DRG cells in vivo.

Purinergic Receptors or Purinoceptors

Adenosine 5'-triphosphate (ATP) and its breakdown product adenosine are neurotransmitters or neuromodulators (Burnstock and Wood, 1996). Receptors for these ligands are divided into those that respond to adenosine (A receptors, originally P1 receptors) and those that respond to ATP (P2 receptors).

Adenosine Receptors

Adenosine acts through adenosine (A) receptors, which are at present divided into three subtypes with a splice variant (A1, A2a, A2b, and A3) (Kaelin-Lang et al., 1998). A significant number of trigeminal ganglion neurons are immunostained for the A1 receptor (Schindler et al., 2001), and 5–10% of large DRG cell profiles in rats contain mRNA for A2a receptors (Kaelin-Lang et al., 1998).

ATP (P) Receptors

ATP (P) receptors are divided into ionotropic (P2X) and metabotropic (P2Y) categories with seven P2X and eight P2Y subtypes (Burnstock and Wood, 1996; Burnstock, 2000).

P2X Purinoceptors

DRG cells express $P2X_{1-6}$ receptors (Collo et al., 1996; Xiang et al., 1998). $P2X_2$ receptors also are present on satellite cells (Vulchanova et al., 1997). Attention focuses on the $P2X_3$ subunit because it is found in many DRG cells, particularly small ones, and only DRG cells express the mRNA for these receptors (Chen et al., 1995; Lewis et al., 1995; Bradbury et al., 1998; Llewellyn-Smith and Burnstock, 1998; Vulchanova et al., 1998; Xiang et al., 1998; Guo, A. et al., 1999; Novakovic et al., 1999). Evidence that $P2X_3$ DRG cells are nociceptive is they: (1) are small (see above references); (2) co-localize with peripherin, a marker for small dark cells (Chen et al., 1995); (3) bind the lectin IB4, a label for nociceptive cells whose central processes end in lamina IIi (Bradbury et al., 1998; Vulchanova et al., 1998; Guo, A. et al., 1999); and (4) $P2X_3$ mRNA and numbers of $P2X_3$ cells are considerably reduced after capsaicin treatment in newborns (Chen et al., 1995; Vulchanova et al., 1998). In addition, small DRG cells from tooth pulp, presumably nociceptors, express $P2X_3$ receptors, whereas large cells from stretch receptors, presumably non-nociceptors, do not (Cook et al., 1997). Physiological evidence, not considered here, is in accord (Burnstock and Wood, 1996). Accordingly the evidence is quite strong that the majority of $P2X_3$-expressing DRG cells are nociceptors.

There is little co-localization with somatostatin, calcitonin gene-related peptide, or substance P, providing evidence that $P2X_3$ receptors identify primarily non-peptidergic nociceptors (Bradbury et al., 1998; Vulchanova et al., 1998). Coexpression of $P2X_2$ and $P2X_3$ occurs and is necessary for the ATP-activated currents that characterize DRG cells (Lewis et al., 1995; Vulchanova et al., 1996; Petruska et al., 2000). $P2X_1$ receptors co-localize with both peptides and IB4, and there is a subgroup of $P2X_2$/$P2X_3$ neurons that do not co-localize with IB4 but do with NF200 (large light cells) (Petruska et al., 2000). Axons in muscle nerves contain significantly lower proportions of $P2X_3$ receptors than cutaneous or visceral nerves (Bradbury et al., 1998; Vulchanova et al., 1998). $P2X_3$ receptors are completely absent from DRG cells in $P2X_3$ knock-out mice, but there is no change in $P2X_2$ immunostaining in this condition compared to wild-type animals (Zhong et al., 2001).

Sciatic nerve section reduces the number of $P2X_3$ labeled cells and staining in the dorsal horn (Bradbury et al., 1998) whereas chronic constriction of the sciatic nerve increases numbers of labeled cells and dorsal horn staining (Novakovic et al., 1999), and either transection or constriction of the inferior alveolar nerve increases numbers of cells expressing the $P2X_3$ receptor (Eriksson et al., 1998). A related finding is that DRG cells with transected peripheral axons decrease production of $P2X_3$ mRNA whereas undamaged neurons in the same ganglion increase expression of this receptor (Tsuzuki et al., 2001), implicating undamaged neighboring neurons as the major source of the $P2X_3$ increase.

Use of antisense oliogonucleotides reduces both pain-related behavior and $P2X_3$ in sensory neurons (Barclay et al., 2002). Administration of GDNF reverses the $P2X_3$ changes after sciatic nerve section and increases numbers of $P2X_3$ cell profiles as does NGF (Bradbury et al., 1998; Ramer et al., 2001). It is not clear whether the two trophic factors are acting on the same or different populations of $P2X_3$ cells. $P2X_3$-deficient mice lose all staining for $P2X_3$ in dorsal root ganglia and spinal cord, and these mice also lose rapidly desensitizing ATP currents ion DRG cells and develop severe bladder hyporeflexia, indicating a possible role for these receptors in a clinically important situation (Cockayne et al., 2000).

$P2X_3$ receptors are expressed early in embryonic DRG cells (Cheung and Burnstock, 2002).

P2Y Purinoceptors

P2Y1 receptors are expressed primarily by large DRG cells, and on the basis of several lines of evidence the conclusion is that these cells are involved in the transduction of light touch (Nakamura and Strittmatter, 1996).

Vanilloid Receptors (VRs)

VRs respond to capsaicin, temperature (heat), and protons (acid). DRG cells bind radioactive resiniferatoxin (RTX), a potent capsaicin analog, thus showing VRs on these cells

(Winter et al., 1993; Szallasi et al., 1994a, b; Szallasi, 1995; Szallasi and Blumberg, 1999), and capsaicin activation of VRs in the presence of cobalt ions results in 10% of small DRG neuron profiles labeling (Nagy et al., 1993b). Immunocytochemical and mRNA analyses show VR1 in small and some medium-sized DRG cells (Helliwell et al., 1998; Guo et al., 1999; Michael and Priestley, 1999), and in somewhat larger trigeminal ganglion cells (Fig. 4.5; Ichikawa and Sugimoto, 2000; 2001). Many VR cells co-localize IB4 (Tominaga et al., 1998; Guo, A. et al., 1999; Michael and Priestley, 1999), as well as SP or CGRP (Tominaga et al., 1998; Michael and Priestley, 1999) or P2X3 (Guo, A. et al., 1999; Ichikawa and Sugimoto, 2001), although with another antibody, VR1 cells co-localize minimally with SP and CGRP (Guo et al., 1999). VR1 cells are not 200-kDa neurofilament-positive, so they are not large light cells (Michael and Priestley, 1999). Because of their location on DRG cells, these receptors presumably exert presynaptic effects in the dorsal horn.

The VRs are transported in dorsal root and peripheral sensory axons as evidenced by piling up of resiniferatoxin binding or VR1 immunostaining proximal to a nerve ligature and loss of staining in the dorsal horn after dorsal rhizotomy (Szallasi, 1995; Guo, A. et al., 1999), and mRNA for VR1 is increased proximal to a ligature after peripheral inflammation (Tohda et al., 2001). Treatment with capsaicin or resiniferatoxin reduces resiniferatoxin binding and VR1 mRNA in the ganglion, which may represent desensitization or destruction of neurons (Winter et al., 1993; Szallasi et al., 1994a; Farkas-Szallasi et al., 1996; Mezey et al., 2000). Peripheral nerve ligation has no effect on central resiniferatoxin binding (Farkas-Szallasi et al., 1996), but reduces mRNA for VR1 in the ganglion, suggesting

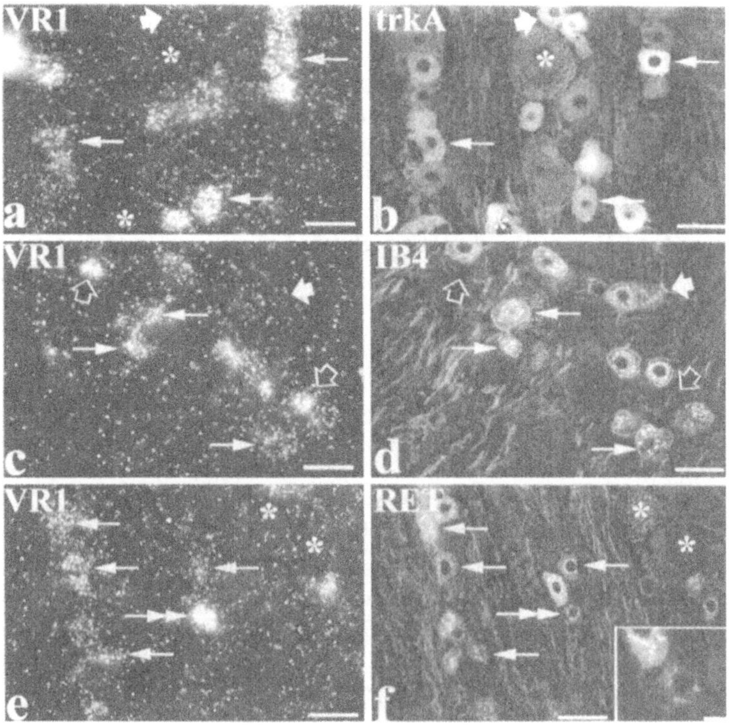

FIGURE 4.5. Cells expressing VR1 mRNA showing co-localizations with trkA, IB4, and retinin. (From Michael and Priestley, 1999.)

undiscovered types of VRs (Michael and Priestley, 1999). DRG cells whose peripheral processes are transected have reduced VR1 expression whereas those with untransected processes have an increase, suggesting that increased VR1 expression is in undamaged DRG cells (Hudson et al., 2001). Intraplantar carrageenan reduces VR1 mRNA in the ganglion and increases it in cord and periphery, suggesting that inflammation increases RNA transport rather than the receptor, implying that the receptors are made locally rather than being transported from the cell body (Tohda et al., 2001). The increases may bear on the increased peripheral sensitivity that accompanies inflammation.

CONCLUSIONS

1. Small dark DRG cells express many different neuroactive substances and receptors. These are almost always considered in relation to nociception, but it should be remembered that there are many small dark DRG cells that are not nociceptive. Reasons for this extraordinary concentration of neuroactive substances and receptors in small DRG cells are not clear, but they must go far beyond classic neural signaling.
2. Numerous functionally meaningful α isoforms of the voltage-gated sodium channel are expressed by DRG cells. Of particular interest are the $Na_V1.3$ isoforms for they are expressed by DRG cells after peripheral axotomy and thus may underlie the development of ectopic activity that sets up central sensitization, a major factor in neuropathic pain.
3. Non-voltage-gated sodium channels are expressed by cells and axons that have an important role in mechano-reception and these channels may also be involved in pH and thermal reception.
4. The fluoride-resistant acid phosphatase FRAP is found in small DRG cells and is used as a marker for the "non-peptide" population of small dark primary afferent neurons. There is controversy about whether it localizes with peptides, however.
5. CA is found in large sensory neurons involved with proprio- and mechanosensation.
6. The enzymes NOS and NADPHd indicate NO neurons. These increase after peripheral axotomy and inflammation.
7. CGRP is expressed by more DRG cells than other peptides, and as far as the cord is concerned, it is present essentially only in primary afferent axons, so it is used as a marker for peptidergic primary afferents in the cord and periphery. Release of CGRP, along with SP, is a major initiator of neurogenic inflammation.
8. GAL, NPY, and VIP are markedly upregulated in DRG cells after peripheral axotomy. Thus, they may play a role in neuropathic pain.
9. Unlike other peptides expressed by DRG cells, somatostatin is not under the control of NGF but rather of GDNF.
10. SP is expressed by many DRG cells and peripheral axons. It is a major factor underlying neurogenic inflammation.
11. The calcium-binding proteins calbindin and calretinin are concentrated primarily in propriospinal and exteroceptive DRG cells.
12. Expression of α-1 and α-2 adrenergic receptors by DRG cells increases after nerve lesions suggesting that these receptors may underlie aspects of sympathetically maintained neuropathic pain.

13. Numerous 5-HT receptor subtypes are found on DRG cells and are combined in different ways suggesting abundant possibilities for integration.
14. Essentially every DRG cell expresses glutamate NMDA receptors, and numerous DRG cells express GABA receptors, so these amino acids can presumably control afferent activity at the level of the cell body.
15. Essentially every DRG cell expresses estrogen receptors, raising issues about the effects of estrogen on primary afferent transmission and suggesting that primary afferent, particularly nociceptive input might be differentially affected during the estrous cycle.
16. Numerous DRG cells express MORs and DORs, which are thus positioned to exert presynaptic control of primary afferent input when transported to the cord.
17. Numerous DRG cells express $P2X_3$ receptors, and the majority of these cells are nociceptors.
18. VRs respond to capsaicin, temperature (heat), and protons (acid), and are found on significant numbers of DRG cells.

Chapter 5

Structure of the Dorsal Horn

This chapter will take as its primary mission the description of laminae I–VI (Fig. 5.1), first presented by Rexed (1952; 1954), who related this organizational pattern to earlier work. To reduce the size of this chapter, we will focus on primary afferent input and only briefly treat the equally important descending input and consider only mammalian studies.

LAMINA I

Lamina I is the classical marginal zone, "a thin veil of gray substance, forming the dorsal-most part of the spinal gray matter" (Rexed, 1952). Bundles of myelinated axons penetrate this lamina, giving it a reticular appearance, and sometimes obscuring the boundary between it and overlying white matter. Early investigators regarded this lamina as the superficial part of the substantia gelatinosa (Ramón y Cajal, 1909; Ranson, 1913), but it is now clear that lamina I is a separate structural and functional entity (Cervero and Iggo, 1980).

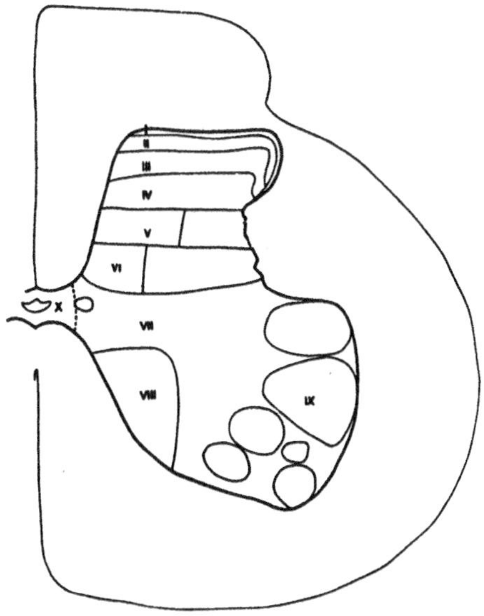

FIGURE 5.1. Schematic indicating the laminae in the L7 segment of cat spinal cord. Laminae I–VI make up the dorsal horn (From Rexed, 1952; reprinted with permission.)

The classic features of this region are large horizontal neurons (the marginal cells of Waldeyer) and a plexus of numerous horizontally arranged fine axons and dendrites that contrast with the tightly packed, radially arranged, small neurons and longitudinally oriented processes in the underlying gelatinosa.

Cell Types

Classic Types: Marginal and Smaller Cells

Neuronal cell bodies in lamina I range widely in size. The largest, the marginal (Waldeyer, 1888) or giant cells (Fig. 5.2) are prominent but sparse. They overlie the gelatinosa with horizontal dendrites (Clarke, 1859; Waldeyer, 1888; Ramón y Cajal, 1909; Scheibel and Scheibel, 1968), so their dendritic fields were depicted as discoids, but reconstructions show that dendritic trees of many lamina I cells are elongated rostrocaudally (Light et al., 1979; Woolf and Fitzgerald, 1983; Lima and Coimbra, 1986, 1989). The giant cells presumably do not form a separate population, because they fit into the same categories as smaller lamina I cells (Molony et al., 1981; Lima and Coimbra, 1986). Marginal cells have been subdivided into projection (P) and association (A) neurons to reflect the axonal destinations of these cells (Narotzky and Kerr, 1978). Some Waldeyer or giant lamina I cells express NK1 receptors and are gephryin-rich (Brown et al., 1995; Puskar et al., 2001). Another classification that applies to these cells is that there are two types of projection neurons, type 1 whose axons proceed out of the medullary dorsal horn with few collaterals, and type 2 whose axons make extensive local collaterals before projecting farther, and these two types have distinctive physiologic characteristics (Li et al., 2000a, b).

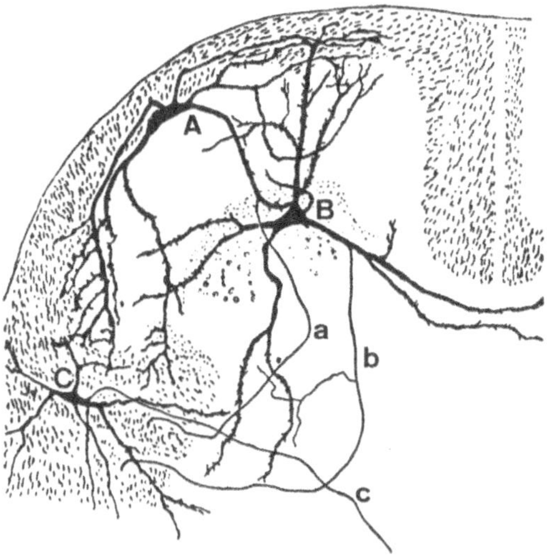

FIGURE 5.2. Drawing of a Golgi-stained section of spinal cord. Cell A is a large marginal cell. Note that majority of the dendrites of this cell pass over the surface of the dorsal horn, but that some pass ventrally into the grey matter. The axon (a) of cell A arises from a ventral dendrite and passes into the lateral funiculus. Cell B is a large cell in the center of the dorsal horn (approximately lamina IV). The dendrites pass medially, laterally, and dorsally. The dorsal dendrites are characteristic of antenna-type neurons. The axon of this cell goes to the lateral funiculus. Cell C does not belong to the dorsal horn but appears to be in the lateral spinal nucleus. (From Ramón y Cajal, 1909.)

Modern Types

Within medullary lamina I, two types of pyramidal cells, with and without numerous dendritic spines, and two types of multipolar neurons, with compact and loose dendritic arbors, all of whose dendrites remain superficial, have been described (Gobel, 1978a). A more complicated subdivision with four groups of neurons, three of which are split into A and B subtypes, is also described (Beal et al., 1981). The subdivision generally accepted today follows that of Gobel except that there are either three or four subtypes: fusiform, multipolar, and pyramidal (Zhang, E. T. et al., 1996; Zhang, E. T. and Craig, 1997) or fusiform, multipolar, pyramidal, and flattened (Fig. 5.3; Lima and Coimbra, 1983, 1986; Galhardo and Lima, 1999). These cell types are clearly seen in slice preparations where rotations allow precise visualizations of the cells (Chéry et al., 2000). Different proportions of these cell types project to different regions (e.g., medulla, parabrachial region, nucleus of the tractus solitarius, thalamus, etc. (Coimbra and Lima, 1989; Lima, 1998; Yu, X. H. et al., 1999), and different proportions express various chemical markers (Chapter 6). No particular cell type fitted with functional categories in earlier studies (Light et al., 1979; Woolf and Fitzgerald, 1983; Steedman et al., 1985; Hylden et al., 1986a; Rethelyi et al., 1989b), and the major physiological correlation was receptive field type with laminar location of dendrites but not cell body or shape of dendritic tree (Light et al., 1979; Bennett et al., 1981). A recent important correlation, however, is pyramidal cells are thermoceptive-specific, fusiform cells nociceptive-specific, and multipolar cells are either polymodal, responsive to heat, pinch, and cold, or nociceptive-specific (Han et al., 1998). NK1 receptors

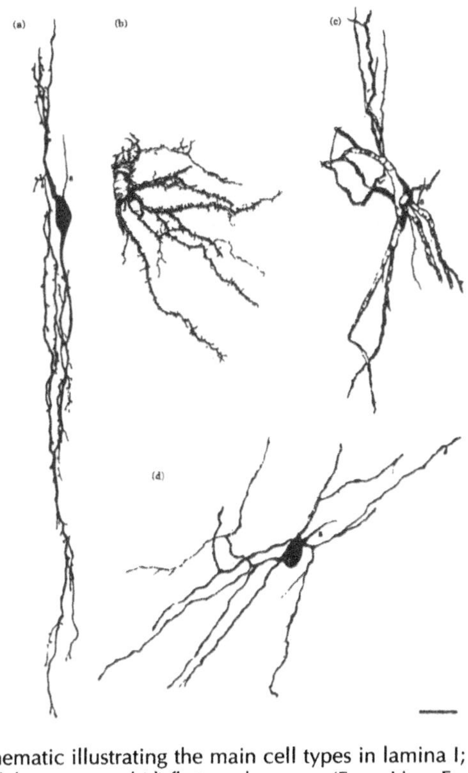

FIGURE 5.3. A schematic illustrating the main cell types in lamina I; (a) Fusiform neuron, (b) multipolar neuron, (c) pyramidal neuron, and (d) flattened neuron. (From Lima Eur J. Pain, 1998.)

are expressed by fusiform and multipolar cells carrying nociceptive information and not by pyramidal cells (Yu et al., 1999), but others find NK1 receptor expression correlated with cell function and not with cell morphology (Todd et al., 2002).

In summary, the neurons of lamina I have classically been divided into large marginal cells characterized by wide-ranging horizontal dendrites and smaller neurons. These categories are simplistic, however, and lamina I cells are now placed into pyramidal, fusiform, multipolar, and sometimes flattened categories. There are correlations of these categories with the destinations of projection neurons and their chemical content (Chapters 6 and 7), but the most precise correlations are that functional types of afferent input are restricted to certain morphologic categories. Progress in this regard will give further insight into lamina I organization.

Primary Afferent Input into Lamina I

Large primary afferent fibers segregate by size either in the proximal dorsal root or as they enter the cord, with the large fibers entering the dorsal columns and fine fibers the tract of Lissauer, and both types then bifurcate into a cranial and a caudal branch, with the input to gray matter of the cord coming from collaterals of these branches (Ramón y Cajal, 1909; Brown, 1981; Fyffe, 1984; Rethelyi, 1984; Fitzgerald, 1989). This is a very useful construct with a few caveats. For example, unmyelinated primary afferents are found in the dorsal columns (Langford and Coggeshall, 1981a; Chung and Coggeshall, 1983; Chung et al., 1985; 1987; Patterson et al., 1989; 1990) and dorsolateral funiculus (Chung et al., 1987). In addition, many hair follicle afferents do not bifurcate (Fig. 10.20; Brown et al., 1977c; Brown, 1981).

The Marginal Plexus

Large numbers of fine primary afferent and propriospinal fibers enter lamina I from the tract of Lissauer and surrounding white matter and form a horizontal plexus, the marginal plexus (Ramón y Cajal 1909; Szentagothai, 1964a; Scheibel and Scheibel, 1968). Quantitative evaluation shows approximately 50% of the synapses in lamina I of the rat dorsal horn are lost following dorsal rhizotomy and thus presumably arise from primary afferents (McNeill et al., 1988; Chung et al., 1989a). The remaining synapses presumably arise from propriospinal and possibly descending fibers. A transverse plexus superficial to the longitudinal plexus is described that is thought to arise from Aδ fibers (Beal and Bicknell, 1981). A complicated longitudinal plexus is also described, and it was suggested that fibers in it less than 1 μm in diameter are C fibers, whereas larger fibers are mostly Aδ fibers (Cruz et al., 1987). Functional correlates of these plexuses are not known.

Large versus Small Fibers

The primary afferent component of the fine primary afferent fibers that enter lamina I from the tract of Lissauer and surrounding white matter make up much of the input into the marginal plexus (Ramón y Cajal, 1909), as confirmed by retrograde transport studies after selective surgery or by reconstructions of individually labeled, functionally identified axons (LaMotte, 1977; Light and Perl, 1977; 1979a) and labeling of lamina I fibers and terminals by wheat germ agglutinin conjugated horseradish peroxidase (WGA-HRP) (Robertson and Grant, 1985; Plenderleith et al., 1988). These fibers arise from Aδ and C fibers (below), and there is no evidence that collaterals from large primary afferent fibers enter lamina I (Brown, 1981).

Aδ versus C Fibers

On the basis of differential primary afferent degeneration in lamina I as opposed to lamina II and finding of many ultrafine primary afferent fibers and terminals from individually labeled C fibers in lamina I, it is clear that much primary afferent input into lamina I is from C fibers (Gobel and Binck, 1977; Gobel et al., 1981; Sugiura et al., 1988; 1989). But relatively large axons also give rise to terminals in lamina I (Beal and Fox, 1976; Beal and Bicknell, 1981), afferent endings in lamina I degenerate more slowly than in lamina II (LaMotte, 1977), and functional studies and individual filling of high-threshold cutaneous mechanoreceptors conducting in the Aδ range (Light and Perl, 1979b; Hayashi, 1985) show a significant Aδ input as well. Thus, lamina I is a major Aδ and C fiber primary afferent termination site (Fyffe, 1984; Mense and Prabhakar, 1986).

Propriospinal versus Primary Afferent Fibers

Pioneers in the study of the tract of Lissauer and the marginal layer of the dorsal horn regarded most of the fine fibers as primary afferents (Lissauer, 1886; Ramón y Cajal, 1909; Ranson, 1913), but lack of change following dorsal rhizotomy or spinal isolation suggested that most of the fibers were propriospinal, particularly laterally in the tract (Earle, 1952; Szentagothai, 1964), and thus this part of the tract was cut to remove propriospinal fibers (Kirk and Denny-Brown, 1970; Narotzky and Kerr, 1978). Fine primary afferent fibers are distributed throughout the tract, however (Chung et al., 1979; Light and Perl, 1979a; Rethelyi et al., 1979; Coggeshall et al., 1981; Chung and Coggeshall, 1982), so a clean separation between propriospinal and primary afferents cannot be achieved by cutting or stimulating a specific part of the tract of Lissauer. Propriospinal connections are shown by localized injections of a retrograde marker into specific parts of laminae I–IV (Petko and Antal, 2000).

Cutaneous, Muscular, and Visceral Afferent Input

There are laminar differences in the termination patterns of the primary afferents in different types of peripheral nerves (Fyffe and Light, 1984; McMahon and Wall, 1985; Fitzgerald, 1989). Visceral nerve projections are reported to occur particularly to laminae I and V (de Groat et al., 1978; Morgan et al., 1981; Nadelhaft et al., 1983; Kuo and de Groat 1985; Cervero and Connell, 1984a, b). By contrast, terminals of cutaneous afferent fibers, although also in lamina I, are prominent in laminae II–IV (Grant and Ygge, 1981; Abrahams et al., 1984b; Cervero and Connell, 1984b; Molander and Grant, 1986). For primary afferents in muscle nerves, the story is more complicated. Early studies in which muscles were injected with tracer show terminals in laminae I–V (Brushart and Mesulam,

1980; Kalia et al., 1981); others report terminals mainly in lamina I (Craig and Mense, 1983; Nyberg and Blomqvist, 1984; Abrahams and Swett, 1986; Mense and Prabhakar, 1986; Mense and Craig, 1988). Others find essentially no or very sparse labeling in lamina I (Ammann et al., 1983; Abrahams et al., 1984; Bakker et al., 1984; Molander and Grant, 1987). Some investigators believe that the intramuscular injection results are the result of spread of label, but lamina I terminals were found after injecting individual fibers (Mense and Prabhakar, 1986). Thus, it seems likely that in some instances muscle afferents terminate in lamina I, but whether this occurs for all muscle nerves in all species is unclear. This work is extended by using various morphologic markers to indicate different types of afferent input (Chapters 4 and 6), and noteworthy is the finding that cross-anastomosing muscle and cutaneous nerves resulted in chemical changes indicating that different peripheral territories change the expression of markers in the sensory cells that project in those nerves (McMahon and Moore, 1988; Lewin and McMahon, 1993).

In summary, primary afferent input into lamina I derives from the marginal plexus which in turn comes from fine fibers in the tract of Lissauer and surrounding spinal white matter. Both Aδ and C fibers provide the primary afferent input to lamina I, and propriospinal fibers also make a major contribution. There are laminar differences in the distribution of terminals from various types of peripheral nerves.

Neuropil Organization

The majority of axons in lamina I are horizontal, parallel to the surface of the cord (Ralston, 1965; 1968a; 1979; Kerr, 1966; 1970a, b; Gobel et al., 1977; Narotzky and Kerr, 1978). These fibers give rise to terminals in lamina I and the terminals can be categorized as containing: (1) round clear vesicles intermixed with only a few large granular vesicles; (2) large granular vesicles intermixed with round clear vesicles; (3) flattened vesicles; and (4) large numbers of small clear vesicles as the central elements of glomeruli (Kerr 1970a, b; Narotzky and Kerr, 1978; Ralston 1979). Round vesicle terminals predominate in lamina I, but large granular vesicle terminals are also common (Ralston, 1979; Snyder, 1982). Flat vesicle terminals are relatively rare except in the peripheral parts of glomeruli, so the round vesicle terminals predominate in lamina I and give way to flat vesicle terminals in deeper laminae (Ralston, 1979; Snyder, 1982). This presumably indicates a predominance of excitation in lamina I changing to a predominance of inhibition in deeper laminae (Ralston, 1979). One group reports that primary afferent input is on distal dendrites and propriospinal input proximal on cell bodies and proximal dendrites, of projection neurons in lamina I (Kerr, 1970a, b; 1975a, b; Narotzky and Kerr, 1978). These findings could not be confirmed (Chung et al., 1979; Light and Perl, 1979b; Ralston, 1979; Gobel et al., 1981).

Glomeruli (complex synaptic arrays, large synaptic complexes) are the most striking synaptic arrangements in laminae I–III. They consist of a prominent central terminal surrounded by dendrites and peripheral terminals, the whole being surrounded by glial processes. Glomeruli will be described in the lamina II section. There seem to be species differences in the lamina I glomeruli. Thus, glomeruli are not reported in lamina I of rat dorsal horn or cat trigeminal nucleus caudalis (Kerr, 1970a; Ribeiro-da-Silva and Coimbra, 1982), but they are found in monkey and cat lamina I (Ralston, 1968a, b; 1979; Gobel et al., 1981). In the monkey many central terminals are characterized by large dense core vesicles, but otherwise they are relatively simple in that they do not seem to have axoaxonic endings on them (Ralston, 1979). By contrast, the glomeruli are very complex in lamina I of the cat, with ample possibilities for presynaptic interaction (Gobel, 1974; 1976). The central terminals of glomeruli are rare compared to the other terminal types.

Nonglomerular primary afferent presynaptic terminals in lamina I are primarily axodendritic (Fig. 5.14), but occasional axosomatic and axoaxonal synapses are seen (Ralston, 1968a; 1979; Ribeiro-da-Silva and Coimbra, 1982). Reports of large numbers of primary afferent synapses on lamina I somata (Kerr, 1970b; Narotzky and Kerr, 1978) are not accepted (Ralston and Ralston, 1979; Gobel et al., 1981).

LAMINA II (THE SUBSTANTIA GELATINOSA)

The "substantia gelatinosa" is a term coined by Rolando (Cervero and Iggo, 1980) for a grossly distinguishable layer of the superficial dorsal horn. The gelatinous appearance of this area is due to the concentration of small neurons and their processes plus a striking absence of myelinated axons (Fig. 5.4). There was debate as to whether this region corresponded to lamina II or both laminae II–III (Rexed, 1952; 1954; Szentagothai, 1964a; Wall, 1967), but it has become clear that there are important differences between lamina II and III in types of primary afferent input and distribution of neuronal markers. Accordingly, most investigators use the term "substantia gelatinosa" as a synonym for lamina II. A reassessment of the relation of the grossly described gelatinosa to laminae has recently been presented, however, with particular reference to locations of different types of primary afferent input (Woodbury et al., 2000) and will be considered below, but here the point is that the laminae are not static definitions, and their significance will change as knowledge increases.

Cell Types

As preliminary comments, an important distinction is between projection neurons (relay neurons, Golgi type I neurons), which convey information to distant regions, and local interneurons (intrinsic neurons, local circuit neurons, Golgi type II neurons), which are

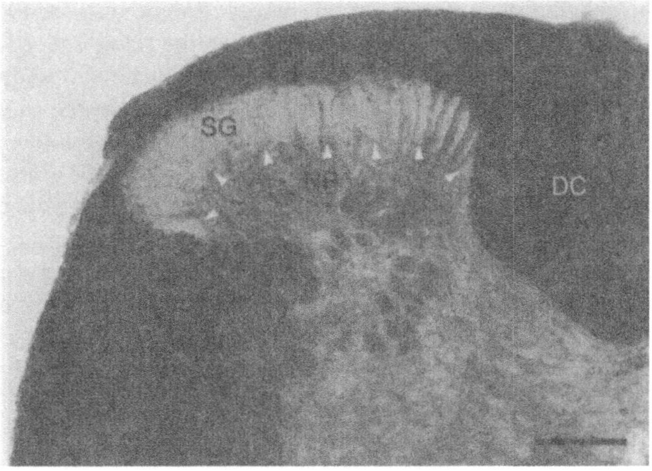

FIGURE 5.4. A plastic embedded section showing the precise lamina II–III boundary (arrow heads) that results from the absence of myelin in lamina II and its presence in lamina III. Width of the substantia gelatinosa (SG) is approximately 2 mm.

enclosed within a structure and have a local integrative function (Rakic, 1975). In many areas of the nervous system this is a clear dichotomy, but not in lamina II of the dorsal horn. First, lamina II neurons have local axonal arbors, so no matter how far the axon projects, there is local integration. Also axonal trees of "short-axoned cells" are not particularly short, in that even the shortest can project to many laminae and over significant distances in the cord. Thus, the distinction between projection (Golgi type I cells) and interneurons (Golgi type II) cells is blurred because most cells do both.

Classic Cell Types

The neurons of the gelatinosa were divided into limiting (limitrophe) and central cells by Ramón y Cajal (1909). This terminology was widely accepted until the mid-1970s.

Limiting (Limitrophe, Border) Cells

Limiting cells are larger than central cells, and their cell bodies arc located in the outer parts of lamina II, particularly at the lamina I–II border. It was on this basis that Rexed (1952) divided lamina II into an outer part with large (limiting) cell bodies and an inner part with smaller (central) cell bodies. The dendrites of these cells pass anteriorly and laterally as well as tangentially in the upper part of the dorsal horn (Ramón y Cajal, 1909; Scheibel and Scheibel, 1968). Thus, these cells have characteristics of dendritic organization that seem to be a mixture of marginal (lamina I) and central cell organizations (Scheibel and Scheibel, 1968). Axons of stalked cells, which are the modern equivalent of limiting cells, end in lamina I (Gobel, 1975; 1978b).

Central Cells

The name "central cell" is a misnomer because these cells are found throughout the gelatinosa and not just centrally. These are the common cells of the gelatinosa. The cell bodies are small with little cytoplasm, and the proximal dendrites usually arise from the dorsal and ventral poles of the cell, giving a radial appearance to lamina II in transverse sections. The dendrites of these cells branch and rebranch, and the richness of these arborizations contributes to the gelatinous appearance of this region (Ramón y Cajal, 1909). These dendritic branches were depicting as interdigating between the flame-shaped arbors of large primary afferent axons (Ramón y Cajal, 1909), but the arbors are now known to be largely confined to lamina III (Brown, 1981). The dendritic trees of central cells are flattened mediolaterally and spread widely in a craniocaudal direction (Ramón y Cajal, 1909; Pearson, 1952; Szentagothai, 1964a; Scheibel and Scheibel, 1968), so these dendritic domains are depicted as thin rectangular sheets (Fig. 5.6), but if central cells are reconstructed from serial Golgi sections, although still flattened, the dendritic trees are much more extensive than previously depicted (Sugiura, 1975). It has been a general pattern to find dendritic and axonal trees more extensive when reconstructed or marked by tracers such as HRP than was indicated in earlier studies, which used primarily single sections from young animals. There are two types of central cells; funicular and short-axoned.

Funicular Cells

Funicular cells are central cells whose axons travel in the spinal white matter. Such cells were assumed to be propriospinal implying that their axons re-enter the gelatinosa after traveling a variable distance in the white matter, so transmitted information remains within the gelatinosa (Szentagothai, 1964b). This is part of the consideration that the gelatinosa is a closed system (see below). The proportions of funicular as opposed to short-axoned cells are higher in young (Ramón y Cajal, 1909; Sugiura, 1975) than adult animals (Gobel, 1979). There are developmental differences between lamina II cells whose axons project into the white matter, and those whose axons stay within the gray matter. The dendrites of the projection cells develop prenatally and increase by elongation and branching. These cells are subdivided into limiting, large and small central, and transverse categories. By contrast, the dendrites of the intrinsic cells arise postnatally as short, beaded processes that arise from the cell body, and these cells are divided into islet, stalked, inverted stalklike, and vertical categories (see below).

Short-Axoned Cells

The axons of short-axoned central cells ramify within the dendritic tree and end near the cell body (Golgi type II cells). Golgi thought that most of the neurons in the gelatinosa fell into this category, but most cells are funicular in other studies (Ramón y Cajal, 1909). In addition, it would now be accepted that the axons of these cells ramify more widely than classically appreciated (see above).

The Gelatinosa as a Closed System?

Most early investigators found gelatinosa neurons whose axons ended in the gelatinosa (short-axon cells), but axons of most gelatinosa cells enter the white matter (funicular cells). The axons of the funicular cells were then claimed to re-enter the gelatinosa so the conclusion was that the "substantia gelatinosa proper, i.e., laminae II and III, form a self-contained neuron system, all axons of which either remain in the (gelatinous) substance or return to it . . .", the axons that never enter the white matter forming a "short-range intrinsic system" and the axons that re-enter the white matter forming a "longer-range longitudinal system." The result would be a lack of "forward transmission," so the gelatinosa is a "closed" system. The only output of this system would be by influencing dorsal dendrites that arise from neurons in deeper laminae (antenna-type neurons) (Szentagothai, 1964a). The above idea, although seminal, needs to be modified. For example, cells in lamina II project to the thalamus or brainstem (Giesler et al., 1978; Willis et al., 1978), and individually reconstructed or microelectrode recorded cells project outside the gelatinosa (Sugiura, 1975; Light and Kavookjian, 1988). However, the enormous concentration of ligands and receptors in the superficial dorsal horn, of which a main component is lamina II, and the dramatic demonstration of dorsal dendrites from antenna cells forming a major part of this plexus, as exemplified by NK1 receptor cells in deeper laminae (Chapter 6), shows that a substantial output of the substantia gelatinosa is influencing dendrites from deeper cells (antenna cells) is a real contribution.

Modern Cell Types

The classification of gelatinosa cells was changed in the mid-1970s with the proposal that the two predominant cell types in lamina II are stalked and islet cells (Gobel, 1975; 1978a) (Figs. 5.5 and 5.6) (Gobel, 1978b). As a terminological issue, lamina II was subdivided into lamina II and III (Gobel, 1975; 1978b), rather than IIo and IIi, as is now done. This

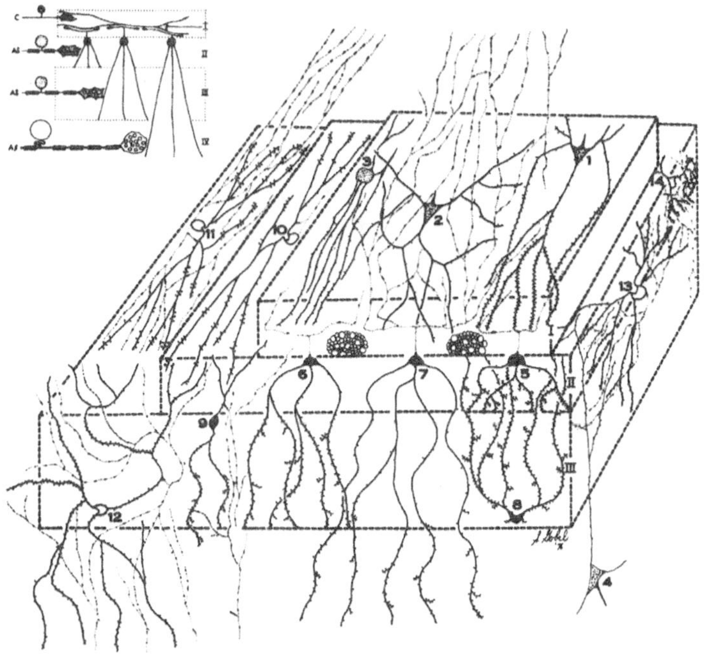

FIGURE 5.5. Schematic diagram of the major cell types of laminae I–IV of the medullary dorsal horn. Lamina 1 neurons are spiny (1) and smooth (2) pyramids and multipolar (3) neurons. Cell 4 is a lamina IV neuron which sends dendrites dorsally into laminae I–III (antenna type neuron). Cells 5–9 are stalked cells, cells 10 and 11 islet cells, cell 12 a spiny cell, cell 13 a lamina II/III border cell, and cell 14 an arboreal cell. (From Gobel, 1978; reprinted with permission.)

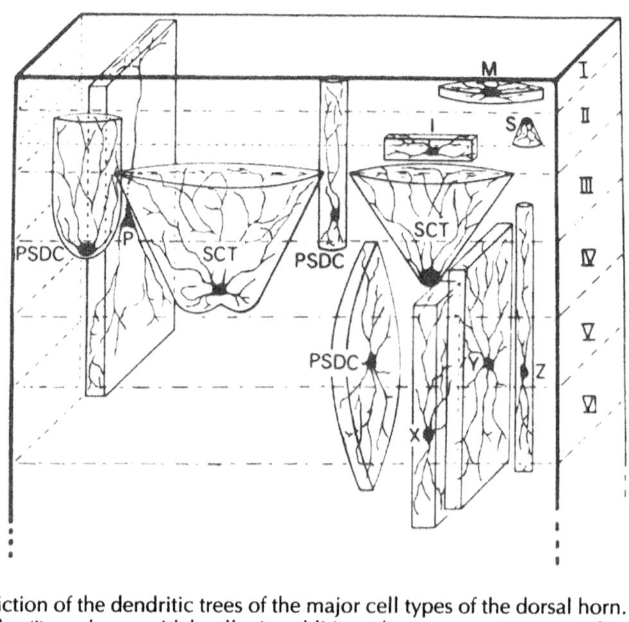

FIGURE 5.6. Depiction of the dendritic trees of the major cell types of the dorsal horn. Note the marginal (M), stalked (S), islet (I), and pyramidal cells. In addition, there are postsynaptic dorsal column (PSDC) and spinocervical tract (SCT) cells in laminae III–V and interneurons (X, Y, and Z) in laminae V and VI. (From Brown, A. G., 1981.)

causes confusion (Brown, 1981), and the terminology was changed (Gobel, 1979; Gobel et al., 1980). Here we will use the IIo and IIi terminology.

Stalked Cells

Stalked cells (Figs. 5.6) are limiting cells (Ramón y Cajal, 1909). They are named because of their short stalk-like spines. The cell bodies are located in lamina IIo, especially at the I–IIo junction. The main characteristic of the dendrites of the stalked cells is that they form a cone emanating from the cell body and passing ventrally through laminae II–IV (Fig 5.5 and 5.6), although some describe these cells as having a fanlike shape (Bennett et al., 1981). The major axonal destination of the stalked cells is into laminae I, but axonal ramifications into deeper laminae have been shown (Schoenen, 1982; Light and Kavookjian, 1988). Thus, these cells are interlaminar interneurons and since their axons ramify outside their dendritic arbors, they are not Golgi type II cells (Gobel, 1979). The lamina I long transmission neurons involved in nociception are presumably major targets for these cells (Gobel, 1978b). The suggestion is that these are excitatory interneurons whose dendrites receive input from primary afferent central terminals in glomeruli but some

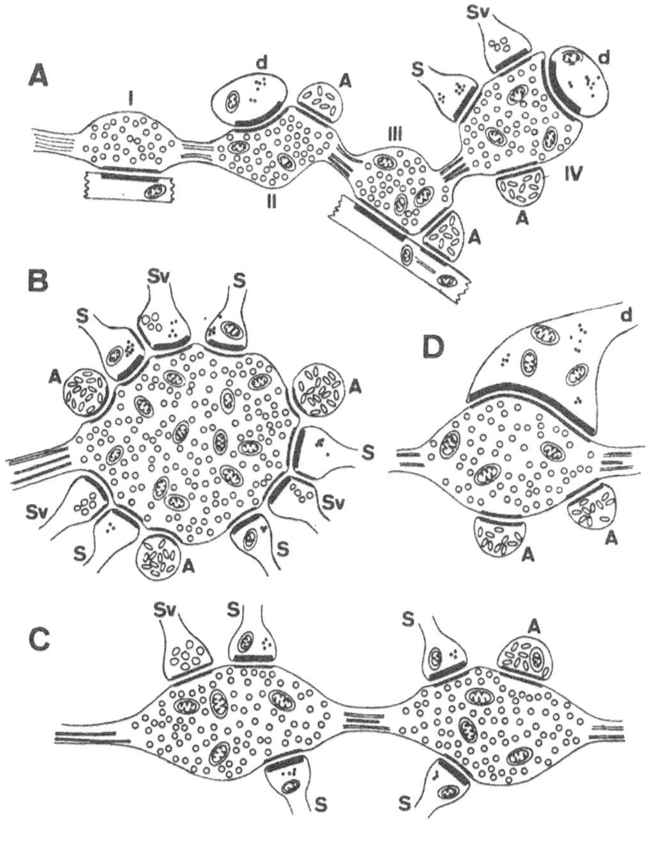

FIGURE 5.7. Schematic drawing of the different synaptic organizations that surround primary afferent fibers. Note in particular the numerous axoaxonic, dendrodendritic, and dendroaxonic terminals. A = Axon; d = dendrite; and S = dendritic spine. The complex in panel B is a typical glomerulus. (From Maxwell and Réthelyi, 1987.)

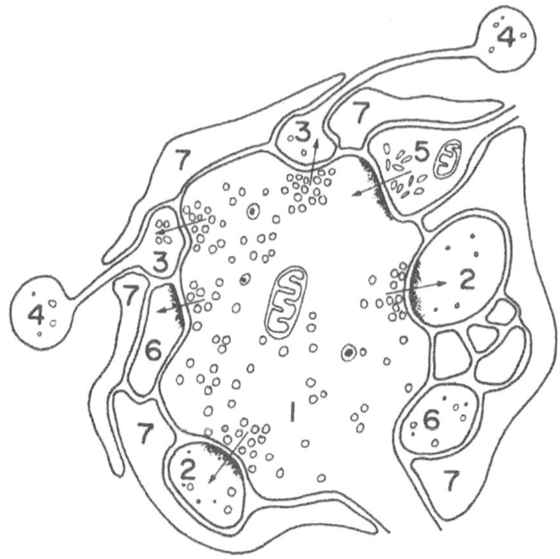

FIGURE 5.8. Diagram of a glomerulus. The components are as follows: (1) the central terminal; (2) dendrites, probably from antenna-type neurons; (3) and (4) dendritic spine heads presumably from substantia gelatinosa neurons; (5) a peripheral terminal filled with flattened vesicles; (6) an unknown dendrite; and (7) glial processes. (From Kerr, 1975.)

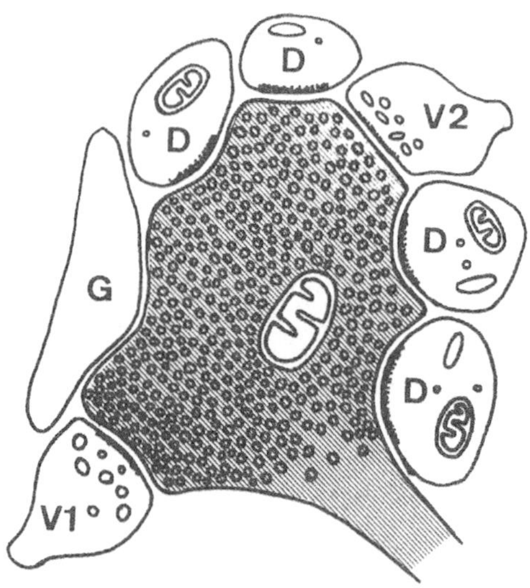

FIGURE 5.9. Diagram of glomerular organization according to the study of Coimbra et al. (1976). The central terminal is darkened, processes D are "ordinary dendrites," V1 is a peripheral terminal that contains clear vesicles of various sizes, V2 is a peripheral terminal with mostly flattened vesicles, and G is a glial process. (From Coimbra et al., 1974.)

stalked cells contain the inhibitory endogenous opioids dynorphin and enkephalin (Bennett et al., 1982; Cruz and Basbaum, 1985). Recordings indicate that they are either nociceptive-specific or wide dynamic range cells (Bennett et al., 1980). Stalked cells do not possess presynaptic dendrites, which is an important difference from islet cells (Gobel et al., 1980).

Islet Cells

(Figures 5.5 and 5.6.) The name "islet cells" (Gobel, 1979) was chosen because small central cells in the medullary trigeminal nucleus are located in clusters or islands (Ramón y Cajal, 1909). Presumably these are central cells, although other types of interneurons are probably also central cells (Gobel, 1979). The dendritic trees of islet cells extend for the full thickness of lamina II and are oriented in the rostrocaudal plane (i.e., flattened mediolaterally) (Gobel, 1978b; Schoenen, 1982) (Figs. 5.5 and 5.6). Islet cells whose dendrites are located largely in lamina IIo are reported to be nociceptive-specific, and those whose dendrites are in lamina IIi mechanoreceptive (Bennett et al., 1980). Other investigators, however, are unable to make such clear correlations of structure with function (Light and Perl, 1979b; Woolf and Fitzgerald, 1983). The axons of islet cells are reported to end largely in lamina II, so these are thought to be Golgi type II cells (Gobel, 1978b; 1979; Schoenen, 1982; Bicknell and Beal, 1984). These cells are thought to be inhibitory (Gobel 1978b) and islet cells contain the inhibitory amino acids GABA and glycine (Chapter 6), thus confirming this suggestion. The dendrites of the GABA-containing islet cells contain vesicles and form presynaptic endings on nearby dendrites and axonal terminals (Spike and Todd, 1992; Chapter 6).

Other Cell Types

The subdivision into stalked and islet cells is generally accepted (Todd and Spike, 1993). Other classification schemes have been proposed, either for cells that do not fall into islet or stalked categories or as a generally different scheme. These have generally not been correlated with chemical markers or functional variables (see below), but the morphological types exist and hopefully will have functional correlates eventually. These classifications are listed for reference.

The Gobel classification: Gobel (1975; 1978b; 1979) subdivided the cells that remained after islet and stalked cells into three subtypes.

Arboreal Cells

Arboreal cells are interneurons predominantly in lamina IIo. They have extensive dendritic branching in lamina IIo with branches extending to laminae I and IIi (Gobel, 1978b). The axons of these cells arborize primarily in the vicinity of the dendrites, but some axonal branches also extend into laminae I and IIi.

II–III Border Cells

Border cells are the rarest of the lamina II interneurons. They give rise to dendrites from their medial and lateral poles, and these dendrites arborize in laminae IIo and IIi. The axon is extensively branched and, on the basis of frequency of spines, is regarded as having a maximum effect in IIo and the outer half of IIi.

Spiny Cells

Spiny cells have a spherical or slightly elliptical dendritic domain, and are reported to be the only ones to enter the magnocellular part of the nucleus caudalis. Their axons distribute to laminae I–III.

The hypothesis (Gobel and Binck, 1977) is that islet cells, arboreal cells, II–III border cells, and spiny cells are Golgi type II interneurons. There is, however, an important difference between islet cells and the others, in that the islet cell axon is confined to the lamina in which the cell body is located whereas the axons of the other cells span laminar boundaries (Gobel, 1979). These are the equivalent of Cajal's central cells, but a difference is that axons of most central cells enter white matter, whereas the modern studies do not find this, presumably because early investigators used tissue from embryonic or young animals (Gobel, 1979).

Another classification, proposed by Schoenen (1982) is that there are islet cells, which are Golgi type II cells, and filamentous, curly, and stellate cells. These cells are described primarily by dendritic domains, which are longitudinally oriented.

Filamentous Cells

Filamentous cells have fine expansions on their terminal dendrites, thus giving a filamentous appearance to the dendritic trees. The dendritic trees for cells in IIo are ventrally oriented and those in IIi dorsally oriented. These cells represent about 20% of the total in lamina II.

Curly Cells

The dendrites of curly cells curve toward each other, giving a curly appearance to the dendritic trees. These are the rarest cells of lamina II, and their axonal trees give relatively few branches in the vicinity of the dendrites.

Stellate Cells

Stellate cells have a simple dendritic tree that spreads in all directions front the cell body, thus leading to a stellate appearance of the cells.

Yet another classification is that the cells in laminae I–III are divided into 10 types. The four types in the gelatinosa are labeled II1–II4. Types II3 and II4 are thought to be Golgi type II cells. In addition, some II3 cells are thought to be short propriospinal neurons (Abdel-Maguid and Bowsher, 1984a, b). In addition there are the categories proposed by the Beal group mentioned above (Beal and Bicknell, 1981; Beal, 1983; Bicknell and Beal, 1984).

Adequacy of Dendritic Classifications for Lamina II Cells

Many lamina II cells do not fit any category of published cell types (Beal and Cooper, 1978; Light et al., 1979; Woolf and Fitzgerald, 1983; Todd and Lewis, 1986; Todd, 1988), and dendritic morphology of lamina II cells can be regarded as a spectrum so any subdivision into a limited number of specific types may be misleading (Beal and Cooper, 1978; Cervero and Iggo, 1980; Molony et al., 1981). Furthermore, the shape of the dendritic tree and the synaptic ultrastructure around the individual cell do not correlate with cell function (Light et al., 1979; Woolf and Fitzgerald, 1983; Steedman et al., 1985; Light and Kavookjian, 1988; Rethelyi et al., 1989), except that dendrites are found primarily in the laminae of the particular afferent input that excites the cell (Light et al., 1979; Rethelyi et al., 1989). Also the extraordinarily wide spread of the dendrites of individually identified and filled cells in the substantia gelatinosa of the medullary dorsal horn may invalidate any simple classification of cell types based on dendritic patterns in lamina II (Li, Y. Q. et al., 1999), again emphasizing that when cells are filled with marker and reconstructed, their axons and dendrites are seen to spread in much greater abundance than when seen in Golgi stains in single sections. On the other hand, correlations have been established for islet and stalked cells, particularly in the arena of chemical markers (Chapter 6).

Axonal Projections

Early studies were concerned primarily with the question of whether or not cells of the substantia gelatinosa sent their axons into the spinal white matter. Then stalked cells were stated to project to lamina I and islet cells to lamina II, arboreal cells to laminae I and II, II–III border cells to laminae IIo and IIi, and spiny cells to laminae I–III (Gobel 1978b; 1979). These then are local interneurons, but some lamina II cells project to thalamus and brainstem (Willis et al., 1978). Later studies agree with the idea that most lamina II cells are local interneurons (Golgi type II cells), although these studies did not provide exact axonal terminations. Reconstructions of individually labeled cells show even wider distributions of the axonal projections, and essentially all neurons have enough local axon branching to provide integration at this level (Li et al., 1999).

In summary, gelatinosa cells were classically divided into limiting and central categories, with the central cells further divided into funicular and short-axoned cells. These classifications were replaced by stalked and islet categories, that have functional and chemical correlates, and other subtypes have been proposed. Physiologic correlates are not well established, however, except that laminae into which dendrites project determine the types of primary afferent input that excites the cell. The axonal projections of lamina II cells have turned out to be more diverse than was classically thought.

Primary Afferent Input

The primary afferent input into lamina II comes primarily from collaterals from fine sensory fibers, which are usually unmyelinated, but the issue of coarse sensory input has recently been raised. The early descriptions were primarily of these fibers as groups, but the profusion of chemical markers, particularly for the C fibers (Chapter 6) has added considerable refinement.

Coarse Primary Afferents

Collaterals from large primary afferent fibers curve ventrally through the medial part of the dorsal horn and then turn dorsally to end in large flame-shaped arbors in the substantia gelatinosa (Ramón y Cajal, 1909; Szentagothai, 1964a; Scheibel and Scheibel, 1968). Those studies were done before the laminae were known or when the gelatinosa was regarded as both lamina II and III. Thus, although the flame-shaped arbors clearly exist (Beal and Fox, 1976; Brown et al., 1977; Hamano et al., 1978; Beal, 1979; Brown, 1981; Fyffe, 1984; Ralston et al., 1984; Woolf, 1987), it has been questioned whether they enter lamina II (Hamano et al., 1978; Fyffe, 1984). Several studies report that distal parts of flame-shaped arbors enter lamina IIi (Proshansky and Egger, 1977; Beal, 1979; Light and Perl, 1979b; Woolf, 1987; Shortland et al., 1989), but this is species-dependent in that it is clear in rats and monkeys but not seen or very limited in cats (Beal and Fox, 1976; Brown et al., 1977; Proshansky and Egger, 1977; Rethelyi, 1977; Hamano et al., 1978; Ralston et al., 1984; Woolf, 1987). The flame-shaped arbors arise from hair follicle afferents (Brown, 1981). Thus, information from hair follicle afferents enters lamina IIi, but most of the above studies would say that the main input is into laminae III–IV. On the other hand, a recent study finds a very considerable input of myelinated afferents, identified by transport of BHRP, into the inner part of the substantia gelatinosa (Woodbury et al., 2000). The study also points out that there are mediolateral and segmental differences with a hemilamina interposed between laminae I and II medially in the lumbosacral region of the cord, which corresponds to the peptide positive region with fine primary afferent endings. From the point of view of laminar organization, the main point is that lamina IIi or the ventral part of the substantia gelatinosa contains a considerable large myelinated afferent fiber input.

Fine Primary Afferents

Fine fibers enter the gelatinosa from the tract of Lissauer and marginal plexus (Ramón y Cajal, 1909; Pearson, 1952; Szentagothai, 1964a; Scheibel and Scheibel, 1968). Early opinion was that myelinated primary afferent input dominated since the flame-shaped arbors were thought to enter the gelatinosa. Then came the realization that the arbors were restricted largely to lamina III (Brown, 1981), although this is now disputed (Woodbury

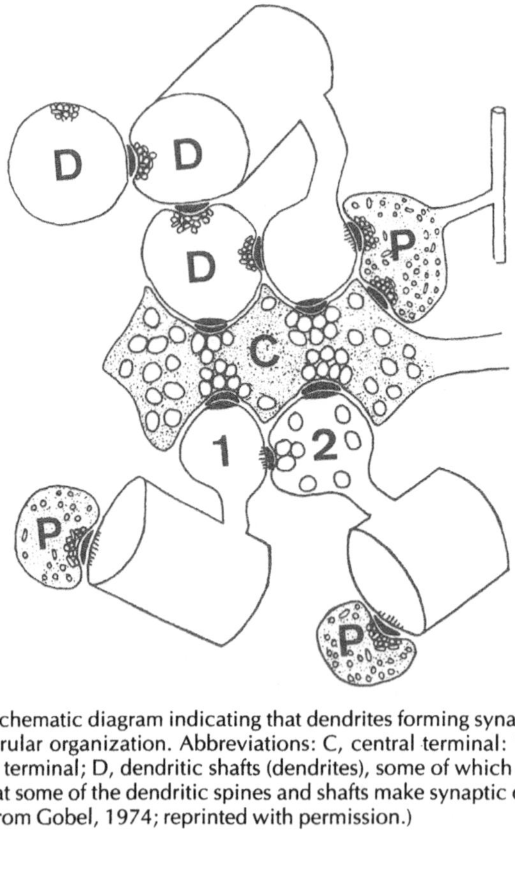

FIGURE 5.10. Gobel's schematic diagram indicating that dendrites forming synapses on each other are an important part of glomerular organization. Abbreviations: C, central terminal: 1 and 2, dendritic spines contacted by the central terminal; D, dendritic shafts (dendrites), some of which are also contacted by the central element (note that some of the dendritic spines and shafts make synaptic contacts with each other); P, peripheral endings. (From Gobel, 1974; reprinted with permission.)

et al., 2000), and there was lack of degeneration in lamina II following dorsal rhizotomy (Ralston, 1965; 1968b; Sterling and Kuypers, 1967; Carpenter et al., 1968; Shriver et al., 1968). Thus, propriospinal fibers were thought to dominate. It has become clear, however, much fine afferent input enters lamina II (Sprague and Ha, 1964a; Szentagothai, 1964a; Heimer and Wall, 1968; Gobel and Binck, 1977; LaMotte, 1977; Light and Perl, 1977; 1979a; Proshansky and Egger, 1977; Rethelyi, 1977; Gobel and Falls, 1979; Ralston and Ralston, 1979; Gobel et al., 1981; Chung et al., 1989), with single afferent fiber-labeling providing the most direct evidence (Light and Perl, 1979a; Rethelyi et al., 1982; Ralston et al., 1984; Mense and Prabhakar, 1986; Sugiura et al., 1986; 1989). Differences in degeneration time of different types of primary afferents were probably responsible for earlier disagreements (Heimer and Wall, 1968). This does not imply a paucity of propriospinal terminals, however, for approximately 50% of the synapses in lamina II survive dorsal rhizotomy, these presumably being a mixture of propriospinal and descending input (Chung et al., 1989). Dorsal rhizotomy in the cat is reported not to lead to fewer synapses, even though primary afferents are removed, presumably because of sprouting (Murray and Goldberger, 1986).

Organizational insights are that presumed propriospinal fibers form synapses *en passant* whereas presumed primary afferents form *boutons terminaux* from showers of boutons that form sagittally oriented sheets (Scheibel and Scheibel, 1968; Rethelyi, 1977; Rethelyi and Capowski, 1977). Of particular interest are studies based on individual fibers. Thus, one unmyelinated high-threshold mechanoreceptor, two polymodal nociceptors, two mechanical and cold nociceptors, and two low-threshold mechanoreceptors principally ended in lamina IIo but endings were also seen in laminae I, IIi, III, and IV and the termination patterns of a functional type of fiber are characteristic (Sugiura et al., 1986). The authors stress that lamina II appears to be the main projection area for unmyelinated primary afferent fibers from the skin. In a second study, unmyelinated visceral afferents terminated in laminae I and II, as well as in deeper lamina (Sugiura et al., 1989). The numbers of terminals for each fiber are much sparser for visceral than for cutaneous fibers, but the visceral fibers distribute to a wider area of the cord, which may be related to the poor localization of visceral sensation. Clearly, more work of this type is needed, but the technical obstacles are formidable.

A major distinction has been that peptidergic (CGRP, SP, and trkA) fine afferent input is distributed to lamina IIo and non-peptidergic (FRAP and IB4) input to lamina IIi. Some studies clearly show this (Silverman and Kruger, 1988a, b; Swett et al., 1991; Molliver et al., 1995), but others show considerable overlap of these populations (Wang et al., 1994) (Chapters 3 and 4). In addition, the inner part of lamina II has recently been reported to contain a major myelinated afferent input (see above), and IB4 axons may predominate in the outer part of lamina II (Woodbury et al., 2000). Many of the references on this issue report diametrically opposed findings (Chapter 4 and 6), so consensus has not been reached.

Fine Myelinated (Aδ) Primary Afferents

Fine myelinated (as defined by conduction velocity) mechanical nociceptive fibers end primarily in the marginal zone (lamina I) and lamina V (Light and Perl, 1979a; Rethelyi et al., 1982), as do fine myelinated nociceptors from deep tissues (Mense and Prabhakar, 1986). That fine myelinated primary afferents do not end in lamina II is in accord with other morphologic work (LaMotte, 1977; Ralston and Ralston, 1979), although some investigators report them here, based on differential degeneration times (Gobel and Binck, 1977) or on diameters of parent axons (Gobel and Falls, 1979). The latter observations, although undoubtedly correct, are more indirect than the single-fiber work. Thus, the conclusion is fine myelinated Aδ primary afferent input into lamina II is sparse or nonexistent.

Neuropil Organization

The most prominent structures in the neuropil of lamina II are glomeruli, so these will be considered first. This will be followed by a consideration of the "nonglomerular" neuropil (Fig. 5.7).

Glomeruli

A glomerulus consists of a central terminal, which is a primary afferent ending, in synaptic contact with several peripheral dendrites and axonal terminals (Figs. 5.7–5.11). Because the peripheral structures (mostly dendrites) indent the central terminal, the name "scalloped ending" is sometimes used. The whole complex is partially separated from the surrounding neuropil by glial processes. These synaptic arrangements were previously given other names "complex synaptic arrays" (Ralston, 1965; 1968a); "central synaptic complexes" or "large synaptic complexes" (Rethelyi and Szentagothai, 1969). The glomeruli are regarded as key structures of the dorsal horn because they offer a morphologic basis for a presynaptic modulation of information transfer. These structures appear very early in primate development (Knyihar-Csillik et al., 1999a).

Central Terminals

Dense or Dark Terminals

Dense terminals contain electron-dense axoplasm and numerous round clear vesicles whose average diameter is approximately 50 run. Numerous mitochondria and some large dense-core vesicles are also seen (Fig. 5.9; Ralston 1965; Kerr, 1966; 1970a; 1975; Ralston, 1968a; 1971a; 1979; Coimbra et al., 1974). They are called dense sinusoidal terminals (Knyihar-Csillik et al., 1982a). The electron density of the axoplasm appears to be intrinsic to the axoplasm, but some investigators also note fine axoplasmic filaments (Knyihar and Gerebtzoff, 1973).

Light (Electron-lucent or Regular Synaptic Vesicle) Terminals

Many central terminals have light axoplasmic matrix with round, clear, uniform vesicles in smaller relative numbers than in the dark terminals (Ralston, 1979; Knyihar-Csillik et al., 1982a, b; Ribeiro-da-Silva and Coimbra, 1982). The light terminals are more prevalent in deeper laminae (Fig. 5.8 Ralston, 1979; Knyihar-Csillik et al., 1982a, b). The light terminals were subdivided into those that contain numerous fine filaments (type IIa) and those that do not (type IIb) with type IIa terminals in lamina IIi and type IIb in lamina III (Ribeiro-da-Silva and Coimbra, 1982; Ribeiro-da-Silva et al., 1985).

Large Dense-Core Vesicle Terminals

Many electron-lucent central terminals contain numerous large dense-core vesicles, particularly in primates (Ralston, 1979; Knyihar-Csillik et al., 1982a).

Although there was early disagreement (Ralston, 1968b; Rethelyi and Szentagothai, 1969), central terminals degenerate following dorsal rhizotomy (Kerr, 1970b; Coimbra et al., 1974; 1984; Gobel, 1974; Knyihar et al., 1974; Knyihar and Csillik, 1976; Ralston and Ralston, 1979; Knyihar-Csillik et al., 1982a), but some find a few surviving fibers (Knyihar et al., 1974; Ralston and Ralston, 1979). Surviving central terminals may come from distant uncut roots, however (Coimbra et al., 1984), and there are different degeneration times for different types of terminals (Heimer and Wall, 1968; Ralston and Ralston, 1979; Knyihar-Csillik et al., 1982b; Coimbra et al., 1984), which may confuse the issue. Thus all or almost all central terminals arise from primary afferent fibers. This conclusion is strengthened by the observation that DRG cell chemical markers are found in central terminals (Chapter 6). The large dense-core vesicle central terminals are found predominantly in lamina 1, dense terminals in lamina II, and light terminals in lamina III (Knyihar-Csillik et al., 1982b). In the rat, dense terminals are found in the ventral part of IIo, light terminals without filaments in IIi, and light terminals with filaments in lamina III (Coimbra et al., 1984). Hypotheses are that fine fibers give rise to both dense and large dense-core vesicle terminals (Knyihar-Csillik et al., 1982b), or that dense terminals arise from unmyelinated fibers, light terminals without filaments from fine myelinated fibers, and light terminals with filaments from large recurrent myelinated fibers (Coimbra et al., 1984).

Peripheral Dendrites

The majority of glomerular processes in synaptic contact with the central terminal are dendrites (Figs. 5.7–5.11). Cells that give rise to these are large "antenna" cells from laminae III–V (Rethelyi and Szentagothai, 1969; 1973) and neurons in lamina II. There two types of dendrites: (1) "typical" dendrites receiving input from presynaptic terminals, and (2) "vesicle-containing dendrites" that, in addition to receiving presynaptic terminals, form synapses with surrounding dendrites and the central terminal (Fig. 5.10; Gobel 1974; 1976; Gobel et al., 1980). Typical dendrites are reported to arise from stalked cells and atypical dendrites from islet cells, but many arise from deeper antenna neurons. Chemical markers, particularly GABA, are found in the vesicle-containing dendrites (Chapter 6).

Peripheral Terminals

Peripheral terminals are axonal endings in synaptic contact with central terminals in glomeruli. Axoaxonic terminals are found in the general neuropil (Zhu et al., 1981), but they are most prominent in glomeruli. Peripheral terminals contain flat vesicles and make symmetric synapses with the central terminal, so they have the morphologic characteristics of inhibitory synapses. Peripheral terminals are referred to as V2 endings (Coimbra et al., 1984), P endings (Gobel, 1974), or F endings (Knyihar-Csillik et al., 1982a). They are not affected by dorsal rhizotomy, and they contain GABA and glycine (Chapter 6; Barber et al., 1978; Ribeiro-DaSilva and Coimbra, 1980; Carlton and Hayes, 1990; Todd and Lochhead, 1990). Care must be taken not to confuse peripheral terminals with presynaptic dendrites.

Further complexities of glomerular circuitry are "triads" where a central terminal makes contact with two dendrites, one of which is presynaptic to the other (Fig. 5.11; Knyihar-Csillik et al., 1982a, b; Ribeiro-da-Silva et al., 1985). Presynaptic dendritic interactions seem to link primary afferent central terminals (Knyihar-Csillik et al., 1982b).

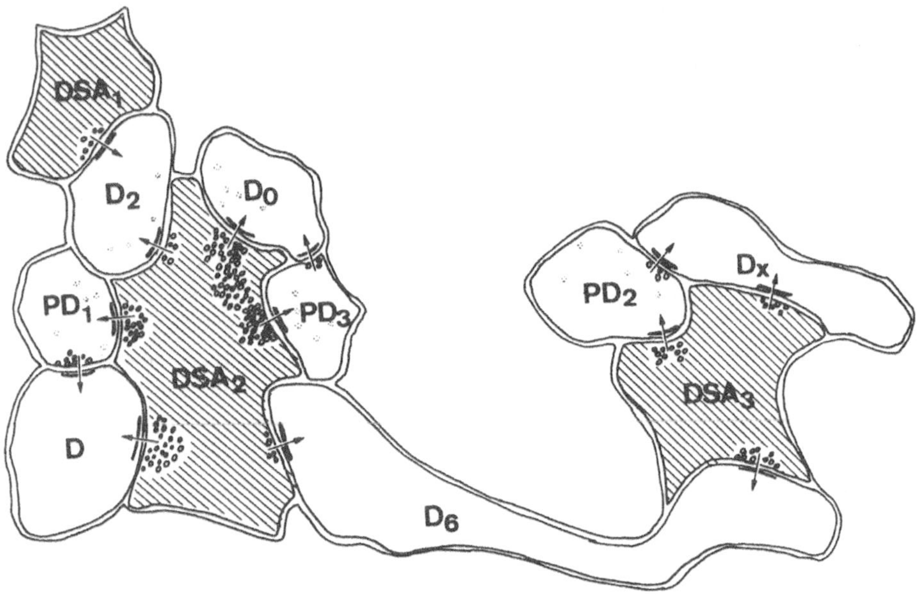

FIGURE 5.11. Schematic drawing of the synaptic arrangement around two primary sensory terminals in lamina II of the monkey. In particular, note the "triads" where two dendrites in synaptic contact with one another are contacted by the terminal. (From Knyihar-Csillik et al., 1982; reprinted with permission.)

These triads could be a system that blocks or shortens transmission at the first primary afferent terminal in the spinal cord (Knyihar-Csillik et al., 1982b), thus modulating primary afferent input in the glomeruli. Further hypotheses are that primary afferent impulses arriving in a central ending would depolarize surrounding dendrites, some of which are from antenna-type projection neurons in laminae IV and V and some of which are from central neurons of the gelatinosa (Kerr, 1975). Gelatinosa neurons would then discharge back onto the central terminal presumably to produce inhibition of primary afferent excitation, thus shutting off and sharpening primary afferent information flow to the projection neurons. The glomerular presynaptic dendrites that synapse back onto the central terminal (dendroaxonic synapses) might be excitatory (Gobel, 1974; 1976), but later work showed such synapses in islet cell dendrites, and islet cells contain GABA and are thought to be inhibitory (Chapter 6). If this is true and if peripheral terminals are also inhibitory, as usually postulated, excitation of a primary afferent central terminal would activate a complicated inhibitory system.

Chemical Markers

The central terminals contain glutamate and express glutamate receptors, with type I terminals expressing GluR1 and type II terminals GluR2. The peripheral terminals contain GABA, often co-localized with glycine, and express GABA and glycine receptors. In addition, CGRP, SP, somatostatin, cholecystokinin, serotonin, and FRAP appear in central terminals and every peptide expressed by small DRG cells presumably could appear in central synapses (Chapter 6).

Nonglomerular Synapses

Although glomeruli are the most prominent elements of the lamina II neuropil, they are in the minority in terms of total numbers of terminals. Several studies mention, for example, that glomerular terminals make up less than 5% of the total synapses in lamina II (Ralston, 1971; Duncan and Morales, 1973; 1978; Ralston and Ralston, 1979). Thus, the large majority of synapses in the neuropil of lamina II are simple and axodendritic in nature. The majority of these terminals contain round, clear, similar-sized vesicles approximately 50 nm in diameter clustered against asymmetric cytoplasmic thickenings. Some large dense-core vesicles are also seen. Other axodendritic synapses contain flattened or pleomorphic vesicles and have symmetric cytoplasmic thickenings. Such synapses are usually thought to be inhibitory. It has been pointed out that the round-vesicle terminals predominate in more superficial laminae and the flat-vesicle terminals predominate deeper (Ralston, 1971).

LAMINA III

Laminae III–VI make up the dorsal horn deep to the substantia gelatinosa. Different cytoarchitectonic subdivisions of this part of the dorsal horn have been proposed and are summarized in excellent historical reviews (Rexed, 1952; 1954). A major factor in the general acceptance of Rexed's laminae is that the receptive fields of dorsal horn neurons change in a predictable way from lamina to lamina (Wall, 1967).

Lamina III forms a broad band across the dorsal horn. In cytoarchitectonic studies, lamina III is distinguished from lamina II by having slightly larger and more widely spaced cells (Rexed, 1952), which is an imprecise boundary, but in myelin stained material, the boundary is sharp because lamina II contains relatively few myelinated fibers and lamina III contains many (Fig. 5.4).

Cell Types

Cells in lamina III are spindle-shaped and have relatively little cytoplasm (Ramón y Cajal, 1909; Rexed, 1952), thus resembling but being somewhat larger than lamina II central cells. Dendritic patterns are slightly more extensive than for lamina II cells in early studies (Ramón y Cajal, 1909; Pearson, 1952; Szentagothai, 1964a; Scheibel and Scheibel, 1968), the dendritic trees are flattened mediolaterally and extend craniocaudally (Scheibel and Scheibel, 1968). Reconstructions of three lamina III cells showed more complicated dendritic fields, however, spanning many segments and only slightly flattened (Mannen and Sugiura, 1976). Similarly 16 individually labeled cells in laminae III–IV had massive dendritic fields with several of the cells sending dendrites to the opposite dorsal horn (Gladfelter et al., 1993). Such reconstructions suggest that the classifications discussed below are based on only proximal parts of the dendritic tree, which does not limit their usefulness, but the fact that these dendritic trees are much more widely distributed should be remembered. The dendrites of lamina III cells form three longitudinally orientated plexuses, in IIo, in IIii and the outer part of III, and in the inner part of III, and some of the cells in ventral lamina III sent dendrites dorsally so that these cells, as well as cells in deeper laminae, are antenna-type neurons (Beal and Cooper, 1978).

Two particular cell types, spinocervical tract cells and postsynaptic dorsal column cells, were demonstrated in this lamina (Fig. 5.7; Brown et al., 1977; Brown, 1981; Brown and Fyffe, 1981), but there has been disagreement about their dendritic organization.

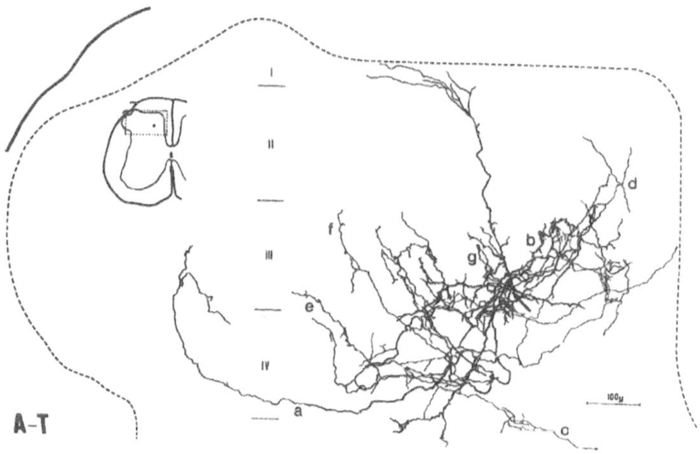

FIGURE 5.12. Reconstruction from serial sections of a lamina III cell. The dendritic and axonal arborizations are more complex than the depictions by earlier investigators. (From Mannen and Sugiura, 1976; reproduced with permission.)

One group shows that the dorsal dendrites of spinocervical tract cells do not travel into lamina I or IIo, so that they would not receive direct fine primary afferent input. But another group says that they do (Bennett et al., 1984). Dendrites of postsynaptic dorsal column cells travel dorsally into laminae I and II, and their dendritic trees are not restricted mediolaterally so they are depicted as dorsoventrally oriented cylinders or cones, and because of their dorsal dendrites they would be antenna-type neurons that could have monosynaptic contacts from fine afferent fibers (Fig. 5.6; Brown et al., 1977; 1986; Brown, 1981; Brown and Fyffe, 1981; Maxwell et al., 1982, 1983).

Pyramidal neurons in lamina III that are not spinocervical or postsynaptic dorsal column neurons have dorsal dendrites into laminae I–II and ventral dendrites into laminae IV–V that are flattened in the transverse plane and so the cells are longitudinally oriented (Scheibel and Scheibel, 1968). These cells could receive direct synaptic input from all types of primary afferent fibers. Another scheme is that there are two types of cells, one with dorsal dendrites (antenna-type cells) and one with radial dendrites (Schoenen, 1982; Bowsher and Abdel-Maguid, 1984). Another plan is that there are two cell types, one whose dendrites arborize dorsoventrally with dorsal dendrites penetrating lamina II, and one whose dendrites are confined to lamina III (Maxwell, 1985). During development, there are two cell types, one that develops before birth and sends axons into the white matter and one that develops after birth and is regarded as a local circuit neuron (Beal et al., 1988). For cells that are not spinocervical and postsynaptic dorsal column neurons, the shape of dendritic trees did not correlate with the function of the cell (Maxwell et al., 1983). Furthermore the extraordinary complexity of the complete dendritic tree (Fig. 5.12) shows that these depictions are only of the proximal dendritic trees.

Axonal Projections

In early studies, projections of lamina III cells were thought to be essentially the same as lamina II cells, thus either to the substantia gelatinosa or to the white matter around the dorsal horn (Ramón y Cajal 1909; Szentagothai, 1964; Scheibel and Scheibel, 1968), with the caveats that axons were shown to cross the midline and terminate in the opposite gelatinosa and that axons that enter the white matter re-enter the gray matter and terminate in the gelatinosa, defined as laminae II–III (Szentagothai, 1964). Thus cells in the gelatinosa were thought to project only to the gelatinosa. But many lamina III cells project to deeper laminae, and reconstructions show a wider axonal arborization than previously appreciated, including to the spinocervical nucleus and into the dorsal columns (Matsushita, 1969; 1970; Mannen and Sugiura, 1976; Brown et al., 1977; Brown, 1981; Brown and Fyffe, 1981; Bennett et al., 1984). Also, an individually visualized cell in cat lamina III had axonal ramifications in laminae IV–VI plus branches in the ipsilateral dorsolateral funiculus and contralateral ventral funiculus (Light and Kavookjian, 1988), and although Golgi type II cells with axons restricted to lamina III are described (Schoenen, 1982; Maxwell et al., 1983; Bowsher and Abdel-Maguid, 1984), most lamina III cells have axonal trees that arborize outside lamina III. It should be emphasized, however, all cells have local branches that mediate activity previously ascribed to Golgi type II cells. A final point is that when cells in various dorsal horn laminae are considered from the point of view of where they project, they cluster in groups that are not laminar-specific, as, for example, cells from various clusters that project to the periaqueductal gray (Vanderhorst et al., 1996; Mouton and Holstege, 2000).

Primary Afferent Input

Lamina III has a major primary afferent input from large primary afferent fibers that enter from the medial division of the dorsal root, curve ventrally through the medial part of the dorsal horn, and then recurve to end in "collaterales grosses et profondes de la substance de Rolando" (Ramón y Cajal, 1909) or "flame shaped arbors" (Scheibel and Scheibel, 1968). This work has been abundantly confirmed (Szentagothai, 1964; Beal and Fox, 1976; Hamano et al., 1978), except that the arbors extend craniocaudally and so are flame-shaped only on transverse section (Scheibel and Scheibel, 1968). Early work did not recognize laminae or regarded the gelatinosa as both laminae II and III, so the question of whether the flame-shaped arbors are found in laminae II is recent. Most reports find flame-shaped arbors primarily in lamina III, e.g. Fyffe (1984), although in the mouse, rat, and monkey they enter lamina IIi (see discussion in **Lamina II** above). However, individually identified, cutaneous myelinated afferent axons remained in laminae III–IV where they ramified extensively (Ritz et al., 1989).

Fine afferent terminals are restricted largely to laminae I and II, but a plexus of presumed fine afferent fine axons arborizes at the base of the flame-shaped arbors (the basal capping plexus) (Scheibel and Scheibel, 1968), selective surgery shows fine-fiber degeneration in laminae I–III with coarse-fiber degeneration in laminae IV–V (LaMotte, 1977), and single-fiber work shows some umyelinated axon terminals in lamina III (Sugiura et al., 1986), although another degeneration study found that removing fine primary afferent fibers caused degeneration mainly in laminae I and II (Light and Perl, 1977). Thus, there is some fine primary afferent input into lamina III, but it seems much less important than the coarse-fiber input. With the advent of intracellular marking techniques, the distribution of single, functionally identified large afferent axons has been determined (Brown, 1981). These reconstructions are described in Chapter 12.

Neuropil Organization

The neuropil of lamina III contains a predominance of axodendritic synapses, a few axoaxonic and axosomatic synapses, and glomeruli similar to those described for lamina II, except that the proportion of flat vesicle terminals with symmetric synaptic thickenings is higher in lamina III than in laminae I and II and in rats the central terminals have a higher proportion of filaments than those in superficial laminae (Ralston, 1979; Ribeiro-da-Silva and Coimbra, 1982).

Fine Structure of Terminals of Identified Axons

Terminals that arise from the individually identified (hair follicle, Pacinian corpuscle, rapidly adapting type I, slowly adapting type I, slowly adapting type II) afferent axons contain spherical granular vesicles and make asymmetric synapses primarily with surrounding dendrites (Ralston et al., 1984; Maxwell and Rethelyi, 1987). Thus they appear to be excitatory axons, which would correlate with their physiology. These terminals are contacted by terminals that contain flattened or pleomorphic vesicles and form symmetric junctions onto the afferent terminals (Fig. 5.13; Maxwell and Rethelyi, 1987) including hair follicle afferents (Maxwell et al., 1982; Ralston et al., 1984; Maxwell and Noble, 1987), Pacinian corpuscle afferents (Maxwell et al., 1984a; Ralston et al., 1984; Semba et al., 1984), rapidly adapting afferents (Maxwell et al., 1984a; Semba et al., 1985) and slowly adapting afferents (Semba et al., 1983; Bannatyne et al., 1984). Many studies show that these presynaptic terminals contain GABA or glycine (Chapter 6).

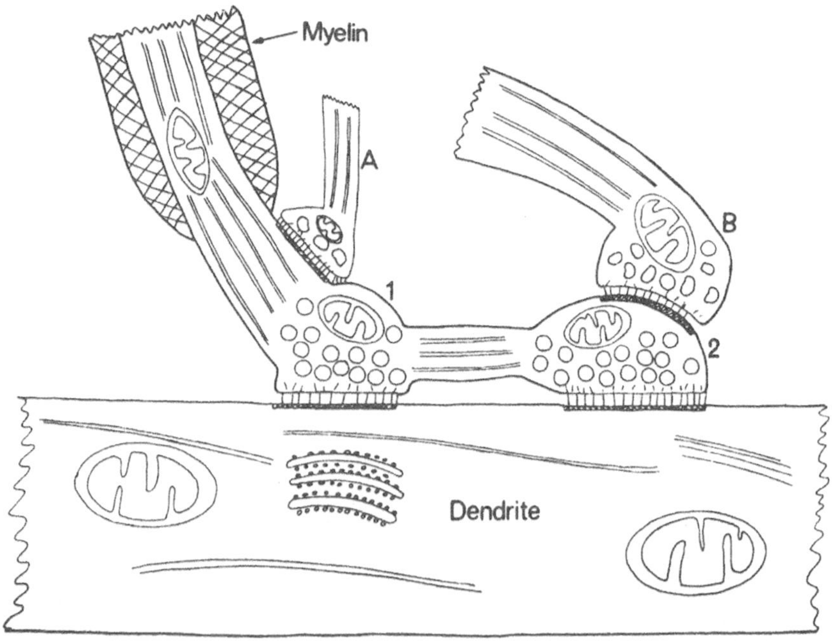

FIGURE 5.13. Schematic diagram of the synaptic organization around a hair follicle afferent terminal. Note the axoaxonic synapses at A and B. The endings of the presynaptic fiber may be *en passant* (1) or terminal (2). (From Maxwell et al., 1982.)

In summary, lamina III is distinguished from lamina II in having slightly larger cells and a neuropil that contains myelinated axons. Two types of cells have been identified in lamina III, postsynaptic dorsal column cells and spinocervical tract cells. Although there is not complete agreement, the postsynaptic dorsal column cells are characterized by dorsal dendrites that penetrate laminae I and II, whereas the dendrites of spinocervical tract cells do not. The unidentified cells in this lamina are of various types. One type is a pyramidal neuron that is flattened in the frontal plane and has dendrites passing through essentially all the laminae of the dorsal horn. Local interneurons (Golgi type II cells) and antenna-type neurons (neurons with prominent dorsal dendrites) are described in abundance, and the axons of the interneurons usually extend beyond the dendritic arbors of these cells.

The classic primary afferent input into this lamina comes from flame-shaped arbors which carry information from hair follicles. Other types of coarse primary afferents that enter lamina III arise from Pacinian corpuscles and rapidly and slowly adapting fibers. The axonal arborizations of each of these fiber types are distinctive (Chapter 12). This is a major finding in somatic sensory mechanisms. Some fine-fiber primary afferent input has been demonstrated in lamina III.

Large primary afferent endings contain round clear vesicles and make asymmetric synapses on surrounding dendritic profiles. These terminals receive terminals, which may be axonal or dendritic in origin, and which resemble inhibitory synapses and contain GABA or glycine (Chapter 6). In a few cases large axons are central endings in glomeruli.

LAMINA IV

Lamina IV is a relatively thick layer, the neurons of which are of various sizes, ranging from approximately 8 by 11 μm to large 35 by 45 μm in size. The largest cells are relatively infrequent but are so prominent that the impression is that this is a layer with large cells, although heterogeneity of cell sizes is seen as most characteristic of this layer (Rexed, 1952). Lamina IV can be distinguished from lamina III by the heterogeneity of neuronal sizes and the presence of some very large cells and from lamina V because the neurons of lamina V are even more heterogeneous than those of lamina IV and there are many longitudinally oriented myelinated axons in the lateral part of lamina V.

Cell Types

Most of the large neurons of laminae IV and V, which make up much of the nucleus proprius, have long spiny dendrites that pass dorsally, laterally, and ventrally and an axon that passes into the white matter (Ramón y Cajal, 1909; Szentagothai, 1964). The dorsal dendrites are particularly emphasized (Schoenen, 1982) for they penetrate the substantia gelatinosa to be contacted by axons from central cells and fine primary afferents. These antenna cells were thought to be essentially the only output from the gelatinosa (Szentagothai, 1964a), which is not the case, but they are a major output (see **lamina II** and Chapter 6). The dendrites of larger lamina IV cells radiate medially, laterally, and dorsally with a single ventral dendrite and so form a cone with a ventral apex (Scheibel and Scheibel, 1968). On the basis of dendritic patterns, the lamina III neurons are divided into three types: (1) antenna-type neurons, smaller in lamina III and larger in laminae IV and V;

(2) central cells with longitudinally oriented dendrites; and (3) cells, primarily in lamina V but also in IV, with transverse dendrites (Rethelyi and Szentagothai, 1973). On the other hand, individually recorded and filled cells in laminae IV and V had very large dendritic fields that often spanned the width of the laminae and extended for long distances cranio-caudally but did not reach dorsally into lamina I and II (Cao et al., 1993). Another pattern is that neurons in laminae IV, V, and VI can be distinguished on the basis of differences in dendritic spread, which is least for the lamina IV cells, intermediate for lamina V cells, and greatest for lamina VI cells (Proshansky and Egger, 1977) and thus correlates with the increasing size of the receptive fields as measured physiologically (Wall, 1967). If these cells are reconstructed, their dendrites extend more widely with dorsal dendrites that spread through laminae I–V of the ipsilateral side and laminae III–IV of the contralateral side and ventral dendrites that enter lamina VII (Mannen, 1975; Mannen and Sugiura, 1976). Spinocervical cells and postsynaptic dorsal column cells have been described as in lamina III, and, except for the location of the cell body, the cells in lamina IV are the same.

Axonal Projections

The classic view is that cells in the deep dorsal horn send their axons into the lateral white column or across the midline (Ramón y Cajal, 1909; Szentagothai, 1964a) presumably to the opposite spinothalamic tract through the anterior white commissure. These axons have a tortuous course and then often bifurcate before they get to the white matter and collaterals of these axons ramify in laminae III–V and deeper laminae including the contralateral side and lamina X (Scheibel and Scheibel, 1968; Matsushita, 1969; 1970; Mannen, 1975).

Primary Afferent Input

Dorsal dendrites of lamina IV antenna cells receive fine afferent input in laminae I and II. Many primary afferent axons directly enter lamina IV. Numerous axons come from the dorsal parts of the posterior columns, thought to be terminal ramifications of large myelinated primary afferent fibers, ramify in the head of the dorsal horn which is located approximately in lamina IV, and form a longitudinal plexus (the longitudinal plexus of the dorsal horn), with apparent synaptic contacts on lamina IV cells (Ramón y Cajal, 1909). They are separate from the coarse primary afferent fibers that form the flame-shaped arbors (Ramón y Cajal, 1909; Scheibel and Scheibel, 1968), although Szentagothai (1964) disagreed. Otherwise the primary afferent input into lamina IV is immersed in descriptions of very complex plexuses (e.g., the longitudinal plexus of the dorsal horn) (Ramón y Cajal, 1909; Sprague and Ha, 1964; Sterling and Kuypers, 1967; Scheibel and Scheibel, 1968; Rethelyi and Szentagothai, 1973; Szentagothai and Rethelyi, 1973). These have largely been supplanted with single afferent fiber reconstructions (Chapter 12).

Experimental degeneration and single fiber studies do not show much fine primary afferent input into lamina IV (LaMotte, 1977; Light and Perl, 1977). At the electron-microscopic level, the degeneration of primary afferent fibers and terminals in lamina IV is characterized by neurofilamentous and electron-lucent degeneration, in contrast to the electron-dense degeneration that characterizes laminae I and II. The neurofilamentous degeneration is thought to characterize large myelinated afferents, the electron-lucent degeneration fine myelinated afferents (Ralston and Ralston, 1982).

Neuropil Organization

Synapses in lamina IV are axodendritic, axosomatic, and axoaxonal (Ralston, 1968b; Kerr, 1970b). Glomeruli are seen in lamina IV, although they are not as prominent as in the superficial laminae (Ralston and Ralston, 1982). Large dense core vesicles are much less prominent in presynaptic terminals than in superficial laminae and do not arise from primary afferents (Ralston, 1982; Ralston and Ralston, 1982). Synaptic terminals in the cord are characterized by round clear vesicles superficially and flattened vesicles more deeply so that in superficial lamina IV about two thirds of the terminals contain round clear vesicles with this proportion decreasing down into laminae V and VI (Ralston, 1982). Thus, inhibitory mechanisms may be more prominent in lamina IV than in more superficial laminae but less than in laminae V or VI. Axoaxonic terminals and presynaptic dendrites are not common. Thus, most presynaptic "synaptic" control of primary afferent input is located more superficially, but volume transmission can occur in the absence of axoaxonic or dendroaxonic synapses and could presynaptically modulate the afferent input.

In summary, lamina IV is a thick layer consisting of nerve cells of various sizes. The large cells have dendrites that pass laterally, dorsally, and medially. Special emphasis is given to cells with prominent dorsal dendrites (antenna-type cells) because they can receive direct fine afferent input in superficial laminae. Postsynaptic dorsal column and spinocervical tract cells are found in this lamina. Afferent input comes from the fine primary afferent input onto the dorsal dendrites of antenna cells and from collaterals of large primary afferent fibers, each type having a characteristic ending (Chapter 12). Axoaxonic, dendrodendritic, or dendroaxonic synapses are relatively rare.

LAMINA V

Lamina V extends as a thick band across the narrowest part of the dorsal horn (Fig. 5.1), in what is often called the neck of the dorsal horn. Its lateral boundary is indistinct because of the many bundles of myelinated fibers coursing longitudinally through this area, giving the lateral part of this lamina a reticulated appearance. Lamina V can be distinguished cytoarchitectonically from lamina IV, since the cells are even more varied than in lamina IV, and from lamina VI, which consists of a medial zone with small, packed, compact cells and a lateral zone with slightly larger cells that are nevertheless more regular than those in lamina V.

Cell Types

The cell bodies in this region vary from 8–10 to 30–45 microns in diameter. (Rexed, 1952). The cells are large and have clear, distinct nucleoli. Nissl substance is relatively sparse but can be clearly seen, particularly in the large cells. Early work did not separate the cells in this region on the basis of their dendritic patterns, so the description of lamina IV cells applies equally well to lamina V (Ramón y Cajal, 1909), but the dendrites of lamina V cells have little longitudinal extent so they are shaped like flattened discs in contrast to lamina IV cells (Scheibel and Scheibel, 1968). By contrast, dendrites of reconstructed or individually visualized cells spread widely and, although flattened craniocaudally are not disk-like (Fig. 5.14; Mannen, 1975; Brown, 1981). Dendritic fields are more widely spread in lamina V than in lamina IV (Proshansky and Egger, 1977). In a correlation of structure and function, multireceptive cells were large and had a dendritic tree that extended in all

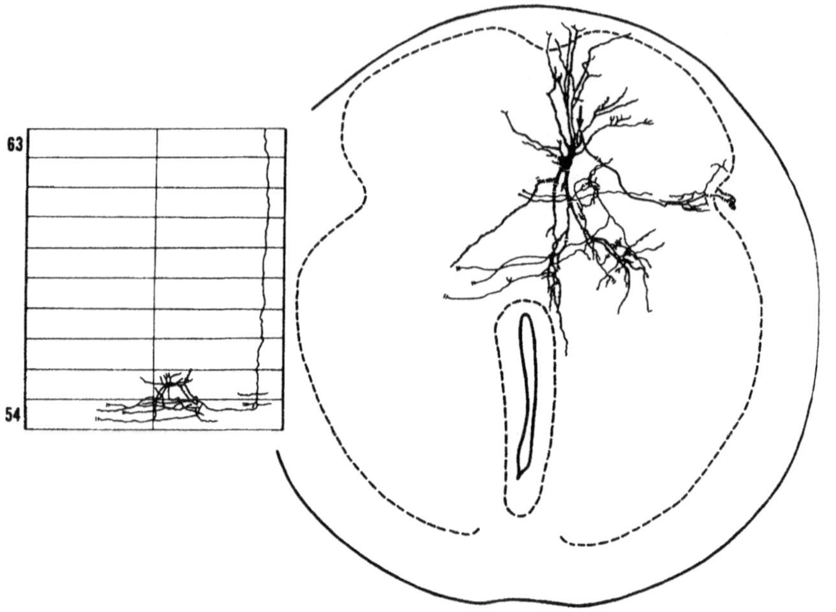

FIGURE 5.14. Large neuron reconstructed from serial sections located at the boundary of laminae IV and V. Note the dorsal dendrites and how widespread the whole dendritic tree is. (From Mannen, 1975.)

directions, whereas cells that responded only to noxious stimuli were smaller, but the dendritic fields were approximately the same, and cells that responded only to innocuous stimuli were smallest and had least extensive dendritic trees (Ritz and Greenspan, 1985).

Axonal Projections

Main projection sites for lamina V neurons are thalamus, lateral cervical nuclei, and dorsal column nuclei. There are relatively more spinothalamic and fewer spinocervical and postsynaptic dorsal column neurons. A number of lamina V neurons send axons to anterior funiculi, and so may be spinothalamic cells, and to motoneurons (Matsushita, 1969; 1970). Synaptic patterns have been mapped on identified lamina V spinothalamic neurons (Carlton et al., 1989).

Primary Afferent Input

There seems to be a dorsoventral orientation of the primary afferent input into lamina V, in contrast to the primarily longitudinal orientation of the primary afferent plexus in lamina IV (Sprague and Ha, 1964; Sterling and Kuypers, 1967; Scheibel and Scheibel, 1968). Terminals in laminae V and VI are not somatotopically organized as they seem to be in laminae III and IV (Sterling and Kuypers, 1967). A more complicated scheme is that the afferent collaterals to the center of the dorsal horn have a horizontal course before breaking up into dense neuropil in the medial part of the neck and bend of the dorsal horn, which they equate with laminae IV and V. For the dorsal horn proper, there seems to be a longitudinally orientated axonal plexus, mainly of primary afferent origin in laminae III, IV, and the upper part of V (Rethelyi and Szentagothai, 1973; Szentagothai and Rethelyi, 1973). As for lamina IV, study of these plexuses has been replaced by individual fiber reconstructions.

In summary, the cells of lamina V are more heterogeneous than those of lamina IV, and the lateral third of lamina V is heavily myelinated and thus has a reticulated appearance. The dendritic organization of lamina V resembles that of lamina IV cells, except that the general orientation of the dendrites changes to vertical, in contrast to the generally longitudinal orientation in lamina IV. The axonal projections of these cells is to the thalamus, the dorsal column nuclei, the lateral cervical nucleus, and various local destinations in the cord. There is a major shift in primary afferent organization in this lamina which has dorsoventral orientation in contrast to the longitudinal orientation of the plexus in lamina IV. The neuropil is similar to that of lamina IV except that there seem to be almost no glomeruli.

LAMINA VI

Lamina VI exists only in the cervical and lumbosacral enlargements of the spinal cord (Fig. 5.1). This is a part of the cord where primary afferent fibers do not predominate as in the rest of the dorsal horn. Cells in this layer are smaller and more regular in their arrangement than those in lamina V. There is more Nissl substance in these cells than in the cells of laminae V and VII. Thus, lamina VI has a darker appearance than laminae V and VII in Nissl sections. Lamina VI is divided into halves, the medial half consisting of a compact group of small or medium-sized heavily staining cells and the lateral half consisting of larger well-stained triangular or star-shaped cells. In this layer are the major parts of the internal and external basal nuclei of Ramón y Cajal.

Cell Types

Lamina VI cells are arranged much like those in lamina V, with dendrites that radiate in dorsoventral and mediolateral planes but with almost no extension in the longitudinal axis of the cord (Scheibel and Scheibel, 1968). These neurons are isodendritic, with long, straight, relatively unbranched dendrites, and the dendritic spread is wider than the dendritic spread of laminae IV and V neurons (Proshansky and Egger, 1977). Neurons in lamina VI often send dendrites across the width of the dorsal horn, but the dorsal dendrites never penetrate laminae I and II, these cells would sample input from wide areas from the base of the dorsal horn, but not fine primary afferent input (Brown, 1981). The variable dendritic patterns of lamina VI cells and patterns of synapses on them are shown in individually marked cells (Maxwell et al., 1997). Neurons in the lateral part of lamina VI tend to be somewhat more variable in size than in medial parts (Tredici et al., 1985). Finally, the dendritic organization of many cells in the middle parts of laminae V–VIII forms a cylinder, which is the basis for an interesting dendritic and cellular organization plan of the entire gray matter that is too extensive to be considered in a chapter devoted to only the dorsal horn (Rethelyi and Szentagothai, 1973; Szentagothai and Rethelyi, 1973; Rethelyi, 1984).

Axonal Projections

Although little has been done to determine connections, lamina VI cells presumably project to the thalamus or lateral cervical nucleus, but the majority are probably propriospinal. Matsushita (1969) points out that small cells in the medial part of lamina VI possess the most abundantly ramifying axon collaterals in the spinal cord which end primarily in laminae IV–VII, and recurrent collaterals end near the parent cell (Matsushita, 1969).

Primary Afferent Input

The primary afferent input to lamina VI is complex. Many collaterals from primary afferent axons destined to reach ventral horn cells end in this area with a particularly prominent group of terminals from Ia fibers in the center of lamina VI (Ramón y Cajal, 1909; Scheibel and Scheibel, 1968). The fine structure of primary afferent endings is described (Maxwell and Bannatyne, 1983). There are several systems of descending fibers ending in this area (Tredici et al., 1985). Morphologically this lamina appears to be distinct from the superficial laminae of the dorsal horn and the motor areas of the ventral horn.

CONCLUSIONS

1. Lamina I neurons were classically divided into marginal and smaller cells. The marginal cells have wide-ranging dendrites that travel tangentially over the surface of the cord. The smaller cells were not as well characterized.
2. The above classifications have been replaced by subdividing lamina I cells into pyramidal, fusiform, multipolar, and sometimes flattened categories. An important correlation is that pyramidal cells are thermoceptive-specific, fusiform cells nociceptive-specific and multipolar cells are either polymodal- or nociceptive-specific.
3. Another major correlation is that receptive field type correlates with laminar distribution of dendrites.
4. Primary afferent input into lamina I comes from fine fibers, both Aδ and C fibers, entering from the marginal plexus and surrounding white matter.
5. Simple axodendritic terminals predominate in lamina I with presumed excitatory synapses predominating, but glomeruli are also seen.
6. Lamina II is synonymous with the substantia gelatinosa, so named for the absence of myelinated fibers and a gelatinous appearance grossly due to intertwined axonal and dendritic branches. The substantia gelatinosa was suggested to be a closed system with the output of this region being only onto dendrites of antenna cells from deeper laminae. This view is too simple, but it has become clear that many deeper neurons send their dendrites into lamina II to receive both primary afferent and intrinsic input.
7. Lamina II cells were classically divided into central and limiting cells, but these categories have been replaced by stalked and islet cells with several other cell types identified by different investigators.
8. Stalked cells were suggested to be excitatory interneurons but some contain the inhibitory opioids dynorphin and enkephalin. These cells are either

nociceptive-specific or wide dynamic range neurons, and they do not possess presynaptic dendrites, an important difference from islet cells.

9. Islet cells, especially small islet cells, contain the inhibitory amino acids GABA and glycine and their dendrites form presynaptic endings on nearby dendrites and axonal terminals.
10. Classically neurons have been divided into local interneurons and projection cells, with islet cells falling in the former category. This is too simple, however, for individual reconstructions reveal that essentially all neurons in the gelatinosa give rise to both extensive local arbors and project relatively long distances. A corollary is that individually reconstructed cells in all laminae have much more extensive axonal trees and dendritic arbors than shown by classic techniques.
11. Unmyelinated fibers were thought to form the primary afferent input into lamina II, but a recent reassessment reveals that many large afferent fiber terminals enter lamina IIi. In addition, there seems to be an extra hemi-lamina containing much fine peptidergic afferent input in rat lumbosacral cord where so many studies have been done. Accordingly, these organizational features must be remembered when assaying sprouting, for example, and the precise description of laminar boundaries will be necessary for good assessment of morphologic data.
12. The neuropil of lamina II consists largely of axodendritic synapses and glomeruli. The axodendritic synapses are not remarkable, but proportionately fewer of them resemble excitatory synapses than in lamina I.
13. Glomeruli consist of a primary afferent central terminal, in synaptic contact with a number of surrounding dendrites and several peripheral axon terminals, the whole being somewhat isolated by glial processes. The central processes fall into light (electron-lucent) or dark (electron-dense) categories, and the light central processes are further subdivided by the presence or absence of large numbers of dense core vesicles. The dark and light dense core vesicle terminals are thought to arise from unmyelinated axons, the latter being peptidergic, and the other light terminals, which occur deeper in lamina II, are thought to arise from myelinated afferents. Although these precise identifications need further verification, the primary afferent origin of the central terminal is regarded as pathognomonic.
14. The peripheral terminals that end on the central processes in glomeruli contain GABA and often glycine.
15. The dendrites that are in contact with the central process in a glomerulus are of two types, presynaptic or regular. The presynaptic dendrites arise from islet cells, are usually GABAergic and make synaptic contacts with surrounding dendrites and sometimes the central terminal. Thus there is both pre- and postsynaptic glomerular integration of primary afferent input.
16. Lamina III contains larger neurons than lamina II, and two specific cell types have been identified, spinocervical and postsynaptic dorsal column neurons.
17. Antenna cells whose dorsal dendrites extend into superficial laminae are major outputs for lamina II in they receive both fine primary afferent and propriospinal input on these distal dendrites. Some of the most dramatic examples of antenna neurons express NK-1 receptors over essentially their entire surface.
18. The primary afferent input into lamina III comes largely as collaterals of large sensory axons. The different types of large myelinated input have distinctive termination patterns as detailed in other sections of this book.
19. Axon terminals, presumably inhibitory, are found on many of the large primary afferent terminals in this lamina.

20. The neuropil in lamina II largely resembles that in lamina III.
21. The same cell types are in lamina IV as in lamina III, with the exception of some spinothalamic neurons, and antenna cells are in abundance.
22. Primary afferent input is as in lamina III, except that there is a longitudinal orientation to the fibers.
23. The neuropil of lamina IV has few glomeruli and many presumed inhibitory synapses.
24. The neurons and neuropil in lamina V are as in lamina IV except that there are numerous spinothalamic neurons and a more vertical orientation of the neuropil.
25. Lamina VI is a transition between the primary-afferent dominated dorsal horn and the motor ventral horn. Descending input predominates, but there are extensive collaterals from Ia afferent fibers.

Chapter 6

Chemical Anatomy of the Dorsal Horn

AMINES

Catecholamines (CA)

Catecholamines are highly reactive compounds derived from tyrosine, which proceed via tyrosine hydroxylase (TH) to DOPA, then through DOPA decarboxylase to dopamine (DA), then through dopamine β-oxidase (DβH) to noradrenaline (NA), and finally through phenylethanolamine N-methyltransferase (PNMT) to adrenaline. These compounds have many biologic functions, for example, neurotransmitters in the sympathetic nervous system, but for spinal sensory functions they are considered in relation to nociceptive transmission.

Dopa

There are to our knowledge no reports of DOPA in cells or fiber systems of the dorsal horn.

Dopamine (DA)

No specific (DA) cells are seen in the dorsal horn but dopamine approximately 10-fold less in amount than NA in the cord is found in the dorsal horn (Skagerberg et al., 1982; Kondo et al., 1985). With specific antisera, however, DA fibers and terminals are seen in lamina III and below, with some in lamina I and a remarkable absence in lamina II (Yoshida and Tanaka, 1988; Mouchet et al., 1992; Ridet et al., 1992; Holstege et al., 1996).

Noradrenaline (NA)

NA is the most abundant CA in the cord. NA and TH fibers and terminals, the latter presumably largely representing NA because of its abundance, are concentrated in superficial laminae, particularly laminae I and IIo, but are more numerous in the intermediolateral column and ventral horn (Carlsson et al., 1964; Dahlstrom and Fuxe, 1965; Fuxe, 1965; Westlund and Coulter, 1980; Westlund et al., 1982; 1983; Schroder and Skagerberg, 1985; Pindzola et al., 1988; Mouchet et al., 1992; Westlund, 1992; Rajaofetra et al., 1992a). There are a few CA neurons in the dorsal horn, which presumably represent NA, mostly around the central canal and deep in the dorsal horn (Commissiong, 1983; Singhaniyom et al., 1983; Moore and Card, 1984; Dietl et al., 1985; Mouchet et al., 1986; Zhang, Y. Q. et al., 2001a), especially in young animals (Pindzola et al., 1988; 1990). These cells are too sparse to challenge the idea that aminergic fibers in the dorsal horn descend from brainstem cells.

Noradrenergic terminals as labeled by DBH staining, contact dendrites, and occasional cells (Hagihira et al., 1990; Doyle and Maxwell, 1991a), although a few axoaxonic synapses are seen with permanganate fixation, but lack of cytologic detail imposed by this fixative made further analysis difficult (Satoh et al., 1982). NA terminals do not form central endings in glomeruli, so they presumably do not arise from fine primary afferent fibers (Hagihira et al., 1990). Some find NA terminals with symmetric synaptic thickenings (Doyle and Maxwell, 1991a, b), others asymmetric (Hagihira et al., 1990), and others find few thickenings at all and suggest non-synaptic transmission (Rajaofetra et al., 1992a; Ridet et al., 1993). Catecholamine terminals are found on approximately 60% of neurons that project to the dorsal columns, and on individually identified spinothalamic neurons indicating aminergic control of these systems (Westlund et al., 1991; Doyle and Maxwell, 1993a). CA or NA fibers develop first in the ventral horn early in development (Commissiong, 1983; Rajaofetra et al., 1992b).

Adrenaline

Adrenaline fibers and terminals, outlined by the PNMT reaction, are found in the superficial dorsal horn, but are much more prominent in the intermediolateral cell column (Chung et al., 1989a; Carlton et al., 1991). Adrenaline cells are not described in the dorsal horn (Hokfelt et al., 1984).

Histamine

Histamine-containing cells are found in the tuberoinfundibular nucleus in adult mammals, and fibers from these cells pass widely in the CNS, including the cord. In the cord, histamine fibers have a slight concentration in superficial regions (Inagaki et al., 1988; Panula et al., 1989). During development, histamine systems are seen but are not prominent in the cord (Vanhala et al., 1994).

Serotonin (5-HT)

5-HT is an amine with widespread actions that are useful in psychomimetic and analgesic arenas.

Laminar Localizations and Intrinsic Cells

5-HT axons and terminals are found in all spinal laminae (Ruda, 1988) but are concentrated in the superficial dorsal horn, particularly I and IIo, in lamina X, and around motoneurons (Carlsson et al., 1964; Dahlstrom and Fuxe, 1965; Segu and Calas, 1978; Ruda and Gobel, 1980; Steinbush, 1981; Kojima et al., 1982; 1983; Ruda et al., 1982; LaMotte and de Lanerolle, 1983a; Light et al., 1983; Glazer and Basbaum, 1984; Leah et al., 1988; Pearson and Jennes, 1988; Tashiro and Ruda, 1988; Tork, 1990; Costa et al., 1994; Leger et al., 2001).

Although there are a few intrinsic 5-HT cells in the cord (LaMotte et al., 1982; Bowker, 1986; Newton and Hamill, 1988), the almost complete disappearance of 5-HT below a spinal transection (see below), the absence of 5-HT in most DRG cells (Chapter 4) and the marking of brainstem 5-HT neurons from the cord indicate this is a descending pathway (Ruda et al., 1986; Ruda, 1988). 5-HT fibers destined for the dorsal horn descend through the dorsolateral funiculus (Hylden et al., 1985; Basbaum et al., 1988; Bullitt and Light, 1989; Westlund et al., 1992), although one study suggests that they descend primarily through the gray matter of the cord (Johannessen et al., 1984). Many neurons projecting to the cord from serotonin nuclei of the brainstem do not contain 5-HT, but the function of the "non-5-HT" projection is not clear (Johannessen et al., 1984; Skagerberg and Bjorklund, 1985). The 5-HT transporter SERT1 is found primarily in laminae I, III, and IV (Sur et al., 1996).

Synaptic Organization

5-HT synaptic terminals tend to be small, although some are sizeable, axodendritic, contain numerous small round clear vesicles, some pleomorphic or flattened vesicles, and occasional large dense-cored vesicles, and attach to postsynaptic elements, usually by symmetrical but occasionally by asymmetrical synaptic thickenings (Ruda and Gobel, 1980; Ruda et al., 1982; LaMotte and de Lanerolle, 1983a; Light et al., 1983; Nishikawa et al., 1983; Basbaum et al., 1986a; Marlier et al., 1991a; Costa et al., 1994). Occasional 5-HT scalloped endings in glomeruli suggest a minor primary afferent input (Ruda and Gobel, 1980; Ruda et al., 1982; LaMotte and de Lanerolle, 1983a). Axoaxonic synapses are rare (LaMotte and de Lanerolle, 1983a; Maxwell et al., 1985b; Matthews et al., 1989). 5-HT synaptic complexes can be subdivided morphologically (Ruda and Gobel, 1980; Ruda et al., 1982), but physiological consequences of these types are not clear.

There is debate concerning the issue of classic vs. volume synaptic transmission. Some investigators note that synaptic thickenings are rare, and on this basis suggest the possibility of volume transmission for 5-HT (Maxwell et al., 1985b; Marlier et al., 1991a; Ridet et al., 1993), whereas others find enough synaptic thickenings to suggest classic transmission (Ruda et al., 1982; Light et al., 1983; Costa et al., 1994). Both types of transmission probably occur (Polgar et al., 2002).

5-HT terminals end on enkephalin and neurotensin interneurons implying direct interactions with cells expressing endogenous opioids (Glazer and Basbaum, 1984; Costa et al., 1994; Miller and Salvatierra, 1998) and on NK1 cells in lamina I (Polgar et al., 2002) and III–IV (Stewart and Maxwell, 2000), with the 5-HT innervation denser on NK1 than non-NK1 cells (Polgar et al., 2002). 5-HT terminals contact spinothalamic or spinomesencephalic (projection) cells (Hylden et al., 1986b; Wu and Wessendorf, 1992), but there seems to be no distinction in the 5-HT innervation of projection vs. non-projection neurons (Hylden et al., 1986b). Nociceptive lamina I neurons and lamina II stalked cells have fewer 5-HT terminals than non-nociceptive lamina I cells and lamina II islet cells (Hoffert et al., 1983; Miletic et al., 1984). 5-HT terminals end on fos-expressing laminae I and IIo projection neurons (Li et al., 1997b), post-synaptic dorsal column neurons (Nishikawa et al., 1983; Wu and Wessendorf, 1992), spinocerebellar tract neurons (Jankowska et al., 1995; Maxwell and Jankowska, 1996), and interneurons excited by group II muscle afferents (Jankowska et al., 1997). Thus 5-HT cells could directly influence many intrinsic systems in the dorsal horn. The innervation is considerably heavier on dendrites than somata (Hylden et al., 1986b; Jankowska et al., 1997).

Co-localizations

Many 5-HT axons contain enkephalin (Tashiro et al., 1988). SP and TRH are in 5-HT cells whose axons project to the cord, but these axons are usually in the ventral horn (Johansson et al., 1981; Pearson and Jennes, 1988; Tashiro and Ruda, 1988; Tuchscherer and Seybold, 1989; Arvidsson et al., 1990), although some are seen in the dorsal horn (Wessendorf and Elde, 1987; Pearson and Jennes, 1988; Reddy et al., 1990a). There is little dorsal horn co-localization with TRH (Arvidsson et al., 1990) nor with any of the commonly studied peptides nor, with few exceptions, glutamate decarboxylase (Maxwell et al., 1996). Some nociceptive spinothalamic neurons are contacted by 5-HT/TRH terminals whereas non-nociceptive postsynaptic dorsal column neurons are contacted by 5-HT only terminals, so descending neurons with terminals only containing 5-HT are not exclusively concerned with nociception (Wu and Wessendorf, 1992).

Experimental Manipulations

Adult or neonatal dorsal rhizotomy usually increases dorsal horn 5-HT, either transiently (Hirakawa and Kawata, 1992a) or permanently (Bregman, 1987; Polistina et al., 1990; Wang et al., 1991a, b; Mitchell et al., 2000), as does cauda equina compression (Kawakami and Tamaki, 1992), and 5-HT is unchanged after neonatal capsaicin (Jancso et al., 1981) or expands its input into lamina IIi (Marlier et al., 1990; 1992). There is a report of a decrease after rhizotomy, but this is regarded as secondary and not a direct result of primary afferent axon death (Tuchscherer and Seybold, 1989). Accordingly these results indicate no 5-HT primary afferent terminals (see above), and the increases presumably result from sprouting although they could also indicate appearance of more 5-HT in extant neuronal processes. Spinal transection removes dorsal horn 5-HT staining distally (Bregman, 1987; Bullitt and Light, 1989; Arvidsson et al., 1990; Hirakawa and Kawata, 1992b), although one study finds residual 5-HT immunostaining below a transection (Hadjiconstantinou et al., 1984). Chronic immobilization decreases dorsal horn 5-HT (Kitayama et al., 1989), and chronic arthritis and nitroglycerin (related to migraine) cause an increase (Marlier et al., 1991b; Pardutz et al., 2002). Depletion of 5-HT prevents dorsal horn loss of SP and enkephalin that normally follows peripheral axotomy (DiGiulio et al., 1987). Fine 5-HT axons are killed but coarse ones are spared after systemic *p*-Chloroamphetamine possibly indicating two separate 5-HT systems in the cord (Ridet et al., 1994a).

Development

The 5-HT innervation of the IML and ventral horn precedes development in the dorsal horn, but by shortly after birth the adult pattern is established (Rajaofetra et al., 1989; Wang et al., 1991b).

AMINO ACIDS

Glutamate

The main excitatory amino acid transmitter is glutamate, and the goal of morphologic studies is to locate neuronal systems that use this compound as a transmitter. Since glutamate is also used for general metabolism, however, labeled cells do not necessarily imply glutamate neurotransmission, although the probability is high, but synapses are not involved in most metabolic processes, so glutamate labeling of synaptic terminals would imply neurotransmission.

Laminar Localizations

Numerous cells and terminals are preferentially stained for glutamate, the enzyme glutaminase, or retrogradely transported radioactive aspartate in laminae I and II (substantia gelatinosa, superficial dorsal horn) (Rustioni and Cuenod, 1982; Martinez-Rodriguez and Diaz, 1987; Weinberg et al., 1987; Miller et al., 1988; de Biasi and Rustioni, 1990; Sluka et al., 1992; Sluka and Westlund, 1993b). Other studies find labeled cells in laminae III as well (Antal et al., 1991; Merighi et al., 1991); glutamate cells are concentrated in laminae III–IV of the spinal trigeminal nucleus (Magnusson et al., 1986; 1987). Many of the latter cells are trigeminothalamic neurons (Magnusson et al., 1987). A few neurons that project to the nucleus of the tractus solitarius are immunostained for glutamate, but the percentages are not high (Gamboa-Esteves et al., 2001a).

Synaptic Organization

Numerous presynaptic "dome-shaped" terminals contain high concentrations of glutamate in the dorsal horn. These terminals contain numerous small (~50 nm) clear vesicles and have asymmetric synaptic thickenings that unite the pre- and postsynaptic elements (de Biasi and Rustioni, 1988; 1990; Miller et al., 1988; Maxwell et al., 1990a; 1993; Merighi et al., 1991; Alvarez et al., 1993; Broman and Adahl, 1994; Valtschanoff et al., 1994; Lekan and Carlton, 1995; Irish et al., 1996; Bae et al., 2000). The glutamate is concentrated in regions containing small vesicles, and the amount of glutamate is proportional to vesicle concentration (Broman and Adahl, 1994). Glutamate containing presynaptic terminals have been subdivided cytologically with: (1) an electron-dense axoplasm and numerous clear vesicles of various sizes; (2) a clear axoplasm with a relatively uniform collection of small vesicles; or (3) variable numbers of large dense-core vesicles scattered amongst the small clear vesicles. The dark and dense-core vesicle terminals tend to be in superficial laminae and are thought to be endings of unmyelinated fibers whereas the clear endings are usually deeper and are thought to be from myelinated axons, particularly myelinated sensory axons (Maxwell et al., 1990b).

Central endings in glomeruli, which arise from primary afferents (Chapter 5), are enriched in glutamate. These include electron-dense endings with clear vesicles of various sizes (type I or DSA endings), relatively large clear endings with a dense population of similar-sized clear vesicles (type II or RSV terminals) and terminals which can be either dark or clear but with significant numbers of dense-core vesicles (LDCV terminals), all of these terminals connecting to postsynaptic elements via asymmetric cytoplasmic thickenings (de Biasi and Rustioni, 1988; 1990; Merighi et al., 1991; Alvarez et al., 1993; Maxwell et al., 1993; Broman and Adahl, 1994; Valtschanoff et al., 1994; Bae et al., 2000). Some of the dark scalloped terminals coexpress SP (de Biasi and Rustioni, 1988) and/or CGRP (Merighi et al., 1991), so glutamate and these peptides are presumably co-released (de Biasi and Rustioni, 1990). The central terminals have GABA-containing flat vesicle terminals with symmetric thickenings ending on them (see below). Thus, primary afferent glutamate input is inhibited or sharpened by GABA inhibition arising mostly from local interneurons. Loss of this sharpening, for example by GABA cell loss after nerve lesions, might have a bearing on neuropathic pain. Glutamate labeling persists through early stages of terminal degeneration (Iliakis et al., 1996). There is a considerable increase in glutamate immunostaining after experimentally induced arthritis (Sluka et al., 1992; Sluka and Westlund, 1993b).

Aspartate

The other excitatory amino acid is aspartate. Many neurons in laminae I–III of the dorsal horn and presynaptic terminals in laminae I–III and axons in the dorsal roots are labeled for aspartate (Westlund et al., 1989b; Merighi et al., 1991; Tracey et al., 1991; Valtschanoff et al., 1994; Keast and Stephensen, 2000; Larsson et al., 2001). Aspartate co-localizes with glutamate in many DRG cells (Tracey et al., 1991), but they are not in the same terminals in the dorsal horn (Merighi et al., 1991). There is a question as to whether aspartate is truly a primary afferent neurotransmitter in the dorsal horn (Valtschanoff et al., 1994); and although aspartate labels DRG cells, dorsal root axons, and synapses in the superficial dorsal horn aspartate is spread through the axoplasm and not associated with vesicles in presynaptic terminals, which may mean use in general metabolism (Larsson et al., 2001).

Gamma-Aminobutyric Acid (GABA)

Laminar Localizations

GAD (glutamic acid decarboxylase, the GABA-synthesizing enzyme, a GABA marker) or GABA labeling is concentrated superficially (superficial dorsal horn, substantia gelatinosa, laminae I–III, laminae II, laminae I, IIi, and III, laminae II–III) (Kelly et al., 1973; Ljungdahl and Hökfelt, 1973; McLaughlin et al., 1975; Barber et al., 1978; 1982; Ribeiro-da-Silva and Coimbra, 1980; Hunt et al., 1981a; Perez de la Mora et al., 1981; Price et al., 1984; Fuji et al., 1985; Basbaum et al., 1986b; Ruda et al., 1986; Magoul et al., 1987; Kosaka et al., 1988; Todd and McKenzie, 1989; Carlton and Hayes, 1990; Mark et al., 1990; Todd and Sullivan, 1990; Waldvogel et al., 1990; Antal et al., 1991; Hiura et al., 1991; Ma et al., 1994; Bernardi et al., 1995; Simmons et al., 1995; Eaton et al., 1998; Tillakaratne et al., 2000), although in some cases label is reported in laminae I–IV or scattered evenly over the dorsal horn (Ljungdahl and Hökfelt, 1973; Price et al., 1984). The appellations are slightly disparate, but laminae I–III is the consensus as to where GABA or GAD elements are concentrated (Todd and Spike, 1993).

Intrinsic Cells

GABAergic or GAD cells are found in all dorsal horn laminae, but they are most common in laminae I–III, where their profiles make up of 28, 31, and 46% of total neuron profiles respectively (Todd and Sullivan, 1990). Most GABA cells in lamina II are islet cells (Barber et al., 1982; Todd and McKenzie, 1989), in contrast to stalked cells which are not enriched in GABA (Todd and McKenzie, 1989). Other GABA cells are fusiform in lamina I with restricted dendrites and multipolar in lamina II with widespread dendrites (Coimbra and Lima, 1989) and filamentous and islet GABA cells in lamina II and antenna cells in lamina III (Waldvogel et al., 1990). Visualization of the dendritic tree can be obtained with Golgi stains combined with immunochemical localizations, and these studies show that large islet cells are GABAergic whereas small islet and stalked cells are not and that the

GABAergic cells in lamina III have dendrites arranged horizontally rather than vertically (Todd and McKenzie, 1989; Powell and Todd, 1992; Spike and Todd, 1992; Todd and Spike, 1993). Showing GABA primarily in islet cells is a cornerstone of subsequent work on synaptic organization.

The GABA system arises from intrinsic neurons with little evidence for descending or primary afferent contributions. However, GABA cells in the medulla project to the cord (Millhorn et al., 1987a; Jones et al., 1991; Todd and Spike, 1993) and GAD accumulates proximal to a transection (Fuji et al., 1985), so there are descending GABAergic fibers, but they do not seem play major roles in spinal inhibition. Similarly, there is no indication of significant GABAergic primary afferent input, but there are two reports of GABAergic DRG cells (Chapter 4) that need to be verified.

Synaptic Organization

Axodendritic or axosomatic contacts. The common GABAergic synapses in the dorsal horn are GABAergic endings with pleomorphic and a few dense-core vesicles on dendrites or rarely cell bodies with thin symmetrical synaptic thickenings (McLaughlin et al., 1975; Ribeiro-da-Silva and Coimbra, 1980; Basbaum et al., 1986b; Magoul et al., 1987; Matthews et al., 1988; Waldvogel et al., 1990; Dawe and Maxwell, 1991; Merighi et al., 1991; Todd and Spike, 1993; Ma and Ribeiro-da-Silva, 1995; Iliakis et al., 1996), although a few GABAergic terminals have round clear vesicles and asymmetric thickenings (McLaughlin et al., 1975; Carlton and Hayes, 1990). Dendrites contacted by GABAergic terminals fall into two approximately equal categories; those that contain vesicles and make synaptic contacts with other neuronal elements (presynaptic dendrites) and those that do not (classical dendrites). Many of the presynaptic dendrites are themselves enriched in GABA (Carlton and Hayes, 1990; 1991; Todd and Lochhead, 1990; Hiura et al., 1991; Alvarez et al., 1992; Hayes and Carlton, 1992; Spike and Todd, 1992). GAD immunostaining is often not associated with either synaptic thickenings or $\beta2/\beta3$ expression of $GABA_A$ receptors suggesting both classical and volume synaptic transmission (Alvarez et al., 1996).

Axoaxonic Synapses

GABA terminals can be presynaptic or postsynaptic in axoaxonic synapses. Presynaptic GABA terminals in axoaxonic synapses end either on other terminals in the neuropil, some types of which have been identified, for example muscle afferent terminals (Maxwell et al., 1991) or on central endings in glomeruli (McLaughlin et al., 1975; Ribeiro-da-Silva and Coimbra, 1980; Basbaum et al., 1986b; Magoul et al., 1987; Matthews et al., 1988; Todd and Lochhead, 1990; Waldvogel et al., 1990; Bernardi et al., 1995; Almond et al., 1996). Evidence that many terminals postsynaptic to GABA endings arise from primary afferents is that many: (1) degenerate after dorsal rhizotomy (Barber et al., 1978); (2) are marked by retrograde afferent labeling (Bernardi et al., 1995; Bae et al., 2000); (3) arise from labeled individual afferents of various types (Maxwell and Noble, 1987; Maxwell et al., 1990c; Alvarez et al., 1992); and (4) are glutamatergic, glutamate being the major primary afferent transmitter (Maxwell et al., 1990b; Merighi et al., 1991; Valtschanoff et al., 1994; Iliakis et al., 1996). Thus there is considerable presynaptic GABA control of primary afferent input into the dorsal horn.

Axoaxonic synapses where the GABA terminal is postsynaptic to another terminal are rarer than axoaxonic synapses with the GABA terminal presynaptic, but they occur in some glomeruli where central primary afferent terminals end on GABA terminals

(Basbaum et al., 1986b; Carlton and Hayes, 1990). Some of these central terminals contain SP indicating that peptidergic input can influence the GABA system monosynaptically (Wang, H. et al., 2000). Since GABA is inhibitory, primary afferent input onto GABA terminals could cause inhibition by exciting the GABA terminal, which would then inhibit its postsynaptic target.

Dendrodendritic, Dendrosomatic, and Dendroaxonic Synapses

The spinal GABA inhibitory system is complicated by the presence of GABAergic presynaptic dendrites. Presynaptic dendrites differ from classic dendrites by forming *en passant* synapses on postsynaptic targets (Ralston, 1971a, b). The presynaptic dendrites in the dorsal horn arise largely from islet cells (Gobel, 1976; Gobel et al., 1980; Todd, 1988). Many presynaptic dendrites are enriched in GABA (Carlton and Hayes, 1990; 1991; Todd and Lochhead, 1990; Hiura et al., 1991; Alvarez et al., 1992; Hayes and Carlton, 1992; Spike and Todd, 1992). Thus, the dendritic arborizations of islet cells should have a central role in spinal inhibitory systems. An important nuance is that GABAergic neurons give rise to presynaptic dendrites whereas "non-GABAergic" neurons do not in the superficial dorsal horn (Spike and Todd, 1992), presumably indicating that the presynaptic dendritic system is largely inhibitory.

Reciprocal synapses and triads: complexity of the GABA system in the cord is further indicated by the presence of triads where both terminals in axoaxonic or dendroaxonic synapses end on a common dendrite (triads). In the dorsal horn the presynaptic terminal is often GABAergic and the postsynaptic terminal a primary afferent ending (Carlton and Hayes, 1990; Todd and Spike, 1993). This allows for very complex excitatory and inhibitory interactions at the primary afferent synapse in the dorsal horn. This complexity is greatly increased with the demonstration of reciprocal synapses between GABAergic presynaptic dendrites and primary afferent endings (Carlton and Hayes, 1990). Thus, an understanding of the basic architecture of GABAergic axonal and dendritic systems, important as it is, is just the beginning of insight into the spinal GABA inhibitory system.

Postsynaptic Targets

Identification of postsynaptic targets gives insight into the GABA wiring diagram. GABA-enriched terminals end on 20–60 % of Aδ HTM endings (Alvarez et al., 1992), on 80–100% of individual, physiologically identified, fast and slowly adapting cutaneous afferent axons (Sutherland et al., 2002), on spinothalamic (Carlton et al., 1992; Lekan and Carlton, 1995) and trigeminothalamic neurons, on type I and type II central endings in glomeruli (see above), and on MOR dendrites and cell bodies (Gong et al., 1997), the latter presumably implying that endogenous opioid inhibitory systems are inhibited by the endogenous GABA inhibitory system. An important negative finding is that GABA terminals do not end on peptidergic primary afferent endings (Ribeiro-da-Silva et al., 1989; Hayes and Carlton, 1992; Alvarez et al., 1993; Todd and Spike, 1993; Bernardi et al., 1995), presumably indicating that peptidergic nociceptive afferent input is not under presynaptic GABA control. For non-nociceptive input, GABA terminals end on individual hair follicle and Ia afferents (Maxwell and Noble, 1987; Maxwell et al., 1990c; Watson and Bazzaz, 2001; Watson et al., 2002), on spinocervical neurons (Maxwell et al., 1991), on postsynaptic dorsal column cells (Maxwell et al., 1995a), and on spinal projections of medullary neurons (Maxwell et al., 1995b). GABA and glycine are co-localized in terminals that end on individually identified lamina I cells, but GABA seems to be located extrasynaptically (Chery and De Koninck, 1999). Ia afferents end on and presumably excite GABA cells in lamina VI, which then presynaptically inhibit Ia afferents (Dawe and Maxwell, 1991).

Co-localizations

Glutamate

GABA and glutamate are not co-localized (Merighi et al., 1989; Maxwell et al., 1990b) which is consistent with the idea that GABA is not a primary afferent transmitter.

Glycine

Glycine is the other major inhibitory amino acid in the dorsal horn. In laminae I–III, all glycinergic cells are GABAergic, but some GABAergic cells are not glycinergic, whereas in the deeper dorsal horn there are glycinergic cells, GABAergic cells, and cells enriched in both (Todd and Sullivan, 1990). GABA and glycine also co-localize in superficial layers of the medullary dorsal horn (Wang, D. et al., 2000). GABA and glycine co-localize in peripheral terminals in type II glomeruli whereas the peripheral terminals in type I glomeruli are only GABAergic implying that the two transmitters are concerned with fine myelinated but not unmyelinated primary afferent input (Todd, 1996). GABAergic cells that are not glycinergic co-localize with enkephalin, NPY, TRH, and ChAT whereas those without glycine do not (Todd, 1991; Todd et al., 1992a; Rowan et al., 1993; Fleming and Todd, 1994), indicating a complicated and undoubtedly significant subdivision of glycinergic and GABAergic inhibitory interneurons. Furthermore, since numerous cells and terminals are both GABAergic and glycinergic, this suggests that some synapses release both powerful inhibitory transmitters onto their postsynaptic targets. In this regard, some GABAergic terminals end on dendrites expressing glycinergic receptors or gephryin (Maxwell et al., 1995b; Todd, 1996), and terminals containing both GABA and glycine face post-synaptic targets enriched in both GABA receptors and gephryin (Todd et al., 1996).

ChAT

ChAT is a marker for cholinergic neurons and in the dorsal horn ChAT colocalizes with GAD or GABA (Kosaka et al., 1988; Todd, 1991). ChAT and glycine do not coexist, however, so there are GABAergic cells with glycine but no ChAT and GABAergic cells with ChAT but no glycine (Todd, 1991).

NOS and NADPH Diaphorase

GABA and NOS or NADPH diaphorase largely coexist in dorsal horn neurons, and many of these neurons are glycinergic (Valtschanoff et al., 1992a; Spike et al., 1993). In the neuropil, GABA and NOS are frequently co-localized in simple axodendritic terminals, but GABAergic terminals in glomeruli do not contain NOS. Thus, NOS presumably has little direct effect on sensory transmission by glomerular primary afferent endings (Bernardi et al., 1995). NOS does not co-localize with cyclooxygenases (Maihofner et al., 2000).

CGRP

GABA neuronal profiles do not contain CGRP (Hayes and Carlton, 1992; Alvarez et al., 1993).

Enkephalin

Many enkephalin cells are enriched in GABA but none with glycine (Todd et al., 1992a).

Galanin

Almost all galanin immunoreactive neurons in the dorsal horn are GABAergic and almost none are glycinergic (Simmons et al., 1995).

Neuropeptide Y

Neuropeptide Y cells are GABAergic but not glycinergic (Rowan et al., 1993). GABAergic-NPY cells target dendrites that express NK1 receptors (Polgar et al., 1999a).

Neurotensin

Neurotensin and GABA do not coexist, although both are found in islet cells (Todd et al., 1992b).

Somatostatin

Somatostatin does not co-localize with GABA in the superficial dorsal horn, but does in a few cells in the deep dorsal horn (Proudlock et al., 1993).

SP

SP does not co-localize with GABA in cell bodies (Patel et al., 1997), but some synaptic terminals in the superficial dorsal horn are enriched with GABA and SP (Ma and Ribeiro-da-Silva, 1995) suggesting that both neuroactive substances are released during synaptic transmission.

Thyrotropin-releasing Hormone

The majority of TRH cells in the dorsal horn are GABAergic but not glycinergic (Fleming and Todd, 1994).

Parvalbumin

Approximately two thirds of parvalbumin cell profiles are GABAergic but few calbindin-D28K cells are (Antal et al., 1990).

Glycine Receptor

Numerous GABAergic terminals end on postsynaptic targets that express glycine receptors (Mitchell et al., 1993).

MORs

Most MOR cells are not enriched for either GABA or glycine, so cells inhibited by morphine are probably excitatory interneurons (Kemp et al., 1996), but GABA terminals end on cells expressing MORs (Gong et al., 1997).

Experimental Manipulations and Development

In chronic Freund's adjuvant arthritis, numbers of GAD mRNA and GABA profiles increase and then decrease roughly in concert with the course of the arthritis (Castro-Lopes et al., 1994a, b). There is an increase in numbers of neuronal profiles for GAD_{67} but not GAD_{65} after spinal ligation (Tillakaratne et al., 2000). During development, there is an increase in both GAD_{65} and GAD_{67} between P7 and P14, followed by a decline to normal levels by adulthood (Ma et al., 1994). Finally, there is a considerable decrease of GABA dorsal horn neuronal profiles after nerve lesions, which presumably has a bearing on neuropathic pain (Castro-Lopes et al., 1993; Ibuki et al., 1997; Ralston et al., 1997; Eaton et al., 1998) and after ischemia (Zhang, A. L. et al., 1994a), and a 5-HT transplant prevents GABA loss after transection (Eaton et al., 1998).

Glycine

Laminar Localizations

Glycinergic cells and neuropil elements are sparse in laminae I–II but numerous in deeper laminae (Hökfelt and Ljungdahl, 1971; Ljungdahl and Hökfelt, 1973; Campistron et al., 1986; Ottersen and Storm-Mathisen, 1987; Van Den Pol and Gorcs, 1988; Todd, 1990; Todd and Sullivan, 1990), especially lamina III (Ribeiro-Da-silva and Coimbra, 1980; Carlton et al., 1996). Glycine tends to be in islet and similar cells in lamina III (Todd and Sullivan, 1990; Spike and Todd, 1992).

Synaptic Organization

Numerous glycinergic presynaptic terminals are attached to postsynaptic elements by symmetrical synaptic thickenings. The vesicles in the presynaptic elements are pleomorphic or flattened and are rounder and more regular in monkeys than in rats (Van Den Pol and Gorcs, 1988; Todd, 1990; Carlton et al., 1996b). Presynaptic terminals of unknown origin end on glycinergic dendrites, but unlike GABA dendrites, glycine is not in presynaptic dendrites (Todd, 1990; Carlton et al., 1995b). Some presynaptic endings on glycinergic targets arise from myelinated primary afferents (Todd, 1990; Todd and Spike, 1993), including hair follicle afferents (Maxwell et al., 1990c) and others contain SP (Wang, D. et al., 2000). In monkeys no glycinergic axoaxonic synapses are seen (Carlton et al., 1996b), but in rats, glycinergic terminals that also contain GABA end on myelinated afferent terminals in type II synaptic glomeruli (Todd, 1990; Todd et al., 1991; Todd and Spike, 1993) and on hair follicle afferent axons (Maxwell et al., 1990c). Type I glomeruli are not associated with glycine (Todd et al., 1991).

Co-localizations

Gephryin

Gephryin is a synaptic protein that anchors glycine receptors to the postsynaptic membrane. Most presynaptic profiles facing gephyrin postsynaptic elements are enriched in glycine (Todd et al., 1995). Most presynaptic terminals facing sites labeled for $GABA_A$-receptor β3 and gephyrin are enriched in both GABA and GLY implying dual transmission from many inhibitory presynaptic endings (Todd et al., 1996).

GABA

Most glycinergic cell bodies, axons, and synapses in the dorsal horn co-localize with GABA, but many GABA elements do not contain glycine. Thus, there are two types of GABAergic elements, those that co-localize with glycine and those that do not (Todd and Sullivan, 1990; Wang et al., 2000a).

Glycine Transporter GLYT2

Almost all glycine transporter GLYT2 is found in glycinergic terminals in laminae I–III (Spike et al., 1997).

NADPH Diaphorase

Approximately half of cells labeled for NADPH diaphorase are enriched in glycine, and many of these glycine cells co-localize with GABA (Spike et al., 1993).

ChAT

ChAT and glycine do not coexist, but ChAT and GABA do. Thus, there are two populations of GABAergic cells; those with glycine but no ChAT and those with ChAT but no glycine (Todd, 1991).

Somatostain

Approximately half the cells enriched in somatostatin co-localize with glycine (Proudlock et al., 1993).

Taurine

Taurine immunoreactivity is concentrated in laminae I–II of the dorsal horn with numerous taurine-expressing cells in these laminae (Lee et al., 1992).

CHANNELS

Calcium Channels

Calcium channel $\alpha 2\delta$-1 receptors are expressed by principally small DRG neurons in naïve rats, and after peripheral nerve injury the expression increases with the increase proportionately greatest in large DRG cells (Newton et al., 2001).

Sodium Channels

For terminology see Chapter 4. $Na_V1.1$, $Na_V1.2$, and $Na_V1.3$ (I–III, α1–3), and Na_X(NaG) are expressed by cells scattered throughout the cord, most strongly in anterior horn cells (Felts et al., 1995; 1997), although in one study no Na_X cells were observed (Waxman and Black, 1996). The $Na_V1.6$ isoform is expressed in many neurons in both the dorsal and ventral horns and is present in high abundance at many but not all nodes of Ranvier (Tzoumaka et al., 2000). Cells expressing β1, β2, and β1A subtypes are scattered in the dorsal horn (Blackburn-Munro and Fleetwood-Walker, 1999; Kazen-Gillespie et al., 2000), and in one study β1 increases and β2 decreases in laminae I and II after axotomy (Blackburn-Munro and Fleetwood-Walker, 1999), but β1 decreases in another (Coward et al., 2001c). During development, $Na_V1.1$ (I) is expressed at P2 and increases thereafter, $Na_V1.2$ (II) is seen at all stages, $Na_V1.3$ (III) is prominent at E17 but steadily declines, Na_X (NaG) is seen only at P2, $Na_V1.6$ (Na6) is seen at E17, peaks at P2 but is also found at P15 and P30, and $Na_V1.7$ (hNe) is not expressed (Felts et al., 1997).

ENZYMES

Cholinergic Enzymes (ChAT and ACHE)

Cholinergic systems are demonstrated by histochemical procedures for two enzymes, the relatively specific choline acetyltransferase (ChAT) and the less specific acetylcholinesterase (ACHE).

ChAT

Numerous ChAT cells are seen in the superficial dorsal horn, particularly lamina III (Kimura et al., 1981; Houser et al., 1983; Barber et al., 1984; Phelps et al., 1984; 1988; Borges and Iversen, 1986). Cholinergic dendrites are concentrated in lamina III and fibers and terminals are concentrated a little deeper, in laminae III–V (Phelps et al., 1984; 1988). There is little evidence that ACh is a primary afferent transmitter (Gwyn and Flumerfelt, 1971; Karczmar et al., 1980), for example, spinal cord transection or neonatal capsaicin has little effect on dorsal horn ChAT (Holzer-Petsche et al., 1986), and cholinergic terminals end on primary afferent terminals, but the afferent terminals are not themselves cholinergic (DaSilva and Cuello, 1990). Thus, this is an intrinsic spinal system (Phelps et al., 1984), although there are reports (Phelps et al., 1988) of ChAT DRG cells that need to be confirmed (Chapter 4).

During development, dorsal horn cholinergic cells are "born" on E13, slightly later than ventral horn and IML cholinergic motor cells (Phelps et al., 1988), and many migrate into the dorsal horn from a U-shaped bundle of more ventral cholinergic cells (Phelps et al., 1990). Many ChAT cells coexpress NADPHd (Wetts and Vaughn, 1994), and more than half of ChAT dorsal horn neuron profiles express GAD or GABA (Kosaka et al., 1988). Peripheral axotomy markedly increases ChAT in the cord (Malatova et al., 1985).

AChE

AChE cells and neuropil are seen in laminae I–III of the superficial dorsal horn (Roessmann and Friede, 1967; Navaratnam and Lewis, 1970; Silver and Wolstencroft, 1971; Odutola, 1972). In these regions there is a vertical alternation of AChE staining (Marchand and Barbeau, 1982). AChE staining in the superficial dorsal horn appears relatively late in development (Plantinga et al., 1992). A speculation is that AChE has a role in shaping developmental processes (Oudega et al., 1993).

Cytochrome Oxidase

Cytochrome oxidase is a mitochondrial enzyme indicating oxidative activity (Wong-Riley, 1989). In the dorsal horn, some cells are labeled in lamina I, including Waldeyer cells.

Labeled cells are scattered through deeper laminae with little labeling in lamina II (Wong-Riley and Kageyama, 1986). A useful feature of this reaction is to indicate primary afferent input from single whiskers in the trigeminal system (barrels and barreloids) (Ma and Woolsey, 1984), and it replaced the succinnic dehydrogenase reaction, which was used similarly (Killackey and Belford, 1979).

Fluoride Resistant Acid Phosphatase (FRAP)

Regular or free or common acid phosphatase is a lysosomal hydrolase important in general cell metabolism. A different acid phosphatase, resistant to fluoride ions, fluoride-resistant acid phosphatase (FRAP), is found in high amounts in a subpopulation of small DRG cells and forms a prominent band of staining in the superficial dorsal horn. Exact delineations are the marginal zone (Gerebtzoff and Maeda, 1968), the substantia gelatinosa or Rolando substance or lamina II (Coimbra et al., 1970; Knyihar and Gerebtzoff, 1973; Csillik and Knyihar-Csillik, 1986; McMahon and Moore, 1988), laminae I–II (McNeill and Hulsebosch, 1987), the middle or upper third of lamina II (Nagy and Hunt, 1982; Coimbra et al., 1986), but the usually accepted localization is lamina IIi (Coimbra et al., 1974; 1986; Nagy and Hunt, 1983; Hunt and Rossi, 1985). The area of FRAP localization in the dorsal horn is deep to the termination zones of the peptides SP and CGRP (Nagy and Hunt, 1982; 1983; Hunt and Rossi, 1985; Csillik et al., 1986; Silverman and Kruger, 1988a, b; 1990), although overlap was noted in one study (Carr et al., 1990). FRAP is found exclusively in unmyelinated axons (Coimbra et al., 1974; Hunt and Rossi, 1985), emphasizing its presence in small DRG cells. Numerous FRAP presynaptic terminals are seen in lamina IIi, and the most prominent are the electron dense or dark type 1 sinusoid terminals (Coimbra et al., 1970; Knyihar and Gerebtzoff, 1973; Knyihar et al., 1974; Ribeiro-da-Silva et al., 1986). FRAP staining in the dorsal horn is removed by dorsal rhizotomy or neonatal capsaicin indicating that all or a very large majority of the FRAP terminals arise from primary afferents (Knyihar, 1971; Coimbra et al., 1974; Knyihar et al., 1974; Jancso and Knyihar, 1975; McNeill and Hulsebosch, 1987; Hammond and Ruda, 1991). FRAP terminals are capsaicin-sensitive (Hiura et al., 1997; 1999).

Peripheral nerve section dramatically reduces FRAP staining in the spinal distribution of the fine primary afferents in the damaged nerve (Colmant, 1959; Sanyal and Rustioni, 1974; Knyihar and Csillik, 1976; Bezzegh et al., 1986; Knyihar-Csillik and Torok, 1989;Knyihar-Csillik et al., 1989). Local capsaicin or vinblastine also deplete FRAP centrally (Ainsworth et al., 1981; Csillik et al., 1985). Administration of NGF to the proximal stump of the cut nerve prevents the loss indicating that maintenance of FRAP is dependent on access to this trophic factor (Csillik et al., 1985; Fitzgerald et al., 1985). The FRAP reappears in consort with regeneration after nerve crush (Knyihar-Csillik and Torok, 1989; Knyihar-Csillik et al., 1989), but it also reappears more slowly after nerve cut even if regeneration is prevented (Devor and Claman, 1980). FRAP is so precisely located that the loss after axotomy is used to demonstrate central projections of FRAP-containing afferents in various peripheral nerves (Devor and Claman, 1980, Qui et al., 1984; Kovacs and Ferencsik, 1986; Castro-Lopes and Coimbra, 1991) and to assess sprouting by showing ingrowth of axons from neighboring regions after cord denervation (Fitzgerald and Vrbova, 1985; McMahon and Kett-White, 1991). An increase in acid phosphatase staining is seen after noxious stimuli in capsaicin treated animals (Kantner and Kirby, 1982).

Monoamine Oxidase

Monoamine oxidase A is found in the superficial dorsal horn, where it labels primarily mitochondria in dendrites, but it is also found in synaptic terminals, including some that are primary afferent in origin (Westlund et al., 1993), raising the issue of amine-containing primary afferent neurons (Chapter 4). Binding for monoamine oxidase B is found in the superficial dorsal horn, where it is unchanged in ALS (Jossan et al., 1994).

Nitric Oxide (NO)

Nitric oxide is a gas that relaxes endothelial cells and has diverse neural signaling functions (Vincent, 1994). NO is derived from L-argenine by nitric oxide synthase (NOS), a neuronal nicotinamide adenine dinucleotide phosphate diaphorase (NADPHd) (Blottner et al., 1995). Thus, histochemical reactions for NADPHd are used interchangeably with NOS antibody reactions to indicate NO production (Hama and Sagen, 1994). The reactions do not always coincide, however (Laing et al., 1994; Traub et al., 1994a; Blottner et al., 1995), and there are discrepancies (Vizzard et al., 1994a; 1998; Callsen-Cencic et al., 1999). Whether this is because of differential sensitivity or represents a more fundamental difference such as diaphorases not synthesizing NO (Blottner et al., 1995) is not clear, but here we will regard the reactions as synonymous, and note that future work may modify this position.

Laminar Localizations

Cells labeled for NOS, NADPHd, and cytosolic urea cycle enzymes are found in all dorsal laminae, but they are concentrated in superficial laminae, particularly at the lamina II–III border (Mizukawa et al., 1989; Anderson, 1992; Bruning, 1992; Dun et al., 1992, 1993; Valtschanoff et al., 1992a, b; Spike et al., 1993; Terenghi et al., 1993; Vizzard et al., 1993a; 1994a; 1997; Laing et al., 1994; Rodrigo et al., 1994; Ruda et al., 1994; Saito et al., 1994; Blottner et al., 1995; Nakamura, 1997; Pullen et al., 1997; Marsala et al., 1998; Lukacova et al., 1999). Many NOS or NADPHd superficial dorsal horn cells are bipolar islet cells with long poorly branching or recurved dendrites (Valtschanoff et al., 1992b; Lee, J. H. et al., 1993; Spike et al., 1993; Aimar et al., 1998), so they are presumably inhibitory.

NOS or NADPHd fibers and terminals are concentrated in laminae I and II, with emphasis on lamina IIi, and there are some fibers in lamina III (Bruning, 1992; Valtschanoff et al., 1992a, b; Spike et al., 1993; Vizzard et al., 1993a, b; Laing et al., 1994; Reuss and Reuss, 2001). This superficial plexus arises mostly from intrinsic neurons and is most prominent in the sacral cord (Valtschanoff et al., 1992a; Vizzard et al., 1994a), where it follows visceral afferent pathways (Foster and Phelps, 2000a).

Co-localizations

NADPH and NOS do not co-localize with glutamate or excitatory peptides in dorsal horn neurons (Herdegen et al., 1994; Laing et al., 1994; Aimar et al., 1998; Reuss and Reuss, 2001) but do co-localize with GABA and to a lesser extent glycine (Spike et al., 1993; Laing et al., 1994), so NO-producing cells are mostly inhibitory interneurons. NADPHd does not co-localize or has a very restricted co-localization with induced FOS in dorsal horn cells (Lee, J. H. et al., 1993a; Herdegen et al., 1994), but NOS processes seem to end on FOS cells suggesting a role for NO in FOS expression. There is restricted co-localization with Jun and Krox transcription factors (Herdegen et al., 1994). Most lamina III and IV NADPHd cells express ChAT and so are cholinergic (Spike et al., 1993; Wetts and Vaughn, 1994), and most NOS or NADPHd dorsal horn cells contain GABA and many contain glycine (see below). NADPHd labeled dendrites are occasionally found postsynaptic to a central, glutamate-containing, glomerular axon terminal, suggesting glutamate activation of NO lamina II cells and much non-anatomical evidence favors the hypothesis that NO production is under glutamate control and newly synthesized NO may in turn liberate excitatory peptides (Aimar et al., 1998).

Experimental Manipulations

Dorsal Rhizotomy

Dorsal rhizotomy causes loss of NO immunostaining in the superficial dorsal horn (Vizzard et al., 1993c, d; Reuss and Reuss, 2001), with particular emphasis on primary afferent fibers passing to the sacral parasympathetic nucleus. Numbers of cell profiles expressing NADPHd also decrease after axotomy possibly indicating dependence of NO manufacture on primary afferent input (Reuss and Reuss, 2001).

Peripheral nerve section. NADPHd cell profiles increase after chronic constriction injury (Hama and Sagen, 1994) but overall NADPHd staining decreases after complete transection or TTX applied to the nerve (Carr et al., 1998). NOS profiles decrease (Goff et al., 1998) but overall staining shows no change (Zhang et al., 1993a; Rydh-Rinder et al., 1996) following chronic constriction or transection of rat sciatic nerve. This variability is unexplained, so effects from pathologic perturbations are difficult to compare unless lesions, species, reactions, time windows, etc. are the same (Callsen-Cencic et al., 1999). An increase in NADPHd cell profiles was prevented by adrenal transplants (Hama and Sagen, 1994). Sciatic nerve transection results in dark cells especially in the ipsilateral superficial dorsal horn, presumably due to glutamate toxicity, and blocking NOS prevents this in a dose-dependent manner (Mao et al., 1997). NO donors and glutamate agonists increase cGMP immunostaining in neurons in the superficial dorsal horn, and this is prevented by a NOS inhibitor (Morris et al., 1993). These findings are examples of an increasing body of evidence linking NO formation to glutamate activation (Meller and Gebhart, 1993).

Spinal Damage

As with nerve transection, different investigators find different responses to spinal damage. Thus, a partial spinal lesion results in increases in NOS neuronal profiles on both sides of the lesion (Sharma et al., 1996), and NADPHd profiles in the nucleus dorsalis after complete transection (Wu, 1993; Wu et al., 1994). On the other hand, transection causes a considerable decrease in NADPHd profiles on both sides of a transection (Callsen-Cencic et al., 1999; Trudrung et al., 2000). This may reflect a biphasic response to spinal lesions, with responses depending on survival time.

Inflammation

Intraplantar formalin or carrageenan and intra-articular CFA increase NADPHd or NOS neuronal profiles and NOS neuropil staining bilaterally in the superficial dorsal horn, although there the increase is usually greater ipsilaterally (Solodkin et al., 1992; Meller and Gebhart, 1993; Herdegen et al., 1994; Traub et al., 1994a, b; Lam et al., 1996; Wu, J. et al., 1998b; Tao and Johns, 2002), although NOS does not increase whereas NADPHd does in one study (Traub et al., 1994a). NO-activated soluble guanylate cyclase is also upregulated in the spinal cord following intraplantar formalin (Tao and Johns, 2002). NOS co-localizes with cyclo-oxygenase-2 after inflammation, the latter being important in prostaglandin synthesis (Maihofner et al., 2000). By contrast, numbers of NOS profiles decrease after acute experimental myositis with additional surgery but increase with myositis alone suggesting that the strength of the input influences numbers of labeled profiles, with stronger or longer-lasting stimuli decreasing profile numbers, consistent with the findings that numbers of NOS profiles increase or decrease depending on the frequency and duration of C-fiber stimulation (Hoheisel et al., 1995; 1997; Hoheisel and Mense, 2000).

Development

Several investigators note that NOS or NADPHd does not occur in superficial dorsal horn cells until 1–2 weeks after birth (Liuzzi et al., 1993; Soyguder et al., 1994; Vizzard et al., 1994b; Wetts et al., 1995), but there is a transient expression of NADPHd in almost all DRG cells before birth that fades out in most neurons except in lumbar areas (Wetts and Vaughn, 1993). Labeled neurons in the deep dorsal horn appear much earlier in development than superficial dorsal horn NO neurons (Foster and Phelps, 2000b).

Protein Kinase C

There are many isoforms of protein kinase C, but protein kinase Cγ (PKCγ) leapt into prominence when it was shown that PKCγ knock-out mice have normal responses to acute painful stimuli but increased thresholds to neuropathic pain and reduced NPY and NK1R increases that normally follow nerve lesions (Malmberg et al., 1997a, b). PKCγ cells are found in lamina II, especially IIi, and some co-localize with GAD-67 but not with parvalbumin, NO, or GABA so they are likely to be excitatory interneurons (Martin et al., 1999; Polgar et al., 1999b; Miletic et al., 2000; Li et al., 2001), although in the medullary dorsal horn PKCγ did co-localize with calcium-binding proteins (Ni et al., 2002). The PKCγ cells receive SP and IB4 input so they may be directly excited by nociceptors (Li et al., 2001). Local inflammation increases PKCγ in the superficial dorsal horn and causes translocation to the cell membrane (Martin et al., 1999).

HORMONES

Corticotrophin Releasing Factor (CRF)

Corticotrophin releasing factor is a 41-amino acid peptide that releases ACTH and thus is important in the adrenal secretion that follows stress. In the dorsal horn CRF fibers and terminals are concentrated in laminae I–II (Olschowka et al., 1982; Merchenthaler et al., 1983; Schipper et al., 1983; Merchenthaler, 1984; Petrusz et al., 1985; Foote and Cha, 1988; Mantyh et al., 1989a). No CRF cell bodies are seen in the dorsal horn in normal animals (Foote and Cha, 1988; Puder and Papka, 2001a), but with colchicine or hypophysectomy a few are seen in the deep dorsal horn and occasionally in the superficial dorsal horn (Merchenthaler et al., 1983; Merchenthaler, 1984; Petrusz et al., 1985). This system is thought to be largely intrinsic (Petrusz et al., 1985), but hypothalamic and pontine CRF cells contribute to spinal cord innervation (Puder and Papka, 2001a) and capsaicin and dorsal rhizotomy lessen dorsal horn staining (Skofitsch et al., 1984), particularly at parasympathetic sacral levels (Kawatani et al., 1996). Thus, descending and primary afferent fibers presumably make a contribution. Care must be taken not to have cross-reactions of CRF and SP immunostaining (Berkenbosch et al., 1986).

Thyroid-releasing Hormone (TRH)

Cells and fibers containing TRH are found in laminae I–III, the cells particularly at the II–III border (Coffield et al., 1986; Harkness and Brownfield, 1986; Tsuruo et al., 1987; Ulfhake et al., 1987; Arvidsson et al., 1990; 1992a; Fleming and Todd, 1994). Cells expressing TRH prohormone are located in lamina III (Lechan et al., 1987). TRH fibers are not affected by spinal transection, but they are reduced by a combined spinal transection-dorsal rhizotomy (Arvidsson et al., 1990). The dorsal rhizotomy result would seem to indicate some TRH primary afferents projecting into the cord, but in the absence of evidence for TRH DRG cells, the conclusion is that this is an intrinsic system (Arvidsson et al., 1990). These fibers do not co-localize with 5-HT in the dorsal horn, although they do in the ventral (Arvidsson et al., 1990; 1992a), and most TRH cells in the dorsal horn are GABAergic presumably indicating an inhibitory system (Fleming and Todd, 1994). TRH is mainly degraded by a TRH-degrading ectoenzyme, and cells expressing this enzyme are spread evenly through all laminae (Heuer et al., 1998).

PEPTIDES

Angiotensin

The active form of angiotensin, angiotensin II, controls blood pressure in addition to many other effects. In the spinal cord angiotensin II is concentrated in the superficial dorsal horn (Fuxe et al., 1976; Changaris et al., 1978; Lind et al., 1985), and angiotensin immunolabeled cells are somewhat more numerous in the substantia gelatinosa of neonatal as compared to adult rats (Sood et al., 1990).

Atrial and Brain Natriuretic Peptide

Atrial natriuretic peptide is a cardiac hormone found in a few fibers in the superficial dorsal horn (Skofitsch et al., 1985b; Kawata et al., 1989; Nohr et al., 1989; Saper et al., 1989). Brain natriuretic peptide is a closely related peptide found in large numbers of axons in laminae I and either IIo (Saper et al., 1989) or IIi (Kawata et al., 1989).

Bombesin

Bombesin was isolated from frog skin and is the prototype for a large number of closely related, biologically active peptides, collectively referred to as bombesin-like peptides, of which three are bombesin itself and the closely related mammalian gastrin-releasing factor and Neuromedin B (Ohki-Hamazaki, 2000). These compounds have various GI and CNS effects (Ohki-Hamazaki, 2000), and are proposed as having a role in nociceptive input. A plexus of bombesin-like fibers and terminals is found in laminae I–II and sometimes III (Fuxe et al., 1983; Massari et al., 1983; Panula et al., 1983; 1988; O'Donohue et al., 1984; Decker et al., 1985; Namba et al., 1985; Panula, 1986; Chung et al., 1989b; Westermark et al., 1999). These arise largely from primary afferents because dorsal rhizotomy and neonatal capsaicin cause a marked decrease of these fibers (Panula et al., 1982; O'Donohue et al., 1984; Decker et al., 1985; Massari et al., 1985). No bombesin-containing cells have been described in the dorsal horn (above references), but cells in the gelatinosa display a moderate signal for bombesin-like mRNA (Wada et al., 1990). Adrenalectomy followed by dexamethasone increases dorsal horn immunostaining, suggesting a trophic action of glucocorticoids on the expression of these peptides (Westermark et al., 1999).

Cholycystokinin (CCK)

Fibers, terminals, and cells are concentrated in the superficial dorsal horn (Larsson and Rehfeld, 1979; Loren et al., 1979; Vanderhaeghen et al., 1980; 1981; 1982; Gibson et al., 1981; Jancso et al., 1981; Maderdrut et al., 1982; Priestley et al., 1982a; Schultzberg et al., 1982; Schroder, 1983; Conrath-Verrier et al., 1984; Fuji et al., 1985; Tuchscherer and Seybold, 1985; Micevych et al., 1986; Tuchscherer et al., 1987; Chung et al., 1989b), in small DRG cells (Chapter 4) and in brainstem cells that project to the cord (Skirboll et al., 1983; Ju et al., 1987a). Dorsal rhizotomy, capsaicin administration, ganglionectomy, and peripheral nerve section remove some but not all spinal cord staining (Ainsworth et al., 1981; Jancso et al., 1981; Gibson et al., 1982; Maderdrut et al., 1982; Priestley et al., 1982a; Micevych et al., 1986; Shehab and Atkinson, 1986a, b; Ju et al., 1987b; Tuchscherer et al., 1987), suggesting that much dorsal horn CCK arises from primary afferents, but since not all staining is removed, there are intrinsic and descending CCK sources.

However, when the C-terminal antibody used in many of the above studies is absorbed with CGRP, DRG staining is abolished (Ju et al., 1986) and CGRP and CCK staining coexist suggesting that CGRP and CCK antibodies recognize the same antigen (Ju et al., 1987b). Also no measurable CCK mRNA is found in rat DRGs (Cortes et al., 1990; Schiffmann et al., 1991), immunochemical CCK is not "true" CCK (Zouaoui et al., 1990), dorsal rhizotomy or capsaicin do not reduce radioassayed spinal CCK (Marley et al., 1982; Schultzberg et al., 1982; Pohl et al., 1990) and staining with an antibody that does not cross-react with CGRP labels only in lamina III and below (Williams et al., 1987) or showed essentially no change in the superficial trigeminal nucleus after rhizotomy (Jacquin et al., 1992). This is strong evidence that CCK does not occur to any great extent in rat DRG cells, and that earlier antibodies were cross-reacting with CGRP or a similar substance. The absence of CCK in primary afferents is species-specific, however, in that there are large numbers of CCK mRNA-expressing DRG cells in the guinea-pig and monkey (Cortes et al., 1990; Seroogy et al., 1990; Verge et al., 1993) and significant CCK by radioassay in cat DRGs (Yaksh et al., 1982a; Rehfeld and Lundberg, 1983). Thus although the rat is by far the most commonly used animal in these studies, it is exceptional in having little CCK in normal adult DRG cells. Interestingly the difficulty with the classic stains seems to apply only to primary afferents because in other regions, classic and modern staining is essentially identical. Intrinsic and descending CCK systems exist in rats (Hökfelt et al., 1991; Jacquin et al., 1992).

CCK antagonizes opioids at the spinal level, and since opioids are largely ineffective in neuropathic pain, this might be because peripheral axotomy increases CCK in the dorsal horn (Hökfelt et al., 1991; 1994a; Wiesenfeld-Hallin and Xu, 1996; 2001). In this regard, although axotomy increases numbers of CCK mRNA rat DRG cells (Verge et al., 1993), there is no increase in CCK itself in the dorsal horn (Bras et al., 1999). Another possibility is that CCK receptors become more "effective" since they increase after axotomy (Zhang et al., 2000). Thus, changes in CCK or its receptor may play a role in morphine ineffectiveness in neuropathic pain, but there are disparities in the morphologic findings (Hökfelt et al., 1994b).

Acute morphine administration does not change spinal CCK in normals or after inflammation, and reduces CCK after peripheral axotomy (Zhang et al., 2000), but chronic morphine administration increases spinal CCK (Ding and Bayer, 1993). Thus morphine affects CCK levels in a complicated way, and further study to clarify parameters is warranted.

Calcitonin Gene-related Peptide (CGRP)

CGRP is a peptide expressed by many DRG cells (Chapter 4). Peripherally CGRP is a major vasodilator and plays an important role in neurogenic inflammation, and in the cord CGRP potentiates the action of SP.

Laminar Localizations

CGRP fibers and terminals are concentrated in laminae I–IIo and V (Carlton et al., 1987; 1988; 2001a; Chung et al., 1988; Bennett et al., 1989; Traub et al., 1989a, b; Tuchscherer and Seybold, 1989; Fried et al., 1990; Cameron et al., 1991; Hirakawa and Kawata, 1992a; Alvarez et al., 1993; Christensen and Hulsebosch, 1997b; Belyantseva and Lewin, 1999; Kashiba et al., 2001a), although others report that all of lamina II contains these axons (Rosenfeld et al., 1983; Gibson et al., 1984a; Skofitsch and Jacobowitz, 1985a; Blumenkopf, 1988; Dipette et al., 1988; Harmann et al., 1988b; Weihe et al., 1988a; Sharkey et al., 1989; Jakab et al., 1990; Traub et al., 1990; Bonfanti et al., 1991; Marlier et al., 1991b; Zhang, L. et al., 1993b, c; 2001; Ma et al., 1995; Ribeiro-da-Silva, 1995; Munglani et al., 1996; Miki et al., 1998a; Ritz et al., 2001), and one study found the labeling in laminae I, II, and the outer part of III (Nothias et al., 1993). The distinction between CGRP labeling in IIo as contrasted to IIi can be checked by comparison with IB-4 or phosphatase labeling which are restricted to IIi (Carlton et al., 2001a).

Intrinsic Neurons

A few CGRP cells have been described in the dorsal horn (Conrath et al., 1989; Tie-Jun et al., 2001) and trigeminal nucleus (Kruger et al., 1988a), although others do not see them (Gibson et al., 1984a; Piehl et al., 1992; Zhang et al., 1993b), and no cells with CGRP mRNA are found in the dorsal horn (Rethelyi et al., 1989a). Further, essentially all CGRP staining in the dorsal horn vanishes after dorsal rhizotomy (see below) so CGRP is used as a pure primary afferent marker notwithstanding reports of a few intrinsic CGRP cells.

Synaptic Organization

CGRP presynaptic terminals are most numerous in laminae I–IIo. These terminals are characterized by myriad small round clear vesicles and numerous dense-core vesicles, with the CGRP largely in the cores of the dense-core vesicles but occasionally free in the axoplasm. The synaptic thickenings between CGRP-containing presynaptic terminals and postsynaptic targets are asymmetrical with the postsynaptic thickenings more prominent. CGRP terminals are of two types; the first, more common, are simple dome-shaped endings forming an attachment to a post-synaptic element. The other is as a central terminal in a glomerulus (Carlton et al., 1987; 1988; 1990; Harmann et al., 1988a; McNeill et al., 1988a; 1990; Merighi et al., 1989; 1991; Alvarez et al., 1993; Zhang et al., 1993b; Ribeiro-da-Silva, 1995). Glomerular endings are where much presynaptic control of primary afferent input is located, but terminals on CGRP endings in glomeruli or in the general neuropil are not seen (Hayes and Carlton, 1992; Alvarez et al., 1993). Thus, there is no morphologic evidence for presynaptic control of the CGRP afferent input, compared to the numerous endings on non-peptidergic central terminals. Two individually labeled unmyelinated CGRP afferent axons do not end in glomeruli possibly indicating that myelinated but not unmyelinated CGRP sensory axons give rise to glomerular central endings (Alvarez et al., 1993). CGRP terminals form synapses on DYN cell bodies axons and terminals indicating that CGRP has a postsynaptic effect on DYN transmission (Carlton and Hayes, 1989).

Dorsal Column Organization

Dorsal column axons are classically described as large and myelinated. Unmyelinated axons, many of which contain CGRP, are also found in dorsal columns, however, and make up approximately 30% of the total (Briner et al., 1988; Kruger et al., 1988b; McNeill et al., 1988b; Tamatani et al., 1989; Traub et al., 1989b; Fabri and Conti, 1990; Patterson et al., 1990; Hirakawa and Kawata, 1992a). Retrograde labeling and appropriate denervations show that DRG cells give rise to most or all of these fibers (Patterson et al., 1989; Traub et al., 1989b; Fabri and Conti, 1990). Presumably these fibers bring nociceptive input to dorsal column nuclei, which might bear on dorsal column stimulation for pain relief. In a retrograde double-labeling study, CGRP DRG cells whose axons ascended to the dorsal column nuclei were large, indicating that they were unlikely to give rise to fine nociceptive afferent axons (Giuffrida and Rustioni, 1992), although in another study it was noted that CGRP fibers that project in the dorsal columns are mostly unmyelinated but the cells that were backfilled were large (Tamatani et al., 1989). Thus, although not all points are resolved, there is an unmyelinated CGRP primary afferent pathway in the dorsal columns. Although peripheral axotomy generally decreases CGRP mRNA and peptide (see below), DRG cells whose axons project in the dorsal column upregulate CGRP (Miki et al., 1998a), presumably indicating enhancement of dorsal column CGRP function after nerve damage.

Descending Fibers

Although most investigators find no descending CGRP fibers, one study reports that dorsolateral funiculus (DLF) lesion reduces CGRP caudally, suggesting a descending CGRP input into the dorsal horn (Wang, H. et al., 2000). The DLF contains numerous primary afferent fibers, however (Chung et al., 1987), and damage to these could also explain this loss.

Experimental Manipulations

Deafferentation

Deafferentation by dorsal rhizotomy or neonatal capsaicin removes essentially all CGRP fibers from the dorsal horn, especially if the rhizotomies are multiple and bilateral.

(Gibson et al., 1984a; Skofitsch and Jacobowitz, 1985b; Chung et al., 1988; Traub et al., 1989a; Tuchscherer and Seybold, 1989; Arvidsson et al., 1990; Pohl et al., 1990; Villar et al., 1991; Hirakawa and Kawata, 1992a). A very few CGRP fibers are seen after extensive rhizotomies, but there are a few ectopic CGRP cells in the proximal dorsal root that would be missed by a typical rhizotomy (Jeftinija and Jeftinija, 1990), and occasionally CGRP dorsal horn cells are reported (see above). However, since the loss of CGRP is essentially complete, CGRP is used as a marker for primary afferent axons in the dorsal horn (Chung et al., 1988; Traub et al., 1989b; Tuchscherer and Seybold, 1989).

Peripheral Nerve Lesions

Most reports show a loss of CGRP in those parts of the dorsal horn into which the axons of adult axotomized sensory cells project (Bennett et al., 1989; Sharkey et al., 1989; Noguchi et al., 1990b; Cameron et al., 1991; Villar et al., 1991; Inaishi et al., 1992; Zhang et al., 1993c; Hökfelt et al., 1994c; Carlton and Coggeshall, 1996; Na et al., 2001; Shi et al., 2001). By contrast, however, nerve damage results in no change in CGRP staining in the dorsal horn in two studies (Garrison et al., 1993; Munglani et al., 1996), and although there was an overall decrease, there was an increase in a subpopulation of neurons whose axons projected to laminae III and IV and dorsal column nuclei after segmental axotomy (Miki et al., 1998a). Thus, although responses to axotomy differ, possibly reflecting different experimental paradigms, the consensus is that peripheral axotomy results in a downregulation of CGRP mRNA and a decrease in the amount of CGRP in the DRG and dorsal horn. The reduction in CGRP did not correlate with the development of symptoms in a neuropathic nerve lesion model, however (Na et al., 2001). Neonatal sciatic nerve lesions in rats lead to extensive loss of DRG cells and almost no CGRP in the cord acutely. Then CGRP recovers completely, presumably from DRG cells in adjacent ganglia and from the few cells that survive peripheral axotomy (Nothias et al., 1993).

Spinal Hemisection

Spinal hemisection leads to a chronic increase in CGRP immunostaining in laminae III and IV bilaterally, interpreted as sprouting (see below) (Christensen and Hulsebosch, 1997b).

Inflammation

Increases in dorsal horn CGRP are seen at 3, 8, or 21 days after induction of unilateral arthritis; 1 and 2 months after polyarthritis; 3, 5, or 7 days after inflammation of the paw, 10 days after experimental cystitis, and 2 hours after formalin stimulation (Weihe et al., 1988a; Marlier et al., 1991b; Donnerer et al., 1992; Mapp et al., 1993; Zhang, R. X. et al., 1994; Galeazza et al., 1995; Mulder et al., 1997; Vizzard, 2001). Declines are seen before (Galeazza et al., 1995; Calza et al., 1998) or after the increase (Mapp et al., 1993). Thus there may be a time-limited increase in dorsal horn CGRP following inflammation with declines that precede and/or follow the increase. During the florid phase of inflammation, a considerable increase of CGRP terminals on DYN dendrites and cells is reported (Takahashi et al., 1988). Permanent changes in peptide systems are reported after neonatal inflammation (Ruda et al., 2000), but care must be taken to ensure that the changes are not the result of persistent inflammation.

Morphine and Tolerance

A major drawback in morphine therapy is development of tolerance. Intrathecal morphine increases CGRP in the dorsal horn in concert with the development of tolerance and reductions in CGRP receptor binding (Menard et al., 1995a, b; 1996; Ma, W. et al., 2001). This may indicate a role for CGRP in morphine tolerance.

Electrical Stimulation

Electrical stimulation of peripheral nerves produces significant depletion of CGRP immunostaining in the superficial dorsal horn (Klein et al., 1990) and equivalent parts of the trigeminal nuclear complex, presumably because of release from fine afferents (Klein et al., 1991). The trigeminal depletion may play a role in migraine (Samsam et al., 2001).

MK-801

The NMDA antagonist MK-801 increases the amount of CGRP in the dorsal horn of normal rats (McNeill et al., 1991a; Garrison et al., 1993).

Dietary Calcium

Increases in dietary calcium are correlated with the increases in CGRP staining (Dipette et al., 1988).

Development and Age

NT-3 knock-out animals lose CGRP cells, so the inference is that not all these cells are under control of NGF (Airaksinen and Meyer, 1996). The CGRP system in sheep develops early, before the glutamate system (Nitsos and Rees, 1993), and the CGRP content of the dorsal horn decreases with age (Bergman et al., 1996).

Sprouting

Formation of new axons and terminals (sprouting) may be of great importance in recovery from spinal injury. For CGRP systems, denervation in a spared root paradigm increases CGRP immunostaining from axons in the spared root (McNeill et al., 1990; 1991b), afferent denervation of the cord leads to loss followed by reappearance of CGRP immunostaining from the edges of the denervated area (Piehl et al., 1992), CGRP afferents stain more intensely and spread over a wider area after peripheral nerve lesion accompanied by spinal denervation (Belyantseva and Lewin, 1999), neonatal sciatic nerve lesions cause an almost complete initial loss followed by a return of CGRP staining (Nothias et al., 1993), CGRP axons form pericellular baskets surrounding large DRG cell bodies in the normal (Kuwayama et al., 1987) and following peripheral nerve lesions (McLachlan and Hu, 1998; Michael et al., 1999; Hu and McLachlan, 2000), and there is an increase in the territory occupied by CGRP fibers after MK-801 administration or spinal hemisection (Harris et al., 1990; McNeill et al., 1991b). These findings are interpreted as indicating sprouting, certainly a reasonable possibility, but they could also be CGRP appearing in previously unlabeled but extant axons and synapses. Finally, CGRP immunostaining is used to see if fine afferent axons penetrate or sprout into the cord to assay regeneration (Houle and Reier, 1989; Itoh and Tessler, 1990; Itoh et al., 1999; Romero et al., 2001).

Endothelin (ET)

Endothelin is a 21-amino acid peptide that is found in three forms, ET-1, ET-2, and ET-3. These are vasoconstrictors but have diverse actions including some that relate to nociception. Immunostaining for endothelin is concentrated in lamina III with some staining in laminae I–II (Yoshizawa et al., 1989).

Galanin

Galanin was isolated from intestine and is found in DRG cells and superficial dorsal horn. It is markedly upregulated in the DRG by peripheral axotomy (Bartfai et al., 1993), and levels correlate with behavioral hypersensitivity (Xu et al., 2000) although the effects of galanin are complicated (Kerr et al., 2000).

Laminar Localizations

A dense galanin axonal plexus is seen in superficial dorsal horn (Ch'ng et al., 1985; Skofitsch and Jacobowitz, 1985c, d; Melander et al., 1986; Arvidsson et al., 1991; Kordower et al., 1992; Bartfai et al., 1993; Wiesenfeld-Hallin and Xu, 1998; Simpson et al., 1999) with a few fibers in deeper laminae (Rokaeus et al., 1984). There is more galanin in sacral cord than other areas via RNA analysis (Arvidsson et al., 1991). Intrinsic galanin cells are seen in the superficial dorsal horn (Melander et al., 1986; Arvidsson et al., 1991; Kordower et al., 1992). These are islet cells, some of which co-localize with GABA, enkephalin, and neuropeptide Y (Simmons et al., 1995; Zhang, X. et al., 1995a).

Synaptic Organization

Galanin containing presynaptic terminals are of two types. First is a simple terminal containing small round clear vesicles and large dense-core vesicles, the cores of which contain the peptide, and pleomorphic vesicles (Zhang, X. et al., 1995d). They have asymmetric thickenings and end on dendrites and occasional axons. The second is as a central element in type I or occasionally type II glomeruli (Arvidsson et al., 1991; Zhang et al., 1993b). Rare axoaxonic and dendrodendritic complexes with the galanin element presynaptic are seen (Zhang, X. et al., 1995a, d).

Experimental Manipulations

Dorsal rhizotomy reduces but does not eliminate galanin staining in superficial laminae indicating primary afferent and intrinsic galanin fibers (Tuchscherer and Seybold, 1989; Arvidsson et al., 1991). All galanin central elements in glomeruli degenerate showing that they are primary afferent in origin (Zhang, X. et al., 1993b; 1995c).

Following peripheral nerve transection, some find less galanin immunostaining in superficial laminae (Klein et al., 1991; Munglani et al., 1996), some no change

(Arvidsson et al., 1991; Wiesenfeld-Hallin and Xu, 2001), others an increase (Zhang, X. et al., 1993c; Rydh-Rinder et al., 1996; Ma, W. and Bisby, 1997; Ramer et al., 1998c) or no change superficially and a slight increase deeper (Zhang, X. et al., 1995c). Since there is increased release of galanin in the cord after this lesion (Zhang, X. et al., 1998a), the parsimonious explanation is that axotomy causes an increased production and central transport of galanin in DRG cells (Villar et al., 1991), so when release predominates there is less cord immunostaining and when release is less or production greater, immunostaining is unchanged or increased. Another possibility is that excess galanin is preferentially transported peripherally. Interestingly, partial nerve lesions in neuropathic pain models cause greater changes than simple transections (Ma, W. and Bisby, 1997), although this difference was not as marked in a later study (Shi et al., 1999a). However, numbers of galanin synapses do not change in the superficial dorsal horn after peripheral axotomy (Carlton SM and Coggeshall, 1996).

Spinal transection does not change galanin in the dorsal horn indicating no descending fibers (Arvidsson et al., 1991). Galanin staining decreases early in arthritis, presumably because of increased release, followed by an increase in galanin mRNA in both cord and ganglion and a return of immunostaining to normal levels (Calza et al., 1998).

Hypocretin

Hypocretin is a peptide found in hypothalamic cells that project in considerable numbers to lamina I of the spinal cord (Van Den Pol, 1999). The hypocretins orexin-A and orexin-B are concentrated in lamina I (Date et al., 2000).

Neuropeptide FF (NPFF)

The peptide FMRF-NH2, isolated from clams, shows immunoreactivity in the mammalian CNS, and homology studies led to the identification of related mammalian peptides neuropeptides FF (NPFF) and AF (NPAF) (Panula et al., 1999). NPFF has antinociceptive properties and attenuates opioid analgesia (Panula et al., 1999), although the effects are complicated (Roumy and Zajac, 1998). NPFF neurons, fibers, and terminals are found in the superficial dorsal horn (Weber et al., 1981; Hokfelt et al., 1983; Chronwall et al., 1984; Panula et al., 1987; Chen et al., 1989; Allard et al., 1991; Kivipelto and Panula, 1991; Lee, C. H. et al., 1993). This is species-specific, however, in that there is no labeling in pig dorsal horn (Wasowicz and Panula, 1994). Many NPFF lamina I cells are large with tangentially oriented dendrites and thus resemble Waldeyer cells (Kivipelto and Panula, 1991). There are only small contributions to the laminae I–II NPFF plexus from descending fibers (Ferrarese et al., 1986; Majane et al., 1989), and no NPFF primary afferent neurons are observed in normal rats (Panula et al., 1987; Majane et al., 1989). Thus, although dorsal rhizotomy reduces dorsal horn FMRH-NH2 (Ferrarese et al., 1986), no such reduction is seen when NPFF is measured chemically (Majane et al., 1989). Accordingly, even though a few small NPFF DRG cells are seen after colchicine exposure (Majane et al., 1989), the NPFF plexus seems primarily of intrinsic origin (Kivipelto and Panula, 1991; Panula et al., 1996).

During development, the NPFF dorsal horn system appears relatively late, not until after birth (Majane et al., 1989; Kivipelto et al., 1991). NPFF processes contact spinothalamic cells (Aarnisalo and Panula, 1998). There is a significant increase in NPFF mRNA in the dorsal horn after inflammation but not after nerve lesions (Vilim et al., 1999).

Neuropeptide Y (NPY), Pancreatic Polypeptide (PPY) and Peptide YY

Laminar Localizations

NPY and PPY fibers and terminals are concentrated in superficial horn, primarily lamina II, and the innervation is usually heaviest in lumbosacral cord (Hökfelt et al., 1981; Hunt et al., 1981a, b; Allen et al., 1984; Chronwall et al., 1985; Sasek and Elde, 1985; de Quidt and Emson, 1986; Blessing et al., 1987; Krukoff, 1987; Merighi et al., 1990; Wakisaka et al., 1991; Doyle and Maxwell, 1993b; 1994) except Broome et al. (1985). These fibers come primarily from NPY cells that usually do not stain intensely but are somewhat concentrated in the superficial dorsal horn (Hökfelt et al., 1981; Hunt et al; 1981a, b; Gibson et al., 1984b; Sasek and Elde, 1985; de Quidt and Emson, 1986; Krukoff, 1987; Merighi et al., 1990), although some NPY fibers descend from cells in locus coeruleus (Holets et al., 1988). Small NPY mRNA expressing cells are numerous in laminae I–III, with fewer but larger such cells in deeper laminae (Minson et al., 2001). No particular cell types have been identified. There are no primary afferent NPY fibers in normal cords (see below), although this may be because of low synthesis rather than complete absence (Hökfelt et al., 1998). NPY cells appear early in development in human laminae I and II and as development proceeds in deeper laminae, ultimately to become most numerous in the intermediolateral cell column (Shen et al., 1994).

Synaptic Organization

NPY presynaptic terminals can be found throughout the dorsal horn but are most common in laminae I–II. They contain pleomorphic small clear vesicles and large dense core vesicles, with the immunostaining free in the axoplasm or associated with the dense-core vesicles. These terminals usually form symmetrical junctions with unidentified dendrites or cell bodies, but approximately one third end on other axon terminals, many of which are the central endings in glomeruli. Thus NPY acts presynaptically, on primary afferent and presumably other spinal inputs, and postsynaptically on dendrites and cell bodies (Doyle and Maxwell, 1993b; 1994).

Co-localizations

There is considerable co-localization of NPY with other neuroactive substances within the CNS (e.g., Hunt et al., 1981b; Holets et al., 1988), and in the dorsal horn NPY and galanin are co-localized in some neurons (Zhang et al., 1995a), and there is considerable co-localization of NPY and catechol amines in brain stem cells that project to the cord, but these are probably not co-localized in cord axons (Doyle and Maxwell, 1994). The most striking co-localization is that essentially every VIP cell is GABAergic (Rowan et al., 1993), which would imply that VIP is released with GABA, an important inhibitory compound.

Experimental Manipulations

Dorsal Rhizotomy or Spinal Transection

There is essentially no effect on spinal NPY systems after dorsal rhizotomy or spinal transection confirming that there is little primary afferent or descending VIP input in normal rats (Hökfelt et al., 1981; Hunt et al., 1981a; Gibson et al., 1984b).

Peripheral Axotomy

Peripheral axotomy leads to a marked upregulation of NPY mRNA and the peptide itself in DRG cells (Chapter 4) and in laminae I–III of the dorsal horn where the label is both more intense and spreads over a wider area, into laminae III–IV, consistent with finding that large DRG cells express NPY after this lesion (Wakisaka et al., 1991; 1992; Ma, W. and Bisby, 1998; Shi et al., 2001), although in one instance this effect was not observed (Allen et al., 1984).

Inflammation

Inflammation was first thought to have no effect (Wakisaka et al., 1992), but paw inflammation increases NPY mRNA in spinal cord neurons, particularly in laminae I and II, as well as FOS and opioid peptides, with no change in DRG cells (Ji et al., 1994). Thus, the effects of inflammation on NPY seem to be mostly on central neurons whereas the effects of peripheral axotomy are more dramatic in DRG cells.

Neurotensin (NT)

Neurotensin is a peptide with diverse actions, with emphasis in the dorsal horn on its antinociceptive properties (Kalivas et al., 1982). Neurotensin cells, axons, and terminals are concentrated in the superficial dorsal horn (Uhl et al., 1979a; Seybold and Elde, 1980; 1982; Hunt et al., 1981a; Ninkovic et al., 1981; Emson et al., 1982; Jennes et al., 1982; Polak and Bloom, 1982; Gibson et al., 1984b; Papadopoulos et al., 1986a; Mai et al., 1987; Chen et al., 1988; Boissonade et al., 1993; Atoji et al., 1995). Most NT cells are in laminae II and III, but one study noted cells primarily in laminae I and VI (Miller and Seybold, 1987). Axons and terminals are concentrated in laminae I and II and concentrate in two bands, the first in laminae I and IIo and the other in IIi and III (Todd and Spike, 1993; Zhang, X. et al., 1995d). NT cells in lamina II resemble islet cells (Seybold and Elde, 1982; Todd et al., 1992b) and those in laminae III II/III border cells (Seybold and Elde, 1982). By contrast, neurotensin cells in laminae I and II of the trigeminal nucleus are spiny cells (Uhl et al., 1979a). Many neurotensin cells co-localize with calbindin (Yoshida et al., 1990) or glutamate (Todd et al., 1994a) and a few central neurotensin terminals in glomeruli co-localize with CGRP (Zhang et al., 1993b). On the other hand, NT cells do not co-localize with GABA (Todd et al., 1992b), somatostatin (Proudlock et al., 1993), or 5-HT (Maxwell et al., 1996), and NT is not found in dorsal horn cells whose axons ascend into the brain stem (Leah et al., 1988).

The superficial dorsal horn contains numerous NT axodendritic and some axosomatic terminals with asymmetric synaptic thickenings (some cell body synapses have symmetric thickenings), containing small clear vesicles and a few large dense-core vesicles, with NT prominent in the cores of the large vesicles (Hunt et al., 1981a; Ninkovic et al., 1981; DiFiglia et al., 1982a; 1984; Seybold and Malay, 1984). NT terminals end on vesicle-containing dendrites, but dendroaxonic synapses are not seen (DiFiglia et al., 1984; Seybold and Malay, 1984). A few NT central glomerular elements, co-localize with CGRP (Zhang, X. et al., 1993b). The other type of NT terminal in the superficial dorsal horn is axoaxonic, with the NT terminal ending on type I or type II central glomerular endings, but without synaptic thickenings, suggesting NT volume transmission for presynaptic control of primary afferent input (Zhang, X. et al., 1995d). Dorsal rhizotomy, capsaicin administration, and spinal transection have no effect on dorsal horn NT staining or content (Seybold and Elde, 1980; Ainsworth et al., 1981; Hunt et al., 1981a; Ninkovic et al., 1981; Yaksh et al., 1982b; Howe et al., 1987; Tuchscherer and Seybold, 1989), so the NT plexus in the dorsal horn arises almost exclusively from intrinsic dorsal horn NT neurons. Peripheral axotomy decreases NT staining in the dorsal horn, indicating a transganglionic effect after this lesion (Zhang, X. et al., 1995d).

OPIOIDS

Dynorphin (DYN)

DYN is an endogenous κ-opioid ligand that comes in A and B forms, with the A form further subdivided by molecular size (e.g., dynorphin 1-8 or 1-17). Although there are functional differences between the "dynorphins" (Faden, 1990), most anatomical studies do not distinguish between them, so we will describe the anatomy of the dynorphin system associated with the dorsal horn.

Laminar Localizations

DYN cells, fibers, and terminals are concentrated in the superficial dorsal horn and to a lesser extent lamina V (Khachaturian et al., 1982; Vincent et al., 1982; Cruz and Basbaum, 1985; Miller and Seybold, 1987; Nahin, 1988; Ruda et al., 1988; Carlton and Hayes, 1989; Tuchscherer and Seybold, 1989). There is discussion as to whether the label is concentrated only in lamina I–IIo or fills lamina I–II completely (Cruz and Basbaum, 1985; Tuchscherer and Seybold, 1989), which may relate to recognition of different forms of DYN. DYN cell profile numbers are approximately equal in laminae I and V in cats but are more numerous in lamina I in rats (Miller and Seybold, 1987). Several of the above studies note that DYN immunostaining is less intense than that for enkephalin.

Intrinsic Cell Types

In lamina I, dynorphin cells are fusiform, pyriform, or triangular (Cruz and Basbaum, 1985) or pyramidal, flattened, fusiform, or polygonal, the largest being marginal cells of Waldeyer (Miller and Seybold, 1989), and in lamina II, stalked and possibly central cells (Cruz and Basbaum, 1985), so no particular dorsal horn cell type contains dynorphin. Some DYN marginal neurons project to the parabrachial nucleus, as do a different population of enkephalin marginal neurons (Standaert et al., 1986), so these opioids may be involved in supraspinal inhibition by stimulating descending inhibitory systems. Prodynorphin neurons in the superficial dorsal horn project to the thalamus (Li, J. L. et al., 1999), but dynorphin spinothalamic neurons are found only in lamina VII (Nahin, 1988). A few lamina I DYN neurons project to the nucleus of the solitary tract (Leah et al., 1988). The increase in DYN that follows inflammation (see below) is seen in both projection and local circuit neurons in the dorsal horn (Nahin et al., 1989). Dynorphin activity is most intense at sacral levels (Cruz and Basbaum, 1985; Basbaum et al., 1986c; Weihe et al., 1988b; Miller and Seybold, 1989).

Co-localizations

Dynorphin terminals often contain SP (Gibbins et al., 1987b) but not κ opioid receptors, the latter speaking against a dynorphin autoreception system (Arvidsson et al., 1995a).

Synaptic Organization

Dynorphin terminals are usually axodendritic, contain mostly small clear round vesicles and some large dense core vesicles, with symmetric and asymmetric synaptic thickenings, the former more common on cell bodies and the latter on dendrites, which may indicate anatomically segregated inhibitory and excitatory inputs to these cells (Cho and Basbaum, 1989). Few dynorphin terminals are axoaxonic, but opioid receptors are on primary afferent terminals so non-synaptic dynorphin presynaptic control of primary afferent input is possible (Cho and Basbaum, 1989). Some primary afferent terminals and numerous CGRP terminals end on dynorphin dendrites (Carlton and Hayes, 1989; Cho and Basbaum, 1989) indicating primary afferent input onto postsynaptic dynorphin targets.

Experimental Manipulations

Dorsal Rhizotomy and Peripheral Axotomy

Dorsal rhizotomy does not change spinal dynorphin chemically (Botticelli et al., 1981; Pohl et al., 1990), but DYN immunostaining in laminae I–II and V in sacral cord is decreased (Basbaum et al., 1986c; Tuchscherer and Seybold, 1989) presumably indicating a modest amount of DYN primary afferent input. Nerve transection or chronic constriction increase spinal dynorphin immunostaining (Cho and Basbaum, 1988; Bennett et al., 1989; Kajander et al., 1990; Wagner et al., 1993; Rosen et al., 2000) with sometimes bilateral effects extending into areas not directly innervated by afferents from the damaged area (Wagner et al., 1993; Malan et al., 2000). Dorsal rhizotomy increases DYN cell staining, as opposed to general neuropil staining, in laminae I, II, and V (Cho and Basbaum, 1988). These increases presumably imply greater inhibitory control.

Spinal Injury

Transection, weight drop, or injection of toxic doses of kainic acid upregulate dynorphin and prodynorphin mRNA above and below the lesion as measured chemically (Cox et al., 1985; Faden et al., 1985a, b; Przewlocki et al., 1988; Faden, 1990; Yakovlev and Faden, 1994; MacArthur et al., 1999; Abraham et al., 2000; 2001; Abraham and Brewer, 2001), and in cells and terminals in laminae I–II (Tachibana et al., 1998; MacArthur et al., 1999). The increase in dynorphin expression caudal to a transection presumably indicates removal of a descending inhibitory system (MacArthur et al., 1999). Spinal injury activates FOS as well as dynorphin (Messersmith et al., 1998), presumably indicating a role for FOS in the increased dynorphin expression (Abraham and Brewer, 2001). High concentrations of dynorphin are neurotoxic when they activate glutamate receptors, with resulting deficits in motor control, as opposed to the analgesic action of this opioid when it acts at κ-receptors (Caudle and Mannes, 2000).

Inflammation

Inflammation increases dynorphin and prodynorphin mRNA as measured chemically or anatomically (Fig. 6.1; Millan et al., 1985; 1986; 1987; 1988; Hollt et al., 1987;

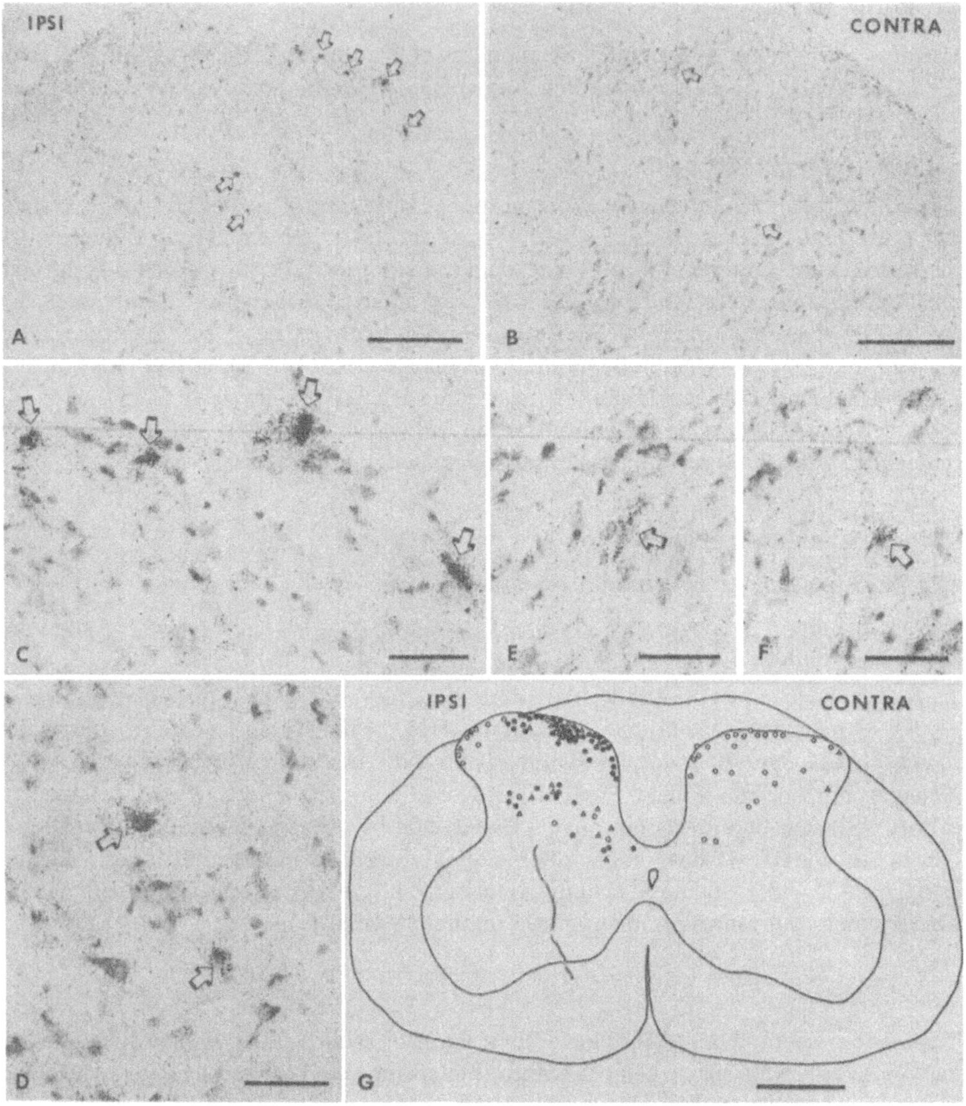

FIGURE 6.1. This figure illustrates increases in preprodynorphin mRNA levels in rat dorsal horn in response to inflammation. In particular notice the sketch (**G**), showing the remarkable increase in the numbers of cells in superficial and deep laminae in the cord ipsilateral to the peripheral inflammation (From Ruda et al., 1988).

Przewlocki et al., 1987; Iadarola et al., 1988a, b; Draisci and Iadarola, 1989; Draisci et al., 1991; Dubner and Ruda, 1992; Persson et al., 1994; Stanfa and Dickenson, 1995; Pohl et al., 1997; Zhang, R. X. et al., 1998; Spetea et al., 2002), in contrast to intermittent acute noxious stimuli where there is a decrease (Millan et al., 1981). These changes are due to increases in numbers of dynorphin or prodynorphin mRNA neuron profiles as well as general increases in intensity of immunostaining or hybridization signals (Fig. 6.1; Ruda et al., 1988; Nahin et al., 1989; Weihe et al., 1989; Noguchi et al., 1991; Przewlocka et al., 1992;

MacArthur et al., 1999). Part of the increase is because cells expressing prodynorphin and proenkephalin mRNAs appear in laminae IV–V (Weihe et al., 1988c). Interestingly opioid receptors do not change in the dorsal horn in these conditions (Iadarola et al., 1988b; Millan et al., 1988; Spetea et al., 2002). Changes in spinal dynorphin considerably exceed changes in the spinal enkephalin, which may be the reason that dynorphin is released into the spinal fluid whereas enkephalin is not after inflammation (Przewlocka et al., 1992; Pohl et al., 1997). CGRP afferent terminals make contact with DYN-expressing cells that appear after inflammation presumably indicating a direct primary afferent input onto these cells (Takahashi et al., 1988). The inflammatory changes are due to fine afferent fiber activity (Hylden et al., 1992), which also leads to FOS expression prior to increases in dynorphin or preprodynorphin mRNA (Noguchi et al., 1991), suggesting a role for FOS expression in the dynorphin increases after inflammation (Dubner and Ruda, 1992; Messersmith et al., 1998). The NMDA antagonist MK-801 reduces the dynorphin increase that follows inflammation presumably indicating a role for glutamate in the increase (Zhang, R. X. et al., 1998). Dynorphin is increased in peripheral immune cells and in cutaneous nerves during inflammation (Hassan et al., 1992).

Endomorphins

Endomoprphin-1 and endomorphin-2 are recently discovered endogenous opioids highly selective for MORs, much more selective than morphine, that cause naloxone-sensitive analgesia (Zadina et al., 1997; Przewlocki and Przewlocka, 2001; Vrinten et al., 2001). Fibers and terminals immunostained for both endomorphins are concentrated in the superficial dorsal horn (substantia gelatinosa, laminae I and IIo, laminae I–II) with endomorphin-2 more prominent (Martin-Schild et al., 1997; 1999; Schreff et al., 1998; Zudina et al., 1999; Dun et al., 2000; Pierce and Wessendorf, 2000; Smith et al., 2001; Mousa et al., 2002). Dorsal rhizotomy removes many of these fibers, and many co-localize with CGRP indicating that many endomorphin-2 fibers in the dorsal horn are central terminals of primary afferents (Pierce et al., 1998). Endomorphin-2 fibers co-localize with SP and MORs, but antisense mapping indicates that responses for endomorphin are somewhat separate from those for morphine (Martin-Schild et al., 1998; Abbadie et al., 2002).

Endomorphin-2 is located in the cores of large dense core vesicles in synaptic terminals, synapses are usually axodendritic with either symmetric or asymmetric synaptic thickenings, and there is a relatively high proportion of axoaxonic synapses with sometimes both terminals being endomorphin labeled (Martin-Schild et al., 1998). Peripheral nerve transection results in loss of endomorphin-2 in superficial laminae consistent with loss of opioids in this condition (Smith et al., 2001). Inflammation causes little change in endomporphin-2 in the cord, but a large number of inflammatory cells containing this peptide migrate into inflamed regions and presumably act on primary afferent terminals (Mousa et al., 2002). Spinal endomorphin-2 is found only postnatally and reaches its adult appearance in adolesence (Barr and Zadina, 1999).

Endorphin

β-endorphin is a classic endogenous ligand for MORs but also activates DORs. It is derived from pro-opiomelanocortin (POMC) as are ACTH and α-melanocyte-stimulating hormone (α-MSH). In adult spinal cord, β-endorphin labeled cells are not reported, although a few β-endorphin and melanocortin cells are seen in neonates (Bloom et al., 1978; Haynes et al., 1982; Mountjoy and Wild, 1998). Descending β-endorphin, ACTH and α-MSH fibers are seen in laminae III–V (Tsou et al., 1986). Some β-endorphin and POMC are found chemically in the cord, and β-endorphin persists below spinal transection suggesting that there may be intrinsic neurons and/or DRG cells that contain POMC and its derivatives (Przewlocki et al., 1983; Gutstein et al., 1992; van der Kraan et al., 1999). The β-endorphin and pro-opiomelanocortin mRNAs are increased by morphine or sciatic nerve crush (Plantinga et al., 1992), and β-endorphin cells are seen in the dorsal horn of obese diabetic mice (MacDonald et al., 1998).

Enkephalin (ENK)

Methionine- and leucine-enkephalin are derived from a precursor molecule, proenkephalin (proenkephalin A, preproenkephalin). Although separate localizations are occasionally done, these compounds are usually considered together.

Laminar Localizations

ENK neurons, fibers, and terminals are found throughout the dorsal horn or spinal trigeminal nucleus but are concentrated in the superficial dorsal horn, most commonly described as laminae I and II (Elde et al., 1976; Hökfelt et al., 1977a, b; Simantov et al., 1977; Uhl et al., 1979b; Del Fiacco and Cuello, 1980; Hunt et al., 1980; 1981a; Seybold and Elde, 1980; Aronin et al., 1981; DiTirro et al., 1981; Finley et al., 1981a; Gibson et al., 1981; Glazer and Basbaum, 1981; 1983; LaMotte and de Lanerolle, 1981; 1983b; Bennett et al., 1982; de Lanerolle and LaMotte, 1982; Haber and Elde, 1982a, b; Sakanaka et al., 1982; Sumal et al., 1982; Arluison et al., 1983a, b; Conrath-Verrier et al., 1983; Bresnahan et al., 1984; Cruz and Basbaum, 1985; Schoenen et al., 1985a; 1986; Ruda et al., 1986; Standaert et al., 1986; Tohyama and Shiotani, 1986; Miller and Seybold, 1987; 1989; Katoh et al., 1988a, b; Nahin, 1988; Tashiro et al., 1988; Chung et al., 1989b; Cheng et al., 1996). There are differences in described distributions, however, with staining reported concentrated in lamina I (Glazer and Basbaum, 1981; 1983; de Lanerolle and LaMotte, 1982; LaMotte and de Lanerolle, 1983b; Standaert et al., 1986), lamina II (Hunt et al., 1981a), lamina IIo (Hunt et al., 1980), met-enkephalin primarily in lamina IIi (Cruz and Basbaum, 1985), laminae I and IIo (Aronin et al., 1981), at the laminae I–III border (Finley et al., 1981a), less in lamina IIo than in laminae I or IIi (Glazer and Basbaum, 1981; 1983; Conrath-Verrier et al., 1983; Bresnahan et al., 1984). These lists seem pedantic, but there are functional differences in primary afferent input into different laminae.

Synkephalin, from which proenkephalin is derived, is distributed as met-enkephalin (Schoenen et al., 1986).

Segmental Levels

Enkephalin immunostaining is most intense at lumbar levels (Miller and Seybold, 1989; Pohl et al., 1990).

Cell Types

Both islet and stalked cells are enkephalinergic (Bennett et al., 1982), and stellate ENK cells are noted in humans (Schoenen et al., 1985a). In lamina I, bipolar and large multipolar spinothalamic cells are enkephalinergic as are large multipolar cells in lamina V (Ruda, 1982; Ruda et al., 1984). Longitudinally oriented spindle-shaped cells are most characteristic in superficial laminae (Hunt et al., 1981a; Katoh et al., 1988a) and some large marginal neurons (Waldeyer cells) are enkephalinergic in lamina I (Glazer and Basbaum, 1981; LaMotte and de Lanerolle, 1981).

Synaptic Organization

ENK terminals in the dorsal horn are dome-shaped, contain numerous small round clear, and fewer but more prominent large dense-core vesicles, the latter intensely stained, and form asymmetric contacts on dendrites and occasional cell bodies mostly in the superficial dorsal horn (Hunt et al., 1980; Aronin et al., 1981; Ruda, 1982; Arluison et al., 1983b; Glazer and Basbaum, 1983; LaMotte and de Lanerolle, 1983b; Bresnahan et al., 1984; Ruda et al., 1984; Katoh et al., 1988a, b; Matthews et al., 1989; Cheng et al., 1995a; 1996). Thus, ENK synapses seem to be excitatory, which is interesting in that ENK is an opioid. Few or no axoaxonic synapses contain ENK (Hunt et al., 1980; Aronin et al., 1981; Sumal et al., 1982; LaMotte and de Lanerolle, 1983b; Matthews et al., 1989), but a few exist where the enkephalinergic terminal is postsynaptic (Sumal et al., 1982; Glazer and Basbaum, 1983). Opioid receptors are on primary afferent terminals, however, so non-synaptic ENK control of primary afferent input is probable.

Although it is difficult to find co-localization of opioids and their receptors in the light microscope (Dado et al., 1993; Arvidsson et al., 1995b), electron-microscopy shows more than half of ENK profiles express DORs and many express MORs (Cheng et al., 1995a; 1996), providing morphologic evidence for presynaptic autocrine interactions. 5-HT terminals end on ENK elements (Miller and Salvatierra, 1998), and ENK and 5-HT are co-localized in axons in the superficial dorsal horn (Reddy et al., 1990a).

Development

The enkephalinergic system develops relatively early throughout the cord in humans and increases steadily in superficial laminae as development proceeds (Charnay et al., 1984; Luo et al., 1988). In opossums and rats, development of this system is relatively late and increases steadily in superficial laminae to reach the adult state after birth (Senba et al., 1982; DiTirro et al., 1983).

Experimental Manipulations

Deafferentation

Dorsal rhizotomy, sensory ganglionectomy, and neonatal capsaicin have no effect on ENK immunostaining in the dorsal horn or spinal nucleus of V (Del Fiacco and Cuello, 1980; Jancso et al., 1981; Basbaum et al., 1986c; Micevych et al., 1986; Pohl et al., 1990; Hirakawa and Kawata, 1992a). Accordingly there are few enkephalin primary afferent fibers in the dorsal horn.

Spinal Injury

Spinal transection does relatively little to enkephalin immunostaining in the dorsal horn, so one view is that there is no descending ENK to the cord (Naftchi et al., 1981; Arvidsson et al., 1992b), but ENK fibers co-localize with 5-HT, a substance found in descending systems, and these double-labeled fibers disappear after transection (Tashiro et al., 1990; Arvidsson et al., 1992b). Brain stem enkephalinergic cells project to the cord (Hokfelt et al., 1979; Reddy et al., 1990b), so enkephalinergic fibers descend into the dorsal horn but presumably not enough to result in a loss of enkephalin immunostaining caudal to a transection. If the cord is transected before inflammation, dorsal horn proenkephalin mRNA increases more than after inflammation alone, suggesting a descending inhibitory effect on opioid manufacture by dorsal horn neurons (MacArthur et al., 1999). Increased numbers of dorsal horn neuron profiles express mRNA for proenkephalin after excitotoxic spinal cord injury (Abraham et al., 2000; 2001).

Nerve Transection

Dorsal horn proenkephalin mRNA shows little change or a considerable reduction after nerve section, with or without a neuroma, or nerve crush (Di Giulio et al., 1985; Draisci et al., 1991; Gorio et al., 1996), but an increase after chronic constriction injury (Sommer and Myers, 1995). The implication is that nerve injury pain paradigms increase ENK in the cord whereas "simple" lesions reduce it. Nerve section prevents the increase in proenkephalin mRNA that follows inflammation, indicating the necessity of axon activity for this change (Faccini et al., 1984). Thalidomide, which reduces CCI pain behaviors, increases enkephalin in the cord (George et al., 2000).

Inflammation

There is a mild rise in proenkephalin mRNA in the dorsal horn and spinal trigeminal nucleus following various inflammations (Faccini et al., 1984; Iadarola et al., 1988a; Draisci and Iadarola, 1989; Noguchi et al., 1989a; 1992; Draisci et al., 1991; Przewlocka et al., 1992; Pohl et al., 1997; MacArthur et al., 1999; Imbe and Ren, 2000; Rosen et al., 2000, reviewed by Dubner and Ruda, 1992; Stanfa and Dickenson, 1995). By contrast, proenkephalin mRNA decreases in dorsal root ganglia in polyarthritis (Pohl et al., 1997).

For enkephalin itself, some report no change after various inflammations (Iadarola et al., 1988a; Millan et al., 1988), others modest increases (Cesselin et al., 1980; Millan et al., 1986; Przewlocka et al., 1992; Pohl et al., 1997; Spetea et al., 2002), others the appearance of enkephalin cells in laminae IV and V (Kar et al., 1991a), and such cells coexpress mRNAs for proenkephalin and prodynorphin (Weihe et al., 1988c), and there is a decrease in peripheral enkephalin fibers (Elhassan et al., 1998). Afferent input is necessary for these changes since nerve transection prevents them (Faccini et al., 1984). In some cases, there is decreased release of enkephalin (Pohl et al., 1997), in others an increase (Przewlocka et al., 1992). Thus, the role of endogenous spinal enkephalins in inflammation is not clear, and attention focuses on dynorphin because of more dramatic changes (Dubner and Ruda, 1992; Stanfa and Dickenson, 1995). There is a large inflammation-induced increase in opioids, including enkephalin and its precursor mRNA, in peripheral, non-neural, inflammatory cells, and these endogenous opioids play a major role in the response of the organism to inflammation (Stein et al., 1990; Przewlocki et al., 1992).

Many enkephalinergic cells express FOS after inflammation so FOS may play a role in the induction of enkephalin synthesis (Draisci and Iadarola, 1989; Dubner and Ruda, 1992; Noguchi et al., 1992). In addition, enkephalinergic terminals seem to end on FOS cells that appear after cutaneous mustard oil (Pretel and Piekut, 1991).

Ovariectomy and Noxious Stimuli

Estrogens cause a quick rise in spinal enkephalin mRNA in ovariectomized rats (Amandusson et al., 1999). Thermal injury results in increased enkephalin in the superficial dorsal horn (de Ceballos et al., 1990).

Co-localizations

Enkephalin co-localizes with DORs or MORs in dorsal horn terminals presumably indicating autocrine opioid synaptic control (Cheng et al., 1995a; 1996). Proenkephalin and prodynorphin co-localize in lamina IV and V cells that become enkephalinergic after inflammation (Weihe et al., 1988c). Some enkephalin terminals and cells contain SP (Katoh et al., 1988a, b) or 5-HT and express 5-HT3 receptors (Tsuchiya et al., 1999). Many enkephalin cells are enriched in GABA but none in glycine (Todd et al., 1992a).

Nociceptin (Orphanin FQ)

Nociceptin, also known as orphanin FQ, is a dynorphin resembling peptide that binds to a novel opioid-like orphan receptor ORL1. The physiologic roles of nociceptin are complex and just beginning to be unraveled (Darland et al., 1998). The mRNA for the nociceptin precursor prepronociceptin is found in numerous cells in laminae I–III (Houtani et al., 1996; 2000), and the mRNA for nociceptin in cells in laminae I–II (Ikeda et al., 1998; Pettersson et al., 2002). Immunolabeling is concentrated in lamina II or I–II (Riedl et al., 1996; Lai et al., 1997; Narita et al., 1999; Pettersson et al., 2002). Since the labeling is not affected by dorsal rhizotomy, nociceptin is presumably not expressed by primary afferents, and there is little co-localization with endogenous opioids (Riedl et al., 1996; Schulz et al., 1996).

Oxytocin (OXY) and Vasopressin (VP)

OXY and VP are hypothalamic peptides that play major roles in uterine contraction, milk ejection, and water balance. There are no OXY or VP cells in the dorsal horn, but descending hypothalamic fibers form a prominent plexus in laminae I and II (Swanson and McKellar, 1979; Swanson and Kuypers, 1980; Kai-Kai et al., 1986; Cechetto and Saper, 1988; Rousselot et al., 1990; Puder and Papka, 2001b). Both symmetric and asymmetric OXY terminals are found in laminae I–II with the OXY being present primarily in cores of large, dense core vesicles, and they do not form central terminals in glomeruli (Rousselot et al., 1990). OXY and VP cells that project to the posterior pituitary are separate from those that project to the cord (Swanson and Kuypers, 1980), which correlates with the observation that dehydration and dexamethasone deplete OXY and VP in the hypothalamus but have no effect in the dorsal horn (Millan et al., 1984). Because of the controversy as to the presence of OXY and VP cells in the DRG (Chapter 4), it would be useful to see if dorsal rhizotomy reduces these peptides in the dorsal horn.

Secretoneurin

Secretoneurin is a product of secretogranin metabolism. Attention usually focuses on the autonomic nervous system, but immunoreactive secretoneurin is present in high concentrations in the superficial dorsal horn, particularly lamina I and IIo (Kirchmair et al., 1993; Marksteiner et al., 1993; 1994a, b; Telser et al., 1995). Numerous neurons in laminae I and II express the mRNA for secretogranin II (Marksteiner et al., 1994b).

Somatostatin

Laminar Localizations and Intrinsic Cells

Somatostatin is a biologically active peptide with widespread effects, but it is predominantly inhibitory in spinal sensory function. Somatostatin fibers and terminals as well as small, spindle-shaped neurons that express somatostatin or its mRNA are found in the superficial dorsal horn, primarily lamina II (Hökfelt et al., 1975a; 1976; Forssmann, 1978; Burnweit and Forssmann, 1979; Forssmann et al., 1979; Seybold and Elde, 1980; DiTirro et al., 1981; 1983; Finley et al., 1981b; Senba et al., 1982; Ho, 1983; 1988; Massari et al., 1983; Ho and Berelowitz, 1984; Johansson et al., 1984; Schroder, 1984; Schoenen et al., 1985a; Vincent et al., 1985; Krukoff et al., 1986; Papadopoulos et al., 1986b; Ruda et al., 1986; Charnay et al., 1987; Leah et al., 1988; Mizukawa et al., 1988a; Chung et al., 1989b; Rosenthal and Ho, 1989; Alvarez and Priestley, 1990), with some indications of a concentration in lamina IIi (Dalsgaard et al., 1981; Hunt et al., 1981a; Merighi et al., 1990; Yin, 1998). Somatostatin is concentrated in the Golgi complex and rough endoplasmic reticulum (Johansson, 1978; Mizukawa et al., 1988a), and somatostatin cells are islet (Schoenen et al., 1985a) or stalked and islet cells (Alvarez and Priestley, 1990).

Deafferentation and Peripheral Axotomy

Dorsal rhizotomy, peripheral axotomy, and capsaicin administration reduce spinal somatostatin staining but do not remove it, and spinal transection has no effect (Seybold and Elde, 1980; Jancso et al., 1981; Massari et al., 1983; Atkinson and Shehab, 1986; Shehab and Atkinson, 1986a, b), indicating significant numbers of primary afferent and intrinsic fibers but no descending input.

Synaptic Organization

Somatostatin is found mostly in unmyelinated axons and presynaptic terminals, the terminals having symmetric or asymmetric thickenings and containing primarily small clear and some large dense-core vesicles, the cores of the latter being intensely stained (Foster and Johansson, 1985; Mizukawa et al., 1988a; Ribeiro-da-Silva and Cuello, 1990b). Some central elements in glomeruli as well as presynaptic dendrites are somatostatin positive (Rosenthal and Ho, 1989; Ribeiro-da-Silva and Cuello, 1990b). Since somatostatin is both a primary afferent transmitter and mostly inhibitory, a possible counterintuitive conclusion is that some primary afferent input is inhibitory.

Development

Somatostatin appears well before birth in rat and human dorsal horn, which is long before primary afferent synaptic input, and shortly after birth in the opossum and then is lost steadily in all regions except the superficial dorsal horn (Senba et al., 1982; DiTirro et al., 1983; Charnay et al., 1987; Marti et al., 1987; Ho, 1988), although the early expression may be very transient (Marti et al., 1987). Thus somatostatin may have a developmental role.

Substance P (SP)

Substance P is a major factor in neurogenic inflammation. It co-localizes with and potentiates CGRP.

Laminar Localizations

Substance P (SP) is concentrated in the superficial dorsal horn, usually laminae I and II, with lesser deep staining, particularly in lamina V (Hökfelt et al., 1975b; 1977b; Pickel et al., 1977; Cuello and Kanazawa, 1978; Gibson et al., 1981; Jancso et al., 1981; de Lanerolle and LaMotte, 1982; Charnay et al., 1983; Bresnahan et al., 1984; Atkinson and Shehab, 1986; Micevych et al., 1986; Vacca et al., 1986; Katoh et al., 1988b; Warden and Young, 1988; Chung et al., 1989b; Arvidsson et al., 1990; Achaval et al., 1991; Bonfanti et al., 1991; Marlier et al., 1991b; Delander et al., 1997; Ishigooka et al., 2002). Careful examination, however, shows the superficial staining concentrated in laminae I and IIo, superficial to FRAP or other markers that are concentrated in lamina IIi (DiTirro et al., 1981; DiFiglia et al., 1982b; Nagy and Hunt, 1982; Priestley et al., 1982b; Hunt, 1983; Schoenen et al., 1985a; Drew et al., 1986; Bennett et al., 1989). Occasionally label is reported in lamina III (Barber et al., 1979; Seybold and Elde, 1980; Kawatani et al., 1985).

Intrinsic Cells

Small SP cells are seen, especially after colchicine treatment, in all laminae of the dorsal horn, with concentrations in superficial laminae, lamina IIi, or lamina III (Ljungdahl et al., 1978; Del Fiacco and Cuello, 1980; Tessler et al., 1980; 1981; 1984; Hunt et al., 1981a; Nagy et al., 1981; Hunt, 1983; Schoenen et al., 1985a; Leah et al., 1988; Arvidsson et al., 1990; Cuello et al., 1993). Many of the SP cells are stalked cells (Ribeiro-da-Silva et al., 1991), and they are responsible for the immunostaining that remains after dorsal rhizotomy (see below). Most SP cells are propriospinal (Leah et al., 1988), but some are projection neurons (Noguchi and Ruda, 1992). Many SP cells are found in the dorsal horn before birth with numbers declining postnatally to adult levels (Senba et al., 1982). Some intrinsic SP cells co-localize with enkephalin (Senba et al., 1988) or galanin (Zhang, X. et al., 1993b).

Synaptic Organization

Most SP-enriched presynaptic terminals are simple with asymmetric thickenings except for axosomatic synapses where some thickenings are symmetric (Barber et al., 1979; de Lanerolle and LaMotte, 1983). Intensely stained terminals are associated with asymmetric thickenings, whereas lighter staining goes with either symmetric or asymmetric thickenings. (Ribeiro-da-Silva et al., 1989). SP glomerular central terminals with large DCVs (type 1) occur but are much less frequent than simple SP endings (de Lanerolle and LaMotte, 1983). Axoaxonic synapses with SP elements are rare (Barber et al., 1979; DiFiglia et al., 1982b; de Lanerolle and LaMotte, 1983; Bresnahan et al., 1984).

SP is concentrated in the cores of large dense-core vesicles (Chan-Palay and Palay, 1977a; Pelletier et al., 1977; Pickel et al., 1977; Barber et al., 1979; DiFiglia et al., 1982b; de Lanerolle and LaMotte, 1983; Bresnahan et al., 1984; Merighi et al., 1989), but may also be found in agranular vesicles (Ribeiro-da-Silva et al., 1989; Ribeiro-da-Silva and Hokfelt, 2000). Many dense-core vesicles are not associated with synapses, so this might indicate volume transmission (Barber et al., 1979). SP co-localizes with glutamate (de Biasi and Rustioni, 1988; Merighi et al., 1991) or CCK (Tuchscherer et al., 1987) in primary afferent terminals, and SP and ENK co-localize in synapses and intrinsic dorsal horn neurons (Ribeiro-da-Silva et al., 1991; Ma et al., 1997). SP primary afferent varicosities co-localized with CCK are largely removed by dorsal rhizotomy and so are presumably of primary afferent origin whereas those with SP alone are largely unaffected suggesting they arise from intrinsic neurons (Tuchscherer et al., 1987).

SP terminals end on marginal (both pyramidal and multipolar), islet, stalked, and presumed spinothalamic cells in the dorsal horn (LaMotte and de Lanerolle, 1981; de Lanerolle and LaMotte, 1982). In addition, SP terminals end on DYN cell bodies and dendrites (Katoh et al., 1988b), and SP and SP+ENK terminals end predominantly on nociceptive-specific or wide dynamic range cells as opposed to non-nociceptive cells (Ma et al., 1997). SP terminals are abundant on intrinsic cells with slow, prolonged excitatory potentials and rare on those that do not have such potentials (De Koninck et al., 1992). Thus, SP terminals seem concerned primarily with nociceptive transmission.

A key issue is whether SP acts by volume transmission or by synaptic connectivity. Volume transmission is favored because: (1) there is a mismatch between SP terminals and NK-1 receptors in lamina II; (2) SP terminals can be found on non-NK-1 elements; and (3) NK-1 receptors are spread over the entire cell surface so SP terminals, even if they synapse on NK-1 cells, are near only a minority of the receptors (Liu et al., 1994a; Nakaya et al., 1994; Littlewood et al., 1995). On the other hand, the mismatch between SP and the NK-1 receptor is attributed in part to the failure of previous antisera to recognize some forms of antigenic presentation, and SP terminals preferentially end on NK-1 postsynaptic elements (McLeod et al., 1998; 1999; Ribeiro-da-Silva and Hokfelt, 2000). Thus, both volume transmission and precise synaptic connectivity may exist for SP, and if so, it will be of interest to sort out resultant functions.

Experimental Manipulations

Deafferentation

Doral rhizotomy, dorsal root ganglionectomy, or neonatal capsaicin greatly but not completely reduce SP staining in the superficial dorsal horn (Chan-Palay and Palay, 1977b;

Barber et al., 1979; Del Fiacco and Cuello, 1980; Seybold and Elde, 1980; Tessler et al., 1980; 1981; 1984; Hunt, 1983; Dalsgaard et al., 1985; Kawatani et al., 1985; Micevych et al., 1986; Shehab and Atkinson, 1986b; Tuchscherer et al., 1987; Tuchscherer and Seybold, 1989; Arvidsson et al., 1990; Knyihar-Csillik et al., 1990; Pohl et al., 1990; Hammond and Ruda, 1991; Villar et al., 1991; Hirakawa and Kawata, 1992a). Thus, much SP staining in the dorsal horn arises from primary afferents, and since there is little evidence for descending SP input, the rest is from intrinsic cells.

Peripheral Nerve Lesions

In adults, nerve lesions by cut, proteolytic enzymes, or cauda equina compression deplete SP in the superficial dorsal horn or subnucleus caudalis of the spinal trigeminal nucleus (Jessell et al., 1979; Barbut et al., 1981; McGregor et al., 1984; Fitzgerald et al., 1985; Atkinson and Shehab, 1986; Shehab and Atkinson, 1986a, b; Sharkey et al., 1987; Rhoades et al., 1988; Bennett et al., 1989; Himes and Tessler, 1989; Villar et al., 1989; Kar et al., 1990; 1991a; Kawakami and Tamaki, 1992; El-Bohy and LaMotte, 1993; Garrison et al., 1993; Nothias et al., 1993; Zhang et al., 1993c; Hökfelt et al., 1994c; Kajander and Xu, 1995; Munglani et al., 1996; Eriksson et al., 1997; Robertson et al., 1999; Na et al., 2001). In neonates, nerve transection or capsaicin leads to more SP depletion than in adults (Himes and Tessler, 1989; Hammond and Ruda, 1991; Nothias et al., 1993), although neonatal infraorbital nerve sections do not cause SP loss (Rhoades et al., 1988). The loss can be prevented by NGF (Fitzgerald et al., 1985; Eriksson et al., 1997) or reversed by NGF after the lesion (Siri et al., 2001). Thus loss of SP after nerve lesions seems to be largely NGF dependent.

The SP depletion seems to be due primarily to lessened production as evidenced by a decline in preprotachyninin mRNA in DRG cells (Jessell et al., 1979; Nielsch and Keen, 1989; Noguchi et al., 1989b; Henken et al., 1990), but loss of cells after the lesion or increased activity that follows nerve transection may play a role. Nerve crush, by contrast, does not cause SP depletion, perhaps because primary afferents still have access to peripheral trophic factors (Barbut et al., 1981; McGregor et al., 1984), but others find a small drop after simple crush and a large drop after "controlled" multiple crushes (Villar et al., 1989; Knyihar-Csillik et al., 1990). Nevertheless, nerve crush is clearly depletes less spinal SP than nerve transection.

The SP loss that follows deafferentation is usually regained over time, but neither repeat dorsal rhizotomy nor capsaicin affects adult repleted SP, so the repletion is in intrinsic neurons (Tessler et al., 1980; 1981; 1984; Micevych et al., 1986; Arvidsson et al., 1990; Hammond and Ruda, 1991). On the other hand, other studies report no repletion of dorsal horn SP after dorsal rhizotomy (Knyihar-Csillik et al., 1990; Hirakawa and Kawata, 1992a), nor after adult peripheral axotomy (Barbut et al., 1981; Hunt et al., 1982; Nothias et al., 1993; Kajander and Xu, 1995; Munglani et al., 1996; Eriksson et al., 1997), although one study reported a slow replenishment (Robertson et al., 1999). There is, however, rapid repletion after neonatal axotomy, and dorsal rhizotomy or repeat axotomy removes much of the repleted stain, interpreted as showing primary afferent sprouting (Rhoades et al., 1988; Knyihar-Csillik et al., 1990; El-Bohy and LaMotte, 1993; Nothias et al., 1993; Munglani et al., 1996; Na et al., 2001). Phenotypic changes or excess production of SP in unsprouted neurons may also play a role. In summary, SP is depleted after nerve lesion, and some find no repletion and others find repletion but not from primary afferents except in newborns.

The role of SP in neuropathic pain is not clear. For example, SP depletion after nerve lesions remains after the behavioral changes resolve (Munglani et al., 1996), and spinal SP is the same in animals with and without dramatic neuropathic pain behaviors (Na et al., 2001). Thus SP changes are not directly linked to neuropathic pain. A hypothesis is that axotomy-induced peptide changes indicate a shift in peptide function from nociception to regeneration (Hökfelt et al., 1994c).

Spinal Lesions

SP increases below a spinal transection (Naftchi et al., 1978; 1981; Arvidsson et al., 1990) although the changes are not prominent (Seybold and Elde, 1980). Dorsolateral funiculus lesions increase SP staining after intraplantar formalin caudal to the lesion, possibly implying a descending inhibitory pathway on SP elements through the DLF (Zhang et al., 1993d).

Inflammation

After monoarthritis, SP rises acutely and falls chronically (Mapp et al., 1993), falls acutely and rises chronically (Sluka and Westlund, 1993b; Calza et al., 1998), falls acutely and returns to normal (Galeazza et al., 1995) or is increased at all measured times (Schoenen et al., 1985b; Marlier et al., 1991b). SP rises after polyarthritis (Kar et al., 1991a), nematode infestation (De Giorgio et al., 2001), bladder inflammation (Vizzard, 2001), and in concert with both phases of formalin inflammation (Kantner et al., 1986; Holland and Goldstein, 1990), but falls mildly after myositis (Hoheisel et al., 1998), significantly after thermal injury (de Ceballos et al., 1990), and markedly at the height of ultraviolet inflammation (Polgár et al., 1998). To explain these apparently divergent results, a hypothesis is that the response to an inflammatory stimulus is early release from spinal primary afferent terminals due to inflammation-induced activity followed by SP mRNA upregulation in DRG and intrinsic cells (Noguchi and Ruda, 1992) and then an SP increase at least to normal and usually beyond in the dorsal horn. It should be noted, however, that only increases were seen on occasion even though the SP was measured minutes after stimulation (Holland and Goldstein, 1990), and in another case, an increase preceded a decrease (Mapp et al., 1993). Thus, further work will be necessary to see if a common SP response to inflammation exists.

Nerve Growth Factor

NGF prevents loss of SP in the dorsal horn after nerve lesions (Fitzgerald et al., 1985; Eriksson et al., 1997), and reverses the loss if given after the lesion (Siri et al., 2001). Furthermore, SP stains a wider area more intensely in transgenic NGF over-expressing mice (Ma, W. et al., 1995). AntiNGF reduces SP in DRG cells (Shadiack et al., 2001), and blocking axoplasmic transport reduces immunostaining for SP in the dorsal horn by 50%, presumably because of lack of retrograde transport of this trophic factor (Kantner et al., 1986).

Local Anesthesia

Local anesthesia stops the phasic part of the SP increase that follows formalin injection in the foot (Holland and Goldstein, 1990), and tetrodotoxin, which stops axon activity, reduces SP immunostaining by 50% (Kantner et al., 1986). Thus the SP increase is related to neuron activity.

Capsaicin

Local capsaicin to an adult nerve depletes SP in the cord (Gibson et al., 1982), as does systemic capsaicin in young adults or neonates (Jancso et al., 1981; Nagy et al., 1981; Priestley et al., 1982a; Salt et al., 1982; Skofitsch and Jacobowitz, 1985b; Pohl et al., 1990).

MK-801

Systemic MK-801 prevents the decrease in dorsal horn SP that follows the CCI partial nerve transection (Garrison et al., 1993), thus indicating a role for NMDA receptors in SP production.

Thiamine Deficiency

Thiamine deficiency decreases spinal SP (Tadano et al., 1995).

Electrical Stimulation

Electrical stimulation of the trigeminal ganglion at high intensity depletes SP in the superficial layers of the cord or spinal trigeminal nucleus (Samsam et al., 2001).

Prostate Stimulation

SP is increased in laminae I–II after noxious prostate stimulation, and the increase is partially blocked by an α-adrenergic antagonist (Ishigooka et al., 2002).

SP Antisera

SP is decreased in adult laminae I–IIo after neonatal administration of antiSP (Achaval et al., 1991).

Hereditary Sensory Neuropathy

SP is decreased in hereditary sensory neuropathy (Scaravilli, 1983; Cummings et al., 1984).

Age and Development

SP is present prenatally but takes its adult distribution postnatally in rats (Senba et al., 1982) and humans (Charnay et al., 1983; Chigr et al., 1991), although it may take its adult distribution in humans by 27 weeks fetal age (Luo et al., 1988). In rats, it appears prenatally and increases rapidly after birth (Marti et al., 1987). SP declines in DRG and cord in senescence (Johnson et al., 1993; Bergman et al., 1996).

Vasointestinal Polypeptide (VIP)

VIP and the closely related compounds PHI and PACAP belong to the glucagon/secretin superfamily and have multiple functions.

Laminar Localizations

In the spinal cord VIP, PHI, and PACAP are concentrated in axons and terminals in the superficial dorsal horn, particularly lamina I, and VIP is most prominent in the sacral cord (Gibson et al., 1981; 1984c; Anand et al., 1983; Basbaum and Glazer, 1983; de Groat et al., 1983; Fuji et al., 1983; Honda et al., 1983; Kawatani et al., 1983; 1985; McGregor et al., 1984; de Groat, 1986; 1987; LaMotte and deLanerolle, 1986; Shehab and Atkinson, 1986a; Triepel et al., 1987; Sasek et al., 1991; Moller et al., 1993; Knyihar-Csillik et al., 1995; Dun, E. C. et al., 1996; Dun, N. J. et al., 1996; Morgan et al., 1999). VIP fibers and terminals are distributed in the same pattern as the pelvic afferents, and this plus the demonstration that many VIP primary afferent neurons project to pelvic viscera presumably indicates a key role for VIP in pelvic visceral afferent function (Basbaum and Glazer, 1983; de Groat et al., 1983; Honda et al., 1983; Kawatani et al., 1983; 1985; de Groat, 1986; 1987; Morgan et al., 1999).

Intrinsic Cells

Dorsal horn VIP neurons are usually not seen, but after colchicine VIP cells are reported in the lateral spinal nucleus and around the central canal up into the deep dorsal horn (Fuji et al., 1983; 1985; Gibson et al., 1984c; LaMotte and de Lanerolle, 1986; Triepel et al., 1987; Leah et al., 1988). Small PACAP neurons can be seen without colchicine in superficial laminae (Jongsma et al., 2000).

Synaptic Organization

VIP terminals contain large dense-core vesicles, the cores of which are intensely labeled, and more numerous small clear round vesicles, and usually make asymmetrical contacts with dendrites but occasionally with the central axon terminals in scalloped glomeruli (Honda et al., 1983; LaMotte and deLanerolle, 1986; Knyihar-Csillik et al., 1995; Morgan et al., 1999).

Experimental Manipulations

Dorsal Rhizotomy

Dorsal rhizotomy results in removal of all or almost all VIP immunostaining in the superficial dorsal horn indicating that VIP is in primary afferent axons and terminals (Honda et al., 1983; Gibson et al., 1984c; Kawatani et al., 1985; Shehab and Atkinson, 1986b).

Peripheral Nerve Section or Crush

VIP and related PHI and PCACP are considerably increased in the dorsal horn following peripheral axotomy (McGregor et al., 1984; Atkinson and Shehab, 1986; Shehab et al.,

1986; Shehab and Atkinson, 1986a, b; Nielsch and Keen, 1989; Knyihar-Csillik et al., 1995; Zhang, Y. Z. et al., 1996). This increase is due to increased production by the DRG cells that have had their peripheral processes cut, apparently both by cells that express VIP normally and by cells that normally do not express VIP or related compounds (Shehab and Atkinson, 1986a, b; Nielsch and Keen, 1989; Noguchi et al., 1989b; Nahin et al., 1994). More VIP fibers appear in deeper laminae consistent with the expression of VIP in large DRG cells after axotomy, and spinal cells expressing the mRNA for PACAP do not change (Jongsma et al., 2000). Some of the increase may be because of the appearance of VIP in glial processes (Knyihar-Csillik et al., 1995). Other substances that rise after peripheral axotomy are GAL, NPY, CCK, and NOS (Hökfelt et al., 1994c). VIP has a generally excitatory role in spinal information processing (Dickinson and Fleetwood-Walker, 1999), and possibly this compound takes over the functions of SP and CGRP that are diminished after peripheral axotomy (Hökfelt et al., 1994c).

Spinal Transection

Spinal transection leads to a pile-up of VIP in lateral funiculus axons caudal to the lesion, probably from intrinsic VIP cells in the lateral spinal nucleus (Fuji et al., 1983).

PROTEINS

Calcium Binding Proteins

A superfamily of calcium binding proteins, including calmodulin, calbindin-$_{28K}$ (calbindin), parvalbumin, and calretinin influence the biological activity of calcium ions. Calmodulin is not found immunochemically in the dorsal horn (Ren and Ruda, 1994). Numerous calbindin and calretinin cells are found in laminae I–II of the dorsal horn or trigeminal spinal nucleus whereas parvalbumin cells are found in laminae II–III, particularly IIi (Celio and Heizmann, 1981; Garcia-Segura et al., 1984; Fournet et al., 1986; Yamamoto et al., 1989; Antal et al., 1990; Celio, 1990; Sequier et al., 1990; Arai et al., 1991; Resibois and Rogers, 1992; Rogers and Resibois, 1992; Heppelmann and Emson, 1993; Ren et al., 1993a; Ren and Ruda, 1994; Li, J. L. et al., 1999b). The parvalbumin and calbindin cells are separate populations (Celio, 1990; Bennett-Clarke et al., 1992; Ren and Ruda, 1994), but there is some co-localization of calretinin and calbindin (Rogers and Resibois, 1992), although a later group reported only about 5% of the neurons co-localized these compounds and approximately 1% of the parvalbumin cells also co-localize with the other compounds (Li, Y. Q. et al., 2000c). Thus, there is some co-localization, but the calcium binding proteins are mostly in separate neuronal populations. The trigeminal calbindin cells do not co-localize with GABA whereas more than 50% of the parvalbumin cells in laminae II–III do, indicating that the parvalbumin cells contain this inhibitory transmitter (Polgar and Antal, 1995). Some calbindin neurons in the dorsal horn colocalize with protein kinase Cγ (Ni et al., 2002).

Many calcium binding protein cells are intrinsic neurons, but many are also projection neurons, as shown by being retrogradely labeled from the thalamus, parabrachial region, hypothalamus, tractus solitarius, and other parts of the spinal cord (Aronin et al., 1991; Bennett-Clarke et al., 1992; Menetrey et al., 1992; Li, J. L. et al., 1999c; Gamboa-Esteves et al., 2001a). Furthermore, most lamina I spinothalamic cells contain calbindin and tracing the axons of these cells showed them in the classic spinothalamic pathway, as opposed to suggestions that only lamina V spinothalamic cells project in this manner (Craig et al., 2002). In another projection study, lamina I spinal neurons projecting to the nucleus tractus solitarius contain calbindin and express FOS after painful stimuli providing evidence that they are nociceptors (Gamboa-Esteves et al., 2001b). Calbindin and calretinin plexuses are in laminae I–II and parvalbumin in laminae II–III (see above references). These fibers arise primarily from intrinsic neurons in that dorsal rhizotomy, ganglionectomy, and neonatal capsaicin do not change the immunostaining (Yamamoto et al., 1989; Ren et al., 1993a; Ren and Ruda, 1994), and there is no evidence for descending fibers containing these proteins. The plexus seems to be mostly dendritic although labeled terminals are seen (Antal et al., 1990; Yoshida et al., 1990; Aronin et al., 1991), particularly during development (Knyihar-Csillik et al., 1999a).

During development neurons expressing calbindin or parvalbumin appear early in prenatal development especially in the superficial dorsal horn and then fade out by birth, remaining only in large DRG cells and their fibers (Zhang, J. H. et al., 1990; Solbach and Celio, 1991). The adult pattern of calcium binding protein localization, as described above, develops primarily postnatally, and the appearance of immunostaining is hypothesized to be related to developing neural activity (Clowry et al., 1997; 2000). In developmental studies on primates, primary afferent parvalbumin neurons form synapses first on cells in Clarke's column, next in the motoneuron pool and finally on dorsal horn neurons, and when synaptic contact is established, the postsynaptic cells begin to express parvalbumin (Knyihar-Csillik et al., 1999b). Then the staining in the axons subsides, even though the synapses remain, presumably explaining why dorsal rhizotomy does not change dorsal horn immunostaining. The development of characteristic dorsal horn synapses, including those with elements stained for parvalbumin and calbindin is relatively much earlier in primates than other species (Knyihar-Csillik et al., 1999a).

Glutamate Transporters

Glutamate is a major excitatory neurotransmitter that becomes toxic if it is not removed from the extracellular space, and sodium-dependent, high-affinity transporter proteins and sodium vesicular glutamate transporters do this. The sodium-dependent, high-affinity transporters are subdivided into those located in glial cells (EEAT1 or GLAST and EEAT2 or GLT-1) and those located in neurons (EEAT3 or EEAC1 and EEAT4, the latter not located in cord or DRG). The glial transporters are distributed throughout the dorsal horn, with some concentration in superficial layers and with little change in the dorsal horn in ALS or motoneuron disease (Schmitt et al., 1996; Furuta et al., 1997; Sasaki et al., 2000; Banner et al., 2002), although GLT-1 is transiently expressed by axons in superficial laminae in early development (Yamada et al., 1998). After spinal injury, EEAC1 and EEAC2 increase presumably indicating greater capacity for absorbing excess glutamate (Vera-Portocarrero et al., 2002). The neuron-specific transporter is seen throughout the dorsal horn (Shashidharan et al., 1997), but is localized mainly in laminae I–II in normal animals with a considerable increase in laminae I–IV after spinal injury (Vera-Portocarrero et al., 2002). The vesicular glutamate transporter 1 and a neutral and basic amino acid transporter are also concentrated in laminae I–II but with significant staining throughout the dorsal horn (Nirenberg et al., 1995; Kaneko et al., 2002).

RECEPTORS

Aminergic Receptors

Adrenergic Receptors

Adrenergic receptors are divided into α_1, α_2, and β categories.

β Adrenergic Receptors

There are no morphologic papers to our knowledge showing β adrenergic receptors in dorsal horn or DRG.

α_1 Receptors

There are three subtypes of α_1 receptors: α_{1A}, α_{1B}, and α_{1D}. The mRNAs for α_1 receptors in general are expressed by a few or no dorsal horn neurons and binding of a radioactive α_1 ligand is uniform and light in rat dorsal horn (Young III and Kuhar, 1980; Roudet et al., 1993), but concentrated in lamina II in cats (Giroux et al., 1999). Binding increased caudal to a spinal transection or after chemical adrenergic destruction, indicating development of hypersensitivity (Roudet et al., 1993; Giroux et al., 1999). Less attention has been paid to these receptors than to the α_2 receptors in regard to nociception and neuropathic pain, but these receptors are expressed by DRG cells and increase after peripheral axotomy (Chapter 4), so they may play some role (Millan, 2002).

α_2 Receptors

There are three subtypes of α_2 receptors, α_{2A}, α_{2B}, and α_{2C}. Binding of α_2 receptors in general is concentrated in the superficial dorsal horn (Sullivan et al., 1987; De Vos et al., 1994; Roudet et al., 1994; Ruggiero et al., 1995; 1998; Giroux et al., 1999). Some of these receptors are on primary afferent terminals as shown by the 20% reduction caused by dorsal rhizotomy (Howe et al., 1987), but these are not on capsaicin sensitive fibers since neonatal capsaicin has no effect (Wikberg and Hajos, 1987). The α_2 receptors increase distal to a spinal transection presumably indicating development of sensitivity (Roudet et al., 1994; Giroux et al., 1999).

Both α_{2A} and α_{2C} receptors are concentrated in the superficial dorsal horn, with α_{2A} receptors predominating (Rosin et al., 1993; Talley et al., 1996; Uhlen et al., 1997; Stone et al., 1998) and α_{2C} cells being rare and located in laminae IV and V (Rosin et al., 1996). The mRNA localizations show α_2 receptor cells concentrated in laminae I, II, and V with some in laminae III and IV (Smith et al., 1995), and more specifically, α_{2A} mRNA cells in laminae II and III but few α_{2C} mRNA cells in the dorsal horn (Nicholas et al., 1993). In contrast to the lack of effect of neonatal capsaicin on α_2 receptors in general (Wikberg and Hajos, 1987), many α_{2A} receptors are on capsaicin sensitive, SP containing primary afferent terminals (Stone et al., 1998), but α_{2C} receptors are not affected by neonatal capsaicin, and are not found on primary afferent or descending terminals or on SP or CGRP fibers (Stone et al., 1998; Olave and Maxwell, 2002), but they are found on terminals expressing NT, somatostatin, enkephalin, GABA, or NPY (Olave and Maxwell, 2002). Interestingly neither α_2 subtype was located on noradrenergic terminals, so there is no indication of autoreception, but the α_{2C} receptor cells often contain enkephalin (Stone et al., 1998), and they are on endings containing NT, somatostatin, enkephalin, GABA, and NPY (Olave and Maxwell, 2002), so they are in a position to presynaptically influence intrinsic and perhaps some primary afferent input.

A major thrust for adrenergic receptor research concerns the role of these receptors in neuropathic pain, and when these receptors are examined in various neuropathic models, the α_{2A} receptors decrease and then return to normal correlated with the development of neuropathic behavior whereas α_{2C} receptors increased in only one of three models with a time course that did not correlate with the behavior (Stone et al., 1999). On the other hand, another group found little change in numbers of dorsal horn cells expressing mRNA for either α_{2A} or α_{2C} receptors (Shi et al., 1999b), so further work is necessary. In early development, α_{2B} and α_{2C} receptor mRNA are co-localized in almost every spinal neuron whereas there is little mRNA for α_{2A}, and as development proceeds, the subtypes come to be expressed by different neurons (Huang et al., 2002).

Dopamine Receptors

Dopamine receptors are subdivided into two families, the D_1 group consisting of D_1 and D_5 receptors and the D_2 group consisting of the D_2, D_3 and D_4 receptors. Binding of radioactive dopamine showed radioactivity spread evenly over the dorsal horn (Kondo et al., 1985). Reports of D_1 ligand binding are disparate with one study showing little binding in the dorsal horn (Dubois et al., 1986) and another showing a very considerable signal with what appears to be a concentration in superficial regions (Levant and McCarson, 2001). Binding of D_2 and D_3 ligands is seen throughout the dorsal horn but concentrated in superficial laminae (Dubois et al., 1986; Bouthenet et al., 1987; Yokoyama et al., 1994; Levant and McCarson, 2001). Immunochemistry showed D_2 receptor cells only in lamina I, and cells expressing D_2 receptor mRNA in laminae I and III–IV with a conspicuous absence in lamina II (van Dijken et al., 1996). Presumably, receptors shown by lamina II binding are on processes of cells in other laminae.

5-HT Receptors

Following revised nomenclature (Martin and Humphrey, 1994), there are seven 5-HT receptor classes (5-HT1–7), some with many subtypes. The 5HT-3 receptors are ligand gated ion channels whereas the others interact with G proteins and are linked to second messengers.

5-HT1 and 5-HT1_A Receptors

Binding of radioactive 5-HT1 ligands such as LSD is concentrated in superficial laminae (Seybold and Elde, 1984; Daval et al., 1987; Fischette et al., 1987; Pazos et al., 1987a; Marlier et al., 1991c), although some investigators find the label slightly deeper, in laminae II–III (Pubols et al., 1992) or at the II–III border (Seybold, 1985a), and one study finds it throughout the dorsal horn (Pazos and Palacios, 1985). As to segmental concentrations, one group finds most label at cervical levels (Fischette et al., 1987), others 5HT-1 or 5HT1_A receptors at lumbosacral levels (Marlier et al., 1991c; Thor et al., 1993; Laporte et al., 1996) and yet another finds no segmental differences (Pubols et al., 1992).

Most studies with more specific ligand binding show label concentrated in the superficial dorsal horn (Daval et al., 1987; Pazos et al., 1987a; Gozlan et al., 1988; Manaker et al., 1988; Palacios, J. M. et al., 1990; Marlier et al., 1991c; Radja et al., 1991; Forrest et al., 1996; Laporte et al., 1996) although some investigators note the labeling more deeply (Pazos and Palacios, 1985) including the whole dorsal horn (Thor et al., 1993), or in laminae I–IV (Giroux et al., 1999), or that it is restricted to laminae III–IV (Pompeiano et al., 1992), and one study noted two distinct bands of label in the superficial dorsal horn (Gozlan et al., 1988). If concentrations of 5HT-1 or 5HT-1_A receptors are in the deeper dorsal horn, these receptors could be involved with large fiber primary afferent input. However, since the most reports find binding concentrated in laminae I–II as does an immunocytochemical localization of the 5HT-1_A receptor (Kia et al., 1996), we regard these receptors as largely restricted to superficial laminae. Some cells in the dorsal horn, particularly laminae III–IV, express 5H-$T1_A$ receptor mRNA (Pompeiano et al., 1992).

Approximately 20% of 5H-$T1_A$ receptors are lost after dorsal rhizotomy (Daval et al., 1987; Laporte et al., 1995), indicating a significant number of these receptors on primary afferent terminals. After spinal transection, 5H-$T1_A$ receptors increase transiently in laminae I and III (Giroux et al., 1999).

5-HT1_B Receptors

Reports do not agree. Thus, several investigators find few 5HT1_B receptors in the dorsal horn (Pazos and Palacios, 1985; Pazos et al., 1987a; Palacios et al., 1990; Radja et al., 1991), others find significant binding in laminae I and III, some binding in lamina IV and little binding in lamina II (Thor et al., 1993), or significant binding in lamina II (substantia gelatinosa) (Marlier et al., 1991c; Castro et al., 1997) or superficial dorsal horn (Laporte et al., 1995), and intense immunochemical labeling in laminae I–III using an antibody that recognizes, among other things, 5-HT1_B receptors (Ridet et al., 1994b) but relatively sparse labeling in the human spinal trigeminal nucleus with an antibody specific to 5HT1_B receptors (Longmore et al., 1997). We cannot resolve these discrepancies, but if it is accepted that 5-HT1_B receptors are found in the dorsal horn, some are on primary afferents as evidenced by the 20% loss of binding after dorsal rhizotomy (Laporte et al., 1995). Major interest in the 5HT1_B receptors comes from the migraine-relieving effects of 5-HT1_B and 5-HT1_D ligands, and there is abundant evidence for 5HT1_B receptors on DRG cells and peripheral vasculature (Chapter 4).

5-HT1_C Receptors

These are now referred to as 5-HT2_C receptors.

5-HT1_D Receptors

Several studies note few 5-HT1_D receptors in the dorsal horn (Palacios et al., 1990; Radja et al., 1991; Castro et al., 1997; Bonaventure et al., 1998a, b), but two studies describe these receptors in the superficial dorsal horn (Mills and Martin, 1995; Laporte et al., 1996).

5-HT1_S Receptors

A type 1 receptor labeled 5-HT1_S is concentrated in the superficial dorsal horn (Murphy and Zemlan, 1992).

5-HT2 Receptors

There is little binding of generic 5-HT2 ligands in the dorsal horn (Pazos et al., 1985a, b; 1987b; Fischette et al., 1987; Palacios et al., 1990; Radja et al., 1991), although one study reported binding in laminae I–II (Forrest et al., 1996). More recently, 5HT2 receptors have been subdivided into three subtypes, of which two, 5-HT2_A and 5H-T2_C (formerly known as 5HT1_C) have been studied in the cord. The 5-HT2_A receptors or cells expressing this mRNA are either not present or are present in low numbers in the dorsal horn (Pompeiano et al., 1994; Maeshima et al., 1998; Cornea-Hebert et al., 1999; Fonseca et al., 2001; Zhang, Y. Q. et al., 2001), but profiles expressing this mRNA markedly increase following peripheral inflammation (Zhang, Y. Q. et al., 2001). There is little binding of ligands for 5-HT1_C receptors, which are now known as 5H-T2_C receptors (Pazos and Palacios, 1985; Pazos et al., 1987a; Radja et al., 1991; Kashiba et al., 1992b; Thor et al., 1993). Cells expressing 5HT2_C receptor mRNA are found in laminae V and VII (Molineaux et al., 1989), or evenly distributed throughout the dorsal horn (Pompeiano et al., 1994) or are numerous everywhere but lamina II (Fonseca et al., 2001).

5-HT$_3$ Receptors

Unlike the other 5-HT receptor types, 5-HT$_3$ receptors are ligand gated ion channels. Binding and mRNA studies show 5-HT$_3$ receptors concentrated in the superficial dorsal horn or similar regions of the spinal trigeminal nucleus (Hamon et al., 1989; Waeber et al., 1989; Gehlert et al., 1991; Laporte et al., 1992; 1995; Kidd et al., 1993; Tecott et al., 1993; Fletcher and Barnes, 1999), although one study shows them uniformly distributed over the dorsal horn (Fonseca et al., 2001). Many of these receptors are located on capsaicin-sensitive primary afferent terminals as evidenced by their dramatic reduction following dorsal rhizotomy and capsaicin treatment (Hamon et al., 1989; Kidd et al., 1993). They are also located postsynaptically, presumably on intrinsic neurons expressing the 5-HT$_3$ receptor mRNA (Kia et al., 1995). Many intrinsic enkephalinergic neurons express these receptors (Tsuchiya et al., 1999).

Cloning has shown 2 5-HT$_3$ receptor subunits; 5-HT$_3$A and 5-HT$_3$B. The 5-HT$_3$A receptors can form functional channels whereas the 5-HT$_3$B receptors cannot, but heteromeric complexes of 5-HT$_3$A and 5-HT$_3$B subtypes have different properties from the homomeric 5-HT$_3$A receptors. 5-HT$_3$A receptor subunits are restricted to the CNS, and cells expressing or ligand binding to these receptors concentrate in the superficial dorsal horn (Morales et al., 1996; Doucet et al., 2000; Miquel et al., 2002; Morales and Wang, 2002). 5-HT$_3$-A$_s$ subunits are primarily presynaptic being found mostly on axons and terminals in the superficial dorsal horn (Miquel et al., 2002). Mice missing 5-HT$_3$A receptors have normal responses to acute nociceptive stimuli (Zeitz et al., 2002).

5-HT$_7$ Receptors

There is weak 5-HT$_7$ receptor binding in the dorsal horn (Gustafson et al., 1996a).

Amino Acid Receptors

Glutamate Receptors

Receptors for the excitatory amino acid glutamate are divided into ionotropic and metabotropic categories.

Ionotropic Glutamate Receptors

The ionotropic receptors mediate neural excitation in the cord and are divided into three main classes on the basis of their responses to the selective ligands; N-Methyl-D-Aspartate (NMDA), α-amino-3-hydroxy-5-methyl-4-isoazolepropionate (AMPA) and kainic acid (KA).

NMDA Receptors. Two gene families encode for NMDA receptor (NMDAR) subunits, called N(MDA)R1 and N(MDA)R2 respectively (Zukin and Bennett, 1995). The NR1 gene can undergo alternative splicing at three different exons to give eight splice variants, of which seven have been found in cDNA libraries. Each of the seven can form channels with distinctive functional properties, and the NR1 splice variants can be combined with any of the four different NR2 subunits, again with different functional properties. In addition, the various subtypes and splices have different distributions, so the complexity becomes daunting. A good nomenclature summary has been provided (Zukin and Bennett, 1995). With pan oligonucleotide probes, most or all DRG and trigeminal ganglion cells and dorsal horn neurons express mRNA for NR1 receptors (Shigemoto et al., 1992a, b; Furuyama et al., 1993; Tölle et al., 1993; 1995a; Luque et al., 1994; Pellegrini-Giampietro et al., 1994; Watanabe et al., 1994). Localization of splice variants is incomplete, but most neurons are labeled by type 1b variants, and to a lesser extent the type 2a variant, whereas 3b seems to be restricted to the dorsal horn and 2b showed a few neurons in lamina II (Luque et al., 1994). In another study, the dominant NR1 slice variant in the dorsal horn is NR1-a (Tölle et al., 1993). Thus, all or almost all neurons in the dorsal horn express NR1 receptors, and the splice variants have yet to be worked out.

The NR2 subtype is subdivided into NR2A–D categories. The NR2 subunits do not form functional receptors, instead they combine with NR1 subunits to modify their activity (Zukin and Bennett, 1995). NR2A is present in all laminae of the dorsal horn except lamina II, NR2B is essentially restricted to lamina II and little signal is seen for NR2C and D subunits (Watanabe et al., 1994; Samarasinghe et al., 1996), which is consistent with immunolabeling (Boyce et al., 1999). By contrast, in another study NR2A and 2B were essentially not expressed in the dorsal horn, NR2C only in a few cells in lamina II and NR2D lightly everywhere in the gray matter (Tölle et al., 1993).

AMPA Receptors. GluR1 and GluR2 subtypes (the latter often being studied as GluR2/3) are more numerous in the dorsal horn, and GluR3 and GluR4 subtypes are more numerous in the ventral horn but are sometimes in the dorsal horn, especially in superficial laminae (Furuyama et al., 1993; Henley et al., 1993; Sato et al., 1993b; Tölle et al., 1993; Pellegrini-Giampietro et al., 1994; Jakowec et al., 1995a, b; Virgo et al., 1996; Grossman et al., 1999; Nazli and Morris, 2000). Two studies report that that the intense labeling for GluR2, particularly GluR2 flip (see below), extends into lamina III (Tölle et al., 1993; 1995b).

For AMPA receptor splice variants, one study reports that GluR1 flip and flop variants are concentrated in laminae I and IIo whereas GluR2 flip is expressed at higher levels in laminae I–III (Tölle et al., 1993; 1995b). Similar findings are reported in developing rats, although there is no confirmation of the label in lamina III, and the label is more pronounced in younger animals (Jakowec et al., 1995b). This latter study also reported considerable numbers of both flip and flop GluR3 and GluR4 flip labeled cells throughout the dorsal horn (Jakowec et al., 1995b). In human cord with the focus on motoneurons, labeling in the dorsal horn was primarily GluR1 flop, but some labeling was seen in almost all laminae for flip GluR1–4, with most intense label consistently seen in laminae I–II (Tomiyama et al., 1996).

Kainate Receptors. A few weakly labeled cells for GluR5 are scattered throughout the dorsal horn (Furuyama et al., 1993), and in another analysis, occasional gelatinosa cells express GluR5 and 7 and moderate numbers express KA-2 receptors (Tölle et al., 1993). These findings contrast with the large numbers of neurons that are labeled by cobalt following stimulation with kainate ligands (see below).

In summary, mRNA for NR1 is present in most if not all dorsal horn neurons, and mRNA for GluR1 or GluR2 is found many neurons, particularly in laminae I and II. The GluR3 and GluR4, GluR5–7, and NR2 subtypes and numerous splice variants are more differentially distributed. These gene localizations sometimes imperfectly mirror localizations as seen in binding and immunochemical studies, presumably indicating that some neurons that possess mRNA for various subtypes and splice variants do not express the receptors.

Laminar Localizations and Segmental Differences. Autoradiographic binding of NMDA, AMPA, and kainate ligands is seen in all laminae but concentrated in laminae I and II (substantia gelatinosa, superficial dorsal horn) (Monaghan and Cotman, 1982; 1985; Greenamyre et al., 1984; Jansen et al., 1990; Mitchell and Anderson, 1991; Allaoua et al., 1992; Kalb et al., 1992; Tallaksen-Greene et al., 1992; Albin et al., 1993; Gonzalez et al., 1993; Henley et al., 1993; Krieger et al., 1993; Shaw et al., 1994a; Tomiyama et al., 1994; Croul et al., 1995; Jakowec et al., 1995a; Kus et al., 1995a; Kalb and Fox, 1997; Krenz and Weaver, 1998; Akesson et al., 2000; Gougis et al., 2002). Some studies, however, report that glutamate receptor binding extends into the dorsal part of lamina III (Shaw et al., 1991; 1994b; Chinnery et al., 1993; Shaw and Ince, 1994). Glutamate receptors are also localized immunochemically, and again most investigators find label concentrated in laminae I and II (substantia gelatinosa, superficial dorsal horn) (Martin et al., 1993; Tachibana et al., 1994; Hama et al., 1995; Kondo et al., 1995; Petralia et al., 1996a; 1997; Aicher et al., 1997; Carlton et al., 1998b; Goff et al., 1998; Krenz and Weaver, 1998; Popratiloff et al., 1998a, b; Spike et al., 1998; Boyce et al., 1999), others say laminae I–III (Petralia et al., 1994a, b; Bonnot et al., 1996; Spike et al., 1998; Yung, 1998), others include the dorsal part of III (Furuyama et al., 1993; Kiyama et al., 1993; Popratiloff et al., 1996; Helgren et al., 1999a; Nazli and Morris, 2000), others that GluR2/3 concentrates in lamina III (Yung, 1998), others that GluR1 concentrates in laminae I and IIo whereas GluR2 concentrates in IIi (Popratiloff et al., 1996; Engelman et al., 1999), others that AMPA receptors concentrate in laminae I, IIi, and the outer part of lamina III (Helgren et al., 1999a), and finally that kainate receptors are concentrated in lamina IIi (Hwang et al., 2001b). Closure on precise laminar localizations would be desirable, but as yet this is not possible, in part because of lack of descriptions of laminar boundaries. The neuropil is lightly and evenly stained for GluR4 without cellular staining (Martin et al., 1993). Since nociceptive afferents end in laminae I and II, where glutamate receptors are concentrated, the usual suggestion is that glutamate receptor activation is primarily involved with nociceptive primary afferent transmission, but all primary afferents presumably use glutamate as a transmitter, and the reason that nociceptive primary afferents require relatively more glutamate receptors is not understood.

Several studies note that there is relatively little variation in segmental labeling for glutamate receptors (Pellegrini-Giampietro et al., 1994; Shaw et al., 1994a, b; Akesson et al., 2000). On the other hand MK-801 and CNQX binding is greater in lumbar and sacral segments (Shaw et al., 1991; Chinnery et al., 1993), and GluR2 is highest at lumbar levels (Grossman et al., 1999).

Primary Afferent Input. Anatomical evidence that NMDA and non-NMDA receptors are located presynaptically on primary afferent fibers is the expression of NMDA, AMPA, and kainate receptors on DRG cells, peripheral sensory axons, and central endings in glomeruli, particularly in laminae I and II, the bidirectional transport of glutamate and NR1 in peripheral nerves (Chapter 4) and the co-localization of glutamate receptors with primary afferent peptides in DRG cells and terminals in the dorsal horn (see peptides). In addition there is the reduction of AMPA and kainate glutamate receptors in the dorsal horn following deafferentation by dorsal rhizotomy (Carlton et al., 1998b; Hwang et al., 2001a, b). Thus, all types of inotropic glutamate receptors are found on DRG cells and their afferent terminals. However, NMDA receptor localizations (Croul et al., 1995; Tseng et al., 2000) and GluR1 and GluR2/3 immunostaining in deeper laminae (Carlton et al., 1998b) increase chronically after rhizotomy. These increases must be due to compensatory increases in receptors on intrinsic and/or descending systems or sprouting of primary afferent axons from adjacent undamaged roots.

Intrinsic and Descending Systems. Many small cells express NR1 and GluR1–4 in laminae I and II of the dorsal horn with fewer labeled cells in deeper laminae (Tachibana et al., 1994; Hama et al., 1995; Popratiloff et al., 1996; 1998b; Spike et al., 1998; Benoliel et al., 2000). The cells in deeper laminae tend to be multipolar (Tachibana et al., 1994). Most GluR1-expressing neuron profiles show high concentrations of GABA and often GLY and are thus thought to be inhibitory whereas GluR2/3 cell profiles are never GABAergic and often contain somatostatin or neurotensin, and are thus thought to be excitatory (Spike et al., 1998), although somatostatin is usually inhibitory. Many NR1 cells, especially in lamina I, express the NK1 receptor (Benoliel et al., 2000). The molecular scaffold protein PSD-95/SAP90 is preferentially expressed in laminae I and IIo and interacts with the NMDA receptor in thermal hyperalgesia (Tao et al., 2000a).

Spinothalamic neurons express NR1, GluR1, and GluR2/3 receptors (Fig. 6.2; Kus et al., 1995b; Ye and Westlund, 1996; Zou et al., 2000). By contrast, relatively few trigeminothalamic neurons express GluR1 and GluR2/3 receptors in contrast to the large numbers that express GABA receptors (Kondo et al., 1995). Injection of capsaicin, which activates nociceptors and causes neurogenic inflammation, increases the numbers of spinothalamic neuron profiles with phosphorylated NR1 receptors (Zou et al., 2000). The rapid change from non-phosphorylated to phosphorylated receptors has implications for synaptic plasticity.

A marking technique is to expose neurons to agonists that activate receptors that open Ca^{++} channels, which can be visualized histochemically with Co^{++}. Enough of the cell body and proximal dendrites are outlined to allow the cell to be identified by classic criteria. Receptor-specificity is checked by showing that an antagonist for a particular receptor prevents labeling. In slices, numerous cells are outlined in all dorsal horn laminae after application of NMDA, AMPA, and KA ligands. In lamina II, for example, application of AMPA resulted primarily in stained stalked cells, kainate in stained islet cells, and NMDA in all types of cells. All ligands stained antenna-type cells in deeper laminae. Overall numbers of labeled cell profiles were highest with kainate ligands, intermediate with AMPA ligands, and least with NMDA ligands, which is the reverse of the

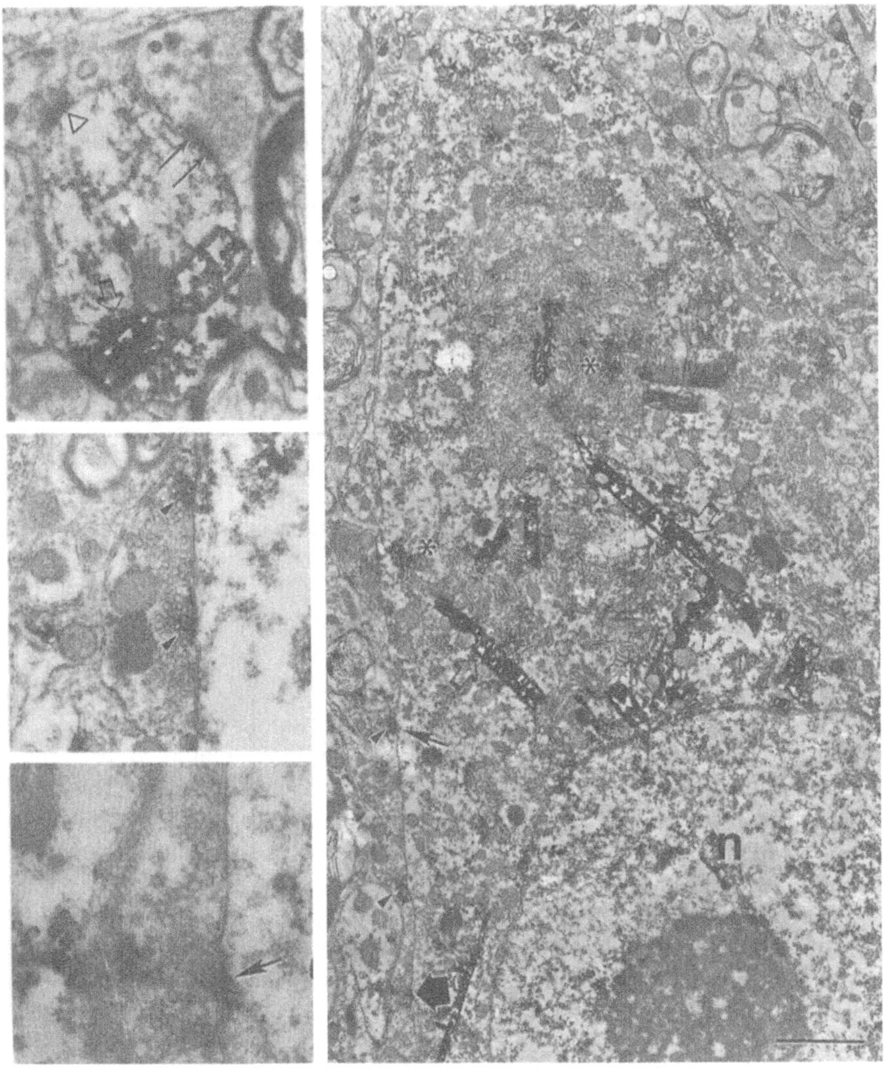

FIGURE 6.2. This figure demonstrates NMDAR1 receptors (diffuse patches of label) expressed by spinothalamic cells (crystalline or rectangular label). (From Ye and Westlund, 1996.)

immunochemical and in situ findings (Nagy et al., 1994). AMPA induced Co^{++} labeling was combined with immuno- and histochemical staining to show that activated neurons are most common in lamina I and IIo and are in the midst of IB4 labeled afferent fibers. The conclusion is that Ca^{++} permeable GluR1 receptors are prominently involved in the postsynaptic integration of nociceptive input into lamina I and IIo. In addition some CO^{++} neurons express NK1 receptors. (Engelman et al., 1999). The CO^{++} labeling method was used to show AMPA responding neurons in early human fetuses (Bardoul et al., 1998).

The m-RNA for a novel glutamate receptor, X-1, is expressed in neurons in laminae II and III of the rat spinal cord (Clabarra et al., 1995).

Synaptic Organization. Synapses classically consist of a presynaptic element attached to a postsynaptic element by synaptic thickenings with receptors in the membrane of the thickening on the postsynaptic side, where they can respond to ligands from presynaptic terminals. Some receptors are not located at active zones, however, but instead are spread over the dendritic or somatic surface without reference to classical synapses. This is evidence for volume transmission. Receptors can also be expressed presynaptically on terminals, where they can control synaptic activity. If a receptor and an appropriate ligand are expressed in the same terminal, this indicates autoreception. Thus key morphologic issues are whether the receptors are clearly localized to postsynaptic thickenings or are spread widely over the cell surface (classic vs. volume transmission), whether they are on axon terminals (presynaptic control), and if the ligand for the receptor is present (autoreception).

NMDA Receptors. NR1 immunostaining is seen in the neuronal cytoplasm and along non-synaptic areas of dendritic membrane, but it is most dramatic in the postsynaptic thickenings that underlie presynaptic terminals (Liu et al., 1994b; Ye and Westlund, 1996; Aicher et al., 1997; Popratiloff et al., 1998a). C1 and C2 primary afferent terminals, dome-shaped terminals of probable intrinsic origin, and SP terminals face postsynaptic thickenings that are enriched in NR1 receptors (Aicher et al., 1997; Popratiloff et al., 1998a; Benoliel et al., 2000). Many NR1 expressing neurons are spinothalamic cells, providing morphologic evidence that the NR1 receptors play an important role in nociceptive relay to higher regions of the nervous system (Ye and Westlund, 1996). NR1 labeling is seen on axon terminals (Liu et al., 1994b), many of which are central elements of glomeruli (Aicher et al., 1997), and these glomerular endings are in turn enriched in glutamate, suggesting autoreception and thus NMDA presynaptic control of fine primary afferent input into the cord (Fig. 6.3; Liu et al., 1994b).

AMPA Receptors. AMPA receptor subunits are found in the cytoplasm of cell bodies and dendrites as well as on synaptic thickenings on the surfaces of the labeled postsynaptic elements (Tachibana et al., 1994; Petralia et al., 1997). Thus these are largely postsynaptic. The presynaptic endings associated with the labeled postsynaptic elements are filled with large round clear vesicles and occasional dense-cored vesicles and the labeled thickenings are asymmetric and so fit the category of excitatory synapses. GluR1 active zones are found largely in type I glomerular terminals, which are thought to arise from unmyelinated primary afferents and the GluR2 is largely associated with type II glomerular endings, which are thought to arise from fine myelinated afferents (Popratiloff et al., 1996). This distinction may have functional import because GluR1 activation results in Ca^{++} flow whereas GluR2 activation prevents Ca^{++} influx.

KA Receptors. Kainate receptors (GluR 5/6/7) are found postsynaptically labeling synaptic thickenings on dendrites and cell bodies, on dendrites and cell bodies not related to postsynaptic thickenings, and on numerous primary afferent terminals, all of these being especially numerous in the superficial dorsal horn (Hwang et al., 2001a, b).

In summary, morphologic evidence indicates that glutamate receptors participate in classical synaptic transmission, volume transmission, presynaptic and postsynaptic control, and autoreception.

Lesions, Diseases, and Excitotoxicity

Dorsal Rhizotomy. Deafferentation by dorsal rhizotomy or ganglionectomy results in increased binding of MK-801, a NMDA receptor ligand, and NR1 immunostaining

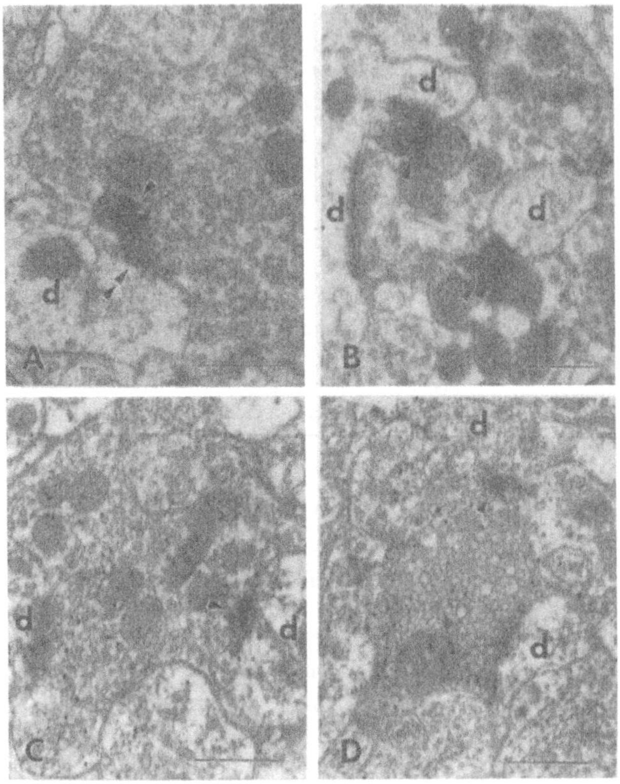

FIGURE 6.3. A figure demonstrating that glutamate-enriched terminals express NMDA receptors. In **A**, NMDAR1 label is expressed by the vesicle containing axon terminal (single arrowhead) and the postsynaptic dendrite (Double arrowhead). In **C** and **D**, glutamate is indicated in the vesicle containing axon terminal by immunogold labeling (black dots) and in the same terminal the NMDAR1 receptor is seen by peroxidase labeling (arrowheads). (From Liu et al., 1994b.)

(Croul et al., 1995; Tseng et al., 2000). The duration of the increase coincides with the duration of autotomy in animals receiving dorsal root ganglionectomies (Tseng et al., 2000). By contrast, deafferentation by dorsal rhizotomy reduces GluR1, GluR2/3, and GluR5/6/7 immunostaining in laminae I and II (Carlton et al., 1998b; Helgren et al., 1999a; Hwang et al., 2001b) but increases it in laminae III–IV (Carlton et al., 1998b). Thus deafferentation by either dorsal rhizotomy or dorsal root ganglionectomy provokes a complicated response with increases in NMDA receptors and AMPA receptors in deeper laminae and a reduction of AMPA and kainate receptors in superficial laminae.

Peripheral Nerve Lesions. Crush of the sciatic nerve leads to a 300% increase in NMDA receptor binding bilaterally in the dorsal horn (Croul et al., 1998), whereas loose ligations of the sciatic nerve, a chronic pain injury model, leads to a significant reduction in NMDA binding and immunocytochemistry (Hama et al., 1995) or no change (Goff et al., 1998). Similarly, transection of the sciatic nerve leads to an increase in immunostaining for AMPA receptors (Harris et al., 1996), whereas damage to the sciatic nerve by pronase injection leads to a decrease (Helgren et al., 1999a). Thus, the response of glutamate receptors in the dorsal horn after nerve lesions is complex, and since changes in these receptors have a great bearing on sensory transmission, more work needs to be done. Implantation of adrenal tissue at the time of the lesion prevents changes in NMDA receptors (Hama et al., 1995). Transection of the sciatic nerve in new-born rats prevents the normal decline of MK-801 (a non-competitive antagonist of NR1) binding on the operated side, which is interpreted as evidence for two populations of NMDA receptors, one affected by surgery and one not (Hughes et al., 1998).

Inflammation. Lipopolysaccharide injection causes dramatic joint swelling, and a bilateral decrease of approximately 25% of GluR1 in neurons of the substantia gelatinosa (lamina II) with no change in the expression of GluR2, GluR3, and NR1 mRNAs (Pellegrini-Giampietro et al., 1994). Similarly, unilateral arthritis caused by injection of Freund's adjuvant into one hindpaw led to bilateral time-limited decreases in mRNA expression and NR1 receptor binding in laminae I and II (Kus et al., 1995a), but in another study an increase in NMDA mRNA (Ohtori et al., 2002). Injection of capsaicin produces a local neurogenic inflammation and morphologic evidence for release of glutamate from laminae I and II, complementing demonstrations of release of glutamate into the cerebrospinal fluid after noxious stimuli (Sasaki et al., 1998). Another approach is to use antisense oligonucleotides. When this is done for NR1, both spinal NR1 immunostaining and the pain of formalin-induced inflammation are reduced (Garry et al., 2000). This would seem to be a promising avenue for future therapy.

Spinal Damage. Many studies determine effects of glutamate antagonists on secondary cell death following spinal trauma and ischemia (Choi, 1994). For changes in these receptors, spinal transection reduces binding of the NR1 ligand MK-801 below the lesion, particularly in lamina II but also in all parts of the cord (Krenz and Weaver, 1998), and NR-1 staining was not changed (Goff et al., 1998; Krenz and Weaver, 1998; Gougis et al., 2002). There is no change in NR1 mRNA in the dorsal horn as contrasted to the ventral horn after spinal injury (Grossman et al., 2000).

Excitotoxicity and Disease. Glutamate excitotoxicity is a large rapidly expanding field (Choi, 1994). As regards the spinal cord, dark or TUNEL cells appear after nerve lesions (Sugimoto et al., 1989b; 1990; Azkue et al., 1998), and the speculation is that these cells die due to glutamate toxicity (Sugimoto et al., 1989b; Bennett, 1991; Dubner, 1991; Azkue et al., 1998). Furthermore, cell loss from various causes can be prevented with both NMDA and non-NMDA antagonists (Sanner et al., 1994; von Euler et al., 1994; Wrathall et al., 1994; Iwasaki et al., 1995; Carriedo et al., 1996), and systemic quisqualate and kainate induce neuronal death in the superficial dorsal horn with relief by appropriate antagonists (Urca and Urca, 1990; Nakamura et al., 1994). Accordingly glutamate excitotoxicity may play a role in various types of spinal cell loss.

A possible mechanism of motor neuron loss in human disease is glutamate toxicity, and glutamate receptors are concentrated in the superficial dorsal horn as well as on motoneurons (Allaoua et al., 1992; Shaw, 1992; Shaw et al., 1992; 1994a, b; Krieger et al., 1993; Tomiyama et al., 1994; Samarasinghe et al., 1996). Thus, superficial dorsal horn cells might also be lost, but there are no reports to our knowledge of such loss in superficial laminae in diseases such as amyotropic lateral sclerosis, and there are no changes in GluR2 receptors in superoxide dismutase mutant mice, a condition related to ALS (Morrison et al., 1998). On the other hand, selective stimulation of large primary afferents leads to superficial dorsal horn cell loss in naïves and more extensively after nerve lesions, and this may represent glutamate toxicity (Coggeshall et al., 2001). The point is that neurons in superficial laminae express high concentrations of glutamate receptors and so may be vulnerable to glutamate toxicity.

A form of diabetes mellitus is inherited in a strain of obese mice, and the presence of this disease is associated with an increase in both AMPA and NMDA receptor binding in the dorsal part of the spinal cord (Li, N. et al., 1999d).

Development. Some find NR1, AMPA, and KA receptors in all laminae in embryonic and early post-natal life that then recede except in superficial dorsal horn, which thus becomes preferentially labeled (Kalb et al., 1992; Watanabe et al., 1994; Jakowec et al., 1995b; Kalb and Fox, 1997; Akesson et al., 2000). Others find glutamate binding always most intense in the substantia gelatinosa (Gonzalez et al., 1993), that NR2A mRNA is in the ventral cord in utero and then expands to the dorsal horn except for lamina II by adulthood, that NR2C mRNA is not seen at any time, that NR2D is seen in every lamina at birth but disappears by P21 (Watanabe et al., 1994) or that GluR2/3 always concentrated in the substantia gelatinosa but Glu4 is always distributed throughout the gray matter except for the substantia gelatinosa (Jakowec et al., 1995a). Affinity binding of AMPA and kainate is not developmentally regulated (does not change), but NMDA binding is, and this may have a bearing on excitotoxicity (Kalb and Fox, 1997).

Co-localizations. Numerous NR1 cells (a splice variant missing exon 22) co-localize with cells expressing the NK1 receptor in laminae I, IV, and V (Benoliel et al., 2000), and lamina I cells with AMPA receptors, as outlined by Co^{++} localizations, express NK1 receptors (Engelman et al., 1999). Thus glutamate and SP can activate many of the same neurons in the superficial dorsal horn. There is an enormous bibliography on the expression of the immediate early gene cfos or its product fos after various types of painful stimuli (Chapman and Besson, 1997). As far as glutamate receptors are concerned, approximately 25% of NR1 neuron profiles in laminae I–II and 55% in lamina V express fos after CFA inflammation (Zhang et al., 1998c), and NMDA receptor antagonists inhibit fos expression after painful stimuli (Chapman and Besson, 1997). Many NMDA receptor-expressing terminals contain high concentrations of glutamate itself (Liu et al., 1994b), indicating autoreception. GluR1 cells co-localize with GABA, glycine, and parvalbumin, and thus are thought to be inhibitory whereas GluR2/3 cells co-localize with neurotensin and somatostatin and thus are thought to be excitatory, whereas NO is expressed by both (Spike et al., 1998).

Metabotropic Glutamate Receptors

Eight metabotrobic glutamate receptors (mGluRs) have been cloned (mGluR1–8), and these are placed in three groups, group 1 consisting of mGluR1 and mGluR5, group 2 of mGluR2 and mGluR3, and group 3 of mGluR4 and mGluR6–8.

Laminar Localizations. Binding of radioactive glutamate in the presence of appropriate blockers shows mGluRs throughout the gray matter but concentrated in the superficial dorsal horn (Tomiyama et al., 1994). For group 1, two studies show little label for mGluR1, including a and b isoforms, in laminae I and II (Romano et al., 1995; Alvarez et al.,

2000; Aronica et al., 2001) whereas others show considerable label in the superficial dorsal horn (Fotuhi et al., 1993; Tang and Sim, 1999). Judgment is difficult, but the pictorial evidence in the first two papers is more convincing. The mGluR5 receptors, including a and b isoforms, are prominent in laminae I and II (Valerio et al., 1997a; Agrawal et al., 1998; Jia et al., 1999; Alvarez et al., 2000; Tao et al., 2000b; Aronica et al., 2001). For group 2, mGluR2 and mGluR2/3 stain intensely in laminae IIi and III with relatively little in laminae I and IIo (Petralia et al., 1996b; Tang and Sim, 1999; Azkue et al., 2000; Aronica et al., 2001; Carlton et al., 2001a). For group 3, mGluRs 4a, 7, 7a, and 7b are concentrated in the superficial dorsal horn (Ohishi et al., 1995a; Li, H. et al., 1997; Kinoshita et al., 1998; Azkue et al., 2001). To estimate relative expression levels, reverse transcriptase-polymerase chain reaction mRNA estimates show mGluR1A and mGluR5A and B to be present in high amounts, mGluRs 2, 3, 4, and 7 in intermediate amounts and mGluRs 1B, 6 and 8 essentially not present (Valerio et al., 1997b). It must be remembered that these are whole cord assays and not just the dorsal horn.

Intrinsic Spinal Systems. Immunostained mGluR1a neurons are primarily in laminae III–VI although one report shows 1a cells in superficial laminae (Tang and Sim, 1999), mGluR1b is in a few weakly stained neurons in lamina II and mGluR5 in numerous neurons concentrated in laminae I and II (Vidnayanszky et al., 1994; Boxall et al., 1998; Jia et al., 1999; Alvarez et al., 2000; Aronica et al., 2001). The mRNA for mGluR1 receptors is present but weak in the substantia gelatinosa (Shigemoto et al., 1992a; Fotuhi et al., 1993). For mGluR1 isoforms, 1b show considerable expression, 1a and 1d moderate expression, and 1c nothing in the superficial dorsal horn (Berthele et al., 1999). A few cells express group 2 mGluRs or their mRNAs, and they are vaguely concentrated in laminae II–III (Ohishi et al., 1993a, b; Boxall et al., 1998; Berthele et al., 1999; Aronica et al., 2001), although sometimes they are not apparent (Boxall et al., 1998; Tomiyama et al., 2001). From their scarcity, it is probable they do not contribute greatly to the group 2 mGluR plexus in the dorsal horn. No cells are immunostained for mGluR4a or mGluR7a and b in the dorsal horn (Kinoshita et al., 1998; Azkue et al., 2001), but dorsal horn cells express mRNA for these receptors (Ohishi et al., 1995b; Berthele et al., 1999) with mGluR7 mRNA in lamina I–II and fewer but still notable in III and IV and deeper laminae (Kinzie et al., 1995; Boxall et al., 1998).

Synaptic Relations. Group 1 mGluR1a is almost exclusively postsynaptic and most of these receptors are extrasynaptic, suggesting volume transmission (Tang and Sim, 1999; Alvarez et al., 2000). The mGluR5 receptors are also predominantly postsynaptic, although a few labeled terminals are seen (Jia et al., 1999; Alvarez et al., 2000; Tao et al., 2000b) and the labeling is mostly on non-specialized dendritic membrane, again suggesting volume transmission, but some synaptic thickenings are specifically labeled (Jia et al., 1999; Alvarez et al., 2000; Tao et al., 2000b). No C1 and few C2 primary afferent glomerular endings are immunostained, suggesting little presynaptic control of primary afferents (Jia et al., 1999), but considerable numbers of primary afferents end on mGluR5 dendrites (Jia et al., 1999; Tao et al., 2000b). Approximately half the terminals immunostained for mGluR1 or mGluR5 contain GABA (Jia et al., 1999; Tao et al., 2000b). If fine primary afferents are labeled with WGA-HRP, many end with asymmetric terminals on postsynaptic elements labeled for mGluR5. GABA-containing symmetric terminals also end on mGluR5 post-synaptic elements (Tao et al., 2000b).

Group 2 mGluRs are found primarily in axons and terminals, mostly at extrasynaptic locations, and concentrated in lamina IIi (Jia et al., 1999; Azkue et al., 2000), although one report emphasized labeled dendrites (Tang and Sim, 1999). Many of the labeled terminals are C2 glomerular endings, and dorsal rhizotomy removes many of these fibers (Jia et al., 1999; Carlton et al., 2001a), so primary afferent endings express group 2 mGluRs indicating a probable presynaptic control of primary afferent input, particularly fine myelinated afferent input.

Group 3 mGluRs are expressed by many presynaptic terminals that make asymmetric contacts with dendrites in laminae I and II, and many of these terminals bind IB-4 and some SP, presumably indicating that they are primary afferent endings concerned with nociception (Ohishi et al., 1995a; Li, H. et al., 1997). The labeling is usually confined to the presynaptic specializations, and when dendrites are labeled, the labeling is usually concentrated in the postsynaptic specializations (Azkue et al., 2001), thus suggesting classical synaptic transmission.

In summary Group 1 and 2 mGluRs seem morphologically to be arranged for volume transmission whereas group 3 is more for classical transmission, and Group 1 is almost exclusively postsynaptic whereas groups 2 and 3 have a large presynaptic component and seem poised to exercise presynaptic control over primary afferent input.

Development. The mGluR3 and 5 mRNAs are more intensely expressed in early development, and particularly mGluR3 expression recedes to reach the adult arrangement (Berthele et al., 1999). Immunostaining for mGluR5 is widespread in neonatal human cords, but concentrated in laminae I and II in adulthood (Valerio et al., 1997a).

Experimental Manipulations. Dorsal rhizotomy causes loss of group 2 mGluRs indicating that many of them are on primary afferent endings (Jia et al., 1999; Carlton et al., 2001a). Sunburn caused by exposure of 9-day-old rat paws to ultraviolet light leads to an increase in mGluR3 mRNA in laminae II–IV for approximately 3 days after exposure and then subsides in concert with the symptoms (Boxall et al., 1998).

GABA and Benzodiazepine Receptors

Receptors for gamma amino butyric acid (GABA), a major inhibitory transmitter, are divided into three subtypes; $GABA_A$, $GABA_B$, and $GABA_C$. The $GABA_A$ receptor subtype mediates rapid ionic transmission through activation of chloride channels, the $GABA_B$ subtype is metabatropic and $GABA_C$ subtypes are not reported in the cord (Bormann and Feigenspan, 1995). Benzodiazapene receptors, which are of considerable clinical importance, will be considered with the $GABA_A$ receptors because the two receptors are associated functionally, although their distributions do not everywhere overlap.

$GABA_A$ Receptors

Laminar and Segmental Localizations. Binding of radioactive muscimol (a $GABA_A$ agonist), or GABA + baclofen (which suppresses $GABA_B$ binding) show relatively even labeling over the gray matter of the cord with somewhat heavier label over the dorsal as compared to the ventral horn (Palacios et al., 1981; Price et al., 1984; Bowery et al., 1987), with a tendency to concentrate in the superficial dorsal horn (Waldvogel et al., 1990; Castro-Lopes et al., 1995). Binding for benzodiazepine receptors is concentrated in laminae II–IV in rats (Young, W. S. and Kuhar, 1979a, b), laminae II–III in humans (Whitehouse et al., 1983; Faull and Villiger, 1986; Waldvogel et al., 1990) and laminae I–II in monkeys (Albin et al., 1993). Immunostaining shows $GABA_A$ and benzodiazepine receptors concentrated in laminae II–IV (Richards et al., 1987; Waldvogel et al., 1990; Bohlhalter et al., 1996). Benzodiazepine receptors are uniformly distributed down the length of the cord during development (Bruning et al., 1990).

$GABA_A$ Subtypes. There are numerous $GABA_A$ receptor subtypes and mRNA localizations are complex. Thus, α2, α3, β3, and γ2 units are found in cord, but α2 subtypes are restricted to the ventral horn (Wisden et al., 1991). A later study reports mRNAs for 13 $GABA_A$ receptor subtypes in the embryonic cord, but only the α2, α3, β3, and γ2 units are expressed in the adult (Ma, W. et al., 1993). Another group finds somewhat more subtypes in the cord, namely α1, α2, α3, and α5 (Poulter et al., 1992), and β2, β3, and γ2 (Poulter et al., 1993). Another group reports eight different subtypes to varying degrees in the cord with γ2 the most ubiquitous (Persohn et al., 1991; 1992), with α3, α5, β3, and γ2 found in laminae II–III (Persohn et al., 1991). Immunostaining of the β3 subunit shows intense label in laminae I–II and some labeling deeper (Todd et al., 1996), and the α2 subunit in lamina II of the dorsal horn (Gutierrez et al., 1996). The α1 and α5 subunits are abundant in laminae III–VI and deeper, whereas the α2 subunit is predominant in the superficial dorsal horn (Fig. 6.4; Bohlhalter et al., 1994; 1996), and the β2 and β3 subunits are particularly prominent in laminae II–III. Thus, there is good but not complete agreement on the distribution of $GABA_A$ receptor subtypes. A major point of these studies is that differences in GABA function may be due to differences in subtype distributions. Trigeminothalamic cells express $GABA_A$ γ1 and γ2 subunits (Kondo et al., 1995).

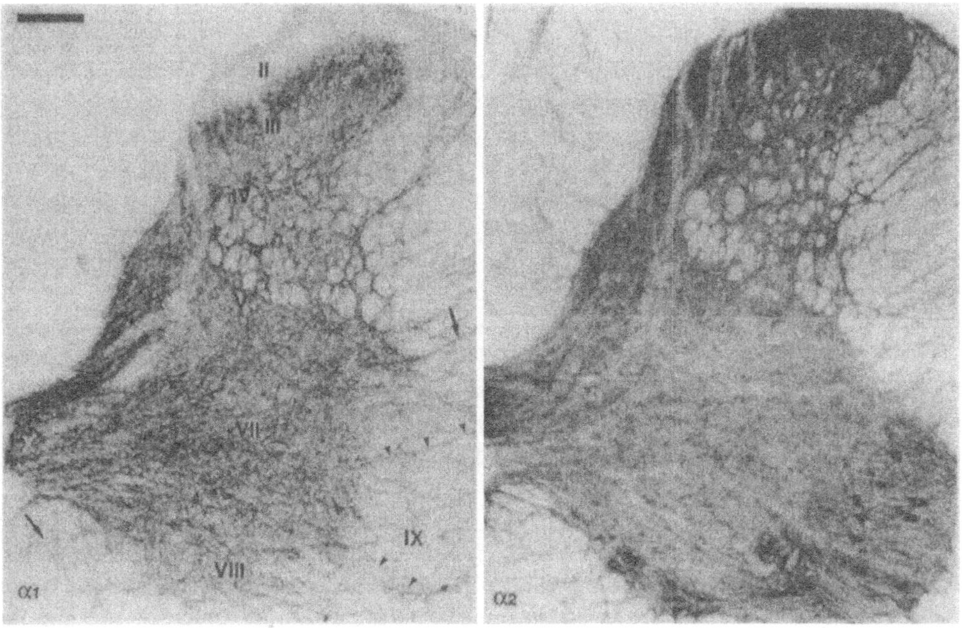

FIGURE 6.4. A pair of micrographs showing the laminar specificity of $GABA_A$ receptors in the dorsal horn. Note in particular the absence of staining for the α1 subunit in lamina II compared to the intense staining of the α2 subunit in this lamina. (From Bohlhalter et al., 1996.)

Synaptic Organization. There is a 20–30% decrease in $GABA_A$ binding sites in the adult superficial dorsal horn after neonatal capsaicin treatment, which is evidence that many of these receptors are presynaptic to fine capsaicin-sensitive primary afferent terminals (Singer and Placheta, 1980). By contrast, however, dorsal rhizotomy does not affect the immunostaining of the β2 and β3 subunits of the $GABA_A$ receptors, even though essentially every dorsal root ganglion cell expresses these subtypes (Alvarez et al., 1996). Since 10–14% of peripheral unmyelinated sensory axons express these receptors (Carlton et al., 1999), $GABA_A$ receptors are on peripheral primary afferent processes, and a possibility to explain the lack of change after rhizotomy is that receptors are shipped only peripherally in DRG cell processes.

Ultrastructural studies show $GABA_A$ immunostaining most often on the postsynaptic side of synaptic complexes where the presynaptic element contains either pleomorphic or flattened clear vesicles or are GABA immunoreactive (Waldvogel et al., 1990; Sur et al., 1995; Alvarez et al., 1996; Todd et al., 1996). The $GABA_A$ immunostaining is not always at synapses and non-synaptic dendritic regions are also stained, suggesting the possibility of volume as well as regular synaptic transmission. Staining is seen in the peripheral axoaxonic synapses in glomeruli so there is an anatomic basis for both pre- and postsynaptic interactions (Waldvogel et al., 1990). Glial labeling is reported (Sur et al., 1995). Glycine, as represented by gephryin localizations, and $GABA_A$ receptors are frequently co-localized in the dorsal horn (Bohlhalter et al., 1994; Todd et al., 1996), possibly indicating the action of two powerful inhibitory substances at the same synapse. Finally gephyrin is a protein that binds $GABA_A$ as well as glycine receptors to their postsynaptic locations, and in accord with this, gephyrin-deficient mice lack the typical GABAergic synaptic clustering on spinal neurons (Kneussel et al., 1999).

$GABA_B$ Receptors

Laminar Localizations. Binding of $GABA_B$ ligands is concentrated in the superficial dorsal horn (laminae I–II, lamina II, substantia gelatinosa), although some label is found in all laminae (Price et al., 1984; 1987; Bowery et al., 1987; Waldvogel et al., 1990; Kroin et al., 1993; Malcangio et al., 1993; Castro-Lopes et al., 1995; Towers et al., 2000). Binding is reduced approximately 50% by dorsal rhizotomy or neonatal capsaicin indicating that approximately half the $GABA_B$ ligands are on primary afferents (Price et al., 1984; 1987). $GABA_B$ receptors are subdivided into $GABA_{B1}$ and $GABA_{B2}$ subtypes, and mRNA for both appears in the dorsal horn (Calver et al., 2000) with $GABA_{B2}$ having a slight concentration in superficial laminae (Durkin et al., 1999). With immunochemistry, $GABA_{B1a}$, $GABA_{B1b}$ and $GABA_{B2}$ are in laminae I–II (Poorkhalkali et al., 2000; Towers et al., 2000; Charles et al., 2001), although the $GABA_{B2}$ labeling is more diffuse throughout the dorsal horn. Most lamina II neurons express $GABA_{B1a}$ and $GABA_{B1b}$ receptor mRNAs, whereas cells expressing $GABA_{B2}$ mRNA are evenly distributed through the gray matter of the cord (Towers et al., 2000).

Experimental Manipulations. A $GABA_B$ agonist reduces and a $GABA_B$ antagonist increases $GABA_B$ binding in the dorsal horn (Kroin et al., 1993; Malcangio et al., 1993). Peripheral nerve lesions decrease $GABA_B$ receptor binding but increase $GABA_A$ binding in the spinal cord (Castro-Lopes et al., 1995). The decrease might arise because of loss of $GABA_B$ expressing of primary afferent endings, and the upregulation of $GABA_A$ receptors presumably indicates development of hypersensitivity. This finding seems to conflict with the observation that L5 peripheral nerve ligation leads to a downregulation of $GABA_A$ $\alpha2$ and $\gamma2$ mRNAs in ipsilateral dorsal root ganglia (Fukuoka et al., 1998). Twenty-four hours after intraplantar formalin, a significant bilateral increase in $GABA_{B1}$ and $GABA_{B2}$ receptors occurs in the cord and unilaterally in the ipsilateral dorsal root ganglion (McCarson and Enna, 1999). By contrast, 3–4 weeks after chronic inflammation caused by intraplantar CFA, there is a significant reduction in $GABA_B$ receptor binding whereas $GABA_A$ binding is not affected (Castro-Lopes et al., 1995).

Glycine Receptors

Glycine is an inhibitory transmitter that increases chloride conductance when it binds to its receptor. Strychnine is the most specific antagonist for the glycine receptor, which in adults is a heterodimer with 3α and 2β subunits, the β subunit being closely linked to the large polypeptide gephyrin, which is used as a marker for glycinergic receptors. There are 4 α isoforms (α_{1-4}) but only 1 β subtype. The $\alpha1$ and $\alpha2$ isoforms are most common in the spinal cord, the $\alpha1$ isoform being the adult form and the $\alpha2$ isoform the neonatal one. Ligands bind to the α isoforms and the β subtype, via its association with gephyrin, has a role in the assembly and expression of the receptor (Betz et al., 1999; Callister et al., 1999).

Laminar Localizations. Ligand (strychnine) binding is concentrated in the superficial dorsal horn (Zarbin et al., 1981; Frostholm and Rotter, 1985; Probst et al., 1986; Bruning et al., 1990; Frost White et al., 1990; Albin et al., 1993), and in Wobbler mice in laminae IIi, III, and IV (Tomiyama et al., 1997). By contrast, immunocytochemical labeling is in deeper laminae, particularly laminae III–IV, but may include lamina IIi (Araki et al., 1988; Basbaum, 1988; Murakami et al., 1988; Van Den Pol and Gorcs, 1988; Mitchell et al., 1993; Bohlhalter et al., 1994). The difference between superficial and deep laminae is so striking that it was hypothesized that spinothalamic cells in lamina I would not receive glycinergic input whereas those in lamina V would (Basbaum, 1988), although it must be remembered that some lamina V cells have dendrites that extend to superficial laminae (antenna-type cells). The discrepancy between binding and immunocytochemical labeling is not resolved. Cells expressing mRNA for the $\alpha1$, $\alpha2$, and β subtypes are scattered throughout the dorsal horn, with some concentration in deeper laminae (Fujita et al., 1991; Malosio et al., 1991; Sato et al., 1991). During embryonic development cells expressing the $\alpha1$ receptor are sparse but steadily increase until they reach maximal levels by P21, $\alpha2$ cells are at a maximum at P0 and then decline to moderate numbers in the adult (Watanabe and Akagi, 1995). A distinction is that α_1 and α_2 isoforms are on neuronal somata in superficial dorsal horn but on both somata and dendrites in the ventral horn (Racca et al., 1998). All glial cells contain the α_1 isoform and about half contain the β subtype (Kirchhoff et al., 1996).

Synaptic Organization. Glycine receptors are localized in synaptic thickenings of axodendritic, axosomatic, and dendrodendritic synapses in laminae I–III of the dorsal horn (Mitchell et al., 1993). Many of the presynaptic profiles at these synapses are GABAergic (Mitchell et al., 1993). GABA and glycine are co-localized in terminals that appose terminals enriched with GABA receptors and GABA and glycine receptors often commingle (Bohlhalter et al., 1994). $GABA_A\beta_3$ subunits are associated with gephyrin immunoreactivity in many synapses, and both GABA and glycine are often co-localized in the presynaptic terminals apposed to the receptors described above (Todd et al., 1996). This is strong evidence that glycine and GABA are often co-transmitters. Glycine receptors are associated with terminals that end on axoaxonic terminals, but these receptors do not seem to be involved in the axoaxonic synapses themselves (Mitchell et al., 1993).

Large lamina I neurons, Waldeyer cells, which project to the parabrachial area and do not express glutamate receptors, have gephyrin puncta on their membranes (Puskar et al., 2001). In the puncta is the α1 glycine subtype isoform, and most of the puncta are adjacent to GABAergic axons and about one third are next to axons that express a glycine transporter. Most of these large cells express fos after noxious stimuli and the cells receive NOS containing synapses (Puskar et al., 2001). Thus these gephyrin-rich cells form a specific population of lamina I projection neurons that convey noxious information and are under very powerful inhibitory control (Puskar et al., 2001). A reduction in glycine receptors was found after CCI, a model of neuropathic pain, on cells in laminae II–IV of the dorsal horn (Simpson Jr. and Huang, 1998). Thus glycine may play a role in deleterious pain states.

Cholinergic Receptors

Cholinergic receptors are divided into nicotinic and muscarinic categories. Nicotinic receptors are ligand-gated ionic channels whose activation leads to rapid increases in permeability to sodium and potassium. Muscarinic receptors are G-protein coupled and activate second-messenger systems.

Nicotinic Receptors

In the dorsal horn, binding for α-bungarotoxin, a nicotinic ligand, is concentrated in laminae I–III (Hunt and Schmidt, 1978; Arimatsu et al., 1981; Ninkovic et al., 1981; Gillberg and Aquilonius, 1985; Gillberg and Wiksten, 1986; Gillberg et al., 1988; Gillberg and Askmark, 1991). The labeled receptors are transported both proximally and distally as shown by piling-up of label in ligated peripheral nerves and dorsal roots (Ninkovic and Hunt, 1983). Labeling in the dorsal horn is significantly reduced by dorsal rhizotomy showing that many of these receptors are on central primary afferent terminals (Ninkovic and Hunt, 1983; Gillberg and Wiksten, 1986; Gillberg and Askmark, 1991). Since primary afferent fibers expressing nicotinic receptors are found in laminae I–III, the implication is that both large and small afferent fibers express these receptors, and nicotinic receptors are found on all sizes of DRG cells (Chapter 4). Another nicotinic ligand, epibatdidine, showed intense binding in laminae I–III and the medial part of lamina IV, and the binding decreased as the blood pressure increased in hypertensive rats (Khan et al., 1996). Nicotinic receptor subtypes have been divided into α and β subtypes. For the nervous system in general, most nicotinic receptors consist of α4 and β2 subunits, but for the spinal cord, with relatively weak localizations, cells expressing α2 and α3 subunits, as well as α4 and β2 subunits, were seen with only α3 being specifically localized to cells of the substantia gelatinosa (Wada et al., 1989).

Muscarinic Receptors

Autoradiographic localizations, principally for quinuclidinyl benzilate but also for other muscarinic ligands, show intense label in laminae I–III of the dorsal horn (Wamsley et al., 1981; Whitehouse et al., 1983; Yamamura et al., 1983; Gillberg et al., 1984; 1988; Scatton et al., 1984; Seybold and Elde, 1984; Gillberg and Aquilonius, 1985; Seybold, 1985a, b; Villiger and Faull, 1985; Gillberg and Wiksten, 1986; Mizukawa et al., 1988b; Gillberg and Askmark, 1991). Reductions in labeling are seen with dorsal rhizotomy, so some of these receptors are located on central primary afferent terminals (Gillberg and Wiksten, 1986; Gillberg and Askmark, 1991). No labeled intrinsic cells are seen so the presumption is that the labeling that persists in laminae I–III after dorsal rhizotomy is due to descending fibers and perhaps intrinsic fibers from cells in other areas of the cord. There are to date five muscarinic receptor subtypes (M1–M5), and M2 and M3 but not M1 receptors are concentrated in laminae I–III in the dorsal horn (Höglund and Baghdoyan, 1997; Yung and Lo, 1997). The antinociceptive properties of intrathecal muscarinic ligands (Naguib and Yaksh, 1997) make this a likely area for further study.

Hormone Receptors

Androgen Receptors

Androgen receptors have been studied most intensively in motoneurons since many muscle groups are responsive to sex steroids, and there is a relation to motoneuron damage in inheritable muscular dystrophies (Li et al., 1995). Sensory systems are also affected however, and radioactive androgen binding and mRNA for androgen receptors show label in the nucleus dorsalis or substantia gelatinosa and nucleus proprius (Sar and Stumpf, 1977; Simerly et al., 1990). Immunostaining shows neurons in unspecified locations in the dorsal horn (Lumbrosa et al., 1996). No gender differences in dorsal horn mRNA are found (Lumbrosa et al., 1996).

Corticosteroid Receptors

Type I corticosteroid receptors (mineralocorticoids) are concentrated in lamina II (Ahima et al., 1991), as are type II glucocorticoid receptors and their mRNAs (Fuxe et al., 1985; Ahima and Harlan, 1990; Marlier et al., 1995).

Corticotrophin Releasing Factor (CRF) Receptors

Binding of radioactive CRF ligands shows CRF receptors concentrated in laminae I–II of the dorsal horn in one study (Skofitsch et al., 1985a), and a generalized label in the dorsal horn with a slight diminution in superficial layers in others (De Souza et al., 1985; Skofitsch et al., 1985c).

Estrogen Receptors

Estrogen is a gonadal steroid hormone with metabolic, developmental, and regenerative activities. Cloning shows two forms of the estrogen receptor, estrogen receptor-α and estrogen receptor-β. Early morphologic localizations, presumably for estrogen receptor-α, show estradiol-concentrating cells in the dorsal horn, principally lamina II, but these cells are relatively sparse (Morrell et al., 1982). In situ and immunolabeling show the same basic pattern (Simerly et al., 1990; Amandusson et al., 1995; 1996; Williams and Papka, 1996; Shughrue et al., 1997), and this distribution characterizes both estrogen receptor-α and estrogen receptor-β (Shughrue et al., 1997). Many immunolabeled estrogen receptor cells contain mRNA for preproenkephalin (Amandusson et al., 1996), administration of estrogen leads to a rapid upregulation of enkephalin (Amandusson et al., 1999), and ovariectomy increases enkephalin expression in the dorsal horn (Amandusson et al., 1999). Thus estrogen may be controlling opioids in the dorsal horn. Estrogen levels correlate with changes in trophic factor expression in that p75 and trkA receptors are upregulated during proestrus (high estrogen) and downregulated after ovariectomy (low estrogen) (Sohrabji et al., 1994).

Insulin Receptors

Insulin receptors belong to the tyrosine kinase receptor family (Reinhardt et al., 1993) and are of interest because of the prevalence of diabetes mellitus. In the nervous system these receptors seem involved in growth-promoting activities rather than glucose metabolism (Karagiannis et al., 1997). Insulin and IGF-1 receptors are found in the cytoplasm of sensory cells in culture (Waldbillig and LeRoith, 1987) and in some lamina V cells (Llewelyn et al., 1988; Karagiannis et al., 1997), but their function here is not known (Sugimoto et al., 2002). Other receptors in this group are the insulin-like growth factors (IGF-1 and IGF-2) and insulin receptor-related receptor (IRR). The IGF receptors, particularly IGF-1, may have a role in amyotropic lateral sclerosis, but they bind in all areas of the cord, not just in motor areas (Adem et al., 1994; Doré et al., 1996).

TRH Receptors

Binding and immunocytochemical studies show TRH receptors concentrated in superficial dorsal horn (lamina I, the substantia gelatinosa, laminae II–III) (Manaker et al., 1985a, b, c; 1988; Mantyh and Hunt, 1985a; Sharif and Burt, 1985; Ulfhake et al., 1987; Wu et al., 1992; Deng et al., 1996). These receptors are in large part on intrinsic spinal neurons, with the cells expressing these receptors largely in lamina III (Tsuruo et al., 1987; Wu et al., 1992). These receptors are considerably decreased in the dorsal horn in amyotrophic lateral sclerosis (Manaker et al., 1985c; 1988) and in the Wobbler mouse (Deng et al., 1996), but the relation of this to pathogenesis is not clear.

Recently another TRH receptor, TRH-R2, has been cloned and cells possessing the mRNA for this receptor in the dorsal horn are located in laminae IV–VI (Cao et al., 1998; Heuer et al., 2000). The function of these cells is not known.

Peptide Receptors

Angiotensin Receptors

Binding of radioactive angiotensin II is concentrated in laminae I–II of the superficial dorsal horn (Gehlert et al., 1986; White et al., 1988; Oldfield et al., 1994), except for diffuse light binding in one study (MacGregor et al., 1995). Binding of the Angiotensin II fragment, angiotensin IV, is diffuse and light in the dorsal horn (Moeller et al., 1995).

Bombesin Receptors

Bombesin-like receptors, as assayed by ligand binding or mRNA analyses, are concentrated in the superficial dorsal horn (O'Donohue et al., 1984; Massari et al., 1985; Zarbin et al., 1985; Wada et al., 1992; 1993). Many of these are lost after dorsal rhizotomy indicating that they are located on primary afferent fibers (Massari et al., 1985).

Bradykinin Receptors

The kinins are peptides that affect smooth muscle and are involved in inflammation and pain. For sensory mechanisms, bradykinin and kallidin, which are plasma kinins formed from α2 globulins in blood, are most intensively studied. There are to date two kinin receptors, B1 and B2. The B2 receptor is the classic bradykinin receptor that selectively binds bradykinin and kallidin and is constitutively present in most tissues. The B1 receptor binds to kinin metabolites and is generally less abundant than the B2 receptor, although the B1 receptor seems to be more robustly upregulated in inflammation. Bradykinin causes considerable pain when injected locally and is a potent sensitizer of nociceptive primary afferent neurons.

Immunolocalizations show B1 receptors concentrated in laminae I and II of the dorsal horn (Raidoo and Bhoola, 1997; Wotherspoon and Winter, 2000). In guinea-pig, rat, and human, bradykinin ligands and B2 immunolocalizations preferentially label the substantia gelatinosa, with some labeling in laminae I and III (Steranka et al., 1988; Lopes et al., 1993; Lopes and Couture, 1997; Murone et al., 1997; Raidoo and Bhoola, 1997; Krstew et al., 1998). Binding is highest in lumbar segments and dorsal rhizotomy, neonatal capsaicin and sciatic nerve section reduce the binding by approximately two thirds, so many of these receptors presumably are on primary afferents (Lopes et al., 1995). Part of the remaining binding decreases after destruction of adrenergic systems, so presumably some of the receptors are on descending adrenergic systems (Lopes et al., 1995).

Cannabinoid Receptors

Cannabinoids are analgesic and addictive compounds. Cannabinoid receptors are divided into CB1 and CB2 subtypes (Pertwee, 2001). Attention focuses on CB1 since CB2 is not yet shown in the dorsal horn or DRG (Hohmann et al., 1999). CB1 binding studies show moderate to light labeling over the entire dorsal horn with no or only slight concentration superficially (Herkenham et al., 1991; Mailleux et al., 1992; Mailleux and Vanderhaeghen, 1992; Glass et al., 1997; Hohmann et al., 1999). By contrast, CB1 immunolocalizations show CB1 concentrated in the superficial dorsal horn (Tsou et al., 1998; Sanudo-Pena et al., 1999), especially laminae I and IIo (Salio et al., 2002), and in one study it is most intense in the DLF and in two bands in the dorsal horn, the first in laminae I–IIo and the second in laminae IIi and III (Farquhar-Smith et al., 2000). Some find CB1 cells throughout the dorsal horn (Ong and Mackie, 1999; Sanudo-Pena et al., 1999), whereas others show them concentrated in the superficial dorsal horn, particularly in lamina IIo (Salio et al., 2002). Many CB1 cells co-localize with GABA, NO, and MORs (Salio et al., 2001; 2002). In neuronal processes, CB1 receptors co-localize with protein kinase Cγ but only in a limited way with primary afferent markers CGRP, SP SOM, and IB4, suggesting that most CB1 in the dorsal horn is in intrinsic neurons (Salio et al., 2002).

The CB1 label is usually cytoplasmic, and when it is found on cell membranes, it is not associated with synaptic thickenings (Ong and Mackie, 1999; Salio et al., 2002). Thus, there are postsynaptic CB1 receptors in the dorsal horn, but they do not seem be associated with synapses, so if an endogenous cannabinoid exists, it would presumably operate by volume transmission. There are also presynaptic receptors since dorsal rhizotomy or capsaicin reduces the amount of spinal CB1 binding (Hohmann and Herkenham, 1998; Hohmann et al., 1999), except one study showed no loss of CB1 staining after dorsal rhizotomy (Farquhar-Smith et al., 2000).

CCK Receptors

CCK receptors are subdivided into CCK_A and CCK_B subtypes (CCK_1R and CCK_2R). Binding of radioactive CCK ligands is concentrated in laminae I–II, and specific blockers show that CCK_B subtypes predominate in rat and rabbit whereas monkey and human are predominantly CCK_A (Hill et al., 1988; Ghilardi et al., 1992), although both are widespread in the dorsal horn (Mercer and Beart, 1997; Mercer et al., 2000). Interestingly DRG cells of all these species express predominantly CCK_B receptors, and the many CCK_A receptors in monkey cord are on intrinsic cells (Ghilardi et al., 1992). Neonatal capsaicin causes a substantial drop in CCK binding so a significant number of CCK receptors are presynaptic on primary afferents.

CGRP Receptors

Although there are several types of CGRP receptors (Quirion et al., 1992), attention focuses on generic CGRP receptors or CGRPα. Some see CGRP receptor binding or immunostaining either of the receptor or a CGRP receptor component protein concentrated in the superficial dorsal horn (Seifert et al., 1985; Tschopp et al., 1985; Kar et al., 1994; Menard et al., 1995a, b; Ye et al., 1999; Pokabla et al., 2002), others report a diffuse distribution or do not indicate laminar concentrations (Skofitsch and Jacobowitz, 1985e; Inagaki et al., 1986; Sexton et al., 1986; Galeazza et al., 1992; Ménard et al., 1994; Kar and Quirion, 1995) and still others show a clear absence from superficial laminae (Kruger et al., 1988b; Garry et al., 1991; Yashpal et al., 1992; Nitsos et al., 1994). These discrepancies, which we cannot resolve, have implications for the type of primary afferent input most affected by CGRP. At the ultrastructural level, most CGRP receptors are located postsynaptically in synaptic thickenings (Ye et al., 1999). Thus, CGRP seems to be involved with classical synaptic transmission in contrast to volume transmission that seems to characterize most peptide synaptic interactions.

During development, receptor binding begins diffusely and then recedes in lamina II (Kar and Quirion, 1995) whereas in sheep binding is absent in superficial laminae from the beginning (Nitsos et al., 1994). Dorsal rhizotomy transiently increases CGRP receptor binding, suggesting that these receptors are not on DRG cells (Garry et al., 1991), but presynaptic endings express this receptor in the cord so there is CGRP presynaptic control in the cord (Ye et al., 1999). Peripheral inflammation and development of tolerance to opioids decrease CGRP binding in superficial laminae (Galeazza et al., 1992; Kar et al., 1994; Ménard et al., 1994), and peripheral nerve injury either decreases or leaves CGRP binding unchanged (Mantyh et al., 1988; Garry et al., 1991). Thus, there is little upregulation of dorsal horn CGRP receptors after the above manipulations.

Endothelin Receptors

Endothelin binding is diffuse in the dorsal horn in one study (Kar et al., 1991b) and concentrated in laminae I–III in another (Niwa et al., 1992).

Galanin Receptors

Radioactive galanin ligand binding is concentrated in laminae I–II (Skofitsch et al., 1986; Arvidsson et al., 1991; Kar and Quirion, 1994; 1995; Zhang, X. et al., 1995c). The binding is intense in embryos and declines to adult levels after birth (Kar and Quirion, 1995). Dorsal rhizotomy and neonatal capsaicin cause an increase in binding and peripheral axotomy a decrease or no change, findings interpreted as indicating that these receptors are primarily on intrinsic neurons, and so galanin acts postsynaptically (Kar and Quirion, 1995; Zhang et al., 1995e). More recently, cloning divides galanin receptors into three subtypes, GAL-R1, GAL-R2, and GAL-R3, and all are found in cord and DRG (Waters and Krause, 2000). GALR-1 mRNA is found in cells in all dorsal horn laminae but concentrated in laminae I–II (Gustafson et al., 1996b; Xu et al., 1996b; O'Donnell et al., 1999), and GALR-2 mRNA cells are spread throughout the laminae (O'Donnell et al., 1999).

Neurokinin Receptors

Neurokinin (Tachykinin) receptors (NKRs) are divided into three categories: (NK-1-3Rs or NK_{1-3}Rs). NK-1R is the most ubiquitous. Substance P was thought to be the most specific NK-1R ligand, (NK-1Rs are sometimes called substance P receptors or SPRs), but Neurokinin A is possibly more specific (Quartara and Maggi, 1998).

Laminar Localizations

Ligand binding and some immunocytochemical localizations show NK-1Rs concentrated in laminae I and II (Maurin et al., 1984; Ninkovic et al., 1984; Charlton and Helke, 1985; Mantyh and Hunt, 1985b; Massari et al., 1985; Buck et al., 1986; Helke et al., 1986; 1990; Dietl et al., 1989; Mantyh et al., 1989b; 1995; Yashpal et al., 1990; 1994; Moussaoui et al., 1992; Stucky et al., 1993; Croul et al., 1995; 1998; Kar and Quirion, 1995; Routh and Helke, 1995; Sann et al., 1995b; Von Banchet and Heppelmann, 1995; Zerari et al., 1995; Dery et al., 1997). Exceptions are binding throughout laminae I–V (Kar et al., 1994), in medial laminae III and IV (Vita et al., 1990; Croul et al., 1998), in laminae I–III (Maguire and Geraghty, 1998) or laminae II–III (Bernau et al., 1993). Most immunolabeling studies, however, show staining concentrated in lamina I, little in lamina II except that provided by dendrites ascending from deep cell bodies to superficial laminae, and a considerable amount in laminae III–V (Bleazard et al., 1994; Brown et al., 1995; Littlewood et al., 1995; Zerari et al., 1995; 1998; Abbadie et al., 1996; Li et al., 1996a; Mantyh et al., 1997; Naim et al., 1998; Todd et al., 1998). Thus, ligand binding shows many receptors in lamina II whereas most immunocytochemical studies show only a few dorsally directed dendrites expressing NK-1Rs in lamina II. A significant number of NK-1R-expressing neurons are found in the white matter surrounding the dorsal horn, and these have a rich GABA input and can respond to peptides in the cerebrospinal fluid (Abbadie et al., 1999).

Segmental Localizations

Substance P binding is greatest at sacral levels (Charlton and Helke, 1985; Massari et al., 1985), but the density of NK-1R-expressing cells seems approximately the same at all spinal levels (Brown et al., 1995).

Intrinsic Neurons

NK-1R neurons are located in lamina I and laminae III–IV or III–V (Bleazard et al., 1994), with essentially none in lamina II (Brown, J. L. et al., 1995; Littlewood et al., 1995), in contrast to young animals (Bleazard et al., 1994; Nakaya et al., 1994; Brown, J. L. et al., 1995; Littlewood et al., 1995; Li, 1999; Polgar et al., 1999c). Early counts show approximately 5% of lamina I neuron profiles are NK-1R cells (Brown, J. L. et al., 1995), and approximately 10% in immature animals (Polgar et al., 1999c), but with the confocal microscope approximately 45% of neurons in lamina I and 25% of neurons in laminae III–IV are NK-1R positive (Todd et al., 1998), which implies that many cells in these laminae respond to SP.

A dramatic finding is that NK-1Rs are spread over the cell surface so that a Golgi-like picture emerges after immunocytochemical staining (Fig. 6.5). Thus, NK-1R cells in lamina I are of various sizes, the majority have a fusiform cell body although some are multipolar with dendrites largely arborizing in lamina I, although many dendrites also pass into lamina IIo and some down through lamina IIi into lamina III (Bleazard et al., 1994; Liu et al., 1994a; Brown, J. L. et al., 1995; Ding et al., 1995a; 1999; Littlewood et al., 1995; Marshall et al., 1996; Polgar et al., 1999c; Cheunsuang and Morris, 2000). The largest are marginal cells of Waldeyer and these have the greatest dendritic expanse (Brown, J. L. et al., 1995; Littlewood et al., 1995). NK-1R cells in laminae III–V are characterized by prominent dorsal dendrites that pass into superficial laminae (Fig. 6.5; Bleazard et al., 1994; Liu et al., 1994a; Brown, J. L. et al., 1995; Ding et al., 1995a; 1999; Littlewood et al., 1995; Marshall et al., 1996; Mantyh et al., 1997; Naim et al., 1997;

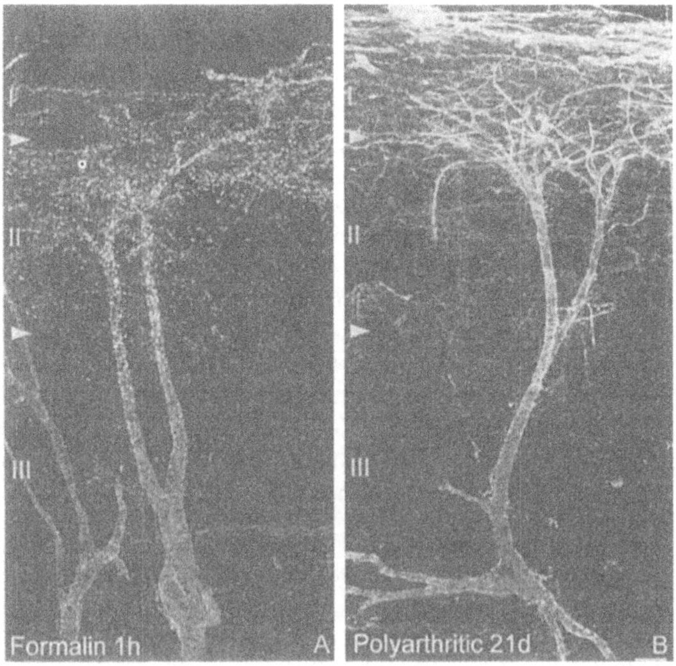

FIGURE 6.5. A figure illustrating two lamina III neurons outlined by NK-1 (SPR) receptor staining whose dendrites extend vertically into lamina I (antenna cells). (From Honore et al., 1999.)

Polgar et al., 1999c), but other dendrites from these cells extend laterally and ventrally. These are antenna or pyramidal cells (Szentagothai, 1964; Szentagothai and Rethelyi, 1973; Brown, A. G., 1981; Maxwell et al., 1983), whose dorsal dendrites, along with those of lamina I NK-1R cells form a dense plexus in lamina I (Liu et al., 1994a; Brown, J. L. et al., 1995; Mantyh et al., 1995; Honore et al., 1999) or I and IIo (Marvizon et al., 1997; 1999a; Naim et al., 1997).

Most NK-1R cells are projection neurons (Littlewood et al., 1995; Marshall et al., 1996; Todd et al., 1998; 2000; Li, 1999; Li et al., 2000b). Targets are the lateral spinal nucleus, the parabrachial nucleus (Ding et al., 1995a; Li and Li, 2000; Todd et al., 2000) the thalamus (Marshall et al., 1996; Li, J. L. et al., 1996b, 1997a), the hypothalamus (Li et al., 1997c); the nucleus of the solitary tract (Guan et al., 1998), the lateral reticular nucleus (Todd et al., 2000), and the periaqueductal gray (Li et al., 1998a), although the latter finding was not confirmed (Todd et al., 2000). The cells that express the NK1 receptor do not sequester GABA or glycine (Littlewood et al., 1995).

Synaptic Organization

Large numbers of SP asymmetric terminals preferentially end on dendrites of NK-1R antenna cells of laminae III–IV (Naim et al., 1997; Sakamoto et al., 1999; Li, J. L. et al., 2000). Since most of the SP terminals contain CGRP, they are presumably of primary afferent origin (Naim et al., 1997). Thus these dendrites integrate afferent information coming into superficial laminae, and they support the idea that fine afferent input to the gelatinosa passes information by influencing dorsal dendrites of antenna cells in deeper laminae (Chapter 5). The SP endings are considerably more common on nociceptive-specific as opposed to wide dynamic range or non-nociceptive neurons (Ma, W. et al., 1996). NK-1R cells in laminae III and IV receive monosynaptic input from larger myelinated primary afferents (Naim et al., 1998). Dendrites of approximately half of lamina III/IV NK-1R neurons receive close contacts from 5-HT axons (Stewart and Maxwell, 2000).

G protein-coupled receptors like NK-1Rs are often internalized. Noxious primary afferent input causes this internalization particularly in the superficial plexus (Mantyh et al., 1995; 1997; Abbadie et al., 1997; Allen et al., 1997; 1999; Liu et al., 1997; Marvizon et al., 1997; 1999a; Honore et al., 1999). The stimuli that induce internalization are intraplantar capsaicin, local CFA, carrageenan, formalin, noxious heat and cold, noxious pinch, intrathecal NMDA, SP and intense electrical stimulation of dorsal root and peripheral nerve. In normal animals, non-noxious stimuli do not lead to internalization. The internalization is stimulus intensity-dependent (Mantyh et al., 1995; Allen et al., 1997; Marvizon et al., 1997) Endocytosis is rapid, within 5–10 minutes (see above references). The receptor then returns to the cell membrane within 1 hour (Mantyh et al., 1995; Abbadie et al., 1997). Some internalization is seen deeper (Marvizon et al., 1997), but much less than in the superficial plexus indicating that most internalized receptor is not retrogradely transported to the cell body (Mantyh et al., 1995). Internalization of NK-1Rs is used to show the time course and extent of SP release after various stimuli (Mantyh et al., 1995; Allen et al., 1997; Marvizon et al., 1997; 1999a) as is NK-1R binding (Yashpal et al., 1994). A caveat is that internalization indicates only the SP bound to the NK-1R, which may explain discrepancies in amounts of SP in fluids that bathe the cord compared with the amounts of NK-1R internalization (Trafton et al., 1999). Since internalization removes receptor from the cell surface, this undoubtedly causes changes in sensitivity of neurons that express NK-1Rs (Allen et al., 1997).

Experimental Manipulations

Prevention of Internalization. Receptor internalization is initiated by SP and prevented by a NK-1R antagonist (Marvizon et al., 1997). A more dramatic prevention is caused by NMDA antagonists whereas non-NMDA antagonists have no effect (Liu et al., 1997; Marvizon et al., 1997). Thus, SP release, as indicated by internalization of NK-1Rs, seems to be controlled by activation of NMDA receptors, primarily by presynaptic mechanisms (Liu et al., 1997; Marvizon et al., 1997). $GABA_B$ but not $GABA_A$ or glycine receptor agonists inhibit SP release after intense dorsal root stimulation, but $GABA_B$ agonists have no effect on NMDA-induced internalization (Marvizon et al., 1999a). Opioids are the most effective control of inflammatory pain, and one opioid action is reduction of SP release by primary afferents. The effects of opioids on NK-1R internalization are complicated, however. Thus, morphine does not influence internalization caused by noxious pinch but does reduce it when combined with a subliminal dose of an NMDA blocker (Trafton et al., 1999). These results are interpreted as indicating that opioid analgesia involves primarily postsynaptic mechanisms and/or presynaptic control of non-SP primary afferents (Trafton et al., 1999). In accord with these observations, chronic administration of either morphine or naloxone does not affect SP binding in the dorsal horn (Gouardéres et al., 1993a). More work needs to done in this area.

Inflammation. Arthritis increases NK-1R mRNA, immunostaining, and binding (Schäfer et al., 1993; Stucky et al., 1993; McCarson and Krause, 1994; 1995; Abbadie et al., 1996; Ohtori et al., 2002), which suggests NK-1R antagonism as possible therapy (McCarson and Krause, 1994), but decreases SP binding (Mantyh et al., 1988; Kar et al., 1994). The reason for the discrepancy is not known but is perhaps related to the time course of the changes (Stucky et al., 1993) A phosphorylated cAMP response element binding protein (pCREB) increases in superficial NK-1R cells after intraplantar formalin, which may implicate pCREB in NK-1R upregulation (Anderson and Seybold, 2000). Inflammation caused by systemic resiniferatoxin does not change NK-1R mRNA in the dorsal horn (Szallasi et al., 1999). After inflammation by intraplantar CFA, which by itself does not cause internalization, internalization caused by noxious heat or pinch was more dramatic, and non-noxious stimuli, which have no effect normally, caused internalization (Abbadie et al., 1997). Chronic inflammations lead to more receptor internalization than do acute inflammations (Honore et al., 1999). Intraplantar CFA-induced inflammation caused an upregulation of NK-1Rs in the dorsal horn, and internalization was driven by Aδ and C fiber activation, and not by Aβ activation (Allen et al., 1999). In summary, the more long-lasting the inflammation, the more extensive the internalization, and NK-1R internalization can be used to indicate SP activity.

Antisense Oligonucleotides. Antisense oligonucleotides for NK-1Rs do not change nociceptive responses to intraplantar formalin or immunocytochemical staining, but intrathecal SP in animals receiving antisense oligonucleotides reduces nociceptive behaviors and immunostaining, essentially the opposite of the effects of SP in normal animals (Hua et al., 1998).

Neurotoxicity and Therapy

A clever use of internalization is to attach a neurotoxic agent to a SP ligand and then stimulate to induce internalization. The result is a destruction of NK-1R-expressing neurons in the dorsal horn (Mantyh et al., 1997; Nichols et al., 1999). The animals so treated are responsive to mildly painful and innocuous stimuli but have great reductions in neuropathic and inflammatory pain behaviors (Mantyh et al., 1997; Benoliel et al., 1999; Nichols et al., 1999). This indicates that NK-1R has an important role in neuropathic and inflammatory pain. The term "noci-toxins" is proposed to describe this type of potential therapy (Benoliel et al., 1999). Another form of neurotoxicity is to administer capsaicin to neonates to remove unmyelinated afferents, which reduces numbers of SP mRNA cells in lamina III (Ohtori et al., 2000).

Neural Injuries

Primary afferent denervation by dorsal rhizotomy rapidly increases SP ligand binding in the superficial dorsal horn followed by a slow return to normal (Massari et al., 1985; Brass et al., 1986; Helke et al., 1986; Yashpal et al., 1991a; Croul et al., 1995). The lack of a change after rhizotomy bears on the conflicting evidence as to whether NK-1Rs are expressed by primary afferent neurons (Chapter 4). Similar changes are seen after CCI, ricin, or pronase application to peripheral nerves, nerve crush, and dorsolateral funiculus lesions (Helke et al., 1986; Aanonsen and Seybold, 1991; Aanonsen et al., 1992; Bernau et al., 1993; Croul et al., 1998; Helgren et al., 1999b). A similar upregulation occurs above and below spinal cord transection (Vita et al., 1990). The exceptions are that capsaicin treatment in a newborn did not change spinal NK-1R mRNA and local capsaicin treatment of a peripheral nerve leads to a reduction in SP binding in the dorsal horn chronically (Sann et al., 1995b). The capsaicin findings notwithstanding, it seems clear that removal of synaptic input into the dorsal horn leads to upregulation of NK-1Rs, which may be a structural basis for development of SP hypersensitivity. In accord with this conclusion, Aδ and C fiber activation led to a greater internalization of NK-1Rs after nerve transection than after similar stimulation in the normal (Allen et al., 1999).

Development

SP ligand binding is diffusely distributed in newborn and early postnatal rats, with the adult distribution emerging as development progresses (Charlton and Helke, 1986; Beresford et al., 1996). This pattern is seen for receptors of many neuroactive substances (Kar and Quirion, 1995). By contrast, dorsal horn binding changes little in old age (Maguire and Geraghty, 1998). In humans, however, the same patterns of NK-1R cells were seen in newborns as in adults except that there were a few more cells in the newborn (Ding et al., 1999). NK-1R receptors are expressed in the embryonic floor plate so they presumably have a role in spinal cord development in addition to their other functions (Heath et al., 1995).

Other Neurokinin Receptors

NK-2 Receptors. Neurokinin A, the main ligand of NK-2Rs, is localized in the superficial dorsal horn (Buck et al., 1986; Yashpal et al., 1990), principally lamina II (Yashpal et al., 1991a), and increases after dorsal rhizotomy, presumably a basis for denervation supersensensitivity (Yashpal et al., 1991a).

NK-3 Receptors. NK-3 binding or immunostaining or binding of eldoisin, a preferred ligand of NK-3Rs, is concentrated in the superficial dorsal horn (Buck et al., 1986;

Danks et al., 1986; Hunter et al., 1987; Yashpal et al., 1990; Croul et al., 1995) and these cells largely co-localize with NO (Yashpal et al., 1990; Seybold et al., 1997). The labeling increases after dorsal rhizotomy, presumably a basis of denervation suprasensitivity (Croul et al., 1995). NK-3R binding was reduced after foot inflammation but mRNA expression was increased (Linden et al., 2000). Neuromedin K and its mRNA are concentrated in lamina II (Ding et al., 1996a).

Neuropeptide Y Receptors (Y-Rs)

Receptors for neuropeptide Y are subdivided pharmacologically into Y1-R and Y2-R subtypes, with Y1-R usually presynaptic and Y2-R usually postsynaptic (Hökfelt et al., 1998). Further subtypes have since been described, but as far as spinal cord and DRG are concerned, focus is on Y1-R and Y2-R (Hökfelt et al., 1998). Binding of radioactive NPY or related ligands in rats and monkeys and immunochemical labeling is most intense in the superficial dorsal horn with lighter labeling in deeper laminae (Kar and Quirion, 1992; Zhang, X. et al., 1995c; Kopp et al., 2002). The binding seems to represent predominantly Y2-Rs (Zhang, X. et al., 1995c), but few (Hökfelt et al., 1998) Y2-R mRNA-expressing neurons are found in the dorsal horn (Gustafson et al., 1997). By contrast, numerous YR-1 neurons and their processes are found in the superficial dorsal horn, particularly lamina IIi, with some in deeper laminae (Kar and Quirion, 1992; Zhang, X. et al., 1994a).

NPY receptor binding is markedly reduced by dorsal rhizotomy or neonatal capsaicin (Kar and Quirion, 1992; Ji et al., 1995a), indicating that many of these receptors are on primary afferent terminals, but dorsal rhizotomy has little effect on spinal YR-1s (Zhang, X. et al., 1994c; 1995c; Brumovsky et al., 2002), and there is no pile-up next to nerve or root ligatures (Zhang et al., 1994c; 1995c) so the loss must be primarily Y2-Rs (Hökfelt et al., 1998). However, many DRG neurons express Y1-Rs, so the lack of effect of dorsal rhizotomy on spinal immunostaining shows relatively little centrifugal transport of this receptor to the cord (Hökfelt et al., 1998).

Many Y1-R spinal neurons express somatostatin, and ultrastructurally Y1-Rs are not related to synaptic thickenings suggesting volume transmission (Zhang, X. et al., 1999a). Peripheral axotomy has little effect on Y1-R mRNA in cord neurons (Ji et al., 1994), but CFA inflammation causes a major upregulation of Y1-R mRNA in cord neurons (Ji et al., 1994). Peripheral axotomy increases receptor binding in laminae III–IV and decreases it in laminae I–II, and blocking with appropriate ligands shows both Y1-Rs and Y2-Rs participate, with emphasis on Y2-Rs (Zhang, X. et al., 1995c). These findings, combined with the observation that many large DRG cells express NPY receptor mRNA after peripheral axotomy (Zhang et al., 1994c), suggest an upregulation and increased central transport of NPY receptors in large primary afferent fibers after peripheral axotomy.

Neurotensin Receptors

Neurotensin binding is concentrated in lamina II (Ninkovic et al., 1981; Young and Kuhar, 1981; Faull et al., 1989) particularly lamina IIi, and neurotensin mRNA is found in large cells at the lamina II–III border (Elde et al., 1990). Neurotensin receptor binding is not affected by dorsal rhizotomy, but a significant number of DRG cells express neurotensin receptor mRNA and peripheral axotomy causes a significant decrease in their numbers as well as in dorsal horn cells expressing this mRNA and neurotensin immunostaining in the superficial laminae (Zhang, X. et al., 1995b).

NeuropeptideFF (NPFF) Receptors

NPFF ligand binding concentrates in laminae I–II (Allard et al., 1992; 1994; Lombard et al., 1995a; Gouardéres et al., 1996; 1997). One group found little change following dorsal rhizotomy (Lombard et al., 1995a), but others with a ligand less susceptible to metabolic breakdown show a considerable loss (Gouardéres et al., 1996; 2000), so these receptors are probably on DRG terminals. These receptors are also found on intrinsic cell processes (Gouardéres et al., 2000). The adult pattern of receptor localization does not appear until after birth in mice (Desprat and Zajac, 1994).

Opioid Receptors

MORs, DORs, and KORs

The main opioid receptors are subdivided into μ, δ, and κ categories. Nomenclature from the cloning literature is MOR, DOR, and KOR, respectively. A recent change in nomenclature from the International Union of Pharmacology was OP3, OP1, and OP2 to indicate μ, δ, and κ categories respectively. We will use the MOR, DOR, and KOR nomenclature here. These are transmembrane G-coupled receptors.

Laminar Localizations

Opioid receptor-expressing neurons and their processes are found throughout the nervous system (Mansour et al., 1988), and many DRG and dorsal horn cells express them. In the cord, they are concentrated superficially (laminae I–II, substantia gelatinosa), with some expression in deeper laminae, around the central canal, and in autonomic areas. The localizations are done by glutamate ligand binding or immunocytochemistry.

Autoradiographic Binding

(Pert et al., 1975; 1976; LaMotte et al., 1976; Atweh and Kuhar, 1977; Goodman et al., 1980; Pearson et al., 1980; Ninkovic et al., 1981; 1982; Herkenham and Pert, 1982; Wamsley et al., 1982; Slater and Patel, 1983; Gillberg and Aquilonius, 1985; Gouardéres et al., 1985; 1986; 1991; 1993a, b; Gillberg and Wiksten, 1986; Waksman et al., 1986; Daval et al., 1987; Faull and Villiger, 1987; Morris and Herz, 1987; Iadarola et al., 1988b; Millan et al., 1988; Sharif et al., 1988; Hunter et al., 1989; Sharif and Hughes, 1989; Zajac et al., 1989; Besse et al., 1990; 1991; 1992a, b, c; Traynor et al., 1990; Besson et al., 1991; Stevens et al., 1991a, b; Kar et al., 1994; Kar and Quirion, 1995; Lombard et al., 1995b; Mansour et al., 1995a; Stevens and Seybold, 1995; Rahman et al., 1998; Hohmann et al., 1999; Ray and Wadhwa, 1999.)

Immunocytochemical Labeling

(Mansour et al., 1988; 1995b; Dado et al., 1993; Mansour et al., 1994; Zerari et al., 1994; Arvidsson et al., 1995a; b; c; Cheng et al., 1995a; 1997; Ding et al., 1995b; 1996b; Elde et al., 1995; Honda and Arvidsson, 1995; Ji et al., 1995b; Kemp et al., 1996; deGroot et al., 1997; Li et al., 1998b; Porreca et al., 1998; Zhang, X. et al., 1998b, c; 2000; Marvizon et al., 1999b; Robertson et al., 1999; Yu et al., 1999; Abbadie et al., 2000a, b.)

MOR labeling is most intense in lamina II, particularly IIi (Ding et al., 1995b; Honda and Arvidsson, 1995; Ji et al., 1995b; Marvizon et al., 1999b). This suggests that MORs are concentrated on endings of unmyelinated sensory fibers, as compared to DORs which are as abundant in lamina I as lamina II and so might also be located on fine myelinated (Aδ) afferent terminals (Honda and Arvidsson, 1995). Lamina IIi appears preferentially labeled after dorsal rhizotomy or nerve section because this region does not lose as much immunoreactivity as laminae I and IIo (Arvidsson et al., 1995a; deGroot et al., 1997). This is presumably because of the concentration of MOR cell bodies and dendrites in lamina II, particularly lamina IIi (Ding et al., 1995b; 1996b; Kemp et al., 1996; Zhang, X. et al., 1998c). IB-4 labeled fine primary afferent fibers end preferentially in lamina IIi, so these non-peptidergic afferents might have particular connections with intrinsic MOR cells here.

Segmental Variability

An early study described KORs concentrated at lumbosacral regions, but other work shows relatively little rostro-caudal variation for the major types of opioid receptors (Besse et al., 1991; Gouardéres et al., 1991; Arvidsson et al., 1995b; Lombard et al., 1995b).

Proportions of Opioid Receptor Subtypes

Opioids modulate pain at the spinal level, so proportions of receptor subtypes here have clinical relevance. One group reported that KORs predominate in the superficial dorsal horn (Gouardéres et al., 1985; 1986), but others report that MORs predominate (54–70% MORs, 23–29% DORs and 7–17% KORs (Besse et al., 1990; 1991; Besson et al., 1991; Gillberg and Askmark, 1991; Stevens et al., 1991b; Rahman et al., 1998). The apparent concentration of KORs was suggested to result from lack of ligand specificity (Besse et al., 1990; 1991). If proportions of receptors are related to effectiveness of ligands, then MOR ligands should be most effective in relieving pain at the spinal level, but all ligands will have some effect.

Primary Afferent Input

All types of opioid receptors are reduced in the dorsal horn following dorsal rhizotomy or capsaicin treatment to the newborn. Some of the studies use general opioid markers while others use markers for specific subtypes (Fig. 6.6; Hohmann et al., 1999). Quantitative estimates vary somewhat, but most estimates are that 50% or more of the binding or staining is lost. The most extensive studies in this regard, taking account of time of survival and numbers of roots sectioned, showed that 76, 61, and 53% of MORs, DORs, and KORs respectively disappear after rhizotomy (Besse et al., 1990; 1992a; Lombard et al., 1995b). The assumption is that the decreases are because of loss of receptors following degeneration of fine afferent fibers rather than because of transsynaptic changes or down-regulation/degradation of postsynaptic opioid receptors. Evidence that the receptors that disappear after rhizotomy are truly located on primary afferent axons and terminals is absence of central degeneration except for primary afferent axons, rapid loss of receptors, faster than post- or transsynaptic changes, temporal correspondence of axon deafferentation and receptor losses, and lack of changes in the affinities of the MOR and DOR binding sites that remain after rhizotomy (Besse et al., 1990; 1992a; Lombard et al., 1995b). KORs are found

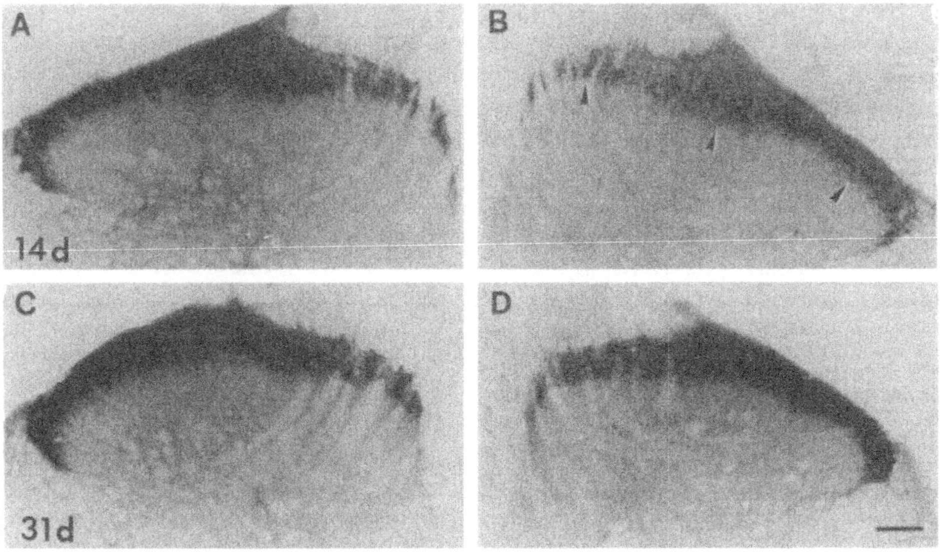

FIGURE 6.6. A figure illustrating loss of MOR receptor staining in the superficial dorsal horn after dorsal rhizotomy. The contralateral side is shown in **A** and **C**, and note the loss of staining on the ipsilateral side 14 (**B**) and 31 (**D**) days after deafferentation (From deGroot et al., 1997.)

postsynaptic to primary afferent input (Arvidsson et al., 1995a) but are also presynaptic since they found on DRG cells and nerve terminals (Elde et al., 1995). Thus, the majority of opioid receptors are presynaptic to primary afferent fibers, but this conclusion is not completely accepted for MORs (Zhang, X. et al., 1998c), and many are postsynaptic as well.

Intrinsic Neurons

Numerous small MOR-expressing neurons are found in lamina II (Ding et al., 1995b; Kemp et al., 1996; Zhang, X. et al., 1998c), particularly in lamina IIi (Ding et al., 1995b). They are presumably central or islet cells.

Synaptic Organization

Presynaptic Elements. Many MOR profiles in superficial laminae are axons and terminals (Cheng et al., 1996), but, most are dendrites (Cheng et al., 1996; 1997; Zhang, X. et al., 1998c). Results of dorsal rhizotomy are not consistent with many reporting a major loss of MORs (see above) but others only a slight loss (Zhang, X. et al., 1998c). On balance a significant number of MORs in the superficial dorsal horn seem to be located on primary afferent terminals. A few MOR presynaptic elements contain pleomorphic vesicles and form symmetric synapses, but most contain round clear and some dense-core vesicles with asymmetric synaptic thickenings (Cheng et al., 1995b; 1996; 1997; Ding et al., 1995b; Li et al., 1998b), suggesting excitatory synapses although MOR transmission is largely inhibitory.

DOR labeled elements are mostly presynaptic (Cheng et al., 1995a; Elde et al., 1995; Zhang, X. et al., 1998b), but one study showed that many labeled profiles are dendritic (Cheng et al., 1995a), and dorsal rhizotomy only removes half the DOR labeling (see above). Thus, many DORs are also postsynaptic. Presynaptic DOR labeling is more punctuate than MOR labeling because there are many intensely labeled large dense-core vesicles (LDCVs) in DOR terminals (Dado et al., 1993; Arvidsson et al., 1995b; Cheng et al., 1995a; 1997; Zhang, X. et al., 1998b). Exocytosis of DOR labeled LDCVs leads to transient labeling of the terminal membrane, presumably indicating a transient presynaptic sensitivity to DOR ligands (Zhang, X. et al., 1998b). This is a secretory process, as contrasted to placement of MORs, which is constitutive (Zhang et al., 1998b). DORs and MORs are expressed on afferent glomerular endings (Zerari et al., 1994; Zhang et al., 1998c), but stained regions are not near active zones suggesting volume transmission.

Synaptic Thickenings—Absence of Opioid Receptors. In classical synapses, a transmitter containing presynaptic element links to a postsynaptic target at receptor containing synaptic thickenings (active zones) on, but MOR and DOR labeling on axons or dendrites is patchy, and synaptic thickenings are almost never labeled (Zerari et al., 1994; Cheng et al., 1995a, b; 1996; 1997; Zhang, X. et al., 1998b, c). Thus active zones do not contain opioid receptors, suggesting volume transmission.

Postsynaptic Elements. MOR and DOR labeled postsynaptic elements (dendrites and cell bodies) are not cytologically distinctive. Some presynaptic endings on MOR postsynaptic elements contain leu-enkephalin (Cheng et al., 1996) and others SP (Aicher et al., 2000a, b; Spike et al., 2002) but without synaptic thickenings suggesting non-synaptic innervation or volume transmission. Some dendrites and somata that express MORs also express DORs and vice versa (Cheng et al., 1997). Numerous MOR-expressing cells contain CCK, a morphologic basis for CCK-opioid interactions (see below). Unlabeled terminals on MOR or DOR post-synaptic elements are diverse with both symmetric and asymmetric synaptic thickenings, which are almost never stained for opioid receptors (Cheng et al., 1997).

Co-localizations

Opioids and Opioid Receptors. At the light microscopic level, no clear examples of ENK or ppDYN presynaptic terminals expressing an opioid receptor are seen (Dado et al., 1993; Arvidsson et al., 1995c), but in the electron microscope 23% of the ENK terminals express DORs and 60% of DOR terminal profiles express ENK (Cheng et al., 1995a). This suggests auto- or paracrine reception in terminals too small to be seen in the light microscope. There is physiologic evidence that MOR and DOR ligands act synergistically, and MORs and DORs are reported on the same terminals (Arvidsson et al., 1995c), but a later report finds dual labeling in somata and dendrites and not in terminals (Cheng et al., 1997). Thus it is not clear whether these interactions are primarily at pre- or postsynaptic levels.

Peptides. CGRP is found in terminals expressing both MORs and DORs and SP in terminals expressing MORs (Dado et al., 1993; Ding et al., 1995b; Li et al., 1998b; Zhang, X. et al., 1998c; Aicher et al., 2000a, b), suggesting that MORs and DORs presynaptically modulate peptide containing primary afferent terminals in the superficial dorsal horn. An important part of the spinal action of morphine (a MOR ligand) is presynaptic inhibition of SP and other peptide release from primary afferent fibers (Ding et al., 1995b; Trafton et al., 1999). However, although morphine does reduce the amount of SP released after painful stimuli, there is no change in the internalization of SP receptors or c-fos expression in the dorsal horn after these stimuli. This implies that although SP release is reduced, there is still enough SP to carry out SP functions. Accordingly the functional significance of the MOR mediated inhibition of SP release is not clear (Trafton et al., 1999). CCK antagonizes the actions of morphine, and 40% of MOR neuron profiles in laminae I and II express CCK and 65.6% of CCK terminals express MORs (Zhang, X. et al., 2000). Thus CCK can presumably affect MOR transmission presynaptically, and MOR ligands can presumably affect CCK transmission similarly.

Lack of Co-localization with Amines and Amino Acids. MORs are not co-localized with GABA or glycine (Kemp et al., 1996), DORs are not co-localized with 5-HT or TH (Arvidsson et al., 1995b), and MOR and ORL-1 are not co-localized (Monteillet-Agius et al., 1998) indicating that these are separate systems.

Experimental Manipulations

Peripheral Nerve Lesions. Neuropathic pain is a devastating sequel to nerve lesions, and opioids are ineffective in alleviating this condition. One possible reason is that peripheral axotomies downregulate opioid receptors, particularly MORs, so that morphine (the most commonly used therapeutic opioid) or other opioid ligands have reduced spinal targets. Many studies show reductions in binding or immunostaining of opioid receptors following nerve lesions (Fig. 6.6; Jessell et al., 1979; Fields et al., 1980; Arvidsson et al., 1995c; Li et al., 1996b; deGroot et al., 1997; Porreca et al., 1998; Zhang, X. et al., 1998c; 2000; Robertson et al., 1999) but the amount of loss varies from study to study. For example, one group shows a reduction in the ratio of binding comparing operated with control sides, but due mostly to an increase on the normal side (Besse et al., 1992c). Another finds binding increased temporarily and then declining towards normal for MORs, a decline for a much longer period for DORs and a complicated time course for KORs (Stevens et al., 1991a). Finally another group reported no change in the binding but a drop in immunostaining (Porreca et al., 1998). The cause of these disparate results is not clear, but they may relate to type of lesion or time of sacrifice.

Morphine. Chronic infusion of morphine causes an approximately 30% increase in both MOR and DOR binding in the superficial dorsal horn of both normal and rhizotomized animals in one study (Besse et al., 1992b), whereas in another, chronic morphine administration caused a 37% increase in DOR binding with no change in MOR binding, and naloxone caused a 50% increase in MOR binding (Gouardéres et al., 1993a). Chronic administration of morphine intensified the loss of MOR immunoreactivity in the superficial dorsal horn following nerve lesions, which may be due to receptor internalization, presumably implying that administration of morphine reduces its receptors and thus its effectiveness (Zhang, X. et al., 2000). Morphine has no effect on reduced foot-flick and tail-flick thresholds that follow an L5/L6 segmental lesion. The lack of effect on foot-flick might be due to a decrease in MORs because this reflex is mediated at the L5/L6 level where there is MOR loss, but tail-flick is mediated at sacral levels, where there is no loss

(Porreca et al., 1998). Thus, there is as yet no proof that opioid ineffectiveness after nerve lesions is because of loss of central opioid receptors.

Ischemic Damage. Neuropathic pain behaviors follow ischemic cord damage induced by laser-irradiation after intravenous erythrosine B, which also causes a reduction in dorsal horn MOR immunostaining that lasts for approximately as long as the allodynia (Yu et al., 1999), suggesting the MOR loss plays a role.

Internalization of Opioid Receptors. MOR activation causes internalization of MORs in transfected cells and brain with concomitant activation of MAP kinases (Marvizon et al., 1999b). Internalization is seen in spinal cord slices in the superficial dorsal horn with a maximum effect 10 minutes after administration of the ligand (Marvizon et al., 1999b). This provides a short-term removal of active MORs in the dorsal horn, but morphine itself does not cause internalization, only less commonly used MOR ligands do (Marvizon et al., 1999b).

Inflammation. Several studies report no or slight changes in opioid binding or receptor staining in the dorsal horn after unilateral or bilateral (polyarthritis) inflammation (Cesselin et al., 1980; Iadarola et al., 1988b; Millan et al., 1988; Delay-Goyet et al., 1989; Kar et al., 1994; Ji et al., 1995b; Zhang, X. et al., 2000). By contrast, KORs decline amid an unchanging overall binding following polyarthritis (Millan et al., 1986), and bilateral MOR and DOR binding increase and KOR binding decreases after unilateral inflammation (Besse et al., 1992d). Thus, there are changes in opioid receptor levels after experimental inflammation, and a tendency for KORs to be diminished, but these responses are not robust. This relative lack of change in spinal opioid receptor organization contrasts with dramatic increases in opioid receptor containing DRG and peripheral inflammatory cells (Chapter 4), and changes in spinal opioids, particularly dynorphin.

Development

MORs are expressed diffusely throughout the dorsal horn at birth in rats and then regress in deeper laminae leaving the receptors concentrated in laminae I and II as adulthood approaches (Allerton et al., 1989b; Kar and Quirion, 1995; Rahman et al., 1998). Interestingly, more large DRG cells express opioid receptors early in development (Beland and Fitzgerald, 2001). In humans, MOR labeling is concentrated in laminae I and II by the 20th week of gestation (Ray and Wadhwa, 1999).

Estrus

Rats express fewer KORs in the superficial dorsal horn during diestrus than in estrus or proestrus suggesting that reproductive hormones have a role in controlling these receptors (Chang et al., 2000).

"Other" Opioid Receptors

Endomorphin Receptors

Endomorphins are the most selective MOR agonists known, and their receptors are the MORs.

Endorphin Receptors

β-endorphin, the main derivative of POMC, is a preferential ligand for MORS but also binds to DORS. The MORs and DORs are present in abundance in the cord (see above). β-endorphin binding, presumably to these receptors, occurs in the dorsal horn and is increased in diabetic obese mice (MacDonald et al., 1998). Another POMC derivative is melanocortin, and the melanocortin-4 receptor is found in the developing dorsal horn (Mountjoy and Wild, 1998) or adult substantia gelatinosa, the latter not being affected greatly by nerve crush (van der Kraan, M. et al., 1999; Vrinten et al., 2001).

ORL1 Receptors

Homology cloning yields a receptor with considerable sequence homology to known opioid receptors. It is referred to as an "orphan" receptor because classic opioid ligands do not bind with high affinity to this receptor and there was no known endogenous ligand. However, an endogenous ligand with considerable sequence homology to prodynorphin, called orphaninFQ or nociceptin, has been found. Nociceptin binding, presumably to ORL1, is found in the dorsal horn (Florin et al., 2000), particularly superficially (Narita et al., 1999; Neal et al., 1999), and is significantly increased in inflammation (Jia et al., 1998) and morphine tolerance (Gouardéres et al., 1999). Part of this binding is by primary afferents as evidenced by a significant reduction after dorsal rhizotomy (Le Cudennec et al., 2002). Cells expressing ORL1 mRNA are concentrated in deeper dorsal horn laminae in one study (Neal et al., 1999) and superficial laminae in another (Pettersson et al., 2002; reviewed by Mollereau and Mouledous, 2000). Immunostaining for ORL1 is concentrated in the superficial dorsal horn (Anton et al., 1996) with no co-localization with MORs (Monteillet-Agius et al., 1998).

Sigma Receptors

Sigma receptors or "non-opiate σ binding sites" are localized to the superficial dorsal horn (Aanonsen and Seybold, 1989; Alonso et al., 2000).

Oxytocin and Vasopressin Receptors

Relatively diffuse binding of radioactive OXY and VP is seen in the dorsal horn, with some studies reporting a concentration in superficial laminae (Brinton et al., 1984; Loup et al., 1989; 1991; Reiter et al., 1994; Tribollet et al., 1997; 1998).

Somatostatin (SRIF) Receptors

Molecular cloning reveals five different G-coupled somatostatin receptors, $sst_{1-5,}$ that are subdivided into two large groups; $SRIF_1$ (sst_2, sst_3, and sst_5, with sst_2 split into sst_{2a} and sst_{2b} subtypes) and $SRIF_2$ (sst_1 and sst_4). General somatostatin ligands are concentrated in the superficial dorsal horn or equivalent part of the spinal trigeminal complex (Reubi and Maurer, 1985; Uhl et al., 1985; Mantyh et al., 1988; Kar and Quirion, 1995; Carpentier et al., 1996; Von Banchet et al., 1999). During development, the same concentration in the superficial dorsal horn exists, but it is emphasized in the adult because labeling in deeper laminae decreases (Charnay et al., 1988; Kar and Quirion, 1995), and one study emphasizes the transient nature of the prenatal binding (Maubert et al., 1994). Inflammation causes a significant decrease in the binding (Mantyh et al., 1988).

More specifically, sst_2 and sst_{2a} are expressed by cells concentrated in the superficial dorsal horn (Holloway et al., 1996; Schindler et al., 1997; 1998; Hervieu and Emson, 1998; Schulz et al., 1998a, b; Todd et al., 1998; Helboe et al., 1999; Helboe and Moller, 2000) although one report shows label in laminae I–III (Von Banchet et al., 1999). This is an intrinsic system in that dorsal rhizotomy and spinal transection have essentially no effect, and labeling is on cell bodies and dendrites (Schindler et al., 1998; Schulz et al., 1998b). The lack of effect of dorsal rhizotomy is in spite of medium-sized DRG cells and cutaneous axons expressing sst_{2a} receptors (Schulz et al., 1998b; Carlton et al., 2001b), perhaps indicating that these receptors are expressed by DRG cells but only transported peripherally. The sst_{2a} receptor-expressing cells do not express SRIF, CGRP, SP, MORs, or NK1 receptors (Todd et al., 1998), but they do express glutamate, GABA transporters, GABA, and glycine (Schindler et al., 1998; Schulz et al., 1998b; Todd et al., 1998). Thus the sst_{2a} system is inhibitory. The sst_{2b} receptors are expressed by neurons throughout the cord with no superficial concentrations (Schulz et al., 1998a), and mRNA (Schindler et al., 1999) for the sst_3 receptors is not found in the superficial dorsal horn but every DRG cell expresses these receptors (Señaris et al., 1995). Immunochemical localizations show a considerable plexus of sst_3 cells and processes in laminae I–III (Von Banchet et al., 1999). An overall review exists (Selmer et al., 2000).

Vasoactive Intestinal Polypeptide (VIP) Receptors

Binding of radioactive VIP and PACAP ligands is concentrated in the superficial dorsal horn (Yashpal et al., 1991b; Kar and Quirion, 1995; Jongsma et al., 2000). The mRNA for the VIP_2 receptor is expressed by a few cells in laminae II and V (Sheward et al., 1995; Dickinson et al., 1999). Following nerve transection, PACAP binding was significantly decreased, and the mRNA for $VPAC_1$ (VIP_1) increased, $VPAC_2$ (VIP_2) decreased, and PAC_1 remained unchanged (Dickinson et al., 1999).

Purinoceptors

Adenosine 5′-triphosphate (ATP) and its breakdown product adenosine function as neurotransmitters or neuromodulators (Burnstock and Wood, 1996). For primary afferents, these compounds are linked primarily to nociceptive transduction. The receptors for these ligands are divided into those that respond to adenosine (originally called P1 receptors, but now often referred to as A receptors) and those that respond to ATP (P2 receptors), which are at present divided into ionotropic (P2X) and metabotropic (P2Y) categories (Burnstock and Wood, 1996). Ligand binding shows adenosine receptors in general and A_1 and A_2 receptors in particular and mRNA for the A_1 receptors in moderate numbers of cells concentrated in the superficial dorsal horn (Goodman and Snyder, 1982; Choca et al., 1988; Glass et al., 1996) and diffusely spread in others (Reppert et al., 1991; Bailey et al., 2002). Immunochemistry of A_1 receptors shows weak diffuse staining in the dorsal horn (Schindler et al., 2001). These receptors are located on intrinsic neurons since dorsal rhizotomy, capsaicin, or spinal transection has no effect (Geiger and Nagy, 1985; Choca et al., 1988).

For P2X purinoceptors, ligand binding (α, β-methylene ATP) shows P2X receptors throughout the cord, but concentrated in the superficial laminae of the dorsal horn, the central gray matter, and the intermediolateral cell column (Bo and Burnstock, 1994; 1995; Tuyau et al., 1997). Immunostaining shows P2X1–X4 fibers and terminals concentrated in laminae I and II (Vulchanova et al., 1996; 1997; Lê et al., 1998; Kanjhan et al., 1999). P2X7 receptors are generally thought not to be in nervous tissue, but this receptor is expressed presynaptically throughout the dorsal horn (Deuchars et al., 2001). P2X3 is a particular focus because it appears only in small DRG cells, and dorsal rhizotomy reduces P2X3 staining in the dorsal horn (Bradbury et al., 1998; Vulchanova et al., 1998), and P2X3 staining piles up proximal to a ligature of the sciatic nerve (Vulchanova et al., 1998). Central terminals in glomeruli, which are primary afferent endings, in lamina IIi are immunostained for P2X3 receptors (Llewellyn-Smith and Burnstock, 1998). Thus a significant number of these receptors are presynaptic to primary afferent input. Neurons that express P2X1–6 receptors are found in all laminae of the dorsal horn except lamina III (Collo et al., 1996). Thus many fibers expressing P2X1–6 receptors in the superficial dorsal horn are primary afferents, but others arise from intrinsic neurons (Collo et al., 1996; Lê et al., 1998; Kanjhan et al., 1999). Descending P2X fibers have not yet been described.

Sciatic nerve section reduces P2X3 staining in the dorsal horn (Bradbury et al., 1998) whereas chronic constriction increases staining (Novakovic et al., 1999). Since a major difference between the two lesions is that some afferents remain intact after constriction, this may imply that activity in untransected fibers plays a role in the increase. Administration of GDNF reverses the changes of sciatic nerve section (Bradbury et al., 1998), and NGF increases P2X3 staining in normal animals (Ramer et al., 2001). Whether this implies two separate populations of P2X3 DRG cells that respond to the different trophic factors is not clear. P2X3-deficient mice lose $P2X_3$ staining in DRG and spinal cord, and these mice also lose rapidly desensitizing ATP currents in DRG cells and develop severe bladder hyporeflexia (Cockayne et al., 2000).

Vanilloid Receptors

Vanilloid receptors (VRs) respond to capsaicin, temperature, and protons (acid). They were first demonstrated in the superficial dorsal horn cord, peripheral nerves, and dorsal root ganglia by autoradiography of resiniferatoxin (RTX), a potent capsaicin analog (Winter et al., 1993; Szallasi et al., 1994a, b; Szallasi, 1995; Szallasi and Blumberg, 1999), and by capsaicin activation of vanilloid receptors in the presence of cobalt ions which outlines the cells that express these receptors (Nagy et al., 1993b). VR cells were noted but not illustrated in the trigeminal nucleus (Mezey et al., 2000). VR1 immunostaining, presumably mostly on sensory fibers, is most prominent in lamina I early in development, but shifts deeper, particularly to lamina IIi as development progresses (Guo et al., 2001).

CONCLUSIONS

1. Cells, axons, and terminals in lamina II express an extraordinary array of neuroactive substances and receptors, inferring a relation to nociceptive integration. Such inferences are reasonable and generally true, but it is not clear why such a plethora of ligands and receptors is necessary for this function. One answer is undoubtedly that these substances and receptors are involved in much more than nociception.

2. 5-HT axons descend from higher centers and terminate on, among other targets, opioid-containing interneurons, indicating that 5-HT can influence endogenous nociception.
3. In keeping with the idea that glutamate is the main excitatory primary afferent transmitter, glutamate-enriched terminals are found throughout the dorsal horn and in central terminals of all glomerular types.
4. The main dorsal horn inhibitory amino acid is GABA. Islet as opposed to stalked cells in lamina II are GABAergic. GABAergic interneurons give rise to presynaptic dendrites. GABA terminals are usually characterized by flattened or pleomorphic vesicles and symmetric synaptic thickenings. GABA-enriched terminals can be presynaptic in typical axodendritic synapses, presynaptic to other terminals (axoaxonic) most prominently as the peripheral terminals in glomeruli, occasionally postsynaptic to other terminals, again usually in glomeruli, and GABA-enriched dendrites make dendrodendritic, dendrosomatic, and dendroaxonic synapses with other neural elements. GABAergic presynaptic terminals end on many of the large fiber primary afferent terminals in the dorsal horn, but they do not end on peptidergic endings in glomeruli presumably indicating that large fiber input is under direct GABAergic control whereas peptidergic primary afferent input is not. GABAergic terminals, either axonic or dendritic, participate in complicated synaptic arrangements such as triads. Thus the possibilities for GABAergic inhibitory control in the dorsal horn are enormous.
5. Glycine, the other inhibitory amino acid in the dorsal horn, co-localizes with GABA in interneurons and their processes. The GABA system is more extensive, however.
6. FRAP intensely stains axons and terminals in lamina IIi. FRAP is so completely and sharply depleted after peripheral nerve section that this loss of staining is used to monitor central projections and sprouting of FRAP afferents.
7. CCK antagonizes the actions of morphine, so changes in this compound following nerve lesions might bear on the failure of opioids to influence neuropathic pain. A difficulty, however, is that some of the antibodies used to visualize CCK may also visualize CGRP.
8. CGRP is expressed by many, mostly small, DRG cells. CGRP fibers and terminals distribute to laminae I, IIo, and V, and there are CGRP fibers in the dorsal columns as well. Because of the absence of intrinsic and descending CGRP fibers, CGRP is used to indicate peprtidergic afferent input to the cord. CGRP distributions are also used to assay sprouting, but controls must be done to assure that increases in CGRP are not the result of changes in extant fibers and terminals.
9. GAL, NPY, and VIP are markedly upregulated in DRG cells after peripheral nerve section. In the cord, NPY and VIP staining are concomitantly increased, but for GAL the results are variable.
10. The major opioids in the dorsal horn are dynorphin and enkephalin, but endomorphin, endorphin, and nociceptin are also found. Dynorphin increases dramatically in the dorsal horn after either inflammation or peripheral axotomy, whereas the changes in enkephalin are much less dramatic.
11. SP is found in many DRG cells as well as intrinsic neurons in the cord. Its peripheral release is a major factor in neurogenic inflammation. Essentially every SP DRG cell coexpresses CGRP.

12. Excess glutamate, which is neurotoxic, is removed from the extracellular space by glutamate transporters. After spinal injury both glial and neuronal transporters increase.
13. Ionotropic glutamate receptors are expressed by fibers and terminals and intrinsic cells in the dorsal horn. Many glutamate receptors, both NMDA and non-NMDA, are also found on primary afferent endings, which since afferent terminals release glutamate, suggests autoreception.
14. Metabotropic glutamate receptors are divided into three groups. Group 1 and 2 mGluRs seem morphologically to be arranged for volume transmission whereas group 3 is more for classical transmission. Group 1 is almost exclusively postsynaptic whereas groups 2 and 3 have a large presynaptic component and seem poised to exercise presynaptic control over primary afferent input.
15. GABA receptors are subdivided into $GABA_A$, $GABA_B$, and $GABA_C$ subtypes, and the $GABA_A$ subtype is subdivided into numerous subunits. Combinations of different $GABA_A$ subunits will result in different responses to GABA, allowing ample integration possibilities. GABA agonists and antagonists increase and decrease $GABA_B$ receptors respectively, suggesting the possibility of use plasticity.
16. Glycine receptors are divided into α and β subunits, the β subunit being closely linked to the large polypeptide gephyrin. Many terminals facing glycine-receptor synaptic thickenings are enriched in GABA, GABA, and glycine are colocalized in terminals facing GABA receptor targets, and GABA and glycine receptors often commingle. Thus there is a considerable synergy in the synaptic organization of the inhibitory amino acids GABA and glycine.
17. Estrogen receptors are found in spinal cells, and there is considerable co-localization with enkephalin, suggesting that estrogens can activate this endogenous opioid. Estrogen levels correlate with changes in trkA and p75 during the estrous cycle.
18. There is controversy about the location of CGRP receptors, with some reporting a concentration in superficial laminae and others reporting a conspicuous absence here. If the latter is true, an explanation as to why so many CGRP terminals end in this area should be provided. CGRP receptors are located in synaptic thickenings, suggesting classic synaptic transmission as opposed to volume transmission, which characterizes most peptides.
19. NK-1 receptors outline large neurons in laminae I and antenna neurons in laminae III–IV. SP terminals, presumably of afferent origin, synapse with the dendrites of these cells in laminae I–II, thus showing that fine primary afferent input will influence antenna cells. NK-1 receptors on these large cells are internalized following noxious stimuli or SP administration, and NMDA antagonists block the internalization. Thus, SP release, as indicated by internalization of NK-1Rs, is controlled by activation of NMDA receptors. Attaching a neurotoxic agent to a SP ligand and then stimulating destroys NK-1R neurons allowing study of the results of removal of this important group of interneurons. These are major advances in our understanding of the organization of the dorsal horn.
20. MORS, DORs, and KORs are expressed by intrinsic neurons and primary afferent terminals. Thus these receptors have pre- and postsynaptic functions. A major feature of opioid receptor synaptic organization is that these receptors are not associated with synaptic thickenings, so volume transmission is probable. Opioid

receptors are expressed by peptidergic terminals indicating presynaptic control of these afferents, but not by amino acid containing processes suggesting no control of this type of input. Opioid receptors are internalized following certain types of stimuli. Peripheral nerve lesions reduce spinal opioid receptors, but it is not clear whether this bears on the failure of opioid receptor activation to relieve neuropathic pain.

21. Purinoceptors, especially P2X3 receptors, are found on primary afferent fibers in the dorsal horn and increase following nerve lesions that result in neuropathic pain behavior.
22. Peripheral transport: it is not uncommon to find that substances expressed by DRG cells do not change in the cord following dorsal rhizotomy or other types of primary afferent removal. Examples are galanin and galanin receptors, and 5-HT 1, $GABA_A$, and somatostatin receptors. A probable explanation is that these substances are transported only peripherally and not to the cord to any extent. A possible functional correlate is that ligands for these receptors can influence DRG cells at peripheral but not central terminals.
23. Volume transmission. There are many instances where receptors are expressed by post-synaptic targets but not at synaptic thickenings. This is generally taken to indicate volume as opposed to classic synaptic transmission. Prominent examples are most of the peptides, opioid and NK-1 receptors and certain metabotropic glutamate receptors. If so, the precise distribution of receptors becomes the key factor in wiring diagrams of the dorsal horn.

Chapter 7

Functional Organization of Dorsal Horn Interneurons

Sensory processing in the spinal cord involves interactions among primary afferent fibers bearing information from sensory receptors, interneurons, ascending tract cells conveying sensory messages to the brain, and descending tract cells that modulate the activity in spinal cord circuits. Such interactions are complex and still poorly understood. In this chapter, the interactions between primary afferent fibers and spinal interneurons will be discussed. In Chapters 8–11, the organization of the ascending tracts will be considered. Chapter 12 includes an overview of the sensory channels and of the descending systems that control somatic sensation.

It can be assumed that electrophysiological recordings from neurons not specifically identified as ascending tract cells are likely to be from propriospinal neurons. The justification for this assumption is that by far the majority of spinal cord neurons are interneurons. Chung, K. et al. (1984) tested this in rats by comparing the number of neurons in the S2 segment of the spinal cord that could be labeled retrogradely 24 hours after horseradish peroxidase (HRP) was applied to the cut surface of the cervical enlargement with the number of unlabeled neurons. Only 1% of the neurons at this level were labeled and so contributed to long ascending tracts. Similar proportions of labeled cells were seen if the HRP was applied at T9 or L1. Since it could be estimated that 2% of the total neurons were motor neurons (few neurons at S2 in rats are preganglionic parasympathetic neurons), this suggests that as many as 97% of the neurons at S2 are propriospinal neurons. Thus, the dominant population of neurons in the spinal cord (like the brain) is the interneuronal component.

ELECTROPHYSIOLOGICAL RECORDINGS FROM INTERNEURONS

Recording Techniques

The response properties of individual interneurons in the spinal cord can be investigated by using extracellular or intracellular (including patch-clamping) recording techniques. The advantage of extracellular recordings is that they can readily be made in the spinal cord in vivo in anesthetized or decerebrate preparations, and experimental conditions often permit stable recordings for hours. This makes it possible to determine the responses of an interneuron to stimulation of peripheral nerves or of its receptive field and to examine alterations in activity due to various experimental manipulations. Intracellular recordings can also be accomplished in vivo, but stable recordings are difficult to maintain for long periods of time. Intracellular or patch-clamp recordings in in vitro preparations offer stable recording conditions, but often the only peripheral input available is stimulation of a dorsal root. However, some in vitro preparations have been developed that allow the analysis of receptive fields.

In Vivo Preparations

Extracellular Recordings from Interneurons

In extracellular recordings from dorsal horn neurons in vivo, interneurons can be distinguished from primary afferent fibers on the basis of their response properties and the configuration of their action potentials. For instance, a primary afferent fiber should respond to stimulation of an individual sensory receptor organ in a predictable way (see Chapter 2), whereas an interneuron can typically be activated by more than one sensory receptor organ, and the effective sensory receptors are often of several different types. The action potentials of primary afferent fibers in extracellular recordings are generally positive in sign, since the recordings are usually made from an internode, and they are very brief in duration (less than 1 ms). The action potentials of interneurons are predominantly negative when recorded in the vicinity of the cell body and proximal dendrites (Fig. 7.1A). The spike potentials may be brief, but they generally have a longer duration than the action potentials of axons. The rising phase of the action potentials may have an inflection if there is a delay in the invasion of the soma-dendritic membrane by an action potential that is initiated in the initial segment of the axon (see Fig. 7.19A). Examples of recordings from spinal cord neurons that show

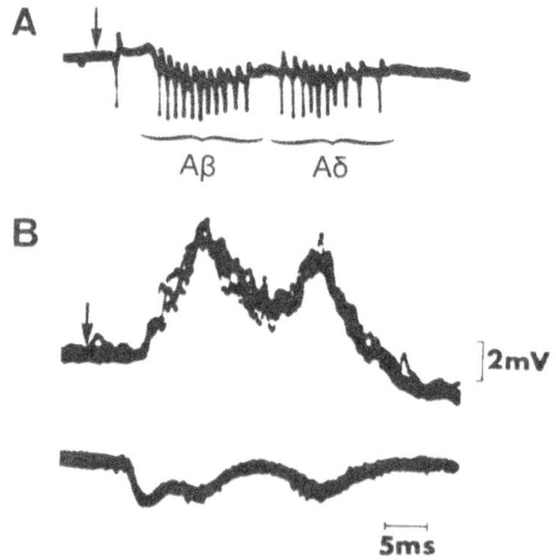

FIGURE 7.1. A, extracellular recording of the responses of a dorsal horn neuron (identified as a spinothalamic tract neuron) to a volley in Aβ and Aδ fibers of the sural nerve. The responses were separated in time because of the distance between the location of the stimulating electrodes on the nerve and that of the recording electrode in the spinal cord (approximately 200 mm). B, the neuron was later impaled by the microelectrode so that intracellular recordings could be made of the excitatory postsynaptic potentials (EPSPs) evoked by the two components of the afferent volley. The arrows represent the time of the stimulus. (From Beall et al., 1977.)

these various features can be found in Brock et al. (1952); Woodbury and Patton (1952); Frank and Fuortes (1955; 1956); Wall (1959); Eccles et al. (1960).

During in vivo experiments, ascending tract cells or motor neurons can be distinguished from interneurons by antidromic activation following stimulation of their axons near projection targets in the brain or of motor axons in a ventral root or peripheral nerve. Criteria for antidromic activation include: (1) the action potential follows the stimulus at a constant latency; (2) there is collision when an orthodromic action potential precedes the antidromic action potential over a predictable range of intervals; and (3) the antidromic action potential can follow high frequencies of stimulation (see Fuller and Schlag, 1976; Trevino et al., 1973).

Direct evidence concerning the morphology of the neural element from which recordings are made can be obtained by marking the unit, either by retrograde transport before the experiment or intracellularly through a microelectrode during the recording session.

Intracellular Recordings from Interneurons

Intracellular recordings can be made from spinal cord interneurons either in vivo (Fig. 7.1B) or in vitro (see next section) in spinal cord slice preparations (Fig. 7.2;

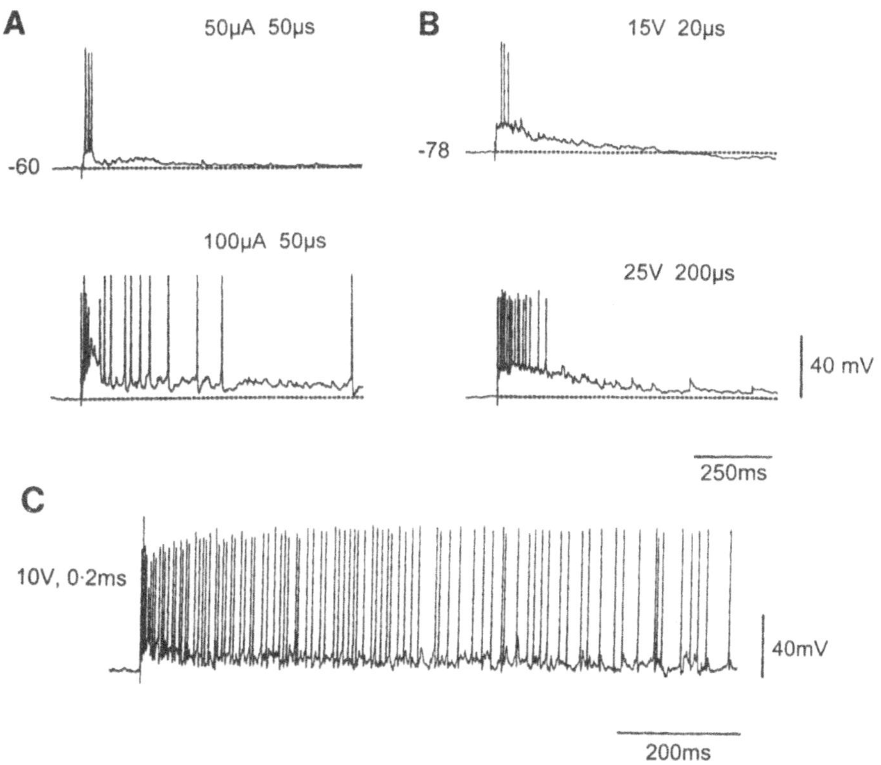

FIGURE 7.2. Intracellular recordings from deep dorsal horn interneurons in a spinal cord slice preparation. A and B, responses of 2 dorsal horn interneurons to two intensities of electrical orthodromic stimulation of a dorsal root. The stimulus parameters are indicated. C, unusually prolonged discharge recorded from another interneuron. (From King et al., 1988.)

Lu et al., 1983; King et al., 1988; see Magnuson et al., 1987; Yoshimura and Jessell, 1989a, b; Bentley and Gent, 1994; Chéry et al., 2000) or from isolated cells (Huang, 1987). Impalement of a neuron with a microelectrode is signaled by the sudden appearance of a negative resting membrane potential and often by large, positive-going action potentials (Fig. 7.2; see Frank and Fuortes, 1955).

Interneurons can be distinguished from primary afferent axons in intracellular recordings in several ways. The action potential of an interneuron is generally longer in duration (up to 2 ms) than is that from an axon (less than 1 ms) (Frank and Fuortes, 1955). Furthermore, interneurons may be identified by the presence of synaptic potentials (Figs. 7.1B, 7.2), in addition to a demonstration of a convergent input from more than one primary afferent fiber or sensory receptor organ (Fig. 7.2A, B). However, care must be taken to distinguish between synaptically evoked action potentials in interneurons and dorsal root reflexes, which may be recorded from primary afferent axons (Frank and Fuortes, 1955). Dorsal root reflexes generally have longer latencies than the orthodromic activity evoked in a primary afferent fiber by peripheral stimulation and may include a variable number (1–4) of action potentials. These impulses tend to fail as the rate of stimulation increases. Action potentials in interneurons often begin after a shorter latency (depending of the conduction velocity of the primary afferents that activate the interneuron), there may be many more than four spikes in a response (Fig. 7.2), and the number of discharges may increase with increases in the rate of stimulation (see later discussion of "wind-up").

Intracellular recordings can be used to demonstrate the membrane properties of interneurons, as well as excitatory postsynaptic potentials (EPSPs; Figs. 7.1B; 7.2) and inhibitory postsynaptic potentials (IPSPs) generated in response to activation of pathways impinging synaptically on these cells (Haapanen et al., 1958; Kolmodin and Skoglund, 1960; Hunt, C. C. and Kuno, 1959a, b; Eccles et al., 1960; Kostiuk, 1960; Hongo et al., 1966; Price, D. D. et al., 1971; Jankowska et al., 1981a; Iggo et al., 1988).

In Vitro Preparations

The absence of movement in spinal cord slices provides an opportunity to maintain very stable recordings, using either penetrations with sharp microelectrodes or whole cell patch-clamp recordings. Such recordings can provide information that can be compared to that obtained in vivo. For example, volleys in large dorsal root afferents often result in an initial, short-lasting EPSP and a brief duration burst of action potentials (Fig. 7.2A, B). In some neurons, volleys in fine afferents result in a train of discharges that can last several seconds (Fig. 7.2C; King et al., 1988).

The membrane properties and the characteristics of the action potentials of interneurons in the deep layers of the dorsal horn have been described based on patch-clamp recordings made in spinal cord slice preparations (Hochman et al., 1997). By using the patch-clamp technique to voltage clamp dorsal horn neurons in vitro (Fig. 7.3; Konnerth et al., 1990; Yoshimura and Nishi, 1993; Yang et al., 2001a), it is possible to study excitatory (EPSCs; Fig. 7.3D) and inhibitory postsynaptic currents (IPSCs; Fig. 7.27). In addition, patch-clamp recordings in spinal cord slices offer the advantage of more stable conditions for observations of miniature EPSCs and miniature IPSCs than do intracellular recordings in vivo (Hori and Endo, 1992; Yoshimura and Nishi, 1993; Yang, K. et al., 1998; Yang and Li, 2001).

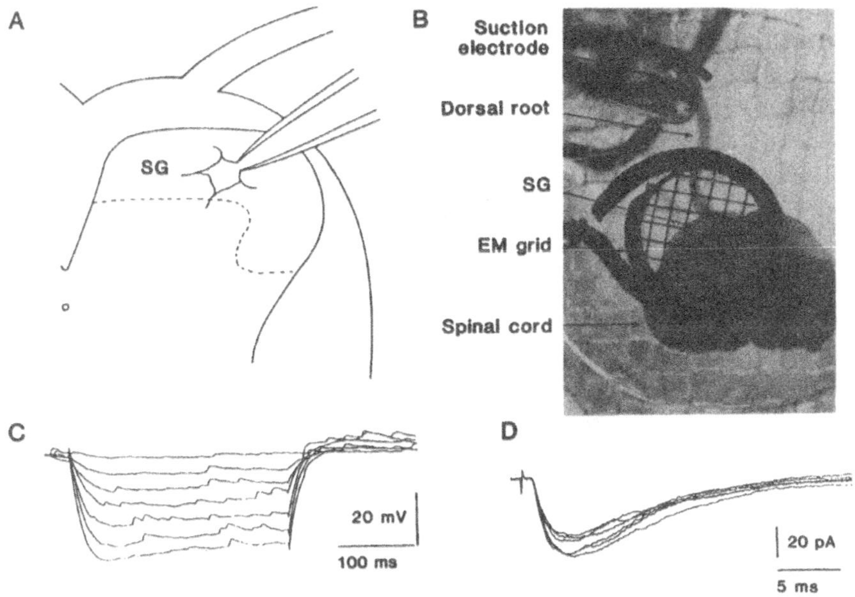

FIGURE 7.3. Recordings from a spinal cord slice preparation. A, diagram of the set-up for whole cell patch-clamp recordings from neurons in the substantia gelatinosa (SG) in a transverse slice of the spinal cord. A dorsal rootlet remains attached to the slice. B, recording chamber with slice placed on nylon mesh and held in place by a silver wire loop. C, hyperpolarizing pulses applied under current clamp. Spontaneous EPSPs are seen as upward-going waveforms. D, monosynaptic EPSCs recorded under voltage clamp conditions in response to stimulation of Aδ fibers in the dorsal root. (From Yoshimura and Nishi, 1993.)

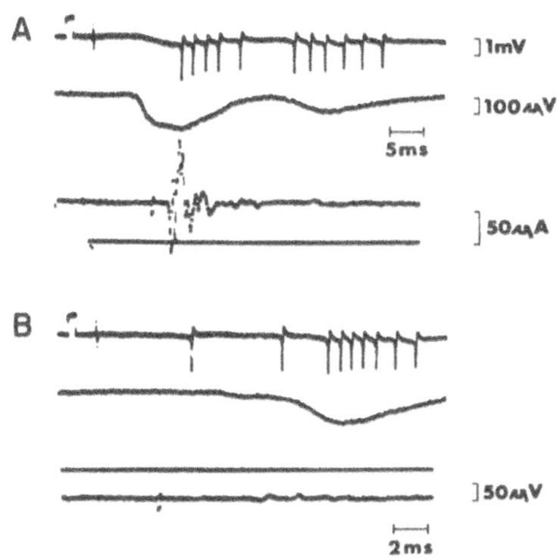

FIGURE 7.4. Responses of a dorsal horn neuron (identified as a spinothalamic tract cell) to volleys in Aβ and Aδ fibers in the sural nerve and the effect of anodal blockade of the Aβ volley. In A and B, the upper traces are extracellular recordings of the action potentials evoked in the neuron by volleys in Aβ and/or Aδ fibers. The second traces are recordings of negative cord dorsum potentials evoked by the sural nerve volleys (see section on Population responses). The third trace in A and the bottom trace in B are neurograms recorded from the sural nerve to show the compound action potentials in Aβ and/or Aδ fibers elicited by the stimulus. The other trace is the monitor of the current used in B (100 μA) for anodal blockade of the Aβ fiber volley. The 5 ms time scale applies to the upper two traces in A and B, while the 2 ms time scale applies to the lower traces. (From Beall et al., 1977.)

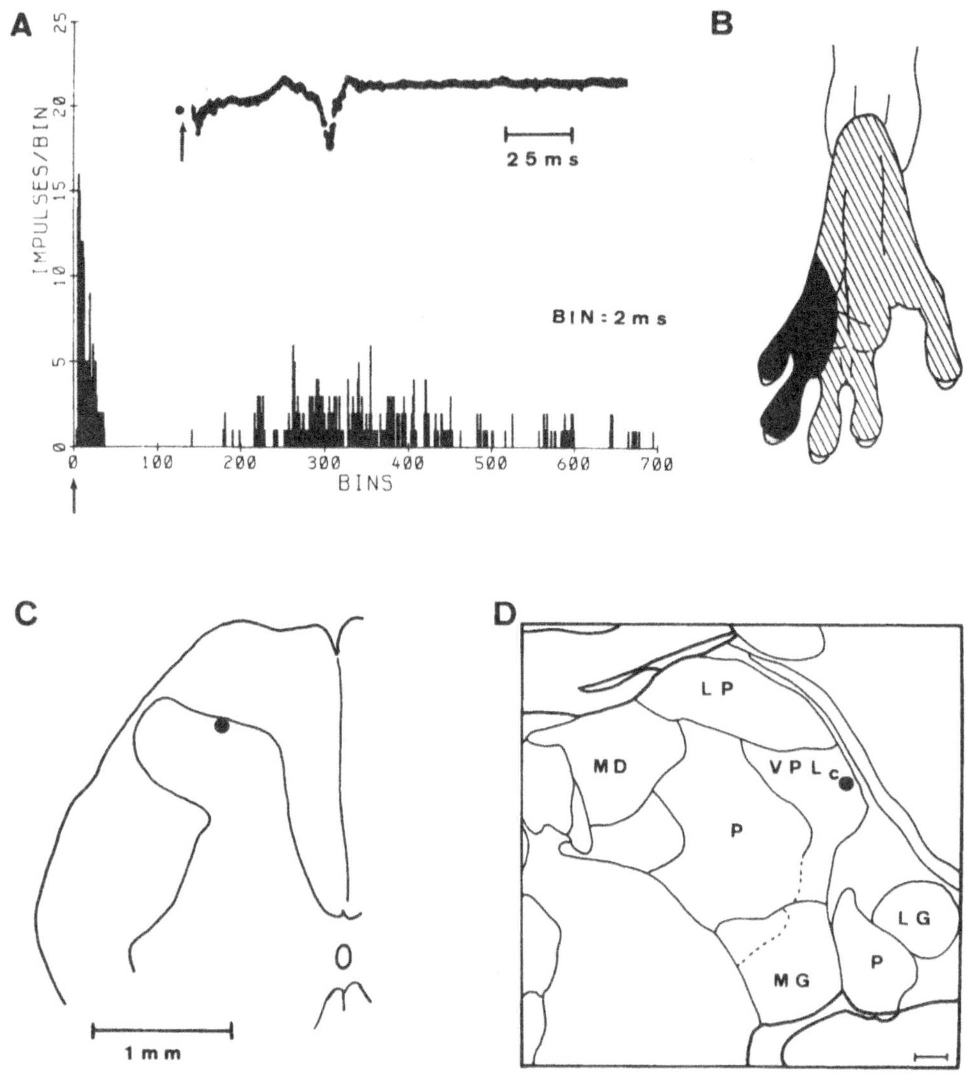

FIGURE 7.5. A, peristimulus time histogram of the responses of a dorsal horn neuron (identified as a spinothalamic tract cell) to volleys in A and C fibers in the sural nerve. The inset shows the compound action potential recorded from the sural nerve 5 cm proximal to the point at which the stimulus was applied. B, receptive field on the foot. The area shown in black was the most sensitive, whereas that shown by hatching was less sensitive. C, location of the recorded neuron in lamina I. D, site in the thalamus from which the cell was activated antidromically. VPL_c = the caudal part of the ventral posterior lateral nucleus. (From Chung et al., 1979.)

Responses of Interneurons to Peripheral Nerve Volleys Evoked by Electrical Stimulation

Volleys in A Fibers

Electrical stimulation of myelinated primary afferent fibers will evoke a synchronous volley of nerve impulses at the point of stimulation, but the volley becomes dispersed as it travels over a considerable length of peripheral nerve to reach the spinal cord. An interneuron in the dorsal horn may respond with separate bursts of discharges to the Aβ and Aδ components of the afferent volley (Beall et al., 1977), which are separated in time when they activate the interneuron (Fig. 7.1A; 7.4A; cf. Fig. 2.2). Anodal current can be used to block conduction in the Aβ fibers, leaving just the Aδ component of the volley. This reveals the activity evoked in the interneuron by just the small myelinated afferents (Fig. 7.4B).

Volleys in A and C Fibers

Similarly, volleys can be evoked in C fibers, as well as in A fibers, if the stimulus strength is increased sufficiently (Fig. 7.5A; inset shows C fiber volley). The response may now include well-separated components due to the successive A and C fiber volleys, which reach the spinal cord at very different times (Chung, J. M. et al., 1979). The initial response shown in the peristimulus time histogram in Fig. 7.5A is due to the arrival of the A fiber volley in the dorsal horn, and the later, more dispersed response is evoked by the C fiber volley. Similar to the experiment shown in Fig. 7.4, the A fiber volley can be eliminated by anodal blockade, leaving just the C fiber volley and the response to it (see Fig. 11.24; Chung, J. M. et al., 1979). The location of the receptive field of the neuron whose response is illustrated in Fig. 7.5A is shown in Fig. 7.5B, and the position of the neuron in the dorsal horn is shown in Fig. 7.5C. This particular neuron could be activated antidromically from the ventral posterior lateral nucleus of the contralateral thalamus (Fig. 7.5D) and so could be identified as a spinothalamic tract neuron (see Chapter 11).

POPULATION RESPONSES

Cord Dorsum Potentials and Intraspinal Field Potentials

Recordings of extracellular potentials can be used to demonstrate the distribution of activity that is evoked in large populations of spinal cord neurons by stimulation of primary afferent fibers (see Figs. 7.1B, lower trace, and Fig. 7.6). The potentials reflect in large part the depolarization of interneurons or of primary afferent fibers in the dorsal horn. These potentials can be recorded from the surface of the spinal cord with a gross electrode ("intermediary cord potentials" or "cord dorsum potentials") or from within the spinal cord ("field potentials") with a microelectrode.

Cord Dorsum Potentials

The potentials that can be recorded from the cord dorsum (Fig. 7.6A) in response to electrical stimulation of myelinated cutaneous afferent fibers in a peripheral nerve include: an afferent volley (V), one or more negative (N) waves, and a positive (P) wave (Gasser and Graham, 1933; Hughes and Gasser, 1934a, b; Bernhard, 1953; Bernhard et al., 1953; Bernhard and Widén, 1953; Lindblom and Ottoson, 1953a, b; Austin and McCouch, 1955; Willis et al., 1973; Beall et al., 1977).

In monkeys, the cord dorsum potentials evoked by stimulation of the A fibers in a cutaneous nerve include a sequence of three negative waves (Fig. 7.6B), the N1, N2, and N3 potentials (Beall et al., 1977). The N1 wave is evoked by activity in Aαβ fibers, the N2 wave by activity in Aαβ and Aδ fibers, and the N3 wave by activity in Aδ fibers.

Similar N and P waves generated in the human spinal cord have been recorded from the epidural space (Shimoji et al., 1977).

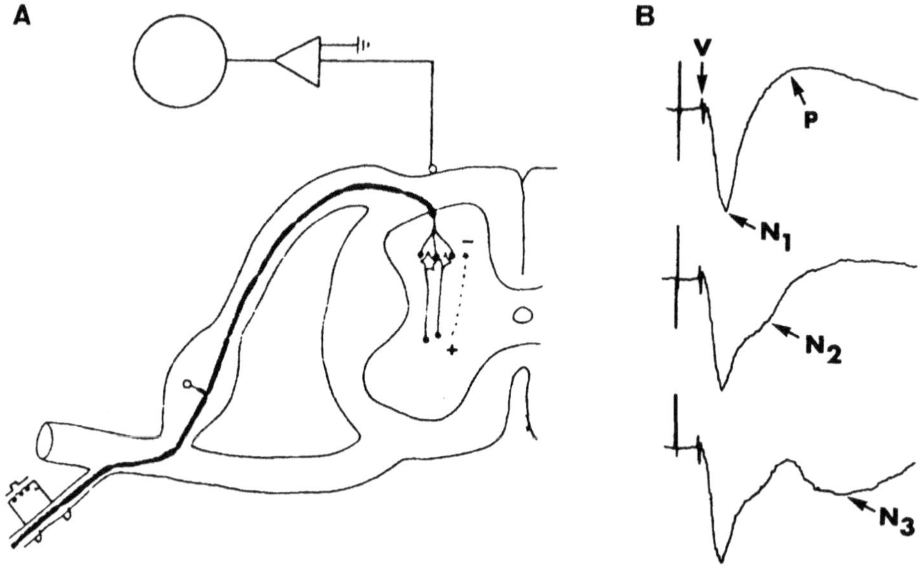

FIGURE 7.6. A, method for recording cord dorsum potentials. An electrode is placed on the dorsal surface of the spinal cord, and a differential recording is made with respect to ground. Stimulating electrodes are placed on a peripheral nerve to activate primary afferent fibers. The arrangement of dorsal horn neurons and their ventrally projecting axons is shown schematically. The interneurons form dipoles, with the negative end of the dipole in the extracellular region in which the interneurons are synaptically depolarized by activity in primary afferent fibers. The positive end of the dipoles is along the ventrally projecting axons, which form current sources during synaptic excitation. (From Willis, 1984.) B, cord dorsum potentials recorded from the spinal cord of a monkey in response to stimulation of a cutaneous nerve. The stimulus strength was increased progressively from a level that activated only Aβ fibers (top trace) to one that also activated some Aδ fibers (middle trace) and finally to a level that activated most A fibers (bottom trace). Abbreviations: V = volley; N_1, N_2, and N_3 = three different negative cord dorsum potentials; P = the positive cord dorsum potential. (From Beall et al., 1977.)

Negative Field Potentials

Recordings from within the spinal cord show that the negative field potentials produced by volleys in A fibers (and that are responsible for the negative cord dorsum potentials) reach a maximum in the dorsal horn. These potentials reverse to become positive in the ventral horn (Fig. 7.7; Howland et al., 1955; Coombs et al., 1956; Fernandez de Molina and Gray, 1957; Willis et al., 1973; Beall et al., 1977). Since the area of peak negativity corresponds to a region that contains many interneurons that are activated by the cutaneous nerve volley, it can be presumed that the negativity reflects excitatory processes (see Fig. 7.1). A negative potential would be expected to develop in the extracellular space because of the shift of cations into dorsal horn interneurons that occurs during excitatory postsynaptic potentials and action potentials (Fig. 7.6A). The positive area in the ventral horn represents current sources distributed along the axons of interneurons that project ventrally (Fig. 7.6A). Neurons that are excited by afferent volleys can also be found in the ventral part of the cord. However, the extracellular potentials represent the summed activity of many neurons, and the negativity around excited cells in the ventral horn of the cord is evidently swamped by the positivity caused by the many, more dorsally situated neurons whose axons in the ventral horn form current sources.

The maximum negative field potential (equivalent to the N1 cord dorsum potential) that is evoked by a volley restricted to the Aβ fibers of a cutaneous nerve in a monkey is centered in laminae IV and V (Fig. 7.7A). Activation of Aβ and some Aδ fibers results in the field potential distribution that is shown in Fig. 7.7B, which corresponds to the N2 cord dorsum potential. The negative wave evoked by Aβ fibers seen in Fig. 7.7A was subtracted from the potential evoked by the entire A fiber volley, and so the negative foci that are illustrated in Fig. 7.7C are due to activation of just Aδ fibers. A volley in Aδ fibers produces a negative field potential that corresponds to the N3 cord dorsum potential. The field potential is centered in two separate loci (Fig. 7.7C), one along the dorsal border of the dorsal horn and the other in laminae V and VI (Beall et al., 1977; cf. Christensen and Perl, 1970).

The distribution of field potentials evoked by Aβ fibers shown in Fig. 7.7A and by Aδ fibers in Fig. 7.7C implies that interneurons should be identified in laminae IV and V that are strongly excited by Aβ fibers and that there should be cells in laminae I and V–VI that are powerfully activated by Aδ fibers. This is in fact the case (see section on Responses to electrical stimulation).

In rats, Schouenborg (1984) has demonstrated that negative field potentials can also be recorded in the dorsal horn in response to stimulation of C fibers in the sural nerve (Fig. 7.8A). The field potentials are maximal in laminae II and V (Fig. 7.8B). There is a close relationship between the size of the C-fiber volley and the negative field potential in lamina II. The negative field potential in lamina II extends only for about 2 mm longitudinally and the sign of the potential reverses at levels beyond 2 mm. The field is undetectable at distances exceeding 6 mm from the central zone of negativity. Anodal blockade of

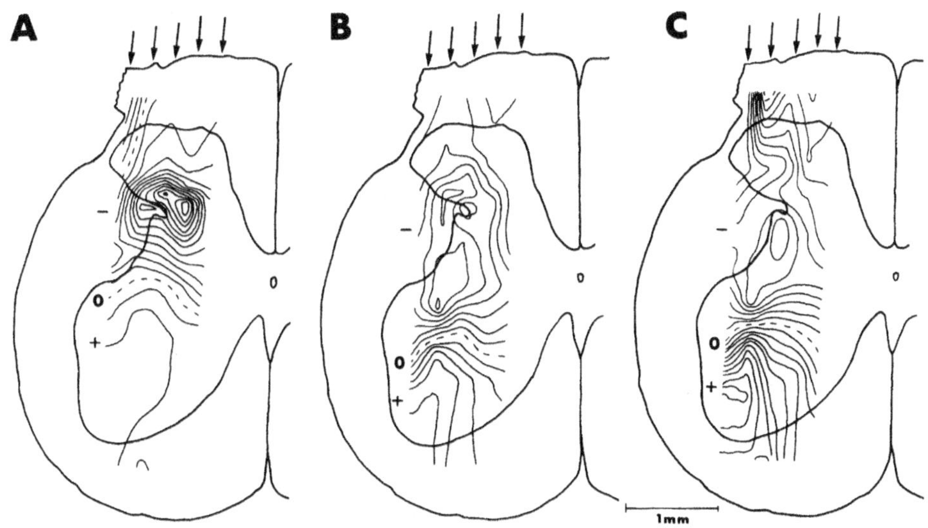

FIGURE 7.7. Isopotential contour plots of the field potentials evoked in the monkey spinal cord by cutaneous afferent volleys in the sural nerve. The stimulus strength was graded in A–C, so that the field potentials produced by the smaller myelinated afferent fibers could be determined by subtraction of the field potential produced by stimulation of larger afferent fibers. The measurements were made at the times of the latencies of the peaks of the cord dorsum N_1, N_2, and N_3 waves. The field potentials were recorded at intervals along five tracks, beginning at the locations indicated by the arrows. A, contour plot showing the spatial distribution in a transverse section of the spinal cord of the field potential that is recorded from the cord dorsum as the N_1 wave. B, field potential corresponding to the N_2 cord dorsum potential. C, field potential corresponding to the N_3 cord dorsum potential. (From Beall et al., 1977.)

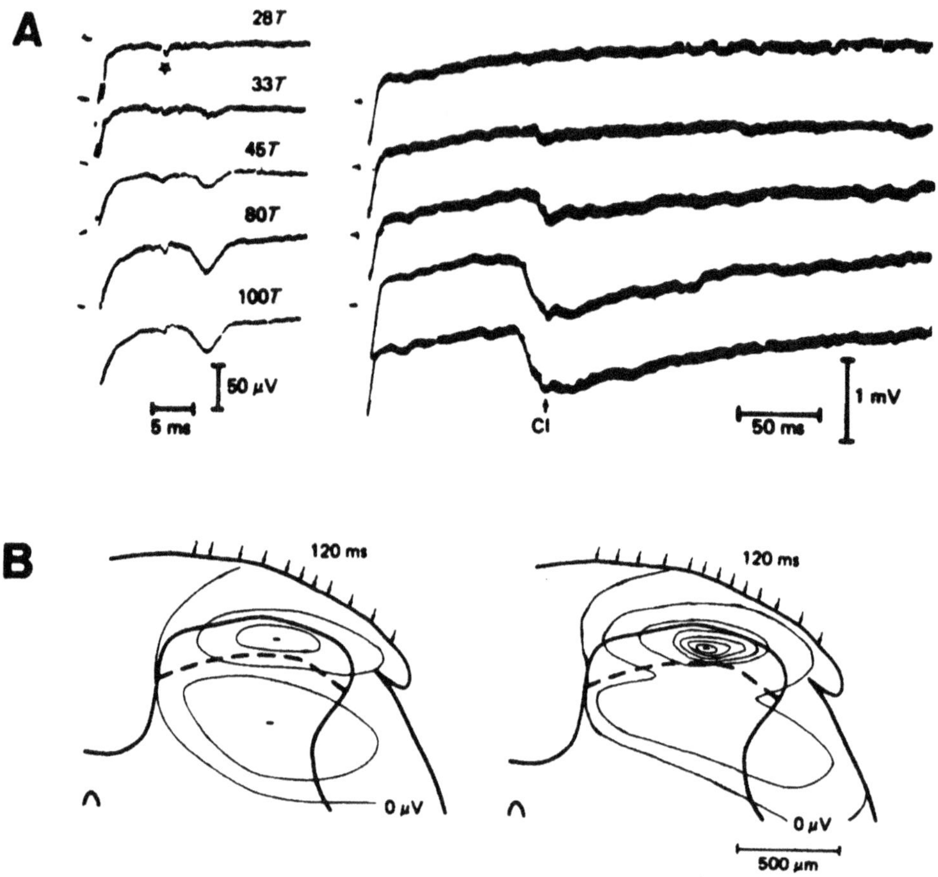

FIGURE 7.8. A (left), compound action potentials in C fibers of the rat sural nerve produced by different stimulus strengths, indicated as multiples of the threshold for the largest afferent fibers in the nerve. A (right), field potentials recorded from lamina II in response to C-fiber volleys evoked in the sural nerve by the same stimuli. B (left), isopotential contour map of the field potential evoked by C-fiber volleys in the sural nerve that were repeated at 0.1 Hz. B (right), similar contour map in the same experiment, but the rate of stimulation was increased to 1 Hz. (From Schouenborg, 1984.)

conduction in A fibers or spinal cord transection does not substantially change the potentials. The field potential in lamina II is enhanced when the stimulus frequency is increased from 0.1 to 1 Hz (Fig. 7.8B), although the size of the C-fiber volley remains constant. The increase in the field potential presumably reflects the same process of EPSP summation that causes "wind-up" of the discharges of dorsal horn interneurons (see below). The field potential produced by C fiber volleys could be explained by an action of nociceptors but not of thermoreceptors or mechanoreceptors. These findings correlate with the evidence of Lynn and Carpenter (1982) that 78% of C afferent fibers in cutaneous nerves supplying the hairy skin of the rat are nociceptors.

Population Responses to Muscle Afferent Volleys

Volleys in muscle afferent fibers can also evoke field potentials. A volley in Group I muscle afferent fibers in cats produces only a small negative cord dorsum potential

(Eccles et al., 1954; Coombs et al., 1956; Lucas and Willis, 1974), presumably because a muscle afferent volley excites fewer interneurons than does a cutaneous afferent volley, and the affected interneurons are located in deeper layers of the gray matter. The maximum negative field potential due to a group I volley is located deep in the dorsal horn, and there is a smaller negative peak in the ventral horn centered over parts of laminae VII and IX (Fig. 7.9; Lucas and Willis, 1974). These areas of peak negativity presumably reflect the monosynaptic excitation of group I interneurons and motor neurons.

Larger N waves are produced in cats by stimulation of high-threshold muscle afferent fibers (Bernhard, 1953; Mendell, 1972). These potentials have also been described in monkeys, in which an afferent volley in all of the myelinated axons of a muscle nerve produces a series of N waves termed the cord dorsum NI, NII, and NIII waves (Foreman et al., 1979). The NI wave is evoked by group I fibers, the NII wave by group II fibers, and the NIII wave by group III fibers. The negative field potential that is equivalent to the NII cord dorsum potential in monkeys is largest in laminae IV–VI, and that equivalent to the NIII wave is largest in laminae V–VI (Foreman et al., 1979). The negative field potentials produced by muscle afferent volleys reverse in polarity in the ventral horn.

The cord dorsum and field potentials are largest when repetitive stimulation is used, especially when nerves to flexor muscles are stimulated (Fig. 7.10C; Eccles et al., 1962b) because of temporal summation.

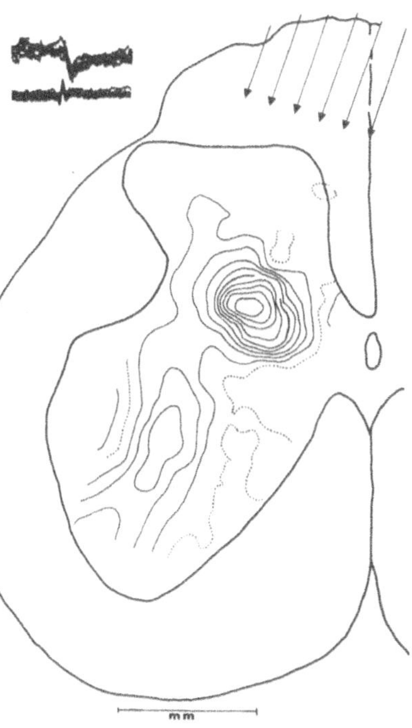

FIGURE 7.9. Isopotential contour plot of the negative field potential evoked by electrical stimulation of the group I afferent fibers in the lateral gastrocnemoius-soleus nerve in a cat. The recordings were made at 200 μm intervals along six tracks oriented as shown by the arrows. The contours were calculated from measurements made at the time of the peak negative field potential. Negative isopotentials are continuous lines, and the zero isopotential line is dotted. The inset shows the maximum potential recorded in the intermediate nucleus. (From Lucas and Willis, 1974.)

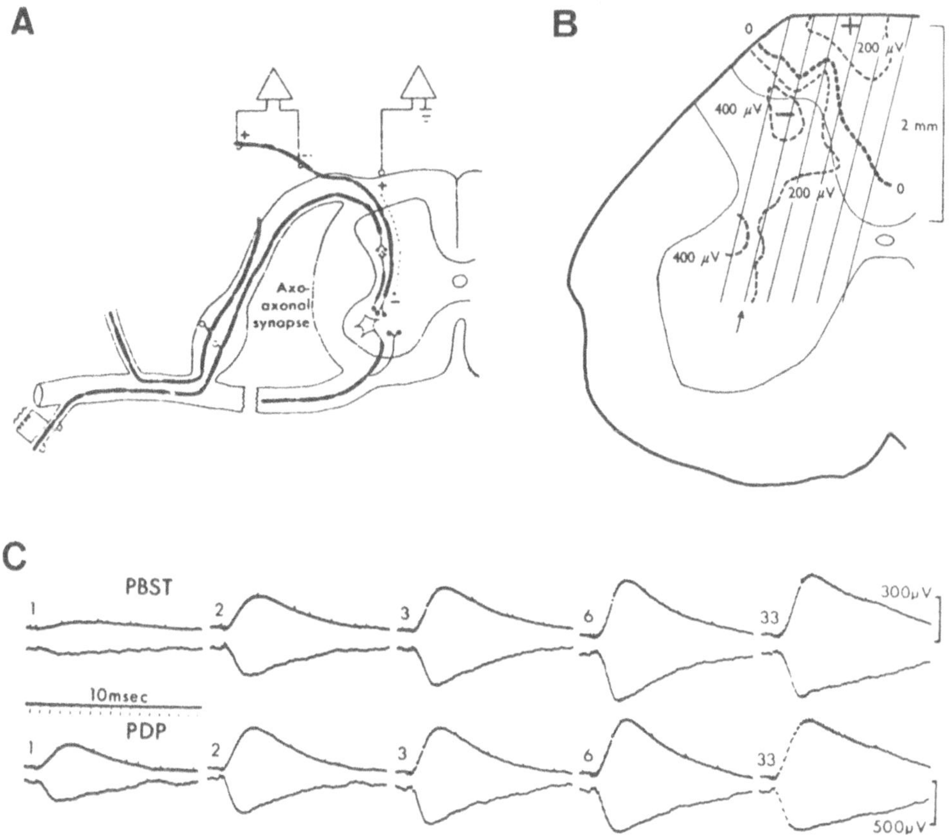

FIGURE 7.10. A, sample recordings of dorsal root potentials and the recording method. The main component of the dorsal root potential (DRP V) is often referred to as the negative dorsal root potential. The drawing shows that the dorsal root potentials can be recorded by a pair of electrodes placed on a cut filament of dorsal root. One electrode is very near the spinal cord, and the sign of the potential at this point is used as a reference in naming the potentials. An electrode for recording cord dorsum potentials is also shown. The orientation of large primary afferent fibers (group Ia muscle afferent fibers) that project to the ventral horn is also indicated. When the synaptic terminals of the afferents are depolarized (PAD), the extracellular space becomes negative in the ventral part of the cord because of ion shifts across the terminal synaptic membranes. In this case, there is an efflux of Cl^- through Cl^- channels opened by $GABA_A$ receptors. Positivity develops dorsally because the primary afferent fibers become current sources. The dorsal root potential is negative, because this reflects the electrotonic spread of the depolarization along the axons in the dorsal root; however, the cord dorsum potential is positive because this reflects the current sources near the dorsal surface of the cord. (From Willis, 1984.) B, isopotential contour plot constructed from measurements of the negative field potential made at 40 ms after a cutaneous afferent volley. The contour map shows the distribution of positivity and negativity in the spinal cord during PAD. (From Eccles, J. C. et al., 1962a.) C, series of recordings of dorsal root potentials (top traces) and cord dorsum potentials (lower traces) recorded in response to repetitive stimulation of the nerves to two sets of flexor muscles (posterior biceps and semitendinosus, PBST, and peronei and deep peroneal, PDP) operating at the knee and ankle joints. The number of stimuli used for each pair of recordings is indicated. The potentials become larger with greater numbers of volleys because of temporal summation. (From Eccles, J. C. et al., 1963c.)

Population Responses to Joint Afferent Volleys

A volley in joint afferent fibers also evokes a series of negative cord dorsum potentials (Schaible et al., 1986). These have been called the NI, NII, and NIII cord dorsum potentials, and they are attributed to volleys in the corresponding sets of group I, II, and III afferent fibers, which have successively lower conduction velocities and higher thresholds. The equivalent negative field potentials are maximal in the dorsal horn and reverse sign in the ventral horn.

Population Responses to Visceral Afferent Volleys

Similarly, stimulation of visceral afferent fibers evokes negative cord dorsum potentials (see, e.g., Selzer and Spencer, 1969; Yates and Thompson, 1985). The negative field potential that results from stimulation of visceral afferent fibers is maximal in lamina V (Selzer and Spencer, 1969).

Positive Cord Dorsum Potential and Primary Afferent Depolarization

Primary Afferent Depolarization

The positive cord dorsum potential (P wave) that follows the N waves evoked by stimulation of cutaneous (Fig. 7.6), muscle (Fig. 7.10), or other types of nerves reflects a long-lasting depolarization of primary afferent fibers (primary afferent depolarization, PAD). This part of the cord dorsum potential corresponds to the negative dorsal root potential, which can be recorded from a disconnected filament of dorsal root (Fig. 7.10A and C, upper rows of traces; Barron and Matthews, 1938; Lloyd and McIntyre, 1949; Lloyd, 1952; Koketsu, 1956; Eccles and Krnjevic, 1959; Eccles et al., 1962b; 1963a).

PAD is considered to be one of the mechanisms responsible for the inhibitory process known as presynaptic inhibition (Eccles, 1964; Schmidt, R. F., 1971; Rudomin and Schmidt, 1999; Willis, 1999). PAD is believed to cause presynaptic inhibition by reducing the amount of excitatory transmitter that is released by afferent fiber terminals when they are invaded by a nerve impulse (Eccles, 1964; Schmidt, R. F., 1971; Rudomin et al., 1998; Rudomin and Schmidt, 1999). PAD has been proposed to be due to axoaxonal synapses made by the terminals of interneurons on primary afferent fiber terminals (Fig. 7.10A; Gray, E. G., 1963; Khattab, 1968; Ralston, 1965; 1968a, b; 1979; Conradi, 1969; Réthelyi and Szentágothai, 1969; Barber et al., 1978; Ralston and Ralston, 1979; 1982; Zhu et al., 1981; Knyhar-Csillik et al., 1982b). However, some of the synaptic contacts observed on primary afferent terminals in the superficial dorsal horn may actually represent dendroaxonal connections (Carlton and Hayes, 1990). Functionally, axoaxonal and dendroaxonal synapses would be equivalent. On the other hand, axoaxonal synapses have been demonstrated on the endings of hair follicle, FA I and SA I afferents in the dorsal horn (Sutherland et al., 2002) and on the terminals of group Ia and group II muscle afferent fibers in lamina VI (Maxwell et al., 1990a; Maxwell and Riddell, 1999).

The cord dorsum P wave evoked by stimulation of cutaneous nerves is also recorded as a field potential that is positive in the dorsal funiculus and that reverses to negativity in the dorsal horn and deeper parts of the spinal cord gray matter (Fig. 7.10B). The presynaptic inhibitory synapses on primary afferent fiber terminals would produce a negative area in the extracellular space because of shifts of anions out of the axon terminals (Fig. 7.10A; see later section on Pharmacology of PAD). Current sources would develop along the course of the primary afferent fibers as they enter the spinal cord and cross the dorsal funiculus, producing a positive potential that is recorded at the dorsal surface of the spinal cord and in the dorsal funiculus (Fig. 7.10A, B; see Eccles et al., 1962a, b).

PAD can be demonstrated not only by recordings of cord dorsum P waves, field potentials, and the dorsal root potential, but also by intracellular recordings directly from primary afferent fibers (Koketsu, 1956; Eccles and Krnjevic, 1959; Eccles et al., 1962b; 1963a), by tests for an increase in the excitability of primary afferent terminals (Wall, 1958; Eccles et al., 1962b; 1963a) and by recordings of dorsal root reflexes (Fig. 7.11B; Toennies, 1938; Eccles et al., 1961; Lin et al., 2000).

A rather specific organization has been found for some of the types of primary afferent fibers that give or receive PAD (see the review by Schmidt, R. F., 1971; see also Whitehorn and Burgess, 1973; Brink et al., 1984). For instance, among cutaneous afferent fibers, the slowly adapting mechanoreceptors preferentially evoke PAD in the terminals of

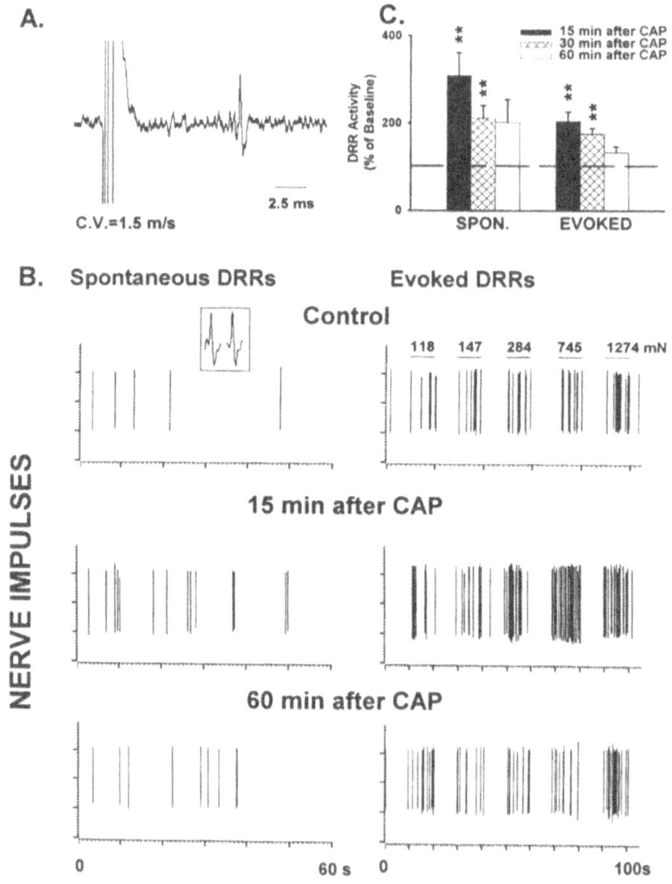

FIGURE 7.11. Dorsal root reflex activity in a C fiber. Recordings were made from the central stump of a cut filament of dorsal root. Other dorsal roots were intact so that discharges in primary afferents could reach the spinal cord. A, action potential in a single primary afferent unit identified as a C fiber by the conduction velocity of the action potential evoked by electrical stimulation of the dorsal root filament (1.5 m/s). The action potential occurred at the same latency on repeated stimulation, ruling out the possibility that dorsal root reflexes elicited by dorsal rootlet stimulation were being confused with the antidromic spike in the afferent unit. B, recordings of spontaneous antidromic activity and dorsal root reflexes (DRRs) evoked by stimulation of the skin with von Frey filaments of graded bending forces (values indicated in mN). The responses were increased following an injection of capsaicin into the skin. C, increased spontaneous and evoked DRR responses of a population of C fibers following capsaicin injection. (From Lin et al., 2000.)

other slowly adapting mechanoreceptors, whereas rapidly adapting receptors tend to produce PAD in the afferent fibers that supply other rapidly adapting receptors (Jänig et al., 1968b).

Excitability testing has also been used to demonstrate changes in the thresholds of the terminals of primary afferent C fibers in the dorsal horn (Hentall and Fields, 1979; Madrid et al., 1979; Fitzgerald and Woolf, 1981; Calvillo et al., 1982). Volleys in A fibers increase the excitability of C-fiber endings. However, the types of receptors connected with the C fibers that generate PAD are unknown, and further evidence is needed concerning the relationship of presynaptic excitability changes to presynaptic inhibition in C-fiber pathways.

Indirect evidence that PAD can be produced in C fibers comes from recordings of dorsal root reflexes in C fibers (Fig. 7.11; Lin et al., 2000). The recording arrangement used for the C-fiber activity shown in Fig. 7.11 was similar to that used for dorsal root potentials (Fig. 7.10). A recording electrode was placed in contact with the proximal part of a cut dorsal root filament, and, in addition, a stimulating electrode was placed on the dorsal root filament near its junction with the spinal cord. Stimulation of the dorsal rootlet was used to determine the conduction velocity of an axon whose activity was isolated by splitting the filament until a single unit was recorded. After this identification procedure, spontaneous or evoked action potentials were recorded from the unit. This activity was conducted away from the spinal cord in the antidromic direction. It was possible for dorsal root reflexes to be evoked by peripheral stimulation because dorsal roots other than the filament that was dissected remained intact. The waveforms of the action potentials representing dorsal root reflexes are shown in the inset in Fig. 7.11B; the vertical lines represent the same action potentials in the dorsal root C fiber but on a slow time base.

Presynaptic inhibition has been investigated not only in vivo but also in vitro. For example, PAD and dorsal root reflexes have been demonstrated in an isolated spinal cord (Bagust et al., 1985a, b). The patch-clamp technique can be used in slice preparations for a quantal analysis of synaptic transmission, at least from dorsal root axons onto dorsal horn neurons. This provides a better opportunity for a quantal analysis of presynaptically mediated events than does in vivo recording (Hori and Endo, 1992; Yang and Li, 2001).

Changes in Metabolic Activity and Intracellular Signaling

The responses of a population of neurons in the spinal cord to peripheral stimulation can also be demonstrated using anatomical methods. Two of these methods are the 2-deoxyglucose technique and immunostaining for immediate-early genes.

2-Deoxyglucose

An increase in the activity of neurons results in an increase in the utilization of glucose. Excited neurons (or their synapses) will take up 2-deoxyglucose in the place of glucose if this substance is introduced into the circulation, but unlike glucose, 2-deoxyglucose is not metabolized. The accumulation of radiolabeled 2-deoxyglucose can be visualized by radioautography in tissue sections. This technique is a valuable way to demonstrate enhanced activity in a number of systems in the brain, and it has also been used to show an increase in glucose metabolism in the dorsal horn in response to electrical stimulation of the sciatic nerve (e.g., Piepmeier et al., 1983; Kadekaro et al., 1985). Since dorsal root ganglia fail to show changes (Kadekaro et al., 1985), it can be concluded that most of the enhanced glucose utilization occurs in afferent terminals or in postsynaptic neurons. Proshansky et al. (1980) and Crockett et al. (1989) demonstrated an enhancement of 2-deoxyglucose uptake in the ipsilateral dorsal horn, as well as in the intermediate region, lamina X, and both ventral horns, in response to innocuous mechanical stimulation of the plantar cushion of the cat's foot. After noxious thermal stimulation of the plantar cushion, Abram and Kostreva (1986) also found increases in the superficial dorsal horn, the intermediate region, lamina X, and both ventral horns. The main difference in the distribution of the enhanced metabolism in these two studies was that there was increased metabolism in the medial part of the dorsal horn in response to innocuous but not to noxious stimulation. This can be explained on the basis that interneurons in the medial dorsal horn participate in the plantar cushion reflex, which is a response to the activation of mechanoreceptors (Egger et al., 1986). Gonzalez-Lima (1986) found that stimulation in the midbrain reticular formation of rats also causes an increase in 2-deoxyglucose in laminae II and III.

The 2-deoxyglucose method has been utilized very effectively for the demonstration of enhanced metabolic activity of neurons that are involved in an animal model of neuropathic pain (Fig. 7.12; Price, D. D. et al., 1991; Mao et al., 1992c, d).

Expression of Immediate, Early Genes

Stimulation, especially noxious stimulation, of peripheral structures, can trigger expression of FOS protein by the c-Fos gene (Hunt, S. P. et al., 1987; Menétrey et al., 1989; Bullitt, 1989). FOS expression can be activated by a substance P agonist, as well as by strong stimulation of a dorsal root, and reduced by administration of an antagonist of NK-1 receptors or inhibitors of PKA or PKC (Badie-Mahdavi et al., 2001). The gene expression appears to be triggered at least in part by calcium influx through voltage-gated channels (Morgan, J. I. and Curran, 1986). The release of nitric oxide in the dorsal horn is also a factor (Wu, J. et al., 2000). FOS protein can be detected immunocytochemically in the superficial dorsal horn within 5–15 min after strong heat stimulation; staining shifts largely to the deeper layers of the dorsal horn by 8 hours after stimulation (Williams, S. et al., 1990; cf., Menétrey et al., 1989). FOS protein is synthesized in the cell body of a neuron and is then transposed to the nucleus, where it can affect the expression of other genes by interacting with DNA (Mugnaini et al., 1989; Herdegen and Leah, 1998).

Immunocytochemical staining for FOS provides a way to survey the distribution of neurons in the spinal cord gray matter that are activated by a particular stimulus (Menétrey et al., 1989; Chi et al., 1993; Munglani and Hunt, 1995). For example, Fig. 7.13 shows the locations of neurons that expressed FOS protein after several types of noxious stimuli were employed (Menétrey et al., 1989). These stimuli included injection of Freund's adjuvant into the plantar foot (Fig. 7.13A), implantation of uric acid crystals in the ankle (Fig. 7.13B), or intraperitoneal injection of acetic acid (Fig. 7.13C). Inflammation of the foot resulted in FOS expression in laminae I and V, the intermediate region, and ventral horn. Inflammation of the ankle had much less of an effect on neurons of lamina I; otherwise, the pattern of staining resembled that produced by the injection of Freund's adjuvant into the foot. Visceral stimulation was particularly effective in eliciting FOS staining in neurons of laminae I, V, and X. The longitudinal distribution of FOS staining in lamina I extended over many segments, whereas staining in laminae V and X was limited to only a few segments (Menétrey et al., 1989; evidence not shown). In another study, when mustard oil was injected into muscle, neurons that stained for FOS protein were found to be chiefly in laminae I and IV–VII (Hunt, S. P. et al., 1987).

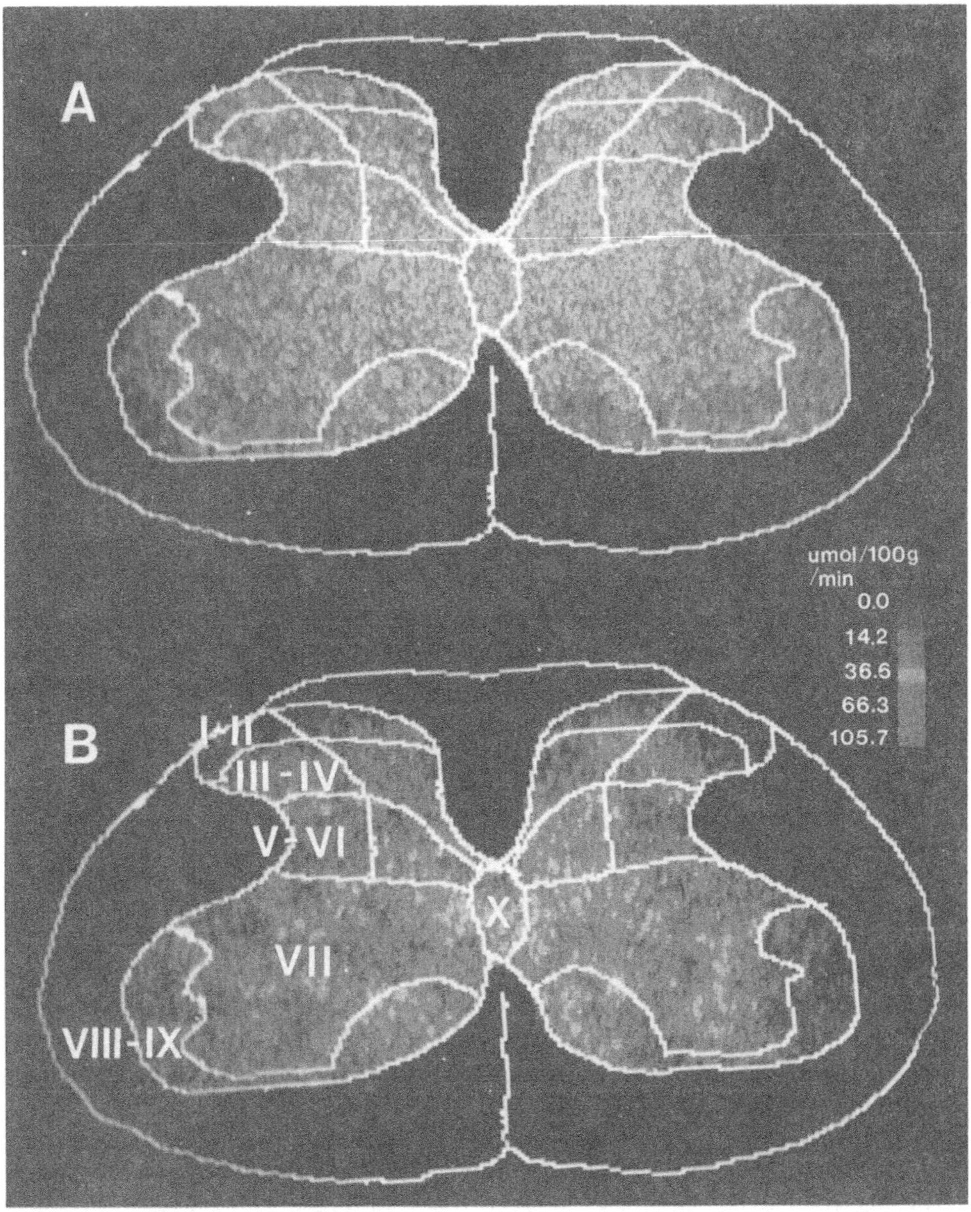

FIGURE 7.12. Pseudocolor images of transverse sections of the L4 spinal cord showing the distribution of 2-deoxyglucose in the gray matter of the spinal cord of rats. A, metabolic activity 10 days following application of four loose ligations of the sciatic nerve (chronic constriction injury model of neuropathic pain; see section on *Models of Peripheral Neuropathic Pain* later in this chapter). B, metabolic activity after sham surgery. (From Price, D. D. et al., 1991.) (A color version of this figure appears between pages 290 and 291.)

FOS staining is also a useful means for demonstrating the effects of therapeutic manipulations that reduce pain (Morgan et al., 1994; Munglani and Hunt, 1995). For example, treatment with morphine produced a naloxone-reversible reduction in the number of neurons stained for FOS protein following noxious stimulation, especially in the

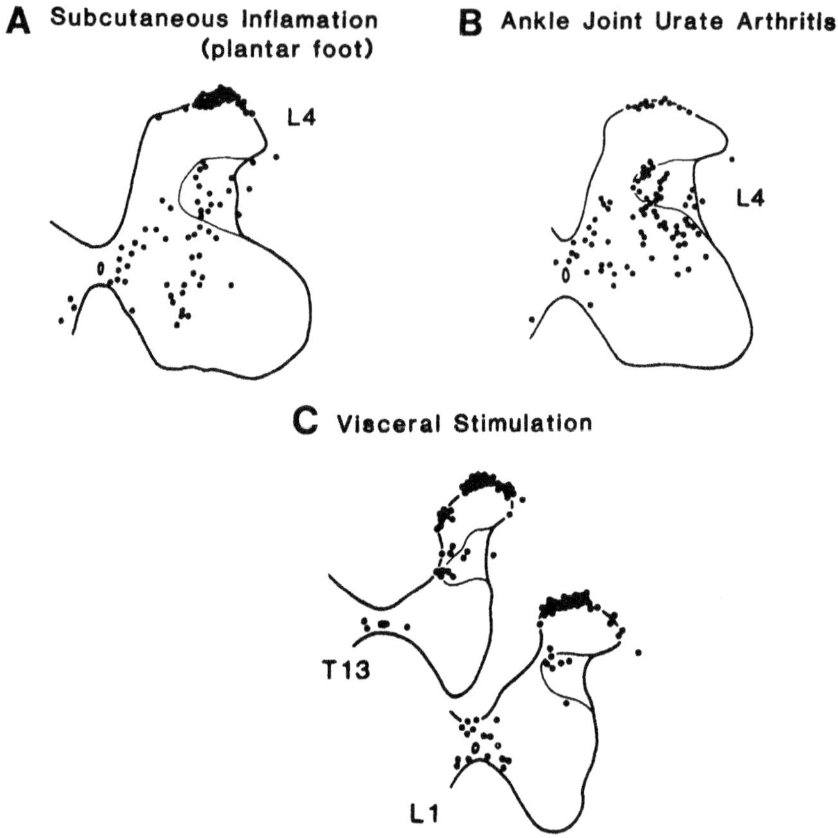

FIGURE 7.13. Locations of neurons whose nuclei immunostain for FOS protein following noxious stimulation in anesthetized rats. A, injection of Freund's adjuvant into the plantar foot. B, implantation of uric acid crystals in the ankle. C, intraperitoneal injection of acetic acid. (From Menétrey et al., 1989, reproduced with permission.)

deep layers of the dorsal horn (Fig. 7.14; Presley et al., 1990; Hammond et al., 1992). Intrathecal pretreatment with fentanyl could also be shown to reduce FOS staining in the formalin test (Nakamura and Takasaki, 2001).

In another example, electrical stimulation in the nucleus raphe magnus, a part of the endogenous analgesia system, reduced the number of dorsal horn neurons that stained for FOS following noxious heating of the foot (Jones and Light, 1990). Thus, FOS staining can be used not only to survey the distribution of neurons that are activated by a particular stimulus, but in addition this procedure can help demonstrate the action of drugs or other manipulations that interfere with responses to the stimulus.

Interneurons in the superficial dorsal horn may be excitatory or inhibitory. Many of the inhibitory interneurons use GABA as their neurotransmitter (see below). After an intradermal capsaicin injection, some interneurons can be immunostained for both GABA and FOS (Todd et al., 1994b; Zou et al., 2001). For example, in Fig. 7.15A, GABA-immunoreactive neurons and FOS-labeled neurons were stained with different fluorescent antibodies. A double-labeled neuron is shown in the right panel. Fig. 7.15B and C show the locations of GABA-labeled neurons (open circles), FOS-labeled neurons (squares),

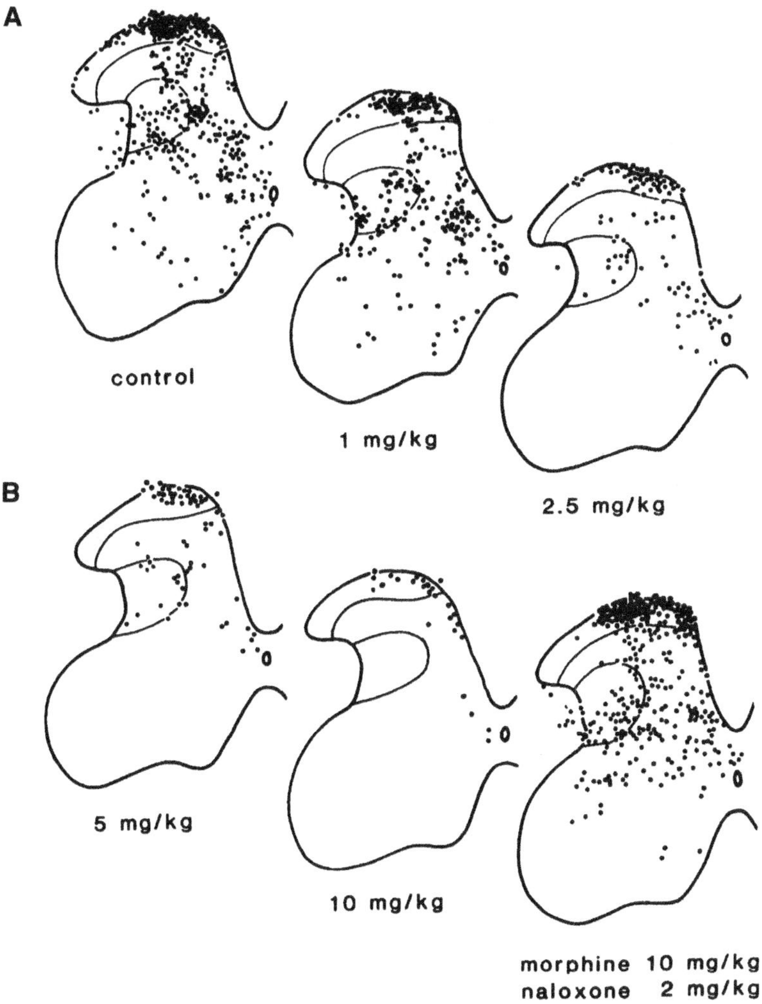

FIGURE 7.14. Reduction in FOS expression following pretreatment with morphine. Drawings are shown of transverse sections from the L4/L5 segments of the rat spinal cord. The dots indicate neurons that were immunostained for FOS protein after a formalin injection into the ipsilateral foot. Prior to the injection, one of several different doses of morphine was administered. The morphine produced a dose-related reduction of FOS-labeling that was especially noticeable for the deep layers of the dorsal horn. Administration of naloxone prevented this effect of morphine. (From Presley et al., 1990.)

and double-labeled neurons (filled circles). After an injection of vehicle into a foot (Fig. 7.15B), several FOS-positive neurons were observed on both sides of the spinal cord, but only a few cells were double-labeled. When capsaicin was injected intradermally into the foot, as expected, many more neurons were immunostained for FOS protein than after vehicle injection (Fig. 7.15C). Most of the neurons that developed FOS-like immunoreactivity were in the superficial dorsal horn on the side ipsilateral to the injection, but some were in the deep dorsal horn and intermediate region. Many of the dorsal horn neurons were double-labeled immunocytochemically for both FOS and GABA (Fig. 7.15C).

Since some of the neurons that stained for GABA also expressed FOS protein, it is evident that inhibitory, as well as excitatory, dorsal horn neurons can respond to

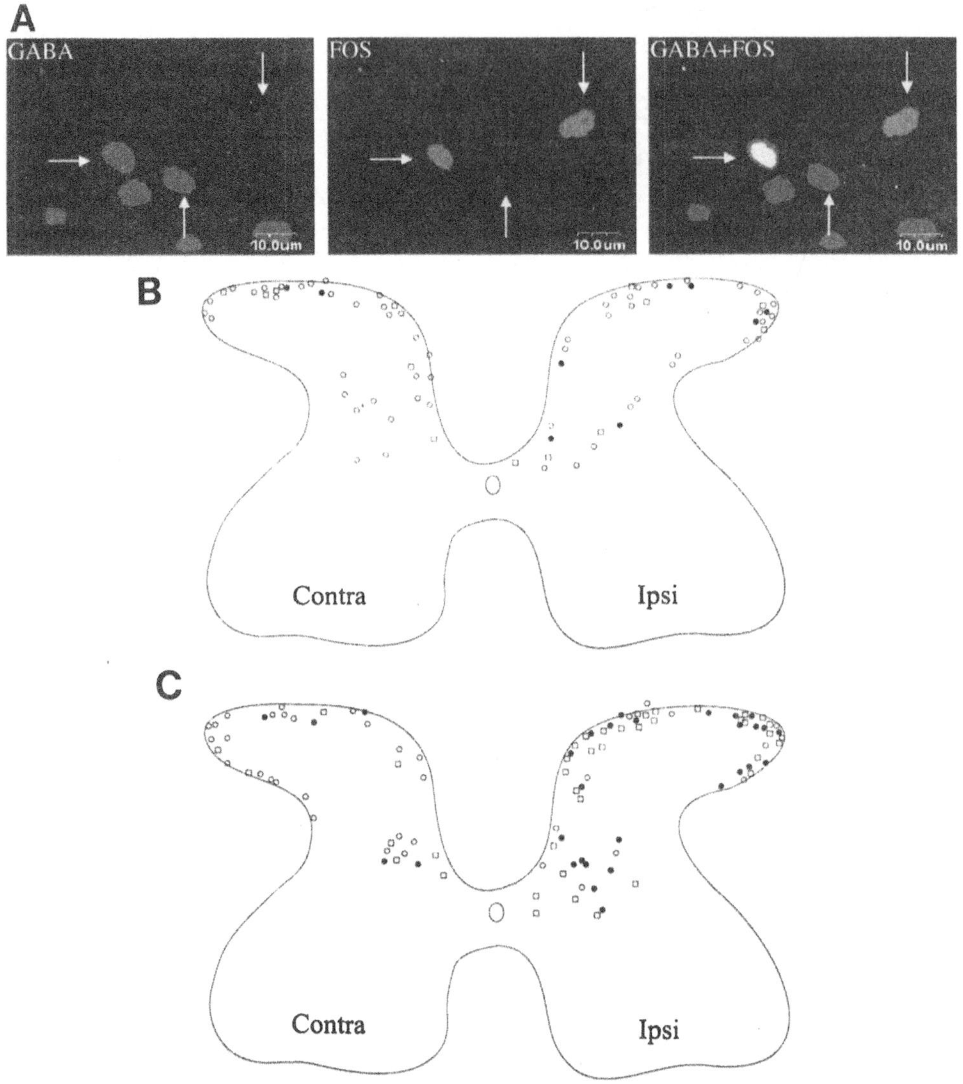

FIGURE 7.15. A, double immunofluorescence labeling of neurons in the rat superficial dorsal horn for GABA and FOS half an hour after an intradermal injection of capsaicin. The images were made using a confocal microscope. The upward vertical arrow shows a neuron that was positive only for GABA; the downward vertical arrow indicates a neuron that had a FOS labeled nucleus but that was not GABAergic; the horizontal arrow points to a neuron that was positive for both GABA and FOS. The drawings show the location and number of neurons immunolabeled for GABA and/or FOS protein half an hour after an intradermal injection of vehicle, B, or of capsaicin, C, into the foot of rats. Open circles indicate neurons with GABA-like immunoreactivity; squares neurons with FOS-like immunoreactivity; and filled circles neurons with both. (From Zou et al., 2001.) (A color version of this figure appears facing page 291.)

noxious stimulation by upregulating FOS. Therefore, mapping the distribution of FOS-immunoreactive neurons is not equivalent to mapping the distribution of neurons belonging to excitatory pathways in the central nervous system, since the distribution of FOS-labeled neurons can also reveal the distribution of inhibitory neurons that are activated by peripheral stimulation.

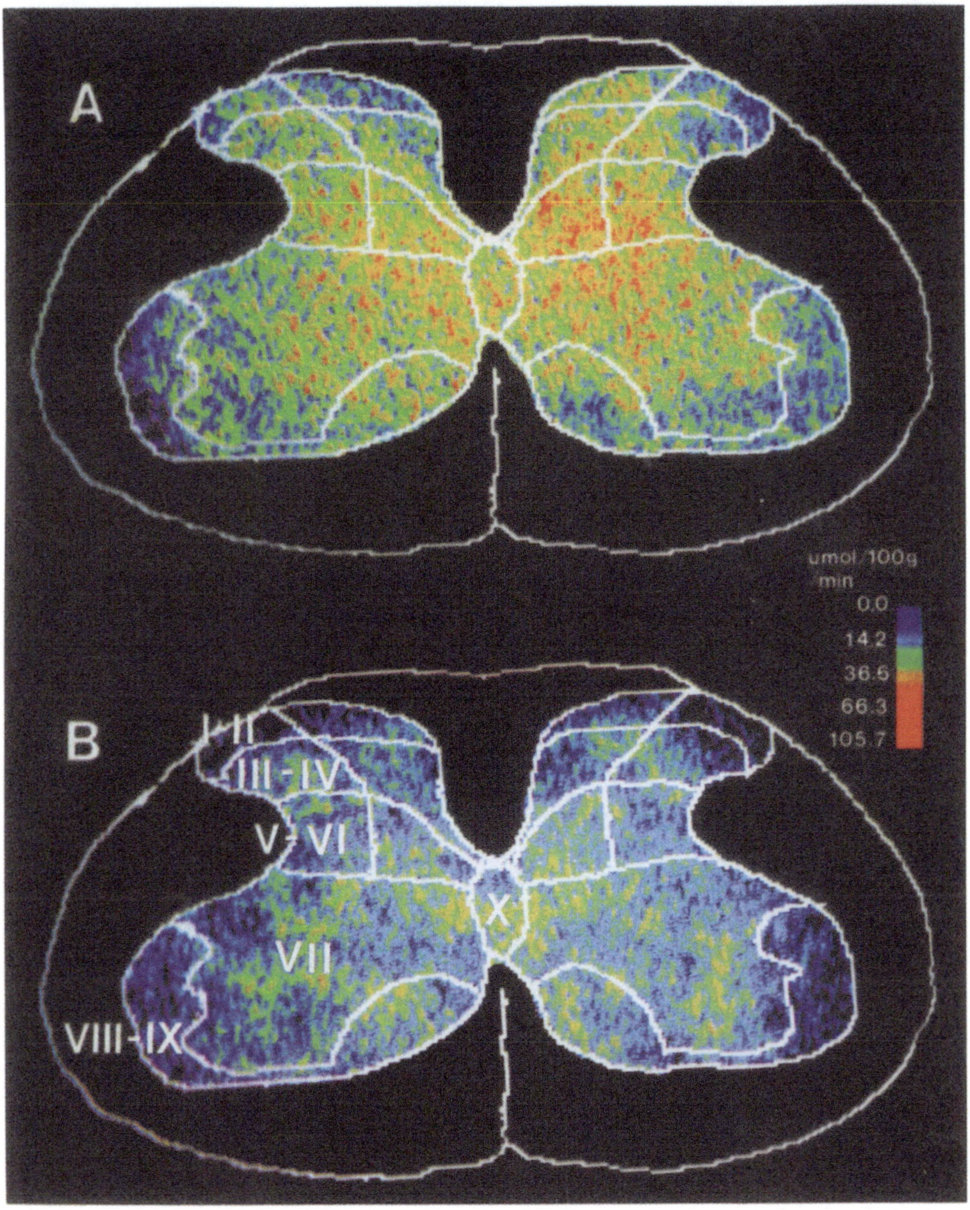

Figure 7.12. Pseudocolor images of transverse sections of the L4 spinal cord showing the distribution of 2-deoxyglucose in the gray matter of the spinal cord of rats. A, metabolic activity 10 days following application of four loose ligations of the sciatic nerve (chronic constriction injury model of neuropathic pain; see section on *Models of Peripheral Neuropathic Pain* later in this chapter). B, metabolic activity after sham surgery. (From Price, D. D. et al., 1991.)

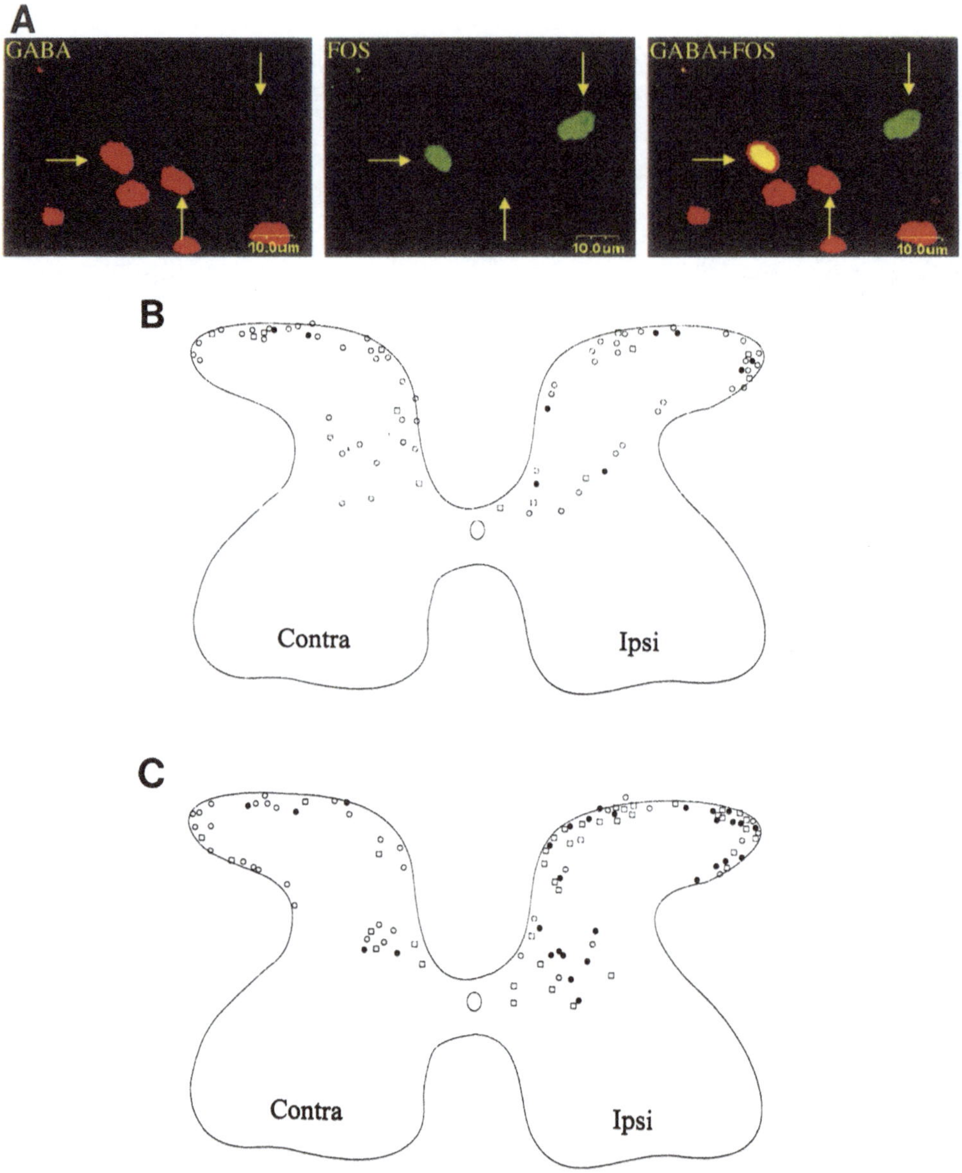

Figure 7.15. A, double immunofluorescence labeling of neurons in the rat superficial dorsal horn for GABA and FOS half an hour after an intradermal injection of capsaicin. The images were made using a confocal microscope. The upward vertical arrow shows a neuron that was positive only for GABA; the downward vertical arrow indicates a neuron that had a FOS labeled nucleus but that was not GABAergic; the horizontal arrow points to a neuron that was positive for both GABA and FOS. The drawings show the location and number of neurons immunolabeled for GABA and/or FOS protein half an hour after an intradermal injection of vehicle, B, or of capsaicin, C, into the foot of rats. Open circles indicate neurons with GABA-like immunoreactivity; squares neurons with FOS-like immunoreactivity; and filled circles neurons with both. (From Zou et al., 2001.)

Besides FOS expression, changes in a number of other immediate, early genes and transcription factors could also be used to investigate primary afferent or spinal neurons that are affected by peripheral stimulation or injury. Some of these include inducible transcription factors, such as c-Jun, JunB, JunD, Fos B, FRA-1, FRA-2, Krox-20, and Krox-24, and constitutive transcription factors, such as CREB, CREM, ATF-2, SRF, and NFκB (Munglani and Hunt, 1995; Doyle and Hunt, 1997; Herdegen and Leah, 1998; Wu, J. et al., 2002).

Summary

The electrophysiological activity of individual interneurons can be recorded extracellularly or intracellularly. The recordings can be done in vivo or in vitro. Each approach has its advantages and limitations. The activity of interneurons must be distinguished from that of primary afferent fibers and of projection cells and motor neurons. The response of interneurons to volleys in primary afferent Aβ, Aδ, and C fibers can be recognized, based on the latencies of the volleys following peripheral stimulation and the stimulus strengths required to evoke the volleys. Receptive fields can be mapped and the types of sensory receptors that activate or inhibit dorsal horn neurons can be evaluated in in vivo preparations and in special in vitro preparations.

Activity in populations of neurons in the dorsal horn can be demonstrated in several ways. For example, field potentials can be recorded electrophysiologically. Synchronous volleys of nerve impulses in primary afferent fibers, evoked by electrical stimulation of peripheral nerves, produce a series of potentials that can be recorded from the surface of the spinal cord or from within the spinal cord. Following the nerve volley are several slow potentials. The first waves are negative in sign when recorded from the dorsal surface of the cord or in the dorsal horn; these largely reflect the excitation of interneurons. Following the negative waves is a later, positive cord dorsum potential that reflects primary afferent depolarization (PAD), which is associated with presynaptic inhibition. These slow potentials reverse in sign more ventrally in the cord. Details of the time course and distribution of the slow potentials depend on the nature of the afferent volley (activity in large or small myelinated fibers versus unmyelinated fibers; activity in cutaneous, muscle, joint, or visceral nerves).

Activity in a population of dorsal horn neurons can also be visualized anatomically. One approach employs 2-deoxyglucose. This technique takes advantage of the fact that active neurons take up 2-deoxyglucose from the circulation, as if it were glucose. However, since neurons cannot metabolize 2-deoxyglucose, this substance accumulates in active neurons and can be demonstrated by radioautography. Activated neurons or their synaptic endings can be mapped with this method. Another approach is to immunostain for FOS protein or other immediate-early genes and transcription factors that may be expressed in neurons after strong stimulation. FOS can be expressed in inhibitory, as well as in excitatory, neurons.

AFFERENT PROJECTIONS TO THE DORSAL HORN

Laminae I–IV

The synaptic connections formed by different-sized primary afferent fibers supplying a variety of sensory receptors on neurons in the superficial dorsal horn are summarized in Fig. 7.16 (Cervero and Iggo, 1980).

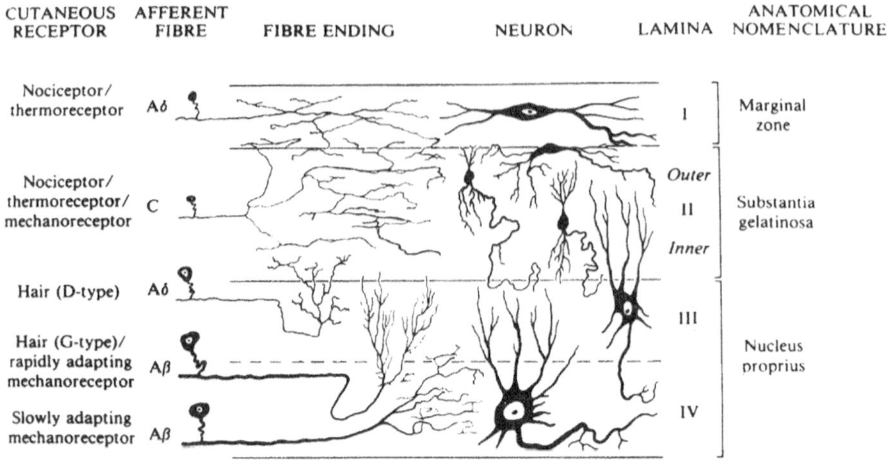

FIGURE 7.16. Schematic drawing of the afferent input to the upper layers of the dorsal horn. The types of afferent fibers are indicated at the left, and examples of their projections from different receptor types onto neurons that are typical of laminae I–IV are shown to the right. (From Cervero and Iggo, 1980; reproduced with permission.)

Afferent fibers of Aδ size from nociceptors and thermoreceptors terminate chiefly in lamina I on neurons that are mostly oriented longitudinally with respect to the spinal cord (cf. Light and Perl, 1979b). Fine muscle, joint, and visceral afferent fibers also project to lamina I (Craig and Mense, 1983; Craig et al., 1988; Sugiura et al., 1989). In addition to lamina I, some Aδ nociceptors also have projections to laminae V and X (Light and Perl, 1979b).

Unmyelinated cutaneous C-afferent fibers from nociceptors, thermoreceptors, and sensitive mechanoreceptors project mainly to lamina II. C-nociceptors and thermoreceptors also terminate in lamina I. The afferent terminals of C-fibers in lamina II are on limiting cells (stalked cells) at the border between laminae II and I or on central cells (islet cells) within the substantia gelatinosa (cf. Sugiura et al., 1986). The neurons of the substantia gelatinosa, like those of lamina I, are oriented longitudinally. Those in the outer part of the substantia gelatinosa are likely to receive inputs from C nociceptive or thermoreceptive afferent fibers, and those in the inner part receive inputs from C mechanoreceptors.

Information from lamina II can reach deeper layers of the dorsal horn by ventrally directed axonal projections (Light and Kavookjian, 1988). Neurons in laminae III and IV with dorsally directed dendrites can also receive direct input from primary afferent fibers that project to lamina II (Fig. 7.16; Todd, A. J., 1989; cf. Wall, 1965).

Aδ fibers from down hair follicles distribute to lamina III (the unmyelinated tips of these fibers may also end in the inner part of lamina II; Fig. 7.16). Other sensitive mechanoreceptors supplied by Aαβ fibers end in laminae III and IV. Neurons in lamina III tend to be oriented longitudinally, but those in lamina IV have a more dorso-ventral orientation (Scheibel and Scheibel, 1968).

Laminae V, VI, and X

Aδ fibers that supply mechanical nociceptors (Light and Perl, 1979b), as well as sensitive mechanoreceptors, project to lamina V, as do many group III and IV muscle, joint, and visceral afferent fibers (Craig and Mense, 1983; Craig et al., 1988; Sugiura et al., 1989). However, it is likely that cells in lamina V also receive input from other kinds of afferent fibers that terminate in the superficial layers of the dorsal horn, since many of these cells have dendrites that extend into more superficial layers (Ritz and Greenspan, 1985).

Proprioceptive afferent fibers project to lamina VI in the dorsal horn, as well as to the ventral horn (see review by Brown, A. G., 1981).

Aδ mechanical nociceptors and unmyelinated visceral afferent fibers also project to lamina X (Light and Perl, 1979b; Sugiura et al., 1989).

RESPONSES TO ELECTRICAL STIMULATION

Neurons in the various laminae of the dorsal horn show the monosynaptic responses that would be predicted by the connections from primary afferent fibers just described. However, they may also have polysynaptic responses that cannot be predicted based on these direct connections.

Lamina I

Neurons in lamina I are activated by volleys in cutaneous Aδ fibers and sometimes by volleys in C fibers as well (Christensen and Perl, 1970; Kumazawa et al., 1975; Cervero et al., 1976; 1979b; Light et al., 1979; Fitzgerald and Wall, 1980; Bennett et al., 1981; Woolf and Fitzgerald, 1983). There have also been reports of responses by units in lamina I to input from Aβ fibers (Christensen and Perl, 1970; Woolf and Fitzgerald, 1983). These may be polysynaptic responses. Recordings have been made from neurons in lamina I of the thoracic spinal cord that respond to cutaneous volleys and to volleys in visceral afferent fibers of the greater splanchnic nerve (Cervero and Tattersall, 1987; see also Alarcon and Cervero, 1990). Responses of lamina I neurons have also been recorded following stimulation of the inferior cardiac nerve (Takahasi, M. and Yokota, 1983).

Lamina II

There have been relatively few studies in which recordings have been made from neurons in the substantia gelatinosa in vivo. However, there are now a number of reports of such recordings in spinal cord slices.

Kumazawa and Perl (1976, 1977, and 1978) found that cells in the substantia gelatinosa are activated primarily by C fibers. However, Aδ fibers may also activate these neurons. Other investigators confirm that fibers of both A and C caliber can often excite neurons of lamina II (Wall et al., 1979a; Bennett et al., 1979; 1980; Fitzgerald, 1981). Similarly to the situation for cells in lamina I, the predominant A-fiber input to lamina II is from Aδ fibers, although there is some Aβ input as well. The latter may be polysynaptic.

Neurons that respond to electrical stimulation of fine visceral afferent fibers in the greater splanchnic nerve have been reported not to occur in lamina II (Alarcon and Cervero, 1990). However, M. Takahashi and Yokota (1983) found neurons in the outer part of lamina II that responded to stimulation of afferent fibers in the inferior cardiac nerve, as well as to stimulation of the skin.

Laminae III–VI

A population of cells can be found along the medial aspect of laminae III–VI of the lumbar spinal cord that are monosynaptically excited by stimulation of Aβ afferent fibers of the medial plantar nerve (Armett et al., 1961).

Mendell and Wall (1965) and Mendell (1966) found that 60% of the neurons of the nucleus proprius that they investigated could be activated by both A and C fibers, whereas the other 40% responded only to stimulation of A fibers. The cell bodies of such neurons appear to be intermingled throughout the dorsal horn (Gregor and Zimmermann, 1972; Menétrey et al., 1977). Volleys evoked by graded stimulus strengths often demonstrate a convergence of Aβ, Aδ, and C fibers upon the same neurons (Wagman and Price, 1969).

Repetitive stimulation of C fibers at rates of 1 per 3 sec or greater produces a progressive enhancement of the responses of many dorsal horn neurons, a phenomenon called "wind-up" (Fig. 7.17B) (Mendell and Wall, 1965; Mendell, 1966; Wagman and Price, 1969; Price and Wagman, 1970; Fitzgerald and Wall, 1980; Woolf and King, 1987; Sivillotti et al., 1993). The basis for wind-up is the very prolonged EPSP that is evoked by a C-fiber volley (Fig. 7.17A). The C-fiber evoked EPSP shows temporal summation when repetitive stimulation is used within a suitable range of rates (Fig. 7.17B). Other neurons

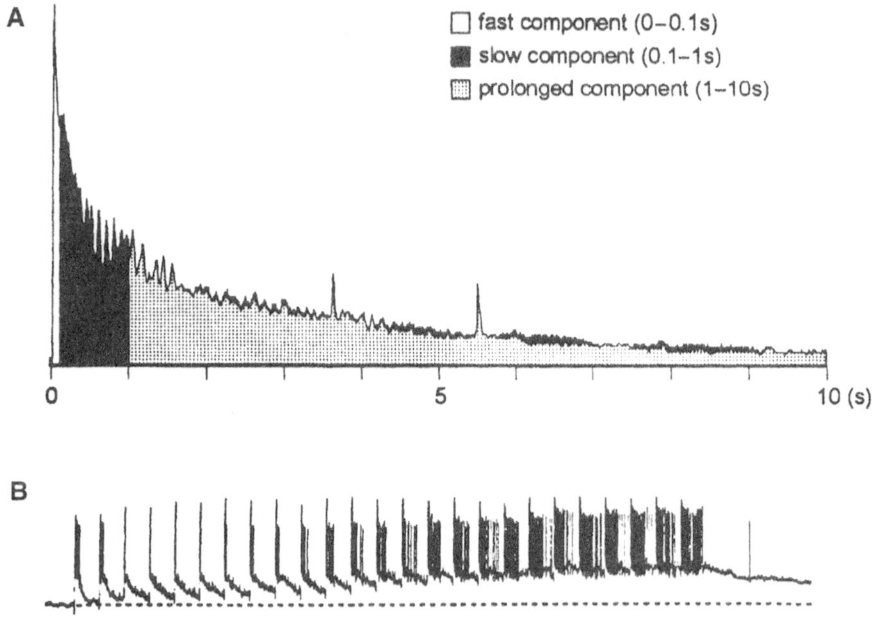

FIGURE 7.17. A, prolonged EPSP evoked by a C-fiber volley. The EPSP was recorded as an electrotonically propagated EPSP from a ventral root in an in vitro preparation of the spinal cord in response to a C-fiber volley in the dorsal root. Thus, the EPSP was generated in motor neurons, rather than in interneurons. However, similar potentials have been observed in interneurons. The initial part of the EPSP is white and is due to the action of glutamate on non-NMDA glutamate receptors; the black part of the EPSP is due to the action of glutamate or aspartate on NMDA receptors; and the shaded part is due to activation of neurokinin receptors. B, intracellular recording of wind-up or temporal summation of EPSPs evoked by repetitive C-fiber volleys and recorded in vitro from a large neuron in the ventral horn of the spinal cord. (A is from Nagy and Woolf, 1996; B is from Sivilotti et al., 1993.)

show a decreasing response to C-fiber volleys, "wind-down" (Fitzgerald and Wall, 1980; Woolf and King, 1987; Alarcon and Cervero, 1990).

In the thoracic and sacral spinal cord, cells in lamina V receive a convergent excitatory input from cutaneous afferent fibers and from visceral afferent fibers of the splanchnic nerve, sympathetic chain, inferior cardiac nerve, or pelvic and pudendal nerves (Pomeranz et al., 1968; Selzer and Spencer, 1969; Foreman, 1977; Foreman and Ohata, 1980; McMahon, S. B. and Morrison, 1982; Takahashi, M. and Yokota, 1983; Alarcon and Cervero, 1990). In the lumbosacral enlargement, comparable cells are activated by cutaneous afferent fibers (Pomeranz et al., 1968; Hancock et al., 1973; Gokin et al., 1977), by group III muscle afferent fibers (Pomeranz et al., 1968), and by joint afferents (Schaible et al., 1986).

Many neurons in the intermediate region (laminae VI and VII) receive monosynaptic excitatory connections from group I muscle afferent fibers (Eccles et al., 1956; 1960; Eccles, R. M., 1965; Lucas and Willis, 1974). The same interneurons may also be excited polysynaptically by cutaneous afferent fibers (Eccles et al., 1960; Eccles, R. M., 1965).

RESPONSES TO NATURAL STIMULATION

Lamina I

Christensen and Perl (1970) recorded responses to several kinds of natural stimuli from neurons of lamina I of the cat dorsal horn. One group of cells could be activated only by intense mechanical stimulation (Fig. 7.18A–C). A second group of cells was excited by intense mechanical stimulation of the skin and by noxious heat. A third group of cells could be activated by innocuous temperature changes. Most of these also responded to intense mechanical stimuli. More units were found that responded to cooling than to warming. Light et al. (1979) marked several lamina I neurons intracellularly in cats. The cells were nociceptive or thermoreceptive.

Kumazawa et al. (1975) extended the findings of Christensen and Perl (1970) to the monkey dorsal horn and also demonstrated that some lamina I cells projected contralaterally

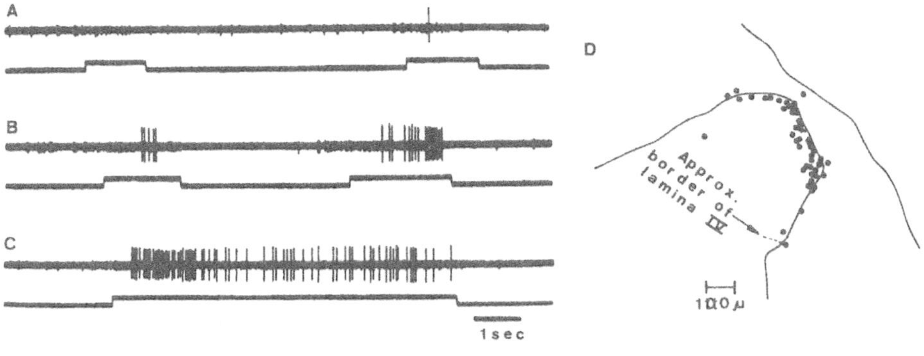

FIGURE 7.18. A, minimal response of a neuron in lamina I to stroking the skin with a glass rod. B, response to squeezing the skin with smooth-surfaced forceps. C, results of squeezing the skin with serrated forceps. D, location of lamina I neurons that responded to strong mechanical stimuli or to strong mechanical and noxious heat stimuli. (From Christensen and Perl, 1970.)

to at least the upper cervical cord. This observation is consistent with a contribution of lamina I to the spinothalamic, spinoreticular, and/or spinomesencephalic tracts (see Chapter 11). Many of the cells responded to noxious mechanical stimuli, but some were specifically activated by innocuous cooling or warming, Kumazawa and Perl (1978) subdivided their population of lamina I cells in the monkey into three categories: (1) cells activated only by intense mechanical stimulation; (2) cells excited by innocuous skin cooling; and (3) a few cells that discharged in response to noxious mechanical and thermal stimuli.

Cervero et al. (1976) confirmed that a high proportion of cells in lamina I of cats can be excited just by intense stimuli. However, some lamina I cells in their sample had a wide dynamic range of responsiveness, indicating that the population of neurons in this lamina is not homogeneous. Wide-dynamic-range cells could usually also be activated by group III and IV muscle afferent fibers.

Although there is general agreement that a large proportion of the neurons in lamina I are activated just by noxious intensities of stimulation, several groups besides that of Cervero et al. (1976) have recorded the responses of wide-dynamic-range neurons in this lamina in rats, cats, and monkeys (Menétrey et al., 1977; Price, D. D. et al., 1979; Bennett et al., 1981; McMahon, S. B. and Wall, 1983b; Woolf and Fitzgerald, 1983; Cervero and Tattersall, 1987; Ferrington et al., 1987d; Owens et al., 1992). Furthermore, some neurons in the marginal zone of the thoracic spinal cord are mechanoreceptive (Cervero and Tattersall, 1987). Several investigations have confirmed the observation of Christensen and Perl (1970) that there are thermoreceptive neurons in lamina I (Hellon and Misra, 1973; Iggo and Ramsey, 1976; Kanui, 1988; Andrew and Craig, 2001; Craig et al., 2001). Mitchell and Hellon (1977) were able to activate several neurons in lamina I of the rat spinal cord by applying noxious heat to the tail. It would be important to know whether these cells play a role in the tail-flick reflex.

Some lamina I neurons receive input from joint receptors (Schaible et al., 1986), and other lamina I neurons have input from visceral afferent fibers (Takahasi, M. and Yokota, 1983; Cervero and Tattersall, 1987; Ness and Gebhart, 1989; Alarcon and Cervero, 1990). Such cells generally also receive a convergent input from cutaneous afferent fibers.

Lamina II

Kumazawa and Perl (1976, 1977, 1978) recorded from a number of neurons at recording sites that were shown by dye marks to be in the substantia gelatinosa. None of the cells could be activated antidromically from the cervical cord (cf. Kumazawa et al., 1975). The neurons were excited either by noxious or by innocuous mechanical stimuli.

Wall et al. (1979a) recorded from neurons in the substantia gelatinosa that were either spontaneously active or that could be antidromically or orthodromically activated by stimulation of Lissauer's fasciculus. Many of the cells had small receptive fields, although other cells located deeper in the substantia gelatinosa had somewhat larger fields, which could encompass those of the more dorsal cells. Repeated stimuli often revealed habituation of the responses. Many cells showed a prolonged discharge, lasting seconds to minutes, following a single brush stimulus or, for other cells, following a noxious stimulus. The prolonged discharge was blocked by barbiturate anesthesia. Most cells responded to hair movement or other weak mechanical stimuli, but some cells required stronger stimuli, such as intense pressure or pinch, before they responded.

Cervero et al. (1977b; 1979a, c; see also the review by Cervero and Iggo, 1980) also used a volley in the isolated Lissauer tract as a search stimulus for neurons of the substantia gelatinosa in the cat spinal cord. Cells could be activated either antidromically or orthodromically by such a volley. One criterion that was used to identify neurons intrinsic to the substantia gelatinosa was that the responses had to differ from those known to characterize cells in laminae I and IV, so that recordings from dendrites of neurons in nearby laminae could be ruled out (cf. Wall, 1965). This restriction would of course eliminate from the sample any substantia gelatinosa neurons that behaved like the cells of neighboring laminae. Another restriction was that the extracellular spikes should show an inflection on the rising phase. This criterion argues against the identification of the units as axons (Fig. 7.19A). Neurons meeting these criteria had a high background discharge, were usually inhibited by cutaneous stimulation (Fig. 7.19D), and sometimes showed prolonged discharges that lasted seconds after brief periods of stimulation. The classification system used for the cells was inverse to that applied to most other dorsal horn neurons (see the section on **Classification of Dorsal Horn Neurons**). Inhibition was produced in some neurons by tactile stimuli (inverse class 1), in others by innocuous and noxious stimuli (inverse class 2), and still others by noxious stimuli (inverse class 3). Inverse classes 1 and 3 were excited by noxious and by innocuous stimuli, respectively. Molony et al. (1981) marked several lamina II interneurons intracellularly. These cells also belonged to the "inverse" categories.

Hentall (1977) found units in lamina II that were inhibited and then excited by weak mechanical stimuli. Habituation was often seen. Noxious stimuli could excite the cells, provided that background activity was present.

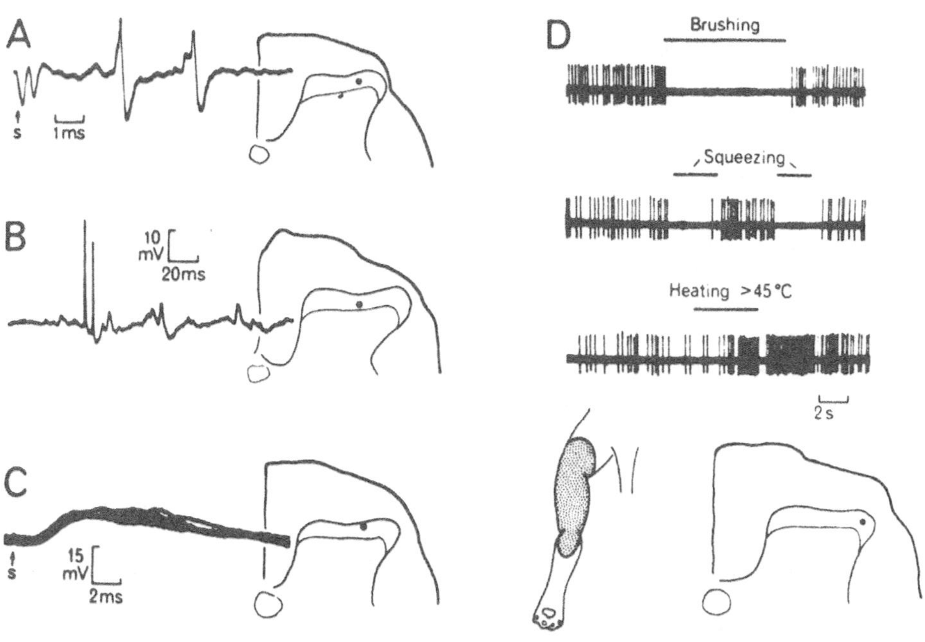

FIGURE 7.19. Response properties of neurons of the substantia gelatinosa. A–D, extra- and intracellular recordings from neurons of the substantia gelatinosa, along with the locations of the recording sites, and in D, a receptive field. The action potentials in A and the EPSP in C were evoked by stimulation of the tract of Lissauer. The activity in B was spontaneous. The responses in D include inhibition due to brushing and squeezing the skin, but excitation due to noxious heating. (From Cervero et al., 1977b.)

Laminae III–VI

Wall (1960) surveyed the responses of neurons in the neck and base of the dorsal horn of the cat to natural forms of stimulation. Cells were typically activated both by rapidly adapting tactile receptors and also by slowly adapting pressure receptors. Noxious stimuli added to the responses. Cells of this kind were later described as having a "wide dynamic range of response" (Mendell, 1966). Convergence of input from a number of types of sensory receptors upon dorsal horn neurons was confirmed by Wall and Cronly-Dillon (1960).

Wall (1967) later found that the response properties of neurons in different laminae could be distinguished. Cells in lamina IV received a convergent input from mechanoreceptors. Hair movement, touch, and stimulation of tactile domes could often excite them. Some cells, but not all, could also be excited by pressure and pinch. Thus, lamina IV was found to contain narrow- and wide-dynamic-range cells. There was no input from deep receptors. Neurons in lamina V responded to touch and pressure. Again, no input was noticed from deep receptors. Blocking descending projections in the spinal cord lowered the threshold for cutaneous stimulation and expanded the receptive fields. Neurons in lamina VI responded to both cutaneous stimulation and joint movement. Spinal cord block increased the excitability to cutaneous stimuli and reduced it to proprioceptive stimuli; thus, activity descending from the brain could act as a switch to shift the responsiveness of cells in lamina VI between cutaneous and proprioceptive inputs.

Other investigators have found neurons in laminae III and IV (and also deeper layers) that are activated just by mechanical stimuli applied to the skin. A convergent input from several types of receptor is typical, although neither invariable nor random (Brown, P. B., 1969; Heavner and DeJong, 1973; Tapper et al., 1973). Wide-dynamic-range cells have also been found in laminae IV–VI by various investigators, and several groups have found interneurons in laminae IV–VI, in addition to those in lamina I, that respond just to noxious stimuli (Kolmodin and Skoglund, 1960; Armett et al., 1962; Fetz, 1968; Pomeranz et al., 1968; Wagman and Price, 1969; Fields et al., 1970a; Besson et al., 1972; Gregor and Zimmermann, 1972; Heavner and DeJong, 1973; Tapper et al., 1973; Price, D. D. and Browe, 1973; Price, D. D. and Mayer, 1974; Handwerker et al., 1975; Menétrey et al., 1977; Egger et al., 1986; Sorkin et al., 1986; Cervero et al., 1988).

Spike-triggered averaging has been used to show that some interneurons in laminae V and VI with cutaneous input can evoke EPSPs or IPSPs in motor neurons and hence that these are premotor interneurons in reflex pathways (Hongo et al., 1989a, b).

Several groups have confirmed the presence of cells in the deepest part of the dorsal horn that receive a proprioceptive input (Kolmodin, 1957; Pomeranz et al., 1968; Heavner and DeJong, 1973; Fields et al., 1977; Menétrey et al., 1977; Jankowska and Riddell, 1994; Jankowska et al., 1981a). However, such neurons are not confined to lamina VI, but rather extend from laminae IV to VIII. Neurons with input from group II muscle afferent fibers have been reported to lie as far dorsally as lamina IV (Fukushima and Kato, 1975). Neurons that receive input from group I muscle afferent fibers are known to exert synaptic actions directly on motor neurons (Jankowska and Roberts, 1972a, b; Jankowska and Lindström, 1972; Czarkowska et al., 1981; Brink et al., 1983a). It is not clear whether the same neurons also play a sensory role. A number of studies have been done to define the spectrum of inputs to these interneurons (see, e.g., Hultborn et al., 1976a, b; Jankowska et al., 1981a; Brink et al., 1983b; Harrison and Jankowska, 1985a, b; Rudomin et al., 1987; 1990).

Cells that respond to visceral input have been found in laminae IV–V, as well as in lamina I and in the ventral horn at thoracic, lumbar, and sacral levels (Pomeranz et al., 1968;

Selzer and Spencer, 1969; Fields et al., 1970a, b; Hancock et al., 1973; Foreman, 1977; Guilbaud et al., 1977; Gokin et al., 1977; Foreman and Ohata, 1980; Price, D. D. et al., 1981; Cervero, 1982a; 1983; McMahon and Morrison, 1982; Takahashi, M. and Yokota, 1983; Cervero and Tattersall, 1985; Kanui, 1985; Ammons, 1986; Tattersall et al., 1986a; Knuepfer et al., 1988; Ness and Gebhart, 1987; 1988; 1989; Alarcon and Cervero, 1990). These cells also receive cutaneous input and can be classified as low-threshold, high-threshold, or wide-dynamic-range cells on this basis. The strength of interneuronal responses to visceral stimulation depends upon the activation of a supraspinal loop through the brain stem (Cervero and Wolstencroft, 1984; Cervero et al., 1985; Tattersall et al., 1986b).

Laird et al. (1996) recorded responses of dorsal horn neurons in the T12-L1 segments of the rat spinal cord to distension of the ureter. Neurons that showed responses to ureter input also responded to noxious somatic stimuli, often from bilateral receptive fields. In some cells, ureter distension affected the somatic receptive fields. Roza et al. (1998) did similar experiments in rats that had an experimental uteric stone implanted in the ureter (cf., Laird et al., 1997). A higher proportion of the neurons sampled had input from the ureter than in the previous study, and the somatic receptive fields could be to innocuous stimulation, suggesting the development of central sensitization.

Neurons in laminae IV–VI (as well as in lamina VIII) can often be excited by articular input (Schaible et al., 1986). These cells commonly have a convergent input from cutaneous and muscle receptors (Schaible et al., 1987a).

Lamina X

A population of neurons that respond to noxious cutaneous stimuli has been found in the vicinity of the central canal of rats (Nahin et al., 1983). The cells were of the high-threshold type and were excited by noxious mechanical and thermal stimuli.

Comparable cells have also been found in the cat (Honda and Perl, 1985; Honda, 1985). However, cells near the central canal of cats include both low-threshold and wide-dynamic-range cells, as well as high-threshold ones (Honda and Perl, 1985). Many of the cells have a convergent input from visceral afferent fibers, and a few respond only to visceral stimuli (Honda, 1985).

Ness and Gebhart (1987) recorded from neurons in the vicinity of the central canal in the L6-S1 segments of the spinal cord in rats. Most of the neurons were projection cells, since they could be activated antidromically from the contralateral white matter at a cervical level or from a supraspinal site, such as the ventrobasal thalamus or the midbrain. Many of the cells responded to noxious colorectal distension, either by excitation or inhibition. They commonly had somatic receptive fields in the perineal region, but many responded to movement of the tail. The latter were thought to be comparable to the primate spinothalamic tract cells described by Milne et al. (1982) in Stilling's nucleus.

Olivar et al. (2000) recorded from neurons in and near lamina X in rats. Many of the neurons responded to colorectal distension (CRD), as well as to noxious somatic stimulation. Colon inflammation by acetic acid resulted in enhanced responses to CRD, but not in changes in responses to somatic stimuli.

Laird et al. (2001c) have developed a behavioral model in mice of visceral-pain-related behaviors in response to colon inflammation with mustard oil or capsaicin. The model is expected to be useful for studies using transgenic animals.

Wall et al. (2002) did systematic tracks through the dorsal horn of the lumbosacral spinal cord and into lamina X in rats. They found that the cutaneous receptive fields of dorsal horn neurons were small and ipsilateral, as expected, but that the receptive fields expanded substantially when the microelectrode entered lamina X, where more than half of the cells had bilateral receptive fields.

CLASSIFICATION OF DORSAL HORN INTERNEURONS

Several attempts have been made to develop a system for classifying dorsal horn interneurons, but none of the approaches suggested to date is completely satisfactory. A major difficulty is that in electrophysiological studies the projections of interneurons are generally unknown, and so the only information available about the cells may be their locations in the spinal cord and their response properties. Another important limitation is that it is generally not known if the interneuron has an excitatory or an inhibitory action on the neurons with which it connects. Based on the results of a study of the responses of dorsal horn neurons to sudden changes in the intensity of a noxious mechanical stimulus, Laird and Cervero (1991) came to the conclusion that the role of the neuron in nociceptive transmission depends more on its projection target than on its input.

Monosynaptic Input

Eccles et al. (1960) described the responses of interneurons in the region of the intermediate nucleus of the spinal cord. The nomenclature they introduced was based on monosynaptic excitatory synaptic connections to the interneurons from group Ia, group Ib, or cutaneous afferent fibers. Interneurons receiving monosynaptic excitation from these different sources were termed type A, B, or C interneurons, respectively. The inference could be drawn that there were no other significant peripheral inputs to these neurons, yet many received strong polysynaptic effects from afferent fiber types other than those making monosynaptic connections. Furthermore, it is difficult to determine whether or not inputs from fine afferent fibers are in fact monosynaptic. It is not at all clear that monosynaptic connections reflect the most important connections for the normal operation of interneuronal pathway, although presumably monosynaptic pathways are subject to minimal interference by convergent inputs (except by means of presynaptic inhibition). Another objection to this nomenclature is that the C-type interneurons would include cells that receive an input from a wide variety of cutaneous receptors (Tapper et al., 1973); thus, classification of interneurons based on a generic cutaneous input would tend to obscure a diversity of functions. Eccles et al. (1960) did subdivide the C class of interneurons into C and CT, the latter being tract cells (presumed spinocervical tract cells).

PAD-related

The alphabetical list of interneuronal classes was expanded by Eccles et al. (1962a) to include a set of interneurons, which they named Type D, that had properties suited to the interneurons responsible for evoking PAD. Type D interneurons were not activated monosynaptically by primary afferent fibers, and so the taxonomy of these cells was based on other criteria, including convergence onto the interneurons of certain kinds of inputs and the time course of the discharges of these neurons in response to single and repetitive volleys in primary afferent fibers. The pattern of convergence has been used as a major criterion in a number of studies of the organization of interneurons in reflex pathways, including those responsible for PAD, and in descending control systems (Jankowska et al., 1981b; Rudomin et al., 1983; Brink et al., 1984; Solodkin et al., 1984; Edgley and Jankowska, 1987; Jankowska, 1992; Davies and Edgley, 1994; Jankowska and Riddell, 1995; Riddell et al., 1995; Quevedo et al., 1997).

Laminar Position

Wall (1967) found that interneurons in each of laminae IV–VI tend to have different properties. Because of this observation, an informal nomenclature was developed in relation to the response properties of the neurons. Interneurons that responded exclusively to tactile stimuli were called "lamina IV neurons" and those that responded to both tactile and noxious stimuli were called "lamina V" cells (see Hillman and Wall, 1969; Besson et al., 1975), irrespective of the actual location of the interneurons. This terminology is no longer used.

A or A Plus C Input

Gregor and Zimmermann (1972) determined whether dorsal horn interneurons were excited just by myelinated cutaneous afferent fibers or by both myelinated and unmyelinated afferent fibers. They also determined whether the excitatory connections from the A fibers were monosynaptic or polysynaptic. Interneurons could then be classified as activated by just A fibers or by A and C fibers, and then they could be subclassified according to whether the inputs were monosynaptic or polysynaptic. However, it is difficult to be sure whether or not inputs from C fibers are monosynaptic.

Bandwidth

Mendell (1966) introduced the scheme that is still the one that is most commonly used. This scheme takes into account the intensity "bandwidth" of stimuli that activate an interneuron (see Fig. 7.20A). Cells that respond just to innocuous mechanical stimuli were called "narrow-dynamic-range cells" or "low-threshold cells." Interneurons activated by both innocuous and noxious mechanical stimuli were labeled "wide-dynamic-range cells." Although "high-threshold cells" were not generally recognized in 1966, they were later accommodated in this scheme as another form of narrow dynamic range cell. Some investigators call these "nociceptive-specific" cells. However, this terminology assumes that wide-dynamic-range neurons do not specifically signal nociceptive events, an assumption that may not be warranted.

Price, D. D. and Browe (1973) and Price, D. D. and Mayer (1974) evaluated the responses of dorsal horn neurons in cats and monkeys to touch, pressure, and pinch stimuli. The cells were then subdivided into five classes, based on the "band-width" of their responses: (1) touch units; (2) touch-pressure units; (3) touch-pressure-pinch units; (4) pressure-pinch units; and (5) pinch units. The third of these classes is essentially

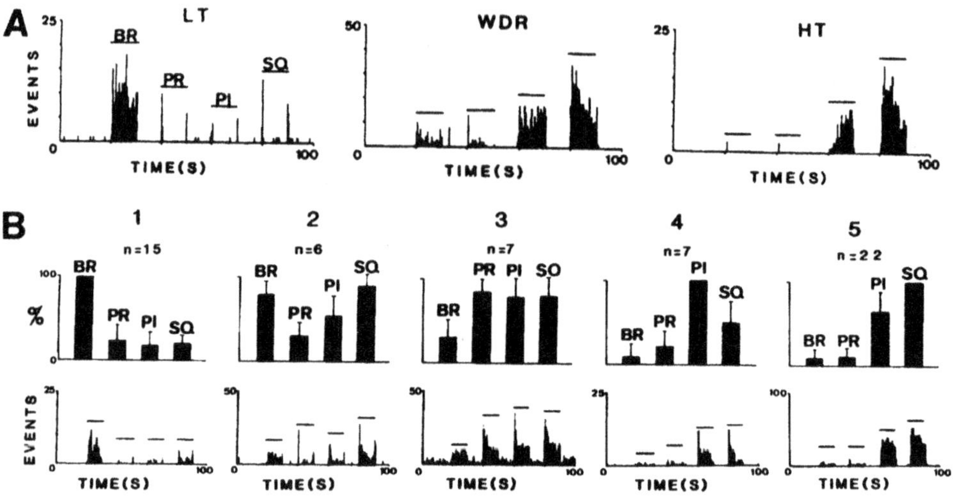

FIGURE 7.20. Classification of dorsal horn neurons. A, the upper row of peristimulus time histograms show responses typical of low-threshold (LT), wide-dynamic-range (WDR), and high-threshold (HT) dorsal horn neurons. The responses of these cells are to brushing (BR), pressure (PR), pinching (PI), and squeezing (SQ) the skin. The PR and PI stimuli were delivered with arterial clips with two different grip forces, and SQ with serrated forceps. The horizontal lines show when the stimuli were applied. B, the bar graphs in 1–5 (upper row) show the response profiles of the five classes of cervical cord interneurons determined by k means cluster analysis, using the same four mechanical stimuli. The histograms (lower row) show responses typical of interneurons in the cervical cord that were placed in the five cell classes. (From Sorkin et al., 1986.)

equivalent to the wide-dynamic-range category, whereas the other classes represent cells with narrow or intermediate dynamic ranges.

Heavner and DeJong (1973) pursued the classification problem by identifying in detail the types of stimuli that activate dorsal horn interneurons. Using the responses to a variety of stimuli, including hair movement, prick, pinch, pressure, vibration, and proprioception, they were able to identify 19 response classes. Tapper et al. (1973) examined the patterns of convergence of input from four different cutaneous receptor types that they could activate selectively and found cells that represent 16 combinations of these inputs.

Handwerker et al. (1975) suggested a simplified terminology. They proposed that cells excited just by large myelinated afferent fibers be called class 1 cells and that cells excited by large myelinated and unmyelinated fibers be designated class 2 cells. The effective receptor organs for class 1 cells could be either sensitive mechanoreceptors of the skin or proprioceptors. Class 2 cells are excited by a variety of sensitive mechanoreceptors, but also by nociceptors. Class 1 cells are equivalent to "low-threshold cells" and class 2 neurons to "wide-dynamic-range cells."

Cervero et al. (1976) expanded the terminology of Handwerker et al. (1975) to include another kind of cell, class 3. Class 3 neurons respond only to activity in fine afferent fibers and to noxious stimuli and are equivalent to "high-threshold cells." Two subtypes were recognized: class 3a cells, which receive input chiefly from Aδ mechanical nociceptors and generally do not respond to noxious heat stimuli, and class 3b cells, which are excited by Aδ and C nociceptors and are activated by noxious heat as well as by noxious mechanical stimuli. Group III and IV muscle afferent fibers also excite some class 3b cells.

Menétrey et al. (1977) elaborated on this scheme. They subdivided dorsal horn interneurons into four classes. Class 1 cells are activated by innocuous stimuli. They may be divided into two subclasses, 1A (activated by hair movement and/or touch only) and 1B (excited by hair movement and/or touch and pressure or pressure only). Class 2 cells are excited by innocuous and noxious stimuli. They can also be divided into two subclasses, 2A (activated by hair movement and/or touch, pressure, pinch, and/or pinprick) and 2B (excited by pressure, pinch, and/or pinprick). Class 3 cells respond only to noxious stimuli (pinch and/or pinprick). Class 4 cells respond just to joint movement or deep pressure.

Cluster Analysis

Another approach to the classification of dorsal horn neurons makes use of multivariate statistical analyses of the responses to several stimuli (Chung, J. M. et al., 1986a; Surmeier et al., 1986a, b; 1988; Sorkin et al., 1986; Owens et al., 1992). With this approach, the responses of interneurons of the cervical spinal cord of the cat to mechanical stimulation of the skin could be subdivided into five classes, based on a k means cluster analysis of the responses to four mechanical stimuli of progressively increasing intensity (Sorkin et al., 1986). The response profiles of these classes are shown in Fig. 7.20B and are compared with the response profiles of cells that were classified as low-threshold, wide-dynamic-range or high-threshold cells (Fig. 7.20A).

There are several difficulties with using cluster analysis for classifying spinal interneurons. A technical reservation is that the number of spatial dimensions needed for the analysis corresponds to the number of different inputs. Another is that a large number of neurons must be sampled to obtain a clear separation of neuronal types. However, once a large population is classified, new neurons can be placed with confidence into a suitable cluster, using discriminant analysis (Owens et al., 1992).

DORSAL HORN NEURONS IN UNANESTHETIZED, BEHAVING ANIMALS

The activity of interneurons is influenced by the presence or absence of anesthesia and also by the level of activity in pathways that descend from the brain. Before it is possible to relate particular classes of interneurons to behavioral events, it is necessary to evaluate the discharge patterns of such neurons in unanesthetized animals. However, the usual experimental preparations available that avoid anesthesia (decerebrate or spinalized animals) have the drawback that there is an abnormal set in the activity of the descending control systems. For example, decerebrate animals have an increase in the activity of the brain stem pathways that produce a tonic inhibition of flexion reflex pathways (see e.g., Eccles, R. M. and Lundberg, 1959), whereas spinalized animals completely lack such descending control.

One solution to this problem is to record from unanesthetized, behaving animals. Such an approach has been very successful in studies of the activity of neurons in the brain. Technical difficulties have slowed the application of this approach to the spinal cord. However, several groups have described methods that have been used successfully to record from spinal cord neurons in awake, behaving rats, cats, and monkeys (Wall et al., 1967; Courtney and Fetz, 1973; Bromberg and Fetz, 1977; Collins, J. G., 1985; 1987;

Sorkin et al., 1988). Initial studies have suggested that the activity of spinal cord interneurons in an awake, behaving animal is highly regulated by inhibitory pathways (Wall et al., 1967; Courtney and Fetz, 1973; Collins, J. G., 1983; 1987; Collins, J. G. and Ren, 1987; Sorkin et al., 1988). This modulation can be abolished by anesthesia (Collins, J. G. and Ren, 1987; Collins, J. G. et al., 1990).

Summary

The projection patterns of different types of primary afferent fibers influence the kinds of responses that are evoked in dorsal horn neurons in different laminae. Cutaneous Aδ nociceptors project to laminae I and V. Aδ thermoreceptors also terminate in lamina I. Cutaneous C fibers, whether mechanoreceptive, nociceptive, or thermoreceptive, end chiefly in lamina II, although some C fibers also terminate in lamina I. Cutaneous Aδ mechanoreceptors distribute mostly to lamina III, whereas Aβ fibers from other types of mechanoreceptors project to laminae III and IV. Proprioceptive afferent fibers from muscle end chiefly in laminae V, VI (as well as VII and IX). Visceral afferent fibers project to several laminae of the dorsal horn, as well as to lamina X.

Lamina I neurons often respond to volleys in cutaneous Aδ and sometimes also C fibers. Cutaneous Aβ volleys are occasionally effective, and, in the thoracic cord, visceral afferent fibers often activate lamina I cells. Natural stimuli that excite lamina I neurons may include noxious mechanical or thermal, innocuous thermal, or sometimes both innocuous and noxious mechanical cutaneous stimuli.

Lamina II neurons are excited by C fibers, but often also by Aδ and even by Aβ fibers. Effective natural stimuli include noxious mechanical and thermal, or innocuous mechanical cutaneous stimuli. Unusual responses (compared to responses of neurons in adjacent laminae) include marked habituation to repeated stimuli, prolonged discharges following brief stimuli, and predominantly inhibitory responses.

Neurons in laminae III–VI are excited by various combinations of cutaneous Aβ, Aδ, and C fibers. Repetitive C-fiber volleys at rates of about 1 Hz may result in incrementing responses ("wind-up"). Volleys in visceral or fine muscle afferent fibers often excite neurons of lamina V. Large muscle afferent fibers activate interneurons in lamina VI (and VII). Some of the neurons in laminae V and VI are premotor interneurons in reflex pathways, since they are directly connected to motor neurons; whether the same interneurons participate in sensory processing, as well, is not clear. Effective natural stimuli include innocuous and/or noxious mechanical stimulation of the skin (depending on whether the cell is of the low-threshold, wide-dynamic-range, or high-threshold class), visceral stimuli, and joint movement for cells in deep layers.

Neurons in lamina X respond to cutaneous inputs and many also have visceral inputs.

Interneurons can be classified according to their response properties. However, no completely satisfactory classification system has been devised. Experimenters interested in pain mechanisms often refer to neurons that are activated just by innocuous mechanical stimuli as low-threshold (class 1) cells, those excited both by innocuous and noxious stimuli as wide-dynamic-range (class 2) cells, and cells that discharge only if noxious stimuli are used as high-threshold (class 3) cells.

In a few studies, it has been possible to record the activity of dorsal horn interneurons in awake, behaving animals. The evidence obtained so far suggests that these neurons are strongly regulated by descending control systems.

PHARMACOLOGY OF DORSAL HORN INTERNEURONS

Excitatory amino acids are thought to produce the fast component of synaptic transmission between many primary afferent fibers and dorsal horn neurons (see reviews by Puil, 1981; Watkins and Evans, 1981; Salt and Hill, 1983; Mayer, M. L. and Westbrook, 1987; Evans, 1989; Dingledine et al., 1999). Another candidate fast excitatory transmitter in the dorsal horn is adenosine triphosphate (ATP; see review by Sawynok and Sweeney, 1989). One or more peptides that are also contained in primary afferents may serve as slow neurotransmitters or modulators (see Chapter 4). Dorsal horn interneurons may release excitatory neurotransmitters or modulators, including excitatory amino acids and a number of peptides, or inhibitory transmitters, such as inhibitory amino acids or inhibitory peptides. Interneurons may participate in local excitatory circuits or in multisynaptic circuits that result in activity that is transmitted to the brain by way of ascending sensory pathways. Other interneurons contribute to local inhibitory circuits or to inhibitory pathways formed by projections that descend from the brain.

A large number of studies have (1) demonstrated the presence of various neuroactive chemicals in synaptic terminals in the spinal cord, (2) shown that these substances are released following the activation of neural pathways, (3) demonstrated pharmacologic actions of these substances on spinal neurons, and (4) provided evidence that specific blocking agents prevent such actions. These studies go far towards meeting the criteria required to prove that a substance is a neurotransmitter (Werman, 1966). However, the evidence may not be complete for a given substance or for a particular synaptic connection.

Pharmacological Techniques

Pharmacological studies of spinal cord neural circuits can be accomplished by using electrophysiological techniques, such as extracellular, intracellular, or patch-clamp recordings, to assay the effects of a drug. Alternatively, pharmacological studies can depend on changes in behavioral responses to provide clues about drug actions on sensory phenomena. Routes of drug administration include (1) systemic, (2) superfusion, (3) intrathecal, (4) microdialysis, or (5) microiontophoresis. Each approach has its advantages and limitations.

Systemic Drug Administration

Drugs can be given systemically, but in this case the site of drug action is frequently unclear. Actions seen could be mediated very indirectly, for example through neural circuits that involve structures in the brain or in remote parts of the spinal cord. However, the effects of a drug given systemically offer an overview of how a dorsal horn interneuron will react to that drug, whatever the circuitry.

Superfusion of the Spinal Cord

Superfusion experiments require an open laminectomy, generally in an anesthetized or decerebrate preparation (Eccles et al., 1963b; Liu, X. G. and Sandkühler, 1998). The drug solution can be confined to a short length of the spinal cord by construction of a suitable barrier (Liu, X. G. and Sandkühler, 1998). This approach permits electrophysiological recordings in the part of the spinal cord on which the drug acts.

Intrathecal Administration

For intrathecal drug administration, a catheter is placed in the intrathecal space, usually through an opening in the meninges at an upper cervical level during brief anesthesia (Yaksh and Rudy, 1976). The catheter is threaded through the spinal canal and the tip of the catheter placed at the desired segmental level. After the animal recovers, drugs can be injected or infused through the catheter and their effects studied. Since there is no open laminectomy, studies can be done in awake, behaving animals. However, it is difficult to make electrophysiological recordings in such experiments. An intrathecal catheter also offers the possibility of removing cerebrospinal fluid to measure substances that are released from the spinal cord (Yaksh, 1984).

Microdialysis

Microdialysis involves diffusion of a drug or other substance across a dialysis membrane (Ungerstedt, 1984; Sorkin et al., 1991; 1992; see Robinson and Justice, 1991). A microdialysis fiber can be implanted during surgery under anesthesia and then left in place for several days after the animal has recovered from the anesthesia so that experiments can be done on an awake, behaving animal (Skilling et al., 1988; Sluka and Westlund, 1992). Alternatively, one or more microdialysis fibers can be implanted during an acute electrophysiological experiment on an anesthetized animal (Lin et al., 1997). The microdialysis fiber can be placed in the dorsal horn by drawing it across the spinal cord with the aid of an attached insect pin. Placement is assisted by marks that are made on the fiber. The extent of spread of a dye included in the dialysis fluid can provide an estimate of the region affected by a drug. For experiments on awake, behaving animals, the microdialysis fiber is drawn across the spinal cord through holes made in the pedicles of the lamina of a vertebra during surgery. Most of the dialysis membrane is sealed before the experiment, except for a short length that is to be exposed to nervous tissue. A solution containing the drug can be pumped through the fiber at a constant rate. The amount of drug that diffuses across the dialysis membrane depends on the concentration of the drug in the dialysate, the speed of the infusion, and the porosity of the dialysis membrane. The ability of a particular drug to diffuse across the dialysis membrane can be measured before the experiment. The reduction in drug concentration across the dialysis membrane is often in the order of 10-fold. A further reduction in drug concentration can be expected within nervous tissue because of the tortuous pathway through the extracellular space and also because of the presence of uptake mechanisms or enzymatic activity that can further reduce the concentration of the drug. Usually, only a rough estimate of drug concentration at the point of recording can be obtained. However, this is also true for the other in vivo techniques. Microdialysis can also be used to investigate the release of substances in the spinal cord (Sorkin et al., 1991; Robinson and Justice, 1991). A substance of interest, such as a neurotransmitter, can be measured in the dialysis fluid.

Microiontophoresis

Microiontophoresis is a technique that allows the release of a drug in an ionized form into the extracellular space near a cell whose activity is recorded (Curtis and Crawford, 1969; see Stone, 1985). Microiontophoresis has the advantage that the site of drug action is either on the neuron being studied or very nearby. A disadvantage is that the drug does not affect a large population of neurons, and so the responses of many individual neurons need to be examined. A different drug may be placed in each of several barrels of a multibarrelled microelectrode array, and recordings are typically made through a central barrel filled with electrolyte or that contains a carbon fiber (Armstrong-James and Millar, 1979; Millar and Pelling, 2001). Each drug barrel is connected to a channel in an electronic device that controls the passage of a predetermined amount and duration of current of a chosen polarity through that barrel. When no drug release is desired, a restraining current of the opposite sign can keep the drug within its barrel. The effective spread of a drug that is released iontophoretically is a sphere with a diameter in the order of several hundred μm (Stone, 1985).

Limitations of Electrophysiological Recordings in Drug Studies

Pharmacological studies are more difficult to do when using intracellular recordings than when recordings are made extracellularly because of the greater likelihood that the recording conditions will deteriorate before the experiment is complete. Small movements of the preparation can occur, because of physiological or other factors. For example, a drug given systemically may change the blood pressure, and this in turn can alter the relative positions of the recording electrode and the impaled neuron. However, some investigators have been successful in making intracellular recordings during microiontophoresis. The microelectrode be used for intracellular recording is often mounted with its tip in advance of the drug-containing electrodes (Stone, 1985). This arrangement allows the assessment of the effects of a drug on membrane properties, as well as on synaptic events.

In general, it is more straightforward to do pharmacological analyses in vitro, for example in spinal cord slices, than in vivo. The concentrations of drugs in the bathing solution are known, and the bathing solution can be changed rapidly when an experimental manipulation is performed. One problem in common with in vivo experiments is that the site of drug action is not known with certainty. Although the site of action is within the slice, the drug action may not be on the recorded neuron. This problem can be avoided if the drug can be included in the electrolyte solution that fills the microelectrode used to patch-clamp the neurons in the slice. A disadvantage of slice preparations is that the synaptic pathways from the periphery and the brain have usually been detached, and response properties may change because of disruption of the normal circuitry. On the other hand, projection neurons can be identified in experiments using in vitro preparations by labeling these neurons retrogradely, for instance with a fluorescent substance, before the experiment (Huang et al., 1985; Huang, 1987; Hori and Endo, 1992; Li, P. et al., 1999; Chéry et al., 2000).

Excitatory Transmitters in the Dorsal Horn

Excitatory Amino Acids: Glutamate and Aspartate

Dorsal root ganglion cells, dorsal root axons and the terminals of primary afferent fibers in the dorsal horn contain glutamate and aspartate (Graham et al., 1967; Duggan and Johnston, 1970; Wanaka et al., 1987; Battaglia and Rustioni, 1988; De Biasi and Rustioni, 1988; Westlund et al., 1989a, b; Kai-Kai and Howe, 1991; Merighi et al., 1991;

Tracey et al., 1991; Broman et al., 1993; Broman and Adahl, 1994; Valtschanoff et al., 1994). For example, Maxwell et al. (1993) have shown that the terminals of hair follicle afferents in the dorsal horn are enriched with glutamate-immunoreactivity. However, Larsson et al. (2001) provide evidence that, in contrast to glutamate, aspartate is not contained in synaptic vesicles in primary afferent terminals (Broman and Adahl, 1994). This observation argues against a specific synaptic release of aspartate from primary afferent fibers.

Although both of these excitatory amino acids are released into the dorsal horn after stimulation of primary afferents (see below), glutamate and aspartate are also found in dorsal horn interneurons (Rizzoli, 1968; Fagg and Foster, 1983; Cotman et al., 1987; Storm-Mathison and Otterson, 1987; Miller et al., 1988; Antal et al., 1991; Kai-Kai and Howe, 1991; Merighi et al., 1991). Therefore, the source of the excitatory amino acids that are released by stimulation of primary afferent fibers could include interneurons activated by the afferent input, as well as the terminals of primary afferent fibers.

Although glutamate is released from dorsal root ganglia (Roberts, P. J., 1974), dorsal rhizotomy does not change the concentration of glutamate in the dorsal horn (Roberts, P. J. and Keen, 1974). This may reflect the large glutamate stores that are present in dorsal horn neurons, which may be unaltered by dorsal rhizotomy.

For details about early work on excitatory amino acids and synaptic transmission in the dorsal horn, see Curtis and Watkins, 1960; Curtis, et al., 1959a; Graham, et al., 1967; Duggan and Johnston, 1970; Johnson, J. L. and Aprison, 1970; Johnson, J. L., 1972; Roberts, P. J., et al., 1973; Roberts, P. J., 1974; Roberts, P. J. and Keen, 1974; Zieglgänsberger and Puil, 1973; Duggan, 1974; Biscoe et al., 1976; McLennan and Lodge, 1979; Watkins and Evans, 1981; McLennan and Liu, 1982; Davies, J. and Watkins, 1983; McLennan, 1983; Peet, et al., 1983; Schneider, S. P. and Perl, 1985; 1988; Jessell, et al., 1986; Childs, et al., 1988; Magnuson, et al., 1988; Sillar and Roberts, 1988.

Release of Excitatory Amino Acids in the Spinal Cord

A number of studies using micorodialysis have shown that activation of primary afferents (especially of nociceptors) can result in an increase in the concentration of excitatory amino acids in the dorsal horn of the spinal cord in vivo (Skilling et al., 1988; Paleckova et al., 1992; Sluka and Westlund, 1992; Sorkin et al., 1992; Sorkin and McAdoo, 1993). However, it has not been established that the increase in excitatory amino acid concentration is due to an accumulation of the excitatory amino acids that are directly released from primary afferent terminals. Rather, the increase may be the result of interneuronal activity, since it is blocked by administration of the non-NMDA receptor antagonist, CNQX (Paleckova et al., 1992). Presumably, the action of CNQX is at least in part on postsynaptic non-NMDA receptors on interneurons (see below). Consistent with this interpretation is the finding that the release of the inhibitory amino acid, glycine, presumably from inhibitory interneurons, is also blocked by CNQX. On the other hand, glutamate receptor antagonists, such as CNQX, could be acting on glutamate receptors on primary afferent terminals (e.g., see Hwang et al., 2001b).

Glutamate and aspartate are also released from spinal cord slices in response to stimulation of primary afferent fibers or to exposure of the slice to SP, CGRP, capsaicin, or $K+$ (Kangrga et al., 1990; Kangrga and Randic, 1991; Ueda et al., 1993; 1994; Teoh et al., 1996b).

Excitatory Action of Glutamate and Aspartate

Curtis et al. (1959a; 1960) and Curtis and Watkins (1960) were the first investigators to demonstrate the excitatory actions of glutamate and aspartate (and other amino acids, such as homocysteic acid) on spinal interneurons, as well as on other neurons of the central nervous system. The technique that they employed was microiontophoretic release of these amino acids into the vicinity of the neurons. Many investigators have since observed that spinal cord interneurons, in vivo or in vitro, are generally, although not invariably, responsive to glutamate or its agonists, such as α-amino-3-hydroxy-5-methylisoxazoleproprionic acid (AMPA), kainate, and N-methyl-D-aspartate (NMDA) (Millar and Armstrong-James, 1982; Davies and Watkins, 1983; Jessell et al., 1986; Nässtrom et al., 1994; Miller and Woolf, 1996).

Schneider and Perl (1985; 1988), who surveyed interneurons in the dorsal horn in a slice preparation from the hamster spinal cord, found in their sample that only about 30% of the units were excited by iontophoretically or bath-applied glutamate or aspartate. Neurons that did respond to excitatory amino acids were concentrated in laminae I and II, whereas insensitive units were distributed throughout laminae I–V (see also Yoshimura and Jessell, 1990). These observations suggest that other substances besides excitatory amino acids may play a role in excitatory transmission within the spinal cord. Alternatively, uptake or breakdown of the excitatory amino acids or desensitization of receptors may have prevented a detectable action of the amino acids on dorsal horn neurons, or the bath-applied glutamate may have activated inhibitory interneurons, and inhibition may have interfered with the expected effects of glutamate (cf. Nässtrom et al., 1994).

Excitatory Amino Acid Receptors

Excitatory amino acid receptors are generally referred to as "glutamate receptors," even though the natural ligand for some may actually be aspartate or some other excitatory amino acid. Glutamate receptors can be subdivided into ionotropic and metabotropic glutamate receptors (these terms were introduced by McGeer et al., 1978; application of these terms to glutamate receptors is attributed to Sugiyama et al., 1987). The effects of ionotropic glutamate receptors are associated with the opening of ion channels, whereas the effects of metabotropic glutamate receptors depend on the activation of G-proteins and intracellular signaling pathways.

Ionotropic Glutamate Receptors

Ionotropic glutamate receptors are subdivided into three types (Dingledine et al., 1999), based on their excitatory responses to three selective agonists, AMPA, kainate, and NMDA. The AMPA and kainate receptors are often grouped together as "non-NMDA receptors." The general stucture of kainate/AMPA receptors is shown diagrammatically in Fig. 7.21A and that of NMDA receptors in Fig. 7.21B. Molecular cloning studies have determined that these different ionotropic glutamate receptors are composed of various combinations of subunits, depending on the particular location of the receptors and sometimes on the history of their functional activity (Hollman and Heinemann, 1994; Dingledine et al., 1999). AMPA receptor subunits include GluR1–4; kainate receptor subunits include GluR5–7, as well as KA-1 and KA-2; NMDA receptor subunits include NR1, NR2A–D, and NR3A.

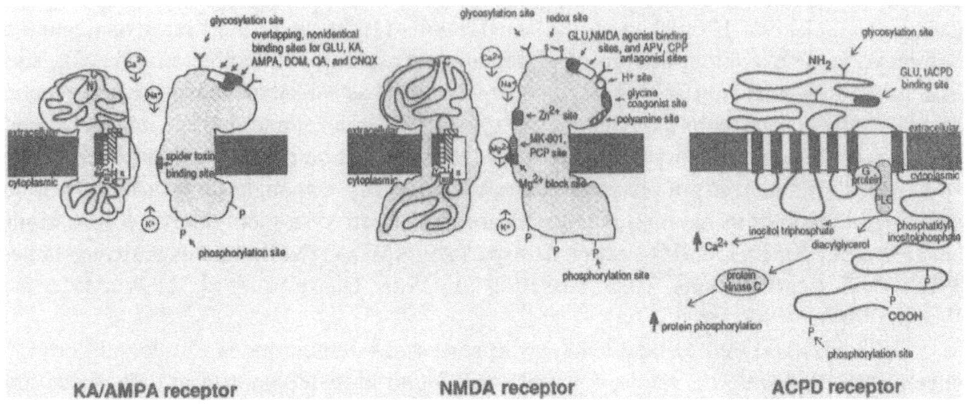

FIGURE 7.21. Diagrammatic representations of the types of glutamate receptors. A, kainate/AMPA receptor. B, NMDA receptor. C, metabotropic glutamate receptor. Sites for interactions with agonists, antagonists, and modulators are shown, as well as phosphorylation sites. (From Hollman and Heinemann, 1994.)

The distribution of binding sites for AMPA, kainate, and NMDA in the dorsal horn has been mapped, using autoradiography (Mitchell and Anderson, 1991; Albin et al., 1993; Henley et al., 1993). All three types of binding sites are in highest concentration in laminae I and II, although they are also distributed elsewhere in the gray matter (Mitchell and Anderson, 1991). Studies using in situ hybridization have shown that mRNA for GluR1–4 is found throughout the dorsal horn, but GluR2 is strongly expressed and GluR1 moderately expressed in laminae II and III (Furuyama et al., 1993). GluR5 and NR1 mRNAs are also found in dorsal horn neurons.

Immunostaining at the electron-microscopic level for AMPA receptor subtypes in the superficial dorsal horn has shown that GluR1 subunits are more likely to be in postsynaptic relationship to the glomerular terminals of C fibers and that GluR2–3 subunits are more likely to be in apposition to the glomerular endings of fine myelinated afferents (Popratiloff et al., 1996). Since AMPA receptors that lack GluR2 are calcium permeable (Hollman and Heinemann, 1994), this suggests that unmyelinated afferent fibers may be positioned to activate calcium-permeable AMPA channels on interneurons.

Spike et al. (1998) report that most (78%) interneurons in laminae I–III that were GluR1 immunoreactive contained GABA (and some also glycine), whereas most interneurons (96%) that were GluR2–3 immunoreactive lacked GABA or glycine. Therefore, C fibers are likely to activate inhibitory and some excitatory interneurons in the superficial dorsal horn by means of calcium-permeable AMPA receptors, whereas Aδ fibers presumably activate excitatory interneurons through AMPA receptors that lack calcium permeability. It will be of interest to determine if the influx of $Ca2^{+}$ through AMPA receptors lacking GluR2 subunits contributes to FOS expression in both inhibitory and excitatory interneurons of the superficial dorsal horn following the activation of C fibers (see the section on **Expression of Immediate, Early Genes**).

A possible implication of the higher concentration of GluR1 subunits of AMPA receptors on GABAergic inhibitory interneurons under pathological circumstances is an excitotoxic action that might result in cell death of inhibitory interneurons that normally help to limit nociceptive transmission from C fibers (Sugimoto et al., 1990; Coggeshall et al.,

2001). Loss of inhibitory interneurons could lead to a chronic pain state (see section on **Neuropathic Pain** in this chapter).

Popratiloff et al. (1998b) also examined the relationship between the same two types of glomerular endings, as well as non-glomerular endings, to postsynaptic sites that contained NR1 subunits of NMDA receptors. Although not so abundant as AMPA receptors, all three types of endings contacted NR1-immunoreactive postsynaptic sites.

Functions of Ionotropic Glutamate Receptors

Yoshimura and Jessell (1990) recorded intracellularly from substantia gelatinosa neurons in spinal cord slices from adult rats. They found that monosynaptic EPSPs evoked by Aδ and/or C primary afferent fibers were mediated by excitatory amino acids. Aδ monosynaptic EPSPs were produced mainly by AMPA/kainate receptors, as shown by the strong blocking action of CNQX, which is an antagonist of non-NMDA glutamate receptors (Fig. 7.22A, left). By contrast, AP5, an antagonist of NMDA glutamate receptors,

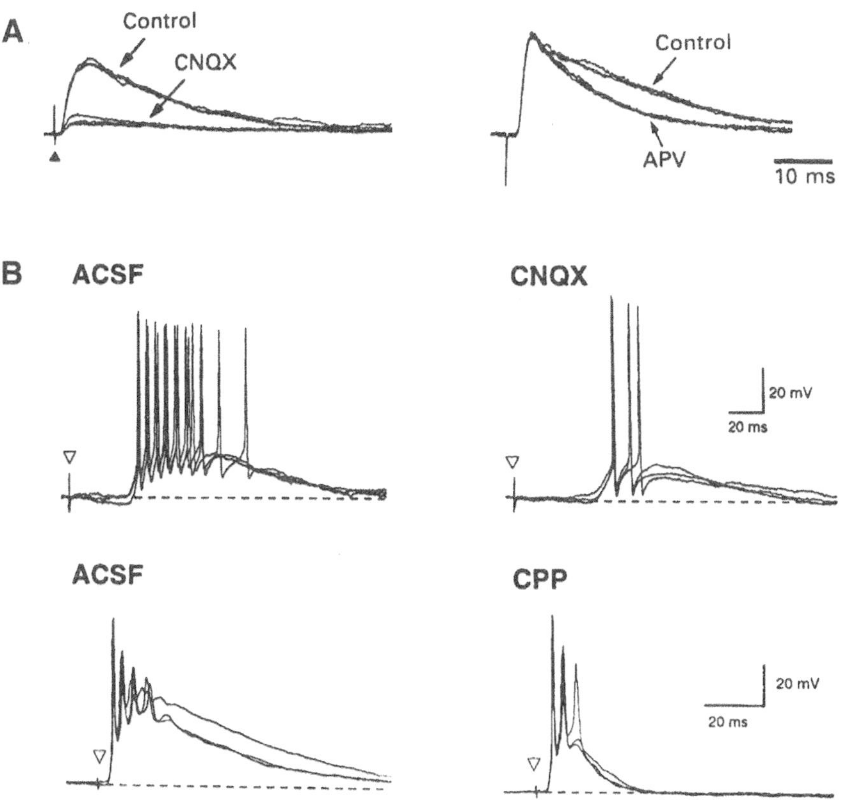

FIGURE 7.22. Effects of CNQX and AP5 on monosynaptic EPSPs recorded intracellularly in spinal cord slice preparations. A, CNQX (10 μM) reduced the monosynaptic EPSP evoked by stimulation of Aδ fibers in the dorsal root, whereas AP5 reduced only the late component of the EPSP in another neuron. (From Yoshimura and Jessell, 1990.) B, upper row, CNQX reduced the EPSP and action potentials evoked by stimulation of Aδ fibers in the dorsal root. Lower row, the NMDA receptor antagonist, CPP (50 μM), also reduced the EPSP, but its greatest effect was on the late component. (From Nässtrom et al., 1994.)

produced only a small reduction in the rate of rise and in the amplitude of Aδ evoked EPSPs in about half the neurons sampled. In almost all of the cells, AP5 reduced the late component of the monosynaptic EPSPs (Fig. 7.22A, right). AP5 also abolished polysynaptic EPSPs in many neurons. CNQX was less effective in reducing monosynaptic EPSPs evoked by C fibers. Bath application of glutamate caused a depolarization of only about half of the substantia gelatinosa neurons in the presence of tetrodotoxin. However, the same neurons generated EPSPs that could be blocked by CNQX, suggesting that glutamate did not affect the neurons when applied in the bath, perhaps because of uptake mechanisms in the slice or desensitization of glutamate receptors. Bath application of AMPA or kainate, for which there are no uptake systems, caused marked CNQX-sensitive depolarizations.

Nässtrom et al. (1994) obtained very similar results when recording intracellularly from neurons of the superficial dorsal horn (laminae I and II) in slices of hamster spinal cord. EPSPs and superimposed action potentials recorded from two neurons are shown in Fig. 7.22B. The non-NMDA glutamate receptor antagonist, CNQX, reduced both the early and late components of the EPSP evoked in one of the neurons by stimulation of Aδ and C fibers in the dorsal root, resulting in fewer discharges (Fig. 7.22B, upper row). An NMDA receptor antagonist, CPP, reduced chiefly the late part of the EPSP in the other neuron (Fig. 7.22B, lower row).

Yoshimura and Nishi (1993) investigated further the excitation of neurons in the substantia gelatinosa by primary afferent fibers, using blind patch-clamp recordings from thick slices of the adult rat spinal cord that had an attached dorsal root. Stimulation of Aβ fibers in the dorsal root failed to evoke short-latency EPSCs in substantia gelatinosa neurons, although Aβ fibers could evoke long and variable-latency, presumably polysynaptic, EPSCs in some neurons. Stimulation of Aδ fibers in the dorsal root evoked monosynaptic EPSCs in half of the neurons examined. These EPSCs were reduced by administration of the non-NMDA antagonist, CNQX. However, the NMDA antagonist, AP5, affected them only minimally. This difference from the results shown in Fig. 7.22 may have been the consequence of an effective Mg^{++} block of NMDA receptors in this patch-clamp study (Mayer et al., 1984; see Kerchner et al., 2001).

Miller and Woolf (1996) obtained similar results in blind patch-clamp recordings from neurons of the deep dorsal horn in horizontal slices of the neonatal rat spinal cord. Fig. 7.23A and B show the effects of AP5, CNQX, or both antagonists on the early and late components of the EPSCs evoked by stimulation of a dorsal root. AP5 had only a small effect on the monosynaptic EPSC, whereas CNQX reduced it markedly. On the other hand, AP5 was very effective in reducing the late components of the EPSC, and the combination of AP5 and CNQX essentially eliminated the EPSCs. Thus, although it has been suggested that polysynaptic EPSPs are mediated by NMDA receptors, this and other studies indicate that both non-NMDA and NMDA receptors contribute to polysynaptic excitation of dorsal horn neurons (Davies and Watkins, 1983; Peet et al., 1983; Schneider and Perl, 1985; Schouenborg and Sjölund, 1986; Morris, 1989; Gerber and Randic, 1989a, b; Dickenson and Sullivan, 1990; Nässtrom et al., 1994; Miller and Woolf, 1996; Yajiri et al., 1997).

An important issue is what glutamate receptors are utilized by neurons that have particular sensory functions. This question is presumably addressed best by using in vivo preparations in which natural forms of stimuli can be applied to receptive fields. In tests involving microiontophoretic application of several amino acid agonists, Aanonsen et al. (1990) observed that NMDA and AMPA preferentially activate nociceptive dorsal horn

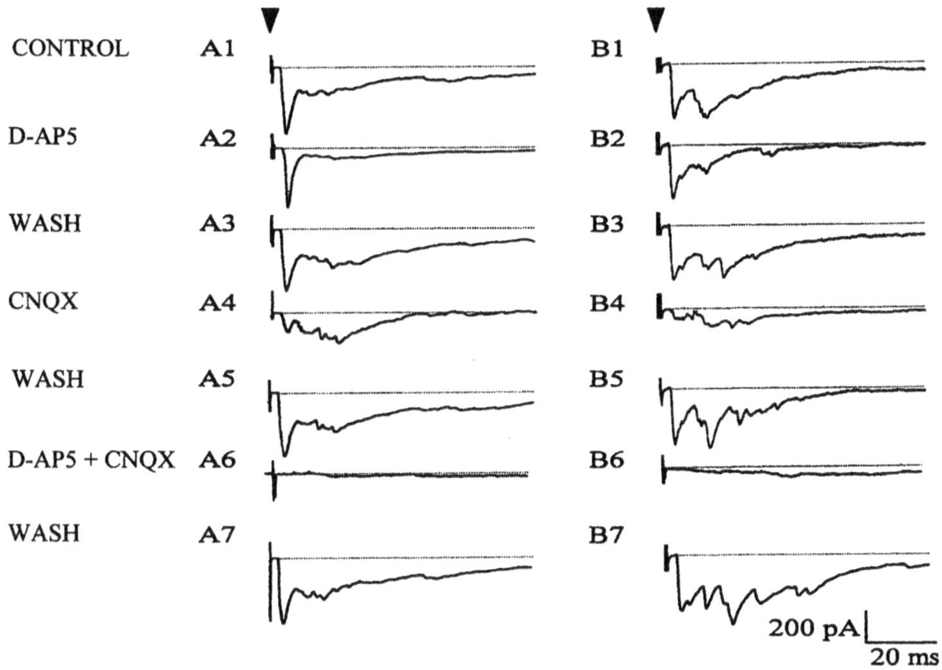

FIGURE 7.23. Effects of AP5 and CNQX on slow EPSCs recorded from two neurons in the deep dorsal horn in a spinal cord slice preparation using the blind patch-clamp technique. The effects of AP5, CNQX, or both on the polysynaptic EPSCs recorded from the neurons are shown in A1–7 and B1–7. (From Miller and Woolf, 1996.)

neurons, whereas kainate preferentially excites low threshold cells. Radhakrishnan and Henry (1993) observed that the responses of dorsal horn neurons to stimulation of hair follicle afferents in decerebrate cats were blocked by CNQX and by AP5. By contrast, responses to noxious heat were blocked by AP5 but not by CNQX. However, the presence of an anesthetic made an important difference, since in chloralose-anesthetized cats, responses to stimulation of hair follicle afferents were blocked by CNQX but not by AP5, and responses to noxious heat were unaffected by either antagonist. Li, P. et al. (1999) found that large afferents form synapses that utilize just AMPA receptors, whereas fine afferents have synapses that activate both AMPA and kainate receptors. Hopefully, the discrepencies in such experiments from different laboratories will be clarified as selective antagonists of AMPA and kainate receptors come into wider use.

Capsaicin increased the rate of spontaneous miniature EPSCs recorded in neurons of the substantia gelatinosa in spinal cord slices in which primary afferent activity was blocked by tetrodotoxin (Fig. 7.24; Yang et al., 1998; Yang and Li, 2001). The amplitudes of the miniature EPSCs were unaltered (Fig. 7.24B2), showing that the site of action of capsaicin was presynaptic. The spontaneous miniature EPSCs could be blocked by CNQX and so were presumably mediated by release of glutamate that acted on non-NMDA glutamate receptors. Capsaicin did not change the frequency of spontaneous IPSCs. This is understandable in the absence of primary afferent transmission, since the capsaicin would act directly on the terminals of capsaicin-sensitive primary afferents, but not on inhibitory interneurons.

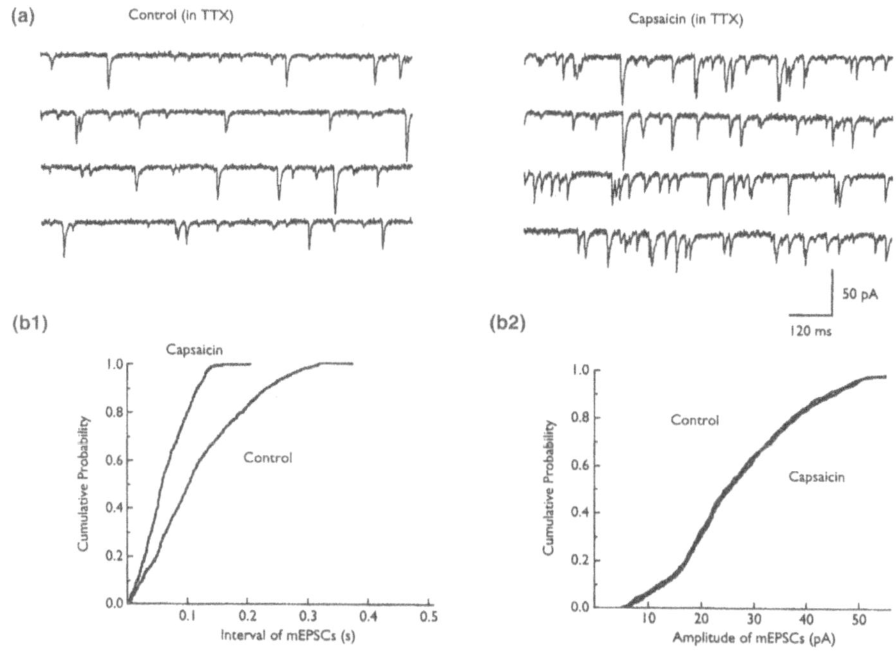

FIGURE 7.24. Effect of capsaicin on miniature EPSCs recorded from neurons of the substantia gelatinosa in a spinal cord slice preparation. The bathing medium contained tetrodotoxin (1 μM) to block action potential activity. (a), control spontaneous miniature EPSCs are shown at the left and the increase in frequency evoked by capsaicin at the right. The capsaicin concentration was 2 μM and it was applied for 60 s. (b1), cumulative frequency plots show the decrease in interevent intervals caused by capsaicin. (b2), the absence of an effect of capsaicin on the amplitudes of the miniature EPSCs. The concentration of capsacin used for b was 1 μM, and the capsaicin was applied for 60 s. Holding potential was −70 mV. (From Yang and Li, 2001.)

Metabotropic Glutamate Receptors

Metabotropic glutamate receptors can be divided into groups I, II, and III (see reviews by Nakanishi, 1992; Pin and Duvoisin, 1995; Conn and Pin, 1997; Schoepp et al., 1999; Schoepp, 2001). Molecular cloning studies have shown that the group I metabotropic glutamate receptors are composed of combinations of mGluR1 and mGluR5 subunits, Group II metabotropic glutamate receptors of mGluR2 and mGluR3 subunits, and group III receptors of mGuR4, 6, 7, and 8 subunits. Metabotropic receptors are coupled to several different second-messenger pathways and they can produce either excitatory or inhibitory modulation of transmission in neural circuits (Pin and Duvoisin, 1995; Saugstad et al., 1995; Conn and Pin, 1997). Group II and III metabotropic receptors activate second-messenger pathways that tend to oppose the actions initiated by group I metabotropic receptors. Fig. 7.21C shows the general structure of a metabotropic glutamate receptor of the type that responds to the non-selective agonist, ACPD.

Several groups have examined the expression of mRNAs for metabotropic glutamate receptors in the spinal cords of rats. Valerio et al. (1997b) found a high level of mRNA for mGluR1a but not mGluR1b. Expression was strong for mGluR5a and 5b. However, mRNAs for mGluR2–4 and 7 were at lower levels, and mGluR8 mRNA was not detected. Berthele et al. (1999) observed mGluR1 and mGluR5 mRNA in the superficial dorsal horn but did not detect mGluR2. The levels of mRNA for different receptors were a function of developmental age. The mRNA for mGluR7 has been found in presynaptic terminals of asymmetric synapses in laminae I and II (Ohishi et al., 1995a–c). Many of these terminals were shown to label for IB4 or substance P (Li, H. et al., 1997).

The distribution of metabotropic glutamate receptors in the spinal cord has also been investigated in immunohistochemical studies. These studies show that mGluR1α (Tang and Sim, 1999) and mGluR5a (Vidnyanszky et al., 1994; Jia et al., 1999; Alvarez et al., 2000) are chiefly found in laminae I–II, whereas there is a concentration of mGluR2/3 in the inner part of lamina II (Jia et al., 1999; Tang and Sim, 1999). However, Alvarez et al. (2000) found that mGluR1a is distributed mainly in the deep dorsal horn, and mGluR1b is found in scattered neurons, including a number in lamina X. Immunoreactivity for mGluR4 appears both at the light and the electron-microscopic levels to be presynaptic; small and medium-sized DRG cells, but not large ones, show immunoreactivity to this subunit (Azkue et al., 2001). At an ultrastructural level, mGluR1α, mGluR2/3 and mGluR5 have been found in dendrites (Vidnyanszky et al., 1994; Jia et al., 1999; Tang and Sim, 1999; see also Alvarez et al., 2000), and mGluR2/3 also in synaptic terminals (Jia et al., 1999; cf. Carlton et al., 2001c). The location of mGluR2/3 in axonal endings is well away from active zones (Azkue et al., 2000). Some of the terminals that contain metabotropic glutamate receptors also stain for GABA, suggesting the possibility that these metabotropic receptors may reduce GABA release, which would result in disinhibition (Jia et al., 1999). Neurons with dendrites that contain mGluR5 have also been observed to receive synaptic contacts from both presumed nociceptive primary afferent terminals and from GABAeregic interneuronal terminals (Tao et al., 2000b).

Functions of Metabotropic Glutamate Receptors

One effect of metabotropic glutamate receptors is to potentiate the action of ionotropic glutamate receptors (Aniksztejn et al., 1991; Bleakman et al., 1992; Cerne and Randić, 1992; Meller et al., 1993; 1996; Bond and Lodge, 1995; however, cf. Jones and Headley, 1995). For example, Meller et al., (1993; 1996) found that acute mechanical hyperalgesia can result from co-activation of AMPA and metabotropic glutamate receptors. Another action is to reduce the effects of inhibitory interneurons (Katsurabayashi et al., 2001).

Several studies have implicated metabotropic glutamate receptors in the following: (1) nociceptive transmission in the spinal cord of normal preparations (Budai and Larson, 1998; Fundytus et al., 1998; Dolan and Nolan, 2000); (2) synaptic plasticity (Young et al., 1998; J. Chen et al., 2000; Gerber et al., 2000; Zhong et al., 2000); (3) models of inflammation (Young et al., 1994; 1995; 1997; 1998; Neugebauer et al., 1994a; 1999; 2000; Fisher and Coderre, 1996a, b; Silva et al., 1997; Boxall et al., 1998; Stanfa and Dickenson, 1998; Karim et al., 2001; Yashpal et al., 2001; L. Zhang et al., 2002); (4) peripheral neuropathic pain (Fundytus et al., 1998; 2001; Yashpal et al., 2001); and (5) spinal cord injury pain (Abraham et al., 2001; Mills et al., 2001a, b; 2002).

ATP

ATP has been proposed as another molecule that serves as a fast primary afferent and interneuronal excitatory neurotransmitter in the dorsal horn (see reviews by Sawynok and Sweeney, 1989, and Salter et al., 1993). There is recent evidence for an action of ATP on P2X receptors on sensory ganglion cells belonging to nociceptors and to muscle stretch receptors in normal animals (Cook et al., 1997). Furthermore, purinergic receptors are upregulated after chronic inflammation (Xu, G. Y. and Huang, 2002).

Purinergic Receptors

ATP-gated ion channels are composed of P2X subunits. P2X3 subunits are nearly exclusively found on sensory neurons, specifically on small and medium-sized dorsal root ganglion cells (Vulchanova et al., 1998). Most of these DRG cells in rats can be eliminated by neonatal treatment with capsaicin. Almost all P2X-containing DRG cells label for IB4; only a few contain peptides. P2X3-receptor immunoreactivity is observed in the inner part of lamina II and this immunoreactivity is reduced following dorsal rhizotomy. Immunostained vanilloid (VR-1) receptors are co-localized with P2X3 purinoceptors and IB4 (Guo et al., 1999). Curiously, Guo et al. (1999) observed few peptidergic neurons that had VR-1 receptors. On the other hand, application of ATP to DRG cells causes depletion of SP, and in spinal cord slice preparations, ATP applications lead to NK-1 receptor internalization, suggesting that P2X receptors cause the release of SP from primary afferents (Nakatsuka et al., 2001; see the section on *SP Release* later in this chapter).

Functions of Purinergic Receptors

ATP has been found to depolarize some dorsal horn neurons in an in vitro preparation (Jahr and Jessell, 1983). Fyffe and Perl (1984) have shown that ATP excites neurons in laminae II and III that receive mechanoceptive or mechanoceptive and nociceptive input (low-threshold and wide-dynamic-range neurons), but not neurons that receive only nociceptive input (high-threshold neurons). It was proposed that ATP might mediate the central actions of mechanoceptive afferent fibers. Salter and Henry (1985) confirmed that most low-threshold neurons, but only a small fraction of wide-dynamic-range neurons, are excited by ATP, supporting the suggestion that ATP mediates input from mechanoceptors. On the other hand, ATP has been proposed as a transmitter in nociceptive pathways that synapse in lamina II (Bardoni et al., 1997).

Li, P. et al. (1998) tested if P2X receptors were responsible for fast transmitter action in neurons of the superficial dorsal horn. Using the whole cell patch-clamp technique in spinal cord slice preparations, they found that monosynaptic EPSCs were completely blocked by a combination of antagonists of non-NMDA and NMDA glutamate receptors, indicating the transmitter for the EPSCs was an excitatory amino acid. However, a P2X receptor antagonist caused a reduction in the EPSCs, suggesting a presynaptic action of P2X receptors. The antagonist did not have any antinociceptive action in behavioral tests.

Jo and Schlichter (1999) provide evidence that many dorsal horn neurons use ATP as a fast excitatory transmitter that acts on P2X receptors. ATP is not co-released with glutamate, but is co-released with GABA. Adenosine, formed from ATP, modulates GABAergic postsynaptic currents. Hugel and Schlichter (2000) examined neurons of laminae I–III in culture and found that ATP produced a fast cation current in these neurons, a postsynaptic effect. In addition, ATP increased the frequency of miniature GABAergic or glycinergic IPSCs and increased the size of evoked IPSCs in some of the neurons, although it had no effect on glutamatergic transmission. These actions were apparently mediated by P2X receptors.

Nakatsuka and Gu (2001) recorded from dorsal horn neurons of lamina V in spinal cord slices, using the patch-clamp technique. Activation of P2X receptors by alpha, beta-methyelene-ATP, which activates P2X receptors, caused an increase in the frequency of spontaneous EPSCs and miniature EPSCs, as well as an increase in the sizes of EPSCs evoked by dorsal root stimulation. A P2X-receptor antagonist blocked this effect. Conversely, Barclay et al. (2002) were able to reduce mechanical hyperalgesia in neuropathic rats and in rats that received an intraplantar injection of complete Freund's adjuvant by infusing an antisense oligonucleotide intrathecally to downregulate P2X3 receptors.

Excitatory Neuropeptides

Substances that are likely to serve as slow excitatory neurotransmitters or modulators in the spinal cord dorsal horn include neuropeptides (see reviews by Otsuka and Takahashi, 1977, and Emson, 1979). Unlike many fast neurotransmitters, peptides must be synthesized in the cell bodies of neurons and then transported to the axon terminals for storage and later release. Furthermore, there is no re-uptake mechanism in the synaptic endings, and so replenishment after release depends on new synthesis and transport of the peptide. On the other hand, only small quantities of peptides are needed since they are quite potent. Peptides may be released from primary afferent fibers, interneurons and/or pathways that descend from the brain.

The terminals of some small-diameter primary afferent fibers in the dorsal horn may contain one or more peptides, such as substance P (SP), calcitonin gene-related peptide (CGRP), somatostatin, or vasoactive intestinal polypeptide (VIP), as well as an excitatory amino acid neurotransmitter (Merighi et al., 1992; Ribeiro-da-Silva, 1995). In addition, many spinal cord interneurons contain peptides (Hunt, S. P. et al., 1981a; Schroder, 1983; Fuji et al., 1985; Shimogesawa et al., 1986; Leah and Menétrey, 1989; Ribeiro-da-Silva et al., 1991), as do the terminals of some of the pathways that descend from the brain (Johansson, O. et al., 1981). Peptides are likely to play a modulatory role by enhancing or inhibiting synaptic transmission through any of several mechanisms (Emson, 1979).

Substance P (SP)

SP is thought to be a slow neurotransmitter or modulator that is released by nociceptive primary afferent fibers (see review by Nicoll et al., 1980). SP has been demonstrated to occur within the dorsal horn not only in primary afferent neurons but also in interneurons (Takahashi and Otsuka, 1975; Ribeiro-da-Silva and Hökfelt, 2000).

Localization of SP. Immunohistochemical studies have demonstrated SP-like immunoreactivity in many small DRG cells, as well as in synaptic terminals in the dorsal horn, where it may be co-localized with glutamate (see e.g., Hökfelt et al., 1975b, c; 1976; Chan-Palay and Palay, 1977a, b; Pickel et al., 1977; Barber et al., 1979; Battaglia and Rustioni, 1988; De Biasi and Rustioni, 1988; see Chapter 4). Dorsal rhizotomy causes a marked decrease in the level of SP in the dorsal horn. Furthermore, SP accumulates distal to a ligature of a dorsal root, indicating that it is transported to the spinal cord from the DRG. SP-containing endings are especially concentrated in laminae I and II (DeLanerolle and LaMotte 1983). There is also a concentration of SP terminals in lamina X (LaMotte and Shapiro, 1991). In addition, SP is found in intrinsic neurons of the spinal cord (Tessler et al., 1980; 1981) and in the terminals of projections to the spinal cord ventral horn from the brain stem (Johansson, O. et al., 1981). The SP in descending projections can be co-localized with serotonin (Chan-Palay et al., 1978; Hökfelt et al., 1978; Johansson, O. et al., 1981).

SP is stored in dense-core synaptic vesicles (Cuello et al., 1977; Pickel et al., 1977; Plenderleith et al., 1990), although immunoreactivity is also found in association with other structures in the synaptic terminals (DeLanerolle and LaMotte, 1983).

It is not certain if primary afferent fibers that contain SP are necessarily all nociceptors (Leah et al., 1985a, b), although many clearly are (Lawson et al., 1997). Furthermore, it has been shown that nociceptive dorsal horn neurons receive SP-containing synapses, whereas non-nociceptive neurons do not (De Koninck et al., 1992). SP-containing endings tend to be on the cell body and dendrites of nociceptive-specific dorsal horn neurons, whereas they tend to be on the proximal and distal dendrites of wide-dynamic-range neurons (Ma et al., 1996).

One problem in studies that use immunohistochemistry to map the distribution of SP-containing neurons or processes is that antibodies to SP may cross-react with a related tachykinin, such as one of the other neurokinins (Kimura, S. et al., 1983; Kanazawa et al., 1984). An alternative approach to mapping neurons that synthesize SP is to use in situ hybridization to localize the mRNA for SP or other tachykinins. In rats, mRNA for SP is contained in DRG cells and in neurons of laminae I and II, the lateral spinal nucleus, and laminae V, VII, and X (Warden and Young, 1988). The concentration of mRNA for SP is very high in many DRG cells in rats and rabbits when compared with central neurons, despite low levels of SP as demonstrated by immunohistochemistry. This may reflect a high turnover of SP in DRG cells (Boemer et al., 1989).

Neurokinin Receptors. Receptors for SP and other tachykinins have been demonstrated in the spinal cord (Ninkovic et al., 1985). The distributions of three different tachykinin receptors have been mapped in the rat spinal cord (Yashpal et al., 1990). Tachykinin receptors appear to be located postsynaptically, since they are upregulated, rather than reduced, following dorsal rhizotomy (Yashpal et al., 1991a).

Neurons that contain NK-1 receptors (which respond preferentially to SP) are concentrated in laminae I, IV–VI, and X. Fewer neurons in laminae II and III have NK-1 receptors (Littlewood et al., 1995; Todd et al., 1998; cf. Zerari et al., 1995; Ding et al., 1999). In fact, very few neurons in lamina II either contain NK-1 receptors or respond to an NK-1 agonist (Bleazard et al., 1994). Neurons whose cell bodies are in laminae III or IV (including spinothalamic tract cells) often have dorsally directed dendrites that branch in the superficial dorsal horn; these dendrites typically possess NK-1 receptors, and they serve as important targets of SP-containing primary afferent terminals (Naim et al., 1997). Few NK-1 receptor positive neurons contain GABA or glycine, and so most NK-1-immunoreactive neurons in the dorsal horn are likely to be excitatory neurons (Littlewood et al., 1995).

In the superficial dorsal horn, NK1 receptors are associated chiefly with dendrites, whereas in the deep layers of the dorsal horn they are associated primarily with cell bodies (Moussaoui et al., 1992). Commonly, an SP-containing synaptic ending is apposed to NK-1 receptors on the membrane of an adjacent dendrite (McLeod et al., 1998; see also McLeod et al., 1999). NK1 receptors are often co-localized with NMDA receptors on second-order neurons in lamina I (Aicher et al., 1997; Yu, X. H. et al., 1999; Benoliel et al., 2000). In addition, NMDA receptors are found on the presynaptic terminals of some primary afferent fibers in the dorsal horn (Liu, H. et al., 1997). The geometric arrangement of SP-containing terminals, NK-1 receptors and glutamate receptors may help account for the synergic effects of SP and NMDA (Randic, M. et al., 1990; 1995; Dougherty and Willis, 1991b; Mjellem-Joly et al., 1991; Budai et al., 1992a; Rusin et al., 1992; 1993a, b; Dougherty et al., 1993; Song and Zhao, 1994; Cumberbatch et al., 1995). NK-1 receptors are found on projection neurons of the rat dorsal horn, including spinothalamic, spinoreticular, and spinoparabrachial neurons (Marshall et al., 1996; Yu, X. H. et al., 1999; Todd et al., 2000).

Heppenstall and Fleetwood-Walker (1997a, b) have suggested that the release of glycine in the dorsal horn by substance P (Smullin et al., 1990) may help account for the observation that NK-1 receptor antagonists are ineffective against transient noxious stimuli but do suppress responses to persistent noxious stimuli. For example, strychnine administration enabled an NK-1 antagonist that had previously been ineffective in blocking the responses of dorsal horn neurons to mustard oil application to prevent the responses (Heppenstall and Fleetwood-Walker, 1997b). Thus, postsynaptic inhibition by glycine suppresses the pronociceptive activity of NK-1 receptors in response to the acute effects of mustard oil on the skin. However, glycine can bind not only to strychnine-sensitive glycine receptor sites but also to a site on NMDA receptors that facilitates the responses of these receptors (Fig. 7.21B; Dickenson and Aydar, 1991; Budai et al., 1992b; see also Heppenstall and Fleetwood-Walker, 1997a). These very different glycine-binding sites should not be confused. During prolonged noxious stimulation, the action of the glycine site on NMDA receptors would help produce a pronociceptive action, especially if the NMDA receptors become sensitized during the prolonged noxious stimuli. Heppenstall and Fleetwood-Walker (1997b) hypothesize that during brief noxious stimulation, glycine inhibition masks the excitatory effect of NK-1 receptors, whereas during prolonged noxious stimulation, sensitization of NMDA receptors by PKC allows glycine acting on NMDA receptors to enhance nociceptive responses.

SP Release. SP is released from the spinal cord following stimulation of small myelinated and unmyelinated primary afferent fibers, noxious mechanical, thermal, or chemical stimulation of the skin, or administration of capsaicin (Fig. 7.25; Angelucci,

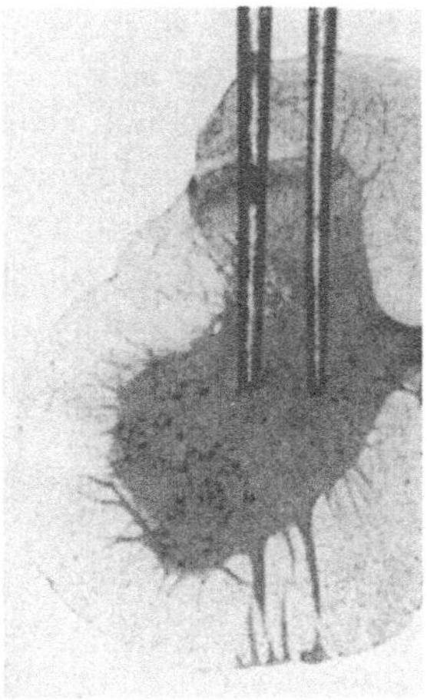

FIGURE 7.25. Release of SP from the spinal cord. Two antibody microprobes are shown in relation to a transverse section of the cat spinal cord. The microprobe at the right shows no release of SP, since no stimulus was given before the microprobe was removed from the cord. The microprobe at the left has two dark areas that did not bind the radioactive SP because SP released in the dorsal horn occupied the antibody sites in this zone. One dark region shows that SP had been released in the region of lamina II in response to noxious stimulation. The other dark region suggests an accumulation of SP at the surface of the spinal cord. Calibration bar = 1 mm. (From Duggan et al., 1988b.)

1956; Otsuka and Konishi, 1976; Gamse et al., 1979; Bergstrom et al., 1983; Brodin et al., 1987; Go and Yaksh, 1987; Duggan and Hendry, 1986; Duggan et al., 1987; 1988b; McCarson and Goldstein, 1991; Teoh et al., 1996b; Zachariou et al., 1997). SP release by capsaicin is in part dependent on an action of NMDA receptors (Afrah et al., 2001). SP can also be released by stimulation in the brain stem (Takano et al., 1984). Since SP and serotonin are co-localized in the terminals of many raphe-spinal axons, in fact in the same dense-core vesicles (Pelletier et al., 1977), it is instructive that SP can be released independently of serotonin (Bergstrom et al., 1983).

NK-1 receptor expression increases following an injection of formalin into the rat hindpaw (McCarson and Krause, 1996). Administration of an NK-1 receptor antagonist blocks this increase in mRNA for NK-1 receptors. This indicates that the activation of NK-1 receptors can result in positive feedback involving the release of SP and upregulation of NK-1 receptors. Similarly, NK1-1 receptors are upregulated in postsynaptic dorsal column neurons following inflammation of the colon by mustard oil (Palecek et al., 2002).

Duggan and his colleagues have introduced the use of antibody microprobes, which can determine the sites of release of peptides, such as SP, with considerable spatial resolution (Fig. 7.25; Duggan and Hendry, 1986; Duggan et al., 1987; 1988a, b, c; 1995). The antibody microprobe is a glass micropipette coated with antibody molecules. When an SP antibody microprobe is introduced into the spinal cord, SP released in the vicinity of the microprobe binds to SP antibodies on the microprobe. The microprobe is withdrawn from the spinal cord and reacted with radiolabeled SP. The distribution of the unlabeled zone to which neurally released SP has bound is determined by autoradiography and related to the layers of the dorsal horn that were adjacent to the microprobe during SP release. A major finding is that noxious stimuli cause the release of SP into the substantia gelatinosa (Fig. 7.25; Zhao et al., 1992). SP is also released in the dorsal horn in response to movements of the knee joint after the development of an acute experimental arthritis (Schaible et al., 1990). Microinjection of peptidase inhibitors into the dorsal horn result in an increased spread of released SP to include all of the dorsal horn and dorsal funiculus (Duggan et al., 1992), indicating that peptide spread is normally restricted by peptidases.

An indirect method for demonstrating the release of SP is by demonstrating reduced SP binding in the spinal cord after noxious stimuli (Yashpal et al., 1994). Another is by visualizing the internalization of SP receptors in histological sections. For example, noxious heating and cooling of the rat hindpaw results in the internalization of SP receptors in the dorsal horn (Allen et al., 1997; 1999). The internalization of NK-1 receptors is blocked by an NK-1 receptor antagonist (as expected) and also by AP5, but not by CNQX, suggesting that the release of SP is regulated by NMDA but not by non-NMDA glutamate receptors (Marvizon et al., 1997). Activation of $GABA_B$ receptors (but not of $GABA_A$ or glycine receptors) prevents SP release, as demonstrated by the effect of $GABA_B$ agents on NK-1 receptor internalization (Marvizon et al., 1999a).

Mantyh et al. (1997) have reported a clever antinociceptive use of the internalization of NK-1 receptors. SP was conjugated with saporin, a cytotoxin. Intrathecal administration of the SP-saporin conjugate led to internalization not only of NK-1 receptors but also the conjugate, leading to the death of exposed neurons in the dorsal horn that contained NK-1 receptors. Many neurons that contained NK-1 receptors were lost in lamina I, but fewer neurons in laminae III–V and X were lost, presumably because the conjugate did not diffuse sufficiently far ventrally (cf. Fig. 7.26A and B). However, there was some reduction in NK-1 receptor staining in the deep layers of the dorsal horn (Fig. 7.26B). In fact, the proportion of such neurons that were lost was comparable to the proportion lost in lamina

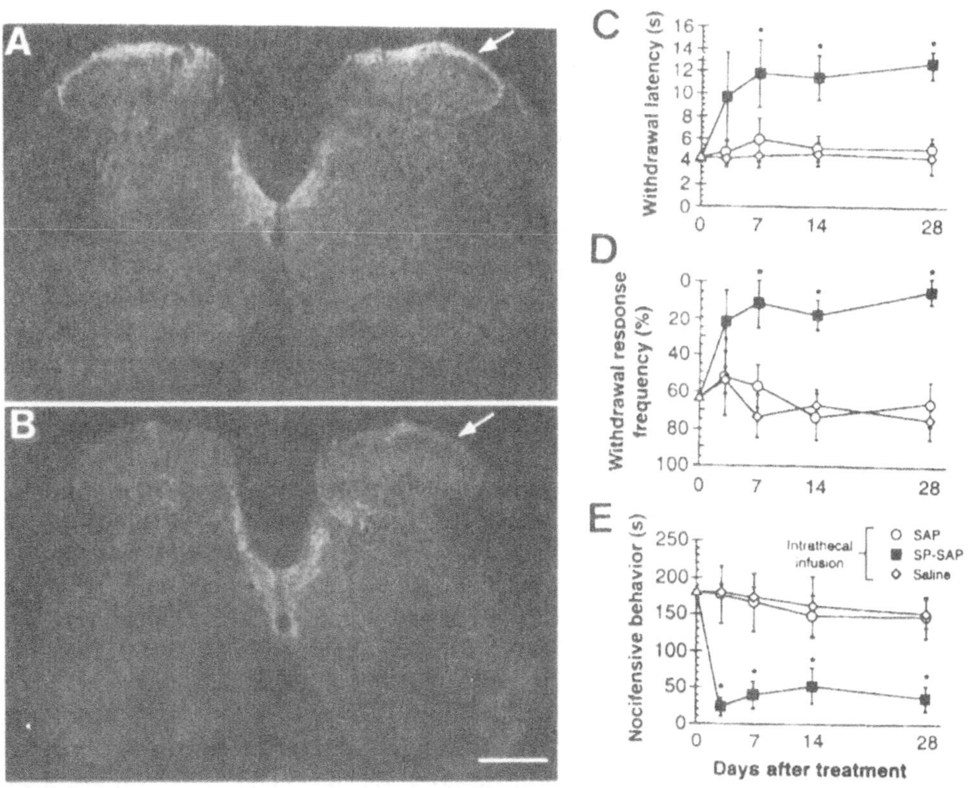

FIGURE 7.26. Effects of cytotoxic action of a SP-saporin conjugate that was internalized following interaction with NK-1 receptors on dorsal horn neurons. A, distribution of immunoreactivity for the NK-1 receptor in a transverse section of the spinal cord. B, loss of immunoreactivity after treatment with SP conjugated to a cytotoxin, saporin. C, nocifensive behavior in response to noxious heat stimuli was reduced after intrathecal treatment with SP-saporin, as shown by a prolongation of the paw withdrawal latency (filled squares). D, the frequency of paw withdrawals from noxious mechanical stimuli was reduced after SP-saporin treatment (filled squares). E, frequency of responses to an intraplantar injection of capsaicin was reduced after treatment with SP-saporin (filled squares). (From Mantyh et al., 1997.) (A color version of this figure appears between pages 322 and 323.)

I (Khasabov et al., 2002). The loss of neurons with NK-1 receptors prevented the primary mechanical and thermal hyperalgesia that is expected following injection of capsaicin into the plantar surface of the rat hindpaw, although the responses to noxious mechanical and thermal stimuli were normal prior to the capsaicin injection (Fig. 7.26C–E). The behavioral changes were attributed to the effect of the saporin on neurons of lamina I that contain NK-1 receptors (Mantyh et al., 1997). However, it is appropriate to consider the possibility that the loss of neurons in the deep dorsal horn and near the central canal might also have played a role, especially considering the high concentration of spinothalamic tract cells in this region of the rat spinal cord (Burstein et al., 1990; see Chapter 9) and the fact that many of these have NK-1 receptors (Marshall et al., 1996).

Functions of SP: Studies at the Cellular Level. When SP is iontophoresed into the vicinity of nociceptive dorsal horn neurons, there is a prolonged excitation, often lasting several minutes (Henry et al., 1975; Henry, 1976; Randic and Miletic, 1977; Sastry, 1979b; Zieglgänsberger and Tulloch, 1979a; Davies and Dray, 1980; Miletic and Randic, 1982).

High doses of SP can sometimes depress the responses of dorsal horn neurons to glutamate (Henry et al., 1975; Davies and Dray, 1980). SP receptor (NK-1) antagonists have been reported to block the action of SP (Urban and Randic, 1984; King et al., 1997a).

Murase et al. (1982) examined the responses of dorsal horn neurons to SP using intracellular recordings in a spinal cord slice preparation. SP caused a depolarization of the neurons. The depolarization had a slow time course and it was often associated with an increase in membrane resistance. A similar slow depolarization could be evoked in about half of the dorsal horn neurons following repetitive stimulation of a dorsal root (Urban and Randic, 1984). Adding SP or a SP antagonist to the bathing solution prevented the depolarization produced by dorsal root stimulation, suggesting that the depolarization was due to the action of SP (Urban and Randic, 1984). Similarly, antibodies to SP blocked the depolarizations produced either by SP or by dorsal root stimulation (Randic et al., 1986).

The mechanism of action of SP has been investigated. One proposal is that SP inhibits the M-current in dorsal horn neurons (Hösli et al., 1981; Nowak and MacDonald, 1982), as it does in sympathetic ganglion cells (Adams et al., 1983). The M-current is a voltage-sensitive K^+ current, and so inhibition of this current would cause a depolarization accompanied by an increased membrane resistance. Another proposal is that SP blocks inward rectification (Stanfield et al., 1985). Alternatively, SP may cause a depolarization through an increase in Na^+ or Ca^{++} conductance (Murase and Randic, 1984). Voltage clamp experiments on dorsal horn neurons in slice preparations suggest that the application of SP augments an inward Ca^{++} current and decreases the M-current (Murase et al., 1986).

MacDermott and her colleagues (Womack et al., 1988) found that application of SP to cells isolated from the dorsal horn cause an increase in cytoplasmic Ca^{++} concentration. Since this effect was independent of extracellular Ca^{++}, the change was due to mobilization of Ca^{++} from intracellular stores. However, Heath et al. (1994) found that these cells were glial and that SP caused an increase in Ca^{++} influx across the cell membrane in isolated neurons. The increase in intracellular Ca^{++} concentration produced by SP application can be localized within a dendrite (Marvizon et al., 1998).

Murase et al. (1989a) have confirmed that SP enhances a voltage-dependent inward Ca^{++} current. Other conductances were also affected, including an increase in the voltage-gated K^+ conductance and the Ca^{++}-activated K^+ conductance, as well as a calcium-sensitive, voltage-independent inward current carried by Na^+ and K^+ ions. There was no change in the inward (anomalous) rectifier current. Neurokinin A had similar complex actions. These multiple effects can account for both the depolarization and variable alterations in membrane resistance that are produced when SP is applied to dorsal horn neurons. According to Murase et al. (1989a), the conductance changes evoked by SP are likely to be mediated by the phosphoinositide second-messenger system (cf. Mantyh et al., 1984b). Second messengers that would be produced following binding of SP to its receptor are inositol 1,4,5-triphosphate and diacylglycerol. Inositol 1,4,5-triphosphate causes release of Ca^{++} from intracellular stores, and diacylglycerol activates protein kinase C, which in turn phosphorylates ion channels, including voltage-dependent Ca^{++} channels (Nowycky et al., 1985).

The amplitude of the EPSP evoked by stimulation of a dorsal root in a spinal cord slice preparation was unaffected by an NK-1 antagonist, but the duration of the EPSP was reduced (King et al., 1997a).

Badie-Mahdavi et al. (2001) recorded nociceptive responses to dorsal root stimulation and also examined FOS expression in isolated spinal cords. The prolonged dorsal root-evoked EPSP was reduced by administration of an NK-1 receptor antagonist, a PKA inhibitor, or a PKC inhibitor. An NK-1 receptor agonist was then used to excite dorsal horn neurons. An NK-1 receptor antagonist, as well as PKA and PKC inhibitors, all reduced this excitation. The effects of the different drugs were not additive, suggesting a convergence of signal transduction pathways onto a common target. The NK-1 receptor antagonist and the PKA and PKC inhibitors also attenuated FOS expression in the dorsal horn following either dorsal root stimulation or administration of an NK-1 receptor agonist (in the presence of tetrodotoxin to prevent indirect actions of the drugs). The effects of the inhibitors on FOS expression were most prominent for neurons in the deep dorsal horn.

Functions of SP: Studies at the Systems Level. Many behavioral and electrophysiological studies, including several that involved the use of transgenic mice, suggest an important role for SP in pain.

SP has been reported to exert presynaptic actions (Randic et al., 1982). Electrical pulses were used to test the excitability of primary afferent axons of Aβ and C fiber caliber within the spinal cord by activating the fibers antidromically. Changes in excitability were often found for both Aβ and C fibers, but the changes could be either reductions or increases in excitability, or there could be biphasic changes. There was no change in SP release when the spinal cord was blocked more rostrally by cold, indicating that tonic descending inhibition (see Chapter 12) is not mediated by presynaptic inhibition of SP release (Duggan et al., 1988c).

Kellstein et al. (1990) have found that superfusion of the spinal cord with SP has little effect on the background activity or responses of dorsal horn neurons to A-fiber volleys. However, the responses to C-fiber volleys are enhanced. The changes in the responses to C-fiber volleys can be prevented with a SP antagonist. Salter and Henry (1991) found that iontophoretic application of SP can activate most nociceptive dorsal horn neurons, whereas non-nociceptive neurons are unaffected by SP.

Antagonists of NK-1 receptors reduce the responses of dorsal horn neurons to C-fiber volleys and to noxious cutaneous stimuli (Kellstein et al., 1990; De Koninck and Henry, 1991; Radhakrishnan and Henry, 1991; 1995; Toda and Hayashi, 1993; Neugebauer et al., 1994c; Radhakrishnan et al., 1998). Superfusion of the spinal cord with a solution containing SP causes enhancement of the background activity and enlargement of the receptive fields of wide-dynamic-range neurons in the superficial dorsal horn, as well as increased responses of the neurons to noxious mechanical stimulation of the skin (Liu and Sandkühler, 1995a). Responses to noxious heat or to C-fiber volleys can be either increased or decreased. Intrathecal co-administration of a high dose of SP with NMDA results in an increase in responses to volleys in Aδ fibers and wind-up in wide-dynamic-range dorsal horn neurons (Chapman et al., 1994). Correspondingly, co-administration of high doses of SP and NMDA enhances nociceptive caudally directed licking and biting behavior, although a low dose of SP is antinociceptive (Masuyama et al., 1996; Masuyama and Shimizu, 1997). The responses of wide-dynamic-range dorsal horn neurons to C-fiber volleys are reduced by administration of an NK-1 receptor antagonist (Budai and Larsson, 1996).

Systemic administration of resiniferotoxin, a vanilloid receptor (VR-1) agonist, to rats causes a decrease in mRNA for SP in dorsal root ganglia but no change in mRNA for NK-1 receptors (Szallasi et al., 1999). A reduced synthesis of SP was proposed as a mechanism for resiniferotoxin-mediated analgesia.

Mice with a deletion of the preprotachykinin A gene have intact responses to mildly noxious stimuli but show a reduction in responses to moderate or intense noxious stimuli and an almost complete absence of neurogenic inflammation (Cao et al., 1998). Similarly, mice with a deletion of the gene responsible for the production of NK-1 receptors are able to respond to acutely painful stimuli (Laird et al., 2001b). However, they lack intensity coding for pain and also wind-up, and they do not show a full response to stress (De Felipe et al., 1998). Weng et al. (2001) obtained comparable results by using NK-1 receptor knock-out mice. Nociceptive neurons in the deep dorsal horn of the knock-out mice had normal responses to mechanical, thermal, and electrical stimuli and to the acute effects of mustard oil. However, wind-up to repetitive C-fiber volleys was decreased, as was central sensitization following mustard oil application. NK-1 receptor knock-out mice show reduced responses to noxious visceral stimuli, both acutely and in animals with visceral inflammation (Laird et al., 2000).

Other Neurokinins: Neurokinin A (NKA) and Neurokinin B (NKB)

Localization of NKA and NKB. In addition to SP, neurokinin A (NKA) is found in small DRG cells (Bakhle and Bell, 1995) and in the spinal cord dorsal horn (McLeod et al., 2000). Neurokinin B (NKB) is found in synaptic terminals and dendrites in laminae I–III, with its highest density in inner lamina II (McLeod et al., 2000). NKB-containing profiles are independent of the SP-containing elements. The mRNA for NKB is chiefly in neurons of lamina III, and it is also present in some neurons of laminae I, II, and X, but not in DRG cells. The presence of mRNA for SP and the lack of mRNA for NKB in the DRG is consistent with the observation that the concentration of SP, but not of NKB, in the spinal cord is reduced by dorsal rhizotomy (Ogawa et al., 1985).

Release of NKA. The antibody microprobe technique has shown that noxious stimuli cause the release of neurokinin A in the spinal cord (Duggan et al., 1990; Hope et al., 1990a, b). However, there are major differences in the release of SP and of NKA. There is a substantial basal level of released NKA, but not of SP, in the absence of stimulation. SP is released focally in the superficial dorsal horn, whereas NKA is released diffusely in the cord gray matter. SP remains in the extracellular space for only a brief period, whereas NKA persists for at least 30 min after stimulation.

NK-2 and NK-3 Receptors. NK-2 receptors (which are particularly responsive to neurokinin A) are apparently not found in the dorsal horn (Trafton et al., 2001). NK-3 receptors (which have their greatest affinity for neurokinin B and eledoisin) are found in the most superficial part of the dorsal horn. The mRNA for NK-3 receptors is also found in the dorsal horn (Shughrue et al., 1996).

Whereas SP selectively activates NK-1 receptors, NKA is usually thought to have a preferred action on NK-2 receptors and NKB on NK-3 receptors. However, NKA can also activate NK-1 receptors (Radhakrishnan and Henry, 1997). In fact, Trafton et al. (2001) showed that NKA is as powerful as SP in causing the internalization of NK-1 receptors, and an NK-1 receptor antagonist can block this action. Curiously, the antagonist that they used does not block the effects of SP on internalization of NK-1 receptors.

Functions of NKA and NKB. In behavioral experiments, intrathecal injections of NKA in mice resulted in nocifensive activity, including hindlimb scratching, licking, and biting, and NKA decreased the tail-flick latency (Gamse and Saria, 1986). Iontophoretic application of NKA was found to activate both nociceptive and non-nociceptive dorsal horn neurons, unlike SP, which only excited nociceptive neurons (Salter and Henry, 1991).

The effects of NKA include a facilitation of thermal nociceptive responses in behavioral tests, and NK-2 receptor antagonists block such responses (Fleetwood-Walker, et al., 1990; 1992). Consistent with this, an NK-2 receptor antagonist blocked the responses of dorsal horn neurons to noxious heat but not to mechanical stimuli (Fleetwood-Walker et al., 1993). An NK-2 receptor antagonist was also able to inhibit the activity of dorsal horn neurons generated in response to peripheral application of mustard oil, whereas NK-1 receptor antagonists were nearly ineffective (Munro, F. E. et al., 1993). Furthermore, NK-2 but not NK-1 antagonists were able to prevent an increase in expression of mRNA for preprodynorphin that is expected after intraplantar carrageenan injection (Parker et al., 1993). NKA also enhances the responses of dorsal horn neurons to noxious heat (Cumberbatch et al., 1995). Radhakrishnan and Henry (1997) found that the responses of dorsal horn neurons to innocuous mechanical or noxious mechanical and thermal stimuli were unaffected by an NK-2 receptor antagonist. However, King et al. (1997b) observed a reduction in the duration of the dorsal root-evoked EPSP in a slice preparation when an NK-2 antagonist was administered. Li, P. and Zhuo (2001b) provide evidence that both SP and NKA mediate slow EPSCs in dorsal horn neurons recorded using the patch-clamp technique.

Neugebauer et al. (1996a) found that an NK-2 receptor antagonist reduced the effects of NKA and also the responses of spinal cord neurons to pressure applied to the knee joint. During acute arthritis, the NK-2 antagonist produced an even more powerful reduction of the responses to noxious and innocuous pressure than in normal animals. It was concluded that NKA plays a role in pain arising from normal and arthritic joints.

The processing of visceral nociceptive responses also depends on the activation of NK-1 and NK-2 receptors, as shown by the actions of neurokinin receptor agonists and antagonists (Julia et al., 1994; Okano et al., 2002). A role of NK-1 receptors in acute and inflammatory visceral pain has been confirmed in knock-out mice lacking NK-1 receptors (Laird et al., 2000). An NK-2 receptor antagonist was found to prevent the enhancement of responses of neurons in the lumbosacral spinal cord to colorectal distension after inflammation of the colon with acetic acid but did not affect the responses to distension of the normal colon (Laird et al., 2001a).

Calcitonin Gene-related Peptide (CGRP)

Localization of CGRP. CGRP is a peptide that is found in many small DRG cells, as well as in some larger ones (Rosenfeld et al., 1983; Skofitsch and Jacobowitz, 1985a; Yokokawa et al., 1986; Ju et al., 1987b; Molander et al., 1987). The CGRP found in the dorsal horn is thought to be restricted to the terminals of primary afferent fibers (Chung, K. et al., 1988).

Frequently, CGRP is co-localized with SP (Wiesenfeld-Hallin et al., 1984; Gibbins et al., 1985; 1987b; Skofitsch and Jacobowitz, 1985c; Merighi et al., 1991; Ribeiro-da-Silva, 1995). In fact, CGRP and SP can be found in the same dense-core vesicles (Gulbenkian et al., 1986; Merighi et al., 1988; Plenderleith et al., 1990), and as many as four different peptides can be found in the same DRG cells (Leah et al., 1985b; Cameron et al., 1988). CGRP is also often co-localized in the same neurons that express the nerve growth factor receptor, trkA (Fried et al., 1990). Carr et al. (1990) have reported that capsaicin treatment causes a substantial depletion of CGRP in the dorsal horn and that 30% of small and 44% of medium-sized CGRP-containing DRG neurons stain positively for fluoride-resistant acid phosphatase (cf. Guo, A. et al., 1999).

CGRP is found in Aδ and C fibers in the dorsal roots and in the tract of Lissauer, as well as in the terminals of primary afferent fibers in the dorsal horn in several species

(Gibson et al., 1984b; Carlton et al., 1987; Harmann et al., 1988b; McNeill, D. L. et al., 1988a, b; Traub et al., 1990; Hayes and Carlton, 1992). Few CGRP-containing terminals remain in the dorsal horn after bilateral multiple dorsal rhizotomies and transection of the spinal cord (Chung, K. et al., 1988), suggesting that the only source of CGRP in the lumbar dorsal horn is DRG cells. The greatest concentration of CGRP-containing terminals is in laminae I–II and V, although there are a few CGRP-containing fibers in other laminae (Carlton et al., 1987; McNeill, D. L. et al., 1988a; Harmann et al., 1988b; Traub et al., 1989b; Yashpal et al., 1992; Chung, K. et al., 1993a). CGRP-containing afferents supply not only somatic structures, but also viscera (Sharkey et al., 1989; Chung, K. et al., 1993a). CGRP-containing primary afferents can project as many as five segments from the dorsal root through which they enter the spinal cord, and, while most end ipsilaterally, some terminate on the contralateral side (Traub et al., 1989b). Some CGRP-containing terminals are presynaptic to other terminals, including other CGRP-containing endings (Carlton et al., 1988).

CGRP or mRNA for CGRP has also been observed in dorsal root ganglia and the dorsal horn of the human spinal cord, as well as in neurons of the ventral horn (Gibson et al., 1984b; 1988; Harmann et al., 1988b; Jakab et al., 1990).

Unlike SP, CGRP is not normally observed in interneurons of the dorsal horn (Gibson et al., 1984b; Kawai et al., 1985; Skofitsch and Jacobowitz, 1985b; Traub et al., 1989a, b; however, see Conrath et al., 1989) or in axons descending from the brain. The mRNA for CGRP is found in DRG cells and motor neurons, but normally not in interneurons of the dorsal horn (Gibson et al., 1988). CGRP levels are reduced dramatically by dorsal rhizotomy, especially if multiple, bilateral rhizotomies are performed (Gibson et al., 1984b; Chung, K. et al., 1988; Traub et al., 1989a, b). Thus, CGRP is a marker for primary afferent terminals in the dorsal horn (Carlton et al., 1988). However, it should be noted that CGRP is contained in the cell bodies of motor neurons and preganglionic autonomic neurons (Gibson et al., 1984b; Senba and Tohyama, 1988). Furthermore, CGRP and the mRNA for CGRP are observed in neurons of lamina III in mice after dorsal rhizotomy (Tie-Jun et al., 2001), suggesting a postsynaptic role for CGRP in the plastic changes that occur in the dorsal horn after injury.

CGRP-containing terminals have been shown to synapse on retrogradely labeled primate spinothalamic tract cells in both the superficial and deep dorsal horn (Carlton et al., 1990; see Chapter 11).

CGRP-containing afferent terminals are often found in relationship to GABA-immunoreactive neurons in the dorsal horn (Hayes and Carlton, 1992). CGRP endings are commonly presynaptic to GABA-containing dendrites. However, GABAergic presynaptic dendrites are also found on postsynaptic CGRP-containing primary afferent terminals. This could be a substrate for presynaptic inhibition. CGRP-containing synapses are in contact with neurons in lamina V that contain dynorphin A (1–8)-like immunoreactivity (Carlton and Hayes, 1989; Takahashi et al., 1990).

There are two forms of CGRP in DRG cells: α-CGRP and β-CGRP (Amara et al., 1985; Steenbergh et al., 1985). Small and medium-sized DRG cells may contain the mRNA for either or both types of CGRP, whereas large DRG cells contain mRNA chiefly for α-CGRP (Noguchi et al., 1990a, b). SP and both forms of CGRP may coexist in small DRG cells.

Release of CGRP. The release of CGRP in the spinal cord has been demonstrated by radioimmunoassay (Saria et al., 1986; Diez Guerra et al., 1988). Using the antibody microprobe technique, Morton and Hutchison (1989) showed that noxious mechanical or thermal stimulation of the skin caused the release of CGRP in the substantia gelatinosa. CGRP was also released when fine afferent fibers in the tibial nerve were stimulated electrically, but not when only large myelinated fibers were stimulated. Concurrent release of SP and CGRP following noxious thermal stimulation could also be demonstrated.

Klein et al. (1990b) found a reduction in CGRP immunostaining in the dorsal horn following stimulation of the sciatic nerve at an intensity that would activate C fibers. Presumably, this reduction reflected the depletion of CGRP from synaptic terminals after its release. Schaible et al. (1994) also found that CGRP is released when C fibers are stimulated electrically before the development of experimental arthritis, but not in response to innocuous stimulation of the knee joint. However, during arthritis the basal level of CGRP in the dorsal horn is increased and CGRP is released in response to weak mechanical stimulation of the knee joint.

The release of CGRP (but not SP) from spinal cord slices by exposure to K^+ was inhibited by GABA and by muscimol, but not by baclofen (Bourgoin et al., 1992). This effect was prevented by bicuculline. These observations suggest that GABA exerts a presynaptic inhibitory control of CGRP release from primary afferent fibers in the dorsal horn.

Dymshitz and Vasco (1994) found that the release of CGRP and SP from cultured DRG cells by capsaicin is increased by sodium nitroprusside, but not by other NO-donors. Pretreatment with a nitric oxide synthase inhibitor did not affect the release of these peptides by capsaicin, although it prevented an increase in cAMP. Garry et al. (1994) observed that sodium nitroprusside also caused release of CGRP and SP from rat spinal cord slices. Garry et al. (2000) report that the release of CGRP in the spinal cord in response to capsaicin administration is reduced when nitric oxide synthase is inhibited and also when antagonists of NMDA and AMPA/kainate receptors are given. However, CGRP release did not depend on changes in cGMP. Nitric oxide synthase is not co-localized with CGRP, and so the effect of NO is indirect.

CGRP Receptors. CGRP-binding sites are concentrated in lamina I and in deep layers of the dorsal horn, as well as around the central canal (Skofitsch and Jacobowitz, 1985a). There is less binding in lamina II and in the ventral horn (Yashpal et al., 1992). Ye et al. (1999) observed the distribution of immunostained CGRP receptors at the light and electron-microscopic levels. CGRP receptors were found on postsynaptic membranes of cell bodies and dendrites of neurons in the superficial dorsal horn. There were also presynaptic CGRP receptors on some synaptic endings within the dorsal horn.

Functions of CGRP: Studies at the Cellular Level. Functions of CGRP include enhancement of the action of SP (Wiesenfeld-Hallin et al., 1984; Biella et al., 1991). Several mechanisms for such an action have been proposed. CGRP has been found to inhibit the enzymatic degradation of SP (Greves et al., 1985), and it may also potentiate the release of SP (Oku et al., 1987), perhaps by increasing Ca^{++} influx into synaptic terminals (Oku et al., 1988). Ryu et al. (1988b) have shown that CGRP increases both the amplitude and duration of Ca^{++} spikes in DRG cells. They propose that enhanced Ca^{++} currents in primary afferent fibers help explain the enhancement of synaptic transmission that is produced in the dorsal horn by CGRP.

In addition to presynaptic modulatory effects, CGRP has a postsynaptic action. Miletic and Tan (1988) applied CGRP iontophoretically onto dorsal horn neurons and found that this peptide activates low-threshold and wide-dynamic-range neurons, but not high-threshold neurons. The time course of the excitation is slow, with an onset after 30 sec to 3 min. The duration of the excitation can be up to 10 min. Yu, Y. et al. (2002) were also able to excite wide-dynamic-range dorsal horn neurons by iontophoretic release of CGRP. The excitation was prevented by iontophoretic application of the CGRP receptor antagonist, $CGRP_{8\text{-}37}$. The antagonist also reduced the responses of the neurons evoked by electrical stimulation of the hindpaw (Yu, L. C. et al., 1999).

Ryu et al. (1988a) applied CGRP to slice preparations of the dorsal horn and DRG of rat spinal cord while recording intracellularly from dorsal horn neurons. The peptide often caused a slow, dose-dependent depolarization (sometimes preceded by a transient hyperpolarization) that was not prevented when voltage-gated Na^+ and K^+ currents were blocked by tetrodotoxin and tetraethylammonium, respectively. Membrane conductance was reduced during the depolarization, and excitability was enhanced. The action of SP was stronger than that of CGRP when the two were compared on the same neurons.

Functions of CGRP: Studies at the Systems Level. Kawamura et al. (1989) observed that intrathecal administration of an antiserum against CGRP returned the behavioral responses of rats to radiant heat or paw pressure to normal in animals made hyperalgesic by injection of Freund's adjuvant or carrageenan. Kuraishi et al. (1988) reported a similar effect of CGRP antibody in reducing the hyperalgesia seen in rats with adjuvant arthritis.

Neugebauer et al. (1996b) have demonstrated an involvement of CGRP in the responses of dorsal horn neurons to knee joint stimulation before and during experimental arthritis. The CGRP receptor antagonist $CGRP_{8\text{-}37}$ reduced the responses of dorsal horn neurons to both innocuous and noxious pressure applied to the knee joint. The antagonist was also effective in reducing the responses to mechanical stimulation during arthritis.

The injection of acetic acid into the colon increased the number of abdominal contractions observed in response to colorectal distension, presumably because of sensitization of visceral afferents (Plourde et al., 1997). These effects were prevented by intravenous administration of a CGRP receptor antagonist. The CGRP antagonist also reduced the number of abdominal contractions in response to colorectal distension, and CGRP injection increased the number of contractions. Intrathecal administration of a CGRP receptor antagonist reversed the visceral hypersensitivity caused by acetic acid or CGRP.

Combined Action of Excitatory Amino Acids and Peptides

At some synapses on dorsal horn interneurons, stimulation of primary afferent fibers causes a fast excitation followed by a slow excitation with two components (Gerber and Randic, 1989b). The initial excitation is due to monosynaptic transmission, often as a result of an action of an excitatory amino acid on non-NMDA receptors. The first part of the slow excitation lasts several seconds and appears to result from the activation of NMDA receptors, perhaps in addition to non-NMDA receptors, by excitatory amino acids released from interneurons (Gerber and Randic, 1989b). The later part of the slow excitation that lasts for several minutes seems to be peptidergic (Urban and Randic, 1984; Randic et al., 1986; Budai and Larson, 1996).

Application of SP or CGRP to an in vitro spinal cord preparation causes an increased release of glutamate and aspartate (Kangrga and Randic, 1990; Kangrga et al., 1990). Neonatal capsaicin treatment prevents the release of excitatory amino acids by dorsal root stimulation and by SP, but does not modify the release produced by CGRP (Kangrga and Randic, 1990; Kangrga et al., 1990).

Co-application of excitatory amino acids and SP (or a selective NK-1 agonist) to dorsal horn neurons (including spinothalamic tract cells) can produce a long-lasting enhancement of the responses of these cells to later applications of an excitatory amino acid or to peripheral input (see Fig. 11.39; Dougherty and Willis, 1991b; Dougherty et al., 1993; see also, Randic et al., 1990; Rusin et al., 1992; 1993a, b; Cumberbatch et al., 1995; see review by Randic et al., 1995). As already mentioned the synergistic effects of the responses to NMDA and SP may relate to a close geometric relationship between NK-1 and NMDA receptors on the postsynaptic membranes of dendrites of dorsal horn neurons (Aicher et al., 1997).

The responses of dorsal horn neurons in vitro to NMDA are also enhanced by CGRP, as well as by glycine (Murase et al., 1989b). The responses of dorsal horn neurons with input from the knee joint to NMDA and to AMPA are both enhanced by iontophoretic application of CGRP. The effect on NMDA responses is prevented by the CGRP antagonist, $CGRP_{8\text{-}37}$, but this antagonist enhances the responses to AMPA (Ebersberger et al., 2000). It was suggested that different signal transduction pathways mediate the different effects of $CGRP_{8\text{-}37}$ on the responses to NMDA and to AMPA.

NKA increases the responses of dorsal horn neurons to NMDA without affecting those to kainate (Cumberbatch et al., 1995). NKB has the opposite effect. A selective agonist for NK-2 receptors reduces the responses to AMPA and kainate, without affecting those to NMDA, and a selective agonist of NK-3 receptors increases the responses to AMPA and kainate.

Other Excitatory Peptides

Besides CGRP and neurokinins, such as SP and NKA, several other peptides have been reported to have an excitatory action on dorsal horn neurons. These include vasoactive intestinal polypeptide (VIP), neurotensin, cholecystokinin (CCK), thyrotropin-releasing hormone (TRH), corticotropin-releasing hormone (CRH), pituitary adenylate cyclase-activating polypeptide (PACAP), and islet amyloid polypeptide (IAPP or amylin). Much less is known about the roles of these substances in the dorsal horn than about SP and CGRP.

Vasoactive Intestinal Polypeptide (VIP)

VIP is found in the terminals of small-diameter primary afferent fibers in the dorsal horn, especially of the sacral spinal cord (Gibson et al., 1981; Basbaum and Glazer, 1983; Honda et al., 1983; Morgan et al., 1999; see also Romualdi et al., 1990). Many VIP-containing afferent fibers are likely to supply pelvic visceral organs (Morgan et al., 1981; Nadelhaft et al., 1983). However, VIP-containing afferent fibers are also found at other levels of the spinal cord (LaMotte, C. C. and DeLanerolle, 1986; Sasek et al., 1991). Terminals that contain VIP are especially concentrated in laminae I, II, V, VII, and X (Morgan et al., 1999). VIP-containing neurons have been observed in laminae II–IV, X, and the intermediolateral cell column (Fuji et al., 1985; LaMotte, C. C. and DeLamerolle, 1986; Sasek et al., 1991). VIP binding sites have their highest density in the lumbar and sacral segments of the spinal cord. Binding is greatest in laminae I and II, around the central canal, and in the sacral parasympathetic nucleus (Yashpal et al., 1991b).

Iontophoretic application of VIP causes excitation of nociceptive dorsal horn neurons (Jeftinija et al., 1982; Salt and Hill, 1983; Dickinson et al., 1997). An antagonist of VIP and pituitary adenylate cyclase-activating polypeptide (PACAP) receptors inhibits the responses of wide-dynamic-range neurons to innocuous brushing and also to mustard oil

(Dickinson et al., 1997). VIP applied to spinal cord slices causes a depolarization of neurons in the superficial dorsal horn, as well as an increase in spontaneous EPSPs (Liu, X. H. and Morris, 1999).

Neurotensin

Neurotensin has not been observed in primary afferent neurons. However, it is found in interneurons located in the inner part of the substantia gelatinosa (Uhl et al., 1979a; Seybold and Elde, 1982; Yaksh et al., 1982) and also in lamina I (Miller and Seybold, 1989). Immunostaining of neurotensin in synaptic terminals in the superficial dorsal horn may demonstrate the axons of neurotensin-containing neurons intrinsic to the dorsal horn (Seybold and Elde, 1980). Glutamate is co-localized with neurotensin in these terminals (Todd et al., 1994a). Neurotensin-containing dorsal horn neurons do not have co-localized somatostatin, glycine, or GABA (Todd et al., 1992a; Proudlock et al., 1993), consistent with electrophysiological evidence that neurotensin-containing neurons are excitatory. Neurotensin binding sites are present in the dorsal horn (Ninkovic et al., 1981), and mRNA for neurotensin receptors is found in neurons of the dorsal horn and the area around the central canal (Elde et al., 1990). In the human spinal cord, neurotensin receptors are especially dense in the inner part of lamina II (Faull et al., 1989).

Neurotensin is released from the spinal cord following stimulation of the sciatic nerve (Yaksh et al., 1982). Iontophoretic application of neurotensin produces an excitation of nociceptive and non-nociceptive neurons in laminae I–III, but not in laminae IV–VII (Miletic and Randic, 1979; Stanzione and Zieglgänsberger, 1983). The excitatory actions of neurotensin and glutamate are additive.

Cholecystokinin (CCK)

Cholecystokinin-like immunoreactivity has been reported in a population of DRG cells and in neurons in several laminae of the dorsal horn (Schroder, 1983; Fuji et al., 1985). It has been questioned whether CCK-immunoreactivity actually represents cross-reactivity with CGRP (Ju et al., 1986). However, the precursor peptide for CCK (pro-cholecystokinin) is present in the spinal cord (Beinfeld, 1985), and in monkeys the mRNA for CCK is found in DRG cells and in neurons in the dorsal horn (Verge et al., 1993). CCK released from the spinal cord has been collected by microdialysis and measured by radioimmunoassay (Gustafsson et al., 1999). CCK binding sites are in highest concentration in laminae I and II. CCK-B is the main receptor subtype found on DRG cells, but within the monkey spinal cord, CCK-A is the main receptor subtype (Ghilardi et al., 1992).

Jeftinija et al. (1981) found that many of the neurons tested in laminae I–VII could be excited by iontophoretic application of CCK in both the intact rat spinal cord and in an in vitro preparation. All classes of neurons were affected. Since CCK was still effective in the absence of Ca^{++} in the bathing solution, it was concluded that CCK has a direct postsynaptic excitatory action. However, there is evidence that CCK can also have an antinociceptive action. This may be through an indirect inhibitory action of CCK by causing the release of GABA, perhaps because of an excitatory action on GABAergic inhibitory interneurons (Rodriguez et al., 1987). Intrathecal administration of a CCK-B receptor antagonist reduced flinching during both Phase 1 and Phase 2 in the formalin test, whereas a CCK-A antagonist had no effect (Yamamoto and Nozaki-Taguchi, 1996). The role of CCK in antagonizing the effects of opiates is discussed below.

Thyrotropin-releasing Hormone (TRH)

TRH is found not only in the ventral horn but also in neuronal cell bodies and terminals in the dorsal horn (Coffield et al., 1986; Harkness and Brownfield, 1986; Ulfhake et al., 1987). Although evidence is given below that TRH-containing neurons are excitatory, most TRH-immunoreactive neurons in the dorsal horn also contain GABA (Fleming and Todd, 1994). There are TRH-binding sites in the dorsal horn (Manaker et al., 1985a, b; Mantyh and Hunt, 1985a), although the mRNA for TRH receptors is too low for detection by in situ hybridization (Wu, W. et al., 1992).

Iontophoretically applied TRH facilitates the responses of wide-dynamic-range and high-threshold neurons in laminae II–V to glutamate (Jackson and White, 1988). Often, facilitation is preceded by inhibition of wide-dynamic-range neurons. It is unclear if the inhibition and facilitation both result from an action of interneurons that co-localize TRH and GABA or if different interneurons are involved. The TRH facilitation is NMDA-specific, since AMPA responses are not enhanced; TRH also facilitates wind-up and responses to noxious stimuli (Chizh and Headley, 1994; 1996).

Corticotropin-releasing hormone (CRH)

CRH-immunoreactive axons have been described in the sacral spinal cord (Kawatani et al., 1996). They terminate in the dorsal commissure region, in the sacral parasympathetic nucleus, as well as in the superficial dorsal horn (see also, Schipper et al., 1983; Puder and Papka, 2001a). CRH- and VIP-immunostaining are eliminated by dorsal rhizotomy, but are unaffected by transection of the spinal cord at a lower thoracic level. CRH fibers are also eliminated following capsaicin treament in newborn rats (Skofitsch et al., 1984). In addition to lamina I, CRH is also found in laminae V–VII, X, and the intermediolateral column. Hypophysectomy results in CRH immunostaining of neuronal cell bodies in laminae V–VII and X, as well as in the IML column (Merchenthaler et al., 1983). CRH binding sites and CRH receptor protein are found in the superficial dorsal horn (Skofitsch et al., 1985a; Bell and de Souza, 1988). Evidence concerning the functional role of CRH is needed.

Pituitary Adenylate Cyclase-activating Polypeptide (PACAP)

PACAP resembles VIP and has been demonstrated in small DRG cells (Mulder et al., 1999) and in many axons in the superficial dorsal horn (Narita et al., 1996). PACAP is released from the spinal cord in response to intrathecal administration of capsaicin (Zhang, Y. et al., 1997). Intrathecal administration of PACAP causes nocifensive behaviors and a reduction in tail-flick latency (Narita et al., 1996). Iontophoretic application of PACAP excites nociceptive dorsal horn neurons (Dickinson et al., 1997). Conversely, antagonists of VIP/PACAP receptors inhibit the responses of dorsal horn neurons to mustard oil, but not to brush, in animals with a neuropathic pain state due to chronic constriction injury of the sciatic nerve (Dickinson et al., 1999).

Islet Amyloid Polypeptide (IAPP or Amylin)

IAPP resembles CGRP and has been observed in small DRG cells (Mulder et al., 1995; 1999), including ones that co-localize CGRP and/or SP (Mulder et al., 1995;

Gebre-Medhin et al., 1998). IAPP knock-out mice have a reduced response in the late phase of the formalin test, suggesting that IAPP has a pronociceptive role (Gebre-Medhin et al., 1998; Mulder et al., 1999). However, joint inflammation induced by injection of adjuvant was unaltered, presumably because the CGRP and SP content of sensory neurons remained intact (Gebre-Medhin et al., 1998).

Inhibitory Transmitters in the Dorsal Horn

Inhibitory Amino Acids

Both GABA and glycine are regarded as fast inhibitory neurotransmitters in the spinal cord. These amino acids and their synthetic enzymes are distributed widely in the spinal cord (Aprison and Werman, 1965; Curtis and Watkins, 1965; Davidoff et al., 1967; Graham et al., 1967; Johnston, G. A. R., 1968; Aprison et al., 1969; Graham and Aprison, 1969; Johnston and Vitali, 1969; McLaughlin et al., 1975). GABA is concentrated more dorsally, whereas glycine is concentrated more ventrally in the spinal cord gray matter (Graham et al., 1967). However both are found in the dorsal horn.

Iontophoresis of GABA or glycine results in the inhibition of the activity of dorsal horn interneurons (Curtis et al., 1959b; 1967a, b; 1968; 1977; Curtis and Watkins, 1960; 1965; Bruggecate and Engberg, 1968; Werman et al., 1968; Willcockson et al., 1984), including those in the substantia gelatinosa (Zieglgänsberger and Sutor, 1983).

GABA and glycine are both candidate transmitters to account for inhibitory postsynaptic potentials (IPSPs) in dorsal horn interneurons that result from stimulation of primary afferent fibers (Graham et al., 1967; Bruggencate and Engberg, 1968; Curtis et al., 1968; 1971a, b; Davidoff et al., 1969; Game and Lodge, 1975; Rudomin et al., 1990). These amino acids would be released by inhibitory interneurons to act at postsynaptic sites to generate the IPSPs. The ionic mechanism responsible for the IPSPs is an increased chloride conductance (Barker, J. L. and Nicoll, 1973). The role of GABA in presynaptic inhibition will be discussed in the section on **Pharmacology of PAD**.

Gamma-aminobutyric Acid (GABA)

Localization of GABA. GABA is synthesized by the action of glutamic acid decarboxylase (GAD). The mRNA for GAD is found throughout the gray matter of the spinal cord, except in the motor nucleus (Castro-Lopes et al., 1994b). The mRNAs of two isoforms of GAD, GAD65 and GAD67, have been localized by in situ hybridization and also by RT-PCR (Feldblum et al., 1995). GAD67 is more abundant than GAD65 in the dorsal horn.

GABA occurs in a large fraction (24–33%) of the interneurons in laminae I–IV of the dorsal horn and also in lamina X; GABA is also found in axons of the dorsal and lateral funiculi and tract of Lissauer (Hunt, S. P. et al., 1981a; Barber et al., 1982; Kaduri et al., 1987; Magoul et al., 1987; Todd, A. J. and McKenzie, 1989; Carlton and Hayes, 1990; Waldvogel et al., 1990). Neurons that can take up tritiated GABA are found in laminae I–III (Ribeiro-DaSilva and Coimbra, 1980).

Todd, A. J. and McKenzie (1989) and Spike and Todd (1992) employed the Golgi stain in combination with immunohistochemistry to show that many central (islet) cells of the substantia gelatinosa contain GABA, whereas some central cells and also limiting (stalked) cells do not. Neurons in lamina III were also observed that contain GABA (and some also glycine; Powell and Todd, 1992). These neurons received synapses from axons that were likely to have supplied sensitive mechanoreceptors.

Terminals that contain CGRP have been shown to synapse on GABA-immunoreactive processes in the dorsal horn (Hayes and Carlton, 1992). GABAergic interneurons can also be postsynaptic to SP-containing terminals (Wang, D. et al., 2000). GABA-immunoreactive synaptic endings are found on spinothalamic tract neurons in rats (Lekan and Carlton, 1995) and monkeys (Carlton et al., 1992). A detailed analysis of synaptic endings immunocytochemically stained for GABA has demonstrated numerous GABAergic presynaptic dendrites in the superficial laminae of the primate dorsal horn (Carlton and Hayes, 1990; see also, Todd and Lochhead, 1990; Todd, 1996). GABAergic dendrites that receive synaptic connections from the central terminals of glomeruli often participate in reciprocal synapses.

GABAergic synaptic terminals more often contact C2 (type 2) glomeruli, which probably originate from small myelinated primary afferent axons, than C1 (type 1) glomeruli, which are likely to originate from unmyelinated afferents (Bernardi et al., 1995). Some GABAergic neurons contain nitric oxide synthase (NOS). The GABAergic terminals that are presynaptic to primary afferent terminals are negative for NOS. However, type 2 terminals often synapse on NOS-positive dendrites. Valtschanoff et al. (1992a) found that all NOS-positive synaptic endings in the substantia gelatinosa also contain GABA. These terminals did not belong to primary afferents but probably to interneurons.

GABAergic interneurons appear to be of several different phenotypes, since different combinations of other active substances can be co-localized with GABA in different populations of neurons (Todd et al., 1991; 1992a, b; Spike et al., 1993; Laing et al., 1994; Jo et al., 1998). These substances include glycine, acetylcholine, enkephalin, galanin, neuropeptide Y (NPY), nitric oxide synthase (NOS), and parvalbumin. GABAergic neurons that contain parvalbumin also contain glycine, but not NPY or NOS. GABAergic neurons that lack glycine-immunoreactivity may contain NPY, enkephalin, acetylcholine, or NOS. GABAergic interneurons in the dorsal horn that contain NPY may produce presynaptic inhibition of nociceptive afferents, since NPY is known to cause analgesia when injected intrathecally (Rowan et al., 1993). However, NPY-containing GABAergic neurons often make synaptic contacts with neurons that have NK-1 receptors (Polgar et al., 1999b), and so the analgesic effects of NPY may involve a postsynaptic action. GABAergic interneurons also often contain galanin (Simmons et al., 1995). Neurotensin, which is an excitatory neuropeptide, does not co-localize with GABA (Todd et al., 1992a).

Release of GABA. Activation of primary afferents by perineural application of capsaicin results in a reduction in GABA-immunoreactivity in the superficial dorsal horn (Wei and Zhao, 1996), presumably because the strong nociceptive input causes GABA release and depletion. This action is partially blocked by a SP receptor antagonist. It is suggested that noxious stimuli activate inhibitory interneurons, which in turn release GABA to modulate spinal nociception. This is consistent with the observation mentioned previously that noxious stimuli can cause FOS expression in GABAergic interneurons (see Fig. 7.15).

Spinal cord stimulation is often helpful for the treatment of pain in patients. In rats, spinal cord stimulation increases the release of GABA (Linderoth et al., 1994). It was suggested that a similar effect in humans could help explain pain relief by this procedure.

GABA Receptors. The best-studied GABA receptors include $GABA_A$ and $GABA_B$ receptors. In addition, there are $GABA_C$ receptors.

The $GABA_A$ receptor is ionotropic and involves a chloride channel (Fig. 7.27A; Tyndale et al., 1995). Muscimol and isoguvacine are agonists of this receptor, and a number of drugs, including barbiturates and benzodiazepines, can modify the activity of the receptors by binding to different sites on the receptor (Fig. 7.27A; Watling et al., 1995). The $GABA_A$ receptor antagonists, picrotoxin and bicuculline, can block the inhibitory actions of GABA that are mediated by this receptor (Fig. 7.27B1–3; Curtis et al., 1969; 1971a; Engberg and Thaller, 1970; Straughan et al., 1971; Benoist et al., 1972; Game and Lodge, 1975). $GABA_A$ receptors are heteropentameric and are formed from combinations of more than a dozen subunits (Bohlhalter et al., 1996; Pirker et al., 2000).

$GABA_A$ receptors can be found presynaptically on primary afferents or postsynaptically on dorsal horn neurons. The distribution of various subunits has been examined by immunohistochemistry or by in situ hybidization for mRNA and related to neurons in particular laminae of the spinal cord dorsal horn or to different size ranges of DRG cells (Persohn et al., 1991; Alvarez et al., 1996; Bohlhalter et al., 1996). Immunoreactivity in the dorsal part of the cord was observed chiefly on the cell bodies and dendrites of neurons in laminae II, III, and X. Large DRG cells contain alpha 2 subunits, and a subpopulation of large and small DRG neurons contain gamma 2 subunits (Persohn et al., 1991).

By contrast, the $GABA_B$ receptor is metabotropic and is a G-protein coupled receptor (Watling et al., 1995). Baclofen is an agonist of $GABA_B$ receptors, and selective antagonists include phaclofen and 2-hydroxysaclofen (Watling et al., 1995). The locations of $GABA_B$ binding sites have been mapped. They are found throughout the dorsal horn, although the highest concentrations are in lamina II (Price, G. W. et al., 1984; 1987; Albin et al., 1993). Like $GABA_A$ receptors, $GABA_B$ receptors are found in both pre- and postsynaptic locations (Charles et al., 2001; Yang et al., 2001c). Dorsal rhizotomy or neonatal capsaicin treatment produces a substantial reduction in binding sites (Price, G. W. et al., 1984; 1987), which supports the idea that many $GABA_B$ receptors are on primary afferent terminals. Also consistent with this, mRNA for $GABA_B$ receptors is found in DRG cells, as well as in dorsal horn neurons (Towers et al., 2000). $GABA_B$ receptors are composed of two different subunits that are expressed by the $GABA_{(B1)}$ and $GABA_{(B2)}$ genes (Charles et al., 2001). $GABA_B$ receptors are heterodimers of these subunits.

A third class of GABA receptor is the $GABA_C$ receptor. This receptor is an iontropic receptor that involves a chloride channel, is resistant to bicuculline, but is antagonized by picrotoxin (Watling et al., 1995). $GABA_C$ receptors respond to cis-4-aminocrotonic acid, and they can be antagonized by imidazole-4-acetic acid (Park et al., 1999).

Functions of GABA: Studies at the Cellular Level. Recordings from neurons of the substantia gelatinosa in spinal cord slice preparations have revealed that stimulation of Aδ fibers in the dorsal root result in inhibitory postsynaptic currents (IPSCs) that can be blocked by bicuculline or by strychnine (Fig. 7.27B, C; Yoshimura and Nishi, 1993). In intracellular recordings, an EPSP is seen, followed by short and/or long monosynaptic IPSPs (Yoshimura and Nishi, 1995). Strychnine blocks short-lasting IPSPs and bicuculline long-lasting ones, indicating that the IPSPs are produced by actions at glycine and $GABA_A$ receptors, respectively. The IPSPs are disynaptic, and so it is not unexpected that they can also be blocked by CNQX; AP5 is less effective.

Patch-clamp recordings have been made in lamina I of the spinal cord of IPSCs that involve $GABA_B$ but not $GABA_A$ receptors (Chery and De Koninck, 2000). However,

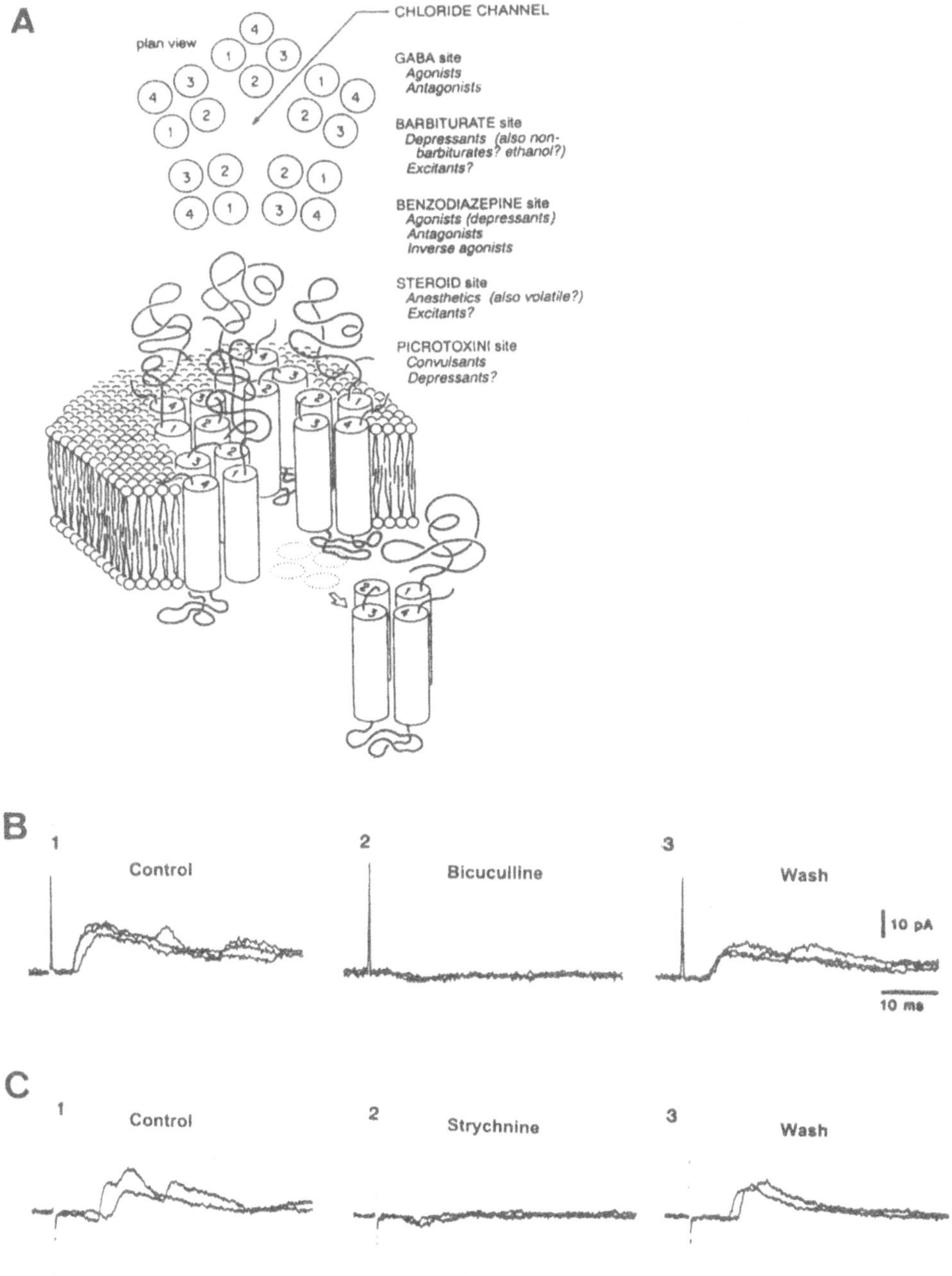

FIGURE 7.27. A, Structure and binding sites of the $GABA_A$ receptor chloride channel complex. The channel is pentameric and is formed from five families of subunits. Different classes of binding sites are indicated. (From Tyndale et al., 1995.) B, C, IPSCs evoked by stimulation of Aδ fibers and recorded from neurons of the substantia gelatinosa, using the blind patch-clamp technique. The holding potential was −40 mV. B,1–3 show the effect of bicuculline (10 μM) on an IPSC presumed to be due to GABA release, and C, 1–3 show the effect of strychnine (0.5 μM) in blocking an IPSC due to glycine release. The latencies and shapes of the IPSCs were variable, presumably because of the disynaptic pathway through an inhibitory interneuron. (From Yoshimura and Nishi, 1993.)

in vivo patch-clamp recordings from neurons of the substantia gelatinosa reveal IPSCs in response to tacile stimulation of the skin that could be blocked by either bicuculline or strychnine (Narikawa et al., 2000).

The $GABA_B$ receptor agonist, baclofen, hyperpolarizes neurons in laminae I–IV in spinal cord slices and decreases EPSPs evoked by stimulation of dorsal root fibers (Allerton et al., 1989a; Kangrga et al., 1991). The hyperpolarization is associated with a decreased membrane resistance. These observations suggest that baclofen causes both pre- and postsynaptic inhibition of synaptic transmission to dorsal horn neurons.

The action of baclofen on miniature EPSCs and IPSCs has been examined in neurons of the substantia gelatinosa in spinal cord slices, using the whole cell patch-clamp technique (Iyadomi et al., 2000). The frequencies of the miniature EPSCs and IPSCs are reduced without any change in their amplitudes, indicating a presynaptic inhibitory action of baclofen. Yang et al. (2001b, c) have obtained further evidence that baclofen produces both postsynaptic and presynaptic inhibition. For example, bath application of baclofen produces a slow IPSC in dorsal horn neurons, but it also decreases monosynaptic EPSCs evoked in dorsal horn neurons by Aβ and Aδ primary afferent fibers.

AMPA receptors can be either Ca^{++} permeable or impermeable. Spike et al. (1998) have found that GABA-immunoreactive interneurons in the dorsal horn generally express GluR1, but not GluR2/3, subunits, suggesting that the glutamate receptors on these GABAergic interneurons are likely to have calcium permeability. Albuquerque et al. (1999) confirmed this in work that showed that most dorsal horn neurons that express either GABA or NK-1 receptors contain Ca^{++} permeable AMPA receptors, whereas several types of excitatory interneurons that contain calretinin or calbindin do not.

Functions of GABA: Studies at the Systems Level. EPSPs have been recorded from dorsal horn neurons in response to stimulation of the DRG neurons belonging to individual Aβ or Aδ axons that supply hair follicle afferents. Administration of bicuculline, but not strychnine, revealed a late inhibitory component of the response to stimulation of Aδ hair follicle afferents (De Koninck and Henry, 1994).

Activation of $GABA_A$, $GABA_B$, and glycine receptors produces the IPSCs that can be recorded from neurons in laminae I and II in spinal cord slices (Chery and De Koninck, 1999). In agreement with Yoshimura and Nishi (1995), those due to $GABA_A$ receptors had slower decays than IPSCs due to activation of glycine receptors. Lamina I neurons seemed to lack IPSCs due to activation of $GABA_A$ receptors, but the same neurons responded to bath applications of both GABA and glycine. Administration of a benzodiazepine drug revealed the presence of monosynaptic GABAergic IPSCs. The slower time course of the GABAergic mIPSCs was attributed to an extrasynaptic location of $GABA_A$ receptors.

Intravenous administration of enantiomers of baclofen, a $GABA_B$ receptor agonist, inhibits nociceptive responses of wide-dynamic-range dorsal horn neurons (Henry, 1982). Intrathecal administration of baclofen also inhibits the responses of wide-dynamic-range dorsal horn neurons to C fibers and to pinch stimuli without having much effect on responses to A fibers or to tactile stimuli (Dickenson et al., 1985). This inhibition is not affected by bicuculline. Instead, the depression of EPSCs recorded from dorsal horn neurons in response to dorsal root stimulation by baclofen is reduced in the presence of a $GABA_B$ receptor antagonist (Blake et al., 1993). Baclofen causes a reduction in the release of GABA and glutamate from the spinal cord, and several $GABA_B$ receptor antagonists increase the release of these amino acids (Teoh et al., 1996a).

A model of tactile allodynia can be produced in decerebrate, spinalized rats by intrathecal injection of strychnine or bicuculline, but not a $GABA_B$ receptor antagonist (Sivilotti and Woolf, 1994). Tactile allodynia induced in mice by intrathecal administration of bicuculline is blocked by the non-NMDA receptor antagonist, GAMS, and by the metabotropic glutamate receptor antagonists, L-AP3 and L-AP4, but not by AP5 or CNQX (Onaka et al., 1996). Microdialysis adminstration of bicuculline into the dorsal horn results in increased activity of nociceptive dorsal horn neurons (Sorkin et al., 1998). Responses to tactile input are most affected, but responses to strong mechanical stimuli are also enhanced. The increased responses are blocked by CNQX but not by AP7. These effects contrast with the allodynia and hyperexcitability of dorsal horn neurons that result from strychnine administration, which are relieved by NMDA receptor antagonists (Onaka et al., 1996; Sorkin and Puig, 1996).

Glycine

Localization of Glycinergic Neurons. Neurons and synaptic terminals that immunostain for glycine are found in both the dorsal and ventral horns (van den Pol and Gorcs, 1988), including the superficial laminae of the dorsal horn (Todd, A. J., 1990). Spinal cord slices take up tritiated glycine (Neal, 1971). Presumably, the uptake of glycine in these studies corresponds to the presence of a glycine transporter (GLYT2) in glycinergic neurons of the dorsal horn (Spike et al., 1997).

Synapses made by HRP-labeled hair follicle afferents have been observed on glycine-immunoreactive dendrites (Todd et al., 1991). In addition, hair follicle afferent terminals that are postsynaptic to glycine-containing axon endings were demonstrated. Glycine-containing interneurons can be postsynaptic to SP-containing synaptic terminals (Wang et al., 2000). Some glycinergic neurons in the deep dorsal horn also contain somatostatin (Proudlock et al., 1993).

Release of Glycine. Glycine, like GABA, is released in a calcium-dependent fashion following stimulation (Aprison, 1970; Hopkin and Neal, 1970; Hammerstad et al., 1971; Collins, G. G. S., 1974). Substance P administration in the dorsal horn results in the release of glycine (Smullin et al., 1990).

Glycine Receptors. Glycine receptor immunoreactivity (van den Pol and Gorcs, 1988) and strychnine binding sites (Albin et al., 1993) are in all laminae, although the concentration is highest in the dorsal horn (Basbaum, 1988; Albin et al., 1993). Glycine receptors are found at axodendritic and axosomatic synapses and in lamina II at dendro-dendritic synapes (Mitchell et al., 1993). Axoaxonic synapses do not have glycine receptor-like immunoreactivity, even though glycine may be co-localized with GABA in synaptic terminals. Gephyrin is a membrane protein that is usually associated with glycinergic synapses (Todd et al., 1995). The glycine receptor antagonist, strychnine, prevents the activation of glycine receptors (Fig. 7.26C; Curtis, 1962; Curtis et al., 1969; 1971b; Davidoff, R. A. et al., 1969; Game and Lodge, 1975).

Acetylcholine (ACh)

Localization of ACh. Many small DRG cells and a few large ones have been found to contain choline acetyltransferase (ChAT), indicating that these are cholinergic primary afferent neurons (Sann et al., 1995a). Commonly, these DRG cells and their projections to the superficial dorsal horn have co-localized CGRP. Dorsal rhizotomy reduces ChAT immunoreactivity in the dorsal horn.

In addition, ChAT is found in some neurons in the dorsal horn and lamina X, as well as in motor neurons (Borges and Iversen, 1986; Ribeiro-da-Silva and Cuello, 1990a; Sann et al., 1995a). Both myelinated and unmyelinated primary afferent fibers synapse directly on cholinergic dorsal horn interneurones (Olave et al., 2002). Conversely, ChAT-positive terminals can apparently make contacts with primary afferent terminals in the dorsal horn (Barber et al., 1984; Ribeiro-da-Sliva and Cuello, 1990a; Todd, 1991).

ACh Receptors. ACh receptors include nicotinic and muscarinic receptors. Both classes of cholinergic receptors have been found on primary afferent neurons (Morita and Katayama, 1984) and on dorsal horn neurons (Khan et al., 1994).

Nicotinic ACh receptors are ligand-gated ion channels with a pentameric structure (Watling, 2001). Ten α and four β subunits are assembled in various combinations to form subtypes of nicotinic ACh receptors. The properties of nicotinic ACh receptors depend on the subunit combination. The mRNA for nicotinic ACh receptors is found in DRG cells in the chick, and many of these neurons are sensitive to ACh (Boyd et al., 1991). Nicotine binding is observed on rat DRG cells, and this is reduced in animals treated neonatally with capsaicin, suggesting an association between nicotinic ACh receptors and capsaicin-sensitive primary afferents (Roberts, R. G. et al., 1995).

The mRNA for nicotinic ACh receptors has been examined in rats using RT-PCR analysis (Genzen et al., 2001). Subunits α2–7 and β2–4 have been analyzed. Many rat DRG neurons have α7 subunits (77% of large and 32% of small DRG cells). These cells respond to choline, as well as to ACh, and the responses are blocked with α7 antagonists. A second category of DRG cells is thought to have α3, β4 subunits. These cells are activated better by epibatidine than by ACh or cytosine, and the responses are blocked by mecamylamine. A few DRG cells have $\alpha4\beta2$ subunits; they are not sensitive to cytosine but are to nicotine or epibatidine. Dihydro-β-erythroidin, an agent that is selective for $\alpha4\beta2$ receptors, is an effective antagonist. A few DRG cells could not be classified.

Nicotinic cholinergic binding sites have been mapped in the spinal cord (Gillberg et al., 1988). They are found in laminae II–III and X, as well as in the motor nucleus.

Immunostaining and the presence of mRNA show that rat DRG cells also contain M1–M4 muscarinic ACh receptors. According to Bernardini et al. (1999), M2 and M4 muscarinic receptors are chiefly on small DRG cells, whereas M1 and M3 receptors are on all sizes of DRG cells. Haberberger et al. (1999) and Tata et al. (1999) also found immunoreactivity and the mRNA for M2 receptors in small to medium-sized DRG cells in rats. The receptors are expressed on IB-4 containing DRG cells, suggesting that these DRG neurons are unmyelinated nociceptors. Tata et al. (2000) did a Northern analysis of rat dorsal root ganglia and found a strong mRNA signal for M2 receptors, faint signals for M3 and M4 receptors, but no signal for M1 receptors. The mRNA for M2–M4 receptors was confirmed using RT-PCR and in situ hybridization. With RT-PCR, there was evidence of M1 and M5 receptors, as well.

Muscarinic receptors are found chiefly in small to medium-sized DRG cells. Haberberger and Bodenbenner (2000) have immunostained the terminals of nerve fibers at the epidermal–dermal border that contain M2 receptors and that also label for IB4; these fibers are distinct from others that contain SP-immunoreactivity. Beneath the epidermis are myelinated axons that label for M2 receptors but for neither SP nor IB4.

Functions of ACh: Studies at the Cellular Level. Nicotinic acetylcholine receptors cause the release of a number of transmitters, including inhibitory amino acids (Kiyosawa et al., 2001; see reviews by Wonnacott, 1997; MacDermott et al., 1999). For example, nicotine increases the frequency without altering the amplitudes of glycinergic mIPSCs in neurons of the substantia gelatinosa recorded in spinal cord slices using the patch-clamp technique (Kiyosawa et al., 2001). Sometimes nicotine also evokes a slow postsynaptic current. Recordings from neurons dissociated from the substantia gelatinosa with intact synaptic boutons demonstrate glycinergic mIPSCs due to glycine release, and nicotine enhances the frequency of these mIPSCs. As in slices, some dissociated neurons have postsynaptic responses to nicotine. The action of nicotine could be blocked by mecamylamine or by dihydro-β-erythroidine, but not by α-bungarotoxin. This suggests that the receptors contain α4 but not α7 subunits (Kiyosawa et al., 2001). Cd^{++} does not block transmitter release in response to nicotine. However, transmitter release does depend on the presence of extracellular Ca^{++}, suggesting that the Ca^{++} required enters through the ACh receptor channel.

Dissociated DRG cells exposed to muscarine show an increase in intracellular Ca^{++}; this could be blocked by atropine or an M3 receptor antagonist, but not by an M2 receptor antagonist (Haberberger et al., 2000). The Ca^{++} response depended on the presence of extracellular Ca^{++}. Immunostaining demonstrated the presence of both M2 and M3 receptors in small and medium-sized DRG cells. Most of these DRG cells could be co-labeled for IB4. The mRNA for M2 and M3 receptors could also be demonstrated in DRG cells by RT-PCR (Haberberger et al., 2000).

Muscarinic receptors can also have antinociceptive effects, either by reducing excitatory transmission or by enhancing inhibitory transmission. For example, bath application of the muscarinic receptor agonist carbachol reduces EPSCs recorded from dorsal horn neurons in spinal cord slice preparations, and this effect is blocked by atropine (Li, P. and Zhuo, 2001a). Carbachol also causes an increase in the frequency, but not amplitude, of GABAergic mIPSCs in neurons of the substantia gelatinosa of spinal cord slices (Baba et al., 1998). Neostigmine has a similar effect. The actions of both carbachol and neostigmine are blocked by atropine. On the other hand, ACh has been reported to facilitate the late, presumably NMDA receptor-mediated, component of EPSCs recorded with the blind patch-clamp technique from dorsal horn neurons in spinal cord slice preparations (Garraway and Hochman, 2001). It is possible that these dorsal horn neurons were inhibitory interneurons.

Functions of ACh: Studies at the Systems Level. Acetylcholine (ACh) contributes to sensory mechanisms of pain. ACh can be either pronociceptive or antinociceptive, depending on its site of action. ACh applied to human skin causes pain (Steen and Reeh, 1993). Administration of nicotine in the skin also causes irritation or pain (Jansco et al., 1961; Keele and Armstrong, 1964; Magerl et al., 1990), presumably due to activation of nicotinic cholinergic receptors. Cutaneous C-nociceptors are activated by both nicotinic and muscarinic receptors (Steen and Reeh, 1993). ACh may also have an action on the central terminals of primary afferent fibers (Genzen et al., 2001). This could presumably either cause transmitter release or, conversely, presynaptic inhibition.

Spinal administration of cholinergic agents results in antinociception in animal studies (Yaksh et al., 1985; Zhuo and Gebhart, 1991; Bleazard and Morris, 1993; Abram and Winne, 1995; Naguib and Yaksh, 1997; Ghelardini et al., 2000) and in analgesia and enhanced opiate analgesia in human patients (Eisenach, 1999). In rats, intrathecal administration of antagonists of nicotinic or muscarinic cholinergic receptors reduces the analgesic effects of systemically administered morphine (Chen, S. R. and Pan, 2001). Conversely, morphine administration results in an increased release of ACh in the spinal cord, as demonstrated by positron emission tomography and by microdialysis in monkeys (Gage et al., 2001).

Wide-dynamic-range dorsal horn neurons respond to intracutaneous injections of nicotine in a concentration-dependent manner (Jinks and Carstens, 1999). Repeated injections lead to tachyphylaxis. The nicotinic receptor antagonist, mecamylamine, reduces the responses to nicotine injection. However, nicotine is also antinociceptive through an action on nicotinic cholinergic receptors in the pain system (Aceto et al., 1983; Bannon et al., 1998; Marubio et al., 1999). Nicotine analgesia is reduced in animals lacking $\alpha 4$ and $\beta 2$ nicotinic ACh receptor subunits (Marubio and Changeux, 2000; Marubio et al., 1999).

Epibatidine, an alkaloid derived from the skin of an Ecuadorian species of frog, has a strong analgesic action (Spande et al., 1992; Bonhaus et al., 1995; Rao et al., 1996; Bannon et al., 1998; Lawand et al., 1999). Badio and Daly (1994) found that the nicotinic receptor antagonist, mecamylamine, blocks this action. Nicotine and other nicotinic receptor agonists, including epibatidine, enhance ion flux from F11 DRG cells, which are considered to be peptidergic, nociceptive neurons (Puttfarcken et al., 1997). This effect is antagonized by mecamylamine and less effectively by dihydro-β-erythroidin. Epibatidine releases SP from these cells, and this action is blocked by mecamylamine.

Peripheral release of ACh in primary afferent fibers may contribute to neurogenic inflammation. Evidence for this is that the responses to antidromic stimulation of peripheral nerves can be partially blocked by atropine (Lembeck and Holzer, 1979; see also Lawand et al., 1999).

Inhibitory Peptides in the Dorsal Horn

Opioid Peptides

The role of opiates and endogenous opioids in the central nervous system has been reviewed on numerous occasions (see, e.g., Frederickson and Geary, 1982; Duggan and North, 1984; Martin, W. R., 1984; Yaksh and Noueihed, 1985; Olson et al., 1989; Zadina et al., 1986; Liu, J. G. and Anand, 2001; Pasternak, 2001; Przewlocki and Przewlocka, 2001; Vaccarino and Kastin, 2001; Williams et al., 2001).

Localization of Opioids. Several opioid peptides are known to be present in the dorsal horn. These include leu- and met-enkephalin, dynorphin (Hökfelt et al., 1977a, b; Hunt, S. P. et al., 1980; Aronin et al., 1981; Glazer and Basbaum, 1981; 1982; LaMotte and DeLanerolle, 1983a; Charnay et al., 1984; Schoenen et al., 1985a), and the recently discovered endomorphins 1 and 2 (Zadina et al., 1997; Wu, S. Y. et al., 1999). Many enkephalin-containing neurons are found in laminae I and II (Glazer and Basbaum, 1981; Bennett et al., 1982). Dynorphin-containing interneurons are found in laminae I, II, and V (Ruda et al., 1988; Cho and Basbaum, 1988; Carlton and Hayes, 1989). CGRP-containing terminals, presumably from primary afferent fibers, make synaptic contact with dynorphin-containing cells in lamina II (Carlton and Hayes, 1989). Some enkephalin-containing terminals belong to axons that descend from the ventral medulla; the same endings may also contain somatostatin (Millhorn et al., 1987b). Immunostaining for endomorphin-2 has been reported in the terminals of primary afferent neurons in the superficial dorsal horn of rats and monkeys, as well as in dorsal root ganglion cells (Martin-Schild et al., 1997; Pierce et al., 1998). Endomorphin-2-like immunoreactivity is co-localized in the primary afferents with substance P and μ-opiate receptors, suggesting that this peptide could be important in pain modulation (Martin-Schild et al., 1998). At the electron microscopic level, immunostaining for endomorphin-2 has been seen in axon terminals and also in postsynaptic elements (Wang, Q. P. et al., 2002).

Opiate Receptors. Specific binding studies have described the distribution of μ, δ, and κ opiate receptors in the spinal cord (Atweh and Kuhar, 1977; Czlonkowski et al., 1983; Gouarderes and Cros, 1984; Gouarderes et al., 1985; 1986; Zarr et al., 1986; Faull and Villiger, 1987; Hunter et al., 1989; Sales et al., 1989). Opiate receptors are especially concentrated in the superficial dorsal horn.

Opiate receptor binding in the superficial dorsal horn decreases significantly following dorsal rhizotomy (LaMotte, C. C. et al., 1976). This is consistent with other evidence that there are opiate receptors on primary afferent fibers, although loss of opiate receptor binding secondary to postsynaptic changes cannot be ruled out. Such presynaptic receptors could account for the ability of opioids to prevent SP release from primary afferent terminals (Jessell and Iversen, 1977; Yaksh et al., 1980; Hirota et al., 1985). These receptors appear to be of the μ and δ subtypes, but not the κ subtype (Fields et al., 1980; Hirota et al., 1985).

Endomorphin-1 and -2 both activate cloned human μ-opiate receptors in a dose-dependent, naloxone reversible manner. Like morphine, they inhibit the cAMP/adenylyl cyclase signal transduction pathway and activate G-protein-coupled inwardly rectifying potassium channels; however, they do not act on delta receptors (Gong et al., 1998; Monory et al., 2000). Curiously, when morphine binds to μ-opiate receptors, the receptors are not internalized; however, these receptors are internalized after they are bound by DAMGO or by endomorphin-1 (Burford et al., 1998; McConalogue et al., 1999).

Release of Opioids. Met-enkephalin is released from the spinal cord in response to stimulation of fine afferent fibers and to perfusion with K^+ or SP (Yaksh and Elde, 1980; 1981; Cesselin et al., 1985; 1989). Le Bars et al. (1987a) found that met-enkephalin is not released from a particular region of the spinal cord by stimulation of a related area of the body, but instead is released when noxious stimuli are applied to distant areas of the body. They suggest that endogenous opioids may be involved in the mechanisms underlying "diffuse noxious inhibitory controls" (DNIC; see later section of this chapter), a proposal that is supported by the observations that DNIC is naloxone reversible (Le Bars et al., 1987b) and that the release of met-enkephalin in the lumbar cord is blocked by lesions of the dorsal lateral funiculi (Le Bars et al., 1987b; see Chapter 12 and also Woolf, 1984). However, noxious heat has the opposite effect, causing release of met-enkephalin from the segmentally related region of the spinal cord and not from a remote region (Cesselin et al., 1989). This is unexpected, since stimulation with noxious heat evokes DNIC, just as does noxious mechanical stimulation. Interestingly, morphine can reduce the amount of met-enkephalin that is released from the spinal cord (Jhamandas et al., 1984). This effect is prevented by naloxone.

Using antibody microprobes, Hutchison et al. (1990) showed that stimulation of C fibers, but not of A fibers, in a peripheral nerve causes the release of dynorphin A in lamina I of the dorsal horn. Dynorphin release could be prevented by transection of the spinal cord, suggesting that the release is under supraspinal control.

Release of endomorphin-2 from the spinal cord during electrical stimulation of the dorsal root entry zone has been demonstrated by the antibody microprobe technique (Williams et al., 1999; Dun et al., 2000).

Functions of Opioid Peptides. Iontophoretic application of enkephalin (or the stable analog, met-enkephalinamide) can produce a selective inhibition of nociceptive dorsal horn interneurons (Duggan et al., 1976; 1977; 1981; Randić; and Miletić, 1978; Zieglgänsberger and Tulloch, 1979b; Sastry and Goh, 1983). This inhibition can be blocked by the opiate receptor antagonist, naloxone.

Duggan et al. (1977) were able to inhibit the responses of dorsal horn neurons by iontophoretic application of met-enkephalin near neuronal cell bodies in laminae IV and V, but the action was relatively nonspecific, affecting responses to both noxious and innocuous stimuli, like the action of GABA or glycine. However, application of met-enkephalin in the substantia gelatinosa produced a more selective inhibition of the responses to noxious stimuli. It was suggested that enkephalin has a postsynaptic inhibitory effect when applied to the cell bodies of dorsal horn neurons. The more selective action of enkephalin released in the substantia gelatinosa could be explained by a presynaptic action. If the opioid action is due to a decrease in the release of SP from fine afferent terminals, it would be anticipated that iontophoresis of SP at the same site as the release of enkephalin would prevent the inhibitory action of the latter; however, simultaneous release of SP does not change the inhibitory action of enkephalin (Duggan et al., 1979). On the other hand, the release of SP from the spinal cord by capsaicin is prevented by opioids acting at either μ or δ (but not κ) receptors (Aimone and Yaksh, 1989).

The responses of the dorsal horn neurons to excitatory amino acids were also blocked by opioids, suggesting that the inhibition produced by opioids is at least partially postsynaptic. Davies, J. and Dray (1978) found the enkephalins (and morphine) had little or no effect when applied iontophoretically onto dorsal horn neurons, although morphine released in the substantia gelatinosa caused a decrease in nociceptive responses in deeper-lying neurons. Sastry and Goh (1983) found that interneurons of the substantia gelatinosa are excited by enkephalin (and also by morphine), whereas deeper-lying nociceptive cells are inhibited; they suggest that the lamina II cells are inhibitory interneurons that mediate part of the depressant action on the deeper-lying dorsal horn neurons.

Zieglgänsberger and Tulloch (1979b) found that iontophoretic application of enkephalin reduced the background activity of dorsal horn neurons and also their responses to excitatory amino acids, presumably indicating a postsynaptic inhibition. The responses to noxious and innocuous stimuli were also nonselectively inhibited. Further evidence of a postsynaptic action of enkephlin has been obtained by use of spinal cord slice preparations (Murase et al., 1982; Yoshimura and North, 1983; Jeftinija et al., 1987). Many dorsal horn neurons are hyperpolarized by enkephalin and their membrane resistance is reduced.

Dickenson et al. (1986) provide evidence that the inhibitory actions of enkephalin on dorsal horn neurons are mediated by δ opiate receptors. Jeftinija (1988), on the other hand, offers evidence to support a role for μ receptors.

Evidence for a presynaptic action of enkephalin has been reported (Sastry, 1978; 1979a; see also Jeftinija, 1988). Sastry suggests that enkephalin potentiates presynaptic inhibition. Mudge et al. (1979) find the enkephalin prevents the release of SP from DRG cells in culture. This effect is accompanied by a reduction in the duration of the action potentials of these neurons. Naloxone reduces the ability of volleys in A fibers to increase the excitability of C-fiber terminals (Woolf and Fitzgerald, 1982). β-endorphin and met-enkephalinamide have been found to depress dorsal root potentials in an in vitro preparation of rat spinal cord (Suzue and Jessell, 1980).

Dynorphin applied to the surface of the rat spinal cord can produce either excitatory or inhibitory actions on the responses of nociceptive dorsal horn neurons to C-fiber volleys (Knox and Dickenson, 1987). The κ-receptor agonist, U50488H, has a similar effect. Knox and Dickenson suggest that the excitatory effects of dynorphin may be mediated through SP release.

Fleetwood-Walker et al. (1988) examined the effects of iontophoretically released opioids on wide-dynamic-range spinocervical tract cells. Agonists of κ-receptors, but not of μ- or δ-receptors, produced a selective reduction in nociceptive responses of the neurons when the drugs were released near the cell body. However, when the same drugs were released in the substantia gelatinosa, it was found that μ- but not δ- or κ-opiate agonists were selectively antinociceptive. Hope et al. (1990c) had similar results when κ-opiate agonists were applied iontophoretically in laminae III–V. However, the same agonists had no effect when released near wide-dynamic-range or high-threshold neurons in lamina I. On the other hand, a μ-opiate agonist inhibited the nociceptive responses of WDR and HT cells in lamina I and a δ-opiate agonist inhibited WDR cells but had mixed effects on HT cells.

The endomorphins have been shown to have antinociceptive effects when administered intrathecally (Stone et al., 1997; Przewlocka et al., 1999; Tseng, L. F. et al., 2000; Ohsawa et al., 2001). The mechanism of action of endomorphin-1 appears to be different from that of endomorphin-2 (Tseng, L. F. et al., 2000; Ohsawa et al., 2001; Sakurada, S. et al., 2001). The action of endomorphin-1 is through μ-opiate receptors. Endomorphin-2 has a similar action on these receptors, but in addition causes the release of dynorphin A (1–17), which in turn acts on spinal cord κ-opiate receptors. Endomorphin-2 also causes the release of enkephalin, which acts on δ-opiate receptors (Sakurada, S. et al., 2001).

The endomorphins are less effective than morphine in counteracting acute and formalin-induced nociception. Interestingly, endomorphins are effective in blocking allodynia in a model of neuropathic pain, in contrast to morphine, which is ineffective (Przewlocka et al., 1999). Intraplantar injection of endomorphin-1 was found to produce a dose-dependent reduction in FOS expression in the dorsal horn following inflammation by injection of carrageenan into the foot (Jin et al., 1999).

In electrophysiological experiments, the endomorphins cause inhibition of the responses of nociceptive dorsal horn neurons to peripheral nerve stimulation (Chapman et al., 1997). In an in vitro preparation, endomorphins caused a reduction in the excitatory and inhibitory postsynaptic potentials evoked by stimulation of the dorsal root entry zone and a hyperpolarization of neurons in the substantia gelatinosa (Wu, S. Y. et al., 1999; Dun et al., 2000). Yajiri and Huang (2000) found that endomorphins blocked EPSCs evoked by stimulation of Aδ afferent fibers and that they reduced the frequency but not the amplitude of miniature EPSCs, suggesting an action on presynaptic terminals leading to a reduction in transmitter release.

Other Inhibitory Peptides

Nocistatin and Nociceptin

Recently described peptides found in the dorsal horn include nocistatin (Okuda-Ashitaka et al., 1998) and nociceptin (orphanin FQ) (Meunier et al., 1995; Reinscheid et al., 1995).

Both inhibitory and excitatory actions of these peptides have been reported. Nocistatin has been shown to block the release of glycine and GABA in the dorsal horn (Zeilhofer et al., 2000). Nociceptin is thought to be an agonist of the opioid receptor-like (ORL1) receptor (Mollereau et al., 1994). However, nocistatin acts at another, so far unidentified receptor (Ahmadi et al., 2001a, b). Nociceptin receptors are upregulated during morphine tolerance (Gouardéres et al., 1999).

Nociceptin has a similar cellular action to that of opioid drugs in that it activates an inwardly rectifying K^+ conductance and blocks transmitter release by inhibiting Ca^{++} channels (Moran et al., 2000). Nociceptin causes presynaptic inhibition of glutamate release in the spinal cord (Liebel et al., 1997; Zeilhofer et al., 2000). It, like morphine, reduces FOS expression in the superficial dorsal horn following formalin injection (Yamamoto et al., 2000).

However, Sakurada, T. et al. (1999a, b) found that intrathecal injections of nociceptin result in scratching, biting, and licking behavior in mice. NK-1 receptor antagonists and morphine reduced these behavioral responses, but an NK-2 antagonist, NMDA receptor antagonists, and a NOS inhibitor failed to do so. Minami et al. (2000) found that intrathecally administered nociceptin results in heat hyperalgesia and mechanical allodynia in mice. These responses were prevented by neonatal capsaicin treatment. The heat hyperalgesia was blocked by the NK-1 antagonists, CP96,345 and CP99,994, but the mechanical allodynia was not. Conversely, in GluRepsilon subunit knock-out mice, the allodynia was not present but the heat hyperalgesia was retained. Morphine reduced both the hyperalgesia and the allodynia. It was concluded that SP is involved in the heat hyperalgesia and NMDA receptors in the tactile allodynia.

Somatostatin

Somatostatin is found in primary afferent fibers and in interneurons within the dorsal horn (Hökfelt et al., 1976; Seybold and Elde, 1980; Hunt, S. P. et al., 1981a), as well as in the terminals of axons that descend from the medulla (Millhorn et al., 1987b). Noxious thermal stimuli, but not noxious mechanical stimuli, applied to the skin cause release of somatostatin in the dorsal horn (Kuraishi et al., 1985; Morton et al., 1988; 1989). However, SP is released by both noxious mechanical and thermal stimuli (see previous section), suggesting that somatostatin is released from heat nociceptors, whereas SP is released from polymodal nociceptors. There may also be release of somatostatin from interneurons (Morton et al., 1989). As for SP, somatostatin release is mainly in the region of the substantia gelatinosa, as shown by the antibody microprobe technique (Morton et al., 1989).

Somatostatin has been reported to cause a selective inhibition of nociceptive dorsal horn interneurons following its iontophoretic application (Randić and Miletić, 1978). It is not evident if release of this substance from primary afferent fibers causes inhibition; equally likely is that the inhibitory effects are related to actions at synapses formed by interneurons containing this peptide.

Galanin

Galanin is a peptide that is found in a subpopulation of small dorsal root ganglion cells and that may play an important role in antinociception (Xu, X. J. et al., 2000). It acts through G-protein coupled receptors. Binding sites for galanin have been found on dorsal root ganglion cells (Zhang, X. et al., 1995c). There are at least three types of galanin receptors: GalR1, GalR2, and GalR3 (Smith, K. E. et al., 1997; 1998; Wang, S. et al., 1998; Watling, 2001). These have actions through a variety of signal transduction systems, including adenylyl cyclase (GalR1) and phospholipase C (GalR2). GalR3 opens inwardly rectifying potassium channels.

Intrathecal administration of galanin has been shown to reduce the hyperactive reflex excitability seen during inflammation (Xu, I. S. et al., 1998) and to alleviate the mechanical and cold allodynia that follow peripheral nerve injury (Hao et al., 1999). Mice that overexpress galanin were shown to have elevated thresholds for behavioral responses to noxious heat stimuli (Blakeman et al., 2001).

Superfusion of the spinal cord with galanin causes an inhibition of wide-dynamic-range neurons, both in normal animals and after sciatic nerve ligation; the effect is stronger after nerve injury (Xu, S. et al., 2000; see also Yu, L. C. et al., 2001). On the other hand, intrathecal galanin has been reported to facilitate mechanical, but not thermal, nociceptive transmission (Kuraishi et al., 1991). Furthermore, galanin knock-out mice display reduced second-phase responses in the formalin test and no increase in nociceptive reflex excitability following carrageenan inflammation (Kerr et al., 2001), implying that an action of galanin is necessary for the development of central sensitization.

Intracerebroventricular administration of galanin potentiates the analgesic action of morphine (Przewlocka et al., 1995), and intrathecal administration of a galanin antagonist abolishes the antinociceptive action of morphine (Reimann et al., 1994).

These diverse results suggest an important, but complex, role of galanin in nociceptive processing.

Bombesin

Bombesin and the related peptides, neuromedin B and neuromedin C, also have inhibitory actions. Bombesin-like immunoreactivity has been found in small DRG cells and in the superficial layers of the dorsal horn, and the immunoreactivity in the dorsal horn decreases following dorsal rhizotomy (Fuxe et al., 1983; Panula et al., 1983; Massari et al., 1985). Iontophoretic release of these peptides near interneurons of the dorsal horn inhibits the responses of these cells, especially responses to noxious stimuli (De Koninck and Henry, 1989).

Neuropeptide Y (NPY)

Neuropeptide Y (NPY) is normally not found in DRG neurons or is present at only very low levels (Gibson et al., 1984a). However, Y1 and Y2 NYP receptors are found in DRG cells (Brumovsky et al., 2002). The DRG cells that express Y1 receptors generally also express CGRP and to a lesser extent SP. Y2 receptors are found in large CGRP-containing DRG cells. Peripherally, terminals containing Y1 receptors can be seen in the dermis and even epidermis, and these terminals double-label for CGRP. NYP receptors have also been found on neurons in laminae I–VI. Dorsal rhizotomy eliminates immunostaining for CGRP in the dorsal horn, but immunostaining for Y1 receptors is not clearly changed. This may reflect a substantial staining of Y1 receptors in dorsal horn neurons. Both Y1 and Y2 receptors could contribute to presynaptic inihibition of the effects of primary afferent neurons (Brumovsky et al., 2002). Mice lacking an NPY receptor show less antinociception and plasma extravasation (Naveilhan et al., 2001).

Pharmacology of PAD

The neural circuit that is responsible for the generation of the PAD triggered by the activation of primary afferent fibers is polysynaptic. Several different neurotransmitters and receptors may therefore be involved, including excitatory linkages in the initial part of the pathway and finally an inhibitory linkage that involves the terminals of the primary afferent fibers that are affected by the presynaptic inhibition.

Role of Excitatory Amino Acids

Evans and Long (1989) examined the effect of excitatory amino acid receptor antagonists on the generation of dorsal root potentials following stimulation of an adjacent dorsal root in an in vitro preparation of rat spinal cord. The NMDA antagonist, AP5, had little effect, but the non-NMDA antagonist, CNQX, caused a reduction in the dorsal root potential. Presumably, the reduction was due to an action at the initial synapse in the pathway responsible for evoking the dorsal root potential. Excitatory synapses in the pathway appeared not to use NMDA receptors.

Role of GABA

There is considerable evidence that the synaptic transmitter acting on primary afferent terminals to produce PAD and presynaptic inhibition is GABA. For example, PAD is at least partly blocked by the $GABA_A$ receptor antagonists, picrotoxin and bicuculline (Eccles et al., 1963b; Besson et al., 1971; Curtis et al., 1971a; 1977; Levy et al., 1971; Benoist et al., 1972; 1974; Davidoff, 1972; DeGroat et al., 1972; Levy and Anderson, 1972; Barker, J. L. and Nicoll, 1973; Levy, 1974; 1975; McLaughlin et al., 1975; Gmelin and Cerletti, 1976; Repkin et al., 1976; Barber et al., 1978; Gallagher et al., 1978; Hackman et al., 1982; Mokha et al., 1983; Rudomin et al., 1990; see reviews by Schmidt, R. F., 1971; Nicoll and Alger, 1979; Rudomin and Schmidt, 1999; Willis, 1999). PAD is also reduced by depletion of GABA by semicarbazide (Bell and Anderson, 1972; Repkin et al., 1976), and PAD is enhanced when GABA-transaminase is inhibited (Davidoff et al., 1973). Finally, group Ia afferent fibers are depolarized by iontophoretically applied GABA (Curtis and Lodge, 1982; Curtis et al., 1982; however, cf. Curtis and Ryall, 1966).

Ionic Mechanism

The depolarization of primary afferent fibers by GABA appears to involve an ionic mechanism different from that which produces IPSPs. Dorsal root ganglion cells contain a different chloride transporter than do postsynaptic dorsal horn neurons. The chloride transporter in DRG cells pumps Cl^- into the cells by simultaneously moving 2 Cl^- ions, 1 Na^+ ion and 1 K^+ ion across the surface membrane and into the cytoplasm (Fig. 7.28; Alvarez-Leefmans et al., 1988; 1998; 2001). The Na^+ and K^+ concentration gradients are readjusted by the Na^+-K^+ ATPase pump. When GABA acting at $GABA_A$ receptors opens Cl^- channels, Cl^- leaves the cytoplasm of DRG cells, depolarizing them. A similar change can be presumed to occur at the presynaptic terminals of the same primary afferent neurons.

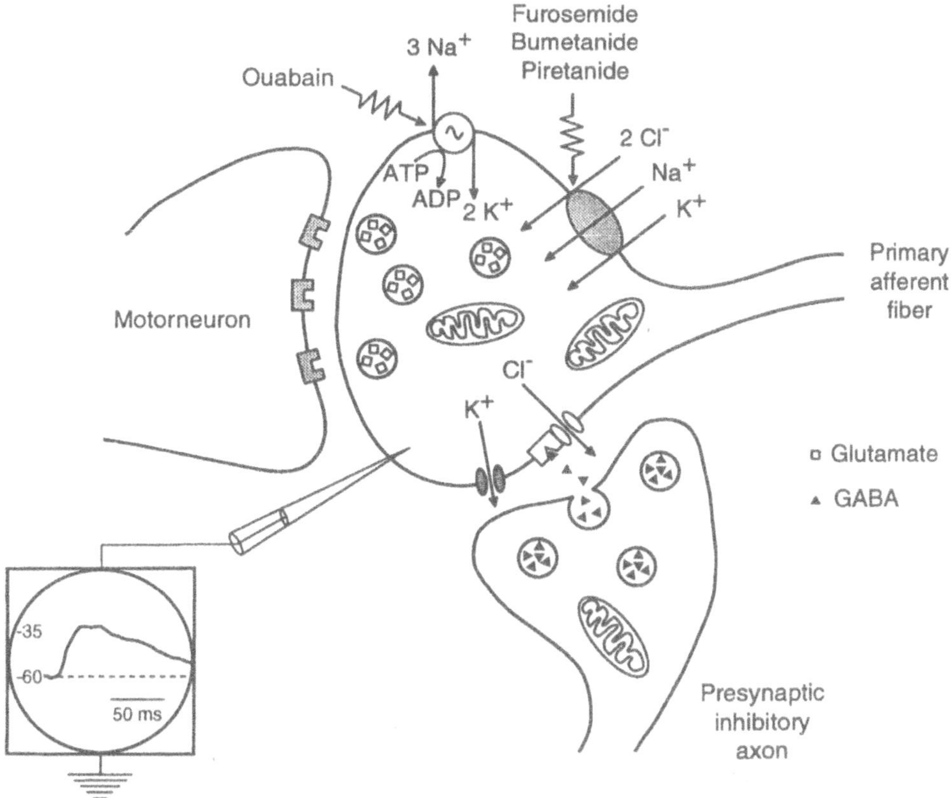

FIGURE 7.28. Mechanism of PAD. Drawing of a synapse formed by a group Ia muscle afferent fiber (center) on a motor neuron (left). The synaptic vesicles in the afferent terminal contain glutamate (open squares). The chloride co-transporter is shown in the terminal membrane by the dark oval. Nearby is the sodium pump. The co-transporter causes Cl^- to be concentrated in the cytoplasm of the afferent neuron. Below, the ending of a GABAergic interneuron is shown releasing GABA (filled triangles). The GABA opens chloride channels, resulting in the exodus of Cl^-, which depolarizes the group Ia fiber (recording at the lower left). K^+ leaves the terminal passively. Several drugs that can affect the chloride co-transporter are indicated. (From Alvarez-Leefmans et al., 1998.)

Other Transmitters

Two other neurotransmitters that may also affect ion channel-coupled receptors on primary afferent terminals to produce a depolarization are serotonin acting on 5-HT3 receptors, and glutamate, acting on kainate receptors (Agrawal and Evans, 1986; Evans and Long, 1989; Peng et al., 2001). A direct action of glutamate on kainate receptors on primary afferents terminals would be an alternative explanation of the effectiveness of CNQX in blocking PAD.

Evidence that the activation of kainate receptors can produce presynaptic inhibition comes from in vitro experiments (Kerchner et al., 2001). In one set of experiments, isolated DRG cells were co-cultured with dorsal horn neurons. Recordings were made from dorsal horn neurons in response to direct stimulation of a synaptically coupled DRG neuron. AMPA responses were blocked with the AMPA receptor-selective antagonist, SYM2206, and inhibitory responses to GABA and glycine were blocked with bicuculline and strychnine. Mg^{++} was not included in the bathing solution, and so fixed latency

NMDA-mediated EPSCs could be recorded when the DRG cell was activated. Addition of kainate to the bathing solution produced a dose-dependent reduction in the EPSCs. Antagonists of $GABA_B$ or metabotropic glutamate receptors did not affect this reduction. The GluR5 agonist, APTA, mimicked the effect of kainate. APTA has an action on kainate receptors on primary afferent neurons, which express more mRNA for GluR5 subunits than do dorsal horn neurons. Kainate also reduces the amplitude of NMDA-mediated EPSCs evoked by dorsal root stimulation in a spinal cord slice preparation.

Other agents can produce presynaptic inhibition without PAD. These substances interact with presynaptic receptors to decrease Ca^{++} influx in response to the invasion of the presynaptic action potential and consequently cause a reduction in the release of excitatory transmitter from the terminals. An example of this is GABA, or an agonist such as baclofen, acting on $GABA_B$ receptors (Redman, 1998). Other examples are peptides, such as endogenous opioids, that act on presynaptic peptide receptors (see section on Inhibitory Peptides in the Dorsal Horn).

Potassium Release

An alternative mechanism for producing PAD is the release of K^+ into the extracellular space (Vyklický et al., 1972; 1975; 1976; Krnjevic and Morris, 1972; 1974; Kríz et al., 1974; 1975; Bruggencate et al., 1974; Somjen and Lothman, 1974; Lothman and Somjen, 1975; Syková et al., 1976; Syková and Vyklický, 1977; 1978; Deschenes and Feltz, 1976; Deschenes et al., 1976; Nicoll, 1979; Davidoff et al., 1980; 1988; Evans, 1980; Syková et al., 1980; Czéh et al., 1981; Bagust et al., 1982; Urban et al., 1985; Svoboda et al., 1988). The main region of spinal cord gray matter that undergoes large changes in K^+ concentration is the intermediate region, rather than the superficial dorsal horn or ventral horn.

An increase in extracellular K^+ concentration does not account for the specific organization of PAD (Rudomin et al., 1981). Nor does such an increase in $[K^+]o$ account for the susceptibility of PAD to $GABA_A$ receptor antagonists, such as picrotoxin and bicuculline, and to CNQX (see above). Thus, there are two independent processes for generating PAD: an early, synaptically generated component that can be decreased by GABA antagonists and CNQX, and a late component that is secondary to K^+ release and is unaffected by GABA antagonists. K^+ release would also have important postsynaptic effects (Czéh et al., 1981).

The uptake of the excess K^+ seems to be through active transport and may produce primary afferent hyperpolarization by electrogenic pumping of Na^+ (Davidoff and Hackman, 1980; Czéh et al., 1981).

Activation of motor neurons is not an important mechanism of K^+ release in the spinal cord, since antidromic activation of large numbers of motor neurons fails to cause a detectable change in K^+ level (Nicoll, 1979). Using the isolated frog spinal cord, Davidoff, R. A. et al. (1988) showed that K^+ release was reduced by 85% when the Mg^{++} concentration in the bathing solution was increased to a level that prevented synaptic transmission from primary afferent fibers. They concluded that 85% of the K^+ release is from synaptically activated interneurons and 15% from primary afferent fibers. K^+ release could also be produced by administration of excitatory amino acids and, to a lesser extent, tachykinin peptides. A nonselective glutamate receptor antagonist prevented most of the K^+ release caused by excitatory amino acids; the NMDA antagonist AP5 was less effective; and a SP antagonist had no effect. Since release of K^+ by excitatory amino acids and tachykinins was prevented by tetrodotoxin, it was concluded that the K^+ release by these agents depended on the excitation of interneurons.

Acute nociceptive stimuli cause a transient increase in extracellular K^+ concentration that lasts 5–30 sec, but chemical or thermal injury causes a prolonged increase, beginning 5–15 min after injury and lasting more than 2 hr (Svoboda et al., 1988). This change in extracellular K^+ concentration could play a role in sensory changes that follow injury.

Summary

A number of candidate neurotransmitters and modulators have been identified that may play important roles in synaptic transmission in pathways that involve dorsal horn interneurons. Excitatory transmitters include excitatory amino acids (glutamate, aspartate) and ATP, as well as several peptides, including SP and CGRP. Excitatory amino acid receptors include ionotropic glutamate receptors (AMPA, kainate, and NMDA receptors) and metabotropic glutamate receptors (groups I, II, and III). Cloning of each of these has revealed that they are made up of various combinations of subunits. The kinetic properties of the various excitatory amino acid receptors differ. AMPA and kainate receptors have fast kinetics that account for fast synaptic transmission. NMDA receptors have slower kinetics because Mg^{++} blocks the receptor channel until the membrane is depolarized to some extent. Metabotropic glutamate receptors and peptide receptors have still slower kinetics, because these receptors are coupled to intracellular metabolic pathways through G-proteins.

Primary afferent nociceptors contain co-localized glutamate and such peptides as SP and CGRP. SP is also found in dorsal horn interneurons and in the axons of some bulbospinal axons. Excitation of dorsal horn neurons by primary afferent fibers may involve early and late components as a result of successive actions by excitatory amino acids and peptides. The membrane effects of SP include a slow depolarization due to a complex action on several conductance mechanisms and activation of second-messenger pathways. CGRP appears to enhance the action of SP by inhibiting its breakdown and by promoting its release. It can also contribute to the slow depolarization of dorsal horn neurons. Other excitatory peptides include neurokinin A, neurotensin, cholecystokinin, thyrotropin-releasing hormone and perhaps corticotropin-releasing hormone, pituitary adenylate cyclase-activating polypeptide, and islet amyloid polypeptide.

Fast-acting inhibitory substances in the dorsal horn include the inhibitory amino acids, GABA and glycine, and acetylcholine. These substances are found in small interneurons, including many cells in the superficial layers of the dorsal horn. Acetylcholine is also found in some primary afferent neurons. GABA and glycine both produce postsynaptic inhibition through an increase in the membrane conductance for Cl^- ions. Chloride is pumped out of interneurons, and so an increase in chloride conductance in these postsynaptic neurons results in their hyperpolarization. GABA also produces PAD by an action of $GABA_A$ receptors that increases Cl^- conductance in the primary afferent terminals. However, the chloride gradient is reversed in primary afferent neurons, and so an increase in Cl^- conductance leads to a depolarization, since anions leave the primary afferent terminals when the chloride channels are opened. A chloride co-transporter is responsible for the concentration of Cl^- in these neurons. PAD can also be produced by release of K^+ ions into the extracellular space, at least in the intermediate nucleus.

Inhibitory peptides in the dorsal horn are found in interneurons or in primary afferent terminals. Dorsal horn inhibitory peptides include enkephalin, which may be released when noxious stimuli are applied. Other opioid peptides in the dorsal horn include dynorphin and the endomorphins. Endomorphins act on μ opiate receptors, enkephalin on δ receptors, and dynorphin on κ receptors. Opiate receptors are located both on primary afferent terminals and on the membranes of postsynaptic neurons. A presynaptic opioid action results in presynaptic inhibition of neurotransmitter release. The postsynaptic action of opioids includes a hyperpolarization due to an increase in K^+ conductance. It is unclear if dynorphin is inhibitory or excitatory.

Other inhibitory peptides in the dorsal horn are nocistatin and nociceptin, somatostatin, galanin, bombesin, and neuropeptide Y.

MODULATION OF THE RESPONSES OF DORSAL HORN NEURONS

Receptive Fields and Alterations in their Sizes

Dorsal horn neurons often have both excitatory and inhibitory receptive fields. Wall (1960) found that the excitatory receptive fields of cells in the dorsal horn are ovoid and larger than the receptive fields of primary afferent fibers. The central part of a receptive field was the most sensitive. Strong stimuli applied to the skin adjacent to the excitatory field could produce inhibition. Hillman and Wall (1969) investigated further the receptive fields of cells located in the region of lamina V. The cells had larger receptive fields than did cells characteristic of lamina IV. The receptive fields consisted of three concentric zones (Fig. 7.29). In the center of the receptive field, the movement of hair was sufficient to activate the neuron, and stronger stimuli caused stronger excitation. Pressure or pinch stimuli were required to activate the cell when stimuli were applied more peripherally. In the surrounding area of skin, tactile stimuli produced inhibition. Electrical stimulation of the dorsal columns or dorsal roots other than those carrying the excitatory afferent fibers also produced inhibition of neurons in lamina V.

Excitatory Receptive Fields

Cascade Theory

Wall (1967) suggested that neurons in lamina IV excite those in lamina V and that cells in lamina V excite those in lamina VI. The evidence was that neurons in lamina V had larger receptive fields than did those in lamina IV, and the latency for activation was longer by an average of 1.5 msec. For neurons in lamina VI, the receptive fields on the skin were similar to those of cells in lamina V, but response latencies were longer still. Thus, Wall hypothesized that the excitation of neurons in the deep layers of the dorsal horn is produced by a cascade of excitation that descends from lamina IV to lamina V and then to lamina VI.

The progressive increase in receptive field size that Wall (1967) described for neurons at progressively greater depths in the dorsal horn was not confirmed by Brown, P. B. et al. (1975; see also Brown, P. B., 1969), nor was the progressively greater latency of response. Brown, P. B. et al. (1975), therefore, dispute the idea of a dorsal to ventral cascade of excitation of interneurons in the deep layers of the dorsal horn (however, cf. Egger et al., 1986).

Somatotopic Organization

Wall (1967) described a somatotopic organization of the dorsal horn. The toes are represented medially and the lateral aspect of the foot and more proximal regions are

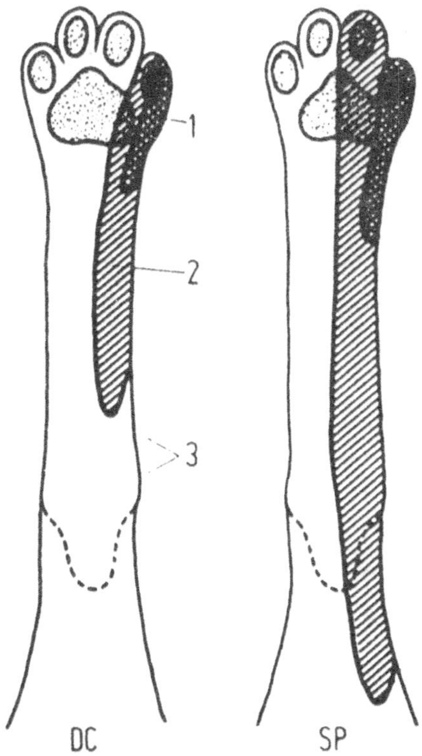

FIGURE 7.29. Organization of the receptive field of a wide-dynamic-range dorsal horn neuron. In the decerebrate state (DC, left), the receptive field included three concentric zones. The cell responded to innocuous and noxious stimuli that were applied in zone 1, but only to noxious stimuli applied in zone 2 (innocuous stimuli here produced inhibition). Mechanical stimuli applied in the surrounding zone 3 (boundary not shown) resulted in inhibition. In the spinalized state (SP, right) produced by cold block of the spinal cord, zones 1 and 2 expanded but zone 3 disappeared. (From Hillman and Wall, 1969.)

represented laterally in the dorsal horn. This description of the somatotopic arrangement of the dorsal horn is similar to that described later by Brown, P. B. and Fuchs (1975). Brown and Fuchs found both disto-proximal and ventro-dorsal gradients in the receptive fields of interneurons located across the medio-lateral extent of the dorsal horn (Fig. 7.30). The observation by Devor and Wall (1976) that cells placed laterally in the dorsal horn may have very proximal receptive fields is consistent with this scheme. Several groups have confirmed and extended these observations (Brown, P. B. and Koerber, 1978; Pubols and Goldberger, 1980; Light and Durkovic, 1984; Sorkin et al., 1986; Wilson, P. et al., 1986; Hylden et al., 1987). The adult pattern is already present in the neonatal kitten (Wilson, P. and Snow, 1988).

P. B. Brown and his colleagues proposed that the somatotopic organization of dorsal horn interneurons is a consequence of the organization of the cutaneous afferent projections to the dorsal horn. Transganglionic labeling has been used to demonstrate the distribution of the afferent projections of various cutaneous nerves. The findings are generally consistent with the somatotopic organization determined from recordings from postsynaptic neurons of the dorsal horn (Brown, P. B. and Culberson, 1981; Koerber and Brown, 1980; 1982; Molander and Grant, 1985; 1986; Swett and Woolf, 1985; Woolf and Fitzgerald, 1986). However, the afferent projections appear to extend to some degree into

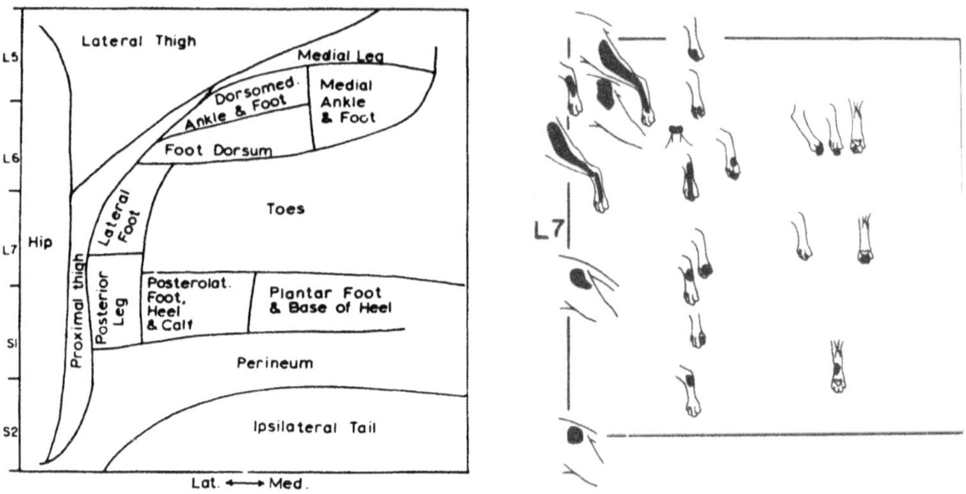

FIGURE 7.30. Dorsal horn somatotopic organization. **Left**, scheme of the somatotopic organization of the dorsal horn in a horizontal plane through the lumbosacral enlargement of the cat. The abscissa indicates the lateral-to-medial extent of the dorsal horn, while the ordinate shows the segmental level. **Right**, receptive fields of dorsal horn neurons in the L7 segment. (From Brown, P. B. and Fuchs, 1975.)

non-somatotopically appropriate regions (Koerber and Brown, 1982; Meyers and Snow, 1984; 1986; Meyers et al., 1984; LaMotte, C. C. et al., 1989; Shortland et al., 1989). Such projections could form the basis of plastic changes that can be produced in receptive fields (see below) and that have been attributed to changes in the effectiveness of "relatively ineffective synapses" in exciting dorsal horn neurons (Basbaum and Wall, 1976; Devor and Wall, 1981a). Other concepts that may help account for such plastic changes are "long-ranging" afferent fibers (Mendell et al., 1978; Wall and Werman, 1976), the "low-probability firing fringe" of the receptive fields of some dorsal horn interneurons (Woolf and King, 1989) and more recently, "silent glutamatergic synapses" (Li, P. and Zhuo, 1998; see also Malenka and Nicoll, 1997; Isaac et al., 1999).

The somatotopic organization of primary afferent projections and of interneurons in the cervical enlargement of the cat is similar to that seen in the lumbosacral enlargement (Koerber, 1980; Nyberg and Blomqvist, 1985; Sorkin et al., 1986). The same statement can be made for the cervical spinal cord of monkeys (Florence et al., 1988; Brown, P. B. et al., 1989). There is a simpler somatotopic organization in the thoracic cord (Grant, G. and Ygge, 1981; Smith, C. L., 1983; Ygge and Grant, 1983).

Bryan et al. (1973; 1974) have suggested that the somatotopic organization of the dorsal horn reflects the embryological development of the innervation of the limbs (cf. Sherrington, 1893). The embryonic spinal cord initially has a simple somatotopic arrangement, in which neurons in the lateral part of the dorsal horn receive projections of afferent fibers supplying receptive fields in the dorsal parts of the corresponding dermatomes and neurons in the medial dorsal horn from the ventral parts of the dermatomes. This arrangement is maintained in the adult thoracic cord (Grant, G. and Ygge, 1981; Smith, C. L., 1983; Ygge and Grant, 1983; cf. Cervero and Tattersall, 1985). However, in the enlargements, as the limb bud elongates and rotates, the simple dermatomal pattern is distorted (Fig. 7.31A, B). Afferent fibers from more rostral dermatomes continue to project to more rostral segments, and afferent fibers from the dorsal parts of the embryonic

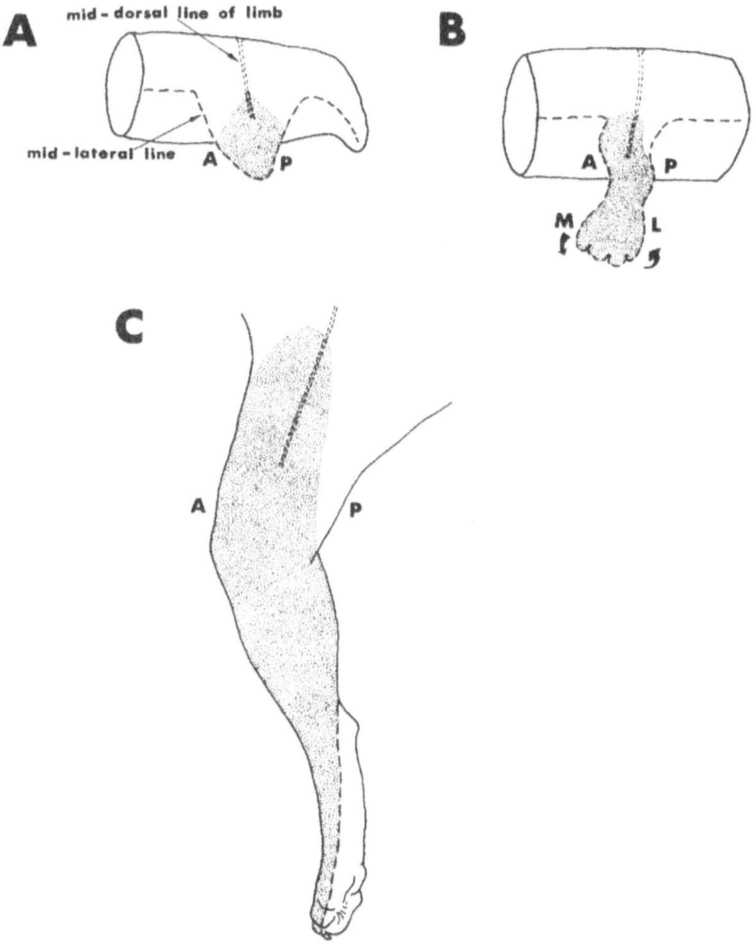

FIGURE 7.31. Relationship of the developing limb-bud to the adult dermatomes in the cat hindlimb. The mid-lateral line is indicated in A and B. The dorsal surface of the L6 dermatome is stippled. A–C, progressive elongation and medial rotation of the distal limb. (From Bryan et al., 1973.)

dermatomes maintain their projection to the lateral dorsal horn, but the location of the embryonic dorsal surface of the limb shifts. For example, in the hindlimb of the adult cat, the dorsal embryonic surface forms the lateral aspect of the thigh and leg, but the dorsal surface of the foot (Fig. 7.31C). Therefore, it would be predicted that afferent fibers from the dorsal skin of the foot would project more laterally and afferent fibers from the plantar surface of the foot more medially in the dorsal horn (Brown, P. B. and Fuchs, 1975; Sorkin et al., 1986).

Xian-Min and Mense (1990) have reported that neurons in the deep layers of the dorsal horn with input from deep tissues are also somatotopically organized. Neurons that are excited by inputs from more distal parts of the limb are located medially to neurons with input from more proximal parts. If there is convergent input from the skin, the cutaneous input to these neurons is not somatotopically organized, suggesting that these neurons function to signal deep rather than superficial input.

For further discussion of somatotopy with respect to the ascending tracts, see Chapters 9–11.

Changes in Receptive Fields

The receptive fields of at least some dorsal horn neurons are modifiable. This implies that the sensations evoked by stimulation of these receptive fields can change under certain conditions. Many sensory abnormalities are known to occur in pathologic conditions, including paresthesias, dysesthesias, pain, hyperalgesia, and allodynia. Presumably, many of these examples of sensory pathology are reflected in the plasticity of the receptive fields of dorsal horn neurons.

Dubuisson et al. (1979) described spontaneous alterations in the sizes of the cutaneous receptive fields of neurons of the substantia gelatinosa. They referred to these changeable receptive fields as "ameboid." Devor and Wall (1981a) mapped the excitatory receptive fields of neurons in laminae IV and V of decerebrate, spinalized cats. The neurons in lamina IV had stable receptive fields, but those in lamina V had receptive fields that varied spontaneously by as much as 30% in size. These changes resembled the ameboid receptive fields in the superficial dorsal horn, except that the changes were not as large. Tetanic electrical stimulation often caused a reduction in receptive field size, followed by gradual recovery over several minutes. Blocking or cutting a peripheral nerve could cause loss of the entire receptive field, loss of part of the receptive field, or no change. In a few cases, a new field developed. By contrast, stimulation of C fibers caused an increase in responsiveness of dorsal horn neurons (Cervero et al., 1984; Cook, A. J. et al., 1987).

Pharmacological evidence for plasticity of the receptive fields of dorsal horn neurons comes from the experiments of Zieglgänsberger and Herz (1971). These investigators showed that the iontophoretic release of excitatory amino acids in the vicinity of dorsal horn neurons causes an increase in the sizes of their receptive fields (see also Dougherty and Willis, 1991a).

Noxious stimuli have been found to produce a long-lasting alteration of the receptive fields of dorsal horn neurons. For example, injury to the skin proximal to the receptive field of a dorsal horn neuron causes the expansion of that receptive field toward the area of damage (McMahon and Wall, 1983a). Noxious heat causes an enhancement of the responses of spinothalamic neurons to repeated noxious heat stimuli (Kenshalo et al., 1979; Ferrington et al., 1987b) and to innocuous mechanical stimuli (Kenshalo et al., 1982). Simone et al. (1989b, 1991) showed that intradermal injections of capsaicin cause an expansion of the receptive fields of dorsal horn neurons, including spinothalamic neurons, to innocuous and noxious mechanical stimulation. This change parallels the development of secondary allodynia and hyperalgesia in human subjects given similar injections (Simone et al., 1989a; 1991). A similar change is produced by application of mustard oil to the skin (Woolf and King, 1989), intradermal injection of Freund's adjuvant (Hylden et al., 1989), or noxious stimulation of deep tissue (Hoheisel and Mense, 1989). Experimentally induced acute arthritis also leads to increases in the responsiveness of dorsal horn neurons (Schaible et al., 1987b; Calvino et al., 1987; Neugebauer and Schaible, 1990).

Cervero et al. (1988) found that repeated noxious pinching of the skin of the tail causes a change in the responses of nociceptive neurons in the dorsal horn of rats. Further exploration of this (Laird and Cervero, 1989) revealed that wide-dynamic-range neurons often showed a dramatic increase in receptive field size, whereas there were either no changes or only small increases in the receptive fields of high-threshold neurons (Fig. 7.32). Based on

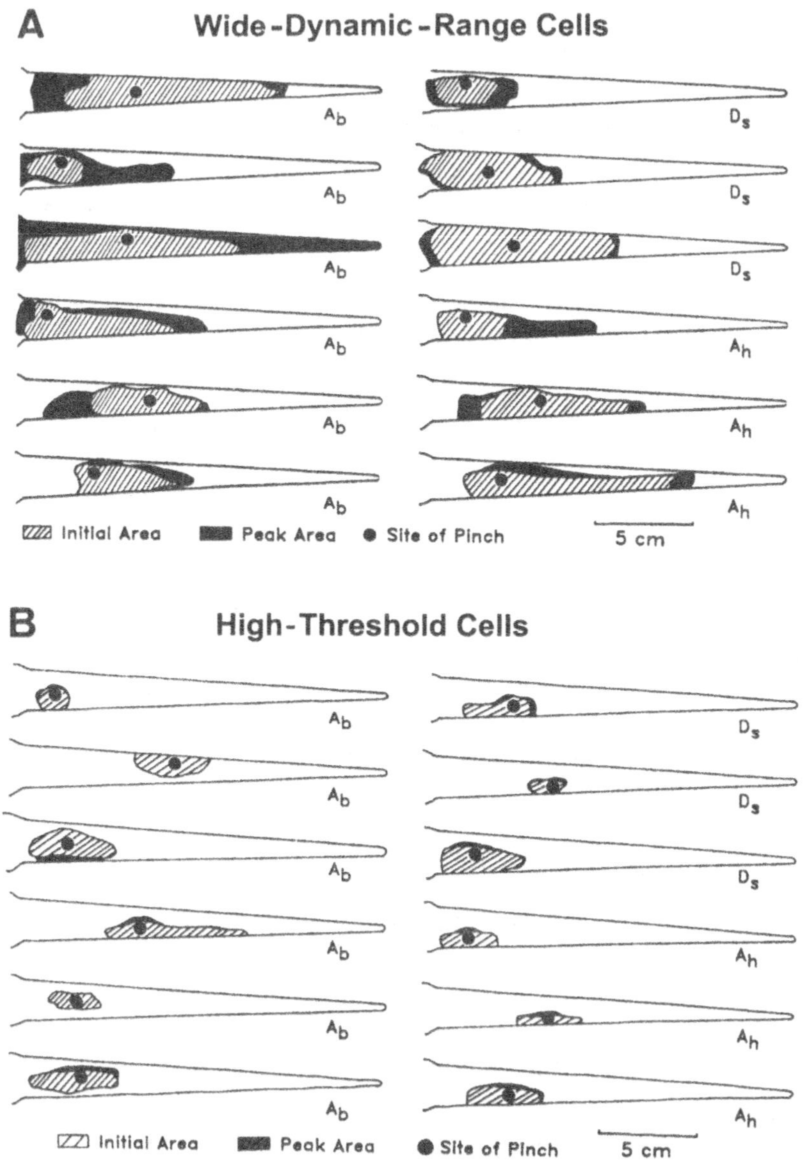

FIGURE 7.32. A, expansion of receptive fields of wide-dynamic range dorsal horn neurons following prolonged noxious pinch of the skin. B, minimal changes in the receptive fields of high-threshold dorsal horn neurons after noxious pinch. Abbreviations refer to the anesthetic state: A_b, barbiturate-anesthetized; A_h, halothane-anesthetized; D_s, decerebrate-spinalized. (From Laird and Cervero, 1989.)

this evidence, it was proposed that the responses of wide-dynamic-range neurons are more readily changed than are those of high-threshold neurons.

Cervero et al. (1992) mapped the cutaneous receptive fields of dorsal horn neurons in anesthetized cats. Noxious stimulation of the biliary system by distension of the bladder not only activated some of the dorsal horn neurons but also increased the sizes of their cutaneous receptive fields. If there were no input from the gall bladder, the cutaneous receptive fields were unaffected.

Long-term Potentiation and Long-term Depression

Evidence that spinal cord neurons can undergo long-lasting changes in excitability was obtained in experiments done in vitro on spinal cord neurons, and these changes were designated as long-term potentiation (LTP) and long-term depression (LTD) (Randic et al., 1993).

Several groups have since confirmed that LTP can be produced in the spinal cord (Svendsen et al., 1997; Sandkühler and Liu, 1998; Rygh et al., 1999). LTP has been demonstrated in recordings of C-fiber-evoked field potentials (Fig. 7.33; Liu and Sandkühler, 1995b) or the discharges of individual nociceptive dorsal horn neurons (Svendsen, F. et al., 1997; 1998; 1999; Svendsen, L. J. et al., 1999). To trigger LTP or LTD by electrical stimulation of primary afferent fibers, it is necessary to use the appropriate stimulation parameters. For example, according to Liu and Sandkühler (1997), high-frequency, high-intensity stimulation of C fibers always induces LTP in dorsal horn neurons (four 1 s trains of 100 Hz stimulus pulses of 30–40 V and 0.5 ms duration). LTP can also be produced by noxious stimulation (Rygh et al., 1999). However, Sandkühler et al. (1997) found that low frequency stimulation of Aδ fibers instead evokes LTD (nine hundred 10 V, 0.1 ms pulses at 1 Hz).

LTP in the spinal cord is blocked by administration of an antagonist of AMPA or NMDA receptors (Liu and Sandkühler, 1995b; Svendsen et al., 1998) or of NK-1 or NK-2 receptors (Liu and Sandkühler, 1997). LTP is initiated if NMDA or neurokinin receptors are activated (Liu and Sandkühler, 1998).

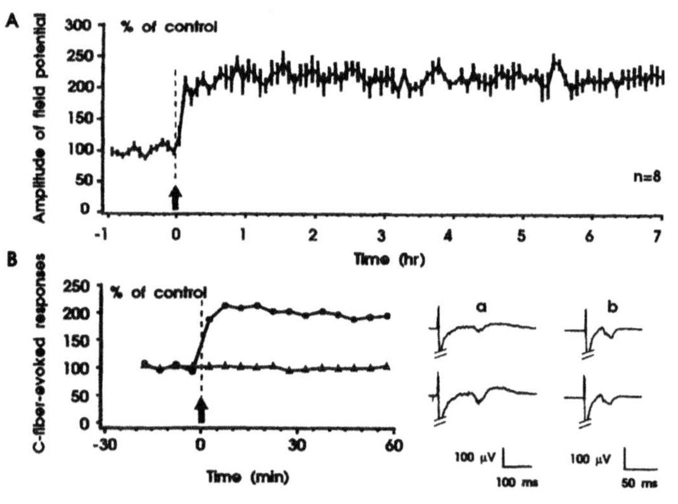

FIGURE 7.33. A, LTP of C-fiber evoked field potentials in rat spinal cord. The sciatic nerve was stimulated at the time indicated by the arrow. The mean size of the field potential ± SE is shown relative to control size over time. B, measurements of the mean amplitudes of C-fiber-evoked field potentials (filled circles) and C-fiber volleys (filled triangles) recorded at 1-min intervals before and after stimulation. Sciatic nerve stimulation resulted in LTP but did not affect the size of the C-fiber volley. To the right are recordings of the C-fiber field potentials (a) and C-fiber volleys (b) before (top) and 25 min after (bottom) stimulation. (From Liu and Sandkühler, 1997.)

INHIBITORY RECEPTIVE FIELDS

As mentioned earlier, it is well known that dorsal horn neurons have inhibitory, as well as excitatory, receptive fields (Wall, 1960; Hillman and Wall, 1969). Inhibition can be produced by natural stimulation of the skin or other organs, by electrical stimulation of peripheral nerves or of the dorsal column, and by electroacupuncture (Hongo et al., 1966; Hillman and Wall, 1969; Besson et al., 1974; Handwerker et al., 1975; Pomeranz and Cheng, 1979; Woolf and Wall, 1982; Cadden et al., 1983; Fitzgerald, 1983; Tomlinson et al., 1983; Woolf, 1983; Kanui, 1987; Morris, R., 1987; Paik et al., 1988; Tsuruoka et al., 1990). It should not be surprising that inhibition from stimulation of peripheral structures is often employed as a maneuver to relieve pain, using such procedures as counterirritation, acupuncture, transcutaneous electrical nerve stimulation, vibratory stimulation, and dorsal column stimulation (Gammon and Starr, 1941; Wall and Sweet, 1967; Shealy et al., 1970; Nashold and Friedman, 1972; Kaada, 1974; Long, D. M. and Hagfors, 1975; Melzack, 1975; Lundeberg et al., 1984).

Gate Theory and Diffuse Noxious Inhibitory Controls

Inhibition of dorsal horn neurons following peripheral stimulation appears to result from activation of either of two different inhibitory mechanisms. One inhibitory system depends on stimulation of large myelinated afferent fibers and the other on the activation of finely myelinated and unmyelinated afferent fibers. The characteristics of the inhibitions that are produced by these two inhibitory systems are quite different. The inhibition produced by large afferent fibers tends to be of the "surround" type, and it is produced by weak mechanical stimuli. It depends on segmental mechanisms and is the type of inhibition on which the "gate theory of pain" is based (see next section). The inhibition evoked by fine afferent fibers can be produced by stimulation anywhere on the body surface and results from the activation of nociceptors. This system involves both a segmental and a supraspinal mechanism. This second inhibitory system is termed "diffuse noxious inhibitory controls" or DNIC (see below).

Gate Theory of Pain

In 1965, Melzack and Wall published a model of the dorsal horn circuitry responsible for pain transmission. They called this model the "gate control system." The gate theory has stimulated a great deal of research and has had a major impact on clinical thought. Nathan (1976) has criticized this theory, and Wall (1978) has rebutted some of the criticisms. Cervero and Iggo (1980) have advanced additional criticisms.

A major feature of the gate theory of Melzack and Wall (1965) is that activity in large afferent fibers inhibits synaptic transmission in a system otherwise activated by small afferent fibers carrying the signal for pain. A similar proposal, based on clinical findings, had already been made by Noordenbos (1959). Melzack and Wall illustrated a specific hypothetical circuit for the organization of a spinal cord gating mechanism (Fig. 7.34). They proposed that inhibitory neurons in the substantia gelatinosa act as a gating mechanism to control afferent input to nociceptive transmission (T) cells. The output of the T cells results in activation of neural mechanisms in the brain that allow the perception of pain and that trigger other pain responses.

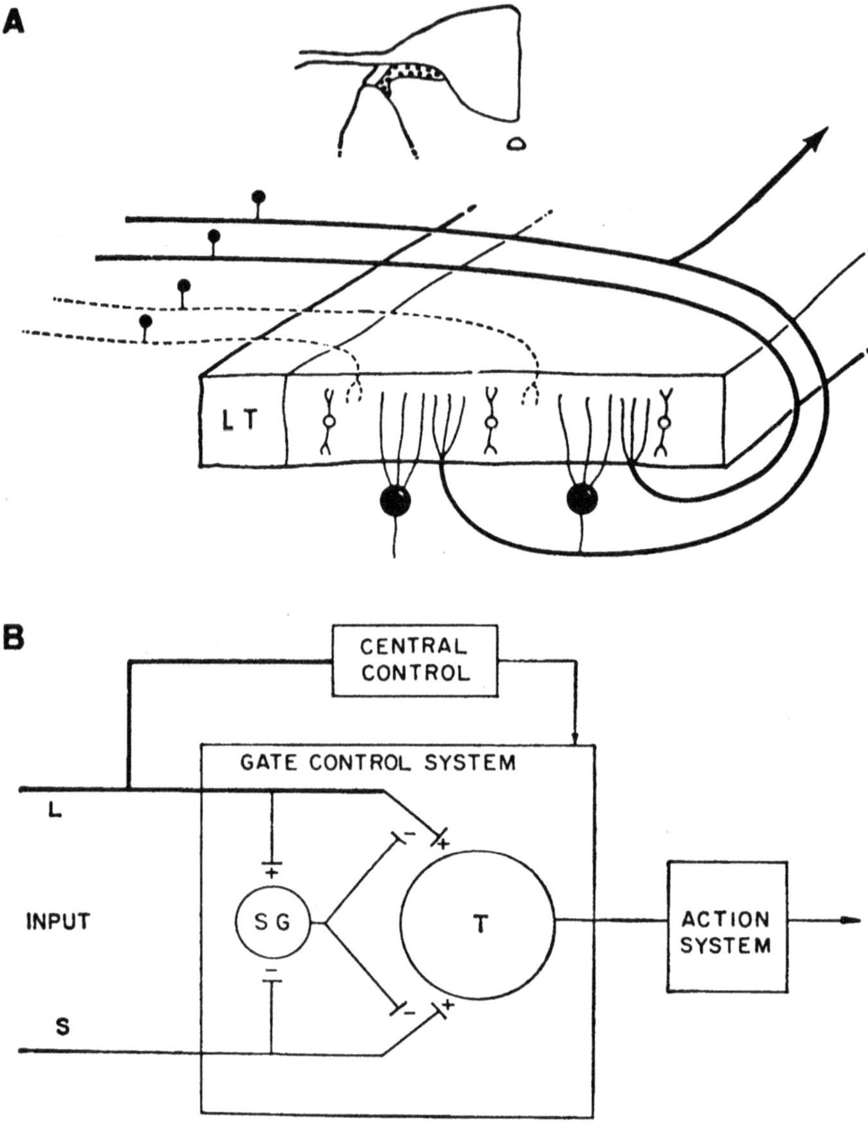

FIGURE 7.34. Gate theory of pain. A, representation of the terminations of large (heavy lines) and small afferent fibers (dashed lines) in the superficial dorsal horn. Some of the large afferents give off collaterals that ascend in the dorsal column (see line ending in arrowhead). The substantia gelatinosa contains interneurons whose cell bodies are shown by the open circles. Dendrites of neurons in deeper laminae are shown extending dorsally into the substantia gelatinosa. LT, Lissauer tract. B, circuit proposed for a gating system that might determine the output of the "action system" for pain. L, large; S, small; SG, substantia gelatinosa neuron; T, transmission cell. (From Melzack and Wall, 1965; copyright 1965 by the AAAS.)

Evidence for the involvement of the substantia gelatinosa in the control of nociceptive afferent input came from the experiments of Wall (1962) and Mendell and Wall (1964) on dorsal root potentials. Recordings from wide-dynamic-range dorsal horn neurons (candidate T-cells) showed that large afferent fibers produce a short-lasting excitation, whereas the additional activation of small fibers results in greatly increased responses (Wall, 1958; 1959;

Mendell and Wall, 1965; Mendell, 1966). Negative and positive feedback loops were proposed to account for the different responses (Melzack and Wall, 1965). The feedback was thought to occur at a presynaptic level, although postsynaptic actions were not excluded.

For the gate to operate properly there would have to be: (1) ongoing activity in small myelinated and unmyelinated fibers, (2) stimulus-evoked activity, and (3) a relative balance of activity in large and small afferent fibers. The tonic activity in small fibers would keep the gate partly open, whereas an input over large fibers would tend to close the gate, limiting the discharge of the T-cells. An increase in stimulus intensity would shift the balance of large- and small-fiber inputs, thereby increasing the output through the T cells. A prolonged stimulus would result in adaptation of the large afferent fibers, and the small fibers would get the upper hand, opening the gate further. The gate could be returned toward a closed position by additional large-fiber input.

An important aspect of the gate theory is the role of descending control systems, which can alter the gate. The activity in descending pathways must be appropriate to the situation, and so a central control trigger was proposed. The dorsal column pathway and the spinocervicothalamic pathway were considered as likely candidates to provide the discriminative information needed by the brain to decide whether or not to alter the sensitivity of the gating mechanism.

Since the theory was proposed, criticisms of the gate theory center have been based on findings in three areas: (1) the properties of cutaneous receptors, (2) the dorsal root potentials that are produced by large and small afferent fibers, and (3) the response characteristics of dorsal horn neurons.

Although the proportion of primary afferent nerve fibers that could be called nociceptors was thought to be small in 1965, later investigations have shown that there are in fact large numbers of nociceptors in peripheral nerves to skin, muscle, joints, and viscera. Nociceptive afferent axons are generally in the size ranges of small myelinated and unmyelinated fibers, although a few nociceptors are supplied by large myelinated axons (see Chapter 2). However, in the absence of damage or inflammation, nociceptive afferent fibers are silent, in contradiction to one of the requirements of the gate theory.

The circuit proposed to explain the negative and positive feedback in the operation of the gate control mechanism is shown in Fig. 7.34B. The neurons of the substantia gelatinosa are supposed to be excited by large afferent fibers, but they then produce presynaptic inhibition of the transmission from the large afferent fibers and so limit the excitation of the T cells. The PAD produced in large cutaneous fibers by the activity in other large afferent fibers is in keeping with this model (Koketsu, 1956; Eccles and Krnjevic, 1959; Eccles et al., 1963a; Hodge, 1972; Whitehorn and Burgess, 1973). However, Melzack and Wall (1965) suggested that small afferent fibers inhibit the tonic activity of inhibitory interneurons in the substantia gelatinosa and thus cause presynaptic disinhibition. The dorsal root potential associated with such an event would be positive, instead of the usual negative wave. Positive dorsal root potentials have been described by a number of investigators (Lloyd, 1952; Wall, 1964; Dawson et al., 1970; Mendell, 1970; 1972; 1973; Burke, R. E. et al., 1971; Hodge, 1972). However, in contrast to the prediction of the gate theory, the activation of unmyelinated afferent fibers by electrical stimuli is associated with the generation of negative dorsal root potentials in large fibers (Franz and Iggo, 1968; Zimmermann, 1968; Jänig and Zimmermann, 1971; Gregor and Zimmermann, 1973). Significantly, the activation of C fibers by a noxious thermal stimulus also produces a negative dorsal root potential (Burke, R. E. et al., 1971), even though positive dorsal root potentials could be evoked in the same animal, ruling out the condition of the animal as a factor. Since many small myelinated and unmyelinated fibers are connected with sensitive mechanoreceptors in cats, the finding that negative dorsal root potentials result when a natural noxious stimulus is used appears to be more meaningful than the finding of positive dorsal root potentials following electrical stimulation of small fibers.

Further evidence concerning the presynaptic modulation of afferent fibers comes from the investigation by Whitehorn and Burgess (1973), who showed that the excitability of the terminals of large afferent fibers was increased whether stimulation was at an innocuous or noxious level of intensity. Primary afferent hyperpolarization was not seen except as an off-response. Noxious mechanical stimuli produced primary afferent depolarization of small myelinated nociceptive afferent fibers, although noxious heat had little effect.

Another major finding since the introduction of the gate theory has been the demonstration of a population of dorsal horn neurons that respond specifically to noxious stimuli (Christensen and Perl, 1970; Kumazawa et al., 1975). Thus, specific nociceptors occur in both peripheral nerves and in the spinal cord.

Nathan (1976) cautions that the gate theory is based not upon the results of noxious stimuli but upon the potentials that are seen when nerves are stimulated electrically. Furthermore, stimulation of primary afferent fibers by microneurography fails to support the gate theory (see Chapter 12). Evidence that is consistent with the gate theory has been reviewed by Wall (1978). The fact that many cells in the dorsal horn receive convergent input from mechanoreceptor and nociceptors has already been discussed in the section on **Responses to Natural Stimulation**.

The therapeutic value of stimulating large afferent fibers for the relief of pain has received considerable attention (Shealy et al., 1967; 1970; Wall and Sweet, 1967; Nashold and Friedman, 1972). Wall (1978) admits that the weight of the evidence obtained since 1965 requires a modification of some of the details of the gate theory, but he believes that it is still a useful model. The most important point is that there is a mechanism in the dorsal horn that allows activity in large myelinated afferent fibers to produce an inhibition of the responses of dorsal horn neurons to noxious stimuli.

Diffuse Noxious Inhibitory Controls

Le Bars et al. (1979a, b) observed that wide-dynamic-range neurons in rats anesthetized with a mixture of halothane, nitrous oxide, and oxygen are subject to a powerful and prolonged inhibition when noxious stimuli are applied to any part of the body and face outside of the excitatory receptive field (Fig. 7.35C). However, narrow dynamic range high-threshold and low-threshold, as well as proprioceptive, neurons are unaffected (Fig. 7.35A, B). This form of inhibition was attributed to a system that they called "diffuse noxious inhibitory controls" (DNIC). The amount of inhibition is graded according to stimulus intensity (Le Bars et al., 1981). DNIC are dependent on activation of a supraspinal loop, since interruption of the spinal cord at a more rostral level abolishes DNIC (Le Bars et al., 1979b; see also Morton et al., 1987). However, a weaker, short-lasting segmental inhibition can still be observed after spinal transection (Cadden et al., 1983). The ascending pathway that triggers the DNIC mechanism is in the ventrolateral quadrant, contralateral to the noxious input (Villanueva et al., 1986b). The descending pathway is in the ipsilateral dorsal lateral funiculus (Villanueva et al., 1986a).

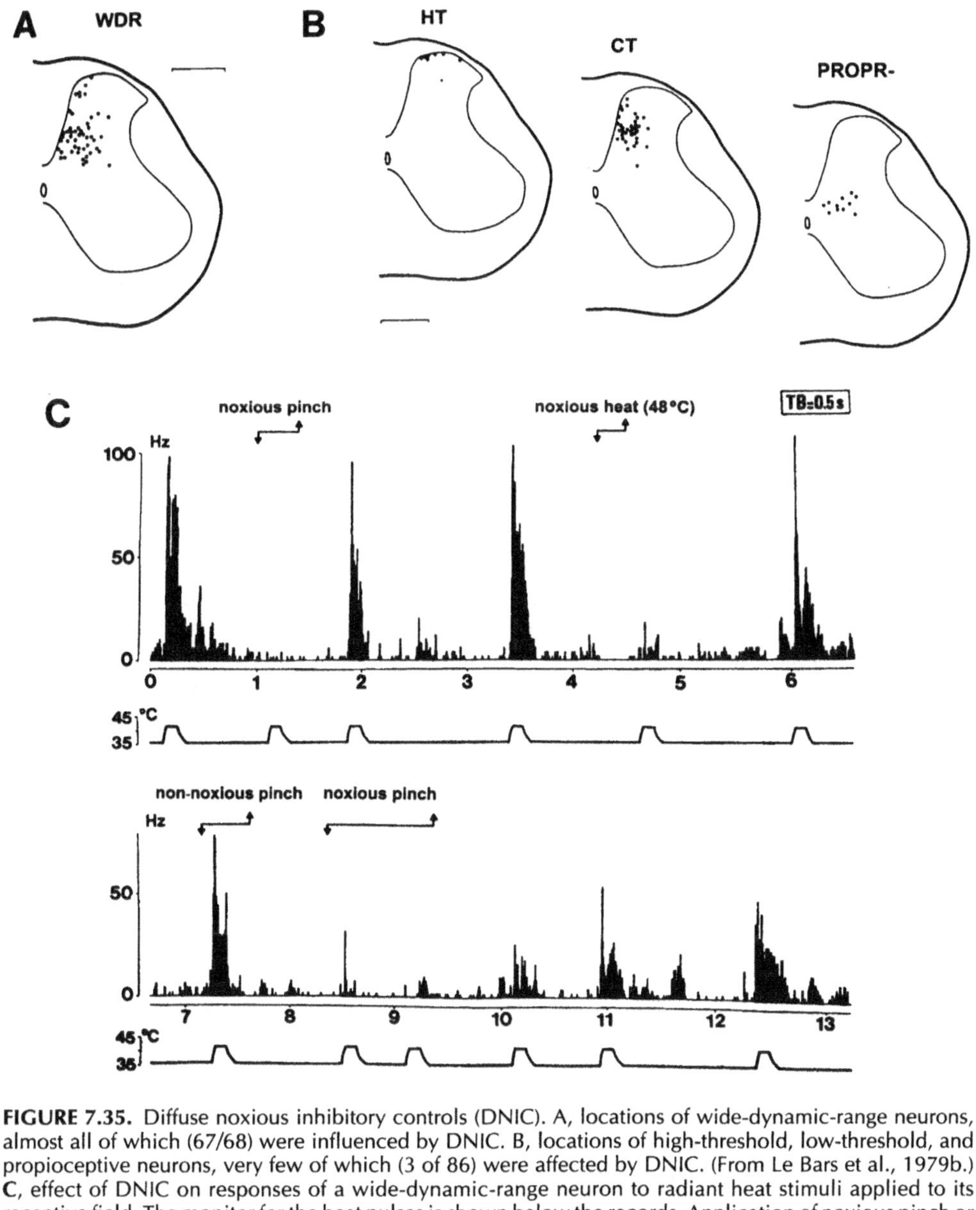

FIGURE 7.35. Diffuse noxious inhibitory controls (DNIC). A, locations of wide-dynamic-range neurons, almost all of which (67/68) were influenced by DNIC. B, locations of high-threshold, low-threshold, and propioceptive neurons, very few of which (3 of 86) were affected by DNIC. (From Le Bars et al., 1979b.) C, effect of DNIC on responses of a wide-dynamic-range neuron to radiant heat stimuli applied to its receptive field. The monitor for the heat pulses is shown below the records. Application of noxious pinch or noxious heat to the tail inhibited the response. (From Le Bars et al., 1979a.)

Le Bars and Chitour (1983) discuss the possible role of DNIC in nociceptive signaling. They point out that wide-dynamic-range neurons are likely to signal pain, since they respond best to noxious stimuli, and yet these cells are also responsive to innocuous stimuli. One possibility is that an innocuous stimulus applied to the body surface would activate only a few WDR cells, whereas a noxious stimulus would activate many because of the structure of their receptive fields. Pain might result when the population response exceeded some threshold centrally. However, another possibility is that noxious stimuli activate the receptive fields of a limited set of WDR cells while simultaneously inhibiting all other WDR cells through the DNIC mechanism. If the central nervous system can recognize the contrast between the activated and the silenced populations of WDR cells, this contrast could serve as a nociceptive signal (Le Bars and Chitour, 1983; see also Talbot, J. D. et al., 1989).

A number of studies have examined the mechanisms of DNIC. A postsynaptic inhibitory action is suggested by the observation that the responses of wide-dynamic-range dorsal horn neurons to pulses of iontophoretically released pulses of glutamate are reduced by DNIC, whereas those of low-threshold, high-threshold, and proprioceptive neurons are not (Villanueva et al. 1984).

Willer et al. (1989) have provided evidence that there is a DNIC-like mechanism in humans (see also Willer et al., 1984; Roby-Brami et al., 1987; Talbot, J. D. et al., 1987; 1989).

Summary

Dorsal horn interneurons often have both excitatory and inhibitory receptive fields. It has been suggested that the excitatory receptive fields of neurons located more dorsally in the dorsal horn are smaller than those of more ventrally placed cells and that the latencies of the initial response are progressively greater for deeper-lying neurons. This led to the idea of an excitatory cascade in which cells in a given lamina excite those in the adjacent deeper lamina. However, this idea is disputed.

Dorsal horn neurons have a somatotopic organization. The dorsal part of a dermatome projects to the lateral dorsal horn and the ventral part to the medial dorsal horn. For the enlargements, input from the distal extremity predominates. The ventral (glabrous) surface of the toes is represented most medially, the dorsal (hairy) surface more laterally, and the dorsal aspect of the proximal limb most laterally.

The responses of dorsal horn neurons are plastic; that is, they can often be shown to change. Alterations can result from stimulation or from lesions such as dorsal rhizotomy or peripheral nerve section. Some of the changes in responsiveness have been attributed to the unmasking of previously "ineffective" synapses. Mechanisms of such changes are still poorly understood.

Inhibitory receptive fields are of two general types. Weak mechanical stimuli may produce a "surround" type of inhibition through a segmental mechanism. This inhibition can also be evoked by electrical stimulation of large myelinated fibers. The original description of the "gate theory of pain" includes a possible circuit that has been shown to be incorrect. Another inhibitory mechanism is activated by noxious stimuli applied anywhere on the body surface or by volleys in fine myelinated and unmyelinated afferent fibers. This inhibition is powerful, long-lasting, and involves both segmental and supraspinal pathways. The supraspinal component is called "diffuse noxious inhibitory controls" (DNIC).

MODELS OF HUMAN DISEASE

Since there are often alterations of sensation in disease states, many experimental studies have been designed to explore the mechanisms of such changes, as well as to develop suitable test systems for exploring potential new therapies for the treatment of disordered sensation. Of course, most of this work is directed at pain mechanisms and pain relief.

Inflammatory Pain

Experimental Arthritis

Several different methods for inducing experimental arthritis have been proposed (Schaible and Grubb, 1993). However, most studies have used the intra-articular injection of kaolin and carrageenan to produce an acute arthritis or of complete Freund's adjuvant to produce a chronic arthritis.

Kaolin and Carrageenan Model

In the kaolin and carrageenan model, the arthritis develops within about 2 hours, is fully developed by 4 hours, and lasts for about 15 hours (Schaible and Schmidt, 1985; 1988a). Nociceptive articular afferent fibers develop background activity and become responsive to weak mechanical stimuli that had previously been ineffective in activating them (Coggeshall et al., 1983; Schaible and Schmidt, 1985). The increased input from joint afferents to the spinal cord leads to increased responses of spinal cord neurons that have joint input (Neugebauer and Schaible, 1990), including ascending tract cells (Fig. 7.36A–C; Schaible et al., 1987b; cf. Fig. 11.31; Dougherty et al., 1992b). However, the increased responsiveness of spinal cord neurons also reflects changes in central excitability, since these neurons develop enhanced responses not only to stimulation of the inflamed knee joint but also to stimulation of the contralateral, uninflamed knee joint (Schaible et al., 1987b; Neugebauer and Schaible, 1990).

The enhanced excitability of dorsal horn neurons may result in part from increases in the concentrations of glutamate and aspartate in the dorsal horn that occur during the arthritis (Sluka and Westlund, 1992; Sorkin et al., 1992). There are also increases in the content of glutamate in articular primary afferents and their terminals in the dorsal horn, as well as changes in peptide content in synaptic terminals in the dorsal horn (Westlund et al., 1992; Sluka et al., 1992; Sluka and Westlund, 1993b). Behavioral signs of hyperalgesia correlate with the level of glutamate immunoreactivity in the superficial dorsal horn (Sluka and Westlund, 1993b). The increased responses of dorsal horn neurons that are observed during acute arthritis can be prevented by iontophoretic release of either a non-NMDA-receptor antagonist (CNQX) or an NMDA-receptor antagonist (AP5), indicating that these receptors play an important role in the heightened central excitability state that accompanies the arthritis (Neugebauer et al., 1993). Metabotropic glutamate receptors were also found to be involved in the enhanced responsiveness of dorsal horn neurons that occurs during arthritis (Neugebauer et al., 1994a).

Substance P acting at NK-1 receptors has been shown to contribute to nociceptive, but not non-nociceptive, processing of joint input to the spinal cord (Neugebauer et al., 1994b). During acute arthritis, spinal cord release of substance P was demonstrated using the antibody microprobe technique (Schaible et al., 1990), and iontophoretic release of an NK-1 receptor antagonist attenuated the expected enhancement of the responses of dorsal horn neurons. It was concluded that substance P plays a role in the maintenance of the enhanced responses of dorsal horn neurons during acute arthritis (Neugebauer et al., 1995).

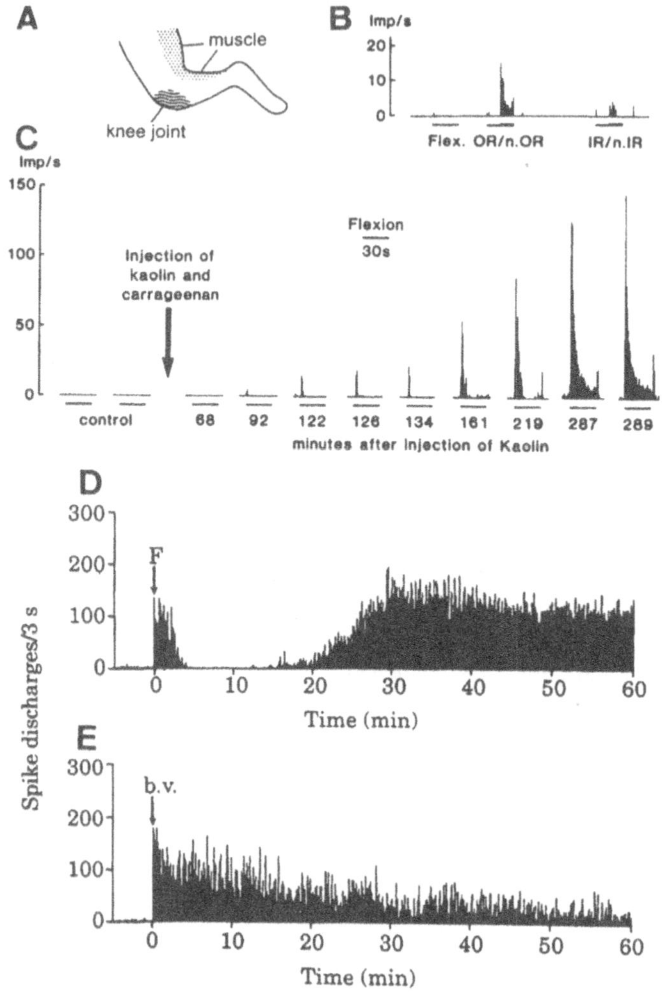

FIGURE 7.36. Increased responses of an ascending tract neuron in the cat spinal cord to flexion of the knee joint following intra-articular injection of kaolin and carrageenan. A, receptive field included the knee joint and muscle. B, initially the cell had almost no response to innocuous knee flexion, although there were responses when the joint was moved into the noxious range. C, responses to flexion of the knee developed in parallel with the onset of arthritis. (From Schaible et al., 1986.) D, biphasic response of a WDR neuron in the rat dorsal horn to injection of formalin into the foot. E, monophasic response of a different WDR neuron to injection of bee venom into the foot. (From You and Chen, 1999.)

Neurokinin A and calcitonin gene-related peptide were also found to be released in the spinal cord as arthritis developed (Hope et al., 1990b; Schaible et al., 1994).

Adjuvant Model

Another form of experimental arthritis can be produced by injection of complete Freund's adjuvant into the tibio-tarsal joint or near the ankle (Butler et al., 1992; Donaldson et al., 1993; Grubb et al., 1993). The inflammation is more chronic than is that from intra-articular injection of kaolin and carrageenan. The inflammation begins within a day, develops over 14 days and is maintained for at least 30 days (Donaldson et al., 1993). Interestingly, after 14 days, a mirror inflammation of the contralateral knee joint develops (Donaldson et al., 1993; 1995). This seems to be associated with the occurrence of dorsal root reflexes in capsaicin-sensitive primary afferent fibers on the contralateral side (Rees et al., 1996).

As in the case of acute arthritis, the responses of wide-dynamic-range neurons in the dorsal horn to stimulation of the inflamed ankle joint were increased, and this hyperexcitability could be counteracted by iontophoretic application of antagonists of either non-NMDA or NMDA receptors (Neugebauer et al., 1994b).

Cutaneous Inflammation

Mustard Oil

In human subjects, application of mustard oil to the skin produces a strong sense of burning pain and mechanical allodynia (Koltzenburg et al., 1992; 1994). There is also heat hyperalgesia because of sensitization of nociceptors to heat. The magnitude of the burning pain correlates with the intensity of the mechanical allodynia. Blocking A fibers in the appropriate peripheral nerve eliminates the mechanical allodynia, whereas blocking C fibers abolishes both mechanical allodynia and the ongoing pain.

In animal experiments, mustard oil inflames the skin and produces a sustained nociceptive input to the dorsal horn and heightened responses of dorsal horn neurons (Woolf and King, 1990; Munro et al., 1993). The prolonged excitation of dorsal horn neurons is associated with increased PKC activity, as shown by the effects of PKC inhibitors and by the translocation of PKC from the cytosol to the membrane fraction in spinal cord tissue (Munro et al., 1994).

Carrageenan

When carrageenan is injected into the plantar surface of the paw in rats, inflammation develops within 3 hours. The latency of paw withdrawal to noxious heat stimuli and the threshold for paw withdrawal from noxious mechanical stimuli are reduced at this time (Meller et al., 1994). The behavioral effects are due, at least in part, to central changes in excitability, since intrathecal injection of a nitric oxide synthase inhibitor reverses the thermal hyperpalgesia (but not the mechanical hyperalgesia). In addition, intrathecal administration of antagonists of NMDA receptors also reduces the heat hyperalgesia (Ren et al., 1992). Administration of an NK-1 receptor antagonist intrathecally delays the development of mechanical and thermal hyperalgesia, but posttreatment with the same agent does not reduce hyperalgesia that is already established (Traub, 1996).

Brain-derived growth factor (BDNF) has been shown to enhance the responses of dorsal horn neurons that are mediated by NMDA receptors following carrageenan injection in rats (Kerr et al., 1999). Conversely, these responses are reduced by trkB-IgG, which acts like a BDNF antagonist. Galan et al. (2002) have found that intraplantar carrageenan injection activates extracellular signaling-regulated kinase-1 and -2 (ERK-kinase 1 and 2) in the spinal cord, and they suggest that this signal transduction path underlies the secondary hyperalgesia seen during the inflammation that follows such an injection.

Complete Freund's Adjuvant

When complete Freund's adjuvant was injected into the paw in neonatal rats, the animals developed inflammation and showed behavioral signs of discomfort (Ruda et al., 2000).

When the same animals reached adulthood, anterograde labeling with wheat germ agglutinin-conjugated horseradish peroxidase demonstrated an increase in the density of terminals of fine primary afferent fibers in laminae I and II of the dorsal horn. There was also an increase in the distribution of these fine afferents to more caudal segments (Ruda et al., 2000). These changes were attributed to sprouting of fine afferent fibers. The affected afferent fibers could be immunostained for CGRP but not for IB4, indicating that the changes were specific for peptidergic afferents. Large primary afferent fibers failed to show similar changes. Behavioral responses of the rats to formalin injection and also the electrophysiologic responses of dorsal horn neurons to brush and pinch stimuli became enhanced on the side of the injection.

Intraplantar injection of complete Freund's adjuvant into the paw of adult rats results in heat hyperalgesia (Ren et al., 1996). The hyperalgesia can be reduced by intrathecal administration of NMDA and/or NK-1 receptor antagonists (Ren and Dubner, 1993; Ren et al., 1992; 1996). The effects of the antagonists of the different receptors are additive. The same antagonists have no effect on the nociceptive thresolds of non-inflamed paws. Thermal and mechanical hyperalgesia still occur after injections of Freund's adjuvant in rats that had been treated neonatally with capsaicin, which eliminates nociceptors with VR1 receptors (Ren et al., 1994).

Guo et al. (2002) investigated the signal transduction pathways and downstream actions of protein kinases in rats given an intraplantar injection of complete Freund's adjuvant. The NR2B (but not the NR2A) subunits of NMDA receptors in the spinal cord became phosphorylated through an action on tyrosine kinase receptors, as shown by the observation that phosphorylation was prevented by genistein, a tyrosine kinase inhibitor. PP2, an inhibitor of the Src family tyrosine kinase, was also effective, indicating a role of Src in the tyrosine kinase signal transduction pathway. In addition, a PKC inhibitor prevented the phophorylation, showing that more than one signal transduction pathway is involved. Blockade of NK-1 or group I metabotropic glutamate receptors was also effective, suggesting that G-protein coupled receptors help initiate the activity that leads to the phosphorylation of the NMDA receptors. Presumably, the change in the NMDA receptors contributes to the hyperexcitability of the dorsal horn neurons. There is also an upregulation and reorganization of AMPA receptor subunit types, which may play a role in the hyperalgesia (Zhou, Q. Q. et al., 2001). In particular, there are changes in the subunit composition of GluR1 and GluR2 subunits. The effect of these changes could result in alterations in the Ca^{++} permeability of the AMPA receptors, which would alter the likelihood of central sensitization and possibly excitotoxicity. The distribution of the changes in excitatory versus inhibitory interneurons is an important issue for understanding the mechanism of the excitability changes.

Formalin Injection

Injection of formalin into the skin of the rat foot leads to behavioral responses and enhanced activity of dorsal horn nociceptive neurons (Dubuisson and Dennis, 1977; Dickenson and Sullivan, 1987; Coderre et al., 1990; 1993; Abbott et al., 1995). Behavioral changes include two phases of increased nociceptive activity (early and late phases), separated by a period of reduced activity (Dubuisson and Dennis, 1977). Dorsal horn neurons also show a biphasic increase in discharge rate (Fig. 7.36D; Dickenson and Sullivan, 1987; You and Chen, 1999), The early phase of nociceptive activity depends largely on changes in primary afferent discharges, whereas the late phase is presumed to result in part from central changes. However, primary afferent activity persists into the late phase, and so peripheral input can contribute to both the early and late phases of response to the formalin test (Puig and Sorkin, 1995).

Central sensitization in the formalin test can be blocked by NMDA receptor antagonists (Coderre and Melzack, 1992a) and by an NK-1 receptor antagonist (Yashpal et al., 1993; Traub, 1996), and so central sensitization in this model appears to be triggered by the synaptic release of excitatory amino acids and SP. Brain-derived growth factor (BDNF) has been shown to enhance the responses of dorsal horn neurons that are mediated by NMDA receptors following formalin injection in rats (Kerr et al., 1999). Conversely, these responses are reduced by trkB-IgG, which acts like a BDNF antagonist.

Following the synaptic excitation of dorsal horn neurons, there is an activation of intracellular signal transduction pathways (Coderre and Yashpal, 1994). This is triggered in part by an influx of Ca^{++} ions into the neurons (Coderre and Melzack, 1992b) through NMDA-activated channels and also through voltage-gated Ca^{++} channels (MacDermott et al., 1986; Murphy and Miller, 1989). In addition, release of Ca^{++} from intracellular stores follows the activation of G-protein coupled receptors, such as NK-1 receptors (Womack et al., 1988). The intracellular signaling that ensues in the formalin model involves the release of nitric oxide and increases in cGMP, activation of phospholipase A2 and production of arachidonic acid metabolites, as well as of inositol triphosphate and diacylglycerol, and the activation of protein kinase C (PKC) (Coderre and Yashpal, 1994).

Bee Venom Injection

Another model of long-lasting cutaneous inflammatory pain is injection of bee venom (lyophilized whole venom of *Apis millifera*) into the skin (Lariviere and Melzack, 1996; Chen, J. et al., 1999a, b). The inflammation that develops is characterized by vasodilation and edema. The behavioral response to bee venom, as in the formalin test, is flinching and lifting/licking of the injected hindpaw. However, in contrast with the behavioral changes in the formalin test, the response to bee venom is monophasic. The behavioral events last 1–2 hr, following which mechanical and heat hyperalgesia can be demonstrated for 72–96 hr. In addition to the ipsilateral changes, heat hyperalgesia (but not mechanical hyperalgesia) also develops contralaterally to the injection. This contralateral heat hyperalgesia was shown to depend on central nervous system changes (Chen, H. S. et al., 2000; 2001).

Recordings from dorsal horn neurons show that bee venom produces a long-lasting, monophasic increase in discharge rate in wide-dynamic-range cells (Fig. 7.36E) (Chen, J. et al., 1998; 1999b; You and Chen, 1999). This is in contrast to the effects of a formalin injection, which results in a biphasic increase in activity (Fig. 7.36D). The activity of WDR cells that was evoked by bee venom could be completely blocked by application of lidocaine to the sciatic nerve, indicating that the activity depends on continous peripheral input (Chen, J. et al., 1998). The activity could also be blocked by systemically administered morphine; this effect was naloxone-reversible (Chen, J. et al., 1998).

Topical application of capsaicin to the sciatic nerve was used to eliminate capsaicin-sensitive afferent fibers (Chen, J. and Chen, 2001). This produced some reduction in the behavioral responses to bee venom injection, but nearly eliminated the behavioral responses to formalin injection. However, capsaicin treatment of the sciatic nerve prevented the development of heat and mechanical hyperalgesia after bee venom injection.

Many dorsal horn neurons develop immunoreactivity for FOS protein after intraplantar injection of bee venom (Luo et al., 1998). Most of the labeled neurons are in the superficial and deep laminae of the dorsal horn at L4 and L5. Labeling is seen by 30 minutes and is maximal at 2 hr. Labeling of some neurons can still be seen even at 96 hr. Morphine reduces the amount of FOS labeling.

The central actions that result in behavioral changes and maintained discharges of WDR neurons that are produced by peripheral input to the spinal cord and that lead to secondary heat hyperalgesia after bee venom injection depend on the activation of NMDA receptors (Chen, J. et al., 1999a; Chen, H. S. and Chen, 2000), as well as neurokinin 1 and 2 receptors (Zheng and Chen, 2001). Interestingly, secondary mechanical hyperalgesia was unchanged by the same drugs that acted on these receptors. The flinching response was reduced by intrathecal administation of a P2X receptor antagonist, suggesting that purinoceptors are involved in this response (Zheng and Chen, 2000). An inhibitor of protein kinase C could block the central sensitization that was responsible for the contralateral heat hyperalgesia (Li, K. C. et al., 2000).

Intradermal Injection of Capsaicin

The intradermal injection of capsaicin offers several advantages as a model of acute inflammatory pain. A major advantage is that it is possible to inject capsaicin in human subjects without permanent effects and therefore to study the consequent human psychophysical changes (Simone et al., 1987; 1989a; LaMotte et al., 1991; 1992; Torebjörk et al., 1992; Baron et al., 1999b). Another advantage from an experimental point of view is that some of the sensory changes last for only a few hours, and so it is possible to examine the effects of more than one capsaicin injection in a single behavioral or electrophysiological experiment. Furthermore, the discharges evoked by capsaicin in primary afferent C fibers last for only about half an hour or less, and sensory changes outlast the primary afferent input (Baumann et al., 1991; LaMotte et al., 1991; 1992). This allows the investigation of long-lasting central changes without continuous afferent input.

The initial effect of a capsaicin injection is severe pain that is felt at the injection site (Simone et al., 1989a; LaMotte et al., 1991; 1992). Depending on the dose of capsaicin, the pain lasts about 15–30 min. Following this, it is possible to demonstrate mechanical and heat hyperalgesia near the injection site (although at the injection site there is a small region of analgesia due to the desensitizing effects of the capsaicin on nociceptors in this area). The hyperalgesia is thought to reflect a peripheral sensitization of nociceptors that have VR1 receptors and are thus capsaicin-sensitive. The heat pain threshold is reduced from the normal level of near 45 °C to about 32 °C.

Perhaps more interesting is the development of mechanical allodynia and hyperalgesia in an area around the injection site. These changes occur in a progressively larger area during a period of about 15–30 min, and then the area shrinks again and disappears after about an hour and a half (cf. Hardy et al., 1952). Primary afferent nociceptors that supply this area do not show any change in responsiveness, presumably because their terminals have not been exposed to the capsaicin (Baumann et al., 1991; LaMotte et al., 1992). Nevertheless, tactile stimulation of the skin or electrical stimulation of tactile afferent fibers by intraneural microstimulation causes pain (mechanical allodynia), and noxious mechanical stimulation of the skin causes more pain than it did previously (mechanical hyperalgesia) (LaMotte et al., 1991; 1992; Torebjörk et al., 1992; Cervero et al., 1994).

The mechanical allodynia and hyperalgesia in the region surrounding the injection site are referred to as secondary mechanical allodynia and hyperalgesia. Curiously, secondary thermal allodynia or hyperalgesia are not observed following intradermal injection of capsaicin (Ali et al., 1996). Behavioral effects that seem to parallel the development of secondary mechanical allodynia and hyperalgesia, without secondary thermal hyperalgesia, can be demonstrated in rats (Sluka and Willis, 1997).

Since the primary afferent nociceptors that supply the area of secondary mechanical allodynia and hyperalgesia are unchanged, there must be a change in the responsiveness of central nociceptive neurons following capsaicin-evoked C-fiber discharges. This change, called "central sensitization," is commonly observed in various pain models.

In experiments in which recordings are made from nociceptive dorsal horn neurons in general or from identified spinothalamic tract (STT) neurons (see Chapter 11), the effects of central sensitization can be demonstrated after intradermal injection of capsaicin. The responses to mechanical stimuli applied to the zone of assumed secondary allodynia and hyperalgesia increase (Simone et al., 1989b; 1991), but the responses to heat stimuli applied to the secondary area are not increased and even decrease (Sluka et al., 1997b). Responses to both mechanical and heat stimuli applied to the area of primary hyperalgesia near the injection site are increased (Simone et al., 1991). Other changes are an increase in background discharges and an expansion of the receptive field. It has been possible to determine if there are changes in the responses of STT cells to iontophoretically released excitatory and inhibitory amino acids. The responses to glutamate, AMPA, and NMDA are increased, whereas those to GABA and glycine are reduced (Dougherty and Willis, 1992; Lin et al., 1996a; 1999b).

Capsaicin injections trigger central sensitization by synaptic actions that involve the release of excitatory amino acids acting at non-NMDA and NMDA receptors (Dougherty et al., 1992a) and at metabotropic glutamate receptors (Palecek et al., 1994b; Neugebauer et al., 1999; 2000), as well as substance P acting at NK-1 receptors (Dougherty and Willis, 1991b; Dougherty et al., 1993; 1994; 1995; Rees et al., 1998). Presumably following influx of Ca^{++} through NMDA and voltage-gated Ca^{++} channels and/or increased intracellular release of Ca^{++} through actions of excitatory transmitters on G-protein coupled metabotropic glutamate receptors and NK-1 receptors, intracellular signaling pathways are activated. These pathways involve a number of second messengers, including nitric oxide, cAMP, cGMP, and several protein kinases, including protein kinase C, protein kinase A, and protein kinase G (Palecek et al., 1994a; Lin et al., 1996b; 1997; Peng et al., 1997; Sluka et al., 1997a, b; Sluka and Willis, 1998; Wu, J. et al., 1998a, b; 2001; Lin et al., 1999a–c; Palecek et al., 1999). Agents that affect these signal transduction pathways also affect the behavioral responses to intradermal capsaicin injection in a predictable way (Sluka and Willis, 1997). Recently, it has been shown that early responses to capsaicin injection involve the upregulation of calcium/calmodulin kinase II (CaMKII; Fang, L. et al., 2002). Some of the proteins that become phosphorylated by the various protein kinases that are activated during central sensitization include NMDA receptors (Zou et al., 2000), as well as transcription factors, such as cyclic AMP response binding element protein (CREB) (Wu, J. et al., 2002).

Because many of these same mechanisms have been found to underlie LTP in the hippocampus, it has been proposed that central sensitization is a spinal cord form of LTP (Willis, 1997; 2001; 2002b).

Cervero and Laird (1996) have proposed an alternative mechanism for the production of mechanical allodynia following intradermal injection of capsaicin. This was based on the observation that capsaicin injection results in central sensitization, during which tactile stimuli in the area of secondary mechanical allodynia can evoke dorsal root reflexes in cutaneous afferents, including Aδ and C fibers (see Lin et al., 1999d; 2000).

Cervero and Laird (1996) point out that such dorsal root reflexes would have central actions that might include the excitation of nociceptive pathways, leading to pain in response to tactile stimuli. This result could be expected, since Aβ fibers are normally thought to inhibit nociceptive pathways in part by presynaptic inhibition of fine primary afferent fibers. Therefore, enhancement of the PAD produced by Aβ fibers in nociceptors could result in dorsal root reflexes in these afferents (see Willis, 1999).

Neuropathic Pain

Deafferentation

A number of laboratories have investigated the influence of deafferentation of the spinal cord on receptive-field properties of dorsal horn neurons and on behavior. Deafferentation is accomplished by interruption of one or more dorsal roots or by sectioning one or more peripheral nerves.

Spared Root Paradigm

Basbaum and Wall (1976) deafferented the lumbosacral spinal cords of cats by cutting all of the dorsal roots below L3 except S1 (spared-root paradigm of Liu, C. N. and Chambers, 1958). At 24 hr after surgery, no cells were found in L4 or L5 that could be excited monosyaptically by cutaneous stimulation. However, by 1 week after surgery, such neurons could be found. After 1 month, the numbers of such cells were stable. The receptive fields were abnormal in the following ways: (1) their locations were on the S1 dermatome or on dermatomes rostral to L4; (2) they were often split; (3) their size was unusually variable; (4) they exhibited less convergence than is normally seen; (5) inhibitory fields were rare; and (6) habituation was common and might be different in one part of the receptive field than in another. It was proposed that the new receptive fields were due either to sprouting of primary afferent fibers or to the unmasking of previously ineffective synapses.

Pubols and Goldberger (1980) also used the spared-root paradigm to examine possible changes in the somatotopic organization of the dorsal horn (see also, Goldberger and Murray, 1974; Murray and Goldberger, 1974). All of the dorsal roots of the lumbosacral enlargement were severed, except L6. Soon after the dorsal rhizotomies, there was a loss of responsiveness of dorsal horn neurons, especially laterally in the dorsal horn, as well as a loss of proximal receptive fields. In chronic preparations, there was recovery of function, with an abnormally high proportion of strictly distal receptive fields. They suggested that this recovery could have been due to presynaptic changes, such as local sprouting or strengthening of synaptic connections, or to postsynaptic changes, such as denervation supersensitivity.

Brinkhus and Zimmermann (1983) also studied the effect of chronic partial deafferentation of dorsal horn neurons in cats by interruption of one or more dorsal roots. Cord dorsum potentials evoked by stimulating peripheral nerves of the hind limb were unchanged when a single dorsal root was cut, but they were smaller after sectioning of L5, L6, and most of L7. The response properties of dorsal horn neurons were little affected by partial deafferentation. The only substantial change was an increase in the background activity of a subpopulation of heat-sensitive neurons. More evidence of enhanced responses was expected, since deafferentation has been proposed as a mechanism of chronic pain (Loeser and Ward, 1967; Lombard et al., 1979; Wall et al., 1979b; Wiesenfeld and Lindblom, 1980; Levitt and Heybach, 1981; Levitt and Levitt, 1981; cf. Wynn Parry, 1980).

Nerve Transection

Devor and Wall (1981a) compared the somatotopic map in animals with intact nerves (controls) with that in animals with an acute or a chronic transection of the sciatic and saphenous nerves (see also, Devor and Wall, 1981b; Ovelman-Levitt et al., 1984; however, cf. Pubols, L. M. and Brenowitz, 1981; Pubols, L. M., 1984). Most of the neurons in the control animals had receptive fields on the toes and foot. Those with receptive fields extending proximal to the ankle were in the lateral part of the dorsal horn. After acute transection of the nerves, cells with proximal receptive fields were still found laterally in the dorsal horn, but most of the cells in the medial dorsal horn were unresponsive. After chronic transection of the nerves (6–105 days), many neurons in the medial part of the dorsal horn had proximal receptive fields. Other neurons were unresponsive. It was concluded that neurons with input exclusively from the toes or foot were in some way converted to neurons that responded to stimulation of proximal skin after chronic transection of a peripheral nerve. The changes could not be attributed to peripheral sprouting of the severed nerves or to the elimination of inhibition. Reasons were given to doubt the development of long-distance central sprouting (cf. Liu, C. N. and Chambers, 1958; Goldberger and Murray, 1974; Devor and Claman, 1980; however, see below for evidence of sprouting). Instead, it was hypothesized that under normal conditions the neurons in the medial dorsal horn have latent proximal receptive fields (Wall and Werman, 1976), but that these are not expressed because the synapses are normally ineffective because of inhibition, low synaptic efficacy, connections to distal dendrites, or other reasons (cf. Merrill and Wall, 1972; Wall, 1977; Markus et al., 1984; Hylden et al., 1987; Markus and Pomeranz, 1987). It was proposed that nerve section causes the ineffective synapses to become effective. Mechanisms might include removal of presynaptic inhibition (cf. Horch and Lisney, 1981; Wall and Devor, 1981), increased quantal release, increased density of postsynaptic receptors, or changes in dendritic geometry or membrane conductance. Alternatively, there might be short-range sprouting, for example with displacement of ineffective synapses to more effective sites.

Woolf and Wall (1982) observed that chronic transection of the sciatic nerve in rats reduces the ability of volleys in A fibers to inhibit the responses of dorsal horn neurons to volleys in either A or C fibers. It was proposed that this reduction in inhibition (see below; also Wall and Devor, 1981; Wall, 1982) plays a role in the unmasking of ineffective synapses and also in the development of chronic pain states.

Lisney (1983) cut the sciatic and saphenous nerves, but then allowed them to regenerate. He found the same receptive field changes that were described by Devor and Wall (1981a) at 1 month. By 9 months neurons in the medial dorsal horn had receptive fields on the toes or foot and not on the proximal limb. However, the somatotopic map had not been restored.

Pubols, L. M. et al. (1986) presented evidence for the presence of relatively ineffective synapses in the cat dorsal horn by using electrical stimulation of the sural nerve to activate the cells. However, it was not possible to determine whether the ineffective synapses were monosynaptic or polysynaptic.

Using intracellular recordings from dorsal horn interneurons in rats, Woolf and King (1989) were able to demonstrate that some neurons had receptive fields that were surrounded by a "low-probability firing fringe" region. Stimulation in this surrounding area evoked EPSPs, but few if any discharges. Such synaptic potentials could be considered to result from the activity of "relatively ineffective synapses."

A number of morphological studies have demonstrated evidence of sprouting of primary afferent fibers following dorsal rhizotomies or peripheral nerve injury (Liu, C. N. and Chambers, 1958; Hulsebosch and Coggeshall, 1981; Goldberger and Murray, 1982; Molander et al., 1988; LaMotte et al., 1989; Fitzgerald et al., 1990; Polistina et al., 1990; however, cf. Rodin et al., 1983; Selzer and Devor, 1984; Pubols, L. M. and Bowen, 1988). In addition, morphological and pharmacological studies have provided evidence of transynaptic changes in dorsal horn interneurons (Gobel and Binck, 1977; Wright and Roberts, 1978; Brown, P. B. et al., 1979; Macon, 1979; Nakata et al., 1979; Gobel, 1984; Sugimoto and Gobel, 1984; Westrum et al., 1984; Kapadia and LaMotte, 1987; Iadarola et al., 1988a, b; Ruda et al., 1988; Kajander et al., 1990; Cameron et al., 1992; see review by Aldskogius et al., 1985). Thus, both sprouting and transynaptic changes in dorsal horn neurons continue to offer explanations for the changes in physiological responses of dorsal horn neurons and presumably of pathological sensations following such injuries.

Of particular interest are recent studies that suggest a reorganization of the terminations of myelinated primary afferent fibers in the dorsal horn (Fitzgerald et al., 1990; Woolf et al., 1992; 1995; Shortland and Woolf, 1993; Chong et al., 1994; Koerber et al., 1994; Doubell and Woolf, 1997; Doubell et al., 1997; Mannion et al., 1998; Kohama et al., 2000; Ma, Q. P. et al., 2000b). Following peripheral nerve transection, the terminals of myelinated primary afferent fibers were labeled transganglionically with either the B-fragment of cholera toxin (B-HRP) or with wheat germ agglutinin conjugated to horseradish peroxidase (WGA-HRP). The assumption was that transport of B-HRP was restricted to myelinated afferent fibers and that transport of WGA-HRP was restricted to unmyelinated afferents (Robertson and Grant, 1985; Rivero-Melian and Grant, 1990). The results were compared to the normal somatotopic arrangement of large myelinated afferents in the dorsal horn (Robertson and Grant, 1985; Swett and Woolf, 1985; Molander and Grant, 1986; LaMotte et al., 1989; Shortland et al., 1989). Normally, large myelinated afferent fibers do not end in lamina II. By 1 week after the peripheral nerve lesion, afferents labeled with B-HRP had begun to invade lamina II, and this invasion progressed for 26 weeks, after which the projection regressed and approached the control pattern by 33–35 weeks. Similar findings were observed when the sciatic nerve was crushed, rather than transected. Ultrastructural observations showed that many of the structures labeled with B-HRP in the dorsal horn were synaptic terminals. In a parallel study, individual Aβ fibers could be traced into lamina II after these were labeled by intra-axonal injection of HRP (Shortland and Woolf, 1993b; see also, Koerber et al., 1999; Kohama et al., 2000). The suggestion was made that these large myelinated, presumably tactile, axons might synapse on neurons in lamina II that are normally responsible for pain sensation and therefore this reorganization could explain mechanical allodynia. Another explanation for the labeling of axons in the superficial dorsal horn by B-HRP after nerve injury is a change in the ability of C fibers to take up and to transport B-HRP (Tong et al., 1999; Bao et al., 2002; see also Doubell and Woolf, 1997).

Popratiloff et al. (1998a) have reported that sectioning the sciatic nerve results in an upregulation of AMPA receptors in primary afferent synapses in the substantia gelatinosa. They suggest that this change may contribute to the increased sensitivity of dorsal horn neurons after nerve injury. The synapses that showed this change were glomerular ones formed by fine afferents; AMPA upregulation did not occur in synapses formed by intrinsic spinal neurons.

A recent study shows that the large myelinated fibers that invade lamina II after a peripheral nerve lesion may release excess amounts of glutamate to produce a neurotoxic action that cause the loss of interneurons in lamina II (Coggeshall et al., 2001; cf. Sugimoto et al., 1990). If the lost interneurons were inhibitory ones, this could result in disinhibition and thus contribute to allodynia (cf. Yaksh, 1989; Sorkin et al., 1998).

Autotomy

In animals with deafferentation of the spinal cord because of lesions of peripheral nerve or dorsal roots, often the only behavioral manifestation is autotomy (self-mutilation). Some investigators regard this behavior as a manifestation of "anaesthesia dolorosa" (Wall et al., 1979b). However, this is disputed and the value of studies of autotomy has been debated (Sweet, 1981; Rodin and Kruger, 1984; Levitt, 1985; Coderre et al., 1986; Neil et al., 1994; Mailis, 1996; see review by Kauppila, 1998). Although autotomy may reflect a response to painful sensations due to activity in primary afferent nociceptors, it is difficult to relate the amount of autotomy to the amount of pain.

Models of Peripheral Neuropathic Pain

Transections of the sciatic nerve or of the dorsal roots are not satisfactory models of neuropathic pain in humans. The reason is that peripheral neuropathic pain patients not only complain of burning pain, but also experience hyperalgesia and allodynia in response to stimulation of the skin in the affected area, implying that at least some of the afferents that supply the affected area are still intact (Mitchell, 1872; Bonica, 1990). Mechanical allodynia in neuropathic pain patients can be eliminated by differential conduction block in Aβ fibers, but ongoing pain continues (Koltzenburg et al., 1994). In some patients, the pain is maintained by the activity of the sympathetic nervous system (Roberts, 1986; Davis et al., 1991). This sympathetically maintained component of neuropathic pain can be relieved by interruption of the sympathetic supply to the painful area (Doupe et al., 1944; Nathan, 1947; Bonica, 1990; Raja et al., 1991; Koltzenburg et al., 1994; Baron et al., 1999a).

Several models of neuropathic pain have been developed for the study of the mechanisms underlying this chronic pain disorder. These include the Bennett, Seltzer, and Chung models (see below). Chung and his colleagues have compared the main features of these models (Kim et al., 1997). Other models have also been proposed (Attal et al., 1990; Sato and Perl, 1991; DeLeo et al., 1994; Na et al., 1994).

Some of the mechanisms that underlie the neuropathic pain state seem to be shared by the various models. These include the development of ectopic discharges in primary afferent fibers, particularly in dorsal root ganglion cells (Wall and Gutnick, 1974; Scadding, 1981; Devor, 1983; Kajander et al., 1992; Sheen and Chung, 1993; Xie et al., 1995; Han et al., 2000). One factor that contributes to the development of ectopic discharges in dorsal root ganglion cells is likely to be an upregulation of sodium channels (cf. Wall and Devor, 1983; Devor et al., 1992; 1993; 1994b; Chaplan et al., 1995; Kim, C. H. et al., 2001). Although tetrodotoxin-resistant sodium channels are a possible source of ectopic discharges (Caffrey et al., 1992; Elliott and Elliott, 1993; Akopian et al., 1996; Sangameswaran et al., 1996; Rush et al., 1998; Lai et al., 2000), tetrodotoxin alleviates the discharges that originate from dorsal root ganglion cells in neuropathic animals (Omana-Zapata et al., 1997; Lyu et al., 2000).

Another cause of ectopic discharges relates to an upregulation of adrenoreceptors in both injured and uninjured primary afferents following a nerve injury (Devor and Jänig, 1981; Korenman and Devor, 1981; Roberts and Elardo, 1985; Häbler et al., 1987; Sanjue and Jun, 1989; Sato and Perl, 1991; Devor et al., 1994a; Bossut and Perl, 1995; Drummond, 1995; Khasar et al., 1995; Kinnman and Levine, 1995; Drummond et al., 1996; Michaelis et al., 1996a; Petersen et al., 1996; O'Halloran and Perl, 1997; Zhang, J.-M. et al., 1997; Ali et al., 1999; Liu, X. et al., 1999; Moon et al., 1999; see also Park et al., 2000; Zhou, J. et al., 2001).

The ectopic discharges result in central sensitization of dorsal horn neurons in neuropathic rats and monkeys (Gracely et al., 1992; Palecek et al., 1992a, b; Laird and Bennett, 1993). Neurons in the dorsal horn that increase their metabolic activity in the neuropathic pain state are presumably undergoing central sensitization (Fig. 7.12; Price, D. D. et al., 1991; Mao et al., 1992c, d). The central sensitization depends on the activation of non-NMDA and NMDA receptors and its maintenance on the activation of NMDA receptors (Davar et al., 1991; Seltzer et al., 1991; Mao et al., 1992a, b).

Another factor that plays a role in supraspinal responses that contribute to mechanical allodynia in neuropathic pain models is activity transmitted to the thalamus through the dorsal column-medial lemniscus pathway (Miki et al., 1998a, b; 2000; Sun et al., 2001). In normal animals, a lesion of the dorsal column or gracile nucleus reduces the responses of neurons in the ventral posterior lateral thalamic nucleus to innocuous mechanical stimuli. However, in neuropathic animals, there is an increase in the number of VPL neurons that respond to noxious stimuli, and a dorsal column lesion reduces the responses of VPL neurons to noxious mechanical stimuli. Apparently, the increased responsiveness of the dorsal column system to mechanical stimulation in the neuropathic state is due in part to changes in primary afferents or in postsynaptic dorsal column neurons that project to the dorsal column nuclei. One change that has been noted in neuropathic animals is an upregulation of peptides, such as substance P and CGRP, in dorsal root ganglion cells that project to the dorsal column nuclei (Noguchi et al., 1995; Miki et al., 1998b). This is in contrast to the downregulation of these same peptides in other dorsal root ganglion cells following axotomy (Nahin et al., 1994).

Bennett Model

Bennett and Xie (1988) introduced a model of neuropathic pain in which the sciatic nerve is chronically compressed by four loose ligatures that are placed around the nerve. The initial tightness of the ligatures is adjusted so that blood flow is only partially occluded (see Mosconi and Kruger, 1996, for improved technique). The sciatic nerve swells, and because of this, many of the axons in the sciatic nerve are interrupted. However, enough survive to allow sensory input to reach the spinal cord and for motor responses to occur. At least part of the effect of this chronic constriction injury (CCI) model has been attributed to inflammation of the sciatic nerve induced by the chromic gut sutures that are employed for the loose ligations (Maves et al., 1993).

Immunocytochemical studies of the spinal cord in rats with CCI show an increase in glutamate levels and a decrease in substance P and CGRP levels in the superficial dorsal horn. Similar changes result when the sciatic nerve is stimulated at C-fiber strength in normal animals, suggesting that the reduced peptide levels are due to peptide depletion because of synaptic activity (Al-Ghoul et al., 1993). Released peptides are metabolized and replacement requires resynthesis and transport. By contrast, glutamate can be synthesized locally in synaptic terminals, and released glutamate is taken up by nerve terminals and glia.

Rats with a CCI develop an altered posture in which the foot remains everted. The animals also develop signs of ongoing pain, as well as mechanical and cold allodynia. The degree of ongoing pain is greater than that seen in the Seltzer or Chung models (Kim et al., 1997).

Membrane-bound protein C is increased in the dorsal horn in the Bennett model, and this increase is correlated with spontaneous pain-related behavior and with thermal hyperalgesia (Mao et al., 1993). Intrathecal treatment with GM1 ganglioside reduces nociceptive behavior and thermal hyperalgesia (Mao et al., 1992d, e).

Seltzer Model

Seltzer et al. (1990) introduced another model of neuropathic pain that retains connections of the spinal cord with peripheral sensory and motor endings. In this model one third to one half of the sciatic nerve is tightly ligated. The animals develop signs of neuropathic pain that are similar to those seen in the Bennett model. Sympathectomy reduces the signs of neuropathic pain (Shir and Seltzer, 1991).

Chung Model

Kim and Chung (1992) proposed still another neuropathic pain model. In the Chung model (spinal nerve ligation or SNL model), axons in spinal nerves L5 and L6 (or sometimes just in L5) in rats are tightly ligated just distal to the corresponding dorsal root ganglia. Axons in spinal nerve L4 provide much of the sciatic nerve innervation of the lower extremity, and so sensory and motor responses are available for tests of the pain state. A similar pain state is induced in monkeys after tight ligation of a single spinal nerve (Palecek et al., 1992b; Carlton et al., 1994; Ali et al., 1999). An advantage of this model is that it is possible to evaluate possible changes in the activity of both injured and intact afferents and their relative contributions to the neuropathic pain state (Yoon et al., 1996; Ali et al., 1999). Dorsal rhizotomy or local anesthetic block at the levels of the injured segments reduce all neuropathic pain behaviors; however, these manipulations at the level of the uninjured segments abolish mechanical and cold allodynia, but the signs of ongoing pain continue (Yoon et al., 1996).

Examination of the L5 and L6 dorsal root ganglia after tight ligation of the corresponding spinal nerves reveals a loss of 50% of the dorsal root ganglion cells (Lekan et al., 1997). However, after a transient reduction in the number of unmyelinated dorsal root axons, the number of dorsal root axons increases again to the normal level by 32 weeks, because of sprouting, mostly of myelinated axons.

Rats with a SNL injury develop signs of ongoing pain and also mechanical and cold allodynia (Kim and Chung, 1992; Choi et al., 1994). The mechanical allodynia is more severe than in the Seltzer or Bennett models (Kim et al., 1997). Sympathectomy produces the most dramatic reduction in the pain state in the Chung model (Fig. 7.37D, E; Kim and Chung, 1991; Kim, S. H. et al., 1993; Choi et al., 1994; Kim, K. J. et al., 1997). This could be correlated with the observation that sympathetic postganglionic nerve fibers sprout and grow into the dorsal root ganglion (Fig. 7.37A–C), beginning within 2 days after SNL (Chung, K. et al., 1993c; 1996; 1997; Chung, K. and Chung, 2001; cf. McLachlan et al., 1993). Changes in the availability of nerve growth factor play a role in this response (Lee, S. E. et al., 1998; Shen et al., 1999).

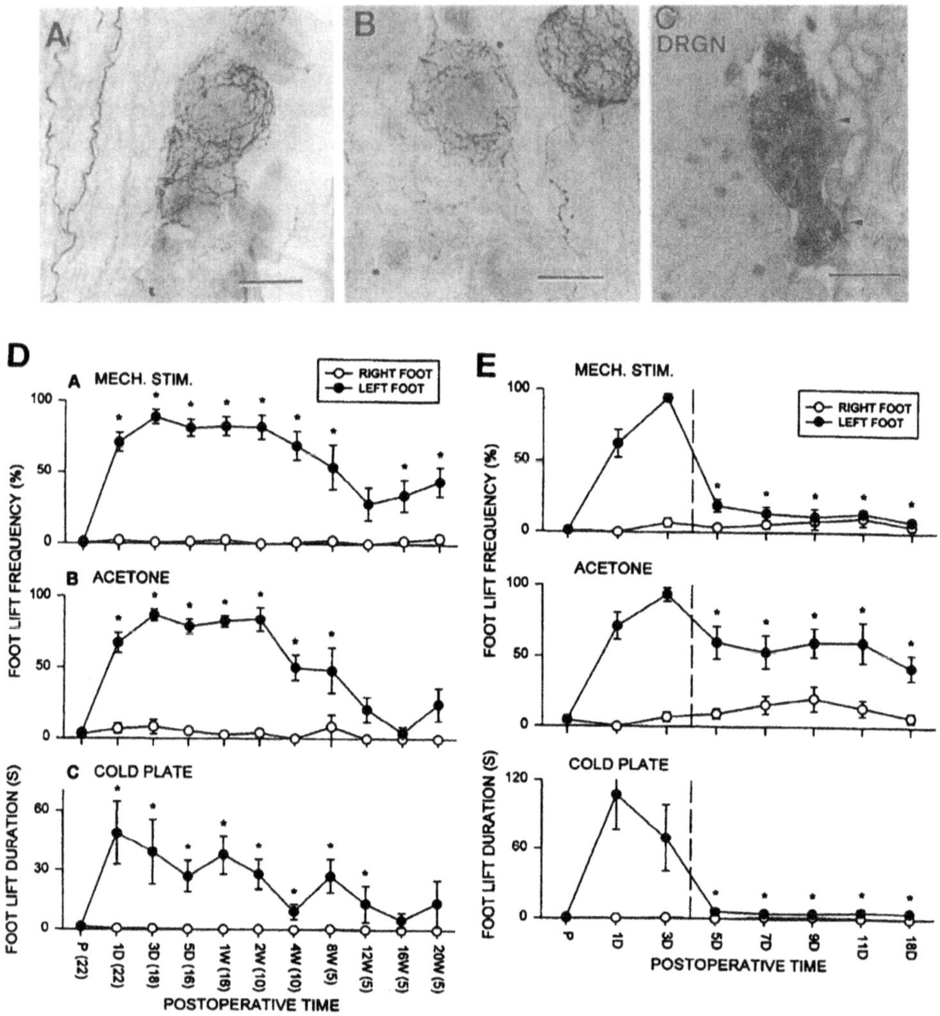

FIGURE 7.37. A, B, light micrographs showing dorsal root ganglion neurons that are surrounded by tyrosine hydroxylase-immunoreactive axons. Scale = 50 μm. C, electron micrograph of a synaptic varicosity that is immunostained for tyrosine hydroxylase and that is in close apposition to a dorsal root ganglion neuron (DRGN). The same varicosity also makes close contacts with a satellite cell (arrowheads). D, time course of neuropathic pain behaviors (increased foot lifts or increased duration of foot lifts in response to mechanical stimulation, acetone, and contact with a cold plate at 5 °C). E, disruption of the neuropathic pain behaviors after surgical sympathectomy. (A, B, D, and E from Chung et al., 1996; C from Chung et al., 1997.)

Primary afferent neurons upregulate their content of adrenoreceptors in neuropathic pain states (Sato and Perl, 1991). This makes primary afferents respond to norepinephrine that is released either in sympathetic ganglia, near their peripheral nerve terminals or in neuromas (Devor and Jänig, 1981; Devor et al., 1994a), and no doubt contributes to the sympathetically maintained component of the neuropathic pain state (Campbell et al., 1992; Devor et al., 1994a). Adrenergic sensitivity and spontaneous discharges develop in uninjured C-fiber nociceptors, presumably contributing to neuropathic pain (Ali et al., 1999).

The adrenergic sensitivity of nerve fibers in the Chung model depends on the strain of rat (Lee, D. H. et al., 1997); in Lewis rats, sympathetic dependency is mediated by $\alpha 1$ adrenoreceptors (Lee, D. H. et al., 1999). In Sprague Dawley rats, adrenoreceptors of different types are upregulated in injured and uninjured primary afferents (Liu, X. et al., 1999; Moon et al., 1999). Purinergic receptors are also involved (Park et al., 2000; Zhou, J. et al., 2001).

Models of Central Neuropathic Pain

Central neuropathic pain is defined as "pain initiated or caused by a primary lesion or dysfunction in the central nervous system" (Merskey and Bogduk, 1994; see Burchiel and Yezierski, 2002). Lesions that interrupt the spinothalamocortical pathway at any level can produce central pain (Leijon et al., 1989; see reviews by Boivie, 1994; Pagni, 1998). Common causes of central neuropathic pain are spinal cord injury and other conditions that affect the spinal cord, including multiple sclerosis, syringomyelia, and neurosurgical interventions, such as anterolateral cordotomy (Tasker, 1990). The pain is projected to a region that is somatotopically related to the damaged area in the central nervous system, and the pain can have any quality, although commonly it is a burning type of pain (Boivie, 1994). Mechanisms that have been proposed to explain central pain include: (1) the presence of an "irritable focus" at the site of damage, perhaps because of central sensitization of nociceptive neurons; (2) disinhibition by interruption of inhibitory pathways in the pain system; (3) substitution of alternative nociceptive pathways; and (4) rewiring of neural circuits (see review by Willis, 2002a). Genetic susceptibility is likely to have an important influence on the chances of developing central neuropathic pain (cf. Levitt and Levitt, 1981).

A number of animal models of spinal cord injury pain have been proposed for the investigation of the mechanisms and potential therapies for this condition.

Cordotomy Model

A cordotomy model was developed (Levitt and Levitt, 1981; Vierck et al., 1990; Ovelmen-Levitt et al., 1995; Vierck and Light, 1999) because of the observation that central pain sometimes follows anterolateral cordotomy (White and Sweet, 1969). In monkeys, a lesion of the lateral funiculus initially causes a reduction in the speed at which escape occurs following electrical stimulation of the skin; however, in time the escape speed not only recovers but may become faster than normal, indicating the development of hyperalgesia (Vierck et al., 1990). Escape responses on the side ipsilateral to the lesion are enhanced if the gray matter at the lesion site is exposed to blood products (Vierck and Light, 1999). Recordings from the ventral posterior lateral nucleus of the thalamus in monkeys following interruption of the spinothalamic tract at a thoracic level reveal hyperactive responses of neurons with inputs from both mechanoreceptors and mechanical nocicpetors (Weng et al., 2000). This activity resembles that seen in human thalamic neurons after spinal cord lesions in patients with central pain (Lenz et al., 1987; 1989). Thalamic neurons with such altered response properties may contribute to central pain.

Contusion Injury Model

Another model of spinal cord injury pain is based on contusion injury of the spinal cord. In this model, a weight drop technique is used in which the spinal cord is damaged by a controlled injury made by dropping a known weight on the cord dorsum from a known elevation (cf. Beattie et al., 1997). In addition to well-characterized motor deficits, a pain state develops (Siddall et al., 1995; Hulsebosch et al., 2000). One of the changes that have been observed in such preparations is an increase in the concentration of excitatory amino acids in the extracellular space of the spinal cord (Liu, D. et al., 1991; 1999; Xu, G.-Y. et al., 1998; McAdoo et al., 2000). As already mentioned, in a model of inflammatory pain, increases in excitatory amino acids in the dorsal horn correlate with hyperalgesia (Sluka and Westlund, 1993b).

Spinal Cord Hemisection Model

Another method for producing a controlled injury to the spinal cord is surgical hemisection at a midthoracic level (Christensen et al., 1996; Christensen and Hulsebosch, 1997a). The animals develop truncal allodynia, as well as mechanical allodynia and thermal hyperalgesia of the hind limbs and fore limbs. In this model, there is a substantial increase in immunostaining for CGRP in the dorsal horn over several segments above and below the lesion (Christensen and Hulsebosch, 1997b). This change may relate to the pain-state, since the increased CGRP staining is in sprouting primary afferent fibers, many of which would be nociceptors.

Quisqualic Acid Model

A model of syringomyelia involves the injection of quisqualic acid into the spinal cord gray matter (Yezierski and Park, 1993; Yezierski et al., 1993; 1998). This excitotoxic substance produces a cavity that resembles the cavities produced by spinal cord injury or in syringomyelia. The animals display overgrooming behavior in an area of the skin that is somatotopically related to the location of the lesion. There is an associated mechanical allodynia. Dorsal horn neurons that respond to the overgroomed region of skin have increased background activity and enhanced responses to mechanical stimulation of the skin.

Ischemia Model

Still another model of spinal cord injury pain involves producing damage to the spinal cord by focusing a laser beam on a region through which blood that contains a photosensitive dye circulates. The result of the interaction between the laser beam and the dye is an ischemic lesion of the spinal cord gray matter. The animals develop truncal mechanical allodynia (Hao et al., 1991a; Xu, X. J. et al., 1992). However, heat responses are not affected (Hao et al., 1992a). The responses of WDR neurons increase after such a lesion (Hao et al., 1991b; 1992b). The number of GABAergic neurons is reduced after an ischemic lesion (Zhang, A. L. et al., 1994), suggesting that the pain state that develops in these animals may relate to disinhibition.

Summary

A number of animal models of human pain conditions have been developed in order to investigate mechanisms underlying different pain states and to explore possible new therapies. Models used to study inflammatory pain include acute arthritis induced by intra-articular injection of kaolin and carrageenan and chronic arthritis induced by injection of complete Freund's adjuvant. Other models of inflammatory pain include injection of carrageenan, complete Freund's adjuvant, formalin, bee venom, or capsaicin into the skin. Behavioral responses and also changes in the responses of nociceptive dorsal horn neurons have been observed during the appropriate timeframe for each model.

Models have also been developed for the study of pain states that are the result of nerve damage. Deafferentation of the spinal cord is accomplished by transection of peripheral nerves or cutting dorsal roots. Some of the animals develop self-mutilation (autotomy) behavior. Changes occur in the receptive fields of dorsal horn neurons that suggest alterations in the strength of synaptic connections. However, deafferentation models do not necessarily mimic human neuropathic pain, which is often accompanied by evoked pain (allodynia and hyperalgesia) and which may depend on sympathetic innervation. Models that closely resemble human neuropathic pain include the Bennett, Seltzer, and Chung models, among others. Central neuropathic pain models that resemble the pain state often seen after spinal cord injury have also been developed. These models involve cordotomy, contusion injury, hemisection, quisqualate injection, or laser-induced spinal cord ischemia.

GENETIC INFLUENCES ON SPINAL CORD SENSORY PROCESSING

As mentioned earlier in this chapter, sensory functions of the spinal cord can be altered by stimuli that affect the activity of immediate, early genes and transcription factors. These in turn can induce changes in gene expression and alterations in the response properties or other functions of spinal cord neurons. Furthermore, constitutive gene expression can help determine the probability that a particular disease or response to a stimulus will occur. In future, genetic engineering may play a role in medical interventions to modify disease processes. An important avenue of research is experiments on genetically altered animals, such as knock-out mice or mice that over-express a gene product. On a smaller scale, gene expression can be altered in a restricted region in order to upregulate or downregulate local gene expression without changing gene expression throughout the organism. Several techniques are available to do this, including the use of antisense oligonucleotides and viral transfection of neurons.

Genetic Basis of Spinal Cord Sensory Functions

Observations have been made of genetic variations that affect a variety of sensory functions. Many of these relate to pain sensation.

Mechanoreceptor-related Genes

As discussed in Chapter 2, one mechanism by which mechanical forces might activate mechanoreceptors is by an action on ion channels that are tethered to the extracellular matrix and to the cytoskeleton (Welsh et al., 2002). Channel proteins belonging to the DEG/ENaC family that have previously been studied in *Caenorhabditis elegans* are candidates for this function (Drummond, H. A. et al., 2000; Price, M. P. et al., 2000), since subunits of these proteins have been demonstrated by immunocytochemistry in several types of cutaneous mechanoreceptors. The role of a member of the DEG/EnaC family known as BNC1 (a non-voltage-dependent Na^+ channel; Price, M. P. et al., 1996) has been examined by a study of knock-out mice that lack BNC1. Both rapidly adapting and slowly adapting cutaneous mechanoreceptors have a reduced sensitivity in the knock-out animals; however, behavioral tests showed that mechanical sensitivity was not abolished in these animals, indicating that other proteins must also be involved in mechanoreception (Price, M. P. et al., 2000). Another cation channel of the DEG/EnaC familty that has been studied with reference to mechanoreception is the DRASIC channel. DRASIC channels have pH sensitivity, as well as mechanosensitivity. In DRASIC-knock-out mice, the sensitivity of mechanoreceptors is actually increased, but nociceptors show reduced responses to pinch and also to acid and noxious heat stimuli (Price, M. P. et al., 2001).

It is known that cold temperatures can affect mechanoreceptor function (see Chapter 2). It is therefore of interest that cold also affects the function of DEG/EnaC channels (Askwith et al., 2001).

Cold-related Gene

Several groups have found ion channels that respond to cold (McKamy et al., 2002; Peier et al., 2002a,b; Viana et al., 2002). One of these is also responsive to menthol, a chemical that is known to activate cold receptors (see Chapter 2).

Nociception-related Genes

A large number of molecules are known to play important roles in pain mechanisms. These include neurotransmitters and modulators, membrane receptors, and signal transduction molecules. Some of these molecules are indicated in Fig. 7.38 in relation to neurons in nociceptors, the ascending pain pathways, or the descending control systems (Mogil et al., 2000). Obviously, many more are likely to be associated with neurons in the pain system found in the brain. Transmitter and modulator molecules can be influenced by genetic alterations in proteins that are involved in the synthesis, storage, release, or degradation of these molecules. Receptor and signal transduction proteins can also be genetically regulated.

TrkA and Congenital Insensitivity to Pain

Cases of congenital insensitivity to painful stimuli have been reported periodically (Dearborn, 1932; Thrush, 1973), although such cases are rare (Chatrian et al., 1975; Low et al., 1978; Manfredi et al., 1981; Matsuo et al., 1981; Dyck et al., 1983; Itoh et al., 1986: Derwin et al., 1994; Rosemberg et al., 1994; Edwards-Lee et al., 1997; Vital et al., 1998; Nolano et al., 2000; Sztriha et al., 2001; Schulman et al., 2001; Shorer et al., 2001). Variations are reported, but in most cases there is an accompanying anhidrosis ("congenital insensitivity to pain with anhidrosis" or CIPA syndrome; another name for this condition is hereditary sensory and autonomic neuropathy, type 4), suggesting a combined defect in sensory and autonomic neurons that are derived from the neural crest. Other findings are generally a consquence of these primary neural defects. These findings include: (1) a reduction in evoked potentials following electric shocks to the skin (Chatrian et al., 1975) and a lack of pain following electrical stimulation of the cornea, dental pulp, or a dorsal root (Manfredi et al., 1981); (2) self-mutilation, Charcot joints, and fractures (Chatrian et al., 1975; Matsuo et al., 1981; Sweet, 1981; Itoh et al., 1986; Derwin et al., 1994;

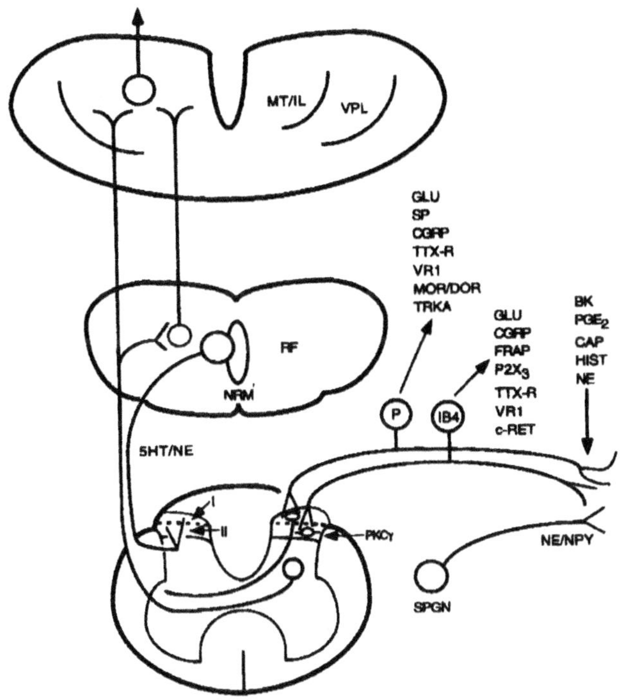

FIGURE 7.38. Diagram showing some of the neural components of the pain system. These include peptidergic (P) and IB4-containing primary afferent nociceptors. Spinal cord laminae I and II contain some spinothalamic tract neurons. Other spinothalamic tract cells and spinoreticular tract neurons originate in the deep dorsal horn and project to ventroposterior lateral (VPL), medial thalamic (MT), and intralaminar (IL) nuclei, and the reticular formation (RF). Descending projections from the nucleus raphe magnus (NRM) and sympathetic postganglionic neurons (SPGN) are also shown. Not indicated are other ascending nociceptive pathways, details about the thalamic, cortical, and other processing areas for pain, and other descending pathways that influence pain signaling. Some important molecules for pain processing are listed. These include substances that act on membrane receptors (BK, bradykinin; CAP, capsaicin; CGRP, calcitonin gene-related peptide; GLU, glutamate; HIST, histamine; 5-HT, 5-hydroxytryptamine; NE, norepinephrine; NPY, neuropeptide Y; PGE_2, prostaglandin E_2; SP, substance P); ion channels (TTX-R, tetrodotoxin-resistance Na^+ channels); membrane receptors (MOR/DOR, mu, and delta opiate receptors; P2X3, purinergic receptor P2X3; TRKA, tyrosine kinase receptor A; VR1, vanilloid receptor 1); signal transduction molecules (PKCγ, protein kinase C gamma); and an enzyme (FRAP, fluoride-resistant acid phosphatase). (From Mogil et al., 2000.)

Nolano et al., 2000; Schulman et al., 2001); (3) loss of flare in response to histamine injection (Manfredi et al., 1981; Nolano et al., 2000); and (4) difficulties with temperature regulation, such as bouts of fever and even death from hyperpyrexia (Matsuo et al., 1981; Itoh et al., 1986; Vital et al., 1998; Sztriha et al., 2001). The problem with thermoregulation is presumably caused by the lack of innervation of sweat glands (Nolano et al., 2000; Sztriha et al., 2001). Mental retardation is common (Rosemberg et al., 1994; Vital et al., 1998; Nolano et al., 2000; Schulman et al., 2001).

One explanation for congenital insensitivity to pain was an overexpression of endogenous opioids, since naloxone lowered the threshold for a reflex response to noxious stimuli in one case of congenital insensitivity to pain (Dehen et al., 1977). However, levels of met-enkephalin in the cerebrospinal fluid were found to be normal in congenital insensitivity to pain, and naloxone was generally ineffective (Cesselin et al., 1984). A more likely explanation is a reduction in the number of primary afferent nociceptors (Larner et al., 1994).

Reports concerning the findings in nerve biopsies often indicate a reduction in small myelinated fibers, but in different studies the number of unmyelinated fibers has been reported as normal or reduced (Low et al., 1978; Manfredi et al., 1981; Matsuo et al., 1981; Dyck et al., 1983; Itoh et al., 1986; Rosemberg et al., 1994; Edwards-Lee et al., 1997; Nolano et al., 2000). C fibers (presumably sympathetic axons) show reduced staining for dopamine-beta-hydroxylase (Dyck et al., 1983), and there is evidently an absence of substance P-containing axons in the knee joint (Derwin et al., 1994). There is also a loss of neurons in sympathetic ganglia (Sztriha et al., 2001) and an absence of skin responses to sympathetic stimulation (Shorer et al., 2001).

The syndrome of congenital insensitivity to pain with anhidrosis (CIPA) is now attributed to any of several mutations of the gene for the trkA receptor (the receptor for nerve growth factor; see Chapter 2) (Greco et al., 1996; 2000; Indo et al., 1996; 2001; Mardy et al., 1999; Yotsumoto et al., 1999; Miura et al., 2000a, b; Shatzky et al., 2000; Toscano et al., 2000; Bodzioch et al., 2001; Houlden et al., 2001; Miranda et al., 2001). The trkA receptor gene is located on chromosome 1 and consists of 17 exons and spans at least 23 kilobases (Indo, 2001). Transfection of COS1 cells with a mutated gene results in an inability of NGF to activate the trkA protein, and transfection of NIHT3 and PC12 cells showed a loss of transforming and differentiating activity (Greco et al., 2000). Thus, the mutations responsible for the CIPA syndrome result in a loss of trkA function. Several mutations have been found that are involved in NGF binding and in intracellular signal-transduction (Mardy et al., 1999).

Mice lacking the gene for trkA also show a loss of responses to painful stimuli, such as pinprick and heat, and they demonstrate self-mutiliation behavior (Smeyne et al., 1994). Anhidrosis is not evident in mice (cf. Indo et al., 1996). There are few small-diameter primary afferents in these animals. The trkA receptor is expressed in adults by peptide-containing primary afferent nociceptors. Non-peptidergic nociceptors that contain IB4 depend on trkA receptors during development, but trkA receptors are not expressed in these neurons in adults (Silos-Santiago et al., 1995). Instead, they express c-Ret, which responds to glial-derived neurotropic factor (GDNF). Corneal sensitivity to noxious stimuli is maintained, suggesting that the cornea is supplied by nociceptors that do not depend on trkA receptors during development (De Castro et al., 1998).

Null mutants for NGF have also been developed. In these animals, there is a developmental loss of primary afferent and sympathetic neurons (Crowley et al., 1994). Primary afferents do not express trkA receptors, and CGRP and SP immunostaining is absent. Most animals die within the first week, but the few that survive more than a week show almost no response to noxious mechanical stimuli. Heterozygous animals have increased tail-flick latencies, suggesting that heat nociceptors are also affected.

In mice that over-express NGF there is a sympathetic hyperinnervation of DRG cells even in the absence of nerve injury (Davis, B. M. et al., 1994). The animals have exaggerated responses to noxious mechanical and thermal stimuli, and neuropathic pain is enhanced. The DGR cells that are targeted by sympathetic sprouts are small ones that should express trkA receptors. It is unclear if the pain state in these animals is sympathetically maintained.

Deletion of Neurokinin-1 Receptors

Mice with a deletion of the gene for neurokinin 1 (NK-1) receptors appear to be normal, and they respond normally to brief noxious mechanical stimuli with von Frey filaments

(Laird et al., 2001b). However, they show a reduction in responses to intradermal injection of capsaicin, and mechanical hyperalgesia and allodynia fails to develop. Presumably, the responses to weak noxious stimuli are mediated by glutamate. Mice with a deletion of the gene for NK-1 receptors also have intact responses in the first phase of the formalin test (De Felipe et al., 1998).

Deletion of CGRP

Mice with a knock-out of the gene for CGRP have normal responses to noxious stimuli (Zhang, L. et al., 2001). However, they fail to develop secondary heat hyperalgesia when the knee joint is inflamed with kaolin and carrageenan.

Changes in Signal Transduction Molecules

Pain responses in mice with a deletion of the R1β subunit of protein kinase A (PKA) or of the gamma isoform of protein kinase C (PKCγ) have been investigated (Malmberg et al., 1997a, b). Responses to acute nociceptive stimuli in the PKA knock-out mice are unaltered, but there is a reduction in the allodynia normally produced by tissue damage. Responses in the second phase of the formalin test are also reduced. Central sensitization that is usually caused by spinal cord injection of PGE2 is also reduced. It was concluded that there is a reduction in both peripheral and central sensitization in these animals. However, there was no change in neuropathic pain.

PKCγ is found in the spinal cord but not in DRG neurons, and it first appears in postnatal animals. PKCγ knock-out mice do not show any change in responsiveness to acute nociceptive stimuli, and they show only minimal changes after nerve injury. In recordings from nociceptive dorsal horn neurons in null mice, the responses to noxious stimuli were normal. However, mustard oil-induced central sensitization was only transient. It was proposed that the interneurons that contain PKCγ are excitatory onto lamina V projection neurons (Martin, W. J. et al., 1999).

The epsilon form of PKC has also been found to play a role in nociception (Khasar et al., 1999b). This depends on an interaction between this isoform of PKC and the effects of NGF (Hundle et al., 1995).

Opiate Receptors

Mice with deletions of the mu opioid receptor gene no longer respond to morphine in tests of pain behavior, although they do respond to other opiates (Matthes et al., 1996; Sora et al., 1997). Their sensitivity to noxious stimuli is normal. This is not surprising, since opiate receptors are not tonically activated under normal circumstances. This is why naloxone does not have any obvious effect when administered in normal volunteers.

Summary

Sensory processing depends on protein molecules and so is subject to genetic regulation. Many kinds of proteins are involved in sensory processing, including proteins that relate to neurotransmitters and modulators, membrane receptors, and signal transduction pathways. Recent observations provide evidence for the proteins that are involved in mechanoreception and cold responsiveness. A number of studies have been directed at "pain genes," which are genes that affect the pain system. These include studies of human patients with congenital insensitivity to pain (usually with anhidrosis) and transgenic mice with alterations in the trkA receptor. Other studies have been made of mice with deletions of genes for the NK-1 receptor, CGRP, certain protein kinases or opiate receptors. Such studies are only beginning to open up a new field of sensory genetics and perhaps gene therapy for sensory malfunction.

CONCLUSIONS

1. Most neurons in the spinal cord are interneurons. Interneurons of the dorsal horn are involved in complex interactions that affect the signaling of somatosensory tract cells (as well as of motor activity).
2. A number of approaches can be used for the electrophysiological study of individual spinal interneurons. These include extracellular and intracellular recordings, as well as patch-clamp recordings, in anesthetized or decerebrate in vivo preparations, in special cases in awake, behaving animals, or in in vitro preparations. Dorsal horn neurons can generally be affected synaptically by volleys in A and/or C fibers. Extracellular recording permits the monitoring of activity of a single neuron for a prolonged period of time while a variety of experimental manipulations are carried out. Intracellular recording allows an appraisal of changes in membrane properties and synaptic events.
3. The activity of populations of dorsal horn interneurons is reflected in slow potentials that can be recorded from the surface of the spinal cord or from within the cord. After an initial afferent volley evoked by an electrical stimulus applied to a peripheral nerve, there may be several slow potentials due to excitatory activity, followed by one with an opposite sign associated with primary afferent depolarization (PAD). When recording from the dorsal surface of the spinal cord, the excitatory field potentials are negative in sign and the one associated with PAD is positive. The details of the field potentials vary with the nerve stimulated and the stimulus strength.
4. Activity in populations of neurons in the dorsal horn can also be demonstrated anatomically by using the 2-deoxyglucose technique or immunocytochemical staining for the presence of FOS protein. An accumulation of 2-deoxyglucose reflects increased glucose metabolism. FOS protein is expressed following induction of the c-fos gene by the activation of several types of neurons, including inhibitory interneurons.
5. The activity of neurons in the different layers of the dorsal horn reflects the input to these layers from particular classes of primary afferent fibers. Neurons in lamina I respond chiefly to inputs conveyed by Aδ and C fibers, although some neurons also respond to Aβ fibers. According to some investigators, cells in lamina II, the substantia gelatinosa, are activated chiefly by C fibers. However, other investigators find that there are also inputs to this lamina from Aδ and even Aβ fibers, as well as from C fibers. Some of these actions must be polysynaptic. Laminae IV–VI contain cells that receive a variety of combinations of inputs.
6. The functions of neurons in different laminae are better reflected by their responses to natural forms of stimulation than to electrically evoked afferent

volleys. Neurons in lamina I may respond chiefly or only to noxious or thermal stimuli, although some wide-dynamic-range cells are also found in this layer. Inputs are from skin, muscle, joints, or viscera. The substantia gelatinosa contains neurons with unusual responses, including habituation, prolonged discharges following brief stimuli, strong inhibitory inputs, and variable or "ameboid" receptive fields. Cells in laminae III–VI include low-threshold, wide-dynamic-range, and high-threshold neurons, based on their responses to mechanical stimulation of the skin. Proprioceptive stimulation can activate neurons in lamina VI (as well as some cells in adjacent laminae). Visceral and joint afferents are often effective in exciting neurons in laminae IV–VI. Cutaneous and visceral nociceptive responses have also been reported for cells of lamina X.

7. A number of schemes for classifying dorsal horn neurons have been proposed. These have been based on a variety of approaches, such as whether particular inputs are mono- or polysynaptic, the possible relationship to PAD, the location of the neurons, the responses to A and/or C fibers and to innocuous and noxious stimuli, and the particular classes of sensory receptors that activate the cells. A commonly used classification is one that designates neurons as low-threshold (class 1) neurons if they respond just to innocuous mechanical stimuli, wide-dynamic-range (class 2) neurons if they respond to both innocuous and noxious stimuli, and high-threshold (class 3) neurons if they respond just to noxious stimuli. An effort has been made to improve on this scheme by a statistical analysis of the responses to defined stimuli, using cluster analysis.
8. Recordings from dorsal horn neurons in unanesthetized, behaving animals are difficult to make. The main finding of the few successful studies is that the activity of dorsal horn neurons in such preparations is under powerful inhibitory regulation.
9. The pharmacology of dorsal horn neurons can be investigated using a number of approaches. These include various routes of drug administration in in vivo experiments (systemic administration, superfusion of the spinal cord, intrathecal administration, microdialysis and microiontophoresis). In addition, the effects of drugs can be examined in in vitro preparations.
10. Many neurotransmitters and modulators affect the activity of dorsal horn neurons. Excitatory substances include the excitatory amino acids, glutamate and aspartate, ATP, and several peptides, such as SP and CGRP. Inhibitory substances include the inhibitory amino acids, GABA and glycine, and acetylcholine, as well as adenosine and several peptides, such as the endogenous opioids.
11. Glutamate and aspartate are likely to serve as the fast excitatory neurotransmitters of many primary afferent terminals, interneurons, and projections that descend from the brain. These excitatory amino acids are found in synaptic endings and are in excess of what is required in the metabolic compartment. When they are released, they excite specific types of receptors. The excitatory amino acid receptors are grouped into ionotropic and metabotropic classes. Ionotropic glutamate receptors are coupled to ion channels and include AMPA, kainate, and NMDA receptors. Metabotropic glutamate receptors are divided into groups I, II, and III and are coupled through G-proteins to intracellular signal transduction pathways. Antagonists are now available that are selective for many of the glutamate receptor types.
12. Another fast excitatory neurotransmitter that can activate some dorsal horn neurons is ATP. The excitation depends on purinergic receptors, such as P2X receptors.

13. Substance P (SP) is a peptide that is contained in many primary afferent neurons, interneurons of the dorsal horn, and in some descending projection fibers. SP is released in the dorsal horn, particularly in the substantia gelatinosa, following noxious mechanical, thermal, or chemical stimulation of the skin. Exposure of nociceptive dorsal horn interneurons to SP results in a prolonged excitation by an action on NK-1 receptors. In vitro studies of the action of SP indicate that this peptide produces a prolonged depolarization due to changes in a number of different conductance mechanisms. It also activates second messenger cascades. Other neurokinins, including NKA, may also contribute to the excitation of dorsal horn neurons.
14. CGRP is another excitatory peptide. In the dorsal horn, CGRP is found only in the terminals of primary afferent fibers. CGRP is released by noxious mechanical or thermal stimulation of the skin. It enhances the action of SP and also has its own postsynaptic action on CGRP receptors.
15. The actions of excitatory amino acids and peptides, such as substance P or CGRP, can be cooperative. For example, co-application of two substances, such as NMDA and substance P, can produce a long-lasting increase in responses to later applications of NMDA.
16. Other excitatory peptides in the dorsal horn include vasoactive intestinal polypeptide, neurotensin, cholecystokinin, thyrotropin-releasing hormone, and corticotropin-releasing hormone, perhaps as well as pituitary adenylate cyclase-activating peptide and islet amyloid polypeptide.
17. Inhibitory neurotransmitters in the dorsal horn include the inhibitory amino acids, GABA and glycine. Both substances inhibit dorsal horn interneurons, and GABA also produces primary afferent depolarization and presynaptic inhibition. $GABA_A$ receptors are antagonized by picrotoxin and by bicuculline, whereas glycine receptors are antagonized by strychnine. GABA also has pre- and postsynaptic inhibitory actions through $GABA_B$ receptors. These can be activated with baclofen and blocked by $GABA_B$ antagonists such as phaclofen. Another inhibitory substance is acetylcholine, acting through nicotinic and muscarinic receptors.
18. Inhibitory peptides include the opioid peptides, such as enkephalin, dynorphin, and the endomorphins, which are present in the dorsal horn. Much of the opioid content is probably in interneurons, although some is in projections that descend from the brain. Endomorphins are in primary afferent terminals. Met-enkephalin is released in the spinal cord in response to noxious stimulation or application of SP to the cord. Opiate receptors include μ-, δ-, and κ-opiate receptors. The endogenous agonists are endomorphins (for μ receptors), enkephalin (for δ receptors), and dynorphin (for κ receptors). Some of the opiate receptors in the spinal cord are on primary afferent terminals, and so opioid effects may be mediated by presynaptic, as well as by postsynaptic, actions. Dynorphin applied topically to the spinal cord can produce excitation or inhibition of dorsal horn neurons.
19. Other inhibitory peptides that are found in the dorsal horn include nocistatin and nociceptin, somatostatin, galanin, bombesin, and neuropeptide Y.
20. Primary afferent depolarization (PAD) can be attributed to an action of GABA released from dorsal horn interneurons on $GABA_A$ receptors on the terminals of primary afferent neurons. An efflux of chloride ions depolarizes the terminals

because chloride is concentrated within primary afferent neurons by a special chloride transporter. Other transmitters, such as serotonin and excitatory amino acids, may also contribute to PAD, but this is less certain. An alternative mechanism for producing PAD is release of K^+ ions into the extracellular space. An increased extracellular K^+ concentration can depolarize not only primary afferent fibers but also dorsal horn interneurons, and so the effects of K^+ may be quite important, especially since damaging stimuli produce increases in K^+ concentration that last for hours. Presynaptic inhibition can also be produced by an action on presynaptic terminals that does not cause PAD but that nevertheless reduces transmitter release. Agents that have such an action include GABA acting on $GABA_B$ receptors and several peptides, including opioid peptides.

21. Dorsal horn neurons typically have both excitatory and inhibitory receptive fields. Wide-dynamic-range neurons often have a central low-threshold excitatory receptive field, surrounded by a high-threshold excitatory field, and outside of that, an inhibitory field.
22. It has been proposed that neurons in lamina IV excite those in lamina V, and that cells in lamina V in turn excite neurons in lamina VI. This "cascade theory" has been disputed.
23. Neurons in the upper layers of the dorsal horn are somatotopically organized. In the lumbar enlargement, the toes are represented medially and the lateral foot and more proximal regions laterally. A similar arrangement is found in the cervical enlargement. This pattern can be related to the embryologic development of the limbs and their dermatomal innervation.
24. Receptive fields are not hard-wired. They can be shown to change, depending on the previous history of sensory activity. Changeable receptive fields have been described as "ameboid." Changes in receptive field size or location and in the responsiveness of dorsal horn neurons following injury or transection of peripheral nerves can help explain the development of secondary allodynia and hyperalgesia.
25. One form of plastic change that has now been demonstrated in the responses of dorsal horn neurons is long-term potentiation. Another is long-term depression. Glutamate or neurokinin receptor antagonists can prevent LTP in dorsal horn neurons.
26. Inhibitory receptive fields are of particular interest in the context of pain mechanisms, since inhibition of nociceptive transmission may lead to analgesia. Inhibitory systems include one that is activated by innocuous stimuli, has a surround organization, and is mediated segmentally. This system is associated with the "Gate Theory of Pain." Another inhibitory system is activated by noxious stimuli applied to most of the body surface and depends on a supraspinal loop. This system is known as "Diffuse Noxious Inhibitory Controls" (DNIC).
27. The "Gate Theory of Pain" was proposed because of the need to account for many features of pain that were difficult to predict from knowledge current at the time. The circuitry suggested involved the substantia gelatinosa as a source of presynaptic inhibition of input to nociceptive transmission cells in the dorsal horn. The inhibition would be evoked by an increase in activity in large afferent fibers, but decreased by inputs over small fibers. There are a number of difficulties with the details of the mechanisms as proposed, but clearly the activation of large mechanoreceptive fibers can lead to inhibition of nociceptive transmission.

Examples are the use of high-frequency, low-intensity transcutaneous electrical nerve stimulation and dorsal column stimulation for the relief of pain.

28. "Diffuse Noxious Inhibitory Controls" (DNIC) produce a powerful inhibition in response to noxious mechanical or thermal stimulation of essentially the entire body and face. The only neurons that are inhibited by such stimuli are wide-dynamic-range neurons. The inhibitory mechanism involves the activation of neurons in the brain stem, which in turn project to the spinal cord. The DNIC mechanism may serve as a contrast-enhancing device.
29. Models of human sensory disease are being developed, particularly in relation to pain mechanisms, in order to learn about possible mechanisms underlying abnormal sensory states and to provide clues for improvements in pain therapy.
30. Useful models of inflammatory pain include the kaolin and carrageenan and the complete Freund's adjuvant models of experimental arthritis, and several models of cutaneous inflammation produced by application of mustard oil to the skin or injection of carrageenan, complete Freund's adjuvant, formalin, bee venom, or capsaicin into the skin.
31. Models of neuropathic pain include deafferentation by dorsal rhizotomy or nerve transections, peripheral nerve damage by loose or tight ligations of peripheral or spinal nerves (Bennett, Selzer, and Chung models), and spinal cord injury (cordotomy, contusion injury, spinal cord hemisection, quisqualic acid injection, or ischemic injury).
32. Genetic studies of sensory mechanisms will no doubt become an important avenue for investigation of sensory mechanisms. There is already progress in identifying molecules that are used by primary afferent neurons as transducers for mechanical, thermal, and chemical stimuli. Pain genes have been identified in humans and in animal models.

References

Aanonsen, L. M. and Seybold, V. S. 1989. Phencyclidine and sigma receptors in rat spinal cord: Binding characterization and quantitative autoradiography. *Synapse* **4**, 1–10.

Aanonsen, L. M. and Seybold, V. S. 1991. 125I-Substance P binding sites in rat spinal cord in a chronic constriction injury and following dorsal rhizotomy. *Ann. NY Acad. Sci.* **632**, 354–357.

Aanonsen, L. M., Lei, S., and Wilcox, G. L. 1990. Excitatory amino acid receptors and nociceptive neurotransmission in rat spinal cord. *Pain* **41**, 309–321.

Aanonsen, L. M., Kajander, K. C., Bennett, G. J., and Seybold, V. S. 1992. Autoradiographic analysis of 125I-substance P binding in rat spinal cord following chronic constriction of the sciatic nerve. *Brain Res.* **596**, 259–268.

Aarnisalo, A. A. and Panula, P. 1998. Neuropeptide FF in the lateral spinal and lateral cervical nuclei: evidence of contacts on spinothalamic neurons. *Exp. Brain Res.* **119**, 159–165.

Abbadie, C. and Basbaum, A. I. 1998. The contribution of capsaicin-sensitive afferents to the dorsal root ganglion sprouting of sympathetic axons after peripheral nerve injury in the rat. *Neurosci. Lett.* **253**, 143–146.

Abbadie, C., Brown, J. L., Mantyh, P. W., and Basbaum, A. I. 1996. Spinal cord substance P receptor immunoreactivity increases in both inflammatory and nerve injury models of persistent pain. *Neuroscience* **70**, 201–209.

Abbadie, C., Trafton, J., Liu, H., Mantyh, P. W., and Basbaum, A. I. 1997. Inflammation increases the distribution of dorsal horn neurons that internalize the neurokinin-1 receptor in response to noxious and non-noxious stimulation. *J. Neurosci.* **17**, 8049–8060.

Abbadie, C., Gultekin, S. H., and Pasternak, G. W. 2000a. Immunohistochemical localization of the carboxy terminus of the novel mu opioid receptor splice variant MOR-1C within the human spinal cord. *NeuroReport* **11**, 1953–1957.

Abbadie, C., Pan, Y. X., and Pasternak, G. W. 2000b. Differential distribution in rat brain of mu opioid receptor carboxy terminal splice variants MOR-1C-like and MOR-1-like immunoreactivity: Evidence for region-specific processing. *J. Comp. Neurol.* **419**, 244–256.

Abbadie, C., Rossi, G. C., Orciuolo, A., Zadina, J. E., and Pasternak, G. W. 2002. Anatomical and functional correlation of the endomorphins with mu opioid receptor splice variants. *Eur. J. Neurosci.* **16**, 1075–1082.

Abbott, F. V., Franklin, K. B. J., and Westbrook, R. F. 1995. The formalin test: Scoring properties of the first and second phases of the pain response in rats. *Pain* **60**, 91–102.

Abdel-Maguid, T. E. and Bowsher, D. 1984a. Classification of neurons by dendritic branching pattern. A categorisation based on Golgi impregnation of spinal and cranial somatic and visceral afferent and efferent cells in the adult human. *J. Anat.* **138**, 689–702.

Abdel-Maguid, T. E. and Bowsher, D. 1984b. Interneurons and proprioneurons in the adult human spinal grey matter and in the general somatic and visceral afferent cranial nerve nuclei. *J. Anat.* **139**, 9–19.

Aberdeen, J., Corr, L., Milner, P., Lincoln, J., and Burnstock, G. 1990. Marked increases in calcitonin gene-related peptide-containing nerves in the developing rat following long-term sympathectomy with guanethidine. *Neuroscience* **35**, 175–184.

Aberdeen, J., Milner, P., Lincoln, J., and Burnstock, G. 1992. Guanethidine sympathectomy of mature rats leads to increases in calcitonin gene-related peptide and vasoactive intestinal polypeptide-containing nerves. *Neuroscience* **47**, 453–461.

Abood, M. E. and Martin, B. R. 1996. Molecular neurobiology of the cannabinoid receptor. *Int. Rev. Neurobiol.* **39**, 197–221.

Abraham, K. E. and Brewer, K. L. 2001. Expression of c-fos mRNA is increased and related to dynorphin mRNA expression following excitotoxic spinal cord injury in the rat. *Neurosci. Lett.* **307**, 187–191.

Abraham, K. E., McGinty, J. F., and Brewer, K. L. 2001. The role of kainic acid/AMPA and metabotropic glutamate receptors in the regulation of opioid mRNA expression and the onset of pain-related behavior following excitotoxic spinal cord injury. *Neuroscience* **104**, 863–874.

Abrahams, V. C. and Swett, J. E. 1986. The pattern of spinal and medullary projections from a cutaneous nerve and a muscle nerve of a forelimb of the cat: A study using the transganglionic transport of horseradish peroxidase. *J. Comp. Neurol.* **246**, 70–84.

Abrahams, V. C., Lynn, B., and Richmond, F. J. R. 1984a. Organization and sensory properties of small myelinated fibres in the dorsal cervical rami of the cat. *J. Physiol.* **347**, 177–187.

Abrahams, V. C., Richmond, F. J., and Keane, J. 1984b. Projections from C2 and C3 nerves supplying muscles and skin of the cat neck: A study using transganglionic transport of horseradish peroxidase. *J. Comp. Neurol.* **230**, 142–154.

Abram, S. E. and Kostreva, D. R. 1986. Spinal cord metabolic response to noxious radiant heat stimulation of the cat hind footpad. *Brain Res.* **386**, 143–147.

Abram, S. E. and Winne, R. P. 1995. Intrathecal acetyl cholinesterase inhibitors produce analgesia that is synergistic with morphine and clonidine in rats. *Anesth. Analg.* **81**, 501–507.

Aceto, M. D., Awaya, H., Martin, B. R., and May, E. L. 1983. Antinociceptive action of nicotine and its methiodide derivatives in mice and rats. *Br. J. Pharmacol.* **79**, 869–876.

Achaval, M., Martinez-Murillo, R., Rodrigo, J., Aguado, F., DeFelipe, M. C., and Del Rio, J. 1991. Lasting loss in substance P following administration of substance P antiserum to newborn rat. An immunohistochemical study. *Neurosci. Lett.* **126**, 75–78.

Ackermann, P. W., Finn, A., and Ahmed, M. 1999. Sensory neuropeptidergic pattern in tendon, ligament and joint capsule. A study in the rat. *NeuroReport* **10**, 2055–2060.

Adams, P. R., Brown, D. A., and Jones, S. W. 1983. Substance P inhibits the M-current in bullfrog sympathetic neurones. *Br. J. Pharmacol.* **79**, 330–333.

Adem, A., Ekblom, J., Gillberg, P.-G., Jossan, S. S., Höög, A., Winblad, B., Aquilonius, S.-M., Wang, L.-H., and Sara, V. 1994. Insulin-like growth factor-I-receptors in human spinal cord: Changes in amyotrophic lateral sclerosis. *J. Neural Trans.* **97**, 73–84.

Adham, N., Bard, J. A., Zgombick, J. M., Durkin, M. M., Kucharewicz, S., Weinshank, R. L., and Brancher, T. A. 1997. Cloning and characterization of the guinea pig 5-HT_{1F} receptor subtype: A comparison of the pharmacological profile to the human species homolog. *Neuropharmacology* **36**, 569–576.

Adler, J. E. 1998. Age-dependent differential regulation of sensory neuropeptides by glial cell line-derived neurotrophic factor. *J. Neurochem.* **71**, 170–177.

Adler, J. E. and Walker, P. D. 2000. Cyclic AMP regulates substance P expression in developing and mature spinal sensory neurons. *J. Neurosci. Res.* **59**, 624–631.

Adriaensen, H., Gybels, J., Handwerker, H. O., and Van Hees, J. 1983. Response properties of thin myelinated (A-δ) fibers in human skin nerves. *J. Neurophysiol.* **49**, 111–122.

Adriaensen, H., Gybels, J., Handwerker, H. O., and Van Hees, J. 1984. Suppression of C-fibre discharges upon repeated heat stimulation may explain characteristics of concomitant pain sensation. *Brain Res.* **302**, 203–211.

Adrian, E. D. 1928. *The Basis of Sensation. The Action of the Sense Organs.* Reprinted by Hafner Publishing Co., New York, 1964.

Adrian, E. D. 1946. *The Physical Background of Perception.* Oxford University Press, Oxford.

Adrian, E. D. and Zotterman, Y. 1926a. The impulses produced by sensory nerve-endings. Part 2. The response of a single end-organ. *J. Physiol.* **61**, 151–171.

Adrian, E. D. and Zotterman, Y. 1926b. The impulses produced by sensory nerve endings. Part 3. Impulses set up by touch and pressure. *J. Physiol.* **61**, 465–483.

Afrah, A. W., Stiller C. O., Olgart, L., Brodin, E., and Gustafsson, H. 2001. Involvement of spinal N-methyl-D-aspartate receptors in capsaicin-induced in vivo release of substance P in the rat dorsal horn. *Neurosci. Lett.* **316**, 83–86.

Agrawal, S. G. and Evans, R. H. 1986. The primary afferent depolarizing action of kainate in the rat. *Br. J. Pharmacol.* **87**, 345–355.

Agrawal, S. K., Theriault, E., and Fehlings, M. G. 1998. Role of group I metabotropic glutamate receptors in traumatic spinal cord white matter injury. *J. Neurotrauma* **15**, 929–941.

Ahima, R. S. and Harlan, R. E. 1990. Charting of type II glucocorticoid receptor-like immunoreactivity in the rat central nervous system. *Neuroscience* **39**, 579–604.

Ahima, R., Krozowski, Z., and Harlan, R. 1991. Type I corticosteroid receptor-like immunoreactivity in the rat CNS: distribution and regulation by corticosteroids. *J. Comp. Neurol.* **313**, 522–538.

Ahmadi, S., Liebel, J. T., and Zeilhofer, H. U. 2001a. The role of the ORL1 receptor in the modulation of spinal neurotransmission by nociceptin/orphanin FQ and nocistatin. *Eur. J. Pharmacol.* **412**, 39–44.

Ahmadi, S., Kotalla, C., Gühring, H., Takeshima, H., Pahl, A., and Zeilhofer, H. U. 2001b. Modulation of synaptic transmission by nociceptin/orphanin FQ and nocistatin in the spinal cord dorsal horn of mutant mice lacking the nociceptin/orphanin FQ receptor. *Mol. Pharmacol.* **59**, 612–618.

Ahmed, M., Srinivasan, G. R., Theodorsson, E., Schultzberg, M., and Kreicbergs, A. 1995a. Effects of surgical denervation on substance P and calcitonin gene-related peptide in adjuvant arthritis. *Peptides* **16**, 569–579.

Ahmed, M., Bjurholm, A., Schultzberg, M., Theodorsson, E., and Kreicbergs, A. 1995b. Increased levels of substance P and calcitonin gene-related peptide in rat adjuvant arthritis. *Arthritis Rheum.* **38**, 699–709.

Ahmed, M., Bjurholm, A., Srinivasan, G. R., Lundeberg, T., Theodorsson, E., Schultzberg, M., and Kreicbergs, A. 1995c. Capsaicin effects on substance P and CGRP in rat adjuvant arthritis. *Regul. Pept.* **55**, 85–102.

Aicher, S. A., Sharma, S., Cheng, P. Y., and Pickel, V. M. 1997. The N-methyl-D-aspartate (NMDA) receptor is postsynaptic to substance P-containing axon terminals in the rat superficial dorsal horn. *Brain Res.* **772**, 71–81.

Aicher, S. A., Sharma, S., Cheng, P. Y., Liu-Chen, L. Y., and Pickel, V. M. 2000a. Dual ultrastructural localization of m-opiate receptors and substance p in the dorsal horn. *Synapse* **36**, 12–20.

Aicher, S. A., Punnoose, A., and Goldberg, A. 2000b. μ-Opioid receptors often colocalize with the substance P receptor (NK1) in the trigeminal dorsal horn. *J. Neurosci.* **20**, 4345–4354.

Aimar, P., Pasti, L., Carmignoto, G., and Merighi, A. 1998. Nitric oxide-producing islet cells modulate the release of sensory neuropeptides in the rat substantia gelatinosa. *J. Neurosci.* **18**, 10375–10388.

Aimi, Y., Fujimura, M., and Vincent, S. R. 1991. Localization of NADPH-diaphorase-containing neurons in sensory ganglia of the rat. *J. Comp. Neurol.* **306**, 382–392.

Aimone, L. D. and Yaksh, T. L. 1989. Opioid modulation of capsaicin-evoked release of substance P from spinal cord in vivo. *Peptides* **10**, 1127–1131.

Ainsworth, A., Hall, P., Wall, P. D., Allt, G., MacKenzie, M., Gibson, S., and Polak, J. M. 1981. Effects of capsaicin applied locally to adult peripheral nerve, II. Anatomy and enzyme and peptide chemistry of peripheral nerve and spinal cord. *Pain* **11**, 379–388.

Airaksinen, M. S. and Meyer, M. 1996. Most classes of dorsal root ganglion neurons are severely depleted but not absent in mice lacking neurotrophin-3. *Neuroscience* **73**, 907–911.

Airaksinen, M. S., Koltzenburg, M., Lewin, G. R., Masu, Y., Helbig, C., Wolf, E., Brem, G., Toyka, K. V., Thoenen, H., and Meyer, M. 1996. Specific subtypes of cutaneous mechanoreceptors require neurotrophin-3 following peripheral target innervation. *Neuron* **16**, 287–295.

Aitkin, S. C. and Lal, S. 1982a. The spatial distribution and functional properties of type I mechanoreceptor units of the sural nerve of the rabbit. *Brain Res. Rev.* **4**, 45–56.

Aitkin, S. C. and Lal, S. 1982b. The functional properties and innervation density of type II mechanoreceptor units of the sural nerve of the rabbit. *Brain Res. Rev.* **4**, 57–64.

Akesson, E., Kjaeldgaard, A., Samuelsson, E. B., Seiger, A., and Sundstrom, E. 2000. Ionotropic glutamate receptor expression in human spinal cord during first trimester development. *Brain Res. Dev. Brain Res.* **119**, 55–63.

Akkina, S. K., Patterson, C. L., and Wright, D. E. 2001. GDNF rescues nonpeptidergic unmyelinated primary afferents in streptozotocin-treated diabetic mice. *Exp. Neurol.* **167**, 173–182.

Akoev, G. N. 1982. The effect of Mg^{2+} and Ca^{2+} on the excitability of Pacinian corpuscles. *Brain Res.* **239**, 391–399.

Akopian, A. N., Sivilotti, L., and Wood, J. N. 1996. A tetrodotoxin-resistant voltage-gated sodium channel expressed by sensory neurons. *Nature* **379**, 257–262.

Akopian, A. N., Souslova, V., England, S., Okuse, K., Ogata, N., Ure, J., Smith, A., Kerr, B. J., McMahon, S. B., Boyce, S., Hill, R., Stanfa, L. C., Dickenson, A. H., and Wood, J. N. 1999. The tetrodotoxin-resistant sodium channel SNS has a specialized function in pain pathways. *Nature Neuroscience* **2**, 541–548.

Akopian, A. N., Chen, C. C., Ding, Y., Cesare, P., and Wood, J. N. 2000. A new member of the acid-sensing ion channel family. *NeuroReport* **11**, 2217–2222.

Alarcon, G. and Cervero, F. 1990. The effects of electrical stimulation of A and C visceral afferent fibres on the excitability of viscerosomatric neurones in the thoracic spinal cord of the cat. *Brain Res.* **509**, 24–30.

Albers, K. M., Perrone, T. N., Goodness, T. P., Jones, M. E., Green, M. A., and Davis, B. M. 1996. Cutaneous overexpression of NT-3 increases sensory and sympathetic neuron number and enhances touch dome and hair follicle innervation. *J. Cell Biol.* **134**, 487–497.

Albin, R. L., Hollingsworth, Z., Sakurai, S. Y., and Gilman, S. 1993. Inhibitory and excitatory amino acid neurotransmitter binding sites in cynomolgus monkey (*Macaca fascicularis*) cervical spinal cord. *Brain Res.* **604**, 354–357.

Albuquerque, C., Lee, C. J., Jackson, A. C., and MacDermott, A. B. 1999. Subpopulations of GABAeregic and non-GABAergic rat dorsal horn neurons expresss Ca^{2+}-permeable AMPA receptors. *Eur. J. Neurosci.* **11**, 2758–2766.

Aldskogius, H. and Risling, M. 1981. Effect of sciatic neurectomy on neuronal number and size distribution in the L7 ganglion of kittens. *Exp. Neurol.* **74**, 597–604.

Aldskogius, H., Arvidsson, J., and Grant, G. 1985. The reaction of primary sensory neurons to peripheral nerve injury with particular emphasis on transganglionic changes. *Brain Res. Rev.* **10**, 27–46.

Aldskogius, H., Elfvin, L. G., and Forsman, C. A. 1986. Primary sensory afferents in the inferior mesenteric ganglion and related nerves of the guinea pig. An experimental study with anterogradely transported wheat germ agglutinin-horseradish peroxidase conjugate. *J. Auton. Nerv. Syst.* **15**, 179–190.

Aley, K. O. and Levine, J. D. 1999. Role of protein kinase A in the maintenance of inflammatory pain. *J. Neurosci.* **19**, 2181–2186.

Aley, K. O., McCarter, G., and Levine, J. D. 1998. Nitric oxide signaling in pain and nociceptor sensitization in the rat. *J. Neurosci.* **18**, 7008–7014.

Aley, K. O., Messing, R. O., Mochly-Rosen, D., and Levine, J. D. 2000. Chronic hypersensitivity for inflammatory nociceptor sensitization mediated by the ε isozyme of protein kinase C. *J. Neurosci.* **20**, 4680–4685.

Aley, K. O., Martin, A., McMahon, T., Mok, J., Levine, J. D., and Messing, R. O. 2001. Nociceptor sensitization by extracellular signal-regulated kinases. *J. Neurosci.* **21**, 6933–6939.

Al-Ghoul, W. M., Volsi, G. L., Weinberg, R. J., and Rustioni, A. 1993. Glutamate immunocytochemistry in the dorsal horn after injury or stimulation of the sciatic nerve of rats. *Brain Res. Bull.* **30**, 453–459.

Ali, Z., Meyer, R. A., and Campbell, J. N. 1996. Secondary hyperalgesia to mechanical but not heat stimuli following a capsaicin injection in hairy skin. *Pain* **68**, 401–411.

Ali, Z., Ringkamp, M., Hartke, T. V., Chien, H. F., Flavahan, N. A., Campbell, J. N., and Meyer, R. A. 1999. Uninjured C-fiber nociceptors develop spontaneous activity and α-adrenergic sensitivity following L_6 spinal nerve ligation in monkey. *J. Neurophysiol.* **81**, 455–466.

Allaoua, H., Chaudieu, I., Krieger, C., Boksa, P., Privat, A., and Quirion, R. 1992. Alterations in spinal cord excitatory amino acid receptors in amyotrophic lateral sclerosis patients. *Brain Res.* **579**, 169–172.

Allard, M., Zajac, J. M., and Simonnet, G. 1992. Autoradiographic distribution of receptors to FLFQPQRFamide, a morphine-modulating peptide, in rat central nervous system. *Neuroscience* **49**, 101–116.

Allard, M., Jordan, D., Zajac, J. M., Ries, C., Martin, D., Monkouanga, D., Kopp, N., and Simonnet, G. 1994. Autoradiographic localization of receptors for neuropeptide FF, FLFQPQRFamide, in human spinal sensory system. *Brain Res.* **633**, 127–132.

Allen, B. J., Rogers, S. D., Ghilardi, J. R., Menning, P. M., Kuskowski, M. A., Basbaum, A. I., Simone, D. A., and Mantyh, P. W. 1997. Noxious cutaneous thermal stimuli induce a graded release of endogenous substance P in the spinal cord: Imaging peptide action in vivo. *J. Neurosci.* **17**, 5921–5927.

Allen, B. J., Menning, P. M., Rogers, S. D., Ghilardi, J., Mantyh, P. W., and Simone, D. A. 1999. Primary afferent fibers that contribute to increased substance P receptor internalization in the spinal cord after injury. *J. Neurophysiol.* **81**, 1379–1390.

Allen, J. M., Gibson, S. J., Adrian, T. E., Polak, J. M., and Bloom, S. R. 1984. Neuropeptide Y in human spinal cord. *Brain Res.* **308**, 145–148.

Allerton, C. A., Boden, P. R., and Hill, R. G. 1989a. Actions of the $GABA_B$ agonist, (-)-baclofen, on neurones in deep dorsal horn of the rat spinal cord in vitro. *Br. J. Pharmacol.* **96**, 29–38.

Allerton, C. A., Smith, J. A. M., Hunter, R. G., Hill, R. G., and Hughes, J. 1989b. Correlation of ontogey with function of [3H]U69593 labelled K opioid binding sites in the rat spinal cord. *Brain Res.* **502**, 149–157.

Alles, A. and Dom, R. N. 1985. Peripheral sensory nerve fibers that dichotomize to supply the brachium and the pericardium in the rat: A possible morphological explanation for referred cardiac pain? *Brain Res.* **342**, 382–385.

Almond, J. R., Westrum, L. E., and Henry, M. A. 1996. Post-embedding immunogold labeling of gamma-aminobutyric acid in lamina II of the spinal trigeminal subnucleus pars caudalis: I. A qualitative study. *Synapse* **24**, 39–47.

Aloisi, A. M., Carli, G., and Rossi, A. 1988. Response of hip joint afferent fibers to pressure and vibration in the cat. *Neurosci. Lett.* **90**, 130–134.

Alonso, G., Phan, V., Guillemain, I., Saunier, M., Legrand, A., Anoal, M., and Maurice, T. 2000. Immunocytochemical localization of the sigma(1) receptor in the adult rat central nervous system. *Neuroscience* **97**, 155–170.

Altman, J. and Bayer, S. A. 1984. The development of the rat spinal cord. *Adv. Anat. Embryol. Cell Biol.* **85**, 1–166.

Alvarez, d. l. R., Zhang, P., Shao, D., White, F., and Canessa, C. M. 2002. Functional implications of the localization and activity of acid-sensitive channels in rat peripheral nervous system. *Proc. Natl. Acad. Sci. USA* **99**, 2326–2331.

Alvarez, F. J. and Priestley, J. V. 1990. Anatomy of somatostatin-immunoreactive fibres and cell bodies in the rat trigeminal subnucleus caudalis. *Neuroscience* **38**, 343–357.

Alvarez, F. J., Cervantes, C., Blasco, I., Villalba, R., Martinez-Murillo, R., Polak, J. M., and Rodrigo, J. 1988. Presence of calcitonin gene-related peptide (CGRP) and substance P (SP) immunoreactivity in intraepidermal free nerve endings of cat skin. *Brain Res.* **442**, 391–395.

Alvarez, F. J., Rodrigo, J., Jessell, T. M., Dodd, J., and Priestley, J. V. 1989a. Morphology and distribution of primary afferent fibres expressing α-galactose extended oligosaccharides in the spinal cord and brainstem of the rat. Light microscopy. *J. Neurocytol.* **18**, 611–629.

Alvarez, F. J., Rodrigo, J., Jessell, T. M., Dodd, J., and Priestley, J. V. 1989b. Ultrastructure of primary afferent fibres and terminals expressing α-galactose extended oligosaccharides in the spinal cord and brainstem of the rat. *J. Neurocytol.* **18**, 631–645.

Alvarez, F. J., Morris, H. R., and Priestley, J. V. 1991. Sub-populations of smaller diameter trigeminal primary afferent neurons defined by expression of calcitonin gene-related peptide and the cell surface oligosaccharide recognized by monoclonal antibody LA4. *J. Neurocytol.* **20**, 716–731.

Alvarez, F. J., Kavookjian, A. M., and Light, A. R. 1992. Synaptic interactions between GABA-immunoreactive profiles and the terminals of functionally defined myelinated nociceptors in the monkey and cat spinal cord. *J. Neurosci.* **12**, 2901–2917.

Alvarez, F. J., Kavookjian, A. M., and Light, A. R. 1993. Ultrastructural morphology, synaptic relationships and CGRP immunoreactivity of physiologically identified C-fiber terminals in the monkey spinal cord. *J. Comp. Neurol.* **329**, 172–190.

Alvarez, F. J., Taylor-Blake, B., Fyffe, R. E., De Blas, A. L., and Zlight, A. R. 1996. Distribution of immunoreactivity for the beta 2 and beta 3 subunits of the $GABA_A$ receptor in the mammalian spinal cord. *J. Comp. Neurol.* **365**, 392–412.

Alvarez, F. J., Villalba, R. M., Carr, P. A., Grandes, P., and Somohano, P. M. 2000. Differential distribution of metabotropic glutamate receptors 1a, 1b, and 5 in the rat spinal cord. *J. Comp. Neurol.* **422**, 464–487.

Alvarez-Leefmans, F. J., Gamiño, S. M., Giraldez, F., and Noguerón, I. 1988. Intracellular chloride regulation in amphibian dorsal root ganglion neurones studied with ion-selective microelectrodes. *J. Physiol.* **406**, 225–246.

Alvarez-Leefmans, F. J., Nani, A., and Marquez, S. 1998. Chloride transport, osmotic balance, and presynaptic inhibition. In P. Rudomin, R. Romo, and L. M. Mendell (eds.), *Presynaptic Inhibition and Neural Control* (pp. 50–79). Oxford University Press, New York.

Alvarez-Leefmans, F. J., León-Olea, M., Mendoza-Sotela, J., Alvarez, F. J., Antón, B., and Garduño, R. 2001. Immunolocalization of the Na^+-K^+-2Cl-cotransporter in peripheral nervous tissue of vertebrates. *Neuroscience* **104**, 569–582.

Amandusson, A., Hermanson, O., and Blomqvist, A. 1995. Estrogen receptor-like immunoreactivity in the medullary and spinal dorsal horn of the female rat. *Neurosci. Lett.* **196**, 25–28.

Amandusson, A., Hermanson, O., and Blomqvist, A. 1996. Colocalization of oestrogen receptor immunoreactivity and preproenkephalin mRNA expression to neurons in the superficial laminae of the spinal and medullary dorsal horn of rats. *Eur. J. Neurosci.* **8**, 2440–2445.

Amandusson, A., Hallbeck, M., Hallbeck, A. L., Hermanson, O., and Blomqvist, A. 1999. Estrogen-induced alterations of spinal cord enkephalin gene expression. *Pain* **83**, 243–248.

Amann, R., Sirinathsinghji, D. J., Donnerer, J., Liebmann, I., and Schuligoi, R. 1996. Stimulation by nerve growth factor of neuropeptide synthesis in the adult rat in vivo: Bilateral response to unilateral intraplantar injections. *Neurosci. Lett.* **203**, 171–174.

Amara, S. G., Arriza, J. L., Leff, S. E., Swanson, L. W., Evans, R. M., and Rosenfeld, M. G. 1985. Expression in brain of a messenger RNA encoding a novel neuropeptide homologous to calcitonin gene-related peptide. *Science* **229**, 1094–1097.

Amaya, F., Decosterd, I., Samad, T. A., Plumpton, C., Tate, S., Mannion, R. J., Costigan, M., and Woolf, C. J. 2000. Diversity of expression of the sensory neuron-specific TTX-resistant voltage-gated sodium ion channels SNS and SNS2. *Mol. Cell Neurosci.* **15**, 331–342.

Ambalavanar, R. and Morris, R. 1992. The distribution of binding by isolectin I-B4 from Griffonia simplicifolia in the trigeminal ganglion and brainstem trigeminal nuclei in the rat. *Neuroscience* **47**, 421–429.

Ambrose, W. W. and McNeill, M. E. 1978. Graphic representation of the distribution of acetylcholinestrase in cat dorsal root ganglion neurons. *Histochem. J.* **10**, 711–720.

Ambrus, A., Kraftsik, R., and Barakat-Walter, I. 1998. Ontogeny of calretinin expression in rat dorsal root ganglia. *Dev. Brain Res.* **106**, 101–108.

Ammann, B., Gottschall, J., and Zenker, W. 1983. Afferent projections from the rat longus capitis muscle studied by transganglionic transport of HRP. *Anat. Embryol.* **166**, 275–289.

Ammons, W. S. 1986. Renal afferent input to thoracolumbar spinal neurons of the cat. *Am. J. Physiol.* **250**, R435–R443.

Anand, P., Gibson, S. J., McGregor, G. P., Blank, M. A., Ghatei, M. A., Bacarese-Hamilton, A. J., Polak, J. M., and Bloom, S. R. 1983. A VIP-containing system concentrated in the lumbosacral region of human spinal cord. *Nature* **305**, 143–145.

Andersen, H. T., Koerner, L., Landgren, S., and Silfvenius, H. 1967. Fibre components and cortical projections of the elbow joint nerve in the cat. *Acta Physiol. Scand.* **69**, 373–382.

Anderson, C. R. 1992. NADPH diaphorase-positive neurons in the rat spinal cord include a subpopulation of autonomic preganglionic neurons. *Neurosci. Lett.* **139**, 280–284.

Anderson, L. E. and Seybold, V. S. 2000. Phosphorylated cAMP response element binding protein increases in neurokinin-1 receptor-immunoreactive neurons in rat spinal cord in response to formalin-induced nociception. *Neurosci Lett.* **283**, 29–32.

Andoh, T., Nagasawa, T., and Kuraishi, Y. 1996. Expression of tachykinin NK1 receptor mRNA in dorsal root ganglia of the mouse. *Mol. Brain Res.* **35**, 329–332.

Andoh, T., Itoh, M., and Kuraishi, Y. 1997. Nociceptin gene expression in rat dorsal root ganglia induced by peripheral inflammation. *NeuroReport* **8**, 2793–2796.

Andreev, N. Y., Dimitrieva, N., Koltzenburg, M., and McMahon, S. B. 1995. Peripheral administration of nerve growth factor in the adult rat produces a thermal hyperalgesia that requires the presence of sympathetic post-ganglionic neurones. *Pain* **63**, 109–115.

Andres, K. H. 1961a. Untersuchungen uber den Feinbau von Spinalganglien. *Z. Zellforsch.* **55**, 1–48.

Andres, K. H. 1961b. Untersuchungen uber Morphologische Veranderungen in Spinalganglien wahrend der retrograden degeneration. *Z. Zellforsch.* **55**, 49–79.

Andres, K. H. 1966. Über die Feinstruktur der Rezeptoren an Sinushaaren. *Z. Zellforsch. Mikrosk. Anat.* **75**, 339–365.

Andres, K. H. and von Düring, M. 1973. Morphology of cutaneous receptors. In A. Iggo (ed.), *Handbook of Sensory Physiology, Vol. II, Somatosensory System* (pp. 3–28). Springer, New York.

Andrew, B. L. 1954. The sensory innervation of the medial ligament of the knee joint. *J. Physiol.* **123**, 241–250.

Andrew, B. L. and Dodt, E. 1953. The deployment of sensory endings at the knee joint of the cat. *Acta Physiol. Scand.* **28**, 287–296.

Andrew, D. and Craig, A. D. 2001. Spinothalamic lamina I neurones selectively responsive to cutaneous warming in cats. *J. Physiol.* **537**, 489–495.

Angelucci, L. 1956. Experiments with perfused frog's spinal cord. *Br. J. Pharmacol.* **11**, 161–170.

Aniksztejn, L., Bregestovski, P., and Ben-Ari, Y. 1991. Selective activation of quisqualate metabotropic receptor potentiates NMDA but not AMPA responses. *Eur. J. Pharmacol.* **205**, 327–328.

Antal, M., Freund, T. F., and Polgar, E. 1990. Calcium-binding proteins, parvalbumin- and calbindin-D 38k-immunoreactive neurons in the rat spinal cord and dorsal root ganglia: A light and electron microscopic study. *J. Comp. Neurol.* **295**, 467–484.

Antal, M., Polgar, E., Chalmers, J., Minson, J. B., Llewellyn-Smith, I., Heizmann, C. W., and Somogyi, P. 1991. Different populations of parvalbumin- and calbindin-D28k-immunoreactive neurons contain GABA and accumulate 3H-D-aspartate in the dorsal horn of the rat spinal cord. *J. Comp. Neurol.* **314**, 114–124.

Anton, B., Fein, J., To, T., Li, X., Silberstein, L., and Evans, C. J. 1996. Immunohistochemical localization of ORL-1 in the central nervous system of the rat. *J. Comp. Neurol.* **368**, 229–251.

Aoki, E., Takeuchi, I. K., Shoji, R., and Semba, R. 1993. Localization of nitric oxide-related substances in the peripheral nervous tissues. *Brain Res.* **620**, 142–145.

Applebaum, M. L., Clifton, G. L., Coggeshall, R. E., Coulter, J. D., Vance, W. H., and Willis, W. D. 1976. Unmyelinated fibres in the sacral 3 and caudal 1 ventral roots of the cat. *J. Physiol.* **256**, 557–572.

Appelberg, B., Bessou, P., and Laporte, Y. 1966. Action of static and dynamic fusimotor fibres on secondary endings of cat's spindles. *J. Physiol.* **185**, 160–171.

Aprison, M. H. 1970. Evidence of the release of [^{14}C]glycine from hemisected toad spinal cord with dorsal root stimulation. *Pharmacologist* **12**, 222P.

Aprison, M. H. and Werman, R. 1965. The distribution of glycine in cat spinal cord and roots. *Life Sci.* **4**, 2075–2083.

Aprison, M. H., Shank, R. P., and Davidoff, R. A. 1969. A comparison of the concentration of glycine, a transmitter suspect, in different areas of the brain and spinal cord in seven different vertebrates. *Comp. Biochem. Physiol.* **28**, 1345–1355.

Arai, R., Winsky, L., Arai, M., and Jacobowitz, D. M. 1991. Immunohistochemical localization of calretinin in the rat hindbrain. *J. Comp. Neurol.* **310**, 21–44.

Arbuthnott, E. R., Boyd, I. A., and Kalu, K. U. 1975. Ultrastructure and conduction velocity of small, myelinated peripheral nerve fibers. In H. H. Kornhuber (ed.), *The Somatosensory System* (pp. 168–175). Thieme, Stuttgart.

Arimatsu, Y., Seto, A., and Amano, T. 1981. An atlas of α-bungarotoxin binding sites and structures containing acetylcholinesterase in the mouse central nervous system. *J. Comp. Neurol.* **198**, 603–631.

Araki, T., Yamano, M., Murakami, T., Wanaka, A., Betz, H., and Tohyama, M. 1988. Localization of glycine receptors in the rat central nervous system: An immunocytochemical analysis using monoclonal antibody. *Neurosci.* **25**, 613–624.

Arluison, M., Conrath-Verrier, M., Tauc, M., Mailly, P., De la Manche, I. S., Cesselin, F., Bourgoin, S., and Hamon, M. 1983a. Different localizations of met-enkephalin-like immunoreactivity in rat forebrain and spinal cord using hydrogen peroxide and triton X-100 light microscopic study. *Brain Res. Bull.* **11**, 555–571.

Arluison, M., Conrath-Verrier, M., Tauc, M., Mailly, P., De la Manche, I. S., Dietl, M., Cesselin, F., Bourgoin, S., and Hamon, M. 1983b. Met-enkephalin-like immunoreactivity in rat forebrain and spinal cord using hydrogen peroxide and Triton X-100: Ultrastructural study. *Brain Res. Bull.* **11**, 573–586.

Armett, C. J., Gray, J. A. B., and Palmer, J. F. 1961. A group of neurones in the dorsal horn associated with cutaneous mechanoreceptors. *J. Physiol.* **156**, 611–622.

Armett, C. J., Gray, J. A. B., Hunsperger, R. W., and Lal, S. 1962. The transmission of information in primary receptor neurones and second-order neurons of a phasic system. *J. Physiol.* **164**, 395–421.

Armstrong-James, M. and Millar, J. 1979. Carbon fibre microelectrodes. *J. Neurosci. Meth.* **1**, 279–287.

Aronica, E., Catania, M. V., Geurts, J., Yankaya, B., and Troost, D. 2001. Immunohistochemical localization of group I and II metabotropic glutamate receptors in control and amyotrophic lateral sclerosis human spinal cord: Upregulation in reactive astrocytes. *Neuroscience* **105**, 509–520.

Aronin, N., Difiglia, M., Liotta, A. S., and Martin, J. B. 1981. Ultrastructural localization and biochemical features of immunoreactive leu-enkephalin in monkey dorsal horn. *J. Neurosci.* **1**, 561–577.

Aronin, N., Chase, K., Folsom, R., Christakos, S., and DiFiglia, M. 1991. Immunoreactive calcium-binding protein (calbindin-D28k) in interneurons and trigeminothalamic neurons of the rat nucleus caudalis localized with peroxidase and immunogold methods. *Synapse* **7**, 106–113.

Arvidsson, U., Cullheim, S., Ulfhake, B., Bennett, G. W., Fone, K. C. F., Cuello, A. C., Verhofstad, A. A. J., Visser, T. J., and Hökfelt, T. 1990. 5-Hydroxytryptamine, substance P, and thyrotropin-releasing hormone in the adult cat spinal cord segment L^7: Immunohistochemical and chemical studies. *Synapse* **6**, 237–270.

Arvidsson, U., Ulfhake, B., Cullheim, S., Bergstrand, A., Theodorsson, E., and Hökfelt, T. 1991. Distribution of ^{125}I-galanin binding sites, immunoreactive galanin, and its coexistence with 5-hydroxytryptamine in the cat spinal cord: Biochemical, histochemical, and experimental studies at the light and electron microscopic level. *J. Comp. Neurol.* **308**, 115–138.

Arvidsson, U., Ulfhake, B., Cullheim, S., Shupliakov, O., Brodin, E., Franck, J., Bennett, G. W., Fone, K. C. F., Visser, T. J., and Hökfelt, T. 1992a. Thyrotropin-releasing hormone (TRH)-like immunoreactivity in the grey monkey (*Macaca fascicularis*) spinal cord and medulla oblongata with special emphasis on the bulbospinal tract. *J. Comp. Neurol.* **322**, 293–310.

Arvidsson, U., Cullheim, S., Ulfhake, B., Ramìrez, V., Dagerlind, Å., Luppi, P.-H., Kitahama, K., Jouvet, M., Terenius, L., Åman, K., and Hökfelt, T. 1992b. Distribution of enkephalin and its relation to serotonin in cat and monkey spinal cord and brain stem. *Synapse* **11**, 85–104.

Arvidsson, U., Riedl, M., Chakrabarti, S., Lee, J.-H., Nakano, A. H., Dado, R. J., Loh, H. H., Wessendorf, M. W., and Elde, R. 1995a. Distribution and targeting of a mu-opioid receptor (MOR-1) in brain and spinal cord. *J. Neurosci.* **15**, 3328–3341.

Arvidsson, U., Dado, R. J., Riedl, M., Lee, J.-H., Law, P. Y., Loh, H. H., Elde, R., and Wessendorf, M. W. 1995b. δ-opioid receptor immunoreactivity: Distribution in brainstem and spinal cord, and relationship to biogenic amines and enkephalin. *J. Neurosci.* **15**, 1215–1235.

Arvidsson, U., Riedl, M., Chakrabarti, S., Vulchanova, L., Lee, J.-H., Nakano, A. H., Lin, X., Loh, H. H., Law, P. Y., Wessendorf, M. W., and Elde, R. 1995c. The k-opioid receptor is primarily postsynaptic: Combined immunohistochemical localization of the receptor and endogenous opioids. *Proc. Natl. Acad. Sci.* **92**, 5062–5066.

Askwith, C. C., Benson, C. J., Welsh, M. J., and Snyder, P. M. 2001. DEG/ENaC ion channels involved in sensory transduction are modulated by cold temperature. *Proc. Natl. Acad. Sci. USA* **98**, 6459–6463.

Atkinson, M. E. and Kenyon, C. 1990. Collateral branching innervation of rat molar teeth from trigeminal ganglion cells shown by double labelling with fluorescent retrograde tracers. *Brain Res.* **508**, 289–292.

Atkinson, M. E. and Shehab, S. A. S. 1986. Peripheral axotomy of the rat mandibular trigeminal nerve leads to an increase in VIP and decrease of other primary afferent neuropeptides in the spinal trigeminal nucleus. *Regul. Pept.* **16**, 69–81.

Atoji, Y., Watanabe, H., Yamamoto, Y., and Suzuki, Y. 1995. Distribution of neurotensin-containing neurons in the central nervous system of the dog. *J. Comp. Neurol.* **353**, 67–88.

Attal, N., Jazat, F., Kayser, V., Guilbaud, G. 1990. Further evidence for "pain-related" behaviours in a model of unilateral peripheral mononeuropathy. *Pain* **41**, 235–251.

Atweh, S. F. and Kuhar, M. J. 1977. Autoradiographic localization of opiate receptors in rat brain: I. Spinal cord and lower medulla. *Brain Res.* **124**, 53–67.

Austin, G. M. and McCouch, G. P. 1955. Presynaptic component of intermediary cord potential. *J. Neurophysiol.* **18**, 441–451.

Averbeck, B. and Reeh, P. W. 2001. Interactions of inflammatory mediators stimulating release of calcitonin gene-related peptide, substance P and prostaglandin E(2) from isolated rat skin. *Neuropharmacology* **40**, 416–423.

Averill, S., McMahon, S. B., Clary, D. O., Reichardt, L. F., and Priestley, J. V. 1995. Immunocytochemical localization of trkA receptors in chemically identified subgroups of adult rat sensory neurons. *Eur. J. Neurosci.* **7**, 1484–1494.

Averill, S., Davis, D. R., Shortland, P. J., Priestley, J. V., and Hunt, S. P. 2002. Dynamic pattern of reg-2 expression in rat sensory neurons after peripheral nerve injury. *J. Neurosci.* **22**, 7493–7501.

Azerad, J., Hunt, C. C., Laporte, Y., Pollin, B., and Thiesson, D. 1986. Afferent fibres in cat ventral roots: Electrophysiological and histological evidence. *J. Physiol.* **379**, 229–243.

Azkue, J. J., Zimmermann, M., Hsieh, T. F., and Herdegen, T. 1998. Peripheral nerve insult induces NMDA receptor-mediated, delayed degeneration in spinal neurons. *Eur. J. Neurosci.* **10**, 2204–2206.

Azkue, J. J., Mateos, J. M., Elezgarai, I., Benitez, R., Osorio, A., Diez, J., Bilbao, A., Bidaurrazaga, A., and Grandes, P. 2000. The metabotropic glutamate receptor subtype mGluR2/3 is located at extrasynaptic loci in rat spinal dorsal horn. *Neurosci. Lett.* **287**, 236–238.

Azkue, J. J., Murga, M., Fernandez-Capetillo, O., Mateos, J. M., Elezgarai, I., Benitez, R., Osorio, A., Diez, J., Puente, N., Bilbao, A., Bidaurrazaga, A., Kuhn, R., and Grandes, P. 2001. Immunoreactivity for the group III metabotropic glutamate receptor subtype mGluR4a in the superficial laminae of the rat spinal dorsal horn. *J. Comp. Neurol.* **430**, 448–457.

Baba, H., Kohno, T., Okamoto, M., Goldstein, P. A., Shimoji, K., and Yoshimura, M. 1998. Muscarinic facilitation of GABA release in substantia gelatinosa of the rat spinal dorsal horn. J. *Physiol.* **508**, 83–93.

Badie-Mahdavi, H., Worsley, M. A., Ackley, M. A., Asghar, A. U., Slack, J. R., and King, A. E. 2001. A role for protein kinase intracellular messengers in substance P- and nociceptor afferent-mediated excitation and expression of the transcription factor Fos in rat dorsal horn neurons in vitro. *Eur. J. Neurosci.* **14**, 426–434.

Badio, B. and Daly, J. W. 1994. Epibatidine, a potent analgesic and nicotinic agonist. *Mol. Pharmacol.* **45**, 563–569.

Bae, Y. C., Ihn, H. J., Park, M. J., Ottersen, O. P., Moritani, M., Yoshida, A., and Shigenaga, Y. 2000. Identification of signal substances in synapses made between primary afferents and their associated axon terminals in the rat trigeminal sensory nuclei. *J. Comp. Neurol.* **418**, 299–309.

Bagust, J., Forsythe, I. D., Kerkut, G. A., and Loots, J. M. 1982. Synaptic and non-synaptic components of the dorsal horn potential in isolated hamster spinal cord. *Brain Res.* **233**, 186–194.

Bagust, J., Forsythe, I. D., and Kerkut, G. A. 1985a. An investigation of the dorsal root reflex using an in vitro preparation of hamster spinal cord. *Brain Res.* **331**, 315–325.

Bagust, J., Forsythe, I. D., and Kerkut, G. A. 1985b. Demonstration of the synaptic origin of primary afferent depolarization (PAD) in the isolated spinal cord of the hamster. *Brain Res.* **341**, 385–389.

Bahns, E., Ernsberger, U., Jänig, W., and Nelke, A. 1986. Functional characteristics of lumbar visceral afferent fibres from the urinary bladder and the urethra in the cat. *Pfluegers Arch.* **407**, 510–518.

Bahns, E., Halsband, U., and Jänig, W. 1987. Responses of sacral visceral afferents from the lower urinary tract, colon and anus to mechanical stimulation. *Pfluegers Arch.* **410**, 296–303.

Bahr, R., Blumberg, H., and Janig, W. 1981. Do dichotomizing afferent fibers exist which supply visceral organs as well as somatic structures? A contribution to the problem of referred pain. *Neurosci. Lett.* **24**, 25–28.

Baik-Han, E. J., Kim, K. J., and Chung, J. M. 1989. Electrophysiological evidence for the presence of looping myelinated afferent fibers in the rat ventral root. *Neurosci. Lett.* **104**, 65–70.

Bailey, A., Matthes, H., Kieffer, B., Slowe, S., Hourani, S. M., and Kitchen, I. 2002. Quantitative autoradiography of adenosine receptors and NBTI-sensitive adenosine transporters in the brains and spinal cords of mice deficient in the mu-opioid receptor gene. *Brain Res.* **943**, 68–79.

Baker, D. G., Coleridge, H. M., Coleridge, L. C. G., and Nerdrum, T. 1980. Search for a cardiac nociceptor: Stimulation by bradykinin of sympathetic nerve endings in the heart of the cat. *J. Physiol.* **306**, 519–536.

Bakhle, Y. S. and Bell, C. 1995. Neurokinin A and substance P vary independently in different regions of rat sensory neurons. *Neuropeptides* **28**, 237–241.

Bakker, D. A., Richmond, F. J. R., and Abrahams, V. C. 1984. Central projections from cat suboccipital muscles: A study using transganglionic transport of horseradish peroxidase. *J. Comp. Neurol.* **228**, 409–421.

Bannatyne, B. A., Maxwell, D. J., Fyffe, R. E., and Brown, A. G. 1984. Fine structure of primary afferent axon terminals of slowly adapting cutaneous receptors in the cat. *Q. J. Exp. Physiol.* **69**, 547–557.

Banner, S. J., Fray, A. E., Ince, P. G., Steward, M., Cookson, M. R., and Shaw, P. J. 2002. The expression of the glutamate re-uptake transporter excitatory amino acid transporter 1 (EAAT1) in the normal human CNS and in motor neurone disease: an immunohistochemical study. *Neuroscience* **109**, 27–44.

Bannon, A. W., Decker, M. W., Holladay, M. W., Curzon, P., Donnelly-Roberts, D., Puttfarcken, P. S., Bitner, R. S., Diaz, A., Dickenson, A. H., Porsolt, R. D., Williams, M., and Arneric, S. P. 1998. Broad spectrum, non-opioid analgesic activity by selective modulation of neuronal nicotinic acetylcholine receptors. *Science* **279**, 77–81.

Bao, L., Wang, H. F., Cai, H. J., Tong, Y. G., Jin, S. X., Lu, Y. J., Grant, G., Hökfelt, T., and Zhang, X. 2002. Peripheral axotomy induces only very limited sprouting of coarse myelinated afferents into inner lamina II of rat spinal cord. *Eur. J. Neurosci.* **16**, 175–185.

Barajon, I., Bersani, M., Quartu, M., Del Fiacco, M., Cavaletti, G., Holst, J. J., and Tredici, G. 1996. Neuropeptides and morphological changes in cisplatin-induced dorsal root ganglion neuropathy. *Exp. Neurol.* **138**, 93–104.

Barakat-Walter, I., Duc, C., and Puymirat, J. 1993. Changes in nuclear 3,5,3″-triiodothyronine receptor expression in rat dorsal root ganglia and sciatic nerve during development: Comparison with regeneration. *Eur. J. Neurosci.* **5**, 319–326.

Barakat-Walter, I., Kraftsik, R., Kuntzer, T., Bogousslavsky, J., and Magistretti, P. 2000. Differential effect of thyroid hormone deficiency on the growth of calretinin-expressing neurons in rat spinal cord and dorsal root ganglia. *J. Comp. Neurol.* **426**, 519–533.

Barbaresi, P., Rustioni, A., and Cuenod, M. 1985. Retrograde labeling of dorsal root ganglion neurons after injection of tritiated amino acids in the spinal cord of rats and cats. *Somatosens. Mot. Res.* **3**, 57–74.

Barber, R. P., Vaughn, J. E., Saito, K., McLaughlin, B. J., and Roberts, E. 1978. GABAergic terminals are presynaptic to primary afferent terminals in the substantia gelatinosa of the rat spinal cord. *Brain Res.* **141**, 35–55.

Barber, R. P., Vaughn, J. E., Slemmon, J. R., Salvaterra, P. M., Roberts, E., and Leeman, S. E. 1979. The origin, distribution and synaptic relationships of substance P axons in rat spinal cord. *J. Comp. Neurol.* **184**, 331–352.

Barber, R. P., Vaughn, J. E., and Roberts, E. 1982. The cytoarchitecture of GABAergic neurons in rat spinal cord. *Brain Res.* **238**, 305–328.

Barber, R. P., Phelps, P. E., Houser, C. R., Crawford, G. D., Salvaterra, P. M., and Vaughn, J. E. 1984. The morphology and distribution of neurons containing choline acetyltransferase in the adult rat spinal cord: An immunocytochemical study. *J. Comp. Neurol.* **229**, 329–346.

Barbut, D., Polak, J. M., and Wall, P. D. 1981. Substance P in the dorsal horn decreases after peripheral nerve injury. *Brain Res.* **205,** 289–298.

Barclay, J., Patel, S., Dorn, G., Wotherspoon, G., Moffatt, S., Eunson, L., Abdel'al, S., Natt, F., Hall, J., Winter, J., Bevan, S., Wishart, W., Fox, A., and Ganju, P. 2002. Functional downregulation of $P2X_3$ receptor subunit in rat sensory neurons reveals a significant role in chronic neuropathic and inflammatory pain. *J. Neurosci.* **22**, 8139–8147.

Bardoni, R., Goldstein, P. A., Lee, C. J., Gu, J. G., and MacDermott, A. B. 1997. ATP P2X receptors mediate fast synaptic transmission in the dorsal horn of the rat spinal cord. *J. Neurosci.* **17**, 5297–5304.

Bardoul, M., Levallois, C., and Konig, N. 1998. Functional AMPA/kainate receptors in human embryonic and foetal central nervous system. *J. Chem. Neuroanat.* **14**, 79–85.

Barker, D. 1962. The structure and distribution of muscle receptors. In D. Barker (ed.), *Symposium on Muscle Receptors* (pp. 227–240). Hong Kong University Press, Hong Kong.

Barker, D., Hunt, C. C., and McIntyre, A. K. 1974. *Handbook of Sensory Physiology: III/2. Muscle Receptors.* Springer, Berlin.

Barker, J. L. and Nicoll, R. A. 1973. The pharmacology and ionic dependency of amino acid responses in the frog spinal cord. *J. Physiol.* **228**, 259–277.

Baron, R., Levine, J. D., and Fields, H. L. 1999a. Causalgia and reflex sympathetic dystrophy: Does the sympathetic nervous system contribute to the generation of pain? *Muscle Nerve* **22**, 678–695.

Baron, R., Wasner, G., Borgstedt, R., Hastedt, E., Schulte, H., Binder, A., Kopper, F., Rowbotham, M., Levine, J. D., and Fields, H. L. 1999b. Effect of sympathetic activity on capsaicin-evoked pain, hyperalgesia, and vasodilatation. *Neurology* **52**, 923–932.

Barr, G. A. and Zadina, J. E. 1999. Maturation of endomorphin-2 in the dorsal horn of the medulla and spinal cord of the rat. *NeuroReport* **10**, 3857–3860.

Barron, D. H. and Matthews, B. H. C. 1938. The interpretation of potential changes in the spinal cord. *J. Physiol.* **92**, 276–321.

Bartfai, T., Hökfelt, T., and Langel, Ü. 1993. Galanin—A neuroendocrine peptide. *Critical Reviews in Neurobiology* **7**, 229–274.

Basbaum, A. I. 1988. Distribution of glycine receptor immunoreactivity in the spinal cord of the rat: Cytochemical evidence for a differential glycinergic control of lamina I and V nociceptive neurons. *J. Comp. Neurol.* **278**, 330–336.

Basbaum, A. I. and Glazer, E. J. 1983. Immunoreactive vasoactive intestinal polypeptide is concentrated in the sacral spinal cord: A possible marker for pelvic visceral afferent fibers. *Somatosensory Res.* **1**, 69–82.

Basbaum, A. I. and Wall, P. D. 1976. Chronic changes in the response of cells in the adult cat dorsal horn following partial deafferentation: The appearance of responding cells in a previously non-responsive region. *Brain Res.* **116**, 181–204.

Basbaum, A. I., Cruz, L., and Weber, E. 1986a. Immunoreactive dynorphin B in sacral primary afferent fibers of the cat. *J. Neurosci.* **6**, 127–133.

Basbaum, A. I., Ralston, D. D., and Ralston, H. J. 1986b. Bulbospinal projections in the primate: A light- and electron-microscopic study of a pain modulating system. *J. Comp. Neurol.* **250**, 311–323.

Basbaum, A. I., Glazer, E. J., and Oertel, W. 1986c. Immunoreactive glutamic acid decarboxylase in the trigeminal nucleus caudalis of the cat: A light- and electron-microscopic analysis. *Somatosens. Mot. Res.* **4**, 77–94.

Basbaum, A. I., Zahs, K., Lord, B., and Lakos, S. 1988. The fiber caliber of 5-HT immunoreactive axons in the dorsolateral funiculus of the spinal cord of the rat and cat. *Somatosens. Mot. Res.* **5**, 177–185.

Battaglia, G. and Rustioni, A. 1988. Coexistence of glutamate and substance P in dorsal root ganglion neurons of the rat and monkey. *J. Comp. Neurol.* **277**, 302–312.

Battaglia, G., Rustioni, A., Altschuler, R. A., and Petrusz, P. 1987. Glutamic acid coexists with substance P in some primary sensory neurons. In R. F. Schmidt, H. G. Schaible, and C. Vahle-Hinz (eds.), *Fine Afferent Nerve Fibers and Pain* (pp. 77–84). VCH, Weinheim.

Baumann, T. K., Simone, D. A., Shain, C. N., and LaMotte, R. H. 1991. Neurogenic hyperalgesia: The search for the primary cutaneous afferent fibers that contribute to capsaicin-induced pain and hyperalgesia. *J. Neurophysiol.* **66**, 212–227.

Baxendale, R. H. and Ferrell, W. R. 1983. Discharge characteristics of the elbow joint nerve of the cat. *Brain Res.* **261**, 195–203.

Bayliss, W. M. 1901. On the origin from the spinal cord of the vaso-dilator fibres of the hindlimb, and on the nature of these fibres. *J. Physiol.* **26**, 173–209.

Beal, J. A. 1979. Serial reconstruction of Ramon y Cajal's large primary afferent complexes in laminae II and III of the adult monkey spinal cord: A Golgi study. *Brain Res.* **166**, 161–165.

Beal, J. A. 1983. Identification of presumptive long axon neurons in the substantia gelatinosa of rat lumbosacral spinal cord: A Golgi study. *Neurosci. Lett.* **41**, 9–14.

Beal, J. A. and Bicknell, H. R. 1981. Primary afferent distribution pattern in the marginal zone (Lamina I) of adult monkey and cat lumbosacral spinal cord. *J. Comp. Neurol.* **202**, 255–263.

Beal, J. A. and Cooper, M. H. 1978. The neurons in the gelatinosal complex (laminae II and III) of the monkey (*Macaca mulatta*): A Golgi study. *J. Comp. Neurol.* **179**, 89–122.

Beal, J. A. and Fox, C. A. 1976. Afferent fibers in the substantia gelatinosa of the adult monkey (*Macaca mulatta*): A Golgi study. *J. Comp. Neurol.* **168**, 113–144.

Beal, J. A., Penny, J. E., and Bicknell, H. R. 1981. Structural diversity of marginal (lamina I) neurons in the adult monkey (*Macaca mulatta*) lumbosacral spinal cord: A Golgi study. *J. Comp. Neurol.* **202**, 237–254.

Beal, J. A., Russell, C. T., and Knight, D. S. 1988. Morphological and development characterization of local-circuit neurons in lamina III of rat spinal cord. *Neurosci. Lett.* **86**, 1–5.

Beall, J. E., Applebaum, A. E., Foreman, R. D., and Willis, W. D. 1977. Spinal cord potentials evoked by cutaneous afferents in the monkey. *J. Neurophysiol.* **40**, 199–211.

Beattie, M. S., Bresnahan, J. C., Mawe, G. M., and Finn, S. 1987. Distribution and ultrastructure of ventral root afferents to lamina I of the cat sacral spinal cord. *Neurosci. Lett.* **76**, 1–6.

Beattie, M. S., Bresnahan, J. C., Komon, J. 1997. Endogenous repair after spinal cord contusion injuries in the rat. *Exp. Neurol.* **148**, 453–463.

Beck, P. W. and Handwerker, H. O. 1974. Bradykinin and serotonin effects on various types of cutaneous nerve fibres. *Pfluegers Arch.* **347**, 209–222.

Beck, P. W., Handwerker, H. O., and Zimmermann, M. 1974. Nervous outflow from the cat's foot during noxious radiant heat stimulation. *Brain Res.* **67**, 373–386.

Beinfeld, M. C. 1985. Cholecystokinin (CCK) gene-related peptides: Distribution and characterization of immunoreactive pro-CCK and an amino-terminal pro-CCK fragment in rat brain. *Brain Res.* **344**, 351–355.

Beland, B. and Fitzgerald, M. 2001. Mu- and delta-opioid receptors are downregulated in the largest diameter primary sensory neurons during postnatal development in rats. *Pain* **90**, 143–150.

Bell, C. 1811. *Idea of a New Anatomy of the Brain.* Strahan and Preston, London. Reprinted in Cranefield, P. F. (ed.), 1974.

Bell, J. and Holmes, M. 1992. Model of the dynamics of receptor potential in a mechanoreceptor. *Math. Biosci.* **110**, 139–174.

Bell, J., Bolanowski, S., and Holmes, M. H. 1994. The structure and function of Pacinian corpuscles: A review. *Prog. Neurobiol.* **42**, 79–128.

Bell, J. A. and Anderson, E. G. 1972. Semicarbazide induced depletion of γ-aminobutyric acid and blockade of presynaptic inhibition. *Brain Res.* **43**, 161–169.

Bell, J. A. and de Souza, E. B. 1988. Functional corticotropin-releasing factor receptors in neonatal rat spinal cord. *Peptides* **9**, 1317–1322.

Belmonte, C. and Gallego, R. 1983. Membrane properties of cat sensory neurones with chemoreceptor and baroreceptor endings. *J. Physiol.* **342**, 603–614.

Belyantseva, I. A. and Lewin, G. R. 1999. Stability and plasticity of primary afferent projections following nerve regeneration and central degeneration. *Eur. J. Neurosci.* **11**, 457–468.

Bennett, D. L. H., Averill, S., Clary, D. O., Priestley, J. V., and McMahon, S. B. 1996a. Postnatal changes in the expression of the trkA high-affinity NGF receptor in primary sensory neurons. *Eur. J. Neurosci.* **8**, 2204–2208.

Bennett, D. L. H., Dmietrieva, N., Priestley, J. V., Clary, D., and McMahon, S. B. 1996b. TrkA, CGRP and IB4 expression in retrogradely labelled cutaneous and visceral primary sensory neurones in the rat. *Neurosci. Lett.* **206**, 33–36.

Bennett, D. L. H., Michael, G. J., Ramachandran, N., Munson, J. B., Averill, S., Yan, Q., McMahon, S. B., and Priestley, J. V. 1998a. A distinct subgroup of small DRG cells express GDNF receptor components and GDNF is protective for these neurons after nerve injury. *J. Neurosci.* **18**, 3059–3072.

Bennett, D. L. H., Koltzenburg, M., Priestley, J. V., Shelton, D. L., and McMahon, S. B. 1998b. Endogenous nerve growth factor regulates the sensitivity of nociceptors in the adult rat. *Eur. J. Neurosci.* **10**, 1282–1291.

Bennett, G. J. 1991. Evidence from animal models on the pathogenesis of painful peripheral neuropathy: Relevance for pharmacotherapy. In A. I. Basbaum and J. M. Besson (eds.), *Towards a New Pharmacotherapy of Pain* (pp. 365–379). Wiley, Baltimore.

Bennett, G. J. and Xie, Y.-K. 1988. A peripheral mononeuropathy in rat that produces disorders of pain sensation like those seen in man. *Pain* **33**, 87–107.

Bennett, G. J., Hayashi, H., Abdelmoumene, M., and Dubner, R. 1979. Physiological properties of stalked cells of the substantia gelatinosa intracellularly stained with horseradish peroxidase. *Brain Res.* **164**, 285–289.

Bennett, G. J., Abdelmoumene, M., Hayashi, H., and Dubner, R. 1980. Physiology and morphology of substantia gelatinosa neurons intracellularly stained with horseradish peroxidase. *J. Comp. Neurol.* **194**, 809–827.

Bennett, G. J., Abdelmoumene, M., Hayashi, H., Hoffert, M. J., and Dubner, R. 1981. Spinal cord layer I neurons with axon collaterals that generate local arbors. *Brain Res.* **209**, 421–426.

Bennett, G. J., Ruda, M. A., Gobel, S., and Dubner, R. 1982. Enkephalin immunoreactive stalked cells and lamina IIb islet cells in cat substantia gelatinosa. *Brain Res.* **240**, 162–166.

Bennett, G. J., Nishikawa, N., Lu, G. W., Hoffert, M. J., and Dubner, R. 1984. The morphology of dorsal column postsynaptic (DCPS) spino-medullary neurons in the cat. *J. Comp. Neurol.* **224**, 568–578.

Bennett, G. J., Kajander, K. C., Saraha, Y., Iadarola, M. J., and Sugimoto, T. 1989. Neurochemical and anatomical changes in the dorsal horn of rats with an experimental peripheral neuropathy In F. Cervero, G. J. Bennett, and P. M. Headley (eds.), *Processing of Sensory Information in the Superficial Dorsal Horn of the Spinal Cord* (pp. 463–471). Plenum Press, New York.

Bennett-Clarke, C. A., Chiaia, N. L., Jacquin, M. F., and Rhoades, R. W. 1992. Parvalbumin and calbindin immunocytochemistry reveal functionally distinct cell groups and vibrissa-related patterns in the trigeminal brainstem complex of the adult rat. *J. Comp. Neurol.* **320**, 323–338.

Benoist, J. M., Besson, J. M., Conseiller, C., and Le Bars, D. 1972. Action of bicuculline on presynaptic inhibition of various origins in the cat's spinal cord. *Brain Res.* **43**, 672–676.

Benoist, J. M., Besson, J. M., and Boissier, J. R. 1974. Modifications of presynaptic inhibition of various origins by local application of convulsant drugs on cat's spinal cord. *Brain Res.* **71**, 172–177.

Benoliel, R., Eliav, E., Mannes, A. J., Caudle, R. M., Leeman, S., and Iadarola, M. J. 1999. Actions of intrathecal diphtheria toxin-substance P fusion protein on models of persistent pain. *Pain* **79**, 243–253.

Benoliel, R., Tanaka, M., Caudle, R. M., and Iadorola, M. J. 2000. Co-localization of *N*-methyl-D-aspartate receptors and substance P (neurokinin-1) receptors in rat spinal cord. *Neurosci. Lett.* **291**, 61–64.

Bentley, G. N. and Gent, J. P. 1994. Electrophysiological properties of substantia gelatinosa neurons in a novel adult spinal slice preparation. *J. Neurosci. Methods* **53**, 157–162.

Berberich, P., Hoheisel, U., and Mense, S. 1988. Effects of a carrageenan-induced myositis on the discharge properties of Group III and IV muscle receptors in the cat. *J. Neurophysiol.* **59**, 1395–1409.

Beresford, I. J., Ireland, S. J., Stables, J., and Hagan, R. M. 1996. Ontogeny and characterization of [125]bolton Hunter-eldoisin sites in rat spinal cord by quantitative autoradiography. *Neuroscience* **46**, 225–232.

Berger, U. V. and Hediger, M. A. 2000. Distribution of the glutamate transporters GLAST and GLT-1 in rat circumventricular organs, meninges, and dorsal root ganglia. *J. Comp. Neurol.* **421**, 385–399.

Bergman, E. and Ulfhake, B. 1998. Loss of primary sensory neurons in the very old rat: Neuron number estimates using the disector method and confocal optical sectioning. *J. Comp. Neurol.* **396**, 211–222.

Bergman, E., Johnson, H., Zhang, X., and Hökfelt, T. 1996. Neuropeptides and neurotrophin receptor mRNAs in primary sensory neurons of aged rats. *J. Comp. Neurol.* **375**, 303–320.

Bergman, E., Carlsson, K., Liljeborg, A., Manders, E., Hokfelt, T., and Ulfhake, B. 1999. Neuropeptides, nitric oxide synthase and GAP-43 in B4-binding and RT97 immunoreactive primary sensory neurons: Normal distribution pattern and changes after peripheral nerve transection and aging. *Brain Res.* **832**, 63–83.

Bergmann, I., Reiter, R., Toyka, K. V., and Koltzenburg, M. 1998. Nerve growth factor evokes hyperalgesia in mice lacking the low-affinity neurotrophin receptor p75. *Neurosci. Lett.* **255**, 87–90.

Bergstrom, L., Hammond, D. L., Go, V. L. W., and Yaksh, T. L. 1983. Concurrent measurement of substance P and serotonin in spinal superfusates: Failure of capsaicin and p-chloroamphetamine to co-release. *Brain Res.* **270**, 181–184.

Berkenbosch, F., Schipper, J., and Tilders, F. J. 1986. Corticotropin-releasing factor immunostaining in the rat spinal cord and medulla oblongata: An unexpected form of cross-reactivity with substance P. *Brain Res.* **399**, 87–96.

Berkley, K. J., Robbins, A., and Sato, Y. 1988. Afferent fibers supplying the uterus in the rat. *J. Neurophysiol.* **59**, 142–163.

Berkley, K. J., Hotta, H., Robbins, S. A., and Sato, Y. 1990. Functional properties of afferent fibers supplying reproductive and other pelvic organs in pelvic nerve of female rat. *J. Neurophysiol.* **63**, 256–272.

Berkley, K. J., Robbins, A., and Sato, Y. 1993. Functional differences between afferent fibers in the hypogastric and pelvic nerves innervating female reproductive organs in the rat. *J. Neurophysiol.* **69**, 533–544.

Bernard, C. 1858. In *Leçons sur la physiologie et la pathologie du systeme nerveux* (pp. 20–112). J.B. Baillière et Fils, Paris.

Bernardi, P. S., Valtschanoff, J. G., Weinberg, R. J., Schmidt, H. H., and Rustioni, A. 1995. Synaptic interactions between primary afferent terminals and GABA and nitric oxide-synthesizing neurons in superficial laminae of the rat spinal cord. *J. Neurosci.* **15**, 1363–1371.

Bernardini, N., Levey, A. I., and Augusti-Tocco, G. 1999. Rat dorsal root ganglia express m1-m4 muscarinic receptor proteins. *J. Peripher. Nerv. Syst.* **4**, 222–232.

Bernau, N. A., Dawson, S. D., Kane, L. A., and Pubols, L. M. 1993. Changes in substance P and 5-HT binding in the spinal cord dorsal horn and lamina 10 after dorsolateral funiculus lesions. *Brain Res.* **613**, 106–114.

Bernhard, C. G. 1953. The spinal cord potentials in leads from the cord dorsum in relation to peripheral source of afferent stimulation. *Acta Physiol. Scand.* **29**(Suppl. 106), 1–29.

Bernhard, C. G. and Widén, L. 1953. On the origin of the negative and positive spinal cord potentials evoked by stimulation of low threshold cutaneous fibres. *Acta Physiol. Scand.* **29**(Suppl. 106), 42–54.

Bernhard, C. G., Lindblom, U. F., and Ottosson, J. O. 1953. The longitudinal distribution of the negative cord dorsum potential following stimulation of low threshold cutaneous fibres. *Acta Physiol. Scand.* **29**(Suppl. 106), 170–179.

Berthele, A., Boxall, S. J., Urban, A., Anneser, J. M., Ziegelgansberger, W., Urban, L., and Tolle, T. R. 1999. Distribution and developmental changes in metabotropic glutamate receptor messenger RNA expression in the rat lumbar spinal cord. *Dev. Brain Res.* **111**, 39–53.

Berthold, C. H. and Carlstedt, T. 1977. Observations on the morphology at the transition between the peripheral and the central nervous system in the cat: II. *Acta Physiol. Scand.* **446**, 23–43.

Bertrand, A. and Weil-Fugazza, J. 1995. Sympathectomy does not modify the levels of dopa or dopamine in the rat dorsal root ganglion. *Brain Res.* **681**, 201–204.

Besse, D., Lombard, M. C., Zajac, J. M., Roques, B. P., and Besson, J. M. 1990. Pre- and postsynaptic distribution of μ, δ, and κ opioid receptors in the superficial layers of the cervical dorsal horn of the rat spinal cord. *Brain Res.* **521**, 15–22.

Besse, D., Lombard, M. C., and Besson, J. M. 1991. Autoradiographic distribution of μ, δ and κ opioid binding sites in the superficial dorsal horn, over the rostrocaudal axis of the rat spinal cord. *Brain Res.* **548**, 287–291.

Besse, D., Lombard, M. C., and Besson, J. M. 1992a. Time-related decreases in μ and δ opioid receptors in the superficial dorsal horn of the rat spinal cord following a large unilateral dorsal rhizotomy. *Brain Res.* **578**, 115–121.

Besse, D., Lombard, M. D., and Besson, J. M. 1992b. Up-regulation of [^{3}H]DAMGO and [^{3}H]DTLET opioid binding sites in laminae I–II of the spinal cord in intact and deafferented morphine-tolerant rats. *Neurosci. Lett.* **136**, 209–212.

Besse, D., Lombard, M. C., Perrot, S., and Besson, J. M. 1992c. Regulation of opioid binding sites in the superficial dorsal horn of the rat spinal cord following loose ligation of the sciatic nerve: Comparison with sciatic nerve section and lumbar dorsal rhizotomy. *Neuroscience* **50**, 921–933.

Besse, D., Weil-Fugazza, J., Lombard, M. C., Butler, S. H., and Besson, J. M. 1992d. Monoarthritis induces complex changes in mu-, d. *Eur. J. Pharmacol.* **223**, 123–131.

Besson, J. M., Rivot, J. P., and Aleonard, P. 1971. Action of picrotoxin on presynaptic inhibition of various origins in the cat's spinal cord. *Brain Res.* **26**, 212–216.

Besson, J. M., Conseiller, C., Hamann, K. F., and Maillard, M. C. 1972. Modifications of dorsal horn cell activities in the spinal cord, after intra-arterial injection of bradykinin. *J. Physiol.* **221**, 189–205.

Besson, J. M., Catchlove, R. F. H., Feltz, P., and Le Bars, D. 1974. Further evidence for postsynaptic inhibitions on lamina 5 dorsal horn interneurons. *Brain Res.* **66**, 531–536.

Besson, J. M., Guilbaud, G., and Le Bars, D. 1975. Descending inhibitory influences exerted by the brain stem upon the activities of dorsal horn lamina V cells induced by intra-arterial injection of bradykinin into the limbs. *J. Physiol.* **248**, 725–739.

Besson, J. M., Lombard, M. C., Zajac, J. M., Besse, D., Peschanski, M., and Roques, B. P. 1991. Opioid receptors in the dorsal horn of intact and deafferented rats: Autoradiographic and electrophysiological studies. In F. Cervero, G. J. Bennett, and P. M. Headley (eds.), *Proceedings of Sensory Information in the Superficial Dorsal Horn of the Spinal Cord* (pp. 415–428). Plenum Press, New York.

Bessou, P. and Laporte, Y. 1961. Étude des recepteurs musculaires innerves par les fibres afferents du groupe III (fibres myelinisées fines), chez le chat. *Arch. Ital. Biol.* **99**, 293–321.

Bessou, P. and Perl, E. R. 1966. A movement receptor of the small intestine. *J. Physiol.* **182**, 404–426.

Bessou, P. and Perl, E. R. 1969. Response of cutaneous sensory units with unmyelinated fibers to noxious stimuli. *J. Neurophysiol.* **32**, 1025–1043.

Bessou, P., Laporte, Y., and Pages, B. 1968. Frequency grams of spindle primary endings elicited by stimulation of static and dynamic fusimotor fibres. *J. Physiol.* **196**, 47–63.

Bessou, P., Burgess, P. R., Perl, E. R., and Taylor, C. B. 1971. Dynamic properties of mechanoreceptors with unmyelinated (C) fibers. *J. Neurophysiol.* **34**, 116–131.

Bettler, B., Boulter, J., Hermans-Borgmeyer, I., O'Shea-Greenfield, A., Deneris, E. S., Moll, C., Boregmeyer, U., Hollmann, M., and Heinemann, S. 1990. Cloning of a novel glutamate receptor subunit, GluR5: Expression in the nervous system during development. *Neuron* **5**, 583–595.

Betz, H., Kuhse, J., Schmieden, V., Laube, B., Kirsch, J., and Harvey, R. J. 1999. Structure and functions of inhibitory and excitatory glycine receptors. *Ann. NY Acad. Sci.* **868**, 667–676.

Bezzegh, A., Knyihar-Csillik, E., Boti, S., Tajti, J., Zaborszky, Z., and Csillik, B. 1986. A computer-aided analysis of the effect of peripheral nerve transection on TMPase activity of substantia gelatinosa roland. *Zeits. f. Mik. Anat. Forsch.* **100**, 428–432.

Bicknell, H. R. and Beal, J. A. 1984. Axonal and dendritic development of substantia gelatinosa neurons in the lumbosacral spinal cord of the rat. *J. Comp. Neurol.* **226**, 508–522.

Biella, G., Panara, C., Pecile, A., and Sotgiu, M. L. 1991. Facilitatory role of calcitonin gene-related peptide (CGRP) on excitation induced by substance P (SP) and noxious stimuli in rat spinal dorsal horn neurons. An iontophoretic study in vivo. *Brain Res.* **559**, 352–356.

Birder, L. A. and Perl, E. R. 1999. Expression of alpha2-adrenergic receptors in rat primary afferent neurones after peripheral nerve injury or inflammation *J. Physiol.* **515**, 533–542.

Birrell, G. J., McQueen, D. S., Iggo, A., and Grubb, B. D. 1990. The effects of 5-HT on articular sensory receptors in normal and arthritic rats. *Br. J. Pharmacol.* **101**, 715–721.

Birrell, G. J., McQueen, D. S., Iggo, A., and Grubb, B. D. 1993. Prostanoid-induced potentiation of the excitatory and sensitizing effects of bradykinin on articular mechanoreceptors in the rat ankle joint. *Neuroscience* **54**, 537–544.

Bisby, M. A. and Keen, P. 1986. Regeneration of primary afferent neurons containing substance P-like immunoreactivity. *Brain Res.* **365**, 85–95.

Biscoe, T. J., Headley, P. M., Lodge, D., Martin, M. R., and Watkins, J. C. 1976. The sensitivity of rat spinal interneurones and Renshaw cells to L-glutamate and L-aspartate. *Exp. Brain Res.* **26**, 547–551.

Black, J. A., Westenbroek, R., Ransonm, B. R., Catterall, W. A., and Waxman, S. G. 1994. Type II sodium channels in spinal cord astrocytes in situ: Immunocytochemical observations. *Glia* **12**, 219–227.

Black, J. A., Dib-Hajj, S., McNabola, K., Jeste, S., Rizzo, M. A., Kocsis, J. D., and Waxman, S. G. 1996. Spinal sensory neurons express multiple sodium channel alpha-subunit mRNAs. *Mol. Brain Res.* **43**, 117–131.

Black, J. A., Cummins, T. R., Plumpton, C., Chen, Y. H., Hormuzdiar, W., Clare, J. J., and Waxman, S. G. 1999. Upregulation of a silent sodium channel after peripheral, but not central, nerve injury in DRG neurons. *J. Neurophysiol.* **82**, 2776–2785.

Black, J. A., Renganathan, M., and Waxman, S. G. 2002. Sodium channel Na(v)1.6 is expressed along nonmyelinated axons and it contributes to conduction. *Mol. Brain Res.* **105**, 19–28.

Blackburn-Munro, G. and Fleetwood-Walker, S. M. 1999. The sodium channel auxiliary subunits beta1 and beta2 are differentially expressed in the spinal cord of neuropathic rats. *Neuroscience* **90**, 153–164.

Blake, J. F., Cao, C. Q., Headley, P. M., Collingridge, G. L., Brugger, F., Evans, R. H. 1993. Antagonism of baclofen-induced depression of whole-cell synaptic currents in spinal dorsal horn neurones by the potent $GABA_B$ antagonist CGP55845. *Neuropharmacology* **32**, 1437–1440.

Blakeman, K. H., Holmberg, K., Hao, J. X., Xu, X. J., Kahl, U., Lendahl, U., Bartfai, T., Wiesenfeld-Hallin, Z., and Hökfelt, T. 2001. Mice over-expressing galanin have elevated heat nociceptive threshold. *NeuroReport* **12**, 423–425.

Bleakman, D. F., Rusin, K. I., Chard, P. S., Glaum, S. R., and Miller, R. J. 1992. Metabotropic glutamate receptors potentiate ionotropic glutamate responses in the rat dorsal horn. *Mol. Pharmacol.* **42**, 192–196.

Bleazard, L. and Morris, R. 1993. The effects of cholinoreceptor agonists and antagonists on C-fibre evoked responses in the substantia gelatinosa of neonatal rat spinal cord slices. *Br. J. Pharmacol.* **110**, 1061–1066.

Bleazard, L., Hill, R. G., and Morris, R. 1994. The correlation between the distribution of the NK1 receptor and the actions of tachykinin agonists in the dorsal horn of the rat indicates that substance P does not have a functional role on substantia gelatinosa (lamina II) neurons. *J. Neurosci.* **14**, 7655–7664.

Blessing, W. W., Oliver, J. R., Hodgson, A. H., Joh, T. H., and Willoughby, J. O. 1987. Neuropeptide Y-like immunoreactive C1 neurons in the rostral ventrolateral medulla of the rabbit project to sympathetic preganglionic neurons in the spinal cord. *J. Auton. Nerv. Syst.* **18**, 121–129.

Blix, M. 1884. Experimentelle Beiträge zur Losung der Frage über die specifische Energie der Hautnerven. *Z. Biol.* **20**, 141–156.

Bloom, F., Battenberg, E., Rossier, J., Ling, N., and Guillemin, R. 1978. Neurons containing beta-endorphin in rat brain exist separately from those containing enkephalin: immunocytochemical studies. *Proc. Natl. Acad. Sci. USA* **75**, 1591–1595.

Blottner, D., Grozdanovic, Z., and Gossrau, R. 1995. Histochemistry of nitric oxide synthase in the nervous system. *Histochem. J.* **27**, 785–811.

Blumenkopf, B. 1988. Neurochemistry of the dorsal horn. *Appl. Neurophysiol.* **51**, 89–103.

Bo, X. and Burnstock, G. 1994. Distribution of [^{3}H] α, β-methylene ATP binding sites in rat brain and spinal cord. *NeuroReport* **5**, 1601–1604.

Bo, X. and Burnstock, G. 1995. Distribution of [^{3}H] α, β-methylene ATP binding sites in rat brain and spinal cord. *J. Neurosci.* **15**, 2985–2994.

Bodzioch, M., Lapicka, K., Aslanidis, C., Kacinski, M. and Schmitz, G. 2001. Two novel mutant alleles of the gene encoding neurotrophic tyrosine kinase receptor type 1 (NTRK1) in a patient with congenital insensitivity to pain with anhidrosis: A splice junction mutation in intron 5 and cluster of four mutations in exon 15. *Hum. Mutat.* **17**, 72.

Boehmer, C. G., Norman, J., Catton, M., Fine, L. G., and Mantyh, P. W. 1989. High levels of mRNA coding for substance P, somatostatin, and alpha-tubulin are expressed by rat and rabbit dorsal root ganglia neurons. *Peptides* **10**, 1179–1194.

Bohlhalter, S., Mohler, H., and Fritschy, J.-M. 1994. Inhibitory neurotransmission in rat spinal cord: Co-localization of glycine- and $GABA_A$-receptors at GABAergic synaptic contacts demonstrated by triple immunofluorescence staining. *Brain Res.* **642**, 59–69.

Bohlhalter, S., Weinmann, O., Mohler, H., and Fritschy, J. M. 1996. Laminar compartmentalization of $GABA_A$-receptor subtypes in the spinal cord: An immunohistochemical study. *J. Neurosci.* **16**, 283–297.

Bolton, C. F., Winkelmann, R. K., and Dyck, P. J., 1966. A quantitative study of Meissner's corpuscles in man. *Neurology* **16**, 1–9.

Boissonade, F. M., Sharkey, K. A., and Lucier, G. E. 1993. Trigeminal nuclear complex of the ferret: Anatomical and immunohistochemical studies. *J. Comp. Neurol.* **329**, 291–312.

Boivie, J. 1994. Central pain. In P. D. Wall and R. Melzack (eds.), *Textbook of Pain* (3rd ed., pp. 871–902). London, Churchill Livingstone.

Boivie, J. and Perl, E. R. 1975. Neural substrates of somatic sensation. In C. C. Hunt (ed.), *MTP International Review of Science, Physiology Series One, Vol. 3, Neurophysiology* (pp. 303–411). University Park Press, Baltimore.

Bolanowski, S. J. 1984. Intensity and frequency characteristics of pacinian corpuscles: III. Effects of tetrodotoxin on transduction process. *J. Neurophysiol.* **51**, 831–839.

Bolanowski, S. J. and Zwislocki, J. J. 1984a. Intensity and frequency characteristics of pacinian corpuscles: I. Action potentials. *J. Neurophysiol.* **51**, 793–811.

Bolanowski, S. J. and Zwislocki, J. J. 1984b. Intensity and frequency characteristics of pacinian corpuscles: II. Receptor potentials. *J. Neurophysiol.* **51**, 812–830.

Bolanowski, S. J., Schyuler, J. E., and Slepecky, N. B. 1994. Semi-serial electron-micrographic reconstruction of putative transducer sites in Pacinian corpuscles. *Somatosen. Mot. Res.* **11**, 205–218.

Bolden, D. A., Sternini, C., and Kruger, L. 1997. GAP-43 mRNA and calcitonin gene-related peptide mRNA expression in sensory neurons are increased following sympathectomy. *Brain Res. Bull.* **42**, 39–50.

Bonaventure, P., Voorn, P., Luyten, W. H., Jurzak, M., Schotte, A., and Leysen, J. E. 1998a. Detailed mapping of serotonin 5-HT1B and 5-HT1D receptor messenger RNA and ligand binding sites in guinea-pig brain and trigeminal ganglion: Clues for function. *Neuroscience* **82**, 469–484.

Bonaventure, P., Voorn, P., Luyten, W. H., and Leysen, J. E. 1998b. 5HT1B and 5HT1D receptor mRNA differential co-localization with peptide mRNA in the guinea pig trigeminal ganglion. *NeuroReport* **9**, 641–645.

Bond, A. and Lodge, D. 1995. Pharmacology of metabotropic glutamate receptor-mediated enhancement of responses to exctitatory and inhibitory amino acids on rat spinal interneurons in vivo. *Neuropharmacology* **34**, 1015–1023.

Bonfanti, L., Bellaridi, S., Ghidella, S., Gobetto, A., Polak, J. M., and Merighi, A. 1991. Distribution of five peptides, three general neuroendocrine markers, and two synaptic-vesicle-associated proteins in the spinal cord and dorsal root ganglia of the adult and newborn dog: An immunocytochemical study. *Am. J. Anat.* **191**, 154–166.

Bongenhielm, U., Nosrat, C. A., Nosrat, I., Eriksson, J., Fjell, J., and Fried, K. 2000. Expression of sodium channel SNS/PN3 and ankyrin(G) mRNAs in the trigeminal ganglion after inferior alveolar nerve injury in the rat. *Exp. Neurol.* **164**, 384–395.

Bonhaus, D. W., Bley, K. R., Broka, C. A., Fontana, D. J., Leung, E., Lewis, R., Shieh, A., and Wong, E. H. 1995. Characterization of the electrophysiological, biochemical and behavioral actions of epibatidine. *J. Pharmacol. Exp. Ther.* **272**, 1199–1203.

Bonica, J. J. (ed.) 1990. *The Management of Pain* (2nd ed). Lea and Febiger, Philadephia.

Bonnot, A., Corio, M., Tramu, G., and Viala, D. 1996. Immunocytochemical distribution of ionotropic glutamate receptor subunits in the spinal cord of the rabbit. *J. Chem. Neuroanat.* **11**, 267–278.

Boom, A., Mollereau, C., Meunier, J.-C., Vassart, G., Parmentier, M., Vanderhaeghen, J.-J., and Schiffmann, S. N. 1999. Distribution of the nociceptin and nocistatin precursor transcript in the mouse central nervous system. *Neuroscience* **91**, 991–1007.

Borges, L. F. and Iversen, S. D. 1986. Topography of choline acetyltransferase immunoreactive neurons and fibers in the rat spinal cord. *Brain Res.* **362**, 140–148.

Borges, L. F. and Moskowitz, M. A. 1983. Do Intracranial and Extracranial Trigeminal Afferents Represent Divergent Axon Collaterals? *Neurosci. Lett.* **35**, 265–270.

Boring, E. G. 1916. Cutaneous sensation after nerve-division. *Q. J. Exp. Physiol.* **10**, 1–95.

Boring, E. G. 1942. *Sensation and Perception in the History of Experimental Psychology.* Appleton-Century-Crofts, New York.

Bormann, J. and Feigenspan, A. 1995. $GABA_C$ receptors. *TINS* **18**, 515–519.

Bossut, D. F. and Perl, E. R. 1995. Effects of nerve injury on sympathetic excitation of Aδ mechanical nociceptors. *J. Neurophysiol.* **73**, 1721–1723.

Bostock, H. 1981. U-turns in rat ventral roots. *J. Physiol. (Lond.)* **312**, 49P–50P.

Botchkarev, V. A., Welker, P., Albers, K. M., Botchkareva, N. V., Metz, M., Lewin, G. R., Bulfone-Paus, S., Peters, E. M. J., Linder, G., and Paus, R. 1998. A new role for neurotrophin-3: Involvement in the regulation of hair follicle regression (catagen). *Am. J. Pathol.* **153**, 785–799.

Botchkarev, V. A., Botchkarena, N. V., Welker, P., Metz, M., Lewin, G. R., Subramaniam, A., Bulfone-Paus, S., Hagen, E., Braun, A., Lommartzsch, M., Renz, H., and Paus, R. 1999. A new role for neurotrophins: Involvement of brain-derived neurotrophic factor and neurotrophin-4 in hair cycle control. *FASEB J.* **13**, 395–410.

Botchkareva, N. V., Botchkarev, V. A., Albers, K. M., Metz, M., and Paus, R. 2000. Distinct roles for nerve growth factor and brain-derived neurotrophic factor in controlling the rate of hair follicle morphogenesis. *J. Invest. Dermatol.* **114**, 314–320.

Botticelli, L., Cox, B., and Goldstein, A. 1981. Immunoreactive dynorphin in mammalian spinal cord and dorsal root ganglia. *Proc. Natl. Acad. Sci.* **78**, 7783–7786.

Bourgoin, S., Pohl, M., Benoliel, J. J., Mauborgne, A., Collin, E., Hamon, M., and Cesselin, F. 1992. Gamma-aminobutryic acid, through $GABA_A$ receptors, inhibits the potassium-stimulated release of calcitonin gene-related peptide- but not that of substance P-like material from rat spinal cord slices. *Brain Res.* **583**, 344–348.

Bouthenet, M. L., Martres, M.-P., Sales, N., and Schwartz, J.-C. 1987. A detailed mapping of dopamine D-2 receptors in rat central nervous system by autoradiography with $^{[125]}$iodosulpride. *Neuroscience* **20**, 117–155.

Bowery, N. G., Hudson, A. L., and Price, G. W. 1987. $GABA_A$ and $GABA_B$ receptor site distribution in the rat central nervous system. *Neuroscience* **20**, 365–383.

Bowker, R. M. 1986. Intrinsic 5HT-immunoreactive neurons in the spinal cord of the fetal non-human primate. *Dev. Brain Res.* **28**, 137–143.

Bowsher, D. and Abdel-Maguid, T. E. 1984. Superficial dorsal horn of the adult human spinal cord. *Neurosurgery* **15**, 893–899.

Boxall, S. J., Berthele, A., Laurie, D. J., Sommer, B., Zieglgansberger, W., Urban, L., and Tolle, T. R. 1998. Enhanced expression of metabotropic glutamate receptor 3 messenger RNA in the rat spinal cord during ultraviolet irradiation induced peripheral inflammation. *Neuroscience* **82**, 591–602.

Boyce, S., Wyatt, A., Webb, J. K., O'Donnell, R., Mason, G., Rigby, M., Sirinathsinghji, D., Hill, R. G., and Rupniak, N. M. 1999. Selective NMDA NR2B antagonists induce antinociception without motor dysfunction: Correlation with restricted localisation of NR2B subunit in dorsal horn. *Neuropharmacology* **38**, 611–623.

Boyd, I. A. 1954. The histological structure of the receptors in the knee-joint of the cat correlated with their physiological response. *J. Physiol.* **124**, 476–488.

Boyd, I. A. 1962. The structure and innervation of the nuclear bag muscle fibre system and the nuclear chain muscle fibre system in mammalian muscle spindles. *Phil. Trans. Roy. Soc. London B.* **245**, 81–136.

Boyd, I. A. and Roberts, T. D. 1953. Proprioceptive discharges from stretch-receptors in the knee-joint of the cat. *J. Physiol.* **122**, 38–58.

Boyd, R. T., Jacob, M. H., McEachern, A. E., Caron, S., and Berg, D. K. 1991. Nicotinic acetylcholine receptor mRNA in dorsal root ganglion neurons. *J. Neurobiol.* **22**, 1–14.

Bradbury, E. J., Burnstock, G., and McMahon, S. B. 1998. The expression of $P2X_3$ purinoreceptors in sensory neurons: Effects of axotomy and glial-derived neurotropic factor. *Mol. Cell. Neurosci.* **12**, 256–268.

Brain, S. D. and Williams, T. J. 1985. Inflammatory oedema induced by synergism between calcitonin gene-related peptide (CGRP) and mediators of increased vascular permeability. *Br. J. Pharmacol.* **86**, 855–860.

Brain, S. D., Williams, T. J., Tippins, J. R., Morris, H. R., and MacIntyre, I. 1985. Calcitonin gene-related peptide is a potent vasodilator. *Nature* **313**, 54–56.

Bras, J. M. A., Epstein, A. L., Bourgoin, S., Hamon, M., Cesselin, F., and Pohl, M. 1998. Herpes simplex virus 1-mediated transfer of preproenkephalin A in rat dorsal root ganglion. *J. Neurochem.* **70**, 1299–1303.

Bras, J. M. A., Laporte, A. M., Benoliel, J. J., Bourgoin, S., Mauborgne, A., Hamon, M., Cesselin, F., and Pohl, M. 1999. Effects of peripheral axotomy on cholecystokinin neurotransmission in the rat spinal cord. *J Neurochem.* **72**, 858–867.

Bras, J. M. A., Becker, C., Bourgoin, S., Lombard, M., Cesselin, F., Hamon, M., and Pohl, M. 2001. Met-enkephalin is preferentially transported into the peripheral processes of primary afferent fibres in both control and HSV1-driven proenkephalin A overexpressing rats. *Neuroscience* **103**, 1073–1083.

Brass, K. M., Newby, A. C., Wilson, V. S., and Snyder, S. H. 1986. Adenosine-containing neurons in the brain localized by immunocytochemistry. *J. Neurosci.* **6**, 1952–1961.

Braz, J., Beaufour, C., Coutaux, A., Epstein, A. L., Cesselin, F., Hamon, M., and Pohl, M. 2001. Therapeutic efficacy in experimental polyarthritis of viral-driven enkephalin overproduction in sensory neurons. *J. Neurosci.* **21**, 7881–7888.

Bregman, B. S. 1987. Development of serotonin immunoreactivity in the rat spinal cord and its plasticity after neonatal spinal cord lesions. *Dev. Brain Res.* **34**, 245–263.

Bresnahan, J. C., Ho, R. H., and Beattie, M. S. 1984. A comparison of the ultrastructure of substance P and enkephalin-immunoreactive elements in the nucleus of the dorsal lateral funiculus and laminae I and II of the rat spinal cord. *J. Comp. Neurol.* **229**, 497–511.

Brewster, W. J., Diemel, L. T., Leach, R. M., and Tomlinson, D. R. 1994. Reduced sciatic nerve substance P and calcitonin gene-related peptide in rats with short-term diabetes or central hypoxaemia co-exist with normal messenger RNA levels in the lumbar dorsal root ganglia. *Neuroscience* **58**, 323–330.

Briner, R. P., Carlton, S. M., Coggeshall, R. E., and Chung, K. 1988. Evidence for unmyelinated sensory fibres in the posterior columns in man. *Brain* **111**, 999–1007.

Brink, E., Harrison, P. J., Jankowska, E., McCrea, D., and Skoog, B. 1983a. Post-synaptic potentials in a population of motoneurones following activity in single interneurones in the cat. *J. Physiol.* **343**, 341–359.

Brink, E., Jankowska, E., McCrea, D., and Skoog, B. 1983b. Inhibitory interactions between interneurones in reflex pathways from group Ia afferents in the cat. *J. Physiol.* **343**, 361–379.

Brink, E., Jankowska, E., and Skoog, B. 1984. Convergence onto interneurones subserving primary afferent depolarization of group I afferents. *J. Neurophysiol.* **51**, 432–449.

Brinkhus, H. B. and Zimmermann, M. 1983. Characteristics of spinal dorsal horn neurons after partial chronic deafferentation by dorsal root transection. *Pain* **15**, 221–236.

Brinton, R. E., Gee, K. W., Wamsley, J. K., Davis, T. P., and Yamamura, H. I. 1984. Regional distribution of putative vasopressin receptors in rat brain and pituitary by quantitative autoradiography. *Proc. Natl. Acad. Sci.* **81**, 7248–7252.

Broberger, C., Farkas-Szallasi, T., Szallasi, A., Lundberg, J. M., Hökfelt, T., Wiesenfeld-Hallin, Z., and Xu, X. J. 2000. Increased spinal cholecystokinin activity after systemic resiniferatoxin: Electrophysiological and in situ hybridization studies. *Pain* **84**, 21–28.

Broberger, C., Holmberg, K., Shi, T. J., Dockray, G., and Hökfelt, T. 2001. Expression and regulation of cholecystokinin and cholecystokinin receptors in rat nodose and dorsal root ganglia. *Brain Res.* **903**, 128–140.

Brock, L. G., Coombs, J. S., and Eccles, J. C. 1952. The recording of potentials from motoneurones with an intracellular electrode. *J. Physiol.* **117**, 431–460.

Brodin, E., Linderoth, B., Gazelius, B., and Ungerstedt, U. 1987. In vivo release of substance P in cat dorsal horn studied with microdialysis. *Neurosci. Lett.* **76**, 357–362.

Broman, J. and Adahl, F. 1994. Evidence for vesicular storage of glutamate in primary afferent terminals. *NeuroReport* **5**, 1801–1804.

Broman, J., Anderson, S., and Ottersen, O. P. 1993. Enrichment of glutamate-like immunoreactivity in primary afferent terminals throughout the spinal cord dorsal horn. *Eur. J. Neurosci.* **5**, 1050–1061.

Bromberg, M. B. and Fetz, E. E. 1977. Responses of single units in cervical spinal cord of alert monkeys. *Exp. Neurol.* **55**, 469–482.

Broome, M., Hökfelt, T., and Terenius, L. 1985. Peptide YY (PYY)-immunoreactive neurons in the lower brain stem and spinal cord of rat. *Acta Physiol. Scand.* **125**, 349–352.

Brown, A. G. 1981. *Organization in the Spinal Cord: The Anatomy and Physiology of Identified Neurones.* Springer-Verlag, Berlin.

Brown, A. G. and Fyffe, R. E. W. 1981. Form and function of dorsal horn neurons with axons ascending the dorsal columns in cat. *J. Physiol.* **321**, 31–47.

Brown, A. G. and Iggo, A. 1967. A quantitative study of the cutaneous receptors and afferent fibers in the cat and rabbit. *J. Physiol.* **193**, 707–733.

Brown, A. G., Iggo, A., and Miller, S. 1967. Myelinated afferent nerve fibers from the skin of the rabbit ear. *Exp. Neurol.* **18**, 338–349.

Brown, A. G., Rose, P. K., and Snow, P. J. 1977c. The morphology of hair follicle afferent fibre collaterals in the spinal cord of the cat. *J. Physiol.* **272**, 779–797.

Brown, A. G., Noble, R., and Rowe, M. J. 1986. Receptive field profiles and integrative properties of spinocervical tract cells in the cat. *J. Physiol.* **374**, 335–348.

Brown, A. M. 1967. Excitation of afferent cardiac sympathetic fibres during myocardial ischaemia. *J. Physiol.* **190**, 35–53.

Brown, J. L., Liu, H., Maggio, J. E., Vigna, S. R., Mantyh, P. W., and Basbaum, A. I. 1995. Morphological characterization of substance P receptor-immunoreactive neurons in the rat spinal cord and trigeminal nucleus caudalis. *J. Comp. Neurol.* **356**, 327–344.

Brown, P. B. 1969. Response of cat dorsal horn cells to variations of intensity, location, and area of cutaneous stimuli. *Exp. Neurol.* **23**, 249–265.

Brown, P. B. and Culberson, J. L. 1981. Somatotopic organization of hindlimb cutaneous dorsal root projections to cat dorsal horn. *J. Neurophysiol.* **45**, 137–143.

Brown, P. B. and Fuchs, J. L. 1975. Somatotopic representation of hindlimb skin in cat dorsal horn. *J. Neurophysiol.* **38**, 1–9.

Brown, P. B. and Koerber, H. R. 1978. Cat hindlimb tactile dermatomes determined with single unit recordings. *J. Neurophysiol.* **41**, 260–267.

Brown, P. B., Moraff, H., and Tapper, D. N. 1973. Functional organization of the cat's dorsal horn: spontaneous activity and central cell response to single impulses in single type I fibers. *J. Neurophysiol.* **36**, 827–839.

Brown, P. B., Fuchs, J. L., and Tapper, D. N. 1975. Parametric studies of dorsal horn neurons responding to tactile stimulation. *J. Neurophysiol.* **38**, 19–25.

Brown, P. B., Busch, G. R., and Whittington, J. 1979. Anatomical changes in cat dorsal horn cells after transection of a single dorsal root. *Exp. Neurol.* **64**, 453–468.

Brown, P. B., Brushart, T. M., and Ritz, L. A. 1989. Somatotopy of digital nerve projections to the dorsal horn in the monkey. *Somatosens. Motor Res.* **6**, 309–317.

Bruce, A. N. 1910. Über die Beziehung der sensiblen Nervenendigungen zum Entzündsvorgang. *Nauyn-Schmiedebergs Arch. Exp. Path. Pharmacol.* **63**, 424–433 (cited in Lembeck and Holzer, 1979).

Bruggemann, I., Schulz, S., Wiborny, D., and Hollt, V. 2000. Colocalization of the mu-opioid receptor and calcium/calmodulin-dependent kinase II in distinct pain-processing brain regions. *Mol. Brain Res.* **85**, 239–250.

Bruggencate, G. ten and Engberg, I. 1968. Analysis of glycine actions on spinal interneurones by intracellular recording. *Brain Res.* **11**, 446–450.

Bruggencate, G. ten, Lux, H. D., and Liebl, L. 1974. Possible relationships between extracellular potassium activity and presynaptic inhibtion in the spinal cord. *Pfluegers Arch.* **349**, 301–317.

Bruinvels, A. T., Landwehrmeyer, B., Moskowitz, M. A., and Hoyer, D. 1992. Evidence for the presence of 5-HT1B receptor messenger RNA in neurons of the rat trigeminal ganglia. *Eur. J. Pharmacol.* **227**, 357–359.

Brumovsky, P. R., Shi, T. J., Matsuda, H., Kopp, J., Villar, M. J., and Hökfelt, T. 2002. NPY Y1 receptors are present in axonal processes of DRG neurons. *Exp. Neurol.* **174**, 1–10.

Bruning, G. 1992. Localization of NADPH diaphorase, a histochemical marker for nitric oxide synthase, in the mouse spinal cord. *Acta Histochem.* **93**, 397–401.

Bruning, G., Bauer, R., and Baumgarten, H. G. 1990. Postnatal development of [^{3}H]flunitrazepam and [^{3}H]strychnine binding sites in rat spinal cord localized by quantitative autoradiography. *Neurosci. Lett.* **110**, 6–10.

Brunsden, A. M. and Grundy, D. 1999. Sensitization of visceral afferents to bradykinin in rat jejunum in vitro. *J Physiol.* **521**, 517–527.

Brushart, T. M. and Mesulam, M. M. 1980. Transganglionic demonstration of central sensory projections from skin and muscle with HRP-Lectin conjugates. *Neurosci. Lett.* **17**, 1–6.

Bryan, R. N., Trevino, D. L., Coulter, J. D., and Willis, W. D. 1973. Location and somatotopic organization of the cells of origin of the spino-cervical tract. *Exp. Brain Res.* **17**, 177–189.

Bryan, R. N., Coulter, J. D., and Willis, W. D. 1974. Cells of origin of the spinocervical tract in the monkey. *Exp. Neurol.* **42**, 574–586.

Bscheidl, C., Hanesch, U., and Heppelmann, B. 1994. NADPH-diaphorase reactivity in articular afferents of a normal and inflamed knee joint in the cat. *Brain Res.* **668**, 266–270.

Buck, S. H. and Burks, T. F. 1986. The neuropharmacology of capsaicin: Review of some recent observations. *Pharmacol. Rev.* **38**, 179–226.

Buck, S. H., Helke, C. J., Burcher, E., Shults, C. W., and O'Donohue, T. L. 1986. Pharmacologic characterization and autoradiographic distribution of binding sites for iodinated tachykinins in the rat central nervous system. *Peptides* **7**, 1109–1120.

Bucknill, A. T., Coward, K., Plumpton, C., Tate, S., Bountra, C., Birch, R., Sandison, A., Hughes, S. P., and Anand, P. 2002. Nerve fibers in lumbar spine structures and injured spinal roots express the sensory neuron-specific sodium channels SNS/PN3 and NaN/SNS2. *Spine* **27**, 135–140.

Budai, D. and Larson, A. A. 1996. Role of substance P in the modulation of C-fiber-evoked responses of spinal dorsal horn neurons. *Brain Res.* **710**, 197–203.

Budai, D. and Larson, A. A. 1998. The involvement of metabotropic glutamate receptors in sensory transmission in dorsal horn of the rat spinal cord. *Neuroscience* **83**, 571–580.

Budai, D., Wilcox, G. L., and Larson, A. A. 1992a. Modulation of N-methyl-D-aspartate and (R,S)-alpha-amino-3-hydroxy-5-methylisoxazole-4-propionate (AMPA) responses of spinal nociceptive neurons by a N-terminal fragment of substance P. *Eur. J. Pharmacol.* **216**, 441–444.

Budai, D., Wilcox, G. L., and Larson, A. A. 1992b. Enhancement of NMDA-evoked neuronal activity by glycine in the rat spinal cord in vivo. *Neurosci. Lett.* **135**, 265–268.

Bulling, D. G., Kelly, D., Bond, S., McQueen, D. S., and Seckl, J. R. 2001. Adjuvant-induced joint inflammation causes very rapid transcription of beta-preprotachykinin and alpha-CGRP genes in innervating sensory ganglia. *J. Neurochem.* **77**, 372–382.

Bullitt, E. 1989. Induction of c-fos-like protein within the lumbar spinal cord and thalamus of the rat following peripheral stimulation. *Brain Res.* **493**, 391–397.

Bullitt, E. and Light, A. R. 1989. Intraspinal course of descending serotoninergic pathways innervating the rodent dorsal horn and lamina X. *J. Comp. Neurol.* **286**, 231–242.

Burchiel, K. J. and Yezierski, R. P. (eds.) 2002. Spinal cord injury pain: Assessment, mechanisms, management. *Progress in Pain Research and Management*, vol. 23. Seattle, IASP Press.

Burford, N. T., Tolbert, L. M., and Sadee, W. 1998. Specific G protein activation and mu-opioid receptor internalization caused by morphine, DAMGO and endomorphin I. *Eur. J. Pharmacol.* **342**, 123–126.

Burgard, E. C., Niforatos, W., van Biesen, T., Lynch, K. J., Touma, E., Metzger, R. E., Kowaluk, E. A., and Jarvis, M. F. 1999. P2X receptor-mediated ionic currents in dorsal root ganglion neurons. *J. Neurophysiol.* **82**, 1590–1598.

Burgess, P. R. and Clark, F. J. 1969. Characteristics of knee joint receptors in the cat. *J. Physiol.* **203**, 317–335.

Burgess, P. R. and Perl, E. R. 1967. Myelinated afferent fibres responding specifically to noxious stimulation of the skin. *J. Physiol.* **190**, 541–562.

Burgess, P. R. and Perl, E. R. 1973. Cutaneous mechanoreceptors and nociceptors. In A. Iggo (ed.), *Handbook of Sensory Physiology*: Vol. II. *Somatosensory System* (pp. 29–78). Springer, New York.

Burgess, P. R., Petit, D., and Warren, R. M. 1968. Receptor types in cat hairy skin supplied by myelinated fibers. *J. Neurophysiol.* **31**, 833–848.

Burgess, P. R., Howe, J. F., Lessler, M. J., and Whitehorn, D. 1974. Cutaneous receptors supplied by myelinated fibers in the cat: II. Number of mechanoreceptors excited by a local stimulus. *J. Neurophysiol.* **37**, 1373–1386.

Burgess, P. R., Wei, J. Y., Clark, F. J., and Simon, J. 1982. Signalling of kinesthetic information by peripheral sensory receptors. *Ann. Rev. Neurosci.* **5**, 171–187.

Burke, D., Hagbarth, K. E., and Lofstedt, L. 1978. Muscle spindle activity in man during shortening and lengthening contractions. *J. Physiol.* **227**, 131–142.

Burke, R. E., Rudomin, P., Vyklický, L., and Zajac, F. E. 1971. Primary afferent depolarization and flexion reflexes produced by radiant heat stimulation of the skin. *J. Physiol.* **213**, 185–214.

Burnstock, G. 2000. P2X receptors in sensory neurones. *Br. J. Anaesth.* **84**, 476–488.

Burnstock, G. and Wood, J. N. 1996. Purinergic receptors: Their role in nociception and primary afferent neurotransmission. *Curr. Opin. Neurobiol.* **6**, 526–532.

Burnweit, C. and Forssmann, W. G. 1979. Somatostatinergic nerves in the cervical spinal cord of the monkey. *Cell Tissue Res.* **200**, 83–90.

Burstein, R., Dado, R. J., and Giesler, G. J. 1990. The cells of origin of the spinothalamic tract of the rat: A quantitative reexamination. *Brain Res.* **511**, 329–337.

Burstone, M. S. 1961. Modifications of histochemical techniques for the demonstration of cytochrome oxidase. *J. Histochem. Cytochem.* **9**, 59–65.

Burt, A. M. 1971. The histochemical localization of choline acetyltransferase. In O. Eranko (ed.), *Progress in Brain Research* (pp. 327–335). Elsevier, Amsterdam.

Burton, H. and McFarlane, J. J. 1973. The organization of the seventh lumbar spinal ganglion of the cat. *J. Comp. Neurol.* **149**, 215–232.

Burton, H., Terashima, S. I., and Clark, J. 1972. Response properties of slowly adapting mechanoreceptors to temperature stimulation in cats. *Brain Res.* **45**, 401–416.

Bushong, J. A. 1963. A report of the frequency and distribution of the Pacinian corpuscle in a palm of a seven month human fetus. Johns Hopkins University, Baltimore, Maryland, MA thesis.

Butler, S. H., Godefroy, F., Besson, J. M., and Weil-Fugazza, J. 1992. A limited arthritic model for chronic pain studies in the rat. *Pain* **48**, 73–81.

Cabot, P. J., Carter, L., Gaiddon, C., Zhang, Q., Schäfer, M., Loeffler, J. P., and Stein, C. 1997. Immune cell-derived β-endorphin: Production, release and control of inflammatory pain in rats. *J. Clin. Invest.* **100**, 142–148.

Cabot, P. J., Carter, L., Schäfer, M., and Stein, C. 2001. Methionine-enkephalin and dynorphin-A-release from immune cells and control of inflammatory pain. *Pain* **93**, 207–212.

Cadden, S. W., Villanueva, L., Chitour, D., and Le Bars, D. 1983. Depression of activities of dorsal horn convergent neurones by propriospinal mechanisms triggered by noxious inputs: Comparison with diffuse noxious inhibitory controls (DNIC). *Brain Res.* **275**, 1–11.

Caffrey, J. M., Eng, D. L., Black, J. A., Waxman, S. G., and Kocsis, J. D. 1992. Three types of sodium channels in adult rat dorsal root ganglion neurons. *Brain Res.* **592**, 283–297.

Cain, D. M., Khasabov, S. G., and Simone, D. A. 2001. Response properties of mechanoreceptors and nociceptors in mouse glabrous skin: An in vivo study. *J. Neurophysiol.* **85**, 1561–1574.

Calancie, B. M. and Stein, R. B. 1988. Microneurography for the recording and selective stimulation of afferents: An assessment. *Muscle Nerve* **11**: 638–644.

Calas, A., Dupuy, J. J., Gamrani, H., Gonella, J., Mourre, C., Condami, M., Pellissier, F., and Bosch, P. van den 1919. Radioautographic investigation of serotonin cells. In B. Haber, S. Gabary, M. R. Issidorides, and S. G. A. Alivisatos (eds.), *Advances in Exp. Med. & Biol.* (pp. 51–66). Plenum Press, New York.

Calignano, A., La Rana, G., Giufridda, A., and Piomelli, D. 1998. Control of pain initiation by endogenous cannabinoids. *Nature* **394**, 277–281.

Callister, R. J., Schofield, P. R., and Sah, P. 1999. Use of murine mutants to study glycine receptor function. *Clin. Exp. Pharmacol. Physiol.* **26**, 929–931.

Callsen-Cencic, P. and Mense, S. 1997. Expression of neuropeptides and nitric oxide synthase in neurones innervating the inflamed rat urinary bladder. *J. Auton. Nerv. Syst.* **65**, 33–44.

Callsen-Cencic, P., Hoheisel, U., Kaske, A., Mense, S., and Tenschert, S. 1999. The controversy about spinal neuronal nitric oxide synthase: Under which conditions is it up- or downregulated? *Cell Tissue Res.* **295**, 183–194.

Calver, A. R., Medhurst, A. D., Robbins, M. J., Charles, K. J., Evans, M. L., Harrison, D. C., Stammers, M., Hughes, S. A., Hervieu, G., Couve, A., Moss, S. J., Middlemiss, D. N., and Pangalos, M. N. 2000. The expression of GABA(B1) and GABA(B2) receptor subunits in the cNS differs from that in peripheral tissues. *Neuroscience* **100**, 155–170.

Calvillo, O., Madrid, J., and Rudomin, P. 1982. Presynaptic depolarization of unmyelinated primary afferent fibers in the spinal cord of the cat. *Neuroscience* **7**, 1389–1400.

Calvino, B., Villanueva, L., and Le Bars, D. 1987. Dorsal horn (convergent) neurones in the intact anaesthetized arthritic rat: I. Segmental excitatory influences. *Pain* **28**, 81–98.

Calza, L., Pozza, M., Zanni, M., Manzini, C. U., Manzini, E., and Hökfelt, T. 1998. Peptide plasticity in primary sensory neurons and spinal cord during adjuvant-induced arthritis in the rat: An immunocytochemical and in situ hybridization study. *Neuroscience* **82**, 575–589.

Calza, L., Pozza, M., Arletti, R., Manzini, E., and Hökfelt, T. 2000. Long-lasting regulation of galanin, opioid, and other peptides in dorsal root ganglia and spinal cord during experimental polyarthritis. *Exp. Neurol.* **164**, 333–343.

Cameron, A. A., Leah, J. D., and Snow, P. J. 1986. The electrophysiological and morphological characteristics of feline dorsal root ganglion cells. *Brain Res.* **362**, 1–6.

Cameron, A. A., Leah, J. D., and Snow, P. J. 1988. The coexistence of neuropeptides in feline sensory neurons. *Neuroscience* **27**, 969–979.

Cameron, A. A., Cliffer, K. D., Dougherty, P. M., Willis, W. D., and Carlton, S. M. 1991. Changes in lectin, GAP-43 and neuropeptide staining in rat superficial dorsal horn following experimental peripheral neuropathy. *Neurosci. Lett.* **131**, 249–252.

Cameron, A. A., Pover, C. M., Willis, W. D., and Coggeshall, R. E. 1992. Evidence that fine primary afferent axons innervate a wider territory in the superficial dorsal horn following peripheral axotomy. *Brain Res.* **575**, 151–154.

Campbell, J. N., Raja, S. N., Meyer, R. A., and Mackinnon, S. E. 1988. Myelinated afferents signal the hyperalgesia associated with nerve injury. *Pain* **32**, 89–94.

Campbell, J. N., Meyer, R., and Raja, S. N. 1992. Is nociceptor activation by alpha-1 adrenoreceptors the culprit in sympathetically maintained pain? *Am. Pain Soc. J.* **1**, 3–11.

Campistron, G., Buijs, R. M., and Geffard, M. 1986. Glycine neurons in the brain and spinal cord: Antibody production and immunocytochemical localization. *Brain Res.* **376**, 400–405.

Cangro, C. B., Sweetnam, P. M., Wrathall, J. R., Haser, W. B., Curthoys, N. P., and Neale, J. H. 1985. Localization of elevated glutaminase immunoreactivity in small DRG neurons. *Brain Res.* **336**, 158–161.

Cangro, C. B., Namboodiri, M. A. A., Sklar, L. A., Corigliano-Murphy, A., and Neale, J. H. 1987. Immunohistochemistry and biosynthesis of n-acetylaspartylglutamate in spinal sensory ganglia. *J. Neurochem.* **49**, 1579–1588.

Cao, C. Q., Djouhri, L., and Brown, A. G. 1993. Lumbosacral spinal neurons in the cat that are candidates for being activated by collaterals from the spinocervical tract. *Neuroscience* **57**, 153–165.

Cao, Y. Q., Mantyh, P. W., Carlton, E. J., Gillespie, A. M., Epstein, C. J., and Basbaum, A. I. 1998. Primary afferent tachykinins are required to experience moderate to intense pain. *Nature* **392**, 390–394.

Carli, G., Farabollini, F., Fontani, G., and Meucci, M. 1979. Slowly adapting receptors in cat hip joint. *J. Neurophysiol.* **42**, 767–779.

Carli, G., Dontani, G., and Meucci, M. 1981. Static characteristics of muscle afferents from gluteus medius muscle: comparison with joint afferents of hip in cats. *J. Neurophysiol.* **45**, 1085–1095.

Carlsson, A., Falck, B., Fuxe, K., and Hillarp, N. A. 1964. Cellular localization of monoamines in the spinal cord. *Acta Physiol. Scand.* **60**, 112–119.

Carlstedt, T. 1977. Observations on the morphology at the transition between the central and the peripheral nervous system in the cat IV. *Acta Physiol. Scand.* **446**, 61–72.

Carlton, S. M. 2002. Localization of CaMKIIalpha in rat primary sensory neurons: Increase in inflammation. *Brain Res.* **947**, 252–259.

Carlton, S. M. and Coggeshall, R. E. 1996. Stereological analysis of galanin and CGRP synapses in the dorsal horn of neuropathic primates. *Brain Res.* **711**, 16–25.

Carlton, S. M. and Coggeshall, R. E. 1997. Immunohistochemical localization of enkephalin in peripheral sensory axons in the rat. *Neurosci. Lett.* **221**, 121–124.

Carlton, S. M. and Hargett, G. L. 2002. Stereological analysis of Ca(2+)/calmodulin-dependent protein kinase II alpha-containing dorsal root ganglion neurons in the rat: Colocalization with isolectin Griffonia simplicifolia, calcitonin gene-related peptide, or vanilloid receptor 1. *J. Comp. Neurol.* **448**, 102–110.

Carlton, S. M. and Hayes, E. S. 1989. Dynorphin A (1–8) immunoreactive cell bodies, dendrites and terminals are postysynaptic to calcitonin gene-related peptide primary afferent terminals in the monkey dorsal horn. *Brain Res.* **504**, 124–128.

Carlton, S. M. and Hayes, E. S. 1990. Light-microscopic and ultrastructural analysis of GABA-immunoreactive profiles in the monkey spinal cord. *J. Comp. Neurol.* **300**, 162–182.

Carlton, S. M. and Hayes, E. S. 1991. GABAergic vesicle-containing dendrites and spines: A critical element in processing sensory input in the monkey dorsal horn. *Neurosci. Lett.* **121**, 40–42.

Carlton, S. M., McNeill, D. L., Chung, K., and Coggeshall, R. E. 1987. A light- and electron-microscopic level analysis of calcitonin gene-related peptide (CGRP) in the spinal cord of the primate: An immunohistochemical study. *Neurosci. Lett.* **82**, 145–150.

Carlton, S. M., McNeill, D. L., Chung, K., and Coggeshall, R. E. 1988. Organization of calcitonin gene-related peptide-immunoreactive terminals in the primate dorsal horn. *J. Comp. Neurol.* **276**, 527–536.

Carlton, S. M., Westlund, K. N., Zhang, D. X., Sorkin, L. S., and Willis, W. D. 1990. Calcitonin gene-related peptide containing primary afferent fibers synapse on primate spinothalamic tract cells. *Neurosci. Lett.* **109**, 76–81.

Carlton, S. M., Honda, C. N., Willcockson, W. S., Lacrampe, M., Zhang, D., Denoroy, L., Chung, J. M., and Willis, W. D. 1991. Descending adrenergic input to the primate spinal cord and its possible role in modulation of spinothalamic cells. *Brain Res.* **543**, 77–90.

Carlton, S. M., Westlund, K. N., Zhang, D., and Willis, W. D. 1992. GABA-immunoreactive terminals synapse on primate spinothalamic tract cells. *J. Comp. Neurol.* **322**, 528–537.

Carlton, S. M., Lekan, H. A., Kim, S., and Chung, J. M. 1994. Behavioral manifestations of an experimental model for peripheral neuropathy produced by spinal nerve ligation in the primate. *Pain* **56**, 155–166.

Carlton, S. M., Zhou, S., and Coggeshall, R. E. 1996a. Localization and activation of substance P receptors in unmyelinated axons of rat skin. *Brain Res.* **734**, 103–108.

Carlton, S. M., Hargett, G. L., and Coggeshall, R. E. 1996b. Distribution of glycine-immunoreactive profiles in the monkey spinal cord: A light-microscopic and ultrastructural study. *J. Comp. Neurol.* **371**, 589–602.

Carlton, S. M., Zhou, S., and Coggeshall, R. E. 1998a. Evidence for the interaction of glutamate and NK1 receptors in the periphery. *Brain Res.* **790**, 160–169.

Carlton, S. M., Hargett, G. L., and Coggeshall, R. E. 1998b. Plasticity in alpha-amino-3-hydroxy-5-methyl-4-isoxazolepropionic acid receptor subunits in the rat dorsal horn following deafferentation. *Neurosci. Lett.* **242**, 21–24.

Carlton, S. M., Zhou, S., and Coggeshall, R. E. 1999. Peripheral $GABA_A$ receptors: Evidence for peripheral primary afferent depolarization. *Neuroscience* **93**, 713–722.

Carlton, S. M., Du, J., Davidson, E., Zhou, S., and Coggeshall, R. E. 2001a. Somatostatin receptors on peripheral primary afferent terminals: Inhibition of sensitized nociceptors. *Pain* **90**, 233–244.

Carlton, S. M., Du, J., Zhou, S., and Coggeshall, R. E. 2001b. Tonic control of peripheral cutaneous nociceptors by somatostatin receptors. *J. Neurosci.* **21**, 4042–4049.

Carlton, S. M., Hargett, G. L., and Coggeshall, R. E. 2001c. Localization of metabotropic glutamate receptors 2/3 on primary afferent axons in the rat. *Neuroscience* **105**, 957–969.

Carpenter, M. B. and Sutin, J. 1983. *Human Neuroanatomy* (8th ed.). Williams & Wilkins, Baltimore.

Carpenter, M. B., Stein, B. M., and Shriver, J. E. 1968. Central projections of spinal dorsal roots in the monkey: II. Lower thoracic, lumbosacral and coccygeal dorsal roots. *Am. J. Anat.* **123**, 75–118.

Carpentier, V., Vaudry, H., Laquerriere, A., Tayot, J., and Leroux, P. 1996. Distribution of somatostatin receptors in the adult human brainstem. *Brain Res.* **734**, 135–148.

Carr, P. A., Yamamoto, T., Karmy, G., Baimbridge, K. G., and Nagy, J. I. 1989a. Analysis of parvalbumin and calbindin D28k-immunoreactive neurons in dorsal root ganglia of rat in relation to their cytochrome oxidase and carbonic anhydrase content. *Neuroscience* **33**, 363–371.

Carr, P. A., Yamamoto, T., Staines, W. A., Whittaker, M. E., and Nagy, J. I. 1989b. Quantitative histochemical analysis of cytochrome oxidase in rat dorsal root ganglia and its co-localization with carbonic anhydrase. *Neuroscience* **33**, 351–362.

Carr, P. A., Yamamoto, T., Karmy, G., Baimbridge, K. G., and Nagy, J. I. 1989c. Parvalbumin is highly colocalized with calbindin D28k and rarely with calcitonin gene-related peptide in dorsal root ganglia neurons of rat. *Brain Res.* **497**, 163–170.

Carr, P. A., Yamamoto, T., and Nagy, J. I. 1990. Calcitonin gene-related peptide in primary afferent neurons of rat: Co-existence with fluoride-resisant acid phosphatase and depletion by neonatal capsaicin. *Neuroscience* **36**, 751–760.

Carr, P. A., Haftel, V., Alvarez, F. J., Cope, T. C., and Fyffe, R. E. 1998. Effect of sciatic nerve transection or TTX application on enzyme activity in rat spinal cord. *NeuroReport* **9**, 357–361.

Carriedo, S. G., Yin, H.-Z., and Weiss, J. H. 1996. Motor neurons are selectively vulnerable to AMPA/kainate receptor-mediated injury in vitro. *J. Neurosci.* **16**, 4069–4079.

Carroll, P., Lewin, G. R., Koltzenburg, M., Toyka, K. V., and Thoenen, H. 1998. A role for BDNF in mechanosensation. *Nature Neurosci.* **1**, 42–46.

Carvalho, T. L., Hodson, N. P., Blank, M. A., Watson, P. F., Mulderry, P. K., Bishop, A. E., Gu, J., Bloom, S. R., and Polak, J. M. 1986. Occurrence, distribution and origin of peptide-containing nerves of guinea-pig and rat male genitalia and the effects of denervation on sperm characteristics. *J. Anat.* **149**, 121–141.

Casey, K. L. and Bushnell, M. C. (eds.) 2000. *Pain Imaging: Progress in Pain Research and Management*, Vol. 18. IASP Press, Seattle.

Castano, P. and Ventura, R. G. 1978. The Meissner's corpuscle of the green monkey (*Cercopithecus aethiops* L.): The organization of the nervous component. *J. Submicros. Cytol.* **10**, 327–344.

Castro, M. E., Pascual, J., Romon, T., Del Arco, C., del Olmo, E., and Pazos, A. 1997. Differential distribution of [^{3}H] sumatriptan binding sites (5-HT_{1B}, 5-HT_{1D}, 5-HT_{1F} receptors) in human brain: Focus on brainstem and spinal cord. *Neuropharmacology* **36**, 535–542.

Castro-Lopes, J. M. and Coimbra, A. 1991. Spinal cord projections of the rat main forelimb nerves, studied by transganglionic transport of WGA-HRP and by the disappearance of acid phosphatase. *Brain Res.* **542**, 187–192.

Castro-Lopes, J. M., Tavares, I., Tölle, T. R., Coito, A., and Coimbra, A. 1992. Increase in GABAergic cells and GABA levels in the spinal cord in unilateral inflammation of the hindlimb in the rat. *Eur. J. Neurosci.* **4**, 296–301.

Castro-Lopes, J. M., Tavares, I., and Coimbra, A. 1993. GABA decreases in the spinal cord dorsal horn after peripheral neurectomy. *Brain Res.* **620**, 287–291.

Castro-Lopes, J. M., Tavares, I., Tölle, T. R., and Coimbra, A. 1994a. Carrageenan-induced inflammation of the hind foot provokes a rise of GABA-immunoreactive cells in the rat spinal cord that is prevented by peripheral neurectomy or neonatal capsaicin treatment. *Pain* **56**, 193–201.

Castro-Lopes, J. M., Tolle, T. R., Pan, B., and Zieglgansberger, W. 1994b. Expression of GAD mRNA in spinal cord neurons of normal and monoarthritic rats. *Mol. Brain Res.* **26**, 169–176.

Castro-Lopes, J. M., Malcangio, M., Pan, B. H., and Bowery, N. B. 1995. Complex changes of $GABA_A$ and $GABA_B$ receptor binding in the spinal cord dorsal horn following peripheral inflammation or neurectomy. *Brain Res.* **679**, 289–297.

Caterina, M. J. and Julius, D. 2001. The vanilloid receptor: A molecular gateway to the pain pathway. *Annu. Rev. Neurosci.* **24**, 487–517.

Caterina, M. J., Schumacher, M. A., Tominaga, M., Rosen, T. A., Levine, J. D., and Julius, D. 1997. The capsaicin receptor: A heat-activated ion channel in the pain pathway. *Nature* **389**, 816–824.

Caterina, M. J., Rosen, T. A., Tominaga, M., Brake, A. J., and Julius, D. 1999. A capsaicin-receptor homologue with a high threshold for noxious heat. *Nature* **398**, 436–441.

Caterina, M. J., Leffler, A., Malmberg, A. B., Martin, W. J., Trafton, J., Petersen-Zeitz, K. R., Koltzenburg, M., Basbaum, A. I., and Julius, D. 2000. Impaired nociception and pain sensation in mice lacking the capsaicin receptor. *Science* **288**, 306–313.

Caudle, R. M. and Mannes, A. J. 2000. Dynorphin: Friend or foe? *Pain* **87**, 235–239.

Cauna, N. 1956. Nerve supply and nerve endings in Meissner's corpuscles. *Am. J. Anat.* **99**, 315–350.

Cauna, N. and Mannan, G. 1958. The structure of human digital Pacinian corpuscles (corpuscula lamellosa) and its functional significance. *J. Anat.* **92**, 1–24.

Cauna, N. and Mannan, G. 1959. Development and postnatal changes of digital Pacinian corpuscles (Corpuscula lamellosa) in the human hand. *J. Anat.* **93**, 271–286.

Cauna, N. and Naik, N. T. 1963. The distribution of cholinesterases in the sensory ganglia of man and some mammals. *J. Histochem. Cytochem.* **11**, 129–138.

Cauna, N. and Ross, L. L. 1960. The fine structure of Meissner's touch corpuscles of human fingers. *J. Biophys. Biochem. Cytol.* **8**, 467–482.

Cechetto, D. F. and Saper, C. B. 1988. Neurochemical organization of the hypothalamic projection to the spinal cord in the rat. *J. Comp. Neurol.* **272**, 579–604.

Celio, M. R. 1990. Calbindin D-28k and parvalbumin in the rat nervous system. *Neuroscience* **35**, 375–475.

Celio, M. R. and Heizmann, C. W. 1981. Calcium-binding protein parvalbumin as a neuronal marker. *Nature* **293**, 300–302.

Cerne, R. and Randić, M. 1992. Modulation of AMPA and NMDA responses in rat spinal dorsal horn neurons by trans-1-aminocyclopentane-1,3-dicarboxylic acid. *Neurosci. Lett.* **144**, 180–184.

Cervero, F. 1982a. Noxious intensities of visceral stimulation are required to activate viscero-somatic multireceptive neurons in the thoracic spinal cord of the cat. *Brain Res.* **240**, 350–352.

Cervero, F. 1982b. Afferent activity evoked by natural stimulation of the biliary system in the ferret. *Pain* **13**, 137–151.

Cervero, F. 1983. Somatic and visceral inputs to the thoracic spinal cord of the cat: Effects of noxious stimulation of the biliary system. *J. Physiol.* **337**, 51–67.

Cervero, F. 1994. Sensory innervation of the viscera: Peripheral basis of visceral pain. *Physiol. Rev.* **74**, 95–138.

Cervero, F. and Connell, L. A. 1984a. Fine afferent fibers from viscera do not terminate in the substantia gelatinosa of the thoracic spinal cord. *Brain Res.* **294**, 370–374.

Cervero, F. and Connell, L. A. 1984b. Distribution of somatic and visceral primary afferent fibres within the thoracic spinal cord of the cat. *J. Comp. Neurol.* **230**, 88–98.

Cervero, F. and Iggo, A. 1980. The substantia gelatinosa of the spinal cord: A critical review. *Brain* **103**, 717–772.

Cervero, F. and Laird, J. M. A. 1996. Mechanisms of touch-evoked pain (allodynia): A new model. *Pain* **68**, 13–23.

Cervero, F. and Laird, J. M. A. 1999. Visceral pain. *Lancet.* **353**, 2145–2148.

Cervero, F. and Sann, H. 1989. Mechanically evoked responses of afferent fibres innervating the guinea-pig's ureter: An in vitro study. *J. Physiol.* **412**, 245–266.

Cervero, F. and Sharkey, K. A. 1988. An electrophysiological and anatomical study of intestinal afferent fibres in the rat. *J. Physiol.* **401**, 381–397.

Cervero, F. and Tattersall, J. E. H. 1985. Cutaneous receptive fields of somatic and viscerosomatic neurones in the thoracic spinal cord of the cat. *J. Comp. Neurol.* **237**, 325–332.

Cervero, F. and Tattersall, J. E. H. 1987. Somatic and visceral inputs to the thoracic spinal cord of the cat: Marginal zone (lamina I) of the dorsal horn. *J. Physiol.* **383**, 383–395.

Cervero, F. and Wolstencroft, J. H. 1984. A positive feedback loop between spinal cord nociceptive pathways and antinociceptive areas of the cat's brain stem. *Pain* **20**, 125–138.

Cervero, F., Iggo, A., and Ogawa, H. 1976. Nociceptor-driven dorsal horn neurones in the lumbar spinal cord of the cat. *Pain* **2**, 5–14.

Cervero, F., Molony, V., and Iggo, A. 1977b. Extracellular and intracellular recordings from neurons in the substantia gelatinosa Rolandi. *Brain Res.* **136**, 565–569.

Cervero, F., Iggo, A., and Molony, V. 1979a. An electrophysiological study of neurones in substantia gelatinosa Rolandi of the cat's spinal cord. *Q. J. Exp. Physiol.* **64**, 297–314.

Cervero, F., Iggo, A., and Molony, V. 1979b. Ascending projections of nociceptor-driven lamina I neurones in the cat. *Exp. Brain Res.* **35**, 135–149.

Cervero, F., Molony, V., and Iggo, A. 1979c. Supraspinal linkage of substantia gelatinosa neurons: Effects of descending impulses. *Brain Res.* **175**, 351–355.

Cervero, F., Schouenborg, J., Sjölund, B. H., and Waddell, P. J. 1984. Cutaneous inputs to dorsal horn neurones in adult rats treated at birth with capsaicin. *Brain Res.* **301**, 47–57.

Cervero, F., Lumb, B. M., and Tattersall, J. E. H. 1985. Supraspinal loops that mediate visceral inputs to thoracic spinal cord neurons in the cat: Involvement of descending pathways from raphe and reticular formation. *Neurosci. Lett.* **56**, 189–194.

Cervero, F., Handwerker, H. O., and Laird, J. M. A. 1988. Prolonged noxious mechanical stimulation of the rat's tail: Responses and encoding properties of dorsal horn neurones. *J. Physiol.* **404**, 419–436.

Cervero, F., Laird, J. M. A., and Pozo, M. A. 1992. Selective changes of receptive field properties of spinal nociceptive neurones induced by noxious visceral stimulation in the cat. *Pain* **51**, 335–342.

Cervero, F., Meyer, R. A., and Campbell, J. N. 1994. A psychophysical study of secondary hyperalgesia: Evidence for increased pain to input from nociceptors. *Pain* **58**, 21–28.

Cesare, P., Dekker, L. V., Sardini, A., Parker, P. J., and McNaughton, P. A. 1999. Specific involvement of PKC-ε in sensitization of the neuronal response to painful heat. *Neuron* **23**, 617–624.

Cesselin, F., Montastruc, J. L., Gros, C., Bourgoin, S., and Hamon, M. 1980. Met-enkephalin levels and opiate receptors in the spinal cord of chronic suffering rats. *Brain Res.* **191**, 289–293.

Cesselin, F., Bourgoin, S., Hamon, M., Artaud, F., Testut, M. F., Rascol, A., and Montastruc, J. L. 1984. Normal CSF levels of met-enkephalin-like material in a case of naloxone-reversible congenital insensitivity to pain. *Neuropeptides* **4**, 217–225.

Cesselin, F., Le Bars, D., Bourgoin, S., Artaud, F., Gozlan, H., Clot, A. M., Besson, J. M., and Hamon, M. 1985. Spontaneous and evoked release of methionine-enkephalin-like material from the rat spinal cord in vivo. *Brain Res.* **339**, 305–313.

Cesselin, F., Bourgoin, S., Clot, A. M., Hamon, M., and Le Bars, D. 1989. Segmental release of met-enkephalin-like material from the spinal cord of rats, elicited by noxious thermal stimuli. *Brain Res.* **484**, 71–77.

Chambers, M. R., Andres, K. H., Düring, M. von, and Iggo, A. 1972. The structure and function of the slowly adapting type II mechanoreceptor in hairy skin. *Q. J. Exp. Physiol.* **57**, 417–445.

Chang, P. C., Aicher, S. A., and Drake, C. T. 2000. Kappa opioid receptors in rat spinal cord vary across the estrous cycle. *Brain Res.* **861**, 168–172.

Changaris, D. G., Keil, L. C., and Severs, W. B. 1978. Angiotensin II immunohistochemistry of the rat brain. *Neuroendocrinology* **25**, 257–274.

Chan-Palay, V., and Palay, S. L. 1977a. Ultrastructural identification of substance P cells and their processes in rat sensory ganglia and their terminals in the spinal cord by immunocytochemistry. *Proc. Natl. Acad. Sci. USA* **74**, 4050–4054.

Chan-Palay, V. and Palay, S. L. 1977b. Immunocytochemical identification of substance P cells and their processes in rat sensory ganglia and their terminals in the spinal cord: Light microscopic studies. *Proc. Natl. Acad. Sci. USA* **74**, 3597–3601.

Chan-Palay, V., Jonsson, G., and Palay, S. L. 1978. Serotonin and substance P coexist in neurons of the cat's central nervous system. *Proc. Natl. Acad. Sci. USA* **75**, 1582–1586.

Chaplan, S. R., Bach, F. W., Shafer, S. L., and Yaksh, T. L. 1995. Prolonged alleviation of tactile allodynia by intravenous lidocaine in neuropathic rats. *Anesthesiology* **83**, 775–785.

Chapman, V. and Besson, J.-M. 1997. Pharmacological studies of nociceptive systems using the c-Fos immunohistochemical technique: An indicator of noxiously activated spinal neurones. In *Handbook of Pharmacology* (pp. 235–279). Springer Verlag, New York.

Chapman, V., Dickenson, A. H., and Tjolen, A. 1994. Bi-directional effects of intrathecal NMDA and substance P on rat dorsal horn neuronal responses. *Neurosci. Lett.* **178**, 90–94.

Chapman, V., Diaz, A., and Dickenson, A. H. 1997. Distinct inhibitory effects of spinal endomorphin-1 and endomorphin-2 on evoked dorsal horn neuronal responses in the rat. *Br. J. Pharmacol.* **122**, 1537–1539.

Charles, K. J., Evans, M. L., Robbins, M. J., Calver, A. R., Leslie, R. A., and Pangalos, M. N. 2001. Comparative immunohistochemical localization of GABA (B1a), GABA (B1b) and GABA (B2) subunits in rat brain, spinal cord and dorsal root ganglion. *Neuroscience* **106**, 447–467.

Charlton, C. G. and Helke, C. J. 1985. Autoradiographic localization and characterization of spinal cord substance P binding sites: High densities in sensory, autonomic, phrenic, and Onuf's motor nuclei. *J. Neurosci.* **5**, 1653–1661.

Charlton, C. G. and Helke, C. J. 1986. Ontogeny of substance P receptors in rat spinal cord: Quantitative changes in receptor number and differential expression in specific loci. *Dev. Brain Res.* **29**, 81–91.

Charnay, Y., Paulin, C., Chayvialle, J. A., and Dubois, P. M. 1983. Distribution of substance P-like immunoreactivity in the spinal cord and dorsal root ganglia of the human foetus and infant. *Neuroscience* **10**, 41–55.

Charnay, Y., Paulin, C., Dray, F., and Dubois, P. M. 1984. Distribution of enkephalin in human fetus and infant spinal cord: An immunofluorescence study. *J. Comp. Neurol.* **223**, 415–423.

Charnay, Y., Chayvialle, J. A., Pradayrol, L., Bouvier, R., Paulin, C., and Dubois, P. M. 1987. Ontogeny of somatostatin-like immunoreactivity in the human fetus and infant spinal cord. *Dev. Brain Res.* **36**, 63–73.

Charnay, Y., Leroux, P., Epelbaum, J., Enjalbert, A., Vaudry, H., and Dubois, P. M. 1988. Displaceable somatostatin binding sites in the gray matter and pyramidal paths of the human developing spinal cord. *Neurosci. Lett.* **84**, 245–250.

Chatrian, G. E., Farrell, D. F., Canfield, R. C., and Lettich, E. 1975. Congenital insensitivity to noxious stimuli. *Arch. Neurol.* **32**, 141–145.

Chaudhary, P. and Baumann, T. K. 2002. Expression of VPAC2 receptor and PAC1 receptor splice variants in the trigeminal ganglion of the adult rat. *Mol. Brain Res.* **104**, 137–142.

Checkosky, C. M. and Bolanowski, S. J. 1992. Effects of stimulus duration on the response properties of Pacinian corpuscles: Implications for the neural code. *J. Acoust. Soc. Am.* **91**, 3372–3380.

Checkosky, C. M. and Bolanowski, S. J. 1994. The effect of stimulus duration on frequency-response functions in the Pacinian (P) channel. *Somatosens. Mot. Res.* **11**, 47–56.

Chen, C.-C., Akopian, A. N., Sivilotti, L., Colquhoun, D., Burnstock, G., and Wood, J. N. 1995. A P2X purinoceptor expressed by a subset of sensory neurons. *Nature* **377**, 428–431.

Chen, C. C., England, S., Akopian, A. N., and Wood, J. N. 1998. A sensory neuron-specific, proton-gated ion channel. *Proc. Natl. Acad. Sci. USA* **95**, 10240–10245.

Chen, H. S. and Chen J. 2000. Secondary heat, but not mechanical, hyperalgesia induced by subcutaneous injection of bee venom in the conscious rat: effect of systemic MK-801, a non-competitive NMDA receptor antagonist. *Eur. J. Pain* **4**, 389–401.

Chen, H. S., Chen, J., and Sun, Y. Y. 2000. Contralateral heat hyperalgesia induced by unilateral intraplantar bee venom injection is produced by central changes: A behavioral study in the conscious rat. *Neurosci. Lett.* **284**, 45–48.

Chen, H. S., Chen, J., Chen, J., Guo, W. G., and Zheng, M. H. 2001. Establishment of bee venom-induced contralateral heat hyperalgesia in the rat is dependent upon central temporal summation of afferent input from the site of injury. *Neurosci. Lett.* **298**, 57–60.

Chen, J. and Chen, H. S. 2001. Pivotal role of capsaicin-sensitive primary afferents in development of both heat and mechanical hyperalgesia induced by intraplantar bee venom injection. *Pain* **91**, 367–376.

Chen, J., Luo, C., and Li, H. L. 1998. The contribution of spinal neuronal changes to development of prolonged, tonic nociceptive responses of the cat induced by subcutaneous bee venom injection. *Eur. J. Pain* **2**, 359–376.

Chen, J., Li, H., Luo, C., Li, Z., and Zheng, J. 1999a. Involvement of peripheral NMDA and non-NMDA receptors in development of persistent firing of spinal wide-dynamic-range neurons induced by subcutaneous bee venom injection in the cat. *Brain Res.* **844**, 98–105.

Chen, J., Luo, C., Li, H., and Chen, H. 1999b. Primary hyperalgesia to mechanical and heat stimuli following subcutaneous bee venom injection into the plantar surface of hindpaw in the conscious rat: A comparative study with the formalin test. *Pain* **83**, 67–76.

Chen, J., Heinke, B., and Sandkühler, J. 2000. Activation of group I metabotropic glutamate receptors induces long-term depression at sensory synapses in superficial spinal dorsal horn. *J. Neuropharmacol.* **39**, 2231–2243.

Chen, S. R. and Pan, H. L. 2001. Spinal endogenous acetylcholine contributes to the analgesic effect of systemic morphine in rats. *Anesthesiology* **95**, 525–530.

Chen, S. T., Tsai, M. S., and Shen, C. L. 1988. Distribution of neurotensin-like immunoreactivity in the central nervous system of the Formosan monkey. *Proc. Natl. Sci. Counc. Repub. China B* **12**, 163–173.

Chen, S. T., Tsai, M. S., and Shen, C. L. 1989. Distribution of FMRFamide-like immunoreactivity in the central nervous system of the Formosan monkey (*Macaca cyclopsis*). *Peptides* **10**, 825–834.

Chen, X. and Levine, J. D. 1999. NOS inhibitor antagonism of PGE_2-induced mechanical sensitization of cutaneous C-fiber nociceptors in the rat. *J. Neurophysiol.* **81**, 963–966.

Chen, X., Tanner, K., and Levine, J. D. 1999. Mechanical sensitization of cutaneous C-fiber nociceptors by prostaglandin E_2 in the rat. *Neurosci. Lett.* **267**, 105–108.

Cheng, P. Y., Svingos, A. L., Wang, H., Clarke, C. L., Jenab, S., Beczkowska, I. W., Inturrisi, C. E., and Pickel, V. M. 1995a. Ultrastructural immunolabeling shows prominent presynaptic vesicular localization of δ-opioid receptor within both enkephalin- and nonekephalin-containing axon terminals in the superficial layers of the rat cervical spinal cord. *J. Neurosci.* **15**, 5976–5988.

Cheng, P. Y., Moriwaki, A., Wang, J. B., Uhl, G. R., and Pickel, V. M. 1995b. Ultrastructural localization of μ-opioid receptor immunoreactivity and relationship to enkephalin in cervical dorsal horn of the rat spinal cord. *Analgesia* **363**, 363–366.

Cheng, P. Y., Moriwaki, A., Wang, J. B., Uhl, G. R., and Pickel, V. M. 1996. Ultrastructural localization of μ-opioid receptors in the superficial layers of the rat cervical spinal cord: extrasynaptic localization and proximity to Leu5-enkephalin. *Brain Res.* **731**, 141–154.

Cheng, P. Y., Liu-Chen, L.-Y., and Pickel, V. M. 1997. Dual ultrastructural immunocytochemical labeling of mu and delta opioid receptors in the superficial layers of the rat cervical spinal cord. *Brain Res.* **778**, 367–380.

Chéry, N. and De Koninck, Y. 1999. Junctional versus extrajunctional glycine and GABA(A) receptor-mediated IPSCs in identified lamina I neurons of the adult rat spinal cord. *J. Neurosci.* **19**, 7342–7355.

Chéry, N. and De Koninck, Y. 2000. GABA(B) receptors are the first target of released GABA at lamina I inhibitory synapses in the adult rat spinal cord. *J. Neurophysiol.* **84**, 1006–1011.

Chéry, N., Yu, X. H., and De Koninck, Y. 2000. Visualization of lamina I of the dorsal horn in live adult rat spinal cord slices. *J. Neurosci. Methods* **96**, 133–142.

Cheung, K. K. and Burnstock, G. 2002. Localization of P2X3 receptors and coexpression with P2X2 receptors during rat embryonic neurogenesis. *J. Comp. Neurol.* **443**, 368–382.

Chi, S. I., Levine, J. D., and Basbaum, A. I. 1993. Peripheral and central contributions to the persistent expression of spinal cord fos-like immunoreactivity produced by sciatic nerve transection in the rat. *Brain Res.* **617**, 225–237.

Chiba, T., Masuko, S., and Kawano, H. 1986. Correlation of mitochondrial swelling after capsaicin treatment and substance P and somatostatin immunoreactivity in small neurons of dorsal root ganglion in the rat. *Neurosci. Lett.* **64**, 311–316.

Chigr, F., Najimi, M., Leduque, P., Chayvialle, J. A., Bouvier, R., and Kopp, A. 1991. Anatomical distribution of substance P-immunoreactive neurons in human brainstem during the first postnatal year. *Brain Res. Bull.* **26**, 515–523.

Childs, A. M., Evans, R. H., and Watkins, J. C. 1988. The pharmacological selectivity of three NMDA antagonists. *Eur. J. Pharmacol.* **145**, 81–86.

Chizh, B. A. and Headley, P. M. 1994. Thyrotropin-releasing hormone (TRH)-induced facilitation of spinal neurotransmission: A role for NMDA receptors. *Neuropharmacology* **33**, 15–121.

Chizh, B. A. and Headley, P. M. 1996. Thyrotropin-releasing hormone facilitates spinal nociceptive responses by potentiating NMDA receptor-mediated transmission. *Eur. J. Pharmacol.* **300**, 183–189.

Ch'ng, J. L. C., Christofides, N. D., Anand, P., Gibson, S. J., Allen, Y. S., Su, H. C., Tatemoto, K., et al. 1985. Distribution of galanin immunoreactivity in the central nervous system and the responses of galanin-containing neuronal pathways to injury. *Neuroscience* **16**, 343–354.

Cho, H. J. and Basbaum, A. I. 1988. Increased staining of immunoreactive dynorphin cell bodies in the deafferented spinal cord of the rat. *Neurosci. Lett.* **84**, 125–130.

Cho, H. J. and Basbaum, A. I. 1989. Ultrastructural analysis of dynorphin B-immunoreactive cells and terminals in the superficial dorsal horn of the deafferented spinal cord of the rat. *J. Comp. Neurol.* **281**, 193–205.

Cho, J. H., Kim, D. S., Lee, N. H., Kim, J. K., Lee, K. M., Hans, K. S., Kang, Y. N., and Kim, K. J. 1997. Changes in the alpha 2-adrenergic receptor subtypes gene expression in rat dorsal root ganglion in an experimental model of neuropathic pain. *NeuroReport* **8**, 3119–3122.

Choca, J. I., Green, R. D., and Proudfit, H. K. 1988. Adenosine A1 and A2 receptors of the substantia gelatinosa are located predominantly on intrinsic neurons: An autoradiographic study. *J. Pharmacol. Exp. Ther.* **247**, 757–764.

Choi, D. W. 1994. Glutamate receptors and the induction of excitotoxic neuronal death. In F. Bloom (ed.), *Progress in Brain Research* (pp. 47–51). Elsevier Science, Amsterdam.

Choi, Y., Yoon, Y. W., Na, H. S., Kim, S. H., and Chung, J. M. 1994. Behavioral signs of ongoing pain and cold allodynia in a rat model of neuropathic pain. *Pain* **59**, 369–376.

Chong, M. S., Reynolds, M. L., Irwin, N., Coggeshall, R. E., Emson, P. C., Benowitz, L. I., and Woolf, C. J. 1994. GAP-43 expression in primary sensory neurons following central axotomy. *J. Neurosci.* **14**, 4375–4384.

Chou, D. K. H., Dodd, J., Jessell, T. M., Costello, C. E., and Jungalwala, F. B. 1988. Identification of α-galactose (α-fucose)-asialo-G_{ml} glycolipid expressed by subsets of rat dorsal root ganglion neurons. *J. Biol. Chem.* **264**, 3409–3415.

Chouchkov, C. 1978. Cutaneous receptors. *Adv. Anat. Embryol. Cell Biol.* **54**, 1–62.

Chouchkov, H. N. 1971. Ultrastructure of Pacinian corpuscles in men and cats. *Z. Mikrosk. Anat. Forsch.* **83**, 17–32.

Christensen, B. N. and Perl, E. R. 1970. Spinal neurons specifically excited by noxious or thermal stimuli: Marginal zone of the dorsal horn. *J. Neurophysiol.* **33**, 293–307.

Christensen, M. D. and Hulsebosch, C. E. 1997a. Chronic central pain after spinal cord injury. *J. Neurotrauma* **14**, 517–537.

Christensen, M. D. and Hulsebosch, C. E. 1997b. Spinal cord injury and anti-NGF treatment result in changes in CGRP density and distribution in the dorsal horn in the rat. *Exp. Neurol.* **147**, 463–475.

Christensen, M. D., Everhart, A. W., Pickelman, J. T., and Hulsebosch, C. E. 1996. Mechanical and thermal allodynia in chronic central pain following spinal cord injury. *Pain* **68**, 97–107.

Chronwall, B. M., Olschowka, J. A., and O'Donohue, T. L. 1984. Histochemical localization of FMRFamide-like immunoreactivity in the rat brain. *Peptides* **5**, 569–584.

Chronwall, B. M., DiMaggio, D. A., Massari, V. J., Pickel, V. M., Ruggiero, D. A., and O'Donohue, T. L. 1985. The anatomy of neuropeptide-Y-containing neurons in rat brain. *Neuroscience* **15**, 1159–1181.

Chung, J. M., Kenshalo, D. R., Gerhart, K. D., and Willis, W. D. 1979. Excitation of primate spinothalamic neurons by cutaneous C-fiber volleys. *J. Neurophysiol.* **42**, 1354–1369.

Chung, J. M., Lee, K. H., Endo, K., and Coggeshall, R. E. 1983b. Activation of central neurons by ventral root afferents. *Science* **222**, 934–935.

Chung, J. M., Surmeier, D. J., Lee, K. H., Sorkin, L. S., Honda, C. N., Tsong, Y., and Willis, W. D. 1986a. Classification of primate spinothalamic and somatosensory thalamic neurons based on cluster analysis. *J. Neurophysiol.* **56**, 308–327.

Chung, J. M., Kim, J., and Shin, H. K. 1986b. Blood pressure response evoked by ventral root afferent fibres in the cat. *J. Physiol.* **370**, 255–265.

Chung, K. and Chung, J. M. 2001. Sympathetic sprouting in the dorsal root ganglion after spinal nerve ligation: Evidence of regenerative collateral sprouting. *Brain Res.*, **895**, 204–212.

Chung, K. and Coggeshall, R. E. 1982. Quantitation of propriospinal fibers in the tract of Lissauer of the rat. *J. Comp. Neurol.* **211**, 418–426.

Chung, K. and Coggeshall, R. E. 1983. Numbers of axons in lateral and ventral funiculi of rat sacral spinal cord. *J. Comp. Neurol.* **214**, 72–78.

Chung, K. and Coggeshall, R. E. 1984. The ratio of dorsal root ganglion cells to dorsal root axons in sacral segments of the cat. *J. Comp. Neurol.* **225**, 24–30.

Chung, K., Langford, L. A., Applebaum, A. E., and Coggeshall, R. E. 1979. Primary afferent fibers in the tract of Lissauer in the rat. *J. Comp. Neurol.* **184**, 587–598.

Chung, K., Kevetter, G. A., Willis, W. D., and Coggeshall, R. E. 1984. An estimate of the ratio of propriospinal to long tract neurons in the sacral spinal cord of the rat. *Neurosci. Lett.* **44**, 173–177.

Chung, K., Sharma, J., and Coggeshall, R. E. 1985. Numbers of myelinated and unmyelinated axons in the dorsal, lateral, and ventral funiculi of the white matter of the S2 segment of cat spinal cord. *J. Comp. Neurol.* **234**, 117–121.

Chung, K., Langford, L. A., and Coggeshall, R. E. 1987. Primary afferent and propriospinal fibers in the rat dorsal and dorsolateral funiculi. *J. Comp. Neurol.* **263**, 68–75.

Chung, K., Lee, W. T., and Carlton, S. M. 1988. The effects of dorsal rhizotomy and spinal cord isolation on calcitonin gene-related peptide containing terminals in the rat lumbar dorsal horn. *Neurosci. Lett.* **90**, 27–32.

Chung, K., McNeill, D. L., Hulsebosch, C. E., and Coggeshall, R. E. 1989a. Changes in dorsal horn synaptic disc numbers following unilateral dorsal rhizotomy. *J. Comp. Neurol.* **283**, 568–577.

Chung, K., Lee, W. T., and Westlund, K. N. 1989b. Adrenergic fibers in the spinal cord of the monkey: Light- and electron-microscopic study. *J. Auton. Ner. Sys.* **28**, 203–210.

Chung, K., Lee, W. T., and Park, M. J. 1993a. Spinal projection of pelvic visceral afferents of the rat: A calcitonin gene-related peptide (CGRP) immunohistochemical study. *J. Comp. Neurol.* **337**, 63–69.

Chung, K., Park, M. J., Sheen, K., and Chung, J. M. 1993b. Neonatal sciatic nerve lesion triggers the sprouting of fibers in the contralateral ventral root of the rat. *Brain Res.* **632**, 80–85.

Chung, K., Kim, H. J., Na, H. S., Park, M. J., Chung, J. M. 1993c. Abnormalities of sympathetic innervation of the area of injured peripheral nerve in a rat model of neuropathic pain. *Neurosci. Lett.* **162**, 85–88.

Chung, K., Lee, B. H., Yoon, Y. W., and Chung, J. M. 1996. Sympathetic sprouting in the dorsal root ganglia of the injured peripheral nerve in a rat neuropathic pain model. *J. Comp. Neurol.* **376**, 241–252.

Chung, K. J., Yoon, Y. W., and Chung, J. M. 1997. Sprouting sympathetic fibers form synaptic varicosities in the dorsal root ganglion of the rat with neuropathic injury. *Brain Res.* **751**, 275–280.

Cincotta, M., Beart, P. M., Summers, R. J., and Lodge, D. 1989. Bidirectional transport of NMDA receptor and ionophore in the vagus nerve. *Eur. J. Pharmacol.* **160**, 167–171.

Clabarra, A. M., Sullivan, J. S., Gahn, L. G., Pecht, G., Heinemann, S., and Sevarino, K. A. 1995. Cloning and characterization of x-1: A developmentally regulated member of a novel class of the ionotropic glutamate family. *J. Neurosci.* **15**, 6498–6508.

Clark, F. J. 1975. Information signaled by sensory fibers in medial articular nerve. *J. Neurophysiol.* **38**, 1464–1472.

Clark, F. J. and Burgess, P. R. 1975. Slowly adapting receptors in cat knee joint: Can they signal joint angle? *J. Neurophysiol.* **38**, 1448–1463.

Clark, S. L. 1926. Nissl granules of primary afferent neurones. *J. Comp. Neurol.* **41**, 423–451.

Clark, S. L. 1931. Innervation of the pia mater of the spinal cord and medulla. *J. Comp. Neurol.* **53**, 129–145.

Clarke, J. L. 1859. Further researches on the grey substance of the spinal cord. *Phil. Trans. R. Soc. Lond. B.* **149**, 437–467.

Clements, J. R. and Beitz, A. J. 1987. A quantitative light microscopic analysis and ultrastructural description of cholecystokinin-like immunoreactivity in the spinal trigeminal nucleus of the rat. *Neuroscience* **20**, 427–438.

Clifton, G. L., Coggeshall, R. E., Vance, W. H., and Willis, W. D. 1976. Receptive fields of unmyelinated ventral root afferent fibres in the cat. *J. Physiol.* **256**, 573–600.

Clowry, G. J., Fallah, Z., and Arnott, G. 1997. Developmental expression of parvalbumin by rat lower cervical spinal cord neurones and the effect of early lesions to the motor cortex. *Dev. Brain Res.* **102**, 197–208.

Clowry, G. J., Arnott, G. A., Clement-Jones, M., Fallah, Z., Gould, S., and Wright, C. 2000. Changing pattern of expression of parvalbumin immunoreactivity during human fetal spinal cord development. *J. Comp. Neurol.* **423**, 727–735.

Cockayne, D. A., Hamilton, S. G., Zhu, Q. M., Dunn, P. M., Zhong, Y., Novakovic, S., Malmberg, A. B., Cain, G., Berson, A., Kassotakis, L., Hedley, L., Lachnit, W. G., Burnstock, G., McMahon, S. B., and Ford, A. P. 2000. Urinary bladder hyporeflexia and reduced pain-related behaviour in P2X3-deficient mice. *Nature* **407**, 1011–1015.

Coderre, T. J. and Melzack, R. 1992a. The contribution of excitatory amino acids to central sensitization and persistent nociception after formalin-induced tissue injury. *J. Neurosci.* **12**, 3665–3670.

Coderre, T. J. and Melzack, R. 1992b. The role of NMDA receptor-operated calcium channels in persistent nociception after formalin-induced tissue injury. *J. Neurosci.* **12**, 3671–3675.

Coderre, T. J. and Yashpal, K. 1994. Intracellular messengers contributing to persistent nociception and hyperalgesia induced by L-glutamate and substance P in the rat formalin pain model. *Eur. J. Neurosci.* **6**, 1328–1334.

Coderre, T. J., Grimes, R. W., and Melzack, R. 1986. Deafferentation and chronic pain in animals: An evaluation of evidence suggesting autotomy is related to pain. *Pain* **26**, 61–84.

Coderre, T. J., Vaccarino, A. L., and Melzack, R. 1990. Central nervous system plasticity in the tonic pain response to subcutaneous formalin injection. *Brain Res.* **535**, 155–158.

Coderre, T. J., Fundytus, M. E., McKenna, J. E., Dalal, S., and Melzack, R. 1993. The formalin test: A validation of the weighted-scores method of behavioural pain rating. *Pain* **54**, 43–50.

Coffield, J. A., Miletić, V., Zimmermann, E., Hoffert, M. J., and Brooks, B. R. 1986. Demonstration of thyrotropin-releasing hormone immunoreactivity in neurons of the mouse spinal dorsal horn. *J. Neurosci.* **6**, 1194–1197.

Coggeshall, R. E. 1980. Law of separation of function of the spinal roots. *Physiol. Rev.* **60**, 716–755.

Coggeshall, R. E. and Carlton, S. M. 1997. Receptor localization in the mammalian dorsal horn and primary afferent neurons. *Brain Res. Rev.* **24**, 28–66.

Coggeshall, R. E. and Carlton, S. M. 1998. Ultrastructural analysis of NMDA, AMPA, and kainate receptors on unmyelinated and myelinated axons in the periphery. *J. Comp. Neurol.* **391**, 78–86.

Coggeshall, R. E. and Ito, H. 1977. Sensory fibres in ventral roots L7 and Si in the cat. *J. Physiol.* **267**, 215–235.

Coggeshall, R. E., Coulter, J. D., and Willis, W. D. 1973. Unmyelinated fibers in the ventral root. *Brain Res.* **57**, 229–233.

Coggeshall, R. E., Coulter, J. D., and Willis, W. D., Jr. 1974. Unmyelinated axons in the ventral roots of the cat lumbosacral enlargement. *J. Comp. Neurol.* **153**, 39–58.

Coggeshall, R. E., Chung, K., Chung, J. M., and Langford, L. A. 1981. Primary afferent axons in the tract of Lissauer in the monkey. *J. Comp. Neurol.* **196**, 431–442.

Coggeshall, R. E., Hong, K. A. P., Langford, L. A., Schaible, H. G., and Schmidt, R. F. 1983. Discharge characteristics of fine medial articular afferents at rest and during passive movements of inflamed knee joints. *Brain Res.* **272**, 185–188.

Coggeshall, R. E., Zhou, S., and Carlton, S. M. 1997. Opioid receptors on peripheral sensory axons. *Brain Res.* **764**, 126–132.

Coggeshall, R. E., Lekan, H. A., White, F. A., and Woolf, C. J. 2001. A-fiber sensory input induces neuronal cell death in the dorsal horn of the adult rat spinal cord. *J. Comp. Neurol.* **435**, 276–282.

Cohen, L. A. 1955. Activity of knee joint proprioceptors recorded from the posterior articular nerve. *Yale J. Biol. Med.* **28**, 225–232.

Cohen, R. H. and Perl, E. R. 1990. Contributions of arachidonic acid derivatives and substance P to the sensitization of cutaneous nociceptors. *J. Neurophysiol.* **64**, 457–464.

Cohen, R. H. and Vierck, C. J. 1993. Population estimates for responses of cutaneous mechanoreceptors to a vertically indenting probe on the glabrous skin of monkeys. *Exp. Brain Res.* **94**, 105–119.

Coimbra, A. and Lima, D. 1989. Projections and neurochemical specificity of the different morphological types of marginal cells. In F. Cervero, G. J. Bennett, and P. M. Headley (eds.), *Processing of Sensory Information in the Superficial Dorsal Horn of the Spinal Cord* (pp. 199–213). Plenum Press, New York.

Coimbra, A., Magalhaes, M. M., and Sodre-Borges, B. P. 1970. Ultrastructural localization of acid phosphatase in synaptic terminals of the rat substantia gelatinosa Rolandi. *Brain Res.* **22**, 142–146.

Coimbra, A., Sodre-Borges, B. P., and Magalhaes, M. M. 1974. The substantia gelatinosa Rolandi of the rat. Fine structure, cytochemistry (acid phosphatase) and changes after dorsal root section. *J. Neurocytol.* **3**, 199–217.

Coimbra, A., Ribeiro-da-Silva, A., and Pignatelli, D. 1984. Effects of dorsal rhizotomy on the several types of primary afferent terminals in laminae I–III of the rat spinal cord. *Anat. Embryol.* **170**, 279–287.

Coleman, G. T., Bahramali, H., Zhang, H. Q., and Rowe, M. J. 2001. Characterization of tactile afferent fibers in the hand of the marmoset monkey. *J. Neurophysiol.* **85**, 1793–1804.

Collins, G. G. S. 1974. The spontaneous and electrically evoked release of ^{3}H-γ-aminobutyric acid (^{3}H-GABA) from spinal cord. *Br. J. Pharmacol.* **47**, 641P.

Collins, J. G. 1983. Neuronal activity recorded from the spinal cord dorsal horn of physiologically intact, awake, drug-free, restrained cats: A preliminary report. *Brain Res.* **322**, 301–304.

Collins, J. G. 1985. A technique for chronic extracellular recording of neuronal activity in the dorsal horn of the lumbar spinal cord in drug-free, physiologically intact cats. *J. Neurosci. Meth.* **12**, 277–287.

Collins, J. G. 1987. A descriptive study of spinal dorsal horn neurons in the physiologically intact, awake, drug-free cat. *Brain Res.* **416**, 34–42.

Collins, J. G. and Ren, K. 1987. WDR response profiles of spinal dorsal horn neurons may be unmasked by barbiturate anesthesia. *Pain* **28**, 369–378.

Collins, J. G., Ren, K., Saito, Y., Iwasaki, H., and Tang, J. 1990. Plasticity of some spinal dorsal horn neurons as revealed by pentobarbital-induced disinhibition. *Brain Res.* **525**, 189–197.

Collo, G., North, R. A., Kawashima, E., Merlo-Pich, E., Neidhart, S., Surprenant, A., and Buell, G. 1996. Cloning of $P2X_5$ and $P2X_6$ receptors and the distribution and properties of an extended family of ATP-gated ion channels. *J. Neurosci.* **16**, 2495–2507.

Colmant, H. J. 1959. Aktivitatsschwankungen der sauren Phosphatase im Ruvkenmark und den Spinalganglien der Ratte nach Durchschneidung des Nervus ischiadicus. *Arch. f. Psych. u. Zeitsch. f. d. ges. Neurologie.* **199**, 60–71.

Colpaert, F. C. 1987. Evidence that adjuvant arthritis in the rat is associated with chronic pain. *Pain* **28**, 210–222.

Commissiong, J. W. 1983. The development of catecholaminergic nerves in the spinal cord of rat: II. Regional development. *Brain Res.* **313**, 75–92.

Conn, P. J. and Pin, J. P. 1997. Pharmacology and functions of metabotropic glutamate receptors. *Annu. Rev. Pharmacol. Toxicol.* **37**, 205–237.

Conradi, S. 1969. On motoneuron synaptology in adult cats. *Acta Physiol. Scand. Suppl.* **332**, 1–115.

Conrath, M., Taquet, H., Pohl, M., and Carayon, A. 1989. Immunocytochemical evidence for calcitonin gene-related peptide-like neurons in the dorsal horn and lateral spinal nucleus of the rat cervical spinal cord. *J. Chem. Neuroanat.* **2**, 335–347.

Conrath-Verrier, M., Dietl, M., Arluison, M., Cesselin, F., Bourgoin, S., and Hamon, M. 1983. Localization of met-enkephalin-like immunoreactivity within pain-related nuclei of cervical spinal cord, brainstem and midbrain in the cat. *Brain Res. Bull.* **11**, 587–604.

Conrath-Verrier, M., Dietl, M., and Tramu, G. 1984. Cholecystokinin-like immunoreactivity in the dorsal horn of the spinal cord of the rat: A light and electron micorscopic study. *Neuroscience* **13**, 871–885.

Cook, A. J., Woolf, C. J., Wall, P. D., and McMahon, S. B. 1987. Dynamic receptive field plasticity in rat spinal cord dorsal horn following C-primary afferent input. *Nature* **325**, 151–153.

Cook, S. P. and McCleskey, E. W. 2002. Cell damage excites nociceptors through release of cytosolic ATP. *Pain* **95**, 41–47.

Cook, S. P., Vulchanova, L., Hargreaves, K. M., Elde, R., and McCleskey, E. W. 1997. Distinct ATP receptors on pain-sensing and stretch-sensing neurons. *Nature* **387**, 505–508.

Coombs, J. S., Curtis, D. R., and Landgren, S. 1956. Spinal cord potentials generated by impulses in muscle and cutaneous afferent fibres. *J. Neurophysiol.* **19**, 452–467.

Corey, D. P. and García-Añoveros, J. 1996. Mechanosensation and the DEG/EnaC ion channels. *Science* **273**, 323–324.

Cornea-Hebert, V., Riad, M., Wu, C., Singh, S. K., and Descarries, L. 1999. Cellular and subcellular distribution of the serotonin 5-HT2A receptor in the central nervous system of adult rat. *J. Comp. Neurol.* **409**, 187–209.

Corness, J., Shi, T. J., Xu, Z. Q., Brulet, P., and Hokfelt, T. 1996. Influence of leukemia inhibitory factor on galanin/GMAP and neuropeptide Y expression in mouse primary sensory neurons after axotomy. *Exp. Brain Res.* **112**, 79–88.

Corness, J., Stevens, B., Fields, R. D., and Hokfelt, T. 1998. NGF and LIF both regulate galanin gene expression in primary DRG cultures. *NeuroReport* **9**, 1533–1536.

Cortes, R., Arvidsson, U., Schalling, M., Ceccatelli, S., and Hökfelt, T. 1990. In situ hybridization studies on mRNAs for cholycystokinin, calcitonin gene-related peptide and choline acetyl transferase in the lower brian stem, spinal cord and dorsal root ganglia of rat and guinea pig with special reference to motoneurons. *J. Chem. Neuroanat.* **3**, 467–485.

Costa, J. J. L., Averill, S., Saavedra, J. P., and Priestley, J. V. 1994. Serotonin innervation of enkephalin containing neurons in the rat spinal trigeminal nucleus. *Neurosci. Lett.* **168**, 167–171.

Cotman, C. W., Monaghan, D. T., Ottersen, O. P., and Storm-Mathisen, J. 1987. Anatomical organization of excitatory amino acid receptors and their pathways. *Trends Neurosci.* **10**, 273–280.

Coupland, R. E. and Holmes, R. L. 1957. The use of cholinesterase techniques for the demonstration of peripheral nerve structures. *J. Microscopical Sci.* **98**, 327–330.

Courtney, K. R. and Fetz, E. E. 1973. Unit responses recorded from cervical spinal cord of awake monkey. *Brain Res.* **53**, 445–450.

Covenas, R., DeLeon, M., Chadi, G., Cintra, A., Gustafsson, J.-A., Narvaez, J. A., and Fuxe, K. 1994. Adrenalectomy increases the number of substance P and somatostatin immunoreactive nerve cells in the rat lumbar dorsal root ganglia. *Brain Res.* **640**, 352–356.

Coward, K., Aitken, A., Powell, A., Plumpton, C., Birch, R., Tate, S., Bountra, C., and Anand, P. 2001a. Plasticity of TTX-sensitive sodium channels PN1 and brain III in injured human nerves. *NeuroReport* **12**, 495–500.

Coward, K., Mosahebi, A., Plumpton, C., Facer, P., Birch, R., Tate, S., Bountra, C., Terenghi, G., and Anand, P. 2001b. Immunolocalisation of sodium channel NaG in the intact and injured human peripheral nervous system. *J. Anat.* **198**, 175–180.

Coward, K., Jowett, A., Plumpton, C., Powell, A., Birch, R., Tate, S., Bountra, C., and Anand, P. 2001c. Sodium channel beta1 and beta2 subunits parallel SNS/PN3 alpha-subunit changes in injured human sensory neurons. *NeuroReport* **12**, 483–488.

Cox, B. M., Mollineaux, C. J., Jacobs, T. P., Rosenberger, J. G., and Faden, A. I. 1985. Effects of traumatic injury on dynorphin immunoreactivity in spinal cord. *Neuropeptides* **5**, 571–574.

Craig, A. D. and Mense, S. 1983. The distribution of afferent fibers from the gastrocnemius-soleus muscle in the dorsal horn of the cat, as revealed by the transport of horseradish peroxidase. *Neurosci. Lett.* **41**, 233–238.

Craig, A. D., Heppelmann, B., and Schaible, H.-G. 1988. Projection of the medal and posterior articular nerves of the cat's knee to the spinal cord. *J. Comp. Neurol.* **276**, 279–288.

Craig, A. D., Krout, K., and Andrew, D. 2001. Quantitative response characteristics of thermoreceptive and nociceptive lamina I spinothalamic neurons in the cat. *J. Neurophysiol.* **86**, 1459–1480.

Craig, A. D., Zhang, E. T., and Blomqvist, A. 2002. Association of spinothalamic lamina I neurons and their ascending axons with calbindin-immunoreactivity in monkey and human. *Pain* **97**, 105–115.

Cranefield, P. F. 1974. *The Way in and the Way Out.* Futura: Futura Publishing Co., Mount Kisco, New York.

Creutzfelt, O. D. 1995. *Cortex Cerebri: Performance, Structural and Functional Organization of the Cortex.* Oxford University Press, Oxford.

Crockett, D. P., Smith, W. K., Proshansky, E., Kauer, J. S., Stewart, W. B., Woodward, D. J., Schusselberg, D. S., and Egger, M. D. 1989. Computer-assisted three-dimensional reconstruction of [14C]-2-deoxy-D-glucose metabolism in cat lumbosacral spinal cord following cutaneous stimulation of the hindfoot. *J. Comp. Neurol.* **288**, 326–338.

Cronk, K. M., Wilkinson, G. A., Grimes, R., Wheeler, E. F., Jhaveri, S., Fundin, B. T., Silos-Santiago, I., Tessarollo, L., Reichardt, L. F., and Rice, F. L. 2002. Diverse dependencies of developing Merkel innervation on the trkA and both full-length and truncated isoforms of trkC. *Development* **129**, 3739–3750.

Croul, S., Sverstiuk, A., Radzievsky, A., and Murray, M. 1995. Modulation of neurotransmitter receptors following unilateral L1–S2 deafferentation: NK1, NK3 NMDA and 5HT1a receptor binding autoradiography. *J. Comp. Neurol.* **361**, 633–644.

Croul, S., Radzievsky, A., Sverstiuk, A., and Murray, M. 1998. NK1, NMDA, 5HT1a, and 5HT2 receptor binding sites in the rat lumbar spinal cord: Modulation following sciatic nerve crush. *Exp. Neurol.* **154**, 66–79.

Crowley, C., Spencer, S. D., Nishmura, M. C., Chen, K. S., Pitts-Meek, S., Armanini, M. P., Ling, L. H., McMahon, S. B., Shelton, D. I., Levinson, A. D., and Phillips, H. S. 1994. Mice lacking nerve growth factor display perinatal loss of sensory and sympathetic neurons yet develop basal forebrain cholinergic neurons. *Cell* **76**, 1001–1011.

Croze, S., Duclaux, R., and Kenshalo, D. R. 1976. The thermal sensitivity of the polymodal nociceptors in the monkey. *J. Physiol.* **263**, 539–562.

Cruz, F., Lima, D., and Coimbra, A. 1987. Several morphological types of terminal arborizations of primary afferents in laminae I–II of the rat spinal cord, as shown after HRP labeling and Golgi impregnation. *J. Comp. Neurol.* **261**, 221–236.

Cruz, L. and Basbaum, A. I. 1985. Multiple opioid peptides and the modualtion of pain: Immunohistochemical analysis of dynorphin and enkephalin in the trigeminal nucleus caudalis and spinal cord of the cat. *J. Comp. Neurol.* **240**, 331–348.

Csillik, B. and Knyihar-Csillik, E. 1986. The Protean Gate, structure and plasticity of the primary nociceptive analyser. akademai Kaido, Budapest, pp. 1–294.

Csillik, B., Schwab, M. E., and Thoenen, H. 1985. Transganglionic regulation of central terminals of dorsal root ganglion cells by nerve growth factor (NGF). *Brain Res.* **331**, 11–15.

Csillik, B., Knyihar-Csillik, E., and Bezzegh, A. 1986. Comparative electron histochemistry of thiamine monophosphatase and substance P in the upper dorsal horn. *Acta Histochem.* **80**, 125–134.

Cuello, A. C., Jessell, T. M., Kanazawa, I., and Iversen, L. L. 1977. Sustance P: localization in synaptic vesicles in rat central nervous system. *J. Neurochem.* **29**, 747–751.

Cuello, A. C., Del Fiacco, M., and Paxinos, G. 1978. The central and peripheral ends of the substance P-containing sensory neurons in the rat trigeminal system. *Brain Res.* **152**, 499–509.

Cuello, A. C., Ribeiro-da-Silva, A., Ma, W., De Konink, Y., and Henry, J. L. 1993. Organization of substance P primary sensory neurons: Ultrastructural and physiological correlates. *Regul. Pept.* **46**, 155–164.

Cui, M. and Nicol, G. D. 1995. Cyclic AMP mediates the prostaglandin E2-induced potentiation of bradykinin excitation in rat sensory neurons. *Neuroscience* **66**, 459–466.

Cumberbatch, M. J., Chizh, B. A., and Headley, P. M. 1995. Modulation of excitatory amino acid responses by tachyknins and selective tachykinin receptor agonists in the rat spinal cord. *Br. J. Pharmacol.* **115**, 1005–1012.

Cummings, J. F., de Lahunta, A., Simpson, S. T., and McDonald, J. M. 1984. Reduced substance P-like immunoreactivity in hereditary sensory neuropathy of pointer dogs. *Acta Neuropathol. (Berl)* **63**, 33–40.

Cummins, T. R. and Waxman, S. G. 1997. Downregulation of tetrodotoxin-resistant sodium currents and upregulation of a rapidly repriming tetrodotoxin-sensitive sodium current in small spinal sensory neurons after nerve injury. *J. Neurosci.* **15**, 3503–3514.

Curtis, D. R. 1962. The depression of spinal inhibition by electrophoretically administered strychnine. *Int. J. Neuropharmacol.* **1**, 239–250.

Curtis, D. R. and Crawford, J. M. 1969. Central synaptic transmission-microelectrophoretic studies. *Ann. Rev. Pharmacol.* **9**, 209–240.

Curtis, D. R. and Lodge, D. F. 1982. The depolarization of feline ventral horn group Ia spinal afferent terminations by GABA. *Exp. Brain Res.* **46**, 215–233.

Curtis, D. R. and Ryall, R. W. 1966. Pharmacological studies upon spinal presynaptic fibres. *Exp. Brain Res.* **1**, 195–204.

Curtis, D. R. and Watkins, J. C. 1960. The excitation and depression of spinal neurones by structurally related amino acids. *J. Neurochem.* **6**, 117–141.

Curtis, D. R. and Watkins, J. C. 1965. The pharmacology of amino acids related to γ-aminobuyric acid. *Pharmacol. Rev.* **17**, 347–391.

Curtis, D. R., Phillips, J. W., and Watkins, J. C. 1959a. Chemical excitation of spinal neurones. *Nature* **183**, 611–612.

Curtis, D. R., Phillis, J. W., and Watkins, J. C. 1959b. The depression of spinal neurones by γ-amino-n-butyric acid and β-alanine. *J. Physiol.* **146**, 185–203.

Curtis, D. R., Phillis, J. W., and Watkins, J. C. 1960. The chemical excitation of spinal neurones by certain acidic amino acids. *J. Physiol.* **150**, 656–682.

Curtis, D. R., Hösli, L., and Johnston, G. A. R. 1967a. Inhibition of spinal neurons by glycine. *Nature* **215**, 1502–1503.

Curtis, D. R., Hösli, L., Johnston, G. A. R., and Johnston, I. H. 1967b. Glycine and spinal inhibition. *Brain Res.* **5**, 112–114.

Curtis, D. R., Hösli, L., and Johnston, G. A. R. 1968. A pharmacological study of the depression of spinal neurones by glycine and related amino acids. *Exp. Brain Res.* **6**, 1–18.

Curtis, D. R., Duggan, A. W., and Johnston, G. A. R. 1969. Glycine, strychnine, picrotoxin and spinal inhibition. *Brain Res.* **14**, 759–762.

Curtis, D. R., Duggan, A. W., Felix, D., and Johnston, G. A. R. 1971a. Bicuculline, an antagonist of GABA and synaptic inhibition in the spinal cord of the cat. *Brain Res.* **32**, 69–96.

Curtis, D. R., Duggan, A. W., and Johnston, G. A. R. 1971b. The specificity of strychnine as a glycine antagonist in the mammalian spinal cord. *Exp. Brain Res.* **12**, 547–565.

Curtis, D. R., Lodge, D., and Brand, S. J. 1977. GABA and spinal afferent terminal excitability in the cat. *Brain Res.* **130**, 360–363.

Curtis, D. R., Lodge, D., Bornstein, J. C., Peet, M. J., and Leah, J. D. 1982. The dual effects of GABA and related amino acids on the electrical threshold of ventral horn group Ia afferent terminations in the cat. *Exp. Brain Res.* **48**, 387–400.

Czaja, K. 2000. Distribution of primary afferent neurons innervating the porcine oviduct and their immunohistochemical characterization. *Cells Tissues Organs* **166**, 275–282.

Czarkowska, J., Jankowska, E., and Sybirska, E. 1981. Common interneurones in reflex pathways from group Ia and Ib afferents of knee flexors and extensors in the cat. *J. Physiol.* **310**, 367–380.

Czéh, G., Kříž, N., and Syková, E. 1981. Extracellular potassium accumulation in the frog spinal cord induced by stimulation of the skin and ventrolateral columns. *J. Physiol.* **320**, 57–72.

Czlonkowski, A., Costa, T., Przewlocki, R., Pasi, A. S., and Herz, A. 1983. Opiate receptor binding sites in human spinal cord. *Brain Res.* **267**, 392–396.

Dado, R. J., Law, P. Y., Loh, H. H., and Elde, R. 1993. Immunofluorescent identification of a delta (δ)-opioid receptor on primary afferent nerve terminals. *NeuroReport* **5**, 341–344.

Dahlstrom, A. and Fuxe, K. 1965. Evidence for the existence of monoamine neurons in the central nervous system: II. Experimentally induced changes in the intraneuronal amine levels of bulbospinal neuron systems. *Acta Physiol. Scand.* **64**, 7–36.

Dallenbach, K. M. 1927. The temperature spots and end-organs. *Am. J. Psychol.* **39**, 402–427.

Dalsgaard, C. J., Hökfelt, T., Johansson, O., and Elde, R. 1981. Somatostatin immunoreactive cell bodies in the dorsal horn and the parasympathetic intermediolateral nucleus of the rat spinal cord. *Neurosci. Lett.* **27**, 335–339.

Dalsgaard, C. J., Risling, M., and Cuello, C. 1982a. Immunohistochemical localization of substance P in the lumbosacral spinal pia mater and ventral roots of the rat. *Brain Res.* **246**, 168–171.

Dalsgaard, C. J., Vincent, S. R., Hökfelt, T., Lundburg, J. M., Dahlstrom, A., Schultzberg, M., Dockray, G. J., and Cuello, A. C. 1982b. Coexistence of cholecystokinin and substance-P like peptides in neurons of the dorsal root ganglia of the rat. *Neurosci. Lett.* **33**, 159–163.

Dalsgaard, C. J., Ygge, J., Vincent, S., Ohrling, M., Dockray, G., and Elde, R. 1984. Peripheral projections and neuropeptide coexistence in a subpopulation of fluoride-resistant acid phosphatase reactive spinal primary sensory neurons. *Neurosci. Lett.* **51**, 139–144.

Dalsgaard, C. J., Haegerstrand, A., Theodorsson-Norheim, E., Brodin, E., and Hökfelt, T. 1985. Neurokinin A-like immunoreactivity in rat primary sensory neurons: Coexistence with substance P. *Histochemistry* **83**, 37–39.

Danks, J. A., Rothman, R. B., Cascieri, M. A., Chicchi, G. G., Liang, T., and Herkenham, M. 1986. A comparative autoradiographic study of the distributions of substance P and eledoisin binding sites in rat brain. *Brain Res.* **385**, 273–281.

Darian-Smith, I. 1984. The sense of touch: Performance and peripheral neural processes. In J. M. Brookhart and V. B. Mountcastle (eds.), *Handbook of Physiology, Section 1: The Nervous System, Vol. III, Sensory Processes, Part 2* (pp. 739–788). Am. Physiol. Soc., Bethesda.

Darian-Smith, I. and Kenins, P. 1980. Innervation density of mechanoreceptive fibers supplying glabrous skin of the monkey's index finger. *J. Physiol.* **309**, 147–155.

Darian-Smith, I. and Oke, L. E. 1980. Peripheral neural representation of the spatial frequency of a grating moving across the monkey's finger pad. *J. Physiol.* **309**, 117–133.

Darian-Smith, I., Johnson, K. O., and Dykes, R. 1973. "Cold" fiber population innervating palmar and digital skin of the monkey: Response to cooling pulses. *J. Neurophysiol.* **36**, 325–346.

Darland, T., Heinricher, M. M., and Grandy, D. K. 1998. Orphanin FQ/nociceptin: A role in pain and analgesia, but so much more. *Trends Neurosci.* **21**, 215–221.

DaSilva, A. R. and Cuello, A. C. 1990. Choline acetyltransferase-immunoreactive profiles are presynaptic to primary sensory fibers in the rat superficial dorsal horn. *J. Comp. Neurol.* **295**, 370–384.

Date, Y., Mondal, M. S., Matsukura, S., and Nakazato, M. 2000. Distribution of orexin-A and orexin-B (hypocretins) in the rat spinal cord. *Neurosci. Lett.* **288**, 87–90.

Daval, G., Vergé, D., Basbaum, A., Bourgoin, S., and Hamon, M. 1987. Autoradiographic evidence of serotonin$_1$ binding sites on primary afferent fibres in the dorsal horn of the rat spinal cord. *Neurosci. Lett.* **83**, 71–76.

Davar, G., Hama, A., Deykin, A., Vos, B., Maciewicz, R. 1991. MK-801 blocks the development of thermal hyperalgesia in a rat model of experimental painful neuropathy. *Brain Res.* **553**, 327–330.

Davidoff, R. A. 1972. The effects of bicuculline on the isolated spinal cord of the frog. *Exp. Neurol.* **35**, 179–193.

Davidoff, R. A. and Hackman, J. C. 1980. Hyperpolarization of frog primary afferent fibres caused by activation of a sodium pump. *J. Physiol.* **302**, 297–309.

Davidoff, R. A., Graham, L. T., Shank, R. P., Werman, R., and Aprison, M. H. 1967. Changes in amino acid concentrations associated with loss of spinal interneurones. *J. Neurochem.* **14**, 1025–1031.

Davidoff, R. A., Aprison, M. H., and Werman, R. 1969. The effects of strychnine on the inhibition of interneurons by glycine and γ-aminobutyric acid. *Int. J. Neuropharmacol.* **8**, 191–194.

Davidoff, R. A., Grayson, V., and Adair, R. 1973. GABA-transaminase inhibitors and presynaptic inhibition in the amphibian spinal cord. *Am. J. Physiol.* **224**, 1230–1234.

Davidoff, R. A., Hackman, J. C., and Osorio, I. 1980. Amino acid antagonists do not block the depolarizing effects of potassium ions on frog primary afferents. *Neuroscience* **5**, 117–126.

Davidoff, R. A., Hackman, J. C., Holohean, A. M., Vega, J. L., and Zhang, D. X. 1988. Primary afferent activity, putative excitatory transmitters and extracellular potassium levels in frog spinal cord. *J. Physiol.* **397**, 291–306.

Davies, H. E. and Edgley, S. A. 1994. Inputs to group II-activated midlumbar interneurones from descending motor pathways in the cat. *J. Physiol.* **497**, 463–473.

Davies, J. and Dray, A. 1978. Pharmacological and electrophysiological studies of morphine and enkephalin on rat supraspinal neurones and cat spinal neurones. *Br. J. Pharmacol.* **63**, 87–96.

Davies, J. and Dray, A. 1980. Depression and facilitation of synaptic responses in cat dorsal horn by substance P administered into the substantia gelatinosa. *Life Sci.* **27**, 2037–2042.

Davies, J. and Watkins, J. C. 1983. Role of excitatory amino acid receptors in mono- and polysynaptic excitation in the cat spinal cord. *Exp. Brain Res.* **49**, 280–290.

Davis, B. M., Albers, K. M., Seroogy, K. B., and Katz, D. M. 1994. Overexpression of nerve growth factor in transgenic mice induced novel sympathetic projections to primary sensory neurons. *J. Comp. Neurol.* **349**, 464–474.

Davis, B. M., Goodness, T. P., Soria, A., and Albers, K. M. 1998. Over-expression of NGF in skin causes formation of novel sympathetic projections to trkA-positive sensory neurons. *NeuroReport* **9**, 1103–1107.

Davis, C. L., Naeem, S. B., Phagoo, S. B., Campbell, E. A., Urban, L., and Burgess, G. M. 1996. B1 bradykinin receptors and sensory neurones. *Br. J. Pharmacol.* **118**, 1469–1476.

Davis, J. B., Gray, J., Gunthorpe, M. J., Hatcher, J. P., Davey, P. T., Overend, P., Harries, M. H., Latcham, J., Clapham, C., Atkinson, K., Hughes, S. A., Rance, K., Grau, E., Harper, A. J., Pugh, P. L., Rogers, D. C., Bingham, S., Randall, A., and Sheardown, S. A. 2000. Vanilloid receptor-1 is essential for inflammatory thermal hyperalgesia. *Nature* **405**, 183–187.

Davis, K. D., Treede, R.-D., Raja, S. N., Meyer, R. A., and Campbell, J. N. 1991. Topical application of clonidine relieves hyperalgesia in patients with sympathetically maintained pain. *Pain* **47**, 309–317.

Davis, K. D., Meyer, R. A., and Campbell, J. N. 1993. Chemosensitivity and sensitization of nociceptive afferents that innervate the hairy skin of monkey. *J. Neurophysiol.* **69**, 1071–1081.

Dawe, G. S. and Maxwell, D. J. 1991. Contacts between group Ia muscle spindle afferent axons and GABA-immunoreactive perikarya in lamina VI of cat spinal cord. *Exp. Physiol.* **76**, 461–464.

Dawson, G. D., Merrill, E. G., and Wall, P. D. 1970. Dorsal root potentials produced by stimulation of fine afferents. *Science* **167**, 1385–1387.

Dearborn, G. V. N. 1932. A case of congenital pure analgesia. *J. Nerv. Ment. Dis.* **75**, 612–615.

De Biasi, S. and Rustioni, A. 1988. Glutamate and substance P coexist in primary afferent terminals in the superficial laminae of spinal cord. *Proc. Natl. Acad. Sci. USA* **85**, 7820–7824.

De Biasi, S. and Rustioni, A. 1990. Ultrastructural immunocytochemical localization of excitatory amino acids in the somatosensory system. *J. Histochem. Cytochem.* **38**, 1745–1754.

De Castro, F., Silos-Santiago, I., de Armentia, M. L., Barbacid, M., and Belmonte, C. 1998. Corneal innervation and sensitivity to noxious stimuli in trkA knockout mice. *Eur. J. Neurosci.* **10**, 146–152.

De Ceballos, M. L., Jenner, P., and Marsden, C. D. 1990. Increased [met]enkephalin and decreased substance P in spinal cord following thermal injury to one limb. *Neuroscience* **36**, 731–736.

Decker, M. W., Towle, A. C., Bissette, G., Mueller, R. A., Lauder, J. M., and Nemeroff, C. B. 1985. Bombesin-like immunoreactivity in the central nervous system of capsaicin-treated rats: A radioimmunoassay and immunohistochemical study. *Brain Res.* **342**, 1–8.

Decosterd, I., Ji, R. R., Abdi, S., Tate, S., and Woolf, C. J. 2002. The pattern of expression of the voltage-gated sodium channels Na(v)1.8 and Na(v)1.9 does not change in uninjured primary sensory neurons in experimental neuropathic pain models. *Pain* **96**, 269–277.

De Felipe, C., Herrero, J. F., O'Brien, J. A., Palmer, J. A., Doyle, C. A., Smith, A. J., Laird, J. M., Belmonte, C., Cervero, F., and Hunt, S. P. 1998. Altered nociception, analgesia and aggression in mice lacking the receptor for substance P. *Nature* **392**, 394–397.

De Giorgio, R., Barbara, G., Blennerhassett, P., Wang, L., Stanghellini, V., Corinaldesi, R., Collins, S. M., and Tougas, G. 2001. Intestinal inflammation and activation of sensory nerve pathways: A functional and morphological study in the nematode infected rat. *Gut* **49**, 822–827.

DeGroat, W. C., Lalley, P. M. and Saum, W. R. 1972. Depolarization of dorsal root ganglia in the cat by GABA and related amino acids: antagonism by picrotoxin and bicuculline. *Brain Res.* **44**, 273–277.

de Groat, W. C. 1986. Spinal cord projections and neuropeptides in visceral afferent neurons. *Prog. Brain Res.* **67**, 185–187.

de Groat, W. C. 1987. Neuropeptides in pelvic afferent pathways. *Experientia* **43**, 801–813.

de Groat, W. C., Nadelhaft, I., Morgan, C., and Schauble, T. 1978. Horseradish peroxidase tracing of visceral efferent and primary afferent pathways in the sacral cord of the cat using benzidine processing. *Neurosci. Lett.* **10**, 103–108.

de Groat, W. C., Kawatani, M., Hisamitsu, T., Lowe, I., Morgan, C., Roppolo, J., Booth, A. M., Nadelhaft, I., Kuo, D., and Thor, K. 1983. The role of neuropeptides in the sacral autonomic reflex pathways of the cat. *J. Auton. Ner. Sys.* **7**, 339–350.

DeGroot, J. F., Coggeshall, R. E., and Carlton, S. M. 1997. The reorganization of μ opioid receptors in the rat dorsal horn following peripheral axotomy. *Neurosci. Lett.* **233**, 113–116.

Dehen, H., Willer, J. C., Prier, S., Boureau, F., and Cambier, J. 1978. Congenital insensitivity to pain and the "morphine-like" analgesia system. *Pain* **5**, 351–358.

DeKoninck, Y. and Henry, J. L. 1989. Bombesin, neuromedin B and neuromedin C selectively depress superficial dorsal horn neurones in the cat spinal cord. *Brain Res.* **498**, 105–117.

De Koninck, Y. and Henry, J. L. 1991. Substance P-mediated slow excitatory postsynaptic potential elicited in dorsal horn neurons in vivo by noxious stimulation. *Proc. Natl. Acad. Sci. USA* **88**, 344–388.

DeKoninck, Y. and Henry, J. L. 1994. Prolonged $GABA_A$-mediated inhibition following single hair afferent input to single spinal dorsal horn neurones in cats. *J. Physiol.* **476**, 89–100.

De Koninck, Y., Ribeiro-da-Silva, A., Henry, J. L., and Cuello, A. C. 1992. Spinal neurons exhibiting a specific nociceptive response receive abundant substance P-containing synaptic contacts. *Proc. Natl. Acad. Sci. USA* **89**, 5073–5077.

Delander, G. E., Schött, E., Brodin, E., and Fredholm, B. B. 1997. Temporal changes in spinal cord expression of mRNA for substance P, dynorphin and enkephalin in a model of chronic pain. *Acta Physiol. Scand.* **161**, 509–516.

DeLanerolle, N. C. and LaMotte, C. C. 1983. Ultrastructure of chemically defined neuron systems in the dorsal horn of the monkey: I. Substance P immunoreactivity. *Brain Res.* **274**, 31–49.

Delay-Goyet, P., Kayser, V., Zajac, J. M., Guilbaud, G., Besson, J. M., and Roques, B. P. 1989. Lack of significant changes in μ, δ opioid binding sites and neutral endopeptidase EC 3.4.24.11 in the brain and spinal cord of arthritic rats. *Neuropharmacology* **28**, 1341–1348.

Delay-Goyet, P., Satoh, H., and Lundberg, J. M. 1992. Relative involvement of substance P and CGRP mechanisms in antidromic vasodilatation in the rat skin. *Acta Physiol. Scand.* **146**, 537–538.

DeLeo, J. A., Coombs, D. W., Willinbring, S., Colburn, R. W., Fromm, C., Wagner, R., and Twitchell, B. B. 1994. Characterization of a neuropathic pain model: Sciatic cryoneurolysis in the rat. *Pain* **56**, 9–16.

DeLeon, M., Covenas, R., Chadi, G., Narvaez, J. A., Fuxe, K., and Cintra, A. 1994. Subpopulations of primary sensory neurons show coexistence of neuropeptides and glucocorticoid receptors in the rat spinal and trigeminal ganglia. *Brain Res.* **636**, 338–342.

Del Fiacco, M. and Cuello, A. C. 1980. Substance P- and enkephalin-containing neurones in the rat trigeminal system. *Neuroscience* **5**, 803–815.

Deng, Y.-P., Li, X.-S., Zhang, S.-H., and Vacca-Galloway, L.-L. 1996. Changes in receptor levels for thyrotropin releasing hormone, serotonin, and substance P in cervical spinal cord of Wobbler mouse: A quantitative autoradiography study during early and late stages of the motoneuron disease. *Brain Res.* **725**, 49–60.

Deng, Y. S., Zhong, J. H., and Zhou, X. F. 2000. Effects of endogenous neurotrophins on sympathetic sprouting in the dorsal root ganglia and allodynia following spinal nerve injury. *Exp. Neurol.* **164**, 344–350.

Denny-Brown, D., Kirk, E. J., and Yanagisawa, N. 1973. The tract of Lissauer in relation to sensory transmission in the dorsal horn of spinal cord in the macaque monkey. *J. Comp. Neurol.* **151**, 175–200.

de Quidt, M. E. and Emson, P. C. 1986. Distribution of neuropeptide Y-like immunoreactivity in the rat central nervous system: II. Immunohistochemical analysis. *Neuroscience* **18**, 545–618.

Derwin, K. A., Glover, R. A., and Wojtys, E. M. 1994. Nociceptive role of substance-P in the knee joint of a patient with congenital insensitivity to pain. *J. Pediatr. Orthop.* **14**, 258–262.

Dery, O., Frobert, Y., Zerari, F., Creminon, C., Grassi, J., Fischer, J., Conrath, M., and Couraud, J. Y. 1997. A monoclonal antibody to the ligand-binding domain of the neurokinin 1 receptor (NK1-R) for the neuropeptide substance P. *J. Neuroimmunol.* **76**, 1–9.

Deschenes, M. and Feltz, P. 1976. GABA-induced rise of extracellular potassium in rat dorsal root ganglion: An electrophysiological study in vivo. *Brain Res.* **118**, 494–499.

Deschenes, M., Feltz, P., and Lamour, Y. 1976. A model for an estimate in vivo of the ionic basis of presynaptic inhibition: an intracellular analysis of the GABA-induced depolarization of rat dorsal root ganglia. *Brain Res.* **118**, 486–493.

De Souza, E. B., Insel, T. R., Perrin, M. H., Rivier, J., Vale, W. W., and Kuhar, M. J. 1985. Corticotropin-releasing factor receptors are widely distributed within the rat central nervous system: An autoradiographic study. *J. Neurosci.* **5**, 3189–3203.

Desprat, C. and Zajac, J. M. 1994. Ontogeny of neuropeptide FF pharmacology and receptors in mouse brain. *Dev. Brain Res.* **82**, 118–126.

Deuchars, S. A., Atkinson, L., Brooke, R. E., Musa, H., Milligan, C. J., Batten, T. F., Buckley, N. J., Parson, S. H., and Deuchars, J. 2001. Neuronal P2X7 receptors are targeted to presynaptic terminals in the central and peripheral nervous systems. *J. Neurosci.* **21**, 7143–7152.

Devor, M. 1983. Nerve pathophysiology and mechanism of pain in causalgia. *J. Auton. Nerv. Syst.* **7**, 371–384.

Devor, M. 1993. The pathophysiology of damaged peripheral nerves, In P. D. Wall and R. Melzack (eds.), *Textbook of Pain* (pp. 79–100). Churchill Livingstone, New York.

Devor, M. and Claman, D. 1980. Mapping and plasticity of acid phosphatase afferents in the rat dorsal horn. *Brain Res.* **190**, 17–28.

Devor, M. and Jänig, W. 1981. Activation of myelinated afferents ending in a neuroma by stimulation of sympathetic supply in the rat. *Neurosci. Lett.* **24**, 43–47.

Devor, M. and Wall, P. D. 1976. Dorsal horn cells with proximal cutaneous receptive fields. *Brain Res.* **118**, 325–328.

Devor, M. and Wall, P. D. 1981a. Effect of peripheral nerve injury on receptive fields of cells in the cat spinal cord. *J. Comp. Neurol.* **199**, 277–291.

Devor, M. and Wall, P. D. 1981b. Plasticity in the spinal cord sensory map following peripheral nerve injury in rats. *J. Neurosci.* **1**, 679–684.

Devor, M., Wall, P. D., and McMahon, S. B. 1984. Dichotomizing somatic nerve fiber exist in rats but they are rare. *Neurosci. Lett.* **49**, 187–192.

Devor, M., Wall, P. D., and Catalan, N. 1992. Systemic lidocaine silences ectopic neuroma and DRG discharge without blocking nerve conduction. *Pain* **48**, 261–268.

Devor, M., Govrin-Lippman, R., and Angelides, K. 1993. Na^+ channel immunolocalization in peripheral mammalian axons and changes following nerve injury and neuroma formation. *J. Neurosci.* **13**, 1976–1992.

Devor, M., Jänig, W., and Michaelis, M. 1994a. Modulation of activity in dorsal root ganglion neurons by sympathetic activation in nerve-injured rats. *J. Neurophysiol.* **71**, 38–47.

Devor, M., Lomazov, P., and Matzner, O. 1994b. Sodium channel accumulation in injured axons as a substrate for neuropathic pain. In J. Boivie, P. Hansson, and U. Lindblom (eds.), *Touch, Temperature, and Pain in Health and Disease: Mechanisms and Assessments* (pp. 207–230). IASP Press, Seattle.

De Vos, H., Bricca, G., De Keyser, J., Bousquet, P., and Vauquelin, G. 1994. Imidazoline receptors, non-adrenergic idazoxan binding sites and α_2-adrenoreceptors in the human central nervous system. *Neuroscience* **59**, 589–598.

Dib-Hajj, S., Black, J. A., Felts, P., and Waxman, S. G. 1996. Down-regulation of transcripts for Na channel alpha-SNS in spinal sensory neurons following axotomy. *Proc. Natl. Acad. Sci. USA* **93**, 14950–14954.

Dib-Hajj, S. D., Tyrrell, L., Black, J. A., and Waxman, S. G. 1998a. NaN, a novel voltage-gated Na channel, is expressed preferentially in peripheral sensory neurons and down-regulated after axotomy. *Proc. Natl. Acad. Sci. USA* **95**, 8963–8968.

Dib-Hajj, S. D., Black, J. A., Cummins, T. R., Kenney, A. M., Kocsis, J. D., and Waxman, S. G. 1998b. Rescue of alpha-SNS sodium channel expression in small dorsal root ganglion neurons after axotomy by nerve growth factor in vivo. *J. Neurophysiol.* **79**, 2668–2676.

DiCarlo, V. 1983. Serotoninergic fibers in dorsal roots of the spinal cord. *Neurosci. Lett.* **43**, 119–125.

Dickenson, A. H. and Aydar, E. 1991. Antagonism at the glycine site on the NMDA receptor reduces spinal nociception in the rat. *Neurosci. Lett.* **121**, 263–266.

Dickenson, A. H. and Sullivan, A. F. 1987. Subcutaneous formalin-induced activity of dorsal horn neurons in the rat: Differential response to an intrathecal opiate administration pre or post formalin. *Pain* **30**, 349–360.

Dickenson, A. H. and Sullivan, A. F. 1990. Differential effects of excitatory amino acid antagonists on dorsal horn nociceptive neurones in the rat. *Brain Res.* **506**, 31–39.

Dickenson, A. H., Brewer, C. M., and Hayes, N. A. 1985. Effects of topical baclofen on C fibre-evoked neuronal activity in the rat dorsal horn. *Neuroscience* **14**, 557–562.

Dickenson, A. H., Sullivan, A., Feeney, C., Fournie-Zaluski, M. C., and Roques, B. P. 1986. Evidence that endogenous enkephalins produce delta-opiate receptor mediated neuronal inhibitions in rat dorsal horn. *Neurosci. Lett.* **72**, 179–182.

Dickinson, T. and Fleetwood-Walker, S. M. 1999. VIP and PACAP: Very important in pain? *Trends Pharmacol. Sci.* **20**, 324–329.

Dickinson, T., Fleetwood-Walker, S. M., Mitchell, R., and Lutz, E. M. 1997. Evidence for roles of vasoactive intestinal polypeptide (VIP) and pituitary adenylate cyclase activating polypeptide (PACAP) receptors in modulating the responses of rat dorsal horn neurons to sensory inputs. *Neuropeptides* **31**, 175–185.

Dickinson, T., Mitchell, R., Robberecht, P., and Fleetwood-Walker, S. M. 1999. The role of VIP/PACAP receptor subtypes in spinal somatosensory processing in rats with an experimental peripheral mononeuropathy. *Neuropharmacol.* **38**, 167–180.

Diehl, B., Hoheisel, U., and Mense, S. 1988. Histological and neurophysiological changes induced by carrageenan in skeletal muscle of cat and rat. *Agents Actions* **25**, 210–213.

Diehl, B., Hoheisel, U., and Mense, S. 1993. The influence of mechanical stimuli and of acetylsalicylic acid on the discharges of slowly conducting afferent units from normal and inflamed muscle in the rat. *Exp. Brain Res.* **92**, 431–440.

Dietl, M., Arluison, M., Mouchet, P., Feuerstein, C., Manier, M., and Thibault, J. 1985. Immunohistochemical demonstration of catecholaminergic cell bodies in the spinal cord of the rat: Preliminary note. *Histochemistry* **82**, 385–389.

Dietl, M. M., Sanchez, M., Probst, A., and Palacios, J. M. 1989. Substance P receptors in the human spinal cord: Decrease in amyotrophic lateral sclerosis. *Brain Res.* **483**, 39–49.

Diez-Guerra, F. J., Zaidi, M., Bevis, P., MacIntrye, I., and Emson, P. C. 1988. Evidence for release of calcitonin gene-related peptide and neurokinin A from sensory nerve endings in vivo. *Neuroscience* **25**, 839–846.

DiFiglia, M., Aronin, N., and Leeman, S. E. 1982a. Immunocytochemical study of neurotensin localization in the monkey spinal cord. *Ann. NY Acad. Sci.* **400**, 405–407.

DiFiglia, M., Aronin, N., and Leeman, S. E. 1982b. Light microscopic and ultrastructural localization of immunoreactive substance P in the dorsal horn of monkey spinal cord. *Neuroscience* **7**, 1127–1139.

DiFiglia, M., Aronin, N., and Leeman, S. E. 1984. Ultrastructural localization of immunoreactive neurotensin in the monkey superficial dorsal horn. *J. Comp. Neurol.* **225**, 1–12.

DiGiulio, A. M., Tenconi, B., Mannavola, A., Mantegazza, P., Schiavinato, A., and Gorio, A. 1987. Spinal cord interneuron degenerative atrophy caused by peripheral nerve lesions is prevented by serotonin depletion. *J. Neurosci. Res.* **18**, 443–448.

Di Marzo, V. 1998. "Endocannabinoids" and other fatty acid derivatives with cannabimimetic properties: Biochemistry and possible physiopathological relevance. *Biochim. Biophys. Acta* **1392**, 153–175.

Dimsdale, J. A. and Kemp, J. M. 1966. Afferent fibres in ventral nerve roots inthe rat. *J. Physiol. (Lond.)* **187**, 25–26P.

Ding, X. Z. and Bayer, B. M. 1993. Increases of CCK mRNA and peptide in different brain areas following acute and chronic administration of morphine. *Brain Res.* **625**, 139–144.

Ding, Y. Q., Nomura, S., Kaneko, T., and Mizuno, N. 1995a. Co-localization of mu-opioid receptor-like and substance P-like immunoreactivities in axon terminals within the superficial layers of the medullary and spinal dorsal horns of the rat. *Neurosci. Lett.* **198**, 45–48.

Ding, Y. Q., Takada, M., Shigemoto, R., and Mizumo, N. 1995b. Spinoparabrachial tract neurons showing substance P receptor-like immunoreactivity in the lumbar spinal cord of the rat. *Brain Res.* **674**, 336–340.

Ding, Y. Q., Shigemoto, R., Takada, M., Ohishi, H., Nakanishi, S., and Mizuno, N. 1996a. Localization of the neuromedin K receptor (NK3) in the central nervous system of the rat. *J. Comp. Neurol.* **364**, 290–310.

Ding, Y. Q., Kaneko, T., Nomura, S., and Mizuno, N. 1996b. Immunohistochemical localization of μ-Opioid receptors in the central nervous system of the rat. *J. Comp. Neurol.* **367**, 375–402.

Ding, Y. Q., Zheng, H. X., Wang, D. S., Xu, J. Q., Gong, L. W., Lu, Y., Qin, B. Z., Shi, J., Li, H. L., Li, J. S., Shigemoto, R., Kaneko, T., and Mizuno, N. 1999. The distribution of substance P receptor (NK-1)-like immunoreactive neurons in the newborn and adult human spinal cord. *Neurosci. Lett.* **266**, 133–136.

Dingledine, R., Borges, K., Bowie, D., and Traynelis, S. F. 1999. The glutamate receptor ion channels. *Pharmacol. Rev.* **51**, 7–61.

Dipette, D. J., Westlund, K. N., and Holland, O. 1988. Dietary calcium modulates spinal cord content of calcitonin gene-related peptide in the rat. *Neurosci. Lett.* **95**, 335–340.

DiTirro, F. J., Ho, R. H., and Martin, G. F. 1981. Immunohistochemical localization of substance-P, somatostatin, and methionine-enkephalin in the spinal cord and dorsal root ganglia of the North American opossum, *Didelphis virginia. J. Comp. Neurol.* **198**, 351–363.

DiTirro, F. J., Martin, G. F., and Ho, R. H. 1983. A developmental study of substance-P, somatostatin, enkephalin, and serotonin immunoreactive elements in the spianl cord of the North American opossum. *J. Comp. Neurol.* **213**, 241–261.

Dixon, A. D. 1968. Cholinesterase distribution in trigeminal ganglionic neurons. *J. Cell Biol.* **39**, 166A.

Djouhri, L. and Lawson, S. N. 1999. Changes in somatic action potential shape in guinea-pig nociceptive primary afferent neurones during inflammation in vivo. *J. Physiol.* **520.2**, 565–576.

Djouhri, L. and Lawson, S. N. 2001. Differences in the size of the somatic action potential overshoot between nociceptive and non-nociceptive dorsal root ganglion neurones in the guinea-pig. *Neuroscience* **108**, 479–491.

Djouhri, L., Bleazard, L., and Lawson, S. N. 1998. Association of somatic action potential shape with sensory receptive properties in guinea-pig dorsal root ganglion neurones. *J. Physiol.* **513**, 857–872.

Dmitrieva, N. and McMahon, S. B. 1996. Sensitization of visceral afferents by nerve growth factor in the adult rat. *Pain* **66**, 87–97.

Dodd, J. and Jessell, T. M. 1985. Lactoseries carbohydrates specify subsets of dorsal root ganglion neurons projecting to the superficial dorsal horn of rat spinal cord. *J. Neurosci.* **5**, 3278–3294.

Dodd, J., Jahr, C. E., Hamilton, P. N., Heath, M. J. S., Matthew, W. D., and Jessell, T. M. 1983. Cytochemical and physiological properties of sensory and dorsal horn neurons that transmit cutaneous sensation. In *Cold Spring Harbor Symposium on Quantitative Biology* (pp. 685–695).

Dodd, J., Solter, D., and Jessell, T. M. 1984. Monoclonal antibodies against carbohydrate differentiation antigens identify subsets of primary sensory neurones. *Nature* **311**, 469–472.

Dodt, E. 1952. The behaviour of thermoreceptors at low and high temperatures with special reference to Ebbecke's temperature phenomena. *Acta Physiol. Scand.* **27**, 295–314.

Dodt, E. and Zotterman, Y. 1952a. Mode of action of warm receptors. *Acta Physiol. Scand.* **26**, 345–357.

Dodt, E. and Zotterman, Y. 1952b. The discharge of specific cold fibres at high temperatures. *Acta Physiol. Scand.* **26**, 358–365.

Dogiel, A. S. 1892. Die Nervenendigunden in Meissnerschen Tastköperen. *Monthly Int. J. Anat. Physiol.* **9**, 76–85.

Dogiel, A. S. 1908. In *Der Bau der spinalganglien des Menschen und der Saugetiere* (pp. Gustav Fischer, Jena).

Dolan, S. and Nolan, A. M. 2000. Behavioural evidence supporting a differential role for group I and II metabotropic glutamate receptors in spinal nociceptive transmission. *Neuropharmacol.* **39**, 1132–1138.

Donaldson, H. H. 1885. On the temperature sense. *Mind* **10**, 399–416.

Donaldson, L. F., Harmar, A. J., McQueen, D. S., and Seckl, J. R. 1992. Increased expression of preprotachykinin, calcitonin gene-related peptide, but not vasoactive intestinal peptide messenger RNA in dorsal root ganglia during the development of adjuvant monoarthritis in the rat. *Mol. Brain Res.* **16**, 143–149.

Donaldson, L. F., Seckl, J. R., and McQueen, D. S. 1993. A discrete adjuvant-induced monoarthritis in the rat: Effects of adjuvant dose. *J. Neurosci. Meth.* **49**, 5–10.

Donaldson, L. F., McQueen, D. S., and Seckl, J. R. 1994. Local anaesthesia prevents acute inflammatory changes in neuropeptide messenger RNA expression in rat dorsal root ganglia neurons. *Neurosci. Lett.* **175**, 111–113.

Donaldson, L. F., McQueen, D. S., and Seckl, J. R. 1995. Neuropeptide gene expression and capsaicin-sensitive primary afferents: Maintenance and spread of adjuvant arthritis in the rat. *J. Physiol.* **486**, 473–482.

Donnerer, J., Schuligoi, R., and Stein, C. 1992. Increased content and transport of substance P and calcitonin gene-related peptide in sensory nerves innervating inflamed tissue: Evidence for a regulatory function of nerve growth in vivo. *Neurosci.* **49**, 693–698.

Donnerer, J., Schuligoi, R., Stein, C., and Amann, R. 1993. Upregulation, release and axonal transport of substance P and calcitonin gene-related peptide in adjuvant inflammation and regulatory function of nerve growth factor. *Regul. Pept.* **46**, 150–154.

Donnerer, J., Amann, R., Schuligoi, R., and Skofitsch, G. 1996. Complete recovery by nerve growth factor of neuropeptide content and function in capsaicin-impaired sensory neurons. *Brain Res.* **741**, 103–108.

Doré, S., Krieger, C., Kar, S., and Quirion, R. 1996. Distribution and levels of insulin-like growth factor (IGF-1 and IGF-II) and insulin receptor binding sites in the spinal cords of amyotrophic lateral sclerosis (ALS) patients. *Mol. Brain Res.* **41**, 128–133.

Dorn, T., Schaible, H. G., and Schmidt, R. F. 1991. Response properties of thick myelinated group II afferents in the medial articular nerve of normal and inflamed knee joints of the cat. *Somatosens. Mot. Res.* **8**, 127–136.

Doubell, T. P. and Woolf, C. J. 1997. Growth-associated protein 43 immunoreactivity in the superficial dorsal horn of the rat spinal cord is localized in atrophic C-fiber, and not in sprouted A-fiber, central terminals after peripheral nerve injury. *J. Comp. Neurol.* **386**, 111–118.

Doubell, T. P., Mannion, R. J., and Woolf, C. J. 1997. Intact sciatic myelinated primary afferent terminals collaterally sprout in the adult rat dorsal horn following section of a neighboring peripheral nerve. *J. Comp. Neurol.* **380**, 95–104.

Doucet, E., Miquel, M. C., Nosjean, A., Verge, D., Hamon, M., and Emerit, M. B. 2000. Immunolabeling of the rat central nervous system with antibodies partially selective of the short form of the 5-HT3 receptor. *Neuroscience* **95**, 881–892.

Dougherty, P. M. and Willis, W. D. 1991a. Modification of the responses of primate spinothalamic neurons to mechanical stimulation by excitatory amino acids and an N-methyl-D-aspartate antagonist. *Brain Res.* **542**, 15–22.

Dougherty, P. M. and Willis, W. D. 1991b. Enhancement of spinothalamic neuron responses to chemical and mechanical stimuli following combined microiontophoretic application of N-methyl-D-aspartic acid and substance P. *Pain* **47**, 85–93.

Dougherty, P. M. and Willis, W. D. 1992. Enhanced responses of spinothalamic tract neurons to excitatory amino acids accompany capsaicin-induced sensitization in the monkey. *J. Neurosci.* **12**, 883–894.

Dougherty, P. M., Paleck, J., Paleckova, V., Sorkin, L. S., and Willis, W. D. 1992a. The role of NMDA and non-NMDA excitatory amino acid receptors in the excitation of primate spinothalamic tract neurons by mechanical, chemical, thermal and electrical stimuli. *J. Neurosci.* **12**, 3025–3041.

Dougherty, P. M., Sluka, K. A., Sorkin, L. S., Westlund, K. N., and Willis, W. D. 1992b. Neural changes in acute arthritis in monkeys: I. Parallel enhancement of responses of spinothalamic tract neurons to mechanical stimulation and excitatory amino acids. *Brain Res. Rev.* **17**, 1–13.

Dougherty, P. M., Palecek, J., Zorn, S., and Willis, W. D. 1993. Combined application of excitatory amino acids and substance P produces long-lasting changes in responses of primate spinothalamic tract neurons. *Brain Res. Rev.* **18**, 227–246.

Dougherty, P. M., Palecek, J., Paleckova, V., and Willis, W. D. 1994. Neurokinin 1 and 2 antagonists attenuate the responses and NK1 antagonists prevent the sensitization of primate spinothalamic tract neurons after intradermal capsaicin. *J. Neurophysiol.* **72**, 1464–1475.

Dougherty, P. M., Palecek, J., Paleckova, V., and Willis, W. D. 1995. Infusion of substance P or neurokinin A by microdialysis alters responses of primate spinothalamic tract neurons to cutaneous stimuli and to iontophoretically released excitatory amino acids. *Pain* **61**, 411–425.

Doughty, S. E., Atkinson, M. E., and Shehab, S. A. 1991. A quantitative study of neuropeptide immunoreactive cell bodies of primary afferent sensory neurons following rat sciatic nerve peripheral axotomy. *Regul. Pept.* **35**, 59–72.

Douglas, W. W. and Ritchie, J. M. 1957. Non-medullated fibres in the saphenous nerve which signal touch. *J. Physiol.* **139**, 385–399.

Doupe, J., Cullen, C. H., and Chance, G. Q. 1944. Post-traumatic pain and the causalgic syndrome. *J. Neurol. Neurosurg. Psychiat.* **7**, 323–48.

Doyle, C. A. and Hunt, S. P. 1997. Reduced nuclear factor κB (p65) expression in rat primary sensory neurons after peripheral nerve injury. *NeuroReport* **8**, 2937–2942.

Doyle, C. A. and Maxwell, D. J. 1991a. Ultrastructural analysis of noradrenergic nerve terminals in the cat lumbosacral spinal dorsal horn: A dopamine-beta-hydroxylase immunocytochemical study. *Brain Res.* **563**, 329–333.

Doyle, C. A. and Maxwell, D. J. 1991b. Catecholaminergic innervation of the spinal dorsal horn: A correlated light- and electron-microscopic analysis of tyrosine hydroxylase-immunoreactive fibres in the cat. *Neuroscience* **45**, 161–176.

Doyle, C. A. and Maxwell, D. J. 1993a. Direct catecholaminergic innervation of spinal dorsal horn neurons with axons ascending the dorsal columns in cat. *J. Comp. Neurol.* **331**, 434–444.

Doyle, C. A. and Maxwell, D. J. 1993b. Neuropeptide Y-immunoreactive terminals form axo-axonic synaptic arrangements in the substantia gelatinosa (lamina II) of the cat spinal dorsal horn. *Brain Res.* **603**, 157–161.

Doyle, C. A. and Maxwell, D. J. 1994. Light- and electron-microscopic analysis of neuropeptide Y-immunoreactive profiles in the cat spinal dorsal horn. *Neuroscience* **61**, 107–121.

Draisci, G. and Iadarola, M. J. 1989. Temporal analysis of increases in c-fos, preprodynorphin and preproenkephalin mRNAs in rat spinal cord. *Mol. Brain Res.* **6**, 31–37.

Draisci, G., Kajander, K. C., Dubner, R., Bennett, G. J., and Iadarola, M. J. 1991. Up-regulation of opioid gene expression in spinal cord evoked by experimental nerve injuries and inflammation. *Brain Res.* **560**, 186–192.

Dray, A., Bettaney, J., Forster, P., and Perkins, M. N. 1988. Bradykinin-induced stimulation of afferent fibres is mediated through protein kinase C. *Neurosci. Lett.* **91**, 301–307.

Dray, A., Bettaney, J., and Forster, P. 1990. Actions of capsaicin on peripheral nociceptors of the neonatal rat spinal cord-tail in vitro: Dependence of extracellular ions and independence of second messengers. *Br. J. Pharmacol.* **101**, 727–733.

Drew, J. P., Westrum, L. E., and Ho, R. H. 1986. Mapping of the normal distribution of substance P-like immunoreactivity in the spinal trigeminal nucleus of the cat. *Exp. Neurol.* **93**, 168–179.

Droz, B. and Kazimierczak, J. 1987. Carbonic anhydrase in primary sensory neurons of dorsal root ganglia. *J. Biochem. Physiol.* **88B**, 713–717.

Drummond, H. A., Price, M. P., Welsh, M. J., and Abboud, F. M. 1998. A molecular component of the arterial baroreceptor mechanotransducer. *Neuron* **21**, 1435–1441.

Drummond, H. A., Abboud, F. M., and Welsh, M. J. 2000. Localization of β and γ subunits of ENaC in sensory nerve endings in the rat foot pad. *Brain Res.* **884**, 1–12.

Drummond, P. D. 1995. Noradrenaline increases hyperalgesia to heat in skin sensitized by capsaicin. *Pain* **60**, 311–315.

Drummond, P. D., Skipworth, S., and Finch, P. M. 1996. α_1-adrenoreceptors in normal and hyperalgesic human skin. *Clin. Sci.* **91**, 73–77.

Du, J., Koltzenburg, M., and Carlton, S. M. 2001. Glutamate-induced excitation and sensitization of nociceptors in rat glabrous skin. *Pain* **89**, 187–198.

Duarte, I. D. G., dos Santos, I. R., Lorenzetti, B. B., and Ferreira, S. H. 1992. Analgesia by direct antagonism of nociceptor sensitization involves the arginine-nitric oxide-cGMP pathway. *Eur. J. Pharmacol.* **217**, 225–227.

Dubner, R. 1991. Neuronal plasticity and pain following peripheral tissue inflammation or nerve injury. In J. E. Charlton and C. J. Woolf (eds.), *Proceedings of the Vth World Congress on Pain* (pp. 263–276). Elsevier, Amsterdam.

Dubner, R. and Ruda, M. A. 1992. Activity-dependent neuronal plasticity following tissue injury and inflammation. *TINS* **15**, 96–103.

Dubois, A., Savasta, M., Curet, O., and Scatton, B. 1986. Autoradiographic distribution of the D_1 agonist [^{3}H]SKF 38393 in the rat brain and spinal cord. Comparison with the distribution of D_2 dopamine receptors. *Neuroscience* **19**, 125–137.

Dubovy, P. and Bednarova, J. 1999. The extracellular matrix of rat Pacinian corpuscles: An analysis of its fine structure. *Anat. Embryol. (Berl.)* **200**, 615–623.

Dubuisson, D. and Dennis, S. G. 1977. The formalin test: A quantitative study of the analgesic effects of morphine, meperidine, and brain stimulation in rats and cats. *Pain* **4**, 161–174.

Dubuisson, D., Fitzgerald, M., and Wall, P. D. 1979. Ameboid receptive fields of cells in laminae 1, 2 and 3. *Brain Res.* **177**, 376–378.

Duc, C., Barakat-Walter, I., and Droz, B. 1993. Innervation of putative rapidly adapting mechanoreceptors by calbindin- and calretinin-immunoreactive primary sensory neurons in the rat. *Eur. J. Neurosci.* **6**, 264–271.

Duce, I. R. and Keen, P. 1977. An ultrastructural classification of the neuronal cell bodies of rat dorsal root ganglion using zinc iodine-osmium impregnation. *Cell Tis. Res.* **185**, 263–277.

Duclaux, R. and Kenshalo, D. R. 1972. The temperature sensitivity of the type I slowly adapting mechanoreceptors in cats and monkeys. *J. Physiol.* **224**, 647–664.

Duclaux, R. and Kenshalo, D. R. 1980. Response characteristics of cutaneous warm receptors in the monkey. *J. Neurophysiol.* **43**, 1–15.

Duggan, A. W. 1974. The differential sensitivity to L-glutamate and L-aspartate of spinal interneurones and Renshaw cells. *Exp. Brain Res.* **19**, 522–528.

Duggan, A. W. and Hendry, I. A. 1986. Laminar localization of the sites of release of immunoreactive substance P in the dorsal horn with antibody-coated microelectrodes. *Neurosci. Lett.* **68**, 134–140.

Duggan, A. W. and Johnston, G. A. R. 1970. Glutamate and related amino acids in cat spinal roots, dorsal root ganglia, and peripheral nerves. *J. Neurochem.* **17**, 1205–1208.

Duggan, A. W. and North, R. A. 1984. Electrophysiology of opioids. *Pharmacol. Rev.* **35**, 219–281.

Duggan, A. W., Hall, J. G., and Headley, P. M. 1976. Morphine, enkephalin and the substantia gelatinosa. *Nature* **264**, 456–458.

Duggan, A. W., Hall, J. G., and Headley, P. M. 1977. Enkephalins and dorsal horn neurones of the cat: Effects on responses to noxious and innocuous skin stimuli. *Br. J. Pharmacol.* **61**, 399–408.

Duggan, A. W., Griersmith, B. T., Headley, P. M., and Hall, J. G. 1979. Lack of effect by substance P at sites in the substantia gelatinosa where met-enkephalin reduces the transmission of nociceptive impulses. *Neurosci. Lett.* **12**, 313–317.

Duggan, A. W., Johnson, S. M., and Morton, C. R. 1981. Differing distributions of receptors for morphine and met5-enkephalinamide in the dorsal horn of the cat. *Brain Res.* **229**, 379–387.

Duggan, A. W., Morton, C. R., Zhao, Z. Q., and Hendry, I. A. 1987. Noxious heating of the skin releases immunoreactive substance P in the substantia gelatinosa of the cat: A study with antibody microprobes. *Brain Res.* **403**, 345–349.

Duggan, A. W., Hendry, I. A., Green, J. L., Morton, C. R., and Hutchison, W. D. 1988a. The preparation and use of antibody microprobes. *J. Neurosci. Meth.* **23**, 241–247.

Duggan, A. W., Hendry, I. A., Morton, C. R., Hutchison, W. D., and Zhao, Z. Q. 1988b. Cutaneous stimuli releasing immunoreactive substance P in the dorsal horn of the cat. *Brain Res.* **451**, 261–273.

Duggan, A. W., Morton, C. R., Hutchison, W. D., and Hendry, I. A. 1988c. Absence of tonic supraspinal control of substance P release in the substantia gelatinosa of the anaesthetized cat. *Exp. Brain Res.* **71**, 597–602.

Duggan, A. W., Hope, P. J., Jarrott, B., Schaible, H. G., and Fleetwood-Walker, S. M. 1990. Release, spread and persistence of immunoreactive neurokinin A in the dorsal horn of the cat following noxious cutaneous stimulation: Studies with antibody microprobes. *Neuroscience* **35**, 195–202.

Duggan, A. W., Schaible, H. G., Hope, P. J., and Lang, C. W. 1992. Effect of peptidase inhibition on the pattern of intraspinally released immunoreactive substance P detected with antibody microprobes. *Brain Res.* **579**, 261–269.

Duggan, A. W., Riley, R. C., Mark, M. A., MacMillan, S. J., and Schaible, H. G. 1995. Afferent volley patterns and the spinal release of immunoreactive substance P in the dorsal horn of the anaesthetized spinal cat. *Neuroscience* **65**, 849–859.

Dumoulin, F. L., Raivich, G., Streit, W. J., and Kreutzberg, G. W. 1990. Differential regulation of calcitonin gene-related peptide (CGRP) in regenerating rat facial nucleus and dorsal root ganglion. *Eur. J. Neurosci.* **3**, 338–342.

Dun, E. C., Huang, R. L., Dun, S. L., and Dun, N. L. 1996. Pituitary adenylate cyclase activating polypeptide-immunoreactivity in human spinal cord and dorsal root ganglia. *Brain Res.* **721**, 233–237.

Dun, N. J., Dun, S. L., Forstermann, U., and Tseng, L. F. 1992. Nitric oxide synthase immunoreactivity in rat spinal cord. *Neurosci. Lett.* **147**, 217–220.

Dun, N. J., Dun, S. L., Wu, S. Y., Forstermann, U., Schmidt, H. H., and Tseng, L. F. 1993. Nitric oxide synthase immunoreactivity in the rat, mouse, cat and squirrel monkey spinal cord. *Neuroscience* **54**, 845–857.

Dun, N. J., Miyazaki, T., Tang, H., and Dun, E. C. 1996. Pituitary adenylate cyclase activating polypeptide immunoreactivity in the rat spinal cord and medulla: Implication of sensory and autonomic functions. *Neuroscience.* **73**, 677–686.

Dun, N. J., Dun, S. L., Lin, H. H., Hwang, L. L., Saria, A., and Fischer-Colbrie, R. 1997. Secretonneurin-like immunoreactivity in rat sympathetic, enteric and sensory ganglia. *Brain Res.* **760**, 8–16.

Dun, N. J., Dun, S. L., Wu, S. Y., Williams, C. A., and Kwok, E. H. 2000. Endomorphins: Localization, release and action on rat dorsal horn neurons. *J. Biomed. Sci.* **7**, 213–220.

Duncan, D. and Keyser, L. L. 1936. Some determinations of the ratio of nerve fibers to nerve cells in thoracic dorsal roots and ganglion of the cat. *J. Comp. Neurol.* **64**, 303–309.

Duncan, D. and Morales, R. 1973. Location of large cored synaptic vesicles in the dorsal grey matter of the cat and dog spinal cord. *Am. J. Anat.* **136**, 123–127.

Duncan, D. and Morales, R. 1978. Relative numbers of several types of synaptic connections in the substantia gelatinosa of the cat spinal cord. *J. Comp. Neurol.* **182**, 601–610.

Durkin, M. M., Gunwaldsen, C. A., Borowsky, B., Jones, K. A., and Branchek, T. A. 1999. An in situ hybridization study of the distribution of the GABA(B2) protein mRNA in the rat CNS. *Brain Res. Mol. Brain Res.* **71**, 185–200.

Dyck, P. J., Kawamura, Y., Low, P. A., Shimono, M., and Solovy, J. S. 1978. The number and sizes of reconstructed peripheral autonomic, sensory and motor neurons in a case of dysautonomia. *J. Neuropathol. Exp. Neurol.* **37**, 741–754.

Dyck, P. J., Jedrzejowska, H., Karnes, J., Kawamura, Y., Low, P. A., O'Brien, P. C., Offord, K., Ohnishi, A., Ohta, M., Pollock, M., and Stevens, J. C. 1979. Reconstruction of motor, sensory, and autonomic neurons based on morphometric study of sampled levels. *Muscle Nerve* **2**, 399–405.

Dyck, P. J., Mellinger, J. F., Reagan, T. J., Horowitz, S. J., McDonald, J. W., Litchy, W. J., Daube, J. R., Fealey, R. D., Go, V. L., Kao, P. C., Brimijoin, W. S., and Lambert, E. H. 1983. Not "indifference to pain" but varieties of hereditary sensory and autonomic neuropathy. *Brain* **106**(Pt. 2), 373–390.

Dykes, R. W. 1975. Coding of steady and transient temperatures by cutaneous "cold" fibers serving the hand of monkeys. *Brain Res.* **98**, 485–500.

Dymshitz, J. and Vasco, M. R. 1994. Nitric oxide and cyclic guanosine 3′,5′-monophosphate do not alter neuropeptide release from rat sensory neurons grown in culture. *Neuroscience* **62**, 1279–1286.

Earle, K. M. 1952. The tract of Lissauer and its possible relation to the pain pathway. *J. Comp. Neurol.* **96**, 93–109.

Eaton, M. J., Plunkett, J. A., Karmally, S., Martinez, M. A., and Montanez, K. 1998. Changes in GAD- and GABA-immunoreactivity in the spinal dorsal horn after peripheral nerve injury and promotion of recovery by lumbar transplant of immortalized serotonergic precursors. *J. Chem. Neuroanat.* **16**, 57–72.

Ebersberger, A., Charbel Issa, P., Vanegas, H., and Schaible, H. G. 2000. Differential effects of calcitonin gene-related peptide and calcitonin gene-related peptide 8-37 upon responses to *N*-methyl-D-asparate or (R,S)-alpha, amino-3-hydroxy-5-methylisoxazole-4-propionate in spinal nociceptive neurons with knee joint input in the rat. *Neuroscience* **99**, 171–178.

Eccles, J. C. 1964. *The Physiology of Synapses.* Springer, New York.

Eccles, J. C. and Krnjevic, K. 1959. Potential changes recorded inside primary afferent fibres within the spinal cord. *J. Physiol.* **149**, 250–273.

Eccles, J. C., Fatt, P., Landgren, S., and Winsbury, G. J. 1954. Spinal cord potentials generated by volleys in the large muscle afferents. *J. Physiol.* **125**, 590–606.

Eccles, J. C., Fatt, P., and Landgren, S. 1956. Central pathway for direct inhibitory action of impulses in largest afferent nerve fibres to muscle. *J. Neurophysiol.* **19**, 75–98.

Eccles, J. C., Eccles, R. M., and Lundberg, A. 1960. Types of neurone in and around the intermediate nucleus of the lumbosacral cord. *J. Physiol.* **154**, 89–114.

Eccles, J. C., Kozak, W., and Magni, F. 1961. Dorsal root reflexes of muscle group I afferent fibres. *J. Physiol.* **159**, 128–146.

Eccles, J. C., Kostyuk, P. G., and Schmidt, R. F. 1962a. Central pathways responsible for depolarization of primary afferent fibres. *J. Physiol.* **161**, 237–257.

Eccles, J. C., Magni, F., and Willis, W. D. 1962b. Depolarization of central terminals of group I afferent fibres from muscle. *J. Physiol.* **160**, 62–93.

Eccles, J. C., Schmidt, R. F., and Willis, W. D. 1963a. Depolarization of the central terminals of cutaneous afferent fibres. *J. Neurophysiol.* **26**, 646–661.

Eccles, J. C., Schmidt, R. F., and Willis, W. D. 1963b. Pharmacological studies on presynaptic inhibition. *J. Physiol.* **168**, 500–530.

Eccles, J. C., Schmidt, R. F., and Willis, W. D. 1963c. The location and the mode of action of the presynaptic inhibitory pathways on to group Ia afferent fibers from muscle. *J. Neurophysiol.* **26**, 506–522.

Eccles, R. M. 1965. Interneurones activated by higher threshold group I muscle afferents. In D. R. Curtis and A. K. McIntyre (eds.), *Studies in Physiology* (pp. 59–64). Springer, New York.

Eccles, R. M. and Lundberg, A. 1959. Supraspinal control of interneurones mediating spinal reflexes. *J. Physiol.* **147**, 565–584.

Eckert, A., Von Banchet, G. S., Sopper, S., and Petersen, M. 1999. Spatio-temporal pattern of induction of bradykinin receptors and inflammation in rat dorsal root ganglia after unilateral nerve ligation. *Pain* **83**, 487–497.

Edgley, S. A. and Jankowska, E. 1987. An interneuronal relay for group I and II muscle afferents in the midlumbar segments of the cat spinal cord. *J. Physiol.* **389**, 675–690.

Edin, B. B. and Abbs, J. H. 1991. Finger movement responses of cutaneous mechanoreceptors in the dorsal skin of the human hand. *J. Neurophysiol.* **65**, 657–670.

Edwards-Lee, T. A., Cornford, M. E., and Yu, K. T. 1997. Congenital insensitivity to pain and anhidrosis with mitochondrial and axonal abnormalities. *Pediatr. Neurol.* **17**, 356–361.

Egger, M. D., Freeman, N. C. G., Jacquin, M., Proshansky, E., and Semba, K. 1986. Dorsal horn cells in the cat responding to stimulation of the plantar cushion. *Brain Res.* **383**, 68–82.

Eisenach, J. C. 1999. Muscarinic-mediated analgesia. *Life Sci.* **64**, 549–554.

Eklund, G. and Skoglund, S. 1960. On the specificity of the Ruffini-like joint receptors. *Acta Physiol. Scand.* **49**, 184–191.

El-Bohy, A. and LaMotte, C. C. 1993. Deafferentation-induced changes in neuropeptides of the adult rat dorsal horn following pronase injection of the sciatic nerve. *J. Comp. Neurol.* **336**, 545–554.

Elcock, C., Boissonade, F. M., and Robinson, P. P. 2001a. Changes in neuropeptide expression in the trigeminal ganglion following inferior alveolar nerve section in the ferret. *Neuroscience* **102**, 655–667.

Elcock, C., Boissonade, F. M., and Robinson, P. P. 2001b. Neuropeptide expression in the ferret trigeminal ganglion following ligation of the inferior alveolar nerve. *Arch. Oral Biol.* **46**, 729–743.

Elde, R., Hokfelt, T., Johansson, O., and Terenius, L. 1976. Immunohistochemical studies using antibodies to leucine-enkephalin: Initial observations on the nervous system of the rat. *Neuroscience* **1**, 349–351.

Elde, R., Schalling, M., Ceccatelli, S., Nakanishi, S., and Hökfelt, T. 1990. Localization of neuropeptide receptor mRNA in rat brain: Initial observations using probes for neurotensin and substance P receptors. *Neurosci. Lett.* **120**, 134–138.

Elde, R., Arvidsson, U., Riedi, M., Vulchanova, L., Lee, J.-H., Dado, R., Nakano, A., Chakrabarti, S., Zhang, X., Loh, H. H., Law, P. Y., Hökfelt, T., and Wessendorf, M. 1995. Distribution of neuropeptide receptors: New views of peptidergic neurotransmission made possible by antibodies to opioid receptors. *Ann. NY Acad. Sci.* **751**, 390–404.

Eldred, E., Yellin, H., DeSantis, M., and Smith, C. M. 1977. Supplement to bibliography on muscle receptors: Their morphology, pathology, physiology, and pharmacology. *Exp. Neurol.* **55** (No. 3, part 2), 1–118.

Elhassan, A. M., Lindgren, J. U., Hultenby, K., Bergstrom, J., and Adem, A. 1998. Methionine-enkephalin in bone and joint tissues. *J. Bone Miner. Res.* **13**, 88–95.

Elliott, A. A. and Elliott, J. R. 1993. Characterization of TTX-sensitive and TTX-resistant sodium currents in small cells from adult rat dorsal root ganglia. *J. Physiol.* **463**, 39–56.

Emery, D. G., Ito, H., and Coggeshall, R. E. 1977. Unmyelinated axons in thoracic ventral roots of the cat. *J. Comp. Neurol.* **172**, 37–48.

Emson, P. C. 1979. Peptides as neurotranmsitter candidates in the mammalian CNS. *Prog. Neurobiol.* **13**, 61–116.

Emson, P. C., Goedert, M., Williams, B., Ninkovic, M., and Hunt, S. P. 1982. Neurotensin: Regional distribution, characterization, and inactivation. *Ann. NY Acad. Sci.* **400**, 198–215.

Endo, K., Kang, Y., Kayano, F., Kojima, H., and Hori, Y. 1985. Synaptic actions of the ventral root afferents on cat hindlimb motoneurons. *Neurosci. Lett.* **58**, 201–205.

Enfiejian, H. J., Chiaia, N. L., MacDonald, G. J., and Rhoades, R. W. 1989. Neonatal transection alters the percentage of substance-P-positive trigeminal ganglion cells that contribute axons to the regenerate infraorbital nerve. *Somatosens. Mot. Res.* **6**, 537–552.

Engberg, I. and Thaller, A. 1970. On the interaction of picrotoxin with GABA and glycine in the spinal cord. *Brain Res.* **19**, 151–154.

Engelman, H. S., Allen, T. B., and MacDermott, A. B. 1999. The distribution of neurons expressing calcium-permeable AMPA receptors in the superficial laminae of the spinal cord dorsal horn. *J. Neurosci.* **19**, 2081–2089.

England, J. D., Gamboni, F., Ferguson, M. A., and Levinson, S. R. 1994. Sodium channels accumulate at the tips of injured axons. *Muscle Nerve* **17**, 593–598.

England, J. D., Happel, L. T., Kline, D. G., Gamboni, F., Thouron, C. L., Liu, Z. P., and Levinson, S. R. 1996. Sodium channel accumulation in humans with painful neuromas. *Neurology* **47**, 272–276.

England, S., Bevan, S., and Docherty, R. J. 1996. PGE_2 modulates the tetrodotoxin-resistant sodium current in neonatal rat dorsal root ganglion neurones via the cyclic AMP-protein kinase A cascade. *J. Physiol.* **495**, 429–440.

English, K. B. 1977. The ultrastructure of cutaneous type I mechanoreceptors (Haarscheiben) in cats following denervation. *J. Comp. Neurol.* **172**, 137–163.

English, K. B., Harper, S., Stayner, N., Wang, Z. M., and Davies, A. M. 1994. Localization of nerve growth factor (NGF) and low-affinity NGF receptors in touch domes and quantification of NGF mRNA in keratinocytes of adult rats. *J. Comp. Neurol.* **344**, 470–480.

Eriksson, J., Bongenhielm, U., Kidd, E., Matthews, B., and Fried, K. 1998. Distribution of P2X3 receptors in the rat trigeminal ganglion after inferior alveolar nerve injury. *Neurosci. Lett.* **254**, 37–40.

Eriksson, N. P., Aldskogius, H., Grant, G., Lindsay, R. M., and Rivero-Melian, C. 1997. Effects of nerve growth factor, brain-derived neurotrophic factor and neurotrophin-3 on the laminar distribution of transganglionically transported choleragenoid in the spinal cord dorsal horn following transection of the sciatic nerve in the adult rat. *Neuroscience* **78**, 863–872.

Erlanger, J. and Gasser, H. S. 1937. *Electrical Signs of Nervous Activity.* University Pennsylvania Press, Philadelphia.

Evans, R. H. 1980. Evidence supporting the indirect depolarization of primary afferent terminals in the frog by excitatory amino acids. *J. Physiol.* **298**, 25–35.

Evans, R. H. 1989. Pharmacology of segmental transmission in the spinal cord. *Prog. Neurobiol.* **33**, 255–279.

Evans, R. H. and Long, S. K. 1989. Primary afferent depolarization in the rat spinal cord is mediated by pathways utilizing NMDA and non-NMDA receptors. *Neurosci. Lett.* **100**, 231–236.

Fabri, M. and Conti, F. 1990. Calcitonin gene-related peptide-positive neurons and fibers in the cat dorsal column nuclei. *Neuroscience* **35**, 167–174.

Faccini, E., Uzumaki, H., Govoni, S., Missale, C., Spano, P. F., Covelli, V., and Trabucchi, M. 1984. Afferent fibers mediate the increase of met-enkephalin elicited in rat spinal cord by localized pain. *Pain* **18**, 25–31.

Faden, A. I. 1990. Opioid and nonopioid mechanisms may contribute to dynorphin's pathophysiological actions in spinal cord injury. *Ann. Neurol.* **27**, 67–74.

Faden, A. I., Molineaux, C. J., Rosenberger, J. G., Jacobs, T. P., and Cox, B. M. 1985a. Increased dynorphin immunoreactivity in spinal cord after traumatic injury. *Regul. Pept.* **11**, 35–41.

Faden, A. I., Molineaux, C. J., Rosenberger, J. G., Jacobs, T. P., and Cox, B. M. 1985b. Endogenous opioid immunoreactivity in rat spinal cord following traumatic injury. *Ann. Neurol.* **17**, 386–390.

Fagg, G. E. and Foster, A. C. 1983. Amino acid neurotransmitters and their pathways in the mammalian central nervous system. *Neuroscience* **9**, 701–719.

Fang, L., Wu, J., Lin, Q., and Willis, W. D. 2002. Calcium-calmodulin-dependent protein kinase II contributes to spinal cord central sensitization. *J. Neurosci.* **22**, 4196–4204.

Fang, X., Djouhri, L., Black, J. A., Dib-Hajj, S. D., Waxman, S. G., and Lawson, S. N. 2002. The presence and role of the tetrodotoxin-resistant sodium channel Na(v)1.9 (NaN) in nociceptive primary afferent neurons. *J. Neurosci.* **22**, 7425–7433.

Fang, X. B. 1987. The population of the dorsal root ganglion cells which have central processes in ventral root and their immunoreactivity. *Brain Res.* **402**, 393–398.

Farkas-Szallasi, T., Bennett, G. J., Blumberg, P. M., Hökfelt, T., Lundberg, J. M., and Szallasi, A. 1996. Vanilloid receptor loss is independent of the messenger plasticity that follows systemic resiniferatoxin administration. *Brain Res.* **719**, 213–218.

Farkas-Szallasi, T., Lundberg, J. M., Wiesenfeld-Hallin, Z., Hokfelt, T., and Szallasi, A. 1995. Increased levels of GMAP, VIP and nitric oxide synthase, and their mRNAs, in lumbar dorsal root ganglia of the rat following systemic resiniferatoxin treatment. *NeuroReport* **6**, 2230–2234.

Farquhar-Smith, W. P., Egertova, M., Bradbury, E. J., McMahon, S. B., Rice, A. S., and Elphick, M. R. 2000. Cannabinoid CB(1) receptor expression in rat spinal cord. *Mol. Cell Neurosci.* **15**, 510–521.

Faull, R. L. M. and Villiger, J. W. 1986. Benzodiazepine receptors in the human spinal cord: A detailed anatomical and pharmacology study. *Neuroscience* **17**, 791–802.

Faull, R. L. M. and Villiger, J. W. 1987. Opiate receptors in the human spinal cord: A detailed anatomical study comparing the autoradiographic localization of [^{3}H]diprenorphine binding sites with laminar pattern of substance P, myelin and Nissl staining. *Neuroscience* **20**, 395–407.

Faull, R. L. M., Villiger, J. W., and Dragunow, M. 1989. Neurotensin receptors in the human spinal cord: A quantitative autoradiographic study. *Neuroscience* **29**, 603–613.

Feldblum, S., Dumoulin, A., Anoal, M., Sandillon, F., and Privat, A. 1995. Comparative distribution of GAD65 and GAD67 mRNAs and proteins in the rat spinal cord supports a differential regulation of these two glutamate decarboxylases in vivo. *J. Neurosci. Res.* **42**, 742–757.

Feldman, A. G. and Latash, M. L. 1982. Afferent and efferent components of joint position sense: Interpretation of kinaesthetic illusion. *Biol. Cybern.* **42**, 205–214.

Felipe, C. D., Gonzalez, G. G., Gallar, J., and Belmonte, C. 1999. Quantification and immunocytochemical characteristics of trigeminal ganglion neurons projecting to the cornea: Effect of corneal wounding. *Eur. J. Pain* **3**, 31–39.

Felts, P. A., Black, J. A., and Waxman, S. G. 1995. Expression of sodium channel alpha- and beta-subunits in the nervous system of the myelin-deficient rat. *J. Neurocytol.* **24**, 654–666.

Felts, P. A., Yokoyama, S., Dib-Hajj, S., Black, J. A., and Waxman, S. G. 1997. Sodium channel α-subunit mRNAs I, II, III, NaG, Na6, and hNE(PN1): Different expression patterns in developing rat nervous system. *Mol. Brain Res.* **45**, 71–82.

Fernandez de Molina, A. and Gray, J. A. B. 1957. Activity in the dorsal spinal grey matter after stimulation of cutaneous nerves. *J. Physiol.* **137**, 126–140.

Ferrarese, C., Iadarola, M. J., Yang, H. Y., and Costa, E. 1986. Peripheral and central origin of Phe-Met-Arg-Phe-amide immunoreactivity in rat spinal cord. *Regul. Pept.* **13**, 245–252.

Ferreira, S. H. and Lorenzetti, B. B. 1994. Glutamate spinal retrograde sensitization of primary sensory neurons associated with nociception. *Neuropharmacology* **33**, 1479–1485.

Ferrell, W. R. 1980. The adequacy of stretch receptors in the cat knee joint for signalling joint angle throughout a full range of movement. *J. Physiol.* **299**, 85–99.

Ferrell, W. R. and Craske, B. 1992. Contribution of joint and muscle afferents to position sense at the human proximal interphalangeal joint. *Exp. Physiol.* **77**, 331–342.

Ferrell, W. R. and Russell, N. J. 1985. Plasma extravasation in the cat knee-joint induced by antidromic articular nerve stimulation. *Pfluegers Arch.* **404**, 91–93.

Ferrell, W. R. and Russell, N. J. 1986. Extravasation in the knee induced by antidromic stimulation of articular C fibre afferents of the anaesthetized cat. *J. Physiol.* **379**, 407–416.

Ferrell, W. R., Nade, S. D., and Newbold, P. I. 1986. The interrelation of neural discharge, intra-articular pressure, and joint angle in the knee of the dog. *J. Physiol.* **373**, 353–365.

Ferri, G. L., Sabani, A., Abelli, L., Polak, J. M., Dahl, D., and Portier, M. M. 1990. Neuronal intermediate filaments in rat dorsal root ganglia: Differential distribution of peripherin and neurofilament protein immunoreactivity and effect of capsaicin. *Brain Res.* **515**, 331–335.

Ferrington, D. G., Rowe, M. J., and Tarvin, R. P. C. 1987a. Actions of single sensory fibres on cat dorsal column nuclei neurones: Vibratory signaling in a one-to-one linkage. *J. Physiol.* **386**, 292–309.

Ferrington, D. G., Sorkin, L. S., and Willis, W. D. 1987b. Responses of spinothalamic tract cells in the superficial dorsal horn of the primate lumbar spinal cord. *J. Physiol.* **388**, 681–703.

Fetz, E. R. 1968. Pyramidal tract effects on interneurons in the cat lumbar dorsal horn. *J. Neurophysiol.* **31**, 69–80.

Fiallos-Estrada, C. E., Kummer, W., Mayer, B., Bravo, R., Zimmerman, M., and Herdegen, T. 1993. Long-lasting increase of nitric oxide synthase immunoreactivity, NDPH-diaphorase reaction and c-JUN co-expression in rat dorsal root ganglion neurons following sciatic nerve transection. *Neurosci. Lett.* **150**, 169–173.

Fields, H. L., Meyer, G. A., and Partridge, L. D. 1970a. Convergence of visceral and somatic input onto spinal neurons. *Exp. Neurol.* **26**, 36–52.

Fields, H. L., Partridge, L. D., and Winter, D. L. 1970b. Somatic and visceral receptive field properties of fibers in ventral quadrant white matter of the cat spinal cord. *J. Neurophysiol.* **33**, 827–837.

Fields, H. L., Clanton, C. H., and Anderson, S. D. 1977. Somatosensory properties of spinoreticular neurons in the cat. *Brain Res.* **120**, 49–66.

Fields, H. L., Emson, P. C., Leigh, B. K., Gilbert, R. F. T., and Iverssen, L. L. 1980. Multiple opiate receptor sites on primary afferent fibres. *Nature* **284**, 351–353.

Finley, J. C. W., Maderdrut, J. L., Roger, L. J., and Petrusz, P. 1981a. The immunocytochemical localization of somatostatin-containing neurons in the rat central nervous system. *Neuroscience* **6**, 2173–2192.

Finley, J. C. W., Maderdrut, J. L., and Petrusz, P. 1981b. The immunocytochemical localization of enkephalin in the central nervous system of the rat. *J. Comp. Neurol.* **198**, 541–565.

Fischette, C. T., Nock, B., and Renner, K. 1987. Effects of 5,7-dihydroxytryptamine on $serotonin_1$ and $serotonin_2$ receptors throughout the rat central nervous system using quantitative autoradiography. *Brain Res.* **421**, 263–279.

Fisher, K. and Coderre, T. J. 1996a. The contribution of metabotropic glutamate receptors (mGluRs) to formalin-induced nociception. *Pain* **68**, 255–263.

Fisher, K. and Coderre, T. J. 1996b. Comparison of nociceptive effects produced by intrathecal administration of mGluR agonists. *NeuroReport* **7**, 2743–2747.

Fitzgerald, E. M., Okuse, K., Wood, J. N., Dolphin, A. C., and Moss, S. J. 1999. cAMP-dependent phosphorylation of the tetrodotoxin-resistant voltage-dependent sodium channel SNS. *J. Physiol.* **516**, 433–446.

Fitzgerald, M. 1981. A study of the cutaneous afferent input to substantia gelatinosa. *Neuroscience* **6**, 2229–2237.

Fitzgerald, M. 1983. Influences of contralateral nerve and skin stimulation on neurones in the substantia gelatinosa of the rat spinal cord. *Neurosci. Lett.* **36**, 139–143.

Fitzgerald, M. 1989. The course and termination of primary afferent fibers. In P. D. Wall and R. Melzack (eds.), *Textbook of Pain* (pp. 46–62). Churchhill Livingstone, New York.

Fitzgerald, M. and Lynn, B. 1977. The sensitization of high threshold mechanoreceptors with myelinated axons by repeated heating. *J. Physiol.* **265**, 549–563.

Fitzgerald, M. and Vrbova, G. 1985. Plasticity of acid phosphatase (FRAP) afferent terminal fields and of dorsal horn cell growth in the neonatal rat. *J. Comp. Neurol.* **240**, 414–422.

Fitzgerald, M. and Wall, P. D. 1980. The laminar organization of dorsal horn cells responding to peripheral C fibre stimulation. *Exp. Brain Res.* **41**, 36–44.

Fitzgerald, M. and Woolf, C. J. 1981. Effects of cutaneous nerve and intraspinal conditioning on C-fibre afferent terminal excitability in decerebrate rats. *J. Physiol.* **318**, 25–39.

Fitzgerald, M., Wall, P. D., Goedert, M., and Emson, C. 1985. Nerve growth factor counteracts the neurophysiological and neurochemical effects of chronic sciatic nerve section. *Brain Res.* **332**, 131–141.

Fitzgerald, M., Woolf, C. J., and Shortland, P. 1990. Collateral sprouting of the central terminals of cutaneous primary afferent neurons in the rat spinal cord: Pattern, morphology and influence of targets. *J. Comp. Neurol.* **300**, 370–385.

Fjällbrandt, N. and Iggo, A. 1961. The effect of histamine, 5-hydroxytryptamine and acetylcholine on cutaneous afferent fibres. *J. Physiol.* **156**, 578–590.

Fjell, J., Cummins, T. R., Dib-Hajj, S. D., Fried, K., Black, J. A., and Waxman, S. G. 1999a. Differential role of GDNF and NGF in the maintenance of two TTX-resistant sodium channels in adult DRG neurons. *Mol. Brain Res.* **67**, 267–282.

Fjell, J., Cummins, T. R., Fried, K., Black, J. A., and Waxman, S. G. 1999b. In vivo NGF deprivation reduces SNS expression and TTX-R sodium currents in IB4-negative DRG neurons. *J. Neurophysiol.* **81**, 803–810.

Fjell, J., Cummins, T. R., Davis, B. M., Albers, K. M., Fried, K., Waxman, S. G., and Black, J. A. 1999c. Sodium channel expression in NGF-overexpressing transgenic mice. *J. Neurosci. Res.* **57**, 39–47.

Fjell, J., Hjelmström, P., Hormuzdiar, W., Milenkovic, M., Aglieco, F., Tyrrell, L., Dib-Hajj, S., Waxman, S. G., and Black, J. A. 2000. Localization of the tetrodotoxin-resistant sodium channel NaN in nociceptors. *NeuroReport* **11**, 199–202.

Fleetwood-Walker, S. M., Hope, P. J., Mitchell, R., El-Yassir, N., and Molony, V. 1988. The influence of opioid receptor subtypes on the processing of nociceptive inputs in the spinal dorsal horn of the cat. *Brain Res.* **451**, 213–226.

Fleetwood-Walker, S. M., Mitchell, R., Hope, P. J., El-Yassir, N., Molony, V., and Bladon, C. M. 1990. The involvement of neurokinin receptor subtypes in somatosensory processing in the superficial dorsal horn of the cat. *Brain Res.* **519**, 169–182.

Fleetwood-Walker, S. M., Mitchell, R., Hope, P. J., El-Yassir, N., Molony, V., and Bladon, C. M. 1992. The involvement of neurokinin receptor subtypes in somatosensory processing in the superficial dorsal horn of the cat. *Brain Res.* **519**, 169–182.

Fleetwood-Walker, S. M., Parker, R. M., Munro, F. E., Young, M. R., Hope, P. J., and Mitchell, R. 1993. Evidence for a role of tachykinin NK2 receptors in mediating brief nociceptive inputs to rat dorsal horn (laminae III–V) neurons. *Eur. J. Pharmacol.* **242**, 173–181.

Fleischer, E., Handwerker, H. O., and Joukhadar, S. 1983. Unmyelinated nociceptive units in two skin areas of the rat. *Brain Res.* **267**, 81–92.

Fleming, A. A. and Todd, A. J. 1994. Thyrotropin-releasing hormone- and GABA-like immunoreactivity coexist in neurons in the dorsal horn of the rat spinal cord. *Brain Res.* **638**, 347–351.

Fletcher, S. and Barnes, N. M. 1999. Autoradiographic localization of the [3H]-(S)-zacopride labelled 5-HT3 receptor in porcine brain. *Neurosci. Lett.* **269**, 91–94.

Florence, S. L., Wall, J. T., and Kaas, J. H. 1988. The somatotopic pattern of afferent projections from the digits to the spinal cord and cuneate nucleus in macaque monkeys. *Brain Res.* **452**, 388–392.

Flores, C. M., DeCamp, R. M., Kilo, S., Roger, S. W., and Hargreaves, K. M. 1996. Neuronal nicotinic receptor expression in sensory neurons of the rat trigeminal ganglion: Demonstration of $\alpha3\beta4$, a novel subtype in the mammalian nervous system. *J. Neurosci.* **16**, 7892–7901.

Florin, S., Meunier, J., and Costentin, J. 2000. Autoradiographic localization of [3H]nociceptin binding sites in the rat brain. *Brain Res.* **880**, 11–16.

Floyd, K. and Morrison, J. F. B. 1974. Splanchnic mechanoreceptors in the dog. *Quart. J. Exp. Physiol.* **59**, 361–366.

Floyd, K., Hick, V. E., and Morrison, J. F. B. 1976. Mechanosensitive afferent units in the hypogastric nerve of the cat. *J. Physiol.* **259**, 457–471.

Floyd, K., Hick, V. E., Koley, J., and Morrison, J. F. B. 1977. The effects of bradykinin on afferent units in intra-abdominal sympathetic nerve trunks. *Q. J. Exp. Physiol.* **62**, 19–25.

Fock, S. and Mense, S. 1976. Excitatory effects of 5-hydroxytryptamine, histamine and potassium ions on muscular group IV afferent units: A comparison with bradykinin. *Brain Res.* **105**, 459–469.

Fonseca, M. I., Ni, Y. G., Dunning, D. D., and Miledi, R. 2001. Distribution of serotonin 2A, 2C and 3 receptor mRNA in spinal cord and medulla oblongata. *Brain Res. Mol. Brain Res.* **89**, 11–19.

Foote, S. L. and Cha, C. I. 1988. Distribution of corticotropin-releasing factor-like immunoreactivity in brainstem of two monkey species (*Saimiri sciureus* and *Macaca fascicularis*): An immunocytochemical analysis. *J. Comp. Neurol.* **276**, 239–264.

Foreman, R. D. 1977. Viscerosomatic convergence onto spinal neurons responding to afferent fibers located in the inferior cardiac nerve. *Brain Res.* **137**, 164–168.

Foreman, R. D. and Ohata, C. A. 1980. Effects of coronary artery occlusion on thoracic spinal neurons receiving viscerosomatic inputs. *Am. J. Physiol.* **238**, H667–H674.

Foreman, R. D., Kenshalo, D. R., Schmidt, R. F., and Willis, W. D. 1979. Field potentials and excitation of primate spinothalamic neurons in response to volleys in muscle afferents. *J. Physiol.* **286**, 197–213.

Forrest, V., Ince, P., Leitch, M., Marshall, E. F., and Shaw, P. J. 1996. Serotonergic neurotransmission in the spinal cord and motor cortex of patients with motor neuron disease and controls: Quantitative autoradiography for 5-HT1a and 5-HT2 receptors. *J. Neurol. Sci.* Suppl. **139**, 83–90.

Forssmann, W. G. 1978. A new somatostatinergic system in the mammalian spinal cord. *Neurosci. Lett.* **10**, 293–297.

Forssmann, W. G., Burnweit, C., Shehab, T., and Triepel, J. 1979. Somatostatin-immunoreactive nerve cell bodies and fibers in the medulla oblongata et spinalis. *J. Histochem. Cytochem.* **27**, 1391–1393.

Forster, C., Greiner, T., Nischik, M., Schmelz, M., and Handwerker, H. O. 1995. Neurogenic flare responses are heterogeneous in superficial and deep layers of human skin. *Neurosci. Lett.* **185**, 33–36.

Foster, G. A. and Johansson, O. 1985. Ultrastructural morphometric analysis of somatostatin-like immunoreactive neurones in the rat central nervous system after labelling with colloidal gold. *Brain Res.* **342**, 117–127.

Foster, J. A. and Phelps, P. E. 2000a. Neurons expressing NADPH-diaphorase in the developing human spinal cord. *J. Comp. Neurol.* **427**, 417–427.

Foster, J. A. and Phelps, P. E. 2000b. NADPH-diaphorase reveals presumptive sympathetic primary afferents in the developing human spinal cord. *Auton. Neurosci.* **84**, 111–117.

Fotuhi, M., Sharp, A. H., Glatt, C. E., Hwang, P. M., von Krosigk, M., Snyder, S. H., and Dawson, T. M. 1993. Differential localization of phosphoinositide-linked metabotropic glutamate receptor (mGluR1) and the inositol 1,4,5-trisphosphate receptor in rat brain. *J. Neurosci.* **13**, 2001–2012.

Fournet, N., Garcia-Segura, L. M., Norman, A. W., and Orci, L. 1986. Selective localization of calcium-binding protein in human brainstem, cerebellum and spinal cord. *Brain Res.* **399**, 310–316.

Franco-Cereceda, A., Henke, H., Lundberg, J. M., Peterman, J. B., Hökfelt, T., and Fisher, J. A. 1986. Calcitonin gene-related peptide (CGRP) in capsaicin-sensitive substance P-immunoreactive sensory neurons in animals and man: Distribution and release by capsaicin. *Peptides* **8**, 399–410.

Frank, K. and Fuortes, M. G. J. 1955. Potentials recorded from the spinal cord with microelectrodes. *J. Physiol.* **130**, 625–654.

Frank, K. and Fuortes, M. G. F. 1956. Unitary activity of spinal interneurones of cats. *J. Physiol.* **131**, 425–435.

Frankenhaeuser, B. 1949. Impulses from a cutaneous receptor with slow adaptation and low mechanical threshold. *Acta Physiol. Scand.* **18**, 68–74.

Franz, M. and Iggo, A. 1968. Dorsal root potentials and ventral root reflexes evoked by nonmyelinated fibers. *Science* **162**, 1140–1142.

Franz, M. and Mense, S. 1975. Muscle receptors with group IV afferent fibres responding to application of bradykinin. *Brain Res.* **92**, 369–383.

Frederickson, R. C. A. and Geary, L. E. 1982. Endogenous opioid peptides: Review of physiological, pharmacological and clinical aspects. *Prog. Neurobiol.* **19**, 19–69.

Freeman, A. W. and Johnson, K. O. 1982a. Cutaneous mechanoreceptors in macaque monkey: Temporal discharge patterns evoked by vibration, and a receptor model. *J. Physiol.* **323**, 21–41.

Freeman, A. W. and Johnson, K. O. 1982b. A model accounting for effects of vibratory amplitude on responses of cutaneous mechanoreceptors in macaque monkey. *J. Physiol.* **323**, 43–64.

Freeman, M. A. R. and Wyke, B. 1967. The innervation of the knee joint: An anatomical and histological study in the cat. *J. Anat.* **101**, 502–532.

Fried, K., Risling, M., Arvidsson, U., and Paulie, S. 1990. Nerve growth factor receptor-like immunoreactivity in nerve fibers in the spinal and medullary dorsal horn of the adult monkey and cat: Correlation with calcitonin gene-related peptide-like immunoreactivity. *Brain Res.* **536**, 321–326.

Frisen, J., Risling, M., Theodorsson, E., and Fried, K. 1992. NPY-like immunoreactivity in sensory nerve fibers in rat sciatic neuroma. *Brain Res.* **577**, 142–146.

Frost White, W., O'Gorman, S., and Roe, A. W. 1990. Three-dimensional autoradiographic localization of quench-corrected glycine receptor specific activity in the mouse brain using ^{3}H-strychnine as the ligand. *J. Neurosci.* **10**, 795–813.

Frostholm, A. and Rotter, A. 1985. Glycine receptor distribution in mouse CNS: Autoradiographic localization of [3H]strychnine binding sites. *Brain Res. Bull.* **15**, 473–486.

Fruhstorfer, H., Zenz, M., Nolte, H., and Hensel, H. 1974. Dissociated loss of cold and warm sensibility during regional anaesthesia. *Pfluegers Arch.* **349**, 73–82.

Frykholm, R. 1951. Cervical nerve root compression resulting from disc degeneration and root-sleeve fibrosis: A clinical investigation. *Acta Chirurgica Scand.* **160**, 1–149.

Frykholm, R., Hyde, J., Norlen, G., and Skoglund, C. R. 1953. On pain sensations produced by stimulation of ventral roots in man. *Acta Physiol. Scand.* **29**, 455–469.

Fu, L. W. and Longhurst, J. C. 1998. Role of 5-HT_3 receptors in activation of abdominal sympathetic C fibre afferents during ischaemia in cats. *J. Physiol.* **509**, 729–740.

Fu, L. W. and Longhurst, J. C. 1999. Interleukin-1β sensitizes abdominal visceral afferents of cats to ischaemia and histamine. *J. Physiol.* **521**, 249–260.

Fu, L. W., Pan, H. L., and Longhurst, J. C. 1997. Endogenous histamine stimulates ischemically sensitive abdominal visceral afferents through H_1 receptors. *Am. J. Physiol.* **273**, H2726–H2737.

Fuji, K., Senba, E., Ueda, Y., and Tohyama, M. 1983. Vasoactive intestinal polypeptide (VIP)-containing neurons in the spinal cord of the rat and their projections. *Neurosci. Lett.* **37**, 51–55.

Fuji, K., Senba, E., Fujii, S., Nomura, I., Wu, J. Y., Ueda, Y., and Tohyama, M. 1985. Distribution, ontogeny and projections of cholecystokinin-8, vasoactive intestinal polypeptide and γ-aminobutyrate-containing neuron systems in the rat spinal cord: An immunohistochemical analysis. *Neuroscience* **14**, 881–894.

Fujita, M., Sato, K., Sato, M., Inoue, T., Kozuka, T., and Tohyama, M. 1991. Regional distribution of the cells expressing glycine receptor β subunit mRNA in the rat brain. *Brain Res.* **560**, 23–37.

Fukuoka, T., Tokunaga, A., Kondo, E., Miki, K., Tachibana, T., and Noguchi, K. 1998. Change in mRNAs for neuropeptides and the $GABA_A$ receptor in dorsal root ganglion neurons in a rat experimental neuropathic pain model. *Pain* **78**, 13–26.

Fukushima, K. and Kato, M. 1975. Spinal interneurons responding to group II muscle afferent fibers in the cat. *Brain Res.* **90**, 307–312.

Fuller, J. H. and Schlag, J. D. 1976. Determination of antidromic excitation by the collision test: Problems of interpetation. *Brain Res.* **112**, 283–298.

Fundin, B. T., Silos-Santiago, I., Ernfors, P., Fagan, A. M., Aldskogius, H., DeChiara, T. M., Phillips, H. S., Barbacid, M., Yancopoulos, G. D., and Rice, F. L. 1997. Differential dependency of cutaneous mechanoreceptors on neurotrophins, trk receptors, and P75 LNGFR. *Dev. Biol.* **190**, 94–116.

Fundytus, M. E., Fisher, K., Dray, A., Henry, J. L., and Coderre, T. J. 1998. In vivo antinociceptive activity of anti-rat mGluR1 and mGLuR5 antibodies in rats. *NeuroReport* **9**, 731–735.

Fundytus, M. E., Yashpal, K., Chabot, J. G., Osborne, M. G., Lefebvre, C. D., Dray, A., Henry, J. L., and Coderre, T. J. 2001. Knockdown of spinal metabotropic glutamate receptor 1 (mGluR(1)) alleviates pain and restores opioid efficacy after nerve injury in rats. *Br. J. Pharmacol.* **132**, 354–367.

Furuta, A., Rothstein, J. D., and Martin, L. J. 1997. Glutamate transporter protein subtypes are expressed differentially during rat CNS development. *J. Neurosci.* **17**, 8363–8375.

Furuyama, T., Sato, M., Sato, K., Araki, T., Inagaki, S., Takagi, H., and Tohyama, M. 1992. Co-expression of glycine receptor beta subunit and GABAA receptor gamma subunit mRNA in the rat dorsal root ganglion cells. *Brain Res. Mol. Brain Res.* **12**, 335–338.

Furuyama, T., Kiyama, H., Sato, K., Park, H. T., Maeno, H., Takagi, H., and Tohyama, M. 1993. Region-specific expression of subunits of ionotropic glutamate receptors (AMPA-type, KA-type and NMDA receptors) in the rat spinal cord with special reference to nociception. *Mol. Brain Res.* **18**, 141–151.

Fuxe, K. 1965. The distribution of monoamine terminals in the central nervous system. *Acta Physiol. Scand.* **64**, 37–85.

Fuxe, K., Ganten, D., Hökfelt, T., and Bolme, P. 1976. Immunohistochemical evidence for the existence of angiotensin II-containing nerve terminals in the brain and spinal cord in the rat. *Neurosci. Lett.* **2**, 229–234.

Fuxe, K., Agnati, L. F., McDonald, T., Locatelli, V., Hökfelt, T., Dalsgaard, C. J., Battistini, N., Yanaihara, N., Mutt, V., and Cuello, A. C. 1983. Immunohistochemical indications of gastrin releasing peptide bombesin-like immunoreactivity in the nervous system of the rat: Codistribution with substance P-like immunoreactive terminal systems and coexistence with substance P-like immunoreactivity in dorsal root ganglion cell bodies. *Neurosci. Lett.* **37**, 17–22.

Fyffe, R. E. W. 1984. Afferent fibers. In R. A. Davidoff (ed.), *Handbook of the Spinal Cord* (pp. 79–136). Dekker, New York.

Fyffe, R. E. W. and Light, A. R. 1984. The ultrastructure of group Ia afferent fiber synapses in the lumbosacral spinal cord of the cat. *Brain Res.* **300**, 201–209.

Fyffe, R. E. W. and Perl, E. R. 1984. Is ATP a central synaptic mediator for certain primary afferent fibers from mammalian skin? *Proc. Natl. Acad. Sci. USA* **81**, 6890–6893.

Gage, H. D., Gage, J. C., Tobin, J. R., Chiari, A., Tong, C., Xu, Z., Mach, R. H., Efange, S. M., Ehrenkaufer, R. L., and Eisenach, J. C. 2001. Morphine-induced spinal cholinergic activation: In vivo imaging with positron emission tomography. *Pain* **91**, 139–145.

Galan, A., Lopes-Garcia, J. A., Cervero, F., and Laird, J. M. 2002. Activation of spinal extracellular signaling-regulated kinase-1 and -2 by intraplantar carrageenan in rodents. *Neurosci. Lett.* **322**, 37–40.

Galeazza, M. T., Stucky, C. L., and Seybold, V. S. 1992. Changes in [^{125}I]hCGRP binding in rat spinal cord in an experimental model of acute, peripheral inflammation. *Brain Res.* **591**, 198–208.

Galeazza, M. T., Garry, M. G., Yost, H. J., Strait, K. A., Hargreaves, K. M., and Seybold, V. S. 1995. Plasticity in the synthesis and storage of substance P and calcitonin gene-related peptide in primary afferent neurons during peripheral inflammation. *Neuroscience* **66**, 443–458.

Galhardo, V. and Lima, D. 1999. Structural characterization of marginal (lamina I) spinal cord neurons in the cat: A Golgi study. *J. Comp. Neurol.* **414**, 315–333.

Gallagher, J. P., Higashi, H., and Nishi, S. 1978. Characterization and ionic basis of GABA-induced depolarizations recorded in vitro from cat primary afferent neurones. *J. Physiol.* **275**, 263–282.

Gallego, R. and Eyzaguirre, C. 1978. Membrane and action potential characteristics of A and C nodose ganglion cells studies in whole ganglia and in tissue slices. *J. Neurophysiol.* **41**, 1217–1232.

Gamboa-Esteves, F. O., Kaye, J. C., McWilliam, P. N., Lima, D., and Batten, T. F. 2001a. Immunohistochemical profiles of spinal lamina I neurones retrogradely labelled from the nucleus tractus solitarii in rat suggest excitatory projections. *Neuroscience* **104**, 523–538.

Gamboa-Esteves, F. O., Lima, D., and Batten, T. F. 2001b. Neurochemistry of superficial spinal neurones projecting to nucleus of the solitary tract that express c-fos on chemical somatic and visceral nociceptive input in the rat. *Metab Brain Dis.* **16**, 151–164.

Game, C. J. A. and Lodge, D. 1975. The pharmacology of the inhibition of dorsal horn neurones by impulses in myelinated cutaneous afferents in the cat. *Exp. Brain Res.* **23**, 75–84.

Gammon, G. D. and Bronk, D. M. 1935. The discharge of impulses from Pacinian corpuscles in the mesentery and its relation to vascular changes. *Am. J. Physiol.* **114**, 77–84.

Gammon, G. D. and Starr, I. 1941. Studies on the relief of pain by counterirritation. *J. Clin. Invest.* **20**, 13–20.

Gamse, R. and Saria, A. 1986. Nociceptive behavior after intrathecal injections of substance P, neurokinin A and calcitonin gene-related peptide in mice. *Neurosci. Lett.* **70**, 143–147.

Gamse, R. and Saria, A. 1987. Antidromic vasodilation in the rat hindpaw measured by laser Doppler flowmetry: Pharmacological modulation. *J. Auton. Nerv. System* **19**, 105–111.

Gamse, R., Molnar, A., and Lembeck, F. 1979. Substance P release from spinal cord slices by capsaicin. *Life Sci.* **25**, 629–636.

Gandevia, S. C. 1985. Illusory movements produced by electrical stimulation of low-threshold muscle afferents from the hand. *Brain* **108**, 965–981.

Gandevia, S. C., Hall, L. A., McCloskey, D. I., and Potter, E. K. 1983. Proprioceptive sensation at the terminal joint of the middle finger. *J. Physiol.* **335**, 507–517.

Gandevia, S. C., McCloskey, D. I., and Burke, D. 1992. Kinaesthetic signals and muscle contraction. *Trends Neurosci.* **15**, 62–65.

Gangula, P. R., Lanlua, P., Wimalawansa, S., Supowit, S., DiPette, D., and Yallampalli, C. 2000. Regulation of calcitonin gene-related peptide expression in dorsal root ganglia of rats by female sex steroid hormones. *Biol. Reprod.* **62**, 1033–1039.

García-Añoveros, J. and Corey, D. P. 1997. The molecules of mechanosensation. *Annu. Rev. Neurosci.* **20**, 567–594.

García-Añoveros, J., Derfler, B., Neville-Golden, J., Hyman, B. T., and Corey, D. P. 1997. BnaC1 and BnaC2 constitute a new family of human neuronal sodium channels related to degenerins and epithelial sodium channels. *Proc. Natl. Acad. Sci USA* **94**, 1459–1464.

García-Añoveros, J., Samad, T. A., Žuvela-Jelaska, L., Woolf, C. J., and Corey, D. P. 2001. Transport and localization of the DEG/EnaC ion channel BNaC1α to peripheral mechanosensory terminals of dorsal root ganglia neurons. *J. Neurosci.* **21**, 2678–2686.

Garcia-Segura, L. M., Baetens, D., Roth, J., Norman, A. W., and Orci, L. 1984. Immunohistochemical mapping of calcium-binding protein immunoreactivity in the rat central nervous system. *Brain Res.* **296**, 75–86.

Gardner, E. 1944. The distribution and termination of nerves in the knee joint of the cat. *J. Comp. Neurol.* **80**, 11–32.

Gardner, E. 1948. Conduction rates and dorsal root inflow of sensory fibers from the knee joint of the cat. *Am. J. Physiol.* **152**, 436–445.

Garraway, S. M. and Hochman, S. 2001. Modulatory actions of serotonin, norepinephrine, dopamine, and acetylcholine in spinal cord deep dorsal horn neurons. *J. Neurophysiol.* **86**, 2183–2194.

Garrett, N. E., Kidd, B. L., Cruwys, S. C., and Tomlinson, D. R. 1995. Changes in preprotachykinin mRNA expression and substance P levels in dorsal root ganglia of monoarthritic rats: Comparison with changes in synovial substance P levels. *Brain Res.* **675**, 203–207.

Garrison, C. J., Dougherty, P. M., and Carlton, S. M. 1993. Quantitative analysis of substance P and calcitonin gene-related peptide immunohistochemical staining in the dorsal horn of neuropathic MK-801-treated rats. *Brain Res.* **607**, 205–214.

Garry, M. G., Miller, K. E., and Seybold, V. S. 1989. Lumbar dorsal root ganglia of the cat: A quantitative study of peptide immunoreactivity and cell size. *J. Comp. Neurol.* **284**, 36–47.

Garry, M. G., Kajander, K. C., Bennett, G. J., and Seybold, V. S. 1991. Quantitative autoradiographic analysis of [125I]-human CGRP binding sites in the dorsal horn of rat following chronic constriction injury or dorsal rhizotomy. *Peptides* **12**, 1365–1373.

Garry, M. G., Richardson, J. D., and Hargreaves, K. M. 1994. Sodium nitroprusside evokes the release of immunoreactive calcitonin gene-related peptide and substance P from dorsal horn slices via nitric oxide-dependent and nitric oxide-indepedent mechanisms. *J. Neurosci.* **14**, 4329–4337.

Garry, M. G., Walton, L. P., and Davis, M. A. 2000. Capsaicin-evoked release of immunoractive calcitonin gene-related peptide from the spinal cord is mediated by nitric oxide but not by cyclic GMP. *Brain Res.* **861**, 208–219.

Gasser, H. S. 1941. The classification of nerve fibers. *Ohio J. Sci.* **41**, 145.

Gasser, H. S. 1950. Unmedullated fibers originating in dorsal root ganglia. *J. Gen. Physiol.* **3**, 651–690.

Gasser, H. S. and Graham, H. T. 1933. Potentials produced in the spinal cord by stimulation of dorsal roots. *Am. J. Physiol.* **103**, 303–320.

Gazelius, B., Edwall, B., Olgart, Lundberg, J. M., Hökfelt, T., and Fisher, J. A. 1987. Vasodilatory effects and coexistence of calcitonin gene-related peptide (CGRP) and substance P in sensory nerves of cat dental pulp. *Acta Physiol. Scand.* **130**, 33–40.

Gebhart, G. F. 1996. Visceral polymodal receptors. In *The Polymodal Receptor: A Gateway to Pathological Pain, Progr. Brain Res.* **113**, 101–112. Elsevier, Amsterdam.

Gebre-Medhin, S., Mulder, H., Zhang, Y., Sundler, F., and Betsholtz, C. 1998. Reduced nociceptive behavior in islet amyloid polypeptide (amylin) knockout mice. *Mol. Brain Res.* **63**, 180–183.

Gehlert, D. R., Gackenheimer, S. L., Wong, D. T., and Robertson, D. W. 1991. Localization of 5-HT_3 receptors in the rat brain using [^{3}H]LY278584. *Brain Res.* **553**, 149–154.

Gehlert, D. R., Speth, R. C., and Wamsley, J. K. 1986. Distribution of [125]angiotensin II binding sites in the rat brain: A quantitative autoradiographic study. *Neuroscience* **18**, 837–856.

Geiger, J. D. and Nagy, J. I. 1985. Localization of [3H] nitrobenzylthioinosine binding sites in rat spinal cord and primary afferent neurons. *Brain Res.* **347**, 321–327.

Geldard, F. A. 1953. *The Human Senses.* John Wiley & Sons, New York.

Genzen, J. R., Cleve, W. van, McGehee, D. S. 2001. Dorsal root ganglion neurons express multiple nicotinic acteylcholine receptor subtypes. *J. Neurophysiol.* **86**, 1773–1782.

George, A., Marziniak, M., Schafers, M., Toyka, K. V., and Sommer, C. 2000. Thalidomide treatment in chronic constrictive neuropathy decreases endoneurial tumor necrosis factor-alpha, increases interleukin-10 and has long-term effects on spinal cord dorsal horn met-enkephalin. *Pain* **88**, 267–275.

Georgopoulos, A. P. 1976. Functional properties of primary afferent units probably related to pain mechanisms in primate glabrous skin. *J. Neurophysiol.* **39**, 71–83.

Georgopoulos, A. P. 1977. Stimulus-response relations in high-threshold mechanothermal fibers innervating primate glabrous skin. *Brain Res.* **128**, 547–552.

Geppetti, P. and Holzer, P. (eds.) 1996. *Neurogenic Inflammation*, CRC Press, Boca Raton.

Gerber, G. and Randic, M. 1989a. Excitatory amino acid-mediated components of synaptically evoked input from dorsal roots to deep dorsal horn neurons in the rat spinal cord slice. *Neurosci. Lett.* **106**, 211–219.

Gerber, G. and Randic, M. 1989b. Participation of excitatory amino acid receptors in the slow excitatory synaptic transmission in the rat dorsal horn in vitro. *Neurosci. Lett.* **106**, 220–228.

Gerber, G., Youn, D. H., Hsu, C. H., Isaev, D., and Randic, M. 2000. Spinal dorsal horn synaptic plasticity: Involvement of group I metabotropic glutamate receptors. *Prog. Brain Res.* **129**, 115–134.

Gerebtzoff, M. A. 1959. Cholinesterases-modern trends in physical science, In *International Monographs on Pure and Applied Biological Science* (pp. 1–105). Pergamon Press, London, NY, LA.

Gerebtzoff, M. A. and Maeda, T. 1968. Caractères et localisation histochimique d'un isoenzyme fluororésistant de la phosphatase acide dans la moelle épinière du rat. *Comp. Rend.* **162**, 2032–2035.

Gernandt, B. and Zotterman, Y. 1946. Intestinal pain: An electrophysiological investigation on mesenteric nerves. *Acta Physiol. Scand.* **12**, 56–72.

Ghelardini, C., Galeotti, N., and Bartolini, A. 2000. Loss of muscarinic antinociception by antisense inhibition of M-1 receptors. *Br. J. Pharmacol.* **129**, 1633–1640.

Ghilardi, J. R., Allen, C. J., Vigna, S. R., McVrey, D. C., and Mantyh, P. W. 1992. Trigeminal and dorsal root ganglion neurons express CCK receptor binding sites in the rat, rabbit, and monkey: Possible site of opiate-CCK analgesic interactions. *J. Neurosci.* **12**, 4854–4866.

Giacobini, E. and Kerpel-Fronius, S. 1970. Histochemical and biochemical correlations of monoamine oxidase activity in autonomic and sensory ganglia of the cat. *Acta Physiol. Scand.* **78**, 522–528.

Giacobini, E. and Norre, B. 1971. Dopa-decarboxylase in autonomic and sensory ganglia of the cat. *Acta Physiol. Scand.* **82**, 209–217.

Giamberardino, M. A., Valente, R., de Bigontina, P., and Vecchiet, L. 1995. Artificial ureteral calculosis in rats: behavioural charactertization of visceral pain episodes and their relationship with referred lumbar muscle hyperalgesia. *Pain* **61**, 459–469.

Gibbins, I. L., Furness, J. B., Costa, M., MacIntyre, I., Hillyard, C. J., and Girgis, S. 1985. Co-localization of calcitonin gene-related peptide-like immunoreactivity with substance P in cutaneous, vascular and visceral sensory neurons of guinea pigs. *Neurosci. Lett.* **57**, 125–130.

Gibbins, I. L., Wattchow, D., and Coventry, B. 1987a. Immunohistochemically identified population of calcitonin gene-related peptide-immunoreactive (CGRP-IR) axons in human skin. *Brain Res.* **414**, 143–148.

Gibbins, I. L., Furness, J. B., and Costa, M. 1987b. Pathway-specific patterns of the co-existence of substance P, calcitonin gene-related peptide, cholecystokinin and dynorphin in neurons of the dorsal root ganglia of the guinea-pig. *Cell Tissue Res.* **248**, 417–437.

Gibson, S. J., Polak, J. M., Bloom, S. R., and Wall, P. D. 1981. The distribution of nine peptides in rat spinal cord with special emphasis on the substantia gelatinosa and on the area around the central canal (lamina X). *J. Comp. Neurol.* **201**, 65–79.

Gibson, S. J., McGregor, G., Bloom, S. R., Polak, J. M., and Wall, P. D. 1982. Local application of capsaicin to one sciatic nerve of the adult rat induces a marked depletion in the peptide content of the lumbar dorsal horn. *Neuroscience* **7**, 3153–3162.

Gibson, S. J., Polak, J. M., Allen, J. M., Adrian, T. E., Kelly, J. S., and Bloom, S. R. 1984a. The distribution and origin of a novel brain peptide, neuropeptide Y, in the spinal cord of several mammals. *J. Comp. Neurol.* **227**, 78–91.

Gibson, S. J., Polak, J. M., Bloom, S. R., Sabate, I. M., Mulderry, P. M., Ghatei, M. A., McGregor, G. P., Morrison, J. F., Kelly, J. S., Evans, R. M., and Rosenfeld, M. G. 1984b. Calcitonin gene-related peptide immunoreactivity in the spinal cord of man and eight other species. *J. Neurosci.* **4**, 3101–3111.

Gibson, S. J., Polak, J. M., Anand, P., Blank, M. A., Morrison, J. F. B., Kelly, J. S., and Bloom, S. R. 1984c. The distribution and origin of VIP in the spinal cord of six mammalian species. *Peptides* **5**, 201–207.

Gibson, S. J., Polak, J. M., Giaid, A., Hamid, Q. A., Kar, S., Jones, P. M., Denny, P., Legon, S., Amara, S. G., Craig, R. K., et al. 1988. Calcitonin gene-related peptide messenger RNA is expressed in sensory neurones of the dorsal root ganglia and also in spinal motoneurones in man and rat. *Neurosci. Lett.* **91**, 283–288.

Giesler, G. J., Cannon, J. T., Urca, G., and Liebeskind, J. C. 1978. Long ascending projections from substantia gelatinosa Rolandi and the subjacent dorsal horn in the rat. *Science* **202**, 984–986.

Gillardon, F., Morano, I., and Zimmermann, M. 1991. Ultraviolet irradiation of the skin attenuates calcitonin gene-related peptide mRNA expression in rat dorsal root ganglion cells. *Neurosci. Lett.* **124**, 144–147.

Gillberg, P. G. and Aquilonius, S.-M. 1985. Cholinergic, opioid and glycine receptor binding sites localized in human spinal cord by *in vitro* autoradiography: Changes in amyotrophic lateral sclerosis. *Acta Neurol. Scand.* **72**, 299–306.

Gillberg, P. G. and Askmark, H. 1991. Changes in cholinergic and opioid receptors in the rat spinal cord, dorsal root and sciatic nerve after ventral and dorsal root lesion. *J. Neural. Trans.* **85**, 31–39.

Gillberg, P. G. and Wiksten, B. 1986. Effects of spinal cord lesions and rhizotomies on cholinergic and opiate receptor binding sites in rat spinal cord. *Acta Physiol. Scand.* **126**, 575–582.

Gillberg, P. G., Nordberg, A., and Aquilonius, S. M. 1984. Muscarinic binding sites in small homogenates and in autoradiographic sections from rat and human spinal cord. *Brain Res.* **300**, 327–333.

Gillberg, P. G., d'Argy, R., and Aquilonius, S. M. 1988. Autoradiographic distribution of [^{3}H] acetylcholine binding sites in the cervical spinal cord of man and some other species. *Neurosci. Lett.* **90**, 197–202.

Giroux, N., Rossignol, S., and Reader, T. A. 1999. Autoradiographic study of α_1- and α_2-noradrenergic and serotonin$_{1A}$ receptors in the spinal cord of normal and chronically transected cats. *J. Comp. Neurol.* **406**, 402–414.

Giuffrida, R. and Rustioni, A. 1992. Dorsal root ganglion neuons projecting to the dorsal column nuclei of rats. *J. Comp. Neurol.* **316**, 206–220.

Gladfelter, W. E., Millecchia, R. J., Pubols, L. M., Sonty, R. V., Ritz, L. A., Covalt-Dunning, D., Culberson, J., and Brown, P. B. 1993. Crossed receptive field components and crossed dentrites in cat sacrocaudal dorsal horn. *J. Comp. Neurol.* **336**, 96–105.

Glass, M., Faull, R. L., and Dragunow, M. 1996. Localisation of the adenosine uptake site in the human brain: A comparison with the distribution of adenosine A1 receptors. *Brain Res.* **710**, 79–91.

Glass, M., Dragunow, M., and Faull, R. L. 1997. Cannabinoid receptors in the human brain: A detailed anatomical and quantitative autoradiographic study in the fetal, neonatal and adult human brain. *Neuroscience* **77**, 299–318.

Glauser, L. and Walter, I. B. 1997. Differential distribution of thyroid hormone receptor isoform in rat dorsal root ganglia and sciatic nerve *in vivo* and *in vitro*. *J. Neuroendocrin.* **9**, 217–227.

Glazer, E. J. and Basbaum, A. I. 1981. Immunohistochemical localization of leucine-enkephalin in the spinal cord of the cat: Enkephalin-containing marginal neurons and pain modulation. *J. Comp. Neurol.* **196**, 377–389.

Glazer, E. J. and Basbaum, A. I. 1982. Opioid neurons and pain modulation: An ultrastructural analysis of enkephalin in cat superficial dorsal horn. *Neuroscience* **10**, 357–376.

Glazer, E. J. and Basbaum, A. I. 1984. Axons which take up [3H] serotonin are presynaptic to enkephalin immunoreactive neurons in cat dorsal horn. *Brain Res.* **298**, 386–391.

Gmelin, G. and Cerletti, A. 1976. Electrophoretic studies on presynaptic inhibition in the mammalian spinal cord. *Experientia* **32**, 756.

Go, V. L. W. and Yaksh, T. L. 1987. Release of substance P from the cat spinal cord. *J. Physiol.* **391**, 141–167.

Gobel, S. 1974. Synaptic organization of the substantia gelatinosa glomeruli in the spinal trigemial nucleus of the adult cat. *J. Neurocytol.* **3**, 219–243.

Gobel, S. 1975. Golgi studies of the substantia gelatinosa neurons in the spinal trigeminal nucleus of the adult cat. *Brain Res.* **83**, 333–338.

Gobel, S. 1976. Dendroaxonic synapse in the substantia gelatinosa glomeruli of the spinal trigeminal nucleus of the cat. *J. Comp. Neurol.* **167**, 165–175.

Gobel, S. 1978a. Golgi studies of the neurons in layer I of the dorsal horn of the medulla (trigeminal nucleus caudalis). *J. Comp. Neurol.* **180**, 375–394.

Gobel, S. 1978b. Golgi studies of the neurons in layer II of the dorsal horn of the medulla (trigeminal nucleus caudalis). *J. Comp. Neurol.* **180**, 395–414.

Gobel, S. 1979. Neural circuitry in the substantia gelatinosa of Rolando: Anatomical insights. *Advances in Pain Research and Therapy* **3**, 175–195.

Gobel, S. 1984. An electron microscopic analysis of the transynaptic effects of peripheral nerve injury subsequent to tooth pulp extirpations on neurons in laminae I and II of the medullary dorsal horn. *J. Neurosci.* **4**, 2281–2290.

Gobel, S. and Binck, J. M. 1977. Degenerative changes in primary trigeminal axons and in neurons in nucleus caudalis following tooth pulp extirpations in the cat. *Brain Res.* **132**, 347–354.

Gobel, S. and Falls, W. M. 1979. Anatomical observations of horseradish peroxidase-filled terminal primary axonal arborizations in layer II of the substantia gelatinosa of Rolando. *Brain Res.* **175**, 335–340.

Gobel, S., Falls, W. M., and Hockfield, S. 1977. The division of the dorsal and ventral horns of the mammalian caudal medulla into eight layers using anatomical criteria. In D. J. Anderson and B. Matthews (eds.), *Pain in the Trigeminal Region* (pp. 443–453). Elsevier Press, North-Holland.

Gobel, S., Falls, W., Bennett, G. J., Abdelmoumene, M., Hayashi, H., and Humphrey, E. 1980. An EM analysis of the synaptic connections of horseradish peroxidase-filled stalked cells and islet cells in the substantia gelatinosa of adult cat spinal cord. *J. Comp. Neurol.* **194**, 781–807.

Gobel, S., Falls, W. M., and Humphrey, E. 1981. Morphology and synaptic connections of ultrafine primary axons in lamina I of the spinal dorsal horn: Candidates for the terminal axonal arbors of primary neurons with unmyelinated (C) axons. *J. Neurosci.* **1**, 1163–1179.

Goehler, L. E. and Sternini, C. 1996. Calcitonin gene-related peptide innervation of the rat hepatobiliary system. *Peptides* **17**, 209–217.

Goff, J. R., Burkey, A. R., Goff, D. J., and Jasmin, L. 1998. Reorganization of the spinal dorsal horn in models of chronic pain: Correlation with behaviour. *Neuroscience* **82**, 559–574.

Gokin, A. P., Kostyuk, P. G., and Preobrazhensky, N. N. 1977. Neuronal mechanisms of interactions of high-threshold visceral and somatic afferent influences in spinal cord and medulla. *J. Physiol., Paris*, **73**, 319–333.

Gold, M. S., White, D. M., Ahlgren, S. C., Guo, M., and Levine, J. D. 1994. Catecholamine-induced mechanical sensitization of cutaneous nociceptors in the rat. *Neurosci. Lett.* **175**, 166–170.

Gold, M. S., Shuster, M. J., and Levine, J. D. 1996a. Characterization of six voltage-gated K^+ currents in adult rat sensory neurons. *J. Neurophysiol.* **75**, 2629–2646.

Gold, M. S., Reichling, D. B., Schuster, M. J., and Levine, J. D. 1996b. Hyperalgesic agents increase a tetrodotoxin-resistant Na^+ current in nociceptors. *Proc. Natl. Acad. Sci. USA* **93**, 1108–1112.

Gold, M. S., Dastmalchi, S., and Levine, J. D. 1996c. Co-expression of nociceptor properties in dorsal root ganglion neurons from the adult rat *in vitro*. *Neuroscience* **71**, 265–275.

Gold, M. S., Dastmalchi, S., and Levine, J. D. 1997. α_2-adrenergic receptor subtypes in rat dorsal root and superior cervical ganglion neurons. *Pain* **69**, 179–190.

Gold, M. S., Levine, J. D., and Correa, A. M. 1998. Modulation of TTX-R I_{Na} by PKC and PKA and their role in PGE_2-induced sensitization of rat sensory neurons *in vitro*. *J. Neurosci.* **18**, 10345–10355.

Goldberger, M. and Murray, M. 1974. Restitution of function and collateral sprouting in cat spinal cord: The deafferented animal. *J. Comp. Neurol.* **158**, 37–54.

Goldberger, M. and Murray, M. 1982. Lack of sprouting and its presence after lesions of the cat spinal cord. *Brain Res.* **241**, 227–239.

Goldin, A. L., Barchi, R. L., Caldwell, J. H., Hofmann, F., Howe, J. R., Hunter, J. C., Kallen, R. G., Mandel, G., Meisler, M. H., Netter, Y. B., Noda, M., Tamkun, M. M., Waxman, S. G., Wood, J. N., and Catterall, W. A. 2000. Nomenclature of voltage-gated sodium channels. *Neuron* **28**, 365–368.

Goldscheider, A. 1884. The specific energy of the sensory nerves of the skin. Transl. from German by Biederman-Thorsen in H. O. Handwerker and K. Brune (eds.) 1987. *Classical German Contributions to Pain Research* (pp. 47–67). Vth World Congress on Pain, Hamburg. Original German: 3. Die Specifische Energie der Gefühlsnerven der Haut. *Monatschefte für practische Dermatologie*. III Band. Nr. 9 und 10.

Goldstein, M. E., House, S. B., and Gainer, H. 1991. NF-L and peripherin immunoreactivities define distinct classes of rat sensory ganglion cells. *J. Neurosci. Res.* **30**, 92–104.

Goldstein, M. E., Grant, P., House, S. B., Henken, D. B., and Gainer, H. 1996. Developmental regulation of two distinct neuronal phenotypes in rat dorsal root ganglia. *Neuroscience* **71**, 243–258.

Gómez-Barrena, E., Martinez-Moreno, E. and Munuera, L. 1996. Segmental sensory innervation of the anterior cruciate ligament and the patellar tendon of the cat's knee. *Acta Orthop. Scand.* **67**, 545–552.

Gong, J., Strong, J. A., Zhang, S., Yue, X., DeHaven, R. N., Daubert, J. D., Cassel, J. A., Yu, G., Mansson, E., and Yu, L. 1998. Endomorphins fully activate a cloned human mu opioid receptor. *FASEB Lett.* **439**, 152–156.

Gong, L.-W., Ding, Y.-Q., Wang, D., Zheng, H.-X., Qin, B.-Z., Li, J.-S., Kaneko, T., and Mizuno, N. 1997. GABAergic synapses on μ-opioid receptor-expressing neurons in the superficial dorsal horn and electron-microscope study in the cat spinal cord. *Neurosci. Lett.* **227**, 33–36.

Gonzalez, D. L., Fuchs, J. L., and Droge, M. H. 1993. Distribution of NMDA receptor binding in developing mouse spinal cord. *Neurosci. Lett.* **151**, 134–137.

Gonzales-Lima, E. 1986. Activation of substantia gelatinosa by midbrain reticular formation stimulation demonstrated with 2-deoxyglucose in the rat spinal cord. *Neurosci. Lett.* **65**, 326–330.

Goodman, R. R. and Snyder, S. H. 1982. Autoradiographic localization of adenosine receptors in rat brain using [3H] cyclohexyladenosine. *J. Neurosci.* **2**, 1230–1241.

Goodman, R. R., Snyder, S. H., Kuhar, M. J., and Young, W. S. 1980. Differentiation of delta and mu opiate receptor localizations by light-microscopic autoradiography. *Proc. Natl. Acad. Sci.* **77**, 6239–6243.

Goodwin, A. W. and Morley, J. W. 1987a. Sinusoidal movement of a grating across the monkey's fingerpad: Representation of grating and movement features in afferent fiber responses. *J. Neurosci.* **7**, 2168–2180.

Goodwin, A. W. and Morley, J. W. 1987b. Sinusoidal movement of a grating across the monkey's fingerpad: Effect of contact angle and force of the grating on afferent fiber responses. *J. Neurosci.* **7**, 2192–2202.

Goodwin, G. M., McCloskey, D. I., and Matthews, P. B. C. 1972a. The persistence of appreciable kinesthesia after paralyzing joint afferents but preserving muscle afferents. *Brain Res.* **37**, 326–329.

Goodwin, G. M., McCloskey, D. I., and Matthews, P. B. C. 1972b. The contribution of muscle afferents to kinaesthesia shown by vibration induced illusions of movement and by the effects of paralyzing joint afferents. *Brain* **95**, 705–748.

Gordon, C. J. and Heath, J. E. 1986. Integration and central processing in temperature regulation. *Ann. Rev. Physiol.* **48**, 595–612.
Gordon, G. (ed.) 1978. Active Touch. *The Mechanism of Recognition of Objects by Manipulation: A Multidisciplinary Approach.* Pergamon Press, Oxford.
Gorio, A., Vergani, L., Ferro, L., Prino, G., and Di Giulio, A. M. 1996. Glycosaminoglycans in nerve injury: II. Effects on transganglionic degeneration and on the expression of neurotrophic factors. *J. Neurosci. Res.* **46**, 572–580.
Gottschaldt, K. M., Iggo, A., and Young, D. W. 1973. Functional characteristics of mechanoreceptors in sinus hair follicles of the cat. *J. Physiol.* **235**, 287–315.
Gouardéres, C. V. and Cros, J. 1984. Opiate binding sites in different levels of rat spinal cord. *Neuropeptides* **5**, 113–116.
Gouardéres, C., Cros, J., and Quirion, R. 1985. Autoradiographic localization of mu, delta and kappa opioid receptor binding sites in rat and guinea-pig spinal cord. *Neuropeptides* **6**, 331–342.
Gouardéres, C., Kopp, N., Cros, J., and Quirion, R. 1986. Kappa opioid receptors in human lumbo-sacral spinal cord. *Brain Res. Bull.* **16**, 355–361.
Gouardéres, C., Beaudet, A., Zajac, J.-M., Cros, J., and Quirion, R. 1991. High resolution radioautographic localizations of [^{125}I]FK-33-824-labelled mu opioid receptors in the spinal cord of normal and deafferented rats. *Neuroscience* **43**, 197–209.
Gouardéres, C., Jhamandas, K., Cridland, R., Cros, J., Quirion, R., and Zajac, J. M. 1993a. Opioid and substance P receptor adaptations in the rat spinal cord following sub-chronic intrathecal treatment with morphine and naloxone. *Neuroscience* **54**, 799–807.
Gouardéres, C., Tellez, S., Tafani, J. A. M., and Zajac, J.-M. 1993b. Quantitative auroradiographic mapping of delta-opioid receptors in the rat central nervous system using [^{125}I][D.Ala^2]deltorphin-I. *Synapse* **13**, 231–240.
Gouardéres, C., Kar, S., and Zajac, J.-M. 1996. Presence of neuropeptide FF receptors on primary afferent fibres of the rat spinal cord. *Neuroscience* **74**, 21–27.
Gouardéres, C., Tafani, J. A. M., Mazaguil, H., and Zajac, J.-M. 1997. Autoradiographic characterization of rat spinal neuropeptide FF receptors by using [^{125}I][D.Tyr^1, (NMe)Phe^3]NPFF. *Brain Res. Bull.* **42**, 231–238.
Gouardéres, C., Tafani, J. A., Meunier, J. C., Jhamandas, K., and Zajak, J. M. 1999. Nociceptin receptors in the rat spinal cord during morphine tolerance. *Brain Res.* **838**, 85–94.
Gouardéres, C., Roumy, M., Advokat, C., Jhamandas, K., and Zajac, J. M. 2000. Dual localization of neuropeptide FF receptors in the rat dorsal horn. *Synapse* **35**, 45–52.
Gougis, S., Prud'homme, M. J., and Rampin, O. 2002. Presence of the N-methyl-D-aspartic acid R1 glutamatergic receptor subunit in the lumbosacral spinal cord of male rats. *Neurosci Lett.* **323**, 224–228.
Gould, H. J., III, England, J. D., Liu, Z. P., and Levinson, S. R. 1998. Rapid sodium channel augmentation in response to inflammation induced by complete Freund's adjuvant. *Brain Res.* **802**, 69–74.
Gould, H. J., III, Gould, T. N., Paul, D., England, J. D., Liu, Z. P., Reeb, S. C., and Levinson, S. R. 1999. Development of inflammatory hypersensitivity and augmentation of sodium channels in rat dorsal root ganglia. *Brain Res.* **824**, 296–299.
Gozlan, H., Ponchant, M., Daval, G., Verge, D., Menard, F., Vanhove, M. A., Beaucourt, J. P., and Hamon, M. 1988. ^{125}I-bolton-hunter-8-methoxy-2-[N-propyl-N- propylamino]tetralin has a new selective radioligand of 5-HT$_{1A}$ sites in the rat brain: *In vitro* binding and autoradiographic studies. *J. Pharm. Exp. Ther.* **244**, 751–759.
Gracely, R. H., Lynch, S. A., and Bennett, G. J. 1992. Painful neuropathy: Altered central processing maintained dynamically by peripheral input. *Pain* **51**, 175–194.
Graham, L. T. and Aprison, M. H. 1969. Distribution of some enzymes associated with the metabolism of glutamate, aspartate, gamma-aminobutyric acid and glutamine in cat spinal cord. *J. Neurochem.* **16**, 559–566.
Graham, L. T., Shank, R. P., Werman, R. and Aprison, M. H. 1967. Distribution of some synaptic transmitter suspects in cat spinal cord: Glutamic acid, aspartic acid, gamma-aminobutyric acid, glycine, and glutamine. *J Neurochem.* **14**, 465–472.
Grandori, F. and Pedotti, A. 1982. A mathematical model of the Pacinian corpuscle. *Biol. Cybern.* **46**, 7–16.
Grant, G. and Ygge, J. 1981. Somatotopic organization of the thoracic spinal nerve in the dorsal horn demonstrated with transganglionic degeneration. *J. Comp. Neurol.* **202**, 357–364.
Gray, E. G. 1963. Electron microscopy of presynaptic organelles of the spinal cord. *J. Anat.* **97**, 101–106.
Greco, A., Villa, R., and Pierotti, M. A. 1996. Genomic organization of the human NTRK1 gene. *Oncogene* **13**, 2463–2466.
Greco, A., Villa, R., Fusetti, L., Orlandi, R., and Pierotti, M. A. 2000. The Gly571Arg mutation, associated with the autonomic and sensory disorder congenital insensitivity to pain with anhidrosis, causes the inactivation of the NTRK1/nerve growth factor receptor. *J. Cell Physiol.* **182**, 127–133.

Green, T. and Dockray, G. J. 1987. Calcitonin gene-related peptide and substance P in afferents to the upper gastrointestinal tract in the rat. *Neurosci. Lett.* **76**, 151–156.

Green, T. and Dockray, G. J. 1988. Characterization of the peptidergic afferent innervation of the stomach in the rat, mouse and guinea-pig. *Neuroscience* **25**, 181–193.

Greenamyre, T., Young, A. B., and Penney, J. B. 1984. Quantitative autoradiographic distribution of L-[3H] glutamate-binding sites in rat central nervous system. *J. Neurosci.* **4**, 2133–2144.

Greenstein, J., Kavanagh, P., and Rowe, M. J. 1987. Phase coherence in vibration-induced responses of tactile fibres associated with Pacinian corpuscle receptors in the cat. *J. Physiol.* **386**, 263–275.

Gregor, M. and Zimmermann, M. 1972. Characteristics of spinal neurones responding to cutaneous myelinated and unmyelinated fibres. *J. Physiol.* **221**, 555–576.

Gregor, M. and Zimmermann, M. 1973. Dorsal root potentials produced by afferent volleys in cutaneous group III fibres. *J. Physiol.* **232**, 413–425.

Gregory, J. E., McIntyre, A. K., and Proske, U. 1989. Tendon organ afferents in the knee joint nerve of the cat. *Neurosci. Lett.* **103**, 287–294.

Greves, P. L., Nyberg, F., Terenius, L., and Hökfelt, T. 1985. Calcitonin gene-related peptide is a potent inhibitor of substance P degradation. *Eur. J. Pharmacol.* **115**, 309–311.

Grigg, P. 1975. Mechanical factors influencing response of joint afferent neurons from cat knee. *J. Neurophysiol.* **38**, 1473–1484.

Grigg, P. 1976. Response of joint afferent neurons in cat medial articular nerve to active and passive movements of the knee. *Brain Res.* **118**, 482–485.

Grigg, P. and Greenspan, B. J. 1977. Response of primate joint afferent neurons to mechanical stimulation of knee joint. *J. Neurophysiol.* **40**, 1–8.

Grigg, P. and Hoffman, A. H. 1982. Properties of Ruffini afferents revealed by stress analysis of isolated sections of cat knee capsule. *J. Neurophysiol.* **47**, 41–54.

Grigg, P. and Hoffman, S. H. 1984. Ruffini mechanoreceptors in isolated joint capsule: Responses correlated with strain energy density. *Somatosens. Res.* **2**, 149–162.

Grigg, P. and Hoffman, A. H. 1996. Stretch-sensitive afferent neurons in cat knee joint capsule: Sensitivity to axial and compression stresses and strains. *J. Neurophysiol.* **75**, 1871–1877.

Grigg, P., Hoffman, A. H., and Fogarty, K. E. 1982. Properties of Golgi-Mazzoni afferents in cat knee joint capsule, as revealed by mechanical studies of isolated joint capsule *J. Neurophysiol.* **47**, 31–40.

Grigg, P., Schaible, H. G., and Schmidt, R. F. 1986. Mechanical sensitivity of group III and IV afferents from posterior articular nerve in normal and inflamed cat knee. *J. Neurophysiol.* **55**, 635–643.

Grill, S. E. and Hallett, M. 1995. Velocity sensitivity of human muscle spindle afferents and slowly adapting type II cutaneous mechanoreceptors. *J. Physiol.* **489**, 593–602.

Grossman, S. D., Wolfe, B. B., Yasuda, R. P., and Wrathall, J. R. 1999. Alterations in AMPA receptor subunit expression after experimental spinal cord contusion injury. *J. Neurosci.* **19**, 5711–5720.

Grossman, S. D., Wolfe, B. B., Yasuda, R. P., and Wrathall, J. R. 2000. Changes in NMDA receptor subunit expression in response to contusive spinal cord injury. *J. Neurochem.* **75**, 174–184.

Groves, M. J., Martinian, L., An, S. F., and Scaravilli, F. 1999. Expression of three oligosaccharide conjugates by neonatal rat dorsal root ganglion neurons: Comparison with CGRP and GAP43 immunoreactivity. *J. Anat.* **195**, 271–280.

Grubb, B. D., Stiller, R. U., and Schaible, H. -G. 1993. Dynamic changes in the receptive field properties of spinal cord neurons with ankle input in rats with unilateral inflammation in the ankle region. *Exp. Brain Res.* **92**, 441–452.

Gu, J., Polak, J. M., Tapia, F. J., Marangos, P. J., and Pearse, A. G. 1981. Neuron-specific enolase in the Merkel cells of mammalian skin: The use of specific antibody as a simple and reliable histologic marker. *Am. J. Pathol.* **104**, 63–68.

Guan, Z. L., Ding, Y. Q., Li, J. L., and Lu, B. Z. 1998. Substance P receptor-expressing neurons in the medullary and spinal dorsal horns projecting to the nucleus of the solitary tract in the rat. *Neurosci. Res.* **30**, 213–218.

Guenther, S., Reeh, P. W., and Kress, M. 1999. Rises in $[Ca^{2+}]_i$ mediate capsaicin- and proton-induced heat sensitization of rat primary nociceptor neurons. *Eur. J. Neurosci.* **11**, 3143–3150.

Guilbaud, G., Benelli, G., and Besson, J. M. 1977. Responses of thoracic dorsal horn interneurons to cutaneous stimulation and to the administration of algogenic substances into the mesenteric artery in the spinal cat. *Brain Res.* **124**, 437–448.

Guilford, J. P. and Lovewell, E. M. 1936. The touch spots and the intensity of the stimulus. *J. Gen. Psychol.* **15**, 149–159.

Guinard, D., Usson, Y., Guillermet, C., and Saxod, R. 1998. Merkel complexes of human digital skin: Three-dimensional imaging with confocal laser microscopy and double immunofluorescence. *J. Comp. Neurol.* **398**, 98–104.

Gulbenkian, S., Merighi, A., Wharton, J., Varndell, I. M., and Polak, J. M. 1986. Ultrastructural evidence for the coexistence of calcitonin gene-related peptide and substance P in secretory vesicles of peripheral nerves in the guinea pig. *J. Neurocytol.* **154**, 535–542.

Gundlach, A. L., Largent, B. L., and Snyder, S. H. 1986. Autoradiographic localization of sigma receptor binding sites in guinea pig and rat central nervous system with (+)3-H3-(3-hydroxyphenyl)-N-(1-propyl)piperidine. *J. Neurosci.* **6**, 1757–1770.

Guo, A., Vulchanova, L., Wang, J., Li, X., and Elde, R. 1999. Immunocytochemical localization of the vanilloid receptor 1 (VR1): Relationship to neuropeptides, the P2X3 purinoceptor and IB4 binding sites. *Eur. J. Neurosci.* **11**, 946–958.

Guo, A., Simone, D. A., Stone, L. S., Fairbanks, C. A., Wang, J., and Elde, R. 2001. Developmental shift of vanilloid receptor 1 (VR1) terminals into deeper regions of the superficial dorsal horn: Correlation with a shift from TrkA to Ret expression by dorsal root ganglion neurons. *Eur. J. Neurosci* **14**, 293–304.

Guo, W., Zou, S., Guan, Y., Ikeda, T., Tal, M., Dubner, R., and Ren, K. 2002. Tyrosine phosphorylation of the NR2B subunit of the NMDA receptor in the spinal cord during the development and maintenance of inflammatory hyperalgesia. *J. Neurosci.* **22**, 6208–6217.

Guo, Z. L. and Longhurst, J. C. 2000. Role of cAMP in activation of ischemically sensitive abdominal visceral afferents. *Amer. J. Physiol.* **278**, H843–H852.

Guo, Z. L., Fu, L. W., Symons, J. D., and Longhurst, J. C. 1998. Signal transduction in activation of ischemically sensitive abdominal visceral afferents: Role of PKC. *Am. J. Physiol.* **275**, H1024–H1031.

Guo, Z. L., Symons, J. D., and Longhurst, J. C. 1999. Activation of visceral afferents by bradykinin and ischemia: Independent roles of PKC and prostaglandins. *Am. J.Physiol.* **276**, H1884–H1891.

Gustafson, E. L., Durkin, M. M., Bard, J. A., Zgombick, J., and Branchek, T. A. 1996a. A receptor autoradiographic and *in situ* hybridization analysis of the distribution of the 5-HT_7 receptor in rat brain. *Br. J. Pharm.* **117**, 657–666.

Gustafson, E. L., Smith, K. E., Durkin, M. M., Gerald, C., and Branchek, T. A. 1996b. Distribution of a rat galanin receptor mRNA in rat brain. *NeuroReport* **7**, 953–957.

Gustafson, E. L., Smith, K. E., Durkin, M. M., Walker, M. W., Gerald, C., Weinshank, R., and Branchek, T. A. 1997. Distribution of the neuropeptide Y Y2 receptor mRNA in rat central nervous system. *Mol. Brain Res.* **46**, 223–235.

Gustafsson, H., de Araujo, Lucas G., Scott, E., Stiller, C. O., Alster, P., Wiesenfeld-Hallin, Z., and Brodin, E. 1999. Measurement of cholecystokinin release in vivo in the rat spinal dorsal horn. *Brain Res. Protoc.* **4**, 192–200.

Gutierrez, A., Khan, Z. U., and De Blas, A. L. 1996. Immunocytochemical localization of the α_6 subunit of the gamma-aminobutyric $acid_A$ receptor in the rat nervous system. *J. Comp. Neurol.* **365**, 504–510.

Gutstein, H. B., Bronstein, D. M., and Akil, H. 1992. Beta-endorphin processing and cellular origins in rat spinal cord. *Pain* **51**, 241–247.

Gwyn, D. G. and Flumerfelt, B. A. 1971. Acetylcholinesterase in non-cholinergic neurons—a histochemical study of dorsal root ganlgion cells in the rat. *Brain Res.* **34**, 193–198.

Gybels, J. M. and Sweet, W. H. 1989. *Neurosurgical Treatment of Persistent Pain.* Karger, Basel.

Gynther, B. D., Vickery, R. M., and Rowe, M. J. 1995. Transmission characteristics for the 1:1 linkage between slowly adapting type II fibers and their cuneate target neurons in cat. *Exp. Brain Res.* **105**, 67–75.

Haapanen, L., Kolmodin, G. M., and Skoglund, C. R. 1958. Membrane and action potentials of spinal interneurons in the cat. *Acta Physiol. Scand.* **43**, 315–348.

Habelt, C., Kessler, F., Distler, C., Kress, M., and Reeh, P. W. 2000. Interactions of inflammatory mediators and low pH not influenced by capsazepine in rat cutaneous nociceptors. *NeuroReport* **11**, 973–976.

Haber, S. and Elde, R. 1982a. The distribution of enkephalin immunoreactive neuronal cell bodies in the monkey brain: Preliminary observations. *Neurosci. Lett.* **32**, 247–252.

Haber, S. and Elde, R. 1982b. The distribution of enkephalin immunoreactive fibers and terminals in the monkey central nervous system: An immunohistochemical study. *Neuroscience* **7**, 1049–1095.

Haberberger, R. V. and Bodenbenner, M. 2000. Immunohistochemical localization of muscarinic receptors (M2) in the rat skin. *Cell Tissue Res.* **300**, 389–396.

Haberberger, R., Henrich, M., Couraud, J. Y. and Kummer, W. 1999. Muscarinic M2-receptors in rat thoracic dorsal root ganglia. *Neurosci. Lett.* **266**, 177–180.

Haberberger, R., Scholz, A., Kummer, W., and Kress, M. 2000. M2-receptor subtype does not mediate muscarine-induced increases in $[Ca^{2+}]_i$ in nociceptive neurons of rat dorsal root ganglia. *J. Neurophysiol.* **84**, 1934–1941.

Häbler, H. J., Jänig, W., and Koltzenburg, M. 1987. Activation of unmyelinated afferents in chronically lesioned nerves by adrenaline and excitation of sympathetic efferents in the cat. *Neurosci. Lett.* **82**, 35–40.

Häbler, H. J., Jänig, W., and Koltzenburg, M. 1988a. Dichotomizing unmyelinated afferents supplying pelvic viscera and perineum are rare in the sacral segments of the cat. *Neurosci Lett.* **94**, 119–124.
Häbler, H. J., Jänig, W., and Koltzenburg, M. 1988b. A novel type of unmyelinated chemosensitive nociceptor in the acutely inflamed urinary bladder. *Agents Actions* **25**, 219–221.
Häbler, H. J., Jänig, W., Koltzenburg, M., and McMahon, S. B. 1990a. A quantitative study of the central projection patterns of unmyelinated ventral root afferents in the cat. *J. Physiol.* **422**, 265–287.
Häbler, H. J., Jänig, W., and Koltzenburg, M. 1990b. Activation of unmyelinated afferent fibres by mechanical stimuli and inflammation of the urinary bladder in the cat. *J. Physiol.* **425**, 545–562.
Häbler, H. J., Jänig, W., and Koltzenburg, M. 1993a. Myelinated primary afferents of the sacral spinal cord responding to slow filling and distension of the cat urinary bladder. *J. Physiol.* **463**, 449–460.
Häbler, H. J., Jänig, W., and Koltzenburg, M. 1993b. Receptive properties of myelinated primary afferents innervating the inflamed urinary bladder of the cat. *J. Neurophysiol.* **69**, 395–405.
Hackman, J. C., Auslander, D., Grayson, V., and Davidoff, R. A. 1982. GABA 'desensitization' of frog primary afferent fibers. *Brain Res.* **253**, 143–152.
Hadjiconstantinou, M., Panula, P., Lackovic, Z., and Neff, N. H. 1984. Spinal cord serotonin: A biochemical and immunohistochemical study following transection. *Brain Res.* **322**, 245–254.
Hagbarth, K. E. and Vallbo, A. B. 1967. Mechanoreceptor activity recorded percutaneously with semi-microelectrodes in human peripheral nerves. *Acta Physiol. Scand.* **69**, 121–122.
Hagbarth, K. E., Hongell, A., Hallin, R. G., and Torebjörk, H. E. 1970. Afferent impulses in median nerve fascicles evoked by tactile stimuli of the human hand. *Brain Res.* **24**, 423–442.
Hagihira, S., Senba, E., Yoshida, S., Tohyama, M., and Yoshiya, I. 1990. Fine structure of noradrenergic terminals and their synapses in the rat spinal dorsal horn: An immunohistochemical study. *Brain Res.* **526**, 73–80.
Hahn, J. F. and Wall, J. T. 1975. Frequency responses of down- and guard-hair receptors in the cat. *Am. J. Physiol.* **229**, 23–27.
Halata, Z. 1975. The mechanoreceptors of mammalian skin: Ultrastructure and morphological classification. *Adv. Anat. Embryol. Cell Biol.* **50**, 4–77.
Halata, Z. 1977. The ultrastructure of sensory nerve endings in the articular capsule of the knee joint of the domestic cat (Ruffini corpuscles and Pacinian corpuscles). *J. Anat.* **124**, 717–729.
Halata, Z. 1993. Sensory innervation of the hairy skin (light- and electron-microscopic study). *J. Invest. Dermatol.* **101**, 75S–81S.
Halata, Z. and Munger, B. L. 1980. Sensory nerve endings in rhesus monkey sinus hairs. *J. Comp. Neurol.* **192**, 645–663.
Halata, Z. and Munger, B. L. 1981. Identification of the Ruffini corpuscle in human hairy skin. *Cell Tissue Res.* **219**, 437–440.
Halata, Z., Badalamente, M. A., Dee, R., and Propper, M. 1984. Ultrastructure of sensory nerve endings in monkey (*Macaca fascicularis*) knee joint capsule. *J. Orthopedic Res.* **2**, 169–176.
Halata, Z., Rettig, T., and Schulze, W. 1985. The ultrastructure of sensory nerve endings in the human knee joint capsule. *Anat. Enbryol.* **172**, 265–275.
Hall, A. K., Ai, X., Hickman, G. E., MacPhedran, S. E., Nduaguba, C. O., and Robertson, C. P. 1997. The generation of neuronal heterogeneity in a rat sensory ganglion. *J. Neurosci.* **17**, 2775–2784.
Hallbeck, M., Hermanson, O., and Blomqvist, A. 1996. Preprovasopressin mRNA is not present in dorsal root ganglia of the rat. *Neurosci. Lett.* **209**, 125–128.
Hallin, R. G. and Torebjörk, H. E. 1976. Studies on cutaneous A and C fibre afferents, skin nerve blocks and perception. In Y. Zotterman (ed.), *Sensory Functions of the Skin in Primates, with Special Reference to Man.* (pp. 137–148). Pergamon Press, New York.
Hallin, R. G., Torebjörk, H. E., and Wiesenfeld, Z. 1981. Nociceptors and warm receptors innervated by C fibres in human skin. *J. Neurol. Neurosurg. Psychiatry* **44**, 313–319.
Hallin, R. G., Wiesenfeld-Hallin, Z., and Duranti, R. 1986. Percutaneous microneurography in man does not cause pressure block of almost all axons in the impaled nerve fascicle. *Neurosci. Lett.* **68**, 356–361.
Hama, A. T. and Sagen, J. 1994. Induction of spinal NADPH-diaphorase by nerve injury is attenuated by adrenal medullary transplants. *Brain Res.* **640**, 345–351.
Hama, A. T., Unnerstall, J. R., Siegan, J. B., and Sagen, J. 1995. Modulation of NMDA receptor expression in the rat spinal cord by peripheral nerve injury and adrenal medullary grafting. *Brain Res.* **687**, 103–113.
Hämäläinen, H. and Pertovaara, A. 1984. Vibrotactile thresholds in mechanoreceptive afferents innervating the foot pad of the cat: The importance of stimulus frequency and duration. *Acta Physiol. Scand.* **120**, 321–327.
Hamano, K., Mannen, H., and Ishizuka, N. 1978. Reconstruction of trajectory of primary afferent collaterals in the dorsal horn of the cat spinal cord, using Golgi-stained serial sections. *J. Comp. Neurol.* **181**, 1–16.

Hamill, O. P. and Martinac, B. 2001. Molecular basis of mechanotransduction in living cells. *Physiol. Rev.* **81**, 685–740.

Hammerstad, J. P., Murray, J. E., and Cutler, R. W. P. 1971. Efflux of amino acid neurotransmitters from rat spinal cord slices: II. Factors influencing the electrically induced efflux of [^{14}C]glycine and 3H-GABA. *Brain Res.* **35**, 357–367.

Hammond, D. L. and Ruda, M. A. 1991. Developmental alterations in nociceptive threshold, immunoreactive calcitonin gene-related peptide and substance P, and fluoride-resistant acid phosphatase in neonatally capsaicin-treated rats. *J. Comp. Neurol.* **312**, 436–450.

Hammond, D. L., Presley, R., Gogas, K. R., and Basbaum, A. I. 1992. Morphine or U-50,488 suppresses Fos protein-like immunoreactivity in the spinal cord and nucleus tractus solitarii evoked by a noxious visceral stimulus in the rat. *J. Comp. Neurol.* **315**, 244–253.

Hamon, M., Gallissot, M. C., Menard, F., Gozlan, H., Bourgoin, S., and Verge, D. 1989. 5-HT_3 receptor binding sites are on capsaicin-sensitive fibres in the rat spinal cord. *Eur. J. Pharm.* **164**, 315–322.

Han, H. C., Lee, D. H., and Chung, J. M. 2000. Characteristics of ectopic discharges in a rat neuropathic pain model. *Pain* **84**, 253–261.

Han, Z. S., Zhang, E. T., and Craig, A. D. 1998. Nociceptive and thermoreceptive lamina I neurons are anatomically distinct. *Nature Neurosci.* **1**, 218–225.

Hancock, M. B., Rigamonti, D. D. and Bryan, R. N. 1973. Convergence in the lumbar spinal cord of pathways activated by splanchnic nerve and hind limb cutaneous nerve stimulation. *Exp. Neurol.* **38**, 337–348.

Handwerker, H. O. 1996. Sixty years of C-fiber recordings from animal and human skin nerves: Historical notes. In T. Kumazawa, L. K. Kruger, and K. Mizumura (eds.), *The Polymodal Receptor-A Gateway to Pathological Pain. Prog. Brain Res.*, **113**, 39–51. Elsevier, Amsterdam.

Handwerker, H. O. and Neher, K. D. 1976. Characteristics of C-fibre receptors in the cat's foot responding to stepwise increase of skin temperature to noxious levels. *Pfluegers Arch.* **365**, 221–229.

Handwerker, H. O., Iggo, A., and Zimmermann, M. 1975. Segmental and supraspinal actions on dorsal horn neurons responding to noxious and non-noxious skin stimuli. *Pain* **1**, 147–165.

Handwerker, H. O., Anton, F., and Reeh, P. W. 1987. Discharge pattern of afferent cutaneous nerve fibers from the rat's tail during prolonged noxious mechanical stimulation. *Exp. Brain Res.* **65**, 493–504.

Handwerker, H. O., Kilo, S., and Reeh, P. W. 1991. Unresponsive afferent nerve fibres in the sural nerve of the rat. *J. Physiol.* **435**, 229–242.

Hanesch, U., Heppelmann, B., and Schmidt, R. F. 1991. Substance P- and calcitonin gene-related peptide immunoreactivity in primary afferent neurons of the cat's knee joint. *Neuroscience* **45**, 185–193.

Hanesch, U., Heppelmann, B., and Schmidt, R. F. 1992. Neurokinin A-like immunoreactivity in articular afferents of the cat. *Brain Res.* **586**, 332–335.

Hanesch, U., Pfrommer, U., Grubb, B. D., Heppelmann, B., and Schaible, H. G. 1993. The proportion of CGRP-immunoreactive and SP-mRNA containing dorsal root ganglion cells is increased by a unilateral inflammation of the ankle joint of the rat. *Regul. Pept.* **46**, 202–203.

Hanesch, U., Heppelmann, B., and Schmidt, R. F. 1995. Somatostatin-like immunoreactivity in primary afferents of the medial articular nerve and colocalization with substance P in the cat. *J. Comp. Neurol.* **354**, 345–352.

Hanesch, U., Heppelmann, B., and Schmidt, R. F. 1997. Quantification of cat's articular afferents containing calcitonin gene-related peptide or substance P innervating normal and acutely inflamed knee joints. *Neurosci. Lett.* **233**, 105–108.

Hanker, J. S. and Peach, R. 1976. Histochemical and ultrastructural studies of primary sensory neurons in mice with dystonia musculorum: I. Acetylcholinesterase and lysosomal hydrolases. *Neuropath. and App. Neurobiol.* **2**, 79–97.

Hanker, J. S., Kusyk, C. J., Bloom, F. E., and Pearse, A. G. E. 1973. The demonstration of dehydrogenases and monoamine oxidase by the formation of osmium blacks at the sites of Hatchett's brown. *Histochemie* **33**, 205–230.

Hanko, J., Hardebo, J. E., Kahrstrom, J., Owman, C., and Sundler, F. 1986. Existence and coexistence of calcitonin gene-related peptide (CGRP) and substance P in cerebrovascular nerves and trigeminal ganglion cells. *Acta Physiol. Scand.* **552**, 29–33.

Hansen, L. A., Alexander, N., Hogan, M. E., Sundberg, J. P., Dlugosz, A., Threadgill, D. W., Magnuson, T., and Yuspa, S. H. 1997. Genetically null mice reveal a central role for epidermal growth factor receptor in the differentiation of the hair follicle and normal hair development. *Am. J. Pathol.* **150**, 1959–1975.

Hao, J. X., Xu, X. J., Aldskogius, H., Seiger, Å., and Wiesenfeld-Hallin, Z. 1991a. Allodynia-like effects in rat after ischaemic spinal cord injury photochemically induced by laser irradiation. *Pain* **45**, 175–185.

Hao, J. X., Xu, X. J., Yu, Y. X., Seiger, Å., and Wiesenfeld-Hallin, Z. 1991b. Hypersensitivity of dorsal horn wide dynamic range neurons to cutaneous mechanical stimuli after transient spinal cord ischaemia in the rat. *Neurosci. Lett.* **128**, 105–108.

Hao, J. X., Xu, X. J., Aldskogius, H., Seiger, Å., and Wiesenfeld-Hallin, Z. 1992a. Photochemically induced transient spinal ischemia induces behavioral hypersensitivity to mechanical and cold stimuli, but not to noxious-heat stimuli, in the rat. *Exp. Neurol.* **118**, 1876–194.

Hao, J. X., Xu, X. J., Yu, Y. X., Seiger, Å., and Wiesenfeld-Hallin, Z. 1992b. Transient spinal cord ischemia induces temporary hypersensitivity of dorsal horn wide dynamic range neurons to myelinated, but not unmyelinated, fiber input. *J. Neurophysiol.* **68**, 384–391.

Hao, J. X., Shi, T. J., Xu, I. S., Kaupilla, T., Xu, X. J., Hökfelt, T., Bartfai, T., and Wiesenfeld-Hallin, Z. 1999. Intrathecal galanin alleviates allodynia-like behaviour in rats after partial peripheral nerve injury. *Eur. J. Neurosci.* **11**, 427–432.

Hardy, J. D., Wolff, H. G., and Goodell, H. 1952. *Pain Sensations and Reactions.* Williams & Wilkins, New York (reprinted in 1967 by Hafner Publishing Co., New York).

Harkness, D. H. and Brownfield, M. S. 1986. A thyrotrophin-releasing hormone-containing system in the rat dorsal horn separate from setotonin. *Brain Res.* **384**, 323–333.

Harmann, P. A., Carlton, S. M., and Willis, W. D. 1988a. Collaterals of spinothalamic tract cells to the periaqueductal gray: A fluorescent double-labeling study in the rat. *Brain Res.* **441**, 87–97.

Harmann, P. A., Chung, K., Briner, R. P., Westlund, K. N., and Carlton, S. M. 1988b. Calcitonin gene-related peptide (CGRP) in the human spinal cord: A light- and electron-microscopic analysis. *J. Comp. Neurol.* **269**, 371–380.

Harper, A. A. and Lawson, S. N. 1985a. Conduction velocity is related to morphological cell type in rat dorsal root ganglion neurones. *J. Physiol.* **359**, 31–46.

Harper, A. A. and Lawson, S. N. 1985b. Electrical properties of rat dorsal root ganglion neurones with different peripheral nerve conduction velocities. *J. Physiol.* **359**, 47–63.

Harrington, T. and Merzenich, M. M. 1970. Neural coding in the sense of touch: Human sensations of skin indentation compared with the responses of slowly adapting mechanoreceptive afferents innervating the hairy skin of monkeys. *Exp. Brain Res.* **10**, 251–264.

Harris, C. H., Fagan, E. L., Shew, R. L., Kammerlocher, T. C., and McNeill, D. L. 1990. MK-801-induced sprouting by CGRP immunoreactive primary afferent fibers in the dorsal spinal cord of the rat. *Neurosci. Lett.* **115**, 24–28.

Harris, J. A., Corsi, M., Quartarolli, M., Arban, R., and Bentivoglio, M. 1996. Upregulation of spinal glutamate receptors in chronic pain. *Neurosci.* **74**, 7–12.

Harrison, P. J. and Jankowska, E. 1985a. Sources of input to interneurones mediating group I non-reciprocal inhibition and motoneurones in the cat. *J. Physiol.* **361**, 379–401.

Harrison, P. J. and Jankowska, E. 1985b. Organization of input to the interneurones mediating group I non-reciprocal inhibition of motoneurones in cat. *J. Physiol.* **361**, 403–418.

Hartschuh, W. and Weihe, E. 1980. Fine structural analysis of the synaptic junction of Merkel cell-axon-complexes. *J. Invest. Dermatol.* **75**, 159–165.

Hartschuh, W., Weihe, E., Yanaihara, N., and Reinecke, M. 1983. Immunohistochemical localization of vasoactive intestinal polypeptide (VIP) in Merkel cells of various mammals: Evidence for a neuromodulator function of the Merkel cell. *J. Invest. Dermatol.* **81**, 361–364.

Hassan, A. H. S., Pzewlocki, R., Herz, A., and Stein, C. 1992. Dynorphin, a preferential ligand for *K*-opioid receptors, is present in nerve fibers and immune cells within inflamed tissue of the rat. *Neurosci. Lett.* **140**, 85–88.

Hassan, A. H. S., Ableitner, A., Stein, C., and Herz, A. 1993. Inflammation of the rat paw enhances axonal transport of opioid receptors in the sciatic nerve and increases their density in inflamed tissue. *Neuroscience* **55**, 185–195.

Haupt, P., Jänig, W., and Kohler, W. 1983. Response pattern of visceral afferent fibres, supplying the colon, upon chemical and mechanical stimuli. *Pfluegers Arch.* **398**, 41–47.

Hayashi, H. 1985. Morphology of terminations of small and large myelinated trigeminal primary afferent fibers in the cat. *J. Comp. Neurol.* **240**, 71–89.

Hayes, E. S. and Carlton, S. M. 1992. Primary afferent interactions: Analysis of calcitonin gene-related peptide-immunoreactive terminals in contact with unlabeled and GABA-immunoreactive profiles in the monkey dorsal horn. *Neuroscience* **47**, 873–896.

Haynes, L. W., Smyth, D. G., and Zakarian, S. 1982. Immunocytochemical localization of beta-endorphin (lipotropin C- fragment) in the developing rat spinal cord and hypothalamus. *Brain Res.* **232**, 115–128.

He, X., Schepelmann, K., Schaible, H. G., and Schmidt, R. F. 1990. Capsaicin inhibits responses of fine afferents from the knee joint of the cat to mechanical and chemical stimuli. *Brain Res.* **530**, 147–150.

Head, H. 1920. *Studies in Neurology* (Vol. 1, pp. 55–65 and 225–329). Oxford University Press, London.

Heath, D. D., Coggeshall, R. E., and Hulsebosch, C. E. 1986. Axon and neuron numbers after forelimb amputation in neonatal rats. *Exp. Neurol.* **92**, 220–233.

Heath, M. J., Womack, M. D., and MacDermott, A. B. 1994. Substance P elevates intracellular calcium in both neurons and glial cells from the dorsal horn of the spinal cord. *J. Neurophysiol.* **72**, 1192–1198.

Heath, M. J., Lints, T. J., Lee, C. J., and Dodd, J. 1995. Functional expression of the tachykinin NK1 receptor by floor plate cells in the embryonic rat spinal cord and brainstem. *J. Physiol.* **486**, 139–148.

Heavner, J. E. and DeJong, R. H. 1973. Spinal cord neuron response to natural stimuli: A microelectrode study. *Exp. Neurol.* **39**, 293–306.

Heimer, L. and Wall, P. D. 1968. The dorsal root distribution to the substantia gelatinosa of the rat with a note on the distribution in the cat. *Exp. Brain Res.* **6**, 89–99.

Heinbecker, P., Bishop, G. H. and O'Leary, J. L. 1933. Pain and touch fibers in peripheral nerves. *Arch. Neurol. Psychiat.* **29**, 771–789.

Heinbecker, P., Bishop, G. H., and O'Leary, J. L. 1934. Analysis of sensation in terms of the nerve impulse. *Arch. Neurol. Psychiat.* **31**, 34–53.

Helboe, L. and Moller, M. 2000. Localization of somatostatin receptors at the light- and electron-microscopical level by using antibodies raised against fusion proteins. *Prog. Histochem. Cytochem.* **35**, 3–64.

Helgren, M. E., Wolfe, M., Arsenault, K., Kapadia, S. E., and LaMotte, C. C. 1999a. Changes in substance P and calcitonin gene-related peptide binding in the dorsal horn of rat spinal cord following pronase-induced deafferentation. *Somatosens. Mot. Res.* **16**, 31–38.

Helgren, M. E., Arsenault, K., Kapadia, S. E., and LaMotte, C. C. 1999b. Deafferentation-induced regulation of AMPA receptors in the spinal cord of the adult rat. *Somatosens. Mot. Res.* **16**, 39–48.

Helke, C. J., Charlton, C. G., and Wiley, R. G. 1986. Studies on the cellular localization of spinal cord substance P receptors. *Neuroscience* **19**, 523–533.

Helke, C. J., Krause, J. E., Mantyh, P. W., Couture, R., and Bannon, M. J. 1990. Diversity in mammalian tachykinin peptidergic neurons: Multiple peptides, receptors, and regulatory mechanisms. *FASEB* **4**, 1606–1615.

Helliwell, R. J. A., McLatchie, L. M., Clarke, M., Winter, J., Bevan, S., and McIntyre, P. 1998. Capsaicin sensitivity is associated with the expression of the vanilloid (capsaicin) receptor (VR1) mRNA in adult rat sensory ganglia. *Neurosci. Lett.* **250**, 177–180.

Hellon, R. F. and Misra, N. K. 1973. Neurones in the dorsal horn responding to scrotal skin temperature changes. *J., Physiol.* **232**, 375–388.

Hellon, R. F., Hensel, H., and Schaefer, K. 1975. Thermal receptors in the scrotum of the rat. *J. Physiol.* **248**, 349–357.

Henken, D. B. and Martin, J. R. 1992a. Herpes simplex virus infection induces a selective increase in the proportion of galanin-positive neurons in mouse sensory ganglia. *Exp. Neurol* **118**, 195–203.

Henken, D. B. and Martin, J. R. 1992b. The proportion of galanin-immunoreactive neurons in mouse trigeminal ganglia is transiently increased following corneal inoculation of herpes simplex virus type-1. *Neurosci. Lett.* **140**, 177–180.

Henken, D. B., Tessler, A., Chesselet, M. -F., Hudson, A., Baldino, F., and Murray, M. 1988. In situ hybridization of mRNA for beta-preprotachykinin and preprosomatostatin in adult rat dorsal ganglia: Comparison with immunocytochemical localization. *J. Neurocytol.* **17**, 671–681.

Henken, D. B., Battisti, W. P., Chesselet, M. F., Murray, M., and Tessler, A. 1990. Expression of beta-preprotachykinin mRNA and tachykinins in rat dorsal root ganglion cells following peripheral or central axotomy. *Neuroscience* **39**, 733–742.

Henle, J. and Kolliker, A. 1845. On the Pacinian corpuscles in the nerves of men and mammalia. *Br. Foreign Med. Rev.* **19**, 78–83.

Henley, J. M., Jenkins, R., and Hunt, S. P. 1993. Localization of glutamate receptor binding sites and mRNAs to the dorsal horn of the rat spinal cord. *Neuropharmacology* **32**, 37–41.

Henry, J. L. 1976. Effects of substance P on functionally identified units in cat spinal cord. *Brain Res.* **114**, 439–451.

Henry, J. L. 1982. Effects of intravenously administered enantiomers of baclofen on functionally identified units in lumbar dorsal horn of the spinal cat. *Neuropharmacology* **21**, 1073–1083.

Henry, J. L., Krnjevic, K. and Morris, M. E. 1975. Substance P and spinal neurones. *Can. J. Physiol. Pharmacol.* **53**, 423–432.

Hensel, H. 1973. Cutaneous thermoreceptors. In A. Iggo (ed.), *Handbook of Sensory Physiology, Vol. II, Somatosensory System* (pp. 79–110). Springer, New York.

Hensel, H. 1974. Thermoreceptors. *Ann. Rev. Physiol.* **36**, 233–249.

Hensel, H. and Boman, K. K. A. 1960. Afferent impulses in cutaneous sensory nerves in human subjects. *J. Neurophysiol.* **23**, 564–578.

Hensel, H. and Iggo, A. 1971. Analysis of cutaneous warm and cold fibres in primates. *Pfluegers Arch.* **329**, 1–8.

Hensel, H. and Zotterman, Y. 1951a. The response of the cold receptors to constant cooling. *Acta Physiol. Scand.* **22**, 96–105.

Hensel, H. and Zotterman, Y. 1951b. The effect of menthol on the thermoreceptors. *Acta Physiol. Scand.* **24**, 27–34.

Hensel, H., Iggo, A., and Witt, I. 1960. A quantitative study of sensitive cutaneous thermoreceptors with C afferent fibres. *J. Physiol.* **153**, 113–126.

Hensel, H., Andres, K. H., and von Düring, M. 1974. Structure and function of cold receptors. *Pfluegers Arch.* **352**, 1–10.

Hentall, I. 1977. A novel class of unit in substantia gelatinosa of the spinal cat. *Exp. Neurol.* **57**, 792–806.

Hentall, I. D. and Fields, H. L. 1979. Segmental and descending influences on intraspinal thresholds of single C-fibers. *J. Neurophysiol.* **42**, 1527–1537.

Heppelmann, B. and Emson, P. C. 1993. Distribution of calretinin mRNA in rat spinal cord and dorsal root ganglia: A study using non-radioactive in situ hybridization histochemistry. *Brain Res.* **624**, 312–316.

Heppelmann, B. and Pawlak, M. 1997a. Sensitisation of articular afferents in normal and inflamed knee joints by substance P in the rat. *Neurosci. Lett.* **223**, 97–100.

Heppelmann, B. and Pawlak, M. 1997b. Inhibitory effect of somatostatin on the mechanosensivity of articular afferents in normal and inflamed knee joints of the rat. *Pain* **73**, 377–382.

Heppelmann, B. and Pawlak, M. 1999. Peripheral application of cyclo-somatostatin, a somatostatin antagonist, increases the mechanosensivity of rat knee joint afferents. *Neurosci. Lett.* **259**, 62–64.

Heppelmann, B., Heuss, C., and Schmidt, R. F. 1988. Fiber size distribution of myelinated and unmyelinated axons in the medial and posterior articular nerves of the cat's knee joint. *Somatosens. Res.* **5**, 273–281.

Heppelmann, B., Messlinger, K., Neiss, W. F., and Schmidt, R. F. 1990. Ultrastructural three-dimensional reconstruction of group III and group IV sensory nerve endings ("free nerve endings") in the knee joint capsule of the cat: Evidence for multiple receptive sites. *J. Comp. Neurol.* **292**, 103–116.

Heppelmann, B., Messlinger, K., Neiss, W. F., and Schmidt, R. F. 1995. Fine sensory innervation of the knee joint capsule by group III and group IV nerve fibers in the cat. *J. Comp. Neurol.* **351**, 415–428.

Heppelmann, B., Shahbazian, Z., and Hanesch, U. 1997. Quantitative examination of calcitonin gene-related peptide immunoreactive nerve fibres in the cat knee joint capsule. *Anat. Embryol.* **195**, 525–530.

Heppenstall, P. A. and Fleetwood-Walker, S. M. 1997a. The glycine site of the NMDA receptor contributes to neurokinin 1 receptor agonist facilitation of NMDA receptor agonist-evoked activity in rat dorsal horn neurons. *Brain Res.* **744**, 235–245.

Heppenstall, P. A. and Fleetwood-Walker, S. M. 1997b. Glycine receptor regulation of neurokinin$_1$ receptor function in rat dorsal horn neurones. *NeuroReport* **8**, 3109–3112.

Herbert, M. K. and Schmidt, R. F. 1992. Activation of normal and inflamed fine articular afferent units by serotonin. *Pain* **50**, 79–88.

Herdegen, T. and Leah, J. D. 1998. Inducible and constitutive transcription factors in the mammalian nervous system: Control of gene expression by Jun, Fos and Krox, and CREB/ATF proteins. *Brain Res. Rev.* **28**, 370–490.

Herdegen, T., Rudiger, S., Mayer, B., Bravo, R., and Zimmermann, M. 1994. Expression of nitric oxide synthase and colocalisation with jun, fos and krox transcription factors in spinal cord neurons following noxious stimulation of the rat hindpaw. *Mol. Brain Res.* **22**, 245–258.

Herkenham, M. and Pert, C. B. 1982. Light microscopic localization of brain opiate receptors: A general autoradiographic method which preserves tissue quality. *J. Neurosci.* **2**, 1129–1149.

Herkenham, M., Lynn, A. B., Johnson, M. R., Melvin, L. S., de Costa, B. R., and Rice, K. C. 1991. Characterization and localization of cannabinoid receptors in rat brain: A quantitative *in vitro* autoradiographic study. *J. Neurosci.* **11**, 563–583.

Hertel, H. C., Howaldt, B., and Mense, S. 1976. Responses of group IV and III muscle afferents to thermal stimuli. *Brain Res.* **113**, 201–205.

Heuer, H., Ehrchen, J., Bauer, K., and Schafer, M. K. 1998. Region-specific expression of thyrotrophin-releasing hormone-degrading ectoenzyme in the rat central nervous system and pituitary gland. *Eur. J. Neurosci.* **10**, 1465–1478.

Heuer, H., Schafer, M. K. H., O'Donnell, D., Walker, P., and Bauer, K. 2000. Expression of thyrotropin-releasing hormone receptor 2 (TRH-R2) in the central nervous system of rats. *J. Comp. Neurol.* **428**, 319–336.

Heym, C., Braun, B., Klimaschewski, L., and Kummer, W. 1995a. Chemical codes of sensory neurons innervating the guinea-pig adrenal gland. *Cell Tissue Res.* **279**, 169–181.

Heym, C., Braun, B., Shuyi, Y., Klimaschewski, L., and Colombo-Benkmann, M. 1995b. Immunohistochemical correlation of human adrenal nerve fibres and thoracic dorsal root neurons with special reference to substance P. *Histochem. Cell Biol.* **104**, 233–243.

Hildebrand, C., Oqvist, G., Brax, L., and Tuisku, F. 1991. Anatomy of the rat knee joint and fibre composition of a major articular nerve. *Anat. Rec.* **229**, 545–555.

Hill, C. E., Gould, D. J., Strigas, J., Burcher, E., and Vidovic, M. 1996. Sensory nerves play an efferent role in the function of the arterioles, but not the dilator muscle, of the rat iris. *J. Auton. Nerv. Syst.* **58**, 89–100.

Hill, D. R., Shaw, T. M., and Woodruff, G. N. 1988. Binding sites for 125I-cholecystokinin in primate spinal cord are of the CCK-A subclass. *Neurosci. Lett.* **89**, 133–139.

Hillman, P. and Wall, P. D. 1969. Inhibitory and excitatory factors influencing the receptive fields of lamina 5 spinal cord cells. *Exp. Brain Res.* **9**, 284–306.

Himes, B. T. and Tessler, A. 1989. Death of some dorsal root ganglion cells and plasticity of others following sciatic nerve section in adult and neonatal rats. *J. Comp. Neurol.* **284**, 215–230.

Hines, A. E., Birn, H., Teglbjaerg, P. S., and Sinkjaer, T. 1996. Fiber type composition of articular branches of the tibial nerve at the knee joint in man. *Anat. Rec.* **246**, 573–578.

Hirakawa, M. and Kawata, M. 1992a. Changes of chemoarchitectural organization of the rat spinal cord following ventral and dorsal root transection. *J. Comp. Neurol.* **320**, 339–352.

Hirakawa, M. and Kawata, M. 1992b. Influence of spinal cord hemisection on the configurational changes in motor and primary afferent neurons and the chemical messenger alterations in the rat lumbar segments. *J Hirnforsch.* **33**, 419–428.

Hirata, T., Kasugai, T., Hirota, S., Nomura, S., Fujisawa, J. and Kitamura, Y. 1995. Characterization of *c-kit*-positive neurons in the dorsal root ganglion of mouse. *Dev. Brain Res.* **85**, 201–211.

Hirota, N., Kuraishi, Y., Hino, Y., Sato, Y., Satoh, M., and Takagi, H. 1985. Met-enkephalin and morphine but not dynorphin inhibit noxious stimulation-induced release of substance P from rabbit dorsal horn in situ. *Neuropharmacology* **24**, 567–570.

Hirst, R. A., Lambert, D. G., and Notcutt, W. G. 1998. Pharmacology and potential therapeutic uses of cannabis. *Br. J. Anaesth.* **81**, 77–84.

Hiss, E. and Mense, S. 1976. Evidence for the existence of different receptor sites for algesic agents at the endings of muscular group IV afferent units. *Pfluegers Arch.* **362**, 141–146.

Hiura, A. 1982. Do the dorsal efferent and ventral afferent fibers exist in the L6 spinal nerve roots in the rat. *Fukushima J. Med. Sci.* **28**, 77–81.

Hiura, A., Ishizuka, H., and Villalobos, E. L. 1991. GABAergic neurons in the mouse superficial dorsal horn with special emphasis on their relation to primary afferent central terminals. *Arch. Histol. Cytol.* **54**, 195–206.

Hiura, A., Nasu, F., Kuwahara, M., and Ishizuka, H. 1997. Fluoride-resistant acid phosphatase (FRAP)-positive afferent terminals make synaptic contact with interneuronal soma in the substantia gelatinosa of the mouse spinal dorsal horn. *Okajimas Folia Anat. Jpn.* **74**, 109–113.

Hiura, A., Nasu, F., and Ishizuka, H. 1999. FRAP-positive and capsaicin-sensitive terminals in the substantia gelatinosa of the mouse spinal trigeminal nucleus caudalis. *Okajimas Folia Anat. Jpn.* **76**, 33–40.

Hník, P., Hudlická, O., Kucera, J., and Payne, R. 1969. Activation of muscle afferents by nonproprioceptive stimuli. *Am. J. Physiol.* **217**, 1451–1458.

Ho, R. H. 1983. Widespread distribution of substance P- and somatostatin-immunoreactive elements in the spinal cord of the neonatal rat. *Cell Tissue Res.* **232**, 471–486.

Ho, R. H. 1988. Somatostatin immunoreactive structures in the developing rat spinal cord. *Brain Res. Bull.* **21**, 105–116.

Ho, R. H. and Berelowitz, M. 1984. Somatostatin 28 1–14 immunoreactivity in primary afferent neurons of the rat spinal cord. *Neurosci. Lett.* **46**, 161–166.

Hochman, S., Garraway, S. M., and Pockett, S. 1997. Membrane properties of deep dorsal horn neurons from neonatal rat spinal cord in vitro. *Brain Res.* **767**, 214–219.

Hodge, C. J. 1972. Potential changes inside central afferent terminals secondary to stimulation of large- and small-diameter peripheral nerve fibers. *J. Neurophysiol.* **35**, 30–43.

Hoffert, M. J., Miletic, V., Ruda, M. A., and Dubner, R. 1983. Immunocytochemical identification of serotonin axonal contacts on characterized neurons in laminae I and II of the cat dorsal horn. *Brain Res.* **267**, 361–364.

Höglund, A. U. and Baghdoyan, H. A. 1997. M2, M3, and M4, but not M1, muscarinic receptor subtypes are present in rat spinal cord. *J. Pharm. Exp. Ther.* **281**, 470–477.

Hoheisel, U. and Mense, S. 1985. Morphological features of dorsal root ganglion cells with groups III and IV afferent fibers. *Neurosci Lett.* **Suppl. 22**, S29.

Hoheisel, U. and Mense, S. 1986. Non-myelinated afferent fibres do not originate exclusively from the smallest dorsal root ganglion cells in the cat. *Neurosci. Lett.* **72**, 153–157.

Hoheisel, U. and Mense, S. 1989. Long-term changes in discharge behaviour of cat dorsal horn neurones following noxious stimulation of deep tissues. *Pain* **36**, 239–247.

Hoheisel, U. and Mense, S. 2000. The role of spinal nitric oxide in the control of spontaneous pain following nociceptive input. *Prog. Brain Res.* **129**, 163–172.

Hoheisel, U., Reinert, A., and Mense, S. 1995. Changes in NADPH-diaphorase activity in the rat dorsal horn following an acute experimental myositis. *Histochemistry* **103**, 459–462.

Hoheisel, U., Kaske, A., Reinert, A., and Mense, S. 1997. Frequency-dependent expression of diaphorase staining and nNOS-immunoreactivity in rat dorsal horn neurones following C-fibre stimulation. *Neurosci. Lett.* **227**, 181–184.

Hoheisel, U., Kaske, A., and Mense, S. 1998. Relationship between neuronal activity and substance P-immunoreactivity in the rat spinal cord during acute and persistent myositis. *Neurosci. Lett.* **257**, 21–24.

Hohenfellner, M., Nunes, L., Schmidt, R. A., Lampel, A., Thuroff, J. W., and Tanagho, E. A. 1992. Interstitial cystitis: Increased sympathetic innervation and related neuropeptide synthesis. *J. Urol.* **147**, 587–591.

Hohmann, A. G. and Herkenham, M. 1998. Regulation of cannabinoid and mu opioid receptor in rat lumbar spinal cord following neonatal capsaicin treatment. *Neurosci. Lett.* **252**, 13–16.

Hohmann, A. G. and Herkenham, M. 1999a. Localization of central cannabinoid CB1 receptor messenger RNA in neuronal subpopulations of rat dorsal root ganglia: A double-label in situ hybridization study. *Neuroscience* **90**, 923–931.

Hohmann, A. G. and Herkenham, M. 1999b. Cannabinoid receptors undergo axonal flow in sensory nerves. *Neuroscience* **92**, 1171–1175.

Hohmann, A. G., Briley, E. M., and Herkenham, M. 1999. Pre- and postsynaptic distribution of cannabinoid and mu opioid receptors in rat spinal cord. *Brain Res.* **822**, 17–25.

Hökfelt, T. and Ljungdahl, A. 1971. Light- and electron-microscopic autoradiography on spinal cord slices after incubation with labeled glycine. *Brain Res.* **32**, 189–194.

Hökfelt, T., Elde, R., Johansson, O., Luft, R., and Arimura, A. 1975a. Immunohistochemical evidence for the presence of somatostatin, a powerful inhibitory peptide, in some primary sensory neurons. *Neurosci. Lett.* **1**, 231–235.

Hökfelt, T., Kellerth, J. O., Nilsson, C. and Pernow, B. 1975b. Experimental immunohistochemical studies on the localization and distribution of substance P in cat primary sensory neurons. *Brain Res.* **100**, 235–252.

Hökfelt, T., Kellerth, J. O., Nilsson, G., and Pernow, B. 1975c. Substance P: localization in the central nervous system and in some primary sensory neurons. *Science* **190**, 889–890.

Hökfelt, T., Elde, R., Johansson, O., Luft, R., Nilsson, G., and Arimura, A. 1976. Immunohistochemical evidence for separate populations of somatostatin-containing and substance P-containing primary afferent neurons in the rat. *Neuroscience* **1**, 131–136.

Hökfelt, T., Elde, K., Johansson, O., Terenius, L., and Stein, L. 1977a. The distribution of enkephalin immunoreactive cell bodies in the rat central nervous system. *Neurosci. Lett.* **5**, 25–32.

Hökfelt, T., Ljungdahl, Å., Terenius, L., Elde, R., and Nilsson, G. 1977b. Immunohistochemical analysis of peptide pathways possibly related to pain and analgesia: Enkephalin and substance P. *Proc. Natl. Acad. Sci. USA* **74**, 3081–3085.

Hökfelt, T., Ljungdahl, Å., Steinbusch, H., Verhofstad, A., Nilsson, G., Brodin, E., Oernow, B., and Goldstein, M. 1978. Immunohistochemical evidence of substance P-like immunoreactivity in some 5-hydroxytryptamine-containing neurons in the rat central nervous system. *Neuroscience* **3**, 517–538.

Hökfelt, T., Terenius, L., Kuypers, H. G., and Dann, O. 1979. Evidence for enkephalin immunoreactive neurons in the medulla oblongata projecting to the spinal cord. *Neurosci. Lett.* **14**, 55–60.

Hökfelt, T., Lundberg, J. M., Terenius, L., Jancso, G., and Kimmel, J. 1981. Avian pancreatic polypeptide (APP) immunoreactive neurons in the spinal cord and spinal trigeminal nucleus. *Peptides* **2**, 81–87.

Hökfelt, T., Lundberg, J. M., Tatemoto, K., Mutt, V., Terenius, L., Polak, J., Bloom, S., Sasek, C., Elde, R., and Goldstein, M. 1983. Neuropeptide Y (NPY)- and FMRFamide neuropeptide-like immunoreactivities in catecholamine neurons of the rat medulla oblongata. *Acta Physiol. Scand.* **117**, 315–318.

Hökfelt, T., Johansson, O., and Goldstein, M. 1984. Central catecholamine neurons as revealed by immunohistochemistry with special reference to adrenaline neurons. In A. Björklund and T. Hökfelt (eds.), *Handbook of Chemical Neuroanatomy* (pp. 157–276). Elsevier, Amsterdam.

Hökfelt, T., Wiesenfeld-Hallin, Z., Villar, M., and Melander, T. 1987. Increase of galanin-like immunoreactivity in rat dorsal root ganglion cells after peripheral axotomy. *Neurosci. Lett.* **83**, 217–220.

Hökfelt, T., Cortesk, R., Schalling, M., Ceccatelli, M., Pelto-Huikko, M., Persson, H., and Villar, M. J. 1991. Distribution patterns of CCK and CCk mRNA in some neuronal and non-neuronal tissues. *Neuropeptides* **199**, 31–43.

Hökfelt, T., Arvidsson, U., Ceccatelli, S., Cortes, R., Cullheim, S., Dagerlind, A., Johnson, H., Orazzo, C., Piehl, F., Pieribone, V., Schalling, M., Terenius, L., Ulfhake, B., Verge, V. M., Villar, M., Wiesenfeld-Hallin, Z., Xu, X. -J., and Xu, Z. 1992. Calcitonin gene-related peptide in the brain, spinal cord, and some peripheral systems. *Ann. NY Acad. Sci.* **657**, 119–134.

Hökfelt, T., Ceccatelli, S., Gustafsson, L., Hulting, A. L., Verge, V., Villar, M., Xu, X. J., Xu, Z. Q., Wiesenfeld-Hallin, Z., and Zhang, X. 1994a. Plasticity of NO synthase expression in the nervous and endocrine systems. *Neuropharmacology* **33**, 1221–1227.

Hökfelt, T., Morino, P., Verge, V., Castel, M. -N., Broberger, C., Zhang, X., Herrera-Marschitz, M., Meana, J. J., Understedt, U., Xu, Z. -J., Hao, J. -X., Puke, M. J. C., Wiesenfeld-Hallin, Z. S., Seiger, A., Hughes, J., Varro, A., and Dockray, G. 1994b. CCK in cerebral cortex and the spinal level.*Ann. NY Acad. Sci.* 157–163.

Hökfelt, T., Zhang, X., and Wiesenfeld-Hallin, Z. 1994c. Messenger plasticity in primary sensory neurons following axotomy and its functional implications. *TINS* **17**, 22–30.

Hökfelt, T., Broberger, C., Zhang, X., Diez, M., Kopp, J., Xu, Z. -Q., Landry, M., Bao, L., Schalling, M., Koistinaho, J., DeArmond, S. J., Prusiner, S., Gong, J., and Walsh, J. H. 1998. Neuropeptide Y: Some viewpoints on a multifaceted peptide in the normal and diseased nervous system. *Brain Res. Rev.* **26**, 154–166.

Hökfelt, T., Holmberg, K., Shi, T. J., and Broberger, C. 2001. CCK-ergic mechanisms in sensory systems. *Scand. J. Clin. Lab Invest. Suppl.* **234**, 69–74.

Holets, V. R., Hokfelt, T., Rokaeus, A., Terenius, L., and Goldstein, M. 1988. Locus coeruleus neurons in the rat containing neuropeptide Y, tyrosine hydroxylase or galanin and their efferent projections to the spinal cord, cerebral cortex and hypothalamus. *Neuroscience* **24**, 893–906.

Holford, L. C., Case, P., and Lawson, S. N. 1994. Substance P, neurofilament, peripherin and SSEA4 immunocytochemistry of human dorsal root ganglion neurons obtained from post-mortem tissue: A quantitative morphometric analysis. *J. Neurocytol.* **23**, 577–589.

Holland, L. N. and Goldstein, B. D. 1990. Changes of substance P-like immunoreactivity in the dorsal horn are associated with the "phasic" behavioral response to a formalin stimulus. *Brain Res.* **537**, 287–292.

Hollman, M. and Heinemann, S. 1994. Cloned glutamate receptors. *Annu. Rev. Neurosci.* **17**, 31–108.

Holloway, S., Feniuk, W., Kidd, E. J., and Humphrey, P. P. 1996. A quantitative autoradiographical study on the distribution of somatostatin sst2 receptors in the rat central nervous system using [^{125}I]-BIM-23027. *Neuro-pharmacology.* **35**, 1109–1120.

Hollt, V., Haarmann, I., Millan, M. J., and Herz, A. 1987. Prodynorphin gene expression is enhanced in the spinal cord of chronic arthritic rats. *Neurosci. Lett.* **73**, 90–94.

Holmberg, K., Shi, T. J., Albers, K. M., Davis, B. M., and Hokfelt, T. 2001. Effect of peripheral nerve lesion and lumbar sympathectomy on peptide regulation in dorsal root ganglia in the NGF-overexpressing mouse. *Exp. Neurol.* **167**, 290–303.

Holmes, A. 1982. Do the dorsal efferent and ventral afferent fibers exist in the L6 spinal nerve roots in the rat? *Fukushima J. Med. Sci.* **28**, 77–81.

Holstege, J. C., van Dijken, H., Buijs, R. M., Goedknegt, H., Gosens, T., and Bongers, C. M. H. 1996. Distribution of dopamine immunoreactivity in the rat, cat, and monkey spinal cord. *J. Comp. Neurol.* **376**, 631–652.

Holtman, J. R. 1989. Localization of substance P immunoreactivity in phrenic primary afferent neurons. *Peptides* **10**, 53–56.

Holzer, P. 1988. Local effector functions of capsaicin-sensitive sensory nerve endings: Involvement of tachykinins, calcitonin gene-related peptide and other neuropeptides. *Neuroscience* **24**, 739–768.

Holzer, P. 1991. Capsaicin: Cellular targets, mechanisms of action, selectivity for thin sensory neurons. *Pharmacol. Rev.* **43**, 143–201.

Holzer, P. and Maggi, C. A. 1998. Dissociation of dorsal root ganglion neurons into afferent and efferent-like neurons. *Neuroscience.* **86**, 389–398.

Holzer-Petsche, U., Rinner, I., and Lembeck, F. 1986. Distribution of choline acetyltransferase activity in rat spinal cord: Influence of primary afferents? *J. Neural. Trans.* **66**, 85–92.

Hölzl, R., Erasmus, L. P., and Möltner, A. 1996. Detection, discrimination and sensation of visceral stimuli. *Biol. Psychol.* **42**, 199–214.

Homor, G. and Kasa, P. 1978. Acetylcholinesterase resynthesis after DFP poisoning: Histochemical and biochemical study. *Acta Histochem.* **62**, 293–301.

Honda, C. N. 1985. Visceral and somatic afferent convergence onto neurons near the central canal in the sacral spinal cord of the cat. *J. Neurophysiol.* **53**, 1059–1078.

Honda, C. N. 1995. Differential distribution of calbindin-D28k and parvalbumin in somatic and visceral sensory neurons. *Neuroscience* **68**, 883–892.

Honda, C. N. and Arvidsson, U. 1995. Immunohistochemical localization of delta- and mu-opioid receptors in primate spinal cord. *NeuroReport* **6**, 1025–1028.

Honda, C. N. and Perl, E. R. 1985. Functional and morphological features of neurons in the midline region of the caudal spinal cord of the cat. *Brain Res.* **340**, 285–295.

Honda, C. N., Réthelyi, M., and Petrusz, P. 1983. Preferential immunohistochemical localization of vasoactive intestinal polypeptide (VIP) in the sacral spinal cord of the cat: Light- and electron-microscopic observations. *J. Neurosci.* **3**, 2183–2196.

Hongo, T., Jankowska, E., and Lundberg, A. 1966. Convergence of excitatory and inhibitory action on interneurones in the lumbosacral cord. *Exp. Brain Res.* **1**, 338–358.

Hongo, T., Kitazawa, S., Ohki, Y., Sasaki, M., and Xi, M. C. 1989a. A physiological and morphological study of premotor interneurones in the cutaneous reflex pathways in cats. *Brain Res.* **505**, 163–166.

Hongo, T., Kitazawa, S., Ohki, Y., and Xi, M. C. 1989b. Functional identification of last-order interneurones of skin reflex pathways in the cat forelimb segments. *Brain Res.* **505**, 167–170.

Honore, P., Menning, P. M., Rogers, S. D., Nichols, M. L., Basbaum, A. I., Besson, J. M., and Mantyh, P. W. 1999. Spinal substance P receptor expression and internalization in acute, short-term, and long-term inflammatory pain states. *J. Neurosci.* **19**, 7670–7678.

Hope, P. J., Lang, C. W., and Duggan, A. W. 1990a. Persistence of immunoreactive neurokinins in the dorsal horn of barbiturate anaesthetized and spinal cats, following release by tibial nerve stimulation. *Neurosci. Lett.* **118**, 25–28.

Hope, P. J., Jarrott, B., Schaible, H.-G., Clarke, R. W., and Duggan, A. W. 1990b. Release and spread of immunoreactive neurokinin A in the cat spinal cord in a model of acute arthritis. *Brain Res.* **533**, 292–299.

Hope, P. J., Fleetwood-Walker, S. M., and Mitchell, R. 1990c. Distinct antinociceptive actions mediated by different opioid receptors in the region of lamina I and laminae III–V of the dorsal horn of the rat. *Br. J. Pharmacol.* **101**, 477–483.

Hopkin, J. M. and Neal, M. J. 1970. The release of [^{14}C]glycine from electrically stimulated rat spinal cord slices. *Br. J. Pharmacol.* **40**, 136–138P.

Horch, K. W. and Burgess, P. R. 1975. Effect of activation and adaptation on the sensitivity of slowly adapting cutaneous mechanoreceptors. *Brain Res.* **98**, 109–118.

Horch, K. W. and Burgess, P. R. 1976. Responses to threshold and suprathreshold stimuli by slowly adapting cutaneous mechanoreceptors in the cat. *J. Comp. Physiol.* **110**, 307–315.

Horch, K. W. and Lisney, S. J. W. 1981. Changes in primary afferent depolarization of sensory neurones during peripheral nerve regeneration in the cat. *J. Physiol.* **313**, 287–299.

Horch, K. W., Whitehorn, D., and Burgess, P. R. 1974. Impulse generation in type I cutaneous mechanoreceptors. *J. Neurophysiol.* **37**, 267–281.

Horch, K. W., Tuckett, R. P., and Burgess, P. R. 1977. A key to the classification of cutaneous mechanoreceptors. *J. Invest. Dermatol.* **69**, 75–82.

Hori, Y. and Endo, K. 1992. Miniature postsynaptic currents recorded from identified rat spinal dorsal horn projection neurons in thin-slice preparations. *Neurosci. Lett.* **142**, 191–195.

Hösli, L., Hösli, E., Zehntner, C., and Landolt, H. 1981. Effects of substance P on neurones and glial cells in cultured rat spinal cord. *Neurosci. Lett.* **24**, 165–168.

Hosoi, J., Murphy, G. F., Lerner, E. A., Grabbe, S., and Granstein, R. D. 1993. Regulation of Langerhans cell function by nerves containing calcitonin gene-related peptide. *Nature* **363**, 159–163.

Hou, M., Kanje, M., Longmore, J., Tajti, J., Uddman, R., and Edvinsson, L. 2001. 5-HT(1B) and 5-HT(1D) receptors in the human trigeminal ganglion: Co-localization with calcitonin gene-related peptide, substance P and nitric oxide synthase. *Brain Res.* **909**, 112–120.

Houk, J. and Henneman, E. 1967. Responses of Golgi tendon organs to active contractions of the soleus muscle of the cat. *J. Neurophysiol.* **30**, 466–481.

Houlden, H., King, R. H., Hashemi-Nejad, A., Wood, N. W., Mathias, C. J., Reilly, M., and Thomas, P. K. 2001. A novel TRK A (NTRK1) mutation associated with hereditary sensory and autonomic neuropathy type V. *Ann. Neurol.* **49**, 521–525.

Houle, J. D. and Reier, P. J. 1989. Regrowth of calcitonin gene-related peptide (CGRP) immunoreactive axons from the chronically injured rat spinal cord into fetal spinal cord tissue transplants. *Neurosci. Lett.* **103**, 253–258.

Houser, C. R., Crowford, G. D., Barber, R. P., Salvaterra, P. M., and Vaughn, J. E. 1983. Organization and morphological characteristics of cholinergic neurons: An immunocytochemical study with a monoclonal antibody to choline acetyltransferase. *Brain Res.* **266**, 97–119.

Houtani, T., Nishi, M., Takeshima, H., Nukada, T., and Sugimoto, T. 1996. Structure and regional distribution of nociceptin/orphanin FQ precursor. *Biochem. Biophys. Res. Commun.* **219**, 714–719.

Houtani, T., Nishi, M., Takeshima, H., Sato, K., Sakuma, S., Kakimoto, S., Ueyama, T., Noda, T., and Sugimoto, T. 2000. Distribution of nociceptin/orphanin FQ precursor protein and receptor in brain and spinal cord: A study using in situ hybridization and X-gal histochemistry in receptor-deficient mice. *J. Comp. Neurol.* **424**, 489–508.

Howe, J. R., Yaksh, T. L., and Go, V. L. W. 1987. The effect of unilateral dorsal root ganglionectomies or ventral rhizotomies on α_2-adrenoceptor binding to, and the substance P, enkephalin, and neurotensin content of, the cat lumbar spinal cord. *Neuroscience* **21**, 385–394.

Howland, B., Lettvin, J. Y., McCulloch, W. S., Pitts, W., and Wall, P. D. 1955. Reflex inhibition by dorsal root interaction. *J. Neurophysiol.* **18**, 1–17.

Howlett, A. C. 1995. Pharmacology of cannabinoid receptors. *Annu. Rev. Pharmacol. Toxicol.* **35**, 607–634.

Hsieh, S. T., Choi, S., Lin, W. M., Chang, Y. C., McArthur, J. C., and Griffin, J. W. 1996. Epidermal denervation and its effects on keratinocytes and Langerhans cells. *J. Neurocytol.* **25**, 513–524.

Hu, P. and McLachlan, E. M. 2000. Distinct sprouting responses of sympathetic and peptidergic sensory axons proximal to a sciatic nerve transection in guinea pigs and rats. *Neurosci. Lett.* **295**, 59–63.

Hu, P. and McLachlan, E. M. 2001. Long-term changes in the distribution of galanin in dorsal root ganglia after sciatic or spinal nerve transection in rats. *Neuroscience* **103**, 1059–1071.

Hua, X. Y., Chen, P., Polgar, E., Nagy, I., Marsala, M., Phillips, E., Wollaston, L., Urban, L., Yaksh, T. L., and Webb, M. 1998. Spinal neurokinin NK1 receptor down-regulation and antinociception: Effects of spinal NK1 receptor antisense oligonucleotides and NK1 receptor occupancy. *J Neurochem.* **70**, 688–698.

Huang, L. Y. M. 1987. Electrical properties of acutely isolated, identified rat spinal dorsal horn projection neurons. *Neurosci. Lett.* **82**, 267–272.

Huang, L. Y. M., Carlton, S. M., and Willis, W. D. 1985. Identification of spinothalamic tract cells in fresh, unfixed rat spinal cord. *J. Neurosci. Methods* **14**, 91–96.

Huang, Y. -L., Ding, M., and Hansson, H.-A. 1998. Dorsal root ganglion cells transiently express increased immunoreactivity of the calcium-binding protein S-100β after sciatic nerve transection. *Brain Res.* **785**, 351–354.

Huang, Y., Stamer, W. D., Anthony, T. L., Kumar, D. V., St John, P. A., and Regan, J. W. 2002. Expression of alpha(2)-adrenergic receptor subtypes in prenatal rat spinal cord. *Dev. Brain Res.* **133**, 93–104.

Hubbard, S. J. 1958. A study of rapid mechanical events in a mechanoreceptor. *J. Physiol.* **141**, 198–218.

Hudson, P. M., Semenenko, F. M., and Lumb, B. M. 2000. Inhibitory effects evoked from the rostral ventrolateral medulla are selective for the nociceptive responses of spinal dorsal horn neurons. *Neuroscience* **99**, 541–547.

Hugel, S. and Schlichter, R. 2000. Presynaptic P2X receptors facilitate inhibitory GABAergic transmission between cultured rat spinal cord dorsal horn neurons. *J. Neurosci.* **20**, 2121–2130.

Hughes, J. and Gasser, H. S. 1934a. Some properties of the cord potentials evoked by a single afferent volley. *Am. J. Physiol.* **108**, 295–306.

Hughes, J. and Gasser, H. S. 1934b. The response of the spinal cord to two afferent volleys, *Am. J. Physiol.* **108**, 307–321.

Hughes, S., Lowrie, M. B., and Smith, M. E. 1998. Evidence for two populations of *N*-methyl-D-aspartate receptors in neonatal rat spinal cord: The effect of peripheral nerve axotomy. *Dev. Brain Res.* **105**, 209–217.

Hulliger, M., Nordh, E., Thelin, A. E., and Vallbo, A. B. 1979. The responses of afferent fibres from the glabrous skin of the hand during voluntary finger movements in man. *J. Physiol.* **291**, 233–249.

Hulsebosch, C. E. and Coggeshall, R. E. 1981. Sprouting of dorsal root axons. *Brain Res.* **224**, 170–174.

Hulsebosch, C. E., Xu, G. Y., Perez-Polo, J. R. 2000. Rodent model of chronic pain after spinal cord contusion injury and effects of gabapentin. *J. Neurotrauma* **17**, 1205–1217.

Hultborn, H., Illert, M., and Santini, M. 1976a. Convergence on interneurones mediating the reciprocal Ia inhibition of motoneurones: I. Disynaptic Ia inhibition of Ia inhibitory interneurones. *Acta Physiol. Scand.* **96**, 193–201.

Hultborn, H., Illert, M., and Santini, M. 1976b. Convergence on interneurones mediating the reciprocal Ia inhibition of motoneurones: II. Effects from segmental flexor reflex pathways. *Acta Physiol. Scand.* **96**, 351–367.

Hundle, B., McMahon, J., Dadger, J., Messing, R. O. 1995. Overexpression of epsilon-protein kinase C enhances nerve growth factor-induced phosphorylation of mitogen-activated protein kinases and neurite outgrowth. *J. Biol. Chem.* **270**, 30134–30140.

Hunt, C. C. 1954. Relation of function to diameter in afferent fibers of muscle nerves. *J. Gen. Physiol.* **38**, 117–131.

Hunt, C. C. 1961. On the nature of vibration receptors in the hind limb of the cat. *J. Physiol.* **155**, 175–186.

Hunt, C. C. and Kuno, M. 1959a. Properties of spinal interneurones. *J. Physiol.* **147**, 346–363.

Hunt, C. C. and Kuno, M. 1959b. Background discharge and evoked responses of spinal interneurones. *J. Physiol.* **147**, 364–384.

Hunt, C. C. and McIntyre, A. K. 1960a. Characteristics of responses from receptors from the flexor longus digitorum muscle and the adjoining interosseous region of the cat. *J. Physiol.* **153**, 74–87.

Hunt, C. C. and McIntyre, A. K. 1960b. Properties of cutaneous touch receptors in cat. *J. Physiol.* **153**, 88–98.

Hunt, C. C. and McIntyre, A. K. 1960c. An analysis of fibre diameter and receptor characteristics of myelinated cutaneous afferent fibres in cat. *J. Physiol.* **153**, 99–112.

Hunt, S. and Schmidt, J. 1978. Some observations on the binding patterns of α-bungaro-toxin in the central nervous system in the rat. *Brain Res.* **157**, 213–232.

Hunt, S. P. 1983. Cytochemistry of the spinal cord. In P. C. Emson (ed.), *Chemical Neuroanatomy* (pp. 53–84). Raven Press, New York.

Hunt, S. P. and Rossi, J. 1985. Peptide- and non-peptide-containing unmyelinated primary afferents: The parallel processing of nociceptive information. *Phil. Trans. Roy. Soc. Lond. B* **308**, 283–289.

Hunt, S. P., Kelly, J. S., and Emson, P. C. 1980. The electron-microscopic localization of methionine-enkephalin within the superficial layers (I and II) of the spinal cord. *Neuroscience* **5**, 1871–1890.

Hunt, S. P., Kelly, J. S., Emson, P. C., Kimmel, J. R., Miller, R. J., and Wu, J. Y. 1981a. An immunohistochemical study of neuronal populations containing neuropeptides or γ-aminobutyrate within the superficial layers of the rat dorsal horn. *Neuroscience* **6**, 1883–1898.

Hunt, S. P., Emson, P. C., Gilbert, R., Goldstein, M., and Kimmel, J. R. 1981b. Presence of avian pancreatic polypeptide-like immunoreactivity in catecholamine and methionine-enkephalin-containing neurones within the central nervous system. *Neurosci. Lett.* **21**, 125–130.

Hunt, S. P., Rossor, M. N., Emson, P. C., and Clement-Jones, V. 1982. Substance P and enkephalins in spinal cord after limb amputation. *Lancet* **1**, 1023.

Hunt, S. P., Pini, A., and Evan, G. 1987. Induction of c-fos-like protein in spinal cord neurons following sensory stimulation. *Nature* **328**, 632–634.

Hunter, J. C., Kilpatrick, G. J., and Brown, J. R. 1987. Pharmacological analysis of 125I-Bolton and Hunter labelled eledoisin in rat spinal cord by quantitative autoradiography. *Neurosci. Lett.* **78**, 12–16.

Hunter, J. C., Birchmore, B., Woodruff, R., and Hughes, J. 1989. Kappa opioid binding sites in the dog cerebral coretex and spinal cord. *Neuroscience* **31**, 735–743.

Hursh, J. B. 1939. Conduction velocity and diameter of nerve fibers. *Am. J. Physiol.* **127**, 131–139.

Hutchison, W. D., Morton, C. R., and Terenius, L. 1990. Dynorphin A: In vivo release in the spinal cord of the cat. *Brain Res.* **532**, 299–306.

Hwang, S. J., Rustioni, A., and Valtschanoff, J. G. 2001a. Kainate receptors in primary afferents to the rat gracile nucleus. *Neurosci. Lett.* **312**, 137–140.

Hwang, S. J., Pagliardini, S., Rustioni, A., and Valtschanoff, J. G. 2001b. Presynaptic kainate receptors in primary afferents to the superficial laminae of the rat spinal cord. *J. Comp. Neurol.* **436**, 275–289.

Hyden, H., Lovtrup, S., and Pigon, A. 1958. Cytochrome oxidase and succinoxidase in spinal ganglion cells and in glial capsule cells. *J. Neurochem.* **2**, 304–311.

Hylden, J. L. K., Ruda, M. A., Hayashi, H., and Dubner, R. 1985. Descending serotonergic fibers in the dorsolateral and ventral funiculi of cat spinal cord. *Neurosci. Lett.* **62**, 299–304.

Hylden, J. L. K., Hayashi, H., Dubner, R., and Bennett, G. J. 1986a. Physiology and morphology of the lamina I spinomesencephalic projection. *J. Comp. Neurol.* **247**, 505–515.

Hylden, J. L. K., Hayashi, H., Ruda, M. A., and Dubner, R. 1986b. Serotonin innervation of physiologically identified lamina I projection neurons. *Brain Res.* **370**, 401–404.

Hylden, J. L. K., Nahin, R. L., and Dubner, R. 1987. Altered responses of nociceptive cat lamina I spinal dorsal horn neurons after chronic sciatic neuroma formation. *Brain Res.* **411**, 341–350.

Hylden, J. L. K., Nahin, R. L., Traub, R. J., and Dubner, R. 1989. Expansion of receptive fields of spinal lamina I projection neurons in rats with unilateral adjuvant-induced inflammation: The contribution of central dorsal horn mechanisms. *Pain* **37**, 229–243.

Hylden, J. L. K., Noguchi, K., and Ruda, M. A. 1992. Neonatal capsaicin treatment attenuates spinal fos activation and dynorphin gene expression following peripheral tissue inflammation and hyperalgesia. *J. Neurosci.* **12**, 1716–1725.

Iadarola, M. J., Brady, L. S., Draisci, G., and Dubner, R. 1988a. Enhancement of dynorphin gene expression in spinal cord following experimental inflammation: Stimulus specificity, behavioral parameters and opoid receptor binding. *Pain* **35**, 313–326.

Iadarola, M. J., Douglass, J., Civelli, O., and Naranjo, J. R. 1988b. Differential activation of spinal cord dynorphin and enkephalin neurons during hyperalgesia: Evidence using cDNA hybridization. *Brain Res.* **455**, 205–212.

Ibuki, T., Hama, A. T., Wang, Z.-T., Wang, G. D., Pappas, G. D., and Sagen, J. 1997. Loss of GABA-immunoreactivity in the spinal dorsal horn of rats with peripheral nerve injury and promotion of recovery by adrenal medullary grafts. *Neuroscience* **76**, 845–858.

Ichikawa, H. and Sugimoto, T. 2000. Vanilloid receptor 1-like receptor-immunoreactive primary sensory neurons in the rat trigeminal nervous system. *Neuroscience* **101**, 719–725.

Ichikawa, H. and Sugimoto, T. 2001. VR1-immunoreactive primary sensory neurons in the rat trigeminal ganglion. *Brain Res.* **890**, 184–188.

Ichikawa, H., Jacobowitz, D. M., and Sugimoto, T. 1992. Calretinin-immunoreactivity in the oro-facial and pharyngeal regions of the rat. *Neurosci Lett.* **146**, 155–158.

Ichikawa, H., Jacobowitz, D. M., and Sugimoto, T. 1993. Calretinin-immunoreactive neurons in the trigeminal and dorsal root ganglia of the rat. *Brain Res.* **617**, 96–102.

Ichikawa, H., Deguchi, T., Nakago, T., Jacobowitz, D. M., and Sugimoto, T. 1994. Parvalbumin, calretinin and carbonic anhydrase in the trigeminal and spinal primary neurons of the rat. *Brain Res.* **655**, 241–245.

Ichikawa, H., He, Y. F., and Sugimoto, T. 1998. S100 protein-immunoreactive trigeminal neurons innervating the rat molar tooth pulp. *Somatosens. Mot. Res.* **15**, 128–133.

Ichikawa, H., Morgan, J. I., and Sugimoto, T. 1999a. Peptide 19 in the dorsal root ganglion and the mesencephalic trigeminal tract nucleus of the adult rat. *Brain Res.* **821**, 231–235.

Ichikawa, H., Itota, T., Torii, Y., Inoue, K., and Sugimoto, T. 1999b. Osteocalcin-immunoreactive primary sensory neurons in the rat spinal and trigeminal nervous systems. *Brain Res.* **838**, 205–209.

Ide, C. and Hayashi, S. 1987. Specializations of plasma membranes in Pacinian corpuscles: Implications for mechano-electric transduction. *J. Neurocytol.* **16**, 759–773.

Ide, C., Yoshida, Y., Hayashi, S., Takashio, M., and Munger, B. L. 1988. A re-evaluation of the cytology of cat Pacinian corpuscles: II. The extreme tip of the axon. *Cell Tissue Res.* **253**, 95–103.

Iggo, A. 1955. Tension receptors in the stomach and the urinary bladder. *J. Physiol.* **128**, 593–607.

Iggo, A. 1959. Cutaneous heat and cold receptors with slowly conducting C afferent fibres. *Q. J. Exp. Physiol.* **44**, 362–370.

Iggo, A. 1960. Cutaneous mechanoreceptors with afferent C fibres. *J. Physiol.* **152**, 337–353.

Iggo, A. 1961. Non-myelinated afferent fibres from mammalian skeletal muscle. *J. Physiol.* **155**, 52–53P.

Iggo, A. 1963. New specific sensory structures in hairy skin. *Acta Neuroveg.* **24**, 175–180.

Iggo, A. 1969. Cutaneous thermoreceptors in primates and subprimates. *J. Physiol.* **200**, 403–430.

Iggo, A. and Kornhuber, H. H. 1977. A quantitative study of C-mechanoreceptors in hairy skin of the cat. *J. Physiol.* **271**, 549–565.

Iggo, A. and Muir, A. R. 1969. The structure and function of a slowly adapting touch corpuscle in hairy skin. *J. Physiol.* **200**, 763–796.

Iggo, A., and Ogawa, H. 1971. Primate cutaneous thermal nociceptors. *J. Physiol.* **216**, 77–78P.

Iggo, A., and Ogawa, H. 1977. Correlative physiological and morphological studies of rapidly adapting mechanoreceptors in cat's glabrous skin. *J. Physiol.* **266**, 275–296.

Iggo, A. and Ramsey, R. L. 1976. Thermosensory mechanisms in the spinal cord of monkeys. In Y. Zotterman (ed.), *Sensory Functions of the Skin in Primates with Special Reference to Man* (pp. 285–304). Pergamon, Oxford.

Iggo, A., Molony, V., and Steedman, W. M. 1988. Membrane properties of nociceptive neurons in lamina II of lumbar spinal cord in the cat. *J. Physiol.* **400**, 367–380.

Iino, S., Kato, M., Hidaka, H., and Kobayashi, S. 1998. Neurocalcin-immunopositive neurons in the rat sensory ganglia. *Brain Res.* **781**, 236–243.

Ikeda, K., Watanabe, M., Ichikawa, T., Kobayashi, T., Yano, R., and Kumanishi, T. 1998. Distribution of prepronociceptin/orphanin FQ mRNA and its receptor mRNA in developing and adult mouse central nervous systems. *J. Comp. Neurol.* **399**, 139–151.

Iliakis, B., Anderson, N. L., Irish, P. S., Henry, M. A., and Westrum, L. E. 1996. Electron microscopy of immunoreactivity patterns for glutamate and gamma-aminobutyric acid in synaptic glomeruli of the feline spinal trigeminal nucleus (subnucleus caudalis). *J. Comp. Neurol.* **366**, 465–477.

Ilyinski, O. B., Akoev, G. N., Krasnikova, T. L., and Elman, S. I. 1976. K and Na ion content in the Pacininan corpuscle fluid and its role in the activity of receptors. *Pfluegers Arch.* **361**, 279–285.

Imbe, H. and Ren, K. 2000. The up-regulation of preprodynorphin mRNA in trigeminoparabrachial neurons after inflammation. *NeuroReport* **11**, 845–847.

Immke, D.C. and McCleskey, E.W. 2001. Lactate enhances the acid-sensing Na^+ channel on ischemia sensing neurons. *Nature Neurosci.* **4**, 869–870.

Inagaki, S. and Kito, S. 1986. Peptides in the peripheral nervous system. In P. C. Emson, M. N. Rossor and M. Tohyama (eds.), *Progress in Brain Res.* (pp. 269–316). Elsevier, Amsterdam.

Inagaki, S., Kito, S., Kubota, Y., Girgis, S., Hillyard, C. J., and MacIntyre, I. 1986. Autoradiographic localization of calcitonin gene-related peptide binding sites in human and rat brain. *Brain Res.* **374**, 287–298.

Inagaki, N., Yamatodani, A., Ando-Yamamoto, M., Tohyama, M., Watanabe, T., and Wada, H. 1988. Organization of histaminergic fibers in the rat brain. *J. Comp. Neurol.* **273**, 283–300.

Inaishi, Y., Kashihara, M., Sakaguchi, M., Nawa, H., and Kuno, M. 1992. Cooperative regulation of calcitonin gene-related peptide levels in rat sensory neurons via their central and peripheral processes. *J. Neurosci.* **12**, 518–524.

Indo, Y., Tsuruta, M., Hayashida, Y., Karim, M. A., Ohta, K., Kawano, T., Mitsubuchi, H., Tonoki, H., Awaya, Y., and Matsuda, I. 1996. Mutations in the TRKA/NGF receptor gene in patients with congenital insensitivity to pain with anhidrosis. *Nat. Genet.* **13**, 382–383.

Indo, Y., Mardy, S., Miura, Y., Moosa, A., Ismail, E. A., Toscano, E., Andria, G., Pavone, V., Brown, D. L., Brooks, A., Endo, F., and Matsuda, I. 2001. Congenital insensitivity to pain with anhidrosis (CIPA): Novel mutations of the TRKA (NTRK1) gene, a putative uniparental disomy, and a linkage of the mutant TRKA and PKLR genes in a family with CIPA and pyruvate kinase deficiency. *Hum. Mutat.* **18**, 308–318.

Inglis, J. T. and Frank, J. S. 1990. The effect of agonist/antagonist muscle vibration on human position sense. *Exp. Brain Res.* **81**, 573–580.

Inglis, J. T., Frank, J. S., and Inglis, B. 1991. The effect of muscle vibration on human position sense during movements controlled by lengthening muscle contraction. *Exp. Brain Res.* **84**, 631–634.

Ino, H., Ishizuka, T., Chiba, T., and Tatibana, M. 1994. Expression of CDK5 (PSSALRE kinase), a neural cdc-2-related protein kinase, in the mature and developing mouse central and peripheral nervous systems. *Brain Res.* **661**, 196–206.

Inyama, C. O., Wharton, J., Su, H. C., and Polak, J. M. 1986. CGRP-immunoreactive nerves in the genitalia of the female rat originate from dorsal root ganglia T11-L3 and L6-S1: A combined immunocytochemical and retrograde tracing study. *Neurosci. Lett.* **69**, 13–18.

Irish, P. S., Iliakis, B., Anderson, N. L., and Westrum, L. E. 1996. Electron microscopic analysis of lesion-induced changes in synaptic structure and immunogold labeling of neurotransmitters within the feline trigeminal nucleus. *Synapse* **24**, 48–59.

Iriuchijima, J. and Zotterman, Y. 1960. The specificity of afferent cutaneous C fibres in mammals. *Acta Physiol. Scand.* **49**, 267–278.

Isaac, J. T., Nicoll, R. A., and Malenka, R. C. 1999. Silent glutamatergic synapses in the mammalian brain. *Can. J. Physiol. Pharmacol.* **77**, 735–737.

Ishida-Yamamoto, A. and Senba, E. 1990. Cell types and axonal sizes of calcitonin gene-related peptide-containing primary sensory neurons of the rat. *Brain Res. Bull.* **24**, 759–764.

Ishida-Yamamoto, A., Senba, E., and Tohyama, M. 1988. Calcitonin gene-related peptide- and substance P-immunoreactive nerve fibers in Meissner's corpuscles of rats: An immunohistochemical analysis. *Brain Res.* **453**, 362–366.

Ishida-Yamamoto, A., Senba, E., and Tohyama, M. 1989. Distribution and fine structure of calcitonin gene-related peptide-like immunoreactive nerve fibers in the rat skin. *Brain Res.* **491**, 93–101.

Ishigooka, M., Nakada, T., Hashimoto, T., Iijima, Y., and Yaguchi, H. 2002. Spinal substance P immunoreactivity is enhanced by acute chemical stimulation of the rat prostate. *Urology* **59**, 139–144.

Itoh, M., Takasaki, I., Andoh, T., Nojima, H., Tominaga, M., and Kuraishi, Y. 2001. Induction by carrageenan inflammation of prepronociceptin mRNA in VR1-immunoreactive neurons in rat dorsal root ganglia. *Neurosci Res.* **40**, 227–233.

Itoh, Y. and Tessler, A. 1990. Ultrastructural organization of regenerated dorsal root axons within transplants of fetal spinal cord. *J. Comp. Neurol.* **292**, 396–411.

Itoh, Y., Yagishita, S. D., Nakajima, S., Nakano, T., and Kawada, H. 1986. Congenital insensitivity to pain with anhidrosis: Morphological and morphometrical studies on the skin and peripheral nerves. *Neuropediatrics* **17**, 103–110.

Itoh, Y., Mizoi, K., and Tessler, A. 1999. Embryonic central nervous system transplants mediate adult dorsal root regeneration into host spinal cord. *Neurosurgery* **45**, 849–856.

Iwasaki, Y., Ikeda, K., Shiojima, T., and Kinoshita, M. 1995. CNQX prevents spinal motor neuron death following sciatic nerve transection in newborn rats. *J. Neurol. Sci.* **134**, 21–25.

Iyadomi, M., Iyadomi, I., Kumamoto, E., Tomokuni, K., and Yoshimura, M. 2000. Presynaptic inhibition by baclofen of miniature EPSCs and IPSCs in substantia gelatinosa neurons of the adult rat spinal dorsal horn. *Pain* **85**, 385–393.

Jackson, D. A. and White, S. R. 1988. Thyrotropin-releasing hormone (TRH) modifies excitability of spinal cord dorsal horn cells. *Neurosci. Lett.* **92**, 171–176.

Jacquin, M. F., Beinfeld, M. C., Chiaia, N. L., and Zahm, D. S. 1992. Cholecystokinin concentrations and peptide immunoreactivity in the intact and deafferented medullary dorsal horn of the rat. *J. Comp. Neurol.* **326**, 22–43.

Jacobs, J. M., Carmichael. N., and Cavanagh, J. B. 1975. Ultrastructural changes in the dorsal root and trigeminal ganglia of rats poisoned with methyl mercury. *Neuropathol. Appl. Neurobiol.* **1**, 1–19.

Jaggar, S. I., Hasnie, F. S., Sellaturay, S., and Rice, A. S. C. 1998. The antihyperalgesic actions of the cannabinoid anandamide and the putative CB2 receptor agonist palmitoylethanolamide in visceral and somatic inflammatory pain. *Pain* **76**, 189–199.

Jahr, C. E. and Jessell, T. M. 1983. ATP excites a subpopulation of rat dorsal horn neurones. *Nature* **304**, 730–733.

Jakab, G., Salamon, I., Petrusz, P., and Rethelyi, M. 1990. Termination patterns of calcitonin gene-related peptide-immunoreactive nerve fibers in the dorsal horn of the human spinal cord. *Exp. Brain Res.* **80**, 609–617.

Jakowec, M. W., Fox, A. J., Martin, L. J., and Kalb, R. G. 1995a. Quantitative and qualitative changes in AMPA receptor expression during spinal cord development. *Neuroscience* **67**, 893–907.

Jakowec, M. W., Yen, L., and Kalb, R. G. 1995b. *In situ* hybridization analysis of AMPA receptor subunit gene expression in the developing rat spinal cord. *Neuroscience* **67**, 909–920.

Jancsó, G. and Knyihar, E. 1975. Functional linkage between nociception and fluoride-resistant acid phosphatase activity in the Rolando substance. *Neurobiology* **5**, 42–43.

Jancsó, G., Hökfelt, T., Lundberg, J. M., Kiraly, E., Halasz, N., Nilsson, G., Terenius, L., and others, 1981. Immunohistochemical studies on the effect of capsaicin on spinal and medullary peptide and monoamine neurons using antisera to substance P, gastrin/CCK, somatostatin, VIP enkephalin, neurotensin and 5-hydroxytryptamine. *J. Neurocytol.* **10**, 963–980.

Jancsó, N., Jancsó-Gabor, A., and Takats, I. 1961. 1: Pain and inflammation induced by nicotine, acetylcholine and structurally related compounds and their prevention by desensitizing agents. *Acta Physiol.* **19**, 113–132.

Jancsó, N., Jancsó-Gábor, A., and Szolcsányi, J. 1967. Direct evidence for neurogenic inflammation and its prevention by denervation and by pretreatment with capsaicin. *Br. J. Pharmacol.* **31**, 138–151.

Jänig, W. 1971a. The afferent innervation of the central pad of the cat's hind foot. *Brain Res.* **28**, 203–216.

Jänig, W. 1971b. Morphology of rapidly and slowly adapting mechanoreceptors in the hairless skin of the cat's hind foot. *Brain Res.* **28**, 217–231.

Jänig, W. 1996. Neurobiology of visceral afferent neurons: Neuroanatomy, functions, organ regulations and sensations. *Biol. Psychol.* **42**, 29–51.

Jänig, W. and Koltzenburg, M. 1990. On the function of spinal primary afferent fibres supplying colon and urinary bladder. *J. Auton. Nerv. Syst.* **30** Suppl. S89–S96.

Jänig, W. and Koltzenburg, M. 1991. Receptive properties of sacral primary afferent neurons supplying the colon. *J. Neurophysiol.* **65**, 1067–1077.

Jänig, W. and Lisney, S. J. W. 1989. Small diameter myelinated afferents produce vasodilatation but not plasma extravasation in rat skin. *J. Physiol.* **415**, 477–486.

Jänig, W. and Zimmermann, M. 1971. Presynaptic depolarization of myelinated afferent fibres evoked by stimulation of cutaneous C fibres. *J. Physiol.* **214**, 29–50.

Jänig, W., Schmidt, R. F., and Zimmermann, M. 1968a. Single unit responses and the total afferent outflow from the cat's foot pad upon mechanical stimulation. *Exp. Brain Res.* **6**, 100–115.

Jänig, W., Schmidt, R. F., and Zimmermann, M. 1968b. Two specific feedback pathways to the central afferent terminals of phasic and tonic mechanoreceptors. *Exp. Brain Res.* **6**, 116–129.

Jänig, W., Levine, J. D., and Michaelis, M. 1996. Interactions of sympathetic and primary afferent neurons following nerve injury and tissue trauma. *Prog. Brain Res.* **113**, 161–184.

Jankowska, E. 1992. Interneuronal relay in pathways from proprioceptors. *Prog. Neurobiol.* **38**, 335–378.

Jankowska, E., and Lindström, S. 1972. Morphology of interneurones mediating Ia reciprocal inhibition of motoneurones in the spinal cord of the cat. *J. Physiol.* **226**, 805–823.

Jankowska, E. and Riddell, J. S. 1994. Interneurones in pathways from group II muscle afferents in sacral segments of the cat spinal cord. *J. Physiol.* **475**, 455–468.

Jankowska, E. and Riddell, J. S. 1995. Interneurones mediating presynaptic inhibition of group I muscle afferents in the cat spinal cord. *J. Physiol.* **483**, 473–479.

Jankowska, E. and Roberts, W. J. 1972a. An electrophysiological demonstration of the axonal projections of single spinal interneurones in the cat. *J. Physiol.* **222**, 597–622.

Jankowska, E. and Roberts, W. J. 1972b. Synaptic actions of single interneurones mediating reciprocal Ia inhibition of motoneurones. *J. Physiol.* **222**, 623–642.

Jankowska, E., Johannisson, T., and Lipski, J. 1981a. Common interneurones in reflex pathways from group Ia and Ib afferents of ankle extensors in the cat. *J. Physiol.* **310**, 381–402.

Jankowska, E., McCrea, D., Rudomin, P., and Sykova, E. 1981b. Observations on neuronal pathways subserving primary afferent depolarization. *J. Neurophysiol.* **46**, 506–516.

Jankowska, E., Riddell, J. S., and McCrea, D. A. 1993. Primary afferent depolarization of myelinated fibres in the joint and interosseous nerves of the cat. *J. Physiol.* **466**, 115–131.

Jankowska, E., Maxwell, D. J., Dolk, S., Krutki, P., Belichenko, P. V., and Dahlstrom, A. 1995. Contacts between serotoninergic fibres and dorsal horn spinocerebellar tract neurons in the cat and rat: A confocal microscopic study. *Neuroscience* **67**, 477–487.

Jankowska, E., Maxwell, D. J., Dolk, S., and Dahlstrom, A. 1997. A confocal and electron microscopic study of contacts between 5-HT fibres and feline dorsal horn interneurons in pathways from muscle afferents. *J. Comp. Neurol.* **387**, 430–438.

Jansen, K. L. R., Faull, R. L. M., Dragunow, M., and Waldvogel, H. 1990. Autoradiographic localisation of NMDA, quisqualate and kainic acid receptors in human spinal cord. *Neurosci. Lett.* **108**, 53–57.

Jeftinija, S. 1988. Enkephalins modulate excitatory synaptic transmission in the superficial dorsal horn by acting at mu-opioid receptor sites. *Brain Res.* **460**, 260–268.

Jeftinija, S. and Jeftinija, K. 1990. Calcitonin gene-related peptide immunoreactivity in neuronal perikarya in the dorsal root. *Brain Res.* **519**, 324–328.

Jeftinija, S., Miletić, V., and Randić, M. 1981. Cholecystokinin octapeptide excites dorsal horn neurons both in vivo and in vitro. *Brain Res.* **213**, 231–236.

Jeftinija, S., Murase, K., Nedeljikov, V., and Randić, M. 1982. Vasoactive intestinal polypeptide excites mammalian dorsal horn neurons both in vivo and in vitro. *Brain Res.* **243**, 158–164.

Jeftinija, S., Urban, L., Kangrga, I., Ryu, P. D., and Randić, M. 1987. Slow excitatory and inhibitory transmission in the rat spinal dorsal horn in vitro and depressant effect of enkephalins. *Neurol. Neurobiol.* **28**, 271–281.

Jenkins, W. L. 1941a. Studies in thermal sensitivity: 15. Effects of stimulus-temperature in seriatum warm-mapping. *J. Exp. Psychol.* **28**, 517–523.

Jenkins, W. L. 1941b. Studies in thermal sensitivity: 16. Further evidence on the effects of stimulus temperature. *J. Exp. Psychol.* **29**, 413–419.

Jennes, L., Stumpf, W. E., and Kalivas, P. W. 1982. Neurotensin: Topographical distribution in rat brain by immunohistochemistry. *J. Comp. Neurol.* **210**, 211–224.

Jessell, T. M. and Dodd, J. 1985. Structure and expression of differentiation antigens on functional subclasses of primary sensory neurons. *Phil. Trans. R. Soc. B* **308**, 271–281.

Jessell, T. M. and Dodd, J. 1986. Neurotransmitters and differentiation antigens in subsets of sensory neurons projecting to the spinal dorsal horn. In J. B. Martin and J. D. Barchas (eds.), *Neuropeptides in Neurologic and Psychiatric Disease* (pp. 111–131). Raven Press, New York.

Jessell, T. M. and Iversen, L. L. 1977. Opiate analgesics inhibit substance P release from rat trigeminal nucleus. *Nature* **68**, 549–551.

Jessell, T. M., Tsunoo, A., Kanazawa, I., and Otsuka, M. 1979. Substance P: Depletion in the dorsal horn of rat spinal cord after section of the peripheral processes of primary sensory neurons. *Brain Res.* **168**, 247–259.

Jessell, T. M., Yoshioka, K., and Jahr, C. E. 1986. Amino acid receptor-mediated transmission at primary afferent synapses in rat spinal cord. *J. Exp. Biol.* **124**, 139–258.

Jhamandas, K., Yaksh, T. L., and Go, V. L. W. 1984. Acute and chronic morphine modifies the in vivo release of methionine enkephalin-like immunoreactivity from the cat spinal cord and brain. *Brain Res.* **297**, 91–103.

Ji, R.-R., Zhang, X., Wiesenfeld-Hallin, Z., and Hökfelt, T. 1994. Expression of neuropeptide Y and neuropeptide Y (Y1) receptor mRNA in rat spinal cord and dorsal root ganglia following peripheral tissue inflammation. *J. Neurosci.* **14**, 5423–5434.

Ji, R. -R., Zhang, X., Zhang, Q., Dagerlind, A., Nilsson, S., Wiesenfeld-Hallin, Z., and Hökfelt, T. 1995a. Central and peripheral expression of galanin in response to inflammation. *Neuroscience* **68**, 563–576.

Ji, R. -R., Zhang, Q., Law, P.-Y., Low, P. Y., Elde, R., and Hökfelt, T. 1995b. Expression of μ-, δ-, and k-opioid receptor-like immunoreactivities in rat dorsal root ganglia after carageenan-induced inflammation. *J. Neurosci.* **15**, 8156–8166.

Jia, H., Rustioni, A., and Valtschanoff, J. G. 1999. Metabotropic glutamate receptors in superficial laminae of the rat dorsal horn. *J. Comp. Neurol.* **410**, 627–642.

Jia, Y., Linden, D. R., Serie, J. R., and Seybold, V. S. 1998. Nociceptin/orphanin FQ binding increases in superficial laminae of the rat spinal cord during persistent peripheral inflammation. *Neurosci Lett.* **250**, 21–24.

Jiang, C. H., Maziéres, L., and Lindström, S. 2002. Cold- and menthol-sensitive C afferents of cat urinary bladder. *J. Physiol.* **543**.1, 211–220.

Jin, S., Lei, L., Wang, Y., Da, D., and Zhao, Z. 1999. Endomorphin-1 reduces carrageenan-induced fos expression in the rat spinal dorsal horn. *Neuropeptides* **33**, 281–284.

Jinks, S. L. and Carstens, E. 1999. Activation of spinal wide-dynamic-range neurons by intracutaneous microinjection of nicotine. *J. Neurophysiol.* **82**, 3046–3055.

Jo, Y. H. and Schlichter, R. 1999. Synaptic corelease of ATP and GABA in cultured spinal neurons. *Nature Neurosci.* **2**, 241–245.

Jo, Y. H., Stoeckel, M. E., and Schlichter, R. 1998. Electrophysiological properties of cultured neonatal rat dorsal horn neurons containing GABA and met-enkephalin-like immunoreactivity, *J. Neurophysiol.* **79**, 1583–1586.

Johannessen, J. N., Watkins, L. R., and Mayer, D. J. 1984. Non-serotonergic origins of the dorsolateral funiculus in the rat ventral medulla. *J. Neurosci* **4**, 757–766.

Johansson, O. 1978. Localization of somatostatin-like immunoreactivity in the Golgi apparatus of central and peripheral neurons. *Histochemistry* **58**, 167–176.

Johansson, O., Hökfelt, T., Pernow, B., Jeffcoate, S. L., White, N., Steinbusch, H. W. M., Verhofstad, A. A. J., Emson, P. C., and Spindel, E. 1981. Immunohistochemical support for three putative transmitters in one neuron: Coexistence of 5-hydroxytryptamine, substance P- and thyrotropin releasing hormone-like immunoreactivity in medullary neurons projecting to the spinal cord. *Neuroscience* **6**, 1857–1881.

Johansson, O., Hokfelt, T., and Elde, R. 1984. Immunohistochemical distribution of somatostatin-like immunoreactivy in the central nervous system of the adult rat. *Neuroscience* **13**, 265–339.

Johansson, O., Fantini, F., and Hu, H. 1999. Neuronal structural proteins, transmitters, transmitter enzymes and neuropeptides in human Meissner's corpuscles: A reappraisal using immunohistochemistry. *Arch. Dermatol. Res.* **291**, 419–424.

Johansson, R. S. 1978. Tactile sensibility in the human hand: Receptive field characteristics of mechanoreceptive units in the glabrous skin area. *J. Physiol.* **281**, 101–123.

Johansson, R. S. and Vallbo, A. B. 1976. Skin mechanoreceptors in the human hand: An inference of some population properties. In Y. Zotterman (ed.), *Sensory Functions of the Skin in Primates* (pp. 171–184). Pergamon Press, New York.

Johansson, R. S. and Vallbo, A. B. 1979a. Tactile sensibility in the human hand: Relative and absolute densities of four types of mechanoreceptive units in glabrous skin. *J. Physiol.* **286**, 283–300.

Johansson, R. S. and Vallbo, A. B. 1979b. Detection of tactile stimuli: Threshold of afferent units related to psychophysical thresholds of the human hand. *J. Physiol.* **297**, 405–422.

Johnson, H., Ulfhake, B., Dagerlind, A., Bennett, G. W., Fone, K. C. F., and Hökfelt, T. 1993. The serotoninergic bulbospinal system and brainstem-spinal cord content of serotonin-, TRH-, and substance P-like immunoreactivity in the aged rat with special reference to the spinal cord motor nucleus. *Synapse* **15**, 63–89.

Johnson, J. A., Grande, J. P., Windebank, A. J., and Kumar, R. 1996. 1,25-dihydroxyvitamin D_3 receptors in developing dorsal root ganglia of fetal rats. *Dev. Brain Res.* **92**, 120–124.

Johnson, J. L. 1972. Glutamic acid as a synaptic transmitter in the nervous system: A review. *Brain Res.* **37**, 1–19.

Johnson, J. L. and Aprison, M. H. 1970. The distribution of glutamic acid, a transmitter candidate, and other amino acids in the dorsal sensory neuron of the cat. *Brain Res.* **24**, 285–292.

Johnson, K. O. 1974. Reconstruction of population response to a vibratory stimulus in quickly adapting mechanoreceptive afferent fiber population innervating glabrous skin of the monkey. *J. Neurophysiol.* **37**, 48–72.

Johnson, K. O., Darian-Smith, I., and LaMotte, C. 1973. Peripheral neural determinants of temperature discrimination in man: A correlative study of responses to cooling of skin. *J. Neurophysiol.* **36**, 347–370.

Johnston, G. A. R. 1968. The intraspinal distribution of some depressant amino acids. *J. Neurochem.* **15**, 1013–1017.

Johnston, G. A. R. and Vitali, M. V. 1969. Glycine-producing transaminase activity in extracts of spinal cord. *Brain Res.* **12**, 471–472.

Jonakait, G. M., Markey, K. A., Goldstein, M., and Black, I. B. 1984. Transient expression of selected catecholaminergic traits in cranial sensory and dorsal root ganglia of the embryonic rat. *Dev. Biol.* **101**, 51–60.

Jones, B. E., Holmes, C. J., Rodriguez-Veiga, E., and Mainville, L. 1991. GABA-synthesizing neurons in the medulla: Their relationship to serotonin-containing and spinally projecting neurons in the rat. *J. Comp. Neurol.* **313**, 349–367.

Jones, M. G., Munson, J. B., and Thompson, S. W. 1999. A role for nerve growth factor in sympathetic sprouting in rat dorsal root ganglia. *Pain* **79**, 21–29.

Jones, M. W. and Headley, P. M. 1995. Interactions between metabotropic and ionotropic glutamate receptor agonists in the rat spinal cord in vivo. *Neuropharmacology* **34**, 1025–1031.

Jones, S. L. and Light, A. R. 1990. Electrical stimulation in the medullary nucleus raphe magnus inhibits noxious heat-evoked *fos* protein-like immunoreactivity in the rat lumbar spinal cord. *Brain Res.* **530**, 335–338.

Jongsma, H., Danielsen, N., Sundler, F., and Kanje, M. 2000. Alteration of PACAP distribution and PACAP receptor binding in the rat sensory nervous system following sciatic nerve transection. *Brain Res.* **853**, 186–196.

Jørum, E., Lundberg, L. E. R., and Torebjörk, H. E. 1989. Peripheral projections of nociceptive unmyelinated axons in the human peroneal nerve. *J. Physiol.* **416**, 291–301.

Jossan, S. S., Ekblom, J., Aquilonius, S. -M., and Oreland, L. 1994. Monoamine oxidase-B in motor cortex and spinal cord in amyotrophic lateral sclerosis studied by quantitative autoradiography. *J. Neural. Trans.* **41**, 243–248.

Ju, G., Hökfelt, T., Fischer, J. A., Frey, P., Rehfeld, J. F., and Dockray, G. J. 1986. Does cholecystokinin-like immunoreactivity in rat primary sensory neurones represent calcitonin gene-related peptide? *Neurosci. Lett.* **68**, 305–310.

Ju, G., Melander, T., Ceccatelli, S., Hökfelt, T., and Frey, P. 1987a. Immunohistochemical evidence for a spinothalamic pathway co-containing cholecystokinin- and galanin-like immunoreactivities in the rat. *Neuroscience* **20**, 439–456.

Ju, G., Hökfelt, T., Brodin, E., Fahrenkrug, J., Fischer, J. A., Frey, P., Elde, R. P., and Brown, J. C. 1987b. Primary sensory neurons of the rat showing calcitonin gene-related peptide immunoreactivity and their relation to substance P-, somatostatin-, galanin-, vasoactive intestinal polypeptide- and cholecystokinin-immunoreactive ganglion cells. *Cell Tissue Res.* **247**, 417–431.

Julia, V., Morteau, O., and Bueno, L. 1994. Involvement of neurokinin 1 and 2 receptors in viscerosensitive response to rectal distension in rats. *Gastroenterology* **107**, 94–102.

Julia, V., Su, X., Buéno, L., and Gebhart, G. F. 1999. Role of neurokinin 3 receptors on responses to colorectal distension in the rat: Electrophysiological and behavioral studies. *Gastroenterology* **116**, 1124–1131.

Junttila, T., Koistinaho, J., Rechardt, L., Hidaka, H., Okazaki, K., and Pelto-Huikko, M. 1995. Localization of neurocalcin-like immunoreactivity in rat cranial motoneurons and spinal cord interneurons. *Neurosci Lett.* **183**, 100–103.

Kaada, B. 1974. Mechanisms of acupuncture analgesia. *Tidsskr. Norske Laegeforen.* **94**, 422–431.

Kadekaro, M., Crane, A. M., and Sokoloff, L. 1985. Differential effects of electrical stimulation of sciatic nerve on metabolic activity in spinal cord and dorsal root ganglion in the rat. *Proc. Natl. Acad. Sci. USA* **82**, 6010–6013.

Kaduri, A. J., Magoul, R., Lescaudron, L., Campistron, G., and Calas, A. 1987. Immunocytochemical approach of GABAergic innervation of the mouse spinal cord using antibodies to GABA. *J Hirnforsch.* **28**, 349–355.

Kai-Kai, M. A. and Che, Y. M. 1995. Distribution of arginine-vasopressin in the trigeminal, dorsal root ganglia and spinal cord of the rat: Depletion by capsaicin. *Comp. Biochem. Physiol. A Physiol.* **110**, 71–78.

Kai-Kai, M. A. and Howe, R. 1991. Glutamate-immunoreactivity in the trigeminal and dorsal root ganglia, and intraspinal neurons and fibres in the dorsal horn of the rat. *Histochem. J.* **23**, 171–179.

Kai-Kai, M. A. and Keen, P. 1985. Localization of 5-hydroxytryptamine to neurons and endoneurial mast cells in rat sensory ganglia. *J. Neurocytol.* **14**, 63–78.

Kai-Kai, M. A., Swann, R. W., and Keen, P. 1985. Localization of chromatographically characterized oxytocin and arginine-vasopressin to sensory neurones in the rat. *Neurosci. Lett.* **55**, 83–88.

Kai-Kai, M. A., Anderton, B. H., and Keen, P. 1986. A quantitative analysis of the interrelationships between subpopulations of rat sensory neurons containing arginine vasopressin or oxytocin and those containing substance P, fluoride-resistant acid phosphatase or neurofilament protein. *Neuroscience* **18**, 475–486.

Kajander, K. C. and Xu, J. 1995. Quantitative evaluation of calcitonin gene-related peptide and substance P levels in rat spinal cord following peripheral nerve injury. *Neurosci. Lett.* **186**, 184–188.

Kajander, K. C., Sahara, Y., Iadarola, M. J., and Bennett, G. J. 1990. Dynorphin increases in the dorsal spinal cord in rats with a painful peripheral neuropathy. *Peptides* **11**, 719–728.

Kajander, K. C., Wakisaka, S., and Bennett, G. J. 1992. Spontaneous discharge originates in the dorsal root ganglion at the onset of a painful peripheral neuropathy in the rat. *Neurosci. Lett.* **138**, 225–228.

Kalb, R. G. and Fox, A. J. 1997. Synchronized overproduction of AMPA, kainate, and NMDA glutamate receptors during human spinal cord development. *J. Comp. Neurol.* **384**, 200–210.

Kalb, R. G., Lidow, M. S., Halsted, M. J., and Hockfield, S. 1992. N-methyl-D-aspartate receptors are transiently expressed in the developing spinal cord ventral horn. *Proc. Natl. Acad. Sci.* **89**, 8502–8506.

Kalia, M., Mei, S. S., and Kao, F. F. 1981. Central projections from ergoreceptors (C fibers) in muscle involved in cardiopulmonary responses to static exercise. *Circ. Res.* **48**, I48–I62.

Kalina, M. and Bubis, J. J. 1968. Histochemical studies on the distribution of acid phosphatases in neurons of sensory ganglia. *Histochemie* **14**, 103–112.

Kalina, M. and Bubis, J. J. 1969. Ultrastructural localization of acetylcholine esterase in neurones of rat trigeminal ganglia. *Experientia* **25**, 388–389.

Kalina, M. and Wolman, M. 1970. Correlative histochemical and morphological study in the maturation of sensory ganglion cells in the rat. *Histochemie* **22**, 100–108.

Kalivas, P. W., Jennes, L., Nemeroff, C. B., and Prange, A. J. 1982. Neurotensin: Topographical distribution of brain sites involved in hypothermia and antinociception. *J. Comp. Neurol.* **210**, 225–238.

Kallakuri, S., Cavanaugh, J. M., and Blagoev, D. C. 1998. An immunohistochemical study of innervation of lumbar spinal dura and longitudinal ligaments. *Spine* **23**, 403–411.

Kanazawa, I., Ogawa, T., Kimura, S., and Munekata, E. 1984. Regional distribution of substance P, neurokinin α and neurokinin β in rat central nervous system. *Neurosci. Res.* **2**, 111–120.

Kaneko, T., Fujiyama, F., and Hioki, H. 2002. Immunohistochemical localization of candidates for vesicular glutamate transporters in the rat brain. *J. Comp. Neurol.* **444**, 39–62.

Kangrga, I. and Randić, M. 1990. Tachykinins and calcitonin gene-related peptide enhance release of endogenous glutamate and aspartate from the rat spinal dorsal horn slice. *J. Neurosci.* **10**, 2026–2038.

Kangrga, I. and Randic, M. 1991. Outflow of endogenous aspartate and glutamate from the rat spinal dorsal horn in vitro by activation of low- and high-threshold primary afferent fibers. Modulation by μ-opioids. *Brain Res.* **553**, 347–352.

Kangrga, I., Larew, J. S. A., and Randic, M. 1990. The effects of substance P and calcitonin gene-related peptide on the efflux of endogenous glutamate and aspatate from the rat spinal dorsal horn in vitro. *Neurosci. Lett.* **108**, 155–160.

Kangrga, I., Jiang, M. C., and Randic, M. 1991. Actions of (-)-baclofen on rat dorsal horn neurons. *Brain Res.* **562**, 265–275.

Kanjhan, R., Housley, G. D., Burton, L. D., Christie, D. L., Kippenberger, A., Thorne, P. R., Luo, L., and Ryan, A. F. 1999. Distribution of the $P2X_2$ receptor subunit of the ATP-grated ion channels in the rat central nervous system. *J. Comp. Neurol.* **407**, 11–32.

Kantner, R. M. and Kirby, M. L. 1982. Changes in acid phosphatase activity in the substantia gelatinosa in response to pain. *Brain Res.* **238**, 451–456.

Kantner, R. M., Goldstein, B., and Kirby, M. 1986. Regulatory mechanisms for substance P in the dorsal horn during a nociceptive stimulus: Axoplasmic transport vs. electrical activity. *Brain Res.* **385**, 282–290.

Kanui, T. I. 1985. Responses of spinal cord neurones to noxious and non-noxious stimulation of the skin and testicle of the rat. *Neurosci. Lett.* **58**, 315–319.

Kanui, T. I. 1987. Thermal inhibition of nociceptor-driven spinal cord neurones in the cat: A possible neuronal basis for thermal analgesia. *Brain Res.* **402**, 160–163.

Kanui, T. I. 1988. Receptive field organization and electrophysiological responses of spinal cord thermoreactive neurones in the rat. *Exp. Brain Res.* **71**, 508–514.

Kapadia, S. E. and LaMotte, C. C. 1987. Deafferentation-induced alterations in the rat dorsal horn: I. Comparison of peripheral nerve injury vs. rhizotomy effects on presynaptic, postsynaptic, and glial processes. *J. Comp. Neurol.* **266**, 183–197.

Kar, S. and Quirion, R. 1992. Quantitative autoradiographic localization of [^{125}] neuropeptide Y receptor binding sites in rat spinal cord and the effects of neonatal capsaicin, dorsal rhizotomy and peripheral axotomy. *Brain Res.* **574**, 333–337.

Kar, S. and Quirion, R. 1994. Galanin receptor binding sites in adult rat spinal cord respond differently to neonatal capsaicin, dorsal rhizotomy and peripheral axotomy. *Eur. J. Neurosci.* **6**, 1917–1921.

Kar, S. and Quirion, R. 1995. Neuropeptide receptors in developing and adult rat spinal cord: An *in vitro* quantitative autoradiography study of calcitonin gene-related peptide, neurokinins, mu-opioid, galanin, somatostatin, neurotensin and vasointestinal polypeptide receptors. *J. Comp. Neurol.* **354**, 253–281.

Kar, S., Gibson, S. J., Scaravilli, F., Jacobs, J. M., Aber, V. R., and Polak, J. M. 1989. Reduced numbers of calcitonin gene-related peptide-(CGRP) and tachykinin-immunoreactive sensory neurones associated with greater enkephalin immunoreactivity in the dorsal horn of a mutant rat with hereditary sensory neuropathy. *Cell Tisue Res.* **255**, 451–466.

Kar, S., Gibson, S. J., and Polak, J. M. 1990. Origins and projections of peptide-immunoreactive nerves in the male rat genitofemoral nerve. *Brain Res.* **512**, 229–237.

Kar, S., Gibson, S. J., Rees, R. G., Jura, W. G., Brewerton, D. A., and Polak, J. M. 1991a. Increased calcitonin gene-related peptide (CGRP), substance P, and enkephalin immunoreactivities in dorsal spinal cord and loss of CGRP- immunoreactive motoneurons in arthritic rats depend on intact peripheral nerve supply. *J Mol. Neurosci.* **3**, 7–18.

Kar, S., Chabot, J. G., and Quirion, R. 1991b. Quantitative autoradiographic localisation of [125I]endothelin-1 binding sites in spinal cord and dorsal root ganglia of the rat. *Neurosci. Lett.* **133**, 117–120.

Kar, S., Rees, R. G., and Quirion, R. 1994. Altered calcitonin gene-related peptide, substance P and enkephalin immunoreactivities and receptor binding sites in the dorsal spinal cord of the polyarthritic rat. *Eur. J. Neurosci.* **6**, 345–354.

Karagiannis, S. N., King, R. H., and Thomas, P. K. 1997. Colocalisation of insulin and IGF-1 receptors in cultured rat sensory and sympathetic ganglion cells. *J Anat.* **191**, 431–440.

Karczmar, A. G., Nishi, S., Minota, S., and Kindel, G. 1980. Electrophysiology, acetylcholine and acetylcholinesterase of immature spinal ganglia of the rabbit: An experimental study and a review. *Gen. Pharmacol.* **11**, 127–134.

Karim, F., Wang, C. C., and Gereau, R. W. 2001. Metabotropic glutamate receptor subtypes 1 and 5 are activators of extracellular signal-regulated kinase signaling required for inflammatory pain in mice. *J. Neurosci.* **21**, 3771–3779.

Karlsson, M. and Hildebrand, C. 1994. Sensory C-fibers in rat ventral roots are capsaicin-insensitive and they do not mediate extravasation from pial vessels. *Brain Res.* **642**, 244–250.

Karmy, G., Carr, P. A., Yamamoto, T., Chan, S. H. P., and Nagy, J. I. 1991. Cytochrome oxidase immunohistochemistry in rat brain and dorsal root ganglia: Visualization of enzyme in neuronal perikarya and in parvalbumin-positive neurons. *Neuroscience* **40**, 825–839.

Kashiba, H. and Senba, E. 1999. Up- and down-regulation of BDNF mRNA in distinct subgroups of rat sensory neurons after axotomy. *NeuroReport* **10**, 3561–3565.

Kashiba, H. and Senba, E. 2000. Delayed expression of somatostatin mRNA in GDNFs-dependent rat sensory neurons during postnatal development. *Dev. Brain Res.* **125**, 147–152.

Kashiba, H. and Senba, E. 2001. Primary sensory neurons expressing histamine H1-receptor mRNA. *Nippon Yakurigaku Zasshi* **118**, 43–49.

Kashiba, H., Senba, E., Ueda, Y., and Tohyama, M. 1990a. Relative sparing of calcitonin gene-related peptide-containing primary sensory neurons following neonatal capsaicin treatment in the rat. *Peptides* **11**, 491–496.

Kashiba, H., Senba, E., Ueda, Y., and Tohyama, M. 1990b. Calbindin D28k-containing splanchnic and cutaneous dorsal root ganglion neurons of the rat. *Brain Res.* **528**, 311–316.

Kashiba, H., Senba, E., Kawai, Y., Ueda, Y., and Tohyama, M. 1992a. Axonal blockade induces the expression of vasoactive intestinal polypeptide and galanin in rat dorsal root ganglion neurons. *Brain Res.* **577**, 19–28.

Kashiba, H., Senba, E., Ueda, Y., and Tohyama, M. 1992b. Co-localized but target-unrelated expression of vasoactive intestinal polypeptide and galanin in rat dorsal root ganglion neurons after peripheral nerve crush injury. *Brain Res.* **582**, 47–57.

Kashiba, H., Ueda, Y., and Senba, E. 1996. Coexpression of preprotachykinin-A, α-calcitonin gene-related peptide, somatostatin, and neurotrophin receptor family messenger RNAs in rat dorsal root ganglion neurons. *Neuroscience* **70**, 179–189.

Kashiba, H., Fukui, H., Morikawa, Y., and Senba, E. 1999. Gene expression of histamine H1 receptor in guinea pig primary sensory neurons: A relationship between H1 receptor mRNA-expressing neurons and peptidergic neurons. *Mol. Brain Res.* **66**, 24–34.

Kashiba, H., Uchida, Y., and Senba, E. 2001a. Difference in binding by isolectin B4 to trkA and c-ret mRNA-expressing neurons in rat sensory ganglia. *Mol. Brain Res.* **95**, 18–26.

Kashiba, H., Fukui, H., and Senba, E. 2001b. Histamine H1 receptor mRNA is expressed in capsaicin-insensitive sensory neurons with neuropeptide Y-immunoreactivity in guinea pigs. *Brain Res.* **901**, 85–93.

Kato, M. and Hirata, Y. 1968. Sensory neurons in the spinal ventral roots of the cat. *Brain Res.* **7**, 479–482.

Kato, M. and Tanji, J. 1971. Physiological properties of sensory fibers in the spinal ventral roots in the cat. *Jpn. J. Physiol.* **21**, 71–77.

Katoh, S., Hisano, S., and Daikoku, S. 1988a. Ultrastructural localization of immunolabeled substance P and methionine-enkephalin-octapeptide in the surface layer of the dorsal horn of rat spinal cord. *Cell Tissue Res.* **253**, 55–60.

Katoh, S., Hisano, S., Kawano, H., Kgotani, Y., and Daikoku, S. 1988b. Light- and electron-microscopic evidence of costoring of immunoreactive enkephalins and substance P in dorsal horn neurons of rat. *Cell Tissue Res.* **253**, 297–303.

Katsurabayashi, S., Kubota, H., Wang, Z. M., Rhee, J. S., and Akaike, N. 2001. cAMP-dependent presynaptic regulation of spontaneous glycinergic IPSCs in mechanically dissociated rat spinal cord neurons. *J. Neurophysiol.* **895**, 332–340.

Katz, D. M. and Erb, M. J. 1990. Developmental regulation of tyrosine hydroxylase expression in primary sensory neurons of the rat. *Dev. Biol.* **137**, 233–242.

Katz, D. M., Markey, K. A., Goldstein, M., and Black, I. B. 1983. Expression of catecholaminergic characteristics by primary sensory neurons in the normal adult rat in vivo. *Proc. Natl. Acad. Sci.* **80**, 3526–3530.

Kaufman, M. P., Iwamoto, G. A., Longhurst, J. C., and Mitchell, J. H. 1982. Effects of capsaicin and bradykinin on afferent fibers with endings in skeletal muscle. *Circ. Res.* **50**, 133–139.

Kauppila, T. 1998. Correlation between autotomy-behavior and current theories of neuropathic pain. *Neurosci. Biobehav. Reviews* **23**, 111–129.

Kausz, M. and Rethelyi, M. 1985. Lamellar arrangement of neuronal somata in the dorsal root ganglion of the cat. *Somatosens. Mot. Res.* **2**, 193–204.

Kawai, Y., Takami, K., Shiosaka, S., Emson, P. C., Hillyard, C. J., Girgis, S., MacIntyre, I., and Tohyama, M. 1985. Topographic localization of calcitonin gene-related peptide in the rat brain: An immunohistochemical analysis. *Neuroscience* **15**, 7547–7563.

Kawakami, M. and Tamaki, T. 1992. Morphologic and quantitative changes in neurotransmitters in the lumbar spinal cord after acute or chronic mechanical compression of the cauda equina. *Spine* **17**, S13–S17.

Kawamura, M., Kuraishi, Y., Minami, M., and Satoh, M. 1989. Antinociceptive effect of intrathecally administered antiserum against calcitonin gene-related peptide on thermal and mechanical noxious stimuli in experimental hyperalgesic rats. *Brain Res.* **497**, 199–203.

Kawamura, Y. and Dyck, P. J. 1978. Evidence for three populations by size in L5 spinal ganglion in man. *J. Neuropathol. Exp. Neurol.* **37**, 269–272.

Kawata, M., Hirakawa, M., Kumamoto, K., Minamino, N., Kangawa, K., Matsuo, H., and Sano, Y. 1989. Brain natriuretic peptide in the porcine spinal cord: An immunohistochemical investigation of its localization and the comparison with atrial natriuretic peptide, substance P, calcitionin gene-related peptide, and enkephalin. *Neuroscience* **33**, 401–410.

Kawatani, M., Lowe, I. P., Nadelhaft, I., Morgan, C., and de Groat, W. C. 1983. Vasoactive intestinal polypeptide in visceral afferent pathways to the sacral spinal cord of the cat. *Neurosci. Lett.* **42**, 311–316.

Kawatani, M., Erdman, S. L., and de Groat, W. C. 1985. Vasoactive intestinal polypeptide and substance P in primary afferent pathways to the sacral spinal cord of the cat. *J. Comp. Neurol.* **241**, 327–347.

Kawatani, M., Nagel, J., and de Groat, W. C. 1986. Identification of neuropeptides in pelvic and pudendal nerve afferent pathways to the sacral spinal cord of the cat. *J. Comp. Neurol.* **249**, 117–132.

Kawatani, M., Suzuki, T., and de Groat, W. C. 1996. Corticotropin releasing factor-like immunoreactivity in afferent projections to the sacral spinal cord of the cat. *J. Auton. Nerv. Syst.* **61**, 218–226.

Kayahara, T. 1986. Synaptic connections between spinal motoneurons and dorsal root ganglion cells in the cat. *Brain Res.* **376**, 299–309.

Kayahara, T., Takimoto, T., and Sakashita, S. 1981. Synaptic junctions in the cat spinal ganglion. *Brain Res.* **216**, 277–290.

Kayahara, T., Sakashita, S., and Takimoto, T. 1984. Evidence for spinal origin of neurons synapsing with dorsal root ganglion cells of the cat. *Brain Res.* **293**, 225–230.

Kazen-Gillespie, K. A., Ragsdale, D. S., D'Andrea, M. R., Mattei, L. N., Rogers, K. E., and Isom, L. L. 2000. Cloning, localization, and functional expression of sodium channel beta1A subunits. *J. Biol. Chem.* **275**, 1079–1088.

Kazimierczak, J., Sommer, E. W., Philippe, E., and Droz, B. 1986. Carbonic anhydrase activity in primary sensory neurons: I. Requirements for the cytochemical localization in the dorsal root ganglion of chicken and mouse by light and electron microscopy. *Cell Tissue Res.* **245**, 487–495.

Keast, J. R. and Gleeson, R. J. 1998. Androgen receptor immunoreactivity is present in primary sensory neurons of male rats. *NeuroReport* **9**, 4137–4140.

Keast, J. R. and Stephensen, T. M. 2000. Glutamate and aspartate immunoreactivity in dorsal root ganglion cells supplying visceral and somatic targets and evidence for peripheral axonal transport. *J. Comp. Neurol.* **424**, 577–587.

Keates, A. C., Castagliuolo, I., Qiu, B., Nikulasso, S., Sengupta, A., and Pothoulakis, C. 1998. CGRP upregulation in dorsal root ganglion and ileal mucosa during *Clostridium difficile* toxin A-induced enteritis. *Am. J. Physiol.* **274**, G196–G201.

Keegan, J. J. and Garrett, E. D. 1948. The segmental distribution of the cutaneous nerves in the limbs of man. *Anat. Rec.* **102**, 409–437.

Keiger, C. J. and Walker, J. C. 2000. Individual variation in the expression profiles of nicotinic receptors in the olfactory bulb and trigeminal ganglion and identification of alpha2, alpha6, and beta3 transcripts. *Biochem. Pharmacol.* **59**, 233–240.

Keele, C. A. and Armstrong, D. 1964. *Substances Producing Pain and Itch.* London, Edward Arnold.

Kellstein, D. E., Price, D. D., Hayes, R. L., and Mayer, D. J. 1990. Evidence that substance P selectively modulates C-fiber-evoked discharges of dorsal horn nociceptive neurons. *Brain Res.* **526**, 291–298.

Kelly, J. S., Gottesfeld, Z., and Schon, F. 1973. Reduction in GAD I activity from the dorsal lateral region of the deafferented rat spinal cord. *Brain Res.* **62**, 581–586.

Kemp, T., Spike, R. C., Watt, C., and Todd, A. J. 1996. The μ-opioid receptor (MOR1) is mainly restricted to neurons that do not contain GABA or glycine in the superficial dorsal horn of the rat spinal cord. *Neuroscience.* **75**, 1231–1238.

Kenins, P. 1981. Identification of the unmyelinated sensory nerves which evoke plasma extravasation in response to antidromic stimulation. *Neurosci. Lett.* **25**, 137–141.

Kenins, P. 1988. The functional anatomy of the receptive fields of rabbit C polymodal nociceptors. *J. Neurophysiol.* **59**, 1098–1115.

Kenshalo, D. R. and Duclaux, R. 1977. Response characteristics of cutaneous cold receptors in the monkey. *J. Neurophysiol.* **40**, 319–332.

Kenshalo, D. R., Leonard, R. B., Chung, J. M., and Willis, W. D. 1979. Responses of primate spinothalamic neurons to graded and to repeated noxious heat stimuli. *J. Neurophysiol.* **42**, 1370–1389.

Kenshalo, D. R., Leonard, R. B., Chung, J. M., and Willis, W. D. 1982. Facilitation of the responses of primate spinothalamic cells to cold and to tactile stimuli by noxious heating of the skin. *Pain* **12**, 141–152.

Kenton, B., and Kruger, L. 1971. Information transmission in slowly adapting mechanoreceptor fibers. *Exp. Neurol.* **31**, 114–139.

Kerchner, G. A., Wilding, T. J., Li, P., Zhuo, M., and Huettner, J. A. 2001. Presynaptic kainate receptors regulate spinal sensory transmission. *J. Neurosci.* **21**, 59–66.

Kerr, B. J., Bradbury, E. J., Bennett, D. H. L., Trivedi, P. M., Dassan, P., French, J., Shelton, D. B., McMahon, S. B., and Thompson, S. W. N. 1999. Brain-derived neurotrophic factor modulates nociceptive sensory inputs and NMDA-evoked responses in the rat spinal cord. *J. Neurosci.* **19**, 5138–5148.

Kerr, B. J., Cafferty, W. B., Gupta, Y. K., Bacon, A., Wynick, D., McMahon, S. B., and Thompson, S. W. 2000. Galanin knockout mice reveal nociceptive deficits following peripheral nerve injury. *Eur. J. Neurosci.* **12**, 793–802.

Kerr, B. J., Gupta, Y., Pope, R., Thompson, S. W., Wynick, D., and McMahon, S. B. 2001. Endogenous galanin potentiates spinal nociceptive processing following inflammation. *Pain* **93**, 267–277.

Kerr, F. W. L. 1966. The ultrastructure of the spinal tract of the trigeminal nerve and substantia gelatinosa. *Exp. Neurol.* **16**, 359–376.

Kerr, F. W. L. 1970a. The fine structure of the subnucleus caudalis of the trigeminal nerve. *Brain Res.* **23**, 129–145.

Kerr, F. W. L. 1970b. The organization of primary afferents in the subnucleus caudalis of the trigeminal: A light- and electron-microscopic study of degeneration. *Brain Res.* **23**, 147–165.

Kerr, F. W. L. 1975. Pain: A central inhibitory balance theory. *Mayo Clinic Proc.* **50**, 685–690.

Kessler, W., Kirchhoff, C., Reeh, P. W., and Handwerker, H. O. 1992. Excitation of cutaneous afferent nerve endings in vitro by a combination of inflammatory mediators and conditioning effect of substance P. *Exp. Brain Res.* **91**, 467–476.

Kessler, F., Habelt, C., Averbeck, B., Reeh, P. W., and Kress, M. 1999. Heat-induced release of CGRP from isolated rat skin and effects of bradykinin and the protein kinase C activator PMA. *Pain* **83**, 289–295.

Khachaturian, H., Watson, S. J., Lewis, M. E., Coy, D., Goldstein, A., and Akil, H. 1982. Dynorphin immunocytochemistry in the rat central nervous system. *Peptides* **3**, 941–954.

Khalsa, P. S., Hoffman, A. H., and Grigg, P. 1996. Mechanical states encoded by stretch-sensitive neurons in feline joint capsule. *J. Neurophysiol.* **76**, 175–187.

Khalsa, P. S., LaMotte, R. H., and Grigg, P. 1997. Tensile and compressive responses of nociceptors in rat hairy skin. *J. Neurophysiol.* **78**, 492–505.

Khan, I. M., Yaksh, T. L., and Taylor, P. 1994. Ligand specificity of nicotinic acetylcholine receptors in rat spinal cord: Studies with nicotine and cytisine. *J. Pharmacol. Exp. Ther.* **270**, 159–166.

Khan, I. M., Youngblood, K. L., Printz, M. P., Yaksh, T. L., and Taylor, P. 1996. Spinal nicotinic receptor expression in spontaneously hypertensive rat. *Hypertension* **28**, 1093–1099.

Khasabov, S. G., Rogres, S. D., Ghilardi, J. R., Peters, C. M., Mantyh, P. W., and Simone, D. A. 2002. Spinal neurons that possess the substance P receptor are required for the development of central sensitization. *J. Neurosci.* **22**, 9086–9098.

Khasar, S. G., Ho, T., Green, P. G., and Levine, J. D. 1994. Comparison of prostaglandin E1- and prostaglandin E2-induced hyperalgesia in the rat. *Neuroscience* **62**, 345–350.

Khasar, S. G., Green, P. G., Chou, B., and Levine, J. D. 1995. Peripheral nociceptive effects of α2-adrenergic receptor agonists in the rat. *Neuroscience* **66**, 427–432.

Khasar, S. G., Gold, M. S., and Levine, J. D. 1998. A tetrodotoxin-resistant sodium current mediates inflammatory pain in the rat. *Neurosci. Lett.* **256**, 17–20.

Khasar, S. G., McCarter, G., and Levine, J. D. 1999a. Epinephrine produces a β-adrenergic receptor-mediated mechanical hyperalgesia and in vitro sensitization of rat nociceptors. *J. Neurophysiol.* **81**, 1104–1112.

Khasar, S. G., Lin, Y. H., Martin, A., Dadgar, J., McMahon, T., Wang, D., Hundle, B., Aley, K. O., Isenberg, W., McCarter, G., Green, P. G., Hodge, C. W., Levine, J. D., and Messing, R. O. 1999b. A novel nociceptor signaling pathway revealed in protein kinase Cε mutant mice. *Neuron* **24**, 253–260.

Khattab, F. I. 1968. A complex synaptic apparatus in spinal cords of cats. *Experientia* **24**, 690–691.

Khoshbaten, A., and Ferrell, W. R. 1990. Alterations in cat knee joint blood flow induced by electrical stimulation of articular afferents and efferents. *J. Physiol.* **430**, 77–86.

Kia, H. K., Miquel, M. C., McKernan, R. M., Laporte, A.-M., Lombard, M. C., Bourgoin, S., Hamon, M., and Verge, D. 1995. Localization of 5-HT_3 receptors in the rat spinal cord: Immunohistochemistry and *in situ* hybridization. *NeuroReport* **6**, 257–261.

Kia, H. K., Miquel, M. C., Brisorgueil, M. J., Daval, G., Riad, M., El Mestikawy, S., Hamon, M., and Verge, D. 1996. Immunocytochemical localization of serotonin1A receptors in the rat central nervous system. *J. Comp. Neurol.* **365**, 289–305.

Kidd, E. J., Laporte, A. M., Langlois, X., Fattaccini, C.-M., Doyen, C., Lombard, M. C., Gozlan, H., and Hamon, M. 1993. 5-HT_3 receptors in the rat central nervous system are mainly located on nerve fibres and terminals. *Brain Res.* **612**, 289–298.

Killackey, H. P. and Belford, G. R. 1979. The formation of afferent patterns in the somatosensory cortex of the neonatal rat. *J. Comp. Neurol.* **183**, 285–303.

Kilo, S., Harding-Rose, C., Hargreaves, K. M., and Flores, C. M. 1997. Peripheral CGRP release as a marker for neurogenic inflammation: A model system for the study of neuropeptide secretion in rat paw skin. *Pain* **73**, 201–207.

Kim, C. H., Oh, Y., Chung, J. M., and Chung, K. 2001. The changes in expression of three subtypes of TTX sensitive sodium channels in sensory neurons after spinal nerve ligation. *Mol. Brain Res.* **95**, 153–161.

Kim, C. H., Oh, Y., Chung, J. M., and Chung, K. 2002. Changes in three subtypes of tetrodotoxin sensitive sodium channel expression in the axotomized dorsal root ganglion in the rat. *Neurosci. Lett.* **323**, 125–128.

Kim, H. J., Na, H. S., Nam, H. J., Park, K. A., Hong, S. K., and Kang, B. S. 1996. Sprouting of sympathetic nerve fibers into the dorsal root ganglion following peripheral nerve injury depends on the injury site. *Neurosci. Lett.* **212**, 191–194.

Kim, H. J., Na, H. S., Sung, B., and Hong, S. K. 1998. Amount of sympathetic sprouting in the dorsal root ganglia is not correlated to the level of sympathetic dependence of neuropathic pain in rat model. *Neurosci. Lett.* **245**, 21–24.

Kim, H. J., Na, H. S., Back, S. K., and Hong, S. K. 2001. Sympathetic sprouting in sensory ganglia depends on the number of injured neurons. *NeuroReport* **12**, 3529–3532.

Kim, K. J., Yoon, Y. W., and Chung, J. M. 1997. Comparison of three rodent neuropathic pain models. *Exp. Brain Res.* **113**, 200–206.

Kim, J., Shin, H. K., and Chung, J. M. 1987. Many ventral root afferent fibers in the cat are third branches of dorsal root ganglion cells. *Brain Res.* **417**, 304–314.

Kim, J., Shin, H. K., Nam, S. C., and Chung, J. M. 1988. Proportion and location of spinal neurons receiving ventral root afferent inputs in the cat. *Exp. Neurol.* **99**, 296–314.

Kim, S. H., and Chung, J. M. 1991. Sympathectomy alleviates mechanical allodynia in an experimental animal model for neuropathy in the rat. *Neurosci. Lett.* **134**, 131–134.

Kim, S. H. and Chung, J. M. 1992. An experimental model for peripheral neuropathy produced by segmental spinal nerve ligation in the rat. *Pain* **50**, 355–363.

Kim, S. H., Na, H. S., Sheen, K., and Chung, J. M. 1993. Effects of sympathectomy on a rat model of peripheral neuropathy. *Pain* **55**, 85–92.

Kimura, H., McGeer, P. L., Peng, J. H., and McGeer, E. G. 1981. The central cholinergic system studied by choline acetyltransferase immunohistochemistry in the cat. *J. Comp. Neurol.* **200**, 151–201.

Kimura, S., Okada, M., Sugita, Y., Kanazawa, I., and Munekata, E. 1983. Novel neuropeptides, neurokinin α and β, isolated from porcine spinal cord. *Proc. J. Acad., Ser. B* **59**, 101–104.

King, A. E., Thompson, S. W. N., Urban, L., and Woolf, C. J. 1988. The responses recorded *in vitro* of deep dorsal horn neurons to direct and orthodromic stimulation in the young rat spinal cord. *Neuroscience* **27**, 231–242.

King, A. E., Ackley, M. A., and Slack, J. R. 1997a. Profile of neuronal excitation following selective activation of the neurokinin-1 receptor in rat deep dorsal horn in vitro. *Brain Res.* **767**, 55–63.

King, A. E., Slack, J. R., Lopez-Garcia, J. A., and Ackley, M. A. 1997b. Tachykinin actions on deep dorsal horn neurons in vitro: An electrophysiological and morphological study in the immature rat. *Eur. J. Neurosci.* **9**, 1037–1046.

Kinkelin, I., Stucky, C. L., and Koltzenburg, M. 1999. Postnatal loss of Merkel cells, but not of slowly adapting mechanoreceptors in mice lacking the neurotrophin receptor p75. *Eur. J. Neurosci.* **11**, 3963–3969.

Kinkelin, I., Brocker, E. B., Koltzenburg, M., and Carlton, S. M. 2000. Localization of ionotropic glutamate receptors in peripheral axons of human skin. *Neurosci. Lett.* **283**, 149–152.

Kinnman, E. and Levine, J. D. 1995. Sensory and sympathetic contributions to nerve injury-induced sensory abnormalities in the rat. *Neuroscience* **64**, 751–767.

Kinoshita, A., Shigemoto, R., Ohishi, H., Van Der Putten, H., and Mizuno, N. 1998. Immunohistochemical localization of metabotropic glutamate receptors, mGluR7 and mGluR7b, in the central nervous system of the adult rat and mouse: A light- and electron-microscopic study. *J. Comp. Neurol.* **393**, 332–352.

Kinzie, J. M., Saugstad, J. A., Westbrook, G. L., and Segerson, T. P. 1995. Distribution of metabotropic glutamate receptor 7 messenger RNA in the developing and adult rat brain. *Neuroscience* **69**, 167–176.

Kirchhoff, C., Jung, S., Reeh, P. W., and Handwerker, H. O. 1990. Carrageenan inflammation increases bradykinin sensitivity of rat cutaneous nociceptors. *Neurosci. Lett.* **111**, 206–210.

Kirchhoff, F., Mulhardt, C., Pastor, A., Becker, C.-M., and Kettenmann, H. 1996. Expression of glycine receptor subunits in glial cells of the rat spinal cord. *J. Neurochem.* **66**, 1383–1390.

Kirchmair, R., Marksteiner, J., Troger, J., Mahata, S. K., Mahata, M., Donnerer, J., Amann, R., Fischer-Colbrie, R., Winkler, H., and Saria, A. 1993. Human and rat primary C-fibre afferents store and release secretoneurin, a novel neuropeptide. *Eur. J. Neurosci.* **6**, 861–868.

Kirk, E. J., and Denny-Brown, D. 1970. Functional variation in dermatomes in the macaque monkey following dorsal root lesions. *J. Comp. Neurol.* **139**, 307–320.

Kitayama, I., Cintra, A., Janson, A. M., Fuxe, K., Agnati, L. F., Eneroth, P., Aronsson, M., Harfstrand, A., Steinbush, H. W., and Visser, T. J. 1989. Chronic immobilization stress: Evidence for decreases of 5-hydroxy-tryptamine immunoreactivity and for increases of glucocorticoid receptor immunoreactivity in various brain regions of the male rat. *J. Neural. Trans.* **77**, 93–130.

Kivipelto, L. and Panula, P. 1991. Origin and distribution of neuropeptide-FF-like immunoreactivity in the spinal cord of rats. *J. Comp. Neurol.* **307**, 107–119.

Kivipelto, L., Rubenstein, J., Yang, H.-Y. T., and Panula, P. 1991. Ontogeny of the F8Famide-like (morphine-modulating) peptides in the central nervous system of rats. *J. Comp. Neurol.* **304**, 14–33.

Kiyama, H., Maeno, H., and Tohyama, M. 1993. Substance P receptor (NK-1) in the central nervous system: Possible functions from a morphological aspect. *Regul. Pept.* **46**, 114–123.

Kiyosawa, A., Katsurabayashi, S., Akaike, N., Pang, Z. P., and Akaike, N. 2001. Nicotine facilitates glycine release in the rat spinal dorsal horn. *J. Physiol.* **1**, 101–120.

Klein, C. M., Westlund, K. N., and Coggeshall, R. E. 1990a. Percentages of dorsal root axons immunoreactive for galanin are higher than those for calcitonin gene-related peptide in the rat. *Brain Res.* **519**, 97–101.

Klein, C. M., Coggeshall, R. E., Carlton, S. M., Westlund, K. N., and Sorkin, L. S. 1990b. Changes in calcitonin gene-related peptide immunoreactivity in the rat dorsal horn following electrical stimulation of the sciatic nerve. *Neurosci. Lett.* **115**, 149–154.

Klein, C. M., Guillamondegui, O., Krenek, C. D., La Forte, R. A., and Coggeshall, R. E. 1991. Do neuropeptides in the dorsal horn change if the dorsal root ganglion cell death that normally accompanies peripheral nerve transection is prevented? *Brain Res.* **552**, 273–282.

Kneussel, M., Brandstatter, J. H., Laube, B., Stahl, S., Muller, U., and Betz, H. 1999. Loss of postsynaptic GABA(A) receptor clustering in gephyrin-deficient mice. *J. Neurosci.* **19**, 9289–9297.

Knibestöl, M. 1973. Stimulus-response functions of rapidly adapting mechanoreceptors in the human glabrous skin area. *J. Physiol.* **232**, 427–452.

Knibestöl, M. 1975. Stimulus-response functions of slowly adapting mechanoreceptors in the human glabrous skin area. *J. Physiol.* **245**, 63–80.

Knibestöl, M. and Vallbo, A. B. 1970. Single unit analysis of mechanoreceptor activity from the human glabrous skin. *Acta Physiol. Scand.* **80**, 178–195.

Knibestöl, M. and Vallbo, A. B. 1980. Intensity of sensation related to activity of slowly adapting mechanoreceptive units in the human hand. *J. Physiol.* **300**, 251–267.

Kniffki, K. D. 1986. *Muskuläre Nociception.* VCH Verlagsgesellschaft mbH, Weinheim.

Kniffki, K. D., Mense, S., and Schmidt, R. F. 1978. Responses of group IV afferent units from skeletal muscle to stretch, contraction and chemical stimulation. *Exp. Brain Res.* **31**, 511–522.

Knox, R. J. and Dickenson, A. H. 1987. Effects of selective and non-selective kappa-opioid receptor agonists on cutaneous C-fibre-evoked responses of rat dorsal horn neurones. *Brain Res.* **415**, 21–29.

Knuepfer, M. M., Akeyson, E. W., and Schramm, L. P. 1988. Spinal projections of renal afferent nerves in the rat. *Brain Res.* **446**, 17–25.

Knyihar, E. 1971. Fluoride-resistant acid phosphatase system of nociceptive dorsal root afferents. *Experientia* **27**, 1205–1207.

Knyihar, E. and Csillik, B. 1976. Effect of peripheral axotomy on the fine structure and histochemistry of the Rolando substance: Degenerative atrophy of central processes of pseudounipolar cells. *Exp. Brain Res.* **26**, 73–87.

Knyihar, E. and Gerebtzoff, M. A. 1973. Extra-lysosomal localization of acid phosphatase in the spinal cord of the rat. *Exp. Brain Res.* **18**, 383–395.

Knyihar, E., Laszlo, I., and Tornyos, S. 1974. Fine structure and fluoride resistant acid phosphatase activity of electron dense sinusoid terminals in the substantia gelatinosa rolandi of the rat after dorsal root transection. *Exp. Brain Res.* **19**, 529–544.

Knyihar-Csillik, E. and Csillik, B. 1981. FRAP: Histochemistry of the primary nociceptive neuron. *Progress in Histochemistry and Cytochemistry* **14**, 1–137.

Knyihar-Csillik, E. and Torok, A. 1989. Cytochemical restoration in the upper dorsal horn after transganglionic degenerative atrophy: Temporospatial and fine structural attributes. *Neuroscience* **33**, 75–91.

Knyihar-Csillik, E., Csillik, B., and Rakic, P. 1982a. Ultrastructure of normal and degenerating glomerular synaptic terminals of dorsal root axons in the substantia gelatinosa of the Rhesus monkey. *J. Comp. Neurol.* **210**, 357–375.

Knyihar-Csillik, E., Csillik, B., and Rakic, P. 1982b. Periterminal synaptology of dorsal root glomerular terminals in the substantia gelatinosa of the spinal cord in the Rhesus monkey. *J. Comp. Neurol.* **210**, 376–399.

Knyihar-Csillik, E., Bezzegh, A., Boti, S., and Csillik, B. 1986. Thiamine monophosphatase: A genuine marker for transganglionic regulation of primary sensory neurons. *J. Histochem. Cytochem.* **34**, 363–371.

Knyihar-Csillik, E., Kreutzberg, G. W., and Csillik, B. 1989. Enzyme translocation in the course of regeneration of central primary afferent terminals in the substantia gelatinosa of the adult rodent spinal cord. *J. Neurosci. Res.* **22**, 74–82.

Knyihar-Csillik, E., Torok, A., and Csillik, B. 1990. Primary afferent origin of substance-P containing axons in the superficial dorsal horn of the rat spinal cord: Depletion, regeneration and replenishment of presumed nociceptive central terminals. *J. Comp. Neurol.* **297**, 594–612.

Knyihar-Csillik, E., Kreutzberg, G. W., and Csillik, B. 1995. Fine structural correlates of VIP-like immunoreactivity in the upper spinal dorsal horn after peripheral axotomy: Possibilities of a neuro-glial translocation of a neuropeptide. *Acta Histochem.* **94**, 1–12.

Knyihar-Csillik, E., Rakic, P., and Csillik, B. 1999a. Development of glomerular synaptic complexes and immunohistochemical differentiation in the superficial dorsal horn of the embryonic primate spinal cord. *Anat. Embryol. (Berl.)* **199**, 125–148.

Knyihar-Csillik, E., Rakic, P., and Csillik, B. 1999b. Illusive transience of parvalbumin expression during embryonic development of the primate spinal cord. *Int. J. Dev. Neurosci.* **17**, 79–97.

Kocher, L., Anton, F., Reeh, P. W., and Handwerker, H. O. 1987. The effect of carrageenan-induced inflammation on the sensitivity of unmyelinated skin nociceptors in the rat. *Pain* **29**, 363–373.

Koelle, G. B., and Valk, J. 1954. Physiological implications of the histochemical localization of monoamine oxidase. *J. Physiol.* **126**, 434–447.

Koerber, H. R. 1980. Somatotopic organization of cat brachial spinal cord. *Exp. Neurol.* **69**, 481–492.

Koerber, H. R., and Brown, P. B. 1980. Projections of two hindlimb nerves to cat dorsal horn. *J. Neurophysiol.* **44**, 259–269.

Koerber, H. R. and Brown, P. B. 1982. Somatotopic organization of hindlimb cutaneous nerve projections to cat dorsal horn. *J. Neurophysiol.* **48**, 481–489.

Koerber, H. R., and Mendell, L. M. 1988. Functional specialization of central projections from identified primary afferent fibers. *J. Neurophysiol.* **60**, 1597–1614.

Koerber, H. R., Druzinsky, R. E., and Mendell, L. M. 1988. Properties of somata of spinal dorsal root ganglion cells differ according to peripheral receptor innervated. *J. Neurophysiol.* **60**, 1584–1596.

Koerber, H. R., Mirnics, K., Brown, P. B., and Mendell, L. M. 1994. Central sprouting and functional plasticity of regenerated primary afferents. *J. Neurosci.* **14**, 3655–3671.

Koerber, H. R., Mirnics, K., Kavookjian, A. M., and Light, A. R. 1999. Ultrastructural analysis of ectopic synaptic boutons arising from peripherally regenerated primary afferent fibers. *J. Neurophysiol.* **81**, 1636–1644.

Kohama, I., Ishikawa, K., and Kocsis, J. D. 2000. Synaptic reorganization in the substantia gelatinosa after peripheral nerve neuroma formation: Aberrant innervation of lamina II neurons by Aβ afferents. *J. Neurosci.* **20**, 1538–1549.

Kojima, M., Tikeuchi, Y., Goto, M., and Sato, Y. 1982. Immunohistochemical study of the distribution of serotonin fibers on the spinal cord of the dog. *Cell Tissue Res.* **226**, 477–491.

Kojima, M., Takeuchi, Y., Goto, M., and Sano, Y. 1983. Immunohistochemical study on the localization of serotonin fibers and terminals in the spinal cord of the monkey. *Cell Tissue Res.* **229**, 23–26.

Koketsu, K. 1956. Intracellular potential changes of primary afferent nerve fibers in spinal cord of cats. *J. Neurophysiol.* **19**, 375–392.

Kokko, A. 1965. Histochemical and cytochemical observations on esterases in the spinal ganglion of the rat. *Acta Physiol. Scand.* **66**, 1–76.

Kolmodin, G. M. 1957. Integrative processes in single spinal interneurones with proprioceptive connections. *Acta Physiol. Scand.* **40**, Supp. 139, 1–89.

Kolmodin, G. M., and Skoglund, C. R. 1960. Analysis of spinal interneurons activated by tactile and nociceptive stimulation. *Acta Physiol. Scand.* **50**, 337–355.

Koltzenburg, M. and Handwerker, H. O. 1994. Differential ability of human cutaneous nociceptors to signal mechanical pain and to produce vasodilatation. *J. Neurosci.* **14**, 1756–1765.

Koltzenburg, M. and McMahon, S. B. 1986. Plasma extravasation in the rat urinary bladder following mechanical, electrical and chemical stimuli: Evidence for a new population of chemosensitive primary sensory afferents. *Neurosci. Lett.* **72**, 352–356.

Koltzenburg, M., Lundberg, L. E., and Torebjörk, H. E. 1992. Dynamic and static components of mechanical hyperalgesia in human hairy skin. *Pain* **51**, 207–219.

Koltzenburg, M., Handwerker, H. O., and Torebjörk, H. E. 1993. The ability of humans to localise noxious stimuli. *Neurosci. Lett.* **150**, 219–222.

Koltzenburg, M., Torebjörk, H. E., and Wahren, L. K. 1994. Nociceptor modulated central sensitization causes mechanical hyperalgesia in acute chemogenic and chronic neuropathic pain. *Brain* **117**, 579–591.

Koltzenburg, M., Stucky, C. L., and Lewin, G. R. 1997. Receptive properties of mouse sensory neurons innervating hairy skin. *J. Neurophysiol.* **78**, 1841–1850.

Koltzenburg, M., Bennett, D. L. H., Shelton, D. L., and McMahon, S. B. 1999. Neutralization of endogenous NGF prevents the sensitization of nociceptors supplying inflamed skin. *Eur. J. Neurosci.* **11**, 1698–1704.

Komori, N., Matsumoto, H., Cain, S. D., Kahn, E. S., and Chung, K. 1999. Predominant presence of beta-arrestin-1 in small sensory neurons of rat dorsal root ganglia. *Neuroscience* **93**, 1421–1426.

Kondo, E., Kiyama, H., Yamano, M., Shida, T., Ueda, Y., and Tohyama, M. 1995. Expression of glutamate (AMPA type) and gamma-aminobutyric acid (GABA)A receptors in the rat caudal trigeminal spinal nucleus. *Neurosci. Lett.* **186**, 169–172.

Kondo, M., Fujiwara, H., and Chikako, T. 1985. Autoradiographic evidence for dopaminergic innervation in guinea pig spinal cord. *Jpn. J. Pharmacol.* **38**, 442–444.

Konietzny, F. 1984. Peripheral neural correlates of temperature sensations in man. *Hum. Neurobiol.* **3**, 21–32.

Konietzny, F., and Hensel, H. 1975. Warm fiber activity in human skin nerves. *Pfluegers Arch.* **359**, 265–267.

Konietzny, F. and Hensel, H. 1977. Response of rapidly and slowly adapting mechanoreceptors and vibratory sensitivity in human skin. *Pfluegers Arch.* **368**, 39–44.

Konietzny, F., Perl, E. R., Trevino, D., Light, A., and Hensel, H. 1981. Sensory experiences in man evoked by intraneural electrical stimulation of intact cutaneous afferent fibers. *Exp. Brain Res.* **42**, 219–222.

Konnerth, A., Keller, B. U., and Lev-Tov, A. 1990. Patch clamp analysis of excitatory synapses in mammalian spinal cord slices. *Pflugers Arch.* **417**, 285–290.

Konttinen, Y. T., Rees, R., Hukkanen, M., Gronblad, M., Tolvanen, E., Gibson, S. J., Polak, J. M., and Brewerton, D. A. 1990. Nerves in inflammatory synovium: Immunohistochemical observations on the adjuvant arthritis rat model. *J. Rheumatol.* **17**, 1586–1591.

Kopaczyk, F., Zabel, J., Silny, W., and Otulakowski, B. 1974. Some histochemical reactions in the neurocytes of the trigeminal ganglion in rabbits. *Folia Morphol. (Warsz.)* **33**, 157–164.

Kopp, J., Xu, Z. Q., Zhang, X., Pedrazzini, T., Herzog, H., Kresse, A., Wong, H., Walsh, J. H., and Hökfelt, T. 2002. Expression of the neuropeptide Y Y1 receptor in the CNS of rat and of wild-type and Y1 receptor knock-out mice. Focus on immunohistochemical localization (1). *Neuroscience* **111**, 443–532.

Kordower, J. H., Le, H. K., and Mufson, E. J. 1992. Galanin immunoreactivity in the primate central nervous system. *J. Comp. Neurol.* **319**, 479–500.

Korenman, E. M. D. and Devor, M. 1981. Ectopic adrenergic sensitivity in damaged nerve axons in the rat. *Exp. Neurol.* **72**, 63–81.

Korhonen, L. K. and Hyyppa, M. 1967. Histochemical localization of carbonic anhydrase activity in the spinal and coeliac ganglia of the rat. *Acta Histochem.* **26**, 75–79.

Kosaka, T., Tauchi, M., and Dahl, J. L. 1988. Cholinergic neurons containing GABA-like and/or glutamic acid decarboxylase-like immunoreactivities in various brain regions of the rat. *Exp. Brain Rev.* **70**, 605–617.

Kostiuk, P. G. 1960. Electrophysiological characteristics of individual spinal cord neurons. *Sechenov Physiol. J. USSR* **46**, 10–22.

Kovacs, A. and Ferencsik, M. 1986. Mapping of spinal projection of primary nociceptive neurones in the rat. *Acta Morphol. Hung.* **34**, 187–194.

Krause, W. 1859. Über Nervenendigungen. *Z. Rat. Med.* **5**, 28–43.

Krenz, N. R., and Weaver, L. C. 1998. Effect of spinal cord transection on *N*-methyl-D-aspartate receptors in the cord. *J. Neurotrauma* **15**, 1027–1036.

Kress, M., and Guenther, S. 1999. Role of $[Ca^{2+}]_i$ in the ATP-induced heat sensitization process of rat nociceptive neurons. *J. Neurophysiol.* **81**, 2612–2619.

Kress, M., Koltzenburg, M., Reeh, P. W., and Handwerker, H. O. 1992. Responsiveness and functional attributes of electrically localized terminals of cutaneous C-fibers in vivo and in vitro. *J. Neurophysiol.* **68**, 581–595.

Kress, M., Fetzer, S., Reeh, P. W., and Vyklicky, L. 1996a. Low pH facilitates capsaicin responses in isolated sensory neurons of the rat. *Neurosci. Lett.* **211**, 5–8.

Kress, M., Rödl, J., and Reeh, P. W. 1996b. Stable analogues of cyclic AMP but not cyclic GMP sensitize unmyelinated primary afferents in rat skin to heat stimulation but not to inflammatory mediators, *in vitro. Neuroscience* **74**, 609–617.

Kress, M., Reeh, P. W., and Vyklicky, L. 1997. An interaction of inflammatory mediators and protons in small diameter dorsal root ganglion neurons of the rat. *Neurosci. Lett.* **224**, 37–40.

Krieger, C., Wagey, R., and Shaw, C. 1993. Amoyotrophic lateral sclerosis: Quantitative autoradiography of $[^3H]$MK-801/NMDA binding sites in spinal cord. *Neurosci. Lett.* **159**, 191–194.

Kríž, N., Syková, E., Ujec, E., and Vyklický, L. 1974. Changes of extracellular potassium concentration induced by neuronal activity in the spinal cord of the cat. *J. Physiol.* **238**, 1–15.

Kríž, N., Syková, E., and Vyklický, L. 1975. Extracellular potassium changes in the spinal cord of the cat and their relation to slow potentials, active transport and impulse transmission. *J. Physiol.* **249**, 167–182.

Krnjevic, K., and Morris, M. E. 1972. Extracellular K^+ activity and slow potential changes in spinal cord and medulla. *Can. J. Physiol. Pharmacol.* **50**, 1214–1217.

Krnjevic, K. and Morris, M. E. 1974. Extracellular accumulation of K^+ evoked by activity of primary afferent fibres in the cuneate nucleus and dorsal horn of cats. *Can. J. Physiol. Pharmacol.* **52**, 852–871.

Kroin, J. S., Bianchi, G. D., and Penn, R. D. 1993. Intrathecal baclofen down-regulates $GABA_B$ receptors in the rat substantia gelatinosa. *J. Neurosurg.* **79**, 544–549.

Krstew, E., Jarrott, B., and Lawrence A. J. 1998. Bradykinin B_2 receptors in nodose ganglia of rat and human. *Eur. J. Pharmcol.* **348**, 175–180.

Kruger, L., Perl, E. R., and Sedivec, M. J. 1981. Fine structure of myelinated mechanical nociceptor endings in cat hairy skin. *J. Comp. Neurol.* **198**, 137–154.

Kruger, L., Sampogna, S. L., Rodin, B. E., Clague, J., Brecha, N., and Yeh, Y. 1985. Thin-fiber cutaneous innervation and its intraepidermal contribution studied by labeling methods and neurotoxin treatment in rats. *Somatosens. Res.* **2**, 335–356.

Kruger, L., Kumazawa, T., Mizumura, K., Sato, J., and Yeh, Y. 1988a. Observations on electrophysiologically characterized receptive fields of thin testicular afferent axons: A preliminary note on the analysis of fine structural specializations of polymodal receptors. *Somatosens. Res.* **5**, 373–380.

Kruger, L., Mantyh, P. W., Sternini, C., Brecha, N. C., and Mantyh, C. R. 1988b. Calcitonin gene-related peptide (CGRP) in the rat central nervous system: Patterns of immunoreactivity and receptor binding sites. *Brain Res.* **463**, 223–244.

Kruger, L., Silverman, J. D., Mantyh, P. W., Sternini, C., and Brecha, N. C. 1989. Peripheral patterns of calcitonin-gene-related peptide general somatic sensory innervation: Cutaneous and deep terminations. *J. Comp. Neurol.* **280**, 291–302.

Krukoff, T. L. 1987. Neuropeptide Y-like immunoreactivity in cat spinal cord with special reference to autonomic areas. *Brain Res.* **415**, 300–308.

Krukoff, T. L., Ciriello, J., and Calaresu, F. R. 1986. Somatostatin-like immunoreactivity in neurons, nerve terminals, and fibers of the cat spinal cord. *J. Comp. Neurol.* **243**, 13–22.

Kummer, W., and Heym, C. 1986. Correlation of neuronal size and peptide immunoreactivity in the guinea-pig trigeminal ganglion. *Cell Tissue Res.* **245**, 657–665.

Kumamoto, K., Senuma, H., Ebara, S., and Matsuura, T. 1993a. Distribution of Pacinian corpuscles in the hand of the monkey, *Macaca fuscata*. *J Anat.* **183**, 149–154.

Kumamoto, K., Takei, M., Kinoshita, M., Ebara, S., and Matsuura, T. 1993b. Distribution of Pacinian corpuscles in the cat forefoot. *J. Anat.* **182**, 23–28.

Kumazawa, T. and Mizumura, K. 1976. The polymodal C-fiber receptor in the muscle of the dog. *Brain Res.* **101**, 589–593.

Kumazawa, T. and Mizumura, K. 1977a. The polymodal receptors in the testis of dog. *Brain Res.* **136**, 553–558.

Kumazawa, T. and Mizumura, K. 1977b. Thin-fibre receptors responding to mechanical, chemical, and thermal stimulation in the skeletal muscle of the dog. *J. Physiol.* **273**, 179–194.

Kumazawa, T. and Mizumura, K. 1979. Effects of synthetic substance P on unit-discharge of testicular nociceptors of dogs. *Brain Res.* **170**, 553–557.

Kumazawa, T. and Mizumura, K. 1980a. Chemical responses of polymodal receptors of the scrotal contents in dogs. *J. Physiol.* **299**, 219–231.

Kumazawa, T. and Mizumura, K. 1980b. Mechanical and thermal responses of polymodal receptors recorded from the superior spermatic nerve of dogs. *J. Physiol.* **299**, 233–245.

Kumazawa, T. and Mizumura, K. 1983. Temperature dependency of the chemical responses of the polymodal receptor units in vitro. *Brain Res.* **278**, 305–307.

Kumazawa, T. and Perl, E. R. 1976. Differential excitation of dorsal horn marginal and substantia gelatinosa neurons by primary afferent units with fine (A-delta and C) fibers. In Y. Zotterman (ed.), *Sensory Functions of the Skin with Especial Reference to Man* (pp. 67–89) Pergamon, Oxford.

Kumazawa, T. and Perl, E. R. 1977. Primate cutaneous sensory units with unmyelinated (C) afferent fibers. *J. Neurophysiol.* **40**, 1325–1338.

Kumazawa, T. and Perl, E. R. 1978. Excitation of marginal and substantia gelatinosa neurons in the primate spinal cord: Indications of their place in dorsal horn functional organization. *J. Comp. Neurol.* **177**, 417–434.

Kumazawa, T., Perl, E. R., Burgess, P. R., and Whitehorn, D. 1975. Ascending projections from marginal zone (lamina I) neurons of the spinal dorsal horn. *J. Comp. Neurol.* **162**, 1–12.

Kumazawa, T., Mizumura, K., and Sato, J. 1987. Response properties of polymodal receptors using in vitro testis superior spermatic nerve preparations of dogs. *J. Neurophysiol.* **57**, 702–711.

Kuo, D. C. and de Groat, W. C. 1985. Primary afferent projections of the major splanchnic nerve to the spinal cord and gracile nucleus of the cat. *J. Comp. Neurol.* **231**, 421–434.

Kuo, D. C., Oravitz, J. J., Eskay, R., and de Groat, W. C. 1984. Substance P in renal afferent perikarya identified by retrograde transport of fluorescent dye. *Brain Res.* **323**, 168–171.

Kuo, D. C., Kawatani, M., and de Groat, W. C. 1985. Vasoactive intestinal polypeptide identified in the thoracic dorsal root ganglia of the cat. *Brain Res.* **330**, 178–182.

Kuraishi, Y., Hirota, N., Sato, Y., Hino, Y., Satoh, M., and Takagi, H. 1985. Evidence that substance P and somatostatin transmit separate information related to pain in the spinal dorsal horn. *Brain Res.* **325**, 294–298.

Kuraishi, Y., Nanayama, T., Ohno, H., Minami, M., and Satoh, M. 1988. Antinociception induced in rats by intrathecal administration of antiserum against calcitonin gene-related peptide. *Neurosci. Lett.* **92**, 325–329.

Kuraishi, Y., Nanayama, T., Ohno, H., Fujii, N., Otaka, A., Yajima, H., and Satoh, M. 1989. Calcitonin gene-related peptide increases in the dorsal root ganglia of adjuvant arthritic rat. *Peptides* **10**, 447–452.

Kuraishi, Y., Kawamura, M., Yamaguchi, T., Houtani, T., Kawabata, S., Futaki, S., Fujii, N., and Satoh, M. 1991. Intrathecal injections of galanin and its antiserum affect nociceptive response of rat to mechanical, but not thermal, stimuli. *Pain* **44**, 321–324.

Kuramoto, H., Furness, J. B., and Gibbins, I. L. 1990. Calbindin immunoreactivity in sensory and autonomic ganglia in the guinea pig. *Neurosci Lett.* **115**, 68–73.

Kus, L., Sanderson, J. J., and Beitz, A. J. 1995a. *N*-methyl-D-aspartate R1 messenger RNA and [^{125}I]MK-801 binding decrease in rat spinal cord after unilateral hind paw inflammation. *Neuroscience* **68**, 159–165.

Kus, L., Saxon, D., and Beitz, A. J. 1995b. NMDA R1 mRNA distribution in motor and thalamic-projecting sensory neurons in the rat spinal cord and brain stem. *Neurosci. Lett.* **196**, 201–204.

Kuwayama, Y. and Stone, R. A. 1986. Cholecystokinin-like immunoreactivity occurs in ocular sensory neurons and partially co-localizes with substance P. *Brain Res.* **381**, 266–274.

Kuwayama, Y., Terenghi, G., Polak, J. M., Trojanowski, J. Q., and Stone, R. A. 1987. A quantitative correlation of substance P-, calcitonin gene-related peptide- and cholecystokinin-like immunoreactivity with retrogradely labeled trigeminal ganglion cells innervating the eye. *Brain Res.* **405**, 220–226.

Laduron, P. M. 1984. Axonal transport of opiate receptors in capsaicin-sensitive neurones. *Brain Res.* **294**, 157–160.

Lai, C. C., Wu, S. Y., Dun, S. L., and Dun, N. J. 1997. Nociceptin-like immunoreactivity in the rat dorsal horn and inhibition of substantia gelatinosa neurons. *Neuroscience* **81**, 887–891.

Lai, J., Hunter, J. C., Ossipov, M. H., and Porreca, F. 2000. Blockade of neuropathic pain by antisense targeting of tetrodotoxin-resistant sodium channels in sensory neurons. *Meth. Enzmol.* **314**, 201–213.

Laing, I., Todd, A. J., Heizmann, C. W., and Schmidt, H. H. 1994. Subpopulations of GABAergic neurons in laminae I–III of rat spinal dorsal horn defined by coexistence with classical transmitters, peptides, nitric oxide synthase or parvalbumin. *Neuroscience* **61**, 123–132.

Laird, J. M. A. and Bennett, G. J. 1993. An electrophysiological study of dorsal horn neurons in the spinal cord of rats with an experimental peripheral neuropathy. *J. Neurophysiol.* **69**, 2072–2085.

Laird, J. M. A. and Cervero, F. 1989. A comparative study of the changes in receptive-field properties of multireceptive and nocireceptive rat dorsal horn neurons following noxious mechanical stimulation. *J. Neurophysiol.* **62**, 854–863.

Laird, J. M. A. and Cervero, F. 1991. Signalling of a step-like intensity change of noxious mechanical stimuli by dorsal horn neurones in the rat spinal cord. *J. Physiol.* **434**, 561–575.

Laird, J. M. A., Roza, C., and Cervero, F. 1996. Spinal dorsal horn neurons responding to noxious distension of the ureter in anesthetized rats. *J. Neurophysiol.* **76**, 3239–3248.

Laird, J. M. A., Roza, C., and Cervero, F. 1997. Effects of artificial calculosis on rat ureter motility: Peripheral contribution to the pain of ureteric colic. *Am. J. Physiol.* **272**, R1409–1416.

Laird, J. M. A., Olivar, T., Roza, C., De Felipe, C., Hunt, S. P., and Cervero, F. 2000. Deficits in visceral pain and hyperalgesia of mice with a disruption of the tachykinin NK1 receptor gene. *Neuroscience* **98**, 345–352.

Laird, J. M. A., Olivar, T., Lopez-Garcia, J. A., Maggi, C. A., and Cervero, F. 2001a. Responses of rat spinal neurons to distension of inflamed colon: Role of tachykinin NK2 receptors. *Neuropharmacology* **40**, 696–701.

Laird, J. M. A., Roza, C., De Felipe, C., Hunt, S. P., and Cervero, F. 2001b. Role of central and peripheral tachykinin NK1 receptors in capsaicin-induced pain and hyperalgesia in mice. *Pain* **90**, 97–103.

Laird, J. M. A., Martinez-Caro, L., Garcia-Nicas, E., and Cervero, F. 2001c. A new model of visceral pain and referred hyperalgesia in the mouse. *Pain* **92**, 335–342.

Lam, H. H. D., Hanley, D. F., Trapp, B. D., Saito, S., Raja, S., Dawson, T. M., and Yamaguchi, H. 1996. Induction of spinal cord neuronal nitric oxide synthase (NOS) after formalin injection in the rat hind paw. *Neurosci. Lett.* **210**, 201–204.

Lamb, G. D. 1983a. Tactile discrimination of textured surfaces: Psychophysical performance measurements in humans. *J. Physiol.* **338**, 551–565.

Lamb, G. D. 1983b. Tactile discrimination of textured surfaces: Peripheral neural coding in the monkey. *J. Physiol.* **338**, 567–587.

LaMotte, C. 1977. Distribution of the tract of Lissauer and the dorsal root fibers in the primate spinal cord. *J. Comp. Neurol.* **172**, 529–562.

LaMotte, C. C. and DeLanerolle, N. C. 1981. Human spinal neurons: Innervation by both substance P and enkephalin. *Neuroscience* **6**, 713–723.

LaMotte, C. C. and DeLanerolle, N. C. 1983a. Ultrastructure of chemically defined neuron systems in the dorsal horn of the monkey: II. Methionine-enkephalin immunoreactivity. *Brain Res.* **274**, 51–63.

LaMotte, C. C. and DeLanerolle, N. C. 1983b. Ultrastructure of chemically defined systems in the dorsal horn of the monkey. III. Serotonin immunoreactivity. *Brain Res.* **274**, 65–77.

LaMotte, C. C. and DeLanerolle, N. C. 1986. VIP terminals, axons, and neurons: Distribution throughout the length of monkey and cat spinal cord. *J. Comp. Neurol.* **249**, 133–145.

LaMotte, C. C. and Shapiro, C. M. 1991. Ultrastructural localization of substance P, met-enkephalin, and somatostatin immunoreactivity in lamina X of the primate spinal cord. *J. Comp. Neurol.* **306**, 290–306.

LaMotte, C. C., Pert, C. B., and Snyder, S. H. 1976. Opiate receptor binding in primate spinal cord: Distribution and changes after dorsal root section. *Brain Res.* **112**, 407–42.

LaMotte, C. C., Johns, D. R., and DeLanerolle, N. C. 1982. Immunohistochemical evidence of indolamine neurons in monkey spinal cord. *J. Comp. Neurol.* **206**, 359–370.

LaMotte, C. C., Kapadia, S. E., and Kocol, C. M. 1989. Deafferentation-induced expansion of saphenous terminal field labeling in the adult rat dorsal horn following pronase injection of the sciatic nerve. *J. Comp. Neurol.* **288**, 311–325.

LaMotte, R. H. and Campbell, J. N. 1978. Comparison of responses of warm and nociceptive C-fiber afferents in monkey with human judgments of thermal pain. *J. Neurophysiol.* **41**, 509–528.

LaMotte, R. H. and Thalhammer, J. G. 1982. Response properties of high-threshold cutaneous cold receptors in the primate. *Brain Res.* **244**, 279–287.

LaMotte, R. H., Thalhammer, J. G., Torebjörk, H. E., and Robinson, C. J. 1982. Peripheral neural mechanisms of cutaneous hyperalgesia following mild injury by heat. *J. Neurosci.* **2**, 765–781.

LaMotte, R. H., Thalhammer, J. G., and Robinson, C. J. 1983. Peripheral neural correlates of magnitude of cutaneous pain and hyperalgesia: A comparison of neural events in monkey with sensory judgments in human. *J. Neurophysiol.* **50**, 1–26.

LaMotte, R. H., Shain, C. N., Simone, D. A., and Tsai, E. -F. P. 1991. Neurogenic hyperalgesia: Psychophysical studies of underlying mechanisms. *J. Neurophysiol.* **66**, 190–211.

LaMotte, R. H., Lundberg, L. E., and Torebjörk, H. E. 1992. Pain, hyperalgesia and activity in nociceptive C units in humans after intradermal injection of capsaicin. *J. Physiol.*, **448**, 749–764.

Landon, D. N. (ed.) 1976. *The Peripheral Nerve.* John Wiley and Sons, New York.

Landry, M., Holmberg, K., Zhang, X., and Hokfelt, T. 2000. Effect of axotomy on expression of NPY, galanin, and NPY Y1 and Y2 receptors in dorsal root ganglia and the superior cervical ganglion studied with double-labeling in situ hybridization and immunohistochemistry. *Exp. Neurol.* **162**, 361–384.

Lang, E., Novak, A., Reeh, P. W., and Handwerker, H. O. 1990. Chemosensitivity of fine afferents from rat skin in vitro. *J. Neurophysiol.* **63**, 887–901.

Langford, L. A. 1983. Unmyelinated axon ratios in cat motor, cutaneous and articular nerves. *Neurosci. Lett.* **40**, 19–22.

Langford, L. A. and Coggeshall, R. E. 1979. Branching of sensory axons in the dorsal root and evidence for the absence of dorsal root efferent fibers. *J. Comp. Neurol.* **184**, 193–204.

Langford, L. A. and Coggeshall, R. E. 1981. Branching of sensory axons in the peripheral nerve of the rat. *J. Comp. Neurol.* **203**, 745–750.

Langford, L. A. and Schmidt, R. F. 1983. Afferent and efferent axons in the medial and posterior articular nerves of the cat. *Anat. Rec.* **206**, 71–78.

Lanlua, P., Decorti, F., Grangula, P. R. R., Chung, K., Taglialatela, G., and Yallampalli, C. 2001. Female steroid hormones modulate receptors for nerve growth factor in rat dorsal root ganglion. *Biology of Reproduction* **64**, 331–338.

Laporte, A. M., Koscielniak, T., Ponchant, M., Verge, D., Hamon, M., and Gozlan, H. 1992. Quantitative autoradiographic mapping of 5-HT_3 receptors in the rat CNS using [^{125}I]iodo-zacopride and [^{3}H]zacopride as radioligands. *Synapse* **10**, 271–281.

Laporte, A. M., Fattaccini, C.-M., Lombard, M. C., Chauveau, J., and Hamon, M. 1995. Effects of dorsal rhizotomy and selective lesion of serotonergic and noradrenergic systems on 5-HT_{1A}, 5-HT_{1B} and 5-HT_3 receptors in the rat spinal cord. *J. Neural. Trans.* **100**, 207–223.

Laporte, A. M., Doyen, C., Nevo, I. T., Chauveau, J., Hauw, J. J., and Hamon, M. 1996. Autoradiographic mapping of serotonin 5-HT1A, 5-HT1D, 5-HT2A and 5-HT3 receptors in the aged human spinal cord. *J. Chem. Neuroanat.* **11**, 67–75.

Lariviere, W. R. and Melzack, R. 1996. The bee venom test: A new tonic pain test. *Pain* **66**, 271–277.

Larner, A. J., Moss, J., Rossi, M. L., and Anderson, M. 1994. Congenital insensitivity to pain: A 20 year follow up. *J. Neurol. Neurosurg. Psychiatry* **57**, 973–974.

Larsson, L. and Rehfeld, J. F. 1979. Localization and molecular heterogeneity of cholecystokinin in the central and peripheral nervous system. *Brain Res.* **165**, 201–218.

Larsson, M., Persson, S., Ottersen, O. P., and Broman, J., 2001. Quantitative analysis of immunogold labeling indicates low levels and non-vesicular localization of L-asparate in rat primary afferent terminals. *J. Comp. Neurol.* **430**, 147–159.

Laurberg, S. and Sorensen, K. E. 1985. Cervical dorsal root ganglion cells with collaterals to both shoulder skin and the diaphragm: A fluorescent double labelling study in the rat. A model for referred pain? *Brain Res.* **331**, 160–163.

Lawand, N. B., Willis, W. D., and Westlund, K. N. 1997a. Blockade of joint inflammation and secondary hyperalgesia by L-NAME, a nitric oxide synthase inhibitor. *NeuroReport* **8**, 895–899.

Lawand, N. B., Willis, W. D., and Westlund, K. N. 1997b. Excitatory amino acid receptor involvement in peripheral nociceptive transmission in rats. *Eur. J. Pharmacol.* **324**, 169–177.

Lawand, N. B., Lu, Y., and Westlund, K. N. 1999. Nicotinic cholinergic receptors: Potential targets for inflammatory pain relief. *Pain* **80**, 291–299.

Lawson, S. N. 1979. The postnatal development of large light and small dark neurons in mouse dorsal root ganglion: A statistical analysis of cell numbers and size. *J. Neurocytol.* **8**, 275–294.

Lawson, S. N. 1992. Morphological and biochemical cell types of sensory neurons. In S. A. Scott (ed.), *Sensory Neurons Diversity, Development, and Plasticity* (pp. 27–59). Oxford University Press, New York.

Lawson, S. N. 1995. Neuropeptides in morphologically and functionally identified primary afferent neurons in dorsal root ganglia: Substance P, CGRP and somatostatin. *Prog. Brain Res.* **104**, 161–173.

Lawson, S. N. 1996. Peptides and cutaneous polymodal nociceptor neurones, In T. Kumazawa, L. Kruger, and K. Mizumura (eds.), *Progress in Brain Research* (pp. 369–385). Elsevier Science.

Lawson, S. N. 2002. Phenotype and function of somatic primary afferent nociceptive neurones with C-, *Ad. Exp. Physiol* **87**, 239–244.
Lawson, S. N. and Biscoe, T. J. 1979. Development of mouse dorsal root ganglia: An autoradiographic and quantitative study. *J. Neurocytol.* **8**, 265–274.
Lawson, S. N. and Waddell, P. J. 1991. Soma neurofilament immunoreactivity is related to cell size and fibre conduction velocity in rat primary sensory neurons. *J. Physiol.* **435**, 41–63.
Lawson, S. N., Caddy, K. W. T., and Biscoe, T. J. 1974. Development of rat dorsal root ganglion neurones: Studies on changes in cell birthdays and changes in mean cell diameter. *Cell Tissue Res.* **153**, 399–413.
Lawson, S. N., Harper, A. A., Harper, E. I., Garson, J. A., and Anderton, B. H. 1984. A monoclonal antibody against neurofilament protein specifically labels a subpopulation of rat sensory neurones. *J. Comp. Neurol.* **228**, 263–272.
Lawson, S. N., Harper, L. I., Harper, A. A., Garson, J. A., Coakham, H. B., and Randle, B. J. 1985. Monoclonal antibody 2C5: A marker for a subpopulation of small neurons in rat dorsal root ganglia. *Neuroscience* **16**, 365–374.
Lawson, S. N., Perry, M. J., Prabhakar, E., and McCarthy, P. W. 1993. Primary sensory neurones: Neruofilament, neuropeptides, and conduction velocity. *Brain Res. Bull.* **30**, 239–243.
Lawson, S. N., McCarthy, P. W., and Prabhakar, E. 1996. Electrophysiological properties of neurones with CGRP-like immunoreactivity in rat dorsal root ganglia. *J. Comp. Neurol.* **365**, 355–366.
Lawson, S. N., Crepps, B. A., and Perl, E. R. 1997. Relationship of substance P to afferent characteristics of dorsal root ganglion neurones in guinea-pig. *J. Physiol.* **505**, 177–191.
Lazarov, N. E. 2002. Comparative analysis of the chemical neuroanatomy of the mammalian trigeminal ganglion and mesencephalic trigeminal nucleus. *Prog. Neurobiol.* **66**, 19–59.
Lê, I.-T., Villeneuve, P., Ramjaun, A. R., McPherson, P. S., Beaudet, A., and Séguéla, P. 1998. Sensory presynaptic and widespread somatodendritic imminolocalization of central ionotropic P2X ATP receptors. *Neuroscience* **83**, 177–190.
Leah, J., and Menétrey, D. 1989. Neuropeptides in propriospinal neurones in the rat. *Brain Res.* **495**, 173–177.
Leah, J. D., Cameron, A. A., and Snow, P. J. 1985a. Neuropeptides in physiologically identified mammalian sensory neurones. *Neurosci. Lett.* **56**, 257–264.
Leah, J. D., Cameron, A. A., Kelly, W. L., and Snow, P. J. 1985b. Coexistence of peptide immunoreactivity in sensory neurons of the cat. *Neuroscience* **16**, 683–690.
Leah, J., Menétrey, D., and De Pommery, J. 1988. Neuropeptides in long ascending spinal tract cells in the rat: Evidence for parallel processing of ascending information. *Neuroscience* **24**, 195–207.
Le Bars, D. and Chitour, D. 1983. Do convergent neurones in the spinal dorsal horn discriminate nociceptive from non-nociceptive information? *Pain* **17**, 1–19.
Le Bars, D., Dickenson, A. H., and Besson, J.-M. 1979a. Diffuse noxious inhibitory controls (DNIC): I. Effects on dorsal horn convergent neurones in the rat. *Pain* **6**, 283–304.
Le Bars, D., Dickenson, A. H., and Besson, J.-M. 1979b. Diffuse noxious inhibitory controls (DNIC): II. Lack of effect on non-convergent neurones, supraspinal involvement and theoretical implications. *Pain* **6**, 305–327.
Le Bars, D., Chitour, D., and Clot, A. M. 1981. The encoding of thermal stimuli by diffuse noxious inhibitory controls (DNIC). *Brain Res.* **230**, 394–399.
Le Bars, D., Bourgoin, S., Clot, A. M., Hamon, M., and Cesselin, F. 1987a. Noxious mechanical stimuli increase the release of met-enkephalin-like material heterosegmentally in the rat spinal cord. *Brain Res.* **402**, 188–192.
Le Bars, D., Bourgoin, S., Villanueva, L., Clot, A. M., Hamon, M., and Cesselin, F. 1987b. Involvement of the dorsolateral funiculi in the spinal release of met-enkephalin-like material triggered by heterosegmental noxious mechanical stimuli. *Brain Res.* **412**, 190–195.
Lechan, R. M., Wu, P., and Jackson, I. M. 1987. Immunocytochemical distribution in rat brain of putative peptides derived from thyrotropin-releasing hormone prohormone. *Endocrinology* **121**, 1879–1891.
Le Cudennec, C., Suaudeau, C., and Costentin, J. 2002. Evidence for a localization of [(3)H]nociceptin binding sites on medullar primary afferent fibers. *J. Neurosci. Res.* **68**, 496–500.
Lee, C. H., Wasowicz, K., Brown, R., Majane, E. A., Yang, H. T., and Panula, P. 1993. Distribution and characterization of neuropeptide FF-like immunoreactivity in the rat nervous system with a monoclonal antibody. *Eur. J. Neurosci.* **5**, 1339–1348.
Lee, D. H., Chung, K., and Chung, J. M. 1997. Strain differences in adrenergic sensitivity of neuropathic pain behaviors in an experimental rat model. *NeuroReport* **8**, 3453–3456.
Lee, D. H., Liu, X., Kim, H. T., Chung, K., and Chung, J. M. 1999. Receptor subtype mediating the adrenergic sensitivity of pain behavior and ectopic discharges in neuropathic Lewis rats. *J. Neurophysiol.* **81**, 2226–2233.

Lee, J. H., Price, R. H., Williams, F. G., Mayer, B., and Beitz, A. J. 1993. Nitric oxide synthase is found in some spinothalamic neurons and in neuronal processes that appose spinal neurons that express Fos induced by noxious stimulation. *Brain Res.* **608**, 324–333.

Lee, K. H., Chung, K., Chung, J. M., and Coggeshall, R. E. 1986. Correlation of cell body size, axon size, and signal conduction velocity for individually labelled dorsal root ganglion cells in the cat. *J. Comp. Neurol.* **243**, 335–346.

Lee, S. E., Shen, H., Taglialatela, G., Chung, J. M., and Chung. K. 1998. Expression of nerve growth factor in the dorsal root ganglion after peripheral nerve injury. *Brain Res.* **796**, 99–106.

Lee, Y., Takami, K., Kawai, Y., Girgis, S., Hillyard, C. J., MacIntyre, I., Emson, P. C., and Tohyama, M. 1985a. Distribution of calcitonin gene-related peptide in the rat peripheral nervous system with reference to its coexistence with substance P. *Neuroscience* **15**, 1227–1237.

Lee, Y., Kawai, Y., Shiosaka, S., Takami, K., Kiyama, H., Hillyard, C. J., Girgis, S., MacIntyre, I., Emson, P. C., and Tohyama, M. 1985b. Coexistence of calcitonin gene-related peptide and substance P- like peptide in single cells of the trigeminal ganglion of the rat: Immunohistochemical analysis. *Brain Res.* **330**, 194–196.

Leem, J. W., Willis, W. D., Weller, S. C., and Chung, J. M. 1993. Differential activation and classification of cutaneous afferents in the rat. *J. Neurophysiol.* **70**, 2411–2424.

Leger, L., Charnay, Y., Hof, P. R., Bouras, C., and Cespuglio, R. 2001. Anatomical distribution of serotonin-containing neurons and axons in the central nervous system of the cat. *J. Comp. Neurol.* **433**, 157–182.

Lehtosalo, J. 1984. Substance P-like immunoreactive trigeminal ganglion cells supplying the cornea. *Histochemistry* **80**, 273–276.

Leijon, G., Boivie, J., Johansson, I. 1989. Central post-stroke pain-neurological symptoms and pain characteristics. *Pain* **36**, 13–25.

Lekan, H. A., and Carlton, S. M. 1995. Glutamatergic and GABAergic input to rat spinothalamic tract cells in the superficial dorsal horn. *J. Comp. Neurol.* **361**, 417–428.

Lekan, H. A., Chung, K., Yoon, Y. W., Chung, J. M., and Coggeshall, R. E. 1997. Loss of dorsal root ganglion cells concomitant with dorsal root axon sprouting following segmental nerve lesions. *Neuroscience* **81**, 527–534.

Lele, P. P. and Weddell, G. 1956. The relationship between neurohistology and corneal sensibility. *Brain* **79**, 119–154.

Lele, P. P. and Weddell, G. 1959. Sensory nerves of the cornea and cutaneous sensibility. *Exp. Neurol.* **1**, 334–359.

Lembeck, F., and Holzer, P. 1979. Substance P as a mediator of antidromic vasodilatation and neurogenic plasma extravasation. *Naunyn-Schmiedebergs Arch. Pharmacol.* **310**, 175–183.

Lenz, F. A., Tasker, R. R., Dostrovsky, J. O., Kwan, H. C., Gorecki, J., Hirayama, T., and Murphy, J. T. 1987. Abnormal single-unit activity recorded in the somatosensory thalamus of a quadriplegic patient with central pain. *Pain* **31**, 225–236.

Lenz, F. A., Kwan, H. C., Dostrovsky, J. O., and Tasker, R. R. 1989. Characteristics of the bursting pattern of action potentials that occurs in the thalamus of patients with central pain. *Brain Res.* **496**, 357–360.

Levant, B., and McCarson, K. E. 2001. D(3) dopamine receptors in rat spinal cord: Implications for sensory and motor function. *Neurosci Lett.* **303**, 9–12.

Levine, J. D., Moskowitz, M. A., and Basbaum, A. I. 1985. The contribution of neurogenic inflammation in experimental arthritis. *J. Immunol.* **135**, 843s–847s.

Levitt, M. 1985. Dysesthesias and self-mutilation in humans and subhumans: A review of clinical and experimental studies. *Brain Res. Rev.* **10**, 247–290. 1985.

Levitt, M., and Heybach, J. P. 1981. The deafferentation syndrome in blind rats: A model of the painful phantom limb. *Pain* **10**, 67–73.

Levitt, M., and Levitt, J. H. 1981. The deafferentation syndrome in monkeys: Dysesthesias of spinal origin. *Pain* **10**, 129–147.

Levy, R. A. 1974. GABA: A direct depolarizing action at the mammalian primary afferent terminal. *Brain Res.* **76**, 155–160.

Levy, R. A. 1975. The effect of intravenously administered γ-aminobutyric acid on afferent fiber polarization. *Brain Res.* **92**, 21–34.

Levy, R. A. and Anderson, E. G. 1972. The effect of GABA antagonists bicuculline and picrotoxin on primary afferent terminal excitability. *Brain Res.* **43**, 171–180.

Levy, R.A., Repkin, A.H. and Anderson, E.G., 1971, The effect of bicuculline on primary afferent terminal excitability. *Brain Res.* **32**, 261–265.

Lew, W. Y., and Longhurst, J. C. 1986. Substance P, 5-hydroxytryptamine, and bradykinin stimulate abdominal visceral afferents. *Am. J. Physiol.* **250**, R465–473.

Lewin, G. R. and McMahon, S. B. 1993. Muscle afferents innervating skin form somatotopically appropriate connections in the adult rat dorsal horn. *Eur. J. Neurosci.* **5**, 1083–1092.

Lewin, G. R. and Mendell, L. M. 1994. Regulation of cutaneous C-fiber heat nociceptors by nerve growth factor in the developing rat. *J. Neurophysiol.* **71**, 941–949.

Lewin, G. R., Ritter, A. M., and Mendell, L. M. 1992. On the role of nerve growth factor in the development of myelinated nociceptors. *J. Neurosci.* **12**, 1896–1905.

Lewin, G. R., Ritter, A. M., and Mendell, L. M. 1993. Nerve growth factor-induced hyperalgesia in the neonatal and adult rat. *J. Neurosci.* **13**, 2136–2148.

Lewin, G. R., Rueff, A., and Mendell, L. M. 1994. Peripheral and central mechanisms of NGF-induced hyperalgesia. *Eur. J. Neurosci.* **6**, 1903–1912.

Lewis, C., Neidhart, S., Holy, C., North, R. A., Buell, G., and Surprenant, A. 1995. Coexpression of P2X2 and P2X3 receptor subunits can account for ATP-gated currents in sensory neurons. *Nature* **377**, 432–435.

Lewis, S. J., Cincotta, M. C., Verberne, A. J. M., Jarrott, B., Lodge, D., and Beart, P. M. 1987. Receptor autoradiography with [3H]L-glutamate reveals the presence and axonal transport of glutamate receptors in vagal afferent neurones of the rat. *Eur. J. Pharmacol.* **144**, 413–415.

Lewis, T. 1927. *The Blood Vessels of the Human Skin and their Responses.* Shaw & Sons, London.

Lewis, T. 1942. *Pain.* Macmillan, London.

Lewis, T., and Pochin, E. E. 1938a. The double pain response of the human skin to a single stimulus. *Clin. Sci.* **3**, 67–76.

Lewis, T., and Pochin, E. E. 1938b. Effects of asphyxia and pressure on sensory nerves of man. *Clin. Sci.* **3**, 1411–155.

Li, H., and Li, Y. Q., 2000. Collateral projection of substance P receptor expressing neurons in the medullary dorsal horn to bilateral parabrachial nuclei of the rat. *Brain Res. Bull.* **53**, 163–169.

Li, H., Ohishi, H., Kinoshita, A., Shigemoto, R., Nomura, D., and Mizuno, N. 1997. Localization of a metabotropic glutamate receptor, mGluR7. in axon terminals of presumed nociceptive, primary afferent fibers in the superficial layers of the spinal dorsal horn: An electron microscope study in the rat. *Neurosci. Lett.* **223**, 153–156.

Li, H. S., and Zhao, Z. Q. 1998. Small sensory neurons in the rat dorsal root ganglia express functional NK-1 tachykinin receptor. *Eur. J. Neurosci.* **10**, 1292–1299.

Li, J. L., Kaneko, T., and Mizuno, N. 1996a. Effects of peripheral nerve ligation on expression of mu-opioid receptor in sensory ganglion neurons: an immunohistochemical study in dorsal root and nodose ganglion neurons of the rat. *Neurosci. Lett.* **214**, 91–94.

Li, J. L., Ding, Y. Q., Shigemoto, R., and Mizuno, N. 1996b. Distribution of trigeminothalamic and spinothalamic-tract neurons showing substance P receptor-like immunoreactivity in the rat. *Brain Res.* **719**, 207–212.

Li, J. L., Kaneko, T., Shigemoto, R., and Mizuno, N. 1997a. Distribution of trigeminohypothalamic and spinohypothalamic tract neurons displaying substance P receptor-like immunoreactivity in the rat. *J. Comp. Neurol.* **378**, 508–521.

Li, J. L., Kaneko, T., Nomura, S., Li, Y. Q., and Mizuno, N. 1997b. Association of serotonin-like immunoreactive axons with nociceptive projection neurons in the caudal spinal trigeminal nucleus of the rat. *J. Comp. Neurol.* **384**, 127–141.

Li, J. L., Ding, Y. Q., Li, Y. Q., Li, J. S., Nomura, S., Kaneko, T., and Mizuno, N. 1998a. Immunocytochemical localization of μ-opioid receptor in primary afferent neurons containing substance P or calcitonin gene-related peptide. A light- and electron-microscope study in the rat. *Brain Res.* **794**, 347–352.

Li, J. L., Ding, Y. Q., Xiong, K. H., Li, J. S., Shigemoto, R., and Mizuno, N. 1998b. Substance P receptor (NK1)-immunoreactive neurons projecting to the periaqueductal gray: Distribution in the spinal trigeminal nucleus and the spinal cord of the rat. *Neurosci Res.* **30**, 219–225.

Li, J. L., Li, Y. Q., Kaneko, T., and Mizuno, N. 1999a. Preprodynorphin-like immunoreactivity in medullary dorsal horn neurons projecting to the thalamic regions in the rat. *Neurosci Lett.* **264**, 13–16.

Li, J. L., Li, Y. Q., Li, J. S., Kaneko, T., and Mizuno, N. 1999b. Calcium-binding protein-immunoreactive projection neurons in the caudal subnucleus of the spinal trigeminal nucleus of the rat. *Neurosci. Res.* **35**, 225–240.

Li, J. L., Wang, D., Kaneko, T., Shigemoto, R., Nomura, S., and Mizuno, N. 2000. The relationship between neurokinin-1 receptor and substance P in the medullary dorsal horn: A light- and electron-microscopic immunohistochemical study in the rat. *Neurosci. Res.* **36**, 327–334.

Li, K. C., Zheng, J. H., and Chen, J. 2000. Involvement of spinal protein kinase C in induction and maintenance of both persistent spontaneous flinching reflex and contralateral heat hyperalgesia induced by subcutaneous bee venom in the conscious rat. *Neurosci. Lett.* **285**, 103–106.

Li, L., and Zhou, X. F. 2001. Pericellular Griffonia simplicifolia I isolectin B4-binding ring structures in the dorsal root ganglia following peripheral nerve injury in rats. *J. Comp. Neurol.* **439**, 259–274.

Li, M., Sobue, G., Doyu, M., Mukai, E., Hashizume, Y., and Mitsuma, T. 1995. Primary sensory neurons in X-linked recessive bulbospinal neuropathy: Histopathology and androgen receptor gene expression. *Muscle Nerve* **18**, 301–308.

Li, N., Young, M. M., Bailey, C. J., and Smith, M. E. 1999. NMDA and AMPA glutamate receptor subtypes in the thoracic spinal cord in lean and obese-diabetic ob/ob mice. *Brain Res.* **849**, 34–44.

Li, P., and Zhuo, M. 1998. Silent glutamatergic synapses and nociception in mammalian spinal cord. *Nature* **393**, 695–698.

Li, P., and Zhuo, M. 2001a. Cholinergic, noradrenergic, and serotonergic inhibition of fast synaptic transmission in spinal lumbar dorsal horn of rat. *Brain Res. Bull.* **54**, 639–647.

Li, P., and Zhuo, M. 2001b. Substance P and neurokinin A mediate sensory synaptic transmission in young rat dorsal horn neurons. *Brain Res. Bull.* **55**, 521–531.

Li, P., Calejesan, A. A. and Zhuo, M. 1998. ATP P2x receptors and sensory synaptic transmission between primary afferent fibers and spinal dorsal horn neurons in rats. *J. Neurophysiol.* **80**, 3356–3360.

Li, P., Wilding, T. J., Kim, S. J., Calejesan, A. A., Huettner, J. E., and Zhuo, M. 1999. Kainate-receptor-mediated sensory synaptic transmission in mammalian spinal cord. *Nature* **397**, 161–164.

Li, Y. Q., Li, H., Kaneko, T., and Mizuno, N. 1999. Substantia gelatinosa neurons in the medullary dorsal horn: An intracellular labeling study in the rat. *J. Comp. Neurol.* **411**, 399–412.

Li, Y. Q., Li, H., Yang, K., Wang, Z. M., Kaneko, T., and Mizuno, N. 2000a. Intracellular labeling study of neurons in the superficial part of the magnocellular layer of the medullary dorsal horn of the rat. *J. Comp. Neurol.* **428**, 641–655.

Li, Y. Q., Li, H., Yang, K., Kaneko, T., and Mizuno, N. 2000b. Morphologic features and electrical membrane properties of projection neurons in the marginal layer of the medullary dorsal horn of the rat. *J. Comp. Neurol.* **424**, 24–36.

Li, Y. Q., Wu, S. X., Li, J. L., Kaneko, T., and Mizuno, N. 2000c, Co-existence of calcium-binding proteins in neurons of the medullary dorsal horn of the rat. *Neurosci Lett.* **286**, 103–106.

Liang, Y. F., Haake, B., and Reeh, P. W. 2001. Sustained sensitization and recruitment of rat cutaneous nociceptors by bradykinin and a novel theory of its excitatory action. *J. Physiol.* **532**, 229–239.

Liebel, J. T., Swandulla, D., and Zeilhofer, H. U. 1997. Modulation of excitatory synaptic transmission by nociceptin in superficial dorsal horn neurons of the neonatal rat spinal cord. *Br. J. Pharmacol.* **121**, 435–432.

Lieberman, A. R. 1976. Sensory ganglia. In D. N. Landon (ed.), *The Peripheral Nerve* (pp. 188–278). Chapman & Hall, London.

Light, A. R., and Durkovic, R. G. 1984. Features of laminar and somatotopic organization of lumbar spinal cord units receiving cutaneous inputs from hindlimb receptive fields. *J. Neurophysiol.* **52**, 449–458.

Light, A. R., and Kavookjian, A. M. 1988. Morphology and ultrastructure of physiologically identified substantia gelatinosa (lamina II) neurons with axons that terminate in deeper dorsal horn laminae (III–V). *J. Comp. Neurol.* **267**, 172–189.

Light, A. R., and Metz, C. B. 1978. The morphology of the spinal cord efferent and afferent neurons contributing to the ventral roots of the cat. *J. Comp. Neurol.* **179**, 501–516.

Light, A. R., and Perl, E. R. 1977. Differential termination of large-diameter and small-diameter primary afferent fibers in the spinal dorsal gray matter as indicated by labeling the horseradish peroxidase. *Neurosci. Lett.* **6**, 59–63.

Light, A. R., and Perl, E. R. 1979a. Reexamination of the dorsal root projection to the spinal dorsal horn including observations on the differential termination of coarse and fine fibers. *J. Comp. Neurol.* **186**, 117–132.

Light, A. R., and Perl, E. R. 1979b. Spinal termination of functionally identified primary afferent neurons with slowly conducting myelinated fibers. *J. Comp. Neurol.* **186**, 133–150.

Light, A. R., Trevino, D. L., and Perl, E. R. 1979. Morphological features of functionally defined neurons in the marginal zone and substantia gelatinosa of the spinal dorsal horn. *J. Comp. Neurol.* **186**, 151–172.

Light, A. R., Kavookjian, A. M., and Petrusz, P. 1983. The ultrastructure and synaptic connections of serotonin-immunoreactive terminals in spinal laminae I and II. *Somatosens. Mot. Res.* **1**, 33–50.

Lim, R. K. S., Liu, C. N., Guzman, F., and Braun, C. 1962. Visceral receptors concerned in visceral pain and the pseudoaffective response to intra-arterial injection of bradykinin and other algesic agents. *J. Comp. Neurol.* **118**, 269–294.

Lima, D. 1998. Anatomical basis for the dynamic processing of nociceptive input. *Eur. J. Pain* **2**, 195–202.

Lima, D., and Coimbra, A. 1983. The neuronal population of the marginal zone (lamina I) of the rat spinal cord. A study based on reconstructions of serially sectioned cells. *Anat. Embryol.* **167**, 273–288.

Lima, D., and Coimbra, A. 1986. A Golgi study of the neuronal population of the marginal zone (lamina I) of the rat spinal cord. *J. Comp. Neurol.* **244**, 53–71.

Lima, D., and Coimbra, A. 1989. Morphological types of spinomesencephalic neurons in the marginal zone (lamina I) of the rat spinal cord, as shown after retrograde labeling with cholera toxin subunit B. *J. Comp. Neurol.* **279**, 327–339.

Lin, Q., Peng, Y. B., and Willis, W. D. 1996a. Inhibition of primate spinothalamic tract neurons by spinal glycine and GABA is reduced during central sensitization. *J. Neurophysiol.* **76**, 1005–1014.

Lin, Q., Peng, Y. B., and Willis, W. D. 1996b. Possible role of protein kinase C in the sensitization of primate spinothalamic tract neurons. *J. Neurosci.* **16**, 3026–3034.

Lin, Q., Peng, Y. B., Wu, J., and Willis, W. D. 1997. Involvement of cGMP in nociceptive processing by and sensitization of spinothalamic neurons in primates. *J. Neurosci.* **17**, 3293–3302.

Lin, Q., Palecek, J., Paleckova, V., Peng, Y. B., Wu, J., Cui, M. L., and Willis, W. D. 1999a. Nitric oxide mediates the central sensitization of primate spinothalamic tract neurons. *J. Neurophysiol.* **81**, 1075–1089.

Lin, Q., Wu, J., Peng, Y. B., Cui, M. L., and Willis, W. D. 1999b. Nitric oxide-mediated spinal disinhibition contributes to the central sensitization of primate spinothalamic tract neurons. *J. Neurophysiol.* **81**, 1986–1094.

Lin, Q., Wu, J., Peng, Y. B., Cui, M. L., and Willis, W. D. 1999c. Inhibition of primate spinothalamic tract neurons by spinal glycine and GABA is modulated by guanosine 3′, 5′-cyclic monophosphate. *J. Neurophysiol.* **81**, 1095–1103.

Lin, Q., Wu, J., and Willis, W. D. 1999d. Dorsal root reflexes and cutaneous neurogenic inflammation after intradermal injection of capsaicin in rats. *J. Neurophysiol.* **82**, 2602–2611.

Lin, Q., Zou, X. J., and Willis, W. D. 2000. Aδ and C primary afferents convey dorsal root reflexes following intradermal injection of capsaicin in rats. *J. Neurophysiol.* **84**, 2695–2698.

Lind, R. W., Swanson, L. W., and Ganten, D. 1985. Organization of angiotensin II immunoreactive cells and fibers in the rat central nervous system: An immunohistochemical study. *Neuroendocrinology* **40**, 2–24.

Lindblom, U. F. 1965. Properties of touch receptors in distal glabrous skin of the monkey. *J. Neurophysiol.* **28**, 966–985.

Lindblom, U. F., and Ottoson, J. O. 1953a. Localization of the structure generating the negative cord dorsum potential evoked by stimulation of low threshold cutaneous fibres. *Acta Physiol. Scand.* **29** (Suppl. 106), 180–190.

Lindblom, U. F., and Ottoson, J. O. 1953b. Effects of spinal sections on the spinal cord potentials elicited by stimulation of low threshold cutaneous fibres. *Acta Physiol. Scand.* **29** (Suppl. 106), 191–208.

Linden, D. R., Reutter, M. A., McCarson, K. E., and Seybold, V. S. 2000. Time-dependent changes in neurokinin(3) receptors and tachykinins during adjuvant-induced peripheral inflammation in the rat. *Neuroscience* **98**, 801–811.

Linderoth, B., Stiller, C. O., Gunasekera, L., O'Conner, W. T., Ungerstedt, U., and Brodin, E. 1994. Gamma-aminobutryic acid is released in the dorsal horn by electrical spinal cord stimulation: An in vivo microdialysis study in the rat. *Neurosurgery* **34**, 484–488.

Lindh, B., Dalsgaard, C. J., Elfvin, L. G., Hökfelt, T., and Cuello, A. C. 1983. Evidence of substance P immunoreactive neurons in dorsal root ganglia and vagal ganglia projecting to the guinea pig pylorus. *Brain Res.* **269**, 365–369.

Lindh, B., Lundberg, J. M., and Hokfelt, T. 1989. NPY-, galanin-, VIP/PHI-, CGRP- and substance P-immunoreactive neuronal subpopulations in cat autonomic and sensory ganglia and their projections. *Cell Tissue Res.* **256**, 259–273.

Lindsay, R. M., and Harmar, A. J. 1989. Nerve growth factor regulates expression of neuropeptide genes in adult sensory neurons. *Nature* **337**, 362–364.

Lisney, S. J. W. 1983. Changes in somatotopic organization of the cat lumbar spinal cord following peripheral nerve transection and regeneration. *Brain Res.* **259**, 31–39.

Lissauer, H. 1886. Beitrag zum Faserverlauf im Hinterhorn des menschlichen Ruckenmark und zum Verhalten desselben bei Tabes dorsalis. *Arch. Psychiat. Nervenkrankh.* **17**, 377–438.

Littlewood, N. K., Todd, A. J., Spike, R. C., Watt, C., and Shehab, S. A. 1995. The types of neuron in spinal dorsal horn which possess neurokinin-1 receptors. *Neuroscience* **66**, 597–608.

Liu, C. N. and Chambers, W. W. 1958. Intraspinal sprouting of dorsal root axons. *Arch. Neurol. Psychiat.* **79**, 46–61.

Liu, D., Thangnipon, W., and McAdoo, D. J. 1991. Excitatory amino acids rise to toxic levels upon impact injury to the rat spinal cord. *Brain Res.* **547**, 344–348.

Liu, D., Xu, G. Y., Pan, E., and McAdoo, D. J. 1999. Neurotoxicity of glutamate at the concentration released upon spinal cord injury. *Neuroscience* **93**, 1383–1389.

Liu, H., Brown, J. L., Jasmin, L., Maggio, J. E., Vigna, S. R., Mantyh, P. W., and Basbaum, A. I. 1994a. Synaptic relationship between substance P and the substance P receptor: Light- and electron-microscopic characterization of the mismatch between neuropeptides and their receptors. *Proc. Natl. Acad. Sci.* **91**, 1009–1013.

Liu, H., Wang, H., Sheng, M., Jan, L. Y., Jan, Y. N., and Basbaum, A. I. 1994b. Evidence for presynaptic N-methyl-D-asparate autoreceptors in the spinal cord dorsal horn. *Proc. Natl. Acad. Sci.* **91**, 8383–8387.

Liu, H., Mantyh, P. W., and Basbaum, A. I. 1997. NMDA-receptor regulation of substance P release from primary afferent nociceptors. *Nature* **386**, 721–724.

Liu, J. G., and Anand, K. J. 2001. Protein kinases modulate the cellular adaptations associated with opioid tolerance and dependence. *Brain Res. Rev.* **38**, 1–19.

Liu, L., Chang, G. Q., Jiao, Y. Q., and Simon, S. A. 1998. Neuronal nicotinic acetylcholine receptors in rat trigeminal ganglia. *Brain Res.* **809**, 238–245.

Liu, X., Chung, K., and Chung, J. M. 1999. Ectopic discharges and adrenergic sensitivity of sensory neurons after spinal cord injury. *Brain Res.* **849**, 244–247.

Liu, X. G., and Sandkühler, J. 1995a. The effects of extrasynaptic substance P on nociceptive neurons in laminae I and II in rat lumbar spinal dorsal horn. *Neuroscience* **68**, 1207–1218.

Liu, X. G., and Sandkühler, J. 1995b. Long-term potentiation of C-fiber-evoked potentials in the rat dorsal horn is prevented by spinal N-methyl-D-aspartic acid receptor blockage. *Neurosci. Lett.* **191**, 43–46.

Liu, X. G., and Sandkühler, J. 1997. Characterization of long-term potentiation of C-fiber-evoked potentials in spinal dorsal horn of adult rat: Essential role of NK1 and NK2 receptors. *J. Neurophysiol.* **78**, 1973–1982.

Liu, X. G., and Sandkühler, J. 1998. Activation of spinal N-methyl-D-asparate or neurokinin receptors induces long-term potentiation of spinal C-fibre-evoked potentials. *Neuroscience* **86**, 1209–1216.

Liu, X. H., and Morris, R. 1999. Vasoactive intestinal polypeptide produces depolarization and facilitation of C-fibre evoked synaptic responses in superficial dorsal horn neurones (laminae I–IV) of the rat lumbar spinal cord in vitro., *Neurosci. Lett.* **276**, 1–4.

Liuzzi, F. J., Wu, W., Scoville, S. A., and Schinco, F. P. 1993. Development of nitric oxide synthase expression in the superficial dorsal horn of the rat spinal cord. *Exp. Neurol.* **121**, 275–278.

Ljungdahl, A., and Hökfelt, T. 1973. Autoradiographic uptake patterns of [3H]GABA and [3H]-glycine in central nervous tissues with special reference to the cat spinal cord. *Brain Res.* **62**, 587–595.

Ljungdahl, Å., Hökfelt, T., and Nilsson, G. 1978. Distribution of substance P-like immunoreactivity in the central nervous system of the rat: I. Cell bodies and nerve terminals. *Neuroscience* **3**, 861–943.

Llewellyn-Smith, I. J., and Burnstock, G. 1998. Ultrastructural localization of $P2X_3$ receptors in rat sensory neurons. *NeuroReport* **9**, 2545–2550.

Llewelyn, J. G., Patel, N. J., Thomas, P. K., Thomson, C. S., Muddle, J. R., Workman, J. M., and Dashwood, M. R. 1988. Insulin receptors in sensory and sympathetic ganglia and in peripheral nerve. *J. Neurol.* **235**, S16.

Lloyd, D. P. C. 1952. Electrotonus in dorsal nerve roots. *Cold Spring Harbor Symp. Quant. Biol.* **17**, 203–219.

Lloyd, D. P. C., and Chang, H. T. 1948. Afferent fibers in muscle nerves. *J. Neurophysiol.* **11**, 199–208.

Lloyd, D. P. C., and McIntyre, A. K. 1949. On the origins of dorsal root potentials. *J. Gen. Physiol.* **32**, 409–443.

Loeb, G. E. 1976. Ventral root projections of myelinated dorsal root ganglion cells in the cat. *Brain Res.* **106**, 159–165.

Loeser, J. D. and Ward, A. A. 1967. Some effects of deafferentation on neurons of the cat spinal cord. *Arch. Neurol.* **17**, 629–636.

Loewenstein, W. R., and Skalak, R. 1966. Mechanical transmission in a Pacinian corpuscle: An analysis and a theory. *J. Physiol.* **182**, 346–378.

Lombard, M. C., Nashold, B. S., Albe-Fessard, D., Salman, N., and Sakr, C. 1979. Deafferentation hypersensitivity in the rat after rhizotomy: A possible animal model of chronic pain. *Pain* **6**, 163–175.

Lombard, M. C., Simonnet, G., Zajac, J.-M., Besson, J. M., and Allard, M. 1995a. Distribution of neuropeptide FF (FLFQPQRFamide) receptors in the adult rat spinal cord: Effects of dorsal rhizotomy and neonatal capsaicin. *Neuroscience* **68**, 1229–1235.

Lombard, M. C., Besse, D., and Besson, J. M. 1995b. Opioid receptors in the superficial layers of the rat spinal cord: Functional implications in pain processing. *Prog. Brain Res.* **104**, 77–92.

Lombet, A., Laduron, P., Mourre, C., Jacomet, Y., and Lazdunski, M. 1985. Axonal transport of the voltage-dependent Na+ channel protein identified by its tetrodotoxin binding site in rat sciatic nerves. *Brain Res.* **345**, 153–158.

Long, D. M., and Hagfors, N. 1975. Electrical stimulation in the nervous system: The current status of electrical stimulation of the nervous system for relief of pain. *Pain* **1**, 109–123.

Long, R. R. 1977. Sensitivity of cutaneous cold fibers to noxious heat: Paradoxical cold discharge. *J. Neurophysiol.* **40**, 489–502.

Longhurst, J. C., and Dittman, L. E. 1987. Hypoxia, bradykinin, and prostaglandins stimulate ischemically sensitive visceral afferents. *Am. J. Physiol.* **253**, H556–H567.

Longhurst, J. C., Mitchell, J. H., and Moore, M. B. 1980. The spinal cord ventral root: An afferent pathway of the hind-limb pressor reflex in cats. *J. Physiol.* **301**, 467–476.

Longhurst, J. C., Rotto, D. M., Kaufman, M. P., and Stahl, G. L. 1991. Ischemically sensitive abdominal visceral afferents: Response to cyclooxygenase blockade. *Am. J. Physiol.* **261**, H2075–H2081.

Longmore, J., Shaw, D., Smith, D., Hopkins, R., McAllister, G., Pickard, J. D., Sirinathsinghji, D. J., Butler, A. J., and Hill, R. G. 1997. Differential distribution of $5HT_{1D}$- and $5HT_{1B}$-immunoreactivity within the human trigemino-cerebravascular system: Implication for the discovery of new antimigraine drugs. *Cephalalgia* **17**, 833–842.

Looft, F. J. 1996a. Response of monkey glabrous skin mechanoreceptors to random noise sequences: II. Dynamic stimulus state analysis. *Somatosens. Mot. Res.* **13**, 11–28.

Looft, F. J. 1996b. Response of monkey glabrous skin mechanoreceptors to random noise sequences: III. Spectral analysis. *Somatosens. Mot. Res.* **13**, 235–244.

Lopes, P., and Couture, R. 1997. Localization of bradykinin-like immunoreactivity in the rat spinal cord: Effects of capsaicin, melittin, dorsal rhizotomy and peripheral axotomy. *Neuroscience* **78**, 481–497.

Lopes, P., Kar, S., Tousignant, C., Regoli, D., Quirion, R., and Couture, R. 1993. Autoradiographic localization of [^{125}I-TYR^{8}]-bradykinin receptor binding sites in the guinea pig spinal cord. *Synapse* **15**, 48–57.

Lopes, P., Kar, S., Chretien, L., Regoli, D., Quirion, R., and Couture, R. 1995. Quantitative autoradiographic localization of the [^{125}I-TYR^{8}]bradykinin receptor binding sites in the rat spinal cord: Effects of neonatal capsaicin, noradrenergic deafferentation, dorsal rhizotomy and peripheral axotomy. *Neuroscience* **68**, 867–881.

López, S. M., Pérez-Pérez, M., Márquez, J. M., Naves, F. J., Represa, J., and Vega, J. A. 1998. p75 and TrkA neurotrophin receptors in human skin after spinal cord and peripheral nerve injury, with special reference to sensory corpuscles. *Anat. Rec.* **251**, 371–383.

Lopshire, J. C., and Nicol, G. D. 1998. The cAMP transduction cascade mediates the prostaglandin E_2 enhancement of the capsaicin-elicited current in rat sensory neurons: Whole-cell and single-channel studies. *J. Neurosci.* **18**, 6081–6092.

Loren, I., Alumets, J., Hakanson, R., and Sundler, F. 1979. Distribution of gastrin and CCK-like peptides in rat brain. *Histochemistry* **59**, 249–257.

Lothman, E. W., and Somjen, G. G. 1975. Extracellular potassium activity, intracellular and extracellular potential responses in the spinal cord. *J. Physiol.* **252**, 115–136.

Loup, F., Tribollet, E., Dubois-Dauphin, M., Pizzolato, G., and Dreifuss, J. J. 1989. Localization of oxytocin binding sites in the human brainstem and upper spinal cord: An autoradiographic study. *Brain Res.* **500**, 223–230.

Loup, F., Tribollet, E., Dubois-Dauphin, M., and Dreifuss, J. J. 1991. Localization of high-affinity binding sites for oxytocin and vasopressin in the human brain: An autoradiographic study. *Brain Res.* **555**, 220–232.

Low, A., and Westerman, R. A. 1989. Neurogenic vasodilation in the rat hairy skin measured using a laser Doppler flowmeter. *Life Sci.* **45**, 49–57.

Low P. A., Burke, W. J., and McLeod, J. G. 1978. Congenital sensory neuropathy with selective loss of small myelinated fibers. *Ann. Neurol.* **3**, 179–182.

Lu, G. W., Bennett, G. J., Nishikawa, N., Hoffert, M. J., and Dubner, R. 1983. Extra- and intracellular recordings from dorsal column postsynaptic spinomedullary neurons in the cat. *Exp. Neurol.* **82**, 456–477.

Lu, Y., Zheng, H. -X., Ding, Y. -Q., Gong, L. -W., Qin, B. -Z., and Li, J. -S. 1997. Coexistence of mu-opioid receptor-like and substance P-like immunoreactivity in the cat dorsal root ganglionic neurons. *J. Brain Res.* **38**, 243–246.

Lucas, M. E., and Willis, W. D. 1974. Identification of muscle afferents which activate interneurons in the intermediate nucleus. *J. Neurophysiol.* **37**, 282–293.

Lukas, Z., Cech, S., and Burianek, P. 1970. Cholinesterases and biogenic monoamines in ganglion semilunare (Gasseri). *Histochemie* **22**, 163–168.

Lumbrosa, S., Sandillon, F., Georget, V., Lobaccaro, J. M., Brinkmann, A. O., Privat, A., and Sultan, C. 1996. Immunohistochemical localization and immunoblotting of androgen receptor in spinal neurons of male and female rats. *Eur. J. Endocrinol.* **134**, 626–632.

Lundberg, J. M., Hökfelt, T., Nilsson, G., Terenius, L., Rehfeld, J., Elde, R., and Said, S. 1978. Peptide neurons in the vagus, splanchnic and sciatic nerves. *Acta Physiol. Scand.* **104**, 499–501.

Lundberg, J. M., Franco-Cereceda, A., Hua, X., Hokfelt, T., and Fischer, J. A. 1985. Co-existence of substance P and calcitonin gene-related peptide-like immunoreactivities in sensory nerves in relation to cardiovascular and bronchoconstrictor effects of capsaicin. *Eur. J. Pharmcol.* **108**, 315–319.

Lundeberg, T., Nordemar, R., and Ottoson, D. 1984. Pain alleviation by vibratory stimulation. *Pain* **20**, 25–44.

Luo, C., Chen, J., Li, H. -L., and Li, J. -S. 1998. Spatial and temporal expression of c-Fos protein in the spinal cord of anesthetized rat induced by subcutaneous bee venom injection. *Brain Res.* **806**, 175–185.

Luo, X. G., Rush, R. A., and Zhou, X. F. 2001. Ultrastructural localization of brain-derived neurotrophic factor in rat primary sensory neurons. *Neurosci. Res.* **39**, 377–384.

Luque, J. M., Bleuel, Z., Malherbe, P., and Richards, J. G. 1994. Alternatively spliced isoforms of the *N*-methyl-D-aspartate receptor subunit 1 are differentially distributed within the rat spinal cord. *Neuroscience* **63**, 629–635.

Lynn, B. 1969. The nature and location of certain phasic mechanoreceptors in the cat's foot. *J. Physiol.* **201**, 765–773.

Lynn, B. 1971. The form and distribution of the receptive fields of Pacinian corpuscles found in and around the cat's large foot pad. *J. Physiol.* **217**, 755–771.

Lynn, B. 1988. Neurogenic inflammation. *Skin Pharmacol.* **1**, 217–224.

Lynn, B. and Carpenter, S. E. 1982. Primary afferent units from the hairy skin of the rat hind limb. *Brain Res.* **238**, 29–43.

Lynn, B., Carpenter, S. E., and Pini, A. 1984. Capsaicin and cutaneous afferents. In L. A. Chahl, J. Szolcsányi, and F. Lembeck (eds.), *Antidromic Vasodilatation and Neurogenic Inflammation.* Hungarian Academy of Science, Budapest.

Lyu, Y. S., Park, S. W., Chung, K. and Chung, J. M. 2000. Low dose of tetrodotoxin reduces neuropathic pain behaviors in an animal model. *Brain Res.* **871**, 98–103.

Ma, P. M. and Woolsey, T. A. 1984. Cytoarchitectonic correlates of the vibrissae in the medullary trigeminal complex of the mouse. *Brain Res.* **306**, 374–379.

Ma, Q. P. and Hargreaves, R. J. 2000. Localization of *N*-methyl-D-aspartate NR2B subunits on primary sensory neurons that give rise to small-caliber sciatic nerve fibers in rats. *Neuroscience* **101**, 699–707.

Ma, Q. P., Hill, R., and Sirinathsinghji, D. 2000a. Basal expression of bradykinin B1 receptor in peripheral sensory ganglia in the rat. *NeuroReport* **11**, 4003–4005.

Ma, Q. P., Tian, L., and Woolf, C. J. 2000b. Resection of sciatic nerve re-triggers central sprouting of A-fibre primary afferents in the rat. *Neurosci. Lett.* **288**, 215–218.

Ma, W. and Bisby, M. A. 1997. Differential expression of galanin immunoreactivities in the primary sensory neurons following partial and complete sciatic nerve injuries. *Neuroscience* **79**, 1183–1195.

Ma, W. and Bisby, M. A. 1998. Partial and complete sciatic nerve injuries induce similar increases of neuropeptide Y and vasoactive intestinal peptide immunoreactivities in primary sensory neurons and their central projections. *Neuroscience* **86**, 1217–1234.

Ma, W. and Bisby, M. A. 1999a. Partial sciatic nerve transection induced tyrosine hydroxidase immunoreactive axon sprouting around both injured and spared dorsal root ganglion neurons which project to the gracile nucleus in middle-aged rats. *Neurosci Lett.* **275**, 117–120.

Ma, W. and Bisby, M. A. 1999b. Increase of galanin mRNA in lumbar dorsal root ganglion neurons of adult rats after partial sciatic nerve ligation. *Neurosci. Lett.* **262**, 195–198.

Ma, W. and Bisby, M. A. 2000. Partial sciatic nerve ligation induced more dramatic increase of neuropeptide Y immunoreactive axonal fibers in the gracile nucleus of middle-aged rats than in young adult rats. *J. Neurosci. Res* **60**, 520–530.

Ma, W. and Ribeiro-da-Silva, A. 1995. Substance P- and GABA-like immunoreactivities are co-localized in axonal varicosities in the superficial laminae of cat but not rat spinal cord. *Brain Res.* **692**, 99–110.

Ma, W., Saunders, P. A., Somogyi, R., Poulter, M. O., and Barker, J. L. 1993. Ontogeny of $GABA_A$ receptor subunit mRNAs in rat spinal cord and dorsal root ganglia. *J. Comp. Neurol.* **338**, 337–359.

Ma, W., Behar, T., Chang, L., and Barker, J. L. 1994. Transient increase in expression of GAD_{65} and GAD_{67} mRNAs during postnatal development of rat spinal cord. *J. Comp. Neurol.* **346**, 151–160.

Ma, W., Ribeiro-da-Silva, A., Noel, G., Julien, J.-P., and Claudio Cuello, A. 1995. Eptopic substance P and calcitonin gene-related peptide immunoreactive fibres in the spinal cord of transgenic mice over-expressing nerve growth factor. *Eur. J. Neurosci.* **7**, 2021–2035.

Ma, W., Ribeiro-da-Silva, A., De Koninck, Y., Radhakrishnan,. V., Henry, J. L., and Cuello, A. C. 1996. Quantitative analysis of substance P-immunoreactive boutons on physiologically characterized dorsal horn neurons in the cat lumbar spinal cord. *J. Comp. Neurol.* **376**, 45–64.

Ma, W., Ribeiro-da-Silva, A., De Konink, Y., Radhakrishnan, V., Cuello, A. C., and Henry, J. L. 1997. Substance P and enkephalin immunoreactivities in axonal boutons presynaptic to physiologically identified dorsal horn neurons: An ultrastructural multiple-labelling study in the cat. *Neuroscience* **77**, 793–811.

Ma, W., Zheng, W. H., Powell, K., Jhamandas, K., and Quirion, R. 2001. Chronic morphine exposure increases the phosphorylation of MAP kinases and the transcription factor CREB in dorsal root ganglion neurons: An in vitro and in vivo study. *Eur. J. Neurosci.* **14**, 1091–1104.

MacArthur, L., Ren, K., Pfaffenroth, E., Franklin, E., and Ruda, M. A. 1999. Descending modulation of opioid-containing nociceptive neurons in rats with peripheral inflammation and hyperalgesia. *Neuroscience* **88**, 499–506.

MacDermott, A. B., Mayer, M. L., Westbrook, G. L., Smith, S. J., and Barker, J. L. 1986. NMDA-receptor activation increases cytoplasmic calcium concentration in cultured spinal cord neurons. *Nature* **321**, 519–522.

MacDermott, A. B., Role, L. W., and Siegelbaum, S. A. 1999. Presynaptic ionotropic receptors and the control of transmitter release. *Ann. Rev. Neurosci.* **22**, 443–485.

MacDonald, R., Bingham, S., Bond, B. C., Parsons, A. A., and Philpott, K. L. 2001. Determination of changes in mRNA expression in a rat model of neuropathic pain by Taqman quantitative RT-PCR. *Mol. Brain Res.* **90**, 48–56.

MacDonald, S., Smith, M. E., and Bailey, C. J. 1998. Expression of β-endorphin and its receptors in the spinal cord of obese-diabetic ob/ob mice. *Neurosci. Lett.* **248**, 85–88.

Macefield, G., Gandevia, S. C., and Burke, D. 1990. Perceptual responses to microstimulation of single afferents innervating joints, muscles and skin of the human hand. *J. Physiol.* **429**, 113–129.

MacGregor, D. P., Murone, C., Song, K., Allen, A. M., Paxinos, G., and Mendelsohn, F. A. O. 1995. Angiotensin II receptor subtypes in the human central nervous system. *Brain Res.* **675**, 231–240.

MacKenzie, R. A., Burke, D., Skuse, N. F., and Lethlean, A. K. 1975. Fibre function and perception during cutaneous nerve block. *J. Neurol. Neurosurg. Psychiat.* **38**, 865–873.

Macon, J. B. 1979. Deafferentation hyperactivity in the monkey spinal trigeminal nucleus: Neuronal responses to amino acid iontophoresis. *Brain Res.* **161**, 549–554.

Maderdrut, J. L., Yaksh, T. L., Petruz, P., and Go, V. L. W. 1982. Origin and distribution of cholecystokinin-containing nerve terminals in the lumbar dorsal horn and nucleus caudalis of the cat. *Brain Res.* **243**, 363–368.

Madrid, J., Alvarado, J., Dutton, H., and Rudomin, P. 1979. A method for the dynamic continuous estimation of excitability changes of single fiber terminals in the central nervous system. *Neurosci. Lett.* **11**, 253–258.

Maeshima, T., Ito, R., Hamada, S., Senzaki, K., Hamaguchi-Hamada, K., Shutoh, F., and Okada, N. 1998. The cellular localization of 5-HT_{2a} receptors in the spinal cord and spinal ganglia of the adult rat. *Brain Res.* **797**, 118–124.

Magendie, F. 1822. Experience sur les functions des racines des nerfs rachidiens. *J. Physiol. Exp. Pathol.* **2**, 276–279 (reprinted in P. F. Cranefield, 1974).

Magerl, W., Szolcsanyi, J., Westerman, R. A., and Handwerker, H. O. 1987. Laser Doppler measurements of skin vasodilation elicited by percutaneous electrical stimulation of nociceptors in humans. *Neurosci. Lett.* **82**, 349–354.

Magerl, W., Graemer, G., and Handwerker, H. O. 1990. Sensations and local inflammatory responses induced by application of carbachol, dopamine, 5-HT, histamine and mustard oil to the skin in humans. *Pfluegers Arch.* **415**, R107.

Magnuson, D. S., Johnson, R., Peet, M. J., Curry, K., and McLennan, H. 1987. A novel spinal cord slice preparation from the rat. *J. Neurosci. Methods* **19**, 141–145.

Magnuson, D. S. K., Curry, K., Peet, M. J., and McLennan, H. 1988. Structural requirements for activation of excitatory amino acid receptors in the rat spinal cord in vitro. *Exp. Brain Res.* **73**, 541–545.

Magnusson, K. R., Larson, A. A., Madl, J. E., Alschuler, R. A., and Beitz, A. J. 1986. Co-localization of fixative-modified glutamate and glutaminase in neurons of the spinal trigeminal nucleus of the rat: An immunohistochemical and immunoradiological analysis. *J. Comp. Neurol.* **247**, 477–490.

Magnusson, K. R., Clements, J. R., Larson, A. A., Madl, J. E., and Beitz, A. J. 1987. Localization of glutamate in trigeminothalamic projection neurons: A combined retrograde transport-immunohistochemical study. *Somatosens. Mot. Res.* **4**, 177–190.

Magoul, R., Onteniente, B., Geffard, M., and Calas, A. 1987. Anatomical distribution and ultrastructural organization of the GABAergic system in the rat spinal cord: An immunocytochemical study using anti-GABA antibodies. *Neuroscience* **20**, 1001–1009.

Magoun, H. W. 1963. *The Waking Brain* (2nd ed). Thomas, Springfield.

Maguire, C. M. and Geraghty, D. P. 1998. Comparison of [^{125}I]-Bolton-Hunter substance P binding in young and aged rat spinal cord. *Brain Res.* **786**, 263–266.

Mai, J. K., Triepel, J., and Metz, J. 1987. Neurotensin in the human brain. *Neuroscience* **22**, 499–524.

Maihofner, C., Tegeder, I., Euchenhofer, C., deWitt, D., Brune, K., Bang, R., Neuhuber, W., and Geisslinger, G. 2000. Localization and regulation of cyclo-oxygenase-1 and -2 and neuronal nitric oxide synthase in mouse spinal cord. *Neuroscience* **101**, 1093–1108.

Mailis, A. 1996. Compulsive targeted self-injurious behaviour in humans with neuropathic pain: A counterpart of animal autotomy? Four case reports and literature review. *Pain* **64**, 569–578.

Mailleux, P. and Vanderhaeghen, J. J. 1992. Distribution of neuronal cannabinoid receptor in the adult rat brain: A comparative receptor binding radioautography and in situ hybridization histochemistry. *Neuroscience* **48**, 655–668.

Mailleux, P., Parmentier, M., and Vanderhaeghen, J. J. 1992. Distribution of cannabinoid receptor messenger RNA in the human brain: An in situ hybridization histochemistry with oligonucleotides. *Neurosci. Lett.* **143**, 200–204.

Majane, E. A., Panula, P., and Yang, H. Y. 1989. Rat brain regional distribution and spinal cord neuronal pathway of FLFQPQRF-NH2, a mammalian FMRF-NH2-like peptide. *Brain Res* **494**, 1–12.

Malan, T. P., Ossipov, M. H., Gardell, L. R., Ibrahim, M., Bian, D., Lai, J., and Porreca, F. 2000. Extraterritorial neuropathic pain correlates with multisegmental elevation of spinal dynorphin in nerve-injured rats. *Pain* **86**, 185–194.

Malatova, Z., Longauer, F., and Marsala, J. 1985. Choline acetyltransferase and acetylcholinesterase in canine spinal ganglia: Increase of choline acetyltransferase activity following sciatic nerve lesion. *J. Hirnforsch.* **26**, 683–688.

Malcangio, M., DaSilva, H., and Bowery, N. G. 1993. Plasticity of $GABA_B$ receptor in rat spinal cord detected by autoradiography. *Eur. J. Pharmacol.* **250**, 153–156.

Malenka, R. C. and Nicoll, R. A. 1997. Silent synapses speak up. *Neuron* **19**, 473–476.

Malinovský, L. 1966. Variability of sensory nerve endings in foot pads of a domestic cat (*Felis ocreata* L., *F. domestica*). *Acta Anat.* **64**, 82–106.

Malinovský, L., Berková, V., and Páč, L. 1982. The ultrastructure of axon processes in sensory corpuscles. *Z. Mikrosk. Anat. Forsch.* **96**, 844–856.

Malmberg, A. B., Brandon, E. P., Idzerda, R. L., Liu, H., McKnight, G. S., and Basbaum, A. I. 1997a. Diminished inflammation and nociceptive pain with preservation of neuropathic pain in mice with a targeted mutation of the Type I regulatory subunit of cAMP-dependent protein kinase. *J. Neurosci.* **17**, 7462–7470.

Malmberg, A. B., Chen, C., Tonegawa, S., Basbaum, A. I. 1997b. Preserved acute pain and reduced neuropathic pain in mice lacking PKCγ. *Science* **278**, 279–283.

Malosio, M.-L., Marqueze-Pouey, B., Kuhse, J., and Betz, H. 1991. Widespread expression of glycine receptor subunit mRNAs in the adult and developing rat brain. *EMBO* **10**, 2401–2409.

Manaker, S., Winokur, A., Rhodes, C. H., and Rainbow, T. C. 1985a. Autoradiographic localization of thyrotropin-releasing hormone (TRH) receptors in human spinal cord. *Neurology* **35**, 328–332.

Manaker, S., Shulman, L. H., Winokur, A., and Rainbow, T. C. 1985b. Autoradiographic localization of thyrotropin-releasing hormone receptors in amyotrophic lateral sclerosis spinal cord. *Neurology* **35**, 1650–1653.

Manaker, S., Winokur, A., Rostene, W. H., and Rainbow, T. C. 1985c. Autoradiographic localization of thyrotropin-releasing hormone receptors in the rat central nervous system. J. *Neurosci.* **5**, 167–174.

Manaker, S., Caine, S. B., and Winokur, A. 1988. Alterations in receptors for thyrotropin-releasing hormone, serotonin, and acetylcholine in amyotrophic lateral sclerosis. *Neurology* **38**, 1464–1474.

Manfredi, M., Bini, G., Cruccu, G., Accornero, N., Berardelli, A., and Medolago, L. 1981. Congenital absence of pain. *Arch. Neurol.* **38**, 507–511.

Mann, M. D., Kasprzak, H., Hiltz, F. L., and Tapper, D. N. 1972. Activity in single cutaneous afferents: Spinal pathways and cortical evoked potentials. *Brain Res.* **39**, 61–70.

Mannen, H. 1975. Reconstruction of axonal trajectory of individual neurons in the spinal cord using Golgi-stained serial sections. *J. Comp. Neurol.* **159**, 357–374.

Mannen, H. and Sugiura, Y. 1976. Reconstruction of neurons of dorsal horn proper using Golgi-stained serial sections. *J. Comp. Neurol.* **168**, 303–312.

Mannion, R. J., Doubell, T. P., Gill, H., and Woolf, C. J. 1998. Deafferentation is insufficient to induce sprouting of A-fibre central terminals in the rat dorsal horn. *J. Comp. Neurol.* **393**, 135–291.

Mannion, R. J., Costigan, M., Decosterd, I., Amaya, F., Ma, O. P., Holstege, J. C., Ji, R. R., Acheson, A., Lindsay, R. M., Wilkinson, G. A., and Woolf, C. J. 1999. Neurotrophins: Peripheral and centrally acting modulators of tactile stimulus-induced inflammatory pain hypersensitivity. *Proc. Natl. Acad. Sci. USA* **96**, 9385–9390.

Mano, I. and Driscoll, M. 1999. DEG/ENaC channels: A touchy superfamily that watches its salt. *Bioessays* **21**, 568–578.

Manocha, S. L. 1973. Experimental protein malnutrition in primates histochemical studies on the dorsal root ganglion cells of healthy and malnourished squirrel monkeys, *Saimiri sciureus. Acta Histochem.* **47**, 220–232.

Mansour, A., Khachaturian, H., Lewis, M. E., Akil, H., and Watson, S. J. 1988. Anatomy of CNS opioid receptors. *TINS* **11**, 308–314.

Mansour, A., Fox, C. A., Thompson, R. C., Akil, H., and Watson, S. J. 1994a. μ-Opioid receptor mRNA expression in the rat CNS: Comparison to μ-receptor binding. *Brain Res.* **643**, 245–265.

Mansour, A., Fox, C. A., Burke, S., and Watson, S. J. 1994b. Immunohistochemical localization of the kappa1 opioid receptors. *Regul. Pept.* **54**, 181.

Mansour, A., Fox, C. A., Watson, A., and Watson, S. J. 1995a. Opioid-receptor mRNA expression in the rat CNS: Anatomical and functional implications. *TINS* **18**, 22–29.

Mansour, A., Fox, C. A., Burke, S., Akil, H., and Watson, S. J. 1995b. Immunohistochemical localization of the cloned mu opioid receptor in the rat CNS. *J. Chem. Neuroanat.* **8**, 283–305.

Mantyh, C. R., Gates, T., Zimmerman, R. P., Kruger, L., Maggio, J. E., Vigna, S. R., Basbaum, A. I., Levine, J., and Mantyh, P. W. 1988. Alterations n the density of receptor binding sites for sensory neuropeptides in the spinal cord of arthritic rats. In J. M. Besson and G. Guilbaud (eds.), *The Arthritic Rat as a Model of Clinical Pain?* (pp. 139–152). Elsevier, Heidelberg.

Mantyh, P. W. and Hunt, S. P. 1985a. Thyrotropin-releasing hormone (TRH) receptors. *J. Neurosci.* **5**, 551–561.

Mantyh, P. W. and Hunt, S. P. 1985b. The autoradiographic localization of substance P receptors in the rat and bovine spinal cord and cat trigeminal nucleus pars caudalis and the effects of neonatal capsaicin. *Brain Res.* **1985**, 315–324.

Mantyh, P. W., Hunt, S. P., and Maggio, J. E. 1984a. Substance P receptors: Localization by light-microscopic autoradiography in rat brain using [^{3}H]SP as the radioliogand. *Brain Res.* **309**, 147–165.

Mantyh, P. W., Pinnock, R. D., Downes, C. P, Goedert, M., and Hunt, S. P. 1984b. Correlation between inositol phospholipid hydrolysis and substance P receptors in rat CNS. *Nature* **309**, 795–797.

Mantyh, P. W., Catton, M. D., Boehmer, C. G., Welton, M. L., Passaro, E. P., Jr., Maggio, J. E., and Vigna, S. R. 1989a. Receptors for sensory neuropeptides in human inflammatory diseases: Implications for the effector role of sensory neurons. *Peptides* **10**, 627–645.

Mantyh, P. W., Gates, T., Mantyh, C. R., and Maggio, J. E. 1989b. Autoradiographic localization and characterization of tachykinin receptor binding sites in the rat brain and peripheral tissues. *J. Neurosci.* **9**, 259–279.

Mantyh, P. W., Allen, C. J., Rogers, S., DeMaster, E., Ghilardi, J. R., Mosconi, T., Kruger, L., Mannon, P. J., Taylor, I. L., and Vigna, S. R. 1994. Some sensory neurons express neuropeptide Y receptors: Potential paracrine inhibition of primary afferent nociceptors following peripheral nerve injury. *J. Neurosci.* **14**, 3958–3968.

Mantyh, P. W., DeMaster, E., Malhotra, A., Ghilardi, J. R., Rogers, S. D., Mantyh, C. R., Liu, H., Basbaum, A., Vigna, S. R., Maggio, J. E., and Simone, D. A. 1995. Receptor endocytosis and dendrite reshaping in spinal neurons after somatosensory stimulation. *Science* **268**, 1629–1632.

Mantyh, P. W., Rogers, S. D., Honore, P., Allen, B. J., Ghilardi, J. R., Li, J., Daughters, R. S., Lappi, D. A., Wiley, R. G., and Simone, D. A. 1997. Inhibition of hyperalgesia by ablation of lamina I spinal neurons expressing the substance P receptor. *Science* **278**, 275–279.

Mao, J., Price, D. D., Mayer, D. J., Lu, J., and Hayes, R. L. 1992a. Intrathecal MK-801 and local nerve anesthesia synergistically reduce nociceptive behaviors in rats with experimental peripheral mononeuropathy. *Brain Res.* **576**, 254–262.

Mao, J., Price, D. D., Hayes, R. L., Lu, J., and Mayer, D. J. 1992b. Differential roles of NMDA and non-NMDA receptor activation in induction and maintenance of thermal hyperalgesia in rats with painful peripheral mononeuropathy. *Brain Res.* **598**, 271–278.

Mao, J., Price, D. D., Coghill, R. C., Mayer, D. J., and Hayes, R. L. 1992c. Spatial patterns of spinal cord [14C]-2-deoxyglucose metabolic activity in a rodent model of painful peripheral mononeuropathy. *Pain* **50**, 89–100.

Mao, J., Hayes, R. L., Price, D. D., Coghill, R. C., Lu, J., and Mayer, D. J. 1992d. Post-injury treament with GM1 gangliocide reduces nociceptive behaviors and spinal cord metabolic activity in rats with experimental peripheral mononeuropathy. *Brain Res.* **584**, 18–27.

Mao, J., Price, D. D., Hayes, R. L., Lu, J., and Mayer, D. J. 1992e. Intrathecal GM1 ganglioside and local nerve anesthesia reduce nociceptive behaviors in rats with experimental peripheral mononeuropathy. *Brain Res.* **584**, 28–35.

Mao, J., Mayer, D. J., Hayes, R. L., and Price, D. D. 1993. Spatial patterns of increased spinal cord membrane-bound protein kinase C and their relation to increases in ^{14}C-2-deoxyglucose metabolic activity in rats with painful peripheral mononeuropathy. *J. Neurophysiol.* **70**, 470–481.

Mao, J., Price, D. D., Zhu, J., Lu, J., and Mayer, D. J. 1997. The inhibition of nitric oxide-activated poly (ADP-ribose) synthetase attenuates transsynaptic alteration of spinal cord dorsal horn neurons and neuropathic pain in the rat. *Pain* **72**, 355–366.

Mapp, P. I., Kidd, B. L., Gibson, S. J., Terry, J. M., Revell, P. A., Ibrahim, N. B., Blake, D. R., and Polak, J. M. 1990. Substance P-, calcitonin gene-related peptide- and C-flanking peptide of neuropeptide Y-immunoreactive fibres are present in normal synovium but depleted in patients with rheumatoid arthritis. *Neuroscience* **37**, 143–153.

Mapp, P. I., Terenghi, G., Walsh, D. A., Chen, S. T., Cruwys, S. C., Garrett, N., Kidd, B. L., Polak, J. M., and Blake, D. R. 1993. Monoarthritis in the rat knee induces bilateral and time-dependent changes in substance P and calcitonin gene-related peptide immunoreactivity in the spinal cord. *Neuroscience* **57**, 1091–1096.

Marchand, J. E., Cepeda, M. S., Carr, D. B., Wurm, W. H., and Kream, R. M. 1999. Alterations in neuropeptide Y, tyrosine hydroxylase, and Y-receptor subtype distribution following spinal nerve injury to rats. *Pain* **79**, 187–200.

Marchand, R. and Barbeau, H. 1982. Vertically oriented alternating acetylcholinesterase rich and poor territories in laminae VI, VII, VII of the lumbosacral cord of the rat. *Neuroscience* **7**, 1197–1202.

Marchettini, P., Simone, D. A., Caputi, G., and Ochoa, J. L. 1996. Pain from excitation of identified muscle nociceptors in humans. *Brain Res.* **740**, 109–116.

Mardy, S., Miura, Y., Endo, F., Matsuda, I., Sztriha, L., Frossard, P., Moosa, A., Ismail, E. A., Macaya, A., Andria, G., Toscano, E., Gibson, W., Graham, G. E., and Indo, Y. 1999. Congenital insensitivity to pain with anhirdosis: Novel mutations in the TRKA (NTRK1) gene encoding a high-affinity receptor for nerve growth factor. *Am. J. Hum. Genet.* **64**, 1570–1579.

Marinozzi, G., Ferrante, F., Gaudio, E., Ricci, A., and Amenta, F. 1991. Intrinsic innervation of the rat knee joint articular capsule and ligaments. *Acta Anat.* (Basel). **141**, 8–14.

Mark, E. K., Hauge, D. H., Standage, G. P., and Doetsch, G. S. 1990. Immunoreactivity for glutamic acid decarboxylase and several neuropeptides in the spinal cord of the raccoon. *Brain Res. Bulletin* **25**, 787–790.

Marksteiner, J., Kirchmair, R., Mahata, S. K., Mahata, M., Fischer-Colbrie, R., Hogue-Angeletti, R., Saria, A., and Winkler, H. 1993. Distribution of secretoneurin, a peptide derived from secretogranin II, in rat brain: An immunocytochemical and radioimmunological study. *Neuroscience* **54**, 923–944.

Marksteiner, J., Saria, A., and Hinterhuber, H. 1994a. Distribution of secretoneurin-like immunoreactivity in comparison with that of substance P in the human brain stem. *J Chem. Neuroanat.* **7**, 253–270.

Marksteiner, J., Mahata, S. K., Pycha, R., Mahata, M., Saria, A., Fischer-Colbrie, R., and Winkler, H. 1994b. Distribution of secretoneurin immunoreactivity in the spinal cord and lower brainstem in comparison with that of substance P and calcitonin gene-related peptide. *J. Comp. Neurol.* **340**, 243–254.

Markus, H. and Pomeranz, B. 1987. Saphenous has weak ineffective synapses in sciatic territory of rat spinal cord: Electrical stimulation of the saphenous or application of drugs reveal these somatotopically inappropriate synapses. *Brain Res.* **416**, 315–321.

Markus, H., Pomeranz, B., and Krushelnycky, D. 1984. Spread of saphenous projection map in spinal cord and hypersensitivity of the foot after chronic sciatic nerve denervation in adult rat. *Brain Res.* **296**, 27–39.

Marley, P. D., Nagy, J. I., Emson, P. C., and Rehfeld, J. F. 1982. Cholecystokinin in the rat spinal cord: Distribution and lack of effect of neonatal capsaicin treatment and rhizotomy. *Brain Res.* **238**, 494–498.

Marlier, L., Rajaofetra, N., Poulat, P., and Privat, A. 1990. Modification of serotonergic innervation of the rat spinal cord dorsal horn after neonatal capsaicin treatment. *J. Neurosci. Res.* **25**, 112–118.

Marlier, L., Sandillon, F., Poulat, P., Rajaofetra, N., Geffard, M., and Privat, A. 1991a. Serotonergic innervation of the dorsal horn of rat spinal cord: Light- and electron-microscopic immunocytochemical study. *J. Neurocytol.* **20**, 310–322.

Marlier, L., Poulat, P., Rajaofetra, N., and Privat, A. 1991b. Modifications of serotonin-, substance P- and calcitonin gene-related peptide-like immunoreactivities in the dorsal horn of the spinal cord of arthritic rats: A quantitative immunocytochemical study. *Exp. Brain Res.* **85**, 482–490.

Marlier, L., Teilhac, J.-R., Cerruti, C., and Privat, A. 1991c. Autoradiographic mapping of 5-HT_1, 5-HT_{1A}, 5-HT_{1B} and 5- HT_2 receptors in the rat spinal cord. *Brain Res.* **550**, 15–23.

Marlier, L., Poulat, P., Rajaofetra, N., Sandillon, F., and Privat, A. 1992. Plasticity of the serotonergic innervation of the dorsal horn of the rat spinal cord following neonatal capsaicin treatment. *J. Neurosci. Res.* **31**, 346–358.

Marlier, L. N. J. L., Csikós, T., Rebaudengo, N., Borboni, P., Patacchioli, F. R., Angelucci, L., Privat, A., and Lauro, R. 1995. Distribution of glucocorticoid receptor mRNA in the rat spinal cord. *NeuroReport* **6**, 2245–2249.

Marsala, J., Vanicky, I., Marsala, M., Jalc, P., Orendacova, J., and Taira, Y. 1998. Reduced nicotinamide adenine dinucleotide phosphate diaphorase in the spinal cord of dogs. *Neuroscience* **85**, 847–862.

Marshall, G. E., Shehab, S. A., Spike, R. C., and Todd, A. J. 1996. Neurokinin-1 receptors on lumbar spinothalamic neurons in the rat. *Neuroscience* **72**, 255–263.

Marti, E., Gibson, S. J., Polak, J. M., Facer, P., Springall, D. R., Van Aswegen, G., Aitchison, M., and Koltzenburg, M. 1987. Ontogeny of peptide- and amine-containing neurones in motor, sensory, and autonomic regions of rat and human spinal cord, dorsal root ganglia, and rat skin. *J. Comp. Neurol.* **266**, 332–359.

Martin, G. R. and Humphrey, P. P. 1994. Receptors for 5-hydroxytryptamine: Current perspectives on classification and nomenclature. *Neuropharmacology* **33**, 261–273.

Martin, H. A. 1990. Leukotriene B4 induced decrease in mechanical and thermal thresholds of C-fiber mechanoreceptors in rat hairy skin. *Brain Res.* **509**, 273–279.

Martin, H. A. and Murphy, P. R. 1995. Interleukin-2 activates a sub-population of cutaneous C-fibre polymodal nociceptors in the rat hairy skin. *Arch. Physiol. Biochem.* **103**, 136–148.

Martin, H. A., Basbaum, A. I., Kwiat, G. C., Goetzl, E. J., and Levine, J. D. 1987. Leukotriene and prostaglandin sensitization of cutaneous high-threshold C- and A-delta mechanoreceptors in the hairy skin of rat hindlimbs. *Neuroscience* **22**, 651–659.

Martin, H. F. and Manning, J. W. 1969. Rapid thermal cutaneous stimulation: Peripheral nerve responses. *Brain Res.* **16**, 524–526.

Martin, H. F. and Manning, J. W. 1972. Response of A-delta-fibers of peripheral nerve to warming of cutaneous fields. *Brain Res.* **43**, 653–656.

Martin, L. J., Blackstone, C. D., Levey, A. I., Huganir, R. L., and Price, D. L. 1993. AMPA glutamate receptor subunits are differentially distributed in rat brain. *Neuroscience* **53**, 327–358.

Martin, W. J., Liu, H., Wang, H., Malberg, A. B., and Basbaum, A. I. 1999. Inflammation-induced up-regulation of protein kinase Cγ immunoreactivity in rat spinal cord correlates with enhanced nociceptive processing. *Neuroscience* **88**, 1267–1274.

Martin, W. R. 1984. Pharmacology of opioids. *Pharmacol. Rev.* **35**, 283–323.

Martinez-Rodriguez, R. and Diaz, G. 1987. Immunocytochemical and histoenzymological studies of aspartate aminotransferase (AAT) in the spinal cord of rat. *Cell. Mol. Biol.* **33**, 159–165.

Martin-Schild, S., Zadina, J. E., Gerall, A. A., Vigh, S., and Kastin, A. J. 1997. Localization of endomorphin-2-like immunoreactivity in the rat medulla and spinal cord. *Peptides* **18**, 1641–1649.

Martin-Schild, S., Gerall, A. A., Kastin, A. J., and Zadina, J. E. 1998. Endomorphin-2 is an endogenous opioid in primary sensory afferent fibers. *Peptides* **19**, 1783–1789.

Martin-Schild, S., Gerall, A. A., Kastin, A. J., and Zadina, J. E. 1999. Differential distribution of endomorphin 1- and endomorphin 2-like immunoreactivities in the CNS of the rodent. *J. Comp. Neurol.* **405**, 450–471.

Marubio, L. M. and Changeux, J. 2000. Nicotinic acetylcholine receptor knockout mice as animal models for studying receptor function. *Eur. J. Pharmacol.* **393**, 133–121.

Marubio, L. M., Del Mar Arroyo-Jimenez, M., Cordero-Erausquin, M., Lena, C., Le Novere, N., De Kerchove D'Exaerde, A., Huchet, M., Damaj, M. I., and Changeux, J. P. 1999. Reduced antinociception in mice lacking neuronal nicotinic receptor subunits. *Nature*, **398**, 805–810.

Maruhashi, J., Mizuguchi, K., and Tasaki, I. 1952. Action currents in single afferent nerve fibres elicited by stimulation of the skin of the toad and the cat. *J. Physiol.* **117**, 129–151.

Marvizon, J. C., Martinez, V., Grady, E. F., Bunnett, N. W., and Mayer, E. A. 1997. Neurokinin 1 receptor internalization in spinal cord slices induced by dorsal root stimulation is mediated by NMDA receptors. *J. Neurosci.* **17**, 8129–8136.

Marvizon, J. C., Eskandari, S., Ennes, H. S., and Mayer, E. A. 1998. Substance P induces brief, localized increase in $[Ca^{2+}]_I$ in dorsal horn neurons. *NeuroReport* **9**, 3369–3374.

Marvizon, J. C., Grady, E. F., Stefani, E., Bunnett, N. W., and Mayer, E. A. 1999a. Substance P release in the dorsal horn assessed by receptor internalization: NMDA receptors counteract a tonic inhibition by $GABA_B$ receptors *Eur. J. Neurosci.* **11**, 417–426.

Marvizon, J. C. G., Grady, E. F., Waszak-McGee, J., and Mayer, E. A. 1999b. Internalization of μ-opioid receptors in rat spinal cord slices. *NeuroReport* **10**, 2329–2334.

Massari, V. J., Tizabi, Y., Park, C. H., Moody, T. W., Helke, C. J., and O'Donohue, T. L. 1983. Distribution and origin of bombesin, substance P and somatostatin in cat spinal cord. *Peptides* **4**, 673–681.

Massari, V. J., Shults, C. W., Park, C. H., Tizabi, Y., Moody, T. W., Cronwall, B. M., Culver, M., and Chase, T. N. 1985. Deafferentation causes a loss of presynaptic bombesin receptors and supersensivity of substance P receptors in the dorsal horn of the cat spinal cord. *Brain Res.* **343**, 268–274.

Masuyama, T. and Shimizu, T. 1997. Antinociceptive involvement of substance P in the spinal cord of mice: Dose effects of substance P on the behavior elicited by intrathecally administered NMDA. *Brain Res.* **759**, 241–246.

Masuyama, T., Shimizu, T., Iwashita, T., Yoshimura, N., and Fukuda, T. 1996. Spinal antinociceptive effect of substance P on the responses induced by intrathecally injected NMDA in mice. *Brain Res.* **722**, 200–202.

Matsuda, L. A., Lolait, S. J., Brownstein, M. J., Young, A. C., and Bonner, T. I. 1990. Structure of a cannabinoid receptor and functional expression of the cloned cDNA. *Nature*, **346**, 561–564.

Matsuo, M., Kurokawa, T., Goya, N., and Ohta, M. 1981. Congential insensitivity to pain with anhidrosis in a 2-month-old boy. *Neurology* **31**, 1190–1192.

Matsushita, M. 1969. Some aspects of the interneuronal connections in cat's spinal grey matter. *J. Comp. Neurol.* **136**, 57–80.

Matsushita, M. 1970. The axonal pathways of spinal neurons in the cat. *J. Comp. Neurol.* **138**, 391–418.

Matsuura, H. 1967. Histochemical observation of bovine spinal ganglia. *Histochemie* **11**, 152–160.

Matsuura, H., Mori, M., and Kawakatsu, K. 1969. A histochemical and electron-microscopic study of the trigeminal ganglion of the rat. *Arch. Oral Biol.* **14**, 1135–1146.

Matsuura, H., Hirose, I., and Fujita, K. 1970. Electron-microscopic localization of alkaline phosphatase in the trigeminal ganglion of the rat. *Histochemie* **23**, 91–97.

Matsuyama, T., Wanaka, A., Yoneda, S., Kimura, K., Kamada, T., Girgis, S., MacIntyre, I., Emson, P. C., and Tohyama, M. 1986. Two distinct calcitonin gene-related peptide-containing peripheral nervous systems: Distribution and quantitative differences between the iris and cerebral artery with special reference to substance P. *Brain Res.* **373**, 205–212.

Matthes, H. W. D., Maldonado, R., Simonin F., Valverde, O., Slowe, S., *et al.* 1996. Loss of morphine-induced analgesia, reward effect and withdrawal symptoms in mice lacking the μ-opioid-receptor gene. *Nature* **383**, 819–823.

Matthews, M. A., McDonald, G. K., and Hernandez, T. V. 1988. GABA distribution in a pain-modulating zone of trigeminal subnucleus interpolaris. *Somatosens. Mot. Res.* **5**, 205–217.

Matthews, M. A., Hernandez, T. V., Hoffmann, K. D., Romanska, A. I., and Liles, S. L. 1989. Synaptic substrates for enkephalinergic and serotoninergic interactions with dental primary afferent terminals in trigeminal subnucleus interpolaris: An immunocytochemical study using peroxidase and colloidal gold. *Synapse* **4**, 175–195.

Matthews, P. B. C. 1964. Muscle spindles and their motor control. *Physiol. Rev.* **44**, 219–288.

Matthews, P. B. C. 1972. *Mammalian Muscle Receptors and Their Central Actions*. Williams & Wilkins, Baltimore.

Matthews, P. B. C. 1977. Muscle afferents and kinaesthesia. *Br. Med. Bull.* **33**, 137–142.

Matthews, P. B. C. 1981. Evolving views on the internal operation and functional role of the muscle spindle. *J. Physiol.* **320**, 1–30.

Matzner, O. and Devor, M. 1994. Hyperexcitability at sites of nerve injury depends on voltage-sensitive Na^+ channels. *J. Neurophysiol.* **72**, 349–359.

Maubert, E., Slama, A., Ciofl, P., Viollet, C., Tramu, G., Dupouy, J. P., and Epelbaum, J. 1994. Developmental patterns of somatostatin-receptors and somatostatin-immunoreactivity during early neurogenesis in the rat. *Neuroscience* **62**, 317–325.

Maurin, Y., Buck, S. H., Wamsley, J. K., Burks, T. F., and Yamamura, H. I. 1984. Light microscopic autoradiographic localization of [^{3}H] substance P binding sites in rat thoracic spinal cord. *Life Sci.* **34**, 1713–1716.

Maves, T. J., Pechman, P. S., Gebhart, G. F., and Meller, S. T. 1993. Possible chemical contribution from chromic gut sutures produces disorders of pain sensation like those seen in man. *Pain* **54**, 57–69.

Mawe, G. M., Bresnahan, J. C., and Beattie, M. S. 1984. Primary afferent projections from dorsal and ventral roots to autonomic preganglionic neurons in the cat sacral spinal cord: Light and electron microscopic observations. *Brain Res.* **290**, 152–157.

Maxwell, D. J. 1985. Combined light and electron microscopy of Golgi-labelled neurons in lamina III of feline spinal cord. *J. Anat.* **141**, 155–169.

Maxwell, D. J. and Bannatyne, B. A. 1983. Ultrastructure of muscle spindle afferent terminations in lamina VI of the cat spinal cord. *Brain Res.* **288**, 297–301.

Maxwell, D. J. and Jankowska, E. 1996. Synaptic relationships between serotonin-immunoreactive axons and dorsal horn spinocerebellar tract cells in the cat spinal cord. *Neuroscience* **70**, 247–253.

Maxwell, D. J. and Noble, R. 1987. Relationships between hair-follicle afferent terminations and glutamic acid decarboxylase-containing boutons in the cat's spinal cord. *Brain Res.* **408**, 308–312.

Maxwell, D. J. and Rethelyi, M. 1987. Ultrastructure and synaptic connections of cutaneous afferent fibres in the spinal cord. *TINS* **10**, 117–122.

Maxwell, D. J. and Riddell, J. S. 1999. Axoaxonic synapses on terminals of group II muscle spindle afferent axons in the spinal cord of the cat. *Eur. J. Neurosci.* **11**, 2151–2159.

Maxwell, D. J., Fyffe, R. E. W., and Brown, A. G. 1982. Fine structure of spinocervical tract neurones and the synaptic boutons in contact with them. *Brain Res.* **233**, 394–399.

Maxwell, D. J., Fyffe, R. E. W., and Rethelyi, M. 1983. Morphological properties of physiologically characterized lamina III neurones in the cat spinal cord. *Neuroscience* **10**, 1–22.

Maxwell, D. J., Bannatyne, B. A., Fyffe, R. E. W., and Brown, A. G. 1984. Fine structure of primary afferent axon terminals projecting from rapidly adapting mechanoreceptors of the toe and foot pads of the cat. *Q. J. Exp. Physiol.* **69**, 381–392.

Maxwell, D. J., Koerber, H. R., and Bannatyne, B. A. 1985a. Light and electron microscopy of contacts between primary afferent fibres and neurones with axons ascending the dorsal columns of the feline spinal cord. *Neuroscience* **16**, 375–394.

Maxwell, D. J., Leranth, C., and Verhofstad, A. A. 1985b. Synaptic arrangements formed by serotonin-immunoreactive axons in the substantia gelatinosa of the rat spinal cord. *Q. J. Exp. Physiol* **70**, 377–388.

Maxwell, D. J., Christie, W. M., Short, A. D., and Brown, A. G. 1990a. Direct observations of synapses between GABA-immunoreactive boutons and muscle afferent terminals in lamina VI of the cat's spinal cord. *Brain Res.* **530**, 215–222.

Maxwell, D. J., Christie, W. E., Short, A. D., Storm-Mathisen, J., and Ottersen, O. P. 1990b. Central boutons of glomeruli in the spinal cord of the cat are enriched with L-glutamate-like immunoreactivity. *Neuroscience* **36**, 83–104.

Maxwell, D. J., Christie, W. M., Ottersen, O. P., and Storm-Mathisen, J. 1990c. Terminals of group Ia primary afferent fibres in Clarke's column are enriched with L-glutamate-like immunoreactivity. *Brain Res.* **510**, 346–350.

Maxwell, D. J., Christie, W. M., Short, A. D., and Brown, A. G. 1991. Direct observations of synapses between GABA-immunoreactive boutons and identified spinocervical tract neurons in the cat's spinal cord. *J. Comp. Neurol.* **307**, 375–392.

Maxwell, D. J., Christie, W. M., Brown, A. G., Ottersen, O. P., and Storm-Mathisen, J. 1993. Identified hair follicle afferent boutons in the spinal cord of the cat are enriched with L-glutamate-like immunoreactivity. *Brain Res.* **606**, 156–161.

Maxwell, D. J., Ottersen, O. P., and Storm-Mathisen, J. 1995a. Synaptic organization of excitatory and inhibitory boutons associated with spinal neurons which project through the dorsal columns of the cat. *Brain Res.* **676**, 103–112.

Maxwell, D. J., Todd, A. J., and Kerr, R. 1995b. Colocalization of glycine and GABA in synapses on spinomedullary neurons. *Brain Res.* **690**, 127–132.

Maxwell, D. J., Kerr, R., Jankowska, E., and Riddell, J. S. 1997. Synaptic connections of dorsal horn group II spinal interneurons: Synapses formed with the interneurons and by their axon collaterals. *J. Comp. Neurol.* **380**, 51–69.

Maxwell, L., Maxwell, D. J., Neilson, M., and Kerr, R. 1996. A confocal microscopic survey of serotoninergic axons in the lumbar spinal cord of the rat: Co-localization with glutamate decarboxylase and neuropeptides. *Neuroscience* **75**, 471–480.

Mayer, E. A., Naliboff, B., and Munakata, J. 2000. The evolving neurobiology of gut feelings. In E. A. Mayer and C. B. Saper (eds.), *The Biological Basis for Mind Body Interactions* (*Prog. Brain Res.* **122**, 195–206). Elsevier, Amsterdam.

Mayer, M. L. and Westbrook, G. L. 1987. The physiology of excitatory amino acids in the vertebrate central nervous system. *Prog. Neurobiol.* **28**, 197–276.

Mayer, M. L., Westbrook, G. L. and Gurthrie, P. B. 1984. Voltage-dependent block by Mg^{++} of NMDA responses in spinal cord neurons. *Nature* **309**, 261–263.

Mayeaux, V. and Valmier, J. 1995. Skeletal muscle contraction modulates carbonic anhydrase phenotype in adult mouse dorsal root ganglion neurons. *Brain Res.* **694**, 191–199.

Mayeux, V., Pons, F., Baldy-Moulinier, M., and Valmier, J. 1996. Early postnatal muscle contractile activity regulates the carbonic anhydrase phenotype of proprioceptive neurons in young and mature mice: Evidence for a critical period in development. *Neuroscience* **71**, 787–795.

Maynard, C. J., Leonard, R. B., Coulter, J. D., and Coggeshall, R. E. 1977. Central connections of ventral root afferents as demonstrated by the HRP method. *J. Comp. Neurol.* **172**, 601–608.

McAdoo, D. J., Xu, G. Y., Robak, G., Hughes, M. G., and Price, E. M. 2000. Evidence that reversed glutamate uptake contributes significantly to glutamate release following experimental injury to the rat spinal cord. *Brain Res.* **865**, 283–285.

McCall, W. D., Farias, M. C., Williams, W. J., and BeMent, S. L. 1974. Static and dynamic responses of slowly adapting joint receptors. *Brain Res.* **70**, 221–243.

McCarson, K. E. 1999. Central and peripheral expression of neurokinin-1 and neurokinin-3 receptor and substance P-encoding messenger RNAs: Peripheral regulation during formalin-induced inflammation and lack of neurokinin receptor expression in primary afferent sensory neurons. *Neuroscience* **93**, 361–370.

McCarson, K. E. and Enna, S. J. 1999. Nociceptive regulation of GABA(B) receptor gene expression in rat spinal cord. *Neuropharmacology* **38**, 1767–1773.

McCarson, K. E. and Goldstein, B. D. 1991. Release of substance P into the superficial dorsal horn following nociceptive activation of the hindpaw of the rat. *Brain Res.* **568**, 109–115.

McCarson, K. E. and Krause, J. E. 1994. NK-1 and NK-3 type tachykinin receptor mRNA expression in the rat spinal cord dorsal horn is increased during adjuvant or formalin-induced nociception. *J. Neurosci.* **14**, 712–719.

McCarson, K. E. and Krause, J. E. 1995. The formalin-induced expression of tachykinin peptide and neurokinin receptor messenger RNAs in rat sensory ganglia and spinal cord is modulated by opiate preadministration. *Neuroscience* **64**, 729–739.

McCarson, K. E. and Krause, J. E. 1996. The neurokinin-1 receptor antagonist LY306,740 blocks nociception-induced increases in dorsal horn neurokinin-1 receptor gene expression. *Mol. Pharmacol.* **50**, 1189–1199.

McCarter, G. C., Reichling, D. B., and Levine, J. D. 1999. Mechanical transduction by rat dorsal root ganglion neurons in vitro. *Neurosci. Lett.* **273**, 179–182.

McCarthy, P. W. and Lawson, S. N. 1990. Cell type and conduction velocity of rat primary sensory neurons with calcitonin gene-related peptide-like immunoreactivity. *Neuroscience* **34**, 623–632.

McCarthy, P. W. and Lawson, S. N. 1997. Differing action potential shapes in rat dorsal root ganglion neurones related to their substance P and calcitonin gene-related peptide immunoreactivity. *J. Comp. Neurol.* **388**, 541–549.

McCarthy, P. W., Prabhaker, E., and Lawson, S. N. 1995. Evidence to support the peripheral branching of primary afferent C-fibres in the rat: An in vitro intracellular electrophysiological study. *Brain Res.* **704**, 79–84.

McCleskey, E. W. and Gold, M. S. 1999. Ion channels of nociception. *Ann. Rev. Physiol.* **61**, 835–856.

McConalogue, K., Grady, E. F., Minnis, J., Balestra, B., Tonini, M., Brecha, N. C., Bunnett, N. W., and Sternini, C. 1999. Activation and internalization of the mu-opioid receptor by the newly discovered endogenous agonists, endomorphin-1 and endomorphin-2. *Neuroscience* **90**, 1051–1059.

McDougal Jr., D. B., McDougal, S. H., and Johnson, E. M. 1985. Effect of capsaicin upon fluoride sensitive acid phosphatases in selected ganglia and spinal cord and upon neuronal size and number in dorsal root ganglion. *Brain Res.* **331**, 63–70.

McGeer, P. L., Eccles, J. C., and McGeer, E. G. 1978. *Molecular Neurobiology of the Mammalian Brain* (pp. 141–142, 171–172). Plenum Press, New York.

McGregor, G. P., Gibson, S. J., Sabate, I. M., Blank, M. A., Christofides, N. D., Wall, P. D., Polak, J. M., and Bloom, S. R. 1984. Effect of peripheral nerve section and nerve crush on spinal cord neuropeptides in the rat: Increased VIP and PHI in the dorsal horn. *Neuroscience* **13**, 207–216.

McIntyre, A. K., Holman, M. E., and Veale, J. L. 1967. Cortical responses to impulses from single Pacinian corpuscles in the cat's hindlimb. *Exp. Brain Res.* **4**, 243–255.

McIntyre, A. K., Proske, U., and Tracey, D. J. 1978. Afferent fibres from muscle receptors in the posterior nerve of the cat's knee joint. *Exp. Brain Res.* **33**, 415–424.

McKemy, D. D., Neuhausser, W. M., and Julius, D. 2002. Identification of a cold receptor reveals a general role for TRP channels in thermosensation. *Nature* **416**, 52–58.

McLachlan, E. M. and Hu, P. 1998. Axonal sprouts containing calcitonin gene-related peptide and substance P form pericellular baskets around large diameter neurons after sciatic nerve transection in the rat. *Neuroscience* **84**, 961–965.

McLachlan, E. M. and Janig, W. 1983. The cell bodies of origin of sympathetic and sensory axons in some skin and muscle nerves of the cat hindlimb. *J. Comp. Neurol.* **214**, 115–130.

McLachlan, E. M., Jänig, W., Devor, M., and Michaelis, M. 1993. Peripheral nerve injury triggers noradrenergic sprouting within dorsal root ganglia. *Nature* **363**, 543–546.

McLaughlin, B. J., Barber, R., Saito, K., Roberts, E., and Wu, J. Y. 1975. Immunocytochemical localization of glutamate decarboxylase in rat spinal cord. *J. Comp. Neurol.* **164**, 305–322.

McLennan, H. 1983. Receptors for the excitatory amino acids in the mammalian central nervous system. *Prog. Neurobiol.* **20**, 251–271.

McLennan, H. and Liu, J. R. 1982. The action of six antagonists of the excitatory amino acids on neurones of the rat spinal cord. *Exp. Brain Res.* **45**, 151–156.

McLennan, H. and Lodge, D. 1979. The antagonism of amino acid-induced excitation of spinal neurones in the cat. *Brain Res.* **169**, 83–90.

McLeod, A. L., Krause, J. E., Cuello, A. C., and Ribeiro-da-Silva, A. 1998. Preferential synaptic relationship between substance P- immunoreactive boutons and neurokinin 1 receptors sites in the rat spinal cord. *Proc. Natl. Acad. Sci. USA* **95**, 15775–15780.

McLeod, A. L., Julien, J. P., Cuello, A. C., Krause, J. E., Ribeiro-da-Silva, A. 1999. Ectopic substance P-immunoreactive boutons are preferentially presynaptic to neurokinin-1 receptor immunoreactive dendrites in the spinal white matter of transgenic mice. *Brain Res.* **836**, 1–8.

McLeod, A. L., Krause, J. E., and Ribeiro-da-Silva, A. 2000. Immunocytochemical localization of neurokinin B in the rat spinal dorsal horn and its association with substance P and GABA: An electron microscopic study. *J. Comp. Neurol.* **420**, 349–362.

McMahon, S. B. 1986. The localization of fluoride-resistant acid phosphatase (FRAP) in the pelvic nerves and sacral spinal cord of rats. *Neurosci. Lett.* **64**, 305–310.

McMahon, S. B. and Gibson, S. 1987. Peptide expression is altered when afferent nerves reinnervate inappropriate tissue. *Neurosci. Lett.* **73**, 9–15.

McMahon, S. B. and Kett-White, R. 1991. Sprouting of peripherally regenerating primary sensory neurones in the adult central nervous system. *J. Comp. Neurol.* **304**, 307–315.

McMahon, S. B. and Moore, C. E. G. 1988. Plasticity of primary afferent acid phosphatase expression following rerouting of afferents from muscle to skin in the adult rat. *J. Comp. Neurol.* **274**, 1–8.

McMahon, S. B. and Morrison, J. F. B. 1982. Two groups of spinal interneurones that respond to stimulation of the abdominal viscera of the cat. *J. Physiol.* **322**, 21–34.

McMahon, S. B. and Wall, P. D. 1983. A system of rat spinal cord lamina I cells projecting through the contralateral dorsolateral funiculus. *J. Comp. Neurol.* **214**, 217–223.

McMahon, S. B. and Wall, P. D. 1984. Receptive fields of rat lamina I projection cells move to incorporate a nearby region of injury. *Pain* **19**, 235–247.

McMahon, S. B. and Wall, P. D. 1987. Physiological evidence for branching of peripheral unmyelinated sensory afferent fibers in the rat. *J. Comp. Neurol.* **261**, 130–136.

McMahon, S. B., Sykova, E., Wall, P. D., Woolf, C. J., and Gibson, S. J. 1984. Neurogenic extravasation and substance P levels are low in muscle as compared to skin in the rat hindlimb. *Neurosci. Lett.* **52**, 235–240.

McMahon, S. B., Armanini, M. P., Ling, L. H., and Phillips, H. S. 1994. Expression and coexpression of Trk receptors in subpopulations of adult primary sensory neurons projecting to identified peripheral targets. *Neuron* **12**, 1161–1171.

McNeill, D. L. and Burden, H. W. 1986. Convergence of sensory processes from the heart and left ulnar nerve onto a single afferent perikaryon: A neuroanatomical study in the rat employing fluorescent tracers. *Anat. Rec.* **214**, 441–444.

McNeill, D. L. and Burden, H. W. 1987. Neuropeptides in sensory perikarya projecting to the rat ovary. *Am. J. Anat.* **179**, 269–276.

McNeill, D. L. and Hulsebosch, C. E. 1987. Intraspinal sprouting of rat primary afferents after deafferentation. *Neurosci. Lett.* **81**, 57–62.

McNeill, D. L., Coggeshall, R. E., and Carlton, S. M. 1988a. A light- and electron-microscopic study of calcitonin gene-related peptide in the spinal cord of the rat. *Exp. Neurol.* **99**, 699–708.

McNeill, D. L., Chung, K., Carlton, S. M., and Coggeshall, R. E. 1988b. Calcitonin gene-related peptide immunostained axons provide evidence for fine primary afferent fibers in dorsal and dorsolateral funiculi of rat spinal cord. *J. Comp. Neurol.* **272**, 303–308.

McNeill, D. L., Westlund, K. N., and Coggeshall, R. E. 1989. Peptide immunoreactivity of unmyelinated primary afferent axons in rat lumbar dorsal roots. *J. Histochem. Cytochem.* **37**, 1047–1052.

McNeill, D. L., Papka, R. E., and Harris, C. H. 1992a. CGRP immunoreactivity and NADPH-diaphorase in afferent nerves of the rat penis. *Peptides* **13**, 1239–1246.

McNeill, D. L., Traugh, N. E., Jr., Vaidya, A. M., Hua, H. T., and Papka, R. E. 1992b. Origin and distribution of NADPH-diaphorase-positive neurons and fibers innervating the urinary bladder of the rat. *Neurosci Lett.* **147**, 33–36.

McNeill, M. E. and Norvell, J. E. 1978. Acetylcholinesterase activity of primary sensory neurons and dorsal root fibers in the cat. *Anat. Rec.* **190**, 155–160.

McRoberts, J. A., Coutinho, S. V., Marvizon, J. C., Grady, E. F., Tognetto, M., Sengupta, J. N., Ennes, H. S., Chaban, V. V., Amadesi, S., Creminon, C., Lanthorn, T., Geppetti, P., Bunnett, N. W., and Mayer, E. A. 2001. Role of peripheral N-methyl-D-aspartate (NMDA) receptors in visceral nociception in rats. *Gastroenterology* **120**, 1737–1748.

Meissner, G. 1859. Untersuchungen über den Tastsinn. *Z. Rat. Med.* **7**, 92–118.

Melander, T., Hökfelt, T., and Rokaeus, A. 1986. Distribution of galaninlike immunoreactivity in the rat central nervous system. *J. Comp. Neurol.* **248**, 475–517.

Meller, S. T. and Gebhart, G. F. 1993. Nitric oxide (NO) and nociceptive processing in the processing in the spinal cord. *Pain* **52**, 127–136.

Meller, S. T., Dykstra, C., and Gebhart, G. F. 1993. Acute mechanical hyperalgesia in the rat is produced by coactivation of ionotropic AMPA and metabotropic glutamate receptors. *NeuroReport* **4**, 879–882.

Meller, S. T., Cummings, C. P., Traub, R. J., and Gebhart, G. F. 1994. The role of nitric oxide in the development and maintenance of the hyperalgesia produced by intraplantar injection of carrageenan in the rat. *Neuroscience* **60**, 367–374.

Meller, S. T., Dykstra, C., and Gebhart, G. F. 1996. Acute mechanical hyperalgesia in the rat can be produced by coactivation of spinal ionotropic AMPA and metabotropic glutamate receptors, activation of phospholipase A2 and generation of cyclooxygenase products. *Prog. Brain Res.* **110**, 177–192.

Melzack, R. 1975. Prolonged relief of pain by brief, intense transcutaneous somatic stimulation. *Pain* **1**, 357–373.

Melzack, R. and Wall, P. D. 1962. On the nature of cutaneous sensory mechanisms. *Brain* **85**, 331–356.

Melzack, R. and Wall, P. D. 1965. Pain mechanisms: A new theory. *Science* **150**, 971–979.

Ménard, D. P., van Rossum, D., Kar, S., and Quirion, R. 1995a. Alteration of calcitonin gene-related peptide and its receptor binding sites during the development of tolerance to μ and δ opioids. *Can. J. Physiol. Pharmacol.* **73**, 1089–1095.

Ménard, D. P., van Rossum, D., Kar, S., Jolicoeur, F. B., Jhamandas, K., and Quirion, R. 1995b. Tolerance to the antinociceptive properties of morphine in the rat spinal cord: Alteration of calcitonin gene-related peptide-like immunostaining and receptor binding sites. *J. Pharmacol. Exp. Ther.* **273**, 887–894.

Mendell, L. M. 1966. Physiological properties of unmyelinated fiber projection to the spinal cord. *Exp. Neurol.* **16**, 316–332.

Mendell, L. M. 1970. Positive dorsal root potentials produced by stimulation of small diameter muscle afferents. *Brain Res.* **18**, 375–379.

Mendell, L. 1972. Properties and distribution of peripherally evoked presynaptic hyperpolarization in cat lumbar spinal cord. *J. Physiol.* **226**, 769–792.

Mendell, L. M. 1973. Two negative dorsal root potentials evoke a positive dorsal root potential. *Brain Res.* **55**, 198–202.

Mendell, L. M. and Wall, P. D. 1964. Presynaptic hyperpolarization: A role for fine afferent fibres. *J. Physiol.* **172**, 274–294.

Mendell, L. and Wall, P. D. 1965. Responses of single dorsal cord cells to peripheral cutaneous unmyelinated fibres. *Nature* **206**, 97–99.

Mendell, L. M., Sassoon, E. M., and Wall, P. D. 1978. Properties of synaptic linkage from long-ranging afferents onto dorsal horn neurones in normal and deafferented cats *J. Physiol.* **285**, 299–310.

Mendelson, B., Albers, K. M., Goodness, T. P., and Davis, B. M. 1996. Overexpression of nerve growth factor in epidermis of transgenic mice preserves excess sensory neurons but does not alter the somatotopic organization of cutaneous nerve projections. *Neurosci. Lett.* **211**, 68–72.

Menétrey, D., Giesler, G. J., and Besson, J. M. 1977. An analysis of response profiles of spinal cord dorsal horn neurones to nonnoxious and noxious stimulion in the spinal rat. *Exp. Brain Res.* **27**, 15–33.

Menétrey, D., Gannon, J. D., Levine, J. D., and Basbaum, A. I. 1989. Expression of c-fos protein in interneurons and projection neurons of the rat spinal cord in response to noxious somatic, articular, and visceral stimulation. *J. Comp. Neurol.* **285**, 177–195.

Menétrey, D., de Pommery, J., Baimbridge, K. G., and Thomasset, M. 1992. Calbindin-D28K (CaBP28k)-like immunoreactivity in ascending projections. *Eur. J. Neurosci* **4**, 61–69.

Mense, S. 1977a. Muscular nociceptors. *J. Physiol. (Paris)* **73**, 233–240.

Mense, S. 1977b. Nervous outflow from skeletal muscle following chemical noxious stimulation. *J. Physiol.* **267**, 75–88.

Mense, S. 1981. Sensitization of group IV muscle receptors to bradykinin by 5-hydroxytryptamine and prostaglandin E2. *Brain Res.* **225**, 95–105.

Mense, S. 1996. Group III and IV receptors in skeletal muscle: Are they specific or polymodal? In T. Kumazawa, L. Kruger, and K. Mizumura (eds.), *The Polymodal Receptor—A Gateway to Pathological Pain* (*Prog. Brain Res.* **113**, 83–100). Elsevier, Amsterdam.

Mense, S. and Craig, J. R. 1988. Spinal and supraspinal terminations of primary afferent fibers from the gastrocnemius-soleus muscle in the cat. *Neuroscience* **26**, 1023–1055.

Mense, S. and Meyer, H. 1985. Different types of slowly conducting afferent units in cat skeletal muscle and tendon. *J. Physiol.* **363**, 403–417.

Mense, S. and Meyer, H. 1988. Bradykinin-induced modulation of the response behaviour of different types of feline group III and IV muscle receptors. *J. Physiol.* **398**, 49–63.

Mense, S. and Prabhakar, N. R. 1986. Spinal termination of nociceptive afferent fibres from deep tissues in the cat. *Neurosci. Lett.* **66**, 169–174.

Mense, S. and Schmidt, R. F. 1974. Activation of group IV afferent units from muscle by algesic agents. *Brain Res.* **72**, 305–310.

Mense, S. and Stahnke, M. 1983. Responses in muscle afferent fibres of slow conduction velocity to contractions and ischaemia in the cat. *J. Physiol.* **342**, 383–397.

Mense, S., Light, A. R., and Perl, R. 1980. Spinal terminations of subcutaneous high threshold mechanoreceptors, In A. G. Brown and M. Rethelyi (eds.), Spinal Cord Sensation. Scottish Academic Press, Edinburgh.

Merchenthaler, I. 1984. Corticotrophin releasing factor (CRF)-like immunoreactivity in the rat central nervous system. Extrahypothalamic distribution. *Peptides* **5**, suppl. 53–69.

Merchenthaler, I., Hynes, M. A., Vigh, S., Shally, A. V., and Petrusz, P. 1983. Immunocytochemical localization of corticotropin releasing factor (CRF) in the rat spinal cord. *Brain Res.* **275**, 373–377.

Merighi, A., Polak, J. M., Gibson, S. J., Gulbenkian, S., Valentino, K. L., and Peirone, S. M. 1988. Ultrastructural studies on calcitonin gene-related peptide-, tachykinins- and somatostatin-immunoreactive neurones in rat dorsal root ganglia: Evidence for the colocalization of different peptides in single secretory granules. *Cell Tiss. Res.* **254**, 101–109.

Merighi, A., Polak, J. M., Fumagalli, G., and Theodosis, D. T. 1989. Ultrastructural localization of neuropeptides and GABA in rat dorsal horn: A comparison of different immunogold labeling techniques. *J. Histochem. Cytochem.* **37**, 529–540.

Merighi, A., Kar, S., Gibson, S. J., Ghidella, S., Gobetto, A., Peirone, S. M., and Polak, J. M. 1990. The immunocytochemical distribution of seven peptides in the spinal cord and dorsal root ganglia of horse and pig. *Anat. Embryol.* **181**, 271–280.

Merighi, A., Polak, J. M., and Theodosis, D. T. 1991. Ultrastructural visualization of glutamate and aspartate immunoreactivities in the rat dorsal horn, with special reference to the co-localization of glutamate, substance P and calcitonin gene-related peptide. *Neuroscience* **40**, 67–80.

Merighi, A., Cruz, F., and Coimbra, A. 1992. Immunocytochemical staining of neuropeptides in terminal arborization of primary afferent fibers anterogradely labeled and identified at light- and electron-microscopic levels. *J. Neurosci. Meth.* **42**, 105–113.

Merkel, F. 1875. Tastzellen und Tastkoerperchen bei den Hausthieren und beim Menschen. *Arch. Mikroskop. Anat.* **11**, 636–652.

Merrill, E. G. and Wall, P. D. 1972. Factors forming the edge of a receptive field: The presence of relatively ineffective afferent terminals. *J. Physiol.* **226**, 825–846.

Merskey, H. and Bogduk, N. (eds.) 1994. *Classification of Chronic Pain: Descriptions of Chronic Pain Syndromes and Definitions of Pain Terms.* (2nd ed.) IASP Press, Seattle.

Merzenich, M. M. and Harrington, T. 1969. The sense of flutter-vibration evoked by stimulation of the hairy skin of primates: Comparison of human sensory capacity with the responses of mechanoreceptive afferents innervating the hairy skin of monkeys. *Exp. Brain Res.* **9**, 236–260.

Messersmith, D. J., Kim, D. J., and Iadarola, M. J. 1998. Transcription factor regulation of prodynorphin gene expression following rat hindpaw inflammation. *Mol. Brain Res.* **53**, 239–269.

Messlinger, K., Schepelmann, K., Pawlak, M., and Schmidt, R. F. 1993. Bradykinin B_1 and B_2 receptor antagonists do not change the ongoing activity of slowly conducting articular afferents in the inflamed knee joint of the cat. *Neurosci. Lett.* **164**, 21–24.

Messlinger, K., Pawlak, M., Schepelmann, K., and Schmidt, R. F. 1994. Responsiveness of slowly conducting articular afferents to bradykinin: Effects of an experimental arthritis. *Pain* **59**, 335–343.

Meunier, J. C., Mollereau, C., Toll, L., Suaudeau, C., Moisand, C., Alvinerie, P., Butour, J. L., Guillemot, J. C., Ferrara, P., Monsarrat, B., Mazarguil, H., Vassart, G., Parmentier, M., Costentin, J. 1995. Isolation and structure of the endogenous agonist of opioid receptor-like ORL1 receptor. *Nature* **377**, 532–535.

Meyer, R. A. and Campbell, J. N. 1981. Myelinated nociceptive afferents account for the hyperalgesia that follows a burn to the hand. *Science* **213**, 1527–1529.

Meyer, R. A., Campbell, J. N., and Raja, S. N. 1988. Antidromic nerve stimulation in monkey does not sensitize unmyelinated nociceptors to heat. *Brain Res.* **441**, 168–172.

Meyer, R. A., Davis, K. D., Cohen, R. H., Treede, R. D., and Campbell, J. N. 1991. Mechanically insensitive afferents (MIAs) in cutaneous nerves of monkey. *Brain Res.* **561**, 252–261.

Meyers, D. E. R. and Snow, P. J. 1984. Somatotopically inappropriate projections of single hair follicle afferent fibres to the cat spinal cord. *J. Physiol.* **347**, 59–73.

Meyers, D. E. R. and Snow, P. J. 1986. Distribution of activity in the spinal terminations of single hair follicle afferent fibers to somatotopically identified regions of the cat spinal cord. *J. Neurophysiol.* **56**, 1022–1038.

Meyers, D. E. R., Wilson, P., and Snow, P. J. 1984. Distribution of the central terminals of cutaneous primary afferents innervating a small skin patch: The existence of somatotopically inappropriate projections. *Neurosci. Lett.* **44**, 179–185.

Mezey, E., Toth, Z. E., Cortright, D. N., Arzubi, M. K., Krause, J. E., Elde, R., Guo, A., Blumberg, P. M., and Szallasi, A. 2000. Distribution of mRNA for vanilloid receptor subtype 1 (VR1), and VR1- like immunoreactivity, in the central nervous system of the rat and human. *Proc. Natl. Acad. Sci. USA* **97**, 3655–3660.

Micevych, P. E., Stroink, A., Yaksh, T., and Go, V. L. W. 1986. Immunochemical studies of substance P and cholecystokinin octapeptide recovery in dorsal horn following unilateral lumbosacral ganglionectomy. *Somatosens. Mot. Res.* **3**, 239–260.

Michael, G. J. and Priestley, J. V. 1999. Differential expression of the mRNA for the vanilloid receptor subtype 1 in cells of the adult rat dorsal root and nodose ganglia and its downregulation by axotomy. *J. Neurosci.* **19**, 1844–1854.

Michael, G. J., Averill, S., Nitkuna, A., Rattray, M., Bennett, D. L. H., Yan, Q., and Priestley, J. V. 1997a. Nerve growth factor treatment increases brain-derived neurotrophic factor selectively in TrkA-expressing dorsal root ganglion cells and in their central terminations within the spinal cord. *J. Neurosci.* **17**, 8476–8490.

Michael, G. J., Kaya, E., Averill, S., Rattray, M., Clary, D. O., and Priestley, J. V. 1997b. TrkA immunoreactivity neurons in the rat spinal cord. *J. Comp. Neurol.* **385**, 441–455.

Michael, G. J., Averill, S., Shortland, P. J., Yan, Q., and Priestley, J. V. 1999. Axotomy results in major changes in BDNF expression by dorsal root ganglion cells: BDNF expression in large trkB and trkC cells, in pericellular baskets, and in projections to deep dorsal horn and dorsal column nuclei. *Eur. J. Neurosci.* **11**, 3539–3551.

Michaelis, M., Devor, M., and Jänig, W. 1996a. Sympathetic modulation of activity in rat dorsal root ganglion neurons changes over time following peripheral nerve injury. *J. Neurophysiol.* **76**, 753–763.

Michaelis, M., Häbler, H. J., and Jänig, W. 1996b. Silent afferents: A separate class of primary afferents? *Clin. Exp. Pharmacol. Physiol.* **23**, 99–105.

Migita, K., Loewy, A. D., Ramabhadran, T. V., Krause, J. E., and Waters, S. M. 2001. Immunohistochemical localization of the neuropeptide Y Y1 receptor in rat central nervous system. *Brain Res.* **889**, 23–37.

Mihaly, A., Priestley, J. V., and Molnar, E. 1996. Expression of *raf* serine/threonine protein kinases in the cell bodies of primary sensory neurons of the adult rat. *Cell Tissue Res.* **285**, 261–271.

Miki, K., Fukuoka, T., Tokunaga, A., and Noguchi, K. 1998a. Calcitonin gene-related peptide increase in the rat spinal dorsal horn and dorsal column nucleus following peripheral nerve injury: Up-regulation in a subpopulation of primary afferent sensory neurons. *Neuroscience* **82**, 1243–1252.

Miki, K., Iwata, K., Tsuboi, R., Fukuoka, T., Tachibana, T., Tokunaga, A., and Noguchi, K. 1998b. Responses of dorsal column nuclei neurons in rats with experimental mononeuropathy. *Pain* **76**, 407–415.

Miki, K., Iwata, K., Tsuboi, Y., Morimoto, T., Kondo, E., Dai, Y., Ren, K., and Noguchi, K. 2000. Dorsal column-thalamic pathway is involved in thalamic hyperexcitability following peripheral nerve injury: A lesion study in rats with experimental mononeuropathy. *Pain* **85**, 263–271.

Miletic, V. and Randic, M. 1979. Neurotensin excites cat spinal neurones located in laminae I–III. *Brain Res.* **169**, 600–604.

Miletic, V. and Randic, M. 1982. Neonatal rat spinal cord slice preparation: Postsynaptic effects of neuropeptides on dorsal horn neurons. *Dev. Brain Res.* **2**, 432–438.

Miletic, V. and Tan, H. 1988. Iontophoretic application of calcitonin gene-related peptide produces a slow and prolonged excitation of neurons in the cat lumbar dorsal horn. *Brain Res.* **446**, 169–172.

Miletic, V., Hoffert, M. J., Ruda, M. A., Dubner, R., and Shigenaga, Y. 1984. Serotoninergic axonal contacts on identified cat spinal dorsal horn neurons and their correlation with nucleus raphe magnus stimulation. *J. Comp. Neurol.* **228**, 129–141.

Miletic, V., Bowen, K. K., and Miletic, G. 2000. Loose ligation of the rat sciatic nerve is accompanied by changes in the subcellular content of protein kinase C beta II and gamma in the spinal dorsal horn. *Neurosci Lett.* **288**, 199–202.

Millan, M. J. 2002. Descending control of pain. *Prog. Neurobiol.* **66**, 355–474.

Millan, M. J., Tsand, Y. F., Przewlocki, R., Hollt, V., and Herz, A. 1981. The influence of foot-shock stress upon brain, pituitary and spinal cord pools of immunoreactive dynorphin in rats. *Neurosci. Lett.* **24**, 75–79.

Millan, M. J., Millan, M. H., Czlonkowski, A., and Herz, A. 1984. Vasopressin and oxytocin in the rat spinal cord: Distribution and origins in comparison to [met]enkephalin, dynorphin and related opioids and their irresponsiveness to stimuli modulating neurohypophyseal secretion. *Neuroscience* **13**, 178–187.

Millan, M. J., Millan, M. H., Pilcher, C. W., Czlonkowski, A., Herz, A., and Colpaert, F. C. 1985. Spinal cord dynorphin may modulate nociception via a kappa-opioid receptor in chronic arthritic rats. *Brain Res.* **340**, 156–159.

Millan, M. J., Millan, M. H., Czlonkowski, A., Hollt, V., Pilcher, W. T., Herz, A., and Colpaert, F. C. 1986. A model of chronic pain in the rat: Response of multiple opioid systems to adjuvant-induced arthritis. *J. Neurosci.* **6**, 899–906.

Millan, M. J., Czlonkowski, A., Pilcher, C. W., Almeida, O. F., Millan, M. H., Colpaert, F. C., and Herz, A. 1987. A model of chronic pain in the rat: Functional correlates of alterations in the activity of opioid systems. *J Neurosci.* **7**, 77–87.

Millan, M. J., Czlonkowski, A., Morris, B., Stein, C., Arendt, R., Huber, A., Höllt, V., and Herz, A. 1988. Inflammation of the hind limb as a model of unilateral localized pain: Influence on multiple opioid systems in the spinal cord of the rat. *Pain* **35**, 299–312.

Millar, J. 1973. Joint afferent fibres responding to muscle stretch, vibration and contraction. *Brain Res.* **63**, 380–383.

Millar, J. 1975. Flexion-extension sensitivity of elbow joint afferents in cat. *Exp. Brain Res.* **24**, 209–214.

Millar, J. and Armstrong-James, M. 1982. The responses of neurones of the superficial dorsal horn to iontophoretically applied glutamate ion. *Brain Res.* **231**, 267–277.

Millar, J. and Pelling, C. W. 2001. Improved methods for construction of carbon fibre electrodes for extracellular spike recording. *J. Neurosci. Meth.* **110**, 1–8.

Millard, C. L. and Woolf, C. J. 1988. Sensory innervation of the hairs of the rat hindlimb: A light microscopic analysis. *J. Comp. Neurol.* **277**, 183–194.

Miller, B. A. and Woolf, C. J. 1996. Glutamate-mediated slow synaptic currents in neonatal rat deep dorsal horn neurons in vitro. *J. Neurophysiol.* **76**, 1465–1476.

Miller, K. E. and Salvatierra, A. T. 1998. Apposition of enkephalin-and neurotensin-immunoreactive neurons by serotonin-immunoreactive varicosities in the rat spinal cord. *Neuroscience* **85**, 837–846.

Miller, K. E. and Seybold, V. S. 1987. Comparison of met-Enkephalin-, dynorphin A-, and neurotensin-immunoreactive neurons in the cat and rat spinal cords: I. Lumbar cord. *J. Comp. Neurol.* **255**, 293–304.

Miller, K. E. and Seybold, V. S. 1989. Comparison of met-enkephalin, dynorphin A, and neurotensin immunoreactive neurons in the cat and rat spinal cords: II. Segmental differences in the marginal zone. *J. Comp. Neurol.* **279**, 619–628.

Miller, K. E., Clements, J. R., Larson, A. A., and Beitz, A. J. 1988. Organization of glutamate-like immunoreactivity in the rat superficial dorsal horn: Light- and electron-microscopic observations. *Synapse* **2**, 28–36.

Miller, K. E., Douglas, V. D., and Kaneko, T. 1993. Glutaminase immunoreactive neurons in the rat dorsal root ganglion contain calcitonin gene-related peptide (CGRP). *Neurosci. Lett.* **160**, 113–116.

Millhorn, D. E., Hökfelt, T., Seroogy, K., Ortel, W., Verhofstad, A. A. T., and Wu, J. 1987a. Immunohistochemical evidence for colocalization of gamma-aminobutyric acid and serotonin in neurons of the ventral medulla oblongata projecting to the spinal cord. *Brain Res.* **410**, 179–185.

Millhorn, D. E., Seroogy, K., Hökfelt, T., Schmued, L. C., Terenius, L., Buchan, A., and Brown, J. C. 1987b. Neurons of the ventral medulla oblongata that contain both somatostatin and enkephalin immunoreactivities project to the nucleus tractus solitarii and spinal cord. *Brain Res.* **424**, 99–108.

Mills, A. and Martin, G. R. 1995. Autoradiographic mapping of [^{3}H] sumatriptan binding in cat brain stem and spinal cord. *Eur. J. Pharmacol.* **280**, 175–178.

Mills, C. D., Fullwood, S. D., and Husebosch, C. E. 2001a. Changes in metabotropic glutamate receptor expression following spinal cord injury. *Exp. Neurol.* **170**, 244–257.

Mills, C. D., Xu, G. Y., McAdoo, D. J., and Hulsebosch, C. E. 2001b. Involvement of metabotropic glutamate receptors in excitatory amino acid and GABA release following spinal cord injury in rat. *J. Neurochem.* **79**, 835–848.

Mills, C. D., Johnson, K. M., and Hulsebosch, C. 2002. Role of group II and group III metabotropic glutamate receptors in spinal cord injury. *Exp. Neurol.* **173**, 153–167.

Milne, R. J., Foreman, R. D., and Willis, W. D. 1982. Responses of primate spinothalamic neurons located in the sacral intermediomedial gray (Stilling's nucleus) to proprioceptive input from the tail. *Brain Res.* **234**, 227–236.

Minami, M. and Satoh, M. 1995. Molecular biology of the opioid receptors: Structures, functions and distributions. *Neurosci Res.* **23**, 121–145.

Minami, M., Kuraishi, Y., Kawamura, M., Yamaguchi, T., Masu, Y., Nakanishi, S., and Satoh, M. 1989. Enhancement of preprotachykinin A gene expression by adjuvant-induced inflammation in the rat spinal cord: possible involvement of substance P-containing spinal neurons in nociception. *Neurosci. Lett.* **98**, 105–110.

Minami, M., Maekawa, K., Yabuuchi, K., and Satoh, M. 1995. Double in situ hybridization study on coexistence of μ-, δ- and *k*-opioid receptor mRNAs with preprotachykinin A mRNA in the rat dorsal root ganglia. *Mol. Brain Res.* **30**, 203–210.

Minami, T., Okuda-Ashitaka, E., Mori, H., Sakimura, K., Watanabe, M., Mishina, M., and Ito, S. 2000. Characterization of nociceptin/orphanin FQ-induced pain responses in conscious mice: Neonatal capsaicin treatment and N-methyl-D-aspartate receptor GluRepsilon subunit knockout mice. *Neuroscience* **97**, 133–142.

Minson, J. B., Llewellyn-Smith, I. J., and Arnolda, L. F. 2001. Neuropeptide Y mRNA expression in interneurons in rat spinal cord. *Auton. Neurosci.* **93**, 14–20.

Miquel, M. C., Emerit, M. B., Nosjean, A., Simon, A., Rumajogee, P., Brisorgueil, M. J., Doucet, E., Hamon, M., and Verge, D. 2002. Differential subcellular localization of the 5-HT3-As receptor subunit in the rat central nervous system. *Eur. J. Neurosci.* **15**, 449–457.

Miranda, C., Di Virgilio, M., Selleri, S., Zanotti, G., Pagliardini, S., Pierotti, M. A., and Greco, A. 2001. Novel pathogenic mechanisms of congenital insensitivity to pain with anhidrosis genetic disorder unveiled by functional analysis of neurotrophic tyrosine receptor kinase type 1/nerve growth factor receptor mutations. *J. Biol. Chem.* **277**, 6455–6462.

Mirnics, K. and Koerber, H. R. 1997. Properties of individual embryonic primary afferents and their spinal projections in the rat. *J. Neurophysiol.* **78**, 1590–1600.

Mitchell, D. and Hellon, R. F. 1977. Neuronal and behavioural responses in rats during noxious stimulation of the tail. *Proc. Roy. Soc. Lond. B.* **197**, 169–194.

Mitchell, G. S., Bach, K. B., Martin, P. A., Foley, K. T., Olson, E. B., Brownfield, M. S., Miletic, V., Behan, M., McGuirk, S., and Sloan, H. E. 2000. Increased spinal monoamine concentrations after chronic thoracic dorsal rhizotomy in goats. *J. Appl. Physiol.* **89**, 1266–1274.

Mitchell, J. J. and Anderson, K. J. 1991. Quantitative autoradiographic analysis of excitatory amino acid receptors in the cat spinal cord. *Neurosci. Lett.* **124**, 269–272.

Mitchell, K., Spike, R. C., and Todd, A. J. 1993. An immunocytochemical study of glycine receptors and GABA in laminae I–III of rat spinal dorsal horn. *J. Neurosci.* **13**, 2371–2381.

Mitchell, S. W. 1872. *Injuries of Nerves and Their Consequences*. J. B. Lippincott, Philadelphia.

Miura, Y., Mardy, S., Awaya, Y., Nihei, K., Endo, F., Matsuda, I., and Indo, Y. 2000a. Mutation and polymorphism analysis of the TRKA (NTRK1) gene encoding a high-affinity receptor for nerve growth factor in congenital insensitivity to pain with anhidrosis (CIPA) families. *Hum. Genet.* **106**, 116–124.

Miura, Y., Hiura, M., Torigoe, K., Numata, O., Kuwahara, A., Matsunaga, M., Hasegawa, S., Boku, N., Ino, H., Mardy, S., Endo, F., Matsuda, I., and Indo, Y. 2000b. Complete paternal uniparental isodisomy for chromosome 1 revealed by mutation analysis of the TRKA (NTRK1) gene encoding a receptor tyrosine kinase for nerve growth factor in a patient with congenital insensitivity to pain with anhidrosis. *Hum. Germnet.* **107**, 205–209.

Mizukawa, K., Otsuka, N., McGreer, P. L., Vincent, S. R., and McGeer, E. G. 1988a. The ultrastructure of somatostatin-immunoreactive cell bodies, nerve fibers and terminals in the dorsal horn of rat spinal cord. *Arch. Histol. Cytol.* **51**, 443–452.

Mizukawa, K., Otsuka, N., Haba, K., and Ogawa, N. 1988b. Distribution of muscarinic cholinergic (mACh) receptors in the rat spinal cord: In vitro quantitative autoradiographic investigation. *Okajimas Folia Anat. Jpn.* **65**, 255–265.

Mizukawa, K., Vincent, S. R., McGeer, P., and McGeer, E. 1989. Distribution of reduced-nicotinamide-adenine-dinucleotide-phosphate diaphorase-positive cells and fibers in the cat central nervous system. *J. Comp. Neurol.* **279**, 281–311.

Mizumura, K., Koda, H., and Kumazawa, T. 1997. Evidence that protein kinase C activation is involved in the excitatory and facilitatory effects of bradykinin on canine visceral nociceptors in vitro. *Neurosci. Lett.* **237**, 29–32.

Mjellem-Joly, N., Lund, A., Berge, O. G., and Hole, K. 1991. Potentiation of a behavioral response in mice by spinal coadministration of substance P and excitatory amino acid agonists. *Neurosci. Lett.* **133**, 1, 21–124.

Moeller, I., Chai, S. Y., Oldfield, B. J., McKinley, M. J., Casley, D., and Mendelsohn, F. A. O. 1995. Localization of angiotensin IV binding sites to motor and sensory neurons in the sheep spinal cord and hindbrain. *Brain Res.* **701**, 301–306.

Mogil, J. S., Yu, L., and Basbaum, A. I. 2000. Pain genes?: Natural variation and transgenic mutants. *Ann. Rev. Neurosci.* **23**, 777–811.

Mokha, S. S., McMillan, J. A., and Iggo, A. 1983. Dorsal root potentials in the cat: effects of bicuculline. *Brain Res.* **259**, 313–318.

Molander, C. and Grant, G. 1985. Cutaneous projections from the rat hindlimb foot to the substantia gelatinosa of the spinal cord studied by transganglionic transport of WGA-HRP conjugate. *J. Comp. Neurol.* **237**, 476–484.

Molander, C. and Grant, G. 1986. Laminar distribution and somatotopic organization of primary afferent fibers from hindlimb nerves in the dorsal horn: A study by transganglionic transport of horseradish peroxidase in the rat. *Neuroscience* **19**, 297–312.

Molander, C. and Grant, G. 1987. Spinal cord projections from hindlimb muscle nerves in the rat studied by transganglionic transport of horseradish peroxidase, wheat germ agglutinin conjugated horseradish peroxidase, or horseradish peroxidase with dimethylsulfoxide. *J. Comp. Neurol.* **260**, 246–255.

Molander, C., Ygge, I., and Dalsgaard, C. J. 1987. Substance P-, somatostatin-, and calcitonin gene-related peptide-like immunoreactivity and fluoride resistant acid phosphatase-activity in relation to retrogradely labeled cutaneous, muscular and visceral primary sensory neurons in the rat. *Neurosci. Lett.* **74**, 37–42.

Molander, C., Kinnman, E., and Aldskogius, H. 1988. Expansion of primary sensory afferent projection following combined sciatic nerve resection and saphenous nerve crush: A horseradish peroxidase study in the adult rat. *J. Comp. Neurol.* **276**, 436–441.

Moller, K., Zhang, Y. Z., Hakanson, R., Luts, A., Sjolund, B., Uddman, R., and Sundler, F. 1993. Pituitary adenylate cyclase activating peptide is a sensory neuropeptide: Immunocytochemical and immunochemical evidence. *Neuroscience* **57**, 725–732.

Mollereau, C., Parmentier, M., Mailleux, P., Butour, J. L., Moisand, C., Chalon, P., Caput, D., Vassart, G., and Meunier, J. C. 1994. ORL1, a novel member of the opioid receptor family. Cloning, functional expression and localization. *FEBS Lett.* **341**, 33–38.

Molliver, D. C., Radeke, M. J., Stuart, R., Feinstein, C., and Snider, W. D. 1995. Presence or absence of TrkA protein distinguishes subsets of small sensory neurons with unique cytochemical characteristics and dorsal horn projections. *J. Comp. Neurol.* **361**, 404–416.

Molliver, D. C., Wright, D. E., Leitner, M. L., Parsadanian, A. S., Doster, K., Wen, D., Yan, Q., and Snider, W. D. 1997. IB4-binding DRG neurons switch from NGF to GDNF dependence in early postnatal life. *Neuron* **19**, 849–861.

Molony, V., Steedman, W. M., Cervero, F., and Iggo, A. 1981. Intracellular marking of identified neurons in the superficial dorsal horn of the cat spinal cord. *Q. J. Exp. Physiol.* **66**, 211–223.

Monaghan, D. T. and Cotman, C. W. 1982. The distribution of [^{3}H]kainic acid binding sites in rat CNS as determined by autoradiography. *Brain Res.* **252**, 91–100.

Monaghan, D. T. and Cotman, C. W. 1985. Distribution of N-methyl-aspartate sensitive c-[3H] glutamate binding sites in rat brain. *J. Neurosci.* **5**, 2909–2919.

Monory, K., Bourin, M. C., Spetea, M., Tomboly, C., Toth, G., Matthes, H. W., Kieffer, B. L., Hanoune, J., and Borsodi, A. 2000. Specific activation of the mu opioid receptor (MOR) by endomorphin 1 and endomorphin 2. *Eur. J. Neurosci.* **12**, 577–584.

Monteillet-Agius, G., Fein, J., Anton, B., and Evans, C. J. 1998. ORL-1 and Mu opioid receptor antisera label different fibers in areas involved in pain processing. *J. Comp. Neurol.* **399**, 373–383.

Moon, D. E., Lee, D. H., Han, H. C., Xie, J., Coggeshall, R. E., and Chung, J. M. 1999. Adrenergic sensitivity of the sensory receptors modulating mechanical allodynia in a rat neuropathic pain model. *Pain* **80**, 589–595.

Moore, R. Y. and Card, J. P. 1984. Noradrenaline-containing neuron systems, In A. Bjorklund and T. Hokfelt (eds.), *Handbook of Chemical Neuroanatomy* (IV, pp. 123–156). Elsevier Science Publishers, London.

Morales, M. and Wang, S. D. 2002. Differential composition of 5-hydroxytryptamine3 receptors synthesized in the rat CNS and peripheral nervous system. *J. Neurosci* **22**, 6732–6741.

Morales, M., Battenberg, E., de Lecea, L., Sanna, P. P., and Bloom, F. E. 1996. Cellular and subcellular immunolocalization of the type 3 serotonin receptor in the rat central nervous system. *Brain Res. Mol. Brain Res.* **36**, 251–260.

Morales, M., McCollum, N., and Kirkness, E. F. 2001. 5-HT(3)-receptor subunits A and B are co-expressed in neurons of the dorsal root ganglion. *J. Comp. Neurol.* **438**, 163–172.

Moran, T. D., Abdulla, F. A., and Smith, P. A. 2000. Cellular neurophysiological actions of nociceptin/orphanin FQ. *Peptides* **21**, 969–976.

Morgan, C., Nadelhaft, I., and De Groat, W. C. 1981. The distribution of visceral primary afferents from the pelvic nerve to Lissauer's tract and the spinal gray matter and its relationship to the sacral parasympathetic nucleus. *J. Comp. Neurol.* **201**, 415–440.

Morgan, C., deGroat, W. C., and Nadelhaft, I. 1986. The spinal distribution of sympathetic preganglionic and visceral primary afferent neurons that send axons into the hypogastric nerves of the cat. *J. Comp. Neurol.* **243**, 23–40.

Morgan, C. W., Ohara, P. T., and Scott, D. E. 1999. Vasoactive intestinal polypeptide in sacral primary sensory pathways in the cat. *J. Comp. Neurol.* **407**, 381–394.

Morgan, J. I. and Curran, T. 1986. Role of ion flux in the control of c-fos expression. *Nature* **322**, 552–555.

Morgan, M. M., Gogas, K. R., and Basbaum, A. I. 1994. Diffuse noxious inhibitory controls reduce the expression of noxious stimulus-evoked Fos-like immunoreactivity in the superficial and deep laminae of the rat spinal cord. *Pain* **56**, 347–352.

Morita, K. and Katayama, Y. 1984. Two types of acetylcholine receptors on the soma of primary afferent neurons. *Brain Res.* **290**, 348–352.

Morley, J. W. and Goodwin, A. W. 1987. Sinusoidal movement of a grating across the monkey's fingerpad: Temporal patterns of afferent fiber responses. *J. Neurosci.* **7**, 2181–2191.

Morrell, J. I., Wolinsky, T. D., Krieger, M. S., and Pfaff, D. W. 1982. Autoradiographic identification of estradiol-concentrating cells in the spinal cord of the female rat. *Exp. Brain Res.* **45**, 144–150.

Morris, B. J. and Herz, A. 1987. Distinct distribution of opioid receptor types in rat lumbar spinal cord. *Arch. Pharmacol.* **336**, 240–243.

Morris, C. E. 1990. Mechanosensitive ion channels. *J. Membr. Biol.* **113**, 93–107.

Morris, J. L., Anderson, R. L., and Gibbins, I. L. 2001. Neuropeptide Y immunoreactivity in cutaneous sympathetic and sensory neurons during development of the guinea pig. *J. Comp. Neurol.* **437**, 321–334.

Morris, R. 1987. Inhibition of nociceptive responses of laminae V–VII dorsal horn neurones by stimulation of mixed and muscle nerves, in the cat. *Brain Res.* **401**, 365–370.

Morris, R. 1989. Responses of spinal dorsal horn neurones evoked by myelinated primary afferent stimulation are blocked by excitatory amino acid antagonists acting at kainate/quisqualate receptors. *Neurosci. Lett.* **105**, 79–85.

Morris, R., Southam, E., Gittins, S. R., deVente, J., and Garthwaite, J. 1993. The no-cGMP pathway in neonatal rat dorsal horn. *Eur. J. Neurosci.* **6**, 876–879.

Morrison, B. M., Janssen, W. G. M., and Gordin, J. W. 1998. Light- and electron-microscopic distribution of the AMPA receptor subunit, GluR2, in the spinal cord of control and G86R mutant superoxide dismutase transgenic mice. *J. Comp. Neurol.* **395**, 523–531.

Morrison, J. F. B. 1973. Splanchnic slowly adapting mechanoreceptors with punctate receptive fields in the mesentery and gastrointestinal tract of the cat. *J. Physiol.* **233**, 349–361.

Morrison, J. F. B. 1977. The afferent innervation of the gastrointestinal tract. In F. P. Brooks and P. W. Evers (eds.), *Nerves and the Gut* (pp. 297–322). C. B. Slack, Thorofare, NJ.

Morton, C. R. and Hutchison, W. D. 1989. Release of sensory neuropeptides in the spinal cord: Studies with calcitonin gene-related peptide and galanin. *Neuroscience* **31**, 807–815.

Morton, C. R., Maisch, B., and Zimmermann, M. 1987. Diffuse noxious inhibitory controls of lumbar spinal neurons involves a supraspinal loop in the cat. *Brain Res.* **410**, 347–352.

Morton, C. R., Hutchison, W. D., and Hendry, I. A. 1988. Release of immunoreactive somatostatin in the spinal dorsal horn of the cat. *Neuropeptides* **12**, 189–197.

Morton, C. R., Hutchison, W. D., Hendry, I. A., and Duggan, A. W. 1989. Somatostatin: Evidence for a role in thermal nociception. *Brain Res.* **488**, 89–96.

Mosconi, T. and Kruger, L. 1996. Fixed-diameter polyethylene cuffs applied to the rat sciatic nerve induce a painful neuropathy: Ultrastructural morphometric analysis of axonal alterations. *Pain* **64**, 37–57.

Motavkin, P. A. and Okhotin, V. E. 1980. Histochemistry of choline acetyltransferase in the spinal cord and spinal ganglia of the cat. *Neurosci. Behav. Physiol.* 307–310.

Mouchet, P., Manier, M., Dietl, M., Feuerstein, C., Berod, A., Arluison, M., Denoroy, L., and Thibault, J. 1986. Immunohistochemical study of catecholaminergic cell bodies in the rat spinal cord. *Brain Res. Bull.* **16**, 341–353.

Mouchet, P., Manier, M., and Feuerstein, C. 1992. Immunohistochemical study of the catecholaminergic innervation of the spinal cord of the rat using specific antibodies against dopamine and noradrenaline. *J Chem. Neuroanat.* **5**, 427–440.

Mountcastle, V. B. 1984. Central nervous mechanisms in mechanoreceptive sensibility. In *Handbook of Physiology, Section 1, The Nervous System, Vol. III, Sensory Processes, Part 2* (pp. 789–878). *Am. Physiol.* Soc., Baltimore.

Mountcastle, V. B. 1998. *Perceptual Neuroscience: The Cerebral Cortex.* Harvard University Press, Cambridge, MA.

Mountcastle, V. B., LaMotte, R. H., and Carli, G. 1972. Detection thresholds for stimuli in humans and monkeys: Comparison with threshold events in mechanoreceptive afferent nerve fibers innervating the monkey hand. *J. Neurophysiol.* **35**, 122–136.

Mountjoy, K. G. and Wild, J. M. 1998. Melanocortin-4 receptor mRNA expression in the developing autonomic and central nervous systems. *Develop. Brain Res.* **107**, 309–314.

Mousa, S. A., Machelska, H., Schafer, M., and Stein, C. 2002. Immunohistochemical localization of endomorphin-1 and endomorphin-2 in immune cells and spinal cord in a model of inflammatory pain. *J. Neuroimmunol.* **126**, 5–15.

Moussaoui, S. M., Hermans, E., Mathieu, A. M., Bonici, B., Clerc, F., Guinet. F., Garret, C., and Laduron, P. M. 1992. Polyclonal antibodies against the rat NK1 receptor: Characterization and localization in the spinal cord. *NeuroReport* **3**, 1073–1076.

Mouton, L. J. and Holstege, G. 2000. Segmental and laminar organization of the spinal neurons projecting to the periaqueductal gray (PAG) in the cat suggests the existence of at least five separate clusters of spino-PAG neurons. *J. Comp. Neurol.* **428**, 389–410.

Mudge, A. W., Leeman, S. E., and Fischbach, G. D. 1979. Enkephalin inhibits release of substance P from sensory neurons in culture and decreases action potential duration. *Proc. Natl. Acad. Sci. USA* **76**, 526–530.

Mugnaini, E., Berrebi, A. S., Morgan, J. I., and Curran, T. 1989. Fos-like immunoreactivity induced by seizure in mice is specifically associated with euchromatin in neurons. *Eur. J. Neurosci.* **1**, 46–52.

Mulder, H., Uddman, R., Moller, K., Zhang, Y. Z., Ekblad, E., Alumets, J., and Sundler, F. 1994. Pituitary adenylate cyclase activating polypeptide expression in sensory neurons. *Neuroscience* **63**, 307–312.

Mulder, H., Leckstrom, A., Uddman, R., Ekblad, E., Westermark, P., and Sundler, F. 1995. Islet amyloid polypeptide (amylin) is expressed in sensory neurons. *J. Neurosci.* **15**, 7625–7632.

Mulder, H., Zhang, Y., Danielsen, N., and Sundler, F. 1997. Islet amyloid polypeptide and calcitonin gene-related peptide expression are upregulated in lumbar dorsal root ganglia after unilateral adjuvant-induced inflammation in the rat paw. *Mol. Brain Res.* **50**, 127–135.

Mulder, H., Jongsma, H., Zhang, Y., Gebre-Medhin, S., Sundler, F., and Danielsen, N. 1999. Pituitary adenylate cyclase-activating polypeptide and islet amyloid polypeptide in primary sensory neurons: Functional implications from plasticity to expression on nerve injury and inflammation. *Mol. Neurobiol.* **19**, 229–253.

Mulderry, P. K. 1994. Neuropeptide expression by newborn and adult rat sensory neurons in culture: Effects of nerve growth factor and other neurotrophic factors. *Neuroscience* **59**, 673–688.

Mulderry, P. K., Ghatei, M. A., Spokes, R. A., Jones, P. M., Pierson, A. M., Hamid, Q. A., Kanse, S., Amara, S. G., Burrin, J. M., and Legon, S. 1988. Differential expression of alpha-CGRP and beta-CGRP by primary sensory neurons and enteric autonomic neurons of the rat. *Neuroscience* **25**, 195–205.

Müller, J. 1840–2. *Elements of Physiology*. Transl. from German, with notes by W. Baly, 2nd ed., London. Original German: 1838. *Handbuch der Physiologie des Menschen*, 2nd ed., Coblenz.

Munger, B. L. 1965. The intraepidermal innervation of the snout of the opossum: A light- and electron-microscope study with observations on the nature of Merkel's "Tastzelle." *J. Cell Biol.* **26**, 79–97.

Munger, B. L. 1971. Patterns of organization of peripheral sensory receptors. In W. R. Loewenstein (ed.), *Handbook of Sensory Physiology* (Vol. 1, pp. 523–556). Springer, Berlin.

Munger, B. L., Yoshida, Y., Hayashi, S., Osawa, T., and Ide, C. 1988. A re-evaluation of the cytology of cat Pacinian corpuscles: I. The inner core and clefts. *Cell Tissue Res.* **253**, 83–93.

Munglani, R. and Hunt, S. P. 1995. Molecular biology of pain. *Br. J. Anaesth.* **75**, 186–192.

Munglani, R., Harrison, S. M., Smith, G. D., Bountra, C., Birch, P. J., Elliot, P. J., and Hunt, S. P. 1996. Neuropeptide changes persist in spinal cord despite resolving hyperalgesia in a rat model of mononeuropathy. *Brain Res.* **743**, 102–108.

Munro, F. E., Fleetwood-Walker, S. M., Parker, R. M., and Mitchell, R. 1993. The effects of neurokinin receptor antagonists on mustard oil-evoked activation of rat dorsal horn neurons. *Neuropeptides* **25**, 299–305.

Munro, F. E., Fleetwood-Walker, S. M., and Mitchell, R. 1994. Evidence for a role of protein kinase C in the sustained activation of rat dorsal horn neurons evoked by cutaneous mustard oil application. *Neurosci. Lett.* **170**, 199–202.

Munro, S., Thomas, K. L., Abu-Shaar, M. 1993. Molecular characterization of a peripheral receptor for cannabinoids. *Nature* **365**, 61–65.

Murakami, T., Araki, T., Yamano, M., Wanaka, A., Betz, H., and Tohyama, M. 1988. Localization of the glycine receptors in the rat central nervous system: An immunocytochemical analysis. *Adv. Exp. Med. Biol.* **236**, 71–81.

Murase, K. and Randic, M. 1984. Actions of substance P on rat spinal dorsal horn neurons. *J. Physiol.* **346**, 20–217.

Murase, K., Nedeljkov, V., and Randic, M. 1982. The actions of neuropeptides on dorsal horn neurons in rat spinal cord slice preparation: An intracellular study. *Brain Res.* **234**, 170–176.

Murase, K., Ryu, P. D., and Randic, M. 1986. Substance P augments a persistent slow inward calcium-sensitive current in voltage-clamped spinal dorsal horn neurons of the rat. *Brain Res.* **365**, 369–376.

Murase, K., Ryu, P. D., and Randic, M. 1989a. Tachykinins modulate multiple ionic conductances in voltage-clamped rat spinal dorsal horn neurons. *J. Neurophysiol.* **61**, 854–865.

Murase, K., Ryu, P. D., and Randic, M. 1989b. Excitatory and inhibitory amino acids and peptide-induced responses in acutely isolated rat spinal dorsal horn neurons. *Neurosci. Lett.* **103**, 56–63.

Murone, C., Paxinos, G., McKinley, M. J., Oldfield, B. J., Müller-Esteri, W., Mendelsohn, F. A. O., and Chai, S. Y. 1997. Distribution of bradykinin B_2 receptors in sheep brain and spinal cord visualized by in vitro autoradiography. *J. Comp. Neurol.* **381**, 203–218.

Murphy, P. G., Ramer, M. S., Borthwick, L., Gauldie, J., Richardson, P. M., and Bisby, M. A. 1999. Endogenous interleukin-6 contributes to hypersensitivity to cutaneous stimuli and changes in neuropeptides associated with chronic nerve constriction in mice. *Eur. J. Neurosci.* **11**, 2243–2253.

Murphy, R. M. and Zemlan, F. P. 1992. Quantitative autoradiographic mapping of a novel serotonin receptor-5- HT1S. *NeuroReport* **3**, 837–840.

Murphy, S. N. and Miller, R. J. 1989. Regulation of Ca^{++} influx into striatal neurons by kainic acid. *J. Pharmacol. Exp. Therap.* **249**, 184–193.

Murray, M. and Goldberger, M. E. 1974. Restitution of function and collateral sprouting in the cat spinal cord: The partially hemisected animal. *J. Comp. Neurol.* **155**, 19–36.

Murray, M. and Goldberger, M. E. 1986. Replacement of synaptic terminals in lamina II and Clarke's nucleus after unilateral lumbosacral dorsal rhizotomy in adult cats. *J. Neurosci.* **6**, 3205–3217.

Na, H. S., Han, J. S., Ko, K. H., and Hong, S. K. 1994. A behavioral model for peripheral neuropathy produced in rat's tail by inferior caudal trunk injury. *Neurosci. Lett.* **177**, 50–52.

Na, H. S., Kim, H. J., Sung, B., Back, S. K., Kim, D. Y., Kim, J. S., and Hong, S. K. 2001. Decrease in spinal CGRP and substance P is not related to neuropathic pain in a rat model. *NeuroReport* **12**, 175–178.

Nade, S., Newbold, P. J., and Straface, S. F. 1987. The effects of direction and acceleration of movement of the knee joint of the dog on medial articular nerve discharge. *J. Physiol.* **388**, 505–519.

Nadelhaft, I., Roppolo, J., Morgan, C., and De Groat, W. C. 1983. Parasympathetic preganglionic neurons and visceral primary afferents in monkey sacral spinal cord revealed following application of horseradish peroxidase to pelvic nerve. *J. Comp. Neurol.* **216**, 36–52.

Nafe, J. P. 1927. The psychology of felt experience. *Am. J. Psychol.* **39**, 367–389.

Nafe, J. P. 1929. A quantitative theory of feeling. *J. Gen. Psychol.* **2**, 199–210.

Naftchi, N. E., Abrahams, S. J., St. Paul, H. M., and Vacca, L. L. 1981. Substance P and leucine-enkephalin changes after cordotomy and morphine treatment. *Peptides* **2**, 61–70.

Nagano, I., Shapshak, P., Yoshioka, M., Xin, K., Nakamura, S., and Bradley, W. G. 1996a. Increased NADPH-diaphorase reactivity and cytokine expression in dorsal root ganglia in acquired immunodeficiency syndrome. *J. Neurol. Sci.* **136**, 117–128.

Nagano, I., Shapshak, P., Yoshioka, M., Xin, K.-Q., Nakamura, S., and Bradley, W. G. 1996b. Parvalbumin and calbindin D-28 k immunoreactivity in dorsal root ganglia in acquired immunodeficiency syndrome. *Neuropath. Appl. Neurobiol.* **22**, 293–301.

Naguib, M. and Yaksh, T. L. 1997. Characterization of muscarinic receptor subtypes that mediate antinociception in the rat spinal cord. *Anesth. Analg.* **85**, 847–853.

Nagy, I. and Rang, H. 1999a. Noxious heat activates all capsaicin-sensitive and also a sub-population of capsaicin-insensitive dorsal root ganglion neurons. *Neuroscience* **88**, 995–997.

Nagy, I. and Rang, H. P. 1999b. Similarities and difference between the responses of rat sensory neurons to noxious heat and capsaicin. *J. Neurosci.* **19**, 10647–10655.

Nagy, I., Urban, L., and Woolf, C. J. 1993a. Morphological and membrane properties of young rat lumbar and thoracic dorsal root ganglion cells with unmyelinated axons. *Brain Res.* **609**, 193–200.

Nagy, I., Pabla, R., Matesz, C., Dray, A., Woolf, C. J., and Urban, L. 1993b. Cobalt uptake enables identification of capsaicin- and bradykinin-sensitive subpopulations of rat dorsal root ganglion cells in vitro. *Neuroscience* **56**, 241–246.

Nagy, I., Woolf, C. J., Dray, A., and Urban, L. 1994. Cobalt accumulation in neurons expressing ionotropic excitatory amino acid receptors in young rat spinal cord: Morphology and distribution. *J. Comp. Neurol.* **344**, 321–335.

Nagy, I. and Woolf, C. J. 1996. Lignocaine selectively reduced C fibre-evoked neuronal activity in rat spinal cord in vitro by decreasing N-methyl-D-aspartate and neurokinin receptor-mediated postsynaptic depolarizations: implications for the development of novel centrally acting analgesics. *Pain* **64**, 59–70.

Nagy, I., Dray, A., and Urban, L. 1995. Possible branching of myelinated primary afferent fibres in the dorsal root of the rat. *Brain Res.* **703**, 223–226.

Nagy, J. I. and Daddona, P. E. 1985. Anatomical and cytochemical relationships of adenosine deaminase-containing primary afferent neurons in the rat. *Neuroscience* **15**, 799–813.

Nagy, J. I. and Hunt, S. P. 1982. Fluoride-resistant acid phosphatase-containing neurones in dorsal root ganglia are separate from those containing substance P or somatostatin. *Neuroscience* **7**, 89–97.

Nagy, J. I. and Hunt, S. P. 1983. The termination of primary afferents within the rat dorsal horn: Evidence for rearrangement following capsaicin treatment. *J. Comp. Neurol.* **218**, 145–158.

Nagy, J. I., Hunt, S. P., Iverson, L. L., and Emson, P. C. 1981. Biochemical and anatomical observations on the degeneration of peptide containing primary afferent neurons after neonatal capsaicin. *Neuroscience* **6**, 1923–1934.

Nagy, J. I., Iversen, L. L., Goedert, M, Chapman, D., and Hunt, S. P. 1983. Dose-dependent effect of capsaicin on primary sensory neurons in the neonatal rat. *J. Neurosci.* **3**, 399–406.

Nagy, J. I., Buss, M., LaBella, L. A., and Daddona, P. E. 1984. Immunohistochemical localization of adenosine deaminase in primary afferent neurons of the rat. *Neurosci. Lett.* **48**, 133–138.

Nahin, R. L. 1988. Immunocytochemical identification of long ascending, peptidergic lumbar spinal neurons terminating in either the medial or lateral thalamus in the rat. *Brain Res.* **443**, 345–349.

Nahin, R. L. and Byers, M. R. 1994. Adjuvant-induced inflammation of rat paw is associated with altered calcitonin gene-related peptide immunoreactivity within cell bodies and peripheral endings of primary afferent neurons. *J. Comp. Neurol.* **349**, 475–485.

Nahin, R. L., Madsen, A. M., and Giesler, G. J. 1983. Anatomical and physiological studies of the gray matter surrounding the spinal cord central canal. *J. Comp. Neurol.* **220**, 321–335.

Nahin, R. L., Hylden, J. L. K., Iadarola, M. J., and Dubner, R. 1989. Peripheral inflammation is associated with increased dynorphin immunoreactivity in both projection and local circuit neurons in the superficial dorsal horn of the rat lumbar spinal cord. *Neurosci. Lett.* **96**, 247–252.

Nahin, R. L., Ren, K., De Leon, M., and Ruda, M. 1994. Primary sensory neurons exhibit altered gene expression in a rat model of neuopathic pain. *Pain* **58**, 95–108.

Naim, M., Spike, R. C., Watt, C., Shehab, S. A., and Todd, A. J. 1997. Cells in laminae III and IV of the rat spinal cord that possess the neurokinin-1 receptor and have dorsally directed dendrites receive a major synaptic input from tachykinin-containing primary afferents. *J. Neurosci.* **17**, 5536–5548.

Naim, M. M., Shehab, S. A. S., and Todd, A. J. 1998. Cells in laminae III and IV of the rat spinal cord which possess the neurokinin-1 receptor receive monosynaptic input from myelinated primary afferents. *Eur. J. Neurosci.* **10**, 3012–3019.

Nakamura, A. and Shiomi, H. 1999. Recent advances in neuropharmacology of cutaneous nociceptors. *Jpn. J. Pharmacol.* **79**, 427–431.

Nakamura, F. and Strittmatter, S. M. 1996. $P2Y_1$ purinergic receptors in sensory neurons: contribution to touch-induced impulse generation. *Proc. Natl. Acid. Sci.* **93**, 10465–10470.

Nakamura, H. 1997. NADPH-diaphorase and cytosolic urea cycle enzymes in the rat spinal cord. *J. Comp. Neurol.* **385**, 616–626.

Nakamura, R., Kamakura, K., and Kwak, S. 1994. Late-onset selective neuronal damage in the rat spinal cord induced by continuous intrathecal administration of AMPA. *Brain Res.* **654**, 279–285.

Nakamura, T. and Takasaki, M. 2001. Intrathecal pre-administration of fentanyl effectively suppresses formalin evoked c-Fos expression in spinal cord of rat. *Can. J. Anaesth.* **48**, 993–999.

Nakamura-Craig, M. and Gill, B. K. 1991. Effect of neurokinin A, substance P and calcitonin gene-related peptide in peripheral hyperalgesia in the rat paw. *Neurosci. Lett.* **124**, 49–51.

Nakanishi, S. 1992. Molecular diversity of glutamate receptors and implications for brain function. *Science* **258**, 597–603.

Nakata, Y., Kusaka, Y., and Segawa, T. 1979. Supersensitivity to substance P after dorsal root section. *Life Sci.* **24**, 1651–1654.

Nakatsuka, T. and Gu, J. G. 2001. ATP P2X receptor-mediated enhancement of glutamate release and evoked EPSCs in dorsal horn neurons of the rat spinal cord. *J. Neurosci.* **21**, 6522–6531.

Nakatsuka, T., Mena, N., Ling, J., and Gu, J. G. 2001. Depletion of substance P from rat primary sensory neurons by ATP, an implication of P2X receptor-mediated release of substance P. *Neuroscience* **107**, 293–300.

Nakaya, Y., Kaneko, T., Shigemoto, R., Nakanishi, S., and Mizuno, N. 1994. Immunohistochemical localization of substance P receptor in the central nervous system of the adult rat. *J. Comp. Neurol.* **347**, 249–274.

Nam, S. C., Kim, K. J., Leem, J. W., Chung, K., and Chung, J. M. 1989. Fiber counts at multiple sites along the rat ventral root after neonatal peripheral neurectomy or dorsal rhizotomy. *J. Comp. Neurol.* **290**, 336–342.

Namba, M., Ghatei, M. A., Gibson, S. J., Polak, J. M., and Bloom, S. R. 1985. Distribution and localization of neuromedin b-like immunoreactivity in pig, cat and rat spinal cord. *Neuroscience* **15**, 1217–1226.

Nandy, K. and Bourne, G. H. 1964. The effects of D-lysergic acid diethylamide tartrate (LSD-25) on the cholinesterases and monoamine oxidase in the spinal cord: A possible factor in the mechanism of hallucination. *J. Neurol., Neurosurg. Psychiat.* **27**, 259–267.

Narikawa, K., Furue, H., Kumamoto, E., and Yoshimura, M. 2000. In vivo patch-clamp analysis of IPSCs evoked in rat substantia gelatinosa neurons by cutaneous mechanical stimulation. *J. Neurophysiol.* **84**, 2171–2174.

Narita, M., Dun, S. L., Dun, N. J., and Tseng, L. F. 1996. Hyperalgesia induced by pituitary adenylate cyclase-activating polypeptide in the mouse spinal cord. *Eur. J. Pharmacol.* **311**, 121–126.

Narita, M., Mizoguchi, H., Oji, D. E., Dun, N. J., Hwang, B. H., Nagase, H., and Tseng, L. F. 1999. Identification of the G-protein-coupled ORL1 receptor in the mouse spinal cord by [35S]-GTPgammaS binding and immunohistochemistry. *Br. J Pharmacol.* **128**, 1300–1306.

Narotzky, R. A. and Kerr, F. W. 1978. Marginal neurons of the spinal cord: Types, afferent synaptology and functional considerations. *Brain Res.* **139**, 1–20.

Nashold, B. S. 1987. Introduction to second international symposium on dorsal root entry zone (DREZ) lesions. *Appl. Neurophysiol.* **51**, 76–77.

Nashold, B. S. and Friedman, H. 1972. Dorsal column stimulation for control of pain: Preliminary report on 30 patients. *J. Neurosurg.* **36**, 590–597.

Nässtrom, J., Schneider, S. P., and Perl, E. R. 1994. Differential L-glutamate responsiveness among superficial dorsal horn neurons. *J. Neurophysiol.* **72**, 2956–2965.

Nathan, P. W. 1947. On the pathogenesis of causalgia in peripheral nerve injuries. *Brain* **70**, 145–171.

Nathan, P. W. 1976. The gate-control theory of pain: A critical review. *Brain* **99**, 123–158.

Nauta, H. J. W., Wehman, J. C., Koliatsos, V. E., Terrell, M. A., and Chung, K. 1999. Intraventricular infusion of nerve growth factor as the cause of sympathetic fiber sprouting in sensory ganglia. *J Neurosurg.* **91**, 447–453.

Navaratnam, V. and Lewis, P. R. 1970. Cholinesterase-containing neurones in the spinal cord of the rat. *Brain Res.* **18**, 411–425.

Naveilhan, P., Neveu, I., Arenas, E., and Ernfors, P. 1998. Complementary and overlapping expression of Y1, Y2 and Y5 receptors in the developing and adult mouse nervous system. *Neuroscience* **87**, 289–302.

Naveilhan, P., Hassani, H., Lucas, G., Blakeman, K. H., Hao, J. X., Xu, X. J., Wiesenfeld-Hallin, Z., Thoren, P., and Ernfors, P. 2001. Reduced antinociception and plasma extravasation in mice lacking a neuropeptide Y receptor. *Nature* **409**, 5213–517.

Nazli, M. and Morris, R. 2000. Comparison of localization of the neurokinin 1 receptor with AMPA-type glutamate receptors in the rat spinal cord. *Anat. Histol. Embryol.* **29**, 277–281.

Neal, C. R., Mansour, A., Reinscheid, R., Nothacker, H.-P., Civelli, O., Akil, H., and Watson, S. J. 1999. Opioid receptor-like (ORL1) receptor distribution in the rat central nervous system: Comparison of ORL1 receptor mRNA expression with ^{125}I-[^{14} Tyr]-orphanin FQ binding. *J. Comp. Neurol.* **412**, 563–605.

Neal, M. J. 1971. The uptake of [^{14}C]glycine by slices of mammalian spinal cord. *J. Physiol.* **215**, 103–117.

Neil, A., Attal, N., and Guilbaud, G. 1994. Effects of guanethidine on sensitization to natural stimuli and self-mutilating behavior in rats with a peripheral mononeuropathy. *Brain Res.* **565**, 237–246.

Neil, E. (ed.) 1972. *Enteroceptors*. Springer-Verlag, Berlin.

Ness, T. J. and Gebhart, G. F. 1987. Characterization of neuronal responses to noxious visceral and somatic stimuli in the medial lumbosacral spinal cord of the rat. *J. Neurophysiol.* **57**, 1867–1892.

Ness, T. J. and Gebhart, G. F. 1988. Characterization of neurons responsive to noxious colorectal distension in the T13–L2 spinal cord of the rat. *J. Neurophysiol.* **60**, 1419–1438.

Ness, T. J. and Gebhart, G. F. 1989. Characteristics of superficial T13–L2 dorsal horn neurons encoding for colorectal distension in the rat: Comparison with neurons in deep laminae. *Brain Res.* **485**, 301–309.

Neugebauer, V. and Schaible, H. G. 1990. Evidence for a central component in the sensitization of spinal neurons with joint input during development of acute arthritis in cat's knee. *J. Neurophysiol.* **64**, 299–311.

Neugebauer, V., Schaible, H. G., and Schmidt, R. F. 1989. Sensitization of articular afferents to mechanical stimuli by bradykinin. *Pfluegers Arch.* **415**, 330–335.

Neugebauer, V., Lücke, T., and Schaible, H.-G. 1993. N-methyl-D-asparate (NMDA) and non-NMDA receptor antagonists block the hyperexcitability of dorsal horn neurons during development of acute arthritis in rat's knee joint. *J. Neurophysiol.* **70**, 1365–1377.

Neugebauer, V., Lücke, T., and Schaible, H.-G. 1994a. Requirement of metabotropic glutamate receptors for the generation of inflammation-evoked hyperexcitability in rat spinal cord neurons. *Eur. J. Neurosci.* **6**, 1179–1186.

Neugebauer, V., Lücke, T., Grubb, B., and Schaible, H.-G. 1994b. The involvement of N-methyl-D-aspartate (NMDA) and non-NMDA receptors in the reponsiveness of rat spinal neurons with input from the chronically inflamed ankle. *Neurosci.Lett.* **170**, 237–240.

Neugebauer, V., Schaible, H.-G., Weiretter, F., and Freudenberger, U. 1994c. The involvement of substance P and neurokinin-1 receptors in the responses of rat dorsal horn neurons to noxious but not to innocuous mechanical stimuli applied to the knee joint. *Brain Res.* **666**, 207–215.

Neugebauer, V., Weiretter, F., and Schaible, H.-G. 1995. Involvement of substance P and neurokinin-1 receptors in the hyperexcitability of dorsal horn neurons during development of acute arthritis in rat's knee joint. *J. Neurophysiol.* **73**, 1574–1583.

Neugebauer, V., Rumenapp, P., and Schaible, H. G. 1996a. The role of spinal neurokinin-2 receptors in the processing of nociceptive information from the joint and in the generation and maintenance of inflammation-evoked hyperexcitability of dorsal horn neurons in the rat. *Eur. J. Neurosci.* **8**, 249–260.

Neugebauer, V., Rumenapp, P., and Schaible, H. G. 1996b. Calcitonin gene-related peptide is involved in the spinal processing of mechanosensory input from the rat's knee joint and in the generation and maintenance of hyperexcitability of dorsal horn neurons during development of acute inflammation. *Neuroscience* **71**, 1095–1109.

Neugebauer, V., Chen, P.-S., and Willis, W. D. 1999. Role of metabotropic glutamate receptor subtype mGluR1 in brief nociception and central sensitization of primate STT cells. *J Neurophysiol.* **82**, 272–282.

Neugebauer, V., Chen, P.-S., and Willis, W. D. 2000. Groups II and III metabotropic glutamate receptors differentially modulate brief and prolonged nociception in primate STT cells. *J. Neurophysiol.* **84**, 2998–3009.

Neumann, S., Doubell, T. P., Leslie, T., and Woolf, C. J. 1996. Inflammatory pain hypersensitivity mediated by phenotypic switch in myelinated primary sensory neurons. *Nature* **384**, 360–364.

Newton, B. W. and Hamill, R. W. 1988. The morphology and distribution of rat serotoninergic intraspinal neurons: An immunohistochemical study. *Brain Res.* **20**, 349–360.

Newton, R. A., Bingham, S., Case, P. C., Sanger, G. J., and Lawson, S. N. 2001. Dorsal root ganglion neurons show increased expression of the calcium channel alpha2delta-1 subunit following partial sciatic nerve injury. *Brain Res. Mol. Brain Res.* **95**, 1–8.

Ni, T. S., Wu, S. X., and Li, Y. Q. 2002. Co-existence of protein kinase C gamma and calcium-binding proteins in neurons of the medullary dorsal horn of the rat. *Neurosignals* **11**, 88–94.

Nicholas, A. P., Pieribone, V. A., Elde, R., and Hokfelt, T. 1991. Initial observations on the localization of mRNA for α and β adrenergic receptors in brain and peripheral tissues of rat using in situ hybridization. *Mol. Cell. Neurosci.* **2**, 344–350.

Nicholas, A. P., Pieribone, V., and Hokfelt, T. 1993. Distributions of mRNAs for alpha-2 adrenergic receptor subtypes in rat brain: An in situ hybridization study. *J. Comp. Neurol.* **328**, 575–594.

Nichols, M. L., Allen, B. J., Rogers, S. D., Ghilardi, J. R., Honore, P., Luger, N. M., Finke, M. P., Li, J., Lappi, D. A., Simone, D. A., and Mantyh, P. W. 1999. Transmission of chronic nociception by spinal neurons expressing the substance P receptor. *Science* **286**, 1558–1561.

Nicol, G. D., Vasko, M. R., and Evans, A. R. 1997. Prostaglandins suppress an outward potassium current in embryonic rat sensory neurons. *J. Neurophysiol.* **77**, 167–176.

Nicoll, R. A. 1979. Dorsal root potentials and changes in extracellular potassium in the spinal cord of the frog. *J. Physiol.* **290**, 113–127.

Nicoll, R. A. and Alger, B. E. 1979. Presynaptic inhibition: Transmitter and ionic mechanisms. *Int. Rev. Neurobiol.* **21**, 217–258.

Nicoll, R. A., Schenker, C., and Leeman, S. E. 1980. Substance P as a transmitter candidate. *Ann. Rev. Neurosci.* **3**, 227–268.

Nielsch, U. and Keen, P. 1989. Reciprocal regulation of tachykinin- and vasoactive intestinal peptide-gene expression in rat sensory neurones following cut and crush injury. *Brain Res.* **481**, 25–30.

Nielsch, U., Bisby, M. A., and Keen, P. 1987. Effect of cutting or crushing the rat sciatic nerve on synthesis of substance P by isolated L5 dorsal root ganglia. *Neuropeptides* **10**, 137–145.

Nielsen, H. S., Hannibal, J., and Fahrenkrug, J. 1998. Embryonic expression of pituitary adenylate cyclase-activating polypeptide in sensory and autonomic ganglia and in spinal cord of the rat. *J. Comp. Neurol.* **394**, 403–415.

Nilsson, B. Y. 1969a. Structure and function of the tactile hair receptors on the cat's foreleg. *Acta Physiol. Scand.* **77**, 396–416.

Nilsson, B. Y. 1969b. Hair discs and Pacinian corpuscles functionally associated with the carpal tactile hairs in the cat. *Acta Physiol. Scand.* **77**, 417–428.

Nilsson, B. Y. and Skoglund, C. R. 1965. The tactile hairs on the cat's foreleg. *Acta Physiol. Scand.* **65**, 364–369.

Ninkovic, M. and Hunt, S. P. 1983. α-Bungarotoxin binding sites on sensory neurones and their axonal transport in sensory afferents. *Brain Res.* **272**, 57–69.

Ninkovic, M. and Hunt, S. P. 1985. Opiate and histamine H1 receptors are present on some substance P-containing dorsal root ganglion cells. *Neurosci. Lett.* **53**, 133–137.

Ninkovic, M., Hunt, S. P., and Kelly, J. S. 1981. Effect of dorsal rhizotomy on the autoradiographic distribution of opiate and neurotensin receptors and neurotensin-like immunoreactivity within the rat spinal cord. *Brain Res.* **230**, 111–119.

Ninkovic, M., Hunt, S. P., and Cleave, J. R. W. 1982. Localization of opiate and histamine H1-receptors in the primate sensory ganglia and spinal cord. *Brain Res.* **241**, 197–206.

Ninkovic, M., Beaujouan, J. C., Torrens, Y., Saffroy, M., Hall, M. D., and Glowinski, J. 1984. Differential localization of tachykinin receptors in rats spinal cord. *Eur. J. Pharmacol.* **106**, 463–464.

Ninkovic, M., Beaujouan, J. C., Torrens, Y., Saffroy, M., Hall, M. D., and Glowinski, J. 1985. Differential localization of tachykinin receptors in rat spinal cord. *Eur. J. Pharmacol.* **106**, 463–464.

Ninomiya, T., Vuillemin, M., Walter-Barakat, I., Winking, H., Pexieder, T., and Droz, B. 1993. Mouse fetal trisomy 13 and hypotrophy of the spinal cord: Effect on calbindin-D_{28k} and calretinin expressed by neurons of the spinal cord and dorsal root ganglion. *Neuroscience* **57**, 1109–1120.

Nirenberg, M. J., Tate, S. S., Mosckovitz, R., Udenfriend, S., and Pickel, V. M. 1995. Immunocytochemical localization of the renal neutral and basic amino acid transporter in rat adrenal gland, brainstem and spinal cord. *J. Comp. Neurol.* **356**, 505–522.

Nishi, K., Oura, C., and Pallie, W. 1969. Fine structure of Pacinian corpuscles in the mesentery of the cat. *J. Cell Biol.* **43**, 539–552.

Nishikawa, N., Bennett, G. J., Ruda, M. A., Lu, G. W., and Dubner, R. 1983. Immunocytochemical evidence for a serotoninergic innervation of dorsal column postsynaptic neurons in cat and monkey: light- and electron-microscopic observations. *Neuroscience* **10**, 1333–1340.

Nissinen, M. J. and Panula, P. 1995. Developmental patterns of histamine-like immunoreactivity in the mouse. *J Histochem. Cytochem.* **43**, 211–227.

Nissl, F. 1894. Über die sogenannten Granula der Nervenzellen. *Neurol. Zbl.* **13**, 676–685.

Nitsos, I. and Rees, S. 1993. Development of immunoreactivity for calcitonin gene-related peptide, substance P and glutamate in primary sensory neurons, and for serotonin in the spinal cord of fetal sheep. *Neuroscience* **54**, 239–252.

Nitsos, I., Sexton, P. M., and Rees, S. 1994. The ontogeny of [125I]rat-α-CGRP binding sites in the spinal cord of sheep: A prenatal and postnatal study. *Neuroscience.* **62**, 257–264.

Niwa, M., Kawaguchi, T., Himeno, A., Fujimoto, M., Kurihara, M., Yamashita, K., Kataoka, Y., Shigematsu, K., and Taniyama, K. 1992. Specific binding sites for 125I-endothelin-1 in the porcine and human spinal cord. *Eur. J Pharmacol.* **225**, 281–289.

Noback, C. R. 1951. Morphology and phylogeny of hair. *Ann. NY Acad. Sci.* **53**, 476–491.

Noguchi, K. and Ruda, M. A. 1992. Gene regulation in an ascending nociceptive pathway: inflammation-induced increase in preprotachykinin mRNA in rat lamina I spinal projection neurons. *J. Neurosci.* **12**, 2563–2572.

Noguchi, K., Morita, Y., Kiyama, H., Sato, M., Ono, K., and Tohyama, M. 1989a. Preproenkephalin gene expression in the rat spinal cord after noxious stimuli. *Mol. Brain Res.* **5**, 227–234.

Noguchi, K., Senba, E., Morita, Y., Saito, M., and Tohyama, M. 1989b. Prepro-VIP and preprotachykinin mRNAs in the dorsal root ganglion cells following peripheral axotomy. *Mol. Brain Res.* **6**, 327–330.

Noguchi, K., Senba, E., Morita, Y., Sato, M., and Tohyama, M. 1990a. Co-expression of α-CGRP and β-CGRP mRNAs in the rat dorsal root ganglion cells. *Neurosci. Lett.* **100**, 1–5.

Noguchi, K., Senba, E., Morita, Y., Sato, M., and Tohyama, M. 1990b. α-CGRP and β-CGRP mRNAs are differentially regulated in the rat spinal cord and dorsal root ganglia. *Mol. Brain Res.* **7**, 299–304.

Noguchi, K., Kowalski, K., Traub, R., Solodkin, A., Iadarola, M. J., and Ruda, M. A. 1991. Dynorphin expression and fos-like immunoreactivity following inflammation induced hyperalgesia are colocalized in spinal cord neurons. *Mol. Brain Res.* **10**, 227–233.

Noguchi, K., Dubner, R., and Ruda, M. A. 1992. Preproenkephalin mRNA in spinal dorsal horn neurons is induced by peripheral inflammation and is co-localized with Fos and Fos-related proteins. *Neuroscience* **46**, 561–570.

Noguchi, K., De Leon, M., Nahin, R. L., Senba, E., and Ruda, M. A. 1993. Quantification of axotomy-induced alteration of neuropeptide mRNAs in dorsal root ganglion neurons with special reference to neuropeptide Y mRNA and the effects of neonatal capsaicin treatment. *J. Neurosci. Res* **35**, 54–66.

Noguchi, K., Kawai, Y., Fukuoka, T., Senba, E., and Miki, K. 1995. Substance P induced by peripheral nerve injury in primary afferent sensory neurons and its effect on dorsal column nucleus neurons. *J. Neurosci.* **15**, 7633–7643.

Nohr, D., Weihe, E., Zentel, H., and Arendt, R. M. 1989. Atrial natriuretic factor-like immunoreactivity in spinal cord and in primary sensory neurons of spinal and trigeminal ganglia of guinea-pig: Correlation with tachykinin immunoreactivity. *Cell Tissue Res.* **258**, 387–392.

Nohr, D., Schafer, M. K., Persson, S., Romeo, H., Nyberg, F., Post, C., Ekstrom, G., and Weihe, E. 1999. Calcitonin gene-related peptide gene expression in collagen-induced arthritis is differentially regulated in primary afferents and motoneurons: Influence of glucocorticoids. *Neuroscience* **93**, 759–773.

Nolano, M., Crisci, C., Santoro, L., Barbieri, F., Casale, R., Kennedy, W. R., Wendelschafer-Crabb, G., Provitera, V., Di Lorenzo, N., and Caruso, G. 2000. Absent innervation of skin and sweat glands in congenital insensitivity to pain with anhidrosis. *Clin. Neurophysiol.* **111**, 1596–1601.

Noordenbos, W. 1959. *Pain*. Elsevier, Amsterdam.

Noordenbos, W. and Wall, P. D. 1976. Diverse sensory functions with an almost totally divided spinal cord: A case of spinal cord transection with preservation of part of one anterolateral quadrant. *Pain* **2**, 185–195.

Nordin, M. 1990. Low-threshold mechanoreceptive and nociceptive units with unmyelinated (C) fibres in the human supraorbital nerve. *J. Physiol.* **426**, 229–240.

Nothias, F., Tessler, A., and Murray, M. 1993. Restoration of substance P and calcitonin gene-related peptide in dorsal root ganglia and dorsal horn after neonatal sciatic nerve lesion. *J. Comp. Neurol.* **334**, 370–384.

Novakovic, S. D., Tzoumaka, E., McGivern, J. G., Haraguchi, M., Sangameswaran, L., Gogas, K. R., Eglen, R. M., and Hunter, J. C. 1998a. Distribution of the tetrodotoxin-resistant sodium channel PN3 in rat sensory neurons in normal and neuropathic conditions. *J. Neurosci.* **18**, 2174–2187.

Novakovic, S. D., Levinson, S. R., Schachner, M., and Shrager, P. 1998b. Disruption and reorganization of sodium channels in experimental allergic neuritis. *Muscle Nerve* **21**, 1019–1032.

Novakovic, S. D., Kassotakis, L. C., Oglesby, I. B., Smith, J. A. M., Eglen, R. M., Ford, A. P. D. W., and Hunter, J. C. 1999. Immunocytochemical localization of $P_{2\times3}$ purinoceptors in sensory neurons in naive rats and following neuropathic injury. *Pain* **80**, 273–282.

Novikoff, A. B., Quintana, N., Villanerde, H., and Forschirm, R. 1966. Nucleoside phosphatase and cholinesterase activities in dorsal root ganglia and peripheral nerve. *J. Cell Biol.* **29**, 525–545.

Novikoff, P. M., Novikoff, A. B., Quintana, N., and Hauw, J.-J. 1971. Golgi apparatus, gerl, and lysosomes of neurons in rat dorsal root ganglia, studied by thick section and thin section cytochemistry. *J. Cell Biol.* **50**, 859–886.

Nowak, L. M. and MacDonald, R. L. 1982. Substance P: Ionic basis for depolarizing responses of mouse spinal cord neurons in cell culture. *J. Neurosci.* **2**, 1119–1128.

Nowycky, M. C., Fox, A. P., and Tsien, R. W. 1985. Three types of neuronal calcium channel with different calcium agonist sensitivity. *Nature* **316**, 440–443.

Nyberg, G. and Blomqvist, A. 1982. The termination of forelimb nerves in the feline cuneate nucleus demonstrated by the transganglionic transport method. *Brain Res.* **248**, 209–222.

Nyberg, G. and Blomqvist, A. 1984. The central projections of muscle afferent fibers to the lower medulla and upper spinal cord: an anatomical study in the cat with the transganglionic transport method. *J Comp. Neurol.* **230**, 99–109.

Nyberg, G. and Blomqvist, A. 1985. The somatotopic organization of forelimb cutaneous nerves in the brachial dorsal horn: An anatomical study in the cat. *J. Comp. Neurol.* **242**, 28–39.

O'Brien, C., Woolf, C. J., Fitzgerald, M., Lindsay, R. M., and Molander, C. 1989. Differences in the chemical expression of rat primary afferent neurons which innervate skin, muscle or joint. *Neuroscience* **32**, 493–502.

Ochoa, J. and Torebjörk, E. 1983. Sensations evoked by intraneural microstimulation of single mechanoreceptor units innervating the human hand. *J. Physiol.* **342**, 633–654.

Ochoa, J. and Torebjörk, E. 1989. Sensations evoked by intraneural microstimulation of C nociceptor fibres in human skin nerves. *J. Physiol.* **415**, 583–599.

O'Conner, B. L. and McConnaughey, J. S. 1978. The structure and innervation of cat knee menisci and their relation to a "sensory hypothesis" of meniscal function. *Am. J. Anat.* **153**, 431–442.

O'Conner, B. L., and Woodbury, P. 1982. The primary articular nerves to the dog knee. *J. Anat.* **134**, 563–572.

O'Donnell, D., Ahmad, S., Wahlestedt, C., and Walker, P. 1999. Expression of the novel galanin receptor subtype GALR2 in the adult rat CNS: Distinct distribution from GALR1. *J. Comp. Neurol.* **409**, 469–481.

O'Donohue, T. L., Massari, V. J., Pazoles, C. J., Chronwall, B. M., Shults, C. W., Quirion, R., Chase, T. N., and Moody, T. W. 1984. A role for bombesin in sensory processing in the spinal cord. *J. Neurosci.* **4**, 2956–2962.

Odutola, A. B. 1972. The organization of cholinesterase-containing systems of the monkey spinal cord. *Brain Res.* **39**, 353–368.

O'Flynn, N. M., Helme, R. D., Watkins, D. J., and Burcher, E. 1989. Autoradiographic localization of substance P binding sites in rat footpad skin. *Neurosci. Lett.* **106**, 43–48.

Ogawa, T., Kanazawa, I., and Kimura, S. 1985. Regional distribution of substance P, neurokinin α and neurokinin β in rat spinal cord, nerve roots and dorsal root ganglia, and the effects of dorsal root section or spinal transection. *Brain Res.* **359**, 152–157.

Oh, U. T., Kim, K. J., Baik-Han, E. J., and Chung, J. M. 1989. Electrophysiological evidence for an increase in the number of ventral root afferent fibers after neonatal peripheral neurectomy in the rat. *Brain Res.* **501**, 90–99.

O'Halloran, K. D. and Perl, E. R. 1997. Effects of partial nerve injury on the responses of C-fiber polymodal nociceptors to adrenergic agonists. *Brain Res.* **759**, 233–240.

Ohishi, H., Shigemoto, R., Nakanishi, S., and Mizuno, N. 1993a. Distribution of the mRNA for a metabotropic glutamate receptor (mGluR3) in the rat brain: An in situ hybridization study. *J. Comp. Neurol.* **335**, 252–266.

Ohishi, H., Shigemoto, R., Nakanishi, S., and Mizuno, N. 1993b. Distribution of the messenger RNA for a metabotropic glutamate receptor, mGluR2, in the central nervous system of the rat. *Neuroscience* **53**, 1009–1018.

Ohishi, H., Nomura, S., Ding, Y. Q., Shigemoto, R., Wada, E., Kinoshita, A., Li, J. L., Neki, A., Nakanishi, S., and Mizuno, N. 1995a. Presynaptic localization of a metabotropic glutamate receptor, mGluR7, in the primary afferent neurons: An immunohistochemical study in the rat. *Neurosci Lett.* **202**, 85–88.

Ohishi, H., Akazawa, C., Shigemoto, R., Nakanishi, S., and Mizuno, N. 1995b. Distributions of the mRNAs for L-2-amino-4-phosphonobutyrate-sensitive metabotropic glutamate receptors, mGluR4 and MGluR7, in the rat brain. *J. Comp. Neurol.* **360**, 555–570.

Ohishi, H., Nomura, S., Ding, Y. Q., Shigemoto, R., Wada, E., Kinoshita, A., Li, J. L., Neki, A., Nakanishi, S., and Mizuno, N. 1995c. Presynaptic localization of a metabotropic glutamate receptor, mGluR7, in the primary afferent neuons: An immunohistochemical study in the rat. *Neurosci. Lett.* **202**, 85–88.

Ohki-Hamazaki, H. 2000. Neuromedin B. *Prog. Neurobiol.* **62**, 297–312.

Ohnishi, A. and Ogawa, M. 1986. Preferential loss of large lumbar primary sensory neurons in carcinomatous sensory neuropathy. *Ann. Neurol.* **20**, 102–104.

Ohsawa, M., Mizoguchi, H., Narita, M., Nagase, H., Kampine, J. P., and Tseng, L. F. 2001. Differential antinociception induced by spinally administered endomorphin-1 and endomorphin-2 in the mouse. *J. Pharmacol. Exp. Therap.* **298**, 592–597.

Ohtori, S., Chiba, T., Takahashi, K., Ino, H., Yamagata, M., Sameda, H., Murata, Y., and Moriya, H. 2000. Neonatal capsaicin treatment decreased substance P receptor immunoreactivity in lamina III neurons of the dorsal horn. *Neurosci. Res.* **38**, 147–154.

Ohtori, S., Takahashi, K., Chiba, T., Yamagata, M., Sameda, H., and Moriya, H. 2001. Phenotypic inflammation switch in rats shown by calcitonin gene-related peptide immunoreactive dorsal root ganglion neurons innervating the lumbar facet joints. *Spine* **26**, 1009–1013.

Ohtori, S., Takahashi, K., Ino, H., Chiba, T., Yamagata, M., Sameda, H., and Moriya, H. 2002. Up-regulation of substance P and NMDA receptor mRNA in dorsal horn and preganglionic sympathetic neurons during adjuvant-induced noxious stimulation in rats. *Ann. Anat.* **184**, 71–76.

Okano, S., Ikeura, Y., and Inatomi, N. 2002. Effects of tachykinin NK1 receptor antagonists on the viscerosensory response caused by colorectal distension in rabbits. *J. Pharmacol. Exp. Ther.* **300**, 925–931.

Okazaki, K., Iino, S., Inoue, S., Kobayashi, S., and Hidaka, H. 1994. Differential distribution of neurocalcin isoforms in rat spinal cord, dorsal root ganglia and muscle spindle. *Biochim. et Biophys. Acta* **1223**, 311–317.

Oku, R., Satoh, M., Fujii, N., Otaka, A., Yajima, H., and Takagi, H. 1987. Calcitonin gene-related peptide promotes mechanical nociception by potentiating release of substance P from the spinal dorsal horn in rats. *Brain Res.* **403**, 350–354.

Oku, R., Nanayama, T., and Satoh, M. 1988. Calcitonin gene-related peptide modulates calcium mobilization in synaptosomes of rat spinal dorsal horn. *Brain Res.* **475**, 356–360.

Okuda-Ashitaka, E., Minami, T., Tachibana, S., Yoshihara, Y., Nishiuchi, Y., Kimura, T., and Ito, S. 1998. Nocistatin, a peptide that blocks nociception action in pain transmission. *Nature* **392**, 286–289.

Olave, M. J. and Maxwell, D. J. 2002. An investigation of neurones that possess the alpha(2C)-adrenergic receptor in the rat dorsal horn. *Neuroscience* **115**, 31–40.

Olave, M. J., Puri, N., Kerr, R., and Maxwell, D. J. 2002. Myelinated and unmyelinated primary afferent axons form contacts with cholinergic interneurons in the spinal dorsal horn. *Exp. Brain Res.* **145**, 448–456.

Oldfield, B. J., Allen, A. M., Hards, D. K., McKinley, M. J., Schlawe, I., and Mendelsohn, F. A. O. 1994. Distribution of angiotensin II receptor binding in the spinal cord of the sheep. *Brain Res.* **650**, 40–48.

O'Leary, J. L., Heinbecker, P., and Bishop, G. H. 1932. Dorsal root fibers which contribute to the tract of Lissauer. *Proc. Soc. Exp. Biol.* **30**, 302–303.

Olivar, T., Cervero, F., and Laird, J. M. A. 2000. Responses of rat spinal neurones to natural and electrical stimulation of colonic afferents: Effect of inflammation. *Brain Res.* **866**, 168–177.

Olschowka, J. A., O'Donohue, T. L., Mueller, G. P., and Jacobwitz, D. M. 1982. The distribution of corticotrophin releasing factor-like immunoreactive neurons in rat brain. *Peptides* **3**, 995–1015.

Olson, G. A., Olson, R. D., and Kastin, A. J. 1989. Endogeous opiates: 1988. *Peptides* **10**, 1253–1280.

Olson, T. H., Riedl, M. S., Vulchanova, L., Ortiz-Gonzalez, X. R., and Elde, R. 1998. An acid sensing ion channel (ASIC) localizes to small primary afferent neurons in rats. *NeuroReport* **9**, 1109–1113.

Omana-Zapata, I., Khabbaz, M. A., Hunter, J. C., Clarke, D. E., and Bley, K. R. 1997. Tetrodotoxin inhibits neuropathic ectopic activity in neuromas, dorsal root ganglia and dorsal horn neurons. *Pain* **72**, 41–49.

Onaka, M., Minami, T., Nishihara, I., and Ito, S. 1996. Involvement of glutamate receptors in strychnine-and bicuculline-induced allodynia in conscious mice. *Anesthesiology* **84**, 1215–1222.

Ong, W. Y. and Mackie, K. 1999. A light- and electron-microscopic study of the CB1 cannabinoid receptor in the primate spinal cord. *J. Neurocytol.* **28**, 39–45.

Ongjoco, R. S., Richardson, C. D., Rudner, X. L., Stafford-Smith, M., and Schwinn, D. A. 2000. Alpha2-adrenergic receptors in human dorsal root ganglia: Predominance of alpha2b and alpha2c subtype mRNAs. *Anesthesiology* **92**, 968–976.

Orr, D. and Rows, R. G. 1901. The nerve cells of the human posterior root ganglia and their changes in general paralysis of the insane. *Brain* **24**, 286–309.

Ortiz-Rey, J. A., Álvarez-Álvarez, C., Antón-Badiola, I., San Miguel-Fraile, P., and De la Fuente-Buceta, A. 2002. Human Meissner corpuscles express Bcl-2 but not Bax protein. *Neurosci. Lett.* **329**, 240–242.

O'Shaughnessy, D. J., McGregor, G. P., Ghatei, M. A., Blank, M. A., Springall, D. R., Gu, J., Polak, J. M., and Bloom, S. R. 1983. Distribution of bombesin, somatostatin, substance-P and vasoactive intestinal polypeptide in feline and porcine skin. *Life Sci.* **32**, 2827–2836.

Ositelu, D. O., Morris, R., and Vaillant, V. 1987. Innervation of facial skin but not masticatory muscles or the tongue by trigeminal primary afferents containing somatostatin in the rat. *Neurosci. Lett.* **78**, 271–276.

Otsuka, M. and Konishi, S. 1976. Release of substance P-like immunoreactivity from isolated spinal cord of newborn rat. *Nature* **264**, 83–84.

Otsuka, M. and Takahashi, T. 1977. Putative peptide neurotransmitters. *Ann. Rev. Pharmacol. Toxicol.* **17**, 425–439.

Otten, U. and Lorez, H. P. 1983. Nerve growth factor increases substance P, cholecystokinin and vasoactive intestinal polypeptide immunoreactivities in primary sensory neurones of newborn rats. *Neurosci. Lett.* **34**, 153–158.

Ottersen, O. P. and Storm-Mathisen, J. 1987. Distribution of inhibitory amino acid neurons in the cerebellum with some observations on the spinal cord: An immunocytochemical study with antisera against fixed GABA, glycine, taurine, and β-alanine. *J. Mind Behav.* **8**, 503–518.

Oudega, M., Lakke, E. A. J. F., Marani, E., and Thomeer, R. T. W. M. 1993. In *Development of the Rat Spinal Cord: Immuno- and Enzyme Histochemical Approaches* (pp. 1–165). Springer-Verlag, New York.

Ouseph, A. K., Khasar, S. G., and Levine, J. D. 1995. Multiple second messenger systems act sequentially to mediate rolipram-induced prolongation of prostaglandin E2-induced mechanical hyperalgesia in the rat. *Neuroscience* **64**, 769–776.

Ovelman-Levitt, J., Johnson, B., Bedenbaugh, P., and Nashold, B. S. 1984. Dorsal root rhizotomy and avulsion in the cat: A comparison of long term effects on dorsal horn neuronal activity. *Neurosurgery* **15**, 921–927.

Ovelmen-Levitt, J., Gorecki, J., Nguyen, K., Iskandar, K., and Nashold, B. S. 1995. Spontaneous and evoked dysesthesias observed in the rat after spinal cordotomies. *Stereotact. Funct. Neurosurg.* **65**, 157–160.

Owens, C. M., Zhang, D., and Willis, W. D. 1992. Changes in the response states of primate spinothalamic tract cells caused by mechanical damage of the skin or activation of descending controls. *J. Neurophysiol.* **67**, 1509–1527.

Owman, C. and Santini, M. 1966. Adrenergic nerves in spinal ganglia of the cat. *Acta Physiol. Scand.* **68**, 127–128.

Pacini, F. 1840. *Nuovi Organi Scoperti nel Corpo Umano.* Ciro, Pistoia. Cited in Pease and Quilliam, 1957.

Pagni, C. A. 1998. *Central Pain. A Neurosurgical Challenge.* Turin, Edizioni Minerva Medica.

Paik, K. S., Nam, S. C., and Chung, J. M. 1988. Differential inhibition produced by peripheral conditioning stimulation on noxious mechanical and thermal responses of different classes of spinal neurons in the cat. *Exp. Neurol.* **99**, 498–511.

Paintal, A. S. 1957. Responses from mucosal mechanoreceptors in the small intestine of the cat. *J. Physiol.* **139**, 353–368.

Paintal, A. S. 1960. Functional analysis of group III afferent fibres of mammalian muscles. *J. Physiol.* **152**, 250–270.

Palacios, J. M., Wamsley, J. K., and Kuhar, M. J. 1981. High affinity GABA receptors—autoradiographic localization. *Brain Res.* **222**, 285–307.

Palacios, J. M., Waeber, C., Hoyer, D., and Mengod, G. 1990. Distribution of serotonin receptors. *Ann. NY Acad. Sci.* **600**, 36–52.

Palecek, J., Paleckova, V., Dougherty, P. M., Carlton, S. M., and Willis, W. D. 1992a. Responses of spinothalamic tract cells to mechanical and thermal stimulation of skin in rats with experimental peripheral neuropathy. *J. Neurophysiol.* **67**, 1562–1573.

Palecek, J., Dougherty, P. M., Kim, S. H., Paleckova, V., Lekan, H., Chung, J. M., Carlton, S. M., and Willis, W. D. 1992b. Responses of spinothalamic tract neurons to mechanical and thermal stimuli in an experimental model of peripheral neuropathy in primates. *J. Neurophysiol.* **68**, 1951–1966.

Palecek, J., Paleckova, V., Dougherty, P. M., and Willis, W. D. 1994a. The effects of phorbol esters on the responses of primate spinothalamic neurons to mechanical and thermal stimuli. *J. Neurophysiol.* **71**, 529–537.

Palecek, J., Paleckova, V., Dougherty, P. M., and Willis, W. D. 1994b. The effect of trans-ACPD, a metabotropic excitatory amino acid receptor agonist, on the responses of primate spinothalamic neurons. *Pain* **56**, 261–269.

Palecek, J., Paleckova, V., and Willis, W. D. 1999. The effect of phorbol esters on spinal cord amino acid concentrations and responsiveness of rats to mechanical and thermal stimuli. *Pain* **80**, 597–605.

Palecek, J., Paleckova, V., and Willis, W. D. 2003. Postsynaptic dorsal column neurons express NK1 receptors following colon inflammation. *Neuroscience*, **116**, 565–572.

Paleckova, V., Palecek, J., McAdoo, D. J., and Willis, W. D. 1992. The non-NMDA antagonist CNQX prevents release of amino acids into the rat spinal cord dorsal horn evoked by sciatic nerve stimulation. *Neurosci. Lett.* **148**, 19–22.

Palmer, C. I. and Gardner, E. P. 1990. Simulation of motion on the skin: IV. Responses of Pacinian corpuscle afferents innervating the primate hand to stripe patterns on the OPTACON. *J. Neurophysiol.* **64**, 236–247.

Pan, H. L., Stahl, G. L., Rendig, S. V., Carretero, O. A., and Longhurst, J. C. 1994. Endogenous BK stimulates ischemically sensitive abdominal visceral C fiber afferents through kinin B2 receptors. *Am. J. Physiol.* **267**, H2398–H2406.

Pan, H. L., Stahl, G. L., and Longhurst, J. C. 1995. Differential effect of 5- and 15-lipoxygenase products on ischemically sensitive abdominal visceral afferents. *Am. J. Physiol.* **269**, H96–H105.

Pang, X., Marchand, J., Sant, G. R., Kream, R. M., and Theoharides, T. C. 1995. Increased number of substance P positive nerve fibres in interstitial cystitis. *Br. J. Urol.* **75**, 744–750.

Panula, P. 1986. Histochemistry and function of bombesin-like peptides. *Med. Biol.* **64**, 177–192.

Panula, P., Yang, H.-Y. T., and Costa, E. 1982. Neuronal location of the bombesin-like immunoreactivity in the central nervous system of the rat. *Reg. Pept.* **4**, 275–283.

Panula, P., Hadjiconstantinou, M., Yang, H. Y. T., and Costa, E. 1983. Immunohistochemical localization of bombesin/gastrin-releasing peptide and substance P in primary sensory neurons. *J. Neurosci.* **3**, 2021–2029.

Panula, P., Kivipelto, L., Nieminen, O., Majane, E. A., and Yank, H. T. T. 1987. Neuroanatomy of morphine-modulating peptides. *Med. Biol.* **65**, 127–135.

Panula, P., Nieminen, O., Falkenberg, M., and Auvinen, S. 1988. Localization and development of bombesin/GRP-like immunoreactivity in the rat central nervous system. *Ann. NY Acad. Sci.* **547**, 54–69.

Panula, P., Pirvola, U., Auvinen, S., and Airaksinen, M. S. 1989. Histamine-immunoreactive nerve fibers in the rat brain. *Neuroscience* **28**, 585–610.

Panula, P., Aarnisalo, A. A., and Wasowicz, K. 1996. Neuropeptide FF, a mammalian neuropeptide with multiple functions. *Prog. Neurobiol.* **48**, 461–487.

Panula, P., Kalso, E., Nieminen, M., Kontinen, V. K., Brandt, A., and Pertovaara, A. 1999. Neuropeptide FF and modulation of pain. *Brain Res.* **848**, 191–196.

Papadopoulos, G. C., Karamanlidis, A. N., Antonopoulos, J., and Dinopoulos, A. 1986a. Neurotensinlike immunoreactive neurons in the hedgehog (*Erinaceus europaeus*) and the sheep (*Ovis aries*) central nervous system. *J. Comp. Neurol.* **244**, 193–203.

Papadopoulos, G. C., Karamanlidis, A. N., Dinopoulos, A., and Antonopoulos, J. 1986b. Somatostatinlike immunoreactive neurons in the hedgehog (*Erinaceus europaeus*) and the sheep (*Ovis aries*) central nervous system. *J. Comp. Neurol.* **244**, 174–192.

Papka, R. E. and McNeill, D. L. 1992. Distribution of NADPH-diaphorase-positive nerves in the uterine cervix and neurons in dorsal root and paracervical ganglia of the female rat. *Neurosci. Lett.* **147**, 224–228.

Papka, R. E., Srinivasan, B., Miller, K. E., and Hayashi, S. 1997. Localization of estrogen receptor protein and estrogen receptor messenger RNA in peripheral autonomic and sensory neurons. *Neuroscience* **79**, 1153–1163.

Pardutz, A., Multon, S., Malgrange, B., Parducz, A., Vecsei, L., and Schoenen, J. 2002. Effect of systemic nitroglycerin on CGRP and 5-HT afferents to rat caudal spinal trigeminal nucleus and its modulation by estrogen. *Eur. J. Neurosci.* **15**, 1803–1809.

Paré, M., Elde, R., Mazurkiewicz, J. E., Smith, A. M., and Rice, F. L. 2001. The Meissner corpuscle revised: A multiafferented mechanoreceptor with nociceptor immunochemical properties. *J. Neurosci.* **21**, 7236–7246.

Paré, M., Smith, A. M., and Rice, F. L. 2002. Distribution and terminal arborizations of cutaneous mechanoreceptors in the glabrous finger pads of the monkey. *J. Comp. Neurol.* **445**, 347–359.

Park, J. S., Higashi, H., Nagata, K., and Yoshimura, M. 1999. Bicuculline-resistant, Cl-dependent GABA response in the rat spinal dorsal horn. *Neurosci. Res.* **33**, 261–268.

Park, M. J., Chung, K., and Chung, J. M. 1994. Immunohistochemical evidence for sprouting of ventral root afferents after neonatal sciatic neurectomy in the rat. *Neurosci. Lett.* **165**, 125–128.

Park, S. K., Chung, K., and Chung, J. M. 2000. Effects of purinergic and adrenergic antagonists in a rat model of painful peripheral neuropathy. *Pain* **87**, 171–179.

Parker, R. M., Fleetwood-Walker, S. M., Rosie, R., Munro, F. E., and Mitchell, R. 1993. Inhibition by NK2 but not NK1 antagonists of carrageenan-induced preprodynorphin mRNA expression in rat dorsal horn lamina I neurons. *Neuropeptides* **25**, 213–222.

Pasternak, G. W. 2001. Insights into mu opioid pharmacology: The role of mu opioid receptor subtypes. *Life Sci.* **68**, 2213–2219.

Patel, R., Kerr, R., and Maxwell, D. J. 1997. Absence of co-localized glutamic acid decarboxylase and neuropeptides in noradrenergic axons of the rat spinal cord. *Brain Res.* **749**, 164–169.

Patrone, C., Andersson, S., Korhonen, L., and Lindholm, D. 1999. Estrogen receptor-dependent regulation of sensory neuron survival in developing dorsal root ganglion. *Proc. Natl. Acad. Sci. USA* **96**, 10905–10910.

Patterson, J. T., Head, P. A., McNeill, D. L., Chung, K., and Coggeshall, R. E. 1989. Ascending unmyelinated primary afferent fibers in the dorsal funiculus. *J. Comp. Neurol.* **290**, 384–390.

Patterson, J. T., Coggeshall, R. E., Lee, W. T., and Chung, K. 1990. Long ascending unmyelinated primary afferent axons in the rat dorsal column: Immunohistochemical localizations. *Neurosci. Lett.* **108**, 6–10.

Pazos, A. and Palacios, J. M. 1985. Quantitative autoradiographic mapping of serotonin receptors in the rat brain: I. Serotonin-1 receptors. *Brain Res.* **346**, 205–230.

Pazos, A., Cortes, R., and Palacios, J. M. 1985. Quantitative autoradiographic mapping of serotonin receptors in the rat brain: II. Serotonin-2 receptors. *Brain Res.* **346**, 231–249.

Pazos, A., Probst, A., and Palacios, J. M. 1987a. Serotonin receptors in the human brain: III. Autoradiographic mapping of serotonin-1 receptors. *Neuroscience* **21**, 97–122.

Pazos, A., Probst, A., and Palacios, J. M. 1987b. Serotonin receptors in the human brain: IV. Autoradiographic mapping of serotonin-2 receptors. *Neuroscience* **21**, 123–139.

Peach, R. 1972. Acid phosphatase distribution in the trigeminal ganglion of the rat. *Anat. Rec.* **174**, 239–250.

Pearson, A. A. 1952. Role of gelatinous substance of spinal cord in conduction of pain. *Arch. Neurol. Psychiatry* **68**, 515–529.

Pearson, J., Brandeis, L., Simon, E., and Hiller, J. 1980. Radioautography of binding of tritiated diprenorphine to opiate receptors in the rat. *Life Sci.* **26**, 1047–1052.

Pearson, J. C. and Jennes, L. 1988. Localization of serotonin- and substance P-like immunofluorescence in the caudal spinal trigeminal nucleus of the rat. *Neurosci. Lett.* **88**, 151–156.

Pease, D. C. and Quilliam, T. A. 1957. Electron microscopy of the Pacinian corpuscle. *J. Biophys. Biochem. Cytol.* **3**, 331–357.

Peet, M. J., Leah, J. D., and Curtis, D. R. 1983. Antagonists of synaptic and amino acid excitation of neurones in the cat spinal cord. *Brain Res.* **266**, 83–95.

Peier, A. M., Moqrich, A., Hergarden, A. C., Reeve, A. J., Andersson, D. A., Story, G. M., Earley, T. J., Dragoni, I., McIntyre, P., Bevan, S., and Patapoutian, A. 2002a. A TRP channel that senses cold stimuli and menthol. *Cell* **108**, 705–715.

Peier, A. M., Moqrich, A., Hergarden, A. C., Reeve, A. J., Andersson, D. A., Story, G. M., Earley, T. J., Dragoni, I., McIntyre, P., Bevan, S., and Patapoutian, A. 2002b. A TRP channel that senses cold stimuli and menthol. *Cell* **108**, 705–715.

Pellegrini-Giampietro, D. E., Fan, S., Ault, B., Miller, B. E., and Zukin, R. S. 1994. Glutamate receptor gene expression in spinal cord of arthritic rats. *J. Neurosci.* **14**, 1576–1583.

Pelletier, G., LeClerc, R., and Dupont, A. 1977. Electron microscope immunohistochemical localization of substance P in the central nervous system of the rat. *J. Histochem. Cytochem.* **25 or 12**, 1373–1380.

Peng, Y. B., Lin, Q., and Willis, W. D. 1997. Involvement of protein kinase C in responses of rat dorsal horn neurons to mechanical stimuli and periaqueductal gray descending inhibition. *Exp. Brain Res.* **114**, 561–570.

Peng, Y. B, Wu, J., Willis, W. D., and Kenshalo, D. R. 2001. $GABA_A$ and 5-HT_3 receptors are involved in dorsal root reflexes: Possible role in periaqueductal gray descending inhibition. *J. Neurophysiol.* **86**, 49–58.

Pereira da Silva, J. A. and Carmo-Fonseca, M. 1990. Peptide containing nerves in human synovium: Immunohistochemical evidence for decreased innervation in rheumatoid arthritis. *J. Rheumatol.* **17**, 1592–1599.

Perez de la Mora, M., Possani, L. D., Tapia, R., Teran, L., Palacios, R., Fuxe, K., Hökfelt, T., and Ljungdahl, A. 1981. Demonstration of central y-aminobutyrate-containing nerve terminals by means of antibodies against glutamate decarboxylase. *Neuroscience* **6**, 875–895.

Perl, E. R. 1968. Myelinated afferent fibres innervating the primate skin and their response to noxious stimuli. *J. Physiol.* **197**, 593–615.

Perl, E. R. 1984. Characterization of nociceptors and their activation of neurons in the superficial dorsal horn: First steps for the sensation of pain. In L. Kruger and J. C. Liebeskind (eds.) *Neural Mechanisms of Pain, Adv. Pain Res. Therapy* (Vol. 6, pp. 23–51). Raven Press, New York.

Perry, M. J. and Lawson, S. N. 1998. Differences in expression of oligosaccharides, neuropeptides, carbonic anhydrase and neurofilament in rat primary afferent neurons retrogradely labelled via skin, muscle or visceral nerves. *Neuroscience* **85**, 293–310.

Perry, M. J., Lawson, S. N., and Robertson, J. 1991. Neurofilament immunoreactivity in populations of rat primary afferent neurons: A quantitative study of phosphorylated and non-phosphorylated subunits. *J. Neurocytol.* **20**, 746–758.

Persohn, E., Malherbe, P., and Richards, J. G. 1991. In situ hybridization histochemistry reveals a diversity of $GABA_A$ receptor subunit mRNAs in neurons of the rat spinal cord and dorsal root ganglia. *Neuroscience* **42**, 497–507.

Persohn, E., Malherbe, P., and Richards, J. G. 1992. Comparative molecular neuroanatomy of cloned $GABA_A$ receptor subunits in the rat CNS. *J. Comp. Neurol.* **326**, 193–216.

Persson, J. K., Lindh, B., Elde, R., Robertson, B., Rivero-Melian, C., Eriksson, N. P., Hökfelt, T., and Aldkogius, H. 1995. The expression of different cytochemical markers in normal and axotomised dorsal root ganglion cells projecting to the nucleus gracilis in the adult rat. *Exp. Brain Res.* **105**, 331–344.

Persson, S., Schafer, M. K., Nohr, D., Ekstrom, G., Post, C., Nyberg, F., and Weihe, E. 1994. Spinal prodynorphin gene expression in collagen-induced arthritis: Influence of the glucocorticosteroid budesonide. *Neuroscience* **63**, 313–326.

Pert, C. B., Kuhar, M. J., and Snyder, S. H. 1975. Autoradiographic localization of the opiate receptor in rat brain. *Life Sci.* **16**, 1849–1854.

Pert, C. B., Kuhar, M. J., and Snyder, S. H. 1976. Opiate receptor: Autoradiographic localization in rat brain. *Proc. Natl. Acad. Sci. USA* **73**, 3729–3733.

Pertwee, R. G. 2001. Cannabinoid receptors and pain. *Prog. Neurobiol.* **63**, 569–611.

Peterfreund, R. A., Kosofsky, B. E., and Fink, J. S. 1995. Cellular localization of dopamine D2 receptor messenger RNA in the rat trigeminal ganglion. *Anesth. Analg.* **81**, 1181–1185.

Peters, A. A., Palay, S., and Webster, H. de F. 1976. *The Fine Structure of the Nervous System: The Neurons and Supporting Cells*. W. B. Saunders Co., Philadelphia.

Petersen, M, Zhang, J., Zhang, J.-M., and LaMotte, R. H. 1996. Abnormal spontaneous activity and response to norepinephrine in dissociated dorsal root ganglion cells after chronic nerve constriction. *Pain* **67**, 391–397.

Petersen, M., Eckert, A. S., Von Banchet, G. S., Heppelmann, B., Klusch, A., and Kniffki, K.-D. 1998a. Plasticity in the expression of bradykinin binding sites in sensory neurons after mechanical nerve injury. *Neuroscience* **83**, 949–959.

Petersen, M., Segond von Banchet, G., Heppelmann, B., and Koltzenburg, M. 1998b. Nerve growth factor regulates the expression of bradykinin binding sites on adult sensory neurons via the neurotrophin receptor p75. *Neuroscience* **83**, 161–168.

Peterson, D. F. and Brown, A. M. 1973. Functional afferent innervation of testis. *J. Neurophysiol.* **36**, 425–433.

Petko, M. and Antal, M. 2000. Propriospinal afferent and efferent connections of the lateral and medial areas of the dorsal horn (laminae I–IV) in the rat lumbar spinal cord. *J. Comp. Neurol.* **422**, 312–325.

Petralia, R. S., Wang, Y.-X., and Wenthold, R. J. 1994a. Histological and ultrastructural localization of the kainate receptor subunits, KA2 and GluR6/7, in the rat nervous system using selective antipeptide antibodies. *J. Comp. Neurol.* **349**, 85–110.

Petralia, R. S., Yokotani, N., and Wenthold, R. J. 1994b. Light- and electron-microscope distribution of the NMDA receptor subunit NMDAR1 in the rat nervous system using a selective anti-peptide antibody. *J. Neurosci.* **14**, 667–696.

Petralia, R. S., Wang, Y.-X., and Wenthold, R. J. 1996a. The NMDA receptor subunits NR2A and NR2B show histological and ultrastructural localization patterns similar to those of NR1. *J. Neurosci.* **14**, 6102–6120.

Petralia, R. S., Wang, Y. X., Niedzielski, A. S., and Wenthold, R. J. 1996b. The metabotropic glutamate receptors, mGluR2 and mGluR3, show unique postsynaptic, presynaptic and glial localizations. *Neuroscience* **71**, 949–976.

Petralia, R. S., Wang, Y. X., Mayat, E., and Wenthold, R. J. 1997. Glutamate receptor subunit 2-selective antibody shows a differential distribution of calcium-impermeable AMPA receptors among populations of neurons. *J. Comp. Neurol.* **385**, 456–476.

Petruska, J. C., Cooper, B. Y., Gu, J. G., Rau, K. K., and Johnson, R. D. 2000. Distribution of P2XI, P2X2, and P2X3 receptor subunits in rat primary afferents: relation to population markers and specific cell types. *J. Chem. Neuroanat.* **20**, 141–162.

Petrusz, P., Merchenthaler, I., Maderdrut, J. L., and Heitz, P. U. 1985. Central and peripheral distribution of corticotropin-releasing factor. *Fed. Proc.* **44**, 229–235.

Pettersson, L. M., Sundler, F., and Danielsen, N. 2002. Expression of orphanin FQ/nociceptin and its receptor in rat peripheral ganglia and spinal cord. *Brain Res.* **945**, 266–275.

Peyronnard, J. M., Charron, L., Lavoie, J., and Messier, J. 1986. Motor, sympathetic and sensory innervation of rat skeletal muscles. *Brain Res.* **373**, 288–302.

Peyronnard, J. M., Charron, L., Messier, J., and Zlavoie, J. 1988a. Differential effects of distal and proximal nerve lesions on carbonic anhydrase activity in rat primary sensory neurons, ventral and dorsal root axons. *Exp. Brain Res.* **70**, 550–560.

Peyronnard, J. M., Charron, L., Lavoie, J., Messier, J. P., and Dubreuil 1988b. Carbonic anhydrase and horseradish peroxidase: Double labeling of rat dorsal root ganglion neurons innervating motor and sensory peripheral nerves. *Anat. Embryol.* **177**, 353–359.

Pezet, S., Onteniente, B., Jullien, J., Junier, M. P., Grannec, G., Rudkin, B. B., and Calvino, B. 2001. Differential regulation of NGF receptors in primary sensory neurons by adjuvant-induced arthritis in the rat. *Pain* **90**, 113–125.

Phelps, P. E., Barber, R. P., Houser, C. R., Crawford, G. D., Salvaterra, P. M., and Vaughn, J. E. 1984. Postnatal development of neurons containing choline acetyltransferase in rat spinal cord: An immunocytochemical study. *J. Comp. Neurol.* **229**, 347–361.

Phelps, P. E., Barber, R. P., and Vaughn, J. E. 1988. Generation patterns of four groups of cholinergic neurons in rat cervical spinal cord: A combined tritiated thymidine autoradiographic and choline acetyltransferase immunocytochemical study. *J. Comp. Neurol.* **273**, 459–472.

Phelps, P. E., Barber, R. P., Brennan, L. A., Maines, V. M., Salvaterra, P. M., and Vaughn, J. E. 1990. Embryonic development of four different subsets of cholinergic neurons in rat cervical spinal cord. *J. Comp. Neurol.* **291**, 9–26.

Phillips, L. H., Park, T. S., Shaffrey, M. E., and Shaffrey, C. L. 2000. Electrophysiological evidence for afferent nerve fibers in human ventral roots. *Muscle Nerve* **23**, 410–415.

Pickel, V. M., Reis, D. J., and Leeman, S. E. 1977. Ultrastructural localization of substance P in neurons of rat spinal cord. *Brain Res.* **122**, 534–540.

Piehl, F., Arvidsson, U., Johnson, H., Dagerlind, Å., Hökfelt, T., Terenius, L., and Ulfhake, B. 1992. Reappearance of calcitonin gene-related peptide-like immunoreactivity in the dorsal horn in long-term root transected rat. *Brain Res.* **585**, 400–404.

Piepmeier, J. M., Kauer, J. S. and Greer, C. A. 1983. Laminar distributions of 2-deoxyglucose uptake in the rat spinal cord following electrical stimulation of the sciatic nerve. *Brain Res.* **259**, 167–171.

Pierau, F. K., Torrey, P., and Carpenter, D. O. 1975. Afferent nerve fiber activity responding to temperature changes of scrotal skin of the rat. *J. Neurophysiol.* **38**, 601–612.

Pierau, F. K., Taylor, D. C. M., Abel, W., and Friedrich, B. 1982. Dichotomizing peripheral fibres revealed by intracellular recording from rat sensory neurones. *Neurosci. Lett.* **31**, 123–128.

Pierau, F. K., Fellmer, G., and Taylor, D. C. 1984. Somato-visceral convergence in cat dorsal root ganglion neurones demonstrated by double-labelling with fluorescent tracers. *Brain Res.* **321**, 63–70.

Pierce, P. A., Xie, G.-X., Levine, J. D., and Peroutka, S. J. 1996. 5-hydroxytryptamine receptor subtype messenger RNAs in rat peripheral sensory and sympathetic ganglia: A polymerase chain reaction study. *Neuroscience* **70**, 553–559.

Pierce, P. A., Xie, G.-X., Meuser, T., and Peroutka, S. J. 1997. 5-Hydroxytryptamine receptor subtype messenger RNAs in human dorsal root ganglia: A polymerase chain reaction study. *Neuroscience* **81**, 813–819.

Pierce, T. L. and Wessendorf, M. W. 2000. Immunocytochemical mapping of endomorphin-2-immunoreactivity in rat brain. *J. Chem. Neuroanat.* **18**, 181–207.

Pierce, T. L., Grahek, M. D., and Wessendorf, M. W. 1998. Immunoreactivity for endomorphin-2 occurs in primary afferents in rats and monkey. *NeuroReport* **9**, 385–389.

Pin, J. P. and Duvoisin, R. 1995. The metabotropic glutamate receptors: Structure and functions. *Neuropharmacology* **34**, 1026.

Pindzola, R. R., Ho, R. H., and Martin, G. F. 1988. Catecholaminergic innervation of the spinal cord in the North American opossum, *Didelphis virginiana. Brain Behav. Evol.* **32**, 281–292.

Pindzola, R. R., Ho, R. H., and Martin, G. F. 1990. Development of catecholaminergic projections to the spinal cord in the North American opossum, *Didelphis virginiana. J. Comp. Neurol.* **294**, 399–417.

Pinkus, H. 1964. Pinkus's Haarscheibe and tactile receptors in cats. *Science* **144**, 891.

Pirker, S., Schwarzer, C., Wieselthaler, A., Sieghart, W., and Sperk, G. 2000. GABA(A) receptors: Immunocytochemical distribution of 13 subunits in the adult rat brain. *Neuroscience* **101**, 815–850.

Plantinga, L. C., Verhaagen, J., Edwards, P. M., Schrama, L. H., Burbach, J. P., and Gispen, W. H. 1992. Expression of the pro-opiomelanocortin gene in dorsal root ganglia, spinal cord and sciatic nerve after sciatic nerve crush in the rat. *Mol. Brain Res.* **16**, 135–142.

Plenderleith, M. B., Cameron, A. A., Key, B., and Snow, P. J. 1988. Soybean agglutinin binds to a subpopulation of primary sensory neurones in the cat. *Neurosci. Lett.* **86**, 257–262.

Plenderleith, M. B., Haller, C. J., and Snow, P. J. 1990. Peptide coexistence in axon terminals within the superficial dorsal horn of the rat spinal cord. *Synapse* **6**, 344–350.

Plourde, V., St-Pierre, S., and Quirion, R. 1997. Calcitonin gene-related peptide in viscerosensitive response to colorectal distension in rats. *Am. J. Physiol.* **273**, G191–196.

Poggio, G. F. and Mountcastle, V. B. 1960. A study of the functional contributions of the lemniscal and spinothalamic systems to somatic sensibility. *Bull. Johns Hopkins Hosp.* **106**, 266–316.

Poggio, G. F. and Mountcastle, V. B. 1963. The functional properties of ventrobasal thalamic neurons studied in unanesthetized monkeys. *J. Neurophysiol.* **26**, 775–806.

Pohl, M., Benoliel, J. J., Bourgoin, S., Lombard, M. C., Mauborgne, A., Taquet, H., Carayon, A., and Besson, J. M. 1990. Regional distribution of calcitonin gene-related peptide-, substance P-, cholecystokinin-, met5-enkephalin-, and dynorphin A (1–8)-like materials in the spinal cord and dorsal root ganglia of adult rats: Effects of dorsal rhizotomy and neonatal capsaicin. *J. Neurochem.* **55**, 1122–1130.

Pohl, M., Collin, E., Bourgoin, S., Conrath, M., Benoliel, J. J., Nevo, I., Hamon, M., Giraud, P., and Cesselin, F. 1994. Expression of preproenkephalin A gene and presence of met-enkephalin in dorsal root ganglia of the adult rat. *J. Neurochem.* **63**, 1226–1234.

Pohl, M., Ballet, S., Collin, E., Mauborgne, A., Bourgoin, S., Benoliel, J. J., Hamon, M., and Cesselin, F. 1997. Enkephalinergic and dynorphinergic neruons in the spinal cord and dorsal root ganglia of the polyarthritic rat—in vivo release and cDNA hybridization studies. *Brain Res.* **749**, 18–28.

Pokabla, M. J., Dickerson, I. M., and Papka, R. E. 2002. Calcitonin gene-related peptide-receptor component protein expression in the uterine cervix, lumbosacral spinal cord, and dorsal root ganglia. *Peptides* **23**, 507–514.

Polacek, P. 1961. Differences in the structure and variability of encapsulated nerve endings in the joints of some species of mammals. *Acta Anat.* **47**, 112–124.

Polak, J. M. and Bloom, S. R. 1982. The central and peripheral distribution of neurotensin. *Ann. NY Acad. Sci.* **400**, 75–93.

Polgar, E. and Antal, M. 1995. The colocalization of parvalbumin and calbindin-D28k with GABA in the subnucleus caudalis of the rat spinal trigeminal nucleus. *Exp. Brain Res.* **103**, 402–408.

Polgar, E., Szuca, P., Urban, L., and Nagy, I. 1998. Alterations of substance P immunoreactivity in lumbar and thoracic segments of rat spinal cord in ultraviolet irradiation induced hyperalgesia of the hindpaw. *Brain Res.* **786**, 248–251.

Polgar, E., Fowler, J. H., McGill, M. M., and Todd, A. J. 1999a. The types of neuron which contain protein kinase C gamma in rat spinal cord. *Brain Res.* **833**, 71–80.

Polgar, E., Shehab, S. A., Watt, C., and Todd, A. J. 1999b. GABAergic neurons that contain neuropeptide Y selectively target cells with the neurokinin 1 receptor in laminae III and IV of the rat spinal cord. *J. Neurosci.* **19**, 2637–2646.

Polgar, E., Szucs, P., Urban, L., Matesz, K., and Nagy, I. 1999c. Immunohistochemical localization of neurokinin-l receptor in the lumbar spinal cord of young rats: Morphology and distribution. *Somatosens. Mot. Res.* **16**, 361–368.

Polgar, E., Puskar, Z., Watt, C., Matesz, C., and Todd, A. J. 2002. Selective innervation of lamina I projection neurones that possess the neurokinin 1 receptor by serotonin-containing axons in the rat spinal cord. *Neuroscience* **109**, 799–809.

Polistina, D. C., Murray, M., and Goldberger, M. E. 1990. Plasticity of dorsal root and descending serotonergic projections after partial deafferentation of the adult rat spinal cord. *J. Comp. Neurol.* **299**, 349–363.

Polz-Tejera, G., Hunt, S. P., and Schmidt, J. 1980. Nicotinic receptors in sensory ganglia. *Brain Res.* **195**, 223–230.

Pomeranz, B. and Cheng, R. 1979. Suppression of noxious responses in single neurons of cat spinal cord by electroacupuncture and its reversal by the opiate antagonist naloxone. *Exp. Neurol.* **64**, 327–341.

Pomeranz, B., Wall, P. D., and Weber, W. V. 1968. Cord cells responding to fine myelinated afferents from viscera, muscle, and skin. *J. Physiol.* **199**, 511–532.

Pomonis, J. D., Rogers, S. D., Peters, C. M., Ghilardi, J. R., and Mantyh, P. W. 2001. Expression and localization of endothelin receptors: Implications for the involvement of peripheral glia in nociception. *J. Neurosci* **21**, 999–1006.

Pompeiano, M., Palacios, J. M., and Mengod, G. 1992. Distribution and cellular localization of mRNA coding for 5-HT1A receptor in the rat brain: Correlation with receptor binding. *J. Neurosci.* **12**, 440–453.

Pompeiano, M., Palacios, J. M., and Mengod, G. 1994. Distribution of the serotonin 5-HT_2 receptor family mRNAs: Comparison between 5-HT_{2A} and 5-HT_{2C} receptors. *Mol. Brain Res.* **23**, 163–178.

Poorkhalkali, N., Juneblad, K., Jonsson, A. C., Lindberg, M., Karlsson, O., Wallbrandt, P., Ekstrand, J., and Lehmann, A. 2000. Immunocytochemical distribution of the GABA(B) receptor splice variants GABA(B) R1a and R1b in the rat CNS and dorsal root ganglia. *Anat. Embryol. (Berl.)* **201**, 1–13.

Popratiloff, A., Weinberg, R. J., and Rustioni, A. 1996. AMPA receptor subunits underlying terminals of fine-caliber primary afferent fibers. *J. Neurosci.* **16**, 3363–3372.

Popratiloff, A., Weinberg, R. J., and Rustioni, A. 1998a. AMPA receptors at primary afferent synapses in substantia gelatinosa after sciatic nerve section. *Eur. J. Neurosci.* **10**, 3220–3230.

Popratiloff, S. A., Weinberg, J. R., and Rustioni, A. 1998b. NMDAR1 and primary afferent terminals in the superficial spinal cord. *NeuroReport* **9**, 2423–2429.

Porreca, F., Tang, Q.-B., Bian, D., Ried, M., Elde, R., and Lai, J. 1998. Spinal opioid mu receptor expression in lumbar spinal cord of rats following nerve injury. *Brain Res.* **795**, 197–203.

Poulos, D. A. 1971. Temperature related changes in discharge patterns of squirrel monkey thermoreceptors. In F. F. Kao, K. Koizumi, and M. Vassalle (eds.), *Research in Physiology* (pp. 441–455). A. Gaggi, Bologna.

Poulos, D. A. and Lende, R. A. 1970a. Response of trigeminal ganglion neurons to thermal stimulation of oral-facial regions: I. Steady-state response. *J. Neurophysiol.* **33**, 508–517.

Poulos, D. A. and Lende, R. A. 1970b. Response of trigeminal ganglion neurons to thermal stimulation of oral-facial regions: II. Temperature change response. *J. Neurophysiol.* **33**, 518–526.

Poulter, M. O., Barker, J. L., O'Carroll, A.-M., Lolait, S. J., and Mahan, L. C. 1992. Differential and transient expression of $GABA_A$ receptor α- subunit mRNAs in the developing rat CNS. *J. Neurosci.* **12**, 2888–2900.

Poulter, M. O., Barker, J. L., O'Carroll, A.-M., Lolait, S. J., and Mahan, L. C. 1993. Co-existent expression of $GABA_A$ receptor β_2, β_3 and δ_2 subunit messenger RNAs during embryogenesis and early postnatal development of the rat central nervous system. *Neuroscience* **53**, 1019–1033.

Powell, J. J. and Todd, A. J. 1992. Light- and electron-microscope study of GABA-immunoreactive neurones in lamina III of rat spinal cord. *J. Comp. Neurol.* **315**, 125–136.

Prabhakar, E. and Lawson, S. N. 1995. The electrophysiological properties of rat primary afferent neurones with carbonic anhydrase activity. *J. Physiol.* 482(Pt 3), 609–622.

Presley, R. W., Menétrey, D., Levine, J. D., and Basbaum, A. I. 1990. Systemic morphine suppresses noxious stimulus-evoked Fos protein-like immunoreactivity in the rat spinal cord. *J. Neurosci.* **10**, 323–335.

Pretel, S. and Piekut, D. T. 1991. Enkephalin, substance P, and serotonin axonal input to c-fos-like immunoreactive neurons of the rat spinal cord. *Peptides* **12**, 1243–1250.

Price, D. D. 1999. *Psychological Mechanisms of Pain and Analgesia. Progress in Pain Research and Management*, Vol. 15. IASP Press, Seattle.

Price, D. D. and Browe, A. C. 1973. Responses of spinal cord neurons to graded noxious and non-noxious stimuli. *Brain Res.* **64**, 425–429.

Price, D. D. and Dubner, R. 1977. Neurons that subserve the sensory-discriminative aspects of pain. *Pain* **3**, 307–338.

Price, D. D. and Mayer, D. J. 1974. Physiological laminar organization of the dorsal horn of *M. mulatta. Brain Res.* **79**, 321–325.

Price, D. D. and Wagman, I. H. 1970. Physiological roles of A and C fiber inputs to spinal dorsal horn of *Macaca mulatta. Exp. Neurol.* **29**, 383–399.

Price, D. D., Hull, C. D., and Buchwald, N. A. 1971. Intracellular responses of dorsal horn cells to cutaneous and sural nerve A and C fiber stimuli. *Exp. Neurol.* **33**, 291–309.

Price, D. D., Hayashi, H., Dubner, R., and Ruda, M. A. 1979. Functional relationships between neurons of marginal and substantia gelatinosa layers of primate dorsal horn. *J. Neurophysiol.* **42**, 1590–1608.

Price, D. D., Bushnell, M. C., and Iadarola, M. J. 1981. Primary afferent and sacral dorsal horn neuronal responses to vaginal probing in the cat. *Neurosci. Lett.* **26**, 67–72.

Price, D. D., Mao, J., Coghill, R. C., d'Avella, D., Cicciarello, R., Fiori, M. G., Mayer, D. J., and Hayes, R. L. 1991. Regional changes in spinal cord glucose metabolism in a rat model of painful neuropathy. *Brain Res.* **564**, 314–318.

Price, G. W., Wilkin, G. P., Turnbull, M. J., and Bowery, N. G. 1984. Are baclofen-sensitive $GABA_B$ receptors present on primary afferent terminals of the spinal cord? *Nature* **307**, 71–74.

Price, G. W., Kelly, J. S., and Bowery, N. G. 1987. The location of $GABA_B$ receptor binding sites in mammalian spinal cord. *Synapse* **1**, 530–538.

Price, J. 1985. An immunohistochemical and quantitative examination of dorsal root ganglion neuronal subpopulations. *J. Neurosci.* **5**, 2051–2059.

Price, J. and Mudge, A. W. 1983. A subpopulation of rat dorsal root ganglion neurones is catecholaminergic. *Nature* **301**, 241–245.

Price, M. P., Snyder, P. M., and Welsh, M. J. 1996. Cloning and expression of a novel human brain Na^+ channel. *J. Biol. Chem.* **271**, 7879–7882.

Price, M. P., Lewin, G. R., McIlwrath, S. L., Cheng, C., Xie, J., Heppenstall, P. A., Stucky, C. L., Mannsfeldt, A. G., Brennan, T. J., Drummond, H. A., Qiao, J., Benson, C. J., Tarr, D. E., Hrstka, R. F., Yang, B., Williamson, R. A., and Welsh, M. J. 2000. The mammalian sodium channel BNC1 is required for normal touch sensation. Nature **407**, 1007–1011.

Price M. P., McIlwrath, S. L., Xie, J., Cheng, C., Qiao, J., Tarr, D. E., Sluka, K. A., Brennan, T. J., Lewin, G. R., and Welsh, M. J. 2001. The DRASIC cation channel contributes to the detection of cutaneous touch and acid stimuli in mice. *Neuron* **32**, 1071–1083.

Priestley, J. V., Bramwell, S., Butcher, L. L., and Cuello, A. C. 1982a. Effect of capsaicin on neuropeptides in areas of termination of primary sensory neurones. *Neurochem. Intern.* **4**, 57–65.

Priestley, J. V., Somogyi, P., and Cuello, A. C. 1982b. Immunocytochemical localization of substance P in the spinal trigeminal nucleus of the rat: A light- and electron-microscopy study. *J. Comp. Neurol.* **211**, 31–49.

Probst, A., Cortest, R., and Palacios, J. M. 1986. The distribution of glycine receptors in the human brain: A light micorscopic autoradiographic study using [3H] strychnine. *Neuroscience* **17**, 11–35.

Prochazka, A., Stephens, J. A., and Wand, P. 1979. Muscle spindle discharges in normal and obstructed movements. *J. Physiol.* **287**, 57–66.

Proshansky, E. and Egger, M. D. 1977. Staining of the dorsal root projection to the cat's dorsal horn by anterograde movement of horseradish peroxidase. *Neurosci. Lett.* **5**, 103–110.

Proshansky, E., Kauer, J. S., Stewart, W. B., and Egger, M. D. 1980. 2-deoxyglucose uptake in the cat spinal cord during sustained and habituated activity in the plantar cushion reflex pathway. *J. Comp. Neurol.* **194**, 505–517.

Proudlock, F., Spike, R. C., Todd, A. J. 1993. Immunocytochemical study of somatostatin, neurotensin, GABA and glycine in rat spinal dorsal horn. *J. Comp. Neurol.* **327**, 289–297.

Przewlocka, B., Lason, W., and Przewlocki, R. 1992. Time-dependent changes in the activity of opioid systems in the spinal cord of monoarthritic rats—a release and in situ hybridization study. *Neuroscience* **46**, 209–216.

Przewlocki, R. and Przewlocka, B. 2001. Opioids in chronic pain, *Eur. J. Pharmacol.* **429**, 79–91.

Przewlocki, R., Gramsch, C., Pasi, A., and Herz, A. 1983. Characterization and localization of immunoreactive dynorphin, alpha-neo-endorphin, met-enkephalin and substance P in human spinal cord. *Brain Res.* **280**, 95–103.

Przewlocki, R., Haarmann, I., Nikolarakis, K., Herz, A., and Hollt, V. 1988. Prodynorphin gene expression in spinal cord is enhanced after traumatic injury in the rat. *Mol. Brain Res.* **4**, 37–41.

Przewlocki, R., Hassan, A. H. S., Lason, W., Epplen, C., Herz, A., and Stein, C. 1992. Gene expression and localization of opioid peptides in immune cells of inflamed tissue: Functional role in antinociception. *Neuroscience* **48**, 491–500.

Przewlocka, B., Machelska, H., Rekowski, P., Kupryszewski, G., and Przewlocki, R. 1995. Intracerebroventricular galanin and N-terminal galanin fragment enhance the morphine-induced analgesia in the rat. *J. Neural Transm. Gen. Sect.* **102**, 229–235.

Przewlocka, B., Mika, J., Labuz, D., Toth, G., and Przewlocki, R. 1999. Spinal analgesic action of endomorphins in acute, inflammatory and neuropathic pain in rats. *Eur. J. Pharmacol.* **367**, 189–196.

Pubols, B. H. 1990. Slowly adapting Type I mechanoreceptor discharge as a function of dynamic force versus dynamic displacement of glabrous skin of raccoon and squirrel monkey hand. *Neurosci. Lett.* **110**, 86–90.

Pubols, B. H. and Benkich, M. E. 1986. Relations between stimulus force, skin displacement, and discharge characteristics of slowly adapting type I cutaneous mechanoreceptors in glabrous skin of squirrel monkey hand. *Somatosens. Res.* **4**, 111–125.

Pubols, B. H. and Pubols, L. M. 1976. Coding of mechanical stimulus velocity and indentation depth by squirrel monkey and raccoon glabrous skin mechanoreceptors. *J Neurophysiol.* **39**, 773–787.

Pubols, B. H. and Pubols, L. M. 1983. Tactile receptor discharge and mechanical properties of glabrous skin. *Fed. Proc.* **42**, 2528–2535.

Pubols, L. M. 1984. The boundary of proximal hindlimb representation in the dorsal horn following peripheral nerve lesions in cats: A reevaluation of plasticity in the somatosensory map. *Somatosens. Res.* **2**, 19–32.

Pubols, L. M. and Bowen, D. C. 1988. Lack of central sprouting of primary afferent fibers after ricin deafferentation. *J. Comp. Neurol.* **275**, 282–287.

Pubols, L. M. and Brenowitz, G. L. 1981. Maintenance of dorsal horn somatotopic organization and increased high-threshold response after single root or spared root deafferentation in cats. *J. Neurophysiol.* **47**, 103–112.

Pubols, L. M. and Goldberger, M. E. 1980. Recovery of function in dorsal horn following partial deafferentation. *J. Neurophysiol.* **43**, 102–117.

Pubols, L. M., Pubols, B. H., and Munger, B. L. 1971. Functional properties of mechanoreceptors in glabrous skin of the raccoon's forepaw. *Exp. Neurol.* **31**, 165–182.

Pubols, L. M., Foglesong, M. E., and Vahle-Hinz, C. 1986. Electrical stimulation reveals relatively ineffective sural nerve projections to dorsal horn neurons in the cat. *Brain Res.* **371**, 109–122.

Pubols, L. M., Bernau, N. A., Kane, L. A., Dawson, S. D., Burleigh, A. L., and Polans, A. S. 1992. Distribution of 5-HT_1 binding sites in cat spinal cord. *Neurosci. Lett.* **142**, 111–114.

Puder, B. A. and Papka, R. E. 2001a. Distribution and origin of corticotropin-releasing factor-immunoreactive axons in the female rat lumbosacral spinal cord. *J. Neurosci. Res.* **66**, 1217–1225.

Puder, B. A. and Papka, R. E. 2001b. Hypothalamic paraventricular axons projecting to the female rat lumbosacral spinal cord contain oxytocin immunoreactivity. *J. Neurosci. Res.* **64**, 53–60.

Puig, S. and Sorkin, L. S. 1995. Formalin-evoked activity in identified primary afferent fibers: Systemic lidocaine suppresses phase-2 activity. *Pain* **64**, 345–355.

Puil, E. 1981. S-glutamate: Its interactions with spinal neurons. *Brain Res. Rev.* **3**, 299–322.

Pullen, A. H., Humphreys, P., and Baxter, R. G. 1997. Comparative analysis of nitric oxide synthase immunoreactivity in the sacral cord of the cat, macaque and human. *J. Anat.* **191**, 161–175.

Puskar, Z., Polgar, E., and Todd, A. J. 2001. A population of large lamina I projection neurons with selective inhibitory input in rat spinal cord. *Neuroscience* **102**, 167–176.

Puttfarcken, P. S., Manelli, A. M., Arneric, S. P., and Donnelly-Roberts, D. L. 1997. Evidence for nicotinic receptors potentially modulating nociceptive transmission at the level of the primary sensory neuron: Studies with F11 cells. *J. Neurochem.* **69**, 930–938.

Quartara, L. and Maggi, C. A. 1998. The tachykinin NK1 receptor: Part II: Distribution and pathophysiological roles. *Neuropeptides* **32**, 1–49.

Quartu, M., Serra, M. P., Bachis, A., Lai, M. L., Ambu, R., and Del Fiacco, M. 1999. Glial cell line-derived neurotrophic factor-like immunoreactivity in human trigeminal ganglion and nucleus. *Brain Res.* **847**, 196–202.

Quevedo, J., Eguibar, J. R., Lomeli, J., and Rudomin, P. 1997. Patterns of connectivity of spinal interneurons with single muscle afferents. *Exp. Brain Res.* **115**, 387–402.

Qui, S. H., Dai, S. J., and Wu, H. X. 1984. The segmental and regional projections of the sciatic, tibial and common peroneal nerves to the substantia gelatinosa of the spinal cord in rats—an experimental study by means of an acid phosphatase (ACP) method. *Okajimas Folia Anat. Jpn.* **61**, 245–251.

Quillium, T. A. 1975. Neuro-cutaneous relationships in fingerprint skin. In H. H. Kornhuber (ed.), *The Somatosensory System* (pp. 193–199). Georg Thieme Verlag, Stuttgart.

Quillium, T. A. and Sato, M. 1955. The distribution of myelin on nerve fibres from Pacinian corpuscles. *J. Physiol.* **129**, 167–176.

Quirion R., van Rossum, D., Dumont, Y., St-Pierre, S., and Fournier, A. 1992. Characterization of $CGRP_1$ and $CGRP_2$ receptor subtypes. *Ann. NY Acad. Sci.* **657**, 88–105.

Racca, C., Gardiol, A., and Triller, A. 1998. Cell-specific dendritic localization of glycine receptor α subunit messenger RNAs. *Neuroscience* **84**, 997–1012.

Radhakrishnan, V. and Henry, J. L. 1991. Novel substance P antagonist, CP-96,345, blocks responses of cat spinal dorsal horn neurons to noxious cutaneous stimulation and to substance P. *Neurosci. Lett.* **132**, 39–43.

Radhakrishnan, V. and Henry, J. L. 1993. Excitatory amino acid receptor mediation of sensory inputs to functionally identified dorsal horn neurons in cat spinal cord. *Neuroscience* **55**, 531–544.

Radhakrishnan, V. and Henry, J. L. 1995. Antagonism of nociceptive responses of cat spinal dorsal horn neurons in vivo by the NK-1 receptor antagonists CP-96,345 and CP-99,994, but not by CP-96,344. *Neuroscience* **64**, 943–958.

Radhakrishnan, V. and Henry, J. L. 1997. Electrophysiological evidence that neurokinin A acts via NK-1 receptors in the cat dorsal horn. *Eur. J. Neurosci.* **9**, 1977–1985.

Radhakrishnan, V., Iyengar, S., and Henry, J. L. 1998. The nonpeptide NK-1 receptor antagonists LY303870 and LY306740 block the responses of spinal dorsal horn neurons to substance P and to peripheral noxious stimuli. *Neuroscience* **893**, 1251–1260.

Radja, F., Laporte, A.-M., Daval, G., Verge, D., Gozlan, H., and Hamon, M. 1991. Autoradiography of serotonin receptor subtypes in the central nervous system. *Neurochem. Int.* **18**, 1–15.

Rahman, W., Dashwood, M. R., Fitzgerald, M., Aynsley-Green, A., and Dickenson, A. H. 1998. Postnatal development of multiple opioid receptors in the spinal cord and development of spinal morphine analgesia. *Dev. Brain Res.* **108**, 239–254.

Raidoo, D. M. and Bhoola, K. D. 1997. Kinin receptors on human neurones. *J. Neuroimmunol.* **77**, 39–44.

Raja, S. N., Treede, R.-D., Davis, K. D., and Campbell, J. N. 1991. Systemic alpha-adrenergic blockade with phentolamine: A diagnostic test for sympathetically maintained pain. *Anesthesiology* **74**, 691–698.

Rajaofetra, N., Sandillon, F., Geffard, M., and Privat, A. 1989. Pre- and post-natal ontogeny of serotonergic projections to the rat spinal cord. *J. Neurosci. Res.* **22**, 305–321.

Rajaofetra, N., Ridet, J.-L., Poulat, P., Marlier, L., Sandillon, F., Geffard, M., and Privat, A. 1992a. Immunocytochemical mapping of noradrenergic projections to the rat spinal cord with an antiserum against noradrenaline. *J. Neurocytol.* **21**, 481–494.

Rajaofetra, N., Poulat, P., Marlier, L., Geffard, M., and Privat, A. 1992b. Pre- and postnatal development of noradrenergic projections to the rat spinal cord: An immunocytochemical study. *Brain Res. Dev. Brain Res.* **67**, 237–246.

Rakic, P. 1975. Local circuit neurons. *Neurosci Res. Prog. Bull.* **13**, 295–416.

Ralston, D. D., Behbehani, M., Sehlhorst, S. C., Meng, X. W., and Ralston, H. J. 1997. Decreased GABA immunoreactivity in rat dorsal horn is correlated with pain behavior: A light- and electron-microscopic study. In T. S. Jensen, J. A. Turner, and Z. Wiesenfeld-Hallin (eds.), *Proceedings of the 8th World Congress on Pain* (pp. 547–560). IASP Press, Seattle.

Ralston, H. J. 1965. The organization of the substantia gelatinosa Rolandi in the cat lumbosacral cord. *Z. Zellforsch.* **67**, 1–23.

Ralston, H. J. 1968a. The fine structure of neurons in the dorsal horn of the cat spinal cord. *J. Comp. Neurol.* **132**, 275–302.

Ralston, H. J. 1968b. Dorsal root projections to dorsal horn neurons in the cat spinal cord. *J. Comp. Neurol.* **132**, 303–330.

Ralston, H. J. 1971. The synaptic organization in the dorsal horn of the spinal cord and in the ventrobasal thalamus in the cat. In R. Dubner and Y. Kawamura (eds.), *Oral-Facial Sensory and Motor Mechanisms* (pp. 229–250). Appleton-Century Crofts, New York.

Ralston, H. J. 1979. The fine structure of laminae I, II and III of the macaque spinal cord. *J. Comp. Neurol.* **184**, 619–642.

Ralston, H. J. 1982. The fine structure of laminae IV, V, and VI of the macaque spinal cord. *J. Comp. Neurol.* **212**, 425–434.

Ralston, H. J. and Ralston, D. D. 1979. The distribution of dorsal root axons in laminae I, II and III of the macaque spinal cord: A quantitative electron microscope study. *J. Comp. Neurol.* **184**, 643–684.

Ralston, H. J. and Ralston, D. D. 1982. The distribution of dorsal root axons to laminae IV, V, and VI of the macaque spinal cord: A quantitative electron-microscope study. *J. Comp. Neurol.* **212**, 435–448.

Ralston, H. J., Light, A. R., Ralston, D. D., and Perl, E. R. 1984. Morphology and synaptic relationships of physiologically identified low-threshold dorsal root axons stained with intra- axonal horseradish in the cat and monkey. *J. Neurophysiol.* **51**, 777–792.

Rambourg, A., Clermont, Y., and Beaudet, A. 1983. Ultrastructural features of six types of neurons in rat dorsal root ganglia. *J. Neurocytol.* **12**, 47–66.

Ramer, M. S. and Bisby, M. A. 1997a. Rapid sprouting of sympathetic axons in dorsal root ganglia of rats with a chronic constriction injury. *Pain* **70**, 237–244.

Ramer, M. and Bisby, M. 1997b. Reduced sympathetic sprouting occurs in dorsal root ganglia after axotomy in mice lacking low-affinity neurotrophin receptor. *Neurosci. Lett.* **228**, 9–12.

Ramer, M. S. and Bisby, M. A. 1998. Sympathetic axons surround neuropeptide-negative axotomized sensory neurons. *NeuroReport* **9**, 3109–3113.

Ramer, M. S. and Bisby, M. A. 1999. Adrenergic innervation of rat sensory ganglia following proximal or distal painful sciatic neuropathy: Distinct mechanisms revealed by anti- NGF treatment. *Eur. J. Neurosci.* **11**, 837–846.

Ramer, M. S., French, G. D., and Bisby, M. A. 1997. Wallerian degeneration is required for both neuropathic pain and sympathetic sprouting into the DRG. *Pain* **72**, 71–78.

Ramer, M. S., Kawaja, M. D., Henderson, J. T., Roder, J. C., and Bisby, M. A. 1998a. Glial overexpression of NGF enhances neuropathic pain and adrenergic sprouting into DRG following chronic sciatic constriction in mice. *Neurosci. Lett.* **251**, 53–56.

Ramer, M. S., Murphy, P. G., Richardson, P. M., and Bisby, M. A. 1998b. Spinal nerve lesion-induced mechanoallodynia and adrenergic sprouting in sensory ganglia are attenuated in interleukin-6 knockout mice. *Pain* **78**, 115–121.

Ramer, M. S., Ma, W., Murphy, P. G., Richardson, P. M., and Bisby, M. A. 1998c. Galanin expression in neuropathic pain: Friend or foe? *Ann. NY Acad. Sci.* **863**, 390–401.

Ramer, M. S., Thompson, S. W., and McMahon, S. B. 1999. Causes and consequences of sympathetic basket formation in dorsal root ganglia. *Pain Suppl.* **6**, S111–S120.

Ramer, M. S., Bradbury, E. J., and McMahon, S. B. 2001. Nerve growth factor induces P2X(3) expression in sensory neurons. *J Neurochem.* **77**, 864–875.

Ramón y Cajal, S. 1909. *Histology of the Nervous System of Man and Vertebrates.* 1995 edition translated from the 1909 French edition by N. Swanson and L. W. Swanson, Oxford University Press, New York.

Randic, M. and Miletic, V. 1977. Effects of substance P in cat dorsal horn neurones activated by noxious stimuli. *Brain Res.* **128**, 164–169.

Randic, M. and Miletic, V. 1978. Depressant actions of methionine-enkephalin and somatostatin in cat dorsal horn neurones activated by noxious stimuli. *Brain Res.* **152**, 196–202.

Randic, M., Carstens, E., Zimmermann, M., and Klumpp, D. 1982. Dual effects of substance P on the excitability of single cutaneous primary afferent C- and A- fibers in the cat spinal cord. *Brain Res.*, **233**, 389–393.

Randic, M., Ryu, P. D., and Urban, L. 1986. Effects of polyclonal and monoclonal antibodies to substance P on slow excitatory transmission in rat spinal dorsal horn. *Brain Res.* **383**, 15–27.

Randic, M., Hecimovic, H., and Ryu, P. D. 1990. Substance P modulates glutamate-induced currents in acutely isolated rat spinal dorsal horn neurones. *Neurosci. Lett.* **117**, 74–80.

Randic, M., Jiang, M. C., and Cerne, R. 1993. Long-term potentiation and long-term depression of primary afferent neurotransmission in the rat spinal cord. *J. Neurosci.* **13**, 5228–5241.

Randic, M., Kolaj, M., Kojic, L., Cerne, R., Cheng, G., and Wang, R. A. 1995. Interaction of neuropeptides and excitatory amino acids in the rat superficial spinal dorsal horn. *Prog. Brain Res.* **104**, 225–253.

Ranieri, F., Mei, N., and Crousillat 1973. Les afferences splanchniques provenant des mecanorécepteurs gastrointestinaux et péritoneaux. *Exp. Brain Res.* **16**, 276–290.

Ranson, S. W. 1912. Degeneration and regeneration of nerve fibers. *J. Comp. Neurol.* **22**, 487–537.

Ranson, S. W. 1913. The course within the spinal cord of the non-medullated fibers of the dorsal roots: A study of Lissauer's tract in cat. *J. Comp. Neurol.* **23**, 259–281.

Ranson, S. W. 1914. An experimental study of Lissauer's tract and the dorsal roots. *J. Comp. Neurol.* **24**, 531–545.

Ranson, S. W. and Billingsley, P. R. 1916. The conduction of painful afferent impulses in the spinal nerves. *Am. J. Anat.* **40**, 571–584.

Rao, G. S., Breazile, J. E., and Kitchell, R. L. 1969. Distribution and termination of spinoreticular afferents in the brain stem of sheep. *J. Comp. Neurol.* **137**, 185–196.

Rao, T. S., Correa, L. D., Reid, R. T., and Lloyd, G. K. 1996. Evaluation of antinociceptive effects of neuronal nicotinic acetylcholine receptor (NAChR) ligands in the rat-tail flick assay. *Neuropharmacology* **35**, 393–405.

Rasband, M. N., Park, E. W., Vanderah, T. W., Lai, J., Porreca, F., and Trimmer, J. S. 2001. Distinct potassium channels on pain-sensing neurons. *Proc. Natl. Acad. Sci. USA* **98**, 13373–13378.

Rasmusson, D. D. and Turnbull, B. G. 1986. Sensory innervation of the raccoon forepaw: 2. Response properties and classification of slowly adapting fibers. *Somatosens. Res.* **4**, 63–75.

Rasool, C. G., Schwartz, A. L., Bollinger, J. A., Reichlin, S., and Bradley, W. G. 1981. Immunoreactive somatostatin distribution and axoplasmic transport in rat peripheral nerve. *Endocrinology* **108**, 996–1001.

Ray, R. H., Mallach, L. E., and Kruger, L. 1985. The response of single guard and down hair mechanoreceptors to moving air-jet stimulation. *Brain Res.* **346**, 333–347.

Ray, S. B. and Wadhwa, S. 1999. Mu opioid receptors in developing human spinal cord. *J. Anat.* **195**, 11–18.

Raybould, H. E., Gschossman, J. M., Ennes, H., Lembo, T., and Mayer, E. A. 1999. Involvement of stretch-sensitive calcium flux in mechanical transduction in visceral afferents. *J. Autonom. Nerv. Syst.* **75**, 1–6.

Raymond, S. A., Thalhammer, J. G., Pipitz-Bergez, F., and Strichartz, G. R. 1990. Changes in axonal impulse conduction correlate with sensory modality in primary afferent fibers in the rat. *Brain Res.* **526**, 318–321.

Rebeck, G. W., Maynard, K. I., Hyman, B. T., and Moskowitz, M. A. 1994. Selective 5-HT_{1D} serotonin receptor gene expression in trigeminal ganglia: Implications for antimigraine drug development. *Proc. Natl. Acad. Sci.* **91**, 3666–3669.

Reddy, V. K., Cassini, P., Ho, R. H., and Martin, G. F. 1990a. Origins and terminations of bulbospinal axons that contain serotonin and either enkephalin or substance-P in the North American opossum. *J. Comp. Neurol.* **294**, 96–108.

Reddy, V. K., Fung, S. J., Zhuo, H., and Barnes, C. D. 1990b. Localization of enkephalinergic neurons in the dorsolateral pontine tegmentum projecting to the spinal cord of the cat. *J. Comp. Neurol.* **291**, 195–202.

Redman, S. J. 1998. The relative contributions of $GABA_A$ and $GABA_B$ receptors to presynaptic inhibition of group Ia EPSPs. In P. Rudomin, R. Romo, and L. M. Mendell (eds.), *Presynaptic Inhibition and Neural Control* (pp. 162–177). Oxford University Press, New York.

Reeh, P. W. 1986. Sensory receptors in mammalian skin in an in vitro preparation. *Neurosci. Lett.* **66**, 141–146.

Reeh, P. W., Kocher, L., and Jung., S. 1986. Does neurogenic inflammation alter the sensitivity of unmyelinated nociceptors in the rat? *Brain Res.* **384**, 42–50.

Reeh, P. W., Bayer, J., Kocher, L., and Handwerker, H. O. 1987. Sensitization of nociceptive cutaneous nerve fibers from the rat's tail by noxious mechanical stimulation. *Exp. Brain Res.* **65**, 505–512.

Rees, H., Sluka, K. A., Westlund, K. N., and Willis, W. D. 1994. Do dorsal root reflexes augment peripheral inflammation? *NeuroReport* **5**, 821–824.

Rees, H., Sluka, K. A., Westlund, K. N., and Willis, W. D. 1995. The role of glutamate and GABA receptors in the generation of dorsal root reflexes by acute arthritis in the anaesthetized rat. *J. Physiol.* **484**, 437–445.

Rees, H., Sluka, K. A., Lu, Y., Westlund, K. N., and Willis, W. D. 1996. Dorsal root reflexes in articular afferents occur bilaterally in a chronic model of arthritis in rats. *J. Neurophysiol.* **76**, 4190–4193.

Rees, H., Sluka, K. A., Urban, L., Walpole, C. J., and Willis, W. D. 1998. The effects of SDZ NKT 343, a potent NK1 receptor antagonist, on cutaneous responses of primate spinothalamic tract neurons sensitized by intradermal capsaicin injection. *Exp. Brain Res.* **121**, 355–358.

Rehfeld, J. F. and Lundberg, J. M. 1983. Cholecystokinin in feline vagal and sciatic nerves: Concentration, molecular form and transport velocity. *Brain Res.* **275**, 341–347.

Reid, G. and Flonta, M. L. 2001. Physiology: Cold current in thermoreceptive neurons. *Nature* **413**, 480.

Reimann, W., Englberger, W., Friderichs, E., Selve, N., and Wilffert, B. 1994. Spinal antinociception by morphine in rats is antagonized by galanin receptor antagonists. *Naunyn Schmiedebergs Arch. Pharmacol.* **350**, 380–386.

Reinhardt, R. R., Chin, E., Zhang, B., Roth, R. A., and Bondy, C. A. 1993. Insulin receptor-related receptor messenger ribonucleic acid is focally expressed in sympathetic and sensory neurons and renal distal tubule cells. *Endocrinology* **133**, 3–10.

Reinhardt, R. R., Chin, E., Zhang, B., Roth, R. A., and Bondy, C. A. 1994. Selective coexpression of insulin receptor-related receptor (IRR) and TRK in NGF-sensitive neurons. *J. Neurosci.* **14**, 4674–4683.

Reinscheid, R. K., Nothacker, H. P., Bourson, A., Ardati, A., Henningsen, R. A., Bunzow, J. R., Grandy, D. K., Langren, H., Monsma, F. J., and Civelli, O. 1995. Orphanin FQ a neuropeptide that activates an opioid-like G protein-coupled receptor. *Science* **270**, 792–794.

Reiter, M. K., Kremarik, P., Freund-Mercier, M. J., Stoeckel, M. E., Desaulles, E., and Feltz, P. 1994. Localization of oxytocin binding sites in the thoracic and upper lumbar spinal cord of the adult and postnatal rat: A histoautoradiographic study. *Eur. J. Neurosci.* **6**, 98–104.

Ren, K. and Dubner, R. 1993. NMDA receptor antagonists attenuate mechanical hyperalgesia in rats with unilateral inflammation of the hindpaw. *Neurosci. Lett.* **163**, 22–26.

Ren, K. and Ruda, M. A. 1994. A comparative study of the calcium-binding proteins calbindin-D28K, calretinin, calmodulin and parvalbumin in the rat spinal cord. *Brain Res. Brain Res. Rev.* **19**, 163–179.

Ren, K. and Ruda, M. A. 1995. Nitric oxide synthase-containing neurons in sensory ganglia of the rat are susceptible to capsaicin-induced cytotoxicity. *Neuroscience* **65**, 505–511.

Ren, K., Williams, G. M., Hylden, J. L., Ruda, M. A., and Dubner, R. 1992. The intrathecal administration of excitatory amino acid receptor antagonists selectively attenuated carrageenan-induced behavioral hyperalgesia in rats. *Eur. J. Pharmacol.* **291**, 235–243.

Ren, K., Ruda, M. A., and Jacobowitz, D. M. 1993a. Immunohistochemical localization of calretinin in the dorsal root ganglion and spinal cord of the rat. *Brain Res. Bull.* **31**, 13–22.

Ren, K., Williams, G. M., Ruda, M. A., and Dubner, R. 1994. Inflammation and hyperalgesia in rats neonatally treated with capsaicin: effect on two classes of nociceptive neurons in the superficial dorsal horn. *Pain* **59**, 287–300.

Ren, K., Iadarola, M. J., and Dubner, R. 1996. An isobolographic analysis of the effects of N-methyl-D-aspartate and NK1 tachykinin receptor antagonists on inflammatory hyperalgesia in the rat. *Br. J. Pharmacol.* **117**, 196–202.

Repkin, A. H., Wolf, P., and Anderson, E. G. 1976. Non-GABA mediated primary afferent depolarization. *Brain Res.* **117**, 147–152.

Reppert, S. M., Weaver, D. R., Stehle, J. H., and Rivkees, S. A. 1991. Molecular cloning and characterization of a rat A1-adenosine receptor that is widely expressed in brain and spinal cord. *Mol. Endocrinol.* **5**, 1037–1048.

Resibois, A. and Rogers, J. H. 1992. Calretinin in rat brain: An immunohistochemical study. *Neuroscience* **46**, 101–134.

Rethelyi, M. 1977. Preterminal and terminal axon arborizations in the substantia gelatinosa of cat's spinal cord. *J. Comp. Neurol.* **172**, 511–528.

Rethelyi, M. 1984. Synaptic connectivity in the spinal dorsal horn, In R. A. Davidoff (ed.), *Handbook of the Spinal Cord* (pp. 137–175). Dekker, New York.

Rethelyi, M. and Capowski, J. J. 1977. The terminal arborization pattern of primary afferent fibers in the substantia gelatinosa of the spinal cord in the cat. *J. Physiol.* **73**, 269–277.

Réthelyi, M. and Szentágothai, J. 1969. The large synaptic complexes of the substantia gelatinosa. *Exp. Brain Res.* **7**, 258–274.

Réthelyi, M. and Szentágothai, J. 1973. Distribution and connections of afferent fibres in the spinal cord. In A. Iggo (ed.), *Handbook of Sensory Physiology, Vol. II. Somatosensory System* (pp. 207–252). Springer-Verlag, New York.

Rethelyi, M., Trevino, D. L., and Perl, E. R. 1979. Distribution of primary afferent fibers within the sacrococcygeal dorsal horn: An autoradiographic study. *J. Comp. Neurol.* **185**, 603–622.

Rethelyi, M., Light, A. R., and Perl, E. R. 1982. Synaptic complexes formed by functionally defined primary afferent units with fine myelinated fibers. *J. Comp. Neurol.* **207**, 381–393.

Rethelyi, M., Metz, C. B., and Lund, P. K. 1989a. Distribution of neurons expressing calcitonin gene-related peptide mRNAs in the brain stem, spinal cord and dorsal root ganglia of rat and guinea-pig. *Neuroscience* **29**, 225–239.

Rethelyi, M., Light, A. R., and Perl, E. R. 1989b. Synaptic ultrastructure of functionally and morphologically characterized neurons of the superficial spinal dorsal horn of cat. *J. Neurosci.* **9**, 1846–1863.

Reubi, J. C. and Maurer, R. 1985. Autoradiographic mapping of somatostatin receptors in the rat central nervous system and pituitary. *Neuroscience* **15**, 1183–1193.

Reuss, M. A. and Reuss, S. 2001. Nitric oxide synthase neurons in the rodent spinal cord: Distribution, relation to substance P fibers, and effects of dorsal rhizotomy. *J. Chem. Neuroanat.* **21**, 181–186.

Rexed, B. 1952. The cytoarchitectonic organization of the spinal cord in the rat. *J. Comp. Neurol.* **96**, 415–466.

Rexed, B. 1954. A cytoarchitectonic atlas of the spinal cord in the cat. *J. Comp. Neurol.* **100**, 297–380.

Rexed, B. and Therman, P. 1948. Calibre spectra of motor and sensory nerve fibres to flexor and extensor muscles. *J. Neurophysiol.* **11**, 133–139.

Rhoades, R. W., Chiaia, N. L., Hess, P. R., and Miller, M. W. 1988. Effect of neonatal infraorbital nerve transection on substance P- and leucine enkephalin-like immunoreactivities in trigeminal subnucleus caudalis of the rat. *J. Comp. Neurol.* **8**, 2234–2247.

Ribeiro-da-Silva, A. 1995. Ultrastructural features of the colocalization of calcitonin gene-related peptide with substance P or somatostatin in the dorsal horn of the spinal cord. *Can. J. Physiol. Pharmacol.* **73**, 940–944.

Ribeiro-da-Silva, A. and Coimbra, A. 1980. Neuronal uptake of [^{3}H]GABA and [^{3}H]glycine in laminae I-III (substantia gelatinosa Rolandi) of the rat spinal cord: An autoradiographic study. *Brain Res.* **188**, 449–464.

Ribeiro-da-Silva, A. and Coimbra, A. 1982. Two types of synaptic glomeruli and their distribution in laminae I–III of the rat spinal cord. *J. Comp. Neurol.* **209**, 176–186.

Ribeiro-da-Silva, A. and Cuello, A. C. 1990a. Choline acetyltransferase-immunoreactive profiles are presynaptic to primary sensory fibers in the rat superficial dorsal horn. *J. Comp. Neurol.* **295**, 370–384.

Ribeiro-da-Silva, A. and Cuello, A. C. 1990b. Ultrastructural evidence for the occurrence of two distinct somatostatin-containing systems in the substantia gelatinosa of rat spinal cord. *J. Chem. Neuroanat.* **3**, 141–153.

Ribeiro-da-Silva, A. and Cuello, A. C. 1995. Organization of peptidergic neurons in the dorsal horn of the spinal cord: Anatomical and functional correlates. *Progr. Brain Res.* **104**, 41–59.

Ribeiro-da-Silva, A. and Hökfelt, T. 2000. Neuroanatomical localisation of substance P in the CNS and sensory neurons. *Neuropeptides* **34**, 256–271.

Ribeiro-da-Silva, A., Pignatelli, D., and Coimbra, A. 1985. Synaptic architecture of glomeruli in superficial dorsal horn of rat spinal cord, as shown in serial reconstructions. *J. Neurocytol.* **14**, 203–220.

Ribeiro-da-Silva, A., Pioro, E. P., and Cuello, A. C. 1991. Substance P- and enkephalin-like immunoreactivities are colocalized in certain neurons of the substantia gelatinosa of the rat spinal cord: An ultrastructural double-labeling study. *J. Neurosci.* **11**, 1068–1080.

Rice, A. S. C., McMahon, S. B., and Wall, P. D. 1993. The electrophysiological consequences of electrode impalement of peripheral nerves in the rat. *Brain Res.* **631**, 221–226.

Rice, A. S. C., Andreev, N. Y., and McMahon, S. B. 1994. The consequences of microneurography electrode-induced injury of peripheral nerves observed in the rat and man. *Pain* **59**, 385–393.

Rice, F. L. and Rasmusson, D. D. 2000. Innervation of the digit on the forepaw of the raccoon. *J. Comp. Neurol.* **417**, 467–490.

Richards, J. G., Schoch, P., Haring, P., Takacs, B., and Mohler, H. 1987. Resolving ABA_A/benzodiazepine receptors: cellular and subcellular localization in the CNS with monoclonal antibodies. *J. Neurosci.* **7**, 1866–1886.

Richardson, J. D., Kilo, S., and Hargreaves, K. M. 1998. Cannabinoids reduce hyperalgesia and inflammation via interaction with peripheral CB_1 receptors. *Pain* **75**, 111–119.

Riddell, J. S., Jankowska, E., and Huber, J. 1995. Organization of neuronal systems mediating presynaptic inhibition of group II muscle afferents in the cat. *J. Physiol.* **483**, 461–471.

Ridet, J.-L., Sandillon, F., Rajaofetra, N., Geffard, M., and Privat, A. 1992. Spinal dopaminergic system of the rat: Light- and electron-microscopic study using an antiserum against dopamine, with particular emphasis on synaptic incidence. *Brain Res.* **598**, 233–241.

Ridet, J.-L., Rajaofetra, N., Teilhac, J.-R., Geffard, M., and Pravat, A. 1993. Evidence for nonsynaptic serotonergic and noradrenergic innervation of the rat dorsal horn and possible involvement of neuron-glia interations. *Neuroscience* **52**, 143–157.

Ridet, J.-L., Geffard, M., and Privat, A. X. 1994a. Light- and electron-microscopic studies of the effect of *p*-chloroamphetamine on the monoaminergic innervation of the rat spinal cord. *J. Comp. Neurol.* **343**, 281–296.

Ridet, J.-L., Tamir, H., and Privat, A. 1994b. Direct immunocytochemical localization of 5-hydroxytryptamine receptors in the adult rat spinal cord: A light- and electron-microscopic study using an anti-idiotypic antiserum. *J. Neurosci. Res.* **38**, 109–121.

Riedl, M., Shuster, S., Vulchanova, L., Wang, J., Loh, H. H., and Elde, R. 1996. Orphanin FQ/nociceptin-immunoreactive nerve fibers parallel those containing endogenous opioids in rat spinal cord. *NeuroReport* **7**, 1369–1372.

Riley, D. A., Ellis, S., and Bain, J. L. W. 1984. Ultrastructural cytochemical localization of carbonic anhydrase activity in rat peripheral sensory and motor nerves, dorsal root ganglia and dorsal column nuclei. *Neuroscience* **13**, 189–206.

Riley, D. A., Sanger, J. R., Matloub, H. S., Yousif, N. J., Bain, J. L. W., and Moore, G. H. 1988. Identifying motor and sensory myelinated axons in rabbit peripheral nerves by histochemical staining for carbonic anhydrase and cholinesterase activities. *Brain Res.* **453**, 79–88.

Risling, M. and Hildebrand, C. 1982. Occurence of unmyelinated axon profiles at distal, middle and proximal levels in the ventral root L7 of cats and kittens. *J. Neurol. Sci.* **56**, 219–231.

Risling, M., Hildebrand, C., and Aldskogius, H. 1981. Postnatal increase of unmyelinated axon profiles in the feline ventral root L7. *J. Comp. Neurol.* **201**, 343–351.

Risling, M., Dalsgaard, C. J., and Cuello, A. C. 1984a. Invasion of lumbosacral ventral roots and spinal pia mater by substance P-immunoreactive axons after sciatic nerve lesion in kittens. *Brain Res.* **307**, 351–354.

Risling, M., Dalsgaard, C. J., Cukierman, A., and Cuello, A. C. 1984b. Electron microscopic and immunohistochemical evidence that unmyelinated ventral root axons make U-turns or enter the spinal pia mater. *J. Comp. Neurol.* **225**, 53–63.

Risling, M., Hildebrand, C., and Cullheim, S. 1984c. Invasion of the L7 ventral root and spinal pia mater by new axons after sciatic nerve division in kittens. *Exp. Neurol.* **83**, 84–97.

Risling, M., Dalsgaard, C.-J., Frisen, J., Sjogren, A.-M., and Fried, K. 1994. Substance P-, calcitonin gene-related peptide, growth-associated protein-43, and neurotrophin receptor-like immunoreactivity associated with unmyelinated axons in feline ventral roots and pia mater. *J. Comp. Neurol.* **339**, 365–386.

Rittenhouse, P. A., Marchand, J. E., Chen, J., Kream, R. M., and Leeman, S. E. 1996. Streptozotocin-induced diabetes is associated with altered expression of peptide-encoding mRNAs in rat sensory neurons. *Peptides* **17**, 1017–1022.

Ritter, A. M. and Mendell, L. M. 1992. Somal membrane properties of physiologically identified sensory neurons in the rat: Effects of nerve growth factor. *J. Neurophysiol.* **68**, 2033–2041.

Ritter, A. M., Lewin, G. R., Kremer, N. E., and Mendell, L. M. 1991. Requirement for nerve growth factor in the development of myelinated nociceptors in vivo. *Nature* **350**, 500–502.

Ritter, A. M., Woodbury, C. J., Albers, K., Davis, B. M., and Koerber, H. R. 2000. Maturation of cutaneous sensory neurons from normal and NGF-overexpressing mice. *J. Neurophysiol.* **83**, 1722–1732.

Ritz, L. A., Brown, P. B., and Bailey, S. M. 1989. Crossed and uncrossed projections to cat sacrocaudal spinal cord: I. Axons from cutaneous receptors. *J. Comp. Neurol.* **289**, 284–293.

Ritz, L. A. and Greenspan, J. D. 1985. Morphological features of lamina V neurons receiving input in cat sacrocaudal spinal cord. *J. Comp. Neurol.* **238**, 440–452.

Ritz, L. A., Murray, C. R., and Foli, K. 2001. Crossed and uncrossed projections to the cat sacrocaudal spinal cord: III. Axons expressing calcitonin gene-related peptide immunoreactivity. *J. Comp. Neurol.* **438**, 388–398.

Rivero-Melian, C. and Grant, G. 1990. Distribution of lumbar dorsal root fibers in the lower thoracic and lumbosacral spinal cord of the rat studied with choleragenoid horseradish peroxidase conjugate. *J. Comp. Neurol.* **299**, 470–481.

Rizzoli, A. A. 1968. Distribution of glutamic acid, aspartic acid, γ-aminobutyric acid and glycine in six areas of cat spinal cord before and after transection. *Brain Res.* **11**, 11–18.

Robain, O. and Jardin, L. 1972. Histoenzymologie du ganglion spinal du lapin. *J. Neurol. Sci.* **17**, 419–433.

Robbins, A., Sato, Y., Hotta, H., and Berkley, K. J. 1990. Responses of hypogastric nerve afferent fibers to uterine distension in estrous or metestrous rats. *Neurosci. Lett.* **110**, 82–85.

Robbins, A., Berkley, K. J., and Sato, Y. 1992. Estrous cycle variation of afferent fibers supplying reproductive organs in the female rat. *Brain Res.* **596**, 353–356.

Roberts, P. J. 1974. Amino acid release from isolated dorsal root ganglia. *Brain Res.* **74**, 327–332.

Roberts, P. J and Keen, P. 1974. Effect of dorsal root section on amino acids of rat spinal cord. *Brain Res.* **74**, 333–337.

Roberts, P. J., Keen, P., and Mitchell, J. F. 1973. The distribution and axonal transport of free amino acids and related compounds in the dorsal sensory neuron of the rat, as determined by the dansyl reaction. *J. Neurochem.* **21**, 199–209.

Roberts, R. G., Stevenson, J. E., Westerman, R. A., and Pennefather, J. 1995. Nicotinic acetylcholine receptors on capsaicin-sensitive nerves. *NeuroReport* **6**, 1578–1582.

Roberts, W. J. 1986. A hypothesis on the physiological basis for causalgia and related pains. *Pain* **24**, 297–311.

Roberts, W. J. and Elardo, S. M. 1985. Sympathetic activation of A-delta nociceptors. *Somatosens. Res.* **3**, 33–44.

Roberts, W. J. and Elardo, S. M. 1986. Clustering of primary afferent fibers in peripheral nerve fascicles by sensory modality. *Brain Res.* **370**, 149–152.

Robertson, B. and Arvidsson, J. 1985. Transganglionic transport of wheat germ agglutinin-HRP and choleragenoid-HRP in rat trigeminal primary sensory neurons. *Brain Res.* **348**, 44–51.

Robertson, B. and Grant G. 1985. A comparison between wheat germ agglutinin- and choleragenoid-horseradish peroxidase as anterogradely transported markers in central branches of primary sensory neurones in the rat with some observations in the cat. *Neuroscience* **14**, 895–905.

Robertson, B. and Grant, G. 1989. Immunocytochemical evidence for the localization of the GM1 ganglioside in carbonic anhydrase-containing and RT 97-immunoreactive rat primary sensory neurons. *J. Neurocytol.* **18**, 77–86.

Robertson, B., Schulte, G., Elde, R., and Grant, G. 1999. Effects of sciatic nerve injuries on delta-opioid receptor and substance P immunoreactivities in the superficial dorsal horn of the rat. *Eur. J. Pain* **3**, 115–129.

Robinson, T. E. and Justice, J. B. (eds.) 1991. *Microdialysis in the Neurosciences*. Elsevier, Amsterdam.

Roby-Brami, A., Bussel, B., Willer, J. C., and Le Bars, D. 1987. An electrophysiological investigation into the pain-relieving effects of heterotopic nociceptive stimuli: Probable involvement of a supraspinal loop. *Brain* **110**, 1497–1508.

Rodin, B. E. and Kruger, L. 1984. Deafferentation in animals as a model for the study of pain: An alternative hypothesis. *Brain Res. Rev.* **7**, 213–228.

Rodin, B. E., Sampogna, S., and Kruger, L. 1983. An examination of intraspinal sprouting in dorsal root axons with the tracer horseradish peroxidase. *J. Comp. Neurol.* **215**, 187–198.

Rodrigo, J., Springall, D. R., Uttenthal, O., Bentura, M. L., Abadia-Molina, F., Riveros-Moreno, V., Martinez-Murillo, R., Polak, J. M., and Moncada, S. 1994. Localization of nitric oxide synthase in the adult rat brain. *Philos. Trans. R. Soc. Lond. B Biol. Sci.* **345**, 175–221.

Rodriguez, R. E., Hill, R. G., and Hughes, J. 1987. Cholecystokinin releases [^{3}H]GABA from the perfused subarachnoid space of the anaesthetized rat spinal cord. *Neurosci. Lett.* **83**, 173–178.

Roessmann, U. and Friede, R. L. 1967. The segmental distribution of acetyl cholinesterase in the cat spinal cord. *J. Anat.* **101**, 27–32.

Rogers, J. H. and Resibois, A. 1992. Calretinin and calbindin-D28k in rat brain: Patterns of partial co-localization. *Neuroscience* **51**, 843–865.

Rokaeus, A., Melander, T., Hökfelt, T., Lundberg, J. M., Tatemoto, K., Carlquist, M., and Mutt, V. 1984. A galanin-like peptide in the central nervous system and intestine of the rat. *Neurosci. Lett.* **47**, 161–166.

Romano, C., Sesma, M. A., McDonald, C. T., O'Malley, K., Van den Pol, A. N., and Olney, J. W. 1995. Distribution of metabotropic glutamate receptor mGluR5 inmmunoreactivity in rat brain. *J. Comp. Neurol.* **355**, 455–469.

Romero, M. I., Rangappa, N., Garry, M. G., and Smith, G. M. 2001. Functional regeneration of chronically injured sensory afferents into adult spinal cord after neurotrophin gene therapy. *J. Neurosci.* **21**, 8408–8416.

Romualdi, P., Lesa, G., Cox, B. M., and Ferri, S. 1990. Distribution and characterization of VIP-related peptides in the rat spinal cord. *Neuropeptides* **16**, 219–225.

Rose, R. D., Koerber, H. R., Sedivec, M. J., and Mendell, L. M. 1986. Somal action potential duration differs in identified primary afferents. *Neurosci. Lett.* **63**, 259–264.

Rosemberg, S., Marie, S. K., and Kliemann, S. 1994. Congenital insensitivity to pain with anhidrosis (hereditary sensory and autonomic neuropathy type IV). *Pediatr. Neurol.* **11**, 50–56.

Rosen, A., Lundeberg, T., Bytner, B., and Nylander, I. 2000. Central changes in nociceptin dynorphin B and met-enkephalin-Arg-Phe in different models of nociception. *Brain Res.* **857**, 212–218.

Rosenfeld, M. G., Mermod, J. J., Amara, S. G., Swanson, L. W., Sawchenko, P. E., Rivier, J., Vale, W. W., and Evans, R. M. 1983. Production of a novel neuropeptide encoded by the calcitonin gene via tissue-specific RNA processing. *Nature* **304**, 129–135.

Rosenthal, B. M. and Ho, R. H. 1989. An electron microscopic study of somatostatin immunoreactive structures in lamina II of the rat spinal cord. *Brain Res. Bull.* **22**, 439–451.

Rosin, D. L., Zeng, D., Stornetta, R. L., Norton, F. R., Riley, T., Okusa, M. D., Guyenet, P. G., and Lynch, K. R. 1993. Immunohistochemical localization alpha$_{2a}$-adrenergic receptors in catecholaminergic and other brainstem neurons in the rat. *Neuroscience* **56**, 139–155.

Rosin, D. L., Talley, E. M., Lee, A., Stornetta, R. L., Gaylin, B. D., Guyenet, P. G., and Lynch, K. R. 1996. Distribution of α_{2C}-adrenergic receptor-like immunoreactivity in the rat central nervous system. *J. Comp. Neurol.* **372**, 135–165.

Ross, M. H., Romrell, L. J., and Kaye, G. I. 1995. *Histology; A Text and Atlas* (3rd ed). Williams & Wilkins, Baltimore.

Rossi, A. and Grigg, P. 1982. Characteristics of hip joint mechanoreceptors in the cat. *J. Neurophysiol.* **47**, 1029–1042.

Rossi, A. and Rossi, B. 1985. Characteristics of the receptors in the isolated capsule of the hip in the cat. *Int. Orthop.* **9**, 123–127.

Rössler, W., Gerstberger, R., Sann, H., and Pierau, F.-K. 1993. Distribution and binding sites of substance P and calcitonin gene-related peptide and their capsaicin-sensitivity in the spinal cord of rats and chicken: A comparative study. *Neuropeptides* **25**, 241–253.

Roudet, C., Savasta, M., and Feuerstein, C. 1993. Normal distribution of alpha-1-adrenoceptors in the rat spinal cord and its modification after noradrenergic denervation: A quantitative autoradiographic study. *J. Neurosci. Res.* **34**, 44–53.

Roudet, C., Mouchet, P., Feuerstein, C., and Savata, M. 1994. Normal distribution of alpha 2-adrenoreceptors in the rat spinal cord and its modification after noradrenergic denervation: A quantitative autoradiography study. *J. Neurosci. Res.* **39**, 319–329.

Roumy, M. and Zajac, J. M. 1998. Neuropeptide FF, pain and analgesia. *Eur. J. Pharmcol.* **345**, 1–11.

Rousselot, P., Papadopoulos, G., Merighi, A., Poulain, D. A., and Theodosis, D. T. 1990. Oxytocinergic innervation of the rat spinal cord. An electron-microscopic study. *Brain Res.* **529**, 178–184.

Routh, V. H. and Helke, C. J. 1995. Tachykinin receptors in the spinal cord. *Prog. Brain Res.* **104**, 93–108.

Rowan, S., Todd, A. J., and Spike, R. C. 1993. Evidence that neuropeptide Y is present in GABAergic neurons in the superficial dorsal horn of the rat spinal cord. *Neuroscience* **53**, 537–545.

Roza, C., Laird, J. M. A., and Cervero, F. 1998. Spinal mechanisms underlying persisent pain and referred hyperalgesia in rats with an experimental ureric stone. *J. Neurophysiol.* **79**, 1603–1612.

Rubin, G., Kaspi, T., Rappaport, Z. H., Cohen, S., Ravikovitch, M., Lomazov, P., and Devor, M. 1997. Adrenosensitivity of injured afferent neurons does not require the presence of postganglionic sympathetic terminals. *Pain* **72**, 183–191.

Ruda, M. A. 1982. Opiates and pain pathways: Demonstration of enkephalin synapses on dorsal horn projection neurons. *Science* **215**, 1523–1525.

Ruda, M. A. 1988. Spinal dorsal horn circuitry involved in the brain stem control of nociception. *Prog. Brain Res.* **77**, 129–140.

Ruda, M. A. and Gobel, S. 1980. Ultrastructural characterization of axonal endings in the substantia gelatinosa which take up [3H]serotonin. *Brain Res.* **184**, 57–83.

Ruda, M. A., Coffield, J., and Steinbusch, H. W. M. 1982. Immunocytochemical analysis of serotonergic axons in laminae I and II of the lumbar spinal cord of the cat. *J. Neurosci.* **2**, 1660–1671.

Ruda, M. A., Coffield, J., and Dubner, R. 1984. Demonstration of postsynaptic opioid modulation of thalamic projection neurons by the combined techniques of retrograde horseradish peroxidase and enkephalin immunocytochemistry. *J. Neurosci.* **4**, 2117–2132.

Ruda, M. A., Bennett, G. J., and Dubner, R. 1986. Neurochemistry and neural circuitry in the dorsal horn. *Prog. Brain Res.* **66**, 219–268.

Ruda, M. A., Iadarola, M. J., Cohen, L. V., and Young, W. S. 1988. In situ hybridization histochemistry and immunocytochemistry reveal an increase in spinal dynorphin biosynthesis in a rat model of peripheral inflammation and hyperalgesia. *Proc. Natl. Acad. Sci. USA* **85**, 622–626.

Ruda, M. A., Besse, D., Inagaki, S., and DeLeon, M. 1994. Nitric oxide expression and regulation in the dorsal root ganglion and spinal cord. *Ann. NY Acad. Sci.* **738**, 181–190.

Ruda, M. A., Ling, Q.-D., Hohmann, A. G., Peng, Y. B., and Tachibana, T. 2000. Altered nociceptive neuronal circuits after neonatal peripheral inflammation. *Science* **289**, 628–631.

Rudomin, P. and Schmidt, R. F. 1999. Presynaptic inhibition in the vertebrate spinal cord revisited. *Exp. Brain Res.* **129**, 1–37.

Rudomin, P., Engberg, I., and Jiménez, I. 1981. Mechanisms involved in presynaptic depolarization of group I and rubrospinal fibers in cat spinal cord. *J. Neurophysiol.* **46**, 532–548.

Rudomin, P., Jiménez, I., Solodkin, M., and Dueñas, S. 1983. Sites of action of segmental and descending control of transmission on pathways mediating PAD of Ia- and Ib-afferent fibers in cat spinal cord. *J. Neurophysiol.* **50**, 743–769.

Rudomin, P. Solodkin, M., and Jiménez, I. 1987. Synaptic potentials of primary afferent fibers and motoneurons evoked by single intermediate nucleus interneurons in the cat spinal cord. *J. Neurophysiol.* **57**, 1288–1313.

Rudomin, P., Jiménez, J., Quevedo, J., and Solodkin, M. 1990. Pharmacologic analysis of inhibition produced by last-order intermediate nucleus interneurons mediating nonreciprocal inhibition of motoneurons in cat spinal cord. *J. Neurophysiol.* **63**, 147–160.

Rudomin, P., Romo, R., and Mendell, L. M. (eds.) 1998. *Presynaptic Inhibition and Neural Control.* Oxford University Press, New York.

Rueff, A. and Dray, A. 1993a. Sensitization of peripheral afferent fibres in the in vitro neonatal rat spinal cord-tail by bradykinin and prostaglandins. *Neuroscience* **54**, 527–535.

Rueff, A. and Dray, A. 1993b. Pharmacological characterization of the effects of 5-hydroxytryptamine and different prostaglandins on peripheral sensory neurons in vitro. *Agents Actions* **38**, C13–15.

Rueff, A. and Mendell, L. M. 1996. Nerve growth factor and NT-5 induce increased thermal sensitivity of cutaneous nociceptors in vitro. *J. Neurophysiol.* **76**, 3593–3596.

Ruffini, A. 1894. Sur un nouvel organe nerveux terminal et sur la presence des corpuscles Golgi-Mazzoni dans le conjunctif sous-cutane de la pulpe des doigts de l'homme. *Arch. Ital. Biol.* **21**, 249–265.

Ruggiero, D. A., Regunathan, S., Wang, H., Milner, T. A., and Reis, D. J. 1995. Distribution of imidazoline receptor binding protein in the central nervous system. *Ann. NY Acad. Sci.* **763**, 208–221.

Ruggiero, D. A., Regunathan, S., Wang, H., Milner, T. A., and Reis, D. J. 1998. Immunocytochemical localization of an imidazoline receptor protein in the central nervous system. *Brain Res.* **780**, 270–293.

Rumio, C., Castano, P., Veraldi, S., Morini, M., and Castano, M. 1995. The innervation of human skin studied with confocal scanning laser microscopy: A comparison between PGP 9.5 immunofluorescence and silver impregnation. *Neuroimage* **2**, 102–111.

Rush, A. M., Brau, M. E., Elliott, D. A., and Elliott, J. R. 1998. Electrophysiological properties of sodium current subtypes in small cells from adult rat dorsal root ganglia. *J. Physiol.* **511**, 771–789.

Rusin, K. I., Ryu, P. D., and Randic, M. 1992. Modulation of excitatory amino acid responses in rat dorsal horn neurons by tachykinins. *J. Neurophysiol.* **68**, 265–286.

Rusin, K. I., Bleakman, D., Chard, P. S., Randic, M., and Miller, R. J. 1993a. Tachykinins potentiate N-methyl-D-aspartate responses in acutely isolated neurons from the dorsal horn. *J. Neurochem.* **60**, 952–960.

Rusin, K. I., Jiang, M. C., Cerne, R., and Randic, M. 1993b. Interactions between excitatory amino acids and tachykinins in the rat spinal dorsal horn. *Brain Res. Bull.* **30**, 329–338.

Rustioni, A. and Cuenod, M. 1982. Selective retrograde transport of d-asparate in spinal interneurons and cortical neurons of rats. *Brain Res.* **236**, 143–155.

Ryall, R. W. and Piercey, M. F. 1970. Visceral afferent and efferent fibers in sacral ventral roots in cats. *Brain Res.* **23**, 57–65.

Rydh-Rinder, M., Holmberg, K., Elfvin, L.-G., Wiesenfeld-Hallin, Z., and Hökfelt, T. 1996. Effects of peripheral axotomy on neuropeptides and nitric oxide synthase in dorsal root ganglia and spinal cord of the guinea pig: An immunohistochemical study. *Brain Res.* **707**, 180–188.

Rygh, L. J., Svendsen, F., Hole, K., and Tjølsen, A. 1999. Natural noxious stimulation can induce long-term increase of spinal nociceptive responses. *Pain* **82**, 305–310.

Rymer, W. Z. and D'Almeida, A. 1980. Joint position sense: The effects of muscle contraction. *Brain* **103**, 1–22.

Ryu, P. D., Gerber, G., Murase, K., and Randic, M. 1988a. Actions of calcitonin gene-related peptide on rat spinal dorsal horn neurons. *Brain Res.* **441**, 357–361.

Ryu, P. D., Gerber, G., Murase, K., and Randic, M. 1988b. Calcitonin gene-related peptide enhances calcium current of rat dorsal root ganglion neurons and spinal excitatory synaptic transmission. *Neurosci. Lett.* **89**, 305–312.

Saito, S., Kidd, G. J., Trapp, B. D., Dawson, T. M., Bredt, D. S., Wilson, D. A., Traystman, R. J., Snyder, S. H., and Hanley, D. F. 1994. Rat spinal cord neurons contain nitric oxide synthase. *Neuroscience* **59**, 447–456.

Sakamoto, H., Spike, R. C., and Todd, A. J. 1999. Neurons in laminae III and IV of the rat spinal cord with the neurokinin-1 receptor receive few contacts from unmyelinated primary afferents which do not contain substance P. *Neuroscience* **94**, 903–908.

Sakanaka, M., Inagaki, S., Shiosaka, S., Senba, E., Takagi, H., Takatsuki, K., Kawai, Y., Iida, H., Hara, Y., and Tohyama, M. 1982. Ontogeny of substance P-containing neuron system of the rat: Immunohistochemical analysis: II. Lower brain stem. *Neuroscience* **7**, 1097–1126.

Sakurada, S., Hayashi, T., Yuhki, M., Orito, T., Zadina, J. E., Kastin, A. J., Fujimura, T., Murayama, K., Sakurada, C., Sakurada, T., Narita, M., Suzuki, T., Tan-no, K., and Tseng, L. F. 2001. Differential antinociceptive effects induced by intrathecally administered endomorphin-1 and endomorphin-2 in the mouse. *Eur. J. Pharmacol.* **427**, 203–210.

Sakurada, T., Sakurada, S., Katsuyama, S., Sakurada, C., Tan-No, K. and Terenius, L. 1999a. Nociceptin (1-7) antagonizes nociceptin-induced hyperalgesia in mice. *Br. J. Pharmacol.* **128**, 941–944.

Sakurada, T., Katsuyama, S., Sakurada, S., Inoue, M., Tan-No, K., Kisara, K., Sakurada, C., Ueda, H. and Sasaki, J. 1999b. Nociceptin-induced scratching, biting and licking in mice: involvement of spinal NK1 receptors. *Br. J. Pharmacol.* **127**, 1712–1718.

Saldanha, G., Hongo, J., Plant, G., Acheson, J., Levy, I., and Anand, P. 1999. Decreased CGRP, but preserved Trk A immunoreactivity in nerve fibres in inflamed human superficial temporal arteries. *J. Neurol. Neurosurg. Psychiatry* **66**, 390–392.

Sales, N., Charnay, Y., Zajac, J. M., Dubois, P. M., and Roques, B. P. 1989. Ontogeny of mu and delta opioid receptors and of neural endopeptidases in human spinal cord: An autoradiographic study. *J. Chem. Neuroanat.* **2**, 179–188.

Salio, C., Fischer, J., Franzoni, M. F., Mackie, K., Kaneko, T., and Conrath, M. 2001. CB1-cannabinoid and μ-opioid receptor co-localization on postsynaptic target in the rat dorsal horn. *NeuroReport* **12**, 3689–3692.

Salio, C., Fischer, J., Franzoni, M. F., and Conrath, M. 2002. Pre- and postsynaptic localizations of the CB1 cannabinoid receptor in the dorsal horn of the rat spinal cord. *Neuroscience* **110**, 755–764.

Salo, P. T. and Theriault, E. 1997. Number, distribution and neuropeptide content of rat knee joint afferents. *J. Anat.* **190**, 515–522.

Salt, T. E. and Hill, R. G. 1983. Neurotransmitter candidates of somatosensory primary afferent fibres. *Neuroscience* **10**, 1083–1103.

Salt, T. E., Crozier, C. S., and Hill, R. G. 1982. The effects of capsaicin pre-treatment on the responses of single neurones to sensory stimuli in the trigeminal nucleus caudalis of the rat: Evidence against a role for substance P as the neurotransmitter serving thermal nociception. *Neuroscience* **7**, 1141–1148.

Salter, M. W. and Henry, J. L. 1985. Effects of adenosine 5'-monophosphate and adenosine 5'-triphosphate on functionally identified units in the cat spinal dorsal horn. Evidence for a differential effect of adenosine 5'-triphosphate on nociceptive vs. non-nociceptive units. *Neuroscience* **15**, 815–825.

Salter, M. W. and Henry, J. L. 1991. Responses of functionally identified neurones in the dorsal horn of the cat spinal cord to substance P, neurokinin A and physalaemin. *Neuroscience* **43**, 601–610.

Salter, M. W., De Koninck, Y., and Henry, J. L. 1993. Physiological roles for adenosine and ATP in synaptic transmission in the spinal dorsal horn. *Prog. Neurobiol.* **41**, 125–156.

Samarasinghe, S., Virgo, L., and de Belleroche, J. 1996. Distribution of the *N*-methyl-D-asparate glutamate receptor subunit NR2A in control and amyotrophic lateral sclerosis spinal cord. *Brain Res.* **727**, 233–237.

Sameda, H., Takahashi, Y., Takahashi, K., Chiba, T., Ohtori, S., and Moriya, H. 2001. Primary sensory neurons with dichotomizing axons projecting to the facet joint and the sciatic nerve in rats. *Spine* **26**, 1105–1109.

Samsam, M., Covenas, R., Csillik, B., Ahangari, R., Yajeya, J., Riquelme, R., Narvaez, J. A., and Tramu, G. 2001. Depletion of substance P, neurokinin A and calcitonin gene-related peptide from the contralateral and ipsilateral caudal trigeminal nucleus following unilateral electrical stimulation of the trigeminal ganglion: A possible neurophysiological and neuroanatomical link to generalized head pain. *J. Chem. Neuroanat.* **21**, 161–169.

Samuel, E. P. 1952. The autonomic and somatic innervation of the articular capsule. *Anat. Rec.* **113**, 53–70.

Sandkühler, J. and Liu, X. 1998. Induction of long-term potentiation at spinal synapses by noxious stimulation or nerve injury. *Eur. J. Neurosci.* **10**, 2476–2480.

Sandkühler, J., Chen, J. G., Cheng, G., and Randić, M. 1997. Low-frequency stimulation of afferent Aδ-fibers induces long-term depression at primary afferent synapses with substantia gelatinosa neurons in the rat. *J. Neurosci.* **17**, 6483–6491.

Sangameswaran, L., Delgado, S. G., Fish, L. M., Koch, B. D., Stewart, G. R., Sze, P., Hunter, J. C., Eglen, R. M., and Herman, R. S. 1996. Structure and function of a novel voltage-gated, tetrodotoxin-resistant sodium channel specific to sensory neurons. *J. Biol. Chem.* **271**, 5953–5956.

Sanjue, H. and Jun, Z. 1989. Sympathetic facilitation of sustained discharges of polymodal nociceptors. *Pain* **38**, 85–90.

Sann, H., McCarthy, P. W., Mäder, M., and Schemann, M. 1995a. Choline acetyltransferase-like immunoreactivity in small diameter neurones of the rat dorsal root ganglion. *Neurosci. Lett.* **198**, 17–20.

Sann, H., Jancso, G., Rossler, W., and Pierau, F.-K. 1995b. Reduction of substance P binding sites in the spinal dorsal horn after perineural capsaicin treatment in the rat. *Neurosci. Lett.* **190**, 151–154.

Sanner, C. A., Cunningham, T. J., and Goldberger, M. E. 1994. NMDA receptor blockade rescues Clark's and Red nucleus neurons after spinal hemisection. *J. Neurosci.* **14**, 6472–6480.

Santini, M., Ibata, Y., and Pappas, G. D. 1971. The fine structure of the sympathetic axons within the Pacinian corpuscle. *Brain Res.* **33**, 279–287.

Sanudo-Pena, M. C., Strangman, N. M., Mackie, K., Walker, J. M., and Tsou, K. 1999. CB1 receptor localization in rat spinal cord and roots, dorsal root ganglion, and peripheral nerve. *Zhongguo Yao Li Xue. Bao.* **20**, 1115–1120.

Sanyal, S. and Rustioni, A. 1974. Phosphatases in the substantia gelatinosa and motoneurones: A comparative histochemical study. *Brain Res.* **76**, 161–166.

Saper, C. B., Hurley, K. M., Moga, M. M., Holmes, H. R., Adams, S. A., Leahy, K. M., and Needleman, P. 1989. Brain natriuretic peptides: Differential localization of a new family of neuropeptides. *Neurosci. Lett.* **96**, 29–34.

Sar, M. and Stumpf, W. E. 1977. Androgen concentration in motor neurons of cranial nerves and spinal cord. *Science* **197**, 77–79.

Saria, A., Gamse, R., Petermann, J., Fischer, J. A., Theodorsson-Norheim, E., and Lundberg, J. M. 1986. Simultaneous release of several tachykinins and calcitonin gene-related peptide from rat spinal cord slices. *Neurosci. Lett.* **63**, 310–314.

Sarrat, R. 1970. Zur Chemodifferenzierung des Rückenmarks und der Spinalganglien der Ratte. *Histochemie* **24**, 202–213.

Sasaki, M., Tohda, C., and Kuraishi, Y. 1998. Region-specific increase in glutamate release from dorsal horn of rats with adjuvant inflammation. *NeuroReport* **9**, 3219–3222.

Sasaki, S., Komori, T., and Iwata, M. 2000. Excitatory amino acid transporter 1 and 2 immunoreactivity in the spinal cord in amyotrophic lateral sclerosis. *Acta Neuropathol. (Berl.)* **100**, 138–144.

Sasek, C. A. and Elde, R. P. 1985. Distribution of neuropeptide Y-like immunoreactivity and its relationship to FMRF-amide-like immunoreactivity in the sixth lumbar and first sacral spinal cord segments of the rat. *J. Neurosci.* **7**, 1729–1739.

Sasek, C. A., Baldwin, C., and Zigmond, R. E. 1991. Distribution of vasoactive intestinal peptide- and peptide histidine isoleucine amide-like immunoreactive neurons and fibers in the thoracic spinal cord of the rat. *Brain Res.* **567**, 159–164.

Sastry, B. R. 1978. Morphine and met-enkephalin effects on sural A-delta afferent terminal excitability. *Eur. J. Pharmacol.* **50**, 269–273.

Sastry, B. R. 1979a. Presynaptic effects of morphine and methionine-enkephalin in feline spinal cord. *Neuropharmacology* **18**, 367–375.

Sastry, B. R. 1979b. Substance P effects on spinal nociceptive neurones. *Life Sci.* **24**, 2169–2178.

Sastry, B. R. and Goh, J. W. 1983. Action of morphine and met-enkephaline-amide on nociceptor driven neurones in substantia gelatinosa and deeper dorsal horn. *Neuropharmacology* **22**, 119–122.

Sato, J. and Perl, E. R. 1991. Adrenergic excitation of cutaneous pain receptors induced by peripheral nerve injury. *Science* **251**, 1608–1610.

Sato, J., Mizumura, K., and Kumazawa, T. 1989. Effect of ionic calcium on the responses of canine testicular polymodal receptors to algesic substances. *J. Neurophysiol.* **62**, 119–125.

Sato, K., Zhang, J.-H., Saika, T., Sato, M., Tadao, K., and Tohyama, T. 1991. Localization of glycine receptor α_1 subunit mRNA-containing neurons in the rat brain: An analysis using in situ hybridization histochemistry. *Neuroscience* **43**, 381–395.

Sato, K., Kiyama, H., Park, H. T., and Tohyama, M. 1993a. AMPA, KA and NMDA receptors are expressed in the rat DRG neurons. *NeuroReport* **4**, 1263–1265.

Sato, K., Kiyama, H., and Tohyama, M. 1993b. The differential expression patterns of messenger RNAs encoding non-*N*-methyl-D-asparate glutamate receptor subunits (GluR1–4) in the rat brain. *Neuroscience* **52**, 515–539.

Satoh, K., Kashiba, A., Kimura, H., and Maeda, T. 1982. Noradrenergic axon terminals in the substantia gelatinosa of the rat spinal cord: An electron-microscopic study using glyoxylic acid-potassium permanganate fixation. *Cell Tissue Res.* **222**, 359–378.

Saugstad, J. A., Segerson, T. P., and Westbrook, G. L. 1995. Modulation of ion channels and synaptic transmission by metabotropic glutamate receptors. In H. Wheal and A. Thompson (eds.), *Excitatory Amino Acids and Synaptic Transmission* (pp. 77–88). New York, Academic.

Sawynok, J. and Sweeney, M. I. 1989. The role of purines in nociception. *Neuroscience* **32**, 557–569.

Scadding, J. W. 1981. Development of ongoing activity, mechanosensitivity, and adrenaline sensitivity in severed peripheral nerve axons. *Exp. Neurol.* **73**, 345–364.

Scatton, B., Dubois, A., Javoy-Agid, F., and Camus, A. 1984. Autoradiographic localization of muscarinic cholinergic receptors at various segmental levels of the human spinal cord. *Neurosci. Lett.* **49**, 239–245.

Schady, W. and Torebjörk, H. E. 1983. Projected and receptive fields. A comparison of projected areas of sensations evoked by intraneural stimulation of mechanoreceptive units and their innervation territories. *Acta Physiol. Scand.* **119**, 267–275.

Schady, W. J. L., Torebjörk, H. E., and Ochoa, J. L. 1983. Peripheral projections of nerve fibres in the human median nerve. *Brain Res.* **277**, 249–261.

Schäfer, K. 1987. A quantitative study of the dependence of feline cold receptor activity on the calcium concentration. *Pfluegers Arch.* **409**, 208–213.

Schafer, M. K., Nohr, D., Krause, J. E., and Weihe, E. 1993. Inflammation-induced upregulation of NK1 receptor mRNA in dorsal horn neurones. *NeuroReport* **4**, 1007–1010.

Schäfer, M. K. H., Bette, M., Romeo, H., Schwaeble, W., and Weihe, E. 1994. Localization of *K*-opioid receptor mRNA in neuronal subpopulations of rat sensory ganglia and spinal cord. *Neurosci. Lett.* **167**, 137–140.

Schaible, H. G. and Grubb, B. D. 1993. Afferent and spinal mechanisms of joint pain. *Pain* **55**, 5–54.

Schaible, H. G. and Schmidt, R. F. 1983a. Activation of groups III and IV sensory units in medial articular nerve by local mechanical stimulation of knee joint. *J. Neurophysiol.* **49**, 35–44.

Schaible, H. G. and Schmidt, R. F. 1983b. Responses of fine medial articular nerve afferents to passive movements of knee joint. *J. Neurophysiol.* **49**, 1118–1126.

Schaible, H. G. and Schmidt, R. F. 1985. Effects of an experimental arthritis on the sensory properties of fine articular afferent units. *J. Neurophysiol.* **54**, 1109–1126.

Schaible, H. G. and Schmidt, R. F. 1988a. Time course of mechanosensitivity changes in articular afferent during developing experimental arthritis. *J. Neurophysiol.* **60**, 2180–2195.

Schaible, H. G. and Schmidt, R. F. 1988b. Excitation and sensitization of fine articular afferents from cat's knee joint by prostaglandin E_2. *J. Physiol.* **403**, 91–104.

Schaible, H. G., Schmidt, R. F., and Willis, W. D. 1986. Responses of spinal cord neurones to stimulation of articular afferent fibres in the cat. *J. Physiol.* **372**, 575–593.

Schaible, H. G., Schmidt, R. F., and Willis, W. D. 1987a. Convergent inputs from articular, cutaneous and muscle receptors onto ascending tract cells in the cat spinal cord. *Exp. Brain Res.* **66**, 479–488.

Schaible, H. G., Schmidt, R. F., and Willis, W. D. 1987b. Enhancement of the responses of ascending tract cells in the cat spinal cord by acute inflammation of the knee joint. *Exp. Brain Res.* **66**, 489–499.

Schaible, H. G., Jarrott, B., Hope, P. J., and Duggan, A. W. 1990. Release of immunoreactive substance P in the spinal cord during development of acute arthritis in the knee joint of the cat: A study with antibody microprobes. *Brain Res.* **529**, 214–223.

Schaible, H. G., Freudenberger, U., Neugebauer, V., and Stiller, R. U. 1994. Intraspinal release of immunoreactive calcitonin gene-related peptide during development of inflammation in the joint in vivo—a study with antibody microprobes in cat and rat. *Neuroscience* **62**, 1293–1305.

Scharf, J. H. 1958. Nervensystem Dritter Teil, Sensible Ganglien, In *Handbook der Mikroskopischen Anatomie des Menschen* (pp. 1–485) Springer-Verlag, Berlin.

Schechter, R., Beju, D., Gaffney, T., Schaefer, F., and Whetsell, L. 1996. Preproinsulin I and II mRNAs and insulin electron microscopic immunoreaction are present within the rat fetal nervous system. *Brain Res.* **736**, 16–27.

Scheibel, M. E. and Scheibel, A. B. 1968. Terminal axon patterns in cat spinal cord: II. The dorsal horn. *Brain Res.* **9**, 32–58.

Schenker, M. and Birch, R. 2000. Intact myelinated fibres in biopsies of ventral spinal roots after preganglionic traction injury to the brachial plexus: A proof that Sherrington's "wrong way afferents" exist in man? *J. Anat.* **197**, 383–391.

Schepelmann, K., Messlinger, K., Schaible, H. G., and Schmidt, R. F. 1992. Inflammatory mediators and nociception in the joint: Excitation and sensitization of slowly conducting afferent fibers of cat's knee by prostaglandin I_2. *Neuroscience* **50**, 237–247.

Schepelmann, K., Messlinger, K., and Schmidt, R. F. 1993. The effects of phorbol ester on slowly conducting afferents of the cat's knee joint. *Exp. Brain Res.* **92**, 391–398.

Schicho, R. and Donnerer, J. 1999. Nerve growth factor stimulates synthesis of calcitonin gene-related peptide in dorsal root ganglion cells during sensory regeneration in capsaicin-treated rats. *Neurosci. Res.* **35**, 183–187.

Schicho, R., Skofitsch, G., and Donnerer, J. 1999. Regenerative effect of human recombinant NGF on capsaicin-lesioned sensory neurons in the adult rat. *Brain Res.* **815**, 60–69.

Schiffmann, S. N., Teugels, E., Halleux, P., Menu, R., and Vanderhaeghen, J. J. 1991. Cholecystokinin mRNA detection in rat spinal cord motoneurons but not in dorsal root ganglia neurons. *Neurosci. Lett.* **123**, 123–126.

Schindler, M., Sellers, L. A., Humphrey, P. P. A., and Emson, P. C. 1997. Immunohistochemical localization of the somatostatin $SST_{2(A)}$ receptor in the rat brain and spinal cord. *Neuroscience* **76**, 225–240.

Schindler, M., Holloway, S., Hathway, G., Woolf, C. J., Humphrey, P. P., and Emson, P. C. 1998. Identification of somatostatin sst2(a) receptor expressing neurones in central regions involved in nociception. *Brain Res.* **798**, 25–35.

Schindler, M., Humphrey, P. P. A., Löhrke, S., and Friauf, E. 1999a. Immunohistochemical localization of the somatostatin $sst_{2(b)}$ receptor splice variant in the rat central nervous system. *Neuroscience* **90**, 859–874.

Schindler, M., Selmer, I. S., Helboe, L., Hick, G. A., Jenkins, D., Papotti, M., Feniuk, W., and Humphrey, P. P. A. 1999b. Localization of somatostatin receptors in the CNS and peripheral tissues. *Regul. Pept.* **80**, 136.

Schindler, M., Harris, C. A., Hayes, B., Papotti, M., and Humphrey, P. P. 2001. Immunohistochemical localization of adenosine A1 receptors in human brain regions. *Neurosci. Lett.* **297**, 211–215.

Schipper, J., Steinbuch, W. M., Vermes, I., and Tilders, F. J. 1983. Mapping of CRF-immunoreactive nerve fibers in the medulla oblongata and spinal cord of the rat. *Brain Res.* **267**, 145–150.

Schlaepfer, W. W. 1968. Acetylcholinesterase activity of motor and sensory nerve fibers in the spinal nerve roots of the rat. *Zeitschrift für Zellforschung* **88**, 441–456.

Schmalbruch, H. 1987. The number of neurons in dorsal root ganglia L4–L6 of the rat. *Anat. Rec.* **219**, 315–322.

Schmelz, M., Schmidt, R., Ringkamp, M., Handwerker, H. O., and Torebjörk, H. E. 1994. Sensitization of insensitive branches of C nociceptors in human skin. *J. Physiol.* **480**, 389–394.

Schmelz, M., Forster, C., Schmidt, R., Ringkamp, M., Handwerker, H. O., and Torebjörk, H. E. 1995. Delayed responses to electrical stimuli reflect C-fiber responsiveness in human microneurography. *Exp. Brain Res.* **104**, 331–336.

Schmelz, M., Schmidt, R., Ringkamp, M., Forster, C., Handwerker, H. O., and Torebjörk, H. E. 1996. Limitation of sensitization to injured parts of receptive fields in human skin C-nociceptors. *Exp. Brain Res.* **109**, 141–147.

Schmelz, M., Schmidt, R., Bickel, A. Handwerker, H. O., and Torebjörk, H. E. 1997. Specific C-receptors for itch in human skin. *J. Neurosci.* **17**, 8003–8008.

Schmelz, M., Schmid, R., Handwerker, H. O., and Torebjörk, H. E. 2000. Encoding of burning pain from capsaicin-treated human skin in two categories of unmyelinated nerve fibres. *Brain* **123**, 560–571.

Schmidt, R., Schmelz, M., Forster, C., Ringkamp, M., Torebjörk, E., and Handwerker, H. O. 1995. Novel classes of responsive and unresponsive C nociceptors in human skin. *J. Neurosci.* **15**, 333–341.

Schmidt, R., Schmelz, M., Ringkamp, M., Handwerker, H. O., and Torebjörk, H. E. 1997. Innervation territories of mechanically activated C nociceptor units in human skin. *J. Neurophysiol.* **78**, 2641–2648.

Schmidt, R., Schmelz, M., Torebjörk, H. E., and Handwerker, H. O. 2000. Mechano-insensitive nociceptors encode pain evoked by tonic pressure to human skin. *Neuroscience* **98**, 793–800.

Schmidt, R. F. 1971. Presynaptic inhibition in the vertebrate central nervous system. *Eregeb. Physiol.* **63**, 20–101.

Schmitt, A., Asan, E., Puschel, B., Jons, T., and Kugler, P. 1996. Expression of the glutamate transporter GLT1 in neural cells of the rat central nervous system: Non-radioactive in situ hybridization and comparative immunocytochemistry. *Neuroscience* **71**, 989–1004.

Schneider, S. P. and Perl, E. R. 1985. Selective excitation of neurons in the mammalian spinal dorsal horn by aspartate and glutamate in vitro: Correlation with location and synaptic input. *Brain Res.* **360**, 339–343.

Schneider, S. P. and Perl, E. R. 1988. Comparison of primary afferent and glutamate excitation of neurons in the mammalian spinal dorsal horn. *J. Neurosci.* **8**, 2062–2073.

Schoenen, J. 1978. Histoenzymology of the developing rat spinal cord. *Neuropathol. Appl. Neurobiol.* **4**, 37–46.

Schoenen, J. 1982. The dendritic organization of the human spinal cord: The dorsal horn. *Neuroscience* **7**, 2057–2087.

Schoenen, J., Budo, C., and Poncelet, G. 1968. Effet de la section du sciatique sur l'activité de l'isoenzyme fluororésistant de la phosphatase acide dans la moelle épinière du rat. *Comp. Rend.* **162**, 2035–2037.

Schoenen, J., Lotstra, F., Vierendeels, G., Reznik, M., and Vanderhaeghen, J. J. 1985a. Substance P, enkephalins, somatostatin, cholecystokinin, oxytocin, and vasopressin in human spinal cord. *Neurology* **35**, 881–890.

Schoenen, J., Hees, J. V., Gybels, J., Costa, M. C., and Vanderhaeghen, J. J. 1985b. Histochemical changes of substance P, FRAP, serotonin and succinic dehydrogenase in the spinal cord of rats with adjuvant arthritis. *Life Sciences* **36**, 1247–1254.

Schoenen, J., Lotstra, F., Liston, D., Rossier, J., and Vanderhaeghen, J. J. 1986. Synkephalin in human and bovine spinal cord. *Cell Tissue Res.* **246**, 641–645.

Schoepp, D. D. 2001. Unveiling the functions of presynaptic metabotropic glutamate receptors in the central nervous system. *J. Pharmacol. Exp. Therap.* **299**, 12–20.

Schoepp, D. D., Jane, D. E., and Monn, J. A. 1999. Pharmacological agents acting at subtypes of metabotropic glutamate receptors. *Neuropharmacology* **38**, 1431–1476.

Schouenborg, J. 1984. Functional and topographical properties of field potentials evoked in rat dorsal horn by cutaneous C-fibre stimulation. *J. Physiol.* **356**, 169–192.

Schouenborg, J. and Sjölund, B. H. 1986. First-order nociceptive synapses in rat dorsal horn are blocked by an amino acid antagonist. *Brain Res.* **379**, 394–398.

Schoultz, T. W. and Swett, J. E. 1972. The fine structure of the Golgi tendon organ. *J. Neurocytol.* **1**, 1–26.

Schoultz, T. W. and Swett, J. E. 1974. Ultrastructural organization of the sensory fibers innervating the Golgi tendon organ. *Anat. Rec.* **179**, 147–162.

Schreff, M., Schulz, S., Wilbrony, D., and Höllt, V. 1998. Immunofluorescent identification of endomorphin-2-containing nerve fibers and terminals in the rat brain and spinal cord. *NeuroReport* **9**, 1031–1034.

Schroder, H. D. 1983. Localization of cholecystokinin-like immunoreactivity in the rat spinal cord, with particular reference to the autonomic innervation of the pelvic organs. *J. Comp. Neurol.* **217**, 176–186.

Schroder, H. D. 1984. Somatostatin in the caudal spinal cord: An immunohistochemical study of the spinal centers involved in the innervation of pelvic organs. *J. Comp. Neurol.* **223**, 400–414.

Schroder, H. D. and Skagerberg, G. 1985. Catecholamine innervation of the caudal spinal cord in the rat. *J. Comp. Neurol.* **242**, 358–368.

Schuligoi, R., Donnerer, J., and Amann, R. 1994. Bradykinin-induced sensitization of afferent neurons in the rat paw. *Neuroscience* **59**, 211–215.

Schulman, H., Tsodikow, V., Einhorn, M., Levy, Y., Shorer, Z., and Hertzanu, Y. 2001. Congenital insensitivity to pain with anhidrosis (CIPA): The spectrum of radiological findings. *Pediatr. Radiol.* **31**, 701–705.

Schultz, R. A., Miller, D. G., Kerr, C. S., and Micheli, L. 1984. Mechano-receptors in human cruciate ligaments: A histological study. *J. Bone Joint Surg.* **66A**, 1072–1076.

Schulz, S., Schreff, M., Gramsch, C., and Höllt, V. 1996. Nociceptin/orphanin FQ and opioid peptides show overlapping distribution but not co-localization in pain-modulatory brain regions. *NeuroReport* **7**, 3021–3025.

Schulz, S., Schreff, M., Schmidt, H., Handel, M., Przewlocki, R., and Hollt, V. 1998a. Immunocytochemical localization of somatostatin receptor SST_{2a} in the rat spinal cord and dorsal root ganglia. *Eur. J. Neurosci.* **10**, 3700–3708.

Schulz, S., Schmidt, H., Handel, M., Schreff, M., and Hollt, V. 1998b. Differential distribution of alternatively spliced somatostatin receptor 2 isoforms (sst2A and sst2B) in rat spinal cord. *Neurosci. Lett.* **257**, 37–40.

Schultzberg, M., Dockray, G. J., and Williams, R. G. 1982. Capsaicin depletes CCK-like activity detected by immunohistochemistry, but not that measured by radioimmunoassay in rat dorsal spinal cord. *Brain Res.* **235**, 198–204.

Schumacher, M. A., Jong, B. E., Frey, S. L., Sudanagunta, S. P., Capra, N. F., and Levine, J. D. 2000. The stretch-inactivated channel, a vanilloid receptor variant, is expressed in small-diameter sensory neurons in the rat. *Neurosci. Lett.* **287**, 215–218.

Seabrook, G. R., Bowery, B. J., Heavens, R., Ford, B. H., Sirinathsinghi, D. J. S., Borkowski, J. A., Hess, J. F., Strader, C. D., and Hill, R. G. 1997. Expression of B_1 and B_2 bradykinin receptor mRNA and their functional roles in sympathetic ganglia and sensory dorsal root ganglia neruones from wild-type and B_2 receptor knockout mice. *Neuropharmacology.* **36**, 1009–1017.

Segu, L. and Calas, A. 1978. The topographical distribution of serotoninergic terminals in the spinal cord of the cat: Quantitative radioautographic studies. *Brain Res.* **153**, 449–464.

Seifert, H., Chesnut, J., DeSouza, E., Rivier, J., and Vale, W. 1985. Binding sites for calcitonin gene-related peptide in distinct areas of rat brain. *Brain Res.* **346**, 195–198.

Selmer, I., Schindler, M., Allen, J. P., Humphrey, P. P., and Emson, P. C. 2000. Advances in understanding neuronal somatostatin receptors. *Regul. Pept.* **90**, 1–18.

Selzer, M. and Spencer, W. A. 1969. Convergence of visceral and cutaneous afferent pathways in the lumbar spinal cord. *Brain Res.* **14**, 331–348.

Selzer, Z. and Devor, M. 1984. Effect of nerve section on the spinal distribution of neighboring nerves. *Brain Res.* **306**, 31–37.

Seltzer, Z., Dubner, R., and Shir, Y. 1990. A novel behavioral model of neuropathic pain disorders produced in rats by partial sciatic nerve injury. *Pain* **43**, 205–218.

Seltzer, Z., Cohn, S., Ginzburg, R., and Berlin, B. 1991. Modulation of neuropathic pain behavior in rats by spinal disinhibition and NMDA receptor blockade of injury discharge. *Pain* **45**, 69–75.

Semba, K., Masarachia, P., Malamed, S., Jacquin, M., and Harris, S. 1983. An electron-microscopic study of primary afferent terminals from slowly adapting type I receptors in the cat. *J. Comp. Neurol.* **221**, 466–481.

Semba, K., Masarachia, P., Malamed, S., Jacquin, M., Harris, S., Yang, G., and Egger, M. D. 1985. An electron-microscopic study of terminals of rapidly adapting mechanoreceptive afferent fibers in the cat spinal cord. *J. Comp. Neurol.* **232**, 229–240.

Semenenko, F. M. and Cervero, F. 1992. Afferent fibres from the guinea-pig ureter: Size and peptide content of the dorsal root ganglion cells of origin. *Neuroscience* **47**, 197–201.

Senaris, R. M., Schindler, M., Humphrey, P. P. A., and Emson, P. C. 1995. Expression of somatostatin receptor 3 mRNA in the motorneurones of the rat spinal cord, and the sensory neurones of the spinal ganglia. *Mol. Brain Res.* **29**, 185–190.

Senba, E. and Kashiba, H. 1996. Sensory afferent processing in multi-responsive DRG neurons, In T. Kumazawa, L. Kruger, and K. Mizumura (eds.), *Progress in Brain Research* (**24**, pp. 387–410). Elsevier Science.

Senba, E. and Tohyama, M. 1988. Calcitonin gene-related peptide containing autonomic efferent pathways to the pelvic ganglia of the rat. *Brain Res.* **449**, 386–390.

Senba, E., Shiosaka, S., Hara, Y., Inagaki, S., Shakanaka, M., Takatsuki, K., Kawai, Y., and Tohyama, M. 1982. Ontogeny of the peptidergic system in the rat spinal cord: Immunohistochemical analysis. *J. Comp. Neurol.* **208**, 54–66.

Senba, E., Yanaihara, C., Yanaihara, N., and Tohyama, M. 1988. Co-localization of substance P and Met-enkephalin–Arg–Gly–Leu in the intraspinal neurons of the rat, with special reference to the neurons in the substantia gelantinosa. *Brain Res.* **453**, 110–116.

Sengupta, J. N. and Gebhart, G. F. 1994. Characterization of mechanosensitive pelvic nerve afferent fibers innervating the colon of the rat. *J. Neurophysiol.* **71**, 2046–2060.

Sengupta, J. N., Kauvar, D., and Goyal, R. K. 1989. Characteristics of vagal esophageal tension-sensitive afferent fibers in the opossum. *J. Neurophysiol.* **61**, 1001–1010.

Sengupta, J. N., Saha, J. K. and Goyal, R. K. 1990. Stimulus-response function studies of esophageal mechanosensitive nociceptors in sympathetic afferents of opossum. *J. Neurophysiol.* **64**, 796–812.

Senok, S. S. and Baumann, K. I. 1997. Functional evidence for calcium-induced calcium release in isolated rat vibrissal Merkel cell mechanoreceptors. *J. Physiol.* **500**, 29–37.

Sequier, J. M., Hunziker, W., Andressen, C., and Celio, M. R. 1990. Calbindin D-28k protein and mRNA localization in the rat brain. *Eur. J. Neurosci.* **2**, 1118–1126.

Seroogy, K., Mohapatra, N. K., Lund, P. K., Rethelyi, M., McGehee, D. S., and Perl, E. R. 1990. Species-specific expression of cholycystokinin messenger RNA in rodent dorsal root ganglia. *Mol. Brain Res.* **7**, 171–176.

Sethi, J. S. and Tewari, H. B. 1971. Histochemical studies on the distribution of some hydrolytic and oxidative enzymes in the neurons of spinal and trigeminal ganglia and in the trigeminal of squirrel (*Funambulus palmarum*). *Acta Morphologica Neerlando-Scandinavica* **9**, 101–115.

Sethi, J. S., Tewari, H. B., and Sood, P. P. 1969. On the distributive patterns of alkaline phosphatase activity and their functional significance amongst the spinal ganglion cells of squirrel. *Acta Neurol. Belg.* **69**, 51–57.

Sexton, P. M., McKenzie, J. S., Mason, R. T., Moseley, J. M., Martin, T. J., and Mendelsohn, F. A. O. 1986. Localization of binding sites for gene-related peptide in rat brain by in vitro autoradiography. *Neuroscience* **19**, 1235–1245.

Seybold, V. S. 1985a. Distribution of histaminergic, muscarinic and serotoninergic binding sites in cat spinal cord with emphasis on the region surrounding the central canal. *Brain Res.* **342**, 291–296.

Seybold, V. S. 1985b. Neurotransmitter receptor sites in the spinal cord. In T. Yaksh (ed.), *Spinal Afferent Processing* (pp. 117–139). Plenum Press, New York.

Seybold, V. S. and Elde, R. P. 1980. Immunohistochemical studies of peptidergic neurons in the dorsal horn of the spinal cord. *J. Histochem. Cytochem.* **28**, 367–370.

Seybold, V. S. and Elde, R. P. 1982. Neurotensin immunoreactivity in the superficial laminae of the dorsal horn of the rat: I. Light-microscopic studies of cell bodies and proximal dendrites. *J. Comp. Neurol.* **205**, 89–100.

Seybold, V. S. and Elde, R. 1984. Receptor autoradiography in thoracic spinal cord: Correlation of neurotransmitter binding sites with sympathoadrenal neurons. *J. Neurosci.* **4**, 2533–2542.

Seybold, V. S. and Malay, B. 1984. Ultrastructural study of neurotensin immunoreactivity in the superficial laminae of the dorsal horn of the rat. *Peptides* **5**, 1179–1189.

Seybold, V. S., Grkovic, I., Portbury, A. L., Ding, Y.-Q., Shigemoto, R., Mizuno, N., Furness, J. B., and Southwell, B. R. 1997. Relationship of NK_3 receptor-immunoreactivity to subpopulations of neurons in rat spinal cord. *J. Comp. Neurol.* **381**, 439–448.

Shadiack, A. M., Sun, Y., and Zigmond, R. E. 2001. Nerve growth factor antiserum induces axotomy-like changes in neuropeptide expression in intact sympathetic and sensory neurons. *J. Neurosci.* **21**, 363–371.

Shah, B. S., Stevens, E. B., Gonzalez, M. I., Bramwell, S., Pinnock, R. D., Lee, K., and Dixon, A. K. 2000. Beta3, a novel auxiliary subunit for the voltage-gated sodium channel, is expressed preferentially in sensory neurons and is upregulated in the chronic constriction injury model of neuropathic pain. *Eur. J. Neurosci.* **12**, 3985–3990.

Shah, B. S., Gonzalez, M. I., Bramwell, S., Pinnock, R. D., Lee, K., and Dixon, A. K. 2001. Beta3, a novel auxiliary subunit for the voltage gated sodium channel is upregulated in sensory neurones following streptozocin induced diabetic neuropathy in rat. *Neurosci. Lett.* **309**, 1–4.

Shantha, T. R., Manocha, C. L., and Bourne, G. H. 1967. Enzyme histochemistry of the mesentric and dorsal root ganglion cells of cat and squirrel monkey. *Histochemie* **10**, 238–245.

Sharif, N. A. and Burt, D. R. 1985. Limbic, hypothalamic, cortical and spinal regions are enriched in receptors for thyrotropin-releasing hormone: Evidence from [3H] ultrofilm autoradiography and correlation with central effects of the tripeptide in rat brain. *Neurosci. Lett.* **60**, 337–342.

Sharif, N. A. and Hughes, J. 1989. Discrete mapping of brain mu and delta opioid receptors using selective peptides: Quantitative autoradiography, species differences and comparison with kappa receptors. *Peptides* **10**, 499–522.

Sharif, N. A., Hunter, J. C., Hill, R. G., and Hughes, J. 1988. [^{125}I] Dynorphin (1–8) produces a similar pattern of κ-opioid receptor labelling to [^{3}H]dynorphin(1–8) and [^{3}H]etorphine in guinea pig brain: A quantitative autoradiographic study. *Neurosci. Lett.* **86**, 272–278.

Sharkey, K. S. 1987. The organization of capsaicin-sensitive visceral afferent systems. *Acta Physiol. Hung.* **69**, 447–458.

Sharkey, K. A., Sobrino, J. A., and Cervero, F. 1987. Evidence for a visceral afferent origin of substance P-like immunoreactivity in lamina V of the rat thoracic spinal cord. *Neuroscience* **22**, 1077–1083.

Sharkey, K. A., Sobrino, J. A., Cervero, F., Varro, A., and Dockray, G. J. 1989. Visceral and somatic afferent origin of calcitonin gene-related peptide immunoreactivity in the lower thoracic spinal cord of the rat. *Neuroscience* **32**, 169–179.

Sharma, H. S., Westman, J., Olsson, Y., and Alm, P. 1996. Involvement of nitric oxide in acute spinal cord injury: An immunocytochemical study using light and electron microscopy in the rat. *Neurosci Res.* **24**, 373–384.

Shashidharan, Huntley, G. W., Murray, J. M., Buku, A., Moran, T., Walsh, M. J., and Plaitakis, A. 1997. Immunohistochemical localization of the neuron-specific glutamate transporter EAACI (EAAT3) in rat brain and spinal cord revealed by a novel monoclonal antibody. *Brain Res.* **773**, 139–148.

Shatzky, S., Moses, S., Levy, J., Pinsk, V., Hershkovitz, E., Herzog, L., Shorer, Z., Luder, A., and Parvari, R. 2000. Congenital insensitivity to pain with anhidrosis (CIPA) in Israeli-Bedouins: Genetic heterogeneity, novel mutations in the TRKA/NGF receptor gene, clinical findings, and results of nerve conduction studies. *Am. J. Med. Genet.* **92**, 353–360.

Shaw, P. J. 1992. Excitatory amino acid neurotransmission, excitotoxicity and excitotoxins. *Cur. Opinion Neurol. Neurosurg.* **5**, 383–390.

Shaw, P. J. and Ince, P. G. 1994. A quantitative autoradiographic study of [^{3}H]kainate binding sutes in the normal human spinal cord, brainstem and motor cortex. *Brain Res.* **641**, 39–45.

Shaw, P. J., Ince, P. G., Johnson, M., Perry, E. K., and Candy, J. 1991. The quantitative autoradiographic distribution of [^{3}H]MK-801 binding sites in the normal human spinal cord. *Brain Res.* **539**, 164–168.

Shaw, P. J., Ince, P. G., Johnson, M., Perry, E. K., and Candy, J. M. 1992. The quantitative autoradiographic distribution of [^{3}H]MK-801 binding sites in the normal human brainstem in relation to motor neuron disease. *Brain Res.* **572**, 276–280.

Shaw, P. J., Ince, P. G., Matthews, J. N., Johnson, M., and Candy, J. M. 1994a. N-Methyl-D-aspartate (NMDA) receptors in the spinal cord and motor cortex in motor neuron disease: A quantitative autoradiographic study using [3H]MK-801. *Brain Res.* **637**, 297–302.

Shaw, P. J., Chinnery, R. M., and Ince, P. G. 1994b. Non-NMDA receptors in motor neuron disease (MND): A quantitative autoradiographic study in spinal cord and motor cortex using [^{3}H]CNQX and [^{3}H]kainate. *Brain Res.* **655**, 186–194.

Shealy, C. N., Mortimer, J. T., and Reswick, J. B. 1967. Electrical inhibition of pain by stimulation of the dorsal columns. *Anesth. Anal.* **46**, 489–491.

Shealy, C. N., Mortimer, J. T., and Hagfors, N. R. 1970. Dorsal column electroanalgesia. *J. Neurosurg.* **32**, 560–564.

Sheehan, D. 1932. The afferent nerve supply of the mesentery and its significance in the causation of abdominal pain. *J. Anat.* **67**, 233–249.

Sheen, K. and Chung, J. M. 1993. Signs of neuropathic pain depend on signals from injured nerve fibers in a rat model. *Brain Res.* **610**, 62–68.

Shehab, S. A. S. and Atkinson, M. 1986a. Vasoactive intestinal polypeptide increases in areas of the dorsal horn of the spinal cord from which other neuropeptides are depleted following peripheral axotomy. *Exp. Brain Res.* **62**, 422–430.

Shehab, S. A. S. and Atkinson, M. E. 1986b. Vasoactive intestinal polypeptide (VIP) increases in the spinal cord after peripheral axotomy of the sciatic nerve originate from primary afferent neurons. *Brain Res.* **372**, 37–44.

Shehab, S. A. S., Atkinson, M. E., and Payne, J. N. 1986. The origins of the sciatic nerve and changes in neuropeptides after axotomy: A double labelling study using retrograde transport of true blue and vasoactive intestinal polypeptide immunohistochemistry. *Brain Res.* **376**, 180–185.

Shen, W. Z., Luo, C. B., Dong, L., Chan, W. Y., and Yew, D. T. 1994. Distribution of neuropeptide Y in the developing human spinal cord. *Neuroscience* **63**, 251–256.

Shen, H., Chung, J. M., and Chung, K. 1999. Expression of neurotrophin mRNAs in the dorsal root ganglion after spinal nerve injury. *Mol. Brain Res.* **64**, 186–192.

Sherrington, C. S. 1893. Experiments in examination of the peripheral distribution of the fibres of the posterior roots of some spinal nerves. *Phil. Trans. Roy. Soc. B*, **184**, 641–765.

Sherrington, C. S. 1894. On the anatomical constitution of nerves of skeletal muscles, with remarks on recurrent fibres in the ventral spinal nerve-root. *J. Physiol.* **17**, 211–257.

Sherrington, C. S. 1898. Experiments in examination of the peripheral distribution of the fibres of the posterior roots of some spinal nerves. Part II. *Phil. Trans. B* **190B**, 45–186.

Sherrington, C. S. 1906. *The Integrative Action of the Nervous System.* (2nd ed., 1947) Yale University. Press, New Haven.

Sheward, W. J., Lutz, E. M., and Harmar, A. J. 1995. The distribution of vasoactive intestinal peptide2 receptor messenger RNA in the rat brain and pituitary gland as assessed by in situ hybridization. *Neuroscience* **67**, 409–418.

Shi, T. J., Zhang, X., Holmberg, K., Xu, Z.-Q. D., and Hökfelt, T. 1997. Expression and regulation of galanin-R2 receptors in rat primary sensory neurons: Effect of axotomy and inflammation. *Neurosci. Lett.* **237**, 57–60.

Shi, T. J., Zhang, X., Berge, O.-G., Erickson, J. C., Palmiter, R. D., and Hökfelt, T. 1998. Effect of peripheral axotomy on dorsal root ganglion neuron phenotype and autotomy behaviour in neuropeptide Y-deficient mice. *Regul. Pept.* **75–76**, 161–173.

Shi, T. J., Cui, J. G., Meyerson, B. A., Linderoth, B., and Hökfelt, T. 1999a. Regulation of galanin and neuropeptide Y in dorsal root ganglia and dorsal horn in rat mononeuropathic models: Possible relation to tactile hypersensitivity. *Neuroscience* **93**, 741–757.

Shi, T. J., Winzer-Serhan, U., Leslie, F., and Hökfelt, T. 1999b. Distribution of alpha2-adrenoceptor mRNAs in the rat lumbar spinal cord in normal and axotomized rats. *NeuroReport* **10**, 2835–2839.

Shi, T. S., Winzer-Serhan, U., Leslie, F., and Hökfelt, T. 2000. Distribution and regulation of alpha(2)-adrenoceptors in rat dorsal root ganglia. *Pain* **84**, 319–330.

Shi, T. J., Tandrup, T., Bergman, E., Xu, Z. Q., Ulfhake, B., and Hokfelt, T. 2001. Effect of peripheral nerve injury on dorsal root ganglion neurons in the C57 BL/6J mouse: Marked changes both in cell numbers and neuropeptide expression. *Neuroscience* **105**, 249–263.

Shigemoto, R., Ohishi, H., Nakanishi, S., and Mizuno, N. 1992a. Expression of the mRNA for the rat NMDA receptor (NMDAR1) in the sensory and autonomic ganglion neurons. *Neurosci. Lett.* **144**, 229–232.

Shigemoto, R., Nakanishi, S., and Mizuno, N. 1992b. Distribution of the mRNA for a metabotropic glutamate receptor (mGluR1) in the central nervous system: An in situ hybridization study in the adult and developing rat. *J. Comp. Neurol.* **322**, 121–135.

Shimoji, K., Matsuki, M. and Shimizu, H. 1977. Wave-form characteristics and spatial distribution of evoked spinal electrogram in man. *J. Neurosurg.* **46**, 304–313.

Shimosegawa, T., Koizumi, M., Toyota, T., Goto, Y., Yanaihara, C., and Yanaihara, N. 1986. An immunohistochemical study of methionine-enkephalin-Arg-Gly-Leu-like immunoreactivity-containing liquor-contacting neurones (LCNs) in the rat spinal cord. *Brain Res.* **379**, 1–9.

Shimosegawa, T., Foda, H. D., and Said, S. I. 1990. [Met]enkephalin–Arg6–Gly7–Leu8-immunoreactive nerves in guinea-pig and rat lungs: Distribution, origin, and co-existence with vasoactive intestinal polypeptide immunoreactivity. *Neuroscience* **36**, 737–750.

Shin, H. K., Kim, J., and Chung, J. M. 1985. Flexion reflex elicited by ventral root afferents in the cat. *Neurosci. Lett.* **62**, 353–358.

Shin, H. K., Kim, J., Nam, S. C., Paik, K. S., and Chung, J. M. 1986. Spinal entry route for ventral root afferent fibers in the cat. *Exp. Neurol.* **94**, 714–725.

Shin, J., Cho, H., Hwang, S. W., Jung, J., Shin, C. Y., Lee, S. Y., Kim, S. H., Lee, M. G., Choi, Y. H., Kim, J., Haber, N. A., Reichling, D. B., Khaser, S., Levine, J. D., and Oh, U. 2002. Bradykinin-12-lipoxygenase-VR1 signaling pathway for inflammatory hyperalgesia. *Proc. Natl. Acad. Sci. USA* **99**, 10150–10155.

Shinder, V., Govrin-Lippmann, R., Cohen, S., Belenky, M., Ilin, P., Fried, K., Wilkinson, H. A., and Devor, M. 1999. Structural basis of sympathetic-sensory coupling in rat and human dorsal root ganglia following peripheral nerve injury. *J. Neurocytol.* **28**, 743–761.

Shir, Y. and Seltzer, Z. 1991. Effects of sympathectomy in a model of causalgiform pain produced by partial sciatic nerve injury in rats. *Pain* **45**, 309–320.

Shorer, Z., Moses, S. W., Hershkovitz, E., Pinsk, V., and Levy, J. 2001. Neurophysiologic studies in congenital insensitivity to pain with anhidrosis. *Pediatr. Neurol.* **25**, 397–400.

Shortland, P. and Woolf, C. J. 1993. Chronic peripheral nerve section results in a rearrangement of the central axonal arborizations of axotomized A beta primary afferent neurons in the rat spinal cord. *J. Comp. Neurol.* **330**, 65–92.

Shortland, P., Woolf, C. J., and Fitzgerald, M. 1989. Morphology and somatotopic organization of the central terminals of hindlimb hair follicle afferents in the rat lumbar spinal cord. *J. Comp. Neurol.* **289**, 416–433.

Shriver, J. E., Stein, B. M., and Carpenter, M. B. 1968. Central projections of spinal dorsal roots in the monkey: I. Cervical and upper thoracic dorsal roots. *Am. J. Anat.* **123**, 27–74.

Shughrue, P. J., Lane, M. V., and Merchenthaler, I. 1996. In situ hybridization analysis of the distribution of neurokinin-3 mRNA in the rat central nervous system. *J. Comp. Neurol.* **372**, 395–414.

Shughrue, P. J., Lane, M. V., and Merchenthaler, I. 1997. Comparative distribution of estrogen receptor-alpha and beta mRNA in the rat central nervous system. *J. Comp. Neurol.* **388**, 507–525.

Siddall, P., Xu, C. L., and Cousins, M. 1995. Allodynia following traumatic spinal cord injury in the rat. *NeuroReport* **6**, 1241–1244.

Siegel, S. M., Grove, B. D., and Carr, P. A. 2002. SSeCKS immunolabeling in rat primary sensory neurons. *Brain Res.* **926**, 126–136.

Sillar, K. T. and Roberts, A. 1988. Unmyelinated cutaneous afferent neurons activate two types of excitatory amino acid receptors in the spinal cord of *Xenopus laevis* embryos. *J. Neurosci.* **8**, 1350–1360.

Silos-Santiago, I., Molliver, D. C., Ozaki, S., Smeyne, R. J., Fagan, A. M. et al. 1995. Non-TrkA-expressing small DRG neurons are lost in TrkA deficient mice. *J. Neurosci.* **15**, 5929–5942.

Silva, E., Cleland, C. L., and Gebhart, G. F. 1997. Contributions of glutamate receptors to the maintenance of mustard oil-induced hyperalgesia in spinalized rats. *Exp. Brain Res.* **117**, 379–388.

Silver, A. and Wolstencroft, J. H. 1971. The distribution of cholinesterases in relation to the structure of the spinal cord in the cat. *Brain Res.* **34**, 205–227.

Silverman, J. D. and Kruger, L. 1988a. Lectin and neuropeptide labeling of separate populations of dorsal root ganglion neurons and associate "nociceptor" thin axons in rat testis and cornea whole-mount preparations. *Somatosens. Mot. Res.* **5**, 259–267.

Silverman, J. D. and Kruger, L. 1988b. Acid phosphatase as a selective marker for a class of small sensory ganglion cells in several mammals: Spinal cord distribution, histochemical properties, and relation to fluoride-resistant acid phosphatase (FRAP) of rodents. *Somatosens. Mot. Res.* **5**, 219–246.

Silverman, J. D. and Kruger, L. 1989. Calcitonin-gene-related peptide-immunoreactive innervation of the rat head with emphasis on specialized sensory structures. *J. Comp. Neurol.* **280**, 303–330.

Silverman, J. D. and Kruger, L. 1990. Selective neuronal glycoconjugate expression in sensory and autonomic ganglia: Relation of lectin reactivity to peptide and enzyme markers. *J. Neurocytol.* **19**, 789–801.

Simantov, R., Kuhar, M. J., Uhl, G. R., and Snyder, S. H. 1977. Opioid peptide enkephalin: Immunohistochemical mapping in rat central nervous system. *Neurobiology* **74**, 2167–2171.

Simerly, R. B., Chang, C., Muramatsu, M., and Swanson, L. W. 1990. Distribution of androgen and estrogen receptor mRNA-containing cells in the rat brain: An in situ hybridization study. *J. Comp. Neurol.* **294**, 76–95.

Simmons, D. R., Spike, R. C., and Todd, A. J. 1995. Galanin is contained in GABAergic neurons in the rat spinal dorsal horn. *Neurosci. Lett.* **187**, 119–122.

Simon, E. 2000. The enigma of deep-body thermosensory specificity. *Int. J. Biometeorol.* **44**, 105–120.

Simon, J., Wei, J. Y., Randic, M., and Burgess, P. R. 1984. Signaling of ankle joint position by receptors in different muscles. *Somatosens. Res.* **2**, 127–147.

Simone, D. A. and Kajander, K. C. 1996. Excitation of rat cutaneous nociceptors by noxious cold. *Neurosci. Lett.* **213**, 53–56.

Simone, D. A. and Kajander, K. C. 1997. Responses of cutaneous A-fiber nociceptors to noxious cold. *J. Neurophysiol.* **77**, 2049–2060.

Simone, D. A., Ngeow, J. Y. F., Putterman, G. J., and LaMotte, R. H. 1987. Hyperalgesia to heat after intradermal injection of capsaicin. *Brain Res.* **418**, 201–203.

Simone, D. A., Baumann, T. K., and LaMotte, R. H. 1989a. Dose-dependent pain and mechanical hyperalgesia in humans after intradermal injection of capsaicin. *Pain* **38**, 99–107.

Simone, D. A., Baumann, T. K., Collins, J. G., and LaMotte, R. H. 1989b. Sensitization of cat dorsal horn neurons to innocuous mechanical stimulation after intradermal injection of capsaicin. *Brain Res.* **486**, 185–189.

Simone, D. A., Sorkin, L. S., Oh, U., Chung, J. M., Owens, C., LaMotte, R. H., and Willis, W. D. 1991. Neurogenic hyperalgesia: Central neural correlates in responses of spinothalamic tract neurons. *J. Neurophysiol.* **66**, 228–246.

Simone, D. A., Marchettini, P., Caputi, G., and Ochoa, J. L. 1994. Identification of muscle afferents subserving sensation of deep pain in humans. *J. Neurophysiol.* **72**, 883–889.

Simpson, K. L., Waterhouse, B. D., and Lin, R. C. 1999. Origin, distribution, and morphology of galaninergic fibers in the rodent trigeminal system. *J. Comp. Neurol.* **411**, 524–534.

Sims, T. J. and Vaughn, J. E. 1979. The generation of neurons involved in an early reflex pathway of embryonic mouse spinal cord. *J. Comp. Neurol.* **183**, 707–720.

Sinclair, D. C. 1955. Cutaneous sensation and the doctrine of specific energy. *Brain* **78**, 584–614.

Sinclair, D. C. 1981. *Mechanisms of Cutaneous Sensation.* Oxford University Press, Oxford.

Sinclair, D. C., Weddell, G., and Feindel, W. H. 1948. Referred pain and associated phenomena. *Brain* **71**, 184–211.

Sindou, M., Quoex, C., and Baleydier, C. 1974. Fiber organization at the posterior spinal cord-rootlet junction in man. *J. Comp. Neurol.* **153**, 15–26.

Sindou, M., Mifsud, J. J., Rosaiti, C., and Boisson, D. 1987. Microsurgical selective posterior rhizotomy in the dorsal root entry zone for treatment of limb spasticity. *Acta neurochiurgica* **39**, 99–102.

Singer, E. and Placheta, P. 1980. Reduction of [^{3}H]muscimol binding sites in rat dorsal spinal cord after neonatal capsaicin treatment. *Brain Res.* **202**, 484–487.

Singhaniyom, W., Wreford, N. G., and Guldner, F. H. 1983. Asymmetric distribution of catecholamine-containing neuronal perikarya in the upper cervical spinal cord of rat. *Neurosci. Lett.* **41**, 91–97.

Siri, C. R., Shortland, P. J., Grant, G., and Olivius, N. P. 2001. Delayed administration of NGF reverses nerve injury induced central alterations of primary afferents. *NeuroReport* **12**, 1899–1902.

Sivilotti, L. and Woolf, C. J. 1994. The contribution of $GABA_A$ and glycine receptors to central sensitization: Disinhibition and touch-evoked allodynia in the spinal cord. *J. Neurophysiol.* **782**, 169–179.

Sivilotti, L. G., Thompson, S. W. N., and Woolf, C. J. 1993. The rate of rise of the cumulative depolarization evoked by repetitive stimulation of small-calibre afferents is a predictor of action potential windup in rat spinal neurons in vitro. *J. Neurophysiol.* **69**, 1621–1631.

Sjölander, P., Johansson, H., Sojka, P., and Rehnholm, A. 1989. Sensory nerve endings in the cat cruciate ligaments: A morphological investigation. *Neurosci. Lett.* **102**, 33–38.

Skagerberg, G. and Bjorklund, A. 1985. Topographic principles in the spinal projections of serotonergic and non-serotonergic brainstem neurons in the rat. *Neuroscience* **15**, 445–480.

Skagerberg, G., Bjorklund, A., Lindvall, O., and Schmidt, R. H. 1982. Origin and termination of the diencephalo-spinal dopamine system in the rat. *Brain Res. Bull.* **9**, 237–244.

Skilling, S. R., Smullin, H. D., Beitz, A. J., and Larsson, A. A. 1988. Extracellular amino acid concentrations in the dorsal spinal cord of freely moving rats following veratridine and nociceptive stimulation. *J. Neurochem.* **51**, 127–132.

Skirboll, L., Hokfelt, T., Dockray, G., Rehfeld, J., Brownstein, M., and Cuello, A. C. 1983. Evidence for periaqueductal cholecystokinin-substance P neurons projecting to the spinal cord. *J. Neurosci.* **3**, 1151–1157.

Skofitsch, G. and Jacobowitz, D. M. 1985a. Autoradiographic distribution of ^{125}I-calcitonin gene-related peptide binding sites in the rat central nervous system. *Peptides* **4**, 975–986.

Skofitsch, G. and Jacobowitz, D. M. 1985b. Calcitonin gene-related peptide: Detailed immunohistochemical distribution in the central nervous system. *Peptides* **6**, 721–745.

Skofitsch, G. and Jacobowitz, D. M. 1985c. Calcitonin gene-related peptide coexists with substance P in capsaicin-sensitive neurons and sensory ganglia of the rat. *Peptides* **6**, 747–754.

Skofitsch, G. and Jacobowitz, D. M. 1985d. Immunohistochemical mapping of galanin-like neurons in the rat central nervous system. *Peptides* **6**, 509–546.

Skofitsch, G. and Jacobowitz, D. M. 1985e. Galanin-like immunoreactivity in capsaicin-sensitive sensory neurons and ganglia. *Brain Res. Bull.* **15**, 191–195.

Skofitsch, G., Hamill, G. S., and Jacobowitz, D. M. 1984. Capsaicin depletes corticotropin-releasing factor-like immunoreactive neurons in the rat spinal cord and medulla oblongata. *Neuroendocrinology* **38**, 514–517.

Skofitsch, G., Insel, T. R., and Jacobowitz, D. M. 1985a. Binding sites for corticotropin releasing factor in sensory areas of the rat hindbrain and spinal cord. *Brain Res. Bull.* **15**, 519–522.

Skofitsch, G., Jacobowitz, D. M., Eskay, R. L., and Zamir, N. 1985b. Distribution of atrial natriuretic factor-like immunoreactive neurons in the rat brain. *Neuroscience* **16**, 917–948.

Skofitsch, G., Zamir, N., Helke, C., Savitt, J., and Jacobowitz, D. 1985c. Corticotropin releasing factor-like immunoreactivity in sensory ganglia and capsaicin sensitive neurons of the rat central nervous system: Colocalization with other neuropeptides. *Peptides* **6**, 307–318.

Skofitsch, G., Sills, M. A., and Jacobowitz, D. M. 1986. Autoradiographic distribution of ^{125}I-galanin binding sites in the rat central nervous system. *Peptides* **7**, 1029–1042.

Skoglund, S. 1956. Anatomical and physiological studies of knee joint innervation in the cat. *Acta Physiol. Scand.* **36** (Suppl. 124), 1–101.

Skoglund, S. 1973. Joint receptors and kinesthesia. In A. Iggo (ed.), *Handbook of Sensory Physiology, Vol. II. Somatosensory System* (pp. 111–136). Springer, Berlin.

Slater, P. and Patel, S. 1983. Autoradiographic localization of opiate k receptors in the rat spinal cord. *Eur. J. Pharmacol.* **92**, 159–160.

Sleeper, A. A., Cummins, T. R., Dib-Hajj, S. D., Hormuzdiar, W., Tyrrell, L., Waxman, S. G., and Black, J. A. 2000. Changes in expression of two tetrodotoxin-resistant sodium channels and their currents in dorsal root ganglion neurons after sciatic nerve injury but not rhizotomy. *J. Neurosci.* **20**, 7279–7289.

Sluka, K. A. and Westlund, K. N. 1992. An experimental arthritis in rats: Dorsal horn aspartate and glutamate increases. *Neurosci. Lett.* **145**, 1451–144.

Sluka, K. A. and Westlund, K. N. 1993a. Centrally administered non-NMDA but not NMDA receptor antagonists block peripheral knee joint inflammation. *Pain* **55**, 217–225.

Sluka, K. A. and Westlund, K. N. 1993b. Behavioral and immunohistochemical changes in an experimental arthritis model in rats. *Pain* **55**, 367–377.

Sluka, K. A. and Willis, W. D. 1997. The effects of G-protein and protein kinase inhibitors on the behavioral responses of rats to intradermal injection of capsaicin. *Pain* **71**, 165–178.

Sluka, K. A. and Willis, W. D. 1998. Increased spinal release of excitatory amino acids following intradermal injection of capsaicin is reduced by a protein kinase G inhibitor. *Brain Res.* **798**, 281–286.

Sluka, K. A., Dougherty, P. M., Sorkin, L. S., Willis, W. D., and Westlund, K. N. 1992. Neural changes in acute arthritis in monkeys: III. Changes in substance P, calcitonin gene-related peptide and glutamate in the dorsal horn of the spinal cord. *Brain Res. Rev.* **17**, 29–38.

Sluka, K. A., Willis, W. D., and Westlund, K. N. 1993. Joint inflammation and hyperalgesia are reduced by spinal bicuculline. *NeuroReport* **5**, 109–122.

Sluka, K. A., Lawand, N. B., and Westlund, K. N. 1994. Joint inflammation is reduced by dorsal rhizotomy and not by sympathectomy or spinal cord transection. *Ann. Rheum. Dis.* **53**, 309–314.

Sluka, K. A., Rees, H., Westlund, K. N., and Willis, W. D. 1995. Fiber types contributing to dorsal root reflexes induced by joint inflammation in cats and monkeys. *J. Neurophysiol.* **74**, 981–989.

Sluka, K. A., Rees, H., Chen, P. S., Tsuruoka, M., and Willis, W. D. 1997a. Capsaicin-induced sensitization of primate spinothalamic tract cells is prevented by a protein kinase C inhibitor. *Brain Res.* **772**, 82–86.

Sluka, K. A., Rees, H., Chen, P. S., Tsuruoka, M., and Willis, W. D. 1997b. Inhibitors of G-proteins and protein kinases reduce the sensitization to mechanical stimulation and the desensitization to heat of spinothalamic tract neurons induced by intradermal injection of capsaicin in the primate. *Exp. Brain Res.* **115**, 15–24.

Smeyne, R. J., Klein, R., Schnapp, A., Long, L. K., Bryant, S. et al. 1994. Severe sensory neuropathies in mice carrying a disrupted Trk/NGF receptor gene. *Nature* **368**, 246–249.

Smith, C. L. 1983. Development and postnatal organization of primary afferent projections to the rat thoracic spinal cord. *J. Comp. Neurol.* **220**, 29–43.

Smith, G. D., Harmar, A. J., McQueen, D. S., and Seckl, J. R. 1992. Increase in substance P and CGRP, but not somatostatin content of innervating dorsal root ganglia in adjuvant monoarthritis in the rat. *Neurosci. Lett.* **137**, 257–260.

Smith, K. E., Forray, C., Walker, M. W., Jones, K. A., Tamm, J. A., Bard, J., Branchek, T. A., Linemeyer, D. L., and Gerald, C. J. 1997. Expression cloning of a rat hypothalamic galanin receptor coupled to phosphoinositide turnover. *J. Biol. Chem.* **272**, 24612–24616.

Smith, K. E., Walker, M. W., Artymyshyn, R., Bard, J., Borowsky, B., Tamm, J. A., Yao, W. J., Vaysse, P. J., Branchek, T. A., Gerald, C., and Jones, K. A. 1998. Cloned human and rat GALR3 receptors: Pharmacology and activation of G-protein inwardly rectifying K^+ channels. *J. Biol. Chem.* **273**, 23321–23326.

Smith, K. R. 1968. The structure and function of the Haarscheibe. *J. Comp. Neurol.* **131**, 459–474.

Smith, K. R. 1970. The ultrastructure of the human Haarscheibe and Merkel cell. *J. Invest. Dermatol.* **54**, 150–159.

Smith, M. S., Schambra, U. B., Wilson, K. H., Page, S. O., Hulette, C., Light, A. R., and Schwinn, D. A. 1995. Alpha 2-adrenergic receptors in human spinal cord: Specific localized expression of mRNA encoding alpha 2-adrenergic receptor subtypes at four distinct levels. *Brain Res. Mol. Brain Res.* **34**, 109–117.

Smith, R. R., Martin-Schild, S., Kastin, A. J., and Zadina, J. E. 2001. Decreases in endomorphin-2-like immunoreactivity concomitant with chronic pain after nerve injury. *Neuroscience* **105**, 773–778.

Smullin, D. H., Skilling, S. R., and Larson, A. A. 1990. Interactions between substance P, calcitonin gene-related peptide, taurine and excitatory amino acids in the spinal cord. *Pain* **42**, 93–101.

Snyder, R. 1977. The organization of the dorsal root entry zone in cats and monkeys. *J. Comp. Neurol.* **174**, 47–70.

Snyder, R. L. 1982. Light and electron microscopic autoradiographic study of the dorsal root projections to the cat dorsal horn. *Neuroscience* **7**, 1417–1437.

Sohrabji, F., Miranda, R. C., and Toran-Allerand, C. D. 1994. Estrogen differentially regulates estrogen and nerve growth factor receptor mRNAs in adult sensory neurons. *J. Neurosci.* **14**, 459–471.

Solbach, S. and Celio, M. R. 1991. Ontogeny of the calcium binding protein parvalbumin in the rat nervous system. *Anat. Embryol. (Berl.)* **184**, 103–124.

Solodkin, A., Traub, R. J., and Gebhart, G. F. 1992. Unilateral hindpaw inflammation produces a bilateral increase in NADPH- diaphorase histochemical staining in the rat lumbar spinal cord. *Neuroscience* **51**, 495–499.

Solodkin, M., Jiménez, I., and Rudomin, P. 1984. Identification of common interneurons mediating pre- and post-synaptic inhibition in the cat spinal cord. *Science* **224**, 1453–1456.

Somjen, G. G. and Lothman, E. W. 1974. Potassium, sustained focal potential shifts, and dorsal root potentials of the mammalian spinal cord. *Brain Res.* **69**, 153–157.

Sommer, E. W., Kazimierczak, J., and Droz, B. 1985. Neuronal subpopulations in the dorsal root ganglion of the mouse as characterized by combination of ultrastructural and cytochemical features. *Brain Res.* **346**, 310–326.

Sommer, C. and Myers, R. R. 1995. Neurotransmitters in the spinal cord dorsal horn in a model of painful neuropathy and in nerve crush. *Acta Neuropathol.* **90**, 478–485.

Song, X. J. and Zhao, Z. Q. 1994. Interaction between substance P and excitatory amino acid receptors in modulation of nociceptive responses of cat spinal dorsal horn neurons. *Neurosci. Lett.* **168**, 49–52.

Sood, P. P., Richoux, J. P., Panigel, M., Bouhnik, J., and Wegmann, R. 1990. Differential distribution of immunoreactive angiotensinogen in the hind-brain and spinal cord of neonatal and adult rats. *Acta Anat. (Basel)* **138**, 230–237.

Sora, I., Takahashi, N., Funada, M., Ujike, H., Revay, R. S., et al. 1997. Opiate receptor knockout mice define μ receptor roles in endogenous nociceptive responses and morphine-induced analgesia. *Proc. Natl. Acad. Sci. USA* **94**, 1544–1549.

Sorkin, L. S. and McAdoo, D. J. 1993. Amino acids and serotonin are released into the lumbar spinal cord of the anesthetized cat following intradermal capsaicin injections. *Brain Res.* **607**, 89–98.

Sorkin, L. S. and Puig, S. 1996. Neuronal model of tactile allodynia produced by spinal strychnine: Effects of excitatory amino acid receptor antagonists and a mu-opiate receptor agonist. *Pain* **68**, 283–292.

Sorkin, L. S., Ferrington, D. G., and Willis, W. D. 1986. Somatotopic organization and response characteristics of dorsal horn neurons in the cervical spinal cord of the cat. *Somatosens. Res.* **3**, 323–338.

Sorkin, L. S., Morrow, T. J., and Casey, K. L. 1988. Physiological identification of afferent fibers and post-synaptic sensory neurons in the spinal cord of the intact, awake cat. *Exp. Neurol.* **99**, 412–427.

Sorkin, L. S., Hughes, M. G., Liu, D., Willis, W. D., and McAdoo, D. J. 1991. Release and metabolism of 5-hydroxytryptamine in the cat spinal cord examined with microdialysis. *J. Pharm. Exp. Therap.* **257**, 192–199.

Sorkin, L. S., Westlund, K. N., Sluka, K. A., Dougherty, P. M., and Willis, W. D. 1992. Neural changes in acute arthritis in monkeys: IV. Time-course of amino acid release into the lumbar dorsal horn. *Brain Res. Rev.* **17**, 39–50.

Sorkin, L. S., Xiao, W. H., Wagner, R., and Myers, R. R. 1997. Tumour necrosis factor-α induces ectopic activity in nociceptive primary afferent fibres. *Neuroscience* **81**, 255–262.

Sorkin, L. S., Puig, S., and Jones, D. L. 1998. Spinal bicuculline produces hypersensitivity of dorsal horn neurons: Effects of excitatory amino acid antagonists. *Pain* **77**, 181–190.

Soyguder, Z., Schmidt, H. H. H. W., and Morris, R. 1994. Postnatal development of nitric oxide synthase type 1 expression in the lumbar spinal cord of the rat: A comparison with the induction of *c-fos* in response to peripheral application of mustard oil. *Neurosci. Lett.* **180**, 188–192.

Spande, T. F., Garrafo, H. M., Edwards, M. W., Yeh, H. J. C., Pannel, L., and Daley, J. W. 1992. Epibatidine, a novel (chloropyridyl) azabicyclo-heptane with potent analgesic activity from an Ecuadorian poison frog. *J. Am. Chem. Soc.* **114**, 3475–3480.

Spetea, M., Rydelius, G., Nylander, I., Ahmed, M., Bileviciute-Ljungar, I., Lundeberg, T., Svensson, S., and Kreicbergs, A. 2002. Alteration in endogenous opioid systems due to chronic inflammatory pain conditions. *Eur. J. Pharmacol.* **435**, 245–252.

Spike, R. C. and Todd, S. J. 1992. Ultrastructural and immunocytochemical study of lamina II islet cells in rat spinal dorsal horn. *J. Comp. Neurol.* **323**, 359–369.

Spike, R. C., Todd, A. J., and Johnston, H. M. 1993. Coexistence of NADPH diaphorase with GABA, glycine, and acetylcholine in rat spinal cord. *J. Comp. Neurol.* **335**, 320–333.

Spike, R. C., Watt, C., Zafra, F., and Todd, A. J. 1997. An ultrastructural study of the glycine transporter GLYT2 and its association with glycine in the superficial laminae of the rat spinal dorsal horn. *Neuroscience* **77**, 543–551.

Spike, R. C., Kerr, R., Maxwell, D. J., and Todd, A. J. 1998. GluR1 and GluR2/3 subunits of the AMPA-type glutamate receptor are associated with particular types of neurone in laminae I–III of the spinal dorsal horn of the rat. *Eur. J. Neurosci.* **10**, 324–333.

Spike, R. C., Puskar, Z., Sakamoto, H., Stewart, W., Watt, C., and Todd, A. J. 2002. MOR-1-immunoreactive neurons in the dorsal horn of the rat spinal cord: Evidence for nonsynaptic innervation by substance P-containing primary afferents and for selective activation by noxious thermal stimuli. *Eur. J. Neurosci.* **15**, 1306–1316.

Sprague, J. M. and Ha, H. 1964. The terminal fields of dorsal root fibers in the lumbosacral spinal cord of the cat and the dendritic organization of the motor nuclei. *Prog. Brain Res.* **11**, 120–152.

Spray, D. C. 1986. Cutaneous temperature receptors. *Ann. Rev. Physiol.* **48**, 625–638.

Stacey, M. J. 1969. Free nerve endings in skeletal muscle of the cat. *J. Anat.* **105**, 231–254.

Standaert, D. G., Watson, S. J., Houghten, R. A., and Saper, C. B. 1986. Opioid peptide immunoreactivity in spinal and trigeminal dorsal horn neurons projecting to the parabrachial nucleus in the rat. *J. Neurosci.* **6**, 1220–1226.

Stanfa, L. and Dickenson, A. H. 1995. Spinal opioid systems in inflammation. *Inflam. Res.* **44**, 231–241.

Stanfa, L. C. and Dickenson, A. H. 1998. Inflammation alters the effects of mGlu receptor agonists on spinal nociceptive neurons. *Eur. J. Pharmacol.* **347**, 165–172.

Stanfield, P. R., Nakajima, Y., and Yamaguchi, K. 1985. Substance P raises neuronal membrane excitability by reducing inward rectification. *Nature* **315**, 498–501.

Stanzione, P. and Zieglgänsberger, W. 1983. Action of neurotensin on spinal cord neurons in the rat. *Brain Res.* **268**, 111–118.

Stark, B., Carlstedt, T., Hallin, R. G., and Risling, M. 1998. Distribution of human Pacinian corpuscles in the hand. A cadaver study. *J. Hand Surg.* **23B**, 370–372.

Steedman, W. N., Molony, V., and Iggo, A. 1985. Nociceptive neurones in the superficial dorsal horn of cat lumbar spinal cord and their primary afferent inputs. *Exp. Brain Res.* **58**, 171–182.

Steel, J. H., Terenghi, G., Chung, J. M., Na, H. S., Carlton, S. M., and Polak, J. M. 1994. Increased nitric oxide synthase immunoreactivity in rat dorsal root ganglia in a neuropathic pain model. *Neurosci. Lett.* **169**, 81–84.

Steele, P. A., Aromataris, C. E., and Riederer, B. M. 1996. Endogenous opioid peptides in parasympathetic, sympathetic and sensory nerves in the guinea-pig heart. *Cell Tissue Res.* **284**, 331–339.

Steen, K. H. and Reeh, P. W. 1993. Actions of cholinergic agonists and antagonists on sensory nerve endings in rat skin in vitro. *J. Neurophysiol.* **70**, 397–405.

Steen, K. H., Reeh, P. W., Anton, F., and Handwerker, H. O. 1992. Protons selectively induce lasting excitation and sensitization to mechanical stimulation of nociceptors in rat skin, in vitro. *J. Neurosci.* **12**, 86–95.

Steen, K. H., Steen, A. E., and Reeh, P. W. 1995. A dominant role of acid pH in inflammatory excitation and sensitization of nociceptors in rat skin, in vitro. *J. Neurosci.* **15**, 3982–3989.

Steen, K. H., Steen, A. E., Kreysel, H. W., and Reeh, P. W. 1996. Inflammatory mediators potentiate pain induced by experimental tissue acidosis. *Pain* **66**, 163–170.

Steenbergh, P. H., Hoppener, J. W. M., Zandberg, J., Lips, C. J. M., and Jansz, H. S. 1985. A second human calcitonin/CGRP gene. *FEBS Lett.* **183**, 403–407.

Stein, C. 1995. The control of pain in peripheral tissue by opioids. *N. Engl. J. Med.* **332**, 1685–1690.

Stein, C., Millan, M. J., Shippenberg, T. S., Peter, K., and Herz, A. 1989. Peripheral opiate receptors mediating antinociception in inflammation: Evidence for involvement of *mu, delta* and *kappa* receptors. *J. Pharmacol. Exp. Ther.* **248**, 1269–1275.

Stein, C., Hassan, A. H. S., Przewlocki, R., Gramsch, C., Peter, K., and Herz, A. 1990. Opioids from immunocytes interact with receptors on sensory nerves to inhibit nociception in inflammation. *Proc. Natl. Acad. Sci.* **87**, 5935–5939.

Steinbusch, H. W. M. 1981. Distribution of serotonin-immunoreactivity in the central nervous system of the rat-cell bodies and terminals. *Neuroscience* **6**, 557–618.

Steranka, L. R., Manning, D. C., DeHaas, C. J., Ferkany, J. W., Borosky, S. A., Connor, J. R., Vavrek, R. J., Stewart, J. M., and Snyder, S. H. 1988. Bradykinin as a pain mediator: Receptors are localized to sensory neurons, and antagonists have analgesic actions. *Proc. Natl. Acad. Sci.* **85**, 3245–3249.

Steriade, M., Jones, E. G., and McCormick, D. A. (eds.) 1997. *Thalamus* (2 vols.). Elsevier, Amsterdam.

Sterling, P. and Kuypers, H. G. J. M. 1967. Anatomical organization of the brachial spinal cord of the cat: I. The distribution of dorsal root fibers. *Brain Res.* **4**, 1–15.

Sterne, G. D., Brown, R. A., Green, C. J., and Terenghi, G. 1998. NT-3 modulated NPY expression in primary sensory neurons following peripheral nerve injury. *J. Anat.* **193**, 273–281.

Sternini, C. and Anderson, K. 1992. Calcitonin gene-related peptide-containing neurons supplying the rat digestive system: Differential distribution and expression pattern. *Somatosens. Mot. Res.* **9**, 45–59.

Stevens, C. W. and Seybold, V. S. 1995. Changes of opioid binding density in the rat spinal cord following unilateral dorsal rhizotomy. *Brain Res.* **687**, 53–62.

Stevens, C. W., Kajander, K., Bennett, G. J., and Seybold, V. S. 1991a. Bilateral and differential changes in spinal mu, delta and kappa opioid binding in rats with a painful, unilateral neuropathy. *Pain* **46**, 315–326.

Stevens, C. W., Lacey, C. B., Miller, K. E., Elde, R. P., and Seybold, V. S. 1991b. Biochemical characterization and regional quantification of μ, δ and κ opioid binding sites in rat spinal cord. *Brain Res.* **550**, 77–85.

Stevens, R. T., Hodge, J., and Vaniaapkarian, A. 1983. Catecholine varicosities in cat dorsal root ganglion and spinal ventral roots. *Brain Res.* **261**, 151–154.

Stewart, W. and Maxwell, D. J. 2000. Morphological evidence for selective modulation by serotonin of a subpopulation of dorsal horn cells which possess the neurokinin-1 receptor. *Eur. J. Neurosci.* **12**, 4583–4588.

Stolwijk, J. A. J. and Wexler, I. 1971. Peripheral nerve activity in response to heating the cat's skin. *J. Physiol.* **214**, 377–392.

Stone, L. S., Fairbanks, C. A., Laughlin, T. M., Nguyen, H. O., Bushy, T. M., Wessendorf, M. W., and Wilcox, G. L. 1997. Spinal analgesic actions of the new endogenous opioid peptides endomorphin-1 and -2. *NeuroReport* **8**, 3131–3135.

Stone, L. S., Broberger, C., Vulchanova, L., Wilcox, G. L., Hokfelt, T., Riedl, M. S., and Elde, R. 1998. Differential distribution of alpha2A and alpha2C adrenergic receptor immunoreactivity in the rat spinal cord. *J. Neurosci.* **18**, 5928–5937.

Stone, L. S., Vulchanova, L., Riedl, M. S., Wang, J., Williams, F. G., Wilcox, G. L., and Elde, R. 1999. Effects of peripheral nerve injury on alpha-2A and alpha-2C adrenergic receptor immunoreactivity in the rat spinal cord. *Neuroscience* **93**, 1399–1407.

Stone, T. W. 1985. *Microiontophoresis and Pressure Ejection*, IBRO Handbook Series: Methods in the Neurosciences (Vol. 8). John Wiley & Sons, New York.

Storm-Mathisen, J. and Ottersen, O. P. 1987. Tracing of neurons with glutamate or γ-aminobutyrate as putative transmitters. *Biochem. Soc. Trans.* **15**, 210–213.

Stoyanova, I., Dandov, A., Lazarov, N., and Chouchkov, C. 1998. GABA-and Glutamate-immunoreactivity in sensory ganglia of cat: A quantitative analysis. *Arch. Physiol. Biochem.* **106**, 362–369.

Straile, W. E. 1960. Sensory hair follicles in the mammalian skin: The tylotrich follicle. *Am. J. Anat.* **106**, 133–147.

Straile, W. E. 1961. The morphology of tylotrich follicles in the skin of the rabbit. *Am. J. Anat.* **109**, 1–13.

Straile, W. E. 1969. Encapsulated nerve end-organs in the rabbit, mouse, sheep and man. *J. Comp. Neurol.* **136**, 317–336.

Straughan, D. W., Neal, M. J., Simmonds, M. A., Collinbs, G. G. S., and Hill, R. G. 1971. Evaluation of bicuculline as a GABA antagonist. *Nature* **233**, 352–354.

Stucky, C. L. and Koltzenburg, M. 1997. The low-affinity neurotrophin receptor p75 regulates the function but not the selective survival of specific subpopulations of sensory neurons. *J. Neurosci.* **17**, 4398–4405.

Stucky, C. L. and Lewin, G. R. 1999. Isolectin B(4)-positive and -negative nociceptors are functionally distinct. *J. Neurosci* **19**, 6497–6505.

Stucky, C. L., Galeazza, M. T., and Seybold, V. S. 1993. Time-dependent changes in Bolton-Hunter ^{125}I-substance P binding in rat spinal cord following adjuvant-induced peripheral inflammation. *Neuroscience* **57**, 397–409.

Stucky, C. L., DeChiara, T., Lindsay, R. M., Yancopoulos, G. D., and Koltzenburg, M. 1998. Neurotrophin 4 is required for the survival of a subclass of hair follicle receptors. *J. Neurosci.* **18**, 7040–7046.

Stucky, C. L., Koltzenburg, M., Schneider, M., Engle, M. G., Albers, K. M., and Davis, B. M. 1999. Overexpression of nerve growth factor in skin selectively affects the survival and functional properties of nociceptors. *J. Neurosci.* **19**, 8509–8516.

Su, H. C., Bishop, A. E., Power, R. F., Hamada, Y., and Polak, J. M. 1987. Dual intrinsic and extrinsic origins of CGRP- and NPY-immunoreactive nerves of rat gut and pancreas. *J. Neurosci.* **7**, 2674–2687.

Su, H. Y., Hickford, J. G. H., Bickerstaffe, R., and Palmer, B. R. 1999. Insulin-like growth factor 1 and hair growth. *Dermatol. Online J.* **5**, 1–20.

Sugimoto, K., Murakawa, Y., and Sima, A. A. 2002. Expression and localization of insulin receptor in rat dorsal root ganglion and spinal cord. *J. Peripher. Nerv. Syst.* **7**, 44–53.

Sugimoto, T. and Gobel, S. 1984. Dendritic changes in the dorsal horn following transection of a peripheral nerve. *Brain Res.* **321**, 199–208.

Sugimoto, T. and Takemura, M. 1993. Tooth pulp primary neurons: Cell size analysis, central connection, and carbonic anhydrase activity. *Brain Res. Bull.* **30**, 221–226.

Sugimoto, T., Takemura, M., and Wakisaka, S. 1988. Cell size analysis of primary neurons innervating the cornea and tooth pulp of the rat. *Pain* **32**, 375–381.

Sugimoto, T., Takemura, M., Ichikawa, H., and Akai, M. 1989a. Carbonic anhydrase activity in the trigeminal primary afferent neuronal cell bodies with peripheral axons innervating the mandibular molar tooth pulps of the rat. *Brain Res.* **505**, 354–357.

Sugimoto, T., Bennett, G. J., and Kajander, K. C. 1989b. Strychnine-enhanced transsynaptic degeneration of dorsal horn neurons in rats with an experimental painful peripheral neuropathy. *Neurosci. Lett.* **98**, 139–143.

Sugimoto, T., Bennett, G. J., and Kajander, K. C. 1990. Transynaptic degeneration in the superficial dorsal horn after sciatic nerve injury: Effect of chronic constriction injury, transection and strychnine. *Pain* **42**, 205–213.

Sugiura, Y. 1975. Three-dimensional analysis of the neurons in the substantia gelatinosa Rolandi. *Proc. Jap. Acad.* **51**, 336–341.

Sugiura, Y., Lee, C. L., and Perl, E. R. 1986. Central projections of identified, unmyelinated (C) afferent fibers innervating mammalian skin. *Science* **234**, 358–361.

Sugiura, Y., Hosoya, Y., Ito, R., and Kohno, K. 1988. Ultrastructural features of functionally identified primary afferent neurons with C (unmyelinated) fibers of the guinea pig: Classification of dorsal root ganglion cell type with reference to sensory modality. *J. Comp. Neurol.* **276**, 265–278.

Sugiura, Y., Terui, N., and Hosoya, Y. 1989. Difference in distribution of central terminals between visceral and somatic unmyelinated (C) primary afferent fibers. *J. Neurophysiol.* **62**, 834–840.

Sugiyama, H., Ito, I., and Hirono, C. 1987. A new type of glutamate receptor linked to inositol phospholipid metabolism. *Nature* **325**, 531–533.

Suh, Y. S., Chung, K., and Coggeshall, R. E. 1984. A study of axonal diameters and areas in lumbosacral roots and nerves in the rat. *J. Comp. Neurol.* **222**, 473–481.

Sullivan, A. F., Dashwood, M. R., and Dickenson, A. H. 1987. Alpha2-adrenoceptor modulation of nociception in rat spinal cord: Location, effects and interactions with morphine. *Eur. J. Pharmacol.* **138**, 169–177.

Sumal, K. K., Pickel, V. M., Miller, R. J., and Reis, D. J. 1982. Enkephalin-containing neurons in substantia gelatinosa of spinal trigeminal complex: Ultrastructure and synaptic interaction with primary sensory afferents. *Brain Res.* **248**, 223–236.

Sumino, R. and Dubner, R. 1981. Response characteristics of specific thermoreceptive afferents innervating monkey facial skin and their relationship to human thermal sensitivity. *Brain Res. Rev.* **3**, 105–122.

Sun, H., Ren, K., Zhong, C. M., Ossipov, M. H., Malan, T. P., Lai, J., and Porreca, F. 2001. Nerve injury-induced tactile allodynia is mediated via ascending spinal dorsal column projections. *Pain* **90**, 105–111.

Sun, Y. and Zigmond, R. E. 1996. Leukaemia inhibitory factor induced in the sciatic nerve after axotomy is involved in the induction of galanin in sensory neurons. *Eur. J. Neurosci.* **8**, 2213–2220.

Supowit, S. C., Zhao, H., and Dipette, D. J. 2001. Nerve growth factor enhances calcitonin gene-related peptide expression in the spontaneously hypertensive rat. *Hypertension* **37**, 728–732.

Sur, C., McKernan, R., and Triller, A. 1995. Subcellular localization of the $GABA_A$ receptor δ2 subunit in the rat spinal cord. *Eur. J. Neurosci.* **7**, 1323–1332.

Sur, C., Betz, H., and Schloss, P. 1996. Localization of the serotonin transporter in rat spinal cord. *Eur. J. Neurosci.* **8**, 2753–2757.

Surmeier, D. J., Honda, C. N., and Willis, W. D. 1986a. Responses of primate spinothalamic neurons to noxious thermal stimulation of glabrous and hairy skin. *J. Neurophysiol.* **56**, 328–350.

Surmeier, D. J., Honda, C. N., and Willis, W. D. 1986b. Temporal features of the responses of primate spinothalamic neurons to noxious thermal stimulation of hairy and glabrous skin. *J. Neurophysiol.* **56**, 351–369.

Surmeier, D. J., Honda, C. N., and Willis, W. D. 1988. Natural groupings of primate spinothalamic neurons based on cutaneous stimulation: Physiological and anatomical features. *J. Neurophysiol.* **59**, 833–860.

Sutherland, F. I., Bannatyne, B. A., Kerr, R., Riddell, J. S., and Maxwell, D. J. 2002. Inhibitory amino acid transmitters associated with axons in presynaptic apposition to cutaneous primary afferent axons in the cat spinal cord. *J. Comp. Neurol.* **452**, 154–162.

Sutherland, S. P., Benson, C. J., Adelman, J. P., and McCleskey, E. W. 2001. Acid-sensing ion channel 3 matches the acid-gating current in cardiac ischemia-sensing neurons. *Proc. Nat. Acad. Sci. USA* **98**, 711–716.

Suzue, T. and Jessell, T. 1980. Opiate analgesics and endorphins inhibit rat dorsal root potential in vitro. *Neurosci. Lett.* **16**, 161–166.

Svendsen, F., Tjølsen, A., and Hole, K. 1997. LTP of spinal Aβ and C-fibre evoked responses after electrical sciatic nerve stimulation. *NeuroReport* **8**, 3427–3430.

Svendsen, F., Tjølsen, A., and Hole, K. 1998. AMPA and NMDA receptor-dependent spinal LTP after nociceptive tetanic stimulation. *NeuroReport* **9**, 1185–1190.

Svendsen, F., Tjølsen, J., Gjerstad, J., and Hole, K. 1999. Long-term potentiation of single WDR neurons in spinalized rats. *Brain Res.* **816**, 487–492.

Svendsen, L. J., Rygh, J., Gjerstad, A., Fiskå, K., Hole, K., and Tjølsen, A. 1999. Recording of long-term potentiation in single dorsal horn neurons in vivo in the rat. *Brain Res. Protoc.* **4**, 165–172.

Svoboda, J., Motin, V., Hajek, I., and Sykova, E. 1988. Increase in extracellular potassium level in rat spinal cord dorsal horn induced by noxious stimulation and peripheral injury. *Brain Res.* **458**, 97–105.

Swanson, L. W. and Kuypers, H. G. 1980. The paraventricular nucleus of the hypothalamus: Cytoarchitectonic subdivisions and organization of projections to the pituitary, dorsal vagal complex, and spinal cord as demonstrated by retrograde fluorescence double-labeling methods. *J. Comp. Neurol.* **194**, 555–570.

Swanson, L. W. and McKellar, S. 1979. The distribution of oxytocin- and neurophysin-stained fibers in the spinal cord of the rat and monkey. *J. Comp. Neurol.* **188**, 87–106.

Sweet, W. H. 1981. Animal models of chronic pain: Their possible validation from human experience with posterior rhizotomy and congenital analgesia. *Pain* **10**, 275–295.

Sweetnam, P. M., Wrathall, J. R., and Neale, J. H. 1986. Localization of dynorphin gene product-immunoreactivity in neurons from spinal cord and dorsal root ganglia. *Neuroscience* **18**, 947–955.

Swett, J. E. and Woolf, C. J. 1985. The somatotopic organization of primary afferent terminals in the superficial laminae of the dorsal horn of the rat spinal cord. *J. Comp. Neurol.* **231**, 66–77.

Swett, J. E., Torigoe, Y., Elie, V. R., Bourassa, C. M., and Miller, P. G. 1991. Sensory axons in the rat sciatic nerve. *Exp. Neurol.* **114**, 82–103.

Syková, E. and Vyklický, L. 1977. Changes of extracellular potassium activity in isolated cord of frog under high Mg^{++} concentration. *Neurosci. Lett.* **4**, 161–165.

Syková, E. and Vyklický, L. 1978. Effects of picrotoxin on potassium accumulation and dorsal root potentials in the frog spinal cord. *Neuroscience* **3**, 1061–1067.

Syková, E., Shirayev, B., Kříž, N., and Vyklický, L. 1976. Accumulation of extracellular potassium in the spinal cord of frog. *Brain Res.* **106**, 413–417.

Syková, E., Czeh, G., and Kříž, N. 1980. Potassium accumulation in the frog spinal cord induced by nociceptive stimulation of the skin. *Neurosci. Lett.* **17**, 253–258.

Szabat, E., Soinila, S., Happola, O., Linnala, A., and Virtanen, I. 1992. A new monoclonal antibody against the GABA-protein conjugate shows immunoreactivity in sensory neurons of the rat. *Neuroscience* **47**, 409–420.

Szabo, B. K., Torok, A., Knyihar-Csillik, E., and Csillik, B. 1989. Mapping the distribution of thiamine monophosphatase, fluoride-resistant acid phosphatase, and substance P in the spinal cord with a personal computer compatible program. *Acta Histochemica* **87**, 123–129.

Szallasi, A., Blumberg, P. M., Nilsson, S., Hokfelt, T., and Lundberg, J. M. 1994a. Visualization by [3H]resiniferatoxin autoradiography of capsaicin-sensitive neurons in the rat, pig and man. *Eur. J. Pharmacol.* **264**, 217–221.

Szallasi, A., Nilsson, S., Hökfelt, T., and Lundberg, J. M. 1994b. Visualizing vanilloid (capsaicin) receptors in pig spinal cord by [^{3}H] resiniferatoxin autoradiography. *Brain Res.* **655**, 234–240.

Szallasi, A. 1995. Autoradiographic visualization and pharmacological characterization of vanilloid (capsaicin) receptors in several species, including man. *Acta Physiol. Scand. Suppl.* **629**, 1–68.

Szallasi, A. and Blumberg, P. M. 1999. Vanilloid (capsaicin) receptors and mechanisms. *Pharmacol. Rev.* **51**, 159–212.

Szallasi, A., Farkas-Szallasi, T., Tucker, J. B., Lundberg, J. M., Hökfelt, T., and Krause, J. E. 1999. Effects of systemic resiniferatoxin treatment on substance P mRNA in rat dorsal root ganglia and substance P receptor mRNA in the spinal dorsal horn. *Brain Res.* **815**, 177–184.

Szarijanni, N. and Rethelyi, M. 1979. Differential distribution of small and large neurons in the sacrococcygeal dorsal root ganglia of the cat. *Acta Morph. Acad. Sci. Hung.* **27**, 25–35.

Szentagothai, J. 1964. Neuronal and synaptic arrangement in the substantia gelatinosa Rolandi. *J. Comp. Neurol.* **122**, 219–239.

Szentagothai, J. and Rethelyi, M. 1973. Cyto- and neuropil architecture of the spinal cord. In J. E. Desmedt (ed.), *New Development in Electromyography and Clinical Neurophysiology* (pp. 20–37). Karger, Basel.

Szolcsanyi, J. 1987. Selective responsiveness of polymodal nociceptors of the rabbit ear to capsaicin, bradykinin and ultra-violet irradiation. *J. Physiol.* **388**, 9–23.

Szolcsányi, J. 1988. Antidromic vasodilatation and neurogenic inflammation. *Agents Actions* **23**, 4–11.

Szolcsanyi, J. 1996. Neurogenic inflammation: Reevaluation of axon reflex theory. In P. Geppetti and P. Holzer (eds.), *Neurogenic Inflammation* (pp. 33–42). CRD Press, Boca Raton.

Szolcsanyi, J., Anton, F., Reeh, P. W., and Handwerker, H. O. 1988. Selective excitation by capsaicin of mechano-heat sensitive nociceptors in rat skin. *Brain Res.* **446**, 262–268.

Sztriha, L., Lestringant, G. G., Hertecant, J., Frossard, P. M., and Masouye, I. 2001. Congenital insensitivity to pain with anhidrosis. *Pediatr. Neurol.* **25**, 63–66.

Tachibana, M., Wenthold, R. J., Morioka, H., and Petralia, R. S. 1994. Light and electron microscopic immunocytochemical localization of AMPA-selective glutamate receptors in the rat spinal cord. *J. Comp. Neurol.* **344**, 431–454.

Tachibana, T., Miki, K., Fukuoko, T., Arakawa, A., Taniguchi, M., Maruo, S., and Noguchi, K. 1998. Dynorphin mRNA expression in dorsal horn neurons after traumatic spinal cord injury: Temporal and spatial analysis using in situ hybridization. *J. Neurotrauma* **15**, 485–494.

Tadano, T., Asao, T., Aizawa, T., Sakurada, S., Abe, Y., Yonezawa, A., Ando, R., Arai, Y., Kinemuchi, H., and Kisara, K. 1995. Immunohistochemical determination of rat spinal cord substance P, and antinociceptive effect during development of thiamine deficiency. *Brain Res.* **696**, 21–29.

Taiwo, Y. O. and Levine, J. D. 1990. Effects of cyclooxygenase products of arachidonic acid metabolism on cutaneous nociceptive threshold in the rat. *Brain Res.* **537**, 372–374.

Taiwo, Y. O. and Levine, J. D. 1991. Further confirmation of the role of adenyl cyclase and cAMP-dependent protein kinase in primary afferent hyperalgesia. *Neuroscience* **44**, 131–135.

Tajti, J., Uddman, R., Moller, S., Sundler, F., and Edvinsson, L. 1999. Messenger molecules and receptor mRNA in the human trigeminal ganglion. *J. Auton. Nerv. Syst.* **76**, 176–183.

Takahashi, M. and Yokota, T. 1983. Convergence of cardiac and cutaneous afferents onto neurons in the dorsal horn of the spinal cord in the cat. *Neurosci. Lett.* **38**, 251–256.

Takahashi, O., Traub, R. J., and Ruda, M. A. 1988. Demonstration of calcitonin gene-related peptide immunoreactive axons contacting dynorphin A(1–8) immunoreactive spinal neurons in a rat model of peripheral inflamation and hyperalgesia. *Brain Res.* **475**, 168–172.

Takahashi, O., Shiosaka, S., Traub, R. J., and Ruda, M. A. 1990. Ultrastructural demonstration of synaptic connections between calcitonin gene-related peptide immunoreactive axons and dynorphin A(1–8) immunoreactive dorsal horn neurons in a rat model of peripheral inflammation and hyperalgesia. *Peptides* **11**, 1233–1237.

Takahashi, T. and Otsuka, M. 1975. Regional distribution of substance P in the spinal cord and nerve roots of the cat and the effect of dorsal root section. *Brain Res.* **87**, 1–11.

Takano, Y., Martin, J. E., Leeman, S. E., and Loewy, A. D. 1984. Substance P immunoreactivity released from rat spinal cord after kainic acid excitation of the ventral medulla oblongata: A correlation with increases in blood pressure. *Brain Res.* **291**, 168–172.

Takiguchi-Hayashi, K., Sato, M., Sugo, N., Ishida, M., Sato, K., Uratani, Y., and Arimatsu, Y. 1998. Latexin expression in smaller diameter primary sensory neurons in the rat. *Brain Res.* **801**, 9–20.

Talaat, M. 1937. Afferent impulses in the nerves supplying the bladder. *J. Physiol.* **89**, 1–13.

Talbot, J. D., Duncan, G. H., Bushnell, M. C., and Boyer, M. 1987. Diffuse noxious inhibitory controls (DNICs): Psychophysical evidence in man for intersegmental suppression of noxious heat perception by cold pressor pain. *Pain* **30**, 221–232.

Talbot, J. D., Duncan, G. H., and Bushnell, M. C. 1989. Effects of diffuse noxious inhibitory controls (DNICs) on the sensory-discriminative dimension of pain perception. *Pain* **36**, 231–238.

Talbot, W. H., Darian-Smith, I., Kornhuber, H. H., and Mountcastle, V. B. 1968. The sense of flutter-vibration: Comparison of the human capacity with response patterns of mechanoreceptive afferents from the monkey hand. *J. Neurophysiol.* **31**, 301–334.

Taleghany, N., Sarajari, S., DonCarlos, L. L., Gollapudi, L., and Oblinger, M. M. 1999. Differential expression of estrogen receptor alpha and beta in rat dorsal root ganglion neurons. *J. Neurosci. Res.* **57**, 603–615.

Tallaksen-Greene, S. J., Young, A. B., Penney, J. B., and Beitz, A. J. 1992. Excitatory amino acid binding sites in the trigeminal principal sensory and spinal trigeminal nuclei of the rat. *Neurosci. Lett.* **141**, 79–83.

Talley, E. M., Rosin, D. L., Lee, A., Guyenet, P. G., and Lynch, K. R. 1996. Distribution of α_{2A}-adrenergic receptor-like immunoreactivity in the rat central nervous system. *J. Comp. Neurol.* **372**, 111–134.

Tamatani, M., Senba, E., and Tohyama, M. 1989. Calcitonin gene-related peptide- and substance P-containing primary afferent fibers in the dorsal column of the rat. *Brain Res.* **495**, 122–130.

Tamura, R., Hanesch, U., Schmidt, R. F., Kumazawa, T., and Mizumura, K. 1998. Examination of colocalization of calcitonin gene-related peptide- and substance P-like immunoreactivity in the knee joint of the dog. *Neurosci. Lett.* **254**, 53–56.

Tanaka, M., Cummins, T. R., Ishikawa, K., Dib-Hajj, S. D., Black, J. A., and Waxman, S. G. 1998. SNS Na^+ channel expression increases in dorsal root ganglion neurons in the carrageenan inflammatory pain model. *NeuroReport* **9**, 967–972.

Tandrup, T. 1995. Are the neurons in the dorsal root ganglion pseudounipolar? A comparison of the number of neurons and number of myelinated and unmyelinated fibres in the dorsal root. *J. Comp. Neurol.* **357**, 341–347.

Tang, F. R. and Sim, M. K. 1999. Pre- and post-synaptic localization of metabotropic glutamate receptor 1alpha (mGluR1alpha) and 2/3 (mGluR2/3) in the rat spinal cord. *Neurosci. Res.* **34**, 73–78.

Taniguchi, K., Tomita, M., Kominami, E., and Uchiyama, Y. 1993. Cysteine proteinases in rat dorsal root ganglion and spinal cord, with special reference to the co-localization of these enzymes with calcitonin gene-related peptide in lysosomes. *Brain Res.* **601**, 143–153.

Tao, Y. X. and Johns, R. A. 2002. Activation and up-regulation of spinal cord nitric oxide receptor, soluble guanylate cyclase, after formalin injection into the rat hind paw. *Neuroscience* **112**, 439–446.

Tao, Y. X., Huang, Y. Z., Mei, L., and Johns, R. A. 2000a. Expression of PSD-95/SAP90 is critical for N-methyl-D-aspartate receptor-mediated thermal hyperalgesia in the spinal cord. *Neuroscience* **98**, 201–206.

Tao, Y., Li, Y., Zhao, Z., and Johns, R. A. 2000b. Synaptic relationship of the neurons containing a metabotropic glutamate receptor, mGluR5, with nociceptive primary afferent and GABAergic terminals in rat spinal superficial laminae. *Brain Res.* **875**, 138–143.

Tapper, D. N. 1965. Stimulus–response relationships in the cutaneous slowly-adapting mechanoreceptor in hairy skin of the cat. *Exp. Neurol.* **13**, 364–385.

Tapper, D. N. 1970. Behavioral evaluation of the tactile pad receptor system in hairy skin of the cat. *Exp. Neurol.* **26**, 447–459.

Tapper, D. N. and Mann, M. D. 1968. Single presynaptic impulses evokes postsynaptic discharge. *Brain Res.* **11**, 688–690.

Tapper, D. N., Brown, P. B., and Moraff, H. 1973. Functional organization of the cat's dorsal horn: Connectivity of myelinated fiber systems of hairy skin. *J. Neurophysiol.* **36**, 817–826.

Tashiro, T. and Ruda, M. A. 1988. Immunocytochemical identification of axons containing coexistent serotonin and substance P in the cat lumbar spinal cord. *Peptides* **9**, 383–391.

Tashiro, T., Takahashi, O., Satoda, T., Matsushima, R., and Mizuno, N. 1987. Immunohistochemical demonstration of coexistence of enkephalin- and substance P-like immunoreactivities in axonal components in the lumbar segments of cat spinal cord. *Brain Res.* **424**, 391–395.

Tashiro, T., Satoda, T., Takahashi, O., Matsushima, R., and Mizuno, N. 1988. Distribution of axons exhibiting both enkephalin- and serotonin-like immunoreactivities in the lumbar cord segments: An immunohistochemical study in the cat. *Brain Res.* **440**, 357–362.

Tashiro, T., Satoda, T., Matsushima, R., and Mizuno, N. 1990. Distribution of axons showing both enkephalin- and serotonin-like immunoreactivities in the lumbar cord segments of the Japanese monkey (*Macaca fuscata*). *Brain Res.* **512**, 143–146.

Tasker, R. 1990. Pain resulting from nervous system pathology (central pain). In J. J. Bonica (ed.), *Management of Pain* (Vol. 1, 2nd ed., pp. 264–280). Lea & Febiger, Philadelphia.

Tata, A. M., Vilaro, M. T., Agrati, C., Biagioni, S., Mengod, G., and Augusti-Tocco, G. 1999. Expression of muscarinic m2 receptor mRNA in dorsal root ganglia of neonatal rat. *Brain Res.* **824**, 63–70.

Tata, A. M., Vilaró, M. T., and Mengod, G. 2000. Muscarinic receptor subtypes expression in rat and chick dorsal root ganglia. *Mol. Brain Res.* **82**, 1–10.

Tattersall, J. E. H., Cervero, F., and Lumb, B. M. 1986a. Viscerosomatic neurons in the lower thoracic spinal cord of the cat: Excitations and inhibitions evoked by splanchnic and somatic nerve volleys and by stimulation of brain stem nuclei. *J. Neurophysiol.* **56**, 1411–1423.

Tattersall, J. E. H., Cervero, F., and Lumb, B. M. 1986b. Effects of reversible spinalization on the visceral input to viscerosomatic neurons in the lower thoracic spinal cord of the cat. *J. Neurophysiol.* **56**, 785–796.

Taylor, D. C. M. and Pierau, F.-K. 1982. Double fluorescence labelling supports electrophysiological evidence for dichotomizing peripheral sensory nerve fibres in rats. *Neurosci. Lett.* **33**, 1–6.

Tecott, L. H., Maricq, A. V., and Julius, D. 1993. Nervous system distribution of the serotonin 5-HT_3 receptor mRNA. *Proc. Natl. Acad. Sci. USA* **90**, 1430–1434.

Telser, S., Marksteiner, J., Hinterhuber, H., and Saria, A. 1995. Distribution of secretoneurin-like immunoreactivity in comparison with that of substance P in the human spinal cord. *Neurosci. Lett.* **191**, 83–86.

Tenser, R. B. 1985. Sequential changes of sensory neuron (fluoride-resistant) acid phosphatase in dorsal root ganglion neurons following neurectomy and rhizotomy. *Brain Res.* **332**, 386–389.

Teoh, H., Malcangio, M., and Bowery, N. G. 1996a. GABA, glutamate and substance P-like immunoreactivity release: Effects of novel $GABA_B$ antagonists. *Br. J. Pharmacol.* **118**, 1153–1160.

Teoh, H., Malcangio, M., Fowler, L. J., and Bowery, N. G. 1996b. Evidence for release of glutamic acid, aspartic acid and substance P but not gamma-aminobutyric acid from primary afferent fibres in rat spinal cord. *Eur. J. Pharmacol.* **302**, 27–36.

Terada, M., Yasuda, H., Kogawa, S., Maeda, K., Haneda, M., Hidaka, H., Kashiwagi, A., and Kikkawa, R. 1998. Expression and activity of cyclin-dependent kinase 5/p35 in adult rat peripheral nervous system. *J. Neurochem.* **71**, 2600–2606.

Terenghi, G., Polak, J. M., Ghatei, M. A., Mulderry, P., Butler, J., Unger, W., and Bloom, S. 1985. Distribution and origin of calcitonin gene-related peptide (CGRP) immunoreactivity in the sensory innervation of the mammalian eye. *J. Comp. Neurol.* **233**, 506–516.

Terenghi, G., Polak, J. M., Rodrigo, J., Mulderry, P., and Bloom, S. 1986a. Calcitonin gene-related-immunoreactive nerves in the tongue, epiglottis and pharynx of the rat: Occurence, distribution and origin. *Brain Res.* **365**, 1–14.

Terenghi, G., Zhang, S. Q., Unger, W. G., and Polak, J. M. 1986b. Morphological changes of sensory CGRP-immunoreactive and sympathetic nerves in peripheral tissues following chronic denervation. *Histochemistry* **86**, 89–95.

Terenghi, G., Riveros-Moreno, V., Hudson, L. D., Ibrahim, N. B. N., and Polak, J. M. 1993. Immunohistochemistry of nitric oxide synthase demonstrates immunoreactive neurons in spinal cord and dorsal root ganglia of man and rat. *J. Neurol. Sci.* **118**, 34–37.

Terrado, J., Gerrikagoitia, I., Dominguez, L., Raldua, D., Martinez-Millan, L., and Sarasa, M. 1999. Expression of the genes for alpha-type and beta-type calcitonin gene-related peptide during rat embryogenesis. *Neuroscience* **92**, 713–727.

Tervo, K., Tervo, T., Eranko, L., Eranko, O., and Cuello, A. C. 1981. Immunoreactivity for substance P in the Gasserian ganglion, ophthalmic nerve and anterior segment of the rabbit eye. *Histochem. J.* **13**, 435–443.

Tessler, A., Gazer, E., Artymyshyn, R., Murray, M., and Goldberger, M. E. 1980. Recovery of substance P in the cat spinal cord after unilateral lumbosacral deafferentation. *Brain Res.* **191**, 459–470.

Tessler, A., Himes, B. T., Artymyshyn, R., Murray, M., and Goldberger, M. E. 1981. Spinal neurons mediate return of substance P following deafferentation of cat spinal cord. *Brain Res.* **230**, 263–281.

Tessler, A., Himes, B. T., Soper, K., Murray, M., Goldberger, M. E., and Reichlin, S. 1984. Recovery of substance P but not somatostatin in the cat spinal cord after unilateral lumbosacral dorsal rhizotomy: A quantitative study. *Brain Res.* **305**, 95–102.

Tewari, H. B. and Bourne, G. H. 1962a. Histochemical studies on the distribution of B-gluminidase and succinic dehydrogenase in young and old spinal ganglion cells of the rat. *Zeitschrift für Zellforschung* **58**, 70–75.

Tewari, H. B. and Bourne, G. H. 1962b. The histochemistry of the nucleus and nucleolus with reference to nucleo-cytoplasmic relations in the spinal ganglion neuron of the rat. *Acta Histochem.* **13**, 323–350.

Tewari, H. B. and Bourne, G. H. 1962c. Histochemical evidence of metabolic cycles in rat spinal ganglion cells. *J. Histochem. Cytochem.* **10**, 42–64.

Tewari, N., Chaturvedi, R. P., and Ujwal, Z. S. 1970. A comparative histoenzymological study on the distribution of alkaline phosphatase amongst the neurons of spinal ganglia of rat and guinea pig. *Anat. Auz.* **126**, 411–417.

Thakar, D. S. and Tewari, H. B. 1967. Histochemical studies on the distribution of alkaline and acid phosphatases amongst the neurons of the cerebellum, spinal and trigeminal ganglia of bat. *Acta Histochemica* **28**, 359–367.

Thomas, E. 1972. Contribution of enzyme histochemistry as to the nature and differences among mantle cells (satellite cells), amphicytes, and Schwann cells in Remak fiber bundles and of myelinated fibers in dorsal root ganglia, sympathetic ganglia and in peripheral. *Neuropath. Pol.* **10**, 275–283.

Thomas, P. K., Ochoa, J., Berthold, C. H., Carlstedt, T., and Corneliuson, O. 1993. Microscopic anatomy of the peripheral nervous system. In P. J. Dyck, P. K. Thomas, J. W. Griffin, P. A. Low, and J. F. Poduslo (eds.), *Peripheral Neuropathy* (3rd ed., pp. 28–91). W. B. Saunders, Philadelphia.

Thompson, S. W. N. and Majithia, A. A. 1998. Leukemia inhibitory factor induces sympathetic sprouting in intact dorsal root ganglia in the adult rat in vivo. *J. Physiol.* **506.3**, 809–816.

Thor, K. B., Nickolaus, S., and Helke, C. J. 1993. Autoradiographic localization of 5-hydroxytryptamine$_{1A}$, 5-hydroxytryptamine$_{1B}$ and 5-hydroxytryptamine$_{1C/2}$ binding sites in the rat spinal cord. *Neuroscience* **55**, 235–252.

Thrush, D. C. 1973. Congenital insensitivity to pain: A clinical, genetic and neurophysiological study of four children from the same family. *Brain* **96**, 369–386.

Tie-Jun, S. S., Xu, Z., and Hökfelt, T. 2001. The expression of calcitonin gene-related peptide in dorsal horn neurons of the mouse lumbar spinal cord. *NeuroReport* **12**, 739–743.

Tillakaratne, N. J., Mouria, M., Ziv, N. B., Roy, R. R., Edgerton, V. R., and Tobin, A. J. 2000. Increased expression of glutamate decarboxylase (GAD(67)) in feline lumbar spinal cord after complete thoracic spinal cord transection. *J. Neurosci. Res.* **60**, 219–230.

Tjen-A-Looi, S. C., Pan, H. L., and Longhurst, J. C. 1998. Endogenous bradykinin activates ischaemically sensitive cardiac visceral afferents through kinin B_2 receptors in cats. *J. Physiol.* **510**, 633–641.

Toda, T. and Hayashi, H. 1993. The inhibitory effect of substance P antagonist, CP-96,345, on the later discharges of nociceptive neurons in the rat superficial spinal dorsal horn. *Neurosci. Lett.* **158**, 36–38.

Todd, A. J. 1988. Electron-microscope study of Golgi-stained cells in lamina ii of the rat spinal dorsal horn. *J. Comp. Neurol.* **275**, 145–157.

Todd, A. J. 1989. Cells in laminae III and IV of rat spinal dorsal horn receive monosynaptic primary afferent input in lamina II. *J. Comp. Neurol.* **289**, 676–686.

Todd, A. J. 1990. An electron-microscope study of glycine-like immunoreactivity in laminae I–III of the spinal dorsal horn of the rat. *Neuroscience* **39**, 387–394.

Todd, A. J. 1991. Immunohistochemical evidence that acetylcholine and glycine exist in different populations of GABAergic neurons in lamina III of rat spinal dorsal horn. *Neuroscience* **44**, 741–746.

Todd, A. J. 1996. GABA and glycine in synaptic glomeruli of the rat spinal dorsal horn. *Eur. J. Neurosci.* **8**, 2492–2498.

Todd, A. J. and Lewis, S. G. 1986. The morphology of Golgi-stained neurons in Lamina II of the rat spinal cord. *J. Anat.* **149**, 113–119.

Todd, A. J. and Lochhead, V. 1990. GABA-like immunoreactivtiy in type I glomeruli of rat substantia gelatinosa. *Brain Res.* **514**, 171–174.

Todd, A. J. and McKenzie, J. 1989. GABA-immunoreactive neurons in the dorsal horn of the rat spinal cord. *Neuroscience* **31**, 799–806.

Todd, A. J. and Spike, R. C. 1993. The localization of classical transmitters and neuronpeptides within neurons in laminae I–III of the mammalian spinal dorsal horn. *Prog. Neurobiol.* **41**, 609–645.

Todd, A. J. and Sullivan, A. C. 1990. Light microscope study of the coexistence of GABA-like and glycine-like immunoreactivities in the spinal cord of the rat. *J. Comp. Neurol.* **296**, 496–505.

Todd, A. J., Maxwell, D. J., and Brown, A. G. 1991. Relationships between hair-follicle afferent axons and glycine-immunoreactive profiles in cat spinal dorsal horn. *Brain Res.* **564**, 132–137.

Todd, A. J., Russell, G., and Spike, R. C. 1992a. Immunocytochemical evidence that GABA and neurotensin exist in different neurons in laminae II and III of rat spinal dorsal horn. *Neuroscience* **47**, 685–691.

Todd, A. J., Spike, R. C., Russell, G., and Johnston, H. M. 1992b. Immunohistochemical evidence that Met-enkephalin and GABA coexist in some neurones in rat dorsal horn. *Brain Res.* **584**, 149–156.

Todd, A. J., Spike, R. C., Price, R. F., and Neilson, M. 1994a. Immunocytochemical evidence that neurotensin is present in glutamatergic neurons in the superficial dorsal horn of the rat. *J. Neurosci.* **14**, 774–784.

Todd, A. J., Spike, R. C., Brodbelt, A. R., Price, R. F., and Shehab, S. A. 1994b. Some inhibitory neurons in the spinal cord develop c-fos-immunoreactivity after noxious stimulation. *Neuroscience* **63**, 805–816.

Todd, A. J., Spike, R. C., Chong, D., and Neilson, M. 1995. The relationship between glycine and gephyrin in synapses of the rat spinal cord. *Eur. J. Neurosci.* **7**, 1–11.

Todd, A. J., Watt, C., Spike, R. C., and Sieghart, W. 1996. Colocalization of GABA, glycine and their receptors at synapses in the rat spinal cord. *J. Neurosci.* **16**, 974–982.

Todd, A. J., Spike, R. C., and Polgar, E. 1998. A quantitative study of neurons which express neurokinin-1 or somatostatin sst2a receptor in rat spinal dorsal horn. *Neuroscience* **85**, 459–473.

Todd, A. J., McGill, M. M., and Shehab, S. A. 2000. Neurokinin 1 receptor expression by neurons in laminae I, III and IV of the rat spinal dorsal horn that project to the brainstem. *Eur. J. Neurosci.* **12**, 689–700.

Todd, A. J., Puskar, Z., Spike, R. C., Hughes, C., Watt, C., and Forrest, L. 2002. Projection neurons in lamina I of rat spinal cord with the neurokinin 1 receptor are selectively innervated by substance p-containing afferents and respond to noxious stimulation. *J. Neurosci.* **22**, 4103–4113.

Todd, J. K. 1964. Afferent impulses in the pudendal nerves of the cat. *Q. J. Physiol.* **49**, 258–267.

Toennies, J. F. 1938. Reflex discharge from the spinal cord over the dorsal roots. *J. Neurophysiol.* **1**, 378–390.

Tohda, C., Sasaki, M., Konemura, T., Sasamura, T., Itoh, M., and Kuraishi, Y. 2001. Axonal transport of VR1 capsaicin receptor mRNA in primary afferents and its participation in inflammation-induced increase in capsaicin sensitivity. *J. Neurochem.* **76**, 1628–1635.

Tohyama, K. and Ide, C. 1987. Carbonic anhydrase activity in axon terminals of sensory corpuscles. *Arch. Histol. Jpn.* **50**, 325–333.

Tohyama, M. and Shiotani, Y. 1986. Neuropeptides in spinal cord. In P. C. Emson, M. N. Rossor, and M. Tohyama (eds.), *Progress in Brain Research* (pp. 177–218). Elsevier Science Publishers, Japan.

Toledo-Aral, J. J., Moss, B. L., He, Z. J., Koszowski, A. G., Whisenand, T., Levinson, S. R., Wolf, J. J., Silos-Santiago, I., Halegoua, S., and Mandel, G. 1997. Identification of PN1, a predominant voltage-dependent sodium channel expressed principally in peripheral neurons. *Proc. Natl. Acad. Sci. USA* **94**, 1527–1532.

Tölle, T. R., Berthele, A., Zieglgänsberger, W., Seeburg, P. H., and Wisden, W. 1993. The differential expression of 16 NMDA and non-NMDA receptor subunits in the rat spinal cord and in periaqueductal gray. *J. Neurosci.* **13**, 5009–5028.

Tölle, T. R., Berthele, A., Laurie, D. J., Seeburg, P. H., and Zieglgänsberger, W. 1995a. Cellular and subcellular distribution of NMDAR1 splice variant mRNA in the rat lumbar spinal cord. *Eur. J. Neurosci.* **7**, 1235–1244.

Tölle, T. R., Berthele, A., Zieglgänsberger, W., Seeburg, P. H., and Wisden, W. 1995b. Flip and flop variants of AMPA receptors in the lumbar spinal cord. *Eur. J. Neurosci.* **7**, 1414–1419.

Tominaga, M., Caterina, M. J., Malmberg, A. B., Rosen, T. A., Gilbert, H., Skinner, K., Raumann, B. E., Basbaum, A. I., and Julius, D. 1998. The cloned capsaicin receptor integrates multiple pain-producing stimuli. *Neuron* **21**, 531–543.

Tomiyama, M., Kannari, K., Nunomura, J., Oyama, Y., Takebe, K., and Matsunaga, M. 1994. Quantitative autoradiographic distribution of glutamate receptors in the cervical segment of the wobbler mouse. *Brain Res.* **650**, 353–357.

Tomiyama, M., Rodriguez-Puertas, R., Cortex, R., Christnacher, A., Sommer, B., Pazos, A., Palacios, J. M., and Mengod, G. 1996. Differential regional distribution of AMPA receptor subunit messenger RNAs in the human spinal cord as visualized by in situ hybridization. *Neuroscience* **75**, 901–915.

Tomiyama, M., Kannari, K., and Matsunaga, M. 1997. Strychnine-sensitive and strychnine-insensitive glycine binding sites in the spinal cord of the wobbler mouse. *Tohoku J. Exp. Med.* **183**, 37–43.

Tomiyama, M., Kimura, T., Maeda, T., Tanaka, H., Furusawa, K., Kurahashi, K., and Matsunaga, M. 2001. Expression of metabotropic glutamate receptor mRNAs in the human spinal cord: Implications for selective vulnerability of spinal motor neurons in amyotrophic lateral sclerosis. *J. Neurol. Sci.* **189**, 65–69.

Tomlinson, R. W. W., Gray, B. G., and Dostrovsky, J. O. 1983. Inhibition of rat spinal cord dorsal horn neurons by non-segmental, noxious cutaneous stimuli. *Brain Res.* **279**, 291–294.

Tong, Q., Ma, J., and Kirchgessner, A. L. 2001. Vesicular glutamate transporter 2 in the brain-gut axis. *NeuroReport* **12**, 3929–3934.

Tong, Y. G., Wang, H. F., Ju, G., Grant, G., Hökfelt, T., and Zhang, X. 1999. Increased uptake and transport of cholera toxin B-subunit in dorsal root ganglion neurons after peripheral axotomy: Possible implications for sensory sprouting. *J. Comp. Neurol.* **404**, 143–158.

Tonra, J. R. and Mendell, L. M. 1998. Effects of postnatal anti-NGF on the development of CGRP-IR neurons in the dorsal root ganglion. *J. Comp. Neurol.* **392**, 489–498.

Torebjörk, H. E. 1974. Afferent C units responding to mechanical, thermal and chemical stimuli in human non-glabrous skin. *Acta Physiol. Scand.* **92**, 374–390.

Torebjörk, H. E. and Hallin, R. G. 1973. Perceptual changes accompanying controlled preferential blocking of A and C fibre responses in intact human skin nerves. *Exp. Brain Res.* **16**, 321–332.

Torebjörk, H. E. and Ochoa, J. L. 1980. Specific sensations evoked by activity in single identified sensory units in man. *Acta Physiol. Scand.* **110**, 445–447.

Torebjörk, H. E. and Ochoa, J. L. 1990. New method to identify nociceptor units innervating glabrous skin of the human hand. *Exp. Brain Res.* **81**, 509–514.

Torebjörk, H. E., Ochoa, J. L., and Schady, W. 1984a. Referred pain from intraneural stimulation of muscle fascicles in the median nerve. *Pain* **18**, 145–156.

Torebjörk, H. E., Schady, W., and Ochoa, J. 1984b. Sensory correlates of somatic afferent fibre activation. *Human Neurobiol.* **3**, 15–20.

Torebjörk, H. E., Vallbo, A. B., and Ochoa, J. L. 1987. Intraneural microstimulation in man. Its relation to specificity of tactile sensations. *Brain* **110**, 1509–1529.

Torebjörk, H. E., Lundberg, L. E. R., and LaMotte, R. H. 1992. Central changes in processing of mechanoreceptive input in capsaicin-induced secondary hyperalgesia in humans. *J. Physiol.* **448**, 765–780.

Tork, I. 1990. Anatomy of the serotonergic system. *Ann. NY Acad. Sci.* **600**, 9–35.

Toscano, E., della Casa, R., Mardy, S., Gaetaniello, L., Sadile, F., Indo, Y., Pignata, C., and Andria, G. 2000. Multisystem involvement in congenital insensitivity to pain with anhidrosis (CIPA), a nerve growth factor receptor (TrkA)-related disorder. *Neuropediatrics* **31**, 39–41.

Towers, S., Princivalle, A., Billinton, A., Edmunds, M., Bettler, B., Urban, L., Castro-Lopes, J., and Bowery, N. G. 2000. $GABA_B$ receptor protein and mRNA distribution in rat spinal cord and dorsal root ganglia. *Eur. J. Neurosci.* **12**, 3201–3210.

Tracey, D. J. 1979. Characteristics of wrist joint receptors in cat. *Exp. Brain Res.* **34**, 165–176.

Tracey, D. J., De Biasi, S., Phend, K., and Rustioni, A. 1991. Aspartate-like immunoreactivity in primary afferent neurons. *Neuroscience* **40**, 673–686.

Trafton, J. A., Abbadie, C., Marchand, S., Mantyh, P. W., and Basbaum, A. I. 1999. Spinal opioid analgesia: How critical is the regulation of substance P signaling? *J. Neurosci.* **19**, 9642–9653.

Trafton, J. A., Abbadie, C., and Basbaum, A. I. 2001. Differential contribution of substance P and neurokinin A to spinal cord neurokinin-1 receptor signaling in the rat. *J. Neurosci.* **21**, 3656–3664.

Traub, R. J. 1996. The spinal contribution of substance P to the generation and maintenance of inflammatory hyperalgesia in the rat. *Pain* **67**, 151–161.

Traub, R. J., Iadarola, M. J., and Ruda, M. A. 1989a. Effect of multiple dorsal rhizotomies on calcitonin gene-related peptide-like immunoreactivity in the lumbosacral dorsal spinal cord of the cat: A radioimmunoassay analysis. *Peptides* **10**, 979–983.

Traub, R. J., Solodkin, A., and Ruda, M. A. 1989b. Calcitonin gene-related peptide immunoreactivity in the cat lumbosacral spinal cord and the effects of multiple dorsal rhizotomies. *J. Comp. Neurol.* **287**, 225–237.

Traub, R. J., Allen, B., Humphrey, E., and Ruda, M. A. 1990. Analysis of calcitonin gene-related peptide-like immunoreactivity in the cat dorsal spinal cord and dorsal root ganglia provide evidence for a multisegmental projection of nociceptive C-fiber primary afferents. *J. Comp. Neurol.* **302**, 562–574.

Traub, R. J., Solodkin, A., Meller, S. T., and Gebhart, G. F. 1994a. Spinal cord NADPH-diaphorase histochemical staining but not nitric oxide synthase immunoreactivity increases following carrageenan-produced hindpaw inflammation in the rat. *Brain Res.* **668**, 204–210.

Traub, R. J., Solodkin, A., and Gebhart, G. F. 1994b. NADPH-diaphorase histochemistry provides evidence of a bilateral, somatotopically inappropriate response to unilateral hindpaw inflammation in the rat. *Brain Res.* **647**, 113–123.

Traub, R. J., Hutchcroft, K., and Gebhart, G. F. 1999. The peptide content of colonic afferents decreases following colonic inflammation. *Peptides* **20**, 267–273.

Traynor, J. R., Hunter, J. C., Rodriguez, R. E., Hill, R. G., and Hughes, J. 1990. Delta-opioid receptor binding sites in rodent spinal cord. *Br. J. Pharmacol.* **100**, 319–323.

Tredici, G., Tarelli, L. T., Cavaletti, G., and Marmiroli, P. 1985. Ultrastructural organization of lamina VI of the spinal cord of the cat. *Prog. Neurobiol.* **24**, 293–311.

Treede, R. D., Meyer, R. A., and Campbell, J. N. 1998. Myelinated mechanically insensitive afferents from monkey hairy skin: Heat-response properties. *J. Neurophysiol.* **80**, 1082–1093.

Trevino, D. L., Coulter, J. D., and Willis, W. D. 1973. Location of cells of origin of spinothalamic tract in lumbar enlargement of the monkey. *J. Neurophysiol.* **36**, 750–761.

Tribollet, E., Barberis, C., and Arsenijevic, Y. 1997. Distribution of vasopressin and oxytocin receptors in the rat spinal cord: Sex-related differences and effect of castration in pudendal motor nuclei. *Neuroscience* **78**, 499–509.

Tribollet, E., Arsenijevic, Y., and Barberis, C. 1998. Vasopressin binding sites in the central nervous system: Distribution and regulation. *Prog. Brain Res.* **119**, 45–55.

Triepel, J., Metz, J., Munroe, D., London, S., Sweriduk, S., and Forssmann, W. G. 1987. Vasoactive intestinal polypeptide immunoreactivity in the spinal cord of the guinea pig. *Cell Tissue Res.* **249**, 145–150.

Trotter, W. and Davies, H. M. 1909. Experimental studies in the innervation of the skin. *J. Physiol.* **38**, 134–246.

Troy, C. M., Brown, K., Greene, L. A., and Shelanski, M. L. 1990. Ontogeny of the neuronal intermediate filament protein, peripherin, in the mouse embryo. *Neuroscience* **36**, 217–237.

Trudrung, P., Wirth, U., and Mense, S. 2000. Changes in the number of nitric oxide-synthesizing neurones on both sides of a chronic transection of the rat spinal cord. *Neurosci. Lett.* **287**, 125–128.

Tsai, S. H., Tew, J. M., McLean, J. H., and Shipley, M. T. 1988. Cerebral arterial innervation by nerve fibers containing calcitonin gene-related peptide (CGRP): I. Distribution and Origin of CGRP perivascular innervation in the rat. *J. Comp. Neurol.* **271**, 435–444.

Tsaur, M. L., Wan, Y. C., Lai, F. P., and Cheng, H. F. 1997. Expression of B-type endothelin receptor gene during neural development. *FEBS Lett.* **417**, 208–212.

Tschopp, F. A., Henke, H., Petermann, J., Tobler, P., Janzer, R., Hökfelt, T., Lundberg, J., and Cuello, C. 1985. Calcitonin gene-related peptide and its binding sites in the human central nervous system and pituitary. *Proc. Natl. Acad. Sci. USA* **82**, 248–252.

Tseng, L. F., Narita, M., Suganuma, C., Mizoguchi, H., Ohsawa, M., Nagase, H., and Kampine, J. P. 2000. Differential antinociceptive effects of endomorphin-1 and endomorphin-2 in the mouse. *J. Pharmacol. Exp. Therap.* **292**, 576–583.

Tsou, K., Khachaturian, H., Akil, H., and Watson, S. J. 1986. Immunocytochemical localization of pro-opiomelanocortin-derived peptides in the adult rat spinal cord. *Brain Res.* **378**, 28–35.

Tsou, K., Brown, S., Sanudo-Pena, M. C., Mackie, K., and Walker, J. M. 1998. Immunohistochemical distribution of cannabinoid CB1 receptors in the rat central nervous system. *Neuroscience* **83**, 393–411.

Tsuchiya, M., Yamazaki, H., and Hori, Y. 1999. Enkephalinergic neurons express 5-HT3 receptors in the spinal cord dorsal horn: Single cell RT-PCR analysis. *NeuroReport* **10**, 2749–2753.

Tsuruo, Y., Hökfelt, T., and Visser, T. 1987. Thyrotropin releasing hormone (TRH)-immunoactive cell groups in the rat central nervous system. *Exp. Brain Res.* **68**, 213–217.

Tsuruoka, M., Li, Q. J., Matsui, A., and Matsui, Y. 1990. Inhibition of nociceptive responses of wide-dynamic-range neurons by peripheral nerve stimulation. *Brain Res. Bull.* **25**, 387–392.

Tsuzuki, K., Kondo, E., Fukuoka, T., Yi, D., Tsujino, H., Sakagami, M., and Noguchi, K. 2001. Differential regulation of P2X(3) mRNA expression by the peripheral nerve injury in intact and injured neurons in the rat sensory ganglia. *Pain* **91**, 351–360.

Tuchscherer, M. M. and Seybold, V. S. 1985. Immunohistochemical studies of substance P, cholecytokinin-octapeptide and somatostatin in dorsal root ganglia of the rat. *Neuroscience* **14**, 593–605.

Tuchscherer, M. M. and Seybold, V. S. 1989. A quantitative study of the coexistence of peptides in varicosities within the superficial laminae of the dorsal horn of the rat spinal cord. *J. Neurosci.* **9**, 195–205.

Tuchscherer, M. M., Knox, C., and Seybold, V. S. 1987. Substance P and cholecystokinin-like immunoreactive varicosities in somatosensory and autonomic regions of the rat spinal cord: A quantitative study of coexistence. *J. Neurosci.* **7**, 3984–3995.

Tuckett, R. P. 1982. Innervation pattern of cutaneous hair receptors in cat. *Brain Res.* **249**, 255–263.

Tuckett, R. P. and Wei, J. Y. 1987a. Response to an itch-producing substance in cat: I. Cutaneous receptor populations with myelinated axons. *Brain Res.* **413**, 87–94.

Tuckett, R. P. and Wei, J. Y. 1987b. Response to an itch-producing substance in cat: II. Cutaneous populations with unmyelinated axons. *Brain Res.* **413**, 95–103.

Tuckett, R. P., Horch, K. W., and Burgess, P. R. 1978. Response of cutaneous hair and field mechanoreceptors in cat to threshold stimuli. *J. Neurophysiol.* **41**, 138–149.

Turnbull, B. G. and Rasmusson, D. D. 1986. Sensory innervation of the raccoon forepaw: 1. Receptor types in glabrous and hairy skin and deep tissue. *Somatosens. Res.* **4**, 43–62.

Tuyau, M., Hansen, M. A., Coleman, M. J., Dampney, R. A. L., Balcar, V. J., and Bennett, M. R. 1997. Autoradiography of [^{3}H]alpha, beta-methylene-ATP binding sites in medulla oblongata and spinal cord of the rat. *Neurochem. Int.* **30**, 159–169.

Tyndale, R. F., Olsen, R. W., and Tobin, A. J. 1995. $GABA_A$ receptors: Ligand- and voltage-gated ion channels, In R. A. North (ed.), *Handbook of Receptors and Channels* (pp. 265–290) CRC Press, Boca Raton.

Tzoumaka, E., Tischler, A. C., Sangameswaran, L., Eglen, R. M., Hunter, J. C., and Novakovic, S. D. 2000. Differential distribution of the tetrodotoxin-sensitive rPN4/NaCh6/Scn8a sodium channel in the nervous system. *J. Neurosci. Res.* **60**, 37–44.

Ubink, R., Kopp, J., Wong, H., Walsh, J. H., Pedrazzini, T., and Hokfelt, T. 2001. Transient prenatal expression of NPY-Y1 receptor in trigeminal axons innervating the mystacial vibrissae. *J. Comp. Neurol.* **429**, 183–191.

Uchida, Y. and Murao, S. 1974. Excitation of afferent cardiac sympathetic nerve fibers during coronary occlusion. *Am. J. Physiol.* **226**, 1094–1099.

Uddman, R., Edvinsson, L., Ekman, R., Kingman, T., and McCulloch, J. 1985a. Innervation of the feline cerebral vasculature by nerve fibers containing calcitonin gene-related peptide: Trigeminal origin and co-existence with substance P. *Neurosci. Lett.* **62**, 134–136.

Uddman, R., Luts, A., and Sundler, F. 1985b. Occurrence and distribution of calcitonin gene-related peptide in the mammalian respiratory tract and middle ear. *Cell Tissue Res.* **241**, 551–555.

Ueda, M., Kuraishi, Y., and Satoh, M. 1993. Detection of capsaicin-evoked release of glutamate from spinal dorsal horn slices of rat with on-line monitoring system. *Neurosci. Lett.* **155**, 179–182.

Ueda, M., Kuraishi, Y., Sugimoto, K., and Satoh, M. 1994. Evidence that glutamate is released from capsaicin-sensitive primary afferent fibers in rats: Study with on-line continuous monitoring of glutamate. *Neurosci. Res.* **20**, 231–237.

Uhl, G. R., Goodman, R. R., and Snyder, S. H. 1979a. Neurotensin-containing cell bodies, fibers and nerve terminals in the brain stem of the rat: Immunohistochemical mapping. *Brain Res.* **167**, 77–91.

Uhl, G. R., Goodman, R. R., Kuhar, M. J., Childers, S. R., and Snyder, S. H. 1979b. Immunohistochemical mapping of enkephalin-containing cell bodies, fibers and nerve terminals in the brain stem of the rat. *Brain Res.* **166**, 75–94.

Uhl, G. R., Tran, V., Snyder, S. H., and Martin, J. B. 1985. Somatostatin receptors: Distribution in rat central nervous system and human frontal cortex. *J. Comp. Neurol.* **240**, 288–304.

Uhlen, S., Lindblom, J., Johnson, A., and Wikberg, J. E. 1997. Autoradiographic studies of central alpha 2A- and alpha 2C- adrenoceptors in the rat using [3H]MK912 and subtype-selective drugs. *Brain Res.* **770**, 261–266.

Ulfhake, B., Arvidsson, U., Cullheim, S., Hökfelt, T., and Visser, T. J. 1987. Thyrotropin-releasing hormone (TRH)-immunoreactive boutons and nerve cell bodies in the dorsal horn of the cat L7 spinal cord. *Neurosci. Lett.* **73**, 3–8.

Unger, W. G., Terenghi, G., Ghatei, M. A., Ennis, K. W., Butler, J. M., Zhang, S. Q., Too, H. P., Polak, J. M., and Bloom, S. R. 1985. Calcitonin gene-related polypeptide as a mediator of the neurogenic ocular injury response. *J. Ocul. Pharmacol.* **1**, 189–199.

Unger, W. G., Terenghi, G., Zhang, S. Q., and Polak, J. M. 1988. Alteration in the histochemical presence of tyrosine hydroxylase and CGRP-immunoreactivities in the eye following chronic sympathetic or sensory denervation. *Curr. Eye Res.* **7**, 761–769.

Ungerstedt, U. 1984. Measurement of neurotransmitter release by intracranial dialysis. In C. A. Marsden (ed.), *Measurement of Neurotransmitter Release* In vivo (pp. 81–105). John Wiley & Sons, New York.

Urban, L. and Randić, M. 1984. Slow excitatory transmission in rat dorsal horn: Possible mediation by peptides. *Brain Res.* **290**, 336–341.

Urban, L., Aitken, P. G., and Somjen, G. G. 1985. Interstitial potassium concentration, slow depolarization and focal potential responses in the dorsal horn of the rat spinal slice. *Brain Res.* **331**, 168–171.

Urca, G. and Urca, R. 1990. Neurotoxic effects of excitatory amino acids in the mouse spinal cord: Quisqualate and kainate but not N-methyl-D-aspartate induce permanent neural damage. *Brain Res.* **529**, 7–15.

Vacca, L. L., Abrahams, S. J., and Naftchi, N. E. 1986. A modified peroxidase-antiperoxidase procedure for improved localization of tissue antigens: Localization of substance P in rat spinal cord. *J. Histochem. Cytochem.* **28**, 297–307.

Vaccarino, A. L. and Kastin, A. J. 2001. Endogenous opiates: 2000. *Peptides* **22**, 2257–2328.

Valerio, A., Rizzonelli, P., Paterlini, M., Moretto, G., Knopfel, T., Kuhn, R., Memo, M., and Spano, P. 1997a. mGluR5 metabotropic glutamate receptor distribution in rat and human spinal cord: A developmental study. *Neurosci Res.* **28**, 49–57.

Valerio, A., Paterlini, M., Boifava, M., Memo, M., and Spano, P. 1997b. Metabotropic glutamate receptor mRNA expression in rat spinal cord. *NeuroReport* **8**, 2695–2699.

Vallbo, A. B. 1981. Sensations evoked from the glabrous skin of the human hand by electrical stimulation of unitary mechanosensitive afferents. *Brain Res.* **215**, 359–363.

Vallbo, A. B. and Johansson, R. S. 1984. Properties of cutaneous mechanoreceptors in the human hand related to touch sensation. *Hum. Neurobiol.* **3**, 3–14.

Vallbo, A. B., Olausson, H., and Wessberg, J. 1999. Unmyelinated afferents constitute a second system coding tactile stimuli of the human hairy skin. *J. Neurophysiol.* **81**, 2753–2763.

Valtschanoff, J. G., Weinberg, R. J., Rustioni, A., and Schmidt, H. H. 1992a. Nitric oxide synthase and GABA colocalize in lamina II of rat spinal cord. *Neurosci. Lett.* **148**, 6–10.

Valtschanoff, J. G., Weinberg, R. J., and Rustioni, A. 1992b. NADPH diaphorase in the spinal cord of rats. *J. Comp. Neurol.* **321**, 209–222.

Valtschanoff, J. G., Phend, K. D., Bernardi, P. S., Weinberg, R. L., and Rustioni, A. 1994. Amino acid immunocytochemistry of primary afferent terminals in the rat dorsal horn. *J. Comp. Neurol.* **346**, 237–252.

Van den Pol, A. N. 1999. Hypothalamic hypocretin (orexin): Robust innervation of the spinal cord. *J. Neurosci.* **19**, 3171–3182.

Van den Pol, A. N. and Gorcs, T. 1988. Glycine and glycine receptor immunoreactivity in brain and spinal cord. *J. Neurosci.* **8**, 472–492.

Van der Kraan, M., Tatro, J. B., Entwistle, M. L., Brakkee, J. H., Burbach, J. P., Adan, R. A., and Gispen, W. H. 1999. Expression of melanocortin receptors and pro-opiomelanocortin in the rat spinal cord in relation to neurotrophic effects of melanocortins. *Mol. Brain Res.* **63**, 276–286.

Van Dijken, H., Dijk, J., Voom, P., and Holstege, J. C. 1996. Localization of dopamine D_2 receptor in rat spinal cord identified with immunocytochemistry and in situ hybridization. *Eur. J. Neurosci.* **8**, 621–628.

Van Doren, C. L. 1989. A model of spatiotemporal tactile sensitivity linking psychophysics to tissue mechanics. *J. Acoust. Soc. Am.* **85**, 2065–2080.

Van Hees, J. and Gybels, J. M. 1972. Pain related to single afferent C fibers from human skin. *Brain Res.* **48**, 397–400.

Van Ranst, L. and Lauweryns, J. M. 1990. Effects of long-term sensory vs. sympathetic denervation on the distribution of calcitonin gene-related peptide and tyrosine hydroxylase immunoreactivities in the rat lung. *J. Neuroimmunol.* **29**, 131–138.

Vanderhaeghen, J. J., Lotstra, F., DeMey, J., and Gilles, C. 1980. Immunohistochemical localization of cholecystokinin- and gastrin-like peptides in the brain and hypophysis of the rat. *Proc. Natl. Acad. Sci.* **77**, 1190–1194.

Vanderhaeghen, J. J., Lotstra, F., Vierendeels, G., Gilles, C., Deschepper, C., and Verbanck, P. 1981. Cholecystokinins in the central nervous system and neurohypophysis. *Peptides* Suppl. **2**, 81–88.

Vanderhaeghen, J. J., Deschepper, C., Lotstra, F., Vierendeels, G., and Schoenen, J. 1982. Immunohistochemical evidence for cholecystokinin-like peptides in neuronal cell bodies of the rat spinal cord. *Cell Tissue Res.* **223**, 463–467.

Vanderhorst, V. G., Mouton, L. J., Blok, B. F., and Holstege, G. 1996. Distinct cell groups in the lumbosacral cord of the cat project to different areas in the periaqueductal gray. *J. Comp. Neurol.* **376**, 361–385.

Vanhala, A., Yamatodani, A., and Panula, P. 1994. Distribution of histamine-, 5-hydroxytryptamine-, and tyrosine hydroxylase-immunoreactive neurons and nerve fibers in developing rat brain. *J. Comp. Neurol.* **347**, 101–114.

Vater, A. 1741. Dissertatio de consensu partium corporis humani. In *Haller, Disputationum Anatomicarum Selectarum*, Vol. II. *Gottingae* (pp. 953–972). Cited in Cauna and Mannan (1958).

Vecsernyes, M., Jojart, I., Jojart, J., Laczi, F., and Laszlo, F. A. 1987. Presence of chromatographically identified oxytocin in human sensory ganglia. *Brain Res.* **414**, 153–154.

Vega, J. A., Rodriguez, C., Medina, M., Martinez Telleria, A., Bengoechea, M. E., and Perez-Casa, A. 1989. Acetyl-cholinesterase and fluoride-resistant acid phosphatase activities in dorsal root ganglia. *Cell. Mol. Biol.* **35**, 39–46.

Vega, J. A., Amenta, F., Hernandez, L. C., and Del Valle, M. E. 1991. Presence of catecholamine-related enzymes in a subpopulation of primary sensory neurons in dorsal root ganglia of the rat. *Cell. Mol. Biol.* **37**, 519–530.

Vega, J. A., Del Valle, M. E., Haro, J. J., Calzada, B., Suarez-Garnacho, S., and Malinovsky, L. 1993. Nerve growth factor receptor immunoreactivity in Meissner and Pacinian corpuscles of the human digital skin. *Anat. Rec.* **236**, 730–736.

Vega, J. A., Del Valle, M. E., Haro, J. J., Naves, F. J., Calzada, B., and Uribelarrea, R. 1994a. The inner-core, outer-core and capsule cells of the human Pacinian corpuscles: An immunohistochemical study. *Eur. J. Morphol.* **32**, 11–18.

Vega, J. A., Vazquez, E., Naves, F. J., Calzada, B., del Valle, M. E., and Represa, J. J. 1994b. Expression of epidermal growth factor receptor (EGFr) immunoreactivity in human cutaneous nerves and sensory corpuscles. *Anat. Rec.* **240**, 125–130.

Vega, J. A., Llamosas, M. M., Huerta, J. J. and García-Fernández, J. M. 1996. Study of human cutaneous sensory corpuscles using double immunolabeling and confocal laser scanning microscopy. *Anat. Rec.* **246**, 557–560.

Vera-Portocarrero, L. P., Mills, C. D., Ye, Z., Fullwood, S. D., McAdoo, D. J., Hulsebosch, C. E., and Westlund, K. N. 2002. Rapid changes in expression of glutamate transporters after spinal cord injury. *Brain Res.* **927**, 104–110.

Vergara, I., Oberpaur, B., and Alvarez, J. 1986. Ventral root nonmedullated fibers: Proportion, calibers, and microtubular content. *J. Comp. Neurol.* **248**, 550–554.

Verge, V. M., Xu, Z., Xu, X. J., Wiesenfeld-Hallin, Z., and Hökfelt, T. 1992. Marked increase in nitric oxide synthase mRNA in rat dorsal root ganglia after peripheral axotomy: In situ hybridization and functional studies. *Proc. Natl. Acad. Sci. USA* **89**, 11617–11621.

Verge, V. M., Wiesenfeld-Hallin, Z., and Hökfelt, T. 1993. Cholecystokinin in mammalian primary sensory neurons and spinal cord: In situ hybridization studies in rat and monkey. *Eur. J. Neurosci.* **5**, 240–250.

Verge, V. M. K., Richardson, P. M., Wiesenfeld-Hallin, Z., and Hökfelt, T. 1995. Differential influence of nerve growth factor on neuropeptide expression in vivo: A novel role in peptide suppression in adult sensory neurons. *J. Neurosci.* **15**, 2081–2096.

Viana, F., de La Pena, E., and Belmonte, C. 2002. Specificity of cold thermotransduction is determined by differential ionic channel expression. *Nature Neurosci.* **5**, 254–260.

Vickery, R. M., Gynther, B. D., and Rowe, M. J. 1994. Synaptic transmission between single slowly adapting type I fibres and their cuneate target neurones in cat. *J. Physiol.* **474**, 379–392.

Vidnyanszky, Z., Hamori, J., Negyessy, L., Ruegg, D., Knopfel, T., Kuhn, R., and Gorcs, T. J. 1994. Cellular and subcellular localization of the mGluR5a metabotropic glutamate receptor in rat spinal cord. *NeuroReport* **6**, 209–213.

Vierck, C. J. and Light, A. R. 1999. Effects of combined hemotoxic and anterolateral spinal lesions on nociceptive sensitivity. *Pain* **83**, 447–457.

Vierck, C. J., Greenspan, J. D., and Ritz, L. A. 1990. Long-term changes in purposive and reflexive responses to nociceptive stimulation in monkeys following anterolateral chordotomy. *J. Neurosci.* **10**, 2077–2095.

Vilim, F. S., Aarnisalo, A. A., Nieminen, M. L., Lintunen, M., Karlstedt, K., Kontinen, V. K., Kalso, E., States, B., Panula, P., and Ziff, E. 1999. Gene for pain modulatory neuropeptide NPFF: Induction in spinal cord by noxious stimuli. *Mol. Pharmacol.* **55**, 804–811.

Villanueva, L., Cadden, S. W., and Le Bars, D. 1984. Evidence that diffuse noxious inhibitory controls (DNIC) are mediated by a final post-synaptic inhibitory mechanism. *Brain Res.* **298**, 67–74.

Villanueva, L., Chitour, D., and Le Bars, D. 1986a. Involvement of the dorsolateral funiculus in the descending spinal projections responsible for diffuse noxious inhibitory controls in the rat. *J. Neurophysiol.* **56**, 1185–1195.

Villanueva, L., Peschanski, M., Calvino, B., and Le Bars, D. 1986b. Ascending pathways in the spinal cord involved in triggering of diffuse noxious inhibitory controls in the rat. *J. Neurophysiol.* **55**, 35–55.

Villar, M. J., Cortes, R., Theodorsson, E., and Wiesenfeld-Hallin, Z. 1989. Neuropeptide expression in rat dorsal root ganglion cells and spinal cord after peripheral nerve injury with special reference to galanin. *Neuroscience* **33**, 587–604.

Villar, M. J., Wiesenfeld-Hallin, Z., Xu, X. J., Theodorsson, E., Emson, P. C., and Hökfelt, T. 1991. Further studies on galanin-, substance P-, and CGRP-like immunoreactivities in primary neurons and spinal cord: Effects of dorsal rhizotomies and sciatic nerve lesions. *Exp. Neurol.* **112**, 29–39.

Villiger, J. W. and Faull, R. L. M. 1985. Muscarinic cholinergic receptors in the human spinal cord: Differential localization of [3H] pirenzepine and [3H]quinuclidinylbenzilate binding sites. *Brain Res.* **345**, 196–199.

Vincent, S. B. 1913. The tactile hair of the white rat. *J. Comp. Neurol.* **23**, 1–34.

Vincent, S. R. 1994. Nitric oxide: A radical neurotransmitter in the central nervous system. *Prog. Neurobiol.* **42**, 129–160.

Vincent, S. R., Hökfelt, T., Christensson, I., and Terenius, L. 1982. Dynorphin-immunoreactive neurons in the central nervous system of the rat. *Neurosci. Lett.* **33**, 185–190.

Vincent, S. R., McIntosh, C. H., Bueham, A. M., and Braum, J. C. 1985. Central somatostatin systems revealed with monoclonal antibodies. *J. Comp. Neurol.* **238**, 169–186.

Virgo, L., Samarasinghe, S., and de Belleroche, J. 1996. Analysis of AMPA receptor subunit mRNA expression in control and ALS spinal cord. *NeuroReport* **7**, 2507–2511.

Vita, G., Huan, C. K., Hawkins, E. F., and Engel, W. K. 1990. Effects of experimental spinal cord transection on substance P receptors: A quantitative autoradiography study. *Neuropeptides* **17**, 147–153.

Vital, A., Fontan, D., Julien, J., Talon, P., Heron, B., Routon, M. C., Ponsot, G., and Vital, C. 1998. Congenital insensitivity to pain with anhidrosis. Report of two unrelated cases. *J. Peripher. Nerv. Syst.* **3**, 125–132.

Vitry, G., Chambost, M. G., and Pitel, J. 1960. Étude des phosphatases dans divers territoires du système nerveux chez le chat, le lapin et le chien. *Ann. Histochem.* **5**, 9–18.

Vizzard, M. A. 1997. Increased expression of neuronal nitric oxide synthase in bladder afferent and spinal neurons following spinal cord injury. *Dev. Neurosci.* **19**, 232–246.

Vizzard, M. A. 2001. Alterations in neuropeptide expression in lumbosacral bladder pathways following chronic cystitis. *J. Chem. Neuroanat.* **21**, 125–138.

Vizzard, M. A., Erdman, S. L., Erickson, V. L., Stewart, R. J., Roppolo, J. R., and de Groat, W. C. 1993a. Localization of NADPH diaphorase in the lumbosacral spinal cord and dorsal root ganglia of the cat. *J. Comp. Neurol.* **339**, 62–75.

Vizzard, M. A., Erdman, S. L., and de Groat, W. C. 1993b. Localization of NADPH-diaphorase in pelvic afferent and efferent pathways of the rat. *Neurosci. Lett.* **152**, 72–76.

Vizzard, M. A., Erdman, S. L., and de Groat, W. C. 1993c. Localization of NADPH diaphorase in bladder afferent and postganglionic efferent neurons of the rat. *J. Auton. Nerv. Syst.* **44**, 85–90.

Vizzard, M. A., Erdman, S. L., and de Groat, W. C. 1993d. The effect of rhizotomy on NADPH diaphorase staining in the lumbar spinal cord of the rat. *Brain Res.* **607**, 349–353.

Vizzard, M. A., Erdman, S. L., Forstermann, U., and de Groat, W. C. 1994a. Differential distribution of nitric oxide synthase in neural pathways to the urogenital organs (urethra, penis, urinary bladder) of the rat. *Brain Res.* **646**, 279–291.

Vizzard, M. A., Erdman, S. L., Roppolo, J. R., Forstermann, U., and de Groat, W. C. 1994b. Differential localization of neuronal nitric oxide synthase immunoreactivity and NADPH-diaphorase activity in the cat spinal cord. *Cell Tissue Res.* **278**, 299–309.

Vizzard, M. A., Erdman, S. L., and de Groat, W. C. 1995. Increased expression of neuronal nitric oxide synthase in dorsal root ganglion neurons after systemic capsaicin administration. *Neuroscience* **67**, 1–5.

Vizzard, M. A., Erdman, S. L., and de Groat, W. C. 1996. Increased expression of neuronal nitric oxide synthase in bladder afferent pathways following chronic bladder irritation. *J. Comp. Neurol.* **370**, 191–202.

Vizzard, M. A., Erickson, K., and de Groat, W. C. 1997. Localization of NADPH diaphorase in the thoracolumbar and sacrococcygeal spinal cord of the dog. *J. Auton. Nerv. Syst.* **64**, 128–142.

Vizzard, M. A., Erdman, J., and de Groat, W. C. 1998. Increased expression of neuronal nitric oxide synthase (NOS) in visceral neurons after nerve injury. *J. Neuroscience* **15**, 4033–4045.

Von Banchet, G. S. and Heppelmann, B. 1995. Non-radioactive localization of substance P binding sites in rat brain and spinal cord using peptides labeled with 1.4-nm gold particles. *J. Histochem. Cytochem.* **43**, 821–827.

Von Banchet, G. S. and Schaible, H. G. 1999. Localization of the neurokinin 1 receptor on a subset of substance P-positive and isolectin B4-negative dorsal root ganglion neurons of the rat. *Neurosci. Lett.* **274**, 175–178.

Von Banchet, G. S., Schindler, M., Hervieu, G. J., Beckmann, B., Emson, P. C., and Heppelmann, B. 1999. Distribution of somatostatin receptor subtypes in rat lumbar spinal cord examined with gold-labelled somatostatin and anti-receptor antibodies. *Brain Res.* **816**, 254–257.

Von Euler, M., Seiger, Å., Holmberg, L., and Sundström, E. 1994. NBQX, a competitive non-NMDA receptor antagonist, reduces degeneration due to focal spinal cord ischemia. *Exp. Neurol.* **129**, 163–168.

Von Frey, M. 1896. Treatise on the sensory functions of the human skin. Transl. from German by Biederman-Thorson in H. O. Handwerker and K. Brune, K (eds.), 1987. *Classical German Contributions to Pain Research*, (pp. 69–131). Vth World Congress on Pain, Hamburg. Original German: Untersuchungen über die Sinnesfunctionen der Menschlichen Haut. vol. 23, *Koenigl. Saech. Ges. Wissensch.*, Leipsig, S. Hirzel.

Von Frey, M. 1906. The distribution of afferent nerves in the skin. *JAMA* **47**, 645–648.

Von Frey, M. 1910. Physiologie der Sinnesorgane der Menschlichen Haut. *Ergebnisse Physiol.* **9**, 351–368.

Vrinten, D. H., Kalkman, C. J., Adan, R. A., and Gispen, W. H. 2001. Neuropathic pain: A possible role for the melanocortin system? *Eur. J. Pharmacol.* **429**, 61–69.

Vulchanova, L., Arvidsson, U., Riedl, M., Wang, J., Buell, G., Surprenant, A., North, R. A., and Elde, R. 1996. Differential distribution of two ATP-gated ion channels(P_{2x} receptors) determined by immunocytochemistry. *Proc. Natl. Acad. Sci. USA* **93**, 8063–8067.

Vulchanova, L., Riedl, M. S., Shuster, S. J., Buell, G., Surprenant, A., North, R. A., and Elde, R. 1997. Immunohistochemical study of the $P2X_2$ and $P2X_3$ receptor subunits in rat and monkey sensory neurons and their central terminals. *Neuropharm.* **36**, 1229–1242.

Vulchanova, L., Riedl, M. S., Shuster, S. J., Stone, L. S., Hargreaves, K. M., Buell, G., Surprenant, A., North, R. A., and Elde, R. 1998. P2X3 is expressed by DRG neurons that terminate in inner layer II. *Eur. J. Neurosci.* **10**, 3470–3478.

Vyklický, L., Syková, E., Kříž, N., and Ujec, E. 1972. Post-stimulation changes of extracellular potassium concentration in the spinal cord of the rat. *Brain Res.* **45**, 608–611.

Vyklický, L., Syková, E., and Kříž, N. 1975. Slow potentials induced by changes of extracellular potassium in the spinal cord of the cat. *Brain Res.* **87**, 77–80.

Vyklický, L., Syková, E., and Mellerová, B. 1976. Depolarization of primary afferents in the frog spinal cord under high Mg^{++} concentrations. *Brain Res.* **117**, 153–156.

Vyklický, L., Vlachová, V., Vitásková, Z., Dittert, I., Kabáat, M., and Orkand, R. K. 1999. Temperature coefficient of membrane currents induced by noxious heat in sensory neurones in the rat. *J. Physiol.* **517**, 181–192.

Wada, E., Wada, K., Boulter, J., Deneris, E., Heinemann, S., Patrick, J., and Swanson, L. W. 1989. Distribution of α2, α3, α4, and β2 neuronal nicotinic receptor subunit mRNAs in the central nervous system: A hybridization histochemical study in the rat. *J. Comp. Neurol.* **284**, 314–335.

Wada, E., Way, J., Lebacq-Verheyden, A. M., and Battey, J. F. 1990. Neuromedin B and gastrin-releasing peptide mRNAs are differentially distributed in the rat nervous system. *J. Neurosci.* **10**, 2917–2930.

Wada, E., Wray, S., and Battey, J. 1992. Comparison of gene expression for two distinct bombesin receptor subtypes in postnatal rat central nervous system. *Mol. Cell. Neurosci.* **3**, 446–460.

Wada, E., Battey, J., and Wray, S. 1993. Bombesin receptor gene expression in rat embryos: Transient GRP-R gene expression in the posterior pituitary. *Mol. Cell. Neurosci.* **4**, 13–24.

Waddell, P. J. and Lawson, S. N. 1990. Electrophysiological properties of subpopulations of rat dorsal root ganglion neurons in vitro. *Neuroscience* **36**, 811–822.

Waeber, C., Hoyer, D., and Palacios, J. M. 1989. 5-hydroxytryptamine$_3$ receptors in the human brain: Autoradiographic visualization using [^{3}H]ICS 205–930. *Neuroscience* **31**, 393–400.

Wagman, I. H., and Price, D. D. 1969. Responses of dorsal horn cells of *M. mulatta* to cutaneous and sural A and C fiber stimuli. *J. Neurophysiol.* **32**, 803–817.

Wagner, R. and Meissner, G. 1852. Über Vorhandensein bischer unbekannten eigentümlichen Körperchen (Corpuscula tactus). *Gott. Nachr.* **2**, 17–30.

Wagner, R., Deleo, J. A., Coombs, D. W., Willenbring, S., and Fromm, C. 1993. Spinal dynorphin immunoreactivity increases bilaterally in a neuropathic pain model. *Brain Res.* **629**, 323–326.

Wakisaka, S., Kajander, K. C., and Bennett, G. J. 1991. Increased neuropeptide Y (NPY)-like immunoreactivity in rat sensory neurons following peripheral axotomy. *Neurosci. Lett.* **124**, 200–203.

Wakisaka, S., Kajander, K. C., and Bennett, G. J. 1992. Effects of peripheral nerve injuries and tissue inflammation on the levels of neuropeptide Y-like immunoreactivity in rat primary afferent neurons. *Brain Res.* **598**, 349–352.

Wakisaka, S., Takikita, S., Sasaki, Y., Kato, J., Tabata, M. J., and Kurisu, K. 1993. Cell size-specific appearance of neuropeptide Y in the trigeminal ganglion following peripheral axotomy of different branches of the mandibular nerve of the rat. *Brain Res.* **620**, 347–350.

Waksman, G., Hamel, E., Fournie-Zaluski, M.-C., and Roques, B. P. 1986. Autoradiographic comparison of the distribution of the neutral endopeptidase "enkephalinase" and of μ and δ opioid receptors in rat brain. *Proc. Natl. Acad. Sci. USA* **83**, 1523–1527.

Waldbillig, R. J. and LeRoith, D. 1987. Insulin receptors in the peripheral nervous system: A structural and functional analysis. *Brain Res.* **409**, 215–220.

Waldeyer, H. 1888. *Das Gorilla-Rueckenmark* (147 pp.). Akad. Wissensch., Berlin.

Waldmann, R. and Lazdunski, M. 1998. H^+-gated cation channels: Neuronal acid sensors in the NaC/DEG family of ion channels. *Curr. Opin. Neurobiol.* **8**, 418–424.

Waldvogel, H. J., Faull, R. L., Jansen, K. L., Dragunow, M., Richards, J. G., Mohler, H., and Streit, P. 1990. GABA, GABA receptors and benzodiazepine receptors in the human spinal cord: An autoradiographic and immunohistochemical study at the light and electron microscope levels. *Neuroscience* **39**, 361–385.

Walker, K., Perkins, M., and Dray, A. 1995. Kinins and kinin receptors in the nervous system. *Neurochem. Int.* **26**, 1–16.

Walker, K., Reeve, A., Bowes, M., Winter, J., Wotherspoon, G., Davis, A., Schmid, P., Gasparini, F., Kuhn, R., and Urban, L. 2001. mGlu5 receptors and nociceptive function II: mGlu5 receptors functionally expressed on peripheral sensory neurones mediate inflammatory hyperalgesia. *Neuropharmacology* **40**, 10–19.

Wall, P. D. 1958. Excitability changes in afferent fibre terminations and their relation to slow potentials. *J. Physiol.* **142**, 1–21.

Wall, P. D. 1959. Repetitive discharge of neurons. *J. Neurophysiol.* **22**, 305–320.

Wall, P. D. 1960. Cord cells responding to touch, damage, and temperature of skin. *J. Neurophysiol.* **23**, 197–210.

Wall, P. D. 1962. The origin of a spinal cord slow potential. *J. Physiol.* **164**, 508–526.

Wall, P. D. 1964. Presynaptic control of impulses at the first central synapse in the cutaneous pathway. In J. C. Eccles and J. P. Schade (eds.), *Physiology of Spinal Neurons. Prog. Brain Res.* **12**, 92–115.

Wall, P. D. 1965. Impulses originating in the region of dendrites. *J. Physiol.* **180**, 116–133.

Wall, P. D. 1967. The laminar organization of dorsal horn and effects of descending impulses. *J. Physiol.* **188**, 403–423.

Wall, P. D. 1977. The presence of ineffective synapses and the circumstances which unmask them. *Phil. Trans. Roy. Soc. B.* **278**, 361–372.

Wall, P. D. 1978. The gate control theory of pain mechanisms: A re-examination and re-statement. *Brain* **101**, 1–18.

Wall, P. D. 1982. The effect of peripheral nerve lesions and of neonatal capsaicin in the rat on primary afferent depolarization. *J. Physiol.* **329**, 21–35.

Wall, P. D. and Cronly-Dillon, J. R. 1960. Pain, itch, and vibration. *Arch. Neurol.* **2**, 365–375.

Wall, P. D. and Devor, M. 1981. The effect of peripheral nerve injury on dorsal root potentials and on transmission of afferent signals into the spinal cord. *Brain Res.* **209**, 95–111.

Wall, P. D. and Devor, M. 1983. Sensory afferent impulses originate from dorsal root ganglion cells as well as from the periphery in normal and nerve injured rats. *Pain* **17**, 321–329.

Wall, P. D. and Gutnick, M. 1974. Ongoing activity in peripheral nerves: The physiology and pharmacology of impulses originating from a neuroma. *Exp. Neurol.* **43**, 580–593.

Wall, P. D. and McMahon, S. B. 1985. Microneurography and its relation to perceived sensation: A critical review. *Pain* **21**, 209–229.

Wall, P. D. and Sweet, W. H. 1967. Temporary abolition of pain in man. *Science* **155**, 108–109.

Wall, P. D. and Werman, R. 1976. The physiology and anatomy of long ranging afferent fibres within the spinal cord. *J. Physiol.* **255**, 321–334.

Wall, P. D., Freeman, J., and Major, D. 1967. Dorsal horn cells in spinal and in freely moving rats. *Exp. Neurol.* **19**, 519–529.

Wall, P. D., Merrill, E. G., and Yaksh, T. L. 1979a. Responses of single units in laminae 2 and 3 of cat spinal cord. *Brain Res.* **160**, 245–260.

Wall, P. D., Devor, M., Inbal, R., Scadding, J. W., Schonfeld, D., Seltzer, Z., and Tomkiewicz, M. M. 1979b. Autotomy following peripheral nerve lesions: Experimental anaesthesia dolorosa. *Pain* **7**, 103–115.

Wall, P. D., Kerr, B. J., and Ramer, M. S. 2002. Primary afferent input to and receptive field properties of cells in rat lumbar area X. *J. Comp. Neurol.* **449**, 298–306.

Walsh, G. S. and Kawaja, M. D. 1998. Sympathetic axons surround nerve growth factor-immunoreactive trigeminal neurons: Observations in mice overexpressing nerve growth factor. *J. Neurobiol.* **34**, 347–360.

Walsh, G. S., Krol, K. M., and Kawaja, M. D. 1999. Absence of the p75 neurotrophin receptor alters the pattern of sympathosensory sprouting in the trigeminal ganglia of mice overexpressing nerve growth factor. *J. Neurosci.* **19**, 258–273.

Walshe, F. M. R. 1942. The anatomy and physiology of cutaneous sensibility: A critical review. *Brain* **65**, 48–112.

Walter-Barakat, I. and Droz, B. 1995. Nuclear and cytoplasmic triiodothyronine-binding sites in primary sensory neurons and schwann cells: Radioautographic study during development. *J. Neuroendocr.* **7**, 127–136.

Wamsley, J. K., Lewis, M. S., Young, W. S., III, and Kuhar, M. J. 1981. Autoradiographic localization of muscarinic cholinertic receptors in rat brainstem. *J. Neurosci.* **1**, 176–191.

Wamsley, J. K., Zarbin, M. A., Young, W. S., and Kuhar, M. J. 1982. Distribution of opiate receptors in the monkey brain: An autoradiographic study. *Neuroscience* **7**, 595–613.

Wanaka, A., Matsuyama, T., Yoneda, S., Kimura, K., Kamada, T., Girgis, S., MacIntyre, I., Emson, P. C., and Tohyama, M. 1986. Origins and distribution of calcitonin gene-related peptide-containing nerves in the wall of the cerebral arteries of the guinea pig with special reference to the coexistence with substance P. *Brain Res.* **369**, 185–192.

Wanaka, A., Shiotani, Y., Kuyama, H., Matsuyama, T., Shiosaka, S. and Tohyama, M. 1987. Glutamate-like immunoreactive structures in primary sensory neurons in the rat detected by a specific antiserum against glutamate. *Exp. Brain Res.* **65**, 691–694.

Wang, D., Li, Y. Q., Li, J. L., Kaneko, T., Nomura, S., and Mizuno, N. 2000. Gamma-aminobutryic acid- and glycine-immunoreactive neurons postsynaptic to substance P-immunoreactive axon terminals in the superficial layers of the rat medullary dorsal horn. *Neurosci Lett.* **288**, 187–190.

Wang, G. D., Wang, X. L., and Li, Z. W. 1996. Influence of peripheral process branching in nerve of spinal ganglion on the primary afferent impulse. *Sheng Li Xue Bao* **48**, 31–36.

Wang, H. and Wessendorf, M. W. 2001. Equal proportions of small and large DRG neurons express opioid receptor mRNAs. *J. Comp. Neurol.* **429**, 590–600.

Wang, H., Rivero-Melian, C., Robertson, B., and Grant, G. 1994. Transganglionic transport and binding of the isolectin B4 from Griffonia simplicifolia I in rat primary sensory neurons. *Neuroscience* **62**, 539–551.

Wang, H., Zhang, R. X., and Qiao, J.-T. 1999. Decreased expression of *N*-methyl-D-aspartate (NMDA) receptors in rat dorsal root ganglion following complete Freund's adjuvant-induced inflammation: An immunocytochemical study for NMDA NR1 subunit. *Neurosci. Lett.* **265**, 195–198.

Wang, H., Wang, R., Nie, H., Zhang, R., and Qiao, J. T. 2000. Neurokinin A, calcitonin gene-related peptide, and dynorphin A (1–8) in spinal dorsal horn contribute to descending inhibition evoked by nociceptive afferent pathways: An immunocytochemical study. *Regul. Pept.* **89**, 7–12.

Wang, H. F., Robertson, B., and Grant, G. 1998a. Anterograde transport of horseradish-peroxidase conjugated isolectin B4 from Griffonia simplicifolia I in spinal primary sensory neurons of the rat. *Brain Res.* **811**, 34–39.

Wang, H. F., Shortland, P., Park, M. J., and Grant, G. 1998b. Retrograde and transganglionic transport of horseradish peroxidase-conjugated cholera toxin B subunit, wheatgerm agglutinin and isolectin B4 from *Griffonia simplicifolia I* in primary afferent neurons innervating the rat urinary bladder. *Neuroscience* **87**, 275–288.

Wang, J. F., Khasar, S. G., Ahlgren, S. C., and Levine, J. D. 1996. Sensitization of C-fibres by prostaglandin E_2 in the rat is inhibited by guanosine 5'-O-(2-thiodiphosphate), 2', 5'-dideoxyadenosine and Walsh inhibitor peptide. *Neuroscience* **71**, 259–263.

Wang, Q. P., Zadina, J. E., Guan, J., Kastin, A. J., Funahashi, H., and Shida, S. 2002. Endomorphin-2 immunoreactivity in the cervical dorsal horn of the rat spinal cord at the electron-microscopic level. *Neuroscience* **113**, 593–605.

Wang, S., Clemmons, A., Strader, C., and Bayne, M. 1998. Evidence for hydrophobic interaction between galanin and the GalR1 galanin receptor and GalR1-mediated ligand internalization: Fluorescent probing with a fluorescein-galanin, *Biochemistry* **37**, 9528–9535.

Wang, S. D., Goldberger, M. E., and Murray, M. 1991a. Plasticity of spinal systems after unilateral dorsal rhizotomy in the adult rat. *J. Comp. Neurol.* **304**, 555–568.

Wang, S. D., Goldberger, M. E., and Murray, M. 1991b. Normal development and the effects of early rhizotomy on spinal systems in the rat. *Brain Res. Dev. Brain Res.* **64**, 57–69.

Warden, M. K. and Young, W. S. 1988. Distribution of cells containing mRNAs encoding substance P and neurokinin B in the rat central nervous system. *J. Comp. Neurol.* **272**, 90–113.

Warrington, W. B. and Griffith, F. 1904. On the cells of the spinal ganglia and on the relationship of their histological structure to the axonal distribution. *Brain* **27**, 297–325.

Wasowicz, K. and Panula, P. 1994. Distribution of neuropeptide FF in porcine spinal cord in comparison with other neuropeptides and serotonin. *J. Comp. Neurol.* **346**, 530–540.

Watanabe, M., Mishina, M., and Inoue, Y. 1994. Distinct spatiotemporal distributions of the *N*-methyl-D-aspartate receptor channel subunit mRNAs in the mouse cervical cord. *J. Comp. Neurol.* **345**, 314–319.

Watanabe, E. and Akagi, H. 1995. Distribution patterns of mRNAs encoding glycine receptor channels in the developing rat spinal cord. *Neurosci. Res.* **23**, 377–382.

Waters, S. M. and Krause, J. E. 2000. Distribution of galanin-1, -2 and -3 receptor messenger RNAs in central and peripheral rat tissues. *Neuroscience* **95**, 265–271.

Watkins, J. C., and Evans, R. H. 1981. Excitatory amino acid transmitters. *Ann. Rev. Pharmacol. Toxicol.* **21**, 165–204.

Watling, K. J., Kebabian, J. W. and Neumeyer, J. L. (eds.) 1995. *The RBI Handbook of Receptor Classification and Signal Transduction.* Research Biochemicals International, Natick, MA.

Watling, K. J. (ed.) 2001. *The Sigma-RBI Handbook of Receptor Classification and Signal Transduction.* (4th ed., pp. 4–5) Sigma-RBI, Natick, MA.

Watson, A. H. and Bazzaz, A. A. 2001. GABA and glycine-like immunoreactivity at axoaxonic synapses on 1a muscle afferent terminals in the spinal cord of the rat. *J. Comp. Neurol.* **433**, 335–348.

Watson, A. H., Hughes, D. I., and Bazzaz, A. A. 2002. Synaptic relationships between hair follicle afferents and neurones expressing GABA and glycine-like immunoreactivity in the spinal cord of the rat. *J. Comp. Neurol.* **452**, 367–380.

Waxman, S. G. 2001. Acquired channelopathies in nerve injury and MS. *Neurology* **56**, 1621–1627.

Waxman, S. G. and Black, J. A. 1996. Expression of mRNA for a sodium channel in subfamily 2 in spinal sensory neurons. *Neurochem. Res.* **21**, 395–401.

Waxman, S. G., Kocsis, J. D., and Black, J. A. 1994. Type III sodium channel mRNA is expressed in embryonic but not adult spinal sensory neurons, and is reexpressed following axotomy. *J. Neurophysiol.* **72**, 466–470.

Waxman, S. G., Dib-Hajj, S., Cummins, T. R., and Black, J. A. 1999. Sodium channels and pain. *Proc. Natl. Acad. Sci. USA* **96**, 7635–7639.

Waxman, S. G., Dib-Hajj, S., Cummins, T. R., and Black, J. A. 2000. Sodium channels and their genes: Dynamic expression in the normal nervous system, dysregulation in disease states (1). *Brain Res.* **886**, 5–14.

Weddell, G. 1955. Somesthesis and the chemical senses. *Ann. Rev. Psychol.* **6**, 19–136.

Weddell, G. and Miller, S. 1962. Cutaneous sensibility. *Ann. Rev. Physiol.* **24**, 199–222.

Weddell, G. and Sinclair, D. C. 1953. The anatomy of pain sensibility. *Acta Neuroveg.* **7**, 135–146.

Weddell, G., Sinclair, D. C., and Feindel, W. H. 1948. An anatomical basis for alterations in quality of pain sensibility. *J. Neurophysiol.* **11**, 99–109.

Wee, B. E. F., Emery, D. G., and Blanchard, J. L. 1985. Unmyelinated fibers in the cervical and lumbar ventral roots of the cat. *Am. J. Anat.* **172**, 307–316.

Wei, F. and Zhao, Z. Q. 1996. Blockade of capsaicin-induced reduction of GABA-immunoreactivity by spantide in cat spinal superficial dorsal horn. *Neuroscience* **71**, 277–283.

Wei, J. Y., Simon, J., Randić, M., and Burgess, P. R. 1986. Joint angle signaling by muscle spindle receptors. *Brain Res.* **370**, 108–118.

Weidner, C., Schmelz, M., Schmidt, R., Hansson, B., Handwerker, H. O., and Torebjörk, H. E. 1999. Functional attributes discriminating mechano-insensitive and mechano-responsive C nociceptors in human skin. *J. Neurosci.* **19**, 10184–10190.

Weidner, C., Schmidt, R., Schmelz, M., Hilliges, M., Handwerker, H. O., and Torebjörk, H. E. 2000. Time course of post-excitatory effects separates afferent human C fibre classes. *J. Physiol.* **527**, 185–191.

Weidner, C., Schmelz, M., Schmidt, R., Hammerberg, B., Ørstavik, K., Hilliges, M., Torebjörk, H. E., and Handwerker, H.O. 2002. Neural signal processing: the underestimated contribution of peripheral human C-fibers. *J. Neurosci.* **22**, 6704–6712.

Weihe, E., Hartschuh, W., and Weber, E. 1985. Prodynorphin opioid peptides in small somatosensory primary afferents of guinea pig. *Neurosci Lett.* **58**, 347–352.

Weihe, E., Leibold, A., Norh, D., Fink, T., and Gauweiler, B. 1986. Co-existence of prodynorphin-opioid peptides and substance P in primary sensory afferents of guinea-pigs. *Natl. Inst. Drug Abuse Res.* **75**, 295–298.

Weihe, E., Nohr, D., Millan, M. J., Stein, C., Muller, S., Gramsch, C., and Herz, A. 1988a. Peptide neuroanatomy of adjuvant-induced arthritic inflammation in rat. *Agents and Actions* **25**, 255–259.

Weihe, E., Nohr, D., and Hartschuh, W. 1988b. Immunohistochemical evidence for a co-transmitter role of opioid peptides in primary sensory neurons. *Prog. Brain Res.* **74**, 189–199.

Weihe, E., Millan, M. J., Leibold, A., Nohr, D., and Herz, A. 1988c. Co-localization of proenkephalin- and prodynorphin-derived opioid peptides in laminae IV/V spinal neurons revealed in arthritic rats. *Neurosci. Lett.* **85**, 187–192.

Weihe, E., Millan, M. J., Hollt, V., Nohr, D., and Herz, A. 1989. Induction of the gene encoding pro-dynorphin by experimentally induced arthritis enhances staining for dynorphin in the spinal cord of rats. *Neuroscience* **31**, 77–95.

Weil-Fugazza, J., Onteniente, B., Audet, G., and Philippe, E. 1993. Dopamine as trace amine in the dorsal root ganglia. *Neurochem. Res.* **18**, 965–969.

Weinberg, R. J., Conti, F., Van Eyck, S. L., Petrusz, P., and Rustioni, A. 1987. Glutamate immunoreactivity in superficial laminae of rat dorsal horn and spinal trigeminal nucleus, In T. P. Hicks, D. Lodge, and H. McClennan (eds.), *Excitatory amino acid transmission* (pp. 173–176). Alan R. Liss, New York.

Welsh, M. J., Price, M. P., and Xie, J. 2002. Biochemical basis of touch perception: Mechanosensory function of degenerin/epithelial Na^+ channels. *J. Biol. Chem.* **277**, 2369–2372.

Weng, H. R., Lee, J. I., Lenz, F. A., Schwartz, A., Vierck, C., Rowland, L., and Dougherty, P. M. 2000. Functional plasticity in primate somatosensory thalamus following chronic lesion of the ventral lateral spinal cord. *Neuroscience* **101**, 393–401.

Weng, H. R., Mansikka, H., Winchurch, R., Raja, S. N., and Dougherty, P. M. 2001. Sensory processing in the deep spinal dorsal horn of neurokinin-1 receptor knockout mice. *Anesthesiology* **94**, 1105–1112.

Wenk, H. N. and Honda, C. N. 1999. Immunohistochemical localization of delta opioid receptors in peripheral tissues. *J. Comp. Neurol.* **408**, 567–579.

Werman, R. 1966. Criteria for identification of a central nervous system transmitter. *Comp. Biochem Physiol.* **18**, 745–766.

Werman, R., Davidoff, R. A., and Aprison, M. H. 1968. Inhibitory action of glycine on spinal neurons in the cat. *J. Neurophysiol.* **31**, 81–95.

Werner, G. and Mountcastle, V. B. 1965. Neural activity in mechanoreceptive cutaneous afferents: Stimulus–response relations, Weber functions, and information transmission. *J. Neurophysiol.* **28**, 359–397.

Wessendorf, M. W. and Elde, R. 1987. The coexistence of serotonin- and substance P-like immunoreactivity in the spinal cord of the rat as shown by immunofluorescent double labeling. *J. Neurosci.* **7**, 2352–2363.

Westermark, T., Isaksson, T., Holmberg, P., Kjorell, U., Rantapaa-Dahlqvist, S., and Forsgren, S. 1999. Increase in bombesin-like peptides in the spinal cord after dexamethasone treatment of adrenalectomized rats. *Neurosci. Lett.* **275**, 179–182.

Westlund, K. N. 1992. Anatomy of noradrenergic pathways modulating pain. In J. M. Besson and G. Guilbaud (eds.), *Towards the Use of Noradrenergic Agonists for the Treatment of Pain* (pp. 91–117). Elsevier, Amsterdam.

Westlund, K. N. and Coulter, J. D. 1980. Descending projections of the locus coeruleus and subcoeruelus/medial parabrachial nuclei in monkey: Axonal transport studies and dopamine-β-hydroxylase immunocytochemistry. *Brain Res. Rev.* **2**, 235–264.

Westlund, K., Bowker, R. M., Ziegler, M. G., and Coulter, J. D. 1982. Descending noradrenergic projections and their spinal terminations. *Prog. Brain Res.* **57**, 219–238.

Westlund, K. N., Bowker, R. M., Ziegler, M. G., and Coulter, J. D. 1983. Noradrenergic projections to the spinal cord of the rat. *Brain Res.* **263**, 15–31.

Westlund, K. N., McNeill, D. L., and Coggeshall, R. E. 1989a. Glutamate immunoreactivity in rat dorsal root axons. *Neurosci. Lett.* **96**, 13–17.

Westlund, K. N., McNeill, D. L., Patterson, J. T., and Coggeshall, R. E. 1989b. Aspartate immunoreactive axons in normal rat L4 dorsal roots. *Brain Res.* **489**, 347–351.

Westlund, K. N., Zhang, D., Carlton, S. M., Sorkin, L. S., and Willis, W. D. 1991. Noradrenergic innervation of somatosensory thalamus and spinal cord. *Prog. Brain Res.* **88**, 77–88.

Westlund, K. N., Sun, Y. C., Sluka, K. A., Dougherty, P. M., Sorkin, L. S., and Willis, W. D. 1992. Neural changes in acute arthritis in monkeys. II. Increased glutamate immunoreactivity in the medial articular nerve. *Brain Res. Rev.* **17**, 15–27.

Westlund, K. N., Krakower, T. J., Kwan, S. W., and Abell, C. W. 1993. Intracellular distribution of monoamine oxidase A in selected regions of rat and monkey brain and spinal cord. *Brain Res.* **612**, 221–230.

Westrum. L. E., Johnson, L. R., and Canfield, R. C. 1984. Ultrastructure of transganglionic degeneration in brain stem trigeminal nuclei during normal primary tooth exfoliation and permanent tooth eruption in the cat. *J. Comp. Neurol.* **230**, 198–206.

Wetts, R. and Vaughn, J. E. 1993. Transient expression of beta-NADPH diaphorase in developing rat dorsal root ganglion neurons. *Develop. Brain Res.* **76**, 278–282.

Wetts, R. and Vaughn, J. E. 1994. Choline acetyltransferase and NADPH diaphorase are co-expressed in rat spinal cord neurons. *Neuroscience* **63**, 1117–1124.

Wetts, R., Phelps, P. E., and Vaughn, J. E. 1995. Transient and continuous expression of NADPH diaphorase in different neuronal populations of developing rat spinal cord. *Dev. Dyn.* **202**, 215–228.

White, D. M., Basbaum, A. I., Goetzl, E. J., and Levine, J. D. 1990. The 15-lipoxygenase product, 8R,15S-diHETE, stereospecifically sensitizes C-fiber mechanoheat nociceptors in hairy skin of rat. *J. Neurophysiol.* **63**, 966–970.

White, J. C. and Sweet, W. H. 1969. *Pain and the Neurosurgeon: A Forty-year Experience*. Thomas, Springfield, IL.

White, S. R., Penner, J. D., Speth, R. C., and Chan, J. Y. H. 1988. Angiotensin II receptors in the lumbar spinal cord of the rat. *Brain Res.* **441**, 195–201.

Whitehorn, D. and Burgess, P. R. 1973. Changes in polarization of central branches of myelinated mechanoreceptor and nociceptor fibers during noxious and innocuous stimulation of the skin. *J. Neurophysiol.* **36**, 226–237.

Whitehorn, D., Howe, J. F., Lessler, M. J., and Burgess, P. R. 1974. Cutaneous receptors supplied by myelinated fibers in the cat: I. Number of receptors innervated by a single nerve. *J. Neurophysiol.* **37**, 1361–1372.

Whitehouse, P. J., Wamsley, J. K., Zarbin, M. A., Price, D. L., Tourtellotte, W. W., and Kuhar, M. J. 1983. Amyotrophic lateral sclerosis: Alterations in neurotransmitter receptors. *Ann. Neurol.* **14**, 8–16.

Wiberg, M., Ljungberg, C., O'Byrne, A., Brown, R., Whitworth, I., Liss, A., and Terenghi, G. 1999. Primary sensory neuron survival following targeted administration of nerve growth factor to an injured nerve. *Scand. J. Plast. Reconstr. Surg. Hand Surg.* **33**, 387–392.

Widerberg, A., Kanje, M., and Dahlin, L. B. 2001. C-terminal flanking peptide of neuropeptide Y in DRG following nerve compression. *NeuroReport* **12**, 3193–3196.

Wiesenfeld, Z. and Lindblom, U. 1980. Behavioral and electrophysiological effects of various types of peripheral nerve lesions in the rat: A comparison of possible models for chronic pain. *Pain* **8**, 285–299.

Wiesenfeld-Hallin, Z. and Xu, X. J. 1996. The role of cholecystokinin in nociception, neuropathic pain and opiate tolerance. *Regul. Pept.* **65**, 23–28.

Wiesenfeld-Hallin, Z. and Xu, X. J. 1998. Galanin in somatosensory function. *Ann. NY Acad. Sci.* **863**, 383–389.

Wiesenfeld-Hallin, Z. and Xu, X. J. 2001. Neuropeptides in neuropathic and inflammatory pain with special emphasis on cholecystokinin and galanin. *Eur. J. Pharmacol.* **429**, 49–59.

Wiesenfeld-Hallin, Z., Hökfelt, T., Lundberg, J. M., Firssmann, W. G., Reunecke, M., Tschopp, F. A., and Fischer, J. A. 1984. Immunoreactive calcitonin gene-related peptide and substance P coexist in sensory neurons to the spinal cord and interact in spinal behavioral responses of the rat. *Neurosci. Lett.* **52**, 199–204.

Wikberg, J. E. S. and Hajos, M. 1987. Spinal cord α2-adrenoceptors may be located postsynaptically with respect to primary sensory neurons: Destruction of primary C- afferents with neonatal capsaicin does not affect the number of [^{3}H]clonidine binding sites in mice. *Neuroscience* **76**, 63–68.

Willcockson, W. S., Chung, J. M., Hori, Y., Lee, K. H., and Willis, W. D. 1984. Effects of iontophoretically released amino acids and amines on primate spinothalamic tract cells. *J. Neurosci.* **4**, 732–740.

Willer, J. C., Roby, A., and Le Bars, D. 1984. Psychophysical and electrophysiological approaches to the pain-relieving effects of heterotopic nociceptive stimuli. *Brain* **107**, 1095–1112.

Willer, J. C., De Brouker, T., ands Le Bars, D. 1989. Encoding of nociceptive thermal stimuli by diffuse noxious inhibitory controls in humans. *J. Neurophysiol.* **62**, 1028–1038.

Williams, C. A., Wu, S. Y., Dun, S. L., Kwok, E. H., and Dun, N. J. 1999. Release of endomorphin-2-like substances from the rat spinal cord. *Neurosci. Lett.* **273**, 25–28.

Williams, J. T., Christie, M. J., and Manzoni, O. 2001. Cellular and synaptic adaptations mediating opioid dependence. *Physiol. Rev.* **81**, 299–343.

Williams, S., Evan, G. I., and Hunt, S. P. 1990. Changing patterns of *c-fos* induction following thermal cutaneous stimulation in the rat. *Neuroscience* **36**, 73–81.

Williams, S. J. and Papka, R. E. 1996. Estrogen receptor-immunoreactive neurons are present in the female rat lumbosacral spinal cord. *J. Neurosci. Res.* **46**, 492–501.

Willis, W. D. 1984. Evoked spinal cord potentials in the cat and monkey: Use in the analysis of spinal cord function. In S. Homma and T. Tamaki (eds.), *Fundamentals and Clinical Application of Spinal Cord Monitoring* (pp. 3–19). Saikon Publishing Co. Ltd., Tokyo.

Willis, W. D. 1985. *The Pain System*. Karger, Basel.

Willis, W. D. 1997. Is central sensitization of nociceptive transmission in the spinal cord a variety of long-term potentiation? A commentary on the article by Svendsen, Tjølsen, and Hole. Focus article for *NeuroReport*, **8**, iii.

Willis, W. D. 1999. Dorsal root potentials and dorsal root reflexes: A double-edged sword. *Exp. Brain Res.* **124**, 395–421.

Willis, W. D. 2001. The role of neurotransmitters in sensitization of pain responses. In *Role of Neural Plasticity in Chemical Intolerance. Ann. NY Acad. Sci.* **933**: 42–56.

Willis, W. D. 2002a. Possible mechanisms of central neuropathic pain. In K. J. Burchiel and R. P: Yezierski (eds.), *Spinal Cord Injury Pain: Assessment, Mechanisms, Management. Vol. 23, Pain Research and Management* (pp. 85–115). IASP Press, Seattle.

Willis, W. D. 2002b. Long-term potentiation in spinothalamic neurons. *Brain Res. Rev.*, **40**: 202–214.

Willis, W. D. and Coggeshall, R. E. 1991. *Sensory Mechanisms of the Spinal Cord* (2nd ed.). Plenum Press, New York.

Willis, W. D. and Grossman, R. G. 1981. *Medical Neurobiology* (3rd ed.). C. V. Mosby, St. Louis.

Willis, W. D., Weir, M. A., Skinner, R. D., and Bryan, R. N. 1973. Differential distribution of spinal cord field potentials. *Exp. Brain Res.* **17**, 169–176.

Willis, W. D., Leonard, R. B., and Kenshalo, D. R. 1978. Spinothalamic tract neurons in the substantia gelatinosa. *Science* **202**, 986–988.

Wilson, P. and Snow, P. J. 1988. Somatotopic organization of the dorsal horn in the lumbosacral enlargement of the spinal cord in the neonatal cat. *Exp. Neurol.* **101**, 428–444.

Wilson, P., Meyers, D. E. R., and Snow, P. J. 1986. The detailed somatotopic organization of the dorsal horn in the lumbosacral enlargement of the cat spinal cord. *J. Neurophysiol.* **55**, 604–617.

Windle, W. F. 1931. Neurons of the sensory type in the ventral roots of man and of other mammals. *Arch. Neurol. Psychiat.* **26**, 791–800.

Winter, D. L. 1971. Receptor characteristics and conduction velocities in bladder afferents. *J. Psychiat. Res.* **8**, 225–235.

Winter, J., Walpole, C. S. J., Bevan, S., and James, I. F. 1993. Characterization of resiniferatoxin binding sites on sensory neurons: Co-regulation of resiniferatoxin binding and capsaicin sensitivity in adult rat dorsal root ganglia. *Neuroscience* **57**, 747–757.

Wisden, W., Gundlach, A. L., Barnard, E. A., Seeburg, P. H., and Hunt, S. P. 1991. Distribution of $GABA_A$ receptor subunit mRNAs in rat lumbar spinal cord. *Mol. Brain Res.* **10**, 179–183.

Witt, I. and Hensel, H. 1959. Afferente impulse aus der Extremitätenhaut der Katze bei thermischer und mechanischer Reizung. *Pfluegers Arch.* **268**, 582–596.

Wojtys, E. M., Beaman, D. N., Glover, R. A., and Janda, D. 1990. Innervation of the human knee joint by substance-P fibers. *Arthoscopy* **6**, 254–263.

Womack, M. D., MacDermott, A. B., and Jessell, T. M. 1988. Sensory transmitters regulate intracellular calcium in dorsal horn neurons. *Nature* **334**, 351–353.

Won, M. H., Park, H. S., Jeong, Y. G., and Park, H. J. 1998. Afferent innervation of the rat pancreas: Retrograde tracing and immunohistochemistry in the dorsal root ganglia. *Pancreas* **16**, 80–87.

Wong, J. and Oblinger, M. M. 1991. NGF rescues substance P expression but not neurofilament or tubulin gene expression in axotomized sensory neurons. *J. Neurosci.* **11**, 543–552.

Wong, V., Barrett, C. P., Donati, E. J., Eng, L. F., and Guth, L. 1983. Carbonic anhydrase activity in first-order sensory neurons in the rat. *J. Histochem. Cytochem.* **31**, 293–300.

Wong-Riley, M. T. T. 1989. Cytochrome oxidase: An endogenous metabolic marker for neuronal activity. *Trends Neurosci* **12**, 94–101.

Wong-Riley, M. T. T. and Kageyama, G. H. 1986. Localization of cytochrome oxidase in the mammalian spinal cord and dorsal root ganglia, with quantitative analysis of ventral horn cells in monkeys. *J. Comp. Neurol.* **245**, 41–61.

Wonnacott, S. 1997. Presynaptic nicotinic ACh receptors. *Trends in Neurosci.* **20**, 92–98.

Wood, J. N., Winter, J., James, I. F., Rang, H. P., Yeats, J., and Bevan, S. 1988. Capsaicin-induced ion fluxes in dorsal root ganglion cells in culture. *J. Neurosci.* **8**, 3208–3220.

Woodbury, C. J., Ritter, A. M., and Koerber, H. R. 2000. On the problem of lamination in the superficial dorsal horn of mammals: A reappraisal of the substantia gelatinosa in postnatal life. *J. Comp. Neurol.* **417**, 88–102.

Woodbury, J. W. and Patton, H. D. 1952. Electrical activity of single spinal cord elements. *Cold Spring Harbor Symp. Quant. Biol.* **17**, 185–188.

Woolf, C. J. 1983. C-primary afferent fibre mediated inhibitions in the dorsal horn of the decerebrate-spinal rat. *Exp. Brain Res.* **51**, 283–290.

Woolf, C. J. 1984. A selective effect of naloxone on heterosynaptic C-fibre-mediated inhibitions in the rat dorsal horn. *Neurosci. Lett.* **45**, 169–174.

Woolf, C. J. 1987. Central terminations of cutaneous mechanoreceptive afferents in the rat lumbar spinal cord. *J. Comp. Neurol.* **261**, 105–119.

Woolf, C. J. and Fitzgerald, M. 1982. Do opioid peptides mediate a presynaptic control of C-fibre transmission in the rat spinal cord? *Neurosci. Lett.* **29**, 67–72.

Woolf, C. J. and Fitzgerald, M. 1983. The properties of neurones recorded in the superficial dorsal horn of the rat spinal cord. *J. Comp. Neurol.* **221**, 313–328.

Woolf, C. J. and Fitzgerald, M. 1986. Somatotopic organization of cutaneous afferent terminals and dorsal horn receptive fields in the superficial and deep laminae of the rat lumbar spinal cord. *J. Comp. Neurol.* **251**, 517–531.

Woolf, C. J. and King, A. E. 1987. Physiology and morphology of multireceptive neurons with C-afferent inputs in the deep dorsal horn of the rat lumbar spinal cord. *J. Neurophysiol.* **58**, 460–479.

Woolf, C. J. and King, A. E. 1989. Subthreshold components of the cutaneous mechanoreceptive fields of dorsal horn neurons in the rat lumbar spinal cord. *J. Neurophysiol.* **62**, 907–916.

Woolf, C. J., and King, A. E. 1990. Dynamic alterations in the cutaneous mechanoreceptive fields of dorsal horn neurons in the rat spinal cord. *J. Neurosci.* **10**, 2717–2726.

Woolf, C. J. and Salter, M. W. 2000. Neuronal plasticity: Increasing the gain in pain. *Science* **288**, 1765–1768.

Woolf, C. J. and Wall, P. D. 1982. Chronic peripheral nerve section diminishes the primary afferent A-fibre mediated inhibition of rat dorsal horn neurones. *Brain Res.* **242**, 77–85.

Woolf, C. J., Shortland, P., and Coggeshall, R. E. 1992. Peripheral nerve injury triggers central sprouting of myelinated afferents. *Nature*, **355**, 75–78.

Woolf, C. J., Shortland, P., Reynolds, M., Ridings, J., Doubell, T., and Coggeshall, R. E. 1995. Reorganization of central terminals of myelinated primary afferents in the rat dorsal horn following peripheral axotomy. *J. Comp. Neurol.* **360**, 121–134.

Wotherspoon, G. and Priestley, J. V. 2000. Expression of the 5-HT1B receptor by subtypes of rat trigeminal ganglion cells. *Neuroscience* **95**, 465–471.

Wotherspoon, G. and Winter, J. 2000. Bradykinin B1 receptor is constitutively expressed in the rat sensory nervous system. *Neurosci. Lett.* **294**, 175–178.

Wrathall, J. R., Choiniere, D., and Teng, Y. D. 1994. Dose-dependent reduction of tissue loss and functional impairment after spinal cord trauma with the AMPA/kainate antagonist NBQX. *J. Comp. Neurol.* **14**, 6598–6607.

Wright, D. E. and Snider, W. D. 1995. Neurotrophin receptor mRNA expression defines distinct populations of neurons in rat dorsal root ganglia. *J. Comp. Neurol.* **351**, 329–338.

Wright, D. M. and Roberts, M. H. T. 1978. Supersensitivity to a substance P analogue following dorsal root section. *Life Sci.* **22**, 19–24.

Wu, G., Ekedahl, R., and Hallin, R. G. 1998. Clustering of slowly adapting type II mechanoreceptors in human peripheral nerve and skin. *Brain* **121**, 265–279.

Wu, G., Ekedahl, R., Stark, B., Carlstedt, T., Nilsson, B., and Hallin, R. G. 1999. Clustering of Pacinian corpuscle afferent fibres in the human median nerve. *Exp. Brain Res.* **126**, 399–409.

Wu, J., Lin, Q., McAdoo, D. J., and Willis, W. D. 1998a. Nitric oxide contributes to central sensitization following intradermal injection of capsaicin. *NeuroReport* **9**, 589–592.

Wu, J., Lin, Q., Willis, W. D., and Westlund, K. N. 1998b. Changes in nitric oxide synthase isoforms in the spinal cord of rat following induction of chronic arthritis. *Exp. Brain Res.* **118**, 457–465.

Wu, J., Fang, L., Lin, Q., and Willis, W. D. 2000. Fos expression is induced by increased nitric oxide release in rat spinal cord dorsal horn. *Neuroscience* **96**, 351–357.

Wu, J., Fang, L., Lin, Q., and Willis, W. D. 2001. Nitric oxide synthase in spinal central sensitization following intradermal injection of capsacin. *Pain* **94**, 47–58.

Wu, J., Fang, L., Lin, Q., and Willis, W. D. 2002. The role of nitric oxide in the phosphorylation of cAMP-responsive element-binding protein in the spinal cord after intradermal injection of capsaicin. *J. Pain* **3**, 190–198.

Wu, S. Y., Dun, S. L., Wright, M. T., Chang, J. K., and Dun, N. J. 1999. Endomorphin-like immunoreactivity in the rat dorsal horn and inhibition of substantia gelatinosa neurons in vitro. *Neuroscience* **89**, 317–321.

Wu, W. 1993. Expression of nitric-oxide synthase (NOS) in injured CNS neurons as shown by NADPH diaphorase histochemistry. *Exp. Neurol.* **120**, 153–159.

Wu, W. and Wessendorf, M. W. 1992. Organization of the serotonergic innervation of spinal neurons in rats: I. Neuropeptide coexistence in varicosities innervating some spinothalamic tract neurons but not in those innervating postsynaptic dorsal column neurons. *Neuroscience* **50**, 885–898.

Wu, W., Elde, R., Wessendorf, M. W. and Hökfelt, T. 1992. Identification of neurons expressing thyrotropin releasing-hormone receptor mRNA in spinal cord and lower brainstem of rat. *Neurosci. Lett.* **142**, 143–146.

Wu, W., Liuzzi, F. J., Schinco, F. P., Depto, A. S., Li, Y., Mong, J. A., Dawson, T. M., and Snyder, S. H. 1994. Neuronal nitric oxide synthase is induced in spinal neurons by traumatic injury. *Neuroscience* **61**, 719–726.

Wynn Parry, C. B. 1980. Pain in avulsion lesions of the brachial plexus. *Pain* **9**, 41–53.

Xiang, Z., Bo, X., and Burnstock, G. 1998. Localization of ATP-gated P2X receptor immunoreactivity in rat sensory and sympathetic ganglia. *Neurosci. Lett.* **256**, 105–108.

Xian-Min, Y. and Mense, S. 1990. Somatotopical arrangement of rat spinal dorsal horn cells processing input from deep tissues. *Neurosci. Lett.* **108**, 43–47.

Xie, G. X., Jones, K., Peroutka, S. J., and Palmer, P. P. 1998. Detection of mRNAs and alternatively spliced transcripts of dopamine receptors in rat peripheral sensory and sympathetic ganglia. *Brain Res.* **785**, 129–135.

Xie, J., Lee, Y. H., Wang, C., Mo, C. J., and Chung, K. 2001. Differential expression of alpha1-adrenoceptor subtype mRNAs in the dorsal root ganglion after spinal nerve ligation. *Mol. Brain Res.* **93**, 164–172.

Xie, Y. K., Zhang, J.-M., Petersen, M., and LaMotte, R. H. 1995. Functional changes in dorsal root ganglion cells after chronic nerve constriction in the rat. *J. Neurophysiol.* **73**, 1811–1820.

Xu, G. Y., McAdoo, D. J., Hughes, M. G., Robak, G., and de Castro, R. 1998. Considerations in the determination by microdialysis of resting extracellular amino acid concentrations and release upon spinal cord injury. *Neuroscience* **86**, 1011–1021.

Xu, G. Y. and Huang, L.-Y. M. 2002. Peripheral inflammation sensitizes P2X receptor-mediated responses in rat dorsal root ganglion neurons. *J. Neurosci.* **22**, 93–102.

Xu, I. S., Grass, S., Xu, X. J., and Wiesenfeld-Hallin, Z. 1998. On the role of galanin in mediating spinal flexor reflex excitability in inflammation. *Neuroscience* **85**, 827–835.

Xu, S., Zhang, Y., Lundeberg, T., and Yu, L. 2000. Effects of galanin on wide-dynamic range neuron activity in the spinal dorsal horn of rats with sciatic nerve ligation. *Regul. Pept.* **95**, 19–23.

Xu, X. J., Hao, J. X., Aldskogius, H., Seiger, A., and Wiesenfeld-Hallin, Z. 1992. Chronic pain-related syndrome in rats after ischemic spinal cord lesion: A possible animal model for pain in patients with spinal cord injury. *Pain* **48**, 279–290.

Xu, X. J., Farkas-Szallasi, T., Lundberg, J. M., Hökfelt, T., Wiesenfeld-Hallin, Z., and Szallasi, A. 1997. Effects of the capsaicin analogue resiniferatoxin on spinal nociceptive mechanisms in the rat: Behavioral, electrophysiological and in situ hybridization studies. *Brain Res.* **752**, 52–60.

Xu, X. J., Hökfelt, T., Bartfai, T., and Wiesenfeld-Hallin, Z. 2000. Galanin and spinal nociceptive mechanisms: Recent advances and therapeutic implications. *Neuropeptides* **34**, 137–147.

Xu, Z. Q., Shi, T. J., Landry, M., and Hokfelt, T. 1996a. Evidence for galanin receptors in primary sensory neurones and effect of axotomy and inflammation. *NeuroReport* **8**, 237–242.

Xu, Z. Q., Shi, T.-J., and Hökfelt, T. 1996b. Expression of galanin and a galanin receptor in several sensory systems and bone anlage of rat embryos. *Proc. Natl. Acad. Sci.* **93**, 14901–14905.

Yajiri, Y. and Huang, L. Y. 2000. Actions of endomorphins on synaptic transmission of Adelta fibers in spinal cord dorsal horn neurons. *J. Biomed. Sci.* **7**, 226–231.

Yajiri, Y., Yoshimura, M., Okamoto, M., Takahashi, H., and Higashi, H. 1997. A novel slow excitatory postsynaptic current in substantia gelatinosa neurons of the rat spinal cord in vitro. *Neuroscience* **76**, 673–688.

Yakovlev, A. G. and Faden, A. I. 1994. Sequential expression of *c-fos* protooncogene, TNF-Alpha, and dynorphin genes in spinal cord following experimental traumatic injury. *Mol. Chem. Neuropath.* **23**, 179–190.

Yaksh, T. L. 1984. Spinal superfusion in the rat and cat. In C. A. Marsden (ed.), *Measurement of Neurotransmitter Release* In vivo New York. (pp. 107–124). John Wiley & Sons.

Yaksh, T. L. 1989. Behavioral and autonomic correlates of the tactile evoked allodynia produced by spinal glycine inhibition: Effects of modulatory receptor systems and excitatory amino acid antagonists. *Pain* **37**, 111–123.

Yaksh, T. L. and Elde, R. P. 1980. Release of methionine-enkephalin immunoreactivity from the rat spinal cord in vivo. *Eur. J. Pharmacol.* **63**, 359–362.

Yaksh, T. L., and Elde, R. P. 1981. Factors governing release of methionine enkephalin-like immunoreactivity from mesencephalon and spinal cord of the cat in vivo. *J. Neurophysiol.* **46**, 1056–1075.

Yaksh, T. L. and Noueihed, R. 1985. The physiology and pharmacology of spinal opiates. *Ann. Rev. Pharmacol. Toxicol.* **25**, 433–462.

Yaksh, T. L. and Rudy, T. S. 1976. Chronic catheterization of the spinal subarachnoid space. *Physiol. Behav.* **17**, 1031–1036.

Yaksh, T. L., Jessell, T. M., Gamse, R., Mudge, A. W., and Leeman, S. E. 1980. Intrathecal morphine inhibits substance P release from mammalian spinal cord in vivo. *Nature* **286**, 155–156.

Yaksh, T. L., Schmauss, C., Micevych, P. E., Abay, E. O., and Go, V. L. W. 1982. Pharmacological studies on the application, disposition, and release of neurotensin in the spinal cord. *Ann. NY Acad. Sci.* **400**, 228–243.

Yaksh, T. L., Dirksen, R., and Harty, G. J. 1985. Antinociceptive effects of intrathecally injected cholinomimetic drugs in the rat and cat. *Eur. J. Pharmacol.* **117**, 81–88.

Yamada, H., Honda, T., Yaginuma, H., Kikuchi, S., and Sugiura, Y. 2001. Comparison of sensory and sympathetic innervation of the dura mater and posterior longitudinal ligament in the cervical spine after removal of the stellate ganglion. *J. Comp. Neurol.* **434**, 86–100.

Yamaga, N., Kawasaki, H., Inaizumi, K., Shimizu, M., Nakamura, A., and Kurosaki, Y. 2001. Age-related decrease in calcitonin gene-related peptide mRNA in the dorsal root ganglia of spontaneously hypertensive rats. *Jpn. J. Pharmacol.* **86**, 448–450.

Yamamoto, K. and Ohnishi, A. 1996. Ascending myelinated axons of primary sensory neurons of the sciatic nerve in the posterior column of the rat spinal cord. *Acta Neuropathol.* **92**, 27–34.

Yamamoto, T. and Nozaki-Taguchi, N. 1996. The effects of intrathecally administered FK480, a cholecytokinin-A receptor antagonist, and YM022, a cholecystokinin-B receptor antagonist, on the formalin test in the rat. *Anesth. Analg.* **83**, 107–113.

Yamamoto, T., Takahashi, K., Satomi, H., and Ise, H. 1977. Origins of primary afferent fibers in the spinal ventral roots in the cat as demonstrated by the horseradish peroxidase method. *Brain Res.* **16**, 350–354.

Yamamoto, T., Carr, P. A., Baimbridge, K. G., and Nagy, J. I. 1989. Parvalbumin- and calbindin D28K-immunoreactive neurons in the superficial layers of the spinal cord dorsal horn of rat. *Neuroscience* **23**, 493–508.

Yamamoto, T., Ohtori, S., and Chibsa, T. 2000. Inhibitory effect of intrathecally administered nociceptin on the expression of Fos-like immunoreactivity in the rat formalin test. *Neurosci. Lett.* **284**, 155–158.

Yamamura, H. I., Wamsley, J. K., Deshmukh, P., and Roeske, W. R. 1983. Differential light-microscopic autoradiographic localization of muscarinic receptors in the brainstem and spinal cord of the rat using [3H]pirenzepine. *Eur. J. Pharmacol.* **91**, 147–149.

Yamashita, Y., Akaike, N., Wakamori, M., Ikeda, I. and Ogawa, H. 1992. Voltage-dependent currents in isolated single Merkel cells of rats. *J. Physiol.* **450**, 143–162.

Yang, K. and Li, Y. Q. 2001. Origins of spontaneous and noxious stimuli-evoked miniature EPSCs in substantia gelatinosa. *NeuroReport* **12**, 39–42.

Yang, K., Kumamoto, E., Furue, H., and Yoshimura, M. 1998. Capsaicin facilitates excitatory but not inhibitory synaptic transmission in the substantia gelatinosa of the rat spinal cord. *Neurosci. Lett.* **255**, 135–138.

Yang, K., Li, Y., Furue, H., and Yoshimura, M. 2001a. Voltage-clamp recordings of postsynaptic currents in substantia gelatinosa neurons in vitro and its applications to assess synaptic transmission. *Brain Res. Protoc.* **7**, 235–240.

Yang, K., Feng, Y., and Li, Y. 2001b. Baclofen inhibition of dorsal root-evoked inhibitory postsynaptic currents in substantia gelatinosa neurons of rat spinal cord slice. *Brain Res.* **900**, 320–323.

Yang, K., Wang, D., and Li, Y. Q. 2001c. Distribution and depression of the GABA(B) receptor in spinal dorsal horn of adult rat. *Brain Res. Bull.* **55**, 479–485.

Yang, Y., Ozawa, H., Lu, H., Yuri, K., Hayashi, S., Nihonyanagi, K., and Kawata, M. 1998. Immunocytochemical analysis of sex differences in calcitonin gene-related peptide in the rat dorsal root ganglion, with special reference to estrogen and its receptor. *Brain Res.* **791**, 35–42.

Yashpal, K., Dam, T. V., and Quirion, R. 1990. Quantitative autoradiographic distribution of multiple neurokinin binding sites in rat spinal cord. *Brain Res.* **506**, 259–266.

Yashpal, K., Dam, T. V., and Quirion, R. 1991a. Effects of dorsal rhizotomy on neurokinin receptor sub-types in the rat spinal cord: A quantitative autoradiographic study. *Brain Res.* **552**, 240–247.

Yashpal, K., Sarrieau, A., and Quirion, R. 1991b. [125I] vasoactive intestinal polypeptide binding sites: Quantitative autoradiographic distribution in the rat spinal cord. *J. Chem. Neuroanat.* **4**, 439–446.

Yashpal, K., Kar, S., Dennis, T., and Quirion, R. 1992. Quantitative autoradiographic distribution of calcitonin gene-related peptide (hCGRP alpha) binding sites in the rat and monkey spinal cord. *J. Comp. Neurol.* **322**, 224–232.

Yashpal, K., Radhakrishnan, V., Coderre, T. J., and Henry, J. L. 1993. CP-96,345, but not its stereoisomer, CP-96,344, blocks the nociceptive responses to intrathecally administered substance P and to noxious thermal and chemical stimuli in the rat. *Neuroscience* **52**, 10398–1047.

Yashpal, K., Kar, S., Quirion, R., Hui-Chan, C. W., and Henry, J. L. 1994. Noxious stimulation decreases substance P binding in rat spinal dorsal horn: Competition by endogenous ligand? *NeuroReport* **5**, 2101–2104.

Yashpal, K., Fisher, K., Chabot, J. G., and Coderre, T. J. 2001. Differential effects of NMDA and group I mGluR antagonists on both nociception and spinal cord protein kinase C translocation in the formalin test and a model of neuropathic pain in rats. *Pain* **94**, 17–29.

Yates, B. J. and Thompson, F. J. 1985. Properties of spinal cord processing of femoral venous afferent input revealed by analysis of evoked potentials. *J. Auton. Nerv. System.* **14**, 201–207.

Ye, Z., Wimalawansa, S. J., and Westlund, K. N. 1999. Receptor for calcitonin gene-related peptide: Localization in the dorsal and ventral spinal cord. *Neuroscience* **92**, 1389–1397.

Yeomans, D. C. and Proudfit, H. K. 1996. Nociceptive responses to high and low rates of noxious cutaneous heating are mediated by different nociceptors in the rat: Electrophysiological evidence. *Pain* **68**, 141–150.

Yeomans, D. C., Pirec, V., and Proudfit, H. K. 1996. Nociceptive responses to high and low rates of noxious cutaneous heating are mediated by different nociceptors in the rat: Behavioral evidence. *Pain* **68**, 133–140.

Yezierski, R. P. and Park, S. H. 1993. The mechanosensitivity of spinal sensory neurons following intraspinal injections of quisqualic acid in the rat. *Neurosci. Lett.* **157**, 115–119.

Yezierski, R. P., Santana, M., Park, S. H. and Madsen, P. W. 1993. Neuronal degeneration and spinal cavitation following intraspinal injections of quisqualic acid in the rat. *J. Neurotrauma* **10**, 445–456.

Yezierski, R. P., Liu, S., Ruenes, G. L., Kajander, K. J., and Brewer, K. L. 1998. Excitotoxic spinal cord injury: Behavioral and morphological characteristics of a central pain model. *Pain* **75**, 141–155.

Ygge, J. and Grant, G. 1983. The organization of the thoracic spinal nerve projection in the rat dorsal horn demonstrated with transganglionic transport of horseradish peroxidase. *J. Comp. Neurol.* **216**, 1–9.

Yin, K. J. 1998. Distribution of somatostatin mRNA containing neurons in the primary pain relaying nuclei of the rat. *Anat. Rec.* **241**, 579–584.

Yokokawa, K., Tohyama, M., Shiosaka, S., Shiotani, Y., Sonoda, T., Emson, P. C., Hillyard, C. V., Girgis, S., and MacIntyre, I. 1986. Distribution of calcitonin-gene related peptide-containing fibers in the urinary bladder of the rat and their origin. *Cell Tiss. Res.* **244**, 271–278.

Yokoyama, C., Okamura, H., Nakajima, T., Taguchi, J.-I., and Ibata, Y. 1994. Autoradiographic distribution of [^{3}H]YM-09151-2, a high- affinity and selective antagonist ligand for the dopamine D_2 receptor group, in the rat brain and spinal cord. *J. Comp. Neurol.* **344**, 121–136.

Yoon, Y. W., Na, H. S., and Chung, J. M. 1996. Contributions of injured and intact afferents to neuropathic pain in an experimental rat model. *Pain* **64**, 27–36.

Yoshida, M. and Tanaka, M. 1988. Existence of new dopaminergic terminal plexus in the rat spinal cord: Assessment by immunohistochemistry using anti-dopamine serum. *Neurosci. Lett.* **94**, 5–9.

Yoshida, S. and Matsuda, Y. 1979. Studies on sensory neurons of the mouse with intracellular- recording and horseradish peroxidase injection techniques. *J. Neurophysiol.* **42**, 1134–1146.

Yoshida, S., Matsuda, Y., and Samejima, A. 1978. Tetrodotoxin-resistant sodium and calcium components of action potentials in dorsal root ganglion cells of the adult mouse. *J. Neurophysiol.* **41**, 1096–1106.

Yoshida, S., Senba, E., Kubota, Y., Hagihira, S., Yoshiya, I., Enson, P. C., and Tohyama, M. 1990. Calcium-binding proteins calbindin and parvalbumin in the superficial dorsal horn of the rat spinal cord. *Neuroscience* **37**, 839–848.

Yoshimura, M. and Jessell, T. M. 1989a. Primary afferent-evoked synaptic responses and slow potential generation in rat substantia gelatinosa neurons in vitro. *J. Neurophysiol.* **62**, 96–108.

Yoshimura, M. and Jessell, T. M. 1989b. Membrane properties of rat substantia gelatinosa neurons in vitro. *J. Neurophysiol.* **62**, 109–118.

Yoshimura, M. and Jessell, T. M. 1990. Amino acid-mediated EPSPs at primary afferent synapses with substantia gelatinosa neurones in the rat spinal cord. *J. Physiol.* **430**, 315–335.

Yoshimura, M. and Nishi, S. 1993. Blind patch-clamp recordings from substantia gelatinosa neurons in adult rat spinal cord slices: Pharmacological properties of synaptic currents. *Neuroscience* **53**, 519–526.

Yoshimura, M. and Nishi, S. 1995. Primary afferent-evoked glycine- and GABA-mediated IPSPs in substantia gelatinosa neurones in the rat spinal cord in vitro. *J. Physiol.* **482**, 29–38.

Yoshimura, M., and North, R. S. 1983. Substantia gelatinosa neurons in vitro hyperpolarized by enkephalin. *Nature* **305**, 529–530.

Yoshimura, N., Seki, S., Novakovic, S. D., Tzoumaka, E., Erickson, V. L., Erickson, K. A., Chancellor, M. B., and de Groat, W. C. 2001. The involvement of the tetrodotoxin-resistant sodium channel Na(v)1.8 (PN3/SNS) in a rat model of visceral pain. *J. Neurosci.* **21**, 8690–8696.

Yoshizawa, T., Kimura, S., Kanazawa, I., Uchiyama, Y., Yanagisawa, M., and Masaki, T. 1989. Endothelin localizes in the dorsal horn and acts on the spinal neurones: Possible involvement of dihydropyridine-sensitive calcium channels and substance P release. *Neurosci. Lett.* **102**, 179–184.

Yotsumoto, S., Setoyama, M., Hozumi, H., Mizoguchi, S., Fukumaru, S., Kobayashi, K., Saheki, T., and Kanzaki, T. 1999. A novel point mutation affecting the tyrosine kinase domain of the TRKA gene in a family with congenital insensitivity to pain with anhidrosis. *J. Invest. Dermatol.* **112**, 810–814.

You, H. J. and Chen, J. 1999. Differential effects of subcutaneous injection of formalin and bee venom on responses of wide-dynamic-range neurons in spinal dorsal horn of the rat. *Eur. J. Pain* **3**, 177–180.

Young, M. R., Fleetwood-Walker, S. M., Mitchell, R., and Munro, F. E. 1994. Evidence for a role of metabotropic glutamate receptors in sustained nociceptive inputs to rat dorsal horn neurons. *Neuropharmacology* **33**, 141–144.

Young, M. R., Fleetwood-Walker, S. M., Mitchell, R., and Dickinson, T. 1995. The involvement of metabotropic glutamate receptors and their intracellular signalling pathways in sustained nociceptive transmission in rat dorsal horn neurons. *Neuropharmacology* **34**, 1033–1041.

Young, M. R., Fleetwood-Walker, S. M., Dickinson, T., Blackburn-Munro, G., Sparrow, H., Birch, P. J., and Bountra, C. 1997. Behavioural and electrophysiological evidence supporting a role for group I metabotropic glutamate receptors in the mediation of nociceptive inputs to the rat spinal cord. *Brain Res.* **777**, 161–169.

Young, M. R., Blackburn-Munro, G., Dickinson, T., Johnson, M. J., Anderson, H., Nakalembe, L., and Fleetwood-Walker, S. M. 1998. Antisense ablation of type I metabotropic glutamate receptor mGluR1 inhibits spinal nociceptive transmission. *J. Neurosci.* **18**, 10180–10188.

Young, W. J. and Chang, C. 1998. Ontogeny and autoregulation of androgen receptor mRNA expression in the nervous system. *Endocrine.* **9**, 79–88.

Young, W. S. and Kuhar, M. J. 1979a. Autoradiographic localization of benzodiazepine receptors in the brains of humans and animals. *Nature* **280**, 391–394.

Young, W. S. and Kuhar, M. 1979b. Radiohistochemical localization of benzodiazepine receptors in rat brain. *J. Pharm. Exp. Ther.* **212**, 337–346.

Young, W. S. and Kuhar, M. J. 1980. Noradrenergic alpha 1 and alpha 2 receptors: Light-microscopic autoradiographic localization. *Proc. Natl. Acad. Sci. USA* **77**, 1696–1700.

Young, W. S. and Kuhar, M. J. 1981. Neurotensin receptor localization by light-microscopic autoradiography in rat brain. *Brain Res.* **206**, 273–285.

Young, W. S., Wamsley, J. K., Zarbin, M. A., and Kuhar, M. J. 1980. Opioid receptors undergo axonal flow. *Science* **210**, 76–78.

Yu, L. C., Zheng, E. M., and Lundeberg, T. 1999. Calcitonin gene-related peptide 8–37 inhibits the evoked discharge frequency of wide dynamic range neurons in dorsal horn of the spinal cord in rats. *Regul. Pept.* **83**, 21–24.

Yu, L. C., Xu, S. L., Xiong, W., and Lundeberg, T. 2001. The effect of galanin on wide-dynamic range neuron activity in the spinal cord of rats. *Regul. Pept.* **101**, 179–182.

Yu, W., Hao, J.-X., Xu, X.-J., Hökfelt, T., Elde, R., and Wiesenfeld-Hallin, Z. 1999. Spinal cord ischemia reduces μ-opioid receptors in rats: Correlation with morphine insensitivity. *Neuropharm.* **10**, 87–91.

Yu, X. H., Zhang, E. T., Craig, A. D., Shigemoto, R., Ribeiro-da-Silva, A., De Koninck, Y. 1999. NK-1 receptor immunoreactivity in distinct morphological types of lamina I neurons of the primate spinal cord. *J. Neurosci.* **19**, 3545–3555.

Yu, Y., Lundeberg, T., and Yu, L. C. 2002. Role of calcitonin gene-related peptide and its antagonist on the evoked discharge frequency of wide dynamic range neurons in the dorsal horn of the spinal cord in rats. *Regul. Pept.* **103**, 23–27.

Yung, K. K. L. 1998. Localization of glutamate receptors in dorsal horn of rat spinal cord. *NeuroReport* **9**, 1639–1644.

Yung, K. K. L. and Lo, Y. L. 1997. Immunocytochemical localization of muscarinic m2 receptor in the rat spinal cord. *Neurosci. Lett.* **229**, 81–84.

Zachariou, V., Goldstein, B. D., and Yeomans, D. C. 1997. Low but not high rate noxious radiant skin heating evokes a capsaicin-sensitive increase in spinal cord dorsal horn release of substance P. *Brain Res.* **752**, 143–150.

Zadina, J. E., Martin-Schild, S., Gerall, A. A., Kastin, A. J., Hackler, L., Ge, L. J., and Zarr, G. P., Werling, L. L., Brown, S. R., and Cox, B. M. 1986. Opioid ligand binding sites in the spinal cord of the guinea-pig. *Neuropharmacol.* **25**, 471–480.

Zadina, J. E., Hackler, L., Ge, L. J., and Kastin, A. J. 1997. A potent and selective endogenous agonist for the mu-opiate receptor. *Nature* **386**, 499–502.

Zajac, J. M., Lombard, M. C., Peschanski, M., Besson, J. M., and Roques, B. P. 1989. Autoradiographic study of μ and δ opioid binding sites and neutral endopeptidase-24.11 in rat after dorsal root rhizotomy. *Brain Res.* **477**, 400–403.

Zarbin, M. A., Wamsley, J. K., and Kuhar, M. J. 1981. Glycine receptor: Light-Microscopic autoradiographic localization with [^{3}H] strychnine[1]. *J. Neurosci.* **1**, 532–547.

Zarbin, M. A., Kuhar, M. J., O'Donohue, T. L., Wolf, S. S., and Moody, T. W. 1985. Autoradiographic localization of (125I-TYR4) bombesin-binding sites in rat brain. *J. Neurosci.* **5**, 429–437.

Zarbin, M. A., Wamsley, J. K., and Kuhar, M. J. 1990. Anterograde transport of opioid receptors in rat vagus nerves and dorsal roots of spinal nerves: Pharmacology and sensitivity to sodium and guanine nucleotides. *Exp. Brain Res.* **81**, 267–278.

Zarr, G. P., Werling, L. L., Brown, S. R., and Cox, B. M. 1986. Opioid ligand binding sites in the spinal cord of the guinea-pig. *Neuropharmacology* **25**, 471–480.

Zeilhofer, H. U., Muth-Selbach, U., Guhring, H., Erb, K., and Ahmadi, S. 2000. Selective suppression of inhibitory synaptic transmission by nocistatin in the rat spinal cord dorsal horn. *J. Neurosci.* **20**, 4922–4929.

Zeitz, K. P., Guy, N., Malmberg, A. B., Dirajlal, S., Martin, W. J., Sun, L., Bonhaus, D. W., Stucky, C. L., Julius, D., and Basbaum, A. I. 2002. The 5-HT_3 subtype of serotonin receptor contributes to nociceptive processing via a novel subset of myelinated and unmyelinated nociceptors. *J. Neurosci.* **22**, 1010–1019.

Zelena, J. 1978. The development of Pacinian corpuscles. *J. Neurocytol.* **7**, 71–91.

Zerari, F., Zouaoui, D., Gastard, M., Apartis, E., Fischer, J., Herbrecht, F., Cupo, A., Cucumel, K., and Conrath, M. 1994. Ultrastructural study of d-opioid receptors in the dorsal horn of the rat spinal cord using monoclonal anti-idiotypic antibodies. *J. Chem. Neuroanat.* **7**, 159–170.

Zerari, F., Dery, O., Fischer, J., Frobert, Y., Couraud, J. Y., and Conrath, M. 1995. Ultrastructural study of substance P receptors in the dorsal horn of the rat spinal cord using monoclonal anti-complementary peptide antibody. *J. Chem. Neuroanat.* **9**, 65–77.

Zerari, F., Fisher, J., Sagot, M. A., Frobert, Y., Couraud, J. Y., and Conrath, M. 1998. Substance P receptor immunodetection in the spinal cord: Comparative use of direct anti-receptor antibody and anti-complementary peptide antibody. *Brain Res. Bull.* **46**, 263–268.

Zhang, A. L., Hao, J. X., Seiger, A., Xu, X.-J., Wiesenfeld-Hallin, Z., Grant, G., and Aldskogius, H. 1994. Decreased GABA immunoreactivity in spinal cord dorsal horn neurons after transient spinal cord ischemia in the rat. *Brain Res.* **656**, 187–190.

Zhang, E. T. and Craig, A. D. 1997. Morphology and distribution of spinothalamic lamina I neurons in the monkey. *J. Neurosci.* **17**, 3274–3284.

Zhang, E. T., Han, Z. S., and Craig, A. D. 1996. Morphological classes of spinothalamic lamina I neurons in the cat. *J. Comp. Neurol.* **367**, 537–549.

Zhang, J. H., Morita, Y., Hironaka, T., Emson, P. C., and Tohyama, M. 1990. Ontological study of calbindin-D28k-like and parvalbumin-like immunoreactivities in rat spinal cord and dorsal root ganglia. *J. Comp. Neurol.* **302**, 715–728.

Zhang, J. H., Chandler, M. J., Miller, K. E., and Foreman, R. D. 1997. Cardiopulmonary sympathetic input does not require dorsal column pathways to excite C_1–C_3 spinal cells in rats, *Brain Res.* **771**, 25–30.

Zhang, J. M., Song, X. J., and LaMotte, R. H. 1997. An in vitro study of ectopic discharge generation and adrenergic sensitivity n the intact nerve-injured rat dorsal root ganglion. *Pain* **72**, 52–57.

Zhang, L., Hoff, A. O., Wimalawansa, S. J., Cote, G. J., Gagel, R. F., and Westlund, K. N. 2001. Arthritic calcitonin/α calcitonin gene-related peptide knockout mice have reduced nociceptive hypersensitivity. *Pain* **89**, 265–273.

Zhang, L., Lu, Y., Chen, Y., and Westlund, K. N. 2002. Group I metabotropic glutamate receptor antagonists block secondary thermal hyperalgesia in rats with knee joint inflammation *JPET* **300**, 149–156.

Zhang, Q., Ji, R. R., Lindsay, R., and Hokfelt, T. 1995a. Effect of growth factors on substance P mRNA expression in axotomized dorsal root ganglia. *NeuroReport* **6**, 1309–1312.

Zhang, Q., Shi, T. J., Ji, R. R., Zhang, Y. Z., Sundler, F., Hannibal, J., Fahrenkrug, J., Hokfelt, T., and Zhang, Y. 1995b. Expression of pituitary adenylate cyclase-activating polypeptide in dorsal root ganglia following axotomy: Time course and coexistence. *Brain Res.* **705**, 149–158.

Zhang, R. X., Mi, Z. P., and Qiao, J. T. 1994. Changes of spinal substance P, calcitonin gene-related peptide, somatostatin, Met-enkephalin and neurotensin in rats in response to formalin-induced pain. *Regul. Pept.* **51**, 25–32.

Zhang, R. X., Ruda, M. A., and Qiao, J. T. 1998. Pre-emptive intrathecal Mk-801, a non-competitive N-methyl-D-aspartate receptor antagonist, inhibits the up-regulation of spinal dynorphin mRNA and hyperalgesia in a rat model of chronic inflammation. *Neurosci. Lett.* **241**, 57–60.

Zhang, X. 1999. Endomorphins: Novel endogenous mu-opiate receptor agonists in regions of high mu-opiate receptor density. *Ann. NY Acad. Sci.* **897**, 136–144.

Zhang, X., Bean, A. J., Wiesenfeld-Hallin, Z., and Hokfelt, T. 1995a. Ultrastructural studies on peptides in the dorsal horn of the rat spinal cord—IV. Effects of peripheral axotomy with special reference to neuropeptide Y and vasoactive intestinal polypeptide/peptide histidine isoleucine. *Neuroscience* **64**, 917–941.

Zhang, X., Ji, R.-R., Arvidsson, J., Lundberg, J. M., Bartfai, T., Bedecs, K., and Hökfelt, T. 1996. Expression of peptides, nitric oxide synthase and NPY receptor in trigeminal and nodose ganglia after nerve lesions. *Exp. Brain Res.* **111**, 393–404.

Zhang, X., Verge, V., Wiesenfeld-Hallin, Z., Ju, G., Bredt, D., Snyder, S. H., and Hökfelt, T. 1993a. Nitric oxide synthase-like immunoreactivity in lumbar dorsal root ganglia and spinal cord of rat and monkey and effect of peripheral axotomy. *J. Comp. Neurol.* **335**, 563–575.

Zhang, X., Nicholas, A. P., and Hökfelt, T. 1993b. Ultrastructural studies on peptides in the dorsal horn of the spinal cord: I. Co-existence of galanin with other peptides in primary afferents in normal rats. *Neuroscience* **57**, 365–384.

Zhang, X., Ju, G., Elde, R., and Hökfelt, T. 1993c. Effect of peripheral nerve cut on neuropeptides in dorsal root ganglia and the spinal cord of monkey with special reference to galanin. *J. Neurocytol.* **22**, 342–381.

Zhang, X., Dagerlind, A., Elde, R. P., Castel, M. N., Broberger, C., Wiesenfeld-Hallin, Z., and Hökfelt, T. 1994a. Marked increase in cholecystokinin B receptor messenger RNA levels in rat dorsal root ganglia after peripheral axotomy. *Neuroscience* **58**, 669–670.

Zhang, X., Bao, L., Xu, Z.-Q., Kopp, J., Arvidsson, U., Elde, R., and Hökfelt, T. 1994b. Localization of neuropeptide Y Y1 receptors in the rat nervous system with special reference to somatic receptors on small dorsal root ganglion neurons. *Proc. Natl. Acad. Sci. USA* **91**, 11738–11742.

Zhang, X., Wiesenfeld-Hallin, Z., and Hökfelt, T. 1994c. Effect of peripheral axotomy on expression of neuropeptide Y receptor mRNA in rat lumbar dorsal root ganglia. *Eur. J. Neurosci.* **6**, 43–57.

Zhang, X., Aman, K., and Hökfelt, T. 1995a. Secretory pathways of neuropeptides in rat lumbar dorsal root ganglion neurons and effects of peripheral axotomy. *J. Comp. Neurol.* **352**, 481–500.

Zhang, X., Xu, Z. Q., Bao, L., Dagerlind, Å., and Hökfelt, T. 1995b. Complementary distribution of receptors for neurotensin and NPY in small neurons in rat lumbar DRGs and regulation of the receptors and peptides after peripheral axotomy. *J. Neurosci.* **15**, 2733–2747.

Zhang, X., Ji, R. R., Nilsson, S., Villar, M. J., Ubink, R., Ju, G., Wiesenfeld-Hallin, Z., and Hökfelt, T. 1995c. Neuropeptide Y and galanin binding sites in rat and monkey lumbar dorsal root ganglia and spinal cord and effect of peripheral axotomy. *Eur. J. Neurosci.* **7**, 367–380.

Zhang, X., Kostarczyk, E., and Gieser, G. J. 1995d. Spinohypothalamic tract neurons in the cervical enlargement of rats: Descending axons in the ipsilateral brain. *J. Neurosci.* **15**, 8393–8407.

Zhang, X., Nicholas, A. P., and Hökfelt, T. 1995e. Ultrastructural studies on peptides in the dorsal horn of the rat spinal cord: II. co-existence of galanin with other peptides in local neurons. *Neuroscience* **64**, 875–891.

Zhang, X., Shi, T., Holmberg, K., Landry, M., Huang, W., Xiao, H., Ju, G., and Hökfelt, T. 1997. Expression and regulation of the neuropeptide Y Y2 receptor in sensory and autonomic ganglia. *Proc. Natl. Acad. Sci. USA* **94**, 729–734.

Zhang, X., Xu, Z. O., Shi, T. J., Landry, M., Holmberg, K., Ju, G., Tong, Y. G., Bao, L., Cheng, X. P., Wiesenfeld-Hallin, Z., Lozano, A., Dostrovsky, J., and Hökfelt, T. 1998a. Regulation of expression of galanin and galanin receptors in dorsal root ganglia and spinal cord after axotomy and inflammation. *Ann. NY Acad. Sci.* **863**, 402–413.

Zhang, X., Bao, L., Arvidsson, U., Elde, R., and Hökfelt, T. 1998b. Localization and regulation of the delta-opioid receptor in dorsal root ganglia and spinal cord of the rat and monkey: Evidence for association with the membrane of large dense-core vesicles. *Neuroscience* **82**, 1225–1242.

Zhang, X., Bao, L., Shi, T. J., Ju, G., Elde, R., and Hökfelt, T. 1998c. Down-regulation of μ-opioid receptors in rat and monkey dorsal root ganlion neurons and spinal cord after peripheral axotomy. *Neuroscience* **84**, 223–240.

Zhang, X., Tong, Y. G., Bao, L., and Hökfelt, T. 1999. The neuropeptide Y Y1 receptor is a somatic receptor on dorsal root ganglion neurons and a postsynaptic receptor on somatostatin dorsal horn neurons. *Eur. J. Neurosci.* **11**, 2211–2225.

Zhang, X., de Araujo, L. G., Elde, R., Wiesenfeld-Hallin, Z., and Hökfelt, T. 2000. Effect of morphine on cholecystokinin and mu-opioid receptor-like immunoreactivities in rat spinal dorsal horn neurons after peripheral axotomy and inflammation. *Neuroscience* **95**, 197–207.

Zhang, X., Gokin, A. P., and Giesler, G. J. 2002. Responses of spinohypothalamic tract neurons in the thoracic spinal cord of rats to somatic stimuli and to graded distension of the bile duct. *Somatosens. Mot. Res.* **19**, 5–17.

Zhang, Y., Malmberg, A. B., Yaksh, T. L., Sjölund, B., Sundler, F., and Håkanson, R. 1997. Capsaicin-evoked release of pituitary adenylate cyclase activating peptide (PACAP) and calcitonin gene-related peptide (CGRP) from rat spinal cord in vivo. *Regul. Pept.* **69**, 83–87.

Zhang, Y. Q., Gao, X., Ji, G. C., and Wu, G. C. 2001. Expression of 5-HT2A receptor mRNA in rat spinal dorsal horn and some nuclei of brainstem after peripheral inflammation. *Brain Res.* **900**, 146–151.

Zhang, Y. Z., Hannibal, J., Zhao, Q., Moller, K., Danielsen, N., Fahrenkrug, J., and Sundler, F. 1996. Pituitary adenylate cyclase activating peptide expression in the rat dorsal root ganglia: Up-regulation after peripheral nerve injury. *Neuroscience* **74**, 1099–1110.

Zhao, Z. Q., Yang, H. Q., Zhang, K. M., and Zhuang, X. X. 1992. Release and depletion of substance P by capsaicin in substantia gelatinosa studied with the antibody microprobe technique and immunohistochemistry. *Neuropeptides* **23**, 161–167.

Zheng, F. and Lawson, S. N. 1997. Neurokinin A in rat renal afferent neurons and in nerve fibres within smooth muscle and epithelium of rat and guinea-pig renal pelvis. *Neuroscience* **76**, 1245–1255.

Zheng, J. H. and Chen, J. 2000. Modulatory roles of the adenosine triphosphate P2x-puroinoceptor in generation of the persistent nociception induced by subcutaneous bee venom injection in the conscious rat. *Neurosci. Lett.* **278**, 41–44.

Zheng, J. H. and Chen, J. 2001. Differential roles of spinal neurokinin 1/2 receptors in development of persistent nociception and hyperalgesia induced by subcutaneous bee venom injection in the conscious rat. *Neuropeptides* **35**, 32–44.

Zhong, J., Gerber, G., Kojic, L., and Randic, M. 2000. Dual modification of excitatory synaptic transmission by agonists at group I metabotropic glutamate receptors in the rat spinal dorsal horn. *Brain Res.* **887**, 359–377.

Zhong, Y., Dunn, P. M., Bardini, M., Ford, A. P., Cockayne, D. A., and Burnstock, G. 2001. Changes in P2X receptor responses of sensory neurons from P2X3- deficient mice. *Eur. J Neurosci.* **14**, 1784–1792.

Zhou, J., Chung, K., and Chung, J. M. 2001. Develoment of purinergic sensitivity in sensory neurons after peripheral nerve injury in the rat. *Brain Res.* **915**, 161–169.

Zhou, Q. Q., Imbe, H., Zou, S., Dubner, R., and Ren, K. 2001. Selective upregulation of the flip-flop splice variants of AMPA receptor subunits in the rat spinal cord after hindpaw inflammation. *Mol. Brain Res.* **88**, 186–193.

Zhou, X. F., Gai, W. P., and Rush, R. A. 1993. CGRP immunoreactive neurons in rat dorsal root ganglia do not all contain low-affinity NGF receptor immunoreactivity. *Brain Res.* **612**, 322–325.

Zhou, X. F., Rush, R. A., and McLachlan, E. M. 1996. Differential expression of the p75 nerve growth factor receptor in glia and neurons of the rat dorsal root ganglia after peripheral nerve transection. *J. Neurosci.* **16**, 2901–2911.

Zhou, X. F., Cameron, D., and Rush, R. A. 1998. Endogenous neurotrophin-3 supports the survival of a subpopulation of sensory neurons in neonatal rat. *Neuroscience* **86**, 1155–1164.

Zhou, X. F., Deng, Y. S., Chie, E., Xue, Q., Zhong, J. H., McLachlan, E. M., Rush, R. A., and Xian, C. J. 1999. Satellite-cell-derived nerve growth factor and neurotrophin-3 are involved in noradrenergic sprouting in the dorsal root ganglia following peripheral nerve injury in the rat. *Eur. J. Neurosci.* **11**, 1711–1722.

Zhu, C. G., Sandri, C., and Akert, K. 1981. Morphological identification of axo-axonic and dendro-dendritic synapses in the rat substantia gelatinosa. *Brain Res.* **230**, 25–40.

Zhuo, M. and Gebhart, G. F. 1991. Tonic cholinergic inhibition of spinal mechanical transmission. *Pain* **46**, 211–222.

Zoli, M., Le Novere, N., Hill, Jr. J. A., and Changeux, J.-P. 1995. Developmental regulation of nicotinic ACh receptor subunit mRNAs in the rat central and peripheral nervous systems. *J. Neurosci.* **15**, 1912–1939.

Zouaoui, D., Benoliel, J. J., Conrath, M., and Cesselin, F. 1990. Cholecystokinin-like immunoreactivity in the dorsal horn of the rat spinal cord: An attempt to analyze contradictory results between immunocytochemistry and radioimmunoassay. *Neuropeptides* **17**, 177–185.

Zieglgänsberger, W. and Herz, A. 1971. Changes of cutaneous receptive fields of spino-cervical-tract neurones and other dorsal column neurons by microiontophoretically administered amino acids. *Exp. Brain Res.* **13**, 111–126.

Zieglgänsberger, W. and Puil, E. A. 1973. Actions of glutamic acid on spinal neurones. *Exp. Brain Res.* **17**, 35–49.

Zieglgänsberger, W. and Sutor, B. 1983. Responses of substantia gelatinosa neurons to putative neurotransmitters in an in vitro preparation of the adult rat spinal cord. *Brain Res.* **279**, 316–320.

Zieglgänsberger, W. and Tulloch, I. F. 1979a. Effects of substance P on neurones in the dorsal horn of the spinal cord of the cat. *Brain Res.* **166**, 273–282.

Zieglgänsberger, W. and Tulloch, I. F. 1979b. The effects of methionine- and leucine-enkephalin on spinal neurones of the cat. *Brain Res.* **167**, 53–64.

Zimmermann, M. 1968. Dorsal root potentials after C-fiber stimulation. *Science* **160**, 896–898.

Zimny, M. L., Schutte, M., and Dabezies, E. 1986. Mechanoreceptors in the human anterior cruciate ligament. *Anat. Rec.* **214**, 204–209.

Zochodne, D. W., Verge, V. M., Cheng, C., Sun, H., and Johnston, J. 2001a. Does diabetes target ganglion neurones? Progressive sensory neurone involvement in long-term experimental diabetes. *Brain* **124**, 2319–2334.

Zochodne, D. W., Sun, H., and Li, X. Q. 2001b. Evidence that nitric oxide- and opioid-containing interneurons innervate vessels in the dorsal horn of the spinal cord of rats. *J. Physiol.* **532**, 749–758.

Zotterman, Y. 1939. Touch, pain and tickling: An electrophysiological investigation on cutaneous sensory nerves. *J. Physiol.* **95**, 1–28.

Zou, X., Lin, Q., and Willis, W. D. 2000. Enhanced phosphorylation of NMDA receptor NR1 subunits in spinal cord dorsal horn and spinothalamic tract neurons following intradermal injection of capsaicin in rats. *J. Neurosci.* **20**, 6989–6997.

Zou, X., Lin, Q., and Willis, W. D. 2001. NMDA or non-NMDA receptor antagonists attenuate increased FOS expression in spinal dorsal horn GABAergic neurons after intradermal injection of capsaicin in rats. *Neuroscience* **106**, 171–182.

Zukin, R. S. and Bennett, M. V. L. 1995. Alternatively spliced isoforms of the NMDAR1 receptor subunit. *Trends Neurosci.* **18**, 306–313.

INDEX